Handbook of Sensory Physiology

Volume VII/3

Central Processing
of Visual Information
Part A

Central Processing of Visual Information A: Integrative Functions and Comparative Data

By

H. Autrum · P. O. Bishop · V. Braitenberg · K. L. Chow
R. L. De Valois · R. B. Freeman Jr. · W. A. van de Grind
O.-J. Grüsser · U. Grüsser-Cornehls · R. Jung · W. R. Levick · H.-U. Lunkenheimer
D. M. MacKay · M. Snyder · J. Stone · N. J. Strausfeld · I. Thomas

Edited by

Richard Jung

With 208 Figures

Springer-Verlag Berlin · Heidelberg · New York 1973

ISBN-13: 978-3-642-65354-4 e-ISBN-13: 978-3-642-65352-0
DOI: 10.1007/978-3-642-65352-0

Typesetting, printing and binding: Brühlsche Universitätsdruckerei Gießen

Preface

The present volume covers the physiology of the visual system beyond the optic nerve. It is a continuation of the two preceding parts on the photochemistry and the physiology of the eye, and forms a bridge from them to the fourth part on visual psychophysics. These fields have all developed as independent specialities and need integrating with each other. The processing of visual information in the brain cannot be understood without some knowledge of the preceding mechanisms in the photoreceptor organs. There are two fundamental reasons, ontogenetic and functional, why this is so: 1) the retina of the vertebrate eye has developed from a specialized part of the brain; 2) in processing their data the eyes follow physiological principles similar to the visual brain centres. Peripheral and central functions should also be discussed in context with their final synthesis in subjective experience, i. e. visual perception.

Microphysiology and ultramicroscopy have brought new insights into the neuronal basis of vision. These investigations began in the periphery: HARTLINE's pioneering experiments on single visual elements of Limulus in 1932 started a successful period of neuronal recordings which ascended from the retina to the highest centres in the visual brain. In the last two decades modern electronmicroscopic techniques and photochemical investigations of single photoreceptors further contributed to vision research. The structural analysis of retinal elements, begun by SCHULTZE in 1866 and developed since 1893 first by CAJAL and later by POLYAK using silver methods, has been made more powerful by ultramicroscopic techniques. In recent years this approach has been extended by SZENTÁGOTHAI and COLONNIER to the study of the visual centres. The electron microscope has revealed details of cell membranes and synaptic structures and has complemented the neuronal investigations which, in many laboratories, continue the work begun by HARTLINE and GRANIT.

SVAETICHIN first recorded graded electrical signals in the fish retina in 1952; later similar slow potentials were found in receptors and various retinal cells of higher mammals, too. Recent research has shown that visual signals may be transmitted from photoreceptors to ganglion cells by graded potentials before the traditional mechanism of spike conduction, on which the early work concentrated, sends information to the brain. After absorbing light quanta in their molecules, the photoreceptors paradoxically show membrane hyperpolarization in vertebrates and induce slow membrane potentials in bipolar, horizontal and amacrine cells which form transverse networks. After this the ganglion cells produce action potentials of constant amplitude in two reciprocal systems, B on-centre neurones and D off-centre neurones, which transmit information concerning brightness and darkness respectively in many parallel lines. Such channels of spike-conducting neurones in all visual pathways of the brain do not, however, exclude the operation of slow potentials in brain centres where Golgi-2 cells with short axons are abundant. This is one of the many questions in cerebral visual physiology still awaiting an answer.

Another problem that may be solved by combined neuronal and psychophysical investigations concerns the coding mechanisms and their perceptual evaluation: Do the dual codes which convey information about brightness and darkness and opponent colours in fact correspond to HERING's "Gegenfarben", namely white-black, green-red and yellow-blue? How does this coding operate to integrate contour and surface values in the visual image?

A detailed description of the morphology and function of the complex structures of the central visual system is necessary in order to understand the delicate performance of the visual brain. Consequently, chapters on anatomy including ultrastructure precede those on neurophysiology. Unfortunately in the present state of our knowledge it is not yet possible to integrate morphology and neurophysiology into a synthesis of structure and function which results in visual perception. After HUBEL and WIESEL in 1959/61 discovered orientation-specific neurones and their columnar arrangement in the visual cortex, future research may establish the basic synaptic mechanisms of this order.

Studies of the behavioural aspects of vision will also add to our knowledge. The comparative study of animal behaviour and visual mechanisms shows promise of fruitful cooperation with neurophysiology. Early behavioural experiments concerned lower animal species, as described in chapter 11. Closer parallels with human perception may come from studies of monkeys with implanted microelectrodes by means of which some of the gaps between neurophysiology and perception may be bridged.

The original plan of this Handbook provided for contributions in German and French as well, so that each author could write in his preferred language. This, with the agreement of the German, French, Hungarian, Dutch and Italian authors, was abandoned in favour of English because it finds a larger audience. The decision not to use the German language seemed a difficult one in the case of visual physiology. Most classical discoveries and many syntheses in this field were reported in German for more than 100 years. The terminology, originally coined in German, has gradually been anglicized, first for neurophysiological publications and later for psychophysical work. Visual research in the nineteenth century, inaugurated by YOUNG and PURKINJE, was then dominated by HELMHOLTZ, HERING, KÜHNE and other pioneers of the visual sciences. However, the predominance of Anglo-American laboratories in experimental methods has brought about a change in the last 50 years. Be this as it may, the fundamental discoveries of the last century and the theories developed at that time, e. g. for visual contrast, colour vision and photochemistry, remain the basis for new advances. It must be appreciated that these various fields are complementary and a continuing synthesis with older work is necessary.

There is some advantage in discussing unsolved questions, even when no immediate solutions can be expected. The history of vision research has shown that it is legitimate to erect theories based upon perceptual phenomena before methods are available for precise neurophysiological experiments. Even when problems appear to be "Aporien" in the philosophical sense of KANT and N.

HARTMANN, i. e. insoluble, it is still worthwhile to speculate about them and to build hypotheses for future experiments. Unexpected new methods can open experimental approaches that seemed impossible or illusory before. Recent examples are the microelectrode techniques and microspectrography of single visual elements. Combinations of several research disciplines promise further progress, so that interrelations between different fields are discussed in this volume. In spite of essential differences in method, all visual research, whether by photochemists, neurophysiologists, behaviourists or psychologists, converges towards a common goal: the understanding of visual perception and its mechanisms.

Of course, even the combined effort of all the disciplines of visual research can never explain the totality of phenomena any more than in other sciences. Nevertheless, the more modest task of seeking relationships between perception and neurophysiological results seems to be fruitful. To combine experiment and theory, or to co-ordinate inductive-experimental and deductive-hypothetical approaches, instead of simply enumerating unrelated facts, is still the most efficient form of scientific synthesis, and one that continues the Handbook tradition.

To facilitate attempts at integration it seemed useful to look at the subjective final results of neuronal information processing as expressed in visual perception. For this purpose two chapters on the relationship between neurophysiology and vision precede the other contributions in this volume. Specific psychophysical problems are treated separately in part VII/4 of this Handbook. A metaphor devised by Ewald HERING may explain why visual perception is introduced into neurophysiology. It expresses the need to look for perceptual correlates when our knowledge of the neuronal mechanisms is insufficient and when experimental problems arise in neurophysiology. HERING used the paradigm of a clock, comparing the internal clockwork with its springs and cogwheels to the neuronal mechanisms of the eye and brain, and the hands of the clock to our sensory perceptions and scalings. HERING stated that sensory physiologists must look at the sensations on the dial when the internal clockwork cannot be directly examined. The position of the hands on the dial, i. e. vision, gives exact and quantitative information about the time, yet this depends upon the mechanisms of the inner "neuronal" clockwork. We all rely upon the clock to tell us the time, often without knowing its internal mechanisms. Similarly, we apply quantitative psychophysical scalings without an exact knowledge of their neuronal basis. Nevertheless, these scalings may indicate some properties of the mechanisms of the clock, i. e. the neurophysiological basis of vision.

Thanks and acknowledgement are extended to the Authors and Publishers for permission to reproduce figures from other publications. Sources are cited in the legends and chapter references.

Freiburg, i. Br., October 1972 R. JUNG

Central Processing of Visual Information

Part A

Contents

Integrative Functions

Comparative Data

List of Contributors

AUTRUM, Hansjochem, Zoologisches Institut der Universität, D-8000 München, Germany

BISHOP, Peter O., Department of Physiology, John Curtin School of Medical Research, Australian National University, Canberra City, A. C. T. 2601, Australia

BRAITENBERG, Valentin, Max-Planck-Institut für biologische Kybernetik, D-7400 Tübingen, Germany

CHOW, Kao Liang, Department of Neurology, Stanford University, Stanford, California 94305, USA

DE VALOIS, Russell L., Department of Psychology, University of California, Berkeley, California 94720, USA

FREEMAN, Robert B., Jr., Universität Konstanz, Fachbereich Psychologie, D-7750 Konstanz, Germany

VAN DE GRIND, Willem Alexander, Lab. Psychophysiology, Jan Swammerdam Institute, University of Amsterdam, Amsterdam, The Netherlands

GRÜSSER, Otto-Joachim, Physiologisches Institut der Freien Universität, D-1000 Berlin 33, Germany

GRÜSSER-CORNEHLS, Ursula, Physiologisches Institut der Freien Universität, D-1000 Berlin 33, Germany

JUNG, Richard, Neurologische Klinik mit Abteilung für Neurophysiologie der Universität Freiburg, D-7800 Freiburg, Germany

LEVICK, William R., Department of Physiology, John Curtin School of Medical Research, Australian National University, Canberra City, A. C. T. 2601, Australia

LUNKENHEIMER, Hans-Ullrich, Psychiatrische Universitätsklinik, D-4000 Düsseldorf-Heerdt, Germany

MACKAY, Donald M., Department of Communication, University of Keele, Staffordshire, Great Britain

SNYDER, Marvin, 5221 Tilden Avenue, Brooklyn, New York 11203, USA

STONE, Jonathan, Department of Physiology, John Curtin School of Medical Research, Australian National University, Canberra City, A. C. T. 2601, Australia

STRAUSFELD, Nicholas James, Max-Planck-Institut für biologische Kybernetik, D-7400 Tübingen, Germany

THOMAS, Inge, Zoologisches Institut der Universität, D-8000 München, Germany

Contents of Part B

Visual Centers in the Brain

Integrative Functions

Chapter 1

Visual Perception and Neurophysiology*

By

RICHARD JUNG, Freiburg i. Br. (Germany)

With 16 Figures

Contents

---

* This chapter is dedicated to Professor J. SZENTAGOTHAI, Budapest, on his 60th birthday, in appreciation of his basic anatomical contributions to neuronal physiology.

„Es ist ein unabweisbarer Glaube des Naturforschers, daß einer jeden Modification des Subjectiven innerhalb der Sinnessphäre jedesmal eine im Objectiven entspreche."

JOHANN PURKINJE, 1819 [369]

I. Introduction

When PURKINJE wrote the lines cited above in 1819 for his doctoral dissertation, he initiated a long period of successful visual research. His conviction, that all subjectively observed sensory phenomena should have correlates in objective physiology, remained a theoretical postulate for over 100 years. The electroretinogram, although discovered in 1865 by HOLMGREN [224], contributed little to these correlations, and objective visual physiology did not begin until 60 years later. The first successful recordings obtained from the vertebrate optic nerve by ADRIAN and R. MATTHEWS in 1927 [4] and from single optic fibres by HARTLINE in 1935 [191] mark the pioneer steps. In 1938 followed HARTLINE's classification of the responses from single optic nerve fibres in the frog [192] and of their receptive fields. Thereafter, the physiology of visual neurones made such rapid progress that it can now provide objective correlates for many visual phenomena, such as brightness, contrast and colour, on the basis of single neurone recordings from the retina, the optic nerve, and the visual projections in the brain itself.

It is the purpose of this chapter to show that neurophysiological and psychophysiological research in vision can, and should, be combined. Coordinated studies of visual perception in man and physiological experiments in animals are complementary and mutually stimulating in three ways. First, neuronal investigations of the visual system can be planned more effectively by making use of subjective psychophysical data of vision. Second, hitherto inexplicable visual phenomena in man may be formulated in neuronal terms. Third, some properties of visual neurones can be predicted from the analysis of human perception.

Admittedly, the conditions of human perception and neuronal animal experiments are not strictly comparable. Species differences and restraint procedures in experimental animals complicate the correlations. Recordings from unrestrained animals are difficult to analyze precisely for receptive fields and motion detection because of head and eye movement. For this reason, most animal experiments are made under severely abnormal conditions such as general anaesthesia, immobilization, fixed pupils and exclusion of eye movement. In spite of these limitations, however, several correlates of neuronal activity in cats with subjective visual phenomena in man have been established (see Fig. 1 and Table 2, p. 28).

I wish to discuss below some *correlations of subjective and objective visual physiology*, with typical examples. But first I shall refer briefly to the historical development of the field.

Subjective and Objective Visual Physiology

Early Work on Visual Perception. During the first phase of research in the nineteenth century visual functions were successfully investigated by observing subjective phenomena in human vision under controlled photic stimulation of the eye. Thus, the visual laws of simultaneous and successive contrast, of binocular vision, and most of the principles of colour vision were discovered by analysing visual perception in human subjects by means of psychophysical methods.

In the nineteenth century, the foundations of *subjective visual physiology* were laid by the original contributions of YOUNG, PLATEAU, PURKINJE, BREWSTER, CHEVREUL, WHEATSTONE, PANUM, and JOHANNES MÜLLER. Systematic research developed rapidly around 1860 when physiologists, physicists and philosophers became interested in vision. The great pioneers HELMHOLTZ [207, 208], MAXWELL [318], HERING [211, 214], MACH [304], FECHNER [130], BRÜCKE [68], HOLMGREN [224], AUBERT [11], WUNDT [460], and VON KRIES [281] published their important discoveries between 1850 and 1885. Most fundamental studies used subjective observations of the effects of exact visual stimulation, or developed physical tools such as the ophthalmoscope [207]. Electrical records were restricted to the electroretinogram in animals. A detailed survey of this development, with portraits of the leading scientists, can be found in POLYAK'S book [362]. The scientific battle between the proponents of the physiological and psychological interpretations of visual phenomena and between opposed schools of functional concepts of vision, such as *Empirismus* and *Nativismus*, were fought rather bitterly by HELMHOLTZ, VON KRIES and their school on the one hand and HERING, MACH and their followers on the other. Such opposing theories are not just a historical curiosity since they stimulated further research. These theses and antitheses can be synthesized today with more understanding now that objective visual research has furnished arguments for both concepts, and shown their validity in restricted fields. General aspects and philosophical parallels are discussed on p. 116.

Objective Recordings from the Visual System. As already mentioned, there was a long gap between the discovery of the electroretinogram in 1865 and the recordings from the optic nerve [4] in 1927 and the *Limulus* eye in 1932 which initiated neuronal research. In 1931–33, too, cerebral evoked potentials from the mammalian visual cortex were discovered by BARTLEY and BISHOP [22, 23, 43], FISCHER, KORNMÜLLER, and TÖNNIES [141, 279, 280]. This technique was used to map the *visual projections* of the retina to the cat's cortex by TALBOT and MARSHALL [426] in 1941. The essential advances came later from single neurone recordings in cats and monkeys. Therefore, this chapter treats mainly the neuronal correlates of vision in mammals.

Neuronal Physiology of the Retina. Fifteen years of research on neuronal responses to light in the *retina* preceded the first recordings from the visual brain. HARTLINE initiated the neuronal physiology of the *visual system of vertebrates* by his recordings from optic nerve fibres of the frog in 1935–40. His basic analysis of visual functions beyond the receptors included three types of response to light-on and -off [192] and their receptive fields [193] with the interaction of neighbouring spots of the retina [194]. WILSKA [457], GRANIT and SVAETICHIN [175] in 1939/41 made the first recordings from *mammalian* retinal ganglion cells using the rat, cat and several other species [170]. GRANIT has summarized this early phase of retinal microphysiology [171, 172]. Following HARTLINE's discovery of lateral inhibition in *Limulus* (1949 [195]), KUFFLER in 1952/53 reported the antagonistic *centre–surround organization of the receptive field* of cat retinal ganglion cells [285, 286]. The earlier triple distinction of on, off, and on-off responses by HARTLINE and GRANIT was modified to a *dual* concept of two neuronal types, 1. on-centre neurones, activated by illuminance increment in the field centre and inhibited by light shed on the surround, and 2. off-centre neurones, activated by illuminance decrement in the centre and by light in the surround, and inhibited by illuminance increment in the field centre. KUFFLER was at first reluctant to accept that his two neuronal types had a clear perceptual correlate [286] but later recognized with BARLOW their correlation with "whiteness" and "blackness" [19]. Thus *two neuronal subsystems*, one signalling "brighter" (B) and the other signalling "darker" (D) in their field centre, originate in the mammalian retina and carry visual messages of *achromatic brightness information* to the visual cortex (see p. 20). In addition, in those species which possess a cone retina, *colour information* is transmitted from the three kinds of spectrally sensitive cones to the brain (see Chap. 3 by DE VALOIS and p. 104).

Neuronal Mechanisms in the Visual Brain. In 1952 neuronal investigations of the *visual cortex* were initiated by recordings from the primary visual area 17 of the cat [254]. In contrast to retinal ganglion cells, about half the neuronal population in the visual cortex, called A-neurones, did not respond to changes in diffuse light [254, 255]. In these early studies, the role of *inhibition* for central processing was stressed by JUNG and BAUMGARTNER in 1955 [255]. Reciprocal inhibition and successive postexcitatory inhibition were related to simultaneous and successive contrast vision and after-images (see Fig. 1). The geniculate relay was investigated from 1953 onwards, mainly by P. BISHOP and his group [44]. Detailed studies of lateral geniculate neurones began in 1956 [123, 85b], and later receptive fields similar to KUFFLER'S dual antagonistic organization of retinal neurones were found in the lateral geniculate. From all these experiments it became apparent that the cortical neuronal systems transform the elementary qualities of brightness and darkness into special *information on form and motion*, and that the visual cortex also receives specific, non-specific and multi-sensory convergence [90, 181, 258]. The main discoveries concerning visual cortex functions made by HUBEL and WIESEL during 1959–1965, were selective orientation of receptive fields [231], columnar organization [233] and correlation with areal architectonics [234].

In concentrating their experiments from 1954—1960 mainly on non-specific afferents [90], multisensory convergence [181, 258, 276], optic nerve stimuli and bright–dark contrast [26], the Freiburg group, comprising BAUMGARTNER, CREUTZFELDT, GRÜSSER and KORNHUBER, overlooked the characteristic oriented receptive fields of the neurones in the primary visual cortex. In 1959—1962 HUBEL and WIESEL [231, 233] discovered the orientation selectivity of receptive fields and their columnar organization in the cat's striate cortex (area 17). Explorations of the neighbouring visual areas beyond area 17 followed: in 1961—1963 KORNHUBER and coworkers found visual, vestibular and acoustic convergence [258, 276] in perivisual areas and the surrounding cortex. HUBEL and WIESEL in 1965 made a detailed study of the paravisual areas 18 and 19 and described the hypercomplex and higher-order fields of their neurones [234]: In 1969 they found similar receptive fields in the perivisual area of the suprasylvian gyrus [237]. In continuation of this work and in relation to anatomical studies, the callosal connections and the vertical meridian in the cat were investigated by several authors and related to the topical projections of the retina, which form a mirror image of the half-fields in each adjacent visual area. During the past five years complex visual functions, such as binocular stereopsis (see Chap. 4), the influence of eye movements [461] and visual learning [51, 217] have been studied in the primary visual cortex of cats and monkeys. The main historical points in the investigations of the neuronal physiology of the visual areas are summarized in Chapter 21 of Part B, p. 327.

Neuronal investigations of the visual areas have been extended during the last 5 years also to *primates* [235]. Monkeys possess highly developed *visual areas in the temporal cortex*. GROSS summarizes our present knowledge of the monkey's inferotemporal area in Chapter 23 of Part B. The *optic tectum* was also studied in monkeys, as described in Chapter 14. Although there are many papers on visual commissures (see Chap. 25), relatively little is known about cortico-cortical interaction and the descending pathways to the brain stem. A large gap also remains between the primary visual cortex and the highest inferotemporal area. Several levels of neuronal transformation which lie in between are just beginning to be investigated in the monkey's brain [154, 155, 367, 465, 466].

Primacy of Perception in Visual Research. Not only in the historical development but also in modern visual studies it has been shown that perceptual phenomena provide a good conceptual frame for experiments. *Perception remains the most solid basis for sensory physiology*. Hence, experimenters working in all fields of vision research should not neglect the results and concepts of the visual physiologists of the last century, who derived their theories from perceptual observations. The most successful modern experiments have been those planned on the basis of classical visual physiology, a knowledge of which helps to distinguish the essential from the trivial.

Sceptics may object that to seek complete parallelism between visual mechanisms and visual perception is an Utopian postulate. Clearly, natural science can never embrace the totality of causal connections as KANT has shown [263]. To seek perfect correlation between visual perception and the complex structure and function of specialized neuronal systems composed of millions of visual cells would be a hopeless task. And yet, there are striking parallels between what modern electrophysiological methods have revealed about neuronal processes in the visual system, and visual phenomena (see Table 2). Although we cannot expect neurophysiology to unravel the complete sequence of visual processing from over 100 million photoreceptors towards an even greater number of cerebral neurones, the perceptual end result in vision may be understood as a *final synthesizing act* of these neuronal transmissions, which appears in consciousness and can be measured by psychophysical methods. We should accept any correlates which can be demonstrated in exact experiments using neurophysiological and psychophysical methods. It would be unwise to reject such correlations a priori on the charge of philosophical dualism, simply because there is no scientific explanation for the transformation of sensory stimuli and their neuronal mechanisms into perception.

This introductory chapter will describe some of the neuronal mechanisms that underlie visual perception in its simplest forms. In addition it will portray a few *anschauliche Bilder*, i.e. images, including visual illusions, which lead to intuitive understanding of visual mechanisms and principles. Although the neuronal basis of conscious visual experience is still unknown, the investigation of these phenomena might repay a neurophysiological approach in future. The visual phenomena selected as examples are those of special significance for ***brightness vision*** and elementary patterns. Form and colour vision will be mentioned only briefly, since these are treated in the next two chapters. Some additional psychophysical aspects of motion detection are included to supplement GRÜSSER's Chapter 6.

A good correlation of neuronal mechanisms and vision naturally implies some reductionism. Even a complete knowledge of neuronal systems cannot explain the whole story of brightness vision. The following examples, however, may show that elementary neurophysiology applied to visual perception does not lead to an atomistic view, as is often believed. Neurophysiology also needs *system concepts* and these may be *derived from perceptual phenomena.* A knowledge of both these fields may preserve the sensory scientist from becoming too reductionistic. In ascending from single retinal cells to neuronal systems carrying information, the gap separating neurophysiology and perception still makes itself felt, but a synthesis of the two may come into view. This should be demonstrated in the first part on the dual afferent systems of brightness vision.

The *limitations of neurophysiology* for explaining perception are obviously that it is restricted to partial aspects and special questions that can be answered through physiological methods. This causes difficulties when we come to investigate higher perceptual functions such as visual invariances, multisensory transpositions and Gestalt recognition, by means of experiments on the central visual structures.

Higher Visual Functions and Neurophysiology

Visual Learning and Recognition. A perceptual synthesis of higher order is the close coupling of present sensation and past experience in *visual recognition.* This, however, demands a greater knowledge of visual learning and memory mechanisms than we have at present. Visual learning and recognition are still outside the scope of neurophysiology, although some progress has been made in this direction by visual conditioning experiments combined with neuronal recordings, as reported in Chapter 21.

The main achievement of neuronal analysis over the last ten years has been to classify and establish the columnar order of receptive fields in the visual cortex (HUBEL and WIESEL [233]), but these findings from anaesthetized cats cannot yet furnish an adequate correlate of form perception, as discussed in Chapter 2. It seems more rewarding, although more difficult too, to record visual neuronal activity from non-anaesthetized monkeys able to direct their eyes to certain visual goals, as in WURTZ's experiments [461]. Neuronal recordings from the visual cortex in man have been attempted in neurosurgical patients by MARG et al. [313] (see Chap. 22) but this procedure has many limitations. Further neuronal recordings of the reactions of trained monkeys to pattern stimuli following visual multisensory conditioning may eventually teach us more about the neuronal physiology of the higher visual systems and of visuomotor coordination. When these experiments are combined with behavioural observations and specific deprivation studies, as begun by BLAKEMORE, HIRSCH and coworkers [51, 217] (see Chap. 21), one may be able to attack the major unsolved problems of visual learning and memory.

Visual learning is just beginning to be treated on the neuronal level. We do not yet know how and where visually coded information acquires meaning, how it is stored and retrieved, nor how neuronal receptive fields may be influenced, modified or stabilized by visual experience. Early attempts to attack the neuronal basis of visual learning are described in Chapter 2 on *form vision* and in Chapter 9 on *visual deprivation*. Several observations on visual conditioning and learning in relation to brain structures and neuronal organization are also mentioned on p. 125 and in Part B, Chapters 21, 23, 24 and 25.

The second part of this volume treats *comparative data*, insofar as they are important in visual physiology, from the lower animals to primates. We had planned to include a more extensive treatment of the cerebral visual structures of primates by HASSLER and WAGNER, because the visual systems of monkeys are very similar to those of humans, but this chapter was unfortunately not completed in time.

The third and largest section on the *function and structure of visual brain centres* summarizes an immense number of physiological and anatomical investigations. Its size made it necessary to publish this part separately as Part B of sub-volume VII/3. The essential structural-functional relations are not yet known in detail, but both the anatomists and the physiologists have striven for maximum interrelatedness. The discrepancy between our detailed knowledge of cortical structure and the relatively coarse localization of micro-electrodes in the visual structures does not yet allow a satisfactory synthesis between the two fields The discovery by HUBEL and WIESEL of columnar organization [233] was a first step towards "anatomical-physiological" correlation and their recent findings concerning horizontal layers for monocular representation [236] mark a further advance, supported by histological evidence. Synaptologic correlates of the columns, however, are still imperfectly known. SZENTÁGOTHAI has summarized his own results together with present knowledge in Chapter 20 on visual cortex synaptology.

II. Terminology and Methods

The Terminology of Vision and the Effects of Light Stimulation

General Terms. The term "*perception*" is used throughout this chapter in a general sense to include sensations, thus discarding the distinction between sensation and perception (Empfindung und Wahrnehmung) traditional in German sensory physiology since HELMHOLTZ [208] which has influenced American psychologists such as BORING [53]. Perception usually denotes *subjectively experienced* sensory information, but some authors also use it to mean objective behavioural responses to sensory stimuli. "*Motion detection*" is used in a general way to denote any suprathreshold response to moving stimuli, including sensory and motor reactions. *Perception of movement and velocity* refers to conscious experience.

Once the dichotomy of sensation and perception is eliminated, it is no longer necessary to produce artificial definitions and extensive discussions like those VON KRIES had to use in his General Sensory Physiology [283]. When he introduced temporal order and motion perception, he had to explain at length the transitions between sensations and perceptions, and deal with nativism and empiricism. The ancient philosophical conception of KANT [262] that, while knowledge cannot transcend experience, it also depends upon a priori syntheses which are not empirical, may still be used in a modified form. Visual physiology can perhaps adapt this concept to cover innate and acquired visual mechanisms (see p. 122).

Practical Terminology. The word "vision" is used here in its most general meaning to include both biological and psychological processing of visual information, not just the subjective end result of visual perception. This also applies to some extent to terms like "brightness", "darkness" and "colour". The term "luminance" is reserved for the actual photic stimulus, not its physiological or perceptual effects.

For brevity, the words "bright", "dark" or "red" may be used in psychophysics and physiology instead of repeating the descriptions given in a physical definition of the method of stimulation.

To counter the objections of terminological purists, one may argue that terminologies that are too rigid lead to complicated paraphrases, which can interfere with understanding. A case in point arose when the behaviourists used lengthy objective descriptions in order to avoid introspective psychological terms which, however, would have made the situation clear at once. We prefer to use *simple and practical terms* which facilitate comprehension, and to avoid sophisticated terminologies and clumsy expressions. Thus, a cat — or a monkey or a human being — does not see and process luminance gradients with incremental or decremental components but in terms of *brightness and darkness*, and its response to their significant *patterns* is expressed in behaviour and perception. In this chapter we assume that a cat, a monkey or a human being may be accorded the privilege of responding to controlled laboratory stimuli in the language in which we ordinarily convey our visual experiences in a comparable environmental situation.

The terms "*bright*" and "*brightness*", although normally reserved for subjective sensations, are thus used in a broader sense to denote subjective and objective information in vision and in visual mechanisms resulting from achromatic light increments (without reference to the spectral composition of the stimulus) or from light-dark contrast in time and space which is experienced or signalled by behavioural responses. The terms "*dark*" and "*darkness*" are used for the reverse information produced by light decrements in time and space (diminished brightness). The words "black" or "blackness" (as opposed to "white" or "whiteness") refer to darkness markedly enhanced by contrasting achromatic light areas in the immediate vicinity and to a quasi-colour quality in perception.

The term "*receptive field*" is used here for the retinal projection of the area within which photic stimuli elicit neural responses that can be recorded objectively from neurones at various levels of the visual systems. A "*perceptive field*" is the subjective equivalent of a receptive field, estimated indirectly in human vision and measured psychophysically (see p. 44).

Luminosity, Luminance and Brightness. The different terminologies employed by physicists and physiologists to describe and measure light intensity and brightness are not strictly comparable and have not been standardized in this Handbook to comply with accepted terminology. As in earlier handbooks [169], we refer to *stimulus effects* rather than to stimuli. The editors and authors have encouraged the tendency to use different terms for physical stimuli and physiological effects, for this helps to distinguish between them.

1. Physical *light intensity* as a photoemission, is called *luminosity*, and can be measured by photometry.

2. The corresponding *light stimulus* is called *luminance (Leuchtdichte)* and, when it is projected onto photoreceptor organs having various spectral components, acts by *retinal illuminance* after some dioptric alterations.

3. The *physiological and psychological effect* of luminance increment is called *brightness (Helligkeit);* brightness is influenced by contrast, and its input to the brain produces perceptual and behavioural responses.

When objective and subjective responses are to be compared, both qualitative and quantitative methods may be employed. In quantitative experiments neuronal discharges are usually measured as *impulse rates* (frequency modulation of maintained discharge) and sensations are generally estimated by magnitude *scaling methods* such as those of STEVENS [413] (see p. 13 and 92).

Physical Luminance versus Physiological Brightness. Even when alterations by the dioptric apparatus of the eye are taken into account, there remains an *essential difference* between *stimulus luminance* and the *sensation of brightness* because the latter is modified by *contrast coding* for visual patterns. This is best demonstrated by HERING's example: Coal in sunlight has a higher luminance than white paper in a dimly lit room, yet the *coal always appears "black" and the paper "white", both in the strongest sunlight and in twilight* [213]. The assumption of MACH and HERING, that contrast brightness is caused by physiological interaction in the retina, has been verified by neuronal research (see p. 25). Objective correlates of subjective *visual contrast stimulation* have been recorded from the firing rates of visual neurones, and parallels were found with subjective perception. Experiments on invertebrates and mammals demonstrated similar principles of contrast vision by lateral inhibition. Counts of spike discharges in the *Limulus* eye by HARTLINE and RATLIFF and their quantitative analysis [195a] as well as the studies by KUFFLER and BAUMGARTNER of neuronal receptive field organization in the cat's retina [286] and central visual system [26] have elucidated the *neuronal basis of contrast and contour vision* (for a summary, see RATLIFF [371]). The peripheral mechanisms are described in the preceding volume on the retina, and the central mechanisms in Chapters 2, 14, 18, 21 and 23 of this volume. Some parallels between perception and reciprocal neuronal systems are discussed below with examples (p. 28 and 47).

Retinal Sensitivity. Briefly, the retina responds most readily to *luminance changes in time and space* and the optic nerves transmit these messages to the brain as transient discharges for transformation into perception and behaviour. Visual neurones do not signal objective levels of luminosity as such, but register *brightness distribution* in the visual field. They possess very fine discrimination for temporal and spatial alterations of luminance, which affect the eye in a short sequence and in neighbouring areas, whereas rapid adaptation occurs to overall changes in luminance. This explains why *eye movements* are essential for vision. Besides the transient responses of visual neurones, which are superimposed upon ongoing discharges, there are sustained responses signalling both light distribution and *diffuse background luminance.* The latter may control pupil size.

The afferent mechanisms for the processing of absolute and diffuse luminance gradations and their relations to maintained discharges are not yet clear (as discussed in Chap. 8 by LEVICK). A few specialized neurones with little adaptation have been found in the optic nerve of cats: on-centre neurones of the B system increase their discharge with increasing luminance while others in the D system decrease it. Some of the light-inhibited D neurones are completely silenced for many minutes by illumination of long duration [250]. These neurones resume tonic activity in total darkness after showing phasic responses to light-off during light adaptation [250]. The tonic dark discharges may be a response to removal of light inhibition which also causes disappearance of the antagonistic field organization (see p. 34). Sustained discharges are more common in certain neurones with slower conducting axons, the so-called X-neurones (see p. 16). Most of these, however, have small receptive fields and may signal patterned light distribution better than diffuse background luminance.

Methods and their Limits

Four Methods of Visual Research. All scientific research into sensory systems, including vision, is based on four methods: 1. *Subjective evaluation of perception;* 2. *Behavioural observations* and recording of reactions to sensory stimuli; 3. *Physiological recordings of sensory mechanisms;* 4. *Formal constructs or theoretical models* designed to link the data from perceptual, behavioural and physiological research. *The first method is restricted to man,* who has sensory experience; but it is not limited to qualitative observations; it can comprise rigorous experimental and quantitative procedures. The *second* and *third methods* are experimental procedures applicable both to *animals and man* so long as they are not harmful to human beings. Physiological methods in the widest sense include biochemical and spectrographic methods (e.g. for investigating photopigments) and are not just electrical and mechanical recordings. The *fourth method* is theoretical and has made progress by way of information theory and cybernetics. However, it is not autonomous and for useful mathematical and physical applications depends more upon biological and psychological data obtained by methods one, two and three than do these three approaches depend on the fourth.

The four disciplines from which these research methods derive, psychology, behaviour research, physiology and cybernetics, are complementary in vision research. Physiology can use a broad physicochemical approach which is adequate for the purpose. Neuronal recordings represent a restricted methodical approach and need to be supplemented by other methods. At the receptor level, spectral reflectance may be used even for man, as in RUSHTON's experiments [383, 385]. To understand the neurophysiology of vision it is useful to know about similar experiments in other fields. Their results can be compared and their methods combined for the successful solution of problems. Neuronal mechanisms become more significant and gain in scientific value when they are matched to psychophysical and behavioural results. Although this volume deals with physiological mechanisms, these will be related as far as possible to perceptual experience. Hence, visual perception also maintains its place in visual neurophysiology.

All the examples of perception quoted here belong to BRINDLEY's "class B" of psychophysical phenomena [62], i.e. they are qualitative, and quantification is effected mainly by psychophysical judgements of intensity, such as STEVENS' scaling estimations. With STEVENS [412, 413], I have some doubt whether it is necessary to make a sharp distinction between these and "class A" measurements, which compare two sensations. As visual physiologists, we are willing to accept any correlations which can be demonstrated between neurophysiology and psychophysics, with due regard to differences in species and experimental conditions.

Limitations of Combining Neurophysiology, Psychophysics and Cybernetics. Each method applied in brain research has been primarily designed to answer a particular range of special questions pertinent to the problem in hand. The combined application of different experimental approaches creates fresh difficulties and needs careful planning and critical examination without undue optimism, since neurophysiological results or cybernetic models cannot always be applied directly to psychological phenomena. Neuronal physiology applied to visual systems has revealed some of the mechanisms of sensory transformation, but the actual cognitive processes that construct the inner representation of the outer world remain obscure. Correlations between perceptive experience and behaviour encounter similar basic difficulties, but they are less serious. In due course, neurochemistry may supply further keys for the investigation of cerebral mechanisms, but at present the neurochemical and neuropharmacological approaches have little to offer sensory physiology, though there has been some success in the investigation of sleep and wakefulness. Black-box models, in cybernetic terms, are worthless unless they direct attention towards correlates in neuroanatomical and neurophysiological findings. CRAIK'S [88] postulate of an *internal model of the external world* which we carry in our heads is a good general statement, but it does not tell us how the brain makes these models. Until specific experimental data become available on the neuronal mechanisms of sensory systems, theoretical models of neuronal circuits cannot contribute much to the neurophysiology of the senses (see p. 110). Nor can sensory physiology be advanced only by "stimulus-response" investigations of the type used in classical conditioning experiments. These methods, although experimentally exact, treat the subject as a living black box. The physiological functions and neuronal mechanisms within this black box can be elucidated only when one penetrates into the brain. Even where psychophysical results yield quantitative measurements, the quantification is not sufficient to enable quantitative neuronal models to be constructed.

Limits of Optic Stimulus Conditions. It should be appreciated that many results of visual experiments are restricted to a particular set of stimulus conditions. Hence generalizations about vision and visual mechanisms should be made with caution if based on the results of a single method. It is not easy to compare the results of various research groups using different methods of light stimulation under natural or laboratory conditions, and the applications of reductionistic experiments to natural vision are limited.

Three methods of light stimulation are used for neuronal recordings: 1. rectangular on-and-off stimulation; 2. periodic stimulation (sinusoidal, flicker or spatial frequencies); 3. moving stimuli. All these are highly artificial. The *natural stimulus* of sequential foveal fixations interrupted by saccadic shifts is usually deliberately excluded in animal experiments. Exceptions are recent experiments by WURTZ [461] which reproduce these conditions in monkeys (see Chaps. 5, 6 and 21).

Ad 1: Since HARTLINE'S pioneering work [192] the classical stimulation method of switching a *light on and off* continues to be used in most laboratories, although this *rectangular* stimulus is very different from optic conditions in natural surroundings, apart from the effects of lid blinking.

Ad 2: Sinusoidal and flicker stimulation, i.e. periodic changes of luminance, as used by many authors, does not correspond to normal vision either. Any regular alternation of these rhythmic stimuli introduces complications of temporal conditioning, avoidable by luminance changes at random intervals. These frequency-dependent stimuli are discussed in Chapter 7 by VAN DE GRIND and coworkers. Sinusoidal stimulation of large areas of the visual field may show some similarity to normal vision for background and adaptation effects; it modulates luminance over a certain medium range, roughly comparable to daylight and twilight illumination. Of course, the natural increase or decrease of local retinal illuminance during eye movements is entirely different from on-off flicker, sinusoidal light or spatial frequency-patterned stimuli.

Ad 3: The third method of stimulation, more comparable but not identical with natural vision, is by the *movement of patterns in certain directions and at certain speeds.* This induces moving retinal images when the eyes are held stationary, i.e. stimulation by visual contrasts gliding over neighbouring receptive fields of many visual neurones. In most animal experiments this method introduces another artefact, namely immobilized eyes. Under natural conditions, however, most moving stimuli provoke ocular movements, i.e. optokinetic nystagmus or slow *eye tracking,* initiated and terminated by rapid saccadic eye movements. Hence, eye tracking compensates for moving retinal images and the *saccades control the onset, duration and location of foveal fixation of the visual targets.* Saccades must be preprogrammed whereas slow ocular tracking must be under continuous retinal control (see p. 85).

The more sophisticated methods using *temporal and spatial frequency transfer* are treated in Chapters 2 and 7 (see also p. 69 and Fig. 12). For pattern stimulation and evoked potentials, see Chapter 28 in Part B.

FECHNER's Function, STEVENS' Function and Visual Neuronal Activity. Both logarithmic (FECHNER, 1860 [130]) and power (STEVENS, 1961 [412]) functions have been found to fit the characteristics of different sensory mechanisms under various experimental conditions. MACKAY [306] in 1963 argued that it was logically inadmissible to infer the characteristics of sensory transducers from the subjective psychophysical functions estimated by the method of STEVENS. He pointed out that both FECHNER'S logarithmic law and STEVENS' power law were compatible with a variety of acceptable physiological hypotheses.

STEVENS' method estimates the final perceptive result of complex information processing in the brain, and one would hardly expect all the neuronal stages of this process, which carries signals from the receptors to the higher centers, to obey the same power law. The exponent may well be different at different levels, and it is of neurophysiological interest to investigate the neuronal equivalents at all levels.

In very few cases is it possible to relate STEVENS' power law to the neuronal recordings of the visual system. For motion detection, STRASCHILL and TAGHAVY recorded motion-sensitive neurones in the cat's optic tectum and related their findings to STEVENS' power-functions. They found an exponent of 0.67 for angular velocity responses of 10 neurones [417]. This is lower than DICHGANS' exponent of 1 for man [105], but considerably higher than the exponent of 0.3 which has been found for luminosity variations in man [413] and for recordings of neuronal discharges as a function of light intensity in the visual pathways of the cat. CREUTZFELDT and SAKMANN maintain that the logarithmic and power functions both give an almost equally good fit with neuronal responses following an increase in light intensity [93]. FISCHER and FREUND [137], in a recent study of retinal receptive fields, demonstrated that STEVENS' power function and FECHNER's logarithmic function may be derived from the same physiological mechanisms as RICCO's law of spatial summation [375]. HERING doubted the general validity of FECHNER's law already in 1876 [210].

Visual Illusions and Abnormalities as Research Tools

Visual Illusions Reveal Visual Laws. One hundred and fifty years ago PURKINJE had already remarked that *optical illusions contain visual truth.* He used his illusions mainly for morphological demonstrations, such as visualizing the vascular tree of the human retina by oblique light [369]. In 1838 when WHEATSTONE elicited binocular stereoscopic illusions by laterally displaced retinal images [455], it became apparent that visual illusions can throw light on the *physiological mechanisms* of vision. For the last 100 years psychological interpretations have competed with mechanistic explanations.

Although HELMHOLTZ in 1859/61 used various optical illusions to demonstrate functions of the visual system, he preferred psychological interpretations based on conscious and unconscious "judgements". Both MACH [304] and HERING [213] maintained that *physiological interactions* were responsible for most contrast illusions. HERING, however, did not deny the role of learning and memory; in colour vision, for example, he recognized the visual effects of memory by his „Gedächtnisfarben" [213, 214].

Most of the work done on visual illusions from 1870 to 1920 relied primarily upon psychological rather than physiological explanations. Adherents of *Gestaltspsychologie*, however, such as WERTHEIMER [451], KÖHLER [272], KOFFKA [269, 270] and RUBIN [382], were open to physiology. For illusions of figure-ground interaction, they postulated physiological equivalents in the visual system. GREGORY, in a survey of various illusions [177], again favours "unconscious interpretation" in his size-constancy hypothesis for the Müller-Lyer illusion and others. An interesting visuomotor correlation of this illusion was demonstrated by YARBUS [463]: records of eye movement excursions paralleled the subjective length extimates of the Müller-Lyer figures. Neuronal correlates of visual illusions will be discussed for the Hermann grid and the Ehrenstein pattern (pp. 47 and 54).

TEUBER in 1960 also used visual illusions for demonstrating constancy regulation [431]. He regards certain illusions as *misapplied constancy effects*. These higher-order phenomena, however, cannot yet be correlated with neuronal mechanisms or other neurophysiological regulations. SCHOBER and RENTSCHLER have just published a pictorial survey of the main visual illusions [391].

Visual Illusions and Perceptive Fields. Visual illusions are efficient tools for visual research when used for specific questions concerning subjective correlates of neuronal mechanisms. The examples depicted in Figs. 5–10 are of special interest for the investigation of perceptive fields and for the interaction of the reciprocal neuronal systems B and D.

The *Hermann grid illusion* [215] was first used by BAUMGARTNER in 1960 to demonstrate correlates of receptive field size for foveal and extrafoveal vision in man and animals by estimating the diameter of perceptive fields in human vision [25]. Another brightness illusion induced by linear patterns (Fig. 9), first described by EHRENSTEIN in 1942 [116], may be used to suggest that neurones with oriented receptive fields in the visual cortex not only signal linear patterns but also contribute to brightness sensation. The *Craik-Cornsweet illusion* demonstrates other inductions of brightness from border cues (Fig. 10). The significance of these illusions for perception and neuronal physiology will be discussed in the following sections.

Investigations of Abnormalities in the Visual System. Besides the visual illusions produced when the normal visual system is overloaded, the study of the *defective visual systems* also represents a legitimate approach to visual research. Lesions of visual structures, whether produced by experimental ablation in animals or by disease in man, may answer the questions asked by physiologists, provided these questions are set in the right framework. Chapters 14, 23 and 25 describe results obtained with experimental lesions in the subcortical and cortical visual systems.

Vision research must use all available methods and should extend to *pathological syndromes and malformations.* One example is the Siamese cat with its abnormal visual projections, another is defective colour vision. In normal as well as in diseased visual systems one can compare physical stimuli and perceptual phenomena such as measured wavelengths and perceived colours. Correlations between them can be studied quantitatively, although only a few details are known of the steps between physical stimulus, physiological response and perception. When *abnormal* correlations occur, as in certain forms of colour blindness, the visual scientist looks for structural, photochemical and genetic differences in the abnormal systems. If he finds any, he may use them as tools supplied by visual pathology for the clarification of normal colour mechanisms. Besides colour vision anomalies, the various diseases known in ophthalmology and neurology to disturb the ordered structures of the visual system can be used in vision research as *experiments of nature* (see pp. 42 and 129).

III. General Concepts and Synthetic Views

System Concepts in the Visual System

Visual Theories and Subsystems. Among the multitude of theories which have been proposed to explain vision, only a few are closely related to neurophysiology. Others are more concerned with photochemistry or anatomy. The duplicity theory of rod and cone receptors for photopic and scotopic vision has little importance for central information processing since rod–cone convergence takes place below the ganglion cell level. The recent hypothesis that there are two visual systems, one localized mainly in the cortex and the other in the mid-brain, needs further confirmation from neuronal experiments. Our *subsystem concept of reciprocal neural mechanisms signalling brightness and darkness* is treated extensively because it allows a systematic order and synthesis of physiological, structural and perceptual results. KUFFLER's finding of two antagonistic receptive field organizations of retinal neurones [285, 286] offers the basis for further reciprocal interaction in the brain, and these neuronal mechanisms show close parallels with the visual phenomena of brightness contrast. Without this synthetic view it is difficult to understand the coding processes of achromatic vision. Similarities in the processing of opponent colour information and some links with HERING's theories are more briefly discussed along with recent neuronal recordings in the visual cortex (see p. 103 and Chap. 3).

Definition of Physiological Systems. Systems and subsystems can be defined as *functionally related physiological mechanisms interacting at different loci* and often having a hierarchic organization at several superimposed levels.

In contrast to localized "centres", such systems are not restricted to certain circumscribed regions but concern *different* parts of the body or the nervous system, with horizontal and vertical interrelations of various structures. This concept of physiological and biological systems can be applied to neural, endocrinal or other functions. It is not equivalent to mathematical system theory, which treats equations relating input to output for information transmission. In spite the many difficulties of applying such theories to biological processes, the system-theory approach may well contribute in the future to our knowledge of information transmission by groups of neurones. Physiological system concepts are — and system theories may become — useful guides for organizing visual research. Within the general frame of a "visual system" several *subsystems* may be distinguished.

Various Visual Subsystems and a Third "Accessory Visual System". Since the classical duplicity theory of the eye (photopic vision from cone receptors and scotopic vision from rod receptors) was developed in the nineteenth century, at least three other dual concepts have been proposed. To avoid confusion, they should be clearly distinguished. There is little relation between these concepts, although some parallels have been drawn between the duplicity theory and the hypothesis of two systems having cortical or tectal projections (see DOTY, Part B, Chap. 24, p. 498). For lower animal species see SMIRNOW [396].
The following dual concepts of visual subsystems are found in the literature:

a) *Rod–cone duplicity of photoreceptors* for scotopic and photopic vision, developed between 1866 and 1896 by M. SCHULTZE [391a], J. VON KRIES [281] and others (see Vols. VII/1 and VII/2).

b) *Two afferent neuronal subsystems, B and D,* related to brightness and contrast vision [249] following KUFFLER's description (1952/53) of the antagonistic organization of two classes of on- and off-centre neurones in the retina. Their perceptual correlates with brightness and darkness sensations are discussed at length in this chapter (p. 20—38).

c) *Two systems of retino-cortical and retino-tectal projections* proposed by SCHNEIDER [390], TREVARTHEN [434] and others in 1967—1968 on the basis of anatomical and behavioural observations. They were thought to represent "focal" and "ambient" vision [434] (see p. 99 and Chaps. 2 and 24).

d) *Two groups of X and Y neurones in the afferent visual pathways.* This is a recent dual concept, developed in 1966—1970 by ENROTH-CUGELL, ROBSON [122] and others. 2 or 3 [79] groups of *fast* and *slow* conduction in optic nerve fibres were related to large and small receptive fields of retinal ganglion cells, respectively, and give predominantly *transient* or *sustained* responses to light (see Chapt. 2, p. 163 and Chap. 21 p. 334 in Part B).

In addition to these dual classifications, we should also mention HASSLER's distinction of *four visual projections to the cortex* in primates, having different structures in day- and night-active monkeys [202], as this is a subdivision of the higher visual systems. Unfortunately, the chapter planned by HASSLER and WAGNER, which should have treated the functional anatomy of these systems and the architecture of the visual cortex, was not ready for printing and will probably appear later as a monograph. (see also [203]).

DOTY's observations on the preservation of pattern vision after removal of area 17 in cats and on pattern blindness (after the whole visual cortex 17—19 is lost) are discussed extensively in Chapter 24 with their possible relations to the two visual systems of cortical and tectal projections.

Apart from these dual and other classifications, and in uncertain relation to them, a *third visual system, the "accessory"*, has a somewhat ghostlike existence.

The function of the so-called *accessory visual system* which runs from the optic tract through the transpeduncular tract to the interpeduncular brain stem is unknown (see Part B, Chap. 15). Whether it should be classed as a third visual system is a matter of surmise, although it is evidently fairly independent of all the other visual structures. It appears less closely related to the retino-geniculo-cortical system than are the retino-tectal or retino-pretectal projections. It may be unrelated to perception. Also unclear is its relation to the nucleus interpeduncularis, which in all vertebrates, including man, receives a powerful and constant connection via Meynert's tractus habenulo-interpeduncularis from the nuclei near the pineal body. The non-committal statements in the literature about accessory-system functions are discussed in Chapter 15 by MARG and in Chapter 24 by DOTY. The strong connections of the neighbouring interpeduncular nuclei with the pineal system and habenular nuclei may indicate *links with visually influenced hormonal functions* which, phylogenetically, may have derived from the parietal eye (see DODT, Chap. 16). In lower vertebrates the accessory system may have functioned as a link between the parietal eye and the lateral eyes for light-induced endocrine regulation. All this is speculation. To what extent these structures have acquired different functions in higher mammals remains an open question.

Functional Principles of Reciprocal Neuronal Systems

The visual system is unique among the senses in having a powerful neuronal system which signals the disappearance of light, that is, the *"negative" quality of light decrement*. True, the eye has receptors only for light and not for dark, but the retina contains the *two reciprocal neuronal systems, B and D*, which signal *light increment (B, on-centre neurones)* and *light decrement (D, off-centre neurones)* respectively, by a spike discharge. Their responses to light express a reciprocal relationship and this is different from that of the receptor cells, which always respond to light with the same polarization. On- and off-centre neurones have nearly the same distribution over the retina, and their receptive fields overlap, so that every stimulus must affect both systems. It seems legitimate to ask: "What is the darkness system for? What is the advantage of having two reciprocally interacting channels in the visual afference?"

Clearly, these are teleological questions which cannot be answered through physiology. However, they may well help us to *understand* some functions and mechanisms of the visual systems and their integration. Indirectly, this will facilitate the planing of experiments to reveal causal connections. In this sense, GRANIT [174] spoke "in defence of teleology" as did SHERRINGTON [393] and ECCLES [114] earlier.

What is the Advantage of Two Reciprocal Systems? A single system of light-excited, on-centre neurones would signal the temporal and spatial derivatives of light distribution rather well by means of excitatory transients, and it would have a bias towards light increments in a "positive" coding. "Negative" values arising from light decrements would only be transferred by *inhibition* in the presence of a considerable maintained discharge which, however, is rather low, even under high continuous illumination, when compared to the discharge peaks. This asymmetry is eliminated by having two opposed systems in a reciprocal relationship: the B system excited by light increment, and the D system excited by light decrement. Thus D excitation signalling the information "darker" is the mirror image of the B excitation signalling "brighter". The light-inhibited off-centre neurones are activated by a luminance decrease in their receptive field centre when the on-centre neurones of the same location are inhibited.

Both systems can indicate slight deviations from the mean luminance level more clearly when a discharge increase in one system inhibits activity in the other and, conversely, inhibition in one system provokes disinhibition of the other at the next level (see Figs. 1 and 4). Thus the ideal dual line for transmitting information concerning "brightness" and "darkness" in both directions would be strengthened by the *reciprocal interaction at several levels of two neuronal channels*. Such double-line interaction has, in fact, developed during the phylogenetic evolution of the vertebrate visual system.

Transgression Beyond the Zero Line by the Reciprocal D Subsystem. The spontaneous discharges of visual neurones allow modulations in both directions, yet their limits are not symmetrical. Both light decrement stimulation and neuronal inhibition are limited at the *zero level* by the complete absence of light or the cessation of impulse firing. The lower margin of decrease susceptible to inhibition by ongoing discharges is much narrower than the upper margin for excitation and discharge increase. An antagonistic field organization that induces visual contrast may sharpen bright–dark discrimination but is also limited by the *zero line of inhibition*. The strongest inhibition in the brightness-signalling B system indicates "no light"; it cannot, however, give the positive information "darker" or "*black*".

The "trick" that enables the visual system to pass below the zero limit of B inhibition is the "invention" of the reciprocal D system with its off-centre neurones, whose discharges carry specific information about darkness or, when contrasted with white, blackness. Although contrast mechanisms are present also in other sensory modalities (for examples, see VON BÉKÉSY [33]), the visual system differs from other sense modalities in having a subsystem which specifically signals the *decrement or absence* of the adequate stimulus "light".

The two subsystems may appear more asymmetric in neurophysiology than in perception when *excitation in one system* is compared with *inhibition in the other system*, irrespective of the kind of stimulation. Subjective brightness does not always correspond to light excitation in B or light inhibition in D, and the *balance of discharges of both systems* has to be considered (see Fig. 1). As discussed on p. 32, a *black* sensation appears after decrement or absence of light only in the presence of contrasting light in the *surround* of the dark area. The intensity scale for subjective blackness elicited by a dark area contrasting with a brighter surround of various luminances is essentially different from a luminance-dependent intensity scale for subjective brightness without contrast.

For the B and D neuronal systems the blackness scale of D depends indirectly on the gradations of luminance in the *surround*, whereas the whiteness scale of B depends directly upon gradations of luminance in the *centre*. Whiteness can be modified by a wide range of luminance increments, whereas luminance decrements in the centre cannot continue below zero. For neuronal inhibition, a similar *asymmetry* applies, since there is complete cessation of spikes from an ongoing discharge (often between 10–40 spikes per second) as against excitation on a scale of increasing discharge rates ranging from about 10 to 800 per second. The *presence of the two reciprocal subsystems, B and D, compensates for this asymmetry*. The visual centres are thus informed about blackness intensities *beyond* the zero level of "no light" which are induced by the contrast light–surround, i.e. blackening without any objective physical decrease in areal luminance.

Significance of Reciprocal Systems in Vision. The dual neuronal coding in the two reciprocal subsystems has several advantages for vision. They may be summarized as points scored for neuronal coding and for perception:

A. *Neurophysiological points*

1. Specific neuronal channels exist for coding light decrements and contrasting stimuli by lateral activation in the D system;

2. Double extension of information is achieved below the zero line of inhibition in the reciprocal neuronal system;

3. Rapid neuronal adaptation to overall changes of luminance in both directions is ensured;

4. Contrast transformation on edges and surfaces is controlled at successive levels for contoured forms of opposite bright–dark borders;

5. Cancellation of residual information is achieved in the sequence of visual stimuli.

B. *Perceptual points*

6. Black acquires the specific nature of colour (instead of brightness diminution);

7. A wider scale of white and black is attained, with finer gradients of shading between the extremes (psychophysical parallel to point 2);

8. Contour vision is selective for patterns of bright and dark borders at different background illumination;

9. Border and areal contrast are balanced in the visual image;

10. Residual images of an earlier fixation are cancelled following eye saccades.

The last two points (4, 5 and 9, 10) in the groups A and B were derived from BARLOW's critical remarks about *recoding transformations* of visual messages at successive stages and are to some extent at variance with his reservations about relating neuronal discharges to visual qualities [15]. The scepticism expressed by BARLOW [16] about the effectiveness of lateral inhibition occurring at successive relays (induction of spread and multiplication of fringes) applies only to single-system processing. If, however, contrast mechanisms are processed by two reciprocal systems which control each other in a regulated balance, stabilized at several levels of the visual system, the fringe corruption of repeated lateral inhibition or activation can be avoided. This reciprocally controlled sequential processing of contrast information, resulting in feature extraction of contours and areal filling-in processes [159], is further facilitated by the reciprocal interaction of two parallel visual subsystems in a hierarchic order of successive stages.

The hypothesis of *reciprocal cancellation,* first put forward by BARLOW for the D system at the 1960 Freiburg Symposium [15] can be extended and related to eye-movement cancelling [253]. During a sequence of eye movements, not only do the off-discharges cancel the visual after-effect of the preceding on-discharge and vice versa, but other residuals of the preceding fixation are also wiped out. This cancelling process appears to be related to *saccades* and will be discussed further in the section on eye movements (p. 82). Saccades and blinks may wipe out the residual effects of previous patterns to obtain an integrated perception of the visual world from a sequence of images and different foveal fixations [253].

A consideration of these points will confer a better understanding of how these reciprocal systems achieve the astonishing performances of vertebrate vision and why approximately equal numbers of "negative" lines (signalling light decrements) have developed alongside "positive" lines (signalling light increments) to carry information from the eye to the brain.

It is true that invertebrate animals such as scallops have also developed "*shadow receptors*" which signal darkness, as shown by HARTLINE [191a], but these are not homologous to the

D system of vertebrates. They project to relatively simple information lines, specifically for certain protective reflexes and do not display the intricate interaction of on-centre and off-centre systems found in vertebrates. In other invertebrates, such as *Limulus*, similar off-neurones which signal luminance decrements appear, not in the eye, but much more centrally beyond the first visual ganglion. In the *Limulus* eye, bright–dark contrast is signalled first only by lateral inhibition of light-activated fibres, which may be compared to an isolated B system of vertebrates but lacking interaction with the reciprocal D system. Conversely, in the vertebrate eye (which in evolution and ontogeny corresponds to a peripheral part of the brain), the dual coding lines begin immediately behind the receptor cells at the bipolar level and continue their reciprocal interaction at several stages up to the visual cortex (Fig. 3).

Similar principles, but with different mechanisms of reciprocal information, are found in the *temperature* sensations of warm and cold and in the spinal *motor* systems of agonist and antagonist muscles. To these SHERRINGTON [393] applied his concept of reciprocal innervation after HERING's theories of opponent colours [211] had indicated the principle.

By 1874 HERING had clearly recognized that in vision there is a specific information quality *black* as an opposite "colour" to white, and that this may be coded by reciprocal mechanisms comparable to opponent spectral colours [211]. HARTLINE in 1938 provided the objective neuronal basis for such black signals by the discovery of light-off responses in special fibres of the frog's retina [192]. This made it clear that specific information lines for black, in contrast to white, originate already in the eye behind the receptors and that the coding of dark contrast is not dependent only on cerebral processing. For contrast vision, two reciprocal systems are superior to a simple neuronal system which signals only one visual quality, even when the latter is composed of many parallel lines for different spatial locations.

In conclusion, the reciprocal-system organization has clear advantages for information transmission and perception because it sends signals of opposing visual qualities via specific channels from the eye to the brain and hence *doubles* the sensory range for contrast vision.

IV. Correlations of Brightness and Contrast Vision in Dual Neuronal Pathways

Two Neuronal Subsystems B and D for Brightness and Darkness Information

Retinal Luminance Processing in On- and Off-Centre Neurons. The coding of luminance for brightness perception in the visual system begins in the *retina* and uses *two neuronal subsystems B and D* which transmit information as *on-centre and off-centre neurones*, respectively. The separate channels of these dual systems are reciprocally organized, and signal *opponent qualities of brightness and darkness*. After the receptors have been excited by light increment, the opposite information of light decrement is also coded in retinal circuits which probably involve special bipolar neurones and intermediate horizontal and amacrine cells. The two systems are clearly separated at the *ganglion cell* level where *on-centre neurones (B) and off-centre neurones (D) form two populations of approximately equal numbers*. These

B and D neurones predominate in subcortical visual structures of cats and monkeys and probably also in man. They have an approximately circular *concentric field configuration with antagonistic surround* in the retina and the lateral geniculate, whereas in the cortex most of these fields are transformed to a variety of specialized neurones with selective receptive fields (Fig. 3).

Both subsystems transmit their information from the eye to the brain with some modifications and their interactions involve lateral, collateral and reciprocal inhibition. The *B system* (signalling "*brighter*") consists of *on-centre neurones*, activated when light shines onto their retinal receptive field centre. These neurones are inhibited by light in the surround of their receptive fields *(lateral inhibition)*. The *D system* (signalling "*darker*") consists of *off-centre neurones*, activated by light-off or light decrement in the receptive field centre; activation also results from light shone onto their receptive field surround *(lateral activation)*.

The *retinal mechanisms* which ensure the dual code between the receptor polarization by light and the two lines of ganglion cells are not yet clarified in their interaction of *graded responses of the receptors, horizontal, amacrine and bipolar cells.* The present state of research is treated in Volume VII/2. We cannot discuss here the interesting problem of whether the darkness signals of the vertebrate eye have developed in relation to the *hyperpolarization of visual receptors* by light, which is characteristic of the vertebrate eye, how the graded polarization of intermediate cells intervenes in the darkness coding, and how the dual line splits at the bipolar cell level. Retinal ganglion cells have an ongoing "spontaneous" discharge that may be regulated by electrotonus in the direction of either excitation or inhibition. It is conceivable that the reciprocal coding in B and D lines is controlled more smoothly by electrotonus and/or synaptic transmitters during opposite polarization of retinal cells than it would be by spike conduction between receptors and ganglion cells. Excitation in on-centre neurones appears to be related to *light-induced receptor hyperpolarization,* whereas the declining receptor hyperpolarization (with a return to the membrane potential existing before light activation) corresponds to *relative depolarization at light-off*, or darkening, and leads to off-centre excitation. Both receptor polarizations may be transmitted via two different bipolar neurones to the ganglion cells; there may even be other intermediate cells with additional synaptic transmitters. All this, however, is highly hypothetical, whereas the effects of illumination changes upon on-centre and off-centre discharge are well documented.

Figure 1 demonstrates how the discharges of these dual neuronal channels are reciprocally related under various conditions of darkness, light-on and light-off, and with successive and simultaneous light–dark contrast. Also shown is the neuronal correlate of Charpentier's dark band. This is perceived following the onset of bright light at about the time when B activity is briefly suppressed and the D neurones are momentarily activated.

Neuronal Interaction. In the optic nerve and all subsequent channels transmitting information by spike conduction and frequency modulation over the lateral geniculate to the visual cortex, the two antagonistic parallel lines *B and D interact at each level of synaptic transmission*. Their interaction is essentially *reciprocal inhibition, B versus D and D versus B,* which may result in disinhibition mainly in the B system. Further, within each subsystem the neurones showing overlapping receptive fields inhibit each other in the receptive field centre to avoid redundancy [145]. This inhibitory process seems to be similar to lateral inhibition at the first retinal synapses but may use different *collateral inhibitory mechanisms within the B system* (B versus B) *and the D system* (D versus D). See Fig. 3 p. 40.

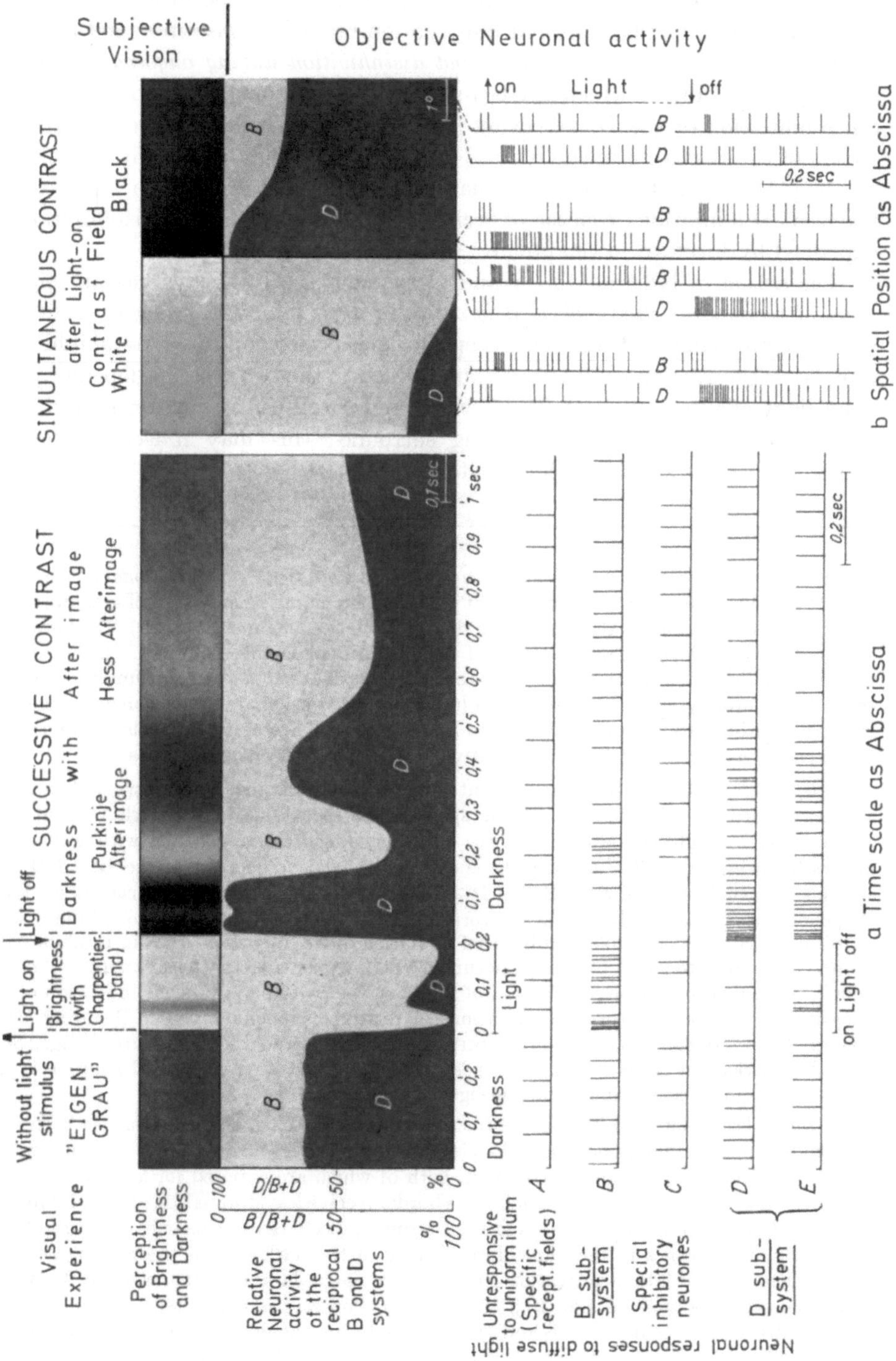

Subjective Vision
Objective Neuronal activity
Visual Experience
Perception of Brightness and Darkness
Relative Neuronal activity of the reciprocal B and D systems
Without light stimulus "EIGEN GRAU"
Light on
Brightness (with Charpentier band)
Light off
Darkness
Purkinje Afterimage
Hess Afterimage
SUCCESSIVE CONTRAST with After image
SIMULTANEOUS CONTRAST after Light-on
Contrast Field
White
Black
D/B+D
B/B+D
Darkness
Light
Darkness
Unresponsive to uniform illum (Specific recept. fields)
B sub-system
Special inhibitory neurones
D sub-system
Neuronal responses to diffuse light
on Light off
0,2 sec
0,1 sec
1 sec
a Time scale as Abscissa
b Spatial Position as Abscissa

The reciprocal activity of the two subsystems does not always deviate symmetrically from the mean level of discharge. Asymmetries occur at various levels of adaptation and of background illumination related to sustained and transient activation or inhibition. *Inhibition and disinhibition during ongoing maintained discharge* is a subject which needs further investigation (see LEVICK, Chap. 8). Since inhibition of the D system by light may have shorter latencies than activation of the B system in the retina [412] and sometimes also in the lateral geniculate [144, 146], it is probable that some B discharges can be triggered by transient D inhibition: Release of reciprocal inhibition causes D disinhibition, as explained by FREUND in Chapter 18.

The action of light on the B and D systems nearly always has an *opposite effect of reciprocal action: when B is activated D is inhibited, and vice versa.* The *centre–surround antagonism* of the receptive fields is also opposite (lateral inhibition in B, lateral activation in D) but surround effects in the two systems depend equally upon light adaptation. Both B on-centre and D off-centre neurones need some centre illumination for an operating surround; this may indeed be the only light effect that is similar in both systems (see p. 34).

Fig. 1. *Scheme of brightness and darkness perception correlated with visual neuronal events.* From JUNG 1961 [249]. The upper row depicts light perception in man. The middle section shows neuronal activity in the cat's visual system (with dual information channels B and D). The lower half summarizes the spike discharge patterns of various types of cortical neurones. The left part of the figure indicates temporal or successive contrast; the right part, spatial or simultaneous contrast. The neurophysiological correlates of luminance perception are activation and reciprocal inhibition within two neuronal subsystems B and D. The B system signals relative brightness by on-centre neurones, the D system relative darkness by off-centre neurones.

a. left: On a time scale showing different conditions of luminance the following phenomena are represented: 1. „Eigengrau", or the spontaneous sensations in the dark without light stimulation; 2. a light-on period of 0.2 sec duration; 3. successive contrast with early after-images following light-off. Just beneath is shown the relative activity of the B and D subsystems. Preponderance of ratio $B/B + D > 0.5$ corresponds to prevailing brightness vision. Preponderance of ratio $D/D + B > 0.5$ corresponds to relative darkness. Strong illumination causes primary activation followed by brief inhibition of the B system. During B inhibition a simultaneous reciprocal D discharge occurs which corresponds to CHARPENTIER's "bande noire". After light-off, strong activation of the D system with simultaneous inhibition of the B system is followed by periodic alterations of relative preponderance of B during after-images. The most constant are the PURKINJE after-image and the HESS after-image, both related to periods of B activation. *Lower left:* Discharge patterns of cortical neurone types, including B and D; same time scale and light stimulus as above.

b. right: To depict *simultaneous contrast* at borders, time on the abscissa has been substituted by a spatial representation of the visual field (in degrees of arc). The bipartite stimulus (above) consists of a light and a dark half, both of which are exposed for a finite period. B neurones with receptive fields overlapping the border receive less stimulation of their inhibitory surrounds than those further, away and their stronger discharge signals enhanced brightness. Conversely, D neurones in the immediate vicinity of the dark side of the border receive a relative increment of lateral activation and therefore signal darkness enhancement. According to BAUMGARTNER's analysis [26, 27], simultaneous contrast is correlated with the sum of neuronal discharges occurring in B and D neurones during the first half-second after stimulus onset or offset. The relative firing rates are schematized in the middle section. *Lower right:* Neuronal firing patterns occurring at various places in the white and black contrast field (dotted lines), when the stimulus is a light flash of given duration (time scale at right)

The neuronal activity which mediates between physical stimulus and visual experience comprises a multitude of mechanisms interacting by complex information processing. Only the first stages from the retina to the cortex have been thoroughly investigated. The higher processes in and beyond the visual cortex are to a great extent unknown, and the transcendence from neuronal activity to visual perception remains an enigma.

Antagonistic Receptive Fields and Reciprocal Systems. Information processing in the lower visual system is based upon two neuronal mechanisms: 1. *antagonistic field organization* of retinal on-centre and off-centre neurones; 2. *dual-system interaction with reciprocal inhibition* between the two neuronal channels B and D.

The first is an essentially *retinal* mechanism [444] and is described in detail in Volume VII/2 by LEVICK [293] and discussed in Chapter 2 of this volume. The second is initiated in the retina but involves mainly *central relays* in the lateral geniculate and visual cortex (see Chaps. 18 and 21). Reciprocal interaction, although it has been discussed in several papers since 1961 [249], is not yet generally recognized as being a principle of dual-system transmission. However, both field organization and reciprocal inhibition appear to be essential to the processing of achromatic vision. Reciprocal interaction is discussed here with several examples in order to demonstrate the validity of this principle in neuronal physiology and in perception. Perceptual correlates can be illustrated by visual experience, including after-images and visual illusions (see Figs. 1–10 and Table 2). Some general principles and teleological problems related to the efficiency of two reciprocal systems were discussed in the preceding section (p. 17).

Contrast Vision and its Neuronal Correlates

Visual Contrast, Historical Sources. *Contrast vision* has been known since antiquity, by experience and common sense, and painters made use of it in the visual arts long before it became the subject of research (see p. 63). The fact that visual contrast between bright and dark areas leads to enhanced perception of whiteness and blackness was first clearly formulated by LEONARDO DA VINCI in his *Treatise on Painting* [441] written around 1500: "Dico che'l bianco che termina con l'oscuro, fà che in essi termini l'oscuro pare più nero, il bianco pare più candido". (For LEONARDO's concept of opponent colours, see p. 104.)

The *terminology and interpretation of contrast phenomena* remained in the realm of psychophysical observations for some 80 years after MACH's fundamental paper [304] of 1865, until HARTLINE's discovery of lateral inhibition in 1949 made contrast vision accessible for neurophysiological experimentation [195]. As RATLIFF has shown in his monograph [371], there are still large gaps in the neuronal explanations and it thus seems useful to discuss the *historical terms* and descriptions of border and areal contrast.

HERING in his early papers (1872/74) first used the word „Lichtkontrast“ as a general term for visual contrast phenomena [211]. HERING followed MACH's proposal [304], which was in opposition to HELMHOLTZ's psychological interpretation, with terms like „Wahrnehmungsurteile“ and „unbewusste Schlüsse“ [208], and explained simultaneous contrast by *retinal interaction.*

In 1874 HERING called the retinal effects „Wechselwirkungen der Netzhautstellen“. In an article which appeared in 1905 in GRAEFE-SAEMISCH's Handbook and in the 1920 edition (to which HERING's notes were added posthumously by C. HESS) [213], he went further, and with the general term *„physiologische Wechselwirkungen der Sehfeldstellen“* extended the concept to include central interaction on the basis of binocular convergence. HERING clearly distinguished *border contrast („Grenzkontrast“)* and *brightness contrast („Helligkeitskontrast“)*, and considered self-regulation of light sensitivity *(„Selbststeuerung der Lichtempfindlichkeit“)* as a controlled process of *local adaptation* which occurs instantaneously *(„Moment-Anpassung“)*. He also discussed the interaction of *simultaneous contrast („Nebenkontrast“)* and *successive contrast („Nachkontrast“)* which appears when eye movements induce a succession of retinal images. In his 1879 article on eye movements and spatial location [212], HERING introduced the concept of *constancy („Raumkonstanz“, „Farbkonstanz“)* and discussed the role of attention in visual localization and goal-directed ocular movenents. For details see also [435, 437].

Visual Contrast and Neuronal Coding in Space and Time. Retinal and geniculostriate neurones respond to *changes in luminance*, whether these are produced in *time* by light-on or light-off, or in *space* by contrast patterns shone upon their receptive fields. Such stimuli are created under natural conditions by the voluntary and involuntary *eye movements* which we make all the time as we view the world around us. There is thus a continuously changing neuronal input from the eye to the visual brain.

The principles of information coding can be demonstrated best in the retina. On-centre and off-centre neurones are reciprocally affected by *luminance and contrast change:* a single spot of light shone on the retina simultaneously affects *both* neurone populations having receptive fields at that location. *Phasic excitation in one system* at a given retinal image location is *related to reciprocal inhibition of the other system*. Spatially, a light–dark border enhances the excitatory responses of on-centre neurones overlapping the dark side, as was shown by BAUMGARTNER [26]. (See Fig. 1, right side.) The psychological correlates are listed in Tables 1 and 2. For contrast stimulation in space and time, the information of subsystems B and D and the resulting visual information can be summarized as follows:

Table 1. Correlation of brightness and darkness perception with discharge of B and D neurones. (Modified from JUNG, 1961 [249])

Subjective Bright–Dark Vision	*Objective Neuronal Correlates of Photic Stimuli*
"brighter" than surround (spatial = simultaneous contrast) than preceding sensation (temporal = successive contrast)	Local B activation with reciprocal D inhibition following spatial and temporal luminance increments in field centre contrasting with surround, or following a lateral luminance decrement in surround contrasting with field centre
"darker" than surround ("black" in simultaneous contrast) than preceding sensation (successive contrast)	Local D activation with reciprocal B inhibition following spatial and temporal luminance decrements in field centre contrasting with surround, or following a lateral luminance increment in surround contrasting with field centre

The information in Table 1 on the B and D subsystems can be expressed as a simple rule: *B signals brighter than surround* (in simultaneous contrast) *or brighter than previous sensation* (in successive stimulation); *D signals darker than surround or than previous sensation* [249]. The perceptual correlates of the relative dominance of the discharges of the two neuronal subsystems are shown schematically for simultaneous and successive contrast in Fig. 1. Whether B or D discharges will predominate may be predicted for the cat from the prevailing sensation of brightness or darkness in man under similar stimulus conditions (cf. Table 2). This perceptual control can be demonstrated in animal experiments when one looks at the stimuli on the perimeter which are exposed to the receptive field of the recorded neurone of the cat, for B discharges correspond to seen temporal and spatial brightness increments and D discharges to darkness increments. Further qualitative correlations of subjective vision in man with objective neuronal recordings in the cat (and probably also in the monkey) can be found in more complex stimulations involving linear patterns and movement. However, prediction for such tests is somewhat less certain. For behavioural correlates in cats see [187].

The neuronal receptive fields of both subsystems, as measured in animals (Fig. 2), and the subjectively estimated perceptive fields in man are found to obey similar laws, though with some species differences (see p. 44 and Figs. 5 and 7). Some of the limits to comparisons between form discrimination and neuronal mechanisms are discussed on p. 127 and in Chapter 2 by STONE and FREEMAN.

When one turns from brightness to *colour* vision, the situation becomes more complex. In perception the simultaneous and successive contrast phenomena that occur for *opponent colours* are very similar to those for white and black, but the neuronal transformations are not yet clear. The flow of information about colour is schematized in HASSENSTEIN's model of data processing (Fig. 16). In spite of some progress in recent years, the central processing of colour information (see pp. 105 ff.) is still far from being understood in terms of neuronal transmission codes. The relationship of colour processing to achromatic processing in the B and D systems is still unclear, and a neuronal scheme like that shown in Fig. 1 cannot yet be made for *chromatic brightness*. The various spectral wavelengths make somewhat different contributions to brightness perception, even with equal energy distribution; e.g. yellow appears brighter than blue. However, the contribution of colour to brightness perception and the nature of the relationship of the broad-band subsystems B and D to the narrow-band neural mechanisms in primates which possess good colour discrimination is difficult to understand. (For a detailed discussion of colour vision, see [433] p. 103 and Chap. 3.)

After-Images and Neuronal Discharge. The *early after-images* can be seen and measured in time by a moving light spot [147] during the first second after a brief illumination. They appear as positive images of the original stimulus in human vision and correspond to periodic *B discharges* in the cat [181, 248]. These are interrupted by pauses and reciprocal *D discharges* which in turn correspond to the *dark intervals* in human perception. The after-images of PURKINJE and HESS (upper left, Fig. 1) and their dark intervals correlate particularly well in time with the periodic after-discharges of the two neuronal subsystems in the retina, the lateral geniculate body, and the visual cortex [181]. Similar correlations have been found for metacontrast [185].

GRÜSSER and GRÜSSER-CORNEHLS in 1962 reinvestigated the neuronal correlates of the early after-images and CHARPENTIER's interval [181]. Their experiments with conditioning light flashes of various durations showed two to four periods of B activation, interrupted by inhibitory phases. The duration of these after-phases increased with stimulus duration. In the sequence of periodic activation, even after the same stimulus, each successive period of B activation was longer than the preceding one. Light stimuli of duration exceeding 1 sec failed to elicit either early after-images or neuronal periods. The earliest after-images of HERING and corresponding neuronal B discharges appear only after illumination periods shorter than 40 msec [181]. The sequence of B and D discharges after such brief stimuli may be compared to the B discharges preceding and following the D burst corresponding to CHARPENTIER's interval during longer stimulus durations (Fig. 1).

The early periodic and positive after-images must be distinguished from the *late after-images*, first called *Blendungsbilder* by PURKINJE [369], which last for several seconds or minutes after intense illumination of long duration. They are also elicited by long fixation of bright contrast patterns and appear mostly *negative* or in colours complementary to those of the stimulus. The neuronal correlates of the late negative after-images remain uncertain but may be related to the flattening of alpha rhythms in the human EEG [240].

Late after-images besides showing slow periodicity (probably related to photochemical changes) are very *sensitive to alterations of background luminance and colour*. Slight changes in brightness and colour will cause late after-images to reverse to opposite colours or from white to blackness. This can easily be verified by lying in the sun with closed eyes: after-images are observed from previous exposures to the sun and the diffused, reddish light penetrating through the lids is altered by closing the lids more tightly or by shading the closed eyes. Simultaneous border contrast, often apparent in sun after-images, also changes with background illumination as a function of successive contrast. The perceptual phenomena may be used as indicators of the *interaction of the opponent colour systems* (red-green, blue-yellow, white-black) within the retina, although they are known to be subject to several transformations in the central pathway. The neuronal correlate is not yet clear. For black-white reversal, a shift of preponderance between B and D due to background light or reciprocal inhibition may be assumed to occur; this is similar to the early after-images in Fig. 1. The strong and enduring inhibition of ongoing D discharge by illumination may favour B preponderance above the 50% mean level. Whether similar shifts occur in neuronal populations signalling opponent colours is a question we cannot yet answer.

There are limits to spatial contrast as determined by the interplay between centre and surround. The mechanisms of contrast vision depend upon the complex relationship between *three factors:* 1. the *absolute intensity of the centre illumination* (below certain luminances the surround is not operational, see pp. 33, 34); 2. the *relative intensity of the centre* illumination in contrast to surround; and 3. the *relative intensity of the surround* illumination in contrast to *centre*. BARLOW et al. [19], POGGIO et al. [361], MAFFEI et al. [311] and FISCHER and KRÜGER [139] have produced experimental evidence for these basic mechanisms. A survey of perceptual phenomena and their neuronal correlates is given in Table 2. This Table and Fig. 1 summarize the neuronal–perceptual correlations and are updated versions of materal published in German in 1961 [249].

For the B subsystem, there is a convincing correspondence between firing patterns in animals and various psychophysical functions in humans, including brightness discrimination, border contrast, integration of spatial and temporal luminance, flicker, after-images and so on. Since similar neuronal responses have been recorded from many mammalian species, one may conclude that transient activation of the on-centre neurones, and their cerebral expression in the B system,

Table 2. Perceptual phenomena and neuronal correlates

Visual Perception	*Neuronal Correlates in Retina (R), Lateral Geniculate (G) or Visual Cortex (Co)*
„*Eigengrau*“ (sensation of grey in complete darkness)	Ongoing or maintained discharge of both B (on-centre) and D (off-centre) subsystems (R, G, Co) with predominant D discharge
Entoptic sensation of clouds and motion in the Eigengrau	Oscillating rates of B and D discharges (R, G) and "propagating activity through neuronal networks" (G)
Relative brightness	B system activation; D system inhibition (activity of on-centre neurones predominates in R, C, Co)
Relative darkness	D system activation; B system inhibition (activity of off-centre neurones predominates in R, G, Co)
Weber-Fechner law as applied to increasing light intensity	Logarithmic increase of discharge rate in the B system with increasing light intensity (R, G)
Charpentier's interval *(bande noire)*	Short pause in discharge of the B system (after primary discharge) with simultaneous "on" discharge of the D system (E neurones in Fig. 1, on-off D)
Successive contrast (early periodical after-images and dark intervals)	Periodically alternating B and D activation and inhibition (R, G, Co) after light-off (Fig. 1)
Simultaneous contrast and contour vision	Lateral inhibition of B system and lateral activation of D system (R, G, Co) resulting in maximal contour activation with B and D enhancement in bright and dark borders (Fig. 1)
Enhancement of contours and distinction of their orientation	Border contrast activation of the B and D systems along edges (R, G, Co, Fig. 3). Activation of simple field neurones with special axis orientation (Co)
Hermann grid illusion: dark spots at grid intersections disappear with foveal fixation	Diminished B discharge at intersections by relative increase of lateral inhibition (Fig. 5) Smaller diameter of receptive field centres in the fovea than in perifoveal ganglion cells (monkey R)
Limited binocular brightness summation and Fechner's paradox	Predominantly monocular connections of the B and D systems from the retina over the lateral geniculate body to the first-order cortex neurones (G, Co)
Binocular rivalry	Inhibition of monocular impulses from the opposite eye (G, Co)
Flicker fusion (CFF)	Maximal flicker frequency for regular neuronal responses with response failures at higher rates, i.e. neuronal CFF (G, Co)
Brücke-Bartley effect	Maximal B system discharge rate with medium flicker frequencies around 10/sec (R, Co)
Similar monocular and binocular CFFs	Predominant monocular neuronal CFF (G, Co) and binocular inhibition
Higher CFF during attention than with fatigue	Increase in neuronal CFF by thalamo-reticular stimuli (Co)

Table 2 (continued)

Visual perception	*Neuronal correlates in retina (R), lateral geniculate (G) or visual cortex (Co)*
Local adaptation with background illumination	Diminished B activation after prolonged light and prolonged D inhibition (R, G)
Binocular stereopsis	Binocular activation of neurones of area 17 by laterally disparate patterns (Co)
Weckblitz (light illusions resulting from an arousing stimulus)	Unspecific activation of cortical neurones, predominantly the B system
Facilitation of visual attention	Convergence of retinal and thalamo-reticular impulses on visual neurones (G, Co)
Galvanic brightness and darkness sensations resulting from anodal and cathodal DC stimuli of the retina	Reversed on- and off-discharges of the B and D systems (R) by a reversal of retinal polarization

are related to brightness perception in mammals including man. For the D subsystem the evidence relating to visual and neuronal correlates of darkness information is less well known. Since scepticism is often expressed concerning the ability of the neuronal channels of the D system to signal darkness, the arguments will be discussed in detail.

Mechanisms and Significance of the D System

Stimulation and Information of Off-Centre Neurones or the D System. The responses to light decrement in off-centre D neurones signalling darkness are equally impressive but less well understood than the responses to light increment in on-centre B neurones signalling brightness. This may be because incremental light stimuli are more easily produced than light decrements. *D neurones are strongly excited by light decrement and inhibited by incremental light stimuli projected onto their field centre;* they are somewhat less strongly *activated by light increment in the field surround* (lateral activation). Both luminance decrement in field centre and luminance increment in surround correspond to the *relative darkness of the field centre in spatial contrast to the surround* (see Fig. 2). *The optimal stimulus is black.* Off-centre neurones further show an excitatory transient to light-off which is proportional to *previous light intensity and duration* [67]; see p. 32. The intensity–response curves for D neurones are very similar to the S-shaped curves for B neurones, except that incremental stimuli of very high contrast result in *prolonged inhibition* and suppression of the off-transient, besides some rhythmic components. The D system responds to extremely short *light flashes*, corresponding to on- and off-stimuli only microseconds apart, with prompt and relatively prolonged *inhibition* [24]. In the retina and lateral geniculate the inhibition of the D system may even precede the excitation of the B system. Flashed spots projected to the off-centre are as effective as diffuse light flashes owing to the *dominance of the field centre* which is primarily inhibited by light and activated by darkness. Such responses are in agreement with the concept that off-centre D neurones signal *information regarding the relative level of darkness,* whether *spatially distributed* on

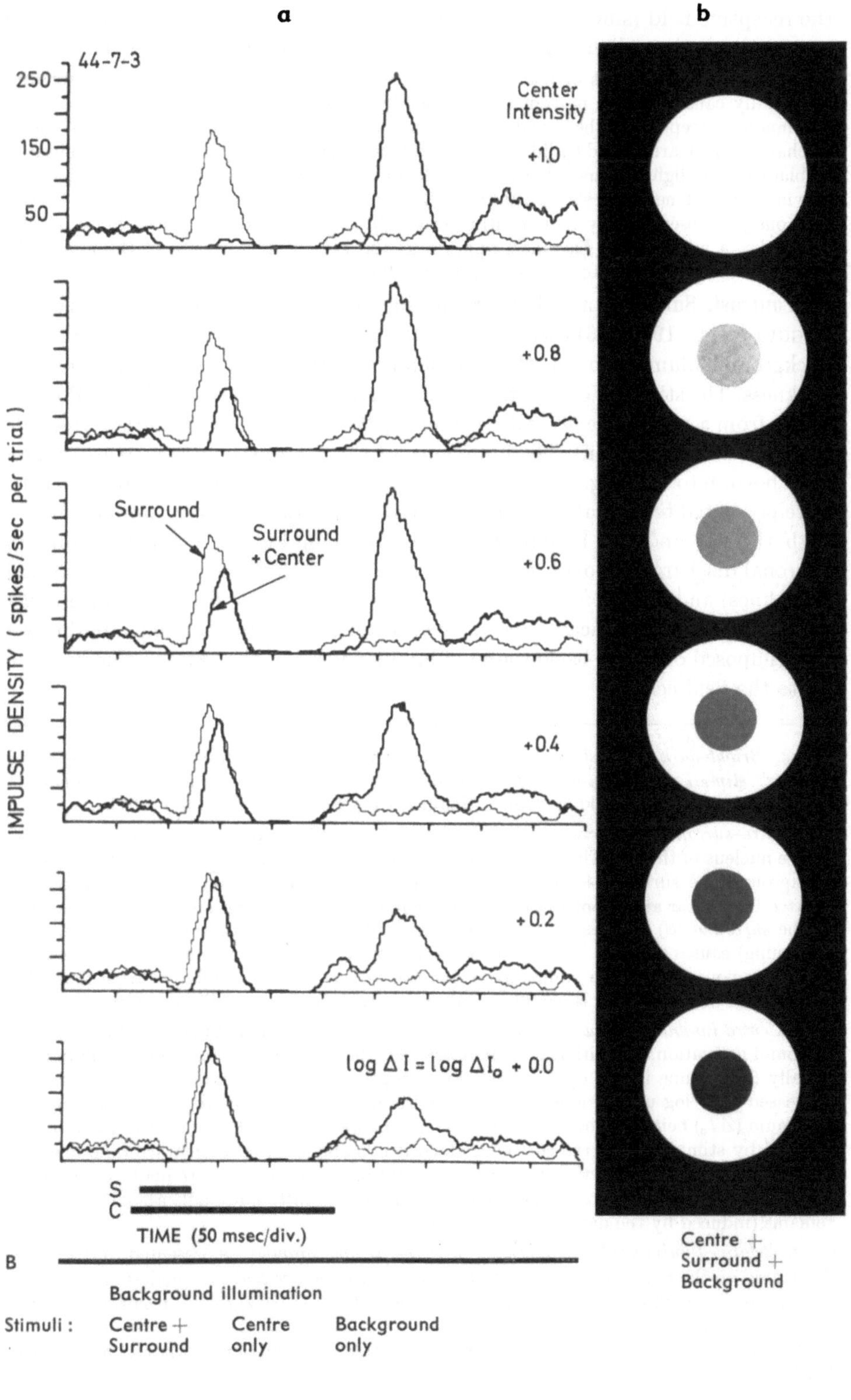
a
b
44-7-3
250
150
50
Center
Intensity
+1.0
+0.8
+0.6
+0.4
+0.2
log ΔI = log ΔI₀ + 0.0
Surround
Surround
+ Center
IMPULSE DENSITY (spikes/sec per trial)
S
C
TIME (50 msec/div.)
B
Background illumination
Stimuli :
Centre +
Surround
Centre
only
Background
only
Centre +
Surround +
Background

the receptive field (simultaneous contrast of centre and surround) or occurring in temporal sequence (light decrement and successive contrast, see Fig. 1).

In daily life sensations of blackness due to successive contrast, as well as off-discharges, are usually caused by eye movements and not by switching the light off. In our stable and continuous perception of the visual world we do not notice the darkness correlates of the off-discharges which are caused by lid blinking and other short diffuse light decrements. Sensations of blackness at light-off may, however, be demonstrated in psychophysical experiments involving contrast near threshold. PIRENNE [357] observed that the sudden interruption of a previously exposed, slowly incrementing test light near threshold is perceived as a "*black flash*", i.e. a sensation of blackness appears as the perceptual correlate of the neuronal D discharge at light-off due to successive and simultaneous contrast of a fixated spot.

Contrast, Surround and Background in the D System. Fig. 2 is adapted from POGGIO et al., 1969 [361] to illustrate the interrelation of centre, surround and background illumination in the D system, and the corresponding perception of darkness. The stimulus configuration on the right (Fig. 2b) was added to POGGIO's traces from a concentric *off-centre neurone of the lateral geniculate nucleus* (recorded from the optic radiation). The luminance of the centre stimulus increases tenfold from bottom to top. Fig. 2b depicts the approximate luminance gradations as they were presented to the cat, showing the bright–dark ratios as they would appear to both the cat and the human eye. The curves in Fig. 2a, computed from the neuronal discharges, show 1. *lateral activation* by the light annulus in the *surround* (thin lines) and 2. *inhibitory–excitatory interaction* resulting from combined *centre–surround* stimulation (heavy lines). The centre disc and surrounding annulus were superimposed on a dim background of luminance below the lowest value projected on to the field centre.

Fig. 2. *Bright–dark contrast and predominance of receptive field centre in the cat's D system for "darker" stimuli at light-on and -off*. Lateral activation is induced by a light annulus in the receptive field surround. Modified from POGGIO et al. (1969) [361].

Centre–surround interaction by light is shown for an off-centre neurone of the lateral geniculate nucleus of the cat. The luminance of the central disc is varied while the illumination of background and surround is kept at a constant level. Stimulation of the *centre* by light (indicated by the bar at the bottom marked *C*) induces primary inhibition, whereas illumination of the *surround* (*S*) induces lateral activation. Switching off the centre stimulus (i.e. centre darkening) causes stronger "off" responses after strong centre illuminance (upper traces) and weaker responses when the luminance of the centre was not much different from that of the background (lower traces). B signifies continuous weak background illumination.

a *Centre inhibition versus surround activation:* Centre inhibition at light-on always precedes surround activation. Inhibition predominates when both centre and surround are illuminated equally (heavy line in the upper rows). From top to bottom the luminance of the centre is decreased by 1 log unit in a series of centre stimuli descending in steps of 0.2 log unit, the minimum (ΔI_0) being 0.3 log unit above the luminance of the background. Lateral activation induced by stimulation of the surround alone is superimposed on each trace (fine line). The suppression of surround activity by centre inhibition (upper row) is gradually reduced with increasing contrast between annulus and disc, until in the bottom row only a small off-response remains (induced by the darker background).

b *Relative luminance levels of centre, surround and background*, as presented to the cat, are illustrated on the right. In the top row the luminance of centre and surround are equal. In the bottom row, the luminance of the centre is twice that of the background, resulting in a slight decrement at light-off.

The traces show that centre–surround interaction is a function of *contrast* and of *centre luminance*

Figure 2 shows that *contrast in space and time* with relative darkness in the region of the field centre is the effective stimulus of the D system and not just "light-off". The D neurone is excited when the light stimulus illuminating the centre, or centre-plus-surround, is *turned off*. It is also excited (lateral activation) when a spatially contrasting stimulus illuminating only the surround (which would appear to man as dark in the centre and light in the surround) is *turned on*. The common feature of these two visual configurations is the transmission of the information "*darker*", located in that part of the visual field which corresponds to the receptive field centre of the recorded neurone. Although the two stimuli occur with different timing and opposite change, i.e. from and towards diffuse background light, the predominance of the receptive field centre explains the constancy of darkness information.

Hence, Fig. 2 demonstrates several principles that are characteristic of the D system and the contrast stimulation of field centre and surround, related to background and that correspond to the perception of *contrast blackness:*

1. D activation corresponds to *relative darkening in the field centre*, not only to luminance decrement in time. Both *spatial* luminance differences and *temporal* luminance decrements interact; thus darkening may appear in time and space.

2. The dark stimulus may be caused either by *light-on* (centre–surround contrast = temporal and spatial darkening) or by *light-off* (temporal darkening following luminance decrement with or without a pattern of spatial contrast).

3. *Luminance increment in the field surround* alone without any luminance decrement in the centre results in D discharge by *lateral activation*, and corresponds to perceptual darkening by simultaneous contrast.

4. *Centre dominance* appears with relatively short latency, both for inhibition by luminance increment and for activation by luminance decrement, and correlates with diminshed or increased darkness seen at the spot which corresponds to the field centre.

5. D activation may be related to the *relative blackness* which the cat or a human observer would see in the area of the field centre.

Recent experiments [139] have shown that these features depend on the presence of a certain amount of *background light* which must affect the *field centre*. Hence, the critical stimulus which maintains surround–centre antagonism is centre illumination and background, not merely general light adaptation (see p. 34).

Many other experiments concerning the retinal and geniculate off-centre neurones of cats and monkeys agree in principle with the results shown in Fig. 2 and the correspondence of relative darkness projected to the receptive field centre with D activation. Some differences between the neurones of the retina and the lateral geniculate (see FREUND, Chap. 18) and species differences between cats and monkeys do not diminish the good correlation with human vision in light and dark adaptation. Light in the field centre of D neurones inhibits their discharge, but it also facilitates in the surround the mechanism of lateral activation by light, thus contributing both to contrast vision and light adaptation (see p. 33).

A peculiarity of the D system is its dependence on the *duration of the preceding test light:* off-activation in retinal and LGN off-centre neurones increases with the previous duration of illuminance up to about 500 msec [67]. It is not yet clear how this peculiarity is related to the facilitation of surround mechanisms by centre illumination, nor is it certain whether there is a corresponding feature in the B system which matches or explains this effect of duration in the D system. This may be one of the asymmetries between the systems. A stimulus capable of producing almost perfectly reciprocal B–D discharges is *sinusoidally modulated light:* B neurones discharge with the phase of light increment, D neurones discharge with light decrement and, if the depth of the temporal modulation is not too great, on-centre and off-centre neurones of both retina and LGN will follow in activity cycles almost "180° out of phase". This method may be of value for systems analysis (see Chap. 8, VAN DE GRIND et al.).

Black Stimuli and Centre Surround Interaction. The mechanisms of on- and off-centre organization are not reciprocally symmetric in all respects as regards the interchange of the stimuli "brighter" and "darker". In the D system, light, instead of darkness is necessary in the field centre for the establishment of the surround mechanism. It seems more satisfactory to regard the interchange as one of "excitation" and "inhibition" when comparing the subsystems. In this sense, the two systems, B and D, do work reciprocally at least under scotopic conditions: Certain photic stimulations activate the neurones of one system while inhibiting the corresponding neurones of the other, and vice versa. This concept leads to some conclusions which may be tested experimentally by applying stimuli which add light or subtract it from different parts of the receptive fields of the two neuronal systems under photopic and scotopic adaptation.

FREUND and coworkers [143a], using a special television set, studied the effects of *black stimuli* that were darker than the general background luminance. They were able to suppress light-induced lateral activation in off-centre neurones by occluding light from the field centre. Moreover, for both B and D systems, surround responses that had been antagonistic with some background illumination were reversed to centre-type responses when the centre was in complete darkness.

Flashing or stationary black areas representing light decrements were displayed on a television screen for various contrasts in the receptive fields of visual neurones. An example is shown in Part B, Chapter 18 (Fig. 10). Sustained blackening of the field centre causes the surround to respond in the same manner as the centre [143a]. Thus, *darkening of the off-centre reverses the typical lateral on-activation* (normally obtained by light onset in the field surround, such as the bright ring in Fig. 2) *into light-off activation from the surround* in off-centre neurones. These "paradoxical" surround responses in the D system (inhibition by light, activation by darkness, normally obtained only from the off-centre itself) result in an extended area responding to light decrement, i.e. *spatial summation in a black area* when it is enlarged beyond the centre. Although the perceptual gain of blackness summation in a dark room under scotopic conditions is less evident than summation of bright areas near threshold, it may be an additional argument for our thesis that off-centre neurones in the D system will signal "darker" under various conditions of light adaptation.

Although darkening of the field centre is the adequate stimulus for phasic D activation, the contrast enhancement for blackness can only be obtained in a single receptive field when some local illumination within the field centre has previously facilitated the surround mechanisms. Hence, the amount of surround antagonism in both B and D systems depends upon the *centre luminance* which also influences the ongoing discharge [139]. Whether different conditions are operational for large field contrasts in neuronal populations remains unclear. Subjectively, areas much larger than receptive field size are seen blacker when bordered by a bright surround (see p. 54).

Neuronal Light Adaptation, Antagonistic Field Organization and Centre-Surround Interaction

Inhibition within the Receptive Field Centre and Interaction with Surround. During the fifteen years following KUFFLER'S work [285, 286], most research was concentrated on lateral, collateral and reciprocal inhibition, whereas the centre mechanisms of inhibition and centre influence on surround were somewhat neglected. However, when one considers that large neuronal populations in the retina are stimulated simultaneously by a small light spot projected to one receptive field centre, it seems evident that inhibitory processes are important for the field centre mechanisms and that this must affect a number of ganglion cells. These retinal interactions precede information processing in brain structures where further inhibitory interactions occur with collateral inhibition within the same neuronal subsystem and reciprocal inhibition between the subsystems B and D (see p. 39). GRÜSSER [178a] and FREUND et al. [145] have made systematic studies of spatial interaction within the field centre. GRÜSSER assumes two types of inhibition: 1. "Subtractive inhibition" which is supposed to be postsynaptic. 2. "Multiplicative shunting inhibition" which is believed to be presynaptic [178a]. Direct proof for post- or presynaptic mechanisms by intracellular recording is still lacking. The mechanisms are discussed in Chapter 7 with reference to temporal transfer. Recent experiments have shown that *centre influence on surround mechanism* is an essential mechanism of neuronal adaptation.

Influence of Field Centre on Surround and Neuronal Adaptation. Since BARLOW, FITZHUGH, and KUFFLER [19] described the disappearance of antagonistic field organization after *dark adaptation*, this was generally accepted as an adaptational regulation which developed slowly with photochemical changes. However, in spite of numerous studies on retinal receptive fields, summarized by LEVICK in Vol. VII/2 [293] of this handbook, the conditions adequate for opposite surround responses in on- and off-centre neurones in the light-adapted eye remained unclear. Quite recent experiments have demonstrated that the essential condition for antagonistic field effects is *not* general light adaptation but some *local illumination within the field centre*. It was shown by MAFFEI and FLORENTINI [311] that the surround inhibition disappears in on-centre neurones after occluding light from the field centre. Recent experiments of FISCHER [139], FREUND [143a], ENROTH [122a] and coworkers have made it clear that the *surround effects in both systems, B and D, depend on centre illumination:* Only when the field centre receives some suprathreshold amount of light do the characteristic effects of the surround mechanisms appear in each system, i.e. lateral inhibition in on-centre neurones and lateral activation in off-centre neurones. These *antagonistic surround responses disappear within a few seconds after the field centre is in complete darkness* and become reversed to the same types of response (although weaker) as those from the field centre. A small light spot above threshold projected somewhere on the field centre is sufficient to reinstate antagonistic surround responses immediately. This excludes photochemical adaptation and shows that a *neurophysiological mechanism induced from the field centre determines the surround effects of lateral activation and inhibition.*

Influence of Surround upon Centre? The centre–surround interaction and centre dominance seem more complex when *sinusoidal stimuli* are used and the influence of a steady surround appears stronger, as shown by the following example. GRÜSSER and coworkers have demonstrated in off-centre neurones, stimulated by *sine-wave flicker* in the centre, strong lateral activation with facilitation of light decrement to responses by a ring of sustained light in the surround which also enhances responses to higher flicker rates (see Chapter 7, Fig. 37). Whether this situation is reversed under scotopic conditions of the centre remains an open question. It is not clear either whether the adaptation level determined by the field centre depends upon the mean intensity or the peak of the flicker luminance, nor whether brightness enhancement at certain flicker frequencies is influenced by the adaptational reversal of lateral activation and inhibition in the two systems.

In both B and D systems the centre–surround interaction is a function of light–dark contrast *and* centre luminance (see p. 34). Without a certain minimal amount of light in the field centre, the antagonistic surround is not operational. This light minimum can be supplied by dim background illumination, as demonstrated in Fig. 2. However, the strength of the surround effects depends on the *amount of luminance shed on the centre.* During combined stimulation of centre and surround in an off-centre neurone, lateral activation at light-on decreases with decreasing contrast *and* with decrease of previous centre luminance. On the other hand, the subsequent *off-discharges* (caused by sudden darkening at light-off towards the low background luminance or by a saccade with fixation of a black object) also depend upon a more or less patterned or uniform background and its stray light. Off-discharges may increase with an increase in the preceding centre luminance when the background remains unchanged (see increase of off-response from lowest row upwards in Fig. 2), and only in this artificial condition are they independent of surround contrast. Similar processing occurs in the B system, but with opposite effects of centre illumination and lateral inhibition.

During adaptation to different absolute levels of luminance the threshold of excitability in B and D neurones may change with increasing background light as the human threshold for brightness changes. Responses to test stimuli are modulated with the amount of steady background light [66] for individual neurones, although maintained discharges may show small differences. Hence the "carrier frequency" of ongoing activity seems to have less importance for the *excitatory transients*, which reach high peak rates, than for threshold responses. Even in the presence of reciprocal inhibition between B and D during maintained and evoked discharge, most visual information is transmitted by transient excitation in each neuronal subsystem B and D. Thus, the asymmetries of excitation and inhibition in one system are balanced by co-operation of the two reciprocal channels, as explained above (p. 18).

Long Distance Interaction in the Retina. McILWAIN in 1964 [322] first described a facilitatory influence of patterned stimuli (mostly moving black discs) from very distant regions of the retina upon retinal ganglion cells and lateral geniculate neurones. The lowering of the threshold to centre and surround stimuli was called the "periphery effect". *Neurones in the central part of the retina were influenced from excentric regions at long distance (up to 90°) from their receptive field centre* [322]. This effect disappears after barbiturates in subanesthetic doses. Psychophysical equivalents of this periphery effect on visual threshold could not be demonstrated by using the same technique in man [406a] but SHARPE obtained positive results with entopic observations of retinal blood vessels [392b]. The effects in human vision might be studied with *large moving patterned field areas* such as used for optokinetic nystagmus.

The pathways and other structural bases of the periphery effects are unknown. Axonal collaterals from excentric ganglion cells to parafoveal ganglion cells which could alter the sensitivity of the latter have not yet been found. Since the latencies of these periphery effects are rather long, transmission through amacrine horizontal cell syncytia with tight junctions appears possible. An efferent central influence returning from the brain to the retina is excluded in the cat by transaction of both optic tracts [322]. Such efferent pathways are only known in birds and lower vertebrates.

Recent experiments of FISCHER and KRÜGER [134] have shown impressive excitatory responses of retinal on- and off-centre neurones by sudden displacements of a grid applied more than 30—50° distant from the receptive field. The response latencies to these stimuli (up to 50 msec) were longer than those of centre responses of comparable discharges. This marked periphery effect which appears in both on- and off-centre neurones as *excitatory response* differs essentially from reciprocal discharges of B- and D-system neurones in the retina. The periphery effect appears more clearly after steady illumination to the centre of on-centre neurones and after black stimuli applied to off-centre neurones. The strength of the periphery effect is comparable to neuronal discharges following field surround stimuli. Long distance interactions in the retina are also of interest for visual processing in scotomas (see p. 42) and in hemianopic fields after lesions in the central visual pathways and for the preferred perception of moving stimuli within these field defects. Residual vision in central field disturbances might be comparable to the poor remnants of vision in cortical blindness. (See p. 77).

Steady Light Perception, „Eigengrau“ and Maintained B-D Discharges

„Eigengrau“ and Sensation of Background Brightness. *Eigengrau* is the sensation of grey that occurs without stimulation by light [113, 249]. Spontaneous dark discharge recorded in the optic nerve and lateral geniculate nucleus shows that off-centre neurones of the D system discharge at a higher rate than on-centre neurones of the B system [20, 250], see Fig. 1a. An increased D discharge over dark periods lasting longer than 5 min was found, even in cortical neurones with concentric or "simple" fields in which a residual preference of the field centre for light or dark stimuli persists (cf. Fig. 6 in [250]). The prevailing dark discharge of the D system agrees well with the subjective *Eigengrau* whose brightness is nearer to black than to white. In the course of 20 min in complete darkness, the Eigengrau brightens somewhat, and diffuse clouds and coloured images appear for which the neuronal correlates are uncertain at present.

With its predominantly rod retina, the cat is behaviourally most active at scotopic luminance levels. There is some evidence that increasing steady illumination augments the tonic discharge of dark-adapted B neurones and simultaneously depresses that of D neurones in the retina [20, 66]. This occurs, however, only over a restricted range of about 2 log units, up to a value where the influence of the receptive-field surround begins to counteract that of the centre. In the lateral geniculate nucleus, B neurones respond transiently but otherwise are only slightly and irregularly affected by increasing luminance. However, geniculate D neurones are strongly depressed when illuminated at scotopic levels, as are retinal off-centre cells [66]. Both the latter D groups show substantial recovery of discharge rate during prolonged exposure.

In the scotopic range, dark-adapted humans find it difficult under certain circumstances to discriminate among steady luminance levels ranging over several log units. Despite this and the lack of a clear correlation between steady luminance and maintained discharge, increasing background luminance regularly and reliably attenuates the transient neuronal response of B neurones to a short test increment of constant intensity [66]. Under the same conditions, humans report that a test spot of constant intensity decreases in brightness as background increases, although the total luminance of the stimulus becomes progressively greater. The capacity of background illumination to reduce both the brightness of incremental test light and the associated phasic neuronal response shows that the retinal "gain control" [384] is operating, even though the average maintained discharge does not reflect absolute luminance levels. Absolute luminance may eventually be shown to be mediated by a relatively few specialized neurones.

Decrease of tonic activity in the D system with increasing luminance may explain the "*dark discharge*" of ARDUINI [8]. He noted a tendency for mass activity in the optic nerve and lateral geniculate to increase in the dark when recording with macroelectrodes. His results may have been due to D inhibition during light exposure. When inhibition by light is removed, the tonic discharge of the D system in darkness would appear as a total increase when mass recordings are made. This may be related to higher retinal metabolism in the dark [393a].

Ongoing Discharge, Steady Illumination, and Brightness Perception. Mass recordings with macroelectrodes have shown changes in ongoing activity (maintained or tonic discharge) in retinal ganglion cells and geniculate neurones with alterations of steady ambient luminance in cats [8] and monkeys [65]. Only a few neurones display a clear relationship between their average discharge rate and the wide range of steady background illumination.

In the dark-adapted cat and during dark test conditions, the maintained discharge rate in off-centre neurones is higher than in on-centre neurones, i.e. D activity exceeds B activity (see Fig. 1 Eigengrau, and LEVICK, Chapter 8). Low levels of steady illumination, however, obscure this difference, causing an increase in mean B discharge and a decrease in D activity, which brings the two groups to an overlapping range. Some off-centre ganglion cells which discharge continuously in the dark may stop firing completely for up to several minutes when exposed to strong light [250]. Some resume firing during light adaptation but do not reach the level of the dark discharge. Above the upper scotopic range there is less change in ongoing discharge of either off-centre or on-centre neurones, probably due to the stronger influence of the antagonistic field surround. Geniculate on-centre neurones show fewer maintained effects of diffuse luminance. In the cortex, the overall level of spontaneous discharge is rather low, at least when recorded from anaesthetized or curarized cats but long diffuse illumination causes a similar depression of D-neurones [250].

The perceptual correlates are less clear. We can recognize the large background light changes of *daylight and dawn* but we cannot distinguish absolute luminances in between. The adaptive ability of the visual system to function well over enormous ranges of light intensity, to ignore absolute luminance levels and to signal small variations around the mean level (brighter and darker) ensures brightness constancy.

Neuronal Adaptation. The source of critical field centre illumination for the centre influence on surround antagonism is of little importance. It may be either *background light*, *stray light* or small *light spots* projected on to the centre. Such background or stray light affecting the field centre is always present during the natural, steady, high ambient luminance of *daylight* conditions. For quantitative studies of centre influence in the laboratory it can be induced or reduced artificially within seconds when the experiment is begun during dark adaptation, and the light is varied only in the field centre of the recorded neurone. These rapid changes demonstrate that the effects occur *independently of photochemical adaptation*. Hence, the facilitation or suppression of surround antagonism induced by centre illumination appears to be a mechanism of *neuronal light adaptation*. It is not yet clear which retinal elements mediate these surround effects; it is probable that horizontal and/or amacrine cells and their graded potentials are involved. In spite of the uncertain classification of horizontal and amacrine cells as neuronal or non-neuronal, such rapid changes in surround organization may be labelled "neuronal adaptation" as opposed to photochemical pigment changes.

Ricco's Law and Background Illumination. In 1877 RICCO, an Italian astronomer, published his law concerning *threshold sensitivity* in relation to visual angle and luminance, and his rules have since been found valuable (within certain limits) for both psychophysical measurements [14] and neuronal responses [140]. RICCO's law suggests that area and luminance are equivalent, at least under threshold conditions. From RICCO's paper [375], however, it is apparent that he did not distinguish clearly enough between stimulus intensity and background luminance and therefore made a number of unnecessary experiments. Both psychophysical and neuronal recordings have shown that *background illumination* is essential for the long-lasting inhibition of off-centre neurones [66]. The effect of early D inhibition by light upon reciprocal B disinhibition in the lateral geniculate depends on a strong background of ongoing or maintained discharge (see p. 36 and Chaps. 8 and 18). Background illumination is also a condition of antagonistic field organization, but it recently became clear that what is relevant for antagonistic surround responses is mainly that part of the background light falling on the receptive field centre [139], and not diffuse background illuminance (see p. 34).

Brightness information is signalled in on-centre neurones by a *transient* increase of impulse rate produced by a luminance increment. Thus the phasic increase in frequency of the B system appears partly independent of absolute luminance and may also show little change with different maintained discharges. BARLOW therefore concluded that the significance of an extra impulse may be relatively constant over the whole range of adaptation levels [17].

Central Information Processing of B–D Input

Diagram of Dual Visual Input to Area 17. The neuronal organization of the central visual system with inhibitory interactions at different levels beyond the retina (lateral geniculate nucleus and area 17) is shown schematically in Fig. 3. On-centre-B neurones are depicted with white centres, off-centre D neurones with black centres. The interneurones are partly hypothetical. The diagram shows *collateral inhibition* (B versus B from surround to centre), *reciprocal inhibition* (B versus D and D versus B in field centre) and *lateral activation* (B versus D from surround to centre on the right). The scheme indicates that the reciprocal organization of the two subsystems B and D for brightness and darkness information continues at least up to the simple field neurones in the visual cortex (second order with contour detectors K). Complex field neurones appear in the third order of visual cortex transformation (upper left) signalling orientation and movement direction (KM).

This scheme, proposed with BAUMGARTNER in 1965 [256], may still be used as a simplified diagram of complex *neuronal interactions beyond the retina*, although BISHOP et al. [46] consider reciprocal inhibition between subsystems B and D as less important, and propose different diagrams for input to the cortex. Our scheme may remain useful for understanding the role of various levels and their inhibitory processes in the visual pathways (right), the receptive field transformation (left), and possible correlates in vision. Stereoscopic mechanisms are not included but can be found in BISHOP'S Chapter 4.

Processing of Projections of B and D Systems. Fig. 3 demonstrates some principles of *transformation of receptive field organization at different stages from the retina to the visual cortex*. The mechanisms of collateral and reciprocal inhibition in the central pathways are depicted with inhibitory interneurones, although this is to some extent hypothetical. The neuronal scheme also shows the *reorganization of geniculo-cortical afferents in area 17*, with transformation of predominantly monocular concentric fields to *binocular* fields of various shape and orientation which correspond to what HUBEL and WIESEL call *simple and complex fields*. During this reorganization the reciprocal antagonism between the B and D systems is gradually lost in favour of specialized information about *contour*, *orientation*, *shape*, and direction or speed of *movement* [27, 256].

This neuronal scheme should be regarded as a working hypothesis and as subject to adjustment if new experimental evidence becomes available. It may be interpreted as a functional guide to central visual structures rather than as a rigid wiring diagram which, in the present state of our knowledge of synaptology, could quite probably be wrong. When this scheme was published in 1965 it was not known that many simple (orientation-selective) neurones may also function as direction-selective motion detectors, as described in the latest work of BISHOP and coworkers [46]. On the basis of these results, the proposed dichotomy of two neuronal populations in the visual cortex for the detection of form and motion [256] has to be modified. There is no clear distinction between a "K system" for vision of contour and form and an "M system" for information on direction and velocity. The role of yet another dichotomy of retino-geniculate afference of X and Y neurones is not yet clear. Another open question is cortico-subcortical coordination and its relation to neurones detecting shape and motion in the higher visual areas. Special form-detecting cells may be found in the inferotemporal cortex of monkeys, as mentioned by GROSS (Chap. 23), and motion detection may develop in relation to cortico-tectal neuronal systems (see Chaps. 14 and 21). It seems premature to propose from these data a hierarchy of visual information processing for shape and movement, since the synaptology is only known for few interneurones [423]. See Chap. 20.

The hierarchy in the various levels of first- to fourth-order cortical neurones, indicated on the right of Fig. 3 with the receptive fields on the left, may not always correspond to a regular progression of integrative transformation at each level, as originally conceived by HUBEL and WIESEL in 1962 [233]. According to recent experiments of STONE (reported in Chaps. 2 and 21), complex visual neurones may also receive direct input from geniculocortical fibers [416] although they often respond with primary inhibition after stimulation of the visual radiation, as shown by DENNEY et al. [97] (see Fig. 19 in Chap. 21). Obviously, the neuronal coding at the various levels of the visual system will be much more complex than can be shown in such a simplified wiring scheme as Fig. 3. There are some higher neuronal transformations which proceed from the visual areas 17—19 to the temporal areas and to the contralateral visual cortex (see vertical meridian p. 71) and which receive multisensory information from other cortical areas (see p. 75), but they are not yet well known.

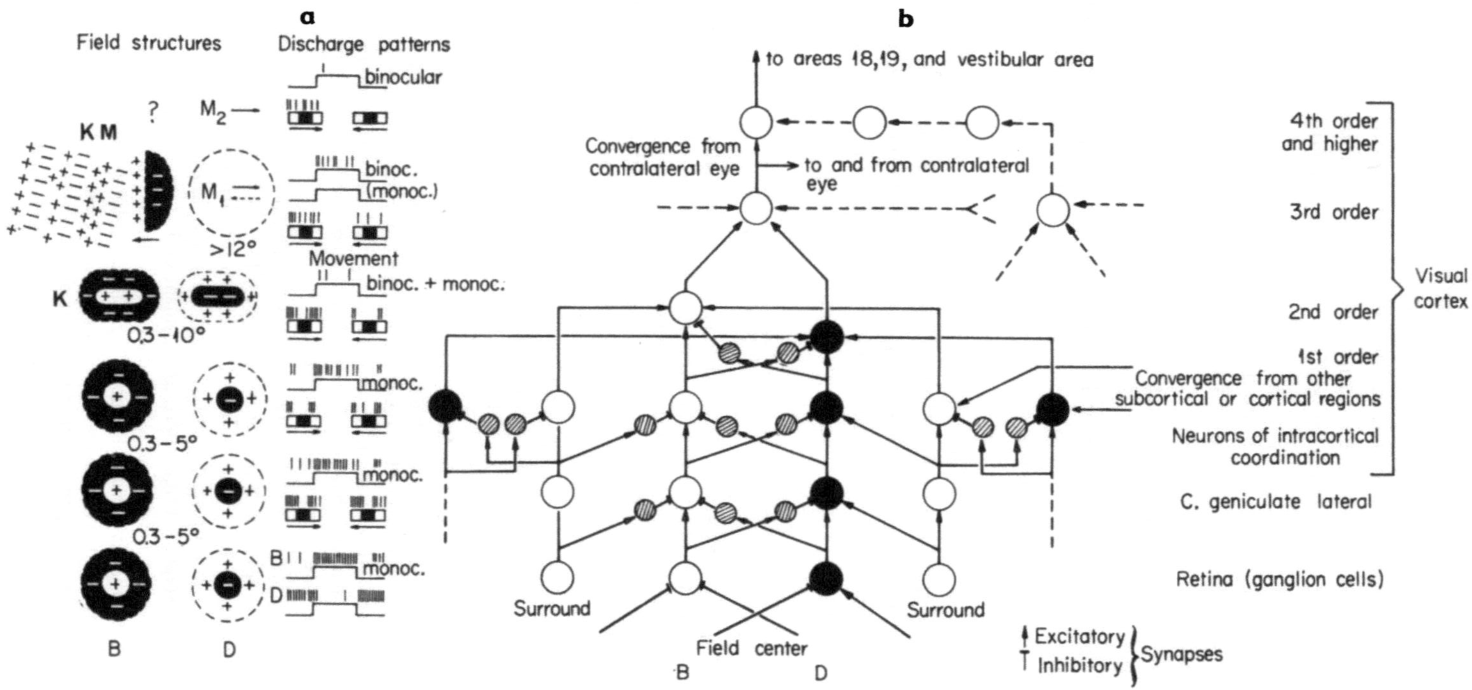

Fig. 3. *Schematic survey of the dual neuronal organisation in the cat's visual system.* Modified from JUNG and BAUMGARTNER (1965) [256].

Brightness Detection versus Contour Discrimination in the Visual Cortex. The dual organization of the reciprocal B and D systems with concentric fields in the retina and lateral geniculate nucleus as a mechanism for mediating brightness perception and contrast is transformed in the visual cortex. Perceptual correlates of cortical cells are less obvious, since neuronal responses are more specialized, and many neurones are not stimulated by diffuse luminance changes (see Part B, Chap. 21)

Besides having highly specialized and different classes of receptive fields, cortical neurones are characterized by additional features which distinguish them from retinal and lateral geniculate neurones: 1. selective adaptation with phasic responses to specific patterns and orientations which decline on repetition of the same stimulus; 2. slower spontaneous discharge rates; 3. binocular convergence with different monocular predominance; 4. topological order in columns representing receptive fields of similar orientation; 5. cortico-cortical connections with paravisual and contralateral areas; 6. convergence with non-specific afferents and other sensory modalities.

The functional significance of these specializations in the visual cortex seems to be a transformation of simple signals on brightness and darkness into more complex neuronal responses carrying information on *contoured patterns, direction-specific movements and stereopsis*. The latter has been investigated in detail by Bishop, Barlow, Blakemore, Pettigrew and coworkers (see Chap. 4).

Fig. 3. a *Receptive fields and neuronal responses of B and D subsystems.* Sequential transformations of field organization occur from retina to cortex. + signifies activation, — inhibition by luminance increment. Field centre responses predominate over surround responses. The corresponding neuronal responses to diffuse light (_⌈⌉_) and moving contrasts (→←) at different levels are drawn schematically. Orientation-selective stimuli are not depicted but would correspond to the receptive field axis.

b *Wiring diagram of the retino-cortical B and D afferents.* Possible interactions and connections within and between the two antagonistic subsystems for brightness (B) and darkness (D) information are schematized, beginning with *on- and off-centre neurones* in the retina. When located in the same retinal area, both subsystems interact by *reciprocal inhibition* (B versus D and D versus B). The left columns show *lateral inhibition* for field centre and surround of the B system, the right columns *lateral activation* from the surround of the D system. This is antagonized by reciprocal inhibition of B neurones having the same receptive field location (crossed connections with hatched interneurones). Thus, beyond the retina lateral activation of D also comes from B neurones whose *field centres* correspond to the *field surround* of the D neurones (right) in the lateral geniculate nucleus.

Up to the fourth synaptic relay from optic nerve to cortex, neuronal transmission *for B and D* occurs in *two parallel channels* which interact with each other. At higher levels of the cortex further convergence is established: this results in larger receptive fields from the second cortical order of simple field neurones upwards, and stronger binocular interaction (third order and complex fields). At this stage, a bright–dark antagonism of B versus D is lost in favour of special information on *motion detection* (M) and oriented *contour selection* (K), elaborated further in areas 18 and 19 to hypercomplex fields signalling contours of a certain length (not shown). Neuronal integration may be achieved in the cortex by second to fourth order transformation, converging at selective target neurones that signal certain directions of movement or pattern configurations *per se*. Oriented contour and motion detection may be combined in the same neurone (KM), as is often found in the complex neurones of Hubel and Wiesel [233] (example at upper left)

All these cortical transformations may involve an alternation of neuronal levels, one processing specific localized patterns, the next non-local patterns. This processing continues from area 17 to 18 and 19 (visual areas II and III), as investigated by HUBEL and WIESEL [234]. The neuronal mechanisms and their "wiring diagrams" are not yet known. HUBEL and WIESEL have proposed a convergence from "simple" field neurones [233]. The stimulation experiments of DENNEY et al. [97] have shown that *inhibition* appears to be essential in the process of transformation from radiation fibres to complex-field neurones.

Pathophysiology of the Dual Systems

Scotomas, Stabilized Images, and Dual B–D Systems. A successful application of the concept of B and D subsystems to *visual pathophysiology* was proposed in 1970 by GERRITS and VENDRIK [159]. It is based upon subjective experience with retinal scotomas and stabilized images. These experiments were made to elucidate the general interaction and adaptation of neuronal activity at various stages of the visual system. The authors assume parallel channels of the reciprocal B and D subsystems from the retina to the visual cortex and postulate that, after this, information processing of brightness and darkness signals proceeds in a hypothetical "higher centre". Whether this centre is located in the visual areas 17, 18, and 19 or in the temporal cortex is still an open question, and the exact neuronal correlations remain speculative. Unlike the lower visual neurones with their ongoing discharges, the higher centre is postulated to have a spreading capacity and neuronal silence after some adaptation time. This was deduced from experiments on the filling-in process of scotomas and from the slow disappearance of stabilized images. In normal vision, small eye movements tend to activate the B and D systems along the contrast borders, and thus preserve continuous perception against local adaptation. GERRITS and VENDRIK believe that the activity of the D system generates a "barrier" and limits the spread of B activity. The authors extend their experiments from brightness to colour. Since colour always disappears before brightness in stabilized images, they postulate a shorter adaptation time for colour than for brightness in the "higher centre".

These hypotheses have not yet been tested by direct recordings from nerve cells in higher visual areas, and they pose interesting problems for study by neuronal experiments. Figure 4 illustrates the hypothesis of GERRITS and VENDRIK concerning the neuronal correlate of contrast and filling-in processes in the retina and the visual centres of the brain.

The authors postulate a marked *spread of B activity* in the cerebral centres which corresponds to *filling-in processes.* It is limited by D activity at contrast borders. The D system shows stronger inhibition by light extending over a wider retinal area than B activity, which is most marked at contoured borders. At higher levels surface brightness is created by a filling-in process indicated at lower right. Although excitation is transmitted solely in the visual pathways, D inhibition correlates with brightness perception, probably by areal B disinhibition.

Amblyopia and Deprivation Studies. Experimental studies on visual deprivation and artificial squint have obvious implications for clinical syndromes of amblyopia.

Amblyopic defects of visual acuity in man may be related to defective organization of receptive fields and interaction of reciprocal systems. This parallels deprivation experiments in newborn cats and monkeys. CHOW (Chap. 9) surveys these studies.

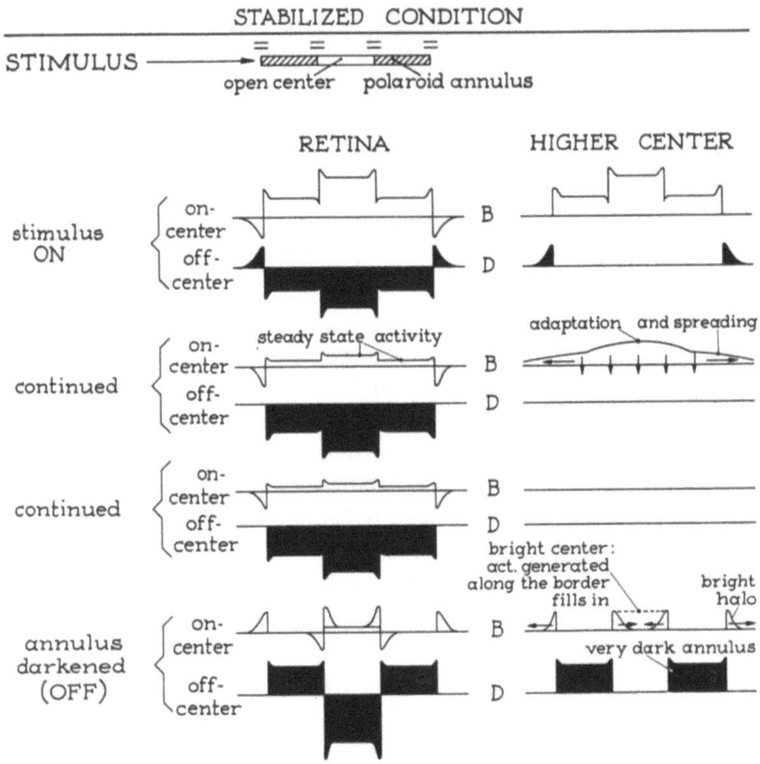

Fig. 4. *The neuronal hypothesis of* GERRITS *and* VENDRIK *concerning simultaneous contrast and filling-in processes in the lower and higher visual systems.* The activities of the retinal cells and higher-centre neurones are not derived from neuronal recordings but inferred from contrast experiments in stabilized images of normal human subjects and from observations in patients with retinal scotoma. From GERRITS and VENDRIK, 1970 [159]. *White area above axis* = excitation of B system (on-centre neurones). *White area below axis* = inhibition of B system. *Black area above axis* = excitation of D system (off-centre neurones). *Black area below axis* = inhibition of D system = stabilized contour

They show that diffuse illumination of the eye prevents only the gross retinal and central atrophy which occurs after complete early light deprivation or damage to the retino-geniculate system, but cannot preserve pattern vision. Early experience of visual contrast stimulation of the retina appears necessary for pattern vision leading to shape recognition. One may assume that the *normal development of the contrast mechanism of the dual systems, their receptive field organization and reciprocal interaction, are dependent on the visual experience of the retina and the visual centres.*

Deprivation experiments of WIESEL and HUBEL [456] also demonstrated central disturbances in the visual cortex. Innate receptive field organizations of simple field neurones, although present in area 17 of newborn kittens before visual experience, partly regress after exclusion of pattern vision during the first months after birth. Thus, the normal development and maintenance of visual perception needs experience of form discrimination for the cerebral substrate of pattern vision.

Nearer to human amblyopia are recent experiments by VON NOORDEN and coworkers [343, 344]. Visually deprived rhesus monkeys (lid closure at different ages after birth) were tested for visual acuity with Landolt rings before and after deprivation. Severe visual impairment without recovery developed only when visual deprivation was begun during the first four weeks of life. Deprivation at a later age, however, caused less or only transitory sight impairment. Good and rapid recovery of visual acuity occurred when lid closure was performed in monkeys aged three months or more. Thus, in monkeys, the critical time when lasting damage is caused, without sight restoration, is about 3–4 months after birth. The corresponding time in man may be much longer.

V. Neuronal Receptive Fields and Perceptive Fields in Human Vision

Receptive Fields and Perceptive Fields

Definition of Receptive and Perceptive Fields. In neurophysiology the response characteristics of receptive fields are measured directly by recording from visual neurones [193]. In human vision, effects analogous to those of receptive fields can be indirectly demonstrated [25, 259] through phenomena which correspond to alterations in the neuronal discharges of the cat's visual system (Fig. 5). Perceptual integration apparently is the end result of co-ordinated organization of receptive fields in neuronal populations. To avoid any confusion between the objective and subjective aspects of information processing, it is proposed to distinguish them by the terms *receptive fields* and *perceptive fields*, respectively [253, 259].

Receptive fields are determined objectively by means of recordings from visual neurones, and their organization is explored by localized visual stimuli.

Perceptive fields are defined as the *psychophysical equivalents of visual receptive field organization in man* [253], and they can be estimated by visual phenomena, including illusions.

Functional Significance of Perceptive Fields and Neuronal Organization. It may be useful to compare receptive and perceptive fields as a bridge between neurophysiology and vision, although it greatly oversimplifies the complex neuronal interaction of cell assemblies at different levels and the unknown perceptive transduction of visual mechanisms. Yet the parallels between single neuronal events and visual contrast phenomena shown in Fig. 5 demonstrate that some elementary laws of nerve cell excitation and inhibition observed in the retina are still effective in higher visual structures and can influence perception.

In human vision perceptive fields are reflected most clearly in the contrast patterns of the *Hermann grid* (Figs. 5–7). Thus they are derived from visual illusions. These subjective phenomena in man have an objective correlate in the neuronal discharges of the cat's visual system (Fig. 5B).

Perceptive fields estimated subjectively in man, and neuronal receptive fields determined by direct recordings in cats have been discussed previously [259] in relation to neuronal populations. Human and animal experiments differ not only in regard to their methods but also in the complexity of their results.

Psychophysical data represent the final stage of integrating numerous neuronal signals, all of which contribute to vision in a way still unknown. Since experiments in animals deal mostly with single-neurone action, the effects of stimulating many neurones simultaneously remain uncertain. Amongst the various mechanisms of interaction and co-ordination, only the reciprocal relationship between the subsystems B and D (for brightness and darkness information) has been thoroughly investigated (see p. 21). The antagonistic responses of these two systems in-

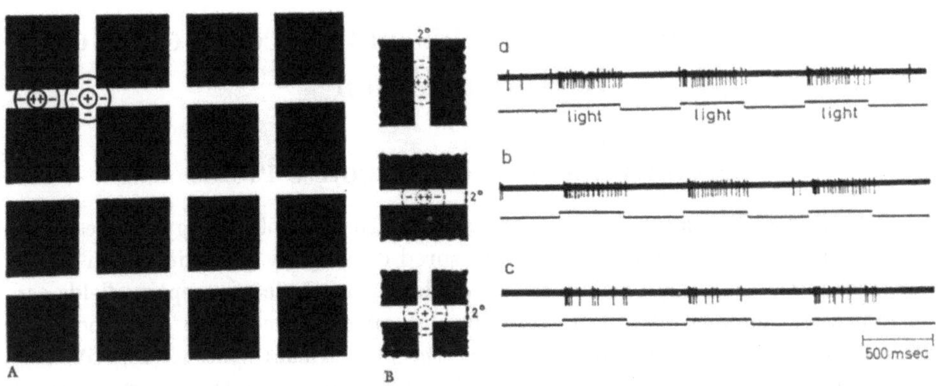

Fig. 5. *Stimulation by the same Hermann grid of perceptive fields in human vision and of receptive fields in the cat's cortex.* Left, subjective effects (A); right, objective recordings of neuronal responses (B). From Jung and Spillmann (1970) [259].

A *Hermann grid* [215] *in man.* Grey spots are seen at the bar intersections except when they are fixated foveally. Projections of perceptive fields are shown in two critical positions relative to the grid. They illustrate how differences in lateral inhibition cause stronger activation (++) of neurones "looking" at a bar than of neurones "looking" at an intersection (+). The disappearance of the illusion with fixation may be explained by the smaller size of the foveal perceptive fields whose entire centres and surrounds are inside the white bars and intersection.

B *Response of a first-order B neurone in the visual cortex of the cat* to various positions of the Hermann grid within its receptive field. In these records by Baumgartner et al. [388] the on-response of the neurone is seen to be diminished when it is stimulated by intersecting horizontal and vertical bars. This reduction in the firing rate is consistent with the attenuation of brightness seen at the grid intersection. With stimulation by one or the other bar (positions a and b), the neuronal response was more than twice as strong as for the intersection (position c). These results are accounted for by differences in the relative amounts of surround illumination and lateral inhibition. The neurone depicted here had a receptive field 6° in diameter, located 20° parcentrally. Its behaviour is representative of neurones having concentric fields in the retina or lateral geniculate and for first-order cortical B-neurones. In the oblong "simple" field neurones of Hubel and Wiesel, the response depends on the orientation of the grid relative to the field axis [388]

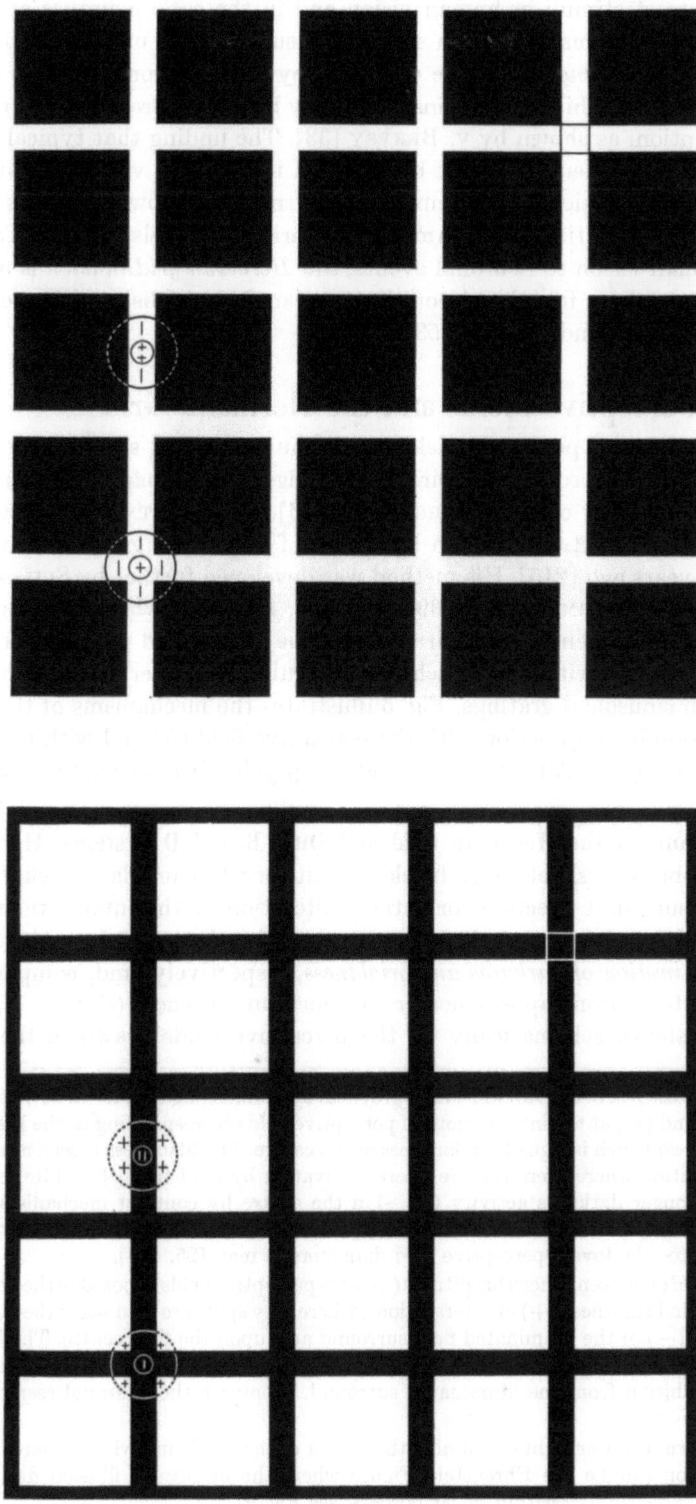

Fig. 6. HERMANN's *contrast illusion* [215] *in black and white grids*. The apparent diminution of brightness and blackness at the intersections as compared with the bars is explained by the weaker surround contrast, i.e. stronger surround influence on centres of perceptive fields. This corresponds to the receptive field organization of the neuronal B and D systems [253]. In the receptive fields + marks neuronal activation by light, and − inhibition by light or activation by darkness.

dicate that contrast stimuli in human vision and in the cat's neurones are processed in a comparable manner. Such similarities suggest that under certain conditions *neuronal mass function* can be described by the same principles as apply to single cells. Lateral inhibition, for instance, may be demonstrated from "neural units" of perception, as shown by v. BEKESY [33]. The finding that typical firing patterns of individual neurones may be reflected in complex visual phenomena implies that psychophysical measurements, in turn, may allow inferences to be made about neuronal action [249]. Among the various parallels listed in Table 2 that relate human vision to neuronal events, the *Hermann grid illusion* is an excellent example of how functional correlations can be established between the visual systems of man and cat [25, 253].

Perceptive Fields and the Hermann Grid

The investigation of perceptive fields in the human visual system relies predominantly on indirect procedures. Direct recordings from human visual neurones were made in only a few cells by MARG et al. [313]. BAUMGARTNER, in 1960, was the first to estimate field size in man by means of a contrast grid, described by HERMANN 100 years ago [215]. His method was developed further by SPILLMANN, KORNHUBER and SINDERMANN [278, 394, 395, 404, 405, 407], who determined the diameters of both field centres and surrounds in the central and peripheral retina. Their results agree well with data which BRYNGDAHL [69, 70] derived from brightness matches in sinusoidal gratings. Fig. 5 illustrates the mechanisms of the Hermann grid illusion in conjunction with the perceptive field (A) and with neuronal correlates in the cat (B). All this corresponds to psychophysical contrast studies [47, 371].

Two Variations of the Hermann Grid and Dual B and D Systems. Hermann grids in reverse brightness relations, black on white or white on black background, show reversed surround effects of brightness alteration at the intersections, but both are caused by the same principle, namely *reduced surround contrast.* This results in a *diminution of darkness and brightness*, respectively, and, compared to the bars, the intersections appear nearer to a medium greyness (Fig. 6a, b). The explanation is shown schematically by the perceptive fields drawn in the grid.

Fig. 6. a The *black grid* evokes an illusion of greyness by diminishing the lateral activation of light in the surround (+) at the intersections of perceptive fields corresponding to the D system (off-centre neurones) which is signalling darkness in the centre. The blackness is seen as strongest at the bar position where there is more lateral activation by light from the white squares, thus inducing stronger darkness activity (— —) in the centre by contrast mechanisms. The grey spots disappear with foveal fixation and reappear with increasing distance when the bar width approximates the foveal perceptive field diameters in man [25, 405].

b The *white grid* evokes enhanced brightness (+ +) in perceptive fields exposed to the bars and a relative decline in brightness (+) at intersections, where grey spots are seen when the stronger lateral inhibition (—) of the illuminated field surround acts upon the field centre. The brightness corresponds to neuronal activation of on-centre receptive fields (B system) but is attenuated by lateral inhibition from the illuminated surround. (Compare the neuronal responses in Fig. 5b).

The lines drawn at lower right continuing the border of the black and white squares do not destroy the illusion (unlike the Ehrenstein Figure where the brightness illusion disappears within linear borders, see Fig. 9)

Perceptive fields having a centre approximately corresponding to the bar diameter and exposed to the bar, receive less light in the surround and, consequently, have less lateral inhibition in the B system and more lateral activation in the D system with stronger brightness (++) or darkness (−−) signals from the centre. Conversely, fields exposed to intersections produce more lateral inhibition in B and less lateral activation in D with the consequence that there is less intense brightness (+) or darkness (−) from the field centre.

The critical values for the perceptive field size are very similar for black and white grids. These subjective phenomena have their objective neuronal correlates in the opposite receptive field organizations of the B and D systems and the similar diameters of their concentric fields in the retina and lateral geniculate. The perceptive field diameters estimated from HERMANN'S grid *increase with retinal eccentricity*, in good agreement with neurophysiological measurements of receptive fields. Fig. 7 shows that the increase of field diameter in man (a) and cats (b) is linear in both, although the slopes are somewhat steeper in cats, which have larger receptive fields in the peripheral parts of the retina. The small diameter of the human perceptive fields corresponding to foveal fixation explains why the grey illusion disappears on fixation. The illusion reappears when the white and black grids are seen from further away, reducing the visual angle of the intersection to approximately 5–10 minutes of arc for white and black grids. Thus, BAUMGARTNER'S explanation of the Hermann grid phenomenon by surround effects [25] is equally valid for the B and D systems having concentric field centres. A tracing of a B neurone in the cat stimulated by the Hermann grid is shown in Fig. 5: at intersections it discharges at about half the spike rate of stimulation in the pure bar position.

BAUMGARTNER and coworkers [388] made neuronal recordings with grid stimulation at different levels of the visual system, most of which are unpublished. Fig. 5b gives an example from the visual cortex.

BAUMGARTNER and coworkers [388] reported that the "simple" field neurones of HUBEL and WIESEL in area 17 of the cat [231] show selective activation when stimulated by a Hermann grid oriented to coincide with their receptive field axes. However, a cortical modification appears to be relevant only for grid orientation and inclination of bars. The brightness illusion as such is sufficiently well explained by the action of concentric fields projecting to retinal, geniculate and first-order cortical neurones. Perhaps the differences in the grid illusions caused by oblique angles [380] indicate a preference for interrelations of simple fields in certain perpendicular orientations.

Another explanation was proposed by the late MOTOKAWA [329] relative to the disappearance of the Hermann grid illusion between oblique rhombic patterns, which had been described by RONCHI and BOTTAI [380]. Without mentioning the relation to receptive field organization, he explained the effect of rhombic bar configurations by differences in retinal induction. It is a moot point whether this corresponds to differences of lateral inhibition in the receptive field surround when it is exposed to rhombic instead of square grids. The intersections of the former differ less from the bars than do the square intersections.

The Hermann grid is the most impressive *demonstratio ad oculos* that similar laws govern the mechanisms of human vision and the neuronal coordinations in the visual system of cats and monkeys.

The subjective brightening and darkening effects at the intersection of a grid may be explained by relative differences of lateral inhibition and activation. A neurone of the B system "looking" at an intersection receives more light in its receptive field surround and produces more lateral inhibition than when it is stimulated by a bar (Fig. 5b). As a consequence, its firing rate is diminished and it signals less brightness. For a neurone of the D system, the same argument holds but with all signs reversed. Thus, D neurones projected to the intersection signal diminished darkness. According to BAUMGARTNER's hypothesis [25], the Hermann grid effect reaches its maximum when the bar width on the retina is approximately equal to the diameter of the receptive field centre. For bars substantially thicker or thinner than this optimal width, brightness enhancement declines until it is cancelled out. An example of this is the disappearance of the contrast illusion at central fixation. The illusionary grey spots become invisible on account of the small size of the foveal perceptive fields but reappear when the visual angle of observation of the grid is decreased [25, 259, 405]. For other variations of the grid see [329, 363a, 380] and p. 51.

Perceptive Fields in the Fovea and the Peripheral Retina. The size of the perceptive field centres in the *human fovea*, as estimated by the Hermann grid illusion was found to be around 5 min of arc [25]. This value corresponds to 25–30 μ on the retina [25]. For centre plus surround, diameters were reported to be in the neighbourhood of 10′ [405], whereas the latest estimates by SPILLMANN [405] are around 18 min of arc. This result agrees well with WESTHEIMER's [453] measurements using increment thresholds for a small stimulus flashed upon a surround of variable size ("lateral sensitization").

In the *peripheral retina*, perceptive fields are much larger. SPILLMANN [404, 259] found that mean diameters increase linearly between 20 and 60° eccentricity. The field centre doubled from 1.5–3° (Fig. 7a). Average values for positive and negative contrast (white grid on a black background and vice versa, shown in Fig. 6a and b) were essentially the same. This suggests that, at a given eccentricity, the B system (signalling brightness) and the D system (signalling darkness) have similar receptive field centres for on- and off-centre neurones. In animals, slightly larger diameters were reported for off-centres than for on-centres [140, 232] (see Fig. 7b).

The increase in the diameter of human perceptive field centres towards the periphery corresponds rather well to HUBEL and WIESEL's measurements [232] of receptive field size of optic nerve fibres in monkeys. It also shows the same trend as the data from the cat's retina recorded by FISCHER and MAY [140]. These investigators obtained a somewhat steeper slope of field increase than was found psychophysically and also described a slightly smaller diameter for on-centre neurones than for off-centre neurones (Fig. 7b). However, the various studies agree sufficiently well to suggest that both objective neuronal measurements and indirect psychophysical estimates are based on the same retinal principles of receptive field organization. It is conceivable that the *invariances* established by FISCHER and MAY for the sensitivity, size and position of receptive fields in the cat [140] hold also for the human eye.

The disappearance of the Hermann grid illusion with direct foveal fixation can be explained by the smaller size of perceptive fields in the fovea [25], but other mechanisms such as after-images and eye movements may also contribute. See discussion by SPILLMANN (1971) [405].

The brightness alteration at intersections varies with differences in contrast configuration (angular orientation and extension of the edges). PAYNE and ANDERSON [350] attempted an objective measurement of brightness diminution by taking reaction times as measures of flash brightness for corner stimulation. Latency increases beyond the bright border as signs of darkening effects were longer for rectangular borders than oblique borders. PAYNE, however, has not related his experiments to BAUMGARTNER's findings of the reappearance of the illusion in the foveal region in grids with a smaller visual angle, corresponding to the diameter of foveal receptive fields [25], and other variations of stimulus conditions such as were used later in quantitative experiments by SPILLMANN [405].

PAYNE's results for oblique borders agree with subjective estimates of brightness. The diminution of the greyish illusory patch at the intersections of bars presented in *oblique instead of rectangular cross-orientation*, first observed by PRANDTL [368] in 1927 and later studied by

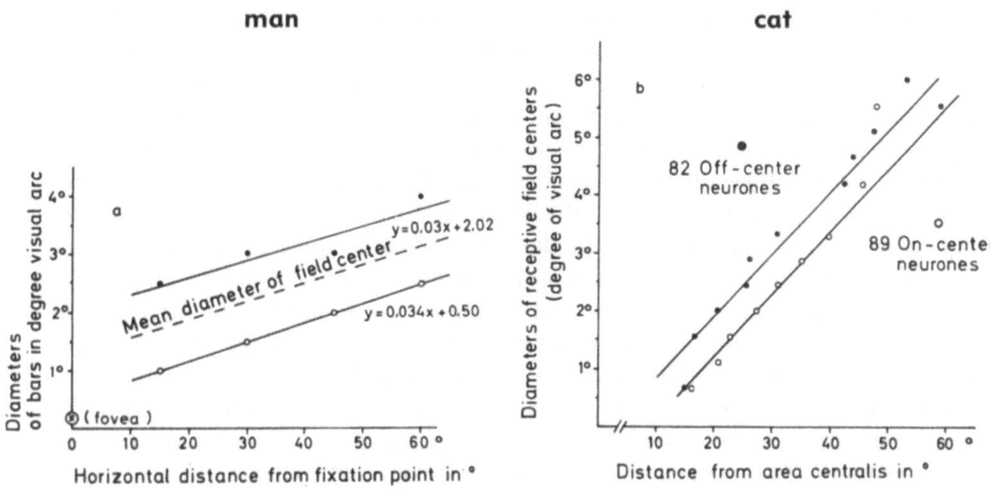

Fig. 7. *Comparison of perceptive fields in man with neuronal receptive fields in cats at the periphery of the visual field.* From JUNG (1972) [253]. In man, the field centres were estimated for bar width and threshold of the Hermann grid illusions by SPILLMANN [259, 404]. In cats, receptive field centres were measured directly from retinal ganglion cell axons in the B and D subsystems by FISCHER and MAY [140].

a *Estimates of human perceptive field centres as a function of retinal eccentricity*, based on the lower and upper thresholds of the Hermann grid illusion [259, 404]. In this experiment by SPILLMANN a black grid was presented on a white background. The occurrence of grey spots at the intersections, caused by a relative decrease of lateral activation in off-centre neurones (D system), was used as a criterion. Grids of various bar width were shown at different horizontal distances from the fixation point, and the critical bar width at which the illusion appeared (circles) or disappeared (dots) was determined. Both types of threshold increase almost linearly from 90 to 180 min of arc between 20 and 60° in the periphery.

b *Measurements of receptive field centres of on-centre and off-centre neurones in the cat as a function of retinal eccentricity.* Data redrawn from FISCHER and MAY [140], who defined the field boundaries by the isosensitive line corresponding to 1/e sensitivity in the central point of the field centre. Towards the periphery of the retina, and for the same range of eccentricities, receptive fields recorded in the cat increase at a greater rate than perceptive fields estimated in man. Off-centre neurones appear to have slightly larger receptive fields than on-centre neurones in the cat, like those in the monkey's optic nerve [232]. However, no essential difference was found in the human visual system between thresholds for white grids (B system on-centre) and black grids (D system off-centre)

Ronchi and Bottai [379, 380] and Motokawa [329], may involve cortical neurones of different axis orientation. These neurones show stronger discharges when the bar borders of a rectangular Hermann grid are projected parallel to the inhibitory flanks of their field axis [388] and weaker discharges at the intersections similar to the concentric field neurone, depicted in Fig. 7. Oblique bars might stimulate a smaller population of simple field neurones at the bar borders than rectangular grids. Such grids of different angles, however, have not yet been studied with neuronal recordings. Field columns with horizontal and vertical orientations may occur more frequently in the human visual cortex (see p. 71) although in cats Hubel and Wiesel found about equal distributions of all orientations in their simple field columns [233].

Contrast Enhancement and Background Luminance. Contrast enhancement changes with varying figure-ground ratio as shown by Spillmann and Levine [407]. They experimented with a modified Hermann grid, graded in two directions, and having fifteen shades of a grey scale viewed against five uniform backgrounds ranging from white to black. These experiments show the importance of *relative contrast*. Spillmann's modification of the Hermann grid (depicted in Fig. 8) demonstrates the conditions under which *maximum and minimum contrast enhancements* occur. His grid consists of stripes of varying luminance superimposed on a bisected background of black and white. In Fig. 8a the illusory dark and bright patches are seen at almost every intersection. However, when the background is rotated through 180 degrees so that the white and black halves are now in opposite positions (Fig. 8b), the illusion is greatly reduced, except for those regions with the highest contrast in the lower left and upper right corners. Obviously, the Hermann-grid illusion depends on the *contrast relation between the grid and its background*. Both brightness and darkness enhancement at an intersection vary with the *difference in luminance between the continuous stripe* (intersecting), *the discontinuous stripe* (intersected), *and the background*. These relationships were quantified by Spillmann and Levine [407] and tentatively explained in terms of lateral inhibition in the B system (on-centre neurones) and lateral activation in the D system (off-centre neurones) related to background contrast.

The essential results can be seen immediately on looking at Fig. 8 in which the vertical lines are continuously intersecting and the horizontal lines discontinuously intersected. Parts A and B are distinguished only by their 180° reversed backgrounds, in opposite bright-dark bisection. The illusion is most distinct when the *contrast ratio between intersecting continuous stripes and background is minimal* while that *between intersecting and intersected stripes is maximal* (upper left and lower right in A). Equi-reflectant stripes similar to the original Hermann grid show a less distinct patch illusion at the intersections (middle part of A and B). The background determines the lighter or darker appearance of the intersections. Brighter patches appear with a white background; similarly, darker patches appear with a black background. The visual angle under which the maximal illusions of Fig. 8 appear is less critical than in ordinary Hermann grids.

Ophthalmological Application

Perceptive Fields, Retinal Interaction and Quantitative Perimetry. Westheimer in 1970 introduced a new method for measuring sensitizing interaction of the human retina [452, 453]. Enoch and coworkers [118–121] applied this technique

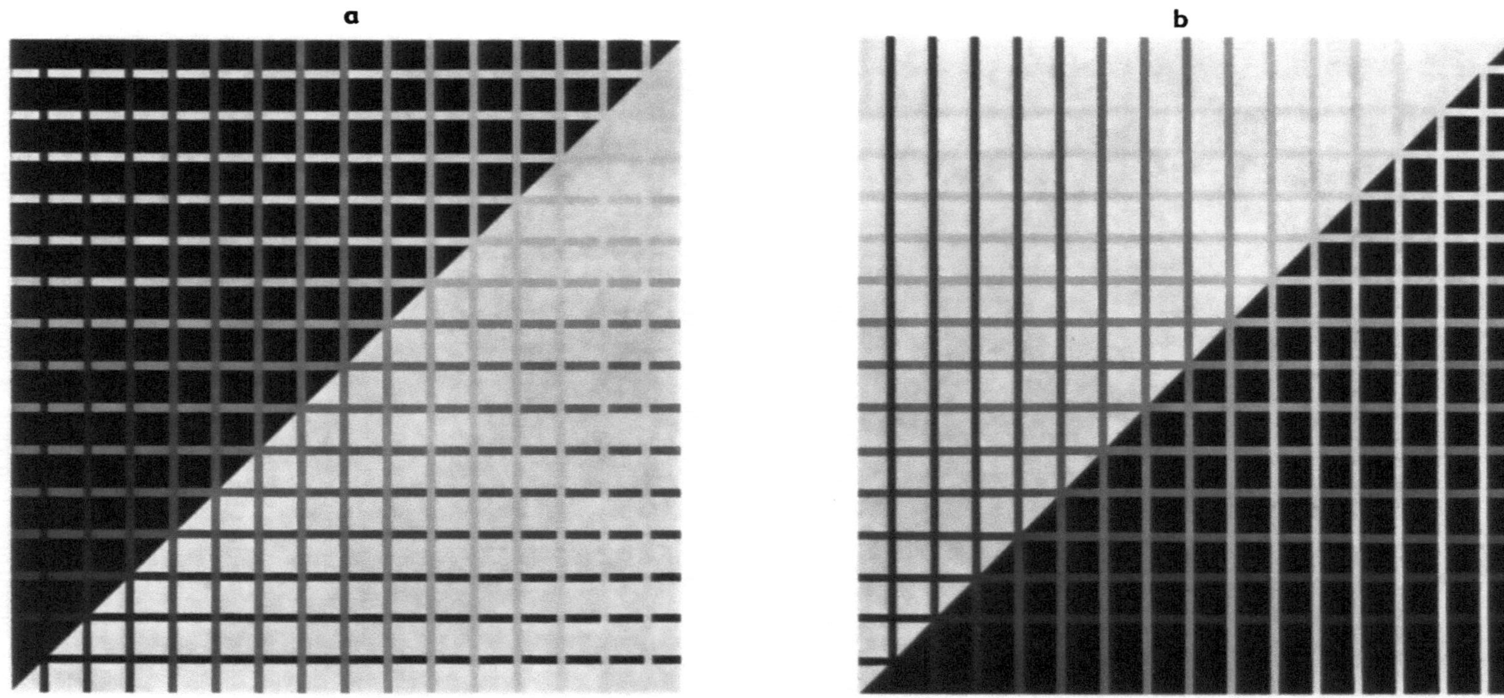

Fig. 8. *Modified Hermann grids for demonstration of maximal and minimal contrast enhancement with white and black backgrounds.* From SPILLMANN and LEVINE, 1971 [407]. Note that the *intersecting bars (vertical)* run continuously across the full length of the grid subdividing the *intersected bars (horizontal)* at regular intervals.

a *Maximal-contrast illusion*, produced by intersecting (vertical) and intersected (horizontal) bars of various grey shades, displays illusory spots at almost every intersection. *Dark patches* occur on the black half of the background and *bright patches* on the white half. The illusion is most distinct where the contrast between the two kinds of bars is greatest (upper left and lower right). Less marked effects are elicited by a pair of equiluminous stripes (as in the orginal Hermann grid of Fig. 5). b *Minimal-contrast illusion* is caused by the reversal of the bisected background relative to the grid. In this configuration the illusion is drastically diminished in strength. However, in some places it may still be seen with rapid eye movements and after change of gaze from the black to the white zone (or vice versa), that is on favouring the formation of after-images to the *quantitative static perimetry* of normal subjects and of patients with visual disturbances. The technique was termed the "Westheimer function" by ENOCH et al. [119]. Spatially interactive properties of human observers were related to data of receptive field dimensions obtained from animals [121]. The results may correspond to perceptive field estimation with by methods such as the Hermann grid in man, although comparative studies of these techniques in the same subjects are lacking. ENOCH et al. claim that receptive field alterations in eye diseases and at the border of visual field defects can be tested by their method. The authors conclude from studies of retinal lesions that the inhibitory component of the Westheimer function has its origin in the inner retina [121].

Before we go on to discuss possible neuronal correlates of contour discrimination in cortical neurones having oriented receptive fields, the visual phenomena of brightness and darkness induced by linear patterns and border cues should be illustrated and viewed. Two illusions, studied by EHRENSTEIN [117] and CRAIK [89], are particularly suited to demonstrate such *interactions of linear patterns and surface brightness or darkness* (Fig. 9 and 10).

VI. Brightness Induction by Linear Patterns and Border Cues to Brightness

Two Kinds of Visual Contrast for Border and Surface

Border and Brightness Contrast. Since MACH and HERING made their observations, contrast phenomena have been divided into: 1. *border contrast* at edges with contour enhancement and Mach bands, and 2. *brightness contrast* of extended surfaces surrounded by contours. HERING called the latter „Helligkeitskontrast" or „Flächenkontrast" [213, 214] to distinguish it from „Grenzkontrast", first systematically investigated by MACH [304]. Hence, VON BÉKÉSY [34] recently proposed two types of lateral inhibition: the first, "*Mach-type inhibition*", for border contrast and the second, "*Hering-type inhibition*", for brightness contrast. At present neuronal correlates are known only for Mach inhibition. For an understanding of Hering inhibition and areal contrast, more neurophysiological investigations are needed (see p. 54).

Evidently border contrast of the Mach type is important for *pattern vision* and the recognition of form boundaries. Some relations to *neuronal contour abstraction* in cortical cells with oriented receptive fields of the simple Hubel-Wiesel type are likely to occur at higher levels (see p. 41 and Chap. 2). Brightness contrast of the Hering type appears related to the *perception of surfaces* having different shades of brightness and colour. Various brightness illusions demonstrate a transition between the two types of contrast phenomena, and this suggests interactions of their neurophysiological mechanisms. The *Hermann-grid illusions*, being related to perceptive field size, may represent mainly the Mach type of lateral inhibition in on-centre neurones of the B system or lateral activation in the D system (Fig. 6), but some contribution of brightness contrast is conceivable. The *Ehrenstein illusion*, which induces surface brightness by lines, may bear witness to possible interactions between contour and surface signalling systems (Fig. 9). In the *Craik-Cornsweet illusion* gradients in the luminance distribution at contours affect surface brightness (Fig. 10).

The Mach and Hering Types of Lateral Inhibition. To explain the different features of visual contrast, VON BÉKÉSY (1968) distinguishes *two types of lateral inhibition in vision* [34]. The first or Mach-type inhibition produces the familiar overshoot and undershoot at boundaries of luminosity and thus causes *edge discontinuites in luminance distribution*, as in the Hermann-grid illusion. The second or Hering-type inhibition covers a much *larger area* and produces *gradients in brightness sensation*. There are special differences between these two inhibitions on exposure to *flicker*, but so far no correlation has been made with the neuronal findings caused by flicker stimulation, and the neuronal mechanisms are not yet clear. BÉKÉSY found that, unlike the Mach-type, the Hering-type inhibition exhibits a gradient which is relatively independent of differences in luminosity. Thus the Hering inhibition near the edge of a brightness step resembles an all-or-none phenomenon. Since the gradient is not dependent upon the magnitude of the brightness difference, it can increase with the number of brightness steps. It remains for further neurophysiological experiments to elucidate the neuronal basis of the Hering inhibition and to distinguish it clearly from the retinal mechanisms of concentric receptive field organization which seem more appropriate for the Mach-type. The possible cerebral mechanisms which could regulate these inhibitory phenomena are not yet clear and their relationship to the Craik-Cornsweet brightness illusion (Fig. 10) also requires further investigation (see p. 57 and 63).

Surface Brightness and Area Contrast. We see and recognize patterns mainly by their contours, which mark the Gestalt borders. Within the contoured surfaces, however, in spite of contour dominance, *areal brightness or colour* appears as a more or less diffuse quality of area contrast. This requires a visual apparatus that can compensate for contour abstraction in favour of surfaces, also at the higher visual levels. Such wide-field effects have not yet been adequately studied at the neuronal level and there are few psychophysical studies besides BÉKÉSY'S work [34]. The homogenous perception of brightness and colour over large surfaces – like most other wide-field effects – cannot yet be correlated with neuronal activity. One may assume that *averaging processes* occur in such wide fields enclosed by contours, similar to the filling-in of scotomas and stabilized images investigated by GERRITS and VENDRIK [159] (see p. 42). Such averaging may either be induced by border configurations, as in the Craik-Cornsweet-illusion (Fig. 10), or occur in the absence of outer borders during diffuse stimulation of large parts of the visual field up to a Ganzfeld divided by a single vertical line. Examples of the latter are depicted in Fig. 2—9 of ZUSNE'S book [468].

Ehrenstein's Brightness Illusion in the Absence of Area Contrast

In 1941 EHRENSTEIN described a brightness illusion elicited by lines only. In several editions of this book [116, 117] he used this demonstration as an argument for his *Gestalt* thesis, that certain patterns (structure configurations), and not just contrast between adjacent areas, determine brightness perception. Ehrenstein's illusion is best seen in a white area within the converging ends of 4–6 radial black lines (Fig. 9a) or in a dark area with white lines. These lines must exceed a certain length and should be four or more in number to induce brightness enhancement.

The central spot is usually seen as round, but changes into the form of a square when the lines are thickened. This change in shape may perhaps be induced by ordinary simultaneous border contrast which appears between neighbouring areas of black and white. Paradoxically, the additional physical contrast does not increase the subjective brightness of the central spot but rather decreases it [259].

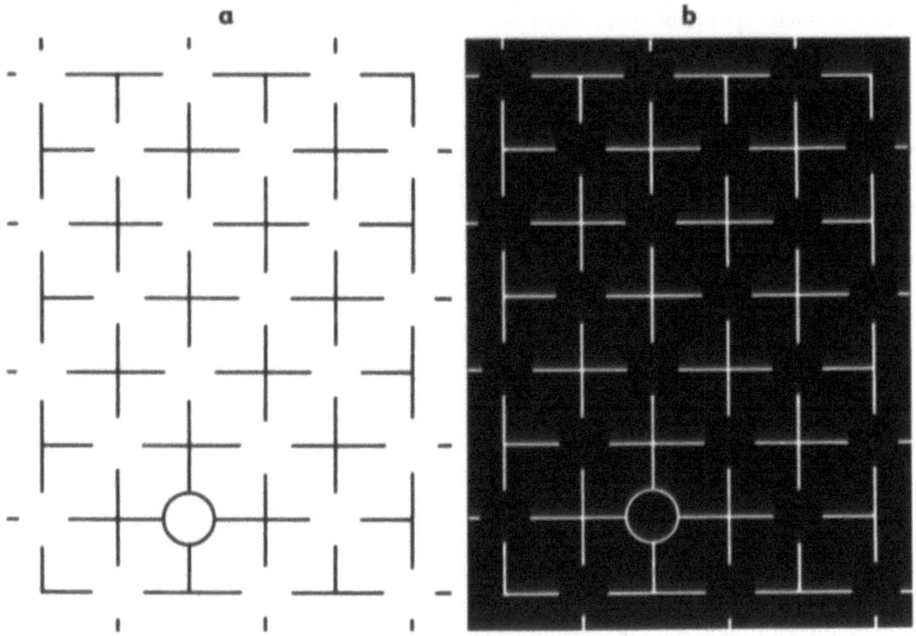

Fig. 9. EHRENSTEIN'S *brightness illusion induced by linear convergence.* Black lines (a) induce brightness enhancement, white lines (b) blackness enhancement in the empty centre towards which the lines converge. To produce this phenomenon, there must be at least four lines of different orientation. The figure was contructed by SPILLMANN and used in his experiments with FULD [406].

a *Brighter spots* appear in a round area induced by the four converging lines. These spots in turn induce an added illusion of *diagonal strips of a bright chain* connecting the enhanced areas, although the background luminance is objectively the same.

b The opposite phenomena of *blacker spots* and dark diagonals of black chains (in the absence of separating lines) are induced by white lines on a dark background. The brightness illusion *disappears within the linear borders,* when the central spot is surrounded by a black or white circle (lower left). The fact that in a and b the illusions correspond may indicate a subjective correlate to the opposite, but in principle similar, receptive field organizations of the visual subsystems B (brightness information) and D (darkness information)

Thus, brightness enhancement in the centre of the pattern can hardly be explained on the basis of circular perceptive fields. The disappearance of the illusion under prolonged steady fixation or when enclosed by a circle rather suggests participation by the "*simple*" *field neurones* described by HUBEL and WIESEL [231, 233]. These explanations become more attractive if "simple" field neurones not only mediate edges, lines and orientation but also contribute to brightness sensation. A neurophysiological basis might be that the ellipsoid receptive fields of these units can be divided into two subclasses carrying brightness and darkness information, respectively: B neurones with oblong on-centres and D neurones with oblong off-centres. Both these organizations could be results of a transformation from concentric fields of the retina and the lateral geniculate nucleus [231, 233].

A somewhat different Ehrenstein figure was reproduced from his book [117] in other papers (Fig. 7b in [259, 253]) in order to demonstrate a *brightness contrast paradox:* the linear brightness enhancement induced by four lines is less effective when these lines are thicker, although the black/white surface contrast is enhanced. This figure also shows another patterned illusion. During foveal fixation the areal brightness enhancement is replaced by a smaller contrast illusion. When the figure is viewed at a suitable distance, a grey *diagonal cross* is seen within the brighter borders of the central area. This cross illusion is best seen with thin or medium line thickness, so it must be more dependent on linear convergence than on black/white contrasting areas [259].

The systematic studies of the Ehrenstein illusion by SPILLMANN and FULD are summarized in Table 3. They show that the receptive field organization of the lower visual system cannot fully explain the Ehrenstein phenomena, although as a model it suffices to explain the Hermann-grid illusion. The binocular effects suggest some participation of the *visual cortex*, and the oriented receptive fields of HUBEL and WIESEL [231] are good candidates for a neurophysiological correlate of the Ehrenstein illusions. Similar illusions are the quasi-perceptive contours of RONCHI and MORI [381] which produce geometrical figures from angular fragments, as described by KANIZSA [260]. These *Gestalt* illusions are probably also related to more complex cortical mechanisms, and possibly to visual learning.

The various effects of different contours on the shape of the Ehrenstein illusions (discs, quadrangles or crosses) are described in detail by SPILLMANN and FULD [406]. They suggest complex interactions of *Gestalt* formation and *area contrast* and possibly a special *pattern contrast (Gestaltkontrast)* which cannot yet be related to neuronal mechanisms.

Table 3. Results of SPILLMANN'S and FULD'S studies of the Ehrenstein-illusion (depicted in Fig. 9)

The Ehrenstein illusion can be seen:

1. Under dichoptic viewing.
2. At brief exposure times.
3. Under stabilized retinal image conditions.
4. In the after-image.
5. In the photopic and scotopic range.
6. At low contrast.
7. After manipulation of component parts of the pattern.
8. Depending on the gap size–line length relationship.

Negative findings:

9. No difference in ΔI as a function of test spot location relative to illusory patch.
10. No colour induction.

Uncertain:

11. Extent of peripheral viewing.
12. Destructive effect of a ring.
13. Effect of defocusing.

Possible Neuronal Basis of Line-Induced Brightness. The Ehrenstein phenomenon suggests that the contour-detecting, orientation-selective "simple-field neurones" may make some contribution to brightness sensation [253, 259, 406]. Unlike the Hermann-grid illusion, Ehrenstein's illusion originates from stimulation by lines in the *absence of contrasting areas*. Thus it cannot be explained by spatial interaction between centre and surround of neurones with concentric field configuration. It could be a suitable psychophysical tool for the investigation of cortical brightness effects of "line detectors".

This would fit in well with recent results of SPILLMANN and FULD [406], who found that brightness and darkness enhancement can be observed in interocular combinations of the four-line patterns. They conclude that the illusory effect is not a purely retinal phenomenon, even though some of its features may be prepared by lateral inhibition or activation of concentric field neurones in the lower visual system before they converge to form the oblong fields of "simple" cortical neurones. As shown by HUBEL and WIESEL, most orientation-specific receptive fields of these neurones have either on-centres or off-centres. Thus, besides signalling oriented contours, they are also higher-level representations of the neuronal B and D subsystems signalling brightness and darkness of oriented shapes in the visual cortex. Accordingly, the bright blot caused by white lines may be correlated to information extracted from B and D afference by the oblong simple field neurones of area 17 [253, 159].

Such field centres should have bar shapes and antagonistic surrounds extending beyond the length of the bar further than is apparent from the schemes of HUBEL and WIESEL. Like most simple field neurones, these neurones should receive *binocular convergence*, since the Ehrenstein illusion can be induced binocularly [406]. The illusion requires also that the lines be of a certain length. The optimal length found in some experiments suggests that there is participation of complex and hypercomplex fields. If we conclude that these visual illusions and their perceptive fields match the receptive fields of certain neuronal populations — which is of course always debatable — some postulates on neuronal information can be made. Simple field neurones, and perhaps also hypercomplex field neurones, not only detect contours, lines or bars, but also participate in brightness sensation. There are several mechanisms which could explain this. The simple field neurones interact with the concentric field neurones which occur in the lower visual system and in the visual cortex and with *disc-shaped fields* surrounded by little or no inhibition (see p. 58). This would compensate contour abstraction for uniform surfaces and generalization *(Flächensehen und Gestaltergänzung)*. Among other phenomena, the Craik-Cornsweet illusion (Fig. 10) demonstrates that the sensation of a uniform area varying in brightness might be induced by border contrast without any objective difference in areal brightness.

Further perceptual correlates of the oblong "simple field neurones", which are arranged in cortical columns of similar orientation, were demonstrated by CAMPBELL and coworkers [50, 75] in their experiments with grating effects and adaptation in man and similar recordings in the cat. Fig. 12, taken from CAMPBELL [75], demonstrates adaptation to oriented gratings in human vision, including movement perception and neuronal population effects (see p. 69—70).

Are Disc-Shaped Receptive Fields Related to Surface Brightness and Colours? Some round, non-oriented receptive fields of cortical neurones are distinguished from the concentric fields of retinal and geniculate neurones by a *weak or missing antagonistic surround*. Such "disc-shaped" fields have been found in cats and monkeys by SPINELLI [408] and others (see Chap 21). Even HUBEL and WIESEL recognized concentric field neurones in the monkey's area 17 [235], although they classify most non-oriented field neurones as geniculate fibres. Such disc-shaped field neurones in the cortex receiving an input from neurones of sustained-response type in the retina and geniculate, seem well suited to signal information about brightness and colour over wide areas, such as would be perceived from surfaces showing different colour and brightness gradations. The colour-specific neurones which GOURAS [166, 167] and ZEKI [467] recently found in the monkey may also belong to this class and represent a spectrally selective population (see p. 107). To be considered as candidates for transmitting colour-specific information, neurones must be stimulated by certain colours of relative spectral selectivity, regardless of the form of the stimulus. Most of the other "colour-opponent" cells, as GOURAS calls them, are more sensitive to contrast. It seems probable that neurones resembling these colour-selective cells exist in the cortical representation of the B and D systems, which signals information about white and black surfaces. However, contrast and contour abstraction in successive stages towards linear contours appears to be stronger in B and D system processing than in the colour systems.

At least *three groups of disc-shaped neurones* which are not spectrally selective occur in the cortex of cats and monkeys:

1. Neurones with strong surround antagonism, resembling geniculate B and D neurones, which show maximal responses to contours of black and white contrast patterns such as HERMANN's grid (Fig. 7).
2. Neurones having little or no antagonistic surround and of two types, responding to white or black stimuli, often with binocular input.
3. Neurones that respond with similar discharges to stimuli of the same form but opposite brightness contrast: white on black or black on white. (For 2. and 3., see Part B, Chap. 21, Fig. 9, p. 352 and Fig. 5B, p. 344.)

Of the neuronal groups whose receptive field is round in shape, only those of the second class are candidates for transmitting brightness or colour of surfaces. The first and third classes probably transmit information about both form and brightness. It is not known whether they are comparable to some geniculate colour cells in the monkey, which are sensitive to both spatial and spectral stimuli. The first group may represent intermediate stages before the cortical neurones are differentiated into oriented simple and complex fields (see Fig. 3). The function of the third class is not yet clear; it is possible they may contribute to pattern vision by signalling spatially contrasting and contoured shapes. The disc-shaped cortical neurones may be associated with the sustained-type responses having slower conduction time found by CLELAND et al. [83], currently attributed to X neurones. Perceptual experience suggests that some interaction may occur between the first and second classes of disc-shaped cortical neurones and that both groups, but mainly the first, may be associated with neurones having oriented fields of the Hubel and Wiesel type, and probably also with direction-specific neurones that contribute to motion detection (p. 39 and 89).

Brightness Contrast and Direction of Movement. Although both B and D systems contribute mainly to brightness contrast, their reciprocal organization may also exert some additional influence on the directional bias in information about motion. This is shown by some recent experiments of MAYHEW on contingent movement after-effects and their adaptation of the B and D systems for movement direction. The opposed black and white contrast of circularly moving patterns results in opposite movement after-effects, one clockwise, the other counterclockwise [320].

Similar after-effects had previously been seen by MAYHEW and ANSTIS [320] after alternating disc rotation with opposite colours, e.g. red clockwise alternating with green counterclockwise. The above observations suggest that the neurones which selectively signal *orientation and movement interact closely with* the neuronal subsystems that are sensitive to *brightness, darkness, and colours,* and that the signals received from these various neuronal populations are co-ordinated at a higher level, giving rise to perceptual phenomena.

Border-Induced Brightness and Darkness

The induction of areal brightness and darkness by graded luminance slopes and the interaction between boundaries and surface brightness is well demonstrated by the Craik-Cornsweet illusion (Fig. 10). It shows clearly how boundaries in-

Addition to Fig. 10: Demonstration of equal luminance of the central areas within the discs A and B. When these are seen without the border ring of luminance gradation by looking through the circular masks they show no brightness difference. After turning this page the areas A and B immediately appear in different brightness: the left is seen brighter, the right darker when surrounded by the borders of opposite slopes of luminance.

The viewing distance should be adjusted for the best effect.

An analogous effect in the domain of spatial frequency has recently been described by MacKay [307b]

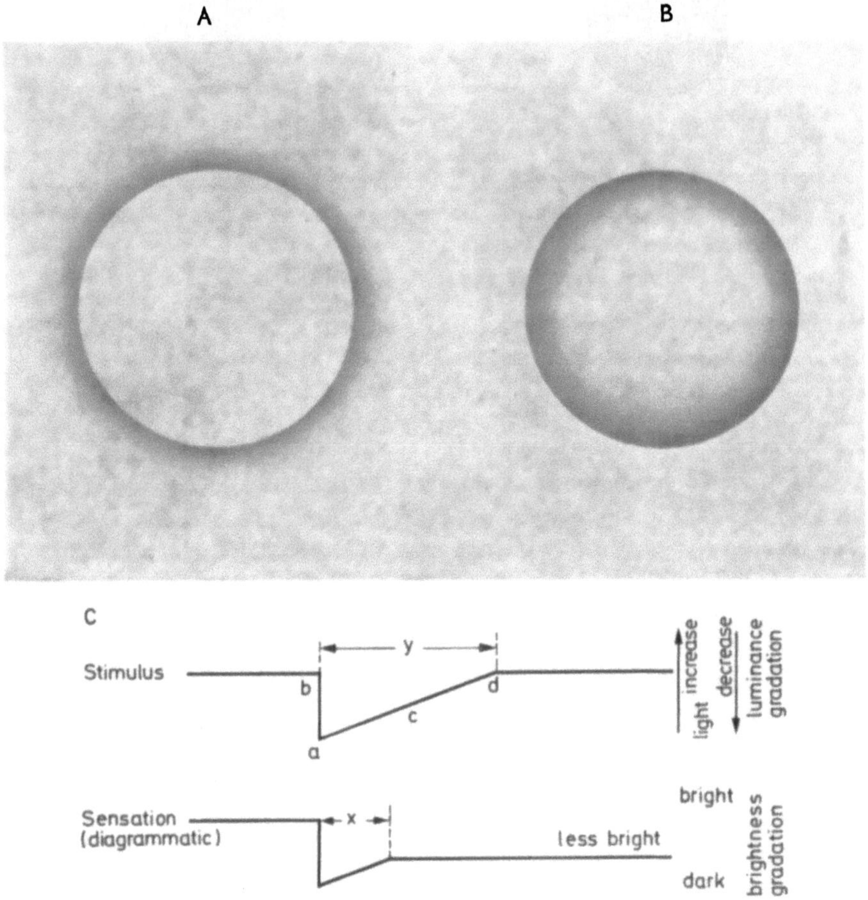

Fig. 10. *The Craik-Cornsweet illusion of border-induced areal contrast in two opposed versions of surface brightness (A, B) with Craik's scheme (C).* The left disc A is seen as brighter and larger than the right B although both are equal in diameter and inner luminance. The fact that the *objective luminance is equal* appears clearly *when the border ring is masked* (turn p. 59).

A The disc-shaped area appears in uniform *surface brightness* when surrounded by a ring of sharp inner-border contrast and gradual outer gradient of increasing luminance. This slowly decreasing darkness induces an illusion of darker surround similar to the centre of the opposite figure B with sharp outer border.

B In contrast to A, the disc area is seen in *surface darkness* when outlined by a gradual border gradient of increasing luminance, inducing decreasing darkness inwards and sharper contrast brightness outwards.

C *Relation of stimulus and sensation for border and diffuse surface brightness.* From K. Craik (1941/1966) [89]. The *objective stimulus* has a gradual border gradient of increasing luminance (a c d) extending from a sharp edge *b—a*. The scheme C corresponds to *a* cross-section of the border slopes of A and B (right edge of disc A, left edge of disc B). The area to the right of *d* corresponds to the central area within the ring B and the surrounding area of A and has the same objective luminance as the area to the left of *b*.

Subjective vision shows a smaller gradient (x) compared to that of the objective stimulus (y) and enlarged areal darkening. Hence, a *continuous diminution in surface brightness* appears subjectively in the area bordered by the slope of diminishing darkness which corresponds to the grey ring of B. Similar stimulus-sensation relations, but with opposite brightness/darkness areal contrast, result from the disc A (subjective brightening by sharp edge inwards)

fluence brightness perception within larger surfaces *beyond* the immediate contrast border. The Ehrenstein illusion induces small spots of brightness or darkness by means of convergent lines. Conversely the Craik-Cornsweet illusion induces brightness or darkness over larger surfaces, depending on either the sharp border or gradual slopes of luminance gradations. The neuronal correlates of these phenomena are less evident than for the Hermann grid. Whereas it is at least probable that oriented simple neurones have a role in the genesis of the Ehrenstein illusion (p. 57), the participation of disc-shaped neurones without antagonistic surround in CRAIK's areal contrast illusion remains a conjecture (p. 58).

Border Cues to Brightness. CRAIK, in 1941 (published posthumously in 1966 [89]), put forward several explanations for areal contrast by eye movements or spatial spread of adaptation and the possible role of the dark and bright fringes at the edges of binocular rivalry. Independently of these physiological explanations, CRAIK considers both phenomena – the uniformly darker appearance of the surface bordered by the gradual slope (Fig. 10B) and the brighter appearance of the surface bordered by the sharp edge (Fig. 10A) – to be a "*function of boundaries* as cues to brightness". This brightness cue for areal surfaces would act also as *compensation for border contrast* and Mach bands near sharp edges. In CRAIK'S words, we see a square of white paper on a black background as "comparatively uniform although the retinal response near its edges is probably considerably greater than at its centre" [89]. The phenomenon may be called the *Craik-Cornsweet effect*, since these two authors have studied it most thoroughly [89, 86]. Similar influences of borders on coloured areas have been known for over a century and were mentioned by PANUM as „*Dominieren der Konturen*" and by HERING as „*Überwiegen der Grenzfarben*" [213, 214].

The changes in surface brightness induced by luminance slopes may be compared to *Hering-type lateral inhibition* over larger areas. BÉKÉSY distinguishes this from Mach-type inhibition, which is restricted to border contrast [34] (see p. 54). The Craik-Cornsweet illusion is not, however, identical with Hering-type inhibition, which is induced by sharp edges.

Border Enhancement versus Area Contrast. Simultaneous contrast phenomena may transcend spatially the border enhancement, as HERING illustrated by several examples [213]. Independent of, but possibly interacting with border contrast is the *brightness contrast in extended areas*. This area contrast may be altered by linear borders, as in KOFFKA'S well-known ring where the line induces area contrast. The striking disappearance of EHRENSTEIN'S brightness illusion (Fig. 9) by a line outlining the central spot is an opposite phenomenon but belongs in the same class of observations. The areal brightness changes induced by border configurations of the Craik type (Fig. 10) are further examples.

It can be assumed that the neuronal events underlying area contrast extend far beyond the region of border contrast and may therefore involve many receptive fields interacting with each other over long distances. Similar interactions between disparate loci have been investigated in the cat's retina by MCILWAIN [322]. In psychological studies, RONCHI and BOTTAI [379] described spatial sequences of contrast enhancement which seem to require comparable mechanisms for their explanation.

It is conceivable that functional interconnections of this kind participate in the visual extinction of retinal scotomata. GERRITS and VENDRIK [159] attempted to relate this filling-in process to neuronal interaction of the B and D systems, and postulated a barrier of D discharge in the central visual system. Both area contrast and filling-in processes seem to generalize areal vision beyond the border of contrast stimulation and to compensate for the progressive abstraction to contour patterns. See p. 42 and Fig. 4. Normally, we perceive images of structured patterns whose borders enclose uniform areas and we do not see areal patterns as linear contours, although contour abstraction facilitates vision of shapes. These visual phenomena indicate interacting processes in large neuronal populations beyond the border of contrast stimulation; the result may be either perceived uniformity of the enclosed area or apparent continuation of contours and contoured areas according to a generalized scheme („Gestaltergänzung"). Obviously, the postulated diffuse spread and far-reaching interaction of neuronal populations lacks precision, but, as a hypothesis for area interaction, it allows predictions of heuristic value which can be tested experimentally.

Brightness contrast appears to be influenced by *Gestalt* factors as in KOFFKA'S ring [269] whereas Mach bands seem less affected by pattern structure. This indicates interactions of simple contrast mechanisms and *higher-order* processes of vision. Brightness illusions induced by linear contours, e.g. the Ehrenstein and Kanisza figures, can be seen under dichoptic viewing conditions (see p. 56) and therefore probably occur at the *cortical* level. GREGORY [177] postulates a still higher process of "cognitive" object hypothesis in perception for these and other illusions of an "intelligent eye". Such cognitive concepts may be used to bridge the old controversy between HELMHOLTZ and HERING in the explanation of contrast phenomena.

In KOFFKA'S *ring*, the Gestalt effect inducing uniform brightness of the ring evidently compensates for retinal mechanisms of simultaneous contrast. A cerebral location of neural mechanisms related to the phenomena of KOFFKA'S ring is suggested by recent experiments of WIST and SUSEN [459], who used variations of retinal disparity between the bisecting line and the central figure of KOFFKA'S pattern. This demonstrated that the perceived depth interval between the line and the ring must influence simultaneous contrast. Such results of stereopsis obviously depend upon processes beyond the retinal level, probably in the visual cortex. If we may conjecture from HUBEL and WIESEL'S experiments in the monkey's area 18 [236a] to cortical mechanisms in man, retinal disparity would be processed mainly *beyond area 17:* Hence, these Gestalt effects may also be localized at a *higher* cortical level, in area 18 or 19, or even in visual areas of the temporal lobe.

GANZ [153] has stressed the essential role of contour and lateral inhibition in KÖHLER and WALLACH'S "figural after-effects" [272a]. The latter cannot be treated here except to mention that MOTOKAWA related it to retinal induction and discussed links with possible neuronal findings and models [328, 329, 331, 332].

Contrast–Contour Interaction and Visual Art

For more than a century, since MACH'S description, it has been known that brightness contrast induces contours. The opposite interaction, contrast formation by contours, has been studied more recently. Both effects were employed in the *art of drawing* long before the laws of contrast were investigated.

Line Drawing and Contour Abstraction. The draughtsman abstracts the essentials of form to linear contours, although there are no lines in nature. This creation of contours may be compared to the opposite process of contour abstraction in vision, which parallels the feature extraction in the neuronal system of the visual cortex and their selective orientation-specific receptive fields [252]. *In sketching, the artist follows the same visual laws, but reverses the abstractive process of vision.* He begins with abstracted contour delineation and later adds hatching or a wash to depict light and plastic form [251, 252].

In perception, gradation of brightness can lead under appropriate conditions to the formation of Mach bands, which lateral inhibition causes to appear as more or less *linear borders* (see RATLIFF [371] for a summary of the vast literature). Besides its physiological interest, contour formation by contrast has practical applications. It is basic to the use of lines in drawing, both in everyday representations of shapes and in works of art. *Line drawings* indicate the extent to which visual perception of forms is dominated by contours. Yet different periods produce different styles which vary in their use of lines and in their tendency to enhance or dissolve contours. This has been illustrated by drawings from various periods in the history of art [252]. Draughtsmen and painters, besides introducing illusions of light, shade and shape by gradations of brightness or by linear hatching, also *distort or deform pure contours* and develop an esthetic value of the pure line (*Eigenwert der Linie* [252]) using various graphical techniques [251].

Thus, *linear structures have an intrinsic value* which attains a higher level of expression and transcends simple contour design. Dynamic lines in a drawing can depict movement and emotion better than non-linear areas of black and white. Despite the variety and relative freedom of his creation, the artist as draughtsman has to obey the fundamental laws of visual abstraction and demonstrates the *dominance of linear contours* [252].

The reverse effect – the creation of brightness contrast by the introduction of contours – demonstrated by the Craik-Cornsweet illusion, also has interesting applications in painting.

Border-Induced Brightness in Art. The fact that contour creates brightness in areal contrast, although very impressive in perception, is often neglected. In the *visual arts*, however, this effect has been used empirically for several centuries by painters for the illusory production of bright and dark areas by means of faint differences beyond contours. RATLIFF depicts good examples of such contour-induced brightness effects in oriental art [372]. The visual impressions created by *Chinese brush-paintings* in ink are similar to those obtained from the *Craik-Cornsweet illusion* (Fig. 10). Surprisingly large areas can be influenced in this fashion by shadings in border luminance at a contour: areal brightness is enhanced beyond the edge where luminance increases sharply at the contour (Fig. 10A) and apparent brightness reversal is seen beyond the border where luminance decreases smoothly, making the whole area appear darker (Fig. 10B).

The first experiments were made by KENNETH CRAIK in 1940 (and published posthumously [89] see p. 61) and later studies came from O'BRIEN [346], so that the designation "Craik-O'Brien effect" was introduced by RATLIFF [372] to refer to the phenomenon as regards its application in the visual arts. CRAIK's experiments correspond in principle to Cornsweet's illusion, induced by a rotating disc causing similar border brightness gradations for concentric areas [86]. For this reason, we prefer the term *Craik-Cornsweet illusion* (see p. 58).

The application of contrast-contour interaction in the visual arts shows extreme variations of linear abstraction and contour dissolution in different styles or in the works of individual artists. These variables demonstrate the relative freedom of the artist's creation [252].

Art and visual perception has been treated in several monographs during recent years: the books of ARNHEIM [9] and GOMBRICH [163] are more concerned with psychological than with physiological aspects and refer mainly to Gestalt concepts. HOMER'S [228] monograph treats the contrast theories of Neo-impressionism which SEURAT derived from sensory physiology; this is also mentioned by RATLIFF [371]. PIRENNE [356, 358] refers mainly to physiological optics and perspective. All this work confirms the impression that artists recognize almost instinctively the significance of visual contour and contrast.

PIRENNE in his book on optics and painting [358] also demonstrates the freedom of the artist and stresses the special attitude of the observer of pictures. This attitude involves the integration of many factors for judging percept, concept, space and depth in a picture, yet the observer is always conscious of the artificial nature of the painting presented on canvas in two dimensions. When viewing a picture, the art connoisseur does not expect an illusion of reality but a *painting* with its specific values of surface, content, style etc. Of course, this sophisticated attitude is entirely different from the naive approach to the world in everyday perception. However, as HELMHOLTZ [209] remarked, the visual arts may give some hints for recognizing which visual impressions are of special interest for our image of the world.

VII. Visual Integration and Cortical Functions

Visual Areas and their Afferences

Areal Architectonics and Visual Functions. Classical procedures of cyto-architectonics and myelo-architectonics have defined three visual areas in the mammalian brain (see BRODMANN [64] and VOGT [442]). These *areas, 17, 18, and 19* of BRODMANN, were determined in the cat by OTSUKA and HASSLER [349] and confirmed physiologically in neuronal recordings by HUBEL and WIESEL [234] and others. The sharp delimitation of areal myelo-architectonics (VOGT called the boundaries „haarscharf") is well seen, even by the inexperienced eye, in myelin-stained sections as demonstrated by Figure 2 in the paper by OTSUKA and HASSLER [349]. As HASSLER'S chapter on the comparative anatomy and architectonics of the visual brain could not be included in this Handbook, his main features of the cat's visual areas are depicted in Fig. 11. This shows the considerable variations in the extent of areas 17, 18 and 19 on the cortical surface in four individual cats. The differences are probably even larger in different breeds of cats, not to mention the definitely abnormal projections in the Siamese cat. Studies of areal variations in man were begun by VOGT and his school, but not finished. VOGT briefly mentioned considerable interindividual differences of areal distribution in the human cortex and suggested relations between various visual faculties in man, including eidetic functions and areal extent or neuronal differentiation [443].

The definition of area 17 or its synonyms (area striata, primary visual cortex, visual I) is usually clear, but the terminology of the *extrastriate* areas which are related to visual function

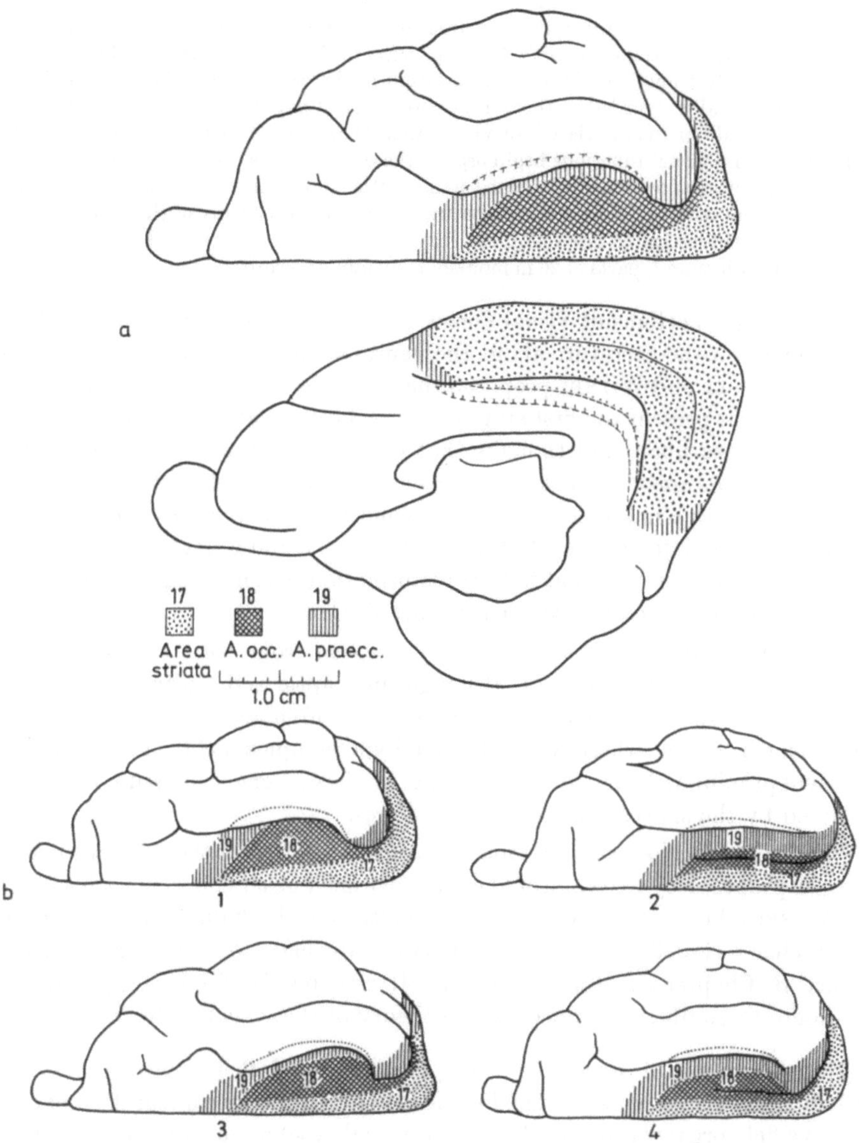

Fig. 11a and b. *Architectonic subdivision of the visual cortex in the cat.* From OTSUKA and HASSLER (1962) [349].

a Aspects of the right cerebral hemisphere in the cat, showing the three main areas determined by cyto-architectonics and myelo-architectonics: *area 17 (striata)*, *area 18 (occipitalis)* and *area 19 (praeoccipitalis)*.

b *Variations of the limits of areas 17, 18, and 19* in four cats showing different extent in the right hemisphere. Anterior and lateral regions of 18 and 19 bordering the suprasylvian gyrus show greater variations.

Retinotopic projections show the vertical meridian located at the border between 17 and 18 and again at the lateral 19 and suprasylvian borders. This topographic arrangement ensures that *each area is a mirror image of the neighbouring visual area.* Area 17 corresponds to the primary visual cortex, 18 and 19 to the paravisual cortex, and the bordering visual area in the suprasylvian gyrus to the perivisual cortex (see Part B, Chap. 21, Fig. 1) is somewhat confusing. BRODMANN'S cyto-architectonic numbers [64] are generally used for areas *17 (striata), 18 (occipitalis) and 19 (praeoccipitalis)* since OTSUKA and HASSLER determined their architectonic boundaries and variation in the cat (Fig. 11). Areas 17, 18, and 19 together are generally regarded as the "*visual cortex*" (see Chap.21). Areas 18 and 19 are often called "paravisual" or "parastriate", or visual areas II and III. Most uncertainties concerning architectonics and visual functions begin *beyond* these three areas. Perivisual cortex [256] is a practical but not architectonically defined name for the adjacent cortical fields extending to the temporal and parietal lobes, from which the inferotemporal cortex of rhesus monkeys has been thoroughly investigated (see Chap. 23). Its architectonic boundaries in BRODMANN'S maps [64] (area 21, 37 in man ?, parts of 20 in monkey ?) remain uncertain, however see also [386].

Neuronal Connections of Visual Cortex Input. As mentioned on p. 42, there is no agreement in the literature about the neuronal circuits of the geniculo-cortical afference. HUBEL and WIESEL proposed that "simple" field neurones with oriented field axes might present the first stage of cortical cells in area 17 which are activated by geniculate fibres and thought that only the monkey's (but not the cat's) visual cortex contains neurones with concentric fields [231, 233]. However, several other authors have found neurones with concentric fields in area 17 in the cat. Examples are depicted in Part B, Chapter 21, Figs 5 and 9. BAUMGARTNER'S view of the various stages of visual field transformation is schematized in Fig. 3, and BISHOP'S wiring scheme is discussed in Part B, Chapter 21.

STONE [416] described *monosynaptic excitation* from optic radiation in most classes of visual cortex neurones, although disynaptic and multisynaptic connections also occurred. Complex field neurones were often activated directly by fast-conducting Y fibres, whereas simple and hypercomplex field neurones received monosynaptic input from slow-conducting X fibres. Longer synaptic delays were explained by the slower conduction of X fibres.

Parallel visual processing by monosynaptic activation of the different neuronal types, as proposed by STONE, would contradict most other concepts such as HUBEL and WIESEL'S hierarchic order [233], the multisynaptic chains in BAUMGARTNER'S wiring scheme (Fig. 3) and DENNEY'S results on visual radiation stimuli, reported in Part B, Chapter 21 (Fig. 19, p. 377). It remains for further experiments to coordinate these different concepts in a theory of the visual cortex.

Transformation of Receptive Fields and Neuronal Cell Types. The synaptic mechanisms which are responsible for the differentiation and integration of special receptive field organizations in the central visual system are still unknown. We are also ignorant about correlations of neuronal receptive fields with the various morphological types of neurones and synapses distinguished by the anatomists and described by SZENTAGOTHAI in Chapters 17 and 20. HUBEL and WIESEL first proposed sequential *levels of integration*. According to them, simple fields are formed in area 17 by synaptic convergence of several concentric fields from lateral geniculate axons and the simple fields in turn converge to complex fields [233]. This may still be accepted in principle but is evidently a more complex process than synaptic convergence at the next neurone. It has to be assumed that there is intermediate processing in networks of interneurones, and intricate coordination of excitation and inhibition. The scheme of B–D interaction we proposed in 1965 (Fig. 3) contains *successive levels of neuronal integration and differentiation* towards selectivity of receptive fields. The circuits of the higher cortical levels are unknown.

Hence, interneurones were depicted only in the lower retino-geniculate structures for reciprocal inhibition and collateral interaction between the B and D subsystems.

This diagram will have to be modified as more precise information becomes available on structure and function of visual cortex neurones and of geniculocortical interaction. The various research groups are not yet in agreement about the role of concentric field neurones and movement-selective cells in the cortex (see p. 88) nor about the input of X and Y afferents to simple and complex field neurones (see Chap. 21). Be this as it may, the existence of some *sequence of levels* for the transformation of various types of receptive fields appears to be a logical postulate and may be accepted in principle. Figure 3 demonstrates the lower levels in a simplified form together with some optimal stimulus configurations to which these neurones respond selectively. A more detailed wiring diagram to explain synaptic mechanisms in the cortex would be conjectural. The latest model of a cortical circuit by BISHOP and coworkers [46] is depicted in Part B, Chapter 21, Fig. 20. This wiring scheme was proposed after analysis of simple field neurones, but the correlation of these neurones and the inhibitory cells with histologically defined cell types of the visual cortex needs verification by detailed studies. BISHOP's identification of simple cells with pyramidal cells in layer V would imply that their axons are cortico-fugal and would reach other neurones within and outside area 17 via the white matter. If this were so, one would expect antidromic excitation of simple neurones to occur rather often following subcortical stimulation, whereas it is exceptional. The correlation of neuronal electrophysiology with structure (i.e. morphology, topography and synaptology of the visual cortex) and the exact marking of the recorded neurones by the modern technique of microstaining at the microelectrode tip is a task for future research.

The relation of the visual system to *extra-geniculate thalamic* nuclei also needs further study in cats and monkeys. Since BUSER et al. [74] described connections of visual areas to the posteromedial thalamus in the cat, the *pulvinar thalami* has been studied in monkeys, which possess elaborate visual areas in the temporal lobe. As is well known to anatomists, the temporal cortex receives many projections from the pulvinar nuclei of the thalamus. These connections probably function as links between the higher visual areas and thalamic nuclei contributing to vision (see GROSS, Chap. 23).

In spite of extensive work over the last 20 years, the visual cortex remains a jungle of undiscovered synaptic connections between the many different neurones. The anatomists have made a preliminary classification of the nerve cells of the various areas and laminas. The cortical connections are not as clear as those in the lateral geniculate, for which SZENTÁGOTHAI has established some basic synaptic relations [424, 425].

Neurophysiologists can only propose hypothetical neuronal circuits but as yet are unable to verify the complex synaptic relations from their recordings. Except in a few cases, it has not been possible to localize and define histologically the neurones recorded in physiological experiments. The anatomists can do no more than stain some nerve cells with all their processes by Golgi methods in unpredictably selected specimens, or select and photograph under the electron microscope small portions of the neuronal membranes with a few synaptic structures. Hence, synaptology is restricted to "circuit fragments", as SZENTÁGOTHAI says (Chap. 20).

All the modern, improved methods, including those of synaptic degeneration and physiological measurements of synaptic delays, have not yet made the field ripe for an ordered synthesis of neuronal networks in the cerebral cortex that could be compared to the clear geometry of the "neuronal machines" in the cerebellum. A promising approach through combined anatomical and physiological experiments was initiated by HUBEL and WIESEL in 1969 [235], disclosing a new order of terminal connections from the ipsilateral and contralateral eye in the monkey's area 17 (Chaps. 20 and 21). Thus we may hope that closer relations between neuronal physiology and synaptology may be obtained in the future.

Cortical Columns and Human Vision

Spatial Frequency Transfer in Man and Columnar Organization. Now that temporal modulation transfer has been extensively studied (see Chap. 7, van de Grind et al.), Campbell's experiments comparing spatial frequency stimuli in cats and humans are of special interest for correlating the neuronal physiology of the visual cortex and visual perception in man. The studies of Campbell and coworkers on spatial adaptation and evoked potentials [49, 50, 75–78] are reported in Chapters 2 and 28. A clear visual demonstration of orientation- and frequency-specific effects in human perception is depicted in Fig. 12. This simple visual experiment shows that *spatial frequency channels of selective orientation* may be operational in the human visual system, and suggests parallels to Hubel and Wiesel's orientation-specific neurones which are located in columns in the visual cortex of cat and monkey [233, 235]. Of course, it has not been directly demonstrated that homologous columns exist in man, but the recordings by Campbell, Maffei, and Blakemore of evoked potentials elicited by spatial frequency stimuli [50, 77, 78] provide good arguments in favour of similar structures in the human visual areas. It may be left open whether these are located only in the primary visual area 17 or also in areas 18 and 19. Areas 18 and 19 extend further into the convexity of the occipital lobe where they would contribute more to visual evoked potentials than area 17, which is buried in the calcarine fissure. Besides a few neuronal recordings from the human visual cortex [313] directe potential recordings are lacking from implants, such as those made by Brindley and Levin [63] for visual substitution in blind subjects. Patients who were blind for many years might have been some cortical atrophy which interfered with response to stimulation.

Another parallel between visual phenomena in man and neuronal recordings in the cat is derived from grating adaptation experiments by Campbell and his group [50]. (Cf. Part B, Chap. 21 on the cortex: Fig. 15 on p. 367 depicts the striking correspondence between the curves for human visual thresholds and neuronal firing rates of cats.)

Gratings, Bars and Form Vision. Experiments on the interaction between visual frequency responses and aperiodic stimuli by Weisstein and Bisaha [449, 450] show how *gratings may mask bars and bars may mask gratings*. This is further discussed in Part B, Chapter 21 (p. 419). As described in Chapters 2 and 6, form vision and motion perception are based upon many complex neuronal interactions, some of which are unkown, and which transcend the laws of contrast for simple patterns. Simplified concepts of dual visual subsystems may help us to understand some of the principles of contrast vision in the brain, but the neurophysiological mechanisms of most phenomena of pattern vision remain unknown. Neither the functional concept of the B–D system antagonism for white and black contrast [249], nor the anatomical postulate of two visual projections by Schneider [390] and

Fig. 12. *Subjective demonstration of spatial frequency-specific "channels", which may correspond to single-cell spatial selectivity in the cat.* From CAMPBELL (1969) [75].

If the reader views the figures from a distance of about two metres, he will see in the centre a person holding an umbrella. The superimposed low-contrast vertical grating represents the "rain".

Now view, for about 60 sec, the high-contrast *vertical grating* of the *same* spatial frequency as the "rain" in the upper left by letting the point of fixation wander around the circle. (This eye movement is required to prevent an after-image of the grating).

After this adapting period, fixate the person's head and it will be noted that the "rain" has stopped, although it returns again after some seconds.

Repeat the experiment, but this time fixate the *horizontal grating* in the upper right. A negative result is obtained. This is also the case if either of the lower vertical gratings of very *different* spatial frequency from that of the "rain" is viewed during the adapting period

TREVARTHEN [434] (the foveo-cortical for discrimination and the extrafoveo-tectal for spatial location), nor again the neurophysiological classification of cortical receptive fields by HUBEL and WIESEL [233—235], allows a satisfactory explanation of visual form perception. See also Chapter 2.

Length and Orientation Perception in Man. CAMPBELL's method of selective adaptation [75, 77] has recently been applied to the perception of line length. NAKAYAMA and ROBERTS [340] found that contrast thresholds for a short-line test grating became higher after the subject had previously viewed short lines of equal length than after he had viewed longer lines. The authors conclude that this may be explained in terms of the hypercomplex neurones of HUBEL and WIESEL [234] or other contour-sensitive neurones with inhibitory flanks. BECHINGER et al. [31, 31a], in their investigations on perception of the length and orientation of lines and endpoints, postulate "neuronal connections which detect simultaneously excited cortical cell columns even across the interhemispheric cleft", and a size constancy mechanism. They also mention a possible role of hypercomplex neurones. A more interesting recent result of BECHINGER's experiments is a clear *orientation preference for horizontal and vertical lines* in human vision as compared to oblique angle orientation in the intermediate range [31]. The situation may be different in the cat, where HUBEL and WIESEL find nearly equal distributions of oriented cells of all angles in the visual cortex [233].

Commissural Transfer, Unity of Visual Half-Fields and Stereopsis

Unity of Vision and the Vertical Meridian. We all see one world despite the separate representations of the two halves of the visual field in the left and right hemispheres. The perfect co-ordination of these half-fields in perception still offers a challenge to neurophysiology. Considerable progress has been made over the last six years in the elucidation of interhemispheric connections and the representation of the *vertical meridian as an interhemispheric link* between the two half-fields. It has become apparent that *callosal connections* between the left and right cortico-visual areas, representing the two half-fields of vision, have a special function in the unification of vision and that the receptive fields on either side of the meridian overlap. This is explained in Chapters 2, 14, and 21. Figure 1 in Chapter 21 shows a synthesis of the structural organization with an elaboration of BERLUCCHI's scheme [39a]. There exist *three different representations of the vertical meridian in the cortex* with a mirror arrangement of the half-fields on either side. Neurones representing the vertical meridian have a disproportionately large share of callosal connections. Moreover, the overlap of meridian receptive fields is larger in the upper and lower regions than in the area centralis or fovea. The physiological significance of the three different cortical projections of the meridian and the mirror arrangements of the central and peripheral parts of these projections may correspond to different levels of visual integration processing from areas 17, 18 and 19 to the suprasylvian convolution in the cat and to the inferotemporal cortex in the monkey. The various neuronal stages of this integration, their relationship to the visuomotor systems and their cortical representation in frontal area 8 are not yet fully understood. NAUTA [341] has recently proposed a scheme for these cortico-cortical connections which will be useful for further neurophysiological experiments. These partly unpublished anatomical findings of NAUTA and his group are summarized in Fig. 13. For callosal connections see [348, 339].

A further step towards the unity of vision, i.e. the bridging of the meridian between the half-fields, is achieved in the inferotemporal cortex. This cortical area contains neuronal receptive fields which are much larger than those of striate neurones at the vertical meridian. They include the fovea and often extend far into the periphery of both visual half-fields [178].

The Inferotemporal Cortex and its Neurones. GROSS's neuronal recordings from the inferotemporal cortex of monkeys are described in Part B, Chapter 23. Briefly, the receptive fields of inferotemporal neurones are essentially different from those of visual-cortex neurones in several respects: they nearly always include the *fovea* as well as large parts of the peripheral retina. More than half the neuronal population extends into *both visual half-fields*. Most receptive fields are very large, often more than 400 degree^2, with diameters of 20–70°. Preferred selective stimuli include moving patterns and coloured forms. Some neurones respond only to *highly complex patterns*, e.g. the shadow of a hand, or rounded shapes. The optimal specific trigger feature was often difficult to determine. It seems highly probable that the inferotemporal cortex is the main centre which processes information coming from the visual cortex; *bilateral input* from visual areas in the left and right hemispheres would then occur at neurones which represent both half-fields of vision. GROSS has shown that the contralateral input is carried over the *hemispheric commissures* (corpus callosum and anterior commissure).

Besides receiving signals from visual areas 17–19, these temporal areas of vision seem to function in close connection with the *pulvinar* of the thalamus, a highly developed nuclear complex in man and higher monkeys. However, inferotemporal neurones in monkeys depend more on the visual input from areas 17–19 and their parietal connections, and the pulvinar afference seems to be an additional source. Pulvinar lesions modify visual responses without eliminating them, and cortico-cortical connections remain the most important information sources for these higher cortical areas, as depicted in Fig. 13.

Visual transfer of pattern information via the corpus callosum and other commissures evidently involves complex processes with interaction of activation and inhibition. The role of *inhibitory mechanisms* is apparent in the neuronal inhibition that is induced in the visual and sensorimotor cortex by electrical stimulation of the symmetrical cortex area of the contralateral hemisphere. In motor cortex neurones this inhibition is limited to low-frequency stimulation of the contralateral sensorimotor area [257]. Shorter inhibitory effects in visual cortex neurones occur after stimulation of the contralateral visual cortex [257]. Although electrical stimuli cause abnormally synchronized discharges, not comparable to the elaborate discharge configurations following patterned visual stimuli, they can be used to determine the timing of callosal volleys.

Callosal and Non-Callosal Commissure Split. The *anterior commissure* in its nonolfactorial parts connects the higher visual areas in both *temporal lobes*, including the inferotemporal cortex (see Part B, Chaps. 23 and 25). Therefore, a callosal section disrupts the commissural connections between the occipital but not the temporal visual areas. Hence, only *two human subjects without cerebral commissures* have shown *two separate visual perceptions from both half-fields*, as described by SPERRY [403] and GAZZANIGA [157]. In these patients, complete section of the corpus callosum, of the anterior and hippocampal commissures and, in one case, of the thalamic massa intermedia was performed to prevent the spread of epilepsy.

Patterns arising from the right visual field and reaching the left hemisphere could be named by *speech*, and patterns from the left field reaching the right hemisphere could be manipulated by the left *hand* but could not be verbally described. These defects, however, were apparent only after short-time exposures in the hemifields. In everyday life, the patients could compensate for their split perception rather well when they viewed the world with both eyes and eye movements. A still better compensation occurs in patients with intact anterior commissure when the section is limited to the corpus callosum. The individual differences in the extent of defects and compensation need further clarification (see Doty and Negrão, Chap. 25, pp. 566–567).

Other Commissural Functions. Since Sperry's original observations [403, 403a], *visual transfer* from one hemisphere to the other has been investigated in split-brain preparations with sectioned optic chiasma. In Chapter 25 Doty and Negrão discuss specific features of transfer in vision together with the role of other forebrain commissures. The preference for visual transfer near the horizontal meridian is related to stereopsis, to high visual acuity in the fovea and to eye fixation. Callosal connections were investigated by Mitchell and co-workers for *stereopsis* [325] and *ocular vergence* [454]. Stereoscopic disparity for depth perception often stems from visual objects projected to the retina near the midline and is represented in the visual cortex of both hemispheres (see Bishop, Chap. 4). These stereoscopic functions are regulated by convergence on nearby objects during ocular fixation. Hence, both depth perception and ocular convergence must obviously depend upon callosal communication. Further experiments in non-anaesthetized monkeys with free and goal-directed eye movements are necessary for analysis of these rather complex mechanisms.

Neuronal Mechanisms of Depth Perception. For 130 years after Wheatstone's pioneer experiments in 1838 [455] on depth perception by binocular disparities stereopsis was investigated only by means of psychophysical methods and assorted stereoscopic apparatus. The neuronal basis of stereopsis in the visual cortex has been elucidated since 1967 by Barlow, Bishop, Pettigrew and co-workers [243]. Their animal experiments yielded most interesting neurophysiological correlates of stereoscopic vision in area 17 and are described in Chapter 4 of this volume by P. O. Bishop. Since subjective aspects of depth perception will not be treated in the psychophysical part of this Handbook (Vol. IV/4), some perceptual correlates are also descibed in Bishop's chapter, such as the horopter, Panum's fusional areas and Julesz's discovery of binocular depth discimination without familarity cues [244]. An account of Julesz's experiments on stereopsis in random-dot accumulations appearing monocularly without regular patterns, and their relations to texture discrimination and depth perception is now available as a monograph [245]. For stereopsy in the monkey's area 18 see [237a].

Temporal aspects of stereopsis, well known since the observation by Fertsch and Pulfrich of depth illusion caused by time delays of monocular stimuli with lower luminance, probably also deserve more attention in neuronal physiology. The paper by Wist and Freund [458] and the last part of Chapter 4, demonstrating that a latency difference of 15 msec might be equivalent to receptive field disparities of 5—31 min of arc, indicate in which direction further neuronal experiments on binocular temporal disparities might be planned.

It remains a neurophysiological enigma, however, how the visual afferent pathways having different conduction velocities can accurately signal small binocular time differences to the brain.

Cortico-Cortical Integration, Intersensory Convergence and Transfer

Cortico-Cortical Connections. Until recently our knowledge about the visual functions of the cerebral cortex beyond areas 17, 18 and 19 was rather scanty, as regards both anatomical connections and neuronal physiology. Studies of evoked potentials and some neuronal recordings had shown how widely the visual input irradiates beyond the visual cortex; nevertheless, the functional significance of this extension of visual signals to other cortical areas, their pathways, and their relationship to the subcortical centres remained uncertain. Yet neurological studies of cortical lesions indicated that higher visual integration must occur outside the primary visual areas in the parietal and temporal cortex.

In 1964 MISHKIN and his group started to investigate the visual functions of the *inferotemporal cortex* by ablation studies in monkeys [287, 324]. A further important step was made by GROSS with his neuronal recordings [178]: they revealed the existence in neurones of the inferotemporal areas of large receptive fields, including the fovea, that are selective for certain visual features as described in Chapter 23. It is not yet clear which are the corresponding areas in carnivores, but the suprasylvian gyrus may be the most likely candidate in the cat's brain, although it seems to overlap with areas which correspond to the parietal cortex in monkeys and man. Judging from neurological experience of human brain lesions, the parietal areas are considered to be representative of visuospatial functions. The *perivisual* areas of the cat's suprasylvian gyrus contain many neurones that respond to diffuse light [258] and to special optic patterns [237]. The lateral part of the suprasylvian cortex, the "Clare-Bishop area", contains a representation of the visual half-fields although it is doubtful whether there is a direct projection from the visual radiation. In this area HUBEL and WIESEL have described complex and hypercomplex receptive fields, often larger than in areas 18 and 19 [237]. However, it is not certain whether they are directly comparable to the inferotemporal neurones which GROSS found in monkeys and to ZEKI's results [465, 466].

The interconnections of the visual areas with other cortical regions of the same hemisphere seem to be rather complex, with involvment of subcortical nuclei. The longer association pathways between occipital, temporal and frontal lobes, postulated in the older literature, are probably multisynaptic. They are less well investigated than the visual projections from subcortical nuclei to paravisual areas and the callosal connection. KUYPERS and co-workers in MISHKIN's group studied the *connections between the visual and inferotemporal cortex* [287], but it remained uncertain how the parietal and frontal cortex received visual messages. A recent study by NAUTA of the frontal lobes [341] contains a survey of cortico-cortical connections from the visual to the temporal, parietal and frontal areas in the monkey although most of these regions are usually regarded as being "non-visual".

These *cortico-cortical connections of the visual system* are depicted in Fig. 13, taken from the studies of NAUTA [341] and his co-workers. They demonstrate that further visual projections are generally *relayed in several stages* from areas 17, 18 and 19 to inferotemporal cortex, parietal cortex and frontal eye fields. These "short" cortico-cortical association pathways are probably supplemented by subcortical relays, but there is as yet little evidence for direct "long" occipito-frontal and fronto-occipital pathways. Gross in Chapter 23 reports the main discoveries of recent years on the temporal areas and their connections with the thalamus and forebrain commissures. Further studies of architectonics, connections and synaptology of these higher visual relays and their neurophysiological correlates are necessary to elucidate the cortico-cortical integration of vision.

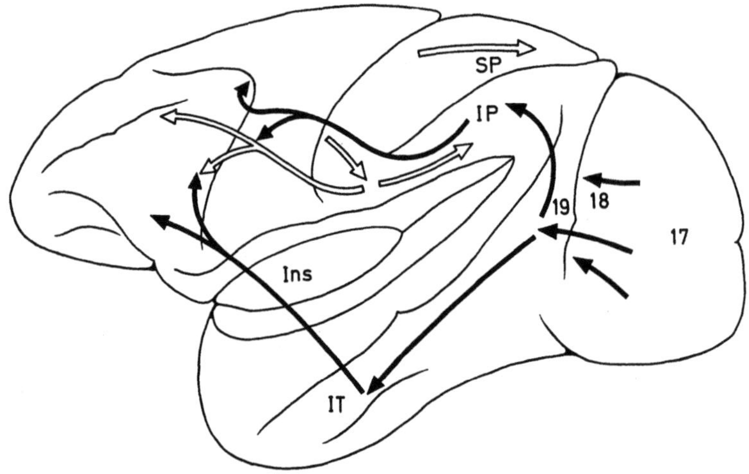

Fig. 13. *Scheme of the major association pathways from the visual areas to the inferotemporal, parietal and frontal cortex in Rhesus monkeys.* Modified from Nauta, 1971 [341].

Cortico-cortical fibre connections conveying *visual* information have solid *black arrows*, other pathways open arrows. Before visual impulses arrive at the frontal eye fields, they are relayed in several stages of synaptic transmission over the paravisual areas 18 and 19, the *superior temporal* and the *inferotemporal cortex* (IT) and the inferior parietal cortex (IP). From the central and parietal areas other connections reach the temporal and frontal regions (white contoured arrows)

The parietal areas also receive somatosensory and acoustic input (the auditory connections are omitted to simplify the picture). Thus, the *parieto-frontal* connection originating from the inferior and superior parietal lobules may convey at least *trisensory* (visual, auditory and somatosensory) information. The inferior parietal lobule (IP) also has efferent connections (not shown) to the cingulate and hippocampal cortex of the limbic systems.

Other essential connections with the contralateral hemisphere which contribute to the unity of vision by commissural fibres are not depicted in this scheme. Callosal connections also contribute to the bilateral hemifield representation of neuronal receptive fields in the inferotemporal cortex (see p. 72).

Multisensory Convergence and Intersensory Transfer. Converging information from sensory organs is essential for behaviour and recognition. In natural surroundings visual input is always accompanied by messages from other sense modalities. *Multisensory responses* are found in neurones of the brain stem and cortex as well as in the primary sensory nuclei and cortical receiving areas. In the visual cortex, acoustic, vestibular and somato-sensory responses can also be recorded from neurones showing bisensory and trisensory input [258, 327]. These findings and their relation to nonspecific systems and to specific visuo-vestibular convergences are discussed in Chapter 21 (see Part B, pp. 401–409).

In brief, nonspecific and specific multimodal effects are distinguished. Neuronal responses showing variable and long latencies (usually activating) are considered to be nonspecific, mediated by the reticular and thalamic integrating systems, and probably related to attentive mechanism and arousal [258]. Other multisensory responses (usually with rather constant and short latencies, which may be excitatory or inhibitory) are signs of intersensory communication [258]. Some results on visuo-vestibular interaction are mentioned in the section on eye movements (p. 98). Special neuronal and psychophysical correlates in relation to head position and the subjective vertical with AUBERT'S and MÜLLER'S phenomena and axis orientation of visual cortex neurones [96, 229] are also discussed in Part B, pp. 407–409. The more complex intersensory transfer related to pattern recognition and arising from various sensory modalities transcends these neuronal investigations. Some interesting transfer observations have recently been obtained from visual substitution.

Intersensory Transfer and Visual Substitution. Transfer between different sensory modalities may be as important for pattern perception as interhemispheric transfer, but has been studied less often. Various attempts to find *sight substitution* for blind people have yielded some interesting results for intersensory pattern perception and visual physiology. BACH-Y-RITA [12a] has just summarized his own and other methods to substitute for visual input by *stimulation of somatosensory systems through the skin*. Eye movements are simulated by a self-induced camera motion which imitates the scan path. BACH-Y-RITA discusses this somatosensory vision substitute in relation to GIBSON'S concepts [160] and HELD'S experiments on sensorimotor linkage in perceptual learning [204, 205]. *Contour abstraction*, which is the basis of visual pattern recognition, also operates in skin stimulation, as shown by BÉKÉSY [33].

These substitutes for other sensory modalities are of special interest for *multisensory convergence* and *intersensory transfer* [12a].

A reverse *transfer from vision to perception* through skin stimulation is apparent in normal humans for visual symbols like numbers and letters. In routine neurological examinations of epicritic skin sensation, subjects with normal skin perception are expected to be able to recognize numbers written on the skin without any prior experience of this stimulation. The subjects could previously have perceived such numbers only as visual or sensorimotor patterns in reading and writing. This recognition of skin writing is an impressive sign of *unintended visuo-somatosensory transfer*. Since this transfer functions normally without being specially learned, it might be expected that the opposite *somatosensory-visual transfer* could be developed to higher perfection with repeated use of skin substitutes and special learning intention. However, the much lower number and density of skin receptors, compared to the retinal mosaic, sets natural limits to this process.

Sequential information by eye movements may also be compared to the sequence of tactile information of a feeling hand which explores the form of an object by moving over its surfaces (see MacKay, Chap. 5 and [307]).

An entirely different visual substitute is Brindley's technique of stimulation of the visual cortex by an implant [63]. Although this procedure avoids intersensory transfer, it has special interest for visuomotor functions (see Part B, Chap. 26).

Vision without Visual Cortex and Visual Functions of other Brain Regions

Visual Cortex Ablations and Tectal Functions. Since Klüver's original experiments on visual cortex ablation in monkeys [266], it seemed to be established that the residual visual abilities were confined to *perception of diffuse brigthness* without shape discrimination. The same was found in human cortical blindness. Such patients usually see only changes in diffuse brightness without any pattern vision and sometimes may show remnants of optokinetic nystagmus [429]. Pupil reactions are largely independent of the visual cortex, although some changes after visual radiation lesions have been observed in man. But what is the contribution of other brain regions to visual perception and behaviour ? The main questions are: *Which visual functions are independent of the visual cortex?* Which brain regions contribute to remnant vision in higher mammals ?

Recent experiments by Weiskrantz and his group [448] have demonstrated that monkeys (as was already known in cats and rabbits) can see more than changes in brightness with the visual connections of their brain stem, including the tectum opticum. Probably these monkeys use the *superior colliculus* for crude perception of shape and movement and for visual localization. In Part B, Chapter 14, the visual functions of the optic tectum are treated extensively by Sprague and co-workers and in Chapter 24 the results of visual cortex ablations are described by Doty. Doty's earlier work [110] demonstrated that cats without visual cortex, when *ablated in early life,* can develop almost normal visual behaviour. The possibility that form discrimination is related to both cortex and superior colliculus is also discussed by Stone and Freeman in Chapter 2. Schneider's "two visual systems", one for visual *orientation located mainly in the superior collicus* and one for visual *discrimination in the visual cortex* [390], may be helpful for a future order, but need further confirmation in man. Sprague, Wurtz and co-workers [410, 462] propose *shifts of visual attention* towards the contralateral half-field as being the main function of the superior colliculus in monkeys. It seems probable that this tectal function cooperates with the cortex in areas 8 and 17–19.

The system for visual localization in space is supposed to have mainly *retino-mesencephalic* connections, and the other system for fine discrimination and pattern vision to use *retinogeniculo-cortical* pathways. Although these systems can be distinguished by their location in the brain, they must function together in close co-ordination and cannot serve independently for the two functions *perception and action.* Their co-operation is apparent from the visuomotor action of eye movements during perception and location within the visual world and by the integration of ocular movements in behavioural attention and motion detection (see p. 99). An integrating function for visuomotor attention may be represented in the frontal eye fields (area 8).

Experiments on the superior colliculi have shown that *in cats and monkeys the tectum functions in close co-ordination with the visual cortex.* Whether this co-ordination is mainly for cortical facilitation of attentive eye movements towards the contralateral visual field, or whether the tectal neurones also need visual discrimination by cortical cells for their visuomotor functions (at least in cats and monkeys) cannot yet be decided. Discrepancies between the findings of different research groups, as to whether tectal neurones signal motion independently from the visual cortex in the cat, may be due to different use of anaesthetics (see Chaps. Chaps. 14 and 21).

KLÜVER's monkey experiments were repeated by WEISKRANTZ and co-workers [447, 448] and additional lesions of the inferotemporal cortex were made. More detailed testing of visual acuity after ablation of area 17 showed a much smaller drop in performance than would have been expected if the lesion had produced absolute scotomas. This seems to indicate that foveal and parafoveal vision is not strictly comparable in man and in rhesus monkeys. The chief incapacity of monkeys deprived of area 17 was the *failure of visual localization in space,* i.e. a function that TREVARTHEN [434] localizes preferentially in the tectum. Recent experiments, however, indicate that even this incapacity is not absolute and that monkeys without visual cortex can be trained to detect and reach out accurately for objects presented visually [448]. These monkeys perform surprisingly well in test situations after some visual training. WEISKRANTZ suggests that the retention of the ability to discriminate after the total removal of the striate cortex in monkeys means they are still able to process information about "total retinal activity" – contour, length, movement and flicker contributing besides luminous flux [447, 448]. Recent experiments by the NORRSELLS [345] on cats with neonatal removal of the neocortex showed visually guided behaviour which was influenced by *brightness rather than form.* An interesting feature was that, during visual conditioning, a worse performance followed a *black* pattern than a similar white pattern.

Visuomotor Behaviour and Visual Cortex. The experiments of HELD and HEIN [204–206] have shown that visually guided behaviour needs a *sensorimotor feedback which is acquired during active movements.* They demonstrated an impressive dissociation between the two eyes of kittens reared with one eye open during active locomotion in the visual world and the other open only during passive transport. This rather differentiated development of visually guided behaviour, with functional blindness of the passive eye, proves that visuomotor co-ordination depends upon spontaneous activity. The mechanisms seem to be partly subcortical, at least for the simpler extension responses. Recent experiments by HEIN [203a] with *ablation of the primary visual cortex* in cats have shown that the extension response remains unaffected and that visually guided behaviour is disturbed but can be *reacquired* in about two weeks. Apparently, this rapid learning is a reactivation of behaviour mechanisms acquired in the early postnatal period. The neuronal basis of all these visuomotor and feedback mechanisms remains unclear.

Cortical Blindness and Eye Movements in Man. Patients having severe bilateral lesions of the visual cortex use *acoustic and tactile* information for orientation and location in the environment. Their preference for acoustic and tactile information is similar to, but not identical with, that of peripherally blind persons. Most patients are also able to direct their eyes in certain directions. Behavioural observations of cortical blindness demonstrate that *multisensory location is preserved.* One of our patients with bilateral occipital cortex lesions saw nothing, but had pupillary reactions to light and could vaguely perceive but not localize the onset of a very bright light. He had a surprisingly *precise faculty for localized pointing with the hand and for stearing goal-directed eye movements towards acoustic stimuli* and towards tactile targets. Thus goal-directed saccades elicited by auditory and somatosensory stimuli may be preserved in man after bilateral striate cortex exclusion with total loss of form discrimination.

Goal-directed saccades contribute to visual spatial localization and are directed by attention (see p. 85). In normal man they depend upon perceived visual stimuli and these are suppressed after bilateral lesions of the visual cortex. Patients with cortical blindness direct their eyes according to acoustic or tactile stimuli, independently of the visual cortex. Optokinetic nystagmus is practically absent in these patients, although some remnants may be seen when the whole visual surround moves [429].

Evidently the visual cortex co-operates with oculomotor mechanisms in the brain stem by tectal and other mesencephalic projections. Experiments in monkeys have elucidated some of the neuronal co-ordinations that precede saccadic eye movements in the brain stem. Several authors have found neuronal discharges in the midbrain tegmentum 5—10 msec before a saccade begins in the oculomotor neurone discharges [148], and in the tectum even 50—100 msec before the saccade [462]. In contrast, neuronal discharges and potentials related to saccadic eye movement appear much later in the lateral geniculate, the visual cortex [85] and the fontal eye fields [48], where they *follow* the saccades. Most observations indicate that the visuomotor locating systems in the midbrain receive inputs from several sense modalities besides the visual system, and use them for saccade programming.

Vision and Limbic System. Projections of the visual system to the temporal lobe are not confined to higher cortical connections from the visual cortex and the pulvinar thalami. Besides the inferotemporal isocortex, which receives these projections, the hippocampal *allocortex* in AMMON's horn is also supplied by the subcortical visual system. Fibres from the lateral geniculate via the temporal loop of the visual radiation project to the posterior hippocampus and the fusiform and lingual gyrus of squirrel monkeys, as shown by MACLEAN and co-workers [308, 309]. The hippocampal areas of the two hemispheres are connected by a powerful commissure. the *psalterium*. Visual neuronal responses in the allocortex and other limbic areas are discussed in Chapter 21 (p. 396). MACLEAN proposes that these visual projections to the limbic system are functionally related to emotional behaviour, neuro-vegetative regulation, visuo-visceral experience, and the influence of light on alertness and sleep [308]. These findings are of some interest for clinical observations in temporal lobe lesions. Complex visual hallucinations and *déjà-vu* feelings in man were described by PENFIELD and co-workers [351, 352] after electrical stimulation of the temporal isocortex, whereas stimulation of the hippocampus shows fewer perceptual correlates. In temporal lobe epilepsy, however, localized discharge of the allocortex and other limbic structures may be related to dreamy states and to visual or multisensory hallucinations, including hearing, smell and taste. Whether these sensory phenomena require activation from the limbic system is not yet clear. The visual input to the limbic brain may be of special interest for the recent hypothesis of O'KEEFE and DOSTROVSKY [348] who postulate that the hippocampus makes an essential contribution to the spatial map of the outer world.

Cortical Functions and Lower Visual Mechanisms in Frog, Monkey and Man. Since LETTVIN, MATURANA, and co-workers [292, 317] described various response types of optic nerve fibres in the frog which signal visual patterns of biological significance to the tectum, the frog has become a useful model for motion detection and pattern vision. In this Handbook, Chapter 6 by GRÜSSER on movement detection contains many references to the frog's tectum. However, if we may adapt LETTVIN's pictorial language for our purpose: *What the frog's eye tells the frog's brain is very different from what the human eye tells the human brain.* In the frog the information processing is already highly organized in retinal neurones for the extraction from the visual world of meaningful signals about prey and enemies. In the rabbit, too, motion detection is processed in retinal ganglion cells at a lower level than in carnivores and monkeys. In man, the encoding and decoding mechanisms for pattern and motion apparently occur at a higher level of the brain, as they do in the monkey and probably also in the cat. In these species the optic nerve fibres and the lateral geniculate neurones contain no specialized "bug detectors" or "newness" or "sameness" neurones, such as described by LETTVIN for frogs [292]. Peripheral information processing in the retina of frog and rabbit may be of general interest for motion detection, but its value for human perception is limited although it does demonstrate some general biological principles relating to vision. In man, to *distinguish meaningful from meaningless patterns* is apparently a function of the *visual cortex* and of other *higher paravisual cortical areas* which resemble the monkey's occipital and temporal cortex. The way they process information for colour and form vision is not yet fully understood. Several research groups such as those of GROSS (Part B, Chap. 23) and ZEKI [465—467], are now working in this field. Their experiments may soon reveal additional functions of higher visual areas in the occipito-temporal cortex, especially since they propose a new anatomical order in cooperation with anatomists such as POWELL and GAREY [154, 155].These anatomists distinguish several *higher-level areas of the visual brain* in the occipito-temporal cortex of monkeys. POWELL has investigated these areas between area 17 and the inferotemporal cortex in the posterior bank of the superior temporal sulcus of rhesus. With some deviation from the classical architectonic areas 17, 18, and 19 of BRODMANN [64] and VOGT [442], he distinguishes at least four visual areas: I, II, III, and IV [367]. For ZEKI's results on colour coding in the paravisual area IV of the rhesus monkey [467] see p. 107.

Physiological and anatomical studies demonstrate that the elaborately organized cortex of the monkey has functions which are localized in the lower visual system of other species like frog and rabbit. It seems probable that the *monkey's cortex* can serve as a useful model of higher visual processing in man.

VIII. Eye Movements, Motion Detection and Visuo-Vestibular Co-ordination

Vision and Eye Movements

It is obvious that we see with moving eyes and that images stabilized on the retina disappear within a few seconds. Although normal vision requires eye movements, visual research has been directed mainly towards afferent messages traveling from the eyes to the brain. The most important visuomotor functions of the brain, the co-operation of vision and eye movement, remained a neglected field until recently. Yet the perceptual *integration of space, time and motion in the visual world* needs a precise correlation of visual messages with ocular movements. In this volume eye movements and vision will be treated only in Chapter 5 by MacKay, while some related problems are briefly mentioned in others: The role of eye movements in form vision is explained by Stone and Freeman in Chapter 2 and the relation of ocular tracking and motion detection is discussed by Grüsser in Chapter 6. Eye movements and the visual cortex are covered in Chapters 21 and 26. However, a comprehensive theory of visuomotor functions is still lacking.

Many open problems of visual and oculomotor coordination cannot be treated fully in this volume. Two chapters in volume VII/4 on eye movements by Alpern [5] and Matin [316] may also be consulted. Several symposia during the last 10 years [13, 36, 101] have brought together important contributions on visuomotor functions. From all this it appears that the field is not yet ripe for a theory and synthesis of vision and eye movements. The most pertinent, still unsolved problems are:

1. *Perceptual continuity* of the *sequence* of eye-movement-induced retinal images in time and space.
2. *Stability of the visual world during eye movements.*
3. *Saccadic functions* of selecting, timing, translocating and cancelling sequential visual messages.
4. *Co-ordination of eye movements and somatomotor activity.*

Perceptual continuity is discussed on p. 128. Visual stability is mentioned on p. 83 and is treated more extensively in Chapter 5 where MacKay explains that the old hypotheses postulating suppression or subtraction of vision during rapid eye movements are unsatisfactory and re-evaluation may operate effectively with less accurate oculomotor information.

Eye Saccades and Vision. Each saccade initiates a new retinal image with further foveal fixation. Saccades determine the duration of fixation since all fixations are intersaccadic intervals (see p. 87). Most saccades are *goal-directed eye movements (Blickzielbewegungen)* and must be pre-programmed to meet the intended visual target. The "Blickzielbewegung" *selects* the next target, *locates* it by foveal fixation, *terminates* the previous fixation period and *cancels* its residual trace to avoid disturbance by multiple images of the visual world [253]. *Thus saccades control the onset, duration and location of retinal images.* By collateral discharge to visual centres the saccades might clear the screen of previous pictures and give additional timing data for new visual messages sent from the retina to cerebral visual structures. But the physiological mechanisms of translocation, timing and cancelling are not yet known. Cancelling may be related to lid blinking [1].

Presynaptic inhibition in the lateral geniculate [6,312], thought to be a suppressive mechanism, has not been confirmed in its morphological substrate, as discussed by SZENTÁGOTHAI ([425]) and Chap. 17). Whereas neuronal discharges in the midbrain tectum [462] and tegmentum [148] occur prior to saccades, no such early or preceding electrical potentials were found in the lateral geniculate dorsal nucleus. Some geniculate neurones receive excitatory or inhibitory impulses *after* the saccade. During the saccade visual messages are transmitted. Not only light flashes but also object movements against background may be seen during the saccade, although the wide-field motion of the whole retinal projection is not perceived. Recently BÜTTNER and FUCHS [74a] localized (in experiments on awake monkeys) the main saccadic input into the *pregeniculate nuclei* which also receive visual afferences. These neurones began their discharge modulation rather late, around 80—200 msec, lasting 100—500 msec after the saccade. In 95% of the neurones in the lateral geniculate dorsal nucleus these authors found *no* change in activity with saccades. On the contrary, most of the pregeniculate neurones were saccade influenced. BÜTTNER and FUCHS conclude that the pregeniculate nucleus receives both an oculomotor and visual input [74a].

Subjective observations of after-images and of other visual phenomena suggest intermittent *cancelling* besides transposition of the retinal image to the new spatial location in the outer world achieved by goal-directed saccades. Although the cancelling–transposition concept differs from the old postulate of saccadic suppression, a brief interruption of the after-image by a saccade may be used as subjective evidence for either suppression or cancellation. In man, the implants in blind patients by BRINDLEY and LEWIN [63] brought new evidence for cortical mechanisms: *localized phosphenes caused by visual cortex stimulation* are translated and wander in the direction of eye movements. This phenomenon has been compared to moving after-images in sighted man, but a brief disappearance of these phosphenes during a saccade has not been described. An interesting observation of BRINDLEY's concerns *vestibular regulation of eye position:* passive turning of the patient's head with compensating eye movement towards a constant spatial direction left the phosphene remaining fixed in space (see Part B, Chap. 26).

For the relation of *eye and body movements*, differences as well as similarities between oculomotor and other sensorimotor functions should be stressed. Eye and body movements are essentially different, because skeletal muscles have to overcome stronger gravitational and inertial forces than do ocular muscles. It is just these differences and some special characteristics of oculomotor neurones, e.g. their regular discharge frequencies, maximal saccadic activations and the lack of stretch reflexes, which should be considered when we investigate the correlation of eye, head and body movement and the interaction of visual and vestibular messages. See p. 94.

General Aspects of Visuomotor Regulation. Three symposia on oculomotor functions appeared in 1964 [36], 1971 [13] and 1972 [101] and mark the increasing interest in research on eye movements and vision. The main advances during the past two decades stem from the following five lines of research which tend to use both psychophysical and neurophysiological methods:

1. Experiments with stabilized retinal images demonstrating the necessity of moving the eyes to overcome local adaption, to regenerate retinal images and so to integrate visual information over time.

2. The precise study of two types of ocular movements
a) rapid saccades representing a spatial location system and
b) slow tracking movements functioning as a motion-control system.

3. Recordings of neuronal activity in the brain stem and cortex during saccades, eye tracking and nystagmus.

4. Visuo-vestibular integration in optokinetic and vestibular nystagmus and the co-operation of eye, head and body movements, investigated in man and animals.

5. Scaling of velocity perception in relation to ocular movements and motion in space.

Eye Movements and Stability of the Visual World. The combination of afferent visual and efferent oculomotor processes during human movement perception calls for a model of sensorimotor co-ordination which modern cybernetic concepts have not yet offered. Only general hypotheses have been discussed since Helmholtz [208] and Hering [212] more than 100 years ago considered spatial constancy during voluntary eye movements as a physiological problem.

To explain visual constancy during eye movements, three theories have been proposed: 1. suppression of visual information during rapid eye movements; 2. compensation of motion information by subtraction of the intended movement; 3. re-evaluation of visual information by matching.

The first explanation, which postulated a visual block during eye saccades, has never been proved by conclusive experiments. The second explanation dates back to Helmholtz [208] and his conception of positive information by voluntary impulses *(Willensimpulse)* of ocular movements. This concept has found a modern formulation in a model by von Holst and Mittelstaedt [225–227]. They postulate an "Efferenzkopie" during voluntary eye movements for cancelling the movement signal caused by the retinal image shift during the same eye movement. A similar model by Sperry [402] uses a corresponding concept of a "central adjustor" or "central kinetic factor", termed the "corollary discharge" [402, 430], as a physiological equivalent of von Holst's "Efferenzkopie". The third theory by MacKay expects and presupposes stability of the outer world during movements of the organism, except for information which is *not* adequate to this movement [307]. MacKay has brought forward good arguments in favour of his theory and against the first explanation by showing that a reduction of visibility similar to that found during eye movements occurs during visual field displacements before a stationary eye. His experiments suggest that the suppressive effects of rapid displacements of retinal images without eye movement are not in principle different from those of moving retinal images during a voluntary eye saccade or a quick phase of nystagmus.

MacKay's *evaluation theory uses a matching concept*. It regards perception as an organizing system during the readiness of the organism. With this theory, the perceived intensity of the sensory stimulus reflects not the impulse frequency from the receptor organ but rather the internal matching effort. Although this internal matching response may counterbalance the afferent impulse frequencies like von Holst's and Mittelstaedt's "Efferenzkopie", MacKay's conception differs from von Holst's by offering a simpler explanation of sensory stability during movement: it presumes the stability of the outer world during movements of the organism and admits external movements only during significant deviations from this hypothesis [307].

According to MacKay the system presupposes stability as a "Null-hypothesis". Like Hess' earlier proposition [216] that perceptions and sensorimotor regulations use identical central regulations, MacKay's postulate is that all transformations occur in the *same* central nervous substrate, e.g. 1) internal representation of the organism in its world, 2) adaptive action and readiness to act in the world, 3) matching for signals received from the world. In his own words: "The motion of the organism in the world is implicitly represented by the transformation which is demanded". For details see MacKay, Chapter 5.

Sequential Order of Ocular Movements and Visual Attention. Since Hering [212], the close association of visual *attention* and eye movements has been recognized, but quantitative investigations are lacking except for a few attempts to *measure* attention towards moving stimuli by optokinetic nystagmus [104, 253]. These investigations used optokinetic nystagmus elicited by large moving fields. Similar attentive selection also determines the *sequence of fixation*, when looking at a stationary picture or during exploration of the visual world. Here attention co-operates with "form vision" in selecting certain marked contours and angles of visual patterns, as first shown by Stratton in 1902 [418a]. Motivation and interest further direct the gaze to certain details of the picture as demonstrated by Yarbus [463]. Optokinetic selection of moving patterns may be used as a model for visual target detection.

Selection of targets during optokinetic nystagmus and involuntary attention changes were investigated under two conditions: *railway nystagmus* and *ambivalent nystagmus*. Ter Braak introduced the latter and summarized the relation to attention with Buis [428]. In his last paper [427a], Ter Braak enlarged the subject with new stimulation methods in relation to Wertheimer's apparent movements [451]. Ter Braak's ambivalent nystagmus shows clearly the limits of voluntary attention since after fatigue the eyes convert involuntarily to the stimulus of opposite direction. Railway nystagmus, often observed since Helmholtz's description in the first edition of his book [208] in 1866, was objectively recorded only recently [104]. The records demonstrated clearly the *attentive selection* and disclosed the faculty of eye tracking in following objects with changing velocities by adequate speed variations. Of course, attention also involves multisensory mechanisms, including opto-acoustic adversion and other sensorimotor coordinations (see p. 79 for cortical blindness).

The many problems of successive integration of vision during eye movements still need experimental investigation and a new system concept. Some work was begun by the research groups of GAARDER [150], HABER [188] and others. GAARDER stresses the discontinuity of the visual flow and the necessity of a level hierarchy for perception of "packaged" visual input [150]. See also p. 128.

Functions of Visual Saccades and Pursuit

Saccades, a Spatial Location System, and Ocular Tracking, a Motion Control System. The coordination of eye movements and vision can be understood only when the two main classes of ocular movements are distinguished and their relation to visual information is defined: 1. *Rapid* eye movements or *saccades* are used in exploring the visual world. These are goal-directed movements *(Blickzielbewegungen)* which precede fixation *before* visual objects are brought onto the area of central vision. 2. *Slow* following movements or *eye tracking* serve to maintain visual fixation of moving objects and to detect motion.

During rapid eye saccades, no motion is perceived but the object envisaged is *localized in space* with anticipation of its appropriate location in the visual surround. At this spatially determined point, which varies in relation to eye and head movement, the goal-directed saccades end in the intended fixation. Objective recordings of eye movements, as well as different subjective sensations elicited by rapid and slow eye movements, suggest the existence of *two different sensorimotor systems* for saccades and eye tracking. They are sensitive to special visual information and assist in processing and controlling position in space and movement perception, respectively.

1. *A spatial-location system* is related to the cerebral mechanism of *rapid saccadic eye movements.*

2. *A motion-control system* is represented in the cerebral mechanism of *slow tracking eye movements.*

The first gives rise to no perception of movement; the second specifically monitors movements in the visual surround. Both are intimately linked in nystagmus by the quick and slow phases. We assume that the two systems cooperate in a cybernetic regulation and that the first is related to the formation of a map of the outer world which the subject carries in his head as a model for sensorimotor coordination. Every movement of the eyes, head and body must affect the relative visual location in the outer world which is continuously being changed by active and environmental movements. The simplest way for the organism to cope with these complex changes is to *direct the eyes to selected targets* in the visual world. This is foveal fixation in human vision and is guided by selective attention [104]. Thus, our *visual perception needs continuous control by eye movements.* Visual tracking and goal-directed saccades not only receive their information from moving and stationary stimuli, but also *control* movement and localization in a well organized co-ordination.

The essential sensorimotor function of the dual system of eye movements is *the selective control of visual input.* The saccadic mechanism is sensitive to localized stimuli and attains the visual target by a rapid eye shift for subsequent fixation.

The ocular *tracking mechanism* is dependent on moving stimuli and constitutes a motion control system. When the visual goal stops moving or a fixed target is viewed, a similar stimulus-locked mechanism maintains *fixation of stationary objects*. During movements of the viewing subject, tracking and fixation are combined in a complex co-ordination of visuo-vestibular-proprioceptive interaction (see p. 95 and [274]).

Slow and Rapid Phases of Optokinetic Nystagmus, Saccades, and Visual Selection. As explained in the following sections, the slow nystagmic phases of optokinetic nystagmus are regulated in relation to the velocity of moving objects in eye pursuit. The *slow* phases are therefore the essential mechanisms for velocity estimation in *oculomotor motion detection*. They reflect the tracking behaviour of the eye. The *rapid* nystagmic phases going in the opposite direction, which contribute to location when they grasp new visual targets, correspond to a *saccade* during intended changes of gaze. Both voluntarily controlled saccades *(Blickzielbewegungen)* as well as quick phases of optokinetic nystagmus serve to visually capture peripheral objects which attract our attention. Saccades and rapid nystagmus phases contribute to *visual target location*. Both slow and rapid phases of nystagmus are directed by *selective attention* (see p. 93).

During optokinetic nystagmus new objects appearing in the peripheral visual field are first captured by the quick phase and then tracked by the slow phase from the periphery to the center of the field of regard. Usually the slow phase does not follow it far beyond this central region. Thus, optokinetic nystagmus usually "beats" on the *field side of the quick phase* and remains in it also in its slow phase. The German expression for this, "*Schlagfeld auf seiten der raschen Phase*" [246, 247], makes it rather clear that optokinetic nystagmus is not simply a passive tracking reflex in a moving visual surround (slow phases) but also an *active search and selection process*, starting anew with each quick phase [109, 247]. The quick phases are visual grasps of new objects which appear in the periphery and therefore "beat" opposite to movement direction towards the side where the moving contrasts first appear when they enter the visual field. This has some biological significance, since every new moving object which appears in the visual field of regard may be of interest to the organism (potentially dangerous or welcome) and this, or the label "uninteresting", can only be decided after the new object has been fixated by the central area of foveal vision. This foveal fixation is the essential part of the *Schaunystagmus* of TER BRAAK [427]. After ocular tracking, the objects which have been fixated and recognized usually lose in interest, and are not followed beyond the central part of the field of regard when new objects appear continuously. The *quick phase* during optokinetic nystagmus, therefore, is a mechanism which *selects, localizes and prepares for visual fixation* during the slow phase. This selection can best be studied during *railway nystagmus* and the voluntary fixation of near and more distant objects in the foreground and background [104].

Visuomotor Scanning. The interesting *scan-path* hypothesis of NOTON and STARK [345a] for correlations of visual and oculomotor information is discussed in Part B, Chapter 21, p. 422. These authors postulate that eye-movement paths are a sensorimotor habit contributing to visual learning and a matching procedure for internal models of the external visual world. Eye scanning is composed of irregular sequences of sacades, fixations and pursuit movements, which are determined by shifts of attention and visual patterns in their sequential order. This appears in a simplified form under laboratory conditions of optokinetic nystagmus as an alternating sequence of quick phases (equivalent to saccades) and slow phases (eye tracking). Slow ocular tracking appears related to oculomotor motion detection *and* recognition of moving objects, whereas the quick saccades contribute to selection and location of visual targets before they are identified. The co-ordination of both with possible neuronal correlates deserves further consideration. The following remarks include some speculations which might be used later for experimental verification.

Fixation, Visual Pursuit and Neuronal Functions

Duration of Fixation and Information Processing. The optimal time of visual fixations should obviously be determined by visual information processing, perceptual demands and attention. Since each fixation is limited by saccadic movements, its duration can be measured as the *intersaccadic interval*, which usually varies between 160 and 400 msec. Similar values around 0.3 sec occur also for the duration of the slow phase of optokinetic nystagmus, for reading movements and during observation of pictures. The information content is more complex and the perceptual processing should be more difficult in reading than in optokinetic nystagmus, elicited by simple black and white stripes. Yet one can find statistical similarities in their durations. Although longer fixations occur under special conditions and for inexperienced readers [463], the duration appears determined not only by the visual information content, but also by inherent visuomotor mechanisms [94].

The perceptual interpretation is somewhat at variance with the concepts on intersaccadic intervals of communication engineers who postulated a saccadic clock and "pulse generator" which need a certain time for their preprogramming and have a "dead time". This duration, however, must not last as long as 200 msec, since shortest intervals for correction saccades or other saccadic sequences give much smaller values of below 100 msec. Whether these short durations can become longer in different and opposite directions is not yet clear. It does seem rather certain that the duration of fixation and slow nystagmic phases is regulated by several factors, including the demands of information processing and perception, as well as oculomotor parameters.

Fixation and Eye Pursuit. In spite of successful recent recordings from the oculomotor neuronal systems of the brainstem, the visual and frontal cortex (area 8), the neuronal mechanisms of visual fixation and oculomotor (efferent) motion perception by eye tracking are practically unknown, although the functions of *selection and fixation* can be studied well in railway nystagmus [104]. GRÜSSER in Chapter 6 reports on neurones responding to moving stimuli with some observations made during eye movements, and postulates an oculomotor control by his movement-sensitive neuronal systems MDS and OS. MACKAY, in Chapter 5, discusses possible *motion-signalling* and *position-signalling* systems with "*drift-sensitive*" neurones. For neuronal correlates of motion perception see also [182, 183].

Possible Neuronal Correlates of Drift, Pursuit and Fixation. Theoretically one may postulate, in addition to the saccadic location system, at least two neuronal subsystems which control motion detection by ocular tracking and guarantee pattern perception of moving objects during fixation: 1. A neuronal population which *signals the velocity of a moving object during eye tracking*, independent of its pattern, and which must be related to *visuomotor* mechanisms such as optokinetic nystagmus and its slow phases. 2. *Fixation neurones* which *during foveal eye pursuit* secure the information continuity of a pursued moving object and signal the significant features of this object undisturbed by any moving retinal images of the background. The second mechanism co-operates with pattern vision and goal-directed saccades in a selective process which operates during railway nystagmus when certain objects that are selected from the visual surroundings show apparent movements of very different angular velocities in the fore-, middle- and background of the viewed world, which appears to move backwards in the frame of the train window [104].

The visual recognition of the moving object and other higher visual functions of feature extraction during fixation may be similar to that during fixation of a stationary object, and it may be left open whether a third group comprising a special motion–pattern signalling system exists for recognition of moving objects. Some speculation about these neuronal functions may be useful for further research.

Stimulus-locked neurones would receive their input from a relatively fixed retinal image showing only minor displacements. They sustain their discharge during adequate stationary stimulation for a remarkably long time (see Part B, Chapter 21, Fig. 16, taken from NODA et al.). Hence, they may be called "*fixation neurones*". It remains undecided whether there are several groups of fixation neurones with visuomotor and purely visual functions: 1. those which secure stimulus-locked fixation, related to the pursuit 2. "drift-sensitive" neurones, postulated by MACKAY, coordinated with oculomotor systems (when the stimulus moves), and 3. those which signal only fixed patterns. Some relation of these neurones to TREVARTHEN'S visual location and identifying systems appears conceivable (see p. 78, 99). Although coordination of the two must be close, one might search for the pursuit neurones in the midbrain and pontine structures and for the fixation neurones in the geniculo-striate projections of the cerebral cortex.

A *drift-sensitive* neuronal subsystem, as postulated by MACKAY, would detect relative movements between different parts of the visual field and serve as feedback generator to prevent retinal image motion during ocular pursuit. Such drift-sensitive neurones should be *visuo-afferent* and different from the visuomotor mechanisms to which they send motion information for control of eye tracking. Psychophysical studies indicate that, besides drift detectors, two neuronal systems with sensorimotor and perceptual functions may exist, namely, *pursuit control neurones and fixation neurones*, which are involved in velocity perception of moving targets by *eye tracking*.

The *pursuit subsystem* has *visuomotor* functions and is supposed to control active visual pursuit during ocular tracking. It may correct slight lags of the eye behind stimulus movement which are signalled partly by "drift-detectors" and partly by neurones influenced by wide-field displacements (see Chap. 21). The final processing of the *pursuit control* system may contribute to *velocity perception of moving targets which are followed by the eyes*. Hence, some *outflow information* must be fed back to afferent movement-detecting systems since we are able to match both the inflow *and* outflow signals in combined visuo-afferent and visuomotor velocity perception during scaling experiments [104, 105]. The frontal eye fields may be involved in these control functions, possibly co-operating with the tectum in visual attention (see p. 77).

The second system has *visuo-perceptive* functions and may help to select and maintain significant *patterns* of the fixated objects, locked to the fovea according to stimulus movement. Such "*fixation neurones*" are apparently undisturbed by global retinal image movements which occur in the background during fixation of certain moving objects, but must be co-ordinated with the pursuit control system. Visual suppression of wide-field movements, described by WURTZ [461] as inhibition during saccades or rapid movements of a patterned background, may provide an additional mechanism of short inhibitory interruptions between subsequent fixation periods.

Besides WURTZ's experiments in monkeys [461], there are very few studies of the *effects of eye movements on visual cortex neurones*, (reported in Part B, Chap. 21, pp. 367–372). An earlier paper on ocular movements of the cat in darkness shows the effects of saccades in the *absence* of contrasting visual stimuli: VALLEALA [438] found that discharge alterations in visual cortex neurones follow the eye saccades, usually 20–40 msec after the onset of the saccadic EMG in eye muscles. Most neurones increase their discharge rate in total darkness, a few decrease it, and only a small fraction of neurones are not influenced. For lambda waves following saccades see p. 102.

Velocity and Grating Pattern Detectors in Man. BREITMEYER investigated stationary gratings of variable spatial frequency and moving patterns of variable velocity [60]. Expanding CAMPBELL'S experiments on human neuronal mechanisms which respond selectively to stimulus size and velocity, BREITMEYER distinguishes two types, *speed* and *direction analyzers*. Results from varying background illumination indicate that the response of a unit to a pulsed stimulus presentation determines neuronal *size-velocity preference*. With reference to previous work of the Freiburg groups on flash stimulation of retinal [180] and cortical neurones [24], BREITMEYER concludes that the interval between the onset of primary and secondary excitatory discharges of the neuronal response is related to preferred velocity and spatial frequency or size of the visual stimuli. The formula $V \cdot fs \cdot i = 1$ is proposed for preferred velocity V, spatial frequency fs and discharge interval i. Whether these intervals are related to the early after-images (p. 26) and to the phenomena of brightness enhancement at certain flicker frequencies [179, 180, 186, 296] may be left open.

Two Mechanisms of Motion Detection, Afferent-Retinal and Efferent-Oculomotor

Inflow and Outflow Mechanisms of Movement Perception. Visual motion detection relies on two different sources of information: 1. a sensory process and 2. a sensorimotor process. The *first* derives object motion from the *shift of retinal images* occurring with eyes held stationary. It represents an afferent modus of velocity information. This type of motion detection corresponds to an *inflow* of visual motion signals which is based on retinal projections of patterns moving with given velocities across the field of view (Fig. 14a). The velocity sensations obtained from this procedure with eyes stationary and retinal shifts are for brevity called *retinal motion perception* (afferent).

The *second* process infers motion from the *pursuit movements of the eyes* following a stimulus and fixating it with fovea. This modus of processing information of velocity corresponds to *efferently controlled oculomotor information*. It represents an *outflow* processing. It computes stimulus motion from the angular speed of the ocular tracking movements which follow the moving object with the fovea (Fig. 14b). The velocity sensations using eye tracking are called briefly *oculomotor motion perception*.

The two modes of motion detection are distinguished by different geometrical conditions of optics besides their entirely different physiological mechanisms.

Figure 14 shows the *physical principles of retinal and oculomotor motion detection* and velocity estimation. *a)* When the eyes are held stationary, optic patterns moving through the visual field produce shifting retinal images that elicit a purely afferent *retinal velocity information*. *b)* When the eyes follow the moving target in order to maintain foveal fixation, no retinal image shift occurs. Instead, only the *fovea is moved* by the rotation of the eyeball. Since the eye rotates around an axis through the centre of the eyeball, the *angular displacement of the fovea* (b) following

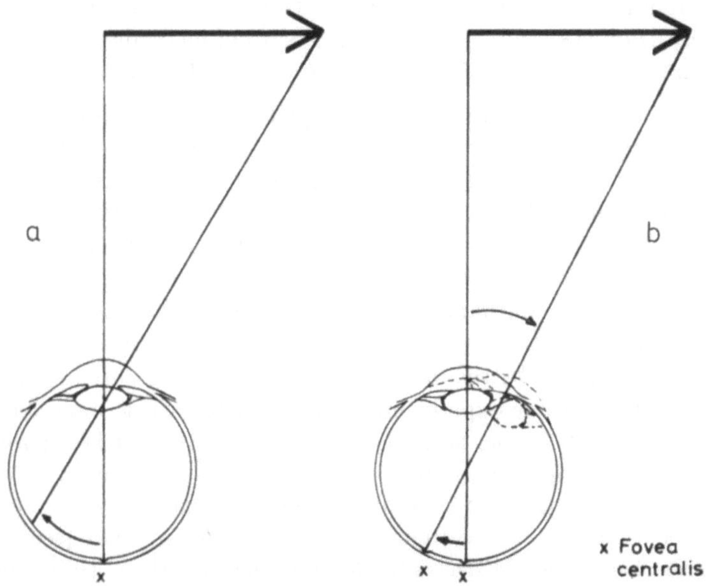

Fig. 14. *Two mechanisms of motion detection and the physical basis of retinal and oculomotor velocity perception:*
a afferent information by moving retinal images and visual inflow, **b** efferently controlled eye tracking and oculomotor outflow. *The visual angle of the retinal image displacement in a corresponds to about twice the visual angle of eye displacement in b.* Under natural conditions both kinds of motion detection may be combined, e.g. when attempted tracking lags behind higher stimulus velocities. **a** *Retinal motion detection* uses visual mechanisms: When the eyes are held stationary, moving patterns are projected with a similar angular velocity to their images on the retina. This movement causes successive stimulation of many receptive fields of retinal ganglion cells; signals proceed to the brain, where further information processing occurs. **b** *Oculomotor motion detection* uses visuomotor mechanisms: Efferently controlled motion detection occurs when the foveal fixation of a moving stimulus is maintained by adequate visual pursuit at the same speed as the angular velocity of the object. This requires a complex sensorimotor coordination and precisely controlled oculomotor innervation to hold the object on the same retinal locus. From JUNG 1972 [253].

the same moving object must be *smaller than* the visual angle of a corresponding *shift of the retinal image* (a), which is projected through the dioptric apparatus at the cornea–lens complex in the frontal part of an eye held stationary (Fig. 14). For the dioptric projection through the lense of external motion along a certain distance with eyes stationary the shift of the retinal image may result in nearly double angular values in relation to the displacement of the fovea for the same moving stimulus. There are considerable changes in this relationship at distances less than 1 m from the cornea.

Similar differences are experienced between retinal and oculomotor velocity perception. It will be discussed later whether the differential physical principles and visual angles would suffice to explain the overestimation of angular velocities of retinal movements, studied by DICHGANS et al. and illustrated in Fig. 15.

Several investigators have found that afferent retinal motion detection causes a considerably greater sensation of velocity for the same stimulus movement than efferent oculomotor motion detection. Rough estimations in the older literature varied around double or triple values for afferent signals with stationary eyes, as compared to efferent tracking velocity perception. However, recent psychophysical measurements by DICHGANS, who used STEVENS' scaling methods, found a ratio of 1.7 : 1 *for afferent and efferent velocity information* [104, 105, 273]. See Fig. 15.

Apparent Movement, Perceptive Fields and Optokinetic Nystagmus. Apparent movement of the WERTHEIMER type [451] occurs without eye movements and may be classified as a special kind of afferent retinal motion perception. SPILLMANN determined the *perceptive fields* for apparent movement [404]. They are about *20–30 times larger than those for the* brightness [259, 404], and also increase their diameters linearly towards the peripheral retina (increase from 15–60° eccentricity by a factor of two). Mechanisms similar to those in human motion perception may be assumed for the cat. Motion-sensitive neurones in the cat's visual cortex may also be stimulated by two successive light stimuli presented at different loci in the visual field. In early experiments SMITH and KAPPAUF observed optokinetic nystagmus to stroboscopic movements in the cat [397]. It is apparent in animals and in man that optokinetic nystagmus depends on attention [104]. Of course, attentive mechanisms which facilitate optokinetic nystagmus and correlate with motion perception can be investigated more precisely in man (see p. 92). TER BRAAK in the last paper [427a] after his untimely death in 1971, made a thorough study of *ambivalent optokinetic stimulation* by repeated displacements of patterns over a small angle below 1°. Subjectively, he found two types of motion perception: *"jump" movement*, and *"flow" movement* of continuous motion, related to, but not caused by optokinetic nystagmus. The close relations of jump and flow sensations with WERTHEIMER'S *β and φ phenomena* [451] are discussed [427a].

Quantification of Velocity Perception

Optokinetic Nystagmus and Oculomotor Velocity Control. A simple method of studying ocular tracking is to record optokinetic nystagmus and measure its slow phase velocity during stimulation by a rotating drum with contrasting patterns projected at various speeds on a white half-cyclindrical screen of 140° of visual arc.

The *angular velocity of the slow phases* of optokinetic nystagmus is measured and compared with subjectively scaled velocities. STEVENS' method of magnitude or ratio estimation is used for the latter.

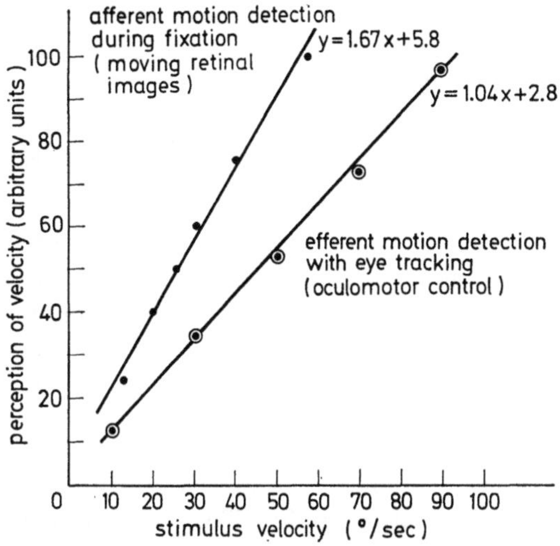

Fig. 15. *Comparison of retinal and oculomotor velocity estimation.* From DICHGANS and BRANDT, 1972 [103]. Mean values of 10 normal subjects. The scaling of velocities by *afferent retinal* motion detection resulting from moving retinal images during fixation (upper line) shows a steeper slope with overestimation when compared to *efferently controlled oculomotor* motion detection. The latter is based on tracking and optimal attention towards the moving stimulus. All estimates were compared to a standard stimulus with eye tracking. Stimulus velocities were: for steady fixation, 13, 20, 26, 30, 40 and 58°/sec: for tracking, 10, 30, 50, 70 and 90°/sec. The ratio of efferent to afferent velocity estimation was 1 : 1.67 in these experiments. Plotted on double logarithmic scales, both estimates would show an exponent of 1 in STEVENS' power functions with a parallel shift between oculomotor and retinal velocity perception

Magnitude Scaling. STEVENS [413,414] since 1961 has developed simple scaling methods with reference to a standard test stimulus for many sensory modalities including brightness, and this has been systematized for psychophysiological measurements [297]. These ratio estimations, however, were not used for higher visual functions until recently. DICHGANS et al. [105, 273] carried out systematic velocity scaling for the two kinds of motion detection by comparing estimates: with eyes stationary, to those with ocular tracking. RACHLIN'S experiments on scaling velocities [370] were not combined with simultaneous recordings of eye movement and may confuse the two modes of motion detection. It remains an open question why this author obtained lower values for the exponent of STEVENS' power function (0.63) than DICHGANS found for both retinal and oculomotor velocity estimation (1.0). Consistent differences appear for subjective velocity when perceived velocities are scaled while seeing moving retinal images with eyes stationary and when scaled during tracking. The results are summarized in Fig. 15. To evaluate these experiments it is necessary to keep in mind the essentially different mechanisms of the two modes of motion detection, the afferent-retinal and the efferent-oculomotor (Fig. 14a and b).

Oculomotor Motion Detection and its Limits. Normal velocity perception without special instructions utilizes cues from tracking movements occurring automatically during optokinetic nystagmus. With attentive ocular tracking, the velocity of the slow nystagmus phase was found to match stimulus velocity up to 50–90°/sec, beyond which the velocity of the slow phase began to lag behind the velocity of the moving stripes [104, 105, 273]. For a comparison with velocity perception, the quickest of the slow phases recorded during attentive eye following were measured. Irregular periods of slow phases with diminished velocity are interspersed between more accurate eye tracking. They indicate periods of diminished attention, often unnoticed by the subjects themselves.

The upper limit of accurate eye tracking varies in different subjects from 50 to 110°/sec. Magnitude estimates of perceived velocity, using a rotating stripe pattern moving at 70°/sec for reference, increase linearly with stimulus speed as long as eye tracking remains accurate. At higher speeds, when eye movements lag behind the moving stimulus, estimated magnitudes markedly exceed stimulus velocities.

Diminished Attention and Velocity Overestimation. During *inattentive states* all subjects produce estimates of higher velocity than during attention and show a larger intra- and inter-individual variability. The curve of velocity estimation obtained from these data shows a progressively steeper slope with increasing stimulus velocity. Therefore, it deviates more and more from the regression line for optimal eye tracking and velocity estimation. Ocular velocity during inattentive states is always much slower than stimulus velocity, often below 15 degrees per second, thus lagging further and further behind with higher speeds of the moving stimulus. The magnitude of estimated velocity during inattention increases proportionally with the velocity of moving retinal images, computed as the difference between ocular velocity during the slow phase and stimulus velocity. Estimated velocity increases with decrease of tracking speed. The reciprocal relation of eye tracking to velocity estimation at higher stimulus speeds is similar to that observed for lower speeds and inattention. This suggests that increased retinal movement velocity accentuates visual velocity perception. The angular velocity of moving retinal images is multiplied for perceived velocity by a factor of 1.7. Thus, the same rule of higher velocity sensation of afferent motion perception from moving retinal images may be assumed for the higher scaling magnitudes during attention at faster stimulus speeds, as during inattentive following movements. The common factor in both conditions is that the *nystagmic slow phase velocity lags behind the stimulus velocity*.

Overrating of Retinal Velocity Estimation. *Velocity perception without eye movement* was obtained by fixating a stationary target projected into the moving field. In this series of experiments stimulus velocities were estimated with controlled suppression of eye following by steady fixation. Under these conditions the movement of retinal images has the same angular velocity as the stimulus. Over the whole range of stimulus speeds with eyes stationary the subjects *overestimated* the stimulus velocities by a factor of approx. 1.7, compared to oculomotor velocity perception during tracking. Thus, the ratio of retinal to oculomotor velocity information is about 1.7 : 1 in magnitude scales (Fig. 15).

A *combination of retinal and oculomotor velocity information* occurs when eye following movements lag behind stimulus velocity. This happens regularly during optokinetic nystagmus at velocities higher than 100–120°/sec, or at lower velocities, when visual attention is diminished [104, 105, 273]. Under these conditions, velocity is overestimated in proportion to the relative part played by the afferent mechanism. A biological interpretation and a tentative teleological explanation of this result may be as follows: to ensure continuous stimuli for eliciting directed optokinetic tracking movements by a moving surround, perfect pursuit movements by foveal fixation are dominant over contrasting afferent motion signals from the peripheral retina. Only when the relative lag results in a retinal image shift, does their stronger excitation with velocity overestimation again elicit tracking. Perfect tracking is further regulated and facilitated by selective visual attention which induces foveal fixation.

Velocity Estimation, Attention and Optokinetic Nystagmus. The principal findings of the experiments on motion detection, optokinetic nystagmus, attention and velocity estimation by DICHGANS and co-workers may be summarized as follows:

1. *The two mechanisms of retinal and oculomotor motion perception may be separated by voluntary attention:* a) afferent inflow by moving retinal images signals movement with eyes during fixation of an immobile object or with lagging slow phases during inattention; b) oculomotor efferent outflow with tracking signals velocity during visual attention and requires an intention to look.

2. *Oculomotor motion perception by eye tracking occurs under natural conditions of visual attention.* Retinal motion perception appears under more artificial fixation or beyond certain stimulus velocities. Both mechanisms may interact by summation of information.

3. *Retinal velocity scaling,* using only retinal shifts exaggerates the velocity information by a factor of approximately 1.7, as compared to oculomotor velocity estimation.

4. The *slow phase of optokinetic nystagmus* is the mechanism for correcting velocity information by monitoring oculomotor output, controlled by foveal fixation.

5. *Attentive facilitation* is necessary for correct tracking of the slow phases of optokinetic nystagmus.

6. *Attentive selection* with foveal fixation determines the tracking of a selected object against a moving background during railway nystagmus.

Visuo-Vestibular Interaction

Eye and Head Movements and Vestibular Information. Eye movements rarely occur in isolation. They are most often coordinated with *head and body movements*. Under natural conditions, we explore the visual world not only by goal-directed eye movements but also by *body motion* with successive changes of our viewpoint. During active movement we collect additional multisensory information that is compared with the motor commands, for example, by reafferent information. Accordingly, the inner model of the outer world must be adapted to this continously changing relation to obtain contingent perception. All movements of eyes, head and body modify this internal model and the position of the eyes relative to the visual surround must be adjusted to each new viewpoint. Fixation targets are mainly selected and localized by vision, but in addition head acceleration is signalled by the labyrinth during head turning. Therefore, *vestibular signals and proprioceptive information* from neck and extremities must be coordinated in the visuo-oculomotor and visuo-graviceptor integrations. Although goal-directed eye movements are usually elicited by visual targets, the *compensatory eye movements during head turning depend upon vestibular information* from the semicircular canals. This correction of the eye position after an associated head turn is regulated almost entirely by the labyrinth, as shown by DICHGANS and BIZZI [102, 326a]. Eye muscle proprioceptors are negligible, and neck proprioceptors are not very important in normal subjects for the coordination of eye and head turning, but may compensate for defective vestibular information after labyrinthine lesions.

Recordings from various brain-stem structures have shown *neuronal correlates of visuo-vestibular interaction* but our knowledge of their synaptic connnections is too scanty to furnish nerve cell diagrams even for nystagmus, although LORENTE DE NÓ had already proposed such neuronal circuits 30 years ago [299, 300]. The most promising approaches seem to be exact timing studies with recordings from tectal and tegmental neurones in rabbits, cats and monkeys, and further investigations of area 8 in the monkey's cerebral cortex. In this area BIZZI described two groups of specific neurones, one activated after saccades and the other by ocular tracking [48]. Similar neurones of the brain stem discharge earlier and *precede* the saccades in the tectum [462] and the reticular formation [148]. See p. 82.

Orienting Behaviour of all organisms moving in contrasting surrounds needs coordination of active sensorimotor effects with passive sensory signals. Both have some influence on perception. The main multisensory interaction is *visuo-vestibular* and *visuo-proprioceptive* and this concerns behavioural adaptation rather than perceptual experience. *Biological models* for this complex interaction may be derived from insects. Just as *Limulus* has taught us some elementary mechanisms of lateral inhibition, so flight behaviour in insects may provide some fundamental information about visuomotor mechanisms [164, 199, 200, 373].

REICHARDT's recent analysis of flight behaviour of flies [372a] promises to become a biological model of visuomotor mechanisms in higher animals which may even be compared to human perception. REICHARDT distinguishes between „Eigenverhalten" in a homogeneous surround and flight behaviour induced by visual patterns caused either by the fly's own movement (active) or by movement in the surroundings (passive). This may be compared to the perceptive effects of egocentric and exocentric localization in man (p. 96) and to the illusory alterations of the subjective vertical, such as the Aubert phenomenon after head tilt [11]. This behaviour is influenced by brain regions outside the visual system, since pathological alterations in perceptions of the vertical after brain damage in man are more often observed in frontal and parietal lesions than in occipital [35].

Egocentric and Exocentric Motion Perception and Illusion during Optokinetic Stimulation. The term "*egocentric localization*" was defined in 1917 by the sensory psychologist G. E. MÜLLER [333, 334]; it was used for spatial location and optokinetic phenomena by HOFMANN [223] and M. H. FISCHER and KORNMÜLLER [142, 143]. It concerns *subjective localization and movement related to our body scheme.* There are two modes of movement perception relating the organism to its surroundings according to the perceived stationary reference point: "*egocentric*" (subject stationary, surround moving) or "*exocentric*" (surround stationary, subject moving). These two motion perceptions interact with vestibular signals and the two visual motion detection mechanisms, the afferent-retinal and the efferent-oculomotor [104], giving rise to various illusions of self-rotation during optokinetic and vestibular nystagmus such as the oculogyric illusion.

The terminology "egocentric" and "exocentric" becomes difficult to understand when applied to motion illusions. Simple descriptive terms such as "*circular vection*" (CV) or apparent self-rotation may then be preferable. The *apparent* self-rotation CV which is experienced by a stationary subject when sitting still and looking into a rotating drum corresponds to an "exocentric" illusion of rotating in a stationary surrounding world (objectively moving, but perceived as stationary), although an outside observer would expect "egocentric" perceptions of surround rotation. BRANDT and DICHGANS [58, 58a, 59, 103] therefore use "*exocentric*" *for subjective movement of the subject* and "*egocentric*" *for any motion perception of the external visual world*, be it real or illusory. Mixed sensations are common (see Table 4). Although one might be inclined to call naively any self-induced movement egocentric, and the equivocal terminology egocentric and exocentric has its drawbacks, it maintains the position of TSCHERMAK's exact subjectivism in sensory physiology [436] and avoids discussion concerning the "relative" and "absolute" relations of subject and surround in spatial and movement perception during optokinetic and vestibular stimulation.

The terms *ego-oriented* (*ichorientiert*, self-related) and *exo-oriented*, (*umwelt-orientiert*, environment-related) motion perception may be preferable to avoid the ambiguous term "centric" and to denote the process of expectation and preprogramming, associated with movements of the organism and in the environment. Ego-oriented motion sensations are experienced in relation to the observer; exo-oriented information occurs, when the observer perceives himself moving by detection of relative motion of the visual environment.

Experiments of BRANDT, WIST, and DICHGANS yielded the following results on motion illusions [59, 103]. The *illusion of self-rotation* known as *circular vection (CV)* [143] is regularly induced by purely visual moving stimuli when the entire visible surroundings are rotating around the subject and optokinetic nystagmus is elicited. This circular vection is one kind of *exocentric motion illusion* (self-rotation is experienced relative to a rotating surround which, however, is perceived to be stable) and cannot be distinguished subjectively from real objective motion of the subject during chair rotation.

When the head is tilted during optokinetic stimulation *pseudo-coriolis effects* are induced which are indistinguishable from true coriolis effects of vestibular stimulation by head tilt during chair rotation. The optokinetic pseudo-coriolis effects outlast the visual stimulation by up to 30 sec whereas optokinetic after-nystagmus lasts much longer, up to 180 sec. In recent experiments BRANDT and DICHGANS [58a] found that exocentric motion perception and illusions such as apparent self-rotation (CV) are induced predominantly from *peripheral* parts of the visual field, whereas motion perception of surroundings (egocentric, subject stationary) is induced mainly from *central* parts of the visual field (foveal and parafoveal). Patterns seen in the foveal field are related rather to active attention and are accompanied by visual tracking (oculomotor motion detection). Stimuli in the peripheral field elicit a reflex-like "passive" attention. Whether the perceptions of self- or surround-movements are influenced by visual motion information of eye movements, fed into the vestibular system, as found in lower animals [103, 389] is unclear.

The following scheme of BRANDT and DICHGANS [58a] summarizes the various movement perceptions related to subject and surroundings that are elicited by optokinetic stimulation (Table 4).

This scheme demonstrates egocentric and exocentric localization under different conditions of optokinetic stimulation. Combinations of both are common, as everybody knows from sitting in the train, and looking at other moving trains and stationary objects during railway nystagmus. Egocentric and exocentric perceptions are less dependent on attention than retinal and oculomotor perception appearing simultaneously during diminished attention [104].

Table 4. *Optokinetic stimuli, motion perception, egocentric and exocentric localization and their interactions indicated by arrows.* From BRANDT, DICHGANS, and KOENIG [58a]

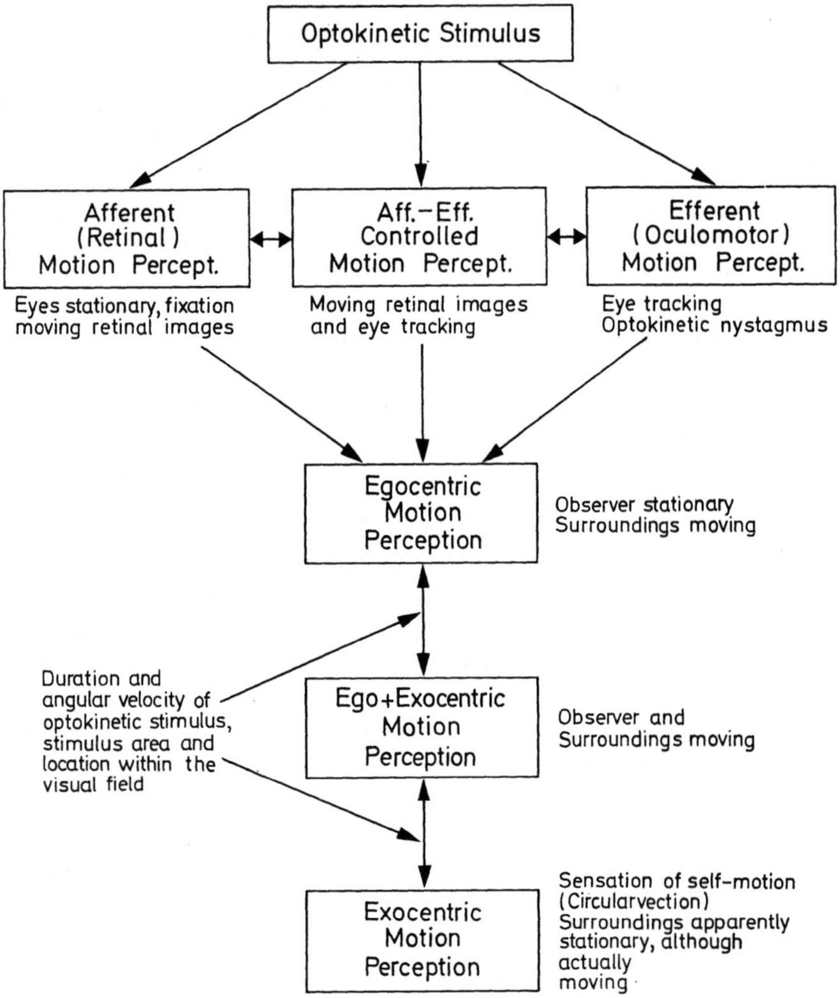

Three modes of observation of optokinetic stimulation occur from the two mechanisms of motion detection and their combinations a) *Afferent* by moving retinal images when the eyes are stationary; b) *Afferent-efferent* with pursuit movement and simultaneously moving retinal images when eye movements lag behind a moving pattern or with tracking of a target that moves in reference to a stationary background; c) *Efferent* with optimal eye tracking.

These stimuli cause three kinds of *motion perception:* 1) *egocentric;* 2) *ego + exocentric;* 3) *exocentric.* Whether perceptions 1, 2 or 3 occur depends upon stimulus duration, velocity, area and location within the visual field, but not on eye movements. During exocentric motion illusion the apparent velocity of selfmotion is not different for eye tracking and eyes stationary.

Specific Visuo-Graviceptor Interaction and the Subjective Vertical. The perception of horizontal and vertical orientation is well developed in man and is modified by the graviceptors of the body, mainly the macular receptors in the labyrinth. Head tilt changes the subjective vertical, as known since AUBERT's experiments [11] in 1861, but non-vestibular afferents also contribute such as rotating visual patterns [102, 103a].

Correlates of apparent shifts of the vertical are the *modifications of receptive field axis orientation of cortical neurones by head tilt*, as described by HORN and co-workers [229] and DENNEY and ADORJANI [96] in area 17. (See Part B, Chapt. 21, Fig. 27, p. 408.) The overshoot in the direction of head tilt observed by DENNEY would correspond to the apparent shift of the subjective vertical in the same direction of head tilt (AUBERT [11]). The neurones showing undershoot in relation to the expected orientation after head tilt would correspond to the E-effect described by MÜLLER [334]. It should be added that, in both phenomena, the *objective* (true) vertical (if not corrected by the observing subject) appears tilted in the opposite direction when compared to the *apparent* subjective vertical. It seems reasonable to correlate the modifications of the neuronal receptive field axes to the apparent vertical and not to the true vertical.

Brain Stem Localization of Optokinetic Functions and Visuo-Vestibular Interaction. Many cerebral structures are involved in the regulation of eye movements, and brain stem systems play a larger role for the sequence of ocular tracking and saccades than for purely afferent visual information. The slow and quick phases of nystagmus are automatic and rhythmic paradigms of tracking-saccade sequences. There seem to be several nodal points in the brain stem where the various visual, vestibular and proprioceptive messages converge for attentive regulation of *eye movements and nystagmus*. Since the classical experiments of LORENTE DE NÒ [298—300] the *reticular formation of pons and midbrain* was considered as the major integrating system of nystagmus. OHM [347], however, in his many papers on nystagmography in man postulated a coordinating function of the *vestibular nuclei* on the basis of older experiments by SPIEGEL and others. Neurological observations on brain stem lesions showing disturbances of eye movements and nystagmus favour the pontine and mesencephalic reticular formation as the main integrating system [36, 37] and marked slowing of saccades and quick nystagmic phases occurs after lesions of the pontine tegmentum. The optic tectum seems to be less important in primates than in lower forms. Experiments in monkeys by BENDER, COHEN and others yielded a more precise localization in the paramedian reticular formation [37, 85a] where classical neurology had localized a "*Blickzentrum*". Recent experiments by POMPEIANO, COHEN, DICHGANS and their groups again demonstrated the role of the vestibular nuclei in eye movement regulation and even a participation of some efferent connections with the labyrinth for the facilitation of optokinetic after-nystagmus [102]. KORNHUBER also postulated an essential role of the *cerebellum* for eye movement regulation as part of sensorimotor behaviour [275, 275a]. The coordination of brain stem structures with cortical neuronal systems, as recorded by BIZZI [48] in area 8 of monkeys, is not yet clear but it seems certain that neuronal discharges preceding the eye saccades and quick nystagmic phases are coordinated in the midbrain *before* the cortical eye fields in area 8 are involved [462]. For visuo-vestibular coordination see [246, 247].

Locating and Identifying Vision in Primates. The conception of two anatomically distinct visual systems was first developed for lower mammals [390] but also applied to the monkey by TREVARTHEN [434]. After experiments with split-brain monkeys, he considered vision of space and vision of object identity as two different brain mechanisms localized in the midbrain and visual cortex, respectively. Vision would then involve two parallel processes, one *ambient* (or extensive) determining space in a larger surround, the other *focal* (or intensive) with detailed vision in small restricted areas. The midbrain visual system, with the representation of the peripheral parts of the visual field, would govern mainly the first process in relation to locomotor behaviour. The cortical visual system, with a magnified representation of the fovea and parafovea, would subserve the second in the central space and control specific identification, fixation and scanning eye movements. The extensive behavioural space around the body is mapped with the aid of eye and head movements. Hence, TREVARTHEN stresses the importance of *velocity patterns* for information in three-dimensional space and its relation to eye and body movements.

In the eye, the first system of "ambient vision" would be located mainly in the *peripheral retina* and the second with "focal vision" in the *fovea* and parafovea. This distinction may be related to differences in visual acuity, especially in primates, including humans, who possess a fovea. It may also be applied to movement perception and to the findings of DICHGANS and his group [58a, 326a] on movement perceived in the environment (elicited from the parafovea) and of self-rotation elicited from peripheral parts of the retina, summarized in Table 4 (p. 97). The focal system would be related to identifying form vision and self-related movement perception, the ambient system to spatial vision and environment-related motion detection.

IX. Visual Evoked Potentials and Electroencephalography

This handbook contains *three chapters on visually evoked cortical potentials:* in Part B of the present volume (VII/3) CREUTZFELDT and KUHNT describe animal experiments (Chap. 27), and MACKAY and JEFFREYS recordings in man (Chap. 28); in Volume VII/4 RIGGS and WOOTEN discuss relations to psychophysical data [378]. Some remarks on neuronal equivalents, on difficulties in comparing animal and human evoked potentials, on readiness potentials and on lambda waves may be added.

Evoked Potentials

Evoked Potentials and Vision. The bulk of the voluminous literature on visual evoked potentials and averaging techniques is somewhat disappointing as a source of reliable perceptual correlates. For human evoked potentials the situation appears similar to that for the EEG in general. Both fields have some practical applications in neurology but their neurophysiological basis remains obscure. This basic ignorance also hampers correlations between evoked potentials and neuronal events in animals.

Some *perceptual correlates of visual evoked potentials* in man have been found. Since SPEHLMANN first studied the influence of patterned stimuli on evoked potentials [400], several laboratories have been active in this field, as described by MACKAY and JEFFREYS in Chapter 28. Their results and those of CAMPBELL and his group on grating patterns [49, 77, 78] seem to be the most interesting for human vision (see also Chap. 2, and in Part B, Chap. 21, p. 366, and Chap. 28, p. 659).

Evoked potentials induced by pattern stimuli have recently found *clinical applications* also. Occasional observations and television experience in epileptics had drawn attention to the effects of pattern stimuli for inducing epileptic attacks. Geometric patterns were found to be more effective than diffuse illumination for provoking photoconvulsive responses in EEG [42, 241]. *Pattern-induced epilepsy*, earlier considered to be a very rare event, is now encountered more often after television experience. It may be explained by the abnormally synchronous excitation of neuronal populations responding to certain spatial orientations in columns of the visual cortex. Similar mechanisms have been discussed for the geometric illusions in migraine (p. 129).

Pattern reversal is also the best stimulus by which the slowed nerve conduction after unilateral optic neuritis (demyelination) can be measured in man using evoked potentials. It consists of the sudden reversal of a black-and-white checkerboard which simultaneously changes small contrasting squared areas of B excitation into D, and D excitation into B. HALLIDAY et al. [189] showed that the resulting evoked potentials in the occipital cortex appear more clearly after these pattern stimuli than after diffuse flashes.

Some abnormal changes of evoked potentials in animals are without perceptual correlates for the simple reason that they are caused by unphysiological stimulus conditions which do not occur in man, i.e. electrical stimuli of both optic nerves after which BREMER described an enormous facilitation of occipital evoked potentials in encéphale isolé cats [61] and stimulation of the optic pathways in blinded cats which elicits higher-amplitude visual evoked potentials [264].

Neuronal Discharge and Evoked Potentials. Inhibitory and excitatory processes interact at all levels of the visual system. Collateral and reciprocal inhibition of the two neuronal subsystems B and D begins in the retina and is elaborated further centrally until another mode of pattern processing with specialized neurones begins in the visual cortex and its columns. The neuronal activity of the visual structures oscillates around a mean level of activity with negative and positive feedback. When one considers the reversal of polarity of evoked potentials in the middle cortical layers and the unknown orientation of the various cortical dipole sources which contribute to cerebral potentials, it seems unlikely that simple relations to neuronal inhibition or excitation would be recognizable in evoked potentials. High-amplitude evoked potentials following diffuse light flashes in animals and man demonstrate that more visual cortex neurones are influenced by diffuse illumination than would appear from HUBEL and WIESEL's studies.

In man the deep foldings of the cerebral cortex and the parasagittal location of the area striata in the calcarine fissure at the median hemispheric surface are reflected in the topography of certain evoked potential components (see Chap. 26 by MACKAY and JEFFREYS). However, it seems improbable that neuronal excitatory or inhibitory postsynaptic potentials of the primary visual cortex can be related to definite positive or negative waves of visual evoked potentials picked up from human scalp recordings. See also [440].

Although recordings from visual areas in animals allow fairly exact electrode positions for exploration of electric fields, all attempts to correlate evoked potentials with postsynaptic excitatory and inhibitory potentials (EPSP and IPSP) have to be interpreted with some reserve. In contrast to his earlier optimistic statements CREUTZFELDT now doubts clear correlations of EPSPs and IPSPs with the positive and negative waves of visual evoked potentials, since he found essential differences for strong and weak stimuli (see Chap. 27, CREUTZFELDT and KUHNT). The origin of sources and sinks of evoked potentials remain unknown, and the role of the laminar and synaptic organization of the visual cortex for brain potentials is not yet fully understood.

It remains uncertain whether evoked potentials can be explained as the synaptic potentials of cortical neurones. The spatial extension of synaptic events at nerve cells would appear too limited except in the large apical dendrites. Nor can summations of spike discharges in axons account for the form of evoked potentials although some temporal synchronicity must be present. According to our present state of knowledge the theory of evoked and slow potentials is highly speculative. Axonal presynaptic depolarization might also provide candidates for slow potentials such as those causing dorsal root potentials by slow membrane changes of axons in the spinal cord. But similar mechanisms of presynaptic inhibition have not yet been convincingly demonstrated in the visual cortex. It is also a matter of surmise whether the *retina* can be used as a *model of brain processes* [439]. If we accept that it can, then slow cell potentials without spikes like those observed in the retinal bipolar and horizontal cells should occur in cortical neurones with short axons, such as Golgi-II cells. However this may be, the spatial distribution of such hypothetical electrotonic changes is unknown and the possible participation of chains of non-neuronal elements such as glial cells with tight junctions and low membrane resistance cannot be excluded.

Hence, the neuronal basis of evoked potentials and brain waves with their relations to the cyclic changes in excitability of the optic pathway, described by G. H. BISHOP in 1933 [43] as accompaniments of rhythmic potentials in the visual cortex, remains nearly as obscure today as it appeared 40 years ago, when the recording of evoked potentials and EEG began [23, 141, 279, 280].

Preparatory Potentials and their Reafferent Cancellation

Readiness Potentials. Recent observations of *human brain potentials preceding movements* may hold more interest for visuomotor research than the ordinary evoked potentials. These early potentials were recorded in man by averaging EEG techniques during conditioning experiments and before motor action. They were related to conditioning processes, to limb movements and to the somatic sensorimotor systems: G. WALTER's [446] *expectancy wave* (CNV = conditioned negative variation) and KORNHUBER's "*Bereitschaftspotential*" [277] (readiness potential and motor potential) *precede conditioned or voluntary movements*. They appear with predominantly *surface-negative* components in the precentral and frontal cortex regions and are *terminated* (cancelled) by a *positive* potential shift, probably triggered from peripheral reafferent messages after the intended movement has begun [277]. To carry out similar experiments on eye movements is more difficult because large electrical corneo-retinal potential shifts, caused by the movement of the eyeball itself, obscure the tracing of cerebral potentials from the human scalp.

J. S. BARLOW and CIGANEK in 1969 first described "anticipatory potentials" which *preceded voluntary eye movements* by 150–200 msec but were absent before involuntary or compensatory ocular movements [21]. Recently BECKER and co-workers [32] studied the various components of readiness and premotor potentials which precede eye movements, and differentiated them from those related to limb movements. No motor potential appears on the contralateral precentral region before eye saccades, whereas with hand movements it does. During experiments on visuomotor tasks, attention, motivation and decision appears to exert an influence similar to WALTER'S [446] expectancy wave in conditioning experiments. The Bereitschaftspotentiale preceding saccades directed toward significant visual patterns were found to be larger than those which were followed by saccades away from the significant object. Also, higher amplitudes occurred in a bright surround than in darkness. BECKER et al. [32] briefly discuss relations to lambda waves which *follow* eye saccades and postulate some non-visual and oculomotor components in the lambda complex.

It appears evident that attentive expectancy and preprogramming play an essential role in the preparation and timing of goal-directed saccades in visuospatial location. In a sequence of eye fixations (during which visual information flows in) each must be prepared by saccades directed towards the *next* fixation target within the visual world. To attain these visual targets the brain uses *inner models* of outer space according to CRAIK's concept [88]. This inner space representation is coordinated with visuomotor activity. Visuomotor intention and its readiness to act should be controlled by feedback and informed about the action by reafferent signals. Even if one does *not* accept HOLST's "Reafferenztheorie", a *cancellation of movement intention by reafferent information of movement effect* would seem necessary. For oculomotor performances the reafferent control is *visual* and not proprioceptive or tactile (as in somato-motor action). Therefore the lambda waves, which are related to visual contrast during saccades and *following* eye movements, should be mentioned here. For latency studies see [326].

Lambda Waves. The so-called *lambda waves*, recorded in the human EEG and from the visual cortex, appear *after* ocular saccades in man [124, 156, 374] and monkey [85], in the latter with latencies of about 35 msec. In man they occur about 50 msec after goal-directed saccades when viewing a picture, or quick nystagmic phases during optokinetic nystagmus and attention. Although lambda waves depend upon visual contrast patterns [21, 178], these patterns are usually *not perceived* during the saccades. Thus, some relation to cortical *inhibition* has to be considered, since WURTZ [461] found neurones in area 17 of the monkey which were inhibited after saccades (see Part B, Chap. 21, Figs. 17c and 18). The failure of many cortical neurones in the awake cat and monkey to respond appropriately to very fast retinal motions as they occur during saccades might be due to a cancellation process, which prevents perception of the moving environment during a saccade and wipes out the previous image and thus cleans the visual table for the next fixation [253]. This cancellation may be related to lambda waves and to blinks which accompany large saccades and provoke off-discharges in the neuronal D system [253]. However, it should not be considered as "saccadic suppression" and recordings in the visual cortex of awake cats and monkeys have not shown any neuronal correlate of HOLST's [225, 226] *Efferenzkopie* (equivalent to the corollary discharge [430]), which still remains a hypothetical postulate (see p. 83). Whether reafferent concellation is a function of lambda waves in the visual cortex and of similar potentials in the lateral geniculate nucleus remains to be decided by further experiments.

Alpha Waves and Vision. GAARDER and co-workers [151] described phase locking of alpha waves and saccadic eye movements. The alpha-like components in evoked responses which resemble lambda waves are *phase-locked both before and after the saccade.* Whether the pacemaker is a component of the occipital alpha rhythm or whether both saccade and alpha component are paced by a third process remains undecided. GAARDER's hypothesis that the saccade causes a "packet of edge information" for short-term memory storage is an interesting concept, but it does not fit in with the lack of perception during saccades. It may be related to G. BISHOP's cyclic excitability [43], mentioned on p. 101.

In the human EEG correlations of trains of alpha waves with the *fading of stabilized images* were described by LEHMANN et al. [290] and JASPER and CRUIKSHANK as early as 1937 [240] related after-image periodicity to alpha periods.

Cortical DC-Potential Shifts Accompanying Arousal, Sleep and Eye Movements. After ARDUINI [7] had described enduring potential changes in the cerebral cortex during stimulation of the non-specific brain stem system, opposite DC shifts were observed in *arousal and sleep.* The negative shifts appearing on arousal may be similar to the readiness potentials which precede motor action and accompany emotion and which are most prominent in the sensorimotor cortex. In the visual cortex, opposite CD shifts occur during different phases of *sleep;* these were found to be surface-positive during slow-wave sleep and surface-negative during REM (rapid-eye-movement) sleep [127, 265]. Rapid eye movements and changes in the discharge of visual cortex neurones in darkness related to eye saccades [438], may be correlates of visual dreams. Another interesting finding is POMPEIANO's distinction of *two types of DC-potential* shifts in non-anaesthetized cats and their relation to *vestibular activation:* 1. type-I rhythmic negative waves persisting after lesions of the vestibular nuclei, although rapid eye movements were diminished; and 2. large type-II phasic negative shifts associated with these eye movements and abolished by vestibular nucleus destruction. The negative type-I shifts were most marked in the visual cortex and absent in the frontal regions [265].

Conclusions. The present state of investigation of visually evoked potentials does not seem very rewarding, either for visual perception or for neuronal correlations. However, this very lack of neuronal correlation might be a valuable feature for demonstrating essential differences from classical neuronal mechanisms. Slow potentials and DC recordings may lead on into a new field or may reveal some little-known neurobiological processes which cannot be explained by spike-discharging nerve cells alone but which may well be important for visual information processing.

X. Colour Vision and Neuronal Mechanisms

Colour Coding in Visual Centres

The Riddle of Colour Information Processing. Our knowledge about the central mechanisms of colour vision is limited to some neuronal correlates of the classical YOUNG-HELMHOLTZ three-colour theory and of HERING's opponent colour concept. Von KRIES' first attempt at a synthesis of these two theories [281, 282] has found a modern form in HASSENSTEIN's information flow model (see p. 107).

DE VALOIS in Chapter 3 summarizes research on colour processing in the brain up to 1971. The photochemical and neuronal mechanisms of colour information at the receptor level and in the retina, described in Vol. VII/1 and VII/2, are better known than the cerebral colour coding. However, some very recent and partly unpublished experiments of GOURAS and ZEKI show new aspects at the cortical level which are interesting for colour vision and deserve a mention along with other neurophysiological and psychophysical results.

The psychophysical approach to colour vision has produced an immense literature in the last 100 years. Several modern aspects of this centennial work are described in the next volume of this Handbook, edited by JAMESON and HURVICH (Vol. VII/4), but a discussion is not possible here.

Opponent Colour Coding. Since LEONARDO DA VINCI about 1500 (p. 24) distinguished six ground colours and their opponent order [441] which correspond to HERING's *Gegenfarben*, red/green, yellow/blue, white/black, the opponent colour theory remained a challenge to the Young-Helmholtz concept, but recently it became apparent that the two theories find their photochemical and neurophysiological substrates at different levels. The specific wavelength absorption curves of three classes of cone receptors in the human and monkey retina are now well established, and the physical equivalents of colour qualities were determined in terms of nm wavelength within the spectrum already in the nineteenth century. HELMHOLTZ's theoretical curves [208] show remarkable parallels with the cone absorption spectra [310, 445]. However, a similar amount of precision has been lacking for neuronal correlates of central colour information processing in the cerebral cortex until recently.

There are two problems needing further study: 1. how the opponent colour coding, which was found by DE VALOIS in the monkey's lateral geniculate neurones, results from the broad spectrum cone information, and 2. how the monochromatic responses in the cortex are related to the retinal and geniculate processes coding for colour.

Retinal Mechanisms. Different *maxima of broad absorption spectra of single cones* (445, 535 and 570 nm) (analogous to the Young-Helmholtz three-colour theory) were found in both monkey and human retina by McNICHOL, WALD, and coworkers [310, 445]. These cones respond well to *white light* and are not only colour-specific. In 1956 SVAETICHIN [420] supplied the first *neurophysiological support for* HERING's *colour concept* in retinal mechanisms. He discovered opposite polarizations of retinal elements in fish for the two pairs of opponent colours, yellow/blue and red/green. Although his original idea that these S potentials originate in cones, or later in glial elements, had to be revised, SVAETICHIN's experiments remain an important demonstration that opponent-colour processing begins early in the eye and may not be restricted to cerebral centres as VON KRIES had thought. In 1965 SVAETICHIN et al. [422] made detailed propositions for the function of horizontal and amacrine cells in retinal colour processing. In cats the story is less clear [95].

Subcortical Colour Coding. There are in the lateral geniculate nucleus special subsystems of spectrally opponent neurones which respond to one waveband of the spectrum with excitation and to another waveband (corresponding to the opponent colour) with inhibition. In addition to these *opponent-colour* neurones, de Valois and coworkers have found neurones that respond to all wavelengths with either excitation or inhibition [99, 100] and which would correspond to our B and D-systems. For the transmission of information in the retina, de Valois postulated bipolar cells with opposing excitatory and inhibitory action. The spectrally opponent cells carry colour information to the cortex and seem to correspond to Hering's *Gegenfarben*. They could be assigned by de Valois and coworkers to red (R), green (G), yellow (Y) and blue (B): +R−G, +G−R, +Y−B, +B−Y cells.

Another neuronal population of broad-band spectrum cells resemble the on-centre and off-centre neurones of the cat retina and may correspond to Hering's white /black achromatic colour system and our B and D subsystems (see p. 20).

From what de Valois writes in Chapter 3 and Autrum and Thomas in Chapter 11 (on animal colour vision and behaviour) it is apparent that there are still serious gaps in our knowledge of what happens between the relatively simple receptor mechanisms of the three classes of colour-sensitive cones showing broad absorption with spectral peaks, and the rich and subjectively clear spectrum of colour perception. Colour theories have only partly bridged this gap.

Hubel and Wiesel's studies of colour-sensitive geniculate neurones and Gouras' recent observations on special cortical neurones in monkeys with selective responses to monochromatic stimuli [165, 166] suggest a complicated processing in several stages and a *recoding* of spectral distribution in area 17. It seems remarkable that the monochromatic cortical neurones should have much *sharper spectral peaks than the retinal receptors*. The neuronal circuits which induce this colour sharpening are unknown. Thus the neuronal information processing of colour remains a neurophysiological enigma in spite of the many successful recordings from the lateral geniculate nucleus and visual cortex.

Retino-Cortical Colour Channels. Gouras' concept of colour vision begins with the midget bipolar cell system of Polyak in the primate retina, and postulates separate channels for cone mechanisms through the lateral geniculate to the striate cortex. It is described in Chapters 3 and 21 (Part B, p. 385) and the possible retino-cortical connections are depicted in Fig. 21 (p. 386). It seems that each channel receives *antagonistic signals from neighbouring cones* which are spectrally different, and consequently each channel exhibits *colour-opponent properties* that could be related to Hering's theory [214]. To determine how both spatial and colour resolution is obtained in co-ordination with the luminance-signalling B and D subsystems and the higher-level colour-insensitive but spatially selective neurones in the cortex remains a task for future research. Gouras proposed that on-centre and off-centre mechanisms can combine their excitatory drive in colour-specific cells, making them especially sensitive to *successive contrast*. In 1971/72 Gouras made extensive recordings from colour specific neurones in area 17 of monkeys [166]. The latest results will now be reported from Gouras' and Zeki's personal communications [168, 467].

Colour Processing in the Monkey's Cortex

Colour Coding in Area 17. I am grateful to Dr. Gouras for communicating to me some of his recent results in addition to those reported in Chapters 3 and 21. Gouras has studied the spatial and chromatic properties of neurones in the foveal representation of *striate cortex in lightly anaesthetized rhesus monkeys* [166, 167]. This work, and recent experiments with DOW [168], tries to identify the three principal cone mechanisms of primate vision in the responses of cortical neurones. Slit and spot stimuli are used and their energy, wavelength, size, velocity and orientation are regulated. Strong chromatic adapting lights help to isolate each cone mechanism. The classes of cells found by means of this approach are: 1. broad spectrum responders, and 2. *colour-opponent neurones* and, among the latter, *colour-specific neurones.* About half of the driveable neurones have the relatively broad action spectrum of the first class, similar to the photopic luminosity function. Most broad-spectrum cells are also direction- and orientation-selective and sensitive to both red and green but more rarely to blue. All respond to white light, and different cone mechanisms seem to act synergistically in such cells; responses to one wavelength can be reproduced by any other wavelength simply by changing the energy of stimulation. Thus, the role of the broad-spectrum neurones in colour vision may not be important. A relation to the complex or hypercomplex cells of HUBEL and WIESEL [233, 234] is possible.

The second class, the *colour-opponent neurones*, have a *narrower* action spectrum than the absorption spectra of cone pigments. Three types are distinguished; one cone mechanism is excitatory, and the opponent-cone mechanism is usually only apparent by its *inhibition*. The neurones are respectively *red-*, *green- or blue-sensitive.* For example: a red-sensitive cell responds well to long wavelengths; it is unresponsive in the presence of a red adapting field; it responds better in a blue-green adapting field than in a neutral adapting field. Most colour-opponent cells of the orientated type also respond to *white* light provided the spatial characteristics of the stimulus are appropriate; this may indicate that the colour-opponent mechanism is stronger in the *surround* of the cell's receptive field. Orientation and direction selectivity occur in many colour-opponent neurones and these resemble the simple or complex categories of HUBEL and WIESEL [233]. Others have *non-oriented* or concentrically organized receptive fields, and some cannot be classified spatially at all.

Colour-Orientation Cells and Psychophysics. The combination of orientation and colour opponency in a group of cortical neurones is of psychophysical interest to explain the *McCollough effect of orientation-specific colour adaptation* [321]. These results as well as those of BISHOP [46] reveal that the class of "simple" field neurones, defined by HUBEL and WIESEL [231] as orientation-specific, may not be a "simple" group and that their limit to complex cells is drawn differently by various authors. The simple class contains cells with wide differences in selectivity, at least in the monkey. Besides the two kinds of on- and off-centre simple cells described by HUBEL and WIESEL [231] and related to the B and D subsystem by us [253], there may be as many as six *colour-opponent subclasses* in the monkey [168], several *direction-specific* in the cat [46] and possibly others yet undistinguished. In order to explain the McCollough effect, one has to assume that GOURAS' six colour-opponent channels formed by red, green, and blue on-centre and off-centre cells go through a stage of *orientation selectivity* before their information culminates in colour vision. This may be related to the "colour-opponent orientation-selective cells" described by GOURAS [168].

Colour-Specific Systems in Area 17. The second class of *colour-specific neurones* among the colour-opponent cells of GOURAS responds to *certain wavelengths* but *not to white light*, regardless of its spatial distribution, and shows *no orientation and direction specificity*. An excitatory response from each of the cone mechanisms converging on these cells becomes manifest as either an *on-discharge or an off-discharge*, depending upon the *selective chromatic adaptation*. Trichromatic colour-opponent neurones are less common than *antagonistic pairs*.

Orientation or direction selectivity is usually lacking. The receptive fields are either *circular* or *elliptical* and overlap for opposing cone mechanisms. The latter seem related to *unresponsiveness to white light* regardless of its spatial distribution. Thus only this special class of colour-opponent cells is truly *colour-specific*.

Colour-specific Neurones Beyond the Visual Cortex. The latest results of ZEKI in the rhesus monkey [467] have demonstrated *monochromatic neurones in the temporal lobe* anterior to visual areas 18 and 19 in a cortical region which ZEKI calls "*visual area IV*" (posterior bank of the superior temporal sulcus above the inferotemporal cortex). Most neurones had specific responses to *one colour* of various areal distribution with little form specificity. *Red*-coloured stimuli, with or without a green surround, caused strong activation in most of the recorded neurones. Green and blue activation effects were less frequently seen. Most of these neurones did not respond to white light or black/white patterns and thus may be regarded as *colour-specific*. ZEKI's experiments in the superior temporal cortex may fill the gap between the visual areas and the inferotemporal cortex where neurones responding to specific patterned stimuli were recorded by GROSS (see Part B, Chap. 23).

Information Transfer from Cones to Colour Vision

A Flow Diagram. HASSENSTEIN's model of data processing of human colour vision [201] is described on p. 113 and illustrated in Fig. 16. It is based upon the spectral absorption curves of three types of cones and on psychophysical data. It also envisages a synthesis of the three-colour theory of YOUNG and HELMHOLTZ with the opponent-colour theory of HERING and develops a clear biocybernetic framework for VON KRIES' *Zonentheorie* of colour vision [282] which cannot yet be correlated with particular anatomical levels. Since HASSENSTEIN's model will not be mentioned in the chapters on colour vision in Volume VII/4 on psychophysics, I report some of its essentials here in as much as they may have relevance for colour processing beyond the receptor level.

HASSENSTEIN, in agreement with the HURVICHS [238], supposes that blue receptors and their channels are antagonistic to yellow and that green receptors are antagonistic to the combined action of blue and yellow. With this formalism, schematized in Fig. 16, it is possible to explain the distribution of colour sensation within the spectrum, the elicitation of red sensations by short wavelengths in the violet band, the wavelengths of opponent colours and the form of spectral hue discrimination, as well as various kinds and symptoms of colour deficiency. The lack of colour vision in the peripheral parts of the retina attested by perimetry can also be interpreted according to this scheme. To explain the achromatic brightness perception of white, HASSENSTEIN postulates that the receptors which signal the opponent colours green and yellow-red are stimulated simultaneously with equivalent intensity. Since this depends upon the relatively broad absorption curves of the three colour receptors, the sensation of white might also determine which wavelength would elicit pure colours. Thus, monochromatic green at the wavelength of approximately 500 nm would not stimulate the antagonistic blue or yellow

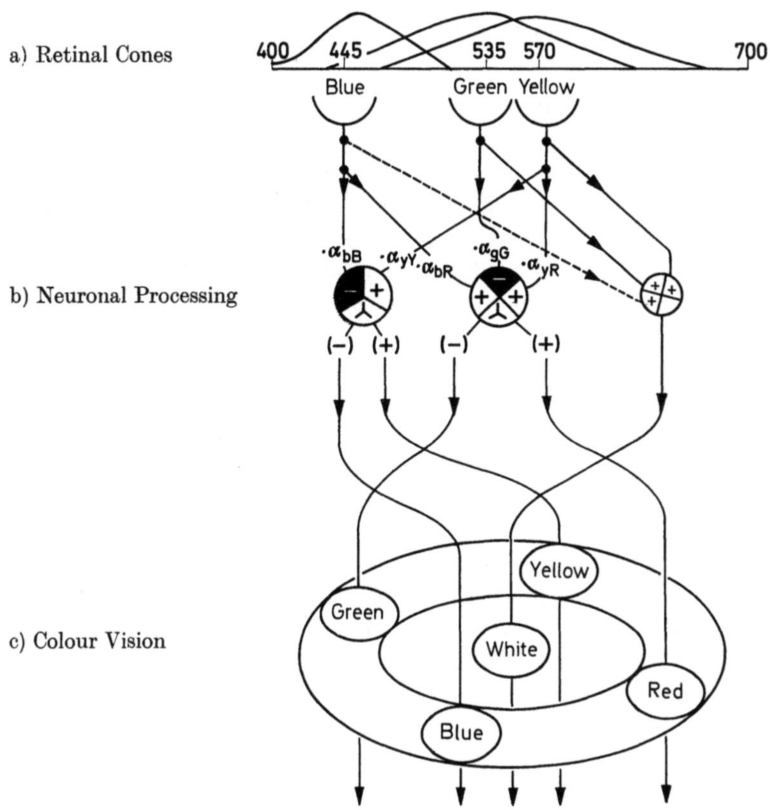

Fig. 16. *Information flow diagram for data processing in colour vision.* From HASSENSTEIN, 1968 [201].

a *Receptor Level:* the messages of the three types of retinal cones, colour-sensitive for blue, green and yellow, are schematized according to their spectral absorption curves from 400 to 680 nm

b *Retinal and Cerebral Information Flow:* the connections from a towards the distributors in b depict antagonistic activating (+) and inhibitory (—) processes. They show that "red" messages can be triggered by short-wave (violet) as well as by middle- and long-wave spectral stimuli. The factors α indicate the relative weight of information of the three classes of spectral messages. These data are processed by summation or subtraction according to the + or — signs in the various sectors of the system. The sign ⅄ indicates that positive and negative results are signalled in different channels. Therefore these two channels do not carry + and — signals simultaneously and only *one* of them can send signals at a given time, so it must be over either the plus or the minus channel

c *Psychophysical Level:* the higher centres receiving the signals are supposed to supply the physiological basis for the various colour sensations, including white at the perceptual level. The substrate for white sensations would receive and summate messages from the green, yellow and red, but would have little or no activation from the blue spectral band. In the two sectors which combine the opponent-colour messages the output would then be zero, i.e. it would correspond to achromatic black

system, although the absorption curve for "blue" and "yellow" cones activated maximally at 450 and 570 nm would overlap in the 500 nm region. In another paper [201 a] HASSENSTEIN discusses his theory in relation to SVAETICHIN's hypothesis of colour processing in horizontal and amacrine cells of the retina [422] but otherwise avoids (with good reason) any commitment as to neuronal connections, since detailed circuits cannot be derived from his block diagram of the data-processing level in Fig. 16b.

Mammalian "Ommatidia" and Colour Vision. M. BOUMAN recently developed a concept of the human eye as an analogue of the ommatidia of insects [55]. BOUMAN goes back to the old picture of the receptor mosaic in the human retina which SCHULTZE [392] described in 1866. SCHULTZE demonstrated a concentric order of photoreceptors, mainly of cones in the fovea and of rods and cones in the peripheral parts of the retina. BOUMAN's "fluctuation concept" of retinal function tries to combine the threshold behaviour of the eye in terms of quantum absorptions with adaptation processes and colour vision in man. He postulates "square-root coincidence scalers" which eliminate irrelevant quantum fluctuations during adaptation. This scaler is compared to a "neural quantum". BOUMAN's ommatidium concept is related to adaptation to light and postulates different receptor types (rods and different chromatic cone systems) for the same stimulus. His "ommatidia" are thought to be discrete *colour units* which contain one blue, two to three red and two to three green cones. Five to seven of these "ommatidia" constitute a recipient unit. Each ommatidium belongs to five to seven different but overlapping units. It produces R + G and R − G signals for signalling brightness and red/greenness which are scaled by the fluctuations of the recipient unit around the ommatidium. *Balanced red/green colour stimulation* leads to *white* perception if only one unit is stimulated at low luminance or to yellow perception if one "ommatidium" is stimulated.

The consequences this ommatidium concept of the eye has for the central transformation of visual signals in the brain are not yet clear. They seem to be more important for colour vision than for pattern vision. BOUMAN has only demonstrated some psychophysical correlates of chromatic and achromatic vision which indicate central transformations of a similar kind. For neurophysiological mechanisms of these transformations, BOUMAN and KOENDERINK [56] have proposed model constructs (see p. 111).

Similar Principles Governing Neuronal Colour and Pattern Selection. Colour processing can be compared to brightness and pattern-feature extraction for the hierarchic principle of *increasing selectivity and specialization at higher levels*. Receptive fields specialized for pattern selectivity may be an analogue to specialized receptive fields of monochromatic selectivity. The signals of broad spectral width which are relatively non-selective at receptor level become differentiated through central colour systems into narrow-band, colour-specific cells of the cortex. At intermediate levels, comparable information processing of bright–dark and colour uses *reciprocal interaction* (opponent colour and B–D antagonism) mainly at the geniculate stage. Although it is probably a crude oversimplification, one could say that colour-opponent neurones representing all *Gegenfarben*, including white and black, are typical for the lateral geniculate, and "pure" colour-specific neurones are more common in the cortex. The latter probably increase in number beyond primary area 17 towards the higher visual areas in the temporal lobe. Some parallels to HASSENSTEIN's information flow diagram [201] may be assumed but need further experimental proof.

When a cortical cell responds to "red" but is not excited by white light of similar intensity (although white light physically includes the same spectral components of red), this response corresponds more to the seen colour "red" than to the spectral stimulus. By analogy, a subjectively seen brightness may differ from objective luminance (see p. 10).

This discrepancy is a common source of misunderstanding between different scientists working in the field of physics and physiology. This occurred in the past, as when GOETHE misunderstood NEWTON in studying colour perception and overrated the Tyndall phenomenon as a basic experiment of colour vision, and it still happens today. GOETHE's theory of perceptual colour phenomena [162] was not only physically but also physiologically false, although it was sometimes psychologically correct.

XI. Models of Visual Mechanisms

The rapid advances in information theory and biological cybernetics, as well as technical progress in computers and computer sciences over the last 20 years, have provoked many attempts to explain information transmission in the nervous system, including vision, on a cybernetic model. However, in visual physiology mathematical theories and physical models have not yet produced any useful biophysical explanation of the central processing of visual information which could lead to an original physiological concept [98]. The many symposia on "biosimulation", "bionics" and "artificial intelligence" were somewhat disappointing for experimental neurophysiology and vision. Biological models taken from *Limulus* and other lower animals, although they stimulated HARTLINE's discoveries concerning the visual physiology of the vertebrate eye, also have obvious limitations.

Models of Retinal Mechanisms

Linear Models. Simple models which use the theory of linear systems were proposed by v. SEELEN [392a] for the visual system of vertebrates. They are based upon MARKO's theory of homogeneous layers [315] for neuronal circuits but they are still far from allowing neurophysiological applications, experimental proofs and synaptic equivalents. These models are easy to handle but have the disadvantage of not accounting for the nonlinearities which are present in the visual system.

Electronic Models. A hardware model for *motion detection* was built by ECKMILLER and GRÜSSER [115] and is discussed in Chapter 7 with several other models. A simple physical steady-state *model for visual contrast* was proposed by MARIMONT in 1961 [314] using photocells, logarithmic amplifiers, averager and differencer. This model can explain how high illuminance increases apparent contrast and how it can induce a decrease in perceived brightness. The model can be further improved to explain the Mach bands and the disappearance of stabilized retinal images.

Nonlinear Shunt Model. A useful *model of shunting inhibition* was described in 1965 by FURMAN [149] and may be equivalent to "forward inhibition" in retinal mechanisms. To explain lateral inhibition in vertebrates, which is different from subtractive inhibition in *Limulus* eyes, FURMAN postulates a shunting inhibition for the vertebrate retina which is supposed to cause a *shunt of excitatory depolarization currents*. Whereas subtractive inhibition results in linear input-output relations and contrast effects which are not dependent on discontinuities and amplitude, FURMAN's shunting inhibition predicts several *non-linear input–output relations* and the dependency of contrast mechanisms on discontinuities and input amplitudes. The model further predicts the Weber-Fechner relations, even when visual receptors are assumed to be linear rather than logarithmic. The model can be correlated with visual intensity discriminations in man and with the experiments of FISCHER and KRÜGER on the surround effects of centre illumination (see p. 34) in the receptive fields of retinal ganglion cells [139].

SPERLING [401] has further developed FURMAN's model of shunting or subtractive inhibition. His model of visual adaptation and contrast detection contains three components: 1. a feedback stage, 2. a feed-forward stage and 3. a threshold detector. The latter is supposed to correspond to physiological processes in retinal ganglion cells. The model is said to make several predictions of physiological findings including the Weber-Fechner law, high luminances masking effects of background luminance, changes of receptive fields on dark and light adaptation, and contrast phenomena with typical Mach bands.

Neuronal correlations with FURMAN's model were described by FISCHER and FREUND in a mathematical formulation based upon experimental data and applicable both to systems B and D by doubling one constant for the off-centre neurones: $2\mathfrak{c}$ (D system) $= \mathfrak{c}$ (B system) [137].

Although this formulation is limited to stationary responses, it has the interesting implication that the laws of RICCO, WEBER, FECHNER, and STEVENS are included as approximations of neuronal summations [137]. The incorporation of thresholds on the basis of linear summation at a given level of adaptation, as presented by FISCHER and MAY [140], differs somewhat from the model of SPERLING. The combination of linear threshold summation with nonlinear suprathreshold summation and adaptation mechanisms may result in a valid but rather complex model of visual information processing in the nervous system of vertebrates.

Field theories of large numbers of inter acting nerve cells, as proposed by BEURLE and GRIFFITH, are discussed for information processing in the retinal layers by FISCHER [136]. In this approach the neuronal activity is ruled by spatio-temporal partial differential equations, as used in several other physical research disciplines, e.g. electrical fields, diffusion or heat conduction.

BOUMAN [55] used SCHULTZE's [391a] classical studies from 1866 onwards on the distribution of retinal receptors to suggest an *ommatidia-like order* in the mammalian retina, including that of man. This hypothesis, already mentioned on p. 109 postulates that a certain number of receptors are grouped together to form a *functional unit*. BOUMAN's concept would cast some doubt on the traditional dichotomy of rod and cone functions in scotopic and photopic vision. BOUMAN suggests that both rods and cones contribute to chromatic and achromatic vision by multiple coincidence and ordered hexagonal patterns of receptors. Several "ommatidia" would presumably be assembled in the outer plexiform layer of the retina. (For consequences relative to colour vision, see p. 109).)

This concept is illustrated by psychophysical results, neuronal recordings and relations to integration models. Theoretically, the scaling ensemble may work as so-called "*De Vries–Rose machines*" or "*Weber machines*" on the basis of light quanta coincidence corresponding to the state of light adaptation. Even with a constant number of neurones forming a scaling ensemble, the corresponding receptive area increases from the central to the peripheral retina. This seems to correlate with recent studies in the cat [134] showing that receptive-field-centre size and ganglion-cell density cancel out in a way that provides a nearly *constant number of ganglion cells* representing the area of a retinal field centre at any given retinal position. It is not yet clear whether the hexagonal order emphasized by BOUMAN is only a general physical property of densely packed elements or whether it has some functional significance, implying a special structure of synaptic connnections between receptors and neuronal and non-neuronal elements of the retina. It seems fairly certain that it has no relationship to special receptive functions for light polarized in different directions, as has been discussed for insect ommatidia.

The integrative nature of single functional units is apparent in the features of their *spatio-temporal receptive fields*, which show spatial and temporal summation at threshold and suprathreshold stimulation, as demonstrated by FISCHER [138].

Models of Central Visual Processing

Mathematical Formalism. Quite different models of the central visual system were proposed by HOFFMAN in 1966 [220]. He considered the first stage of visual perception as a transformation of the visual field to the visual manifold, that is, the two-dimensional representation surface which has its physiological correlate in the cytoarchitecture of the primary visual cortex and its associated receptive-field types, as found by HUBEL and WIESEL in area 17. According to the various visual invariances, special visual forms remain constant under this transformation leading to the basic Lie transformation group over the visual manifold. The prolonged Lie derivations are used to account for higher perceptual mechanisms and their neuronal basis in the paravisual cortex (areas 18 and 19).

These models may allow interesting applications of principles in visual perception and neurophysiology. HOFFMAN's derivatives, although purely hypothetical mathematical formalisms, appeal to the visual scientist for several reasons: 1. his models are applied to several visual illusions, mainly movement illusions and complementary after-images in special patterns, such as those of MACKAY [305]; 2. correlates in higher visual structures of the cortex are discussed in the second paper in relation to the neuronal responses found by HUBEL and WIESEL in areas 18 and 19; 3. applications of his mathematical formalisms for visual constancy and shape recognition seem possible.

Needless to say, no experimental verification of these highly mathematical theories has been attempted; this may well prove more difficult for higher visual information processing than for retinal mechanisms.

Receptive Field Models. MOTOKAWA in his last book [329] published abstracted models of neuronal receptive-field transformations in the visual cortex for concentric, simple, complex and hypercomplex fields, as described by HUBEL and WIESEL. In the proposed circuits backward-inhibition, self-inhibition and cross-inhibition are combined, and progressive integration of several fields at different stages is used to build a hierarchy from concentric to hypercomplex fields. This model is based upon the work of his pupils NAKAGAWA, HIWATASHI, FUKUSHIWA. A simpler model, proposed for movement detectors, uses sequential activations of neighbouring neurones with concentric fields proceeding in one direction [329].

Models of Visuomotor Functions. Following CRAIK [88], *internal models of the external world* have been discussed for conscious perception. Several black-box models have been developed (see Chap. 5) to explain visual constancy and the spatial co-ordination of eye movements and vision. However, the neuronal basis of these models remains obscure. All oculomotor models produced by physiologists and communication scientists are constructed as block diagrams and not as neuronal wiring diagrams. The first attempts were HESS' models in 1941 [216] and VON HOLST's reafference principle in 1950/51 [225, 227]; cybernetic models of eye movement control followed [411]. These and models of visual space [419] cannot be discussed here. Such black-box models are of limited interest to the physiologist as long as the neuronal substrates of these regulations in the central nervous system are unknown. Optic schemes of the two modes of motion detection (Fig. 14) also have obvious limitations.

Information Flow Diagram of Colour Vision. HASSENSTEIN [201] proposes a model in which the spectral absorption curves of colour-sensitive cones form the basis for general and formal interactions of excitation and inhibition of the flow of data processing which results finally in the perception of colours and white sensations. This model is schematized in Fig. 16, and has been discussed in relation to the mechanisms of colour processing on p. 107. The model represents in a simplified way *three levels:* the *receptors* (a), the formal processing channels corresponding to *retinal and cerebral interaction* (b), and the perceptive result of *colour sensation* (c). In contrast to other models which are based on animal experiments or purely theoretical constructs, it is specially designed for *human colour vision* and its colour deficiences. The inputs of the three spectrally selective cones of monkeys are the only animal data from which it proceeds towards the perceptive output of colour sensation in man. For the latter HASSENSTEIN uses various psychophysical data, including colour abnormalities, for control of this flow diagram.

The model cannot yet be correlated with specific retinal and cerebral neuronal mechanisms but at least it suggests some connections in information flow at the processing level (b) of Fig. 16. One neurophysiological difficulty seems to be the postulate of a linear transmission and 1 : 1 translation between the input and the output of the cones and between the data processing systems and colour perception. Of course, this cannot be applied to neuronal transmission or to linear-versus-logarithmic processing in visual pathways. The model neglects relations to neuronal circuits beyond the level of the receptors (see also p. 108).

Although HASSENSTEIN has used a remarkable array of data, which include absorption curves of retinal cones and many psychophysical observations, his model presents similar difficulties for neurophysiologists as do others. Its block diagram is still far from offering any direct application to neuronal structures for colour coding in the retina and the brain.

The models of HASSENSTEIN and BOUMAN show some correlation with psychophysical findings and include cerebral processes, but their neuronal application is limited. Neuronal circuits can only be derived from the shunt models for inhibitory actions to peripheral retinal mechanisms. Even in these, it remains uncertain which elements of the model correspond to which neurones or synaptic transmissions. It is still an unanswered question whether the same or different circuits are used for central visual processing.

BARLOW has thrown doubt on the simple assumption that the central visual system continues to use lateral inhibition for its information processing at different levels [15, 16]. However, similar but more complex mechanisms of collateral inhibition are probably functioning in the lateral geniculate nucleus [319] and visual cortex for the transformation of receptive fields, as discussed on p. 67. Psychophysical observations may also be used to argue in favour of the continuous influence of lateral inhibition and contrast vision from concentric retinal receptive fields upwards in the cerebral levels of the visual systems. The same principles can be demonstrated subjectively in the Hermann grid and similar contrast phenomena which were used to determine perceptive fields in human vision (p. 47, Figs. 5–7). Of course, one cannot yet propose neuronal circuits and models to simulate the end-results in visual perception.

In conclusion, modelmakers working in close contact with electrophysiologists, neuroanatomists and psychophysical laboratories interested in vision may develop useful system theories and models of visual mechanisms. On the other hand, experimental work which is designed to verify or reject predictions derived from special models and theories should be restricted to those cybernetic schemes of vision which are based on neuronal recordings or clear perceptual observations of vision in order to avoid unnecessary refutation of fancy models.

Biological Models and their Limits

Primitive nervous systems of lower forms clearly show certain principles of visual organization. HARTLINE has used the *Limulus* eye as a model for visual research with great success and others have followed his lead in the investigation of lateral inhibition and other visual mechanisms [195a]. However, even „primitive" nervous systems have specializations and species peculiarities which have developed during phylogenesis and adaptation to certain environments. They may obscure general applications and restrict biological models.

The ommatidial eyes of insects, although very different from mammalian eyes, have very clear geometrical arrangements in their nerve-fibre plexus (see Chap. 10). The rather rigid, machine-like behaviour of insects and of the functions of their nervous systems is another advantage. Thus insects can be used as biological models for *motion detection* and *optomotor responses*, as shown by REICHARDT [372a, 373] HASSENSTEIN [199, 200], and GÖTZ [164].

In vertebrate animals the *retina* may also be a useful model for the visual brain, although the retinal mechanisms are specialized more for transducing and transmitting than for elaborating visual information.

Limitations of Limulus as a Biological Model for Visual Inhibition. In spite of many papers on inhibition in the mammalian visual system and its role in visual information processing, very little is known about the membrane mechanisms of neuronal inhibition in the retina. Although HARTLINE's discovery of lateral inhibition in *Limulus* [195] has been a constant challenge to researchers on the mechanisms of contrast vision in the higher forms ever since 1949, the *Limulus* eye is only of limited value as a biological model for the vertebrate visual system. At the mammalian geniculate and cortical levels CREUTZFELDT's quasi-intracellular recordings allowed some inferences to be drawn on synaptic potentials (see Chap. 21). However, the enormously complicated morphological substrates in the synaptic glomeruli of the lateral geniculate, as described by SZENTÁGOTHAI in Chapter 17, are not yet ready for neuronal models and the interpretation of axonal and dendritic elements is still a matter for debate between histologists. There are no analogues from the optic ganglia of lower animals which could be used as biological models. Inhibition is well analysed in the *Limulus* eye and its first nerve nets *but not in the first optic ganglion*, where off-discharges begin to appear, and which might be compared to retinal and central visual circuits in vertebrates. Inhibition in *Limulus* ommatidia was investigated by rigorous quantitative methods, and the dynamics of excitation and inhibition were treated with linear system analyses, as described by HARTLINE and RATLIFF in Vol. VII/2 [195a]. A distinction between self-inhibition and lateral inhibition like that proposed for the *Limulus* eye cannot yet be made in the mammalian visual system.

All visual and nervous systems have a remarkable reliability which has developed by adaptation to special conditions and environments. "Uncertainty", which BURNS [72] ascribes to the nervous system, is not a general rule of its function. What is uncertain is whether there are consistent parallels between different species. These matters have to be tested in experiments before they can be used for the proof of general principles in neurophysiology and perception.

Reliability and Uncertainty of Sensory Systems. After the rise of communication sciences, it became fashionable to compare the nervous system, including its sensory processing, with computers and to stress the probabilistic function and uncertainty of the brain (BURNS [72]). However, in sensory physiology, it is *not* the uncertain but the reliable sensation that counts. The *precision of perception,* related to stimulus conditions, is an impressive performance and this is what needs physiological explanation. A high degree of predictability is achieved by the visual system. Even sensory illusions demonstrate phenomena which may be predicted from general principles. Illusions offer further cues for the precise *functional order* of the sensory neuronal systems in the deviating sensations which result from overloading, as demonstrated by several examples in this chapter (Fig. 5—10). BULLOCK [71] has recently stressed the *reliability of the neuronal order* and contradicted the current trend to overrate probabilistic actions in the brain and indeterminacy. The visual system, too, works with predictable rather than statistic functions and uses a precise *neuronal order at several levels* with multiple safeguards of compensation which are guaranteed, and not merely statistically, by the large number of parallel lines. BULLOCK's conclusion, that many neurones have a high reliability in both connections and discharge, although derived from lower-order animals, can be endorsed for the visual system of higher mammals. Hence, the only models which should be used in visual physiology are those based on precise sensory and perceptual data and reliable experiments.

Visual Modelling and Neurophysiological Experiments. Many neurophysiologists mistrust cybernetics and consider models of sensory and nervous systems to be a waste of time and intellectual effort. They suspect modelmakers of neglecting neurophysiological experiments. On the other hand, some information theorists in mathematical biophysics probably regard neuronal recording as trivial and unlikely to reveal the essential properties and general laws of the nervous system. Of course, both attitudes are unrealistic. However, the conviction that speculative model-making by communication engineers needs to be checked against the physiological equivalents in the living nervous system has been strengthened by the experience of the last 20 years.

There are, of course, no limits to what may be done with models, but only those which have proved useful for visual perception and physiology deserve a report. The fact that models have anatomical correlates is not always a criterion of success. Some interesting models developed on exact histological and synaptic bases, such as PITTS and MCCULLOCH's scanning hypothesis of the visual cortex [359], have never been experimentally confirmed. Conversely, other more formal constructs, such as FURMAN's inhibition model [149] of receptive fields have stimulated successful neuronal experiments in the retina [135—140].

To be useful, models of the visual system should be derived from an analysis of living organisms and tested and confirmed by physiological and psychophysical experiments; moreover, they should lead the experimenter to new conceptions, general principles and theories. A theoretical approach is badly needed for the study of central visual mechanisms.

Some recent models and mathematical formulations of retinal functions mentioned on p. 110 may indicate the direction in which further progress could be expected in central processing of visual information. Of course, any such model must be up-to-date with recent experimental results and with the synaptology of neuronal network connections. Unfortunately the central visual structures are too complex to be simulated by simple models, and as long as the details remain uncertain, more complex models are overwhelmingly likely to be false. To load the literature with improbable models seems of doubtful utility. Up to the time of writing the cybernetic approach has produced only oversimplified neuronal models. This has created a certain prejudice among neurophysiologists. However, a common approach to visual problems by a team of neurophysiologists, communication scientists and experimental psychologists may create useful information-flow models and working links between visual physiology and psychophysics in the future.

XII. General Implications and Philosophy

Since the end of the seventeenth century, visual science has been closely related to philosophy. This can be seen as a continuous line from LOCKE, LEIBNIZ, HUME and KANT in philosophy to JOHANNES MÜLLER, HELMHOLTZ and VON KRIES in visual physiology. KANT's apriorism of space and time was modified by HELMHOLTZ [208] and VON KRIES [283] for vision. It may be useful to discuss philosophical correlations in order to demonstrate some general principles of visual science, although modern trends do not favour this. However, a *need for theory* is felt to counterbalance the increasing fragmentation of science studied by specialists. I realize that this may be a thankless task, since we are as far from a general theory of visual functions as are all neuroscientists. The task may be criticized by physiologists as useless speculation and generalization or by philosophers as a misinterpretation of philosophical principles and a reversal of KANT. In spite of this, I propose a pragmatic application of KANT's apriorisms to neurobiology:

Innate structures and functions of the visual system are considered as a priori conditions which are modified by visual experience in perception.

Kant's Influence on Sensory Physiology

J. MÜLLER [336], HELMHOLTZ [208], and VON KRIES [283, 284] felt the impact of KANT's synthesis of rationalism and empiricism in sensory experience and used his principles with some alterations when they discussed the innate and acquired mechanisms of vision. Although a Kantian in principle, HELMHOLTZ turned against KANT's apriorism of space and time and defended the dependence of space perception upon empirical experience [208]. W. JAMES also denied the a priori-category of space [239]. A "logical apriorism", however, was considered necessary by VON KRIES who tended to distinguish the primary act of sensation from the logical and intellectual effort of perception [284]. However, our concept, in contrast to VON KRIES', may be more rewarding for the sensory scientist when KANT's a priori is freed of its abstract idealism and applied to reality. Then a *concrete a priori* appears as an innate order of sensory and cerebral structure of the perceiving organism. Whether the abstract apriorisms in thought and logic can be derived from this structural and functional order may be left open.

KANT's **Philosophy of Nature and Sensory Sciences.** Although a speculative thinker, KANT was all his life fascinated by mathematics, physics and the natural sciences. This interest first appeared in his early book on theory and origin of the solar system in 1755 and was maintained until his last years. KANT's interest in the experimental sciences was, however, theoretical as he himself "experimented" only in the sphere of the mind. In 1787 he postulated that *philosophy should imitate natural science* to find out in the elements of reason "what can be verified or refuted by experiments" [262]. Of course, KANT's speculative contributions to natural philosophy are limited by the relatively underdeveloped state of eighteenth-century experimental knowledge. However, several of his general conceptions are still interesting for present-day science. In the following we discuss some applications to vision research with our pragmatic simplification of KANT's a priori concepts. For these KANT says that our psychic apparatus orders the world, as experienced by our senses, in space and time.

HELMHOLTZ does not accept KANT's a priori concept of space for visual perception [208] as a primary intuition. He considers spatial vision to be acquired empirically by *learning*. Against HELMHOLTZ's purely empiricist explanation of space perception, VON KRIES maintains that all sensory percepts are primarily experienced in a spatial frame [284]. He says that this *spatial order*, be it of psychological or physiological origin, is an *a priori „Anschauung"*, similar to KANT's intuition of space, and that LOTZE's „Lokalzeichen" of sensations [303] cannot give a sufficient explanation of spatial order. Not all authors who accept HELMHOLTZ's view that spatial perceptions may develop by learning deny the *innate preconditions of these acquired space concepts:* HOFMANN [223] and VON KRIES [283] try to reconcile HERING's nativism and HELMHOLTZ's empiricism by the statement that both are partly correct and that, obviously, Anlage and learning are coordinated processes in perception. VON KRIES, in his General Sensory Physiology [283], gives a synthesis of these theories and points out that innate a priori structures and functions form the essential basis that is further developed by learning. Hence, while avoiding all dogmatic statements about the predominance of innate or acquired conditions of sensations, the neurophysiologist must accept nativism, at least for the structural order of the visual system which is revealed so clearly by the topographic projections and by modern histological and ultramicroscopic techniques, as decsribed in Chapters 9, 17, 19, and 20 of these volumes (see also p. 122).

NICOLAI HARTMANN in his appreciation of KANT's empirical realism and transcendental idealism [196] defines some notions for modern philosophy. Essential points in KANT's reasoning are the *analogies of experience* which are given a priori to the mind and function as preconditions of empirical perception. Although analysis is mostly a priori and synthesis a posteriori, the converse is not true. Kant emphasizes that not all a priori is analytic and that aprioristic syntheses are indeed possible ("synthetische Urteile a priori"). In the following, we analogise these mental faculties with brain functions in perception.

For Kant and Hartmann, similar rules of mathematical logic are valid for nature and mind. Natural laws contain mathematics. This idea may even date back to antiquity when Pythagoras postulated that the principle of numbers was also a principle of real things.

In his philosophy of nature, Kant [261] overrates mathematics and physics as the only "pure" sciences and demonstrates an eighteenth-century bias against the exactness of other research by classifying the biological sciences as empirical "knowledge" (Wissen), and chemistry as a scientific "art". One wonders whether Kant would have disparaged chemistry a hundred years later when the periodic system and atomic weights became known. However, his preference for mathematics, mechanics and quantitative explanations in opposition to metaphysics, natural history and dynamic interpretations would be in agreement with trends of modern science. In sensory physiology, the combination of subjective, perceptual and objective, neuronal data includes qualitative and quantitative, static and dynamic concepts for theory and experiment. The different degree of applicability of mathematics to science may be better understood when one considers the hierarchic order of the sciences.

The *hierarchic concept* of levels is a modern elaboration of these notions, foreshadowed but not yet defined by Kant in his "Critik der Urtheilskraft" [263]. It was clearly expressed by Nicolai Hartmann when he recognized the dependence of the higher levels upon the laws of the lower levels (see p. 119). The validity of the causality principle of the lower levels in relation to teleology at higher levels is summarized in his poignant sentence: "*The causal order is not an obstacle for the teleology of will and action, but just its necessary precondition*" [197, 198].

Kant's natural philosophy is often criticized and its dependence on the Zeitgeist appears obvious. However, his clear insights into essential principles are astonishing when one considers that they were conceived after reading a few descriptions of facts without having done any experiments himself. Contrary to the tendency to overrate Kant's achievement in natural science, Adickes demonstrates the limits of Kant's speculative simplifications without experimental and empirical proof [2]. This criticism, however, is mainly directed to Kant's unpublished notes from the last ten years of his life when he was showing signs of senility and the cortical atrophy that was verified by an autopsy showing reduced brain weight at his death at the age of 80. But even after the age of 75, Kant remained interested in biological experiments and assisted in Ritter's electrophysiological demonstrations of nerve stimulation which elaborated upon Galvani's discoveries.

In spite of many inadequacies, Kant's faculty for seeing the essential in synopsis and theory may still be useful for modern sensory science. In sensory perception, experience a posteriori and reasoning a priori are usually combined into an intricate synthesis. Neurophysiology investigates *partial* aspects of this integration in sensory stimulation experiments with a posteriori conditions but relies on the cerebral order given a priori. Any attempts to investigate the visual system on the basis of randomness are hopeless and sterile tasks.

The Hierarchic Level Concept and Biology

Kant's leading principle of biological science, proposed in his Kritik der Urteilskraft 1790 [263], may still be valuable for sensory research. He says that all products and functions of nature, even the most purposeful ("zweckmäßig", i.e. tending to finality), should be explained mechanically as far as possible, but that at a higher level one should also consider and understand these mechanisms as purposeful and directed towards a goal [263]. All mechanisms follow the laws of causality. Although the higher level processes contain trends of finality, they also depend upon the strict causal laws of the lower levels. Causality as a basic condition of finality was demonstrated by NICOLAI HARTMANN in his "Kategorienlehre", i.e. the laws of dependency of different levels in nature [197]. KANT has developed similar concepts, although his lengthy but carefully formulated sentences were not always understood, and some of his essential terms, such as "Anschauung", "Vorstellung" or "gegeben", are difficult to translate into meaningful English.

Psychologically expressed, "mind" selects and orders the raw material of sensations at a higher level according to its own purposes, but it can influence the lower only *in accordance with* the causal laws, not against them. Here, teleology meets with causality. The "freedom" of will finds its limits in the selection among biological and physical mechanisms, which are not or are only partly represented in conscious experience. However, subjective experience and objective sensory events may be only different aspects of an identical processing of information. The inner and outer world are governed by the *same* natural laws. As NICOLAI HARTMANN put it succinctly in his appreciation of KANT "beyond Idealism and Realism" [196], "*Erkenntnisprinzipien sind zugleich Gegenstandsprinzipien*". According to HARTMANN's *Schichtstruktur*, the higher levels are dependent on the lower but the lower can function without the higher. An ordered dependence is maintained by interrelation of basic causal and higher final processes. These are differentiated and integrated according to special conditions at each level and are limited by an a priori structural order. This results in a *coexistence of causal and teleological laws in the same world* [198].

Pure empiricism appears as insufficient for sensory physiology as pure rationalism, and both supplement each other in percept and concept or in "Anschauung und Idee", to use KANT's terms. In the history of natural science, BACON's emphasis on induction was already superseded by NEWTON who re-established the value of theory and mathematical deduction.

Sensory functions involve three levels of the world, the *inorganic, organic and psychic* realms. Their investigation thus requires a scientific approach at three levels: 1. the *physics and chemistry* of stimuli and transducers, 2. the *neurobiology* of sensory processing and 3. the *psychology* of perceptive experience. Although each level is dependent upon that preceding it, all follow their own natural laws which become increasingly more complex in the higher levels. The lower levels are independent of the higher and any influence from the higher to the elementary level is very limited or non-existent: physicochemical processes proceed whether biological or psychic functions are present or not. But biological and psychological functions are possible only in the basis of physical and chemical events. However, as CLAUDE BERNARD remarked a century ago, biological processes, although dependent upon physical and chemical mechanisms proceed in a direction and towards an order which they would not attain at the lower physico-chemical level alone [40]. The dependence of the higher levels on the lower and the development of a specific order and of finality in the higher levels with appearance of new rules corresponds to the "categorical dependence laws" of NICOLAI HARTMANN [197]. He applied them to causality and finality processes in biology and psychology and extended them up to the highest levels of humanistic accomplishments in culture. Natural laws are apparent in the three levels of inorganic, organic and psychic processes and are further modified and refined at the highest level of social and cultural interaction in human society, again with more complex rules and sophisticated conditions of social communication in ethics.

"Three-World" Interrelation. Another hierarchic concept that has some similarities to HARTMANN's four-level system is discussed by ECCLES [114] in his book "*Facing Reality*". It has been derived from KARL POPPER's "three worlds": 1) the physical, 2) the conscious and 3) that comprising human knowledge [366]. According to ECCLES, one must recognize a certain autonomy of conscious experience in World 2. It follows that perceptual experience in vision is not only a physiological event elicited by physical light, but also a *creation of the individual* which transmutes the patterns of World 1 to the experiencing self in World 2. Of course, the problem of understanding visual perception is only a part of the general problem of understanding the external world and ourselves, which involves the creation of sensory hypotheses to be tested in reality. This theorem is partially related to POPPER's logic of science [365] and his concept of testing and refuting theories. Although it is to some extent based upon KANT and may also be applicable to vision research, POPPER's book cannot be discussed here.

According to ECCLES [114], there is no direct communication between Worlds 2 and 3. World 3 is information coded on records stored in World 1, such as books, pictures, tapes, etc. This basic material must be acquired by learning in Worlds 1 and 2 and information flow needs sensory experience, including memory. In other words, World 3 information is only effective in World 2 after it has been sensed by the ordinary *perceptual process* in a World 1 operation and then, after transmission to the cortex, decoded in a quite unknown and immensely complicated brain process before it becomes a percept through liaison with consciousness in World 2. Likewise, all conscious *actions* have to be communicated by some unknown process from World 2 to the liaison brain and so eventually to the motor cortex, and hence to some desired movement. The whole sequence from liaison brain to movement is, of course, in World 1 although it can be partly experienced as willed action controlled by the inner World 2. Goal-directed eye saccades are marginal cases of such willed actions since they are not experienced as movements but as attention directed towards objects in a stable outer world. Finally, the action instituted by conscious decision and transmitted via the brain to give an effect in the outer world may be in the field of cultural communication and creation so that it becomes coded and assimilated into World 3 [114].

When we apply these concepts to vision, it becomes clear that one cannot expect a complete coincidence of physical, physiological and perceptual processes. Each higher level transcends the lower and each contains several sublevels in itself.

POPPER's and ECCLES' three worlds are not homologous to HARTMANN's levels but some parallels and analogies may be drawn between these two concepts. The inorganic (physicochemical) and vital (biological) levels of HARTMANN combined would correspond to World 1 in the Popper-Eccles concept, whereas conscious perception would occur in World 2. The two highest levels of conscious experience (World 2) and of social organization including science (World 3) would correspond rather well in both orders.

Whether similar principles of dependence of higher levels upon lower ones are valid for the three-world concept, as for HARTMANN's categorical levels was not stated explicitly by ECCLES, but may be inferred from his writings [114].

Pragmatic Apriorism in Vision

Our concept of a priori mechanisms in sensory processing, which is applied to vision, considers the biological structures and functions of the visual system as prerequisites for perception of the visual world. In other words, *our ability to see outer structures in space and time is rendered possible by the intrinsic structural order of the visual system and the brain.*

Structural Apriorism in Biology and Rational Apriorism in Philosophy. The biological structural order of the brain, given a priori to a perceiving organism, and the philosophical rational order of KANT's synthetic judgements, given a priori to perceiving man, certainly differ. However, they both show some conceptual analogies of an *innate order* which can be termed aprioristic. KANT's rational a priori may acquire a biological sense when applied to structure and function, although this may be a reversal of its original meaning. Such analogies of concept can be useful for the theory of sensory sciences. They suggest preperceptive structural apriorisms at the lower levels of the sensory systems, and perceptive rational apriorisms at the higher levels of sensory recognition. Thus some synthesis of nativism and empiricism appears to be as necessary here as in philosophy.

Structural Order a priori and Sensory Experience a posteriori. Expressed in most general terms, innate sensory structures are the formal prerequisites of sensation. They form the substrate of *a priori framework*, whereas acquired sensory data add the empirical *content* of perception *a posteriori*. These parallels of philosophy and science should not be regarded as fortuitious. They may signify a basic neurobiological order indicating that some principles of thinking are also principles of brain function. In the terminology of KANT [262, 263], which unites the empiricist and rationalist conceptions in philosophy, *Anschauung* and *Denken* complement each other. Both contribute to abstractional cognition and are formed by "pure" and formal categories a priori as well as by empirical experiences a posteriori. KANT's succint dictum: "Gedanken ohne Inhalt sind leer, Anschauungen ohne Begriffe sind blind" [262] may be adapted for visual perception and understanding. Applied to vision, it would say: *Visual input without abstractive labelling is blind and formal abstraction without visual images is empty.* It appears that visual information-processing in the brain provides the abstractive order and selection of the inflow of peripheral signals produced at the receptor level and thus renders possible visual recognition.

If a simplification of KANT's "synthetic judgement a priori" for perception is permitted, one may say: *Perceptual judgements a priori become possible through the structural and functional order of neural sensory systems.* Of course, such an a priori order needs some development. It is only innate for the basic structure and is mostly acquired by learning for visual perception and recognition. Hence, it can only be used *a posteriori* after the processing of visual messages induced from the outer world.

Our *pragmatic apriorism* differs from the philosophical and its "pure" a priori categories of time and space, postulated by KANT [262]. It is even possible to reject this neurobiological apriorism as a reversed non-Kantian concept by citing KANT's own writings when one selects those with more idealistic conceptions. Yet KANT's principle, that manifold sensations are ordered formally in perception by use of general a priori categories, remains valid for sensory research. The validity is *not* dependent upon an abstract or concrete interpretation of the categories (as mental faculties or as the cerebral order of structures). These principles can be applied to sensory processes whether one believes in KANT's "thing-in-itself" as an abstraction or regards it only as a physical event. We must also accept conscious perception whether we believe in an abstract "consciousness" or not.

To sum up, the visual scientist encounters *two groups of a priori categories, his own and those inherent in the visual system.* The first influence his thinking and perception, the second are the primary biological strucures and functions, fundamental for vision. He must investigate the visual system as it has developed, with all its astonishing performances and even with its defects. Visual research analyses its action by various appropriate methods in experiment and theory: the visual mechanisms by neurophysiology, photochemistry, and morphology; the final perceptual phenomena by experienced observation and psychophysical quantification.

Innate Structure and Acquired Content of Vision

A few general remarks on biological foundations and historical concepts of innate order and acquired mechanisms of visual perception may be appropriate. Innately determined visual structures form the basis of all visual functions *before* percepts can be stored and labelled for visual recognition. Such innate mechanisms may be defined generally as neurobiological correlates of genetically determined structural-functional relations in the visual systems. This concept of *structural order* dates back to early nineteenth-century biology when the morphological "Bauplan" of organisms was conceived by GOETHE, OKEN, CUVIER and other comparative zoologists. A special *Bauplan of the visual mechanisms* constitutes the anatomical structure and determines the physiological and perceptual frame of vision.

Preperceptive and Structural Order of Vision. All visual structures, peripheral and central, and their mechanisms form the basic conditions for vision. They include *physiological optics of the eyes as well as neuronal circuits of the brain.* Anatomical *projections of visual pathways* to brain stem and cortex constitute the first structural order in the brain. The triple representation of the vertical meridian in three areas of the cerebral cortex (see Part B, Chap. 21, Fig. 1, p. 329) connected by their callosal commissure fibres, constitute an *innate order of the visual fields* which seems to be rather constant in higher mammals with partial chiasmatic decussations. It provides links between the two half-fields of vision. (See p. 71.) Thus the unity of the visual field represented in two cerebral hemispheres is based upon an innate *Bauplan* of the central visual systems.

Invariance in Vision and Biology. Only during the last 20 years have central visual functions and their neuronal mechanisms been investigated in detail. Neurophysiological research is now beginning to build bridges between classical physiological optics and visual perception as conceived by HELMHOLTZ [208]. Objective neuronal correlations with subjective psychophysical studies of vision are beginning to appear. But large gaps remain. Visual thinking and abstraction as conceived by ARNHEIM [10] cannot yet be correlated with neuronal physiology. The *constancies and invariances of visual perception,* stressed by HERING [212, 214] and summarized by TEUBER [430], are still unknown in their neurophysiological basis. Yet these integrating functions of higher levels may also be based on elementary biological structures. They may have parallels in general biological laws and can be developed from Kantian principles as LORENZ used them for his innate releasing mechanism [301, 302], called earlier *angeborenes Schema* in context with his concept of "*angeborene Formen möglicher Erfahrung*" [301]. These may be rooted in evolutionary genetic invariances, as defined by molecular biologists.

Visual structures and their functions may be considered as *biological invariances* in a most general sense, determined in their evolution by a *genetic code*. The intrinsic order of the latter was investigated successfully by modern molecular biology. Molecular investigation of the visual system, however, is still an underdeveloped field although some initial steps have been taken in connection with the photochemical and photoreceptor mechanisms described in Volumes VII/1 and VII/2.

Structural Order and Experience. The old problem of "*Nativismus*" and "*Empirismus*", discussed so ardently in classical visual physiology, can be attacked by neurophysiology with new methods since WIESEL and HUBEL [456] and others have combined visual deprivation experiments with neuronal recordings (see Chaps. 9 and 21). WIESEL and HUBEL's suggestion, that the receptive field organization of orientation-selective neurones in the cat's visual cortex may be *innate*, and that early visual exposure is only necessary for their maturation and maintenance, has recently been questioned but the fundamental concept remains a powerful guide to experiments. Future research has to decide what is the interdependence of "native" order, mostly based on structure and innate neural connections, and "empirical" visual learning with acquired functional relations such as synaptic facilitation and biochemical traces of memory.

Of course, the correlation or orientation-selective neurones with visual phenomena involving lines and edges in perception is a simplification, and the check provided by behavioural incapacity to respond to oriented patterns in deprived kittens [409] are somewhat crude tests. Yet such combined studies of deprivation, behaviour and neuronal recordings give some hope for the emergence of experimental methods to test the relative roles of innate and acquired visual mechanisms and to relate function more closely to structure.

Deduction and Induction in Vision and Visual Science. The modern trend of communication science which uses mathematical and physical models in biocybernetics has re-established the value of *logical deduction* and theory in neurobiology. Deductive methods had been underestimated by the tradition of inductive empirical investigation during the last century. But already in 1863 and 1865 LIEBIG [295] had defended logical theories and deduction as necessary guides for the collection of facts in natural science. LIEBIG contradicted FRANCIS BACON's ancient emphasis on induction (as opposed to deduction), his insufficient appreciation of hypotheses and his overrating of empirical data. The inductive accumulation of many facts in BACON's writings was restricted to crude empirical observations and remained far from the systematic experimentation of modern science which uses a theoretical framework. The present hypothetico-deductive method of science is formulated by POPPER in 1935, [365]. But in stressing the role of deduction one need not exclude the additional use of inductive methods or call induction "superfluous" as POPPER did. However, we agree with POPPER's emphasis on the primacy of deductive methods and the importance of theory in the natural sciences, and with his statement: "*The empirical sciences are systems of theories*" ("Erfahrungswissenschaften sind Theoriensysteme").

Induction and deduction are coordinated in perception. In the brain processes underlying vision, as in other modalities, the stimulus inductively elicits a perceptual matching hypothesis, whose implications are deductively evaluated though not necessarily at a conscious level [307a]. In recognition both inductive and deductive processes of course take account of data stored in memory. Deduction is also used for attentive perception in visuomotor action. The active search for certain biologically significant visual patterns, which is aided by attentive eye movements and instinctive and learned schemes, may be compared to a "*perceptual theory*" of the organism. Hence, by using expectation and attention, we can *select the relevant* from the multitude of irrelevant visual data. In all these perceptual acts, vision also remains dependent upon a priori frames of innate order. This order appears in the structures and functions of sensory organs and of the brain and in fixed patterns of behaviour.

In natural science, single facts are worthless if they are not reproducible, do not follow general laws or are not systematized according to theory. *Facts acquire scientific importance only in context with other facts and theory.* In visual science, they have to be discussed in association with different experiments in various contexts and in their relation to neurophysiology and perception. This causes redundancy, but such *redundancy may be useful.*

It is necessary to follow a certain hypothesis in experimental research until it is shown to be false. Of course, theoretical concepts should always be modified by experimental results and are rarely verified in their original form. Even the application of a false theory as a provisional guide may serve scientific planning better than having no concept at all, accompanied by blind experimentation which leads to uninteresting results and an enumeration of unrelated facts. A paradigm of the role of theory in science may be found in the natural order of perception.

Comment. The messages of the senses are selected, ordered and classified by the brain in a temporal sequence with an end-result of perception. Some redundancy in sensory information facilitates classification and recognition. These quasi-conceptual processes, largely unknown in their neurophysiological mechanisms, may be neurobiological analogues to philosophical postulates of an ordering mind which synthesizes percepts and concepts by comparing and classifying them within a priori categories and with a posteriori experiences. The sensory raw material delivered by the receptors cannot become a percept without information processing over several levels in the brain. This includes feature extraction, spatial and temporal order and memory comparison which involves some reverberation and redundant resonance of the sensory messages. The riddle of the sequential order and unity of vision may be less perplexing for the neurophysiologist when he knows that the philosophy of perception encounters similar unsolved problems. In visual physiology, it is helpful when objective recordings from the visual systems can be correlated with subjective phenomena and when general rules can be applied to neurophysiological results and psychophysical data. This may be stated as apology and explanation for this general section.

XIII. Unsolved Problems

Among the many unsolved questions related to vision and neurophysiology, those of constancy, pathophysiology and plasticity should be discussed. Some problems concerning structure and function, visual learning and deprivation have already been mentioned in preceding sections (pp. 7 and 80). Some recent experiments on the relation of neuronal plasticity to visual memory may be discussed as regards their aspects for future research.

Visual Learning and Plasticity

Visual Learning. The capacity of the higher visual systems to learn and the storage of a visual Gestalt after a single presentation shows that visual learning is closely related to *visual memory*, which underlies perceptual recognition. The bewildering variety in the shaping and reshaping of visual functions is usually grouped under the vague term "*plasticity*". Some plastic effects of specific visual functions, e.g. receptive field alterations in the cortex, are now being investigated by several groups with the aim of determining how malleable such capacities are in the normal life of the behaving organism. However, the neuronal mechanisms of visual learning remain obscure and experiments on visual conditioning and its correlates in the central nervous system are in their infancy. Current work on the effects of early experience mostly concerns the *negative* aspect of visual deprivation, and this is reviewed in Chapter 9 by Chow.

All deprivation experiments indicate that visual learning maintains and facilitates visual function in the cortex most effectively during the *early* periods of life. Short- and long-term visual memory are not only necessary for acquiring form recognition, but are also prerequisites for the normal function and early development of innate neuronal coordination.

Receptive Field Plasticity. The main achievement of neuronal experiments in the visual cortex during the last ten years has been the description of selective receptive-field organizations. This has led to the classifications of Hubel and Wiesel as described in Chapter 21, and to correlations with contour vision as discussed on p. 63 and in Chapter 2 by Stone and Freeman. Recent experiments by Ervin [387], Denney [96] and their coworkers indicate, besides abstraction and selection of stimulus features, more plastic responses in unanaesthetized animals with or without normal eye movements. Receptive fields may also be changed by nonvisual influences both of a specific sensory and non-sensory nature. The significance of multisensory influences is not always clear, but convergence with specific inputs from the vestibular and other proprioceptive modalities indicates that orientations in the visual world have a functional role which needs additional clues of verticality. Modifications of the receptive-field axes, which can be correlated with perceptual phenomena of the Aubert and Müller type of subjective changes in verticality during head tilt, are plastic mechanisms of visuo-graviceptor regulations, as discussed on p. 98 and in Chapter 21.

Other neurones that respond to stimuli, particularly movement, in large parts of the visual field are of great interest for visual perception. Very large fields, up to 45 degrees in diameter and including the fovea, are characteristic of the monkey's inferotemporal cortex (see GROSS, Chap. 23) and the suprasylvian gyrus of the cat. Whether similar large receptive fields also occur in visual areas 18 and 19 has not yet been analysed well enough by experiments without anaesthesia.

Visual Experience and Receptive Fields in Area 17. Recent experiments of BLAKEMORE [51], SPINELLI, HIRSCH [217], and others [353], which are described in Chapter 21, have elucidated a negative aspect of visual learning, the *deprivation* effects on receptive field configurations. These authors demonstrated the influence of special visual deprivation of vertical or horizontal contours on the predominant neuronal responses in orientation-selective cells. The last paper of SPINELLI, HIRSCH, and co-workers [409] described the possible *restoration* of these deprivation effects. After normal visual stimuli were revived, an increasing number of neurones were found which were absent or rare after selective visual experience with deprivation of their adequate selective stimuli. Monocularly activated neurones, however, did not re-acquire binocular responses. The authors concluded that there is a direct, continuing and lasting influence of early visual experience and deprivation. This is of considerable interest for normal visual learning, which still remains a neurophysiological enigma. (See also PIAGET [355]).

Correction of Space Perception and Visual Learning. STRATTON'S [418] classical experiment on vision with re-inversion of the retinal images and more elaborate experiments of KOHLER [271, 271a] on the transformation of the perceptual world and of HELD and HEIN [204–206] on sensory-motor learning have demonstrated the remarkable ability of the human visual system to adapt to spatial inversions and to correct distortions. Amphibia and other lower vertebrates have no such ability of correction, and they never learn behavioural adaptation to an inverted visual field. In frogs, unlike primates, complex visual processing of patterned and moving stimuli occurs already in the retina and the main central projection is to the optic tectum and not to the visual cortex. In primates and man, the higher transformations and the learning processes of the visual system that underlie perception of the visual world, apparently occur mainly in cortical visual systems.

Visual Constancies and Continuity of Vision

Visual Constancies. Besides the temporal order and continuity of vision which awaits neural correlates in the future, the constancy of *temporal and spatial stability* is a major challenge for neurophysiology. These problems are closely related to eye movements and have been discussed since HERING [212] in relation to attention (see p. 93). TEUBER'S Handbook article [430] gives a synthesis of the literature up to 1960 with the theory of corollary discharges derived from the concepts of SPERRY [402] and VON HOLST [225].

The constancies described in the order of retinal neurones by FISCHER [135, 138, 140] are related to receptive and perceptive field size, but these anatomic-physiological constancies are, of course, far from the perceptual constancy of apparent size, which does not correspond to the size of the retinal image. Such findings, however, may point the way to investigation of further correlates of higher-order visual constancies in central processing, e.g. the "magnification" in the visual cortex, discussed in Chapters 2 and 9. FISCHER found that the number of ganglion cells overlapping within receptive fields of increasing diameter towards the periphery shows a remarkable constancy of 30—40 neurones per field in parallel [34]. The field area may constitute a parallel to the extension of dendritic processes and the coordinated synaptic input from the receptors. The physiological basis of the central regulation of constancy is still unknown.

The problem of the temporal and spatial stability of vision has received some new stimuli from neuronal physiology, recordings of eye movements and psychophysical observations.

These are discussed for form vision in Chapter 2 by STONE and FREEMAN and for eye movements in Chapter 5 by MACKAY. Daily experience with fixating eye movements shows that the visual world appears stable in temporal continuity, although neurophysiologically it is built up from sequences of different retinal images, packaged in successive time segments. When viewing moving pictures from a *film* presented in a sequence of 16 frames per second, we perceive continuous movements nearly as well as from a moving object which is pursued by eye tracking. During movement perception we do not know whether one or both of the two mechanisms of motion detection, depicted in Fig. 14, is effective. Either the afferent retinal shift, or the efferent eye tracking, or a combination of both may be responsible for velocity perception [104]. The average duration of fixating period in looking at a picture or in reading (200–300 msec) may correspond to the optimal perception time of single images. It is usually longer than 100 msec, although neuronal recordings show that significant information can be transmitted by retinal ganglion cells within about 80 msec. The reason why we need more time for single fixations apparently lies in the central processing.

The interrelation of eye movement and vision requires an exact interaction between visual-afferent and oculomotor-efferent messages. The *input quantization of time*, postulated by BOYNTON [57], can be made only by exact regulation of eye movements which controls the timing by goal-directed saccades that initiate fixation, and finally by the cancellation of previous visual input during a new saccade and lid blinking. This may be compared to an intrinsic chopper mechanism which coordinates vision and eye movements [253] (see p. 82).

General visual phenomena whose neuronal correlates are unknown are the *abstractive classification* and systematization of forms and the *invariance* of different aspects of similar objects. These orderly representations of the variety of visual images are mostly learned functions, but some may be based upon instinctive, perhaps innate and species-specific mechanisms like LORENZ's innate releasing mechanisms [302]. These "fixed patterns" of the ethologists, insofar as they concern visual images, probably have their foundation in the intrinsic structural and functional order of the visual system. BERITOFF's conception of "image memory" as images stored in the brain and forming the basis of higher vertebrate behaviour [38, 39] may comprise a related function which combines learned and innate visual mechanisms.

The *Prägnanz* of the Gestalt psychologists is a more diffuse concept of generalization and recognition. The tendency towards *homogeneity* is another perceptual phenomenon which might have a neuronal basis. Up to now, the pattern experiments of the Gestalt psychologists have been used mainly to demonstrate the influence of certain *linear patterns on brightness perception* (Fig. 9, 10) and the disappearance of contrast effects within a closed Gestalt such as KOFFKA's ring [269]. This is related to *area contrast* as discussed on p. 63.

Simultaneity and Continuity of Vision. The impressive simultaneity of a visual image and of visual perception of complex patterns exposed for fractions of a second cannot yet be explained in neuronal terms. But simultaneous information is a less formidable question than the *continuous unity of successive information.* To the neurophysiologist, who knows that a million optic nerve fibres can simultaneously transmit signals of retinal images in man, the explanation of simultaneity is more a problem of mass communication and computer design than of principle. The common experience of continuous perception of discontinuous motion pictures at 16 frames/sec in the cinema illustrates the temporal integration of our visual system. Problems of time quantization and visual packaging are also discussed by STONE and FREEMAN in Chapter 2 (p. 195).

Serious questions of principle and experimental difficulties arise for the *sequential integration of images and their visuomotor control in time.* The temporal integration of space in perception, which is derived from many sequences of foveal fixations when a human being looks at a picture, is a remarkable performance of the visual and visuomotor systems. The spatial constancy and temporal continuity of the visual world, HERING's *Raumkonstanz in der Zeit* [214], which is explored by sequential information concerning details drawn from different foveal fixations, cannot yet be explained by neurophysiological mechanisms. HERING's postulate [212], that *visual attention* determines location and eye movements in external space by using an inner perceptive model of the visual world, remains a very general hypothesis, and neuronal correlates are still lacking. To attain neurophysiological explanations of the temporal continuity and stability of visuomotor coordination and visual attention in psychophysics and cerebral physiology, it is probable that new theoretical concepts will be necessary before the appropriate experiments can be instituted.

Visual Thinking. General assertions that all thinking is perceptual in nature would be far beyond the scope of neurophysiology. Yet there are some conceptual relations. ARNHEIM [10] has recently proposed that abstraction always starts from sensory experience, mostly visual. This modern version of philosophical sensualism may have some parallels in neurophysiological concepts. Neurophysiologists working on neuronal recordings in the central visual system have often favoured similar concepts of "visual abstraction" and "feature extraction" by specialized neurones signalling certain abstracted patterns which are stored in the memory. This is of course an empiristic generalization. "Visual thinking" may be accepted only for certain associations experienced in "anschauliches Denken", a sort of conceptual symbolization of thought, but may not be adequate for other, higher forms of abstractive thinking.

Abstract mathematical formulations and most intellectual thinking and reasoning are difficult to classify as "perceptual". The process might be compared with the formal logical inductions of a computer programmed to extract information from data delivered by certain inputs, but it remains doubtful whether these are equivalents of perception. ARNHEIM's assertion that sensory processes are not only passive imprints from the outer world but *active accomplishments of the organism* (e.g. in Gestalt vision) may perhaps have some correlates in modern neurophysiological concepts of the transformation of visual input via several stages of information processing in the central visual systems. A special aspect of visual thinking is visual imagery of numbering and temporal sequences, such as GALTON'S number forms [152] (see p. 130).

Problems of Pathophysiology

Geometric Hallucinations, Scintillating Scotomas, and Visual Cortex. Similar *geometric visual patterns* appear in all *initial stages of hallucinatory states,* such as toxic hallucinations of different origin (mescaline, LSD and other hallucinogenic drugs) and the hallucinations experienced during sensory isolation. In 1967 I proposed an explanation by disinhibition phenomena and disintegration in cortical neuronal populations representing certain receptive-field characteristics [250a].

Some interesting neurophysiological correlates have been discussed for the *scintillating serrated scotomas and "fortification" illusions of migraine patients,* which are elementary organic hallucinations. They may wander slowly from near the fovea to the peripheral hemi-field and show a regular visual pattern with orientated lines, the length of which increase towards the retinal periphery with more or less rectangular figures and flicker. The increase in length approximates the increase of perceptive field sizes in the peripheral and parafoveal retina (Fig. 7, p. 49). After LASHLEY, in 1941, postulated visual cortex disturbances [288], W. RICHARDS in a very recent study [376] suggested a relation to the cortical columns of HUBEL and WIESEL and their line and corner detectors.

RICHARDS postulates a wave front expanding in the cortex with the serrated lines, and estimates a rather constant cortical extension in the visual hemifield of about 1.2 mm of cortical distance for each line. The nature of the abnormal process in the cortex during such optic migraine attacks is not yet clear, but a process similar to spreading depression with vascular and electrolyte changes and surface electronegativity is the most likely cause. In severe attacks this would lead to hemianopsia, in weak episodes to concentric bands of relative blindness separated by the serrated lines. The special properties described by the various authors are suggestive of cortical rather than geniculate or retinal origin.

In contrast to these pathological but relatively simple geometric phenomena, the normal but more complex hallucinations in dreams or the number forms of conscious experience cannot yet be related to particular neurophysiological mechanisms, although some neurobiological theories have been proposed [126, 267].

Visual Imagery, Dreams, and Hallucinations. The visual phenomena in the wider field of visual imagery, that cannot be explained by neurophysiological correlates, range from visual concepts (anschauliche Vorstellungen) and symbolization by images to the phantastic illusions appearing before the onset of sleep, described by JOHANNES MÜLLER [337], to genuine dreaming and cover hallucinations during sensory deprivation, drug ingestion or psychopathological syndromes in psychoses. *Visualized abstracts,* such as the "number forms" of GALTON [152] and similar phenomena of coloured or spatial patterns of time sequences, have been studied in detail by psychologists and defined as "*Diagramme*" by G. E. MÜLLER [333]. [Number-forms are visualised patterns of numerical series or time periods (years, months) in mental space and often in colour which certain persons "see" when counting or thinking of time]. Since GALTON, it has been known that similar number forms occur in the same family and that a disposition to spatial visualization may be hereditary [152]. On the other hand, it seems evident that all mental images are acquired by perception and that visual shapes presuppose visual experience. Number forms and other kinds of *Diagramme* are programmed processes which may be facilitated and finally determined in memory by emotion. This is shown by the prominent place and vivid colour of images pertaining to birthday and Christmas in the year-cycle of children's "number forms" [250a].

As for visual attention and imagery, whether the non-specific reticular activating system in the brain stem has an influence has also been discussed relative to dreaming and REM sleep. However, no direct proof is available for brain-stem activation of mental images or dreams, and their neuronal basis remains obscure. The same lack of knowledge exists for visual hallucinations during sensory deprivation or following hallucinogenic drugs. Clinical experience shows that hallucinations may occur under the influence of subcortical structures, mostly during alterations of consciousness and in epileptic dreamy states. PENFIELD's systematic studies of cortex stimulation [352] which elicited special visual images and remembered scenes by electrical stimuli applied to the occipito-temporal cortex (the interpretive cortex of PENFIELD and PEROT [351]) show that such complex visual experience could be induced electrically only in patients with epilepsy.

Although many EEG studies have been made during drug-induced, *toxic, visual hallucinations* after mescaline or LSD, they show only some EEG flattening, and the neuronal mechanisms of hallucinatory images are unknown.

The synaptic block of the geniculate relay found in cats given LSD [45], cannot define the hallucinations in neuronal terms. The LSD block is *postsynaptic*, and the output of transmitter substance does not seem to be diminished during the blocking action, according to BISHOP et al. [45].

Some neurophysiological characteristics of dreaming have been well investigated in the EEG, but it remains unclear what is the neuronal equivalent of the visual dreams in REM sleep, how it may be related to rapid eye movements with EEG activation, and whether it may be regarded as disinhibition.

In spite of the many neurophysiological and neurochemical theories of sleep and dreaming, our ignorance of the neuronal mechanisms is about the same for visual dream images as for toxic and psychotic hallucinations.

XIV. Conclusion

To discuss the relation of perception and neurophysiology as a prelude to the cerebral mechanisms of vision involves much speculation and generalization. In a border field between physiology and psychology, two sciences that use essentially different methods, the pitfalls of uncontrolled equivocation and the risk of confusion of categories are obvious. The hope that neural modelling would form a link between psychological and physiological data has not yet been fulfilled. The theoretical framework and the quantified data of neurophysiological mechanisms, and their correlation with verified synaptic connections and neuronal circuits are insufficient to permit quantitative models to be constructed. In this state of affairs sceptics may prefer to make a sharp distinction between these two logically different approaches to visual physiology and perception.

However, a clear logical distinction between different methods in two diverse fields of research directed to the same object does not exclude scientific convergence. Methodological incompatibility should not induce an attitude which ignores common aspects and essential correlates. Qualitative parallels, as well as quantitative correlations, between two fields of vision research may be used as stimuli for theory and experiment. To demonstrate this research co-ordination some neuronal and perceptual correlates of vision have been selected as examples. These should be regarded as *two complementary aspects of vision* with due reference to the logical and categorical incompatibilities of physiology and conscious perception.

Qualitative and Quantitative Research in Vision. In this chapter the correlations of vision and neurophysiology are discussed in a general way with little reference to quantification and the mathematical methods applied in psychophysics and physiology. In the age of computer science this may appear rather old-fashioned. However, most basic discoveries in the neuronal physiology of the central visual system were made in the years 1952–1962 with the use of *qualitative or semi-quantitative stimulus conditions* in rather simple experiments. Exploratory approaches such as those used by the Freiburg school [25–29, 248, 249, 254] and by Hubel and Wiesel [230–235] opened up the field, were facilitated by correlates with subjective vision and paved the way for further research. Eventually, however, quantitative measurements and rigidly controlled mathematical methods must confirm the insights obtained through the first qualitative experiments in both psychophysical and neuronal approaches.

Spatial and Temporal Order of Visual Constancy. An organism perceives the world by visually integrating time, space and motion. Despite moving eyes, it sees a continuous sequence of images showing *constancy* and *identity* as prerequisites for experience. This well-integrated perceptual end-product, which gives relative stability to the surrounding world during movements of the subject, is selected by attention and composed from a multitude of visual mechanisms. It is further modified and completed by multisensory messages from non-visual modalities. *Visual attention*, besides evaluating objective movements, steers the saccades towards certain goals and facilitates the tracking eye movements, as demonstrated in optokinetic and railway nystagmus [104]. In addition to this oculomotor space location by various eye movements (including convergence and accommodation), stereopsis also, refines depth localization by neuronal integration in the visual cortex [243]. The neurophysiology of these different mechanisms is only partially known.

Considering the bewildering number of visual and visuomotor mechanisms, one may question whether visual constancy and continuity will ever be explained in physiological terms. As yet, no comprehensive picture has emerged from the neurophysiological data. Even with a more complete knowledge of their interrelationships the visual scientist might still feel lost in the multitude of experimental facts. To introduce *order* into these facts, it is advisable to return to introspective sources of information in perception. Some order may often be derived from psychophysical analyses of visual experience under controlled observation, and such an approach may lead to *concepts and theories* that can be correlated with certain neural mechanisms. The aim of this chapter is to show examples of such perceptual integrations and to point out the unsolved problems in relation to neurophysiology. Partial overlap with other chapters was reduced by treating more extensively those fields which were not dealt with or mentioned en passant by other contributors.

In the vast domain of neurophysiological and perceptual studies of vision it is often difficult to see the wood for the trees, and the leaves. The multitude of single facts and observations calls for *synthesis* in a theoretical framework. Some neuronal correlates of visual perception were discussed here, but there is still no general theory of vision and exact mathematical modelling remains a postulate for future research. In the present state of vision research we can at best ask pertinent questions based on perceptual experience, but we cannot yet provide answers to the most pressing problems of visual physiology in neuronal terms. The development of new methods and theories is necessary before some of these problems can be attacked by well-planned experiments.

The twenty-seven chapters that follow should provide a survey of our present state of knowledge on central visual mechanisms and on possible relations to visual perception.

Acknowledgements

I am grateful to the following authors for permission to reproduce their figures or to use their unpublished results: Prof. G. BAUMGARTNER, Zürich (Switzerland); Dr. BARBARA BROOKS, Seattle (USA); Dr. F. W. CAMPBELL, Cambridge (England); Dr. M. J. M. GERRITS, Nijmegen (Netherlands); Prof. B. HASSENSTEIN, Freiburg (Germany); Prof. R. HASSLER, Frankfurt a. M. (Germany); Prof. W. NAUTA, Cambridge, Mass. (USA); Dr. G. F. POGGIO, Baltimore (USA); Prof. R. RÖHLER, München (Germany); Prof. H. SCHOBER, München (Germany); Dr. L. SPILLMANN, Freiburg (Germany); and Prof. A. J. H. VENDRIK, Nijmegen (Netherlands).

Special thanks are due to Dr. B. BROOKS, Seattle (USA); Dr. B. FISCHER, Freiburg (Germany); Prof. D. M. MACKAY, Keele (England), Dr. L. SPILLMANN, Freiburg (Germany); and Prof. L. TEUBER, Cambridge, Mass. (USA) for criticism of parts of the manuscript. Last but not least, the efficient secretarial help of Miss R. SELBMANN, Mrs. S. BRINCKMANN, Mrs. B. CROOK, and Mrs. G. SINON is gratefully acknowledged.

The preparation of this chapter was supported by the Deutsche Forschungsgemeinschaft in the Sonderforschungsbereich 70 (Hirnforschung und Sinnesphysiologie) at the University of Freiburg.

References

Postscript. The necessity of limiting the cited literature forced many omissions, mainly in psychophysics. Early contributions of classical vision research are cited when they show correlates of modern neuronal recordings and preference is given to brightness, pattern and movement perception versus colour vision.

During the printing of this volume, an account of Symposium on Vision, Canberra, 1972, appeared in an ophthalmological journal [47]. Many contributions contain new material on the visual system in the retina and the brain. Some results such as the experiments of BISHOP and HENRY and their criticism of HUBEL and WIESEL's hierarchic concept of receptive fields are discussed from earlier publications in Part B, Chapter 21. Other contributions related to the subject of our chapter are: recordings from the visual cortex of unanaesthetized monkeys by POGGIO [360]; the development of synapses in the cat's visual cortex by CRAIK; PETTIGREW's concept of early visual experience shaping the developing geniculo-striate system of modifiable synapses and having parallels to astigmatism in humans [353]. Further contributions of this symposium on visual attention, eye movements and the tectum opticum (WURTZ, SCHILLER and others), on stereopsis (JULESZ), light–dark adaption, perceptual constancy and illusion (LAY) are important for visual perception and neurophysiology. Another Symposium on eye movements appeared in 1973 [467a].

1. ADAMS, A.: Nystagmographische Untersuchungen über den Lidnystagmus und die physiologische Koordination von Lidschlag und rascher Nystagmusphase. Arch. Ohr.-, Nas.- u. Kehlk.-Heilk. **170**, 543—585 (1957).
2. ADICKES, E.: KANT als Naturwissenschaftler. Kant Studien **29**, 70—97 (1924).
3. ADRIAN, E. D.: The physical background of perception. Oxford: Clarendon-Press 1947.
4. — MATTHEWS, R.: The action of light on the eye. Part I. The discharge of impulses in the optic nerve and its relation to the electric changes in the retina. J. Physiol. (Lond.) **63**, 378—414 (1927).
5. ALPERN, M.: Eye movements. In: JAMESON, D., HURVICH, L. M. (Eds.): Handbook of Sensory Physiology, Vol. VII/4, pp. 303—330. Berlin-Heidelberg-New York: Springer 1972.
6. ANGEL, F., MAGNI, P., STRATA, P.: The excitability of optic nerve terminal in the lateral geniculate nucleus after stimulation of visual cortex. Arch. ital. Biol. **105**, 104—117 (1967).
7. ARDUINI, A.: Enduring potential changes evoked in the cerebral cortex by stimulation of brain stem reticular formation and thalamus. In: JASPER, H. H. et al. (Eds.): Reticular Formation of the Brain, pp. 333—351. Boston-Toronto: Little, Brown and Co. 1958.
8. — The tonic discharge of the retina and its central effects. Progr. Brain Res. **1**, 184—206 (1963).
9. ARNHEIM, R.: Art and visual perception. London: Faber & Faber 1956.
10. — Visual thinking. Berkeley-Los Angeles-London: Univ. of Calif. Press 1971.
11. AUBERT, H.: Eine scheinbare bedeutende Drehung von Objekten bei Neigung des Kopfes nach rechts oder links. Virchows Arch. path. Anat. **20**, 381—393 (1861).
12. — Physiologie der Netzhaut. Breslau: Morgenstern 1865.
12a. BACH-Y-RITA, P.: Brain mechanisms in sensory substitution. New York-London: Academic Press 1972.
13. — COLLINS, C. A., HYDE, E. (Eds.): The control of eye movements. New York: Academic Press 1971.
14. BARLOW, H. B.: Temporal and spatial summation in human vision at different background intensities. J. Physiol. (Lond.) **141**, 337—350 (1958).
15. — Initial remarks. In: JUNG, R., KORNHUBER, H. H. (Hrsg.): Neurophysiologie und Psychophysik des visuellen Systems, pp. 375—376. Berlin-Göttingen-Heidelberg: Springer 1961.
16. — Possible principles underlying the transformations of sensory messages. Three points about lateral inhibition. In: ROSENBLITH, W. A. (Ed.): Sensory Communication, pp. 217 to 234 and 782—785. Cambridge/Mass.: M.I.T. Press; New York-London: J. Wiley & Sons, Inc. 1961.
17. — The coding of sensory messages. In: THORPE, W. H., ZANGWILL, O. L. (Eds.): Current Problems in Animal Behavior. Cambridge: University Press 1961.
18. — The information capacity of nervous transmission. Kybernetik **2**, 1 (1963/65).
19. — FITZHUGH, R., KUFFLER, S. W.: Change of organisation in the receptive fields of the cat's retina during dark adaptation. J. Physiol. (Lond.) **137**, 338—354 (1957).
20. — LEVICK, W. R.: Changes in the maintained discharge with adaptation level in the cat retina. J. Physiol. (Lond.) **202**, 699—718 (1969).
21. BARLOW, J. S., CIGÁNEK, L.: Lambda responses in relation to visual evoked responses in man. Electroenceph. clin. Neurophysiol. **26**, 183—192 (1969).
22. BARTLEY, S. H., BISHOP, G. H.: The cortical response to stimulation of the optic nerve in the rabbit. Amer. J. Physiol. **103**, 159—172 (1933).
23. — NEWMAN, E. B.: Studies of the dog's cortex. Amer. J. Physiol. **99**, 1—8 (1931).
24. BAUMGARTNER, G.: Reaktionen einzelner Neurone im optischen Cortex der Katze nach Lichtblitzen. Pflügers Arch. ges. Physiol. **261**, 457—469 (1955).
25. — Indirekte Größenbestimmung der rezeptiven Felder der Retina beim Menschen mittels der Hermannschen Gittertäuschung. Pflügers Arch. ges. Physiol. **272**, 21—22 (1960).

26. BAUMGARTNER, G.: Die Reaktionen der Neurone des zentralen visuellen Systems der Katze im simultanen Helligkeitskontrast. In: JUNG, R., KORNHUBER, H. H. (Hrsg.): Neurophysiologie und Psychophysik des visuellen Systems, S. 296—311. Berlin-Göttingen-Heidelberg: Springer 1961.
27. — Neuronale Mechanismen des Kontrast- und Bewegungssehens. Ber. dtsch. ophthal. Ges. **66**, 111—125 (1964).
28. — BROWN, J. L., SCHULZ, A.: Visual motion detection in the cat. Science **146**, 1070—1071 (1964).
29. — — — Responses of single units of the cat visual system to rectangular stimulus patterns. J. Neurophysiol. **28**, 1—18 (1965).
30. — HAKAS, P.: Die Neurophysiologie des simultanen Helligkeitskontrastes. Reziproke Reaktionen antagonistischer Neuronengruppen des visuellen Systems. Pflügers Arch. ges. Physiol. **274**, 489—500 (1962).
31. BECHINGER, D., KONGEHL, G., KORNHUBER, H. H.: Visuelle Informationsübertragung beim Menschen: Richtungsanisotropie im peripheren wie im zentralen Sehfeld. Pflügers Arch. ges. Physiol. **318**, R 157 (1970).
31a. — — — Eine Hypothese für die physiologische Grundlage des Größensehens: Quantitative Untersuchungen der Informationsübertragung für Längen und Richtungen mit Punkten und Linien. Arch. Psychiat. Nervenkr. **215**, 181—189 (1972).
31b. BECKER, W.: The control of eye movements in the saccadic system. Bibl. ophthal. (Basel) **82**, 233—243 (1972).
32. — HOEHNE, O., IWASE, K., KORNHUBER, H. H.: Bereitschaftspotential, prämotorische Positivierung und andere Hirnpotentiale bei sakkadischen Augenbewegungen. Vision Res. **12**, 421—436 (1972).
33. BÉKÉSY, G. VON: Sensory inhibition. Princeton: Princeton Univ. Press 1967.
34. — Mach- and Hering-type lateral inhibition in vision. Vision Res. **8**, 1483—1499 (1968).
35. BENDER, M., JUNG, R.: Abweichungen der subjectiven optischen Vertikalen und Horizontalen bei Gesunden und Hirnverletzten. Arch. Psychiat. Nervenkr. **181**, 193—212 (1948).
36. BENDER, M. B. (Ed.): The oculomotor system. New York-Evanston-London: Hoeber, Harper & Row 1964.
37. — SHANZER, S.: Oculomotor pathways defined by electric stimulation and lesion in the brainstem of monkeys. In: BENDER, M. B. (Ed.): The Oculomotor System, pp. 81—140. New York-Evanston-London: Hoeber, Harper & Row 1964.
38. BERITASHVILI, (BERITOFF), I. S.: Phylogeny of memory development in vertebrates. In: KARCZMAR, A. G., ECCLES, J. C. (Eds.): Brain and Human Behavior, pp. 341—351. Berlin-Heidelberg-New York: Springer 1972.
39. BERITOFF, (BERITASHVILI), J. S.: Neural mechanisms of higher vertebrate behavior (Engl. transl.) Boston: Little, Brown & Co. 1965.
39a. BERLUCCHI, G.: Anatomical and physiological aspects of visual functions of corpus callosum. Brain Res. **37**, 371—392 (1972).
40. BERNARD, C.: Leçons sur les phénomènes de la vie commune aux animaux et aux végétaux. (Cours de physiologie générale du muséum d'histoire naturelle), I/II. Paris: J.-B. Baillière et fils 1878/79.
41. BICKFORD, R. G., KLASS, D. W.: Stimulus factors in the mechanism of television induced seizures. Trans. Amer. neurol. Ass. **87**, 176—178 (1962).
42. — — Eye movement and the electroencephalogram. In: BENDER, M. B. (Ed.): The Oculomotor System, pp. 293—302. New York-Evanston-London: Hoeber, Harper & Row 1964.
43. BISHOP, G. H.: Cyclic changes in excitability of the optic pathway of the rabbit. Amer. J. Physiol. **103**, 213—224 (1933).
44. BISHOP, P. O.: Synaptic transmission. An analysis of the electrical activity of the lateral geniculate nucleus of the cat after optic nerve stimulation. Proc. roy. Soc. B **141**, 362—392, (1953).
45. — BURKE, W., HAYHOW, W. R.: Lysergic acid diethylamide block of lateral geniculate synapses and relief by repetitive stimulation. Exp. Neurol. **1**, 556—568 (1959).
46. — COOMBS, J. S., HENRY, G. H.: Interaction effects of visual contours on the discharge frequency of simple striate neurones. J. Physiol. (Lond.) **219**, 659—687 (1971).

47. Bishop, P.O., Gouras, P. (Eds.): U.S.-Australian Symposium on Vision. Canberra Feb. 7—11, 1972. Invest. Ophthal. **11**, 261—548 (1972).
48. Bizzi, E.: Discharge of frontal eye field neurons during saccadic and following eye movements in unanesthetized monkeys. Exp. Brain Res. **6**, 69—80 (1968).
49. Blakemore, C., Campbell, F.W.: Adaptation to spatial stimuli. J. Physiol. (Lond.) **200**, 11—13 (1968).
50. — — On the existence of neurones in the human visual system selectively sensitive to the orientation and size of retinal images. J. Physiol. (Lond.) **203**, 237—260 (1969).
51. — Cooper, G.F.: Development of the brain depends on the visual environment. Nature (Lond.) **228**, 477—478 (1970).
52. — Nachmias, J., Sutton, P.: The perceived frequency shift: evidence for frequency selective neurones in the human brain. J. Physiol. (Lond.) **210**, 727—750 (1970).
53. Boring, E.G.: Sensation and perception in the history of experimental psychology. New York: Appleton-Century Crofts, Inc. 1942.
54. Bornschein, H.: Der Einfluß zeitlicher Reizgradienten auf die Impulsaktivität retinaler Neurone der Katze. Pflügers Arch. ges. Physiol. **275**, 478—494 (1962).
55. Bouman, M.A.: My image of the retina. Quart. Rev. Biophys. **2**, 25—64 (1969).
56. — Koenderink, J.J.: Psychophysical basis of coincidence mechanisms in the human visual system. Ergebn. Physiol. **65**, 126—172 (1972).
57. Boynton, R.M.: Some temporal factors in vision. In: Rosenblith, W.A. (Ed.): Sensory Communication. Cambridge/Mass.: M.I.T. Press 1961.
57a. — Retinal contrast mechanisms. In: Young, F.A., Lindsley, D.B. (Eds.): Early Experience and Visual Information Processing in Perceptual and Reading Disorders, pp. 95—118. Washington, D. C.: Nat. Acad. Sci. 1970.
58. Brandt, Th., Dichgans, J., Koenig, E.: Perception of self-rotation (circular-vection) induced by optokinetic stimuli. Pflügers Arch. ges. Physiol. **332**, R 98 (1972).
58a. — — — Differential effects of central versus peripheral vision on egocentric and exocentric motion perception. Exp. Brain Res. **16**, 476—491 (1973).
59. — Wist, E., Dichgans, J.: Optisch induzierte Pseudocoriolis-Effekte und Circularvektion. Ein Beitrag zur optisch-vestibulären Interaktion. Arch. Psychiatr. Nervenkr. **214**, 365—389 (1971).
60. Breitmeyer, B.G.: A relationship between the detection of size, rate, orientation, and direction in the human visual system. Vision Res. **13**, 41—58 (1973).
61. Bremer, F.: Etude électrophysiologique de la convergence binoculaire dans l'aire visuelle corticale du chat. Arch. ital. Biol. **102**, 333—371 (1964).
62. Brindley, G.S.: Physiology of the retina and visual pathway. 2nd. Ed. London: Edward Arnold Ltd. 1970.
63. — Lewin, W.S.: The sensations produced by electrical stimulation of the visual cortex. J. Physiol. (Lond.) **196**, 479—493 (1968).
64. Brodmann, K.: Vergleichende Lokalisationslehre der Großhirnrinde in ihren Prinzipien dargestellt auf Grund des Zellenbaues. Leipzig: J. A. Barth 1909.
65. Brooks, B.A.: Neurophysiological correlates of brightness discrimination in the squirrel monkey. Exp. Brain Res. **2**, 1—17 (1966).
66. — Bohn, H.: Activity in the optic tract and lateral geniculate nucleus of the cat during the first moments of light adaptation in the scotopic region. Exp. Brain Res. **11**, 213—228 (1970).
67. — Huber, C.: Evidence for the role of the transient neural "off-response" in perception of light decrement. A psychophysical test derived from neuronal data in the cat. Vision Res. **12**, 1291—1297 (1972).
68. Brücke, E.: Über den Nutzeffekt intermittierender Netzhautreizungen. S.-B. Akad. Wiss. Wien, math.-nat. Kl. **49** (II), 128—153 (1864).
69. Bryngdahl, O.: Visual transfer characteristics from Mach band measurements. Kybernetik **2**, 71—77 (1964).
70. — Eine neue Methode zur Bestimmung der Übertragungseigenschaften des Gesichtssinnes. Naturwissenschaften **51**, 177—180 (1964).
71. Bullock, T.H.: The reliability of neurons. J. gen. Physiol. **55**, 565—584 (1970).
72. Burns, B.D.: The uncertain nervous system. London: Edward Arnold Ltd. 1968.

73. BUSER, P., ANGYAN, L., KITSIKIS, A., MITOVA, L., RICHARD, D., WIESENDANGER, M.: Liaisons fonctionnelles entre cortex visuel et cortex moteur chez le chat: bases neurophysiologiques de la coordination visuomotrice. Rev. Can. Biol. **31**, 103—114 (1972).
74. — BRUNER, J., SINDBERG, R.: Influences of the visual cortex upon posteromedial thalamus in the cat. J. Neurophysiol. **26**, 677—691 (1963).
74a. BÜTTNER, U., FUCHS, A.F.: Influence of saccadic eye movement on unit activity in simian lateral geniculate and pregeniculate nuclei. J. Neurophysiol. **36**, 127—141 (1973).
75. CAMPBELL, F.W.: Trends in physiological optics. In: REICHARDT, W. (Ed.): Processing of Optical Data by Organisms and by Machines, pp. 137—143. Varenna: Academic Press 1969.
76. — COOPER, G.F., ENROTH-CUGELL, CH.: The spatial selectivity of the visual cells of the cat. J. Physiol. (Lond.) **203**, 223—235 (1969).
77. — KULIKOWSKI, J.J.: Orientation selectivity of the human visual system. J. Physiol. (Lond.) **187**, 437—445 (1966).
78. — MAFFEI, L.: Electrophysiological evidence for the existence of orientation and size detectors in the human visual system. J. Physiol. (Lond.) **207**, 635—652 (1970).
79. CHANG, H.T.: Fibre groups in primary optic path ways of cat. J. Neurophysiol. **19**, 224—231 (1956).
80. CHARPENTIER, A.: Réaction oscillatoire de la rétine sous l'influence des excitations lumineuses. Arch. Physiol. (Paris) **24**, 541—553 (1892).
81. CHOW, K.L., RIESEN, A.H., NEWELL, F.W.: Degeneration of retinal ganglion cells in infant chimpanzees reared in darkness. J. comp. Neurol. **107**, 27—42 (1957).
82. CLARE, M.H., BISHOP, G.H.: Responses from an association area secondarily activated from optic cortex. J. Neurophysiol. **17**, 271—277 (1954).
83. CLELAND, B.G., DUBIN, M.W., LEVICK, W.R.: Sustained and transient neurones in the cat's retina and lateral geniculate nucleus. J. Physiol. (Lond.) **217**, 473—496 (1971).
84. — ENROTH-CUGELL, C.: Quantitative aspects of gain and latency in the cat retina. J. Physiol. (Lond.) **206**, 73—91 (1970).
85. COHEN, B., FELDMANN, M.: Potential changes associated with rapid eye movement in the calcarine cortex. Exp. Neurol. **31**, 100—113 (1971).
85a. — HENN, V.: The origin of quick phases of nystagmus in the horizontal plane. Bibl. ophthal. (Basel) **82**, 36—55 (1972).
85b. COHN, R.: Laminar electrical responses in lateral geniculate body of cat. J. Neurophysiol. **19**, 317—324 (1956).
86. CORNSWEET, T.N.: Visual perception. New York-London: Academic Press 1970.
87. COWEY, A., GROSS, C.G.: Effects of foveal prestriate and inferotemporal lesions on visual discrimination by rhesus monkeys. Exp. Brain Res. **11**, 128—144 (1970).
88. CRAIK, K.J.W.: The nature of explanation. Cambridge: Univ. Press 1943
89. — The nature of psychology. A selection of papers, essays and other writings by the late K.J.W. Craik, SHERWOOD, S.L. (Ed.): Cambridge: University Press 1966.
90. CREUTZFELDT, O., AKIMOTO, M.: Konvergenz und gegenseitige Beeinflussung aus der Retina und den unspezifischen Thalamuskernen an einzelnen Neuronen des optischen Cortex. Arch. Psychiat. Nervenkr. **196**, 520—538 (1957/58).
91. — FUSTER, J.M., HERZ, A., STRASCHILL, M.: Some problems of information transmission in the visual system. In: ECCLES, J.C. (Ed.): Brain and Conscious Experience, pp. 138 to 164. Berlin-Heidelberg-New York: Springer 1966.
92. — ITO, M.: Functional synaptic organization of primary visual cortex neurones in the cat. Exp. Brain Res. **6**, 324—352 (1968).
93. — SAKMANN, B.: Neurophysiology of vision. Ann. Rev. Physiol. **31**, 499—544 (1969).
94. CUNITZ, R.J., STEINMAN, R.M.: Comparison of saccadic eye movements during fixation and reading. Vision Res. **9**, 683—693 (1969).
95. DAW, N.W., PEARLMAN, A.L.: Cat colour vision: evidence for more than one cone process. J. Physiol. (Lond.) **211**, 125—137 (1970).
96. DENNEY, D., ADORJANI, C.: Orientation specificity of visual cortical neurons after head tilt. Exp. Brain Res. **14**, 312—317 (1972).
97. — BAUMGARTNER, G., ADORJANI, C.: Responses of cortical neurones to stimulation of the visual afferent radiations. Exp. Brain Res. **6**, 265—272 (1968).
98. DEUTSCH, S.: Models of the nervous system. New York-London: J. Wiley & Sons, Inc. 1967.

99. De Valois, R. L.: Contribution of different lateral geniculate cell types to visual behavior. Vision Res. Suppl. 3, 383—396 (1971).
100. — Jacobs, G. H.: Primate color vision. Science 162, 533—540 (1968).
101. Dichgans, J., Bizzi, E. (Eds.): Cerebral control of eye movements and motion perception. Bibl. ophthal. (Basel) 82, 1—403 (1972).
102. — — Personal communication 1972.
103. — Brandt, Th.: Visual-vestibular interaction and motion perception. Bibl. ophthal. (Basel) 82, 327—338 (1972).
103a. — Held, R., Young, L. R., Brandt, Th.: Moving visual scenes influence the apparent direction of gravity. Science 178, 1217—1219 (1972).
104. — Jung, R.: Attention, eye movements and motion detection: Facilitation and selection in optokinetic nystagmus and railway nystagmus. In: Evans, C. R., Mulholland, T. B. (Eds.): Attention in Neurophysiology, pp. 348—375. London: Butterworth 1969.
105. — Körner, F., Voigt, K.: Vergleichende Skalierung des afferenten und efferenten Bewegungssehens beim Menschen: Lineare Funktionen mit verschiedener Anstiegssteilheit. Psychol. Forsch. 32, 277—295 (1969).
106. — Nauck, B., Wolpert, E.: The influence of attention, vigilance and stimulus area on optokinetic and vestibular nystagmus and voluntary saccades. In: Zikmund (Ed.): The oculomotor system and brain functions. pp. 281—294. Bratislava: Slovak Acad. Sci. 1973.
107. — Schmidt, C. L., Wist, E. R.: Frequency modulation of afferent and efferent unit activity in the vestibular nerve by oculomotor impulses. Progr. Brain Res. 37, 449—456 (1972).
108. Dodge, E.: Visual perception during eye movement. Psychol. Rev. 7, 454—465 (1900).
109. — The participation of the eye movements in the visual perception of motion. Psychol. Rev. 11, 1—14 (1904).
110. Doty, R. W.: Functional significance of the topographical aspects of the retino-cortical discussion. In: Jung, R., Kornhuber, H. H. (Eds.): The Visual System: Neurophysiology and Psychophysics, pp. 228—247. Berlin-Göttingen-Heidelberg: Springer 1971.
111. Duensing, F.: Die Erregungskonstellation im Rautenhirn des Kaninchens bei den Labyrynthstellreflexen (Magnus). Naturwissenschaften 48, 581—690 (1961).
112. Dumont, S., Dell, P.: Facilitations spécifiques et non-spécifiques des réponses visuelles corticales. J. Physiol. (Paris) 50, 261—264 (1958).
113. Ebbecke, U.: Receptorenapparat und entoptische Erscheinungen. In: Bethe, A., Bergmann, G. von, Embden, G., Ellinger, A. (Hrsg.): Handbuch der normalen und pathologischen Physiologie. Receptionsorgane II, pp. 233—265. Berlin: Springer 1929.
114. Eccles, J. C.: Facing reality. Philosophical adventures by a brain scientist. Berlin-Heidelberg-New York: Springer 1970.
114a.— Functional significance of arrangement of neurones in cell assemblies. Arch. Psychiat. Nervenkr. 215, 92—106 (1971).
115. Eckmiller, R., Grüsser, O.-J.: Electronic simulation of the velocity function of movement-detecting neurons. Bibl. ophthal. 82, (Basel) 274—279 (1972).
116. Ehrenstein, W.: Beiträge zur ganzheitspsychologischen Wahrnehmungslehre. Leipzig: J. A. Barth 1942.
117. — Probleme der ganzheitspsychologischen Wahrnehmungslehre. (3. Aufl.). Leipzig: J. A. Barth 1954.
118. Enoch, J. M., Berger, R., Birns, R.: A static perimetric technique believed to test receptive field properties: extension and verification of the analysis. Docum. ophthal. (Den Haag) 29, 127—153 (1970).
119. — — — A static perimetric technique believed to test receptive field properties: responses near visual field lesions with sharp borders. Docum. ophthal. (Den Haag) 29, 154—167 (1970).
120. — Sunga, R. N.: Development of quantitative perimetric tests. Docum. ophthal. (Den Haag) 26, 215—229 (1969).
121. — — Neue Wege der quantitativen Perimetrie. Albrecht v. Graefes Arch. klin. exp. Ophthal. 179, 259—270 (1970).
122. Enroth-Cugell, C., Robson, J. G.: The contrast sensitivity of retinal ganglion cells of the cat. J. Physiol. (Lond.) 187, 517—552 (1966).

122a. ENROTH-CUGELL, C., PINTO, L. H.: Properties of the surround response mechanism of cat retinal ganglion cells and centre-surround interaction. J. Physiol. **220**, (Lond.) 403—439 (1972).
123. ERULKAR, S. D., FILLENZ, M.: The effects of light flashes on single unit activity in the lateral geniculate body of the cat. J. Physiol. (Lond.) **133**, 46 P (1956).
124. EVANS, C. C.: Spontaneous excitation of the visual cortex and association areas — lambda waves. Electroenceph. clin. Neurophysiol. **5**, 69—74 (1953).
125. EVARTS, E. V.: Activity of neurons in visual cortex of cat during sleep with low voltage fast EEG activity. J. Neurophysiol. **25**, 812—816 (1962).
126. — A neurophysiological theory of hallucinations. In: WEST, L. J. (Ed.): Hallucinations, pp. 1—14. New York-London: Grune & Stratton 1962.
127. — Photically evoked responses in visual cortex units during sleep and waking. J. Neurophysiol. **26**, 229—248 (1963).
128. — LANDAU, W., FREYGANG, W., JR., MARSHALL, W. H.: Some effects of lysergic acid diethylamide and bufotenine on electrical activity in the cat's visual system. Amer. J. Physiol. **182**, 594—598 (1955).
129. EYSEL, U. TH., GRÜSSER, O.-J.: Neurophysiological basis of pattern recognition in the cat's visual system. In: GRÜSSER, O.-J., KLINKE, R. (Hrsg.): Zeichenerkennung durch biologische und technische Systeme. Berlin-Heidelberg-New York: Springer 1971.
130. FECHNER, G. T.: Elemente der Psychophysik, Teil 1 und 2. Leipzig: Breitkopf und Härtel 1860.
131. FIDANZATI, G.: Some suggestions about the orientation of oblong shaped fields in the visual system. Atti Fond. G. Ronchi **1**, 97—101 (1972).
132. FIORENTINI, A., BAYLY, E. J., MAFFEI, L.: Peripheral and central contributions to psychophysical spatial interactions. Vision Res. **12**, 253—259 (1972).
133. — MAFFEI, L.: Transfer characteristics of excitation and inhibition in the human visual system. J. Neurophysiol. **33**, 285—292 (1970).
134. FISCHER, B.: Personal communication. 1972.
135. — Optische und neuronale Grundlagen der visuellen Bildübertragung: einheitliche mathematische Behandlung des retinalen Bildes und der Erregbarkeit von retinalen Ganglienzellen mit Hilfe der linearen Systemtheorie. Vision Res. **12**, 1125—1144 (1972).
136. — A neuron field theory. Mathematical approaches to the problem of large numbers of interacting nerve cells. Bull. math. Biophys. **35** (in press 1972).
137. — FREUND, H. J.: Eine mathematische Formulierung für Reiz-Reaktionsbeziehungen retinaler Ganglienzellen. Kybernetik **7**, 160—166 (1970).
138. — KRAUSE, D., MAY, H. U.: Schwellenerregung, zeitliche Summation und Impulsreaktionsfunktion in der Retina der Katze: Temporale rezeptive Felder retinaler Ganglienzellen. Exp. Brain Res. **15**, 212—224 (1972).
139. — KRÜGER, J.: Zentrumsbeleuchtung als Bedingung für die antagonistische Feldorganisation retinaler Ganglienzellen nach Dunkeladaptation der übrigen Retina. Pflügers Arch. ges. Physiol. **335**, R 81 (1972).
140. — MAY, H. U.: Invarianzen in der Katzenretina: Gesetzmäßige Beziehungen zwischen Empfindlichkeit, Größe und Lage receptiver Felder von Ganglienzellen. Exp. Brain Res. **11**, 448—464 (1970).
141. FISCHER, M. H.: Elektrobiologische Erscheinungen an der Hirnrinde. Pflügers Arch. ges. Physiol. **230**, 161—178 (1932).
142. — KORNMÜLLER, A. E.: Optokinetisch ausgelöste Bewegungswahrnehmungen und optokinetischer Nystagmus. J. Psychol. Neurol. **41**, 273—308 (1930).
143. — — Optische egozentrische Richtungslokalisation. Z. Sinnesphysiol. **61**, 87—147 (1930/31).
143a. FREUND, H.-J., Personal communication 1972.
144. — BAUMGARTNER, G.: Beeinflussung der neuronalen Reaktion genikulärer on-Zentrum-Neurone durch retinale off-Zentrum-Neurone. Pflügers Arch. ges. Physiol. **312**, R 137 (1969).
145. — GRÜNEWALD, G., BAUMGARTNER, G.: Räumliche Summation im receptiven Feldzentrum von Neuronen des Geniculatum laterale der Katze. Exp. Brain Res. **8**, 53—65 (1969).
146. — WITA, C. W., BRÜSTLE, R.: Latency differences between inhibitory and excitatory responses of cat optic tract units. Exp. Brain Res. **16**, 60—74 (1972).

147. Fröhlich, F. W.: Die Empfindungszeit. Jena: Gustav Fischer 1929.
148. Fuchs, A. F., Luschei, E. S.: Unit activity in the brainstem related to eye movement. Possible inputs to the motor nuclei. Bibl. ophthal. (Basel) **82**, 17—27 (1972).
149. Furmann, G. G.: Comparison of models for subtractive and shunting lateral inhibition in receptor-neuron fields. Kybernetik **2**, 257—274 (1965).
150. Gaarder, K. R.: Eye movements and perception. In: Young, F. A., Lindsley, D. B. (Eds.): Early Experience and Visual Information Processing in Perceptual and Reading Disorders, pp. 79—94. Washington, D. C.: Nat. Acad. Sci. 1970.
151. Gaarder, K., Koresko, R., Kropfl, W.: The phasic relation of a component of alpha rhythm to fixation saccadic eye movements. Electroenceph. clin. Neurophysiol. **21**, 544—551 (1966).
152. Galton, F.: Inquiries into human faculty and its development. London: Macmillan 1883.
153. Ganz, L.: Lateral inhibition and the location of visual contours: An analysis of figural after effects. Vision Res. **4**, 465—481 (1964).
154. Garey, L. J., Jones, E. G., Powell, T. P. S.: Interrelationships of striate and extrastriate cortex with the primary relay sites of the visual pathway. J. Neurol. Neurosurg. Psychiat. **31**, 135—157 (1968).
155. — Powell, T. P. S.: An experimental study of the termination of the lateral geniculo-cortical pathway in the cat and monkey. Proc. roy. Soc. B **179**, 41—63 (1971).
156. Gastaut, Y.: Un signe électroencéphalographique peu connu: les pointes occipitales survenant pendant l'ouverture des yeux. Rev. Neurol. **84**, 640—643 (1951).
157. Gazzaniga, M. S.: The bisected brain. New York: Appleton-Century-Crofts 1970.
158. Gerrits, H. J. M., Vendrik, A. J. H.: Artificial movements of a stabilized image. Vision Res. **10**, 1443—1456 (1970).
159. — — Simultaneous contrast, filling-in process and information processing in man's visual system. Exp. Brain Res. **11**, 411—430 (1970).
160. Gibson, J.: The perception of the visual world. Boston: Houghton Mifflin 1950.
161. Glezer, V. D.: The receptive fields of the retina. Vision Res. **5**, 497—525 (1965).
162. Goethe, J. W. von: Zur Farbenlehre. (1. Ausgabe 1810.) In: Goethes Werke. Vollständige Ausgabe letzter Hand. 52. Bd. Stuttgart-Tübingen: J. G. Cotta 1833.
163. Gombrich, E. H.: Art and illusion. 2nd Ed. London: Phaidon 1962.
164. Götz, K. G.: Die optischen Übertragungseigenschaften der Komplexaugen von *Drosophila*. Kybernetik **2**, 215—221 (1965).
165. Gouras, P.: Trichromatic mechanisms in single cortical neurons. Science **168**, 489—492 (1970).
166. — The function of the midget cell system in primate color vision. Vision Res. Suppl. **3**, 397—410 (1971).
167. — Colour opponency from fovea to striate cortex. Invest. Ophthal. **7**, 427—434 (1972).
168. — Personal communication. Paper by Daw, B. M., and Gouras, P. submitted to J. Neurophysiol. 1972.
169. Graham, C. H. (Ed.), Bartlett, N. R., Brown, J. L., Hsia, Y., Mueller, C. G., Riggs, L. A.: Vision and visual perception. New York-London-Sidney: J. Wiley & Sons, Inc. 1965.
170. Granit, R.: Isolation of colour-sensitive elements in a mammalian retina. Acta physiol. scand. **2**, 93—109 (1941).
171. — Sensory mechanisms of the retina with an appendix on electroretinography. London-New York-Toronto: G. Cumberledge, Oxford Univ. Press 1947.
172. — The organization of the vertebrate retinal elements. Ergebn. Physiol. **46**, 31—70 (1950).
173. — Receptors and sensory perception. London: Yale Univ. Press 1955.
174. — In defence of teleology. In: Karczmar, A. G., Eccles, J. C. (Eds.): Brain and Human Behavior, pp. 400—408. Berlin-Heidelberg-New York: Springer 1972.
175. — Svaetichin, G.: Principles and technique of the electrophysiological analysis of colour reception with the aid of microelectrodes. Upsala Läk.-Fören. Förh. **65**, 161—177 (1939).
176. Gregory, R. L.: Eye and brain. London: World University Library 1966.
177. — The intelligent eye. London: Weidenfeld and Nicolson 1970.

178. GROSS, C. H., BENDER, D. B., ROCHA-MIRANDA, C. E.: Visual receptive fields of neurons in inferotemporal cortex of the monkey. Science **166**, 1303—1306 (1969).

178a. GRÜSSER, O.-J.: A quantitative analysis of spatial summation of excitation and inhibition within the receptive field of retinal ganglion cells of cats. Vision Res. Suppl. **3**, 103—127 (1971).

179. — CREUTZFELDT, O.: Eine neurophysiologische Grundlage des Brücke-Bartley-Effekts: Maxima der Impulsfrequenz retinaler und corticaler Neurone bei Flimmerlicht mittlerer Frequenzen. Pflügers Arch. ges. Physiol. **263**, 668—681 (1957).

180. — GRÜSSER-CORNEHLS, U.: Mikroelektrodenuntersuchungen zur Konvergenz vestibulärer und retinaler Afferenzen an einzelnen Neuronen des optischen Cortex der Katze. Pflügers Arch. ges. Physiol. **270**, 227—238 (1960).

181. — — Periodische Aktivierungsphasen visueller Neurone nach kurzen Lichtreizen verschiedener Dauer. Beziehungen zu den periodischen Nachbildern und dem Charpentier-Intervall. Pflügers Arch. ges. Physiol. **275**, 292—311 (1962).

182. — — Neurophysiologie des Bewegungssehens. Bewegungsempfindliche und richtungsspezifische Neurone im visuellen System. Ergebn. Physiol. **61**, 178—265 (1969).

183. — — Interaction of vestibular and visual inputs in the visual system. In: BRODAL, A., POMPEIANO, O. (Eds.): Progress in Brain Research, pp. 573—583. Amsterdam-London-New York: Elsevier 1972.

184. — GRÜTZNER, A.: Reaktionen einzelner Neurone des optischen Cortex der Katze nach elektrischen Reizserien des Nervus opticus. Arch. Psychiat. Nervenkr. **197**, 405—432 (1958).

185. — PETERSEN, A., SASOWSKI, R.: Neurophysiologische Grundlagen des Metakontrastes. Pflügers Arch. ges. Physiol. **283**, R 50 (1965).

186. — RABELO, C.: Die Wirkung von Flimmerreizen mit Lichtblitzen an einzelnen corticalen Neuronen. In: BOGAERT, L. v., RADERMECKER, J. (Eds.): Proc. 1st Int. Congr. Neurol. Sci., Vol. 3, Electroencephalography, Clinical Neurophysiol. and Epilepsy, pp. 371—375. London-New York-Paris: Pergamon 1959.

187. — SNIGULA, F.: Vergleichende verhaltensphysiologische und neurophysiologische Untersuchungen am visuellen System von Katzen. II. Simultankontrast. Psychol. Forsch. **32**, 43—63 (1968).

188. HABER, R. N.: How we remember what we see. Sci. Amer. **222**, 104—115 (1970).

188a. HAJOS, A.: Sinnespsychologische Untersuchungen zur Invarianzbildung im visuellen System des Menschen. In: GRÜSSER, O.-J., KLINKE, R. (Hrsg.): Zeichenerkennung durch biologische und technische Systeme, pp. 98—113. Berlin-Heidelberg-New York: Springer 1971.

189. HALLIDAY, A. M., MCDONALD, W. I., MUSHIN, J.: Delayed visual evoked response in optic neuritis. Lancet **1972 II**, 982—985.

190. HARMS, H., AULHORN, E.: Studien über den Grenzkontrast. I. Mitteilung. Ein neues Grenzphänomen. Albrecht v. Graefes Arch. Ophthal. **157**, 3—23 (1955).

191. HARTLINE, H. K.: Impulses in single optic nerve fibres of the vertebrate retina. Amer. J. Physiol. **113**, 59 P (1935).

191a. — The discharge of impulses in the optic nerve of Pecten in response to illumination of the eye. J. cell. comp. Physiol. **11**, 465—477 (1938).

192. — The response of single optic nerve fibres of the vertebrate eye to illumination of the retina. Amer. J. Physiol. **121**, 400—415 (1938).

193. — The receptive fields of optic nerve fibres. Amer. J. Physiol. **130**, 690—699 (1940).

194. — The effects of spatial summation in the retina on the excitation of the fibers of the optic nerve. Amer. J. Physiol. **130**, 700—711 (1940).

195. — Inhibition of activity of visual receptors by illuminating nearby retinal areas in the Limulus eye. Fed. Proc. **8**, 69 (1949).

195a. — RATLIFF, F.: Inhibitory interaction in the retina of Limulus. In: FUORTES, M. G. F. (Ed.): Handbook of Sensory Physiology, Vol. VII/2, pp. 381—447. Berlin-Heidelberg-New York: Springer 1972.

196. HARTMANN, N.: Diesseits von Idealismus und Realismus. Ein Beitrag zur Scheidung des Geschichtlichen und Übergeschichtlichen in der Kantischen Philosophie. Kant Studien **29**, 106—206 (1924).

197. HARTMANN, N.: Der Aufbau der realen Welt. Grundriß der allgemeinen Kategorienlehre. Berlin: W. de Gruyter 1940.
198. — Philosophie der Natur. Abriß der speziellen Kategorienlehre. Berlin: W. de Gruyter 1950.
199. HASSENSTEIN, B.: Funktionsanalyse der Bewegungsperzeption eines Käfers. Naturwissenschaften **38**, 507 (1951).
200. — Wie sehen Insekten Bewegungen? Naturwissenschaften **48**, 207—214 (1961).
201. — Modellrechnung zur Datenverarbeitung beim Farbensehen des Menschen. Kybernetik **4**, 209—223 (1968).
201a. — Überlegungen zur Datenverarbeitung beim Farbensehen. In: DRISCHEL, H., TIEDT, N. (Hrsg.): Biokybernetik III, pp. 242—246. Jena: VEB Gustav Fischer 1971.
202. HASSLER, R.: Comparative anatomy of the central visual systems in day- and night-active primates. In: HASSLER, R., STEPHAN, H. (Eds.): Evolution of the Forebrain, pp. 419—434. Stuttgart: H. Thieme 1966.
203. — WAGNER, A.: Experimentelle und morphologische Befunde über die vierfache corticale Projektion des visuellen Systems. Proc. 8. Int. Congr. Neurol. **3**, 77—96, Wien 1965.
203a. HEIN, A.: L'aquisition de la coordination perceptivomotrice et sa réaquisition après lésion du cortex visuell, pp. 123—136. In: HECAEN, H. (Ed.): Neurophysiologie de la perception visuelle. Paris: Masson et Cie. 1971.
204. HELD, R.: Dissociation of visual functions by deprivation and rearrangement. Psychol. Forsch. **31**, 338—348 (1968).
205. — Two models of processing spatially distributed visual stimulation. In: SCHMITT, F. O. (Ed.): The Neurosciences: Second Study Program, pp. 317—324. New York: Rockefeller Univ. Press 1970.
206. — HEIN, A.: Movement produced stimulation in the development of visually-guided behavior. J. comp. physiol. Psychol. **56**, 872—876 (1963).
207. HELMHOLTZ, H.: Beschreibung eines Augen-Spiegels zur Untersuchung der Netzhaut im lebenden Auge. Berlin: A. Förstner 1851.
208. HELMHOLTZ, H. VON: Handbuch der physiologischen Optik (1. Aufl. 1859/67). 2. Aufl. Hamburg-Leipzig: G. Voss 1896.
209. — Optisches über Malerei (Vorträge 1871—73). In: Vorträge und Reden, Bd. 2, S. 93 bis 135, 5. Aufl. Braunschweig: Vieweg 1903.
210. HERING, E.: Zur Lehre von der Beziehung zwischen Leib und Seele. I. Über Fechner's psychophysisches Gesetz. S.-B. Akad. Wien, math.-naturw. Kl. Abt. III. **72**, 310—348 (1876).
211. — Zur Lehre vom Lichtsinne. Wien: Gerold u. Söhne 1878.
212. — Der Raumsinn und die Bewegungen der Augen. In: HERMANN, L. (Hrsg.): Handbuch der Physiologie, Bd. 3, S. 343—601. Leipzig: F. C. W. Vogel 1879.
213. — Grundzüge der Lehre vom Lichtsinn. Berlin: Springer 1920.
214. — Wissenschaftliche Abhandlungen, Bd. I u. II. Leipzig: G. Thieme 1931.
215. HERMANN, L.: Eine Erscheinung des simultanen Contrastes. Pflügers Arch. ges. Physiol. **3**, 13—15 (1870).
216. HESS, W. R.: Die Motorik als Organisationsproblem. Biol. Zbl. **61**, 545—572 (1941).
217. HIRSCH, H. V. B., SPINELLI, D. N.: Modification of the distribution of receptive field orientation in cats by selective visual exposure during development. Exp. Brain Res. **13**, 509—527 (1971).
218. HOFFMANN, K. P., STONE, J.: Conduction velocity of afferents to cat visual cortex: a correlation with cortical receptive field properties. Brain Res. **32**, 460—466 (1971).
219. HOFFMAN, W. C.: Pattern recognition by the method of isoclines: I. A mathematical model for the visual integrative process. Mathematical Note Nr. 351 (Ed. Mathematics Res. Lab., Boeing Sci. Res. Labs.) DI-82-0351 (1964).
220. — The lie algebra of visual perception. J. math. Psychol. **3**, 65—98 (1966).
221. — Higher visual perception as prolongation of the basic lie transformation group. Math. Biosci. **6**, 437—471 (1970).
222. — Memory grows. Kybernetik **8**, 151—157 (1971).
223. HOFMANN, F. B.: Die Lehre vom Raumsinn des Doppelauges. Ergebn. Physiol. **15**, 238 bis 339 (1915).

224. HOLMGREN, F.: Method att objectivera effecten av ljusintryck pa retina. Upsala Läk.-Fören. Förh. **1**, 177—191 (1865—1866).
225. HOLST, E. VON: Zentralnervensystem und Peripherie in ihrem gegenseitigen Verhältnis. Klin. Wschr. **29**, 97—105 (1951).
226. — Aktive Leistungen der menschlichen Gesichtswahrnehmung. Stud. Gen. **10**, 231—243 (1957).
227. — MITTELSTAEDT, H. v.: Das Reafferenzprinzip (Wechselwirkung zwischen Zentralnervensystem und Peripherie). Naturwissenschaften **37**, 464—476 (1950).
228. HOMER, W. I.: Seurat and the science of painting. Cambridge/Mass.: M. I. T. Press 1964.
229. HORN, G., STECHLER, G., HILL, R. M.: Receptive fields of units in the visual cortex of the cat in the presence and absence of bodily tilt. Exp. Brain Res. **15**, 113—132 (1972).
230. HUBEL, D. H.: Single unit activity in striate cortex of unrestrained cats. J. Physiol. (Lond.) **147**, 226—238 (1959).
231. — WIESEL, T. N.: Receptive fields of single neurones in the cat's striate cortex. J. Physiol. (Lond.) **148**, 574—591 (1959).
232. — — Receptive fields of optic nerve fibres in the spider monkey. J. Physiol. (Lond.) **154**, 572—580 (1960).
233. — — Receptive fields, binocular interaction and functional architecture in the cat's visual cortex. J. Physiol. (Lond.) **160**, 106—154 (1962).
234. — — Receptive fields and functional architecture in two nonstriate visual areas (18 and 19) of the cat. J. Neurophysiol. **28**, 229—289 (1965).
235. — — Receptive fields and functional architecture of monkey striate cortex. J. Physiol. (Lond.) **195**, 215—243 (1968).
236. — — Anatomical demonstration of columns in the monkey striate cortex. Nature (Lond.) **221**, 747—750 (1969).
237. — — Visual area of the lateral suprasylvian gyrus (Clare-Bishop-Area) of the cat. J. Physiol. (Lond.) **202**, 251—260 (1969).
237a. — — Cells sensitive to binocular depth in area 18 of the macaque monkey cortex. Nature (Lond.) **225**, 41—42 (1970).
238. HURVICH-JAMESON, D., HURVICH, L. M.: Opponent-colors theory and physiological mechanisms. In: JUNG, R., KORNHUBER, H. H. (Hrsg.): Neurophysiologie und Psychophysik des visuellen Systems, pp. 152—163. Berlin-Göttingen-Heidelberg: Springer 1961.
239. JAMES, W.: Principles of psychology. New York: H. Holt 1890.
240. JASPER, H. H., CRUIKSHANK, R. M.: Electroencephalography. II. Visual stimulation and after-images as effecting the occipital alpha rhythm. J. gen. Psychol. **17**, 29—48 (1937).
241. JEAVONS, P. M., HARDING, G. F. A., PANAYIOTOPOULOS, C. P., DRASDO, N.: The effect of geometric patterns combined with intermittent photic stimulation in photosensitive epilepsy. Electroenceph. clin. Neurophysiol. **33**, 221—224 (1972).
242. JONES, E. G.: Visual cortex: Structure and connection. U.S.-Australian Symp. on Vision. Invest. Ophthal. **11**, 333—337 (1972).
243. JOSHUA, D. E., BISHOP, P. O.: Binocular single vision and depth discrimination. Receptive field disparities for central and peripheral vision and binocular interaction on peripheral units in cat striate cortex. Exp. Brain Res. **10**, 389—426 (1970).
244. JULESZ, B.: Binocular depth perception without familarity cues. Science **145**, 356—362 (1964).
245. — Foundations of cyclopean perception. Chicago-London: Univ. of Chicago Press 1971.
246. JUNG, R.: Die Registrierung des postrotatorischen und optokinetischen Nystagmus und die optisch-vestibuläre Integration beim Menschen. Acta oto-laryng. (Stockh.) **36**, 199—201 (1948).
247. — Nystagmographie. Zur Physiologie und Pathologie des optisch-vestibulären System beim Menschen. In: v. BERGMANN, G., FREY, W., SCHWIECK, H. (Hrsg.): Handbuch d. inneren Medizin, 4. Aufl., Bd. **5**/1, S. 1325—1379. Berlin-Heidelberg-New York: Springer 1953.
248. — Neuronal integration in the visual cortex and its significance for visual information. In: ROSENBLITH, W. A. (Ed.): Sensory Communication. New York-London: M. I. T. Press 1961.

249. JUNG, R.: Korrelationen von Neuronentätigkeit und Sehen. In: JUNG, R., KORNHUBER, H.H. (Hrsg.): Neurophysiologie und Psychophysik des visuellen Systems, S. 410—435. Berlin-Heidelberg-New York: Springer 1961.
250. — Neuronale Grundlagen des Hell-Dunkelsehens und der Farbwahrnehmung. Ber. dtsch. ophthal. Ges. **66**, 69—111 (1964).
250a. — Neurophysiologie und Psychiatrie. In: GRUHLE, H.W., JUNG, R., MAYER-GROSS, W., MÜLLER, M. (Hrsg.): Psychiatrie der Gegenwart Forschung und Praxis, I/1 A, pp. 335—928. Berlin-Heidelberg-New York: Springer 1967.
251. — Neurophysiologie des Konturensehens und Graphik. In: BAMMER, H.G. (Hrsg.): Zukunft der Neurologie, S. 214—225. Stuttgart: G. Thieme 1967.
252. — Kontrastsehen, Konturbetonung und Künstlerzeichnung. Stud. Gen. **24**, 1536—1565 (1971).
253. — Neurophysiological and psychophysical correlates in vision research. In: KARCZMAR, A.G., ECCLES, J.C., (Eds.): Brain and Human Behavior, pp. 209—258. Berlin-Heidelberg-New York: Springer 1972.
254. — BAUMGARTEN, R. VON, BAUMGARTNER, G.: Mikroableitungen von einzelnen Neuronen im optischen Cortex der Katze: Die lichtaktivierten B-Neurone. Arch. Psychiat. Nervenkr. **189**, 521—538 (1952).
255. — BAUMGARTNER, G.: Hemmungsmechanismen und bremsende Stabilisierung an einzelnen Neuronen des optischen Cortex: Ein Beitrag zur Koordination corticaler Erregungsvorgänge. Pflügers Arch. ges. Physiol. **261**, 434—456 (1955).
256. — — Neuronenphysiologie der visuellen und paravisuellen Rindenfelder. 8. Int. Congr. Neurol. Wien Proc. **3**, 47—75 (1965).
257. — CREUTZFELDT, O., BAUMGARTNER, G.: Microphysiologie des neurones corticaux, processus de coordination et d'inhibition du cortex optique et moteur. Coll. Int. C.N.R.S. **67**, 411—457. In: FESSARD, A. (Ed.): Microphysiologie comparée des éléments excitables. Paris: Gif-sur-Yvette 1955.
258. — KORNHUBER, H.H., DA FONSECA, J.S.: Multisensory convergence on cortical neurons: Neuronal effects of visual, acoustic and vestibular stimuli in the superior convolutions of the cat's cortex. Progr. Brain Res. **1**, 207—240 (1963).
259. — SPILLMANN, L.: Receptive-field estimation and perceptual integration in human vision. In: YOUNG, F.A., LINDSLEY, D.B. (Eds.): Early Experience and Visual Information Processing in Perceptual and Reading Disorders, pp. 181—197. Washington, D. C.: Nat. Acad. Sci. 1970.
260. KANIZSA, G.: Margini quasi-percettivi in campi con stimolazione omogenea. Riv. Psicol. **49**, I, 7—30 (1955).
261. KANT, I.: Metaphysische Anfangsgründe der Naturwissenschaft. Riga: J. F. Hartknoch 1786.
262. — Critik der reinen Vernunft. 2. Aufl. (1. Ed. Riga 1781). Riga: J. F. Hartknoch 1787.
263. — Critik der Urtheilskraft. Berlin-Libau: Lagarde u. Friederich 1790.
264. KASAMATSU, T., KIYONO, S., IWAMA, K.: Electrical activities of the visual cortex in chronically blinded cats. Tohoku J. exp. Med. **93**, 139—152 (1967).
265. KAWAMURA, H., POMPEIANO, O.: Phasic D. C. potential shifts in the sensorimotor and visual cortices during desynchronized sleep. Pflügers Arch. ges. Physiol. **317**, 10—19 (1970).
266. KLÜVER, H.: Visual functions after removal of the occipital lobes. J. Psychol. **11**, 23—45 (1941).
267. — Mechanisms of hallucination. In: Studies in Personality, pp. 175—207. New York: McGraw-Hill Book Comp. 1942.
268. — Behaviour mechanisms in monkeys, 2. Ed. Chicago: University Press 1957.
269. KOFFKA, K.: Zur Grundlegung der Wahrnehmungspsychologie. Eine Auseinandersetzung mit Benussi. Z. Psychol. **73**, 11—90 (1915).
270. — Über Feldbegrenzung und Felderfüllung. Psychol. Forsch. **4**, 176—203 (1923).
271. KOHLER, I.: Über Aufbau und Wandlungen der Wahrnehmungswelt; insbesondere über „bedingte Empfindungen", S.-B. Österr. Akad. Wiss. Phil.-hist. Kl. **227/1**, 1—118. Wien: Rohrer 1951. (Engl. transl. with additions: Psychological Issues **3**, nr. 4 Monogr. **12**. The formation and transformation of the perceptual world. New York: Int. Univ. Press Inc. 1964).

271a. KOHLER, I.: Zentralnervöse Korrekturen in der Wahrnehmung. Naturwissenschaften **48**, 259—264 (1961).
272. KÖHLER, W.: Zur Theorie der stroboskopischen Bewegung. Psychol. Forsch. **3**, 397—407 (1923).
272a. — WALLACH, H.: Figural after-effects: An investigation of visual processes. Proc. Amer. Phil. Soc. **88**, 269—357 (1944).
273. KÖRNER, F., DICHGANS, J.: Bewegungswahrnehmung, optokinetischer Nystagmus und retinale Bildwanderung. Albrecht v. Graefe's Arch. klin. exp. Ophthal. **174**, 34—48 (1967).
274. KORNHUBER, H. H.: Physiologie und Klinik des zentralvestibulären Systems (Blick- und Stützmotorik). In: BERENDES, J., LINK, R., ZÖLLNER, F. (Hrsg.): Hals- Nasen-Ohrenheilkunde, ein kurzgefaßtes Handbuch, Vol. III, pp. 2150—2351. Stuttgart: G. Thieme 1966.
275. — Motor functions of cerebellum and basal ganglia: the cerebello-cortical saccadic (ballistic) clock, the cerebello-nuclear hold regulator, and the basal ganglia ramp (voluntary speed smooth movement) generator. Kybernetik **8**, 157—162 (1971).
275a. — Das vestibuläre System, mit Exkursen über die motorischen Funktionen der Formatio reticularis, des Kleinhirns, der Stammganglien und des motorischen Cortex sowie über die Raumkonstanz der Sehdinge. In: KEIDEL, PLATTIG (Hrsg.): Vorträge der Erlanger Physiologentagung 1970. Berlin-Heidelberg-New York: Springer 1971.
276. — DA FONSECA, J. S.: Optovestibular integration in the cat's cortex: a study of sensory convergence on cortical neurons. In: BENDER, M. B. (Ed.): The Oculomotor System, pp. 239—279. New York-Evanston-London: Hoeber, Harper & Row 1964.
277. — DEECKE, L.: Hirnpotentialänderungen bei Willkürbewegungen und passiven Bewegungen des Menschen: Bereitschaftspotential und reafferente Potentiale. Pflügers Arch. ges. Physiol. **284**, 1—17 (1965).
278. — SPILLMANN, L.: Zur visuellen Feldorganisation beim Menschen: Die rezeptiven Felder im peripheren und zentralen Gesichtsfeld bei Simultankontrast, Flimmerfusion, Scheinbewegung und Blickfolgebewegung. Pflügers Arch. ges. Physiol. **279**, R5—6 (1964).
279. KORNMÜLLER, A. E.: Architektonische Lokalisation bioelektrischer Erscheinungen auf der Großhirnrinde. I. Mitt.: Untersuchungen am Kaninchen bei Augenbelichtung. J. Psychol. Neurol. **44**, 447—459 (1932).
280. — TÖNNIES, J. F.: Registrierung der spezifischen Aktionsströme eines architektonischen Feldes der Großhirnrinde vom uneröffneten Schädel. Psychiat.-neurol. Wschr. **34**, 581 (1932).
281. KRIES, J. VON: Über Farbensysteme. Z. Psychol. Physiol. Sinnesorg. **13**, 241—324 (1897).
282. — Normale und anormale Farbsysteme. Die Theorien des Licht- und Farbensinns. Zusätze zu v. HELMHOLTZ Handbuch der physiologischen Optik, 3. Aufl. II, S. 333—378. Hamburg-Leipzig: L. Voss 1911.
283. — Allgemeine Sinnesphysiologie. Leipzig: F. C. W. Vogel 1923.
284. — Immanuel Kant und seine Bedeutung für die Naturforschung der Gegenwart. Berlin: Springer 1924.
285. KUFFLER, S. W.: Neurones in the retina: organization, inhibition and excitation problems. Cold Spr. Harb. Symp. quant. Biol. **17**, 281—292 (1952).
286. — Discharge patterns and functional organization of mammalian retina. J. Neurophysiol. **16**, 37—68 (1953).
287. KUYPERS, S. G., SZWARCBART, M. K., MISHKIN, M., ROSVOLD, J. E.: Occipitotemporal cortico-cortical connections in the rhesus monkey. Exp. Neurol. **11**, 245—262 (1965).
288. LASHLEY, K. S.: Patterns of cerebral integration indicated by the scotomas of migraine. Arch. Neurol. Psychiat. **46**, 331—339 (1941).
289. LEHMANN, D.: EEG, evoked potentials, and eye and image movements. In: BACH-Y-RITA, P., COLLINS, C. C. (Eds.): The Control of Eye Movements, pp. 149—174. New York-London: Academic Press 1971.
290. — BEELER, G. W., JR., FENDER, D. H.: Changes in patterns of the human electroencephalogram during fluctuations of perception of stabilized retinal images. Electroenceph. clin. Neurophysiol. **19**, 336—343 (1965).
291. LEIBOWITZ, H., MOTE, F. A., THURLOW, W. R.: Simultaneous contrast as a function of separation between test and inducing fields. J. exp. Psychol. **46**, 453—456 (1953).

292. LETTVIN, J. Y., MATURANA, H. R., PITTS, W. H., MCCULLOCH, W. S.: Two remarks on the visual system of the frog. In: ROSENBLITH, W. A. (Ed.): Sensory Communication, pp. 757—776. Cambridge/Mass.: M.I.T. Press. New York-London: J. Wiley & Sons, Inc. 1961.

293. LEVICK, W. R.: Receptive fields of retinal ganglion cells. In: FUORTES, M. G. F. (Ed.): Handbook of Sensory Physiology, Vol. VII/2, pp. 531—566. Berlin-Heidelberg-New York: Springer 1972.

294. LHERMITTE, F., CHAIN, F., ARON-ROSA, D., LEBLANC, M., SOUTY, O.: Enregistrement des mouvements du regard dans un cas d'agnosie visuelle et dans un cas de désorientation spatiale. Rev. neurol. **121**, 121—137 (1969).

295. LIEBIG, J. VON: Über Bacon und die Methode der Naturforschung. 1863. Induction und Deduction, 1865. In: Reden und Abhandlungen. S. 220—254, 296—309. Leipzig-Heidelberg 1874.

296. LINDSLEY, D. B.: Electrophysiology of the visual system and its relation to perceptual phenomena. In: BRAZIER, M. A. (Ed.): Brain and Behaviour, pp. 359—392. Washington D. C.: Nat. Sci. Found. 1961.

297. LIPETZ, L. E.: The relation of physiological and psychological aspects of sensory intensity. In: LOEWENSTEIN, W. R. (Ed.): Handbook of Sensory Physiology, Vol. I, pp. 191—225. Berlin-Heidelberg-New York: Springer 1971.

298. LORENTE DE NÓ, R.: Ausgewählte Kapitel aus der vergleichenden Physiologie des Labyrinthes. Die Augenmuskelreflexe beim Kaninchen und ihre Grundlagen. Ergebn. Physiol. **32**, 73—242 (1931).

299. — Analysis of the activity of the chains of internuncial neurons. J. Neurophysiol. **1**, 207—244 (1938).

300. — Transmission of impulses through cranial motor nuclei. J. Neurophysiol. **2**, 402—464 (1939).

301. LORENZ, K.: Die angeborenen Formen möglicher Erfahrung. Z. Tierpsychol. **5**, 235—409 (1943).

302. — Methods of approach to the problems of behavior. Harvey Lect. **54**, 60—103 (1958/59).

303. LOTZE, R. H.: Medicinische Psychologie (Physiologie der Seele). Leipzig: Weidmannsche Buchhandlung 1852.

304. MACH, E.: Über die Wirkung der räumlichen Verteilung des Lichtreizes auf die Netzhaut. I. S.-Ber. math.-naturw. Kl. d. Kaiserl. Akad. d. Wiss. **52/2**, 303—322 (1865).

304a. — Die Analyse der Empfindungen und das Verhältnis des Physischen zum Psychischen, 2. Aufl. Jena: Gustav Fischer 1900.

305. MACKAY, D. M.: Interactive processes in visual perception. In: ROSENBLITH, W. A. (Ed.): Sensory Communication, pp. 339—355. Cambridge/Mass.: M.I.T. Press. New York-London: J. Wiley & Sons, Inc. 1961.

306. — Psychophysics of perceived intensity: a theoretical basis for Fechner's and Steven's laws. Science **139**, 1213—1216 (1963).

307. — Cerebral organization and the conscious control of action. In: ECCLES, J. C. (Ed.): Brain and Conscious Experience, pp. 442—445. New York-Heidelberg-Berlin: Springer 1966.

307a. — Perception and brain function. In: SCHMITT, F. O. et al. (Eds.): The Neurosciences, Second Study Programme, pp. 303—316. New York: Rockefeller University Press 1970.

307b. — Personal communication. Nature (Lond.) (1973), in press.

308. MACLEAN, P. D.: The limbic system in relation to vision. Int. J. Neurol. (Montevideo) **6**, 297—305 (1968).

309. — CRESWELL, G.: Anatomical connections of visual system with limbic cortex of monkey. J. comp. Neurol. **138**, 265—278 (1970).

310. MCNICHOL, E. F., JR.: Retinal mechanisms of color vision. Vision Res. **4**, 119—133 (1964).

311. MAFFEI, L., FIORENTINI, A.: Retinogeniculate convergence and analysis of contrast. J. Neurophysiol. **35**, 65—72 (1972).

312. MARCHIAFAVA, P. L., POMPEIANO, O.: Excitability changes of the intrageniculate optic tract fibres produced by electrical stimulation of the vestibular system. Pflügers Arch. ges. Physiol. **290**, 275—278 (1966).

313. MARG, E., ADAMS, J. E., RUTKIN, B.: Receptive fields of cells in the human visual cortex. Experientia (Basel) **24**, 348—350 (1968).
314. MARIMONT, R. B.: Model for visual response to contrast. J. Opt. Soc. Amer. **52**, 800—806 (1962).
315. MARKO, H.: Die Systemtheorie der homogenen Schichten. I. Mathematische Grundlagen. Kybernetik **5**, 221—240 (1969).
316. MATIN, L.: Eye movements and perceived visual direction. In: JAMESON, D., HURVICH, L.M. (Eds.): Handbook of Sensory Physiology, Vol. VII/4, pp. 331—380. Berlin-Heidelberg-New York: Springer 1972.
317. MATURANA, H. R., LETTVIN, J. Y., McCULLOCH, W. S., PITTS, W. H.: Anatomy and physiology of vision in the frog (Rana pipiens). J. gen. Physiol. **43**, 129—176 (1960).
318. MAXWELL, J. C.: On the theory of compound colours and the relations of the colours of the spectrum. Phil. Transact. roy. Soc. **150**, 57—84 (1860).
319. MAY, U., FREUND, H.-J., JUNG, R.: Kollaterale Hemmung im Corpus geniculatum laterale der Katze bei Dunkeladaptation. Pflügers Arch. ges. Physiol. **312**, R136 (1969).
320. MAYHEW, J. E. W., ANSTIS, S. M.: Movement aftereffects contingent on color, intensity and pattern. Perception & Psychophysics **12**, 77—85 (1972).
321. McCULLOUGH, C.: Color adaptation of edge-detectors in the human visual system. Science **149**, 1115—1116 (1965).
322. McILWAIN, J. T.: Receptive fields of optic tract axons and lateral geniculate cells: peripheral extent and barbiturate sensitivity. J. Neurophysiol. **27**, 1154—1173 (1964).
323. MINKE, B., AUERBACH, E.: Latencies and correlation in single units and visual evoked potentials in the cat striate cortex following monocular and binocular stimulations. Exp. Brain Res. **14**, 409—422 (1972).
324. MISHKIN, M.: Visual mechanisms beyond the striate cortex. In: RUSSEL, R., (Ed.): Frontiers in Physiological Psychology. New York: Academic Press 1966
325. MITCHELL, D. E., BLAKEMORE, C.: Binocular depth perception and the corpus callosum. Vision Res. **10**, 49—54 (1970).
326. MONNIER, M.: Mesure de la durée d'un processus d'integration corticale: temps d'intégration optomotrice chez l'homme. Helv. physiol. pharmacol. Acta **7**, C52—53 (1949).
326a. MORASSO, P., BIZZI, E., DICHGANS, J.: Adjustment of saccade characteristics during head movements. Exp. Brain Res. **16**, 492—500 (1972/73).
327. MORREL, F.: Integrative properties of parastriate neurones. In: KARCZMAR, A. G., ECCLES, J.C. (Eds.): Brain and Human Behavior. Berlin-Heidelberg-New York: Springer 1972.
328. MOTOKAWA, K.: Field of retinal induction and optical illusion. J. Neurophysiol. **13**, 413—426 (1950).
329. — Color and pattern vision. Tokio: Igaku Shoin Ltd. Berlin-Heidelberg-New York: Springer 1970.
330. — EBE, M.: The physiological mechanism of apparent movement. J. exp. Psychol. **45**, 378—386 (1953).
331. — NAKAGAWA, D., KOHATA, T.: Figural after-effects and retinal induction. J. gen. Psychol. **57**, 121—135 (1957).
332. — OGAWA, T.: The electrical field in the retina and pattern vision. Tohoku J. exp. Med. **78**, 209—221 (1962).
333. MÜLLER, G. E.: Zur Analyse der Gedächtnistätigkeit und des Vorstellungsverlaufes. 3. Bände Z. Psychol. Ergänzungsbände. Leipzig: J.-A. Barth 1911—1917.
334. — Über das Aubertsche Phänomen. Z. Sinnesphysiol. **49**, 109—246 (1916).
335. — Darstellung und Erklärung der verschiedenen Typen der Farbenblindheit. Göttingen: Vandenhoeck and Ruprecht 1924.
336. MÜLLER, J.: Zur vergleichenden Physiologie des Gesichtssinnes des Menschen und der Thiere, nebst einem Versuch über die Bewegungen der Augen und über den menschlichen Blick. Leipzig: Knobloch 1826.
337. — Über die phantastischen Gesichtserscheinungen. Coblenz: Jacob Hölscher 1826.
338. MYERS, R. E.: Commissural connections between occipital lobes of the monkey. J. comp. Neurol. **118**, 1—16 (1962).
339. — Organization of visual pathways. In: ETTLINGER, E. G. (Ed.): Functions of the Corpus Callosum, pp. 133—138. London: Churchill 1965.

340. Nakayama, K., Roberts, D. J.: Line length detectors in the human visual system: evidence from selective adaptation. Vision Res. **12**, 1709—1713 (1972).

341. Nauta, W. J. H.: The problem of the frontal lobe: a reinterpretation. J. psychiat. Res. **8**, 167—187 (1971).

342. Niimi, K., Kawamura, S., Ishimaru, S.: Projections of the visual cortex to the lateral geniculate and posterior thalamic nuclei in the cat. J. comp. Neurol. **143**, 279—312 (1971).

343. Noorden, G. K. von, Dowling, J. E.: Experimental amblyopia in monkeys. II. Behavioral studies in strabismic amblyopia. Arch. ophthal. **84**, 215—220 (1970).

344. — — Ferguson, D. C.: Experimental amblyopia in monkeys. I. Behavioral studies of stimulus deprivation amblyopia. Arch. ophthal. **84**, 206—214 (1970).

345. Norrsell, K., Norrsell, U.: Visual discrimination after total neonatal removal of the neocortex in the cat. Acta physiol. scand. **84**, 29—30 (1972).

345a. Noton, D., Stark, L.: Scanpaths in saccadic eye movements while viewing and recognizing patterns. Vision Res. **11**, 929—942 (1971).

346. O'Brien, V.: Contour perception, illusion and reality. J. Opt. Soc. Amer. **48**, 112—119 (1958).

347. Ohm, J.: Zur Tätigkeit des Augenmuskelsenders. 2Bde. Bottrop: Selbstverlag 1928/29.

348. O'Keefe, J., Dostrovsky, J.: The hippocampus as a spatial map. Preliminary evidence from unit activity in the freely-moving rat. Brain Res. **34**, 171—175 (1971).

349. Otsuka, R., Hassler, R.: Über Aufbau und Gliederung der corticalen Sehsphäre bei der Katze. Arch. Psychiat. Nervenkr. **203**, 212—234 (1962).

350. Payne, W. H., Anderson, D. E.: Border-contrast: corner brightening effects. Vision Res. **9**, 1309—1313 (1969).

351. Penfield, W., Perot, P.: The brain's record of auditory and visual experience. Brain **86**, 595—696 (1963).

352. — Rasmussen, T.: The cerebral cortex of man. New York: Macmillan Comp. 1950.

353. Pettigrew, J. D.: The importance of early visual experience for neurons of the developing geniculostriate system. Invest. Ophthal. **11**, 386—393 (1972).

354. — Nikara, T., Bishop, P. O.: Responses to moving slits by single units in cat striate cortex. Exp. Brain Res. **6**, 373—390 (1968).

355. Piaget, J.: Les notions de mouvement et de vitesse chez l'enfant. Paris: Presses Univ. France 1946.

356. Pirenne, M. H.: Les lois de l'optique et la liberté de l'artiste. J. Psychol. norm. path. **60**, 151—166 (1963).

357. — On the problem of black. J. Physiol. (Lond.) **185**, 64—65P (1966).

358. — Optics, painting and photography. Cambridge: Univ. Press 1970.

359. Pitts, W., McCulloch, W. S.: How we know universals. The perception of auditory and visual forms. Bull. Math. Biophys. **9**, 127—147 (1947).

360. Poggio, G. F.: Spatial properties of neurons in striate cortex of unanesthetized macaque monkey. In: U.S.-Australian Symp. on Vision. Invest. Ophthal. **11**, 368—376 (1972).

361. — Baker, F. H., Lamarre, Y., Sanseverino, E. R.: Afferent inhibition at input to visual cortex of the cat. J. Neurophysiol. **32**, 892—915 (1969).

362. Polyak, S.: The retina. Chicago: University of Chicago Press 1941.

363. — In: The Vertebrate Visual System. Klüver, H (Ed.). Chicago: University of Chicago Press 1957.

363a. Popp, C.: Doppelstrichgitterkontrast. Pflügers Arch. ges. Physiol. **291**, R 81 (1966).

364. Poppele, R. E., Maffei, L.: Retinal responses with different background light and psychophysical correlation. Arch. ital. Biol. **105**, 189—200 (1967).

365. Popper, K. R.: Logik der Forschung. Springer: Wien 1935. (English edit.): The Logic of Scientific Discovery. London: Hutchinson 1959.

366. — Objective knowledge: an evolutionary approach. Oxford: Clarendon Press 1972.

367. Powell, T. P. S.: Personal communication.

368. Prandtl, A.: Über gleichsinnige Induktion und die Lichtverteilung in gitterartigen Mustern. Z. Sinnesphysiol. **58**, 263—307 (1927).

369. PURKINJE, J.: Beiträge zur Kenntnis des Sehens in subjectiver Hinsicht. Prag: J. G. Calve 1819.

370. RACHLIN, H. C.: Scaling subjective velocity, distance and duration. Perception & Psychophysics **1**, 77—82 (1966).

371. RATLIFF, F.: Mach bands: quantitative studies on neural networks in the retina. San Francisco-London-Amsterdam: Holden Day Inc. 1965.

372. — Contour and contrast. Proc. Amer. phil. Soc. **115**, 150—163 (1971).

372a. REICHARDT, W.: Musterinduzierte Flugorientierung (Verhaltensversuche an der Fliege Musca-domestica). Naturwissenschaften **60**, 122—138 (1973).

373. — VARJÚ, D.: Übertragungseigenschaften im Auswertesystem für das Bewegungssehen. Z. Naturforsch. B **14**, 674—689 (1959).

374. RÉMOND, A., LESÊVRE, N., TORRES, F.: Etude chronotopographique de l'activité occipitale moyenne recueillée sur le scalp chez l'homme en relation avec les déplacements du regard (complex lambda). Rev. Neurol. **107**, 177—187 (1965).

375. RICCO, A.: Relazione fra il minimo angolo visuale e l'intensita luminosa. Memorie della Regia Accademia di Scienze, lettere ed Arti in Modena **17**, 47—160 (1877).

376. RICHARDS, W.: The fortification illusions of migraines. Sci. Amer. **224**, 89—96 (1971).

377. RIGGS, L. A.: Progress in the recording of human retinal and occipital potentials. J. opt. Soc. Amer. **59**, 1558—1566 (1969).

378. — WOOTEN, B. R.: Electrical measures and psychophysical data on human vision. In: JAMESON, D., HURVICH, L. M. (Eds.): Handbook of Sensory Physiol, Vol. VII/4, pp. 690 to 731. Berlin-Heidelberg-New York: Springer 1972.

379. RONCHI, L., BOTTAI, G.: On the visual effects produced by a test object consisting of two stripes darker than the background, intersecting with another. Atti Fond. G. Ronchi **18**, 47—70 (1963).

380. — — Simultaneous contrast effects at the center of figures showing different degrees of symmetry. Atti Fond. G. Ronchi **20**, 85—100 (1964).

381. — MORI, G. F.: On the factors which affect the contrast enhancement in a figure with "quasi perceptive contours" and a practical application of such a figure. Atti Fond. G. Ronchi **14**, 495—508 (1959).

382. RUBIN, E.: Visuell wahrgenommene Figuren. Kobenhavn: Gyldendalske Bokhandel 1921.

383. RUSHTON, W. A. H.: The difference spectrum and the photosensitivity of rhodopsin in the living human eye. J. Physiol. (Lond.) **134**, 11—29 (1956).

384. — Visual adaptation: the Ferrier lecture. Proc. roy. Soc. B **126**, 20—46 (1965).

385. — Visual pigments in man. In: DARTNALL, H. J. A. (Ed.): Photochemistry of Vision. Handbook of Sensory Physiology, VII/1, pp. 364—394. Berlin-Heidelberg-New York: Springer 1972.

386. SANIDES, F., HOFFMANN, J.: Cyto- und myeloarchitecture of the visual cortex of the cat and of the surrounding integration cortices. J. Hirnforsch. **11**, 79—104 (1969).

387. SASAKI, H., BEAR, D. M., ERVIN, F. R.: Quantitative characterization of unit response in the visual system. Exp. Brain Res. **13**, 239—255 (1971).

388. SCHEPELMANN, F., ASCHAYERI, H., BAUMGARTNER, G.: Die Reaktionen der "simple field"-Neurone in Area 17 der Katze beim Hermann-Gitter-Kontrast. Pflügers Arch. ges. Physiol. **294**, 57 (1967). „

389. SCHMIDT, C. L., WIST, E. R., DICHGANS, J.: Efferent frequency modulation in the vestibular nerve of goldfish correlated with saccadic eye movements. Exp. Brain Res. **15**, 1—14 (1972).

390. SCHNEIDER, G. E.: Contrasting visuomotor functions of tectum and cortex in the golden hamster. Psychol. Forsch. **31**, 52—62 (1968).

391. SCHOBER, H., RENTSCHLER, I.: Das Bild als Schein der Wirklichkeit. München: Heinz Moos 1972.

391a. SCHULTZE, M.: Zur Anatomie und Physiologie der Retina. Arch. mikr. Anat. **2**, 165—175 (1866).

392. SEELEN, W. VON: Zur Informationsverarbeitung im visuellen System der Wirbeltiere I. Kybernetik **7**, 43—60 (1970).

392a. — Zur Informationsverarbeitung im visuellen System der Wirbeltiere II. Kybernetik **7**, 89—106 (1970).

392b. SHARPE, C. R.: A perceptual correlate of McIlwains "periphery effect". Vision Res. **12**, 519—520 (1972).

393. SHERRINGTON, C. S.: The integrative action of the nervous system. New Haven; Constable: Yale Univ. Press 1906.

393a. SICKEL, W.: Retinal metabolism in dark and light. In: FUORTES, M. G. F. (Ed.): Handbook of Sensory Physiology, Vol. VII/2, pp. 668—727. Berlin-Heidelberg-New York: Springer 1972.

394. SINDERMANN, F., DEECKE, L.: Subjective Intensität des Hermann-Gitter-Phänomens in der Netzhautperipherie. Pflügers Arch. ges. Physiol. **316**, R 95 (1970).

395. — PIEPER, E.: Größenschätzung von fovealen Projektionen receptiver Kontrastfelder (Zentrum und Umfeld) beim Menschen im psycho-physischen Versuch. Pflügers Arch. ges. Physiol. **283**, R 47—48 (1965).

396. SMIRNOV, G. D.: Comparative approach to the neurophysiology of vision. In: BRAZIER, M. A. B. (Ed.): Brain and Behavior, Proc. of the I. Conf., pp. 263—298. Washington, D. C.: Amer. Inst. Biol. Sci. 1961.

397. SMITH, K. U., KAPPAUF, W. E.: A neurological study of apparent movement. J. gen. Psychol. **23**, 315—327 (1940).

398. — SMITH, W. M.: Perception and motion. An analysis of spacestructured behavior. Philadelphia-London: W. B. Saunders Co. 1962.

399. SPEHLMANN, R.: Acetylcholine and prostigmine electrophoresis at visual cortex neurons. J. Neurophysiol. **26**, 127—139 (1963).

400. — The averaged electrical responses to diffuse and to patterned light in the human. Electroenceph. clin. Neurophysiol. **19**, 560—567 (1965).

401. SPERLING, G.: Model of visual adaptation and contrast detection. Perception & Psychophysics 8, 143—157 (1970).

402. SPERRY, R. W.: Neural basis of the spontaneous optokinetic response produced by visual inversion. J. comp. physiol. Psychol. **43**, 482—489 (1950).

403. — Split-brain approach to learning problems. In: QUARTON, G. C., MELNECHUK, T., SCHMITT, F. O. (Eds.): The Neurosciences, pp. 714—722. New York: Rockefeller Univ. Press 1967.

403a. — GAZZANIGA, M. S., BOGEN, J. E.: Interhemispheric relationships: the neocortical commissures: syndromes of hemisphere disconnection. In: VINKEN, P. J., BRUYN, G. W. (Eds.): Handbook of Clinical Neurology, Vol. 4, pp. 273—290. New York: J. Wiley & Sons Inc. 1969.

404. SPILLMANN, L.: Zur Feldorganisation der visuellen Wahrnehmnug beim Menschen. Ph. D. Thesis, Univ. of Münster 1964.

405. — Foveal perceptive fields in the human visual system measured with simultaneous contrast in grids and bars. Pflügers Arch. ges. Physiol. **326**, 281—299 (1971).

406. — FULD, K.: Brightness contrast in the Ehrenstein illusion. A psychophysical and perceptual study. Vision Res. (submitted for publication 1972).

406a. — GAMBONE, G. V.: A test of the McIlwain effect in man. Vision Res. **11**, 751—753 (1971).

407. — LEVINE, J.: Contrast enhancement in a Hermann grid with variable figure-ground ratio. Exp. Brain Res. **13**, 547—559 (1971).

408. SPINELLI, D. N., BARRETT, T. W.: Visual receptive field organization of single units in the cat's visual cortex. Exp. Neurol. **24**, 76—98 (1969).

409. — HIRSCH, H. V. B., PHELPS, R. W., METZLER, J.: Visual experience as a determinant of the response characteristics of cortical receptive fields in cats. Exp. Brain Res. **15**, 289—304 (1972).

410. SPRAGUE, J. M., MEIKLE, T. H.: The role of the superior colliculus in visually guided behavior. Exp. Neurol. **11**, 115—146 (1965).

411. STARK, L.: The control system for versional eye movements. In: BACH-Y-RITA, P., COLLINS, C.C. (Eds.): The Control of Eye Movements, pp. 363—428. New York-London: Academic Press 1971.
412. STEINBERG, R.H.: High-intensity effects on slow potentials and ganglion cell activity in the area centralis of cat retina. Vision Res. **9**, 333—350 (1969).
413. STEVENS, S.S.: The psychophysics of sensory function. In: ROSENBLITH, W.A. (Ed.): Sensory Communication, pp. 1—33. New York—London: John Wiley & Sons 1961.
414. — Sensory power functions and neural events. In: LOEWENSTEIN, W. (Ed.): Handbook of Sensory Physiology, Vol. I, pp. 227—242. Berlin-Heidelberg-New York: Springer 1971.
415. STONE, J.: Quantitative analysis of distribution of ganglion cells in the cat's retina. J. comp. Neurol. **124**, 337—352 (1965).
416. — Morphology and physiology of the geniculocortical synapse in the cat: The question of parallel input to the striate cortex. Invest. Ophthal. **11**, 338—346 (1972).
417. STRASCHILL, M., TAGHAVY, A.: Neuronale Reaktionen im tectum opticum der Katze. Exp. Brain Res. **3**, 353—367 (1967).
418. STRATTON, G.M.: Vision without inversion of the retinal image. Psychol. Rev. **4**, 341—360, 463—481 (1897).
418a. — Eye movements and the aisthesis of visual form. Philos. Stud. (Wundt) **20**, 336—359 (1902).
419. SUTRO, L.L.: A model of visual space. In: BERNARD, E.E., KARE, M.R. (Eds.): Biological Prototypes and Synthetic Systems, (Symp.) pp. 75—87. New York: Plenum Press 1962.
420. SVAETICHIN, G.: Spectral response curves from single cones. Acta physiol. scand. **39**, Suppl. 134, 17—46 (1956).
421. — Receptor mechanisms for flicker and fusion. Acta physiol. scand. **39**, Suppl. **134**, 47—54 (1956).
422. — NEGISHI, K., FATECHAND, R.: Cellular mechanisms of a Young-Hering visual system. In: DE REUCK, A.V.S., KNIGHT, J. (Eds.): Colour Vision. Physiology and Experimental Psychology. CIBA Found.-Symp., pp. 178—207. London: J. & A. Churchill 1965.
423. SZENTÁGOTHAI, J.: The synapses of short local neurons in the cerebral cortex. In: SZENTÁGOTHAI, J. (Ed.): Modern Trends in Neuromorphology, pp. 251—276. Symp. Biol. Hung., Vol. 5. Budapest: Akadémiai Kiadó 1965.
424. — Glomerular synapses, complex synaptic arrangements and their operational significance. In: SCHMITT, F.O. (Ed.): The Neurosciences: Second Study Program, pp. 427—443. New York: Rockefeller University Press 1970.
425. — Lateral geniculate body structure and eye movement. Bibl. opthal. (Basel) **82**, 178 to 188 (1972).
426. TALBOT, S.A., MARSHALL, W.H.: Physiological studies on neural mechanisms of visual localization and discrimination. Amer. J. Ophthal. **24**, 1255—1263 (1941).
427. TER BRAAK, J.W.G.: Untersuchungen über optokinetischen Nystagmus. Arch. néerl. Physiol. **21**, 309—376 (1936).
427a. — Ambivalent optokinetic stimulation and motion detection. Bibl. ophthal. (Basel) **82**, 308—316 (1972).
428. — BUIS, C.: Optokinetic nystagmus and attention. Int. J. Neurol. **8**, 34—42 (1970).
429. — SCHENK, V.W.D., VLIET, A.G.M. VAN: Visual reaction in a case of long-lasting cortical blindness. J. Neurol. Neurosurg. Psychiat. **34**, 140—147 (1971).
430. TEUBER, H.-L.: Perception. In: FIELD, J., MAGOUN, H.W., HALL, E.V (Eds.): Handbook of Physiology, Sect. 1, Neurophysiology **3**, 1595—1688. Washington: Amer. Physiol. Soc. 1960.
431. — Subcortical vision: A prologue. Brain Behav. Evol. **3**, 7—15 (1970).
432. — BATTERSBY, W.S., BENDER, M.B.: Visual field defects after penetrating missile wounds of the brain. Cambridge/Mass.: Harvard Univ. Press 1960.
433. TRENDELENBURG, W.: Der Gesichtssinn. Grundzüge der physiologischen Optik. 2. Auflage überarbeitet von MONNIER, M., SCHMIDT, I., SCHÜTZ, E. Berlin-Heidelberg-New York: Springer 1961.
434. TREVARTHEN, C.B.: Two mechanisms of vision in primates. Psychol. Forsch. **31**, 299—337 (1968).

435. Tschermak, A.: Über Kontrast und Irradiation. Ergebn. Physiol. **2**, 726—798 (1903).
436. — Der exakte Subjektivismus in der neueren Sinnesphysiologie. Pflügers Arch. ges. Physiol. **188**, 1—20 (1921).
437. — Licht- und Farbensinn. In: Bethe, A., Bergmann, G. v., Embden, G., Ellinger, A. (Eds.): Handbuch der normalen und pathologischen Physiologie, Bd. XII/2, S. 295—501. Berlin: Springer 1929.
438. Valleala, P.: The temporal relation of unit discharge in visual cortex and activity of the extraocular muscles during sleep. Arch. ital. Biol. **105**, 1—14 (1967).
439. Vallecalle, E., Svaetichin, G.: The retina as model for the functional organization of the nervous system. In: Jung, R., Kornhuber, H. H. (Hrsg.): Neurophysiologie und Psychophysik des visuellen Systems, pp. 489—492. Berlin-Göttingen-Heidelberg: Springer 1961.
440. Vaughan, H. G., Jr.: The perceptual and physiologic significance of visual evoked responses from the scalp in man. In: Clinical Electroretinography, pp. 203—223. Oxford-New York: Pergamon Press 1966.
441. Leonardo da Vinci: Trattato della pittura di Lionardo da Vinci, novamente dato in luce, con la vita istesso autore, seritta da Raffaelle du Fresne. (Du Fresne, R., Ed.). Paris: G. Langlois 1651.
442. Vogt, C., Vogt, O.: Allgemeinere Ergebnisse unserer Hirnforschung. 1.—4. Mitteilung. J. Psychol. Neurol. (Lpz.) **25**, Erg.heft I, 277—462 (1919).
443. Vogt, O.: Die anatomische Vertiefung der menschlichen Hirnlokalisation. Klin. Wschr. **29**, 111—125 (1951).
444. Wagner, H. G., MacNichol, E. F., Wolbarsht, H. L.: Functional basis for on- and off-center receptive fields in the retina. J. opt. Soc. Amer. **53**, 66—70 (1963).
445. Wald, G.: The receptors of human colour vision. Science **145**, 1007—1017 (1964).
446. Walter, W. G.: Slow potential waves in the human brain associated with expectancy, attention and decision. Arch. Psychiat. Nervenkr. **206**, 309—322 (1964) .
447. Weiskrantz, L.: Contour discrimination in a young monkey with striate cortex ablation. Neuropsychologia **1**, 145—164 (1963).
448. — Cowey, A.: Striate cortex lesions and visual acuity of the rhesus monkey. J. comp. physiol. Psychol. **56**, 225—231 (1963).
449. Weisstein, N.: Neuronal symbolic activity: A psychophysical measure. Science **168**, 1489—1491 (1970).
450. — Bisaha, J.: Gratings mask bars and bars mask gratings: Visual frequency response to aperiodic stimuli. Science **176**, 1047—1049 (1972).
451. Wertheimer, M.: Experimentelle Studien über das Sehen von Bewegung. Z. Psychol. **61**, 161—265 (1912).
452. Westheimer, G.: Spatial interaction in human cone vision. J. Physiol. (Lond.) **190**, 139—154 (1967).
453. — Rod-cone independence for sensitizing interaction in the human retina. J. Physiol. (Lond.) **206**, 109—116 (1970).
454. — Mitchell, D. E.: The sensory stimulus for disjunctive eye movements. Vision Res. **9**, 149—156 (1969).
455. Wheatstone, C.: Contributions to the physiology of vision. Part 1. On some remarkable, and hitherto unobserved, phenomena of binocular vision. Phil. Trans. roy. Soc. **128**, 371—394 (1838).
456. Wiesel, T. N., Hubel, D. H.: Single-cell responses in striate cortex of kittens deprived of vision in one eye. J. Neurophysiol. **26**, 1003—1017 (1963).
457. Wilska, A.: Aktionspotentialentladungen einzelner Netzhautelemente der Katze. Acta Soc. med. Fenn. „Duodecium" A. **12**, 63—71 (1939).
458. Wist, E. R., Freund, H. J.: The neuronal basis of binocular vision. In: Grüsser, J., Klinke, R. (Eds.): Pattern Recognition in Biological and Technical System, pp. 288—300. Berlin-Heidelberg-New York: Springer 1971.
459. — Susen, P.: Evidence for the role of post-retinal processes in simultaneous contrast. Psychol. Forsch. Submitted for publication 1972.
460. Wundt, W.: Theorie der Sinneswahrnehmungen. Leipzig-Heidelberg: C. F. Winter 1862.

461. WURTZ, R.H.: Responses of striate cortex neurones to stimuli during rapid eye movements in the monkey. J. Neurophysiol. **32**, 975—986 (1969).
462. — GOLDBERG, M.E.: The primate superior colliculus and the shift of visual attention. Invest. Ophthal. **11**, 441—450 (1972).
463. YARBUS, A.L.: Eye movements and vision. New York: Plenum Press 1967.
464. YOUNG, T.: On the theory of light and colours. Phil. Trans. roy. Soc. **1802**, 12—48.
465. ZEKI, S.M.: Convergent input from the striate cortex (area 17) to the cortex of the superior temporal sulcus in the rhesus monkey. Brain Res. **28**, 338—340 (1971).
466. — Cortical projections from two prestriate areas in the monkey. Brain Res. **34**, 19—35 (1971).
467. — Personal communication 1972.
467a. ZYKMUND, V. (edit.): The oculomotor system and brain functions. London: Butterworth 1973.
468. ZUSNE, L.: Visual perception of form. New York-London: Academic Press 1970.

Chapter 2

Neurophysiological Mechanisms in the Visual Discrimination of Form

By

JONATHAN STONE, Canberra (Australia), and
ROBERT B. FREEMAN, JR., Konstanz (Germany)

With 9 Figures

Contents

Introduction

Man, in common with many animals, has an extremely well developed ability to detect, discriminate and react to visual stimuli. The determinants of visual behaviour are many and include, for example, physical factors such as ambient illumination and the optical properties of the eye, behavioural or psychological factors such as the subject's past experience, as well as the neurophysiological factors with which this chapter is particularly concerned. It is, of course, a basic assumption of neurophysiologists that the neurophysiological organisation of the visual pathway, and of the associative and efferent pathways involved in visually based behaviour, determines all measures of visual behaviour, from basic measures such as absolute sensitivity and visual acuity to the higher functions of visual memory, discrimination and perception. Yet, although psychophysicists and psychologists have investigated visual performance at all these levels, present understanding of the neurophysiological mechanisms of vision extends very little past the mechanisms involved in processing visual input. Our principal emphasis consequently is on the neurophysiological encoding of form information in the afferent visual pathway, and discussion of behavioural and psychophysical assessments of form vision is included only where relevant neurophysiological mechanisms have been described or sought. For the same reason the term "form discrimination" is taken to denote the behavioural demonstration by an experimental subject (whether animal or human) of the ability to distinguish one form or pattern or shape from another; this excludes from our scope one of the ultimate problems of form vision, viz. how neural activity gives rise to perception. Similarly, no distinction is made between the terms "pattern", "shape" or "form", although there may be grounds for doing so in other contexts.

Organisation of the Chapter. The chapter is in three sections. Section I is concerned primarily with basic processes of form vision, such as visual acuity, contrast sensitivity and contour detection. Section II traces the work of neurophysiologists who, in recent years, have taken advantage of their separate access to the different visual centres of the brain, and to individual cells of each centre, to describe the way in which the neurones of the afferent visual pathway react to, and encode, features of visual stimuli. More complex (but still fundamental) aspects of form vision are considered in Section III. In particular the role of fixational eye movements and the temporal and spatial stability of form vision are discussed to the extent that they have received attention by neurophysiologists.

I. Basic Mechanisms of Form Discrimination: Contrast Sensitivity, Contour Detection and Visual Acuity

1. Contrast Sensitivity

A basic property of the visual system is its sensitivity to local contrast. The eye's ability to detect an edge or contour is determined, to an important degree, by the change in the brightness of the visual field across that contour. Within a wide range the absolute level of illumination is relatively unimportant.

MACH (1865, 1866 a, b, 1868 a, b) first demonstrated local contrast sensitivity by describing and analysing the illusion now known as Mach bands. He showed that our subjective judgement of brightness is markedly affected by changes, particularly sharp changes, in the luminance gradient of a viewed pattern. MACH's conclusion is succinct: "The individual retinal points evaluate the light incident upon them in terms of the average light on their surroundings" (1868 b). "Whatever is near the mean of the surroundings becomes effaced, whatever is above or below is disproportionately brought into prominence" (1868 a). MACH concluded that the bands of increased brightness and darkness near changes of surface luminance "can only be explained on the basis of a reciprocal action (*Wechselwirkung*) of neighboring areas of the retina" (quoted by RATLIFF, 1965, p. 98). He presented a mathematical theory of the effect assuming that the effects of one retinal area upon another were a non-linear, decreasing function of their separation. His observations have been extensively confirmed and are reviewed in detail by RATLIFF (1965).

A different, and less quantitative account of local brightness contrast was given by HERING (1872—1874; cf. HERING, 1964). Like MACH's theory, the HERING theory was based on the notion of reciprocal interaction among elements of the "visual substance". Although HERING discusses at length the applicability of his theory to simultaneous brightness contrast, he also makes explicit its possible relevance to pattern perception: "It is to reciprocal interactions in the somatic visual field that we owe, to a large extent, our visual acuity . . ." (HERING, 1964, p. 124). Like MACH, HERING gave emphasis to the assumed physiological basis of simultaneous contrast, as opposed to the judgment theory of HELMHOLTZ (1867).

Recent investigations distinguish at least *two types of spatial contrast illusion.* One type, now known as Mach bands, appears in the form of narrow light and dark bands which appear, respectively, on the darker and lighter sides of abrupt changes of luminance ("edges"). Mach bands are notable for their small width, being on the order of only a few minutes of arc. A second, and more familiar, illusion is that in which, e.g. a grey patch of paper, when superimposed on a white sheet of paper, appears darker than when placed on a black sheet. This effect, variously called "brightness contrast" or "brightness induction", is distinct from Mach bands in that induction of a test field can be achieved by an inducing field even at considerable distances from the test field (e.g. DIAMOND, 1953; LEIBOWITZ et al., 1953; FRY and ALPERN, 1953). Recently VON BÉKÉSY (1968) has provided evidence that these two types of spatial contrast phenomena may be dependent upon different physiological mechanisms, which he termed Mach- and Hering-type lateral inhibition. Of the two types, phenomena of the Mach type are probably of greater importance in form discrimination, since they concern the discrimination of contours and edges. Moreover, some of the most striking neurophysiological features of visual afferent neurones are clearly related to edge-phenomena of the

Mach type. On the other hand, the physiological basis of Hering-type brightness contrast is still obscure.

Lateral Interaction in the Retina as a Determinant of Contrast Sensitivity. Perhaps the most extensive neurophysiological treatment of lateral interaction is that of RATLIFF and HARTLINE (1959), RATLIFF et al. (1963), and RATLIFF (1965, Chapter 4). They showed that lateral, neural interaction in the eccentric eye of the horseshoe crab, *Limulus polyphemus*, is capable of generating patterns of neural excitation whose spatial distribution is strikingly similar to the spatial distribution of brightness in the visual Mach band effect in humans. Recording from a single ommatidium of the crab eye, they found that the discharge of the ommatidium varies with its position with respect to a dark light border (Fig. 1a). It is maximally excited when just on the bright side of the border and maximally inhibited when just on the dark side. HARTLINE and his co-workers have found no evidence for lateral excitatory interaction in the eye of the crab (HARTLINE, 1949; HARTLINE et al., 1956) and RATLIFF and HARTLINE (1959) showed, moreover, that these effects can be accounted for by mutual lateral inhibition between neighbouring ommatidia. They proposed that lateral inhibition of this type may be the neurophysiological basis of the local contrast sensitivity of the visual system and in particular of phenomena such as Mach bands and the variation of visual threshold near a contrast border (FIORENTINI et al., 1955).

Despite great differences in retinal structure and function between *Limulus* and mammals, KUFFLER (1953) demonstrated that the output cells of the cat retina, the ganglion cells, have receptive fields with analogous properties. Each receptive field consists of two regions, an approximately circular centre and an annular surround, whose influences on a cell are always opposite and antagonistic. Because centre and surround regions are spatially separate, this antagonism is a form of lateral inhibition. As a result, diffuse illumination falling onto both centre and surround excites these cells only weakly, but when the illumination is non-uniform, containing say a border between lighter and darker areas, the cell may respond strongly. For example, when the border runs tangentially to the centre region of a receptive field, leaving the centre region on one side and part of the surround on the other, the centre region becomes predominant in determining the cell's response, generating strong excitation or inhibition. ENROTH-CUGELL and ROBSON (1966) have measured the response of a ganglion cell to a contrast border, as a function of the position of the border on the receptive field. This function (Fig. 1b) closely resembles the analogous function in the *Limulus* eye despite the different neural organisations of the two eyes. BAUMGARTNER et al. (1965) have demonstrated comparable properties of receptive fields in the lateral geniculate nucleus and visual cortex.

The proposition that local contrast sensitivity is determined by the centre-surround organisation of retinal receptive fields has been tested experimentally. BARLOW et al. (1957) showed that as the adaptation level of the retina falls the surround influence of the cat retinal receptive field disappears and predicted that this change should result in a fall in contrast sensitivity. ENROTH-CUGELL and ROBSON (1966) confirmed this prediction, showing that the contrast sensitivity of these receptive fields falls markedly as mean retinal illumination falls. RODIECK and STONE (1965a) analysed the responses of cat retinal ganglion cells to cardboard bars (black or white) moved in front of a grey background. They showed that the sharpest modulation of ganglion cell firing (i.e. excitation or inhibition) occurs when the leading and trailing edges of the bar move between the surround

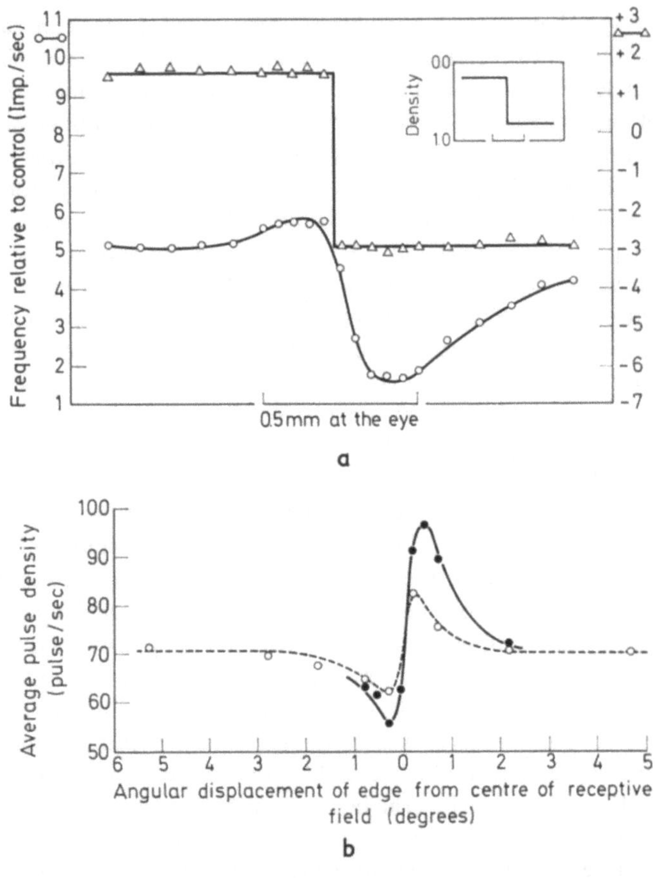

Fig. 1 a and b. Effects of lateral inhibition in *limulus* and cat retina. a: The discharge rate of the eccentric cell of a single ommatidium of the *limulus* eye plotted as a function of the position on the retinal mosaic of a rectangular "step" pattern of illumination. The inset at upper right shows the illumination gradient across this step. The lower (curvilinear) graph shows that the response of the cell is maximal when the step is located so that the ommatidium is just on the bright side of the border, and minimal (in fact inhibited below control level) when the ommatidium is just on the dark side of the border. The scale of the ordinate is at left. The upper (rectilinear) graph shows that when the retinal mosaic is masked, so that only the ommatidium under test was exposed to the step pattern (and hence was not inhibited by its neighbours), the discharge rate is closely related to the incident illumination. Discharge rate is then determined only by whether the ommatidium is on the bright or dark side of the step, and not by its distance from the step. Ordinate scale is on the right. (From Ratliff and Hartline, 1959). b: Response of a cat retinal ganglion cell with an ON-centre receptive field, as a function of the position of contrast edge in its receptive field. The response was measured during the period 10—20 sec after the presentation of the edge. As in the *limulus* eye, the cell's response shows its maximum and minimum when the receptive field is located just on the bright or dark side of the border. Two different contrast levels were used: filled circles 0.4; open circles 0.2 (From Enroth-Cugell and Robson, 1966)

and centre regions of the field. When the background luminance was reduced until the surround influence disappeared, leaving the object-background contrast constant, the sharp modulations in firing were much reduced (Fig. 2a) and similar, more weakly modulated responses were obtained from rare receptive fields which lacked a surround, even in light adapted conditions (Fig. 2b).

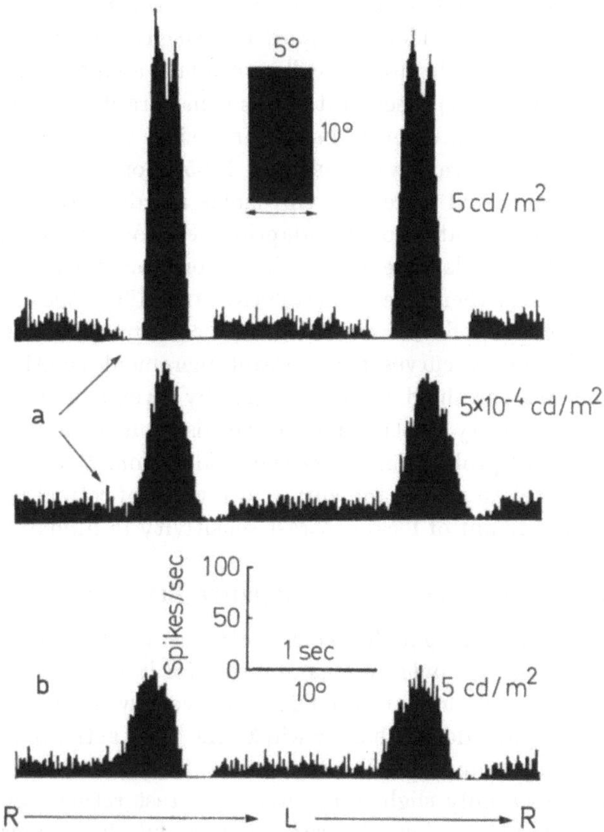

Fig. 2a and b. Effect of receptive field surround on response of a cat retinal ganglion cell to moving figures (From RODIECK and STONE, 1965a). a: The upper histogram shows the response (modulation of firing rate) of an OFF-centre ganglion cell as a black bar moved through its receptive field (against a grey background) at a speed of 10°/sec first from right (R) to left (L), then left to right as indicated at bottom. The response to the two directions of movement (i.e. the left and right halves of the histogram) are essentially identical, showing inhibition as the bar enters the receptive field, a bimodal burst of firing as the bar crosses the field and inhibition as the bar leaves the field. The ambient illumination was 5 cd/m² (mesopic). In the lower histogram the ambient illumination was reduced by 4 log units (i.e. became scotopic). In these conditions the cell's response is weaker, showing no entry inhibition, a weaker, unimodal burst of firing and weaker exit inhibition. RODIECK and STONE argue that these changes result from the loss of the receptive field surround influence at low ambient illumination. b: The response of an OFF-centre unit at mesopic levels of illumination. This unit had no detectable surround. In both cases the absence of surround influence is associated with more weakly modulated discharge patterns in response to stimuli with the same object-background contrast

HUBEL and WIESEL (1960) showed that the receptive fields of monkey retina resemble those of the cat. GOURAS (1967) showed further that, as in the cat, the surround influence of these fields disappears, and the apparent centre region enlarges, with dark adaptation, although the effect of these changes on the contrast sensitivity of individual fields has yet to be tested. Comparable data are not available for the human retina, but because its structure closely resembles that of primate retina (POLYAK, 1957; BOYCOTT and DOWLING, 1969) it is likely that the same basic receptive field organisation exists in the human retina as well. Two lines of psychophysical evidence support this suggestion. First, the contrast sensitivity of human retina falls dramatically with dark adaptation (see GRAHAM, 1965, p. 68 for a brief review). Second, there is considerable psychophysical evidence of inhibitory lateral interaction in human vision, some elements of which are presumably retinal (for a review see GRAHAM, 1965). For example, WESTHEIMER (1967) has shown that visual increment thresholds are dependent upon the size as well as the retinal illumination of the adapting field. As the size of both foveal and parafoveal adapting fields increases, the increment thresholds for small flashes of light first increase and then decrease (WESTHEIMER, 1965, 1967, 1970). This is true also when the adapting field is stabilized on the retina (TELLER et al., 1966). Furthermore area-intensity curves for constant increment threshold levels are very similar to the area-threshold curves obtained by BARLOW et al. (1957), which supports a neural inhibitory explanation of the visual increment effect (WESTHEIMER, 1965). It seems probable from these considerations that the antagonistic lateral interaction between centre and surround regions of retinal receptive fields is an important determinant of local contrast sensitivity in human vision.

2. Contour Detection, and the Coding of Relative Brightness

The responses of individual ganglion cells to contrast borders can be generalised to populations of cells spread over an area of retina (or in the *Limulus*, to an area of the eye comprising many ommatidia). ON-centre cells with their centre regions located just on the bright side of a border will be the most activated, those on the dark side will be the most inhibited (BAUMGARTNER, 1961 a). Cells at some distance from the border will be only slightly affected. Greatest retinal activity is thus associated with the border region; the retina is particularly responsive to sharp contrast borders. This property of "border enhancement" is clearly important in the discrimination of form.

Fifty percent of cat (and monkey) retinal ganglion cells have OFF-centre receptive fields, and (within wide limits) their responses to patterned stimuli are exactly opposite to those of ON-centre cells (RODIECK and STONE, 1965a, b). They are maximally excited when their centre region is just on the dark side of the border (BAUMGARTNER, 1961 a). Several authors (BARLOW et al., 1957; BAUMGARTNER, 1961 b; JUNG, 1961, 1964) have suggested that border contrast is coded by a combination of "darkness" signals from OFF-centre cells on the dark side of the border and "brightness" signals from ON-centre cells on the bright side.

It is interesting to note that these "darkness" and "brightness" signals result from disinhibition, i.e. the active cells respond relatively strongly because they are partially *released from* the lateral inhibitory influence of their receptive field surrounds. CORNSWEET (1970) has argued for example that "there is no way in which neural processing, no matter how elaborate or of what kind, can increase the amount of information about a visual object beyond the amount that is present in the retinal image. The very best that neural processing can do is *not* to *lose* information". Lateral inhibition, in these terms, allows for selective transmission of brightness information at contrast borders.

BARLOW (1961 b) stressed some advantages of border enhancement in the coding of brightness information. He argued that it is more economical for the retina to signal changes in brightness than to code the absolute brightness of each small segment of a pattern. The visual system might then interpret the comparative brightness of two areas in terms of the luminance gradient at the border between them. There is intriguing psychophysical evidence that this is the sort of judgement the visual system makes. For example, equal luminance areas of a disc can be made to appear unequal if an exponential change in brightness, reflected on itself, is created along a short segment of the radius of the disc, leaving the brightness of the rest of the disc unchanged. As a result the equally luminous inner disc and outer annulus appear to differ in brightness, the direction of the difference depending on the direction of change at the false border (see RATLIFF, 1965, Fig. 2.25; CORNSWEET, 1970, Fig. 11.2). O'BRIEN (1958) showed that the subjective judgement of the relative brightness of two areas can be reversed (i.e. a brighter area can be made to appear darker) by a suitably constructed luminance gradient at the border between them. These particular effects are illusions, of course, but they give evidence that the sensitivity of retinal ganglion cells to contrast borders is a property on which higher centres put a lot of reliance in interpreting relative brightness.

3. Visual Acuity

Visual acuity is a measure of the ability of the visual system to resolve patterned stimulation. It is therefore a convenient psychophysical measure of the sensitivity of the neural mechanisms underlying pattern discrimination, at least at the initial stages of neural coding, and it seems important to give some account of the neurophysiological mechanisms which determine acuity.

Acuity as Contrast Detection at High Spatial Frequencies. Visual acuity "involves, near threshold, the discrimination of very small illuminance contrast differences over very small areas of retina" (BOYNTON, 1962). In testing for example the maximal acuity of a subject for a grating pattern, the spatial frequency of the grating (i.e. the number of dark-bright cycles/degree measured along an axis perpendicular to the bars) is typically increased until at about 50 cycles/degree (WESTHEIMER, 1960; CAMPBELL and GREEN, 1965) the subject can no longer distinguish any portion of the grating pattern from a uniform field of the same mean luminance. As spatial frequency is increased there are two changes in the retinal image; predictably, its spatial frequency increases and, importantly, its contrast decreases, even though the contrast of the target grating is constant (WESTHEIMER, 1960). The reduction in image contrast results from the diffraction of light at the pupil, from the spherical and chromatic aberrations of the eye's optical apparatus and from light scatter in the optic media of the eye. These two changes in the image are found quite generally whenever the spatial frequency of some part of the retinal image is increased, e.g. as two edges are brought close together.

Neural Mechanisms in Acuity. It is likely therefore that visual acuity is determined by the *contrast sensitivity* and by the *size* of retinal receptive fields. The importance of receptive field size is clear. At the fovea of the monkey's retina the ganglion cells are maximal in density (POLYAK, 1941; ROLLS and COWEY, 1970) and their receptive fields are minimal in size (4′ in centre diameter or less, HUBEL and WIESEL, 1960). The same high density of ganglion cells is found at the human fovea (POLYAK, 1941), and their receptive fields are very probably as small. Correspondingly the angle of resolution of human vision is minimal (i.e. acuity is maximal) at the fovea (WEYMOUTH, 1958). CAMPBELL and GREEN (1965) reported on the appearance of a wide grating pattern as its spatial frequency is increased. The area of the visual field over which the grating can be distinguished gradually diminishes until, at about 50 cycles/degree, the gratings can be distinguished only over a small central region of about 1/2° diameter. This increase in visual acuity between peripheral and foveal retina probably results entirely from the decrease in receptive field size and the increase in ganglion cell density. There is only a slight change in contrast sensitivity with eccentricity (BLOUGH, 1958), and GREEN (1970; see Fig. 3) has shown directly that elimination of optical degradation of the image does not improve peripheral acuity. For extrafoveal vision, therefore, the size and density of the ganglion cell receptive fields appear to determine acuity.

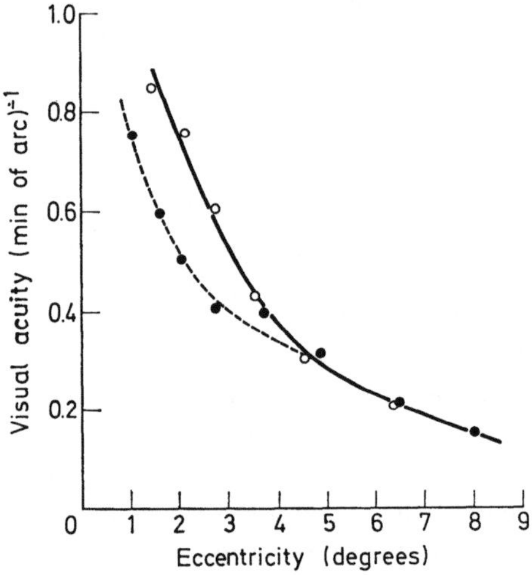

Fig. 3. Visual acuity as a function of eccentricity (From GREEN, 1970). This graph shows the well known gradient of acuity with eccentricity (closed circles). It also shows (open circles) the improvement in acuity achieved by eliminating optical degradation of the image (by forming interference fringes on the retina). A considerable improvement in acuity results, up to 5° from the fovea. At greater eccentricities optical factors do not limit acuity

For foveal vision, however, contrast sensitivity is a major determining factor. Human contrast sensitivity and overall acuity both fall sharply with decreasing ambient illumination although it cannot be assumed from this alone that contrast sensitivity determines acuity. PIRENNE and DENTON (1952) argued that, as background illumination falls, the small receptive fields of the retina become inactive.

In the dark adapted retina, for example, visual thresholds are highest at the fovea (Blough, 1958) where receptive fields are smallest, and acuity is maximal in parafoveal retina (Pirenne and Denton, 1952) where receptive fields are larger. This is probably a major factor in the deterioration of acuity with decreasing ambient illumination. However, the importance of contrast sensitivity for foveal acuity was brought out by Campbell and Green's (1965) elegant study of the contrast transfer function of the optics of the human eye. They presented subjects with grating patterns of variable contrast and spatial frequency, and measured the minimum detectable (threshold) contrast of the grating, over a range of frequencies. By comparing the subjects' contrast thresholds for gratings viewed through the normal optics of the eye with their thresholds for gratings formed on the retina by an interference technique (and hence free of optical degradation), they could estimate the contrast transfer performance of the optics (i.e. image contrast/object contrast) over a range of spatial frequencies. Their results differed from earlier work in indicating that, at frequencies just above the maximum resolvable by a subject viewing normally, there is still significant contrast in the image (about one fifth of object contrast). Campbell and Green's results (their Fig. 9) indicate moreover that when optical degradation of the image contrast was eliminated, their subject could resolve slightly higher frequencies than when viewing normally (60 cycles/degree as against about 46 cycles/degree). This last finding is confirmed by Green's recent (1970) study. It seems likely, therefore, that in normal light-adapted viewing the ability of foveal receptive fields to detect image contrast is the limiting factor for foveal visual acuity.

The higher visual centres seem well organised to preserve the high degree of spatial resolution of foveal vision. In general the proportion of the lateral geniculate nucleus and visual cortex which is devoted to representation of the fovea (area centralis in the cat) is very large (see, for example, Talbot and Marshall, 1941; Hubel and Wiesel, 1962, 1965 a). The proportion seems related to the number of ganglion cells in the fovea (area centralis) and not to the area of retina involved (Bishop et al., 1962; Cowey and Ellis, 1969; Rolls and Cowey, 1970). The cell density in these centres is probably uniform with respect to eccentricity (for example, Chow et al., 1950) so that a large proportion of the cells of each visual centre is concerned with foveal representation. As in the retina, moreover, the foveal receptive fields are small. For example, the excitatory centre regions of area centralis receptive fields of cat visual cortex are similar in size to the centre regions of area centralis retinal fields (about 0.25° in diameter, Hubel and Wiesel, 1962; Stone and Fabian, 1966).

Other Measures of Visual Acuity. So far we have considered only acuity for grating patterns, but a wide variety of other optotypes has been used, including the Landolt C, the Snellen letters, checkerboards and vernier alignment (see LeGrand, 1967). Grating acuity may be determined by the properties of single ganglion cells. The grating acuity of single cells can be tested (Campbell, Cooper, and Enroth-Cugell, 1969; Campbell, Cooper, Robson, and Sachs, 1969), and may approach the acuity of the whole system. However, other types of acuity may not be so simply determined. For example, the vernier acuity of man (the ability to align two halves of a line) is extremely fine, being about 2″ of arc (Riggs, 1965) or even 0.5″ of arc (LeGrand, 1967). LeGrand (1967) has argued in the context of his "continuous theory" that vernier acuity is limited, as is spatial contrast sensitivity, by the contrast transfer of the eye. LeGrand offers as evidence the fact that a black line of 0.5″ thickness may be seen on a white ground, just as a displacement of 0.5″ of vernier offset is the limit of vernier resolution. In neurophysiological terms, it might be argued that, compared with detecting a fine grating, vernier judgment is a relatively complex task, involving the integration of the activity of a number of ganglion cells and hence may be determined by the operations of the higher visual centres. There is evidence that higher-order interactions in the visual system are involved in various other visual acuity tasks. BARTLEY (1941) cites various studies (e.g. FRY and BARTLEY, 1935) showing that the detectability of one contour is a function of the presence and retinal distance of other contours in the visual field. Under experimental conditions, the threshold for detection of one contour decreases with increasing separation from another contour parallel to it. Similar effects have been found in visual masking, in which the visibility of a briefly exposed contour is lowered by a momentary exposure of other contours either preceding (forward masking) or following (backward masking) the test contour. Furthermore, both backward (SEKULER, 1965) and forward (HOULIHAN and SEKULER, 1968) masking decreases (visibility increases) with increasing angular rotation of the masking lines from the orientation of the test lines. These studies suggest interactions among neural units at levels of the visual pathway at which the retinal image has already been coded into lines, presumably at the cortical level.

II. Neurophysiological Encoding of Stimulus Features

1. Retinal Coding

The retina encodes visual information in the impulse activity of retinal ganglion cells, whose axons form the optic nerve. The organisation and mechanisms of ganglion cell receptive fields are considered in more detail elsewhere in this Handbook. Here we are concerned with the way in which these fields code features of a stimulus pattern.

Selective and Non-Selective Retinal Fields. Retinal receptive fields have been studied in a considerable number of species, and a great variety of types has been described. Only part of this variety can be attributed to experimental method (c.f. KEATING and GAZE's (1970) reassessment of MATURANA et al.'s (1960) classification of frog retinal fields). In the present context we suggest classifying the various types of field into two broad groups, those which are *selective* for some spatial or temporal feature of the stimulus pattern, and those which are *non-selective.*

The concentrically arranged, centre-surround receptive fields of the cat's retina are non-selective. They require a minimum brightness contrast between the pattern and background, but they respond to almost any stimulus shape of suprathreshold contrast. For example, they respond to contrast edges or bars of any orientation, and of any length or width above the minimum resolvable, as well as to moving spots, discs and triangles. They respond well to any direction of stimulus movement, and over a wide range of speeds, and are equally responsive to stationary flashing patterns, following any rate of flashing up to the maximum resolvable (KUFFLER, 1953; RODIECK and STONE, 1965 a; OGAWA et al., 1966). They are relatively unresponsive to diffuse flashing light, due to the antagonism of centre and surround regions, and are unaffected by contrast changes which simultaneously and symmetrically excite and inhibit different areas of the receptive field centre. In a proportion of cat retinal fields (the X-cells of ENROTH-CUGELL and ROBSON, 1966), the same is true for the surround region of the receptive field. But with these limited exceptions, non-selective fields are responsive to virtually any change of light distribution within their receptive fields.

In other species, such as the frog (Maturana et al., 1960; Keating and Gaze, 1970), the rabbit (Barlow et al., 1964; Levick, 1967), the ground squirrel (Michael, 1966), the pigeon (Maturana and Frenk, 1963; Holden, 1969), retinal receptive fields have been described whose responses show considerable selectivity for spatial or temporal features of the stimulus pattern. Preferred stimulus features have been termed "trigger features" by Barlow et al. (1964). Typical trigger features are the direction or velocity of stimulus movement, the orientation of a straight edge in the stimulus pattern, and stimulus size. The retinal mechanisms which determine trigger features are considered elsewhere (Levick, Vol. VII/1, of this Handbook).

Coding by Selective Fields. From the simplest point of view, single selective receptive fields might be considered able to code two main facts about the visual environment, viz. the occurrence of a trigger feature, and the visual direction (position in the visual field) of that occurrence. The receptive field properties of each cell determine its selectivity for the trigger feature; the retinal position and limited size of the receptive field determine its ability to specify visual direction.

This view is almost certainly too simple, however, for many types of selective fields respond to a considerable range of stimulus patterns. The directionally-selective fields of rabbit retina, for example, respond maximally to stimulus movement in one "preferred" direction and are inhibited by movement in the opposite, "null" direction. But they give strong though graded responses to movement in directions up to 135° away from the preferred direction (Barlow et al., 1964). Consequently a single field cannot unambiguously signal direction of movement (Oyster and Barlow, 1967). Analogously the "dimming detectors" of the frog retina (Maturana et al., 1960; Keating and Gaze, 1970) are named after their responses to dimming of ambient illumination, but respond well to moving discs, more strongly the larger the disc. The selectivity of most interest for form discrimination is probably orientation selectivity. This property provides a mechanism for 'filtering' the retinal image for straight contours of various orientations, which might be used in "piecing together" a complex shape (Hubel, 1963; Barlow 1961 b). Selectivity of this sort has been clearly described in some rabbit retinal fields (Levick, 1967). Here again, the fields are responsive optimally to edges of one orientation, but respond over a range of orientations. Moreover the range increases as the length of the edge is reduced.

Coding by Non-Selective Fields. In the cat, monkey and probably man, retinal receptive fields are all or, in the case of the cat, almost all (Stone and Fabian, 1966; Rodieck, 1967) non-selective. The problem of how such fields can code the orientation, size, direction and speed of a stimulus pattern has been analysed by Rodieck and Stone (1965 a, b). They studied the responses of non-selective cat retinal units to bars and discs (black or white) moved through their receptive fields, against a grey background. Individual receptive fields respond well to a wide range of stimuli, i.e. no trigger feature selectivity is apparent. There is moreover considerable similarity among different fields in the range of effective stimuli, and in the response pattern (i.e. the modulation of impulse activity) evoked by a particular stimulus. Rodieck and Stone showed that the response of a given cell nevertheless varies in a consistent way with the contrast of the bar or disc (i.e. whether it is brighter or darker than the background), with its width and with its speed of movement. On the other hand, individual cells did not appear to respond differently to different directions of movement or to different orientations or lengths of a bar.

RODIECK and STONE further argued that a population of non-selective fields, distributed over an area across which the stimulus is moving, can provide the brain with considerably more information. Consider, for example, a population of OFF-centre receptive fields distributed over an area of retina across which there is moving the image of a black bar (or disc, or any other shape). At some point in time the image is crossing the centre regions of some receptive fields, exciting them strongly, and it is leaving the centre regions of others, inhibiting them strongly (Fig. 4). The area of excited fields lies ahead of the area of inhibited fields, in the direction of movement, and this asymmetry can provide the brain with precise information about the direction of movement. The area of excited fields will clearly be related to the size and configuration of the stimulus image, i.e. to its outline and orientation. If the receptive fields are assumed to be homogeneous in properties (which is an approximation), the distribution of firing at some instant in time among the population of OFF-centre fields can be derived directly from the response of one OFF-centre field as the stimulus image moves across it

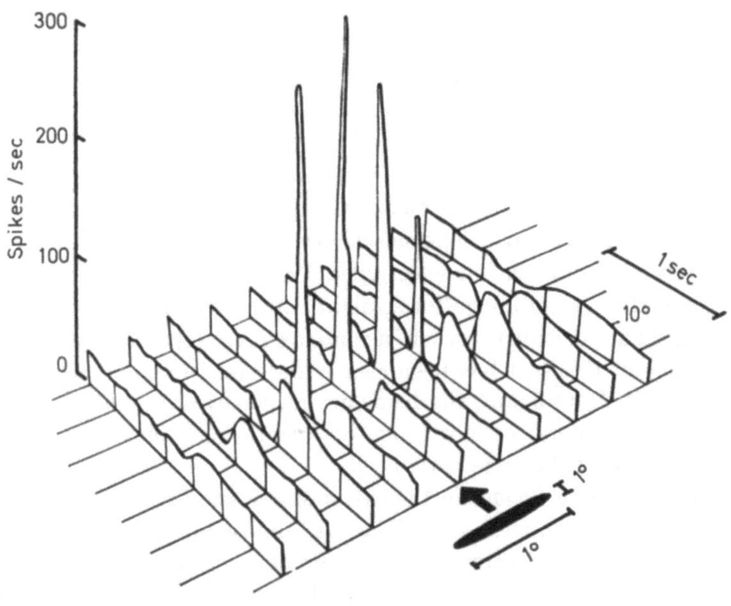

Fig. 4. Schematic representation of the responses of an OFF-centre unit to a black disc moved along different chords through its receptive field. The disc moved 20° along each chord (i.e. through 10° each side of the field), at a velocity of 10°/second. The arrow points to the response obtained as the disc moved horizontally through the receptive field centre. The next histogram to the upper right shows the response obtained when the disc moved along the chord 1/2° higher etc. Time in each histogram is directed to the lower right. This schematic diagram may be alternatively viewed as a map of the local response, at a certain instant, of similar OFF-centre units spatially distributed over a region of retina across which the disc is moving in the direction indicated by the arrow. The angular distance between histograms was made much greater than the angular distance along them, in order that each may be seen. The disc illustrated is drawn with a similar linear distortion (From RODIECK and STONE, 1965 a)

(Fig. 4), and hence will contain information about the contrast and velocity of the stimulus. Moreover these arguments can be quickly extended to ON-centre fields and to white stimulus shapes.

RODIECK and STONE used only constant velocity stimulus movements (usually 10°/sec), but this analysis almost certainly can be extended to saccade-like movements, and to stationary flashing patterns. Stationary flashing patterns consist of one or more contrast borders, each of which can be assumed to generate a localised pattern of firing, like that discussed in Section I above. BAUMGARTNER'S (1961 a) plots of the peak firing rates of cat retinal ganglion cells as a function of the position of a contrast border relative to their receptive fields confirm this idea. Receptive fields located near the contrast borders of the retinal image are particularly excited or inhibited so that the configuration of borders in the stimulus pattern determines the distribution of excitation and inhibition within the retina. Some basic similarities between saccade-like image movements and stationary flashing patterns, as visual stimuli, are discussed in Section III.

In conclusion, spatially distributed populations of non-selective cells seem able to code in their firing rate important features of a stimulus, including its shape and contrast and the direction and speed of any movement. It is necessary to assume that higher visual centres can interpret the simultaneous firing of a population of cells, and can resolve the spatial distribution of their receptive fields in the retina.

Which is Important in Form Discrimination? Selective receptive fields are responsive to more specific features of the visual image than non-selective fields, though it must be stressed again that they are far from completely selective in their responses. Nevertheless even partial selectivity presumably makes for less redundancy (BARLOW'S (1961 a, b) term) in the coding of information; which is to say that fewer receptive fields and/or fewer impulses are needed to code a stimulus feature, provided that feature is a trigger feature. It is initially surprising, therefore, that this more specific and efficient form of coding is uncorrelated, perhaps negatively correlated, with good form discrimination. Form discrimination is best developed in higher primates, which have non-selective retinal fields, and least developed in lower vertebrates like the frog, which has principally selective retinal fields.

In tentative explanation, it can be pointed out that the trigger features of most selective retinal fields do not seem particularly useful for form discrimination. Specification of the direction or velocity of image movement is no doubt useful for visual localisation; the movement-gated convex edge detectors of frog retina might serve as "bug-detectors" (MATURANA et al., 1960), but can code little about the shape of the bug; the direction selective fields of rabbit retina may (OYSTER and BARLOW, 1967) serve as detectors for controlling eye movements. Even the orientation-selective fields of rabbit retina are selective for either horizontally or vertically oriented edges (LEVICK, 1967), and since the rabbit has no trouble discriminating obliquely oriented edges, LEVICK suggests that the rabbit must use other retinal receptive fields for this discrimination (perhaps the 40% of rabbit retinal fields which are non-selective; BARLOW et al., 1964). Moreover species differences in the selectivity of retinal fields seem inversely related to the phylogenetic development of neocortical visual centres. Selective fields predominate in the retinae of the pigeon and frog, in which the optic tectum is the dominant visual centre, the visual cortex being absent or rudimentary. Non-selective fields predominate in cat and monkey (and probably in man) in which species the visual cortex is highly developed. Between these extremes the rabbit and ground squirrel have mixtures of selective and non-selective retinal fields, and their visual cortex is at an intermediate stage of development.

Why have non-selective receptive fields become predominant in the retinae of higher mammals and primates? Our present understanding of retinal coding and central visual processing is probably too rudimentary to provide a definite answer. One possibility emerges from the observation (Hubel and Wiesel, 1962, 1965 a, 1968) that all, or the great majority, of the receptive fields of visual cortical neurones of the cat and monkey are selective in the present sense. The development of selectivity seems to have been "delayed" to the cortical level in higher mammals. Two possible reasons for this can be suggested. First, stimulus selectivity or specificity implies a loss of generality. A selective field is more efficient in coding its trigger feature but makes less contribution to coding non-trigger features. Many selective fields are required to cover each variant of a trigger feature (such as contour orientation) which might occur at any retinal point. The employment of many cells with slightly different trigger features, but congruent receptive fields, is indeed found in the visual cortex (see below). It requires a great proliferation in the number of cells processing visual input, a proliferation which is present in, but not before, the visual cortex. Second, the processing of form information in the cortex can coincide with other processing, in particular the establishment of binocular vision and stereopsis (see below and with Bishop, Chapter 4).

2. Coding Transformation in the Lateral Geniculate Nucleus

The visual image is coded by the retina as the modulation of ganglion cell firing. This activity can reach the visual cortex through a number of pathways, most directly via the dorsal division of the lateral geniculate nucleus (LGN). The axons of retinal ganglion cells reach the LGN via the optic tract and monosynaptically activate LGN neurones, whose axons project to the visual cortex. The structure and neurophysiological properties of the LGN are considered in detail in later chapters by Szentágothai and Freund. Here we are concerned with the transformations which take place in visual coding in the LGN. This question is best approached, as in the retina, in terms of receptive field organisation. How do LGN receptive fields differ from retinal fields? Retina-LGN differences have been best described for two receptive field types, viz. the non-selective fields in cat retina and LGN, and direction-selective fields in rabbit retina and LGN.

Hubel and Wiesel (1961) noted that cat LGN fields are concentric in organisation, with antagonistic centre and surround regions. Like retinal fields they respond less to diffuse illumination, which activates both centre and surround simultaneously, but are strongly excited by flashing spot stimuli which stimulate just the centre region. The fields they studied showed no direction-selectivity or other trigger feature specificity. The principal difference from retinal fields which they noted was the greater suppressive effect of stimulation of the surround on responses evoked from the centre. Kozak et al. (1965) studied the responses of

LGN fields to moving stimulus patterns. Most of their fields (80%) were concentrically organised, and showed no trigger feature specificity. The other 20% differed in various ways, but only 5% showed trigger feature specificity (they were direction-selective).

A more extensive comparison of retinal and LGN fields has recently been made by CLELAND et al. (1970 a, b). They recorded simultaneously from LGN neurones and from retinal neurones with coextensive receptive fields. Their analysis (1970 b) shows that many LGN neurones receive their entire excitatory input from just one ganglion cell; others receive excitatory input from two or several ganglion cells. They show, moreover (1970 a), that LGN fields have a three-component organisation. For example, a LGN cell driven by a single ON-centre ganglion cell has a circular ON-centre and annular OFF-surround, whose response characteristics and size appear to be determined entirely by the ganglion cell. In addition, each LGN cell has a much wider "suppressive field"; i.e. it is inhibited by any ON- or OFF-flash or by movement of a pattern, occurring in a fairly wide region (about 9° in diameter) outside the annular surround of the field. They suggest that the suppressive field component of LGN fields results from the convergence on each LGN cell of activity originating in a wide area of retina. This convergence almost certainly takes place within the LGN, and is probably mediated by inhibitory interneurones of the LGN. As these workers point out, this suppressive field is no doubt the basis of the strong "surround inhibition" of LGN cells reported by earlier workers.

As already discussed, the antagonism of the centre and surround of non-selective retinal fields makes them unresponsive to changes in diffuse illumination but very responsive to contrast borders moved or flashed over their centre regions. Functionally, the suppressive field constitutes an additional element of lateral inhibition in LGN fields. It presumably makes them even more sensitive than retinal fields to the position of a contrast border relative to their receptive fields. It presumably also makes LGN fields less responsive than retinal fields to long contrast edges, which extend out into the suppressive fields, i.e. LGN fields are probably selective, at least to some degree, for short contrast edges.

In the rabbit, LEVICK et al. (1969) have shown that directionally selective LGN fields, while resembling directionally selective retinal fields, show greater specificity for the preferred axis of movement, and are directly inhibited by stimulus movements which do not directly inhibit the retinal fields. These properties constitute an enhanced selectivity for direction on the part of LGN fields, which, these workers suggest, may result from the convergence of two or several direction-selective retinal cells onto one LGN cell, one or some being entirely inhibitory in influence.

In summary, LGN cells receive excitatory influence from one or a few retinal cells and therefore resemble ganglion cell fields in many ways, but they receive inhibitory retinal input which modifies their receptive field properties. These modifications seem, at least in the two examples best documented, to make LGN fields more selective for stimulus features than retinal fields.

In quantitative terms, the retinal afferents form only a minority of all afferents to the LGN (GUILLERY, 1969), but CLELAND et al.'s (1970 a, b) work shows clearly that retinal afferents are the predominant excitatory input to LGN cells. The effects of non-visual inputs on visually-evoked LGN responses has been studied by a number of authors (see, for example, BIZZI, 1966 a, b; KAWAMURA and MARCHIAFAVA, 1968; FELDMAN and COHEN, 1968). While it is clear that these inputs generally modulate the responsiveness of LGN neurones, a change in receptive field organisation has not been shown. These inputs may be important in visual attention and eye movements but are probably not more than sensitivity-modifying influences on basic mechanisms of form discrimination.

3. The Role of the Midbrain Visual Centres

The dorsal portion of the midbrain contains important visual centres in all vertebrates. In fish, amphibia, reptiles and birds the optic tectum of the midbrain is the most prominent visual centre of the brain. In the higher vertebrates in which the forebrain visual centres are highly developed, the roof of the midbrain still contains important visual centres, including principally the superior colliculus and the pretectum.

What role do these structures play in form vision? It is difficult to formulate a precise answer at present, but the problem can be approached in the following way. It is a reasonable assumption, though it has not been tested, that form vision in lower vertebrates is mediated by the optic tectum. In higher vertebrates form discrimination involves the visual cortex, interacting perhaps with other cortical areas and with the superior colliculus (see below and SPRAGUE, BERLUCCHI and RIZZOLATTI, chapter 14). Is there some basic difference in form discrimination between "tectum-based" and "cortex-based" species?

Some observations suggest that form discrimination may be rudimentary or non-existent in some tectum-based species. In the frog, for example, the retinal and tectal receptive fields appear well suited to detect the location and approximate size of a moving object, but to encode relatively little about its form (LETTVIN et al., 1961; GAZE and KEATING, 1970). Gaze and Keating suggest that "form deprived of movement seems to be behaviourally meaningless" to frogs. On the other hand, controlled behavioural experiments have demonstrated apparent form discrimination in lower vertebrates and invertebrates. SUTHERLAND (1969) has compared the criteria by which the octopus, goldfish and rat distinguish certain shapes. All three species could distinguish the test shapes (a parallelogram vs. a square), but on the basis of a number of tests, SUTHERLAND concluded that their criteria for the distinction were different. The octopus (an invertebrate) relied on the filled shape of the object (and its training did not transfer, for example, when it was tested with outlines, instead of filled shapes). The rat (with a partially developed visual cortex) seemed to judge more in terms of contours and outlines. The goldfish (optic tectum) was intermediate in performance, seeming to rely on both outline and filled shapes.

SUTHERLAND'S approach is clearly an important one, but sufficient data are not yet available to generalise about species differences in the criteria used for form discrimination or, consequently, to answer the present question. In some ways the similarities between species are as striking as the differences. LASHLEY (1938) noted at the end of his detailed study of form discrimination by the rat that "if a series of patterns is ranked in order of the conspicuousness of the figures for the human eye, that order will have a high predictive value for the rate at which the rat can learn the figures. Stimuli to which the rat transfers in equivalence tests are those obviously similar for man." In confirmation, SUTHERLAND and

Williams (1969) have shown that rats appear to discriminate a regular from an irregular checkerboard pattern by abstracting the "regularity"; their training in one discrimination transfers to other tests with different irregularities, sizes and orientations of the checkerboards. As these authors stress, regularity is just one example of the very many abstractions which humans can make. But it is of great interest that the rat, with no central retinal specialisation comparable to the fovea and a far less developed LGN and visual cortex, makes an abstraction also made by humans.

For the present, a conservative conclusion is that the criteria used by lower and higher vertebrates in form discrimination are qualitatively similar and that the higher vertebrates are distinguished by the great range and complexity of the visual abstractions of which they are capable. Teuber (1960, p. 1610) reached a similar conclusion, viz. that there is "no support for assuming any abrupt changes in the evolution of vertebrate pattern vision". Beyond this, however, the present question of the relative form discriminative ability of "tectum-based" vertebrates remains largely unanswered.

Does the Superior Colliculus Play a Role in Normal Form Discrimination? When the visual cortex is ablated in cat or monkey or in man by trauma, some visual function remains (Chapter 14), including the ability to distinguish light from dark, to detect gross movements and probably to distinguish gross patterns. Since the colliculus and pretectum are the most prominent remaining visual centres (the LGN undergoes retrograde degeneration due to severance of many of its axons in the cortex) this remaining visual function is generally attributed to them. Although these observations tell something of the capabilities of these midbrain centres after destruction of the geniculo-striate pathway, the question of what role they play in normal form discrimination remains. Recent studies directed to evaluating this role have yielded conflicting conclusions.

In a series of studies of the golden hamster, Schneider (1968, 1969) drew a distinction between *visual orientation* and *visual discrimination*. He showed that lesions to the superior colliculus selectively destroy the hamster's ability for visual orientation, and lesions of the visual cortex selectively disturb pattern discrimination, leaving visual orientation functions intact. The detailed arguments are beyond our present scope, but substantial support for the idea that visual orientation and visual form discrimination are independently organised has been provided by the reviews of Trevarthen (1968) and Held (1968).

Schneider's clear-cut allocation of visual orientation to the superior colliculus and form discrimination to the visual cortex was very welcome for it seemed to resolve the longstanding enigma of collicular function (summarised by Pasik et al. (1966) and Andersen and Symmes (1969), among others.) More recent evidence seems, however, to challenge this allocation, or at least its generality among different species. Sprague et al. (1970) have, for example, reported severe disturbances of pattern discrimination in cats following lesions which included both the superior colliculus and the adjacent pretectal region but left the visual cortex intact. These deficits were not caused by lesions restricted to the colliculus: Sprague et al. stress the importance of the *pretectum*. It may still be too soon to conclude that midbrain centres are not important in normal form discrimination.

Visual Receptive Fields of Optic Tectum and Superior Colliculus: Are They Suitable for Form Discrimination? Receptive fields have been studied in the optic tectum of the frog (LETTVIN et al., 1961; FITE, 1969; GAZE and KEATING, 1970), goldfish (CRONLY-DILLON, 1964; JACOBSON and GAZE, 1964) and pigeon (HOLDEN, 1969) and in the superior colliculus of the rat (HUMPHREY, 1968), rabbit (HILL, 1966; HORN and HILL, 1966), ground squirrel (MICHAEL, 1967), opossum (HILL and GOODWIN, 1968), cat (STRASCHILL and TAGHAVY, 1967; MCILWAIN and BUSER, 1968; STERLING and WICKELGREN, 1969; STRASCHILL and HOFFMANN, 1969 b) and monkey (HUMPHREY, 1968). STRASCHILL and HOFFMANN (1969 a) have described receptive field properties of cells in the cat pretectum. Recent reviews (e.g. HUMPHREY, 1968) have stressed the similarity of the receptive fields described, despite the great phylogenetic range involved.

Size. There is a great range of field sizes in all species, e.g. ranging from 2° to the entire visual field in the frog, and from 2° to 90° in the cat and rat. In general collicular fields are much larger than those found in the retina, LGN and visual cortex of the same species.

Receptive Field Selectivities. Tectal and collicular fields are almost all "selective" in the sense used above. The most common selectivity is for stimulus movement and various types of movement selectivity have been described. Orientation-selectivity is relatively uncommon, and direction-selectivity is usually less specific than in the retina, LGN or cortex. In the cat, the preferred directions of collicular fields are always directed away from the fixation point, suggesting a role for these fields in the correction of fixation errors (STRASCHILL and HOFFMANN, 1969 b).

Habituation. A tendency for the responses of collicular and tectal units to attentuate as a stimulus is repeated (response habituation) has been consistently reported. Its mechanism is not known; because of it receptive fields seem particularly responsive to newly-appearing stimuli and to slight changes in stimuli to which the cell has habituated. In general, response habituation is far more prominent in the colliculus and tectum than in the retina, LGN or visual cortex of any or these species.

These properties do not rule out a form-coding function for the optic tectum or superior colliculus but the large size of the receptive fields, their tendency to habituate and their lack of orientation specificity would seem to make them unsuited to the coding of edges and contours with high spatial resolution. These receptive field properties certainly give weight to the idea that detailed vision in higher vertebrates is normally mediated by the geniculo-striate pathway.

4. Visual Coding in the Neocortex

Our present knowledge of the functions of the neocortex in form discrimination is derived from three principal sources:

(i) Physiological and anatomical studies have sought to determine the limits, topographical organisation, neural connections and cytoarchitecture of those areas of the cortex which are concerned with vision;

(ii) Ablation studies have sought to define the functions of the visual areas of neocortex in terms of the visual functions lost when part or all of these areas are destroyed;

(iii) Studies of the receptive fields of single cortical neurones have sought to establish the coding properties of individual cells.

The receptive field studies have been very largely confined to the cat and monkey; since these two species have also been intensively studied by mapping and ablation techniques, present knowledge of the neurophysiological mechanisms of cortical processing of form information is most advanced in them. The following discussion will therefore be largely confined to the cat and monkey, but where data are available human properties are also discussed. The structural and physiological organisation of the visual cortex is considered in detail in Chapters 19 to 22. Here we are concerned specifically with the functioning of the neocortex in form discrimination.

The Visual Areas of the Cortex: A Brief Summary. *Monkey.* In the monkey the entire cortical projection of the LGN appears to go to the primary visual (striate) area of the occipital cortex (area 17 of Brodmann) (see, for example, Cowey, 1964; Wilson and Cragg, 1967; Diamond and Hall, 1969; Whitteridge, chapter 19). Surrounding area 17 are the secondary visual areas which, on cytoarchitectonic and mapping grounds, have long been considered to consist of an inner and an outer circumstriate belt (areas 18 and 19 of Brodmann, respectively). Zeki (1969 a) critically challenged this subdivision but has subsequently suggested an analogous subdivision based on new observations (1969 b). Areas 18 and 19 do not receive LGN fibres, but they receive fibres from, and send fibres back to, area 17 and the parts of areas 17 and 18 along their joint border send fibres to, and receive fibres from, the contralateral visual cortex. The projection of the visual field on area 17 is retinotopically organised (Talbot and Marshall, 1941; Hubel and Wiesel, 1968), and there appears to be retinotopic organisation in at least area 18 as well (Cowey, 1964; Hubel and Wiesel, 1970).

Within the ipsilateral hemisphere the circumstriate belt (areas 18 and 19) projects to parietal and prefrontal cortex and to the inferotemporal cortex (Kuypers et al., 1965). This latter region has been shown to function in higher order, exclusively visual tasks (see Gross, chapter 23). Hence, in the monkey cortex, the areas devoted entirely to visual processing can be considered as a hierarchy, extending from area 17, which receives visual input, to the circumstriate cortex (areas 18 and 19), and thence to the inferotemporal cortex.

Cat. Areas 17, 18 and 19 are distinguishable in the cat (Otsuka and Hassler, 1962; Sanides and Hoffman, 1969; Hubel and Wiesel, 1962, 1965 a). There is, however, a major difference between cat and monkey in the pattern of geniculo-cortical projection. In the cat the LGN, instead of projecting to only area 17, projects to four predominantly visual areas of cortex (areas 17, 18 and 19 and the lateral wall of the suprasylvian gyrus) (Garey and Powell, 1967; Wilson and Cragg, 1967; Niimi and Sprague, 1970) and appears to project to non-visual areas of cortex as well (Marty et al., 1969; Niimi and Sprague, 1970). The four visual areas are all retinotopically organised (Hubel and Wiesel, 1962; Wilson and Cragg, 1967; Garey and Powell, 1967; Niimi and Sprague, 1970). Area 17 projects to each of areas 18 and 19 and to the suprasylvian gyrus, and areas 18 and 19 project back to 17 (Wilson, 1968). There are, however, considerable differences in the anatomical connections of these areas. They receive different afferent projections from the sub-units of the LGN (Garey and Powell, 1967;

NIIMI and SPRAGUE, 1970), and only areas 18 and 19 appear to project back to the LGN; area 17 apparently does not (HOLLÄNDER, 1970).

Hence, the cat has some elements of the monkey's hierarchy of visual cortical areas (viz. the projection from the LGN to 17, and from 17 to 18 and 19) but it lacks the highest level, since there is no counterpart in the cat to the inferotemporal cortex (Chapter 23). The parallel projections from the LGN to 18, 19 and the suprasylvian gyrus also run counter to the idea of a hierarchy of cortical visual areas in the cat. DIAMOND and HALL (1969) have persuasively argued that the different organisations of cat and monkey visual cortex represent, not different stages of phylogenetic development, but rather different paths of development from a common ancestor. They consider the primate organisation to be an arboreal specialisation, since several features of it, in particular the exclusive projection of the LGN to area 17, are found in arboreal species such as the tree-shrew and squirrel. By contrast the cat, a ground-dwelling carnivore, resembles pre-mammalian forms, in which the circumstriate area receives a direct projection from the LGN. In these species, Diamond and Hall suggest, the circumstriate area must be regarded as a sensory receiving area in parallel with area 17, rather than as visual association cortex or even secondary visual processing cortex.

Man. The striate visual area of human cortex is situated along the lips of the calcarine fissure of the occipital lobe. Gross features of its organisation and connections are well established (see CROSBY et al., 1962, pp. 454–460, for a review). The area is retinotopically organised, and the macular representation is large. The principal afferent pathway is the optic radiation from the LGN; the main efferent pathway (so far identified) is to circumstriate cortex. Circumstriate cortex has been divided into areas 18 and 19, though the border is not sharp (BAILEY and VON BONIN, 1951, pp. 224–228). Area 18 is connected by association fibres to many non-visual areas of cortex and subcortical areas (CROSBY et al., 1962). According to CROSBY et al., the connections of 18 to temporal cortex are multisynaptic. In most of these features human visual cortex resembles the monkey's but data is not available on many important points, such as the cortical target areas of the LGN.

Effects of Cortical Ablation on Form Discrimination. *Monkey.* Bilateral ablation of the *striate cortex* causes apparent blindness in monkeys (HUMPHREY and WEISKRANTZ, 1967). Nevertheless, KLÜVER (1941) showed that destriate monkeys can respond differentially to the luminous energy of a light stimulus and that in certain conditions "the topographical aspects of stimulus configuration may (also) become effective" in determining behavioural responses. KLÜVER hesitated to conclude from this that destriate monkeys have a rudimentary ability to discriminate patterns and most subsequent workers have assumed that they do not. WEISKRANTZ (1963) showed that a destriate monkey can distinguish a pattern with many contours from one with few. HUMPHREY and WEISKRANTZ (1967) showed that destriate monkeys can also accurately localise objects in visual space. WEISKRANTZ and COWEY (1970) argue that these functions can be readily accounted for by the properties of the receptive fields found in the monkey's superior colliculus (HUMPHREY, 1968), and by the visual-localising functions of the superior colliculus, defined in other species (e.g. SCHNEIDER, 1969). Lesions in the *circumstriate cortex* produce less dramatic results, and many authors have reported no visual deficit. ZEKI (1967, 1969 a, b) has reassessed these findings, concluding that circumstiate ablation always produces deficits in learning or retention of a visual discrimination. Lesions of *inferotemporal cortex* produce a permanent, exclusively visual deficit (see Chapter 23 for a detailed consideration). Sensory visual functions, such as visual acuity, are little affected, but the aquisition and retention of difficult discriminations is severely impaired. The effects of inferotemporal ablation can also be produced by destroying the projection to inferotemporal cortex from circumstriate cortex. Visual input can also reach inferotemporal cortex via the superior colliculus and pulvinar, but the importance of this pathway has still to be demonstrated (see also DIAMOND and HALL, 1969). The severe effects of lesions of area 17 are compatible with the idea that it is the exclusive sensory receiving area of the cortex. The "high-order" disturbances caused by ablation of the inferotemporal cortex are consistent with this being the uppermost level of the hierarchy.

Cat. The effects of cortical ablation on form discrimination in the cat are not well defined. This is only partly a result of the more diffuse cortical projection of the LGN in the cat, since most workers have used lesions which included all of areas 17, 18 and 19. It is also a result of the imprecise use of the terms "form discrimination" and "pattern discrimination".

The question of what constitutes an effective test of pattern discrimination has in fact not been settled; this point has already been made by DODWELL and FREEDMAN (1968). Many of the tests used have contained local differences in luminous flux (see for instance SPEAR and BRAUN'S comments, 1969), or have been too simple. Horizontal vs. vertical stripes (GANZ and FITCH, 1969; SPEAR and BRAUN, 1969), horizontal vs. vertical bar (SMITH, 1938), gratings vs. checkerboards (WETZEL et al., 1965) might all be discriminated by subcortical mechanisms whose normal function does not include form discrimination. This is brought out by the elegant (though isolated) observation of DIAMOND and HALL (1969) that the destriate tree-shrew could learn to distinguish an inverted from an upright luminous triangle as fast post- as pre-operatively. But when the patterns were made slightly more complex by surrounding each triangle with a circle, postoperative performance dropped to chance. It may be very difficult to make a clear distinction between discrimination tasks which test pattern discrimination and those which test simpler visual functions, and it certainly seems unlikely that any single pair of discriminanda could be devised which would assess pattern discrimination. Klüver's (1941) comment still seems relevant: "A survey of the recent literature indicates that the role of the occipital lobes in visually guided behaviour . . . cannot be properly evaluated as long as investigators are satisfied with studying only a few visual functions".

Nevertheless, there seems to be general agreement that in the cat substantial cortical lesions cause substantial and permanent loss of visual discrimination ability. SPEAR and BRAUN (1969) concluded that "while pattern discrimination ability is present when visual cortex is removed, it is profoundly retarded". SMITH'S (1938) decorticate animals were slow to learn or failed to reach criterion on his simple test. MEYER (1963) reported failure of decorticate cats on a stripe vs. checkerboard distinction. WINANS' (1967) cats could, with one exception, discriminate all the pairs of triangles presented, but she did not compare their rate of learning or the limits of their performance with normal cats. The separate effects of ablations of area 17 and of areas 18 and 19 do not appear to have been investigated, however.

Man. Destruction of the striate cortex in man produces a severe loss of visual function over the whole visual field; partial destruction causes loss of function over part of the field (scotoma). Within a cortically blind area it is sometimes held that there is no visual function, except for pupillary and perhaps blinking reflexes (CROSBY et al., 1962). Other investigators have concluded that blindness produced by cortical lesions may never be complete. TEUBER et al. (1960) stress the incompleteness of many reports. Awareness of luminous objects is commonly reported in patients with occipital lobe lesions, and in many cases there is a gradual return of visual function after cortical damage, beginning with an awareness of light, then awareness of movement. Subsequently contours may be seen indistinctly, still later some degree of color sensation may return. Correspondingly a scotoma (limited region of blindness) within which there is severe blindness, is usually smallest when tested with stationary lights. The area within which the subject is blind or largely blind to contours or colours is rather wider. The scotoma appears widest when considered as the area within which contour, shape and colour vision are more or less disturbed. TEUBER et al. conclude that "there is a functional hierarchy of different aspects of vision which can be derived from their relative vulnerability to cerebral lesions". In general, however, the (unavoidable) lack of histological control in these reports makes it difficult to conclude whether the simplest functions, such as brightness and movement awareness, are cortically dependent.

Electrical stimulation or pathological irritation of areas 17, 18 and 19 give rise to only simple visual impressions (flashes, colours and simple objects, CROSBY et al., 1962). By contrast, stimulation or irritation of temporal cortex in man produces complex visual hallucinations, involving complex scenes and temporal sequences (PENFIELD and PEROT, 1963), and ablation of temporal cortex produces defects in higher visual function, such as recognition and memory, but not in simple sensory function (Chapter 23). Destruction of either temporal lobe in man impairs visual function but there appear to be differences in the deficits produced by left and right removal (see DORFF et al., 1965; MILNER, 1968). Left temporal lobectomy severely impairs verbalisation of visual experience. Right lobectomy appears to leave verbal ability intact, but produces a deficit in the ability to discriminate complex patterns, and a severe deficit in visual memory. This lateralisation of function may be unique to man, but in many ways the visual deficits caused by temporal lobectomy in man resemble those seen in the monkey.

Receptive Fields in Cat Visual Cortex. Cortical receptive fields have been studied most intensively in the cat (HUBEL and WIESEL, 1959, 1962, 1965 a, 1969 a; BURNS et al., 1962; BURNS and PRITCHARD, 1964; BAUMGARTNER et al., 1965; BARLOW et al., 1967; PETTIGREW et al., 1968 a, b; HENRY and BISHOP, 1971). HUBEL and WIESEL (1962, 1965 a) described three types of receptive field (simple, complex and hypercomplex), and suggested that they code visual information in successively more complex and specific ways. All these field types have quite complex and selective properties when compared with retinal or LGN fields.

Simple Fields. Like retinal and LGN fields, simple cortical fields respond both to moving and to stationary flashing stimuli, but they show marked selectivity for stimulus parameters. All show a marked orientation-selectivity. Although like retinal fields, many simple fields have antagonistic ON and OFF areas, these are arranged in parallel strips, not concentrically (HUBEL and WIESEL, 1959;

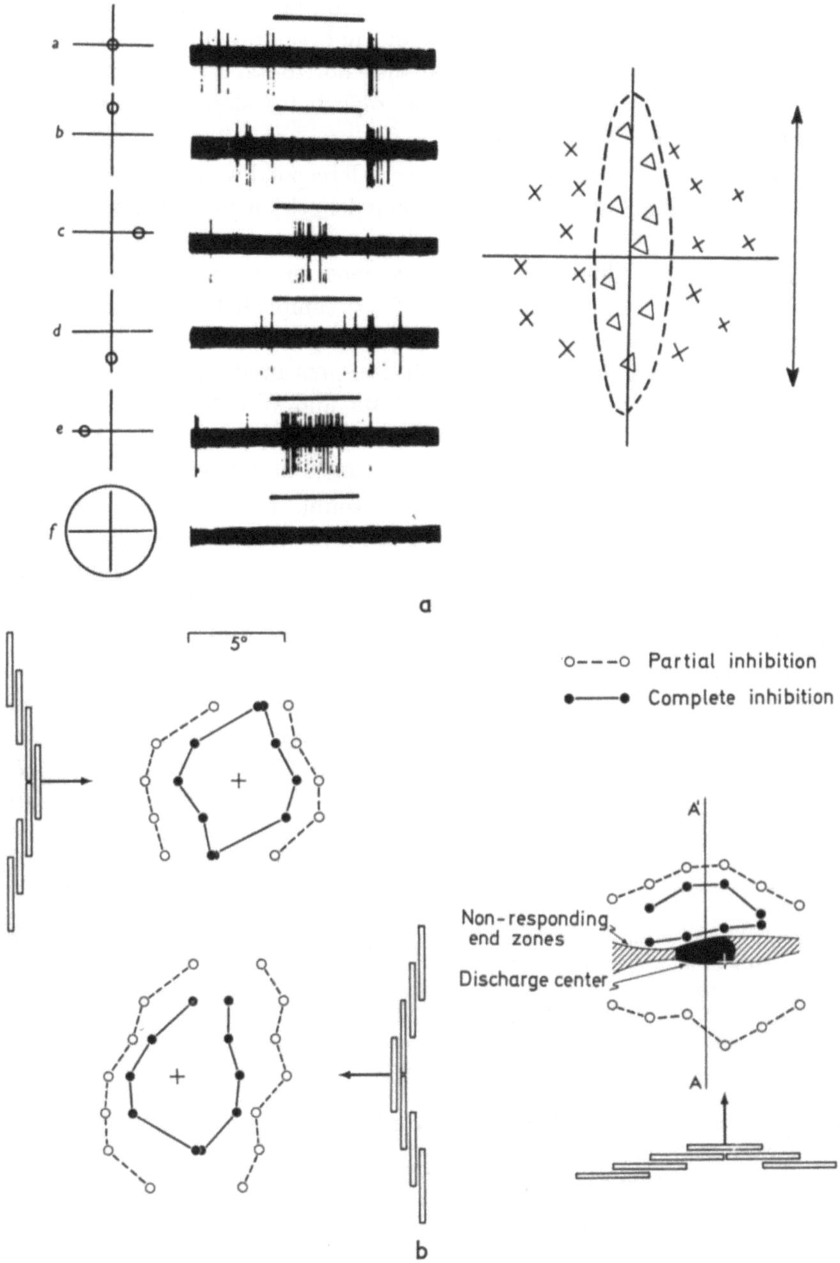

Fig. 5 a and b

Fig. 5a). Consequently the most effective stationary flashing stimulus is a bar of light correctly oriented and positioned to cover either an ON- or an OFF-strip (Fig. 6a). Bars not parallel to the ON- and OFF-strips excite antagonistic areas simultaneously, giving weaker responses. A corresponding orientation-selectivity is seen with moving stimuli (Fig. 6b; Hubel and Wiesel, 1962; Pettigrew et al., 1968 a; Henry and Bishop, 1971). The orientation selectivity is quite sharp, deviations of 10° from the optimal position often abolishing the response. All angles of preferred orientation are found (c.f. the orientation-selective fields of rabbit retina, in which only vertical and horizontal preferences are found, Levick, 1967). Pettigrew et al. (1968 a) reported a slightly higher frequency of horizontal and vertical preference.

Simple fields are often direction-selective. All respond well to correctly oriented bars moving in the direction normal to their length; many fields respond more strongly to one direction of movement than to the opposite. Simple fields also show a preference for slow speeds of movement, most responding optimally to speeds of about 2°/sec (Pettigrew et al., 1968 a). Simple fields have been described only in area 17. Their size varies consistently with position in 17; they are smallest in that part of 17 in which the area centralis is represented. Here the width of their central region may be as little as 0.25° (Hubel and Wiesel, 1962), which is the same size as the smallest retinal fields. They are recorded most frequently in cortical layer IV, which is a principal afferent-receiving layer.

Complex Receptive Fields. These fields resemble simple fields in responding well to moving bar stimuli and in being selective for the orientation of the bar. They differ in giving much weaker responses to flashing stimuli, and in having larger receptive fields (Hubel and Wiesel, 1962). Hubel and Wiesel described complex fields selective for the contrast (dark or bright) and width of a bar stimulus. In these two ways they are more selective than simple fields; on the other hand, complex receptive fields are larger and less selective for stimulus position than simple fields. Complex fields are found commonly in all of areas 17, 18 and 19. They are generally smaller in 17 and 19 than in 18 and within each of 17 and 18 their size has been shown to vary consistently with eccentricity in the visual field, being smallest at the area centralis. They are recorded only rarely in cortical

---

Fig. 5 a and b. Simple receptive fields of cat visual cortex. a: A simple cortical receptive field, as plotted by Hubel and Wiesel (1959) using a stationary flashing 1° spot of light, is shown on the right. The areas marked X gave ON-responses, the areas marked Δ gave OFF-responses. The scale shows 4°. On the left are shown typical ON- and OFF-responses generated at positions in the receptive field shown on the diagrammatic axes at extreme left (a—f). The duration of spot presentation was 1 second, indicated by the horizontal bar over each spike train. b: A simple cortical field plotted with moving slit stimuli, (From Henry and Bishop, 1971). When the slit was optimally (horizontally) oriented (right part of figure) the field is shown to have an excitatory discharge centre and flanking inhibitory regions. The cell is excited when the bar crosses the excitatory region, and inhibited as it crosses inhibitory flanking zones, i.e. the cell is inhibited as the stimulus approaches and leaves the excitatory region. Tested with differently oriented slits (left part of figure), the entire receptive field appears to be inhibitory in influence. These authors stress that the location of the excitatory and inhibitory regions of the receptive field cannot be predicted from the responses evoked by stationary flashing stimuli

layer IV, being more common in deeper and more superficial layers (HUBEL and WIESEL, 1962).

Hypercomplex Fields. Present understanding of hypercomplex fields stems entirely from the analyses of HUBEL and WIESEL (1965 a, 1968). They considered the stimulus selectivity of these fields to be greater than that of simple and

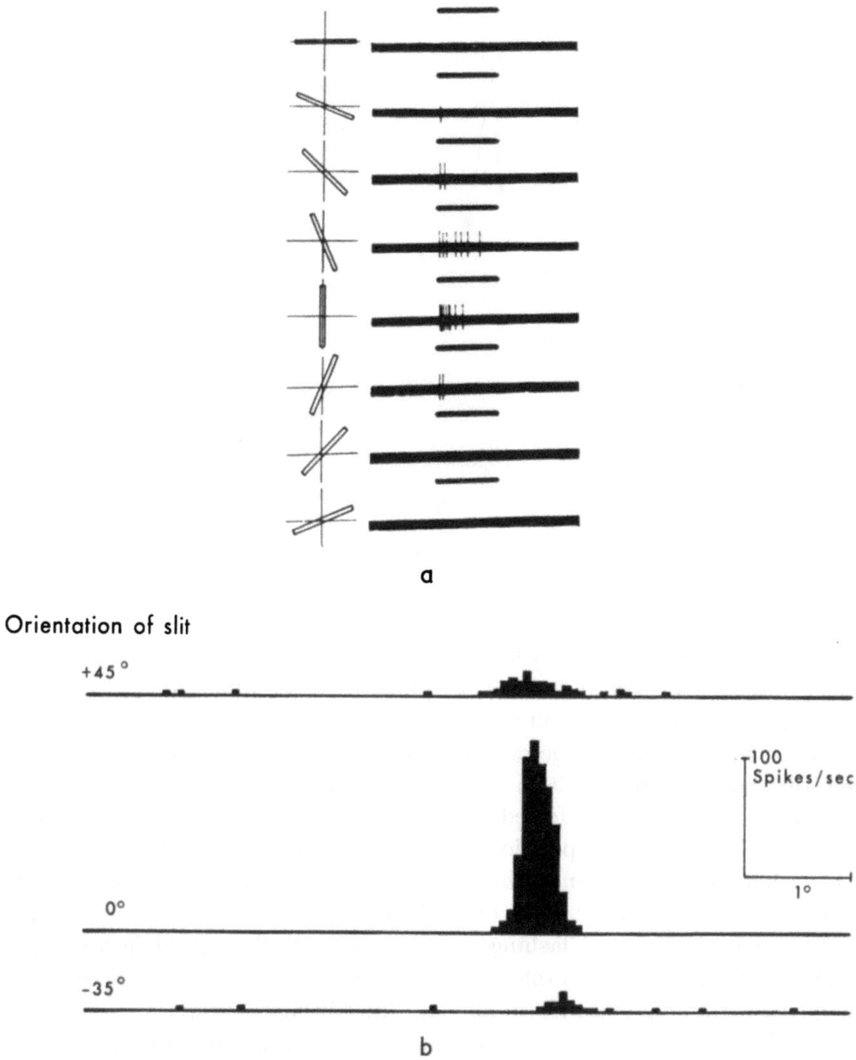

Fig. 6 a, b

Fig. 6a—d. Orientation-selectivity of a simple cortical receptive field. a: Responses of a vertical-selective simple field, to a 1° by 8° bar of light positioned over the receptive field at the orientations shown at left, and flashed ON and OFF. The gradation of response as the orientation deviates from vertical is clearly apparent. (From HUBEL and WIESEL, 1959). b: Averaged responses of a simple field to moving slit stimuli with different orientations. c: The variation of response with slit orientation is shown in the graph. The way in which orientation was varied is shown in d (From HENRY and BISHOP, 1971)

complex fields, and within this group they recognize two types, "lower-"and "higher-order". Lower-order fields are found in all of areas 17, 18 and 19, being specially frequent in 19, higher-order fields only in 19.

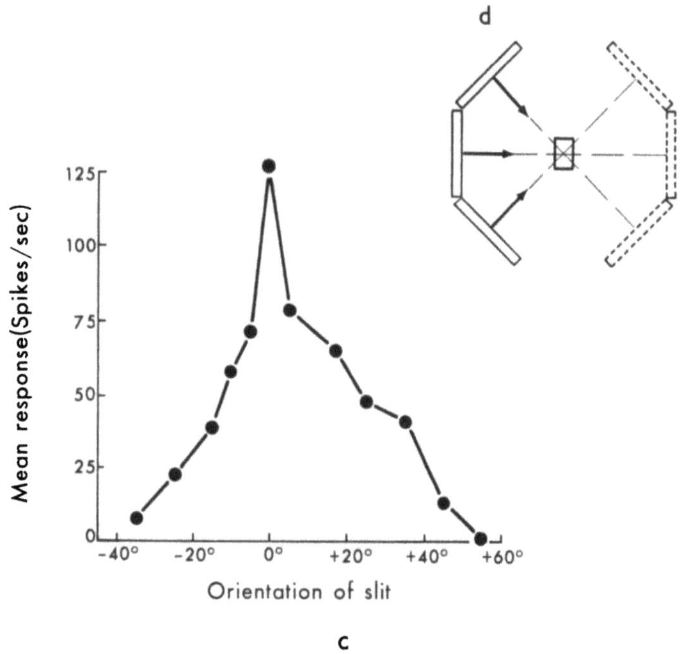

Fig. 6 c, d

Lower-order hypercomplex fields, in addition to being selective for the orientation, width, contrast and often direction of a stimulus, are selective for its length. Excitatory and inhibitory zones could be described for each field, such that a correctly oriented edge or bar would excite the cell as it moved across the excitatory zone and inhibit it as it passed over an inhibitory zone. Since the zones are arranged along the preferred axis (Fig. 7), the optimal stimulus is a bar or edge which ends at such a position that it crosses the full length of the excitatory zone without extending onto the inhibitory zone. Some fields have inhibitory zones at one end, some at both ends. Moving stimuli are generally much more effective than stationary flashing shapes, and stimulus speed is an important parameter. Thus a considerable number of parameters have to be specified for the optimal stimulus and individual cells differ in many ways.

Higher-order hypercomplex cells differ from lower-order in that a correctly oriented and stopped edge or bar could elicit a response by moving in either of two directions oriented at 90° to each other. In the two examples described in Hubel and Wiesel's paper (1965 a) the optimal stimulus was a correctly positioned and oriented, double-stopped edge. The cell responded strongly if the stimulus was moved either in the direction of the edge or normal to it. If the edge was rotated more than 15° from this optimal orientation these movements (relative to the edge) were ineffective.

Non-Orientation Selective Cortical Fields. Persistent but fragmentary reports appear in the literature of cortical receptive fields which are non-selective, most notably for stimulus orientation (Baumgartner et al., 1965; Denney et al.,

1968; BROWN and BAUMGARTNER, 1964; JOSHUA and BISHOP, 1970; HUBEL and WIESEL, 1970). They are uncommon in the cat, though JOSHUA and BISHOP reported them to be more frequent in the area representing peripheral retina. BROWN and BAUMGARTNER (1964) and JOSHUA and BISHOP (1970) concluded that some at least are cortical neurones, and not recordings from afferent fibres from the LGN, since they are binocularly activated, and the responses of some to stationary flashing stimuli are quite different from responses of LGN neurones. Their functional importance has still to be established.

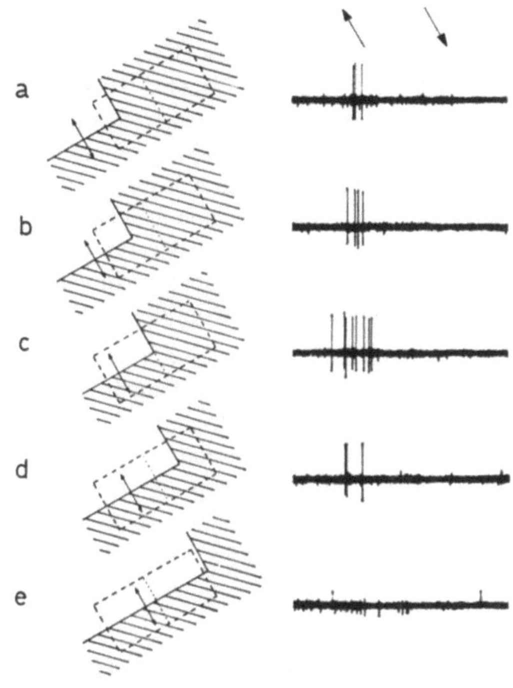

Fig. 7a—e. Plot of hypercomplex cortical field (From HUBEL and WIESEL, 1965 a). The field is approximately 2° × 4°, indicated by the broken-line rectangle. The long axis of the field indicates the preferred orientation of the field. The field can be divided into two halves along this axis, as indicated by the dotted line. The lower-left half is excitatory, the upper-right half inhibitory. Responses a, b, c show that as a bright edge moves across increasing proportions of the excitatory half, an increasingly strong response is elicited from the cell. As the edge extends into the inhibitory half the response is diminished (d, e). The cell is thus optimally responsive to an edge which ends near the border between excitatory and inhibitory halves

Binocularity. Probably all the receptive fields of cat visual cortex can be influenced from both eyes (HENRY et al., 1969). In a minority of cells the influence from one eye is inhibitory or provides only sub-threshold excitation (HENRY et al., 1969). Such fields can be directly excited only from one eye (HUBEL and WIESEL, 1962, 1965 a). The majority of cortical cells can, however, be excited from both eyes. They have very similar receptive fields in the two eyes, and the fields are located at approximately corresponding retinal positions (HUBEL and WIESEL, 1962, 1965 a). The importance of binocular interaction on single cells for the understanding of depth discrimination is considered in detail in Chapter 3. Since form discrimination is as good with monocular vision as with binocular, further consideration is not given here.

Receptive Fields of Monkey Visual Cortex. *Area 17.* The receptive fields of monkey area 17 show remarkable similarities to, and important differences from, cat cortical fields (Hubel and Wiesel, 1968, 1970 a). Among the similarities is the presence of the same basic types of receptive fields (non-orientation selective, simple, complex and hypercomplex). Their distribution among the cortical layers is similar to that seen in the cat but more marked, simple cells being particularly restricted to layer IV. The separate evolution of the cat and monkey from pre-mammalian forms (Diamond and Hall, 1969) makes these basic similarities very remarkable. Among the dissimilarities is the smaller size of monkey cortical fields. They are approximately one quarter the size of cat cortical fields with the same eccentricity. The smallest are found at the area of foveal representation, being less than $1/4° \times 1/4°$ (Hubel and Wiesel, 1968). However, the size of these small fields was not estimated more precisely. Many monkey cortical fields show colour specificity (Hubel and Wiesel, 1968; Gouras, 1970), which is probably not to be found in the cat (Pearlman and Daw, 1970). Finally, most simple fields of monkey cortex can be driven only monocularly whereas in the cat most can be activated from both eyes. Hubel and Wiesel suggest that "in monkey striate cortex impulses from the two eyes probably converge not so much on the simple cell, as in the cat, but chiefly on the complex cell".

Area 18. Hubel and Wiesel (1970 a) describe a new type of receptive field in monkey area 18, the binocular depth cell. Its salient feature is that its activation requires the precise alignment of the two eyes. This type of selectivity may be an extreme case of the facilitatory interaction seen as the right and left eye receptive fields of a cat cortical cell are superimposed (Barlow et al., 1967; Pettigrew et al., 1968 b; Chapter 3). The binocular depth cells constitute about half the cells of area 18. The other half are complex and hypercomplex fields, like those of area 17.

Columnar Organisation of Visual Cortex. One of the striking properties of the visual cortex is its suborganisation into functional columns (Hubel and Wiesel, 1962, 1963, 1965 a, 1968, 1969 b). Each receptive field (simple, complex or hypercomplex) has a certain preferred orientation. Hubel and Wiesel demonstrated that in all of areas 17, 18 and 19 of the cat and in area 17 of the monkey, cells with the same orientation preference are grouped together in columns which extend from the white matter to the surface of the cortex (Fig. 8). Between adjacent columns there is, by definition, a step change in the preferred orientation; within a column there is no change. There is a tendency for neighbouring columns to have only slight and ordered differences in preferred orientations, a tendency more marked in the monkey. The orientation columns appear to be irregular in cross-section (Hubel and Wiesel, 1963; Fig. 9); their width in the cat is typically 0.5 mm. In monkey cortex the orientation columns are apparently narrower (about 0.25 mm across).

Hubel and Wiesel (1968) noted that there appears to be columnar parcellation of monkey area 17 into eye preference columns; i.e. cells with the same eye preference (dominance) are grouped together. Hubel and Wiesel (1969 b) have provided a striking anatomical demonstration of these columns in monkey cortex. A weaker but analogous columnar system is present in the cat visual cortex

(HUBEL and WIESEL, 1965 b). The binocular depth cells of monkey area 18 appear to be grouped into columns according to their requirement of axis alignment (HUBEL and WIESEL, 1970 a).

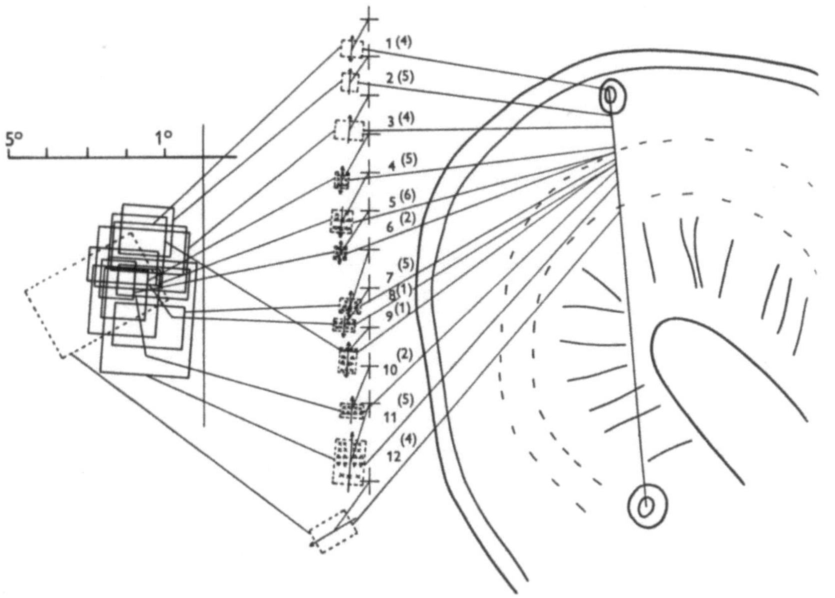

Fig. 8. The spread of receptive field positions within a column (From HUBEL and WIESEL, 1962). At right is a diagrammatic reconstruction of an electrode track through the post-lateral gyrus (area 17) of the cat. Thirteen cells were recorded along the upper one third of the track. Their receptive fields are shown individually in the centre part of the figure. The fields are numbered in the sequence in which they were recorded. Fields 1—3 and 12 were complex, and the remainder simple in organisation. The first twelve have the same orientation preference, i.e. they were recorded from the same orientation column. In the left part of the figure the fields are plotted superimposed, indicating the variation in size and position of the fields

HUBEL and WIESEL (1968, 1969 b) stress that the orientation and eye-preference columns in monkey area 17 appear to be organised independently. Orientation and eye preference columns overlap and there is no evidence of any correlation between the two sets of boundaries. The eye preference columns appear to be about twice as thick as the orientation columns. These workers suggest (1969) the possibility of columnar aggregations of cells according to their direction selective and colour coding properties as well. They suggest that the "analysis of stimulus parameters including eye dominance, stimulus orientation and possibly also direction of movement and color coding, requires a breaking up of the surface (of the cortex) into a number of independent and overlapping systems of mosaics".

The Topographical Organisation of Visual Cortex at the Columnar Level. HUBEL and WIESEL's demonstrations of the columnar parcellation of visual cortex raised the question whether the topographical organisation of visual cortex (which has been described many times from histological and gross physiological studies) can be detected within a small area of cortex, containing a discrete number of columns. HUBEL and WIESEL (1963, 1965 a) provide a detailed answer to this question.

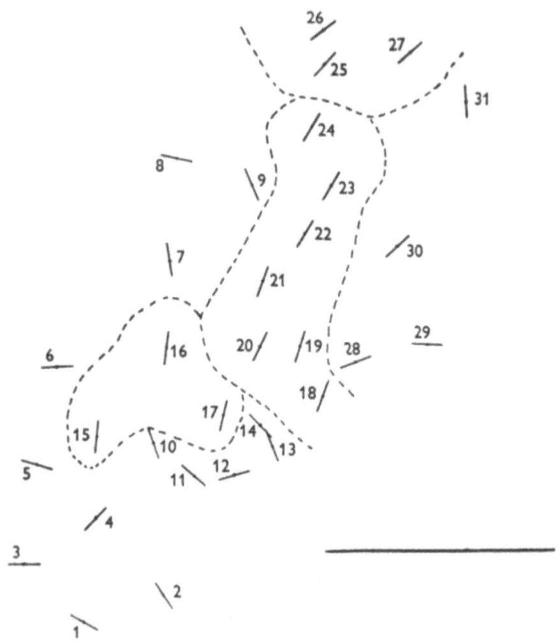

Fig. 9. Map covering the surface of the postero-lateral gyrus of the cat (area 17). The map shows the preferred orientations of cortical fields recorded in 31 penetrations. The interrupted lines indicate regions of relatively constant orientation preference, i.e. they mark the boundaries of orientation columns. The scale marks 1 mm (From HUBEL and WIESEL, 1963)

After recording the receptive field positions of many cells within a column, HUBEL and WIESEL (1965 a) concluded that within a column there is a considerable scatter in receptive field positions, but no systematic variations. Thus, in the example they illustrate (Fig. 8) the positions of 11 receptive fields in one column varied by about 3°, which also was approximately the diameter of the largest of the fields. They never observed a consistent shift in position between the top and bottom of a column, or from one side to another, and concluded that within a column there is no topographical organisation. In a related observation in their 1963 paper, these authors found no perceptible changes in receptive field position between adjacent columns. The regions they considered were in the representation of the area centralis; they extended over 1–2 mm of cortex and spanned several columns (Fig. 9). HUBEL and WIESEL concluded that the changes in mean position between adjacent columns must be small compared to the random variation within a column. HUBEL and WIESEL (1968) report the same basic findings in monkey area 17. Receptive fields are generally smaller in the monkey, and the scatter of field positions within a column is correspondingly smaller.

In describing the mode of termination in the cat LGN of retinal afferents from adjacent areas of the retina, STONE and HANSEN (1966) borrowed the term "partially shifted overlap" from LORENTE DE NÓ (1934). Afferents from adjacent (but separate) areas of retina terminate in volumes of the LGN which *overlap* considerably with each other, but which are *partially shifted* with respect to each other in accordance with the general retinotopic organisation of the LGN. An organisation of this sort in the overall retino-cortex projection would provide a basis both for the scatter of receptive field positions within a column, and for the overlap of positions between adjacent columns, while still being consistent with general retinotopic organisation of visual cortex.

Visual Coding by Cortical Receptive Fields. It seems clear from the above considerations that each small area of retina is represented by a considerable population of cortical receptive fields. Features of a stimulus appear to be coded in terms of differential activity among functional subgroups of that population. For example, cortical cells generally have small receptive fields, especially those cells representing the fovea or area centralis. Consequently, the retinal position of a stimulus feature can be coded in terms of which sub-population of cells is activated. Analogously contours with different orientations will excite different subgroups of cortical fields, even though they are imaged on the same small area of retina. Moreover the positions of the ends of a contour will determine which group of hypercomplex fields it excites. An increasingly sophisticated and detailed body of evidence is accumulating that binocular interaction in cat cortical fields may provide a neurophysiological mechanism for depth discrimination (BARLOW et al., 1967; PETTIGREW et al., 1968 b; HENRY et al., 1969; JOSHUA and BISHOP, 1970; Chapter 3). Here again the discrimination appears to depend on differences in populations of excited cells. Stimuli at different depths relative to the fixation point will optimally excite different populations of cortical neurones. It is of interest that (at least in the cat) the same receptive fields appear to handle both pattern and depth information.

Hence the functional subgroups of the cortical cell population appear able to provide coding of the position, orientation, end-positions and visual depth of a contour. Several limitations of these ideas should be noted however. First, it appears likely (although the point has still to be rigorously tested) that the orientation selectivity of simple fields results from the fact that an incorrectly oriented bar cannot cross the excitatory centre of the field without simultaneously crossing its inhibitory flanks (HUBEL and WIESEL, 1962; HENRY and BISHOP, 1971). However, a bar or edge which is short relative to the width of the field centre may be able to do this. The question is not trivial since fine visual detail contains many short borders, and the relationship between orientation selectivity, border length and receptive field size deserves experimental attention. Second, and more generally, the importance for pattern discrimination of the magnitude and time course of the responses of cortical cells has received relatively little consideration, although these response parameters have been shown to be important in the analysis of binocular interaction (Chapter 3).

Third, relatively little attention has been paid to the responsiveness of cortical fields to image movements, such as those which occur during steady fixation (the most common viewing situation in form discrimination tasks). BURNS et al.

(1962) demonstrated that microsaccade-like stimulus movements do excite some cortical cells, and subsequent experiments (discussed in Section III below) in awake animals (WURTZ, 1969 a; NODA et al., 1971) have suggested an important classification of cortical fields according to their responsiveness during fixation, a classification not apparent in results obtained from paralysed animals.

Neuronal Circuitry of Cortical Receptive Fields. HUBEL and WIESEL (1968) suggested that "the elaboration of simple cortical fields from geniculate concentric fields, complex fields from simple fields and hypercomplex from complex is probably the prime function of the striate cortex". The importance of the cortical field properties described by HUBEL and WIESEL (1962, 1965 a) is clear but some recent evidence does not appear to support the idea of a hierarchy of cortical fields in which complex fields are formed by the convergence of simple fields and hypercomplex fields by the convergence of complex.

Recent work has shown considerable complexity in the organisation of simple cortical fields. First, HUBEL and WIESEL (1962) suggested that the direction-selectivity of simple fields results from the asymmetric arrangement of excitatory and inhibitory portions of the receptive field, but WURTZ (1969 a) and HENRY and BISHOP (1971) could not confirm this. The latter workers suggest that direction-selectivity is a property of the central excitatory region, for which the determining mechanism may be quite complex. Second, when tested with moving stimuli the flanks of simple fields are purely inhibitory in influence; by contrast, with stationary flashing stimuli, the cell can be excited from the flanks. Third, many simple fields have spatially separate discharge centres for leading and trailing edges of a moving bar, whose locations cannot be predicted from responses to stationary flashing stimuli (HENRY and BISHOP, 1971).

PETTIGREW et al. (1968 a) found significant differences between their samples of simple and complex fields in spontaneous discharge rate (mean 3/sec in simple fields, 13/sec in complex), in strength of response (much stronger in complex fields), and in preferred speed (2°/sec for simple fields, 5–6°/sec for complex). The first two differences may perhaps be compatible with the idea of complex fields being determined by a convergence of simple cells, but the different speed preferences are hard to reconcile.

HUBEL and WIESEL (1962, 1965 a, 1968) demonstrated that simple fields are most commonly found in cortical layers III and IV, while complex and hypercomplex fields are rare in IV, but predominant in deeper and more superficial layers. Cortical layers IVB, IVA and the deeper part of layer III are the layers in which geniculo-cortical afferents terminate (CAJAL, 1911; O'LEARY, 1941; POLYAK, 1957; HUBEL and WIESEL, 1970). Moreover, since stellate cells predominate in layer IV, and appear to have oriented dendritic fields, COLONNIER (1964) suggested that simple fields are recorded from stellate cells. Conversely, the layer distribution of cells with complex and hypercomplex fields and of pyramidal cells are comparable, both being common in layers II, III, V and VI, and COLONNIER (1964) suggested that complex fields are recorded from pyramidal cells.

Considerable evidence has accumulated, however, that geniculo-cortical afferents do not terminate exclusively on layer IV stellate cells (see also Chapter 21). GLOBUS and SCHEIBEL (1967) provided evidence of direct geniculocortical termination on pyramidal cells of rabbit visual cortex and they emphasised that although few pyramidal cell bodies are found in the afferent terminal layers, many have apical (and also basal) dendrites in these layers. JONES and POWELL (1970) observed with the electron microscope that thalamic afferents to somatosensory cortex appear to terminate on both the spines of pyramidal cell dendrites and on the spineless dendritic shafts of stellate cells, and GAREY (1970) has reported the same organisation for visual cortex. COLONNIER's (1968) observations in the visual cortex are also clearly consistent with this. Electrophysiologically, TOYAMA and MATSUNAMI (1968) recorded short (0.9 ± 0.17 msec) latency EPSP's in layer III cells of area 18, following stimulation of the optic radiation. The cells of this layer are predominantly pyramidal (O'LEARY, 1941), and have complex or hypercomplex fields (HUBEL and WIESEL, 1965 a). This evidence leads to the idea that both simple and complex fields, and possibly hypercomplex fields as well, are determined, at least partially, by direct LGN input.

Direct support for this idea has come from recent studies which have described some functional correlates of conduction velocity in the retino-cortical pathway of the cat. FUKADA (1971) and CLELAND et al. (1971) have independently shown that the X- and Y-type receptive fields of cat retina (first described by ENROTH-CUGELL and ROBSON, 1966) are recorded from ganglion cells which have, respectively, slow- and fast-conducting axons projecting to the LGN. STONE and HOFFMANN (1971) and CLELAND et al. (1971) have shown that cat LGN cells which receive fast axons from the optic tract have fast axons projecting to visual cortex, and, conversely, LGN cells which receive slow axons have slow axons. (This relationship was first described for the rat, by NODA and IWAMA, 1967). Moreover LGN cells retain, and therefore relay to visual cortex, some of the X- and Y-properties of their retinal input. HOFFMANN and STONE (1971) went on to show that fast and slow optic radiation axons appear to activate different cortical cells. Their evidence seems clear that some complex cortical fields are monosynaptically driven by fast radiation axons and is suggestive that some simple and perhaps hypercomplex receptive fields are driven by the slower axons. The degree to which the X- and Y-properties of retinal fields determine the properties of cortical fields has still to be established, but it has already been shown, for example, that Y-fields are generally bigger than X-fields (FUKADA, 1971; CLELAND et al., 1971) and that complex fields are larger than simple fields (HUBEL and WIESEL, 1962). Considered together these studies suggest that the different cortical fields are processing visual information in parallel, rather than in series. This problem seems basic to our understanding of visual cortex; it requires, and will doubtless receive, continuing experimental attention.

One remarkable conclusion reached by JONES and POWELL (1970) is that (in somatosensory cortex) all extrinsic afferents (thalamo-cortical, commissural and cortico-cortical) end with asymmetric-type synapses on dendritic spines, suggesting that all these afferents are excitatory. By contrast a significant proportion of the terminals of intrinsic afferents (i.e. short axons which terminate close to their cells of origin) appear to terminate in symmetrical synapses on dendritic shafts. These latter synapses are, in other tissues, correlated with physiologically inhibitory synapses (see GRAY, 1969). This, JONES and POWELL conclude, "could lead to the speculation that all inhibition in a subdivision of the cortex is mediated by axons intrinsic to that subdivision". They suggest that short-axon stellate cells may mediate this inhibition.

GAREY (1970) has reported the same basic findings for the visual cortex of cat and monkey, so that JONES and POWELL's suggestion may be equally relevant to visual cortex. Physiologically the idea is well supported, for example, by WATANABE et al. (1966) and TOYAMA et al. (1969) observations that LGN and commissural afferents monosynaptically activate and disynaptically inhibit area 17 and 18 cells (with an inhibitory delay of 0.7–1.0 msec). The work of HENRY et al. (1969) and HENRY and BISHOP (1971) has indicated that the inhibition of a simple field cell by a moving slit is a direct inhibition since the cell's response to a simultaneous excitatory stimulus is sharply suppressed. CREUTZFELDT and ITO's (1968) intracellular recordings from cortical cells confirm the idea of direct inhibition since they provide evidence of unitary IPSP's associated with inhibition of firing caused by flashing stimuli. It is likely that this inhibition is mediated by short-axon cells within the cortex, and not by a withdrawal of tonic excitation from LGN cells or by direct inhibition by LGN cells (the two alternatives considered by HUBEL and WIESEL, 1962). Such intracortical inhibition, interacting in different ways with the direct excitation supplied by LGN afferents, may be important in determining the properties of the various types of cortical fields.

Visual Receptive Fields in Other Cortical Areas. Visual receptive fields have been studied in two other areas of the cat cortex, viz. the lateral wall of the suprasylvian gyrus and the adjacent anterior middle suprasylvian gyrus (AMSS). They are of interest as possible areas for high level form vision, but so far receptive field studies have provided little support for this possibility.

Lateral Wall of Suprasylvian Gyrus. This area receives input from both the LGN and from areas 17, 18 and 19 of both hemispheres (see HUBEL and WIESEL, 1969 a, for a summary of these connections). WRIGHT (1969) and HUBEL and WIESEL (1969 a) have studied the receptive fields of this area. HUBEL and WIESEL describe "clear but rather crude" topographic representation of the visual field, with the vertical meridian situated near the bottom of the suprasylvian sulcus, and more eccentric regions being represented more superficially. They found complex and lower order hypercomplex receptive fields, which were generally larger in area than the fields of areas 17, 18 and 19. No simple fields or lateral geniculate afferents were recognised. WRIGHT's (1969) findings are essentially similar. He found complex and hypercomplex type fields, but in addition described fields with no orientation, edge or length selectivity. HUBEL and WIESEL considered that there is "little evidence that the kinds of form analysis occurring in 19, or even in 17, are carried further in (this) region".

AMSS. This region differs from others so far discussed in being polysensory; it may be genuine "association" cortex. DUBNER and RUTLEDGE (1964) considered, however, that it is predominantly visual in function and DUBNER and BROWN (1968) and DOW and DUBNER (1969) have described visual receptive field properties of AMSS neurones. Dow and Dubner describe three types of receptive fields, all of them large ($>15°$), viz. S-fields which respond only to stationary, flashing stimuli; M-fields which respond well to movement in any direction; E-cells which are selective for oriented edges. Field size varies inversely with eccentricity, being somewhat smaller in peripheral regions. No topographic organization is apparent. The M and E fields were very little affected by KCl-depression of visual cortex. DOW and DUBNER stress the similarities between AMSS fields and the receptive fields of the superior colliculus and conclude that the "AMSS is more likely to be involved in visual attention and orientation than in pattern discrimination".

Inferotemporal Cortex of Monkey. In accord with the exclusively visual deficit caused by ablations of inferotemporal cortex (Chapter 23), the neurones of this cortical area respond only to light stimuli (GROSS et al., 1967). An attempt to characterise their receptive fields has been made by Gross et al. (1969). They found that "virtually all fields were extremely large and included the fovea"; their sizes ranged from 10° × 10° to 70° × 70°. Some of the field properties resembled those of the fields of visual cortex, including orientation and direction preferences, preferences for dark or light edges and binocular interaction. In general, however, the responses of inferotemporal neurones were "less clear" than those of visual cortex neurones and their receptive fields were more difficult to determine. As these authors stress, however, the inferotemporal cortex is involved in high level visual function, and the concept of the "optimal" stimulus for a receptive field, so valuable in the analysis of earlier levels of the visual pathway, may be inappropriate here. The understanding of the neuronal function of the inferotemporal cortex poses a formidable challenge.

Receptive Fields of Human Visual Cortex. Marg and his coworkers (1968) have reported recordings from single neurones in human visual cortex, though the site of the neurones (whether in 17, 18 or 19) could not be established. The receptive fields share some of the properties of monkey and cat fields, including orientation-selectivity and binocular excitatory input. These workers describe some remarkable "plastic" changes in receptive field properties which have not been seen in animal experiments. The experimental demonstrations of these changes are not unequivocal, however, and a conservative conclusion is that the basic properties of human cortical fields seem to be similar to those of cat and monkey (see Chapter 22).

Orientation, Size, Velocity and Direction "Channels" in Human Vision. Additional evidence that human receptive fields basically resemble those of cat and monkey has come from psychophysical descriptions of "channels" in human vision. Campbell and Kulikowski (1966) demonstrated that a background grating with a certain orientation raises the detection threshold for a similarly oriented test grating. This adapting effect of the background decreases as the angle between test and background grating increases, indicating an angular "selectivity" of the adaptation. This result was expected from Hubel and Wiesel's (1962) description of the orientation selectivity of cortical receptive fields. Campbell, Cleland, Cooper, and Enroth-Cugell (1968) showed that the angular "selectivity" of cat cortical fields is rather broader than the selectivity of psychophysical adaptation, but as they point out, monkey cortical fields are more sharply orientation-selective than cat cortical fields (Hubel and Wiesel, 1968), and this is probably true for human cortical fields also. The implication of this work is that the selectively adaptable "orientation channels" of human vision are determined by the properties of single cortical neurones.

Analogous experiments have been performed with the spatial frequency of the gratings as the variable parameter. Pantle and Sekuler (1968 a) and Blakemore and Campbell (1968, 1969) showed that a background grating of a certain frequency raises the detection threshold for a test grating (of the same orientation), over a narrow range of spatial frequencies (centred around the adapting frequency). They conclude that there are selectively-adaptable "spatial frequency" channels or "size channels" in the human visual system. Subsequent workers (Nachmias et al., 1969; Graham and Nachmias, 1971; Campbell and Maffei, 1970) have confirmed and extended this basic observation. Campbell, Cooper, and Enroth-

CUGELL (1969) and CAMPBELL, COOPER, ROBSON, and SACHS (1969) have measured the spatial selectivity of single cells in the cat LGN and visual cortex and in the squirrel monkey LGN. They estimated the high frequency "cut-off" for each cell, which they termed its "characteristic frequency". Within each population of cells there is a range of characteristic frequencies. These frequencies were generally higher for the monkey, compatible with its superior spatial resolution. BLAKEMORE and CAMPBELL (1969) suggest that the "spatial-frequency channels" of human vision "reflect the properties of (individual) neurones".

Cortical receptive fields also show considerable selectivity for stimulus velocity (HUBEL and WIESEL, 1962, 1965 a; PETTIGREW et al., 1968 a; HENRY and BISHOP, 1971) and PANTLE and SEKULER (1968 b) have presented some evidence of velocity-selective channels in human vision, although the selectivity appears weaker than for orientation and size. Again, many cortical receptive fields show direction selectivity (HUBEL and WIESEL, 1962; PETTIGREW et al., 1968a). SEKULER et al. (1968) and PANTLE and SEKULER (1969) have presented psychophysical evidence for direction-selective channels in human vision. The selectivity is fairly weak, however.

These studies are a valuable demonstration of the possible importance in human vision of receptive field properties found in the primary visual pathways of cat and monkey. As a qualification, it should be pointed out that behaviour analogous to psychophysical adaptation has not yet been demonstrated for single receptive fields.

III. Further Considerations: Eye Movements, Visual Coding and Visual Memory

The properties of receptive fields summarised in Section II were established in experiments in which the animal was lightly anaesthetised and, in addition, paralysed so that its eyes could not move. In normal vision, the animal is alert, its head is free to move, and its eyes are constantly in motion. Several questions must now be considered. How important in normal vision are the receptive field properties so far considered? Are other properties apparent in normal vision? What role do eye movements play in visual coding? How do we perceive the visual world to be temporally and spatially stable, when the retinal image is in constant motion?

1. Receptive Field Properties and Visual Behaviour

WURTZ (1969 a, b, c), in an elegant series of experiments, studied the receptive field properties of cells in the visual cortex of awake, unparalysed monkeys. Ten years previously HUBEL (1959, 1960) reported less extensive studies of retinal, LGN and visual cortical fields in awake, unrestrained cats. The properties of receptive fields found in the anaesthetised, paralysed animal (e.g. orientation and direction-selectivity) were also found in awake animals; they do not appear to be artefacts of the experimental situation.

Moreover, these receptive field properties are dependent on visual experience for their full development in the adult animal, and where these properties deteriorate due to deprivation, vision deteriorates. Thus, in three-month-old kittens monocularly deprived (by occlusion or lid suturing) of pattern vision from birth, no cortical receptive fields can be driven from the deprived eye. Virtually all cells can be driven from the normal eye, and their receptive fields for this eye are normal. Correspondingly, such kittens are behaviourally blind with the (opened) deprived eye, and have normal vision with the normal eye (Wiesel and Hubel, 1963). Moreover, the period of susceptibility of cortical fields to deprivation is limited (Hubel and Wiesel, 1970 b) and prolonged deprivation in adults does not disturb either receptive field properties or visual performance. In monocularly deprived cats, Ganz et al. (1969) also found gross abnormalities in the responsiveness of single cortical neurones following monocular deprivation and Ganz and Fitch (1969) demonstrated marked deficits in visual performance with only the deprived eye open.

Unfortunately behavioural studies differ considerably in their evaluations of the effect of deprivation on form discrimination. Meyers and McCleary (1964) and Held (1968) emphasised the severe effects of deprivation on visual orienting behaviour, and the persistence of the ability to discriminate simple patterns, provided the discrimination did not involve visual orientation. By contrast the analysis of Dews and Wiesel (1970) stressed the minor effects of deprivation on visuomotor performance, and the susceptibility of visual acuity. Ganz and Fitch (1969) reported a marked and permanent loss of simple pattern discrimination, but only a slight loss of visual acuity. Despite these differences in emphasis, however, all studies indicate a substantial loss of pattern discrimination ability in animals deprived of visual experience during development.

2. Form Discrimination and Eye Movements

The eyes are constantly in motion during the steadiest fixation. These eye movements largely determine the rate and direction of movement of the retinal image across receptive fields, parameters known to be important in the activation of cortical receptive fields. It is important then to consider the influence of eye movements on form discrimination. During which eye movements is discrimination best? During which is discrimination possible? Form discrimination and/or visual acuity have been tested psychophysically in man in the five following eye movement situations:

(i) *Smooth tracking movements* occur as the eye follows a steadily moving object.

(ii) During *steady fixation* of a stationary object, eye movements have three components (flick, drift and tremor), all of small amplitude.

(iii) *Retinal image movements can be completely eliminated* by various experimental techniques; the retinal image is then said to be "stabilised".

(iv) In all of (i), (ii) and (iii), the retinal image stays approximately or exactly at the same retinal location, the fovea. Acuity has also been tested during *smooth movements of the retinal image across the retina.*

(v) Visual acuity and threshold detection have also been tested during the *fast saccadic jumps* which the eye makes when the direction of fixation is changed. These jumps, or *macrosaccades*, are generally larger and faster than the *microsaccades* that occur during fixation.

Steady fixation of a stationary object is the most common situation in form discrimination tasks, but discrimination is certainly possible in all five situations. We attempt here a brief review of relevant psychophysical findings and of their possible neurophysiological bases.

Psychophysics. *Smooth Tracking Movements.* LUDVIGH and MILLER (1958) showed that small targets (Landolt C-rings) can be seen during the "voluntary ocular pursuit of moving test objects", but that acuity is always poorer than during fixation of stationary test objects. Acuity while following an object moving at 10°/sec was typically one third of acuity during steady fixation, and acuity deteriorated steadily as the velocity increased. WESTHEIMER (1954) showed that ocular pursuit movements are closely matched to the target velocity, errors in velocity being corrected by superimposed saccades (see also ALPERN, 1962, p. 90). LUDVIGH and MILLER argued that the deterioration of acuity with increasing stimulus velocity must be a result of movement of the retinal image relative to the retina, due to "inaccuracy of control of pursuit movements". The movements of the retinal image relative to the retina which occur during pursuit eye movements are thus not exactly defined.

Steady Fixation. Acuity is optimal during steady fixation of a stationary object. The eye movements of fixation then completely determine retinal image movement. These movements are generally classified into three types, involuntary saccades, drifts and tremor (DITCHBURN and GINSBORG, 1953; DITCHBURN et al., 1959). Involuntary saccades, or "flicks", of up to 1/3° occur at irregular intervals (0.03 –5 sec). Between flicks the eye drifts slowly at about 1′/sec. Superimposed upon these movements is a high frequency tremor of varying amplitude (up to 0.5 min of arc), and frequencies up to 150/sec. The same three types of involuntary eye movements are seen in the cat, but are generally coarser than in man, and flicks are rare. Flicks may reach 2° amplitude, drifts are faster (30′/sec), and the tremor movements are lower in frequency (up to 40/sec) and greater in amplitude (PRITCHARD and HERON, 1960; HEBBARD and MARG, 1960). These movements have yet to be investigated in the monkey, but they presumably are present and in magnitude resemble human rather than cat movements.

At least in the human these movements are essential for normal vision. If they are eliminated, and the retinal image thereby stabilised on the retina, acuity and form vision quickly deteriorate (DITCHBURN and GINSBORG, 1952; RIGGS et al., 1953). Several lines of psychophysical evidence suggest that, of the three types of fixational eye movements, the flicks are the most important for maintaining optimal acuity. CORNSWEET (1956) suggested that flicks are triggered by retinal image movement, since they are much less frequent when the image is stabilised, and their direction and amplitude are related appropriately to the eye's deviation from its mean, presumably optimal, position. DITCHBURN et al. (1959) imposed on a stabilised image movements which resembled flicks, drifts and tremors. They concluded that flicks alone are potentially effective in maintaining optimal vision, that drifts are ineffective unless of unusual amplitude, and that tremor is partially effective. CORNSWEET (1956) considered drifts a result of instability of eye position control, but NACHMIAS (1959) and STEINMAN et al. (1967) provide some evidence that drifts may act, like flicks, to keep the eye "on target".

Ditchburn et al. (1959) noted that after the perception of a stabilised retinal image has faded a flick-like movement of image will cause sharp regeneration of the perception. It seems likely, therefore, that in normal fixation flicks provide a constantly repeated regenerative stimulus which is potentially sufficient for the maintenance of optimal acuity (Ditchburn, 1955). What function do the continuous inter-flick movements (drifts and tremor) serve? Ditchburn et al. comment that because the inter-flick periods commonly reach 1 second, drifts and tremor must play some regenerative role, otherwise vision would fade towards the end of such periods. However, their work suggests that it is a minor role and a more positive demonstration of a functional role has not been made.

Stabilised Retinal Image. When the image of a pattern is stabilised on the retina, the perceived outline and contrast in the pattern begin, after a period of about 1 sec, to deteriorate (Ditchburn and Ginsborg, 1952; Riggs et al., 1953; Barlow, 1963). However, Ditchburn and Fender (1955) showed that vision with a stabilised image can be maintained by flashing the image on and off. Moreover, Keesey (1960) showed quantitatively that for a single flash presentation of a target acuity improves with exposure times up to 300 msec and *is equally good whether the retinal image of the target is stabilised or unstabilised* during that time. For the first 200—300 msec eye movements do not improve acuity. It seems that a flash presentation of a stabilised image is capable of generating vision with optimal acuity.

Steady Movement of the Retinal Image. Acuity is generally poor for a target whose image moves quickly across the retina, an effect due largely to "smearing" of the retinal image, but presumably also a result of the fact that fast-moving images remain within the fovea only very briefly. However, with slower image movement, acuity may be considerably better. For example, Ditchburn et al. (1959) showed that the perceived sharpness of a stabilised retinal image does not fade significantly if the image is moved across the retina at a speed of about 0.25°/sec, provided the amplitude is sufficient (greater than about 10′ arc).

Saccadic Eye Movements. The angular velocity of saccadic movements (up to 500°/sec) causes very considerable smearing of the retinal image of a stationary, constant luminance target and acuity for such targets is very poor or non-existent (see, for example, Ludvigh, 1948). This loss of acuity can be largely compensated by presenting the acuity target very briefly. For example, in the experiment of Volkmann (1962), target exposure was limited to 20 μsec and the image fell on or near the fovea. Even under these conditions, however, Volkmann noted in threshold detection tasks a small (0.5 log unit) rise in contrast threshold, and in acuity tasks a small (1.4 times) decrease in the spatial frequency of the finest resolvable grating, i.e. a small loss of acuity. These effects have been termed "saccadic suppression" (Volkmann, 1962; Volkmann et al., 1968; Beeler, 1967). Volkmann et al. (1968) found that the threshold for contrast detection reaches a maximum about 20 msec before the onset of the saccade, suggesting an interpretation of saccadic suppression in terms of partial central inhibition of the sensory afferent pathway by oculomotor centres.

An alternative suggestion was made by MacKay (1970 a), who showed a similar suppression of detection during fixation when the background was shifted over an angular distance and time comparable to a saccadic eye movement.

MacKay found a maximum suppression just prior to the background shift and continuing into it, even without eye movements. This experiment goes against the idea that saccadic suppression is generated by oculomotor centres. Finally, MacKay (1970 b) has subsequently shown that a 10 msec, 4° visual field displacement in one eye causes an almost identical suppression of the frequency-of-seeing curve whether the test flash is presented to the same eye or to the contralateral eye. MacKay suggests "that the surge of neural activity caused by the rapid displacement of the retinal image has side-effects that interfere with processing of the signals generated by the test flash, and so raise the perceptual threshold" (MacKay, 1970 b, p. 873). Furthermore, because of the interoptic transfer of the suppression effect found by MacKay, it seems necessary to assume that this effect must "depend mainly on central rather than purely peripheral physiological interactions" (MacKay, 1970 b, p. 873).

Neurophysiology. Following the above discussion, it is of principal interest (for this Chapter) to account in neurophysiological terms for the fact that visual acuity is good in the presence of (at least) three distinct types of movement of the retinal image, viz.:

(i) The movements, in particular the flick movements, present during steady fixation;

(ii) The absence of any movement (stabilised image), in which situation the image is presented by flashing it;

(iii) Slow, steady movements of the retinal image.

Retinal image movements of the third type are the simplest to consider. Hubel and Wiesel (1962, 1965 a) reported that slow movement of a correctly oriented stimulus bar is a powerful stimulus for all types of cortical fields in the cat; usually a slow speed (1–10°/sec) seemed optimal. Pettigrew et al. (1968 a) measured this speed-selectivity in the cat and found simple fields optimally responsive to speeds of about 2°/sec and complex fields optimally responsive to about 6°/sec. Slow stimulus movement is an effective stimulus for all cortical fields in monkey striate cortex as well (Hubel and Wiesel, 1968).

The flicks or microsaccades of fixation provide a quite complex image movement. Each movement is brief (approximately 25 msec in duration), short (net amplitude 1–20′ arc) and consists of two parts, a primary movement in one direction and a shorter movement in the return direction (Ditchburn and Ginsborg, 1953). In the course of a flick movement any particular contrast border in the retinal image quickly leaves one group of receptive fields on which it has been relatively stable, and about 25 msec later, after accelerating and decelerating twice, comes to rest on another group. In general fast movements of a stimulus which begin or end within a receptive field resemble the flashing of a stationary pattern, especially when the velocity of movement is high. Kozak et al. (1965) demonstrated, for example, that cat LGN cells (which are the principal afferent cells to the visual cortex) respond well to a pattern moved quickly into the receptive field centre and left stationary, giving a response closely resembling their response to a stationary flashing stimulus. Hence the visual stimulus provided by a flick may resemble the flashing of a stationary pattern, a suggestion compatible with the effectiveness of both a flash and flick movement in generating vision with optimal acuity. If a flick movement is of large amplitude a contrast border in the image will also cross a number of receptive fields without stopping. However, DITCHBURN et al. (1959) noted that flicks of very small amplitude (2.5′ arc, i.e. too small to span more than one or two receptive fields) are highly effective in regenerating a faded stabilised image. Regeneration seems to result from the image moving out of some receptive fields into others; it need not cross any fields completely.

Assuming that flashing and flick movement of the retinal image are essentially equivalent visual stimuli, there is still some difficulty in accounting neurophysiologically for their effectiveness. The descriptions of HUBEL and WIESEL (1962, 1965 a) indicate that while many simple cortical fields are responsive to stationary flashing stimuli, complex and hypercomplex fields are more weakly responsive, or else unresponsive. This contrasts with the effectiveness of slow stimulus movements in exciting all field types.

3. Receptive Fields and Eye Movements in Awake Animals

In the discussion of cortical receptive fields presented in Section II, it was implicity assumed that all receptive fields of visual cortex are directly involved in pattern coding. Recent studies of the response properties of single units in the visual cortex of awake cats and monkeys suggest an important classification of receptive fields which goes against this assumption.

WURTZ (1969 a) studied cortical receptive fields in awake monkeys trained to fixate a light spot. He recorded from cortical neurones with receptive fields away from the fixation point and during fixation periods investigated their properties. The receptive fields were then stationary on the tangent screen, except for normal fixational movements. He identified concentric, simple and complex receptive field types. Their size, shape, orientation-specificity and direction-specificity were in accord with previous reports on paralysed animals. Of particular interest is his finding that the fields could be divided into two classes, according to whether their response to an optimally positioned stationary light pattern evoked a phasic (adapting) or tonic (non-adapting) response discharge. Tonic responses have been reported only rarely in the cortex of paralysed animals, and WURTZ cautiously suggests that their frequent occurrence in the alert animal may be a result of fixational eye movements. As Wurtz implied, this tonic activity provides a possible neurophysiological basis for the maintenance of vision during fixation of steady objects, and conversely the absence of tonic activity in paralysed animals provides a basis for the fading of a stabilised image. Moreover the units which gave phasic responses to stationary patterns were those which responded most vigorously to stimulus movements, and which showed direction-selectivity. Wurtz found that about 90% of units could be classified into one group or the other.

NODA et al. (1971) have recently described the response properties of units recorded in the visual cortex of awake cats. In striking parallel with WURTZ's observations, they found a class of units (25% of their sample)which did not respond in the presence of a stationary grating but responded strongly to movement of the grating, and were direction-selective. Conversely, a second class of units (also 25% of the sample) responded continuously in the presence of a stationary grating, were always strongly orientation-selective but were generally unresponsive to large stimulus movements and were not direction-selective. They suggest that the former type of field may be specialised to detect the occurrence and direction of large image movements, and the latter to code pattern information. This suggestion is an important one which cuts across the receptive field classification established in paralysed-animal experiments, and which will be a point of continuing experimental interest.

Moreover, NODA et al. (1972) have described a small number of units (32 out of 357) which responded only during fast (saccadic) eye movements in the presence of a patterned target. Another 7 units responded to such movements in the absence of a visual stimulus (specifically, in the dark). NODA et al. consider the activation of the former to be of retinal origin and the activation of the latter to be of motor origin. It is initially surprising that so many cortical receptive fields seem selectively responsive to large image or eye movements rather than to fixational movements. It is possible, however, that the excitatory and inhibitory responses to eye movements found by NODA et al. in single units in the visual cortex of awake cats are related to the phenomenon of saccadic suppression, discussed earlier. Another possible (and not necessarily conflicting) function of such cortical units is considered below, in relation to the problem of visual stability.

4. Temporal and Spatial Stability of Vision

The eye movements of fixation are essential for the maintenance of vision during fixation. Despite them, our perception of an object is *temporally continuous* or *stable*. We are aware of neither the flick, drift, nor tremor movements which constantly occur during fixation. Moreover, retinal image movements resulting from eye movements do not lead to an impression of object movement. Objects in the visual world appear to be *spatially stable* although if one did move we would detect this movement principally as a result of the movement of its retinal image.

How are these stabilities attained? A separate consideration of the problem is presented in Chapter 5, but a number of aspects have recently been considered neurophysiologically, and deserve mention here.

Temporal Stability. The brief (25 msec) microsaccades or flicks which occur during fixation are not perceived, although we are conscious of voluntary and involuntary macrosaccades, which are larger in amplitude and longer in duration (e.g. 80 msec for a 20° saccade, WESTHEIMER, 1954).

A concept which may be important in understanding temporal stability of vision during fixation is the "input quantisation of time" (BOYNTON, 1961) or "the psychological moment" (see, for example, STROUD, 1956). "Visual input may be packaged in successive time frames and ... therefore any two events that occur within a given frame and that depend upon a temporal discrimination alone for their perception cannot be discriminated" (BOYNTON, 1961); i.e. visual input may consist of a sequence of still pictures (SPERLING, 1963). As long as successive pictures follow at a sufficient rate, temporal continuity is maintained. As BOYNTON (1961) points out, the presentation of 16 still frames per second (i.e. every 62.5 msec), as in motion picture projection, creates an effective illusion of temporal continuity.

Experimental support for the notion of the quantisation of time in tactual, auditory, and visual stimuli is provided by CHEATHAM and WHITE (1952, 1954) and White and Cheatham (1959), who showed that sequences of spatially invariant stimuli could be discriminated temporally up to a presentation rate of 80–90 msec/perceived unit, or 11–12 units/sec. Estimates, based on quite different psychophysical tests, of a duration of 50–100 msec for the "psychological moment" (Eriksen and Collins, 1967) are also in accord with this idea.

Ditchburn and Fender (1955) noted that perception of a stabilised retinal image does not fade if it is flashed on and off, the optimal rate being about 20 cps. Repetitive flashing is necessary because the neurophysiological responses of visual cells are transient, as discussed by Ditchburn (1955). Indeed the duration of neurophysiological responses is in good accord with estimates of the "psychological moment". Levick and Zacks (1970), for example, examined the responses of cat retinal ganglion cells to very short flashes of a spot stimulus. By their estimate the shortest possible responses last 50–70 msec and are evoked by flashes less than 32 msec in duration and less than about 32 × threshold in intensity. Gouras (1967) reported responses of similar duration for monkey retinal ganglion cells responding to a very brief flash and Baker et al. (1969) described responses of comparable duration in LGN cells of the cat. Moreover, the duration of the responses of retinal and LGN cells to *suprathreshold* flashing stimuli show a marked phasic component, so that the maximum firing rate is reached quickly after which the response declines roughly exponentially to a rate only slightly higher than existed prior to the stimulus (see, for example, Rodieck and Stone, 1965 b; Cleland and Enroth-Cugell, 1968). The phasic component is in the order of 100 msec in duration.

Thus the duration of the "psychological moment" may be set by the duration of neurophysiological responses, and temporal continuity may be determined by the occurrence of successive responses at a sufficient rate. In Ditchburn and Fender's (1955) situation, successive responses were generated by successive flashes of the stabilised image. In normal fixation, successive responses are probably generated (as already discussed) by the repeated, small, brief shifts of the retinal image caused by the flick movements of the eye during fixation (c.f. Gaarder, 1966). Boynton, (1961, p. 182) noted that the "quantisation of time" in vision seems to set a limit to the rate at which the human visual system can resolve changes in a viewed scene. He commented that this limited temporal resolution "appears to be imposed by . . . mechanisms (subsequent to the receptors), perhaps to allow the transmission of more spatial information per time frame."

Spatial Stability. "An image moves with respect to the retina both when our eyes move and when the object moves, but in one case we perceive a stationary object, in the other case a moving object. How can we tell the difference?" (Wurtz, 1969 c).

This is the basic puzzle of spatial stability and it has been recognised for many years. Probably all workers who have considered the problem (e.g. Helmholtz, 1910, Vol. III, pp. 246–247; MacKay, 1958; Gregory, 1958; Teuber, 1960, p. 1647) have assumed that spatial stability results from a comparison of the retinal image movements as they are coded in retinal output, with the eye and/or head movements which were "ordered" by the brain an appropriate interval previously. Where a retinal image movement is accountable in terms of eye or head movements, it is suggested, we perceive the object as stable. To the extent that an image movement is not so accountable, we perceive the object to have moved. Recent formulations of this problem by MacKay (1958) and Gregory (1958) have proposed somewhat different logical schemes by which this judgement might be made.

In neurophysiological terms a comparison between eye movements and image movements requires interaction between the oculomotor efferent and visual afferent pathways. A number of workers have found such an interaction at the level of the LGN (the major source of input to the visual cortex). Bizzi (1966 a, b), for example, demonstrated a marked non-visual input to the cat's LGN in synchrony with rapid eye movements of sleep. Kawamura and Marchiafava (1968) have made a comparable observation during eye tracking movements in mid-pontine cats. Feldman and Cohen (1968) have extended the finding to the alert monkey. No assessment has been made, however, of the effects of these inputs on the coding functions of single units. Moreover, Wurtz (1969 c) has demonstrated that in the alert, unparalysed monkey the response characteristics of cortical cells appear to be unaffected by eye movements. Specifically he found that the cell's response as its receptive field swept (during a voluntary eye movement) across a stationary stimulus object did not differ from its response as the stimulus swept, at the same speed, across the receptive field while the eye was stationary (actually fixating). It would seem that if the activity of oculomotor centres does interact with the afferent visual pathway in order to allow judgement to be made about a movement of the retinal image, the interaction must occur at higher levels than the visual cortex.

In this context, however, it is interesting to note again the suggestion by Noda et al. (1971) that different groups of cortical fields code pattern information and information about large image movements. While there is yet no evidence of interaction between these different unit types, their presence in striate cortex provides a basis for their interaction at higher cortical levels.

5. Coding of Complex Forms and Visual Memory

The neurophysiological mechanisms so far considered have been those of only the visual afferent pathway. We have not considered mechanisms for coding patterns or forms more complicated than edges and corners, or the mechanisms underlying visual memory, or those underlying the comparison of visual input with memory. Yet it is here that present knowledge of the mechanisms of form vision essentially stops. We understand only the afferent pathway, and that only incompletely. The following considerations are therefore brief.

On the Coding of Complex Forms. There is a great increase, between the retina and the visual cortex, in the number of cells involved in visual coding. For every retinal ganglion cell there are some hundreds of cells in the visual cortex. The clearest insight we have into how these cells are employed has derived from descriptions of the properties of single cortical fields. As already discussed in Section II, cortical cells with co-extensive receptive fields (i.e. excited from the same small area of retina) are selectively responsive to different orientations of a contrast edge, to the positions of its ends and to differences in its visual depth relative to the fixation point. A change in any one of these parameters, or in retinal position, results in a change in the population of excited cells. Studies in awake animals (discussed above) suggest, moreover, that populations of cells are responsive to different types of image movement, some to fixational movements, some to large saccadic movements, and some to large movements of intermediate velocity. This description is no doubt an oversimplification. The point is that the coding of the features of a stimulus pattern does not appear to converge on a single or several neurones, at least not in the striate cortex. Rather considerable populations of cortical cells are involved in the representation of a small area of retina, and stimulus features determine which sub-groups of that population are excited and which are unaffected or inhibited.

The activity of these populations is presumably "sampled" or "integrated" by the "higher" visual areas of cortex, to provide for the perception of complex forms. It is likely that the circumstriate and inferotemporal areas of cortex are important for this integration (in man and monkey) (PENFIELD and PEROT, 1963, and see Chapter 23). The neurophysiological mechanisms employed are not known, although initial neurophysiological approaches to the problem have been reported (GROSS et al., 1969; Chapter 23). Considerable attention has, however, been paid to the principles by which such sampling of visual input appears to be performed (see, for example, DEUTSCH, 1955; HAKE, 1966; SUTHERLAND, 1968). Most recently SUTHERLAND (1968) in formulating "outlines of a theory of visual pattern recognition", discussed the rules by which information might be abstracted from visual input and the psychophysically established phenomena which must be explained by any abstraction model. The abstraction must be independent of retinal position, size, brightness and small distortions. It must take into account the recognition of shape independent of contrast, the ability to abstract outlines of filled shapes, the speed at which complex scenes can be processed, and the ability to consider segments of a pattern separately. SUTHERLAND'S approach is clearly an important one since it is an attempt to specify, on the basis of many experimental investigations, the basic logic of complex visual processing. Even though this attempt was only preliminary, the challenge posed for neurophysiology is enormous.

Visual Memory. Virtually all tasks which might be termed form discrimination involve visual memory. For example, the recognition of a pattern or form as belonging to some previously learned class, or the discrimination between two patterns in learning tasks, both clearly involve interaction between visual input and memory. Although some characteristics, particularly temporal characteristics, of visual memory have been intensively studied psychophysically (see, for example, HABER, 1969, 1970) little is known of the underlying neuronal mechanisms, except that, in monkey and man, the temporal cortex is intimately involved (Chapter 23).

This review was completed in August 1971.

References

ALPERN, M.: Movements of the eyes: Types of movement. In: DAVSON, H. (Ed.): The Eye. Vol. 3, pp. 63—151. New York: Academic Press 1962.

ANDERSON, K.V., SYMMES, D.: The superior colliculus and higher visual functions in the monkey. Brain Res. **13**, 37—52 (1969).

BAILEY, P., VON BONIN, G.: In: Urbana, Ill.: The Isocortex of Man. Univ. of Illinois Press 1951.

BAKER, F.H., SANSEVERINO, E.R., LAMMARRE, Y., POGGIO, G.F.: Excitatory responses of geniculate neurones of the cat. J. Neurophysiol. **32**, 916—929 (1969).

BARLOW, H. B.: Possible principles underlying the transformations of sensory messages. In: ROSENBLITH, W. A. (Ed.): Sensory Communication. Cambridge, Mass.: M. I. T. Press 1961a.
— Three points about lateral inhibition. In: ROSENBLITH, W. A. (Ed.): Sensory Communication. Cambridge, Mass.: M. I. T. Press 1961 b.
— Slippage of contact lenses and other artefacts in relation to fading and regeneration of supposedly stabilised retinal images. Quart. J. exp. Psychol. **15**, 36—51 (1963).
— BLAKEMORE, C., PETTIGREW, J. D.: The neural mechanism of binocular depth discrimination. J. Physiol. (Lond.) **193**, 327—342 (1967).
— FITZHUGH, R., KUFFLER, S. W.: Change of organisation in the receptive fields of the cat's retina during dark adaptation. J. Physiol. (Lond.) **137**, 338—354 (1957).
— HILL, R. M., LEVICK, W. R.: Retinal ganglion cells responding selectively to direction and speed of image motion in the rabbit. J. Physiol. (Lond.) **173**, 377—407 (1964).
BARTLEY, S. H.: Vision: A study of its basis. New York: Van Nostrand 1941.
BAUMGARTNER, G.: Die Reaktionen der Neurone des zentralen visuellen Systems der Katze im simultanen Helligkeitskontrast. In: JUNG, R., KORNHUBER, H. H. (Eds.): Neurophysiologie und Psychophysik des visuellen Systems. Berlin-Göttingen-Heidelberg: Springer 1961 a.
— Der Informationswert der On-Zentrum und Off-Zentrum Neurone. In: JUNG, R., KORNHUBER, H. H. (Eds.): Neurophysiologie und Psychophysik des visuellen Systems. Berlin-Göttingen-Heidelberg: Springer 1961 b.
— BROWN, J. L., SCHULZ, A.: Responses of single units of the cat visual system to rectangular stimulus patterns. J. Neurophysiol. **38**, 1—18 (1965).
BEELER, G. W., JR.: Visual threshold changes resulting from spontaneous saccadic eye movements. Vision Res. **7**, 769—775 (1967).
BÉKÉSY, G. VON: Mach- and Hering-type lateral inhibition in vision. Vision Res. **8**, 1483—1499 (1968).
BISHOP, P. O., KOZAK, W., LEVICK, W. R., VAKKUR, G. J.: The determination of the projection of the visual field onto the lateral geniculate nucleus in the cat. J. Physiol. (Lond.) **163**, 503—539 (1962).
BIZZI, E.: Changes in orthodromic and antidromic responses of optic tract during the eye movements of sleep. J. Neurophysiol. **29**, 861—870 (1966 a).
— Discharge patterns of single geniculate neurones during the rapid eye movements of sleep. J. Neurophysiol. **29**, 1087—1095 (1966 b).
BLAKEMORE, C., CAMPBELL, F. W.: Adaptation to spatial stimuli. J. Physiol. (Lond.) **200**, 11—13 P (1968).
— — On the existence of neurones in the human visual system selectively sensitive to the orientation and size of retinal images. J. Physiol. (Lond.) **203**, 237—261 (1969).
BLOUGH, P. M.: Difference limen as a function of retinal eccentricity and background brightness. J. opt. Soc. Amer. **48**, 731—735 (1958).
BOYCOTT, B. B., DOWLING, J. E.: Organisation of the primate retina: light microscopy. Phil. Trans. roy. Soc. B **255**, 109—184 (1969).
BOYNTON, R. M.: Some temporal factors in vision. In: ROSENBLITH, W. A. (Ed.): Sensory Communication. Cambridge, Mass.: M. I. T. Press 1961.
— Spatial vision. Annual Rev. Psychol. **13**, 171—200 (1962).
BROWN, J. L., BAUMGARTNER, G.: Discussion on visual encoding. In: The Physiological Basis of Form Discrimination. NIH Symposium, Brown Univ., Providence, R. I. 1964.
BURNS, B. D., HERON, W., PRITCHARD, R.: Physiological excitation of the visual cortex in the cat's unanaesthetised isolated forebrain. J. Neurophysiol. **25**, 165—181 (1962).
— PRITCHARD, R.: Contrast discrimination by neurones in the cat's visual cortex. J. Physiol. (Lond.) **175**, 445—463 (1964).
CAJAL, R. Y.: Le lobe optique des vertébrés inférieurs. In: Histologie du système nerveux de l'homme et des vertébrés. Paris: A. Maloine 1911.
CAMPBELL, F. W., CLELAND, B. G., COOPER, G. F., ENROTH-CUGELL, C.: The angular selectivity of visual cortical cells to moving gratings. J. Physiol. (Lond.) **198**, 237—250 (1968).
— COOPER, G. F., ENROTH-CUGELL, C.: The spatial selectivity of the visual cells of the cat. J. Physiol. (Lond.) **203**, 223—236 (1969).

CAMPBELL, F.W., COOPER, G.F., ROBSON, J.G., SACHS, M.B.: The spatial selectivity of visual cells of the cat and squirrel monkey. J. Physiol. (Lond.) **204**, 120 P (1969).
— GREEN, D.G.: Optical and retinal factors affecting visual resolution. J. Physiol. (Lond.) **181**, 576—593 (1965).
— KULIKOWSKI, J.J.: Orientation selectivity of the human visual system. J. Physiol. (Lond.) **187**, 437—446 (1966).
— MAFFEI, L.: Electrophysiological evidence for the existence of orientation and size detectors in the human visual system. J. Physiol. (Lond.) **207**, 635—652 (1970).
CHEATHAM, P.G., WHITE, C.T.: Temporal numerosity: I. Perceived number as a function of flash number and rate. J. exp. Psychol. **44**, 447—451 (1952).
— — Temporal numerosity: III. Auditory perception of number. J. exp. Psychol. **47**, 425 to 428 (1954).
CHOW, K.-L., BLUM, J.S., BLUM, R.A.: Cell ratios in the thalamo-cortical visual system of Macaca Mulatta. J. comp. Neurol. **92**, 227—239 (1950).
CLELAND, B.G., DUBIN, M.W., LEVICK, W.R.: Simultaneous recording of cat geniculate cells and the retinal ganglion cells driving them. Proc. Aust. Physiol. Pharmacol. Soc. **1/1**, 27 (1970 a).
— — — Field suppression: A functional property of cat lateral geniculate cells. Proc. Aust. Physiol. Pharmacol. Soc. **1/1**, 26 (1970 b).
— — — Sustained and transient neurones in the cat's retina and lateral geniculate nucleus. J. Physiol. (Lond.) **217**, 473—496 (1971).
— ENROTH-CUGELL, C.: Quantitative aspects of sensitivity and summation in the cat retina. J. Physiol. (Lond.) **198**, 17—38 (1968).
COLONNIER, M.: The tangential organisation of the visual cortex. J. Anat. **98**, 327—344 (1964).
— Synaptic patterns on different cell types in the different laminae of the cat visual cortex: an electron microscope study. Brain Res. **9**, 268—287 (1968).
CORNSWEET, T.N.: Determination of the stimuli for involuntary drifts and saccadic eye movements. J. opt. Soc. Amer. **46**, 987—993 (1956).
— Visual Perception. New York: Academic Press 1970.
COWEY, A.: Projection of the retina onto striate and prestriate cortex in monkeys. J. Neurophysiol. **27**, 366—393 (1964).
— ELLIS, C.M.: The cortical representation of the retina in squirrel and rhesus monkeys and its relation to visual acuity. Exp. Neurol. **24**, 374—385 (1969).
CREUTZFELDT, O., ITO, M.: Functional synaptic organisation of primary visual cortex neurones in the cat. Exp. Brain Res. **6**, 324—352 (1968).
CRONLY-DILLON, J.R.: Units sensitive to direction of movement in the goldfish optic tectum. Nature (Lond.) **203**, 214 (1964).
CROSBY, E.C., HUMPHREY, T., LAUER, E.W.: Correlative anatomy of the nervous system. New York: MacMillan 1962.
DENNEY, D., BAUMGARTNER, G., ADORJANI, C.: Responses of cortical neurones to stimulation of the visual afferent radiations. Exp. Brain Res. **6**, 265—272 (1968).
DEUTSCH, J.A.: A theory of shape recognition. Brit. J. Psychol. **46**, 30—37 (1955).
DEWS, P.B., WIESEL, T.N.: Consequences of monocular deprivation on visual behaviour in kittens. J. Physiol. (Lond.) **206**, 437—455 (1970).
DIAMOND, A.L.: Foveal simultaneous brightness contrast as a function of inducing- and test-field luminances. J. exp. Psychol. **45**, 304—314 (1953).
DIAMOND, I.T., HALL, W.C.: Evolution of neocortex. Science **164**, 251—262 (1969).
DITCHBURN, R.W.: Eye movements in relation to retinal action. Optica Acta **1**, 171—176 (1955).
— FENDER, D.H.: The stabilised retinal image. Optica Acta **2**, 128—133 (1955).
— — MAYNE, S.: Vision with controlled movements of the retinal image. J. Physiol. (Lond.) **145**, 98—107 (1959).
— GINSBORG, B.L.: Vision with a stabilised retinal image. Nature (Lond.) **170**, 36—37 (1952).
— — Involuntary eye movements during fixation. J. Physiol. (Lond.) **119**, 1—17 (1953).
DODWELL, P.C., FREEDMAN, N.L.: Visual form discrimination after removal of the visual cortex in cats. Science **160**, 559—560 (1968).

DORFF, J. E., MIRSKY, A. F., MISHKIN, M.: Effects of unilateral temporal lobe removals in man on tachistoscopic recognition in the left and right visual fields. Neuropsychology **3**, 39—51 (1965).

DOW, B. M., DUBNER, R.: Visual receptive fields and responses to movement in an association area of the cat cerebral cortex. J. Neurophysiol. **32**, 773—784 (1969).

DUBNER, R., BROWN, F. J.: Response of cells to restricted visual stimuli in an association area of cat cerebral cortex. Exp. Neurol. **20**, 70—86 (1968).

— RUTLEDGE, L. T.: Recording and analysis of converging input upon neurones in cat association cortex. J. Neurophysiol. **27**, 620—634 (1964).

ENROTH-CUGELL, C., ROBSON, J. G.: The contrast sensitivity of retinal ganglion cells of the cat. J. Physiol. (Lond.) **187**, 517—552 (1966).

ERIKSEN, C. W., COLLINS, J. F.: Some temporal characteristics of visual pattern perception. J. exp. Psychol. **74**, 476—484 (1967).

FELDMAN, M., COHEN, B.: Electrical activity in the lateral geniculate body of the alert monkey associated with eye movements. J. Neurophysiol. **31**, 455—466 (1968).

FIORENTINI, A., JEANNE, M., DI FRANCIA, G. T.: Mésures photométriques sur un champ à gradient d'éclairement variable. Optica Acta **1**, 192—193 (1955).

FITE, K. V.: Single unit analysis of binocular neurones in the frog optic tectum. Exp. Neurol. **24**, 475—486 (1969).

FRY, G. A., ALPERN, M.: The effect of a peripheral glare source upon the apparent brightness of an object.. J opt. Soc. Amer. **43**, 189—195 (1953).

— BARTLEY, S. H.: The effect of one border in the visual field upon the threshold of another. Amer. J. Physiol. **112**, 414—421 (1935).

FUKADA, Y.: Receptive field organisation of cat optic nerve fibres with special reference to conduction velocity. Vision Res. **11**, 209—226 (1971).

GAARDER, K.: Transmission of edge information in the human visual system. Nature (Lond.) **212**, 321—323 (1966).

GANZ, L., FITCH, M.: The effect of visual deprivation on perceptual behaviour. Exp. Neurol. **22**, 638—660 (1969).

— — SATTERBERG, J. A.: The selective effect of visual deprivation on receptive field shape determined neurophysiologically. Exp. Neurol. **22**, 614—637 (1969).

GAREY, L. J.: The termination of thalamo-cortical fibres in the visual cortex of the cat and monkey. J. Physiol. (Lond.) **210**, 15—17 P (1970).

— POWELL, T. P. S.: The projection of the lateral geniculate nucleus upon the cortex in the cat. Proc. Roy. Soc. B **169**, 107—126 (1967).

GAZE, R. M., KEATING, M. J.: Receptive field properties of single units from the visual projection to the ipsilateral tectum in the frog. Quart. J. exp. Physiol. **55**, 143—152 (1970).

GLOBUS, A., SCHEIBEL, A. B.: Synaptic loci on visual cortical neurones of the rabbit: The specific afferent radiation. Exp. Neurol. **18**, 116—131 (1967).

GOURAS, P.: The effects of light-adaptation on rod and cone receptive field organisation of monkey ganglion cells. J. Physiol. (Lond.) **192**, 747—760 (1967).

— Trichromatic mechanisms in single cortical neurones. Science **168**, 489—492 (1970).

GRAHAM, C. H., (Ed.): Vision and visual perception. New York: John Wiley 1965.

GRAHAM, N., NACHMIAS, J.: Detection of grating patterns containing two spatial frequencies: A comparison of single-channel and multi-channel models. Vision Res. **11**, 251—260 (1971).

GRAY, E. C.: Electromicroscopy of excitatory and inhibitory synapses: A brief review. Progr. Brain Res. **31**, 141—155 (1969).

GREEN, D. G.: Regional variations in the visual acuity for interference fringes on the retina. J. Physiol. (Lond.) **207**, 351—356 (1970).

GREGORY, R. L.: Eye movements and the stability of the visual world. Nature (Lond.) **182**, 1214—1216 (1958).

GROSS, C. G., BENDER, D. B., ROCHA-MIRANDA, C. A.: Visual receptive fields of neurones in inferotemporal cortex of the monkey. Science **166**, 1303—1306 (1969).

— SCHILLER, P. H., WELLS, C., GERSTEIN, G. L.: Single unit activity in temporal association cortex of the monkey. J. Neurophysiol. **30**, 833—843 (1967).

GUILLERY, R. W.: A quantitative study of synaptic interconnection in the dorsal lateral geniculate nucleus of the cat. Z. Zellforsch. mikr. Anat. **96**, 39—48 (1969).

Haber, R. N., (Ed.): Information-processing approaches to visual perception. New York: Holt, Rinehart and Winston (1969).
— How we remember what we see. Sci. Amer. **222**/5, 104—112 (1970).
Hake, H. W.: Form discrimination and the invariance of form. In: Uhr, L. (Ed.): Pattern Recognition. pp. 142—173. New York: John Wiley 1966.
Hartline, H. K.: Inhibition of activity of visual receptors by illuminating nearby retinal elements in the *Limulus* eye. Fed. Proc. **8**, 69 (1949).
— Wagner, H. G., Ratliff, F.: Inhibition in the eye of *Limulus*. J. gen. Physiol. **39**, 651—673 (1956).
Hebbard, F. W., Marg, E.: Physiological nystagmus in the cat. J. opt. Soc. Amer. **50**, 151—155 (1960).
Held, R.: Dissociation of visual functions by deprivation and rearrangement. Psychol. Forsch. **31**, 338—348 (1968).
Helmholtz, H. von: Handbuch der physiologischen Optik, Vol. 3. Hamburg: Voss 1867.
— Treatise on physiological optics, 3rd Ed. Southall, J. P. C. (Ed.): Rochester, N. Y.: Opt. Soc. Amer. 1910.
Henry, G. H., Bishop, P. O.: Simple cells of the striate cortex. In: Neff, W. D. (Ed.): Contributions to Sensory Physiology. **5**, 1—46, (1971). New York: Academic Press.
— — Coombs, J. S.: Inhibitory and sub-liminal excitatory receptive fields of simple units in cat striate cortex. Vision Res. **9**, 1289—1296 (1969).
Hering, E.: Zur Lehre vom Lichtsinne. I. II. III. IV. V. Sitzungsberichte der Kaiserlichen Akademie der Wissenschaften in Wien. Mathematisch-naturwissenschaftliche Classe. Abh. III. **66**, 5—24 (1872); **68**, 186—201 (1874); **68**, 229—244 (1874); **69**, 85—104 (1874); **69**, 179—217 (1874).
— Outline of a theory of the light sense. Trans.: Hurvich, L. M., Jameson, D. Cambridge, Mass.: Harvard Univ. Press 1964.
Hill, R. M.: Receptive field properties of the superior colliculus of the rabbit. Nature (Lond.) **211**, 1407—1409 (1966).
— Goodwin, H.: Visual receptive fields from cells of a marsupial (*Didelphis virginiana*) superior colliculus. Experientia (Basel) **24**, 559—560 (1968).
Hoffmann, K. P., Stone, J.: Conduction velocity of afferents to cat visual cortex: a correlation with receptive field properties. Brain Res., **32**, 460—466 (1971).
Holden, A. L.: Receptive properties of retinal cells and tectal cells in the pigeon. J. Physiol. (Lond.) **201**, 56—57 P (1969).
Holländer, H.: The projection from the visual cortex to the lateral geniculate body: An experimental study with silver impregnation methods in the cat. Exp. Brain Res. **10**, 219—235 (1970).
Horn, G., Hill, R. M.: Responsiveness to sensory stimulation of units in the superior colliculus and subadjacent tectotegmental regions of the rabbit. Exp. Neurol. **14**, 199—223 (1966).
Houlihan, K., Sekuler, R. W.: Contour interactions in visual masking. J. exp. Psychol. **77**, 281—285 (1968).
Hubel, D. H.: Single unit activity in striate cortex of unrestrained cats. J. Physiol. (Lond.) **147**, 226—238 (1959).
— Single unit activity in lateral geniculate body and optic tract of unrestrained cats. J. Physiol. (Lond.) **150**, 91—104 (1960).
— The visual cortex of the brain. Sci. Amer. **209**/5, 54—77 (1963).
— Wiesel, T. N.: Receptive fields of single neurones in the cat's striate cortex. J. Physiol. (Lond.) **148**, 574—591 (1959).
— — Receptive fields of optic nerve fibres in the spider monkey. J. Physiol. (Lond.) **154**, 572—580 (1960).
— — Integrative action in the cat's lateral geniculate body. J. Physiol. (Lond.) **155**, 385—398 (1961).
— — Receptive fields, binocular interaction and functional architecture in the cat's visual cortex. J. Physiol. (Lond.) **160**, 106—154 (1962).
— — Shape and arrangement of columns in cat's striate cortex. J. Physiol. (Lond.) **165**, 559 to 568 (1963).

HUBEL, D. H., WIESEL, T. N.: Receptive fields and functional architecture in two non-striate visual areas (18 and 19) of the cat. J. Neurophysiol. **28**, 229—289 (1965 a).
— — Binocular interaction in striate cortex of kittens reared with artificial squint. J. Neurophysiol. **28**, 1041—1059 (1965 b).
— — Receptive fields and functional architecture of monkey striate cortex. J. Physiol. (Lond.) **195**, 215—243 (1968).
— — Visual area of the lateral suprasylvian gyrus (Clare-Bishop area). J. Physiol. (Lond.) **202**, 251—260 (1969 a).
— — Anatomical demonstration of columns in the monkey striate cortex. Nature (Lond.) **221**, 747—750 (1969 b).
— — The period of susceptibility to the physiological effect of unilateral eye closure in kittens. J. Physiol. (Lond.) **206**, 419—436 (1970 a).
— — Stereoscopic vision in macaque monkey. Nature. **225**, 41—42 (1970 b).
HUMPHREY, N. K.: Responses to visual stimuli of units in the superior colliculus of rats and monkeys. Exp. Neurol. **20**, 312—340 (1968).
— WEISKRANTZ, L.: Vision in monkeys after removal of the striate cortex. Nature (Lond.) **215**, 595—597 (1967).
JACOBSON, M., GAZE, R. M.: Types of visual response from single units in the optic tectum and optic nerve of the goldfish. Quart. J. exp. Physiol. **49**, 199—209 (1964).
JONES, E. G., POWELL, T. P. S.: An electron microscopic study of the laminar pattern and mode of termination of different fibre pathways in the somatic sensory cortex of the cat. Phil. Trans. roy. Soc. Lond. B **257**, 45—62 (1970).
JOSHUA, D. E., BISHOP, P. O.: Binocular single vision and depth discrimination. Receptive field disparities for central and peripheral vision and binocular interaction of peripheral single units in cat striate cortex. Exp. Brain Res. **10**, 389—416 (1970).
JUNG, R.: Korrelationen von Neuronentätigkeit und Sehen. In: JUNG, R., KORNHUBER, H. H. (Eds.): Neurophysiologie und Psychophysik des visuellen Systems. Berlin-Göttingen-Heidelberg: Springer 1961.
— Neuronale Grundlagen des Hell-Dunkelsehens und der Farbwahrnehmung. Bericht 66. Zusammenkunft der Dtsch. Ophthalmol. Ges., Heidelberg. München: Bergmann Verlag 1964.
KAWAMURA, H., MARCHIAFAVA, P. L.: Excitability changes along visual pathways during eye tracking movements. Arch. ital. Biol. **106**, 141—156 (1968).
KEATING, M. J., GAZE, R. M.: Observations on the "surround" properties of the receptive fields of frog retinal ganglion cells. Quart. J. exp. Physiol. **55**, 129—142 (1970).
KEESEY, U. T.: Effects of involuntary eye movements on visual acuity. J. opt. Soc. Amer. **50**, 769—773 (1960).
KLÜVER, H.: Visual functions after removal of the occipital lobes. J. Psychol. **11**, 23—45 (1941).
KOZAK, W., RODIECK, R. W., BISHOP, P. O.: Responses of single units in lateral geniculate nucleus of cat to moving visual patterns. J. Neurophysiol. **28**, 19—47 (1965).
KUFFLER, S. W.: Discharge patterns and functional organisation of mammalian retina. J. Neurophysiol. **16**, 37—68 (1953).
KUYPERS, S. G., SZWARCBART, M. K., MISHKIN, M., ROSVOLD, H. E.: Occipitotemporal corticocortical connections in the rhesus monkey. Exp. Neurol. **11**, 245—262 (1965).
LASHLEY, K. S.: The mechanism of vision. XV: Preliminary studies of the rat's capacity for detail vision. J. gen. Psychol. **18**, 123—193 (1938).
LEGRAND, Y.: Form and space vision. Bloomington, Ind.: Indiana Univ. Press (1967).
LEIBOWITZ, H., MOTE, F. A., THURLOW, W. R.: Simultaneous contrast as a function of separation between test and inducing fields. J. exp. Psychol. **46**, 453—456 (1953).
LETTVIN, J. Y., MATURANA, H. R., PITTS, W. H., MCCULLOCH, W. S.: Two remarks on the visual system of the frog. In: ROSENBLITH, W. A. (Ed.): Sensory Communication. Cambridge, Mass.: M. I. T. Press 1961.
LEVICK, W. R.: Receptive fields and trigger features of ganglion cells in the visual streak of the rabbit. J. Physiol. (Lond.) **188**, 285—307 (1967).
— OYSTER, C. W., TAKAHASHI, E.: Rabbit lateral geniculate nucleus: Sharpener of directional information. Science **165**, 712—714 (1969).
— ZACKS, J. L.: Responses of cat retinal ganglion cells to brief flashes of light. J. Physiol. (Lond.) **206**, 677—700 (1970).

LORENTE DE NÓ, R.: Studies on the structure of the cerebral cortex. II. Continuation of the study of the Ammonic system. J. Psychol. Neurol. **46**, 113—117 (1934).
LUDVIGH, E. J.: The visibility of moving objects. Science **108**, 63—64 (1948).
— MILLER, J. W.: Study of visual acuity during the ocular pursuit of moving test objects. I. Introduction. J. opt. Soc. Amer. **48**, 799—802 (1958).
MACH, E.: Über die Wirkung der räumlichen Verteilung des Lichtreizes auf die Netzhaut. I. Sitzungsberichte der mathematisch-naturwissenschaftlichen Klasse der Kaiserlichen Akademie der Wissenschaften **52**/2, 303—322 (1865).*
— Über den physiologischen Effect räumlich vertheilter Lichtreize (Zweite Abhandlung). Sitzungsberichte (see above) **54**/2, 131—144 (1866 a).*
— Über die physiologische Wirkung räumlich vertheilter Lichtreize (Dritte Abhandlung). Sitzungsberichte (see above) **54**/2, 393—408 (1866 b).*
— Über die physiologische Wirkung räumlich vertheilter Lichtreize (Vierte Abhandlung). Sitzungsberichte (see above) **57**/2, 11—19 (1868 a).*
— Über die Abhängigkeit der Netzhautstellen von einander. Vjschr. f. Psychiatrie in ihren Beziehungen zur Morphologie und Pathologie des Central-Nervensystems **2**, 38—51 (1868b)*.
MACKAY, D. M.: Perceptual stability of a stroboscopically lit visual field containing self-luminous objects. Nature (Lond.) **181**, 507—508 (1958).
— Elevation of visual threshold by displacement of retinal image. Nature (Lond.) **225**, 90 to 92 (1970 a).
— Interocular transfer of suppressive effects of retinal image displacement. Nature (Lond.) **225**, 872—873 (1970 b).
MARG, E., ADAMS, J. E., RUTKIN, B.: Receptive fields of cells in the human visual cortex. Experientia (Basel) **24**, 348—350 (1968).
MARTY, R., BENOIT, O., LANGUIER, M. M.: Étude topographique et stratigraphique des projections du corps genouillé latéral sur le cortex cérébral. Arch. ital. Biol. **107**, 723—742 (1969).
MATURANA, H. R., FRENK, S.: Directional movement and horizontal edge detectors in the pigeon retina. Science **142**, 977—979 (1963).
— LETTVIN, J. Y., PITTS, W. H., MCCULLOCH, W. S.: Physiology and anatomy of vision in the frog. J. gen. Physiol. **43** (suppl.), 129—175 (1960).
MCILWAIN, J. T., BUSER, P.: Receptive fields of single cells in the cat's superior colliculus. Exp. Brain Res. **5**, 314—325 (1968).
MEYER, P. M.: Analysis of visual behavior in cats with extensive cortical ablations. J. comp. physiol. Psychol. **56**, 397—401 (1963).
MEYERS, B., MCCLEARY, R. A.: Interocular transfer of a pattern discrimination in pattern deprived cats. J. comp. physiol. Psychol. **57**, 16—21 (1964).
MICHAEL, C. R.: Receptive fields of directionally selective units in the optic nerve of the ground squirrel. Science **152**, 1092—1095 (1966).
— Integration of visual information in the superior colliculus. J. gen. Physiol. **50**, 2485—2486 (1967).
MILNER, B.: Visual recognition and recall after right temporal lobe excision in man. Neuropsychol. **6**, 191—209 (1968).
NACHMIAS, J.: Two-dimensional motion of the retinal image during fixation. J. opt. Soc. Amer. **49**, 901—908 (1959).
— SACHS, M. B., ROBSON, J. G.: Independent spatial frequency channels in human vision. J. opt. Soc. Amer. **59**, 1538 A (1969).
NIIMI, S. K., SPRAGUE, J. M.: Thalamo-cortical organisation of the visual system in the cat. J. comp. Neurol. **138**, 219—250 (1970).
NODA, H., FREEMAN, R. B., JR., CREUTZFELDT, O.: Neuronal correlates of eye movements in the visual cortex of the cat. Science **175**, 661—664 (1972).
— GIES, B., CREUTZFELDT, O.: Neuronal responses in the visual cortex of awake cats to stationary and moving targets. Exp. Brain Res. **12**, 389—405 (1971).
— IWAMA, K.: Unitary analysis of retino-geniculate response times in rat. Vision Res. **7**, 205—213 (1967).

* Translation in RATLIFF (1965).

O'Brien, V.: Contour perception, illusion and reality. J. opt. Soc. Amer. **48**, 112—119 (1958).
Ogawa, T., Bishop, P. O., Levick, W. R.: Temporal characteristics of responses to photic stimulation by single ganglion cells in the unopened eye of the cat. J. Neurophysiol. **29**, 1—30 (1966).
O'Leary, J. L.: Structure of the area striata of the cat. J. comp. Neurol. **75**, 131—164 (1941).
Otsuka, R., Hassler, R.: Über Aufbau und Gliederung der corticalen Sehsphäre bei der Katze. Arch. Psychiat. und Z. ges. Neurol. **203**, 212—234 (1962).
Oyster, C. W., Barlow, H. B.: Direction-selective units in rabbit retina: Distribution of preferred directions. Science **155**, 841—842 (1967).
Pantle, A., Sekuler, R.: Size-detecting mechanisms in human vision. Science **162**, 1146—1148 (1968 a).
— — Velocity-sensitive elements in human vision: Initial psychophysical evidence. Vision Res. **8**, 445—450 (1968 b).
— — Contrast response of human visual mechanisms sensitive to orientation and direction of motion. Vision Res. **9**, 397—406 (1969).
Pasik, T., Pasik, P., Bender, M.: The superior colliculi and eye movements. Arch. Neurol. **15**, 420—436 (1966).
Pearlman, A. C., Daw, N. W.: Opponent color cells in the cat lateral geniculate nucleus. Science **167**, 84—86 (1970).
Penfield, W., Perot, P.: The brain's record of auditory and visual experience. Brain **86**, 595—696 (1963).
Pettigrew, J. D., Nikara, T., Bishop, P. O.: Responses to moving slits by single units in cat striate cortex. Exp. Brain Res. **6**, 373—390 (1968 a).
— — — Binocular interaction on single units in cat striate cortex: Simultaneous stimulation by single moving slits with receptive fields in correspondence. Exp. Brain Res. **6**, 391—410 (1968 b).
Pirenne, M. H., Denton, E. J.: Accuracy and sensitivity of the human eye. Nature (Lond.) **170**, 1039—1042 (1952).
Polyak, S.: The retina. Chicago: Univ. Chicago Press 1941.
— The vertebrate visual system. Chicago: Univ. Chicago Press 1957.
Pritchard, R. M., Heron, W.: Small eye movements of the cat. Canad. J. Psychol. **14**, 131—137 (1960).
Ratliff, F.: Mach bands: Quantitative studies of the neural networks of the retina. San Francisco: Holden-Day 1965.
— Hartline, H. K.: The responses of *Limulus* optic nerve fibres to patterns of illumination of the retinal mosaic. J. gen. Physiol. **42**, 1241—1255 (1959).
— — Miller, W. H.: Spatial and temporal aspects of retinal inhibitory interaction. J. opt. Soc. Amer. **53**, 110—120 (1963).
Riggs, L. A.: Visual acuity. In: Graham, C. H. (Ed.): Vision and visual perception. New York: John Wiley (1965).
— Ratliff, F., Cornsweet, J. C., Cornsweet, T.: The disappearance of steadily fixated visual test objects. J. opt. Soc. Amer. **43**, 495—501 (1953).
Rodieck, R. W.: Receptive fields in the cat's retina: A new type. Science **157**, 90—92 (1967).
— Stone, J.: Response of cat retinal ganglion cells to moving visual patterns. J. Neurophysiol. **28**, 819—832 (1965 a).
— — Analysis of receptive fields of cat retinal ganglion cells. J. Neurophysiol. **28**, 833—849 (1965 b).
Rolls, E. T., Cowey, A.: Topography of the retina and striate cortex and its relationship to visual acuity in rhesus monkeys and squirrel monkeys. Exp. Brain Res. **10**, 298—310 (1970).
Sanides, F., Hoffman, J.: Cyto- and myeloarchitecture of the visual cortex of the cat and of the surrounding integration cortices. J. Hirnforsch. **11**, 79—104 (1969).
Schneider, G. E.: Contrasting visuomotor functions of tectum and cortex in the Golden Hamster. Psychol. Forsch. **31**, 52—62 (1968).
— Two visual systems. Science **163**, 895—902 (1969).

Sekuler, R. W.: Spatial and temporal determinants of visual backward masking. J. exp. Psychol. **70**, 401—406 (1965).

— Rubin, E. L., Cushman, W. H.: Selectivities of human visual mechanisms for direction of movement and contour orientation. J. opt. Soc. Amer. **58**, 1146—1150 (1968).

Smith, K. U.: Visual discrimination in the cat: VI. The relation between pattern vision and visual acuity and the optic projection centres of the nervous system. J. gen. Psychol. **53**, 251—272 (1938).

Spear, P. D., Braun, J. J.: Pattern discrimination following removal of visual neocortex in cat. Exp. Neurol. **25**, 331—348 (1969).

Sperling, G.: A model for visual memory tasks. Human Factors **5**, 19—31 (1963).

Sprague, J. M., Berlucchi, G., di Berardino, A.: The superior colliculus and pretectum in visually guided behaviour and visual discrimination in the cat. Brain, Behav. Evol. **3**, 285—294 (1970).

Steinman, R. M., Cunitz, R. J., Timberlake, G. T., Herman, M.: Voluntary control of microsaccades during maintained monocular fixation. Science **155**, 1577—1579 (1967).

Sterling, P., Wickelgren, B. G.: Visual receptive fields in the superior colliculus of the cat. J. Neurophysiol. **32**, 1—15 (1969).

Stone, J., Fabian, M.: Specialised receptive fields of the cat's retina. Science **152**, 1277—1279 (1966).

— Hansen, S. M.: The projection of the cat's retina on the lateral geniculate nucleus. J. comp. Neurol. **126**, 601—624 (1966).

— Hoffmann, K. P.: Conduction velocity as a parameter in the organisation of the afferent relay in the cat's lateral geniculate nucleus. Brain Res. **32**, 454—459 (1971).

Straschill, M., Hoffmann, K. P.: Response characteristics of movement-detecting neurones in pretectal region of the cat. Exp. Neurol. **25**, 165—176 (1969 a).

— — Functional aspects of localisation in the cat's tectum opticum. Brain Res. **13**, 274—283 (1969 b).

— Taghavy, A.: Neuronale Reaktionen im Tectum Opticum der Katze auf bewegte und stationäre Lichtreize. Exp. Brain Res. **3**, 353—367 (1967).

Stroud, J. M.: The fine structure of psychological time. In: Quastler, H. (Ed.): Information Theory in Psychology. Glencoe, Ill.: Free Press 1966.

Sutherland, N. S.: Outlines of a theoory of visual pattern recognition in animals andman. Proc. roy. Soc. B **171**, 297—317 (1968).

— Shape discrimination in rats, octopus and goldfish: A comparative study. J. comp. physiol. Psychol. **67**, 160—176 (1969).

— Williams, C.: Discrimination of checkerboard patterns by rats. Quart. J. exp. Psychol. **21**, 77—84 (1969).

Talbot, S. A., Marshall, W. H.: Physiological studies on neural mechanisms of visual localisation and discrimination. Amer. J. Ophthal. **24**, 1255—1264 (1941).

Teller, D. Y., Andrews, D. P., Barlow, H. B.: Local adaptation in stabilized vision. Vision Res. **6**, 701—705 (1966).

Teuber, H.-L.: Perception. In: Field, J., (Ed.): Washington, D.C.: Handbook of Physiology: Neurophysiology. Section 1, Vol. 3, pp. 1595—1668. Amer. Physiol. Soc. 1960.

— Battersby, W. S., Bender, M. B.: Visual field defects after penetrating missile wounds of the brain. Cambridge, Mass.: Harvard Univ. Press 1960.

Toyama, K., Matsunami, K.: Synaptic action of specific visual impulses upon cat's parastriate cortex. Brain Res. **10**, 473—476 (1968).

— Tokashiki, S., Matsunami, K.: Synaptic action of commissural impulses upon association efferent cells in cat visual cortex. Brain Res. **14**, 518—520 (1969).

Trevarthen, C. B.: Two mechanisms of vision in primates. Psychol. Forsch. **31**, 299—337 (1968).

Volkmann, F. C.: Vision during voluntary saccadic eye movements. J. opt. Soc. Amer. **52**, 571—578 (1962).

— Schick, A. M., Riggs, L. A.: Time course of visual inhibition during voluntary saccades. J. opt. Soc. Amer. **58**, 562—569 (1968).

WATANABE, S., KONISHI, M., CREUTZFELDT, O.: Postsynaptic potentials in the cat's visual cortex following electrical stimulation of afferent pathways. Exp. Brain Res. **1**, 272—283 (1966).
WEISKRANTZ, L.: Contour discrimination in a young monkey with striate cortex ablation. Neuropsychologia **1**, 145—164 (1963).
— COWEY, A.: Filling in the scotoma: A study of residual vision after striate cortex lesions in monkeys. In: STELLAR, E., SPRAGUE, J.M. (Eds.): Progress in Physiological Psychology, pp. 237—260. New York: Academic 1970.
WESTHEIMER, G.: Eye movement responses to a horizontally moving stimulus. Arch. Ophthal. **52**, 932—941 (1954).
— Modulation thresholds for sinusoidal light distributions on the retina. J. Physiol. (Lond.) **152**, 67—74 (1960).
— Spatial interaction in the human retina during scotopic vision. J. Physiol. (Lond.) **181**, 881—894 (1965).
— Spatial interaction in human cone vision. J. Physiol. (Lond.) **190**, 139—154 (1967).
— Rod-cone independence for sensitizing interaction in the human retina. J. Physiol. (Lond.) **206**, 109—116 (1970).
WETZEL, A. B., THOMPSON, V. E., HOREL, J. A., MEYER, P.M.: Some consequences of perinatal lesions of the visual cortex in the cat. Psychon. Sci. **3**, 381—382 (1965).
WEYMOUTH, F. W.: Visual sensory units and the minimum angle of resolution. Amer. J. Ophthal. **24**, 1255—1264 (1958).
WHITE, C. T., CHEATHAM, P. G.: Temporal numerosity: IV. A comparison of the major senses. J. exp. Psychol. **58**, 441—444 (1959).
WIESEL, T. N., HUBEL, D. H.: Single-cell responses in striate cortex of kittens deprived of vision in one eye. J. Neurophysiol. **26**, 1003—1017 (1963).
WILSON, M. E.: Cortico-cortical connections of the cat visual areas. J. Anat. **102**, 375—387 (1968).
— CRAGG, B. G.: Projections from the lateral geniculate nucleus in the cat and the monkey. J. Anat. **101**, 679—695 (1967).
WINANS, S. S.: Visual form discrimination after removal of the visual cortex in cats. Science **158**, 944—946 (1967).
WRIGHT, M. J.: Visual receptive fields of cells in a cortical area remote from the striate cortex of the cat. Nature (Lond.) **223**, 973—975 (1969).
WURTZ, R. H.: Visual receptive fields of striate cortex neurones in awake monkeys. J. Neurophysiol. **32**, 727—742 (1969 a).
— Responses of striate cortex neurones to stimuli during rapid eye movements in the monkey. J. Neurophysiol. **32**, 975—986 (1969 b).
— Comparison of effects of eye movements and stimulus movements on striate cortex neurones of the monkey. J. Neurophysiol. **32**, 987—994 (1969 c).
ZEKI, S. M.: Visual deficits related to size of lesion in 'pre-striate' cortex of optic chiasm-sectioned monkeys. Life Sci. **6**, 1627—1638 (1967).
— The secondary visual areas of the monkey. Brain Res. **13**, 197—226 (1969 a).
— Representation of central visual fields in prestriate cortex of monkey. Brain Res. **14**, 271 to 291 (1969 b).

Chapter 3

Central Mechanisms of Color Vision

By

Russell L. De Valois, Berkeley, California (USA)

With 17 Figures

Contents

I. Introduction

Color vision has captivated the interest of innumerable people, both amateurs and visual scientists, in the three hundred years since Newton reported his "New Theory about Light and Colours" to the Royal Society. Thousands of psychophysical experiments studying the effects of numerous variables on human color perception have provided us with a rich legacy of knowledge about how the visual system as a whole deals with wavelength information. Despite the fact that one of the principal goals of most visual physiologists is to understand visual behavior on the basis of the physiology of the visual system, there are a surprising number of misconceptions about color vision.

Color vision is often confused with differential spectral sensitivity. An individual who has color vision (or a cell which exhibits color discrimination capabilities) is one which can discriminate between two lights of different spectral composition independent of their relative intensities. Any animal (or plant, for that matter) has differential spectral sensitivity, that is, requires a different amount of light from different spectral regions for a constant response. Therefore a totally color-blind individual (or cell) will give different responses to lights of different wavelengths when they are equated, for instance, for equal energy. But if the relative luminances of the lights are varied for such a colorblind individual, there will be some one point at which the two lights of different spectral composition will be identical. So finding different responses from some visual locus to flashes of light of different wavelength tells one exactly nothing about the presence of color information at that locus. It is only if the response differences are maintained despite changes in the relative luminance of the two lights that one could say that the cells are responding to the color rather than the brightness differences.

A far better way of distinguishing whether a cell was discriminating color rather than some other aspect of the stimulus would be to test its ability to discriminate between two different lights which are equal in luminance (if one has some good behavioral indication of the photopic luminance sensitivity of the organism in question). If a cell could discriminate a shift from a white to a monochromatic light, or from one monochromatic light to another when they were equated in luminance for the animal, that cell could certainly be said to show color discrimination.

Another misconception which one occasionally encounters in the literature is that finding multiple cone inputs to a cell tells one something about the cell's color vision capabilities. A cell in the central visual system which is processing color information must certainly have more than one receptor input, but the presence of more than one cone type does not necessarily indicate that the cell is involved in color vision. A unit might have an input from all the different cone types but be extracting movement information from them, not color information at all. This confusion has its origin, or is at least much facilitated by, the widespread but unfortunate reference to cones as "color receptors" and the use of color names for the different cone types.

Most early hypotheses about visual functions were essentially theories of receptor activity: adaptation, brightness vision, color vision and visual acuity were all accounted for on the basis of the properties of the receptors. The rest of the retinal cells, if mentioned at all, were merely assigned the role of relaying the receptor information intact to the cortex. The discovery by Polyak (1941) of midget bipolar cells which pick up from just one cone and feed into one ganglion cell provided a possible anatomical system (if one were careful to ignore the lateral connections made by the horizontal and amacrine cells) for such a direct, unprocessed feeding of receptor output to the brain. But all recent studies of the physiology of the visual system, and recent investigations of the microanatomy of the paths, indicate that such unprocessed transfer of receptor information cannot and does not occur. Nonetheless, the terminological legacy from such receptor-dominated theories persists in the common reference to "red", "green", and "blue" receptors. This seemingly innocuous terminology has in fact been as important as any factor in hindering an understanding of how the visual system operates in providing us with color information.

A large number of psychophysical, spectrophotometric, and physiological experiments (discussed below) agree in indicating that macaque monkeys and man possess three cone types, containing photopigments which have their peak absorption in the regions of 445, 540, and 570 nm. Furthermore, these studies also agree well in indicating that the 540 and 570 nm pigments, at least, absorb across the whole spectrum, not in just one limited spectral range, see Fig. 4. Thus the 570 pigment cone, so misleadingly called by many the "red receptor", not only is not limited in its sensitivity to that part of the spectrum we see as red (from about 600–700 nm), but does not even absorb maximally in this spectral region. It is in fact more sensitive to that part of the spectrum we see as pure green (c. 510 nm) than to pure red (c. 650 nm). So both on account of its point of maximal absorption, and on account of the breadth of its absorption curve, it is a gross misnomer to term the 570 receptor a "red receptor", or the 540 cone a "green receptor". But worst of all is that calling these receptors "red" or "green" receptors carries a false implication; namely that their activation signals "red" or "green" to the next cells in the visual system.

The various cone types are not "color receptors" but "light receptors", each responding to light of every wavelength, intensity, shape, whether moving or stationary. Information about the color of the stimulus, its shape, and whether it is moving or not is in each case extracted by the later elements in the visual path on the basis of comparing the outputs of different receptors = receptors of different pigment type, or different location, or different activation time. These later units are specifically responsive to certain colors or to certain shapes, or to stimuli moving in particular directions; they might thus quite reasonably be termed "red cells" or "vertical line detectors" or "movement cells". But to term the receptors "red receptors" or "vertical line receptors" or "movement receptors" would be misleading indeed, for no receptors have such specific sensitivities.

In this chapter, therefore, color names will be reserved for those visual units which show a wavelength specificity comparable to that indicated by the color terms. But since short, handy terms are clearly needed to refer to receptors containing different photopigment types, we will use the terms "L cone" (containing the long-wavelength absorbing pigment), "M cone" (medium), and "S cone" (short) to denote the cone types containing the 570, 540 and 445 nm λ_{max} pigment, respectively, in primates or the comparable receptors in other animals.

II. Central Structures Involved in Color Processing

In higher vertebrates the retinal projection to central structures is up four fairly well defined paths: to the lateral geniculate nucleus (LGN) and from there to the striate cortex; to the superior colliculus; to the pretectal area; and to the accessory optic nuclei.

One would like very much to be able to write a chapter on the central neural processing of color information by specifying which of these central paths receive color information, and by describing the way in which color information is sequentially processed in the regions involved. This is unfortunately far from possible at our present stage of knowledge. Almost all of our information about the central mechanisms of color vision comes from studies of the activity of cells at just one non-retinal locus: the LGN. So while a fairly coherent story can be made of the transformations of color information from the receptors to the LGN, very little can be said about what happens to the information past that level. It would appear, however, from the close parallels one finds between the responses of cells at the LGN level in primates and color perception in many visual situations that much of the processing of color information (contrary to the situation with form or movement information) takes place before the cortex.

The ground squirrel, *Citellus mexicanus*, has an all-cone eye and has been shown in psychophysical tests (JACOBS and YOLTON, 1971) to have well developed color vision of a dichromatic variety. MICHAEL (1968) found evidence in optic nerve recording for one class of unit in the ground squirrel which shows opponent responses to different wavelengths of light (discussed further below). In recordings at the next synaptic level, MICHAEL (1971) found similar opponent-color cells in the LGN but not in the superior colliculus.

It would appear that the non-albino rabbit has color vision, but the extent and nature of the color vision have not been well established, the behavioral tests (BROWN, 1937) of its vision being quite inadequate for this purpose. It is therefore hardly the ideal animal in which to investigate color vision. But there are reports (HILL, 1962; HILL and MARG, 1963) of unit responses to flashes of monochromatic light from both the LGN and the transpeduncular path (of the accessory optic system) in rabbit. The results suggest that color processing is found in the LGN but not in the accessory optic system. Their stimulus system suppressed the spontaneous activity, making it difficult to observe inhibition, but when there was a maintained firing rate, spectrally opponent responses were found.

In primates, there is ample evidence for color information going up the geniculo-striate path in Old World monkeys (DE VALOIS et al., 1958, 1966; LENNOX-BUCHTHAL, 1961; ANDERSEN et al., 1962; WIESEL and HUBEL, 1966; HUBEL and WIESEL, 1968). Comparably detailed studies have not been made of the colliculus or other subcortical optic centers. Units in the macaque colliculus have been examined, but primarily in their responses to stimuli moving against a background, with no attempt to control luminance (HUMPHREY, 1968). Many units were found to respond to colored spots, but they gave the same responses to all colors; and because of the lack of luminance control one cannot tell whether they were distinguishing the spots from the background on the basis of color or luminance differences.

In the squirrel monkey, *Saimiri sciureus*, there is again good evidence of the involvement of the LGN in color processing (JACOBS and DE VALOIS, 1965), but few studies of other centers. However, in this species it has been reported (WOLIN et al., 1966) that the superior colliculus does receive color information. Some 14% of the collicular cells were reported (KADOYA and MASSOPUST, 1970) to be color-coded, but how this was defined and determined was not stated. Such cells were not found by these investigators in the pretectal area.

That the geniculo-striate paths is the portion of the primate central nervous system in which much if not all of the processing of color information is accomplished is indicated by the defects in color discrimination which result from cortical lesions in monkeys (KLÜVER, 1941; WEISKRANTZ, 1963). No similar loss has been found to result from destruction of the superior colliculus (ANDERSON and SYMMES, 1969).

The situation in the tree shrew, *Tupaia glis*, appears to be different. This animal, which has been shown in behavioral tests to have dichromatic color vision (POLSON, 1968), has been found to be able to make similar color discriminations before and after a total striate lesion (SNYDER et al., 1969). Lesions in other parts of its optic projections have not been investigated.

In the goldfish, opponent color cells have been found in the optic tectum (JACOBSON, 1964). In this animal, of course, the tectum rather than the geniculo-striate system is the main retinal projection.

III. Lateral Geniculate Nucleus (LGN)

Most of the detailed investigations of the physiology of color vision at post-retinal have been of cells in the LGN of the thalamus. These will be discussed in detail.

1. Types of Color-Specific Cells

Primates. Clear evidence for physiological mechanisms in the central nervous system involved in processing color information came from a study of single units in the LGN of macaque monkeys (DE VALOIS et al., 1958). Cells were found which showed excitation to some wavelengths and inhibition to others when the eye was stimulated with flashes of monochromatic light equated for energy. These different, opposite kinds of responses to light from different spectral regions were maintained over wide intensity ranges so a cell never gave the same response to light from two opposite spectral regions. These cells thus met the criterion of responding differentially to color rather than luminance differences. Such spectrally-opponent cells bear an obvious functional relationship to the opponent S-potentials found by SVAETICHIN (1956) and SVAETICHIN and MACNICHOL (1958) in the fish retina. These early observations have been confirmed by numerous other reports of opponent-cell responses in macaque geniculate cells (DE VALOIS, 1960a, 1965; DE VALOIS et al., 1966; WIESEL and HUBEL, 1966).

Many different locations of peak excitation and inhibition are seen among the various macaque LGN cells. The cell whose responses are shown in Fig. 1 gives its maximum excitation to around 500 nm and maximum inhibition to about 630 nm. These are the spectral regions which a person with normal color vision sees as green and red, respectively; such a cell may thus be termed a green-excitatory, red-inhibitory cell (+ G − R). Other cells show rough mirror-image responses, with maximum excitation to the very long wavelengths and maximum inhibition to the region around 500 nm. They could be termed red-excitatory, green-inhibitory (+ R − G). Although the various opponent cells that are found in great profusion in the dorsal, parvocellular layers of the macaque lateral geniculate peak at virtually every spectral point, and cross over from excitation to inhibition at virtually every spectral region, there is some evidence that they fall into different classes (De Valois et al., 1966). These investigators presented evidence for there being four types of opponent cells in macaque; the two classes mentioned before plus a class of cells which shows maximum excitation to 600 nm and maximum inhibition to 440 nm, and one which shows its maximum excitation to 440 nm and its maximum inhibition to 600 nm. These latter two cell types were

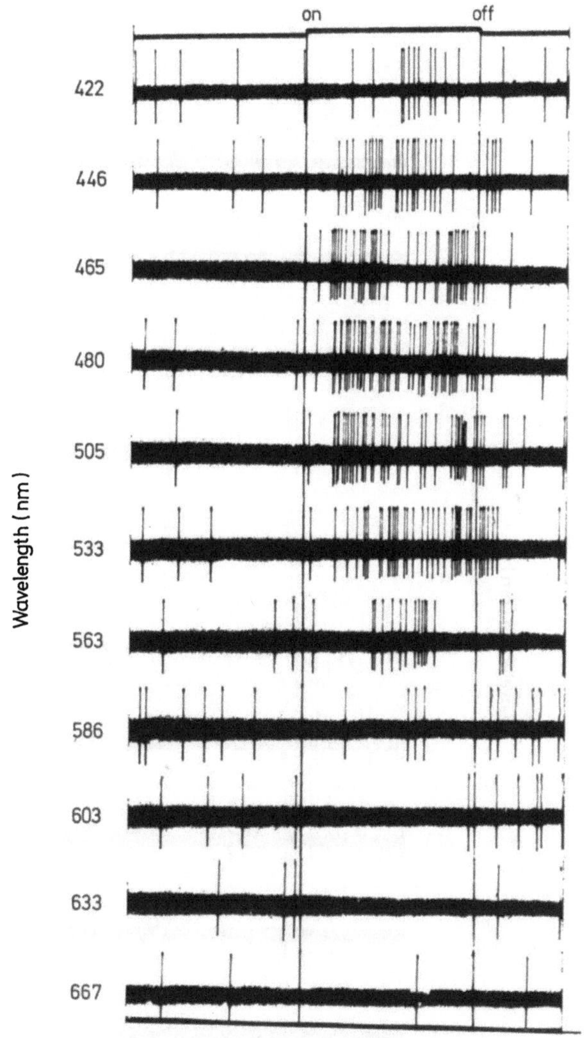

Fig. 1. Responses of a spectrally opponent cell in the macaque LGN to different wavelengths of light, equated for energy. Each record shows the responses during the one second periods before, during, and after the light. It can be seen that light of short wavelengths fires and light of long wavelengths inhibits this cell. Since the wavelengths which produce maximum excitation and inhibition are those we see as green and red, respectively, this is termed a + G — R cell. From De Valois et al. (1966)

termed the yellow-excitatory, blue-inhibitory (+ Y − B) and the blue-excitatory, yellow-inhibitory cells (+ B − Y), respectively, again on the basis of the colors seen at the regions of greatest response. An example of a + B − Y cell is shown in Fig. 2. The evidence for this categorization was based primarily on the distribution of crosspoints from excitation to inhibition, which was found to form a bimodal distribution both among those cells which showed excitation to long wavelengths and among those showing excitation to short wavelengths.

De Valois et al. (1966) concluded that there were, in effect, two separate color vision systems in the primate. One might be called the RG system (+ R − G

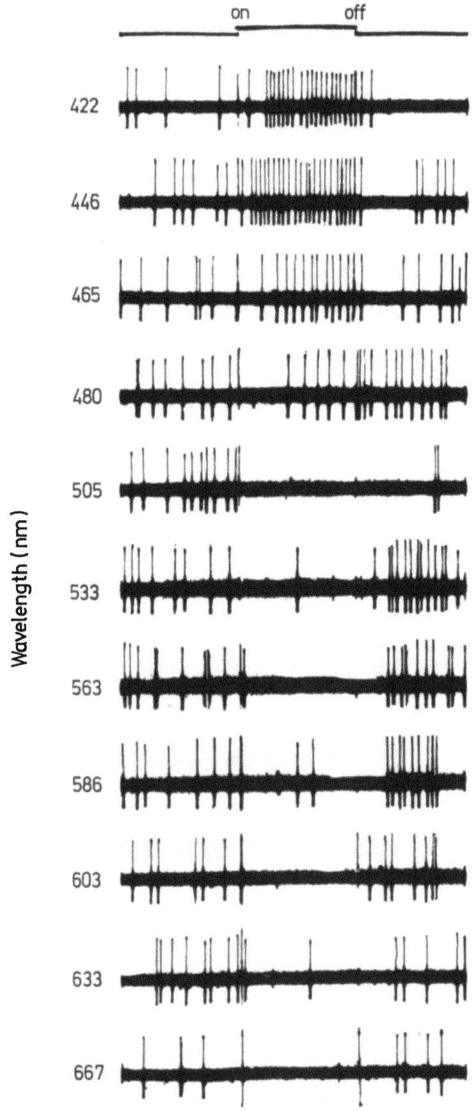

Fig. 2. Responses of a + B — Y cell. Stimulus conditions were identical to those in Fig. 1

and + G − R cells), cells which show peak responses to the red and green and cross from excitation to inhibition above 560 nm. The other, consisting of cells which cross from excitation to inhibition below 560 nm and show peak responses to yellow and blue, might be called the YB system (+ Y − B and + B − Y cells); see Fig. 3.

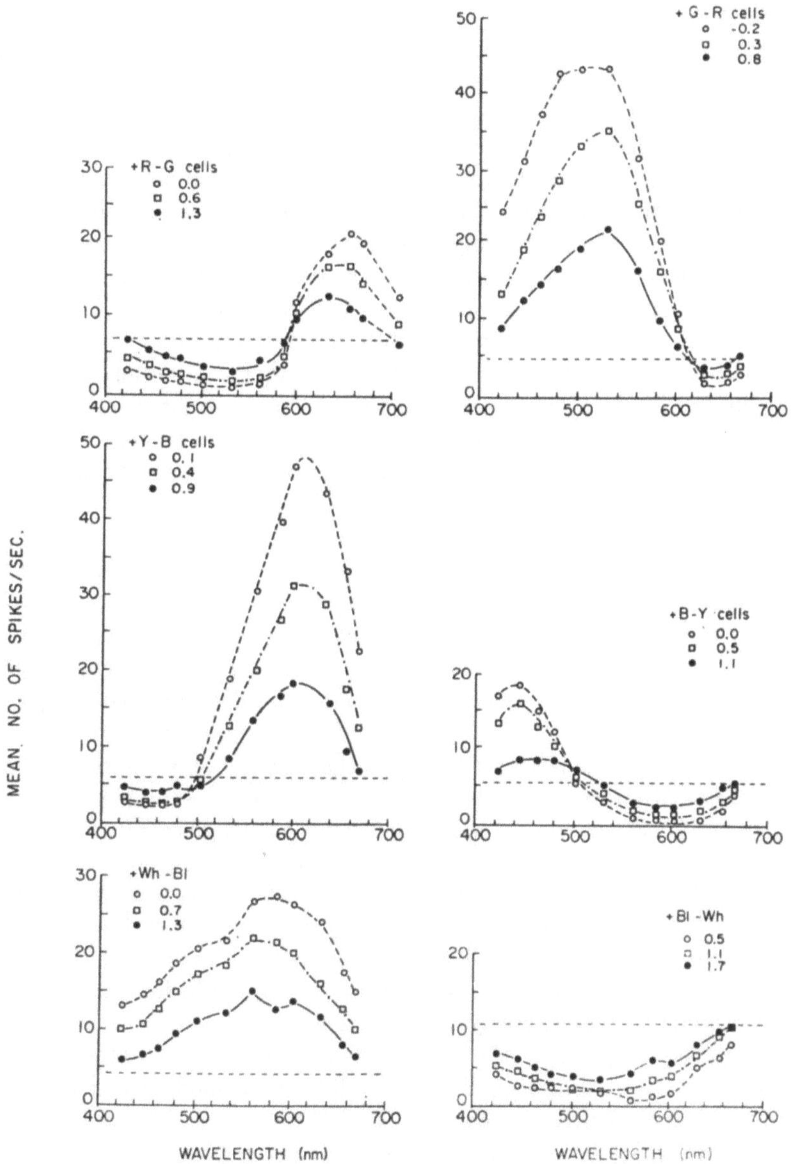

Fig. 3. Average responses of LGN cells. Plots of the average response rates of macaque LGN cells of four spectrally opponent and two non-opponent cell types. The three curves in each case are the responses to different luminance levels, the neutral densities used being indicated. From DE VALOIS et al. (1966)

In a study that was primarily aimed at determining receptive field organization rather than characterizing the optimal stimulus for the cells, WIESEL and HUBEL (1966) confirmed the presence of large numbers of spectrally opponent cells in the macaque LGN. They classified the cells on the basis of receptive field organization and presumed cone inputs (discussed below), rather than on the basis of response to spectral stimuli, but it appears that the same cell types were found as by DE VALOIS et al. (1966).

In addition to these four varieties of spectrally opponent cells, other cells are found in the macaque LGN (DE VALOIS et al., 1958, 1966; WIESEL and HUBEL, 1966) which do not give spectrally opponent responses, but rather respond in the same direction (either excitation or inhibition) to light of all wavelengths. One class of these are the spectrally non-opponent cells which show excitation to flashes of light of all wavelengths. While such cells are maximally responsive to the middle part of the spectrum, they will show exactly the same responses to lights of any wavelengths if they are adjusted appropriately in luminance. Such cells, thus, do not exhibit color discrimination. The other class of spectrally non-opponent cells are those that show inhibition to light of all wavelengths, with maximum inhibition to the middle wavelengths. Again, these cells will be inhibited equally by lights of any wavelength if they are appropriately adjusted in luminance; the cells thus appear not to be involved with color vision. There is thus much evidence for separate chromatic and achromatic channels, the cells picking up from the same receptors but processing the receptor information in different ways; see Fig. 3.

Although the physiological processes involved are quite different from those HERING proposed, it is clear that the cell types seen in the macaque LGN correspond very closely to the types of retinal outputs which HERING (1878) postulated to account for color appearances, and which HURVICH and JAMESON (1957) have developed quantitatively to account for much color discrimination data.

In the initial reports of macaque geniculate cell response (DE VALOIS et al., 1958), evidence was also presented for cells that showed only excitation, but which responded to only narrow spectral regions. These cells resembled the modulators that GRANIT (1947) had reported evidence for in ganglion cell recordings from several different varieties of animals. However, it became apparent from later studies (DE VALOIS, 1965) that these modulator-like cells were in fact opponent cells; the low levels of spontaneous activity and high noise levels in the early recording experiments made it difficult on occasion to detect the inhibition from the opposite spectral region. Clear evidence for such inhibition was that light to a spectrally opposite region always produced a diminution, or even cancellation, of the excitatory response to the narrow spectral region (DE VALOIS, unpublished observations). It would appear likely that the narrow modulator curves obtained by GRANIT in ganglion cell recordings had their origins in a similar failure to detect the inhibitory input from other spectral regions.

In the macaque monkey, the proportion of spectrally opponent units in the LGN is very large. In the quantitative study made by DE VALOIS et al. (1966), approximately 70 % of the cells in the parvocellular layers related to the central retina were found to be spectrally opponent, and the other 30 % spectrally non-opponent cells. The estimate of the two proportions by WIESEL and HUBEL (1966) is 84 % and 16 %, respectively. Thus it can be seen that a large portion of the pathway to the cortex is devoted to cells which are clearly involved with processing color information. The cells in the two magnocellular geniculate layers do not appear to be greatly involved with color (Wiesel and Hubel, 1966), although their function is not clear.

Spectrally opponent cells have also been found in the squirrel monkey, *Saimiri sciureus* (Jacobs and De Valois, 1965). These cells are similar in many respects to opponent cells seen in the macaque monkey, although the proportions to be found in the visual path are far lower: approximately 20 % of all the cells in the squirrel monkey LGN, as opposed to 70 % in the macaque (De Valois and Jacobs, 1968). Furthermore, the relative percentages of cells in the RG as opposed to YB systems of spectrally opponent cells are different in these two varieties of monkeys. Squirrel monkeys have a larger number of yellow-blue cells than red-green cells; in macaque, it is somewhat the reverse. These differences between the squirrel monkey and the macaque monkey have been shown to correspond to differences between the color vision of these two varieties of animals (De Valois and Jacobs, 1968). Thus the macaque monkey has much more highly developed color vision than does the squirrel monkey, and the squirrel monkey is severely protanomalous.

Another New World monkey, the spider monkey, *Ateles*, has also been studied (Hubel and Wiesel, 1960) in optic nerve recordings. Spectrally opponent cells were found, but in far smaller numbers than in macaques. Little is known about the color vision of spider monkeys, but insofar as it resembles that of other New World monkeys it would be much poorer than in the macaque.

Other Species. Spectrally opponent cells have also been found among the ground squirrel, *Citellus mexicanus*, optic nerve fibers and lateral geniculate cells (Michael, 1968, 1971). In this animal, only a single opponent cell system was found, rather than the two sets of opponent cells found in the macaque and squirrel monkey. All of the cells or optic nerve fibers in the ground squirrel are of the YB variety, showing peak excitation around 560 nm and peak inhibition at a 460 nm (+ Y − B) or their mirror-image cells which show their peak excitation in the blue and inhibition in the yellow (+ B − Y). These cells have rather misleadingly been termed blue-green cells by Michael on the basis of the cone types to which they are presumed to be related, the receptors being given color names. How misleading this terminology is can be seen from the firing pattern of these cells. They show, in fact, no response to green light at all: this is the region of the spectrum at which the cells cross over from excitation to inhibition. The point of maximum excitation is not in the green at 510 nm, but around 560 mn, which is the part of the spectrum we see as yellow.

In the ground squirrel, the proportion of opponent cells seen in the optic tract is about 24 % (Michael, 1968). The fact that the ground squirrel has just one mirror-image opponent cell system as opposed to two such systems in the macaque corresponds closely to the nature of its color vision compared to that of the macaque monkey. Jacobs and Yolton (1971) have shown it to be a dichromat with only one opponent color mechanism apparent from its visual behavior. Furthermore, its color vision is not nearly as highly developed as is that of the macaque monkey.

The cat has been widely used as an experimental animal in physiological experimentation, some of the earliest studies of color vision (GRANIT, 1947) being of ganglion cell responses of the cat. Unfortunately these investigations were undertaken before studies of its color vision (MEYER et al., 1954; SECHZER and BROWN, 1964), which made it clear that the cat's color vision is extremely rudimentary, thus making it a very unsuitable subject for investigations of the physiology of color vision. Most of the physiological evidence for color discrimination provided by the early studies (GRANIT, 1947) of cat ganglion cells was quite unconvincing. Most of the modulator functions obtained from this animal were based not on the response properties of the cells, but on the mathematical subtraction of the responses before and after chromatic adaptation. A change in the spectral sensitivity curve due to chromatic adaptation merely shows that the cell under investigation has two different receptor inputs. The presence of two different receptor inputs by no means provides evidence for the cell's being in any way concerned with processing color information, since it is the differential activity of receptors containing different photopigments that provides color information, not just multiple inputs from various cone types. In fact, the macaque spectrally non-opponent cells (the white-black system) are completely color-blind (DE VALOIS and JACOBS, 1968), but they show a shift in spectral sensitivity following chromatic adaptation (WIESEL and HUBEL, 1966), indicating at least two cone inputs. A subtraction of one curve from the other would thus give a "modulator" curve for these cells that cannot discriminate colors.

A recent investigation (PEARLMAN and DAW, 1970) has provided evidence for opponent color cells in the cat LGN. These cells are similar to those reported above for other animals. Something less than three per cent of all the cells found in the cat LGN are spectrally opponent, whereas some 70–80 % are in the macaque monkey. This huge quantitative difference between the amount of the visual system devoted to color processing in the two animals agrees well with the very large difference between the degree of color vision in the two species.

2. Relation to Retinal Structures

Cone Pigments. Overwhelming support has been provided over the past one hundred and fifty years for the notion first put forth clearly by THOMAS YOUNG that three separate varieties of receptors underlie color vision and provide it its basic trichromatic nature. However, specification of the spectral absorption characteristics of the three different cone types has been difficult since it has not been possible to extract cone pigments from the eyes of primates. However, a number of other approaches have been taken to the question of the spectral sensitivities of the different cones, information which is crucial for an understanding of color vision. None of these studies by itself is entirely convincing, but the fact that the same answer has resulted from several different approaches gives one greater confidence.

A great many people have tried over the last century to deduce the spectral sensitivities of the different cone pigments from psychophysical experiments, but extremely diverse answers have resulted, depending upon the assumptions made by the experimenters. It would appear in retrospect that the best answer came from WILLMER (1955) in his studies of the spectral sensitivity of dichromats to small foveal spots of light. The argument here is based on two assumptions. One, that the normal individual is functionally a tritanope (presumably missing the S receptor) when tested with small spots of light in the central fovea (König, 1897); the second is that a dichromat of either the protanopic or deuteranopic variety has the same photopigments as a normal trichromat except for missing one of the long-wavelength pigments — the L cone in the case of a protanope and the M cone in the case of most deuteranopes. If that were the case, such a dichromat, when tested with a small foveal spot of light should be a monochromat, possessing just one cone pigment. Willmer studied both protanopes and deuteranopes, obtaining very precise spectral sensitivity curves for these individuals; the resulting curves peaked at 540 nm for the protanope and 570 nm for the deuteranope. This technique clearly does not allow one to determine the spectral sensitivity of the pigment in the S cone, but in a study of a rare family of monochromats who appear to have only S cones (and rods) in their eyes, Blackwell and Blackwell (1961) obtained a spectral sensitivity function peaking at about 440 nm.

Psychophysical evidence for cone pigment absorption curves in normal trichromats has also come from the use of chromatic adaptation to selectively adapt one cone pigment and thus reveal another in at least partial isolation (Stiles, 1959). That individual cone pigments are completely isolated by this technique is not as convincing as the Willmer experiment, but three of Stiles' five curves (for his π_1, π_4, and π_5 mechanisms) correspond fairly closely to the data obtained by Pitt (1944) and by Willmer (1955). Similar studies have also been done by Wald (1964) with comparable results.

An attempt to answer the question of spectral sensitivity of the cone pigments has also been made by reflection densitometry (Rushton, 1963, 1965). Rushton has coped with the problem of the overlapping absorption curves in normal individuals in the same way as Willmer did, by studying human deuteranopes and protanopes. Two long-wavelength pigment absorptions were found, which showed peaks at about 540 and 570 nm. This technique could not be used to study the S cone because of the high background noise levels produced by scatter of the short wavelengths and the interfering absorption by the rods.

Attempts have also been made to estimate the absorption characteristics of the receptors from unit recordings later in the visual pathway. Since in the macaque there are no pathways to which only one photopigment type contributes (De Valois, 1965), a straightforward answer is not possible. However, selective chromatic adaptation can be used in a manner similar to that of Stiles' studies. Such an experiment, done on various types of opponent cells in the macaque (DeValois, 1965), revealed evidence for three separate cone mechanisms peaking at about 440, 540, and 570 nm. The same procedure has also been used by Wiesel and Hubel (1966), with an additional degree of isolation obtainable by the use of discrete spots projected onto different parts of the receptive field, in the case of opponent cells in which the opposing cone types project to different spatial regions within a receptive field. Their results also indicate the presence of three different cone types with peaks at about the same spectral regions. Michael (1968) has obtained similar evidence for two separate cone mechanisms in the ground squirrel, peaking at about 460 and 525 nm.

All of the above studies aimed at determining cone absorption curves have the sharp limitation that under normal circumstances there are always at least two cone types active; various stratagems have to be employed to minimize the influence of the other pigments and thus reveal one in isolation. These studies also could not answer the question of whether the different photopigment varieties were located in different receptors or intermingled in the same receptors. Both of these limitations were overcome by the use of direct spectrophotometry of individual receptors in extracted monkey, human, and goldfish eyes (MARKS, 1965; MARKS et al., 1964). They provided strong evidence for three different photopigment types in each of these animals, the human and macaque curves peaking at about 445, 535, and 570 nm, and those of the goldfish peaking at about 455, 530, and 625 nm. Although the high noise levels made the exact shapes and peaks of the curves somewhat uncertain, these spectrophotometric studies provide the best evidence for there being three separate photopigments segregated into different receptors and for the rough locations of the absorption peaks.

These diverse attempts to locate the cone pigment peaks and absorption curves agree well on two points. One is that, in the case of the macaque and human, there are three cone types with their absorption peaks at about 445, 540, and 570 nm. The other is that the absorption curves are quite broad, fitting to a first approximation (although slightly narrower than) the Dartnall nomogram; that is, having the shape of the rhodopsin absorption curve when plotted on a frequency scale.

Cone Inputs to LGN Cells. Of principal interest to us here is the relationship between the different receptor types and the varieties of cells found in the central nervous system.

Spectrally opponent cells, obviously, and spectrally non-opponent cells, less obviously, have inputs from more than one receptor type. A question of considerable theoretical interest is which cone types feed into the different types of opponent and non-opponent cells. There is general agreement from a number of studies (DE VALOIS et al., 1963; DE VALOIS, 1965; WIESEL and HUBEL, 1966; ABRAMOV, 1968) that macaque RG cells (+ R − G and + G − R) have their inputs from the L and M cones. One of the cone types is excitatory and the other inhibitory to these opponent cells; for instance, the L cone feeds in an excitatory and the M cone in an inhibitory fashion to the + R − G cells. The + R − G cells therefore show increases or decreases in firing not to the extent to which the L cone produces excitation or the M cone produces inhibition, respectively, but to the extent that there is a difference between the amount of excitation and inhibition produced at each spectral point by the two different cone types. The point at which these cells show maximum excitation is thus not 570 nm, the point at which the L cone is maximally sensitive, but beyond 640, at which the difference between the amount of excitation from the L cone and the amount of inhibition from the M cone is maximum. Correspondingly, maximum inhibition is not to a light of 540 nm, the wavelength to which the M cone is maximally sensitive, but to a light of around 520 nm, at which point the difference between the amount of inhibition from the M cone and the amount of excitation from the L cone is maximal. The effect of the opponent organization is thus to separate the points of maximum responsiveness, compared to that of the underlying receptors; see

Fig. 4. So although the RG cells have inputs from cones maximally sensitive to 570 and 540 nm, they show their maximum responses of opposite types to 640 and 520 nm. The L and M cones are both maximally sensitive in the greenish yellow, but the opponent cells which receive their inputs from these cones show their maximum sensitivity to red and green.

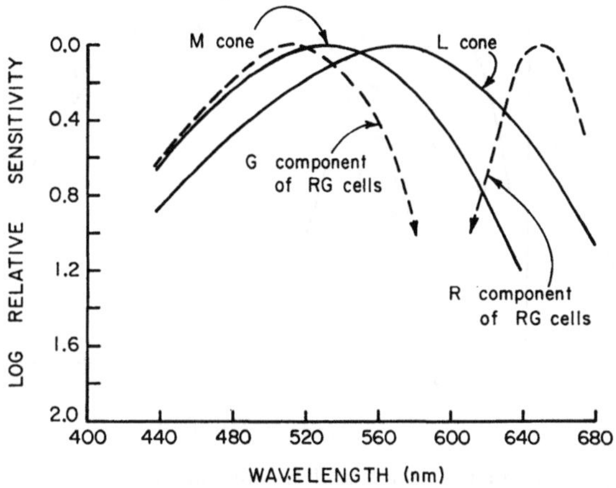

Fig. 4. Cones and opponent color cells. The spectral absorption curves of the macaque and human L and M cones are shown in the solid lines (these curves are based on 535 and 570 peaks and fitted to the Dartnall nomogram. Other estimates would give somewhat narrower curves). The dotted lines represent the spectral sensitivity curves for the R and G components of the macaque RG opponent cells (based on the amount of light required for a certain criterion firing rate under neutral adaptation conditions). Note that the cone peaks are close together, but the peaks of the R and G functions are further apart as a result of the subtractive interactions between the cone inputs to them

Which cones feed into the YB system (+ Y − B and + B − Y cells) is a matter of dispute; one input is clearly from the S cones, but whether the other is from L cones or M cones (or possibly both) is uncertain. WIESEL and HUBEL (1966) report that + Y − B cells receive inputs from S and M cones. They isolated each cone input by chromatic adaptation of the opponent mechanism, and also by taking advantage of the fact that the two different cone types feed into different parts of the receptive fields of some LGN cells. Thus by using small monochromatic test spots in one part of the receptive field in the presence of a chromatic adaptation light, the stimuli could be mainly restricted in effect to a single cone type. DE VALOIS (1965) and ABRAMOV (1968) presented data from a number of YB cells from which they conclude that the inputs to the YB cells are from the S and L receptors, rather than the S and M receptors; see Fig. 5. They used chromatic adaptation to separate the cone types, but, with one exception, did not use selective stimulation with small spots in different parts of the receptive fields. A + Y − B cell, in the absence of chromatic adaptation, shows peak excitation to light of about 600 nm (and peak inhibition at about 440 nm). When one selectively adapts the short-wavelength input to the cell, the peak of the excitation moves towards shorter wavelengths, since it is the opposing short-wavelength inhibition which pushed the peak excitation to longer wavelengths. If one only partially eliminated the short wavelength inhibition, the peak excitation would shift from 600 to say, 590 nm; with a still greater elimination of the short-wavelength input, the peak would shift to still shorter wavelengths. One might argue,

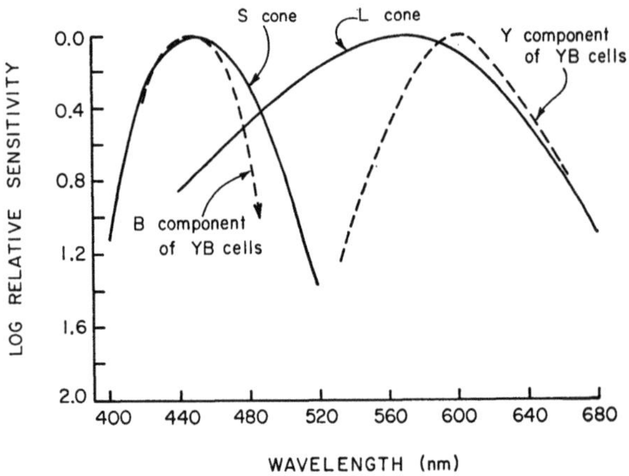

Fig. 5. Same as Fig. 4 except that the S and L cone curves are compared with the B and Y components of the YB cells. Note that since the cones which feed into the YB system lie far apart in the spectrum, the displacement of the opponent-cell peaks is not as great as that seen in the case of the RG cells

therefore, that in the De Valois and Abramov studies the short-wavelength input might not have been completely eliminated, and that they therefore found the long-wavelength input to be peaking at 570 nm (L cone) rather than at 540 nm (M cone). Countering these arguments is the shape of the long-wavelength input to these cells. The 540 and 570 pigments, insofar as they fit a Dartnall nomogram, have quite different long-wavelength absorption curves: the 540 pigment's dropoff would be more rapid in the long-wavelengths than that of the 570 pigment. The spectral sensitivity of a large sample of YB cells at long wavelengths (beyond the point at which the S cone could have influenced the results) was found (Abramov, 1968) to match the drop-off at long wavelengths of a 570, not a 540 nomogram pigment. Another argument for L cones providing the long-wavelength input to YB cells is that the responses of these cells to mixtures of lights of different wavelengths can be predicted (Abramov, 1968) from the 440 and 570 pigment curves, but not if the underlying inputs were assumed to be 440 and 540 pigments. The weight of the evidence perhaps indicates that the YB system has inputs from the S and L cones, but the question must still be considered in doubt. It is, of course, conceivable that both types of combinations occur on occasion.

The fact that the spectrally non-opponent cells have a multiple cone input is not immediately obvious, since the inputs are in the same direction, rather than opposed to each other. However, the spectral sensitivity curve of the non-opponent cells is not smooth and invariable as one would expect if only a single photopigment fed into it, and chromatic adaptation experiments on such cells (Wiesel and Hubel, 1966), and on similar cells found at the ganglion cell level (Gouras,

1968) give good evidence for both the L and M cones (and possibly S cones as well) feeding into these non-opponent cells.

Unit recording experiments thus indicate that there are three cone types underlying the different varieties of macaque (and doubtless human) geniculate cells: the spectrally non-opponent cells add together the (log) outputs of the L and M cones; the spectrally opponent cells subtract the outputs of the L and M cones, in the case of the RG system, and of the L and S cones in the case of the YB system. The comments made earlier about color naming are now seen to have special force, because it is clear that the outputs of the cones feed not just into color systems, as one might suppose if they were to be given color names, but also into the achromatic (black-white) system. Furthermore, the outputs of the L receptors, for instance, appear to feed into every color channel, specifying not just red, but yellow and green and blue and black and white, as well (see Fig. 6).

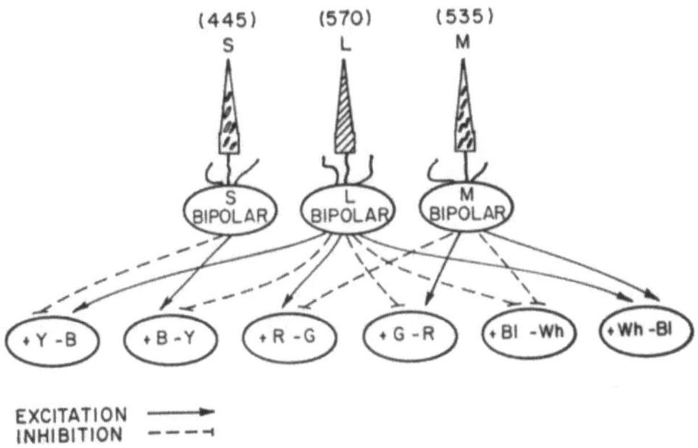

Fig. 6. A model of the connections of the color vision system up to the LGN. No attempt is made to show the spatial organization of the receptive fields or lateral connections. It can be be seen that the light receptors feed into four color and two achromatic channels

As extensive studies have not been made of the inputs to the opponent cells in the squirrel monkey, *Saimiri*; it would be of particular interest to do this since these animals are protanomalous, their color vision being very similar to that of one of the common types of partial human color-blind (JACOBS, 1963). Spectrally opponent cells have been found in the squirrel monkey LGN, and some attempts have been made by chromatic adaptation experiments to determine the spectral sensitivities of the underlying cone types (JACOBS and DE VALOIS, 1965). The results indicate that the M cone has much the same spectral sensitivity as is the case with the macaque, but that the peak sensitivity of the L cone is shifted toward shorter wavelengths, relative to the macaque L cone. Unfortunately, an adequate specification of the spectral sensitivity curves was not provided, but the results suggest that protanomalous color defective individuals have a long wavelength pigment that is different from that of a color-normal, being shifted toward that of the M cone.

In the all-cone ground squirrel, *Citellus tridecemlineatus*, only one photopigment has been isolated by conventional biochemical means (DOWLING, 1968).

He, in fact, suggests that these animals should be color blind because of having only a single cone type. However, JACOBS and YOLTON's (1971) behavioral data on both this animal and the closely related species, *Citellus mexicanus*, indicate that these animals are not color blind, but rather have dichromatic color vision of the protanopic variety; and MICHAEL's recording data from the optic nerve and LGN of *Citellus mexicanus* (1968, 1971) provide evidence for two cone inputs to these cells. One cone pigment would appear to have a peak sensitivity at about 525 nm, and is doubtless the pigment extracted by DOWLING; the other is maximally sensitive in the short wavelengths, with a λ_{max} of about 460 nm. Although both of the photopigments appear to be different from those found in primates, this animal would be most similar to a human protanope with just S and M cones.

Rod-Cone Inputs to LGN Cells. It has been reported (DE VALOIS, 1965; WIESEL and HUBEL, 1966) that not only spectrally non-opponent, but also spectrally opponent cells in the macaque LGN often receive inputs from both rods and cones. In fact, it would appear that most of the cells related to the non-foveal part of the retina receive such combined rod-cone inputs. The evidence for this is that the cells shift from the complex spectral responses given under photopic conditions to a simple rhodopsin-like spectral sensitivity function when tested under scotopic conditions after dark adaptation, see Fig. 7. While it might appear puzzling that there should be such rod inputs to cells that are clearly signalling color information, it should be pointed out that these data are quite consistent both with the anatomy of the visual path and with the psychophysical data on extrafoveal color vision. Anatomical studies (POLYAK, 1957) show no separate rod paths to

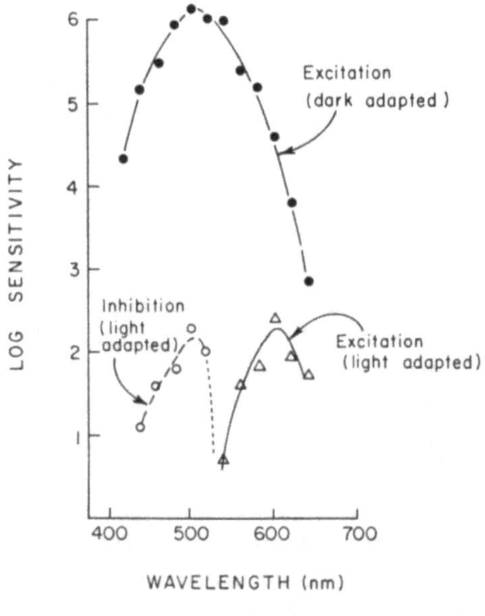

Fig. 7. Rod-cone inputs to opponent cell. Spectral sensitivity curve of an LGN cell which is spectrally opponent when light-adapted (bottom curves) and spectrally non-opponent with a rhodopsin sensitivity curve when dark-adapted (top). Redrawn from WIESEL and HUBEL (1966)

the brain, but rather that rod and cone outputs converge to the same ganglion cells outside the fovea. Psychophysical data also indicate a rod contribution to color vision outside the fovea. Clarke (1963) has found that extrafoveal color vision is tetrachromatic rather than trichromatic, and that this tetrachromacy can be explained on the basis of a fourth receptor input that has the spectral sensitivity characteristics of rods. Lie (1963) found that a foveal chromatic stimulus is seen as much more saturated than the same stimulus presented extrafoveally. This desaturation of extrafoveal lights was found to occur, in a dark-adaptation experiment, at the time at which the rod input first appears, as detected by the dark adaptation sensitivity curve.

Unfortunately, the recording experiments (De Valois, 1965; Wiesel and Hubel, 1966) did not examine rod-cone interactions under mesopic conditions. Studies of macaque ganglion cells (Gouras, 1965; Gouras and Link, 1966), however, indicate that interactions between the rod and cone inputs to a ganglion cell minimize the extent to which rod and cone messages are simultaneously transmitted down the same channel. He found longer latencies to rod than to cone inputs to ganglion cells, and that when both receptor types are activated the cones tend to block the input from the rods.

The anatomical, psychophysical, and physiological data, then, agree on there being both rod and cone inputs to the chromatic as well as to the achromatic systems in the primate. Rods perhaps do not make as much of a contribution under photopic conditions as one might otherwise expect, because of interactions in the pathway. The contribution of rods to the color processing appears to be a negative one in the sense that it tends to desaturate the colors which would otherwise be seen.

Relation to Ganglion Cell Responses. Most of the detailed investigations of color processing in the primate visual system have been at the level of the LGN, but such ganglion cell recordings as have been made indicate that much of the color processing for which evidence is seen at the LGN has already taken place in the retina. Spectrally opponent ganglion cells have been reported in the macaque (Gouras, 1968), and they appear to have much the same properties as opponent LGN cells (Marrocco, 1971). Spectrally opponent cells have also been found in the spider monkey optic nerve (Hubel and Wiesel, 1960), although they are very sparse there. The fact that so few opponent optic nerve fibers were found probably reflects the poor color vision of the spider monkey rather than any change between that level and the LGN.

In the cat, Hubel and Wiesel (1961) noted a tendency for the surrounds of LGN cells to be more powerful than those of ganglion cells, producing a more uniform balance between center and surround. (These cells studied in the cat are similar in their behavior to the spectrally non-opponent cells in the macaque visual system). A similar greater influence of the periphery relative to the center, particularly in non-opponent cells, has been found in the macaque monkey by Marrocco (1971), who also found that the spontaneous activity of LGN cells was more independent of steady state background luminance. However, these findings do not bear directly on color processing.

In the ground squirrel, MICHAEL (1968, 1971) has found the characteristics of spectrally opponent LGN cells to be somewhat different from retinal ganglion cells. The LGN cells have a double-opponent organization (see below), which is not shown by the ganglion cells.

3. Spatial Organization of Receptive Fields

Center-Surround Organization. In most of what has been discussed above, we have been concerned with how the cells in the visual path respond to the wavelength and luminance characteristics of light stimuli presented to them, often in stimulus patterns extending over considerable areas, and with the receptor types involved. But also of interest, principally for an understanding of the physiological organization underlying their responses, is the spatial distribution of the cone inputs: for instance, whether the cones of one type feed just into the center of the cell's receptive field and the cones of the other type into the surround, or whether they are spatially intermingled.

A very common type of receptive field organization, first found by KUFFLER (1953) in the cat and BARLOW (1953) in the rabbit, is the center-surround organization in which a cell is found to have an excitatory center and an inhibitory surround or an inhibitory center and an excitatory surround when tested with spots of white light. Those units studied in the cat and rabbit are comparable to the spectrally non-opponent cells seen in the macaque LGN. Spectrally non-opponent cells found in the macaque also have a center-surround organization (HUBEL and WIESEL, 1966). About half the spectrally non-opponent cells have an excitatory center and an inhibitory surround, and the other half have an inhibitory center and an excitatory surround, with the center usually dominant when both are stimulated.

The situation with regard to the spectrally opponent cells is somewhat more complicated. WIESEL and HUBEL (1966) dichotomized the spectrally opponent cells on the basis of their spatial receptive field organization. The most common variety (Type I, 77 % of all cells found in the dorsal layers) were those that are reported to have an excitatory input from one cone type in the center and an inhibitory input from the other cone-type in the surround, or, vice versa, with an inhibitory input from one cone type in the center and an excitatory input from the other cone type in the surround. WIESEL and HUBEL (1966) report that by far the largest number of these cells are RG cells. All four possibilities of combined spatial and spectral combinations of these cells are found: red excitatory center and green inhibitory surround and, vice versa, green excitatory center and red inhibitory surround; or red inhibitory center and green excitatory surround, and vice versa. Only a few cells were found with blue excitatory center and yellow surround (which they however label "blue-green" on the basis of the presumed cone input – receptors being given color names – rather than the stimuli to which the cell optimally responds), and none of the reverse, with a blue surround.

Found far less frequently (Type II, 7 %) by WIESEL and HUBEL (1966) were spectrally opponent cells that received inputs from two opposing cone types but in which the two cone inputs had identical spatial distributions so that at every point in the receptive field there would be balanced excitatory and inhibitory influences from different spectral regions. About half of these cells were YB cells, and the other half, RG cells.

The clearest evidence for a different type of receptive field organization for Type I and Type II cells was from a study of the responses to spots of different size centered in the receptive field. In the case of the Type I cells, increasing-sized spots of a wavelength to which the center was most sensitive produced an increase in sensitivity which rapidly levelled off, thus indicating a small receptive field center. Spots of a wavelength to which the surround was most sensitive produced no response until they reached a certain large size, at which point the surround response could finally be evoked. In the case of the Type II cells, the area-sensitivity curves for the two opposing wavelengths showed quite the same shape, with the spectrally opposing systems having maximum sensitivity to the same sized spots.

Wiesel and Hubel (1966) attribute the difference between the spatial receptive field organizations of Type I and Type II cells to completely discrete spatial distributions of the opposing receptors feeding into Type I, and overlapping distributions of the receptors feeding into Type II cells, see Fig. 8. A similar dichotomy has also been postulated by Michael (1968) for the optic nerve and LGN cells of the ground squirrel, which he has also found to fall into two different classes on the basis of receptive field organization.

Although there is no question that many spectrally opponent and non-opponent cells have a center-surround organization – that is, exhibit excitation in

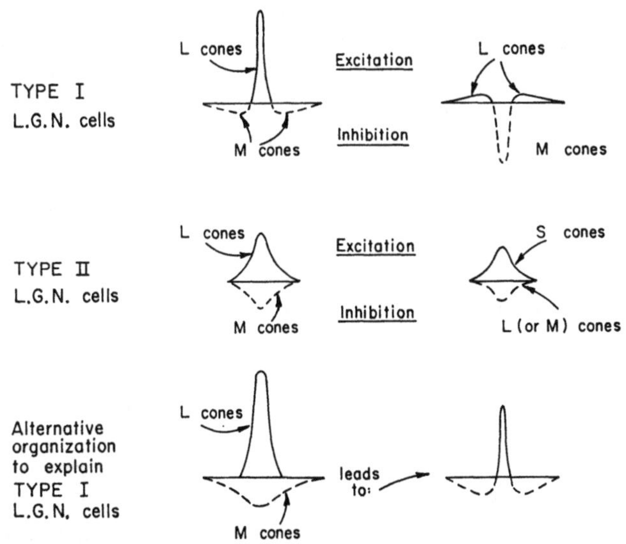

Fig. 8. Schematic representations of possible receptive field organizations of spectrally opponent LGN cells. *Top*: Type I cells, as postulated by Wiesel and Hubel (1966), with different spatial distributions of the opposing cone inputs to produce a center-surround organization. Two of the many possible combinations of cone types are shown. *Middle*: Type II cells (Wiesel and Hubel), with overlapping inputs from the opposing cone types. *Bottom*: an alternative explanation of the inputs to Type I cells. Overlapping distributions of the two cone types as shown at left would lead to a center-surround organization (right) as found by Wiesel and Hubel

the center and inhibition in the surround, or vice versa – there is some question as to whether the two cone inputs are completely discrete spatially or rather overlap with differential sensitivity.

This question was first raised by WAGNER et al. (1963) in studies of goldfish ganglion cells. The spectrally opponent ganglion cells when tested under normal adaptation conditions showed a clear center-surround organization, but WAGNER et al. (1963) showed that when the responsivity of the center mechanism was reduced by chromatically adapting it, the "surround" response could be recorded across the whole receptive field. They found that the cone type producing the surround response was not only distributed across the whole field, but that it was actually most sensitive in the center. These data indicate that the center-surround organization of spectrally opponent cells, then, results from two spatially overlapping cone inputs, the one strongly peaked in the center and the other a flatter function. In the center the one input would predominate, and in the surround the other, as illustrated in the bottom of Fig. 8.

KUFFLER (1953) first reported evidence in the cat ganglion cell for a center-surround organization in which, for instance, the excitatory center was discrete from the inhibitory surround except for an intermediate overlapping region. The studies of RODIECK and STONE (1965), however, provide strong evidence for a complete overlap across the center of the receptive field of the excitatory and inhibitory inputs to these (spectrally non-opponent) cells. They measured the responses to flashes of light or to moving spots and found the best model to quantitatively account for their data to be one of two overlapping Gaussian spatial distributions of input, the center distribution being highly peaked in the center (with a small variance), and the surround distribution being much flatter (larger variance). This model for non-opponent cells, thus, agrees with that put forth for opponent cells by WAGNER et al. (1963). A quantitative examination of squirrel monkey non-opponent cells by JACOBS and YOLTON (1968) also provided evidence for overlapping cone inputs in these cells which show a center-surround organization.

MEAD (1967) examined the receptive field characteristics of macaque LGN cells and concluded that the opposing cone inputs overlapped across the whole center of the receptive field. A cell which, when tested under normal adaptation conditions, showed an inhibitory green center and an excitatory red surround, for example, would when tested in the presence of a short wavelength chromatic adaptation show the excitatory response to red throughout the whole receptive field, and in fact give the largest response in the center; see Fig. 9. Thus his data support the view of WAGNER et al. (1963) that these cells have overlapping excitatory and inhibitory inputs, even in those cases in which there is a clear center-surround relationship.

WIESEL and HUBEL (1966) point out that in the limiting case in which a single cone may feed into the center of the receptive field of a spectrally opponent cell, there obviously must be discrete areas. We find that the receptive field centers of some LGN cells are indeed sufficiently small that they may well have only a single cone as their origin (although many other cones, of course, feed into the surround). But the line spread function of the eye (WESTHEIMER and CAMPBELL,

1962) is such that no stimulus could ever be presented to just one cone, so there would be a functional if not an anatomical overlap even in this case.

So while the issue is perhaps not settled as far as the macaque geniculate cells are concerned, it should be emphasized that in most of the experiments in which quantitative measurements have been made the evidence has supported the mechanism of overlapping spatial distributions rather than spatially discrete inputs to account for the center-surround organization of both spectrally opponent and non-opponent cells. This may appear to be a rather picayune point, but it is not: the discharge patterns, the responses shown to different types of figures, and the amount of successive contrast shown will all differ depending on whether there are discrete or overlapping excitatory and inhibitory inputs.

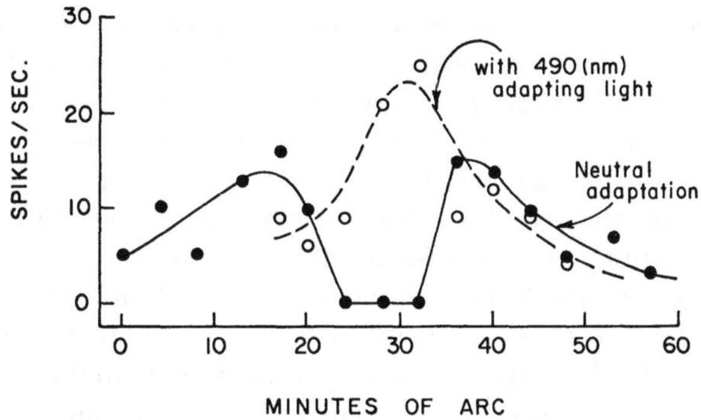

Fig. 9. Effect of chromatic adaptation on the receptive field organization of an opponent cell. This + R — G cell has an inhibitory G center and excitatory R surround. Solid line: response to 650 nm in neutral adaptation state, showing the absence of a response to red in the center. Desensitization of the green "center" component with a 490 nm adapting light, however, reveals that the red "surround" mechanism covers the whole receptive field and is actually most sensitive in the center. This supports the model shown in the bottom of Fig. 8 as an explanation of the cone inputs to a Type I opponent cell. Redrawn from Mead (1967)

Receptive Field Size. Besides varying in center-surround organization, receptive fields of cells in the visual path also vary considerably in size. This point was first brought out clearly by Wiesel (1960) from ganglion cell recordings in the cat. The size of the receptive fields has obvious bearing on the problem of visual acuity and also on consideration of possible size-specific channels in the processing of spatial information. Receptive fields vary in size with retinal location (Wiesel and Hubel, 1966; Mead, 1967), the maximum size increasing with distance from the fovea. However, the fields vary considerably in size within every given region, too. The total size of the receptive field is rather indeterminate since the sensitivity gradually approaches zero with distance. But it is possible to measure the size of the receptive field center, either by examining the concentric circle which gives the largest center response (Wiesel and Hubel, 1966) or by traversing the field with a small spot (Mead, 1967).

For spectrally non-opponent cells, in the case of both on-center (+ Wh – Bl) and off-center (+ Bl – Wh) cells, Wiesel and Hubel (1966) report that the center sizes vary between 8 and 60 min of visual angle, with a median of about

30 min. The receptive field centers of Type I spectrally opponent cells were found to vary between about two minutes and 60 min, with a median of about 8 min of arc. The on-center cells were found to have smaller centers than the off-center cells, but there was some overlap between the two classes. The Type II opponent cells varied in center diameters between 15 and 60 min of arc.

MEAD (1967) found values that correspond in general with those of WIESEL and HUBEL, although he obtained more + Wh − Bl non-opponent cells with small receptive field centers. It is clear, then, that the macaque LGN cells have very small receptive fields (much smaller than those of the cat, for example), that they vary considerably in size within the same field location, and that the opponent and non-opponent cells have receptive fields of much the same size.

Double-Opponent Organization. DAW (1967) pointed out that the center-surround organization of the goldfish spectrally opponent cells found by WAGNER et al. (1963) was just the reverse of that required to account for simultaneous color contrast (the fact that an area adjacent to a chromatic one tends to have its hue shifted in the direction opposite to that of the inducing chromatic field, so that a grey on a red surround appears slightly green or a grey on a green surround appears slightly red). DAW (1967, 1968), examining the responses to very large spots and annuli, found evidence for a surround extending over a much larger area than that which had been found previously in goldfish ganglion cells. This far surround also gave opponent color responses, but of a type opposite to central stimulation, the responses of the far surround would be + G − R. DAW pointed out that such a double-opponent organization could account for simultaneous color contrast.

DE VALOIS and JONES (1961) reported evidence for a cell in the macaque lateral geniculate nucleus which showed simultaneous color contrast. It gave the same direction of response to a grey on a green background as it did to a red light. An occasional cell of this sort has been obesrved in the macaque LGN (MEAD, 1967; DE VALOIS, unpublished observations), but they are not typical of the cells at this point in the visual pathway. WIESEL and HUBEL (1966) reported no evidence for such a double-opponent organization at the macaque LGN, although MICHAEL (1971) found it in the ground squirrel LGN. It would seem that this sort of a neural organization (in common with many others) appears later in the visual path of the primates than in that of lower animals. However, a quite different sort of process may be involved in primate color contrast, as YUND (1970) suggested, with supporting psychophysical evidence.

4. Relation to Visual Behavior

The spectrally opponent cells found in a variety of animals are clearly involved with color vision, as witness the differential responses they give to different wavelengths, independent of the luminance conditions. Thus a + R − G cell shows excitation to long wavelengths and inhibition to short over a wide range of different luminances, never giving the same response to say 500 and 640 nm. However, it would be interesting to look beyond the mere issue of whether the cells have to do with color vision, and examine to what extent one can explain in at least a semi-quantitative way various aspects of color vision on the basis of current physiological knowledge.

A whole host of psychophysical experiments have been carried out in the last one hundred years which specify the discriminative abilities of individuals in a wide variety of different stimulus conditions, or describe the color appearance of objects in different situations. If we compare the responses of cells with the discriminative behavior of individuals in the same situation, or compare the firing patterns of different cell types in a situation with a human observer's reports of the color appearance of objects in that same situation, we can specify the extent to which, and how, a given cell functions in color vision and determine how much color processing has been accomplished at a given stage in the visual system. These are the questions that we explore in this section, with reference to the LGN.

Relation to Colors Seen. Under neutral adaptation conditions an individual with normal color vision sees the different parts of the spectrum to have different colors. Boynton and Gordon (1965) quantified this relation between wavelength and color by giving subjects the opportunity to describe the color of a flash of monochromatic light by one of four color names: red, yellow, green, or blue, modified by another color name chosen from the same four. Thus, a light could be described as pure yellow, or as yellow-red, or as yellow-green, etc. From many such determinations, they obtained the data shown in Fig. 10a in which the extent to which different wavelengths of light are called color names is indicated. As can be seen, wavelengths beyond 620 were called red most of the time, between 600 and 620 they were about equally frequently called red and yellow; light of 580 was called yellow almost all the time, etc.

Not the only conceivable sort of neural organization, but a possible one is that at some point in the visual path there are cells that are selectively sensitive to visual stimuli in such a way as to encode different colors, so that when one sees red it is because a particular variety of cell is active; whereas, when one sees an object as being yellow, another variety of cell is active. One might test this notion by comparing the relative activity rates of different cells with the Boynton and Gordon measurements. The results of such a comparison (De Valois, 1965) can be seen in Fig. 10. All of the responses of the sample of macaque LGN opponent cells to flashes of light of different wavelengths were added together. From such data one can calculate the percentage of all of the opponent cell responses to, say, a light of 640 nm that was exhibited by each of the four varieties of opponent cells. As can be seen, almost all of the opponent cell activity at 640 came from the + R − G cells. Their contribution fell off going toward 600, at which point the + Y − B cells were most active, and so forth, through the spectrum. It is immediately apparent that there is reasonably good agreement between the psychophysical data and the physiological data in Fig. 10, such that for wavelengths beyond 500 nm, at least, one could predict with considerable accuracy the color a human observer would report a flash of monochromatic light to be if one knew the firing rates of each of the four varieties of opponent cells in the macaque LGN. Thus, one could confidently predict that a flash of light that evoked 60 spikes from + G − R cells and 20 spikes from + Y − B cells and essentially no responses from the + R − G and the + B − Y cells would appear a yellowish green to a normal human trichromat.

These data provide good evidence for the notion that there are color-specific cells whose activity rates are closely correlated with the colors seen in the environment. Further, the fact that these close relationships appear at the LGN suggests that there is little further processing of color information *per se* later in the visual pathway.

The agreement between Fig. 10a and 10b is not complete, particularly at shorter wavelengths. The part of the spectrum seen as blue is in the region of 470 nm, shorter wavelengths appearing bluish-red rather than pure blue. There is apparently no secondary rise in the activity of the + R − G cells, and the point at which the + B − Y cells make the maximum relative contribution is in the region of 440, rather than 470 nm.

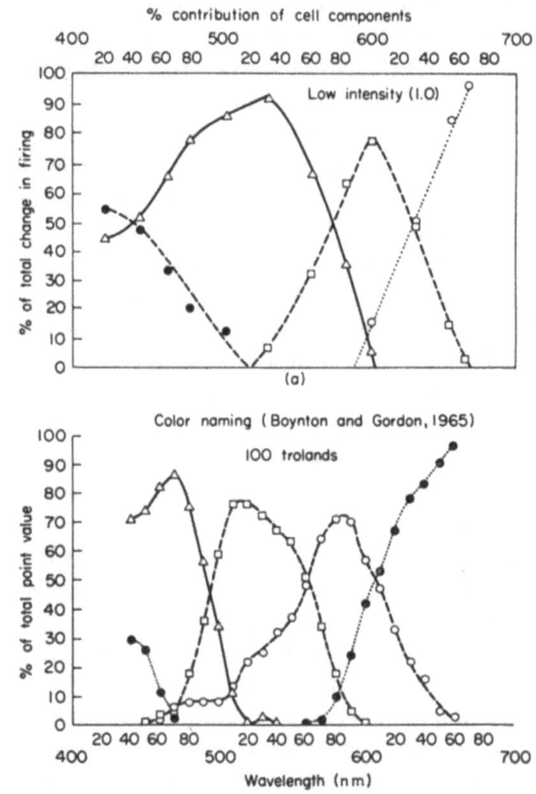

Fig. 10. Color names and opponent cell activity. Comparison of color names given to different wavelengths (bottom. Data from Boynton and Gordon, 1965) with the activity rates of each of the four LGN spectrally opponent cell types (top). Cell types in 10a: ○ - - - ○ : + R — G; □ - - □ : + Y — B; △ — △ : + G — R; ● - - ● : + B — Y. Colornames in 10b: ● - - ● red; ○ - - ○ : yellow; □ - - □: green; △ — △: blue

There are other situations in which the activity rates of the different opponent cells in the macaque LGN and the normal human trichromat's report on the colors seen are in good agreement. One is in a chromatic adaptation experiment: in the presence of a long wavelength adapting light, light of long wavelengths which had previously appeared red now appear greenish, and one sees virtually the whole spectrum as green (Jacobs and Gaylord, 1967). A red chromatic adaptation light changes the cross-point from excitation to inhibition of a + G − R cell toward the longer wavelengths so that wavelengths which previously inhibited the cell (red) now produce excitation (green); the cell in fact responds with excitation across most of the spectrum. See Fig. 11. The same holds true for chromatic adaptation in other spectral regions: the shifts in opponent cell firing patterns are in the direction one would predict from Jacobs and Gaylord's studies of color changes resulting from chromatic adaptation.

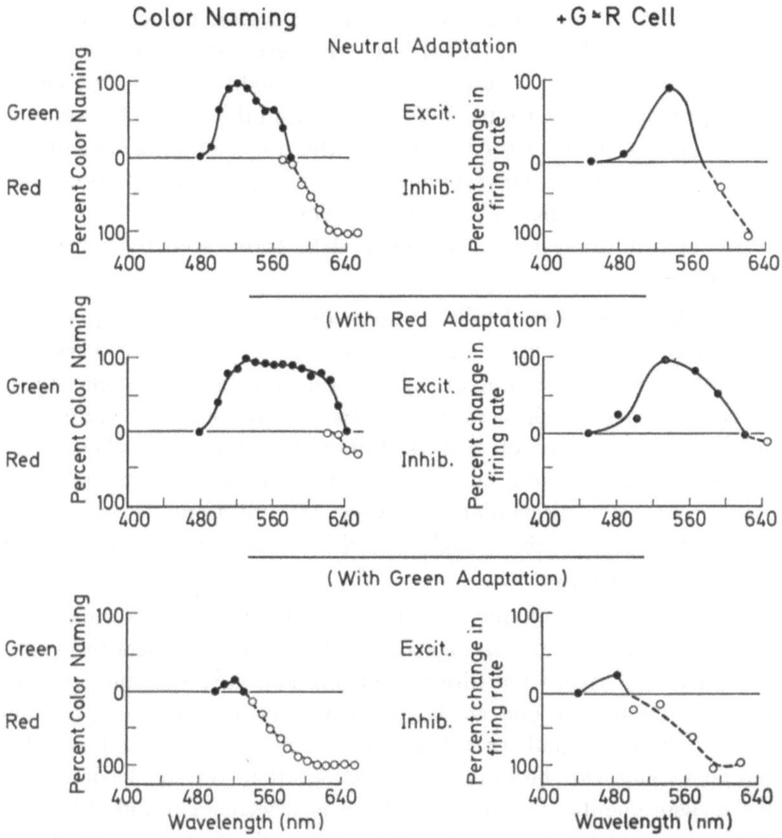

Fig. 11. Effect of chromatic adaptation on color seen and on opponent cell responses. To the left, from Jacobs and Gaylord (1967), are the names given different wavelengths (only the green and red curves are shown). To the right, the responses of a + G — R cell from the macaque LGN. These psychophysical and physiological data are compared under three different adaptation condition: neutral (top pair); red adapting light on (middle); green adapting light on (bottom)

Another situation in which one also finds good agreement between color appearances and firing rates of macaque LGN cells is in the response to spectral lights of varying luminance, the Bezold-Brücke phenomenon. Dim long wavelength lights appear red, for instance, but increasingly brighter ones of the same wavelengths appear more and more yellowish. A similar change is seen (De Valois et al., 1966) in the firing of LGN cells: long wavelength lights of low intensity produce more activity from + R − G than from + Y − B cells, but the balance shifts toward + Y − B cells as one raises the luminance.

Since Jacobs and Yolton (1971) have shown very close relationships between the color vision of ground squirrels and of human dichromats of the protanopic variety, similar comparison between ground squirrel opponent cell activity and the color appearance of stimuli to a protanope might be made. Human protanopes have long been known to perceive the whole spectrum as being either yellow or blue, with a neutral (that is colorless) region in between, at around 500 nm (Judd, 1948). Michael (1968) finds that ground squirrel opponent cells show maximum excitation to the region around 560, a cross-over, that is, no response, to spectral stimuli around 500, and then inhibition to the region around 460 (or, in the case of their mirror-image cells, the reverse points of excitation and inhibition). The cross-point corresponds very closely to that region which we would see as green, but which a protanope cannot discriminate from white, and the points of maximum responses correspond to those seen by a protanope as blue and yellow.

Saturation Discrimination. While comparisons of LGN cell responses to perceptual reports of the appearance of objects is perfectly valid and very meaningful, Brindley (1960) to the contrary, the main body of our psychophysical information comes from discrimination experiments in which the individual does not describe the appearance of objects, but rather judges whether or not two stimuli are different. It would, therefore, be very useful to compare the ability of a human observer to discriminate between stimuli along various dimensions with the ability of cells in the visual pathway to make comparable discriminations by changes in firing rates.

Two such visual situations that have been widely explored in human (Wright, 1947) and monkey (De Valois and Jacobs, 1968) psychophysical tests are wavelength and purity or saturation discrimination. To study discriminability along either of these dimensions, one needs to control the luminance of the lights to make sure that the individual (or cell) is not discriminating on the basis of brightness differences. This can be done by equating them on the basis of photopic luminosity measurements. The question then is whether the individual (or cell) can discriminate a shift from one light to another light of equal brightness but differing in wavelength or purity. It should be emphasized that this is quite different from examining the response to a flash of spectral light, in which both a luminance change and a color change are present. In these discrimination experiments, there is no increment in luminance, but only a shift from one equally bright light to another, to determine how different the lights have to be in wavelength or purity to be discriminably different.

Both behavioral (De Valois and Jacobs, 1968) and physiological studies (De Valois and Marrocco, 1971) of purity or saturation discrimination have been carried out. The behavioral studies of macaque monkeys indicate that their ability to discriminate between a white light and a white to which a small amount of some monochromatic light has been added is quite similar to that of normal human trichromats. The region around 570 nm for both macaques and man is very desaturated relative to the spectral extremes, and the absolute saturation discrimination thresholds are very similar for these two species. In physiological tests, the spectrally non-opponent cells were found to be unable to discriminate among equal luminance lights varying only in purity; they can in fact not discriminate between white light and a pure, monochromatic light (De Valois, 1969; De Valois and Marrocco, 1971). On the other hand, the spectrally opponent cells are very sensitive to purity differences. In Fig. 12 are illustrated the responses of + R − G cells to lights of different purities at each of several different

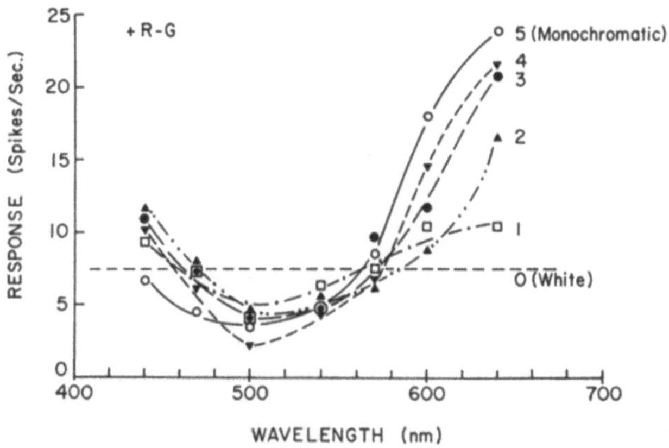

Fig. 12. Responses of + R — G cells to lights of different purity (saturation). At each wavelength, a shift was made from a white light to one of five white-monochromatic mixtures (of equal luminance). Top curve (No. 5) is pure monochromatic light, No. 4 a slightly desaturated spectrum, etc. It can be seen that at long wavelengths the + R — G cell's responses were systematically related to the amount of red in these red-white mixtures. From De Valois and Marrocco, 1971

wavelengths. At each wavelength, five different purity steps, ranging from white through white-monochromatic mixtures to pure monochromatic light, were presented. It can be seen that the responses of these cells are systematically related to the amount of monochromatic light in the monochromatic-white mixture, and that they can discriminate (i.e., alter their firing rates to) very small changes in the purity of stimuli in the long-wavelength end of the spectrum. Similar tests of other cell types indicate that the + B − Y cells can discriminate saturation differences very well in the very short wavelengths, and + G − R cells show relatively good saturation discrimination in the greens; but none of the cell types discriminates saturation differences very well in the yellows. From these data, one can quantitatively account for the saturation discrimination function seen behaviorally (De Valois and Marrocco, 1971); see Fig. 13.

Wavelength Discrimination. A person with normal color vision is easily able to discriminate between different wavelengths of light which are equated for photopic luminance. This wavelength discrimination ability is not uniform across the spectrum. Rather, one can discriminate small wavelength differences in the yellow-orange and in the blue-green parts of the spectrum, but only larger differences in the greens and toward either spectral extreme (WRIGHT, 1947). The same is true for macaque monkeys tested in behavioral experiments (DE VALOIS and JACOBS, 1968). The corresponding ability of various varieties of LGN cells

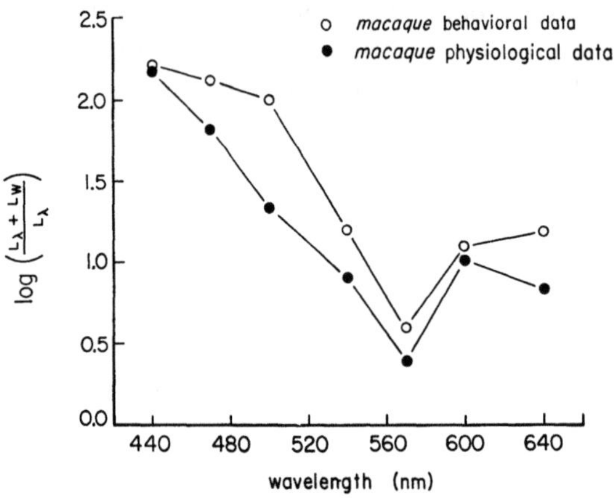

Fig. 13. Comparison of saturation discrimination by macaque monkeys and macaque LGN opponent cells. From data like those in Fig. 11, obtained for each opponent cell type, one can determine the purity level required for a criterion firing level at each wavelength from each cell type. Assuming the animal's behavior is determined by the cell type which best discriminates at each wavelength, one obtains the values plotted in solid dots. The open circles are from psychophysical tests of the ability of monkeys to choose between white and desaturated colored lights. From DE VALOIS and MARROCCO (1971)

in macaque monkeys was examined (DE VALOIS et al., 1964; DE VALOIS et al., 1965). Although spectrally non-opponent cells are, of course, very responsive to a flash of monochromatic light, they cannot discriminate a shift from one monochromatic light to another of equal brightness. On the other hand, the spectrally opponent cells are very responsive in this situation, although to varying degrees in different parts of the spectrum; see Fig. 14. Specifically, the RG cells (+ R − G and + G − R) were found to discriminate very small wavelength differences in the region around 590 nm and to fall off on either side, showing very poor wavelength discrimination in the region of 490 nm, for instance, where the animal can actually discriminate wavelength differences very well. The YB cells (+Y − B and +B − Y), however, show very good wavelength discrimination in the short wavelength region around 490 nm and fall off to either spectral extreme, although these cells also showed a secondary minimum around 580 nm.

The responses of each of these cell types is very reminiscent of the wavelength discrimination curves obtained from dichromatic individuals (WRIGHT, 1947). The curve of the YB cells is quite similar to that seen in protanopes and deuteranopes, and that of the RG cells to that seen in tritanopes. The normal color vision organization in the primate can be considered to consist of two separate dichromatic color vision systems: a red-green system and a yellow-blue system.

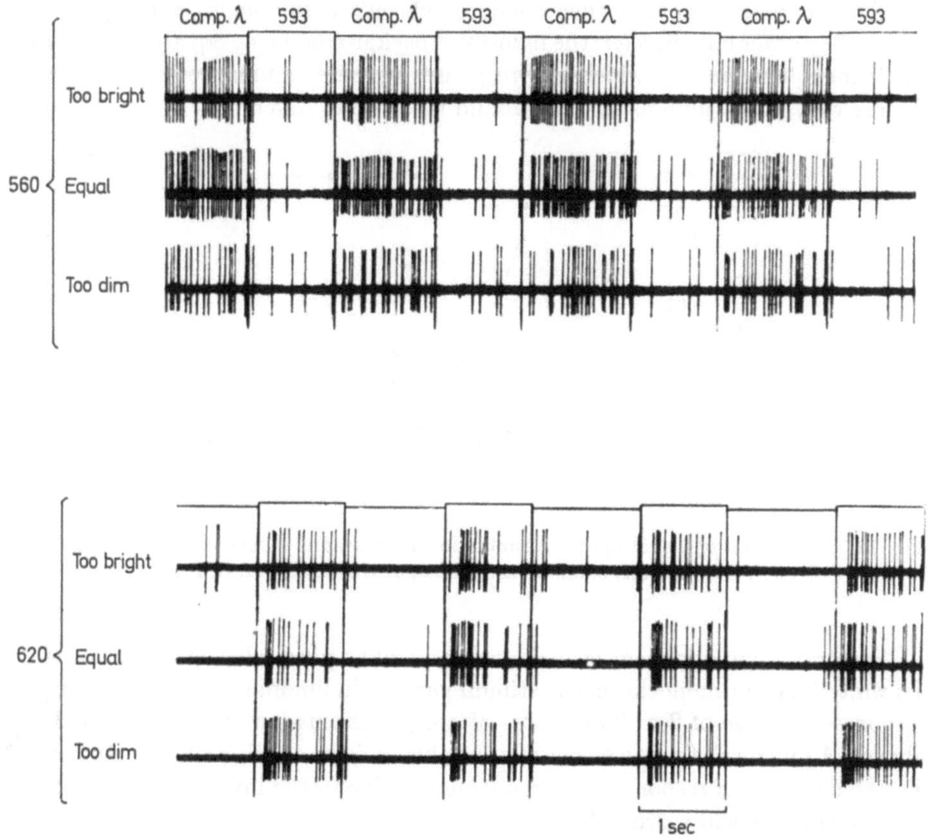

Fig. 14. Wavelength discrimination by a macaque spectrally opponent cell. A shift is made between two wavelengths equated for luminance in the middle line of each pair, and it can be seen that this + G — R cell discriminates (by a change in firing rate) between 593 nm and either 560 nm (top) or 620 nm (bottom). That the discrimination is based on wavelength (that is, is a true color discrimination) is further indicated by the fact that it makes little difference whether the different wavelengths are the same brightness or not. From De Valois et al. (1967)

The overall wavelength discrimination ability of the organism, then, with its double minimum, can be accounted for by the presence of the double dichromatic color vision organization seen in macaque LGN cells; see Fig. 15.

Spatio-Temporal Characteristics. The experiments discussed above, in which the discriminative characteristics of LGN cells were compared to that of the whole organism, were carried out with one-second duration stimuli which covered the whole receptive fields of the cells; the psychophysical experiments used comparably large, long-duration stimuli. Some interesting new relationships arise, however, when the responses of LGN cells are examined to stimuli which vary in size or duration, or to grating patterns of different spatial or temporal frequencies. We (DeValois and Pease, 1971; DeValois, Snodderly, Yund, and Hepler, in preparation) have examined these spatio-temporal characteristics with the two fundamentally different stimuli to which our visual systems respond: one in which the lines or spots differ from their backgrounds only in *luminance*, and the other in which they differ only in *color*, the figure and background being equated in photopic luminance. The way in which the responses to pure luminance and pure color figures vary with stimulus size was found to be quite different.

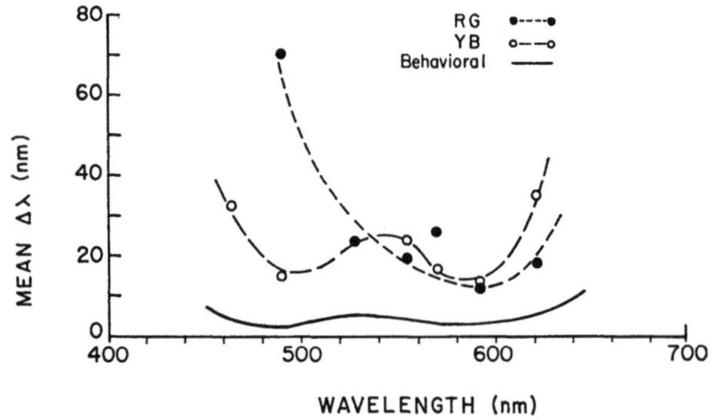

Fig. 15. Comparison of wavelength discrimination by macaque monkeys and macaque LGN cells. Comparisons made on same basis as that discussed in Fig. 17

Spectrally non-opponent cells do not respond to pure color figures of any size. To white-black or monochromatic stimuli varying in luminance, they show largest responses to those of 2 to 30 minutes of arc (see discussion of receptive field size above). A given cell fires to either white or black figures and inhibits to the other. The attenuation of responses to large stimuli is due to the invasion of the antagonistic receptive field surround.

Spectrally opponent cells respond to pure color stimuli (as in the full-field case), but also to black-white figures (as was not clear from the full-field situation). With black-white figures, the maximum response is to some intermediate-sized figure, as was the case for nonopponent cells. With pure color figures, there is little or no response to figures of the size which produce the largest responses in the achromatic case; the response increases with increasing stimulus size with no attenuation. These cells respond optimally, then, to small black-white or large pure color figures. The achromatic information being carried by opponent cells under these conditions (as well as that resulting from rod inputs under scotopic conditions) must be separated out from the chromatic information at some later cortical level.

These results are qualitatively but not quantitatively predictable from the receptive field organization of Type I cells found by Wiesel and Hubel (1966). With achromatic or monochromatic luminance figures which affect both pigment types feeding into a spectrally opponent cell, the center and surround are antagonistic, just as in non-opponent cells. With pure color figures, however, the receptive field center and surround act in the same direction and there is no center-surround antagonism to attenuate the responses to large stimuli. For instance, in a $+G-R$ cell in which the M cone is excitatory in the center and the L cone inhibitory in the surround, a shift from red to green will produce an increase in center excitation and a decrease in surround inhibition. The center and surround thus act in the same direction, so the largest responses are to full-field stimuli.

This absence of center-surround antagonism when pure color figures are involved accounts for the absence of Mach bands or other phenomena dependent on lateral inhibitory interactions when pure color stimuli are examined psychophysically. As a result, acuity is much poorer when based on pure color differences than when based on achromatic or chromatic luminance differences.

Temporal variations have not been examined as thoroughly, but it is clear that an opponent cell cannot follow a pure-color flicker to nearly as high frequencies as the same cell can follow a luminance flicker. The two cone types feeding into a given opponent cell in excitatory and inhibitory directions are in the first case being stimulated out of phase and in the second in phase. Differences in the latency or phase relations between the excitatory and inhibitory paths must be involved in the different temporal characteristics of the responses to these two types of stimuli.

Contrast and Similitude. The color seen at a certain point in space at a particular time depends not only upon the stimulus conditions at that point in space and time, but also on the stimulus configuration in the surrounding regions and that which existed there at a previous time. These interactions are of two different sorts, in two different directions. One type is very well known, and is known as contrast. In the spatial domain, simultaneous contrast refers to the fact that a surround of one color or brightness tends to induce into a neighboring region the opposite color and brightness (opposite in terms of opponent organization). In simultaneous brightness contrast, a grey on a black surround appears whitish, whereas on a white surround it appears black. In simultaneous color contrast, a grey on a red surround appears greenish, on a green surround it appears reddish. Successive contrast, in the temporal domain, is correspondingly an induction of opposite brightness or color into the succeeding time period. After one looks at a green surface for a time, a grey appears red, and after a red surface, one sees the same grey as being green.

These contrast effects, both of color and brightness, have been extensively investigated. Almost totally neglected, however, are effects in the opposite direction, for which we suggest the term "similitude effects" – thus, simultaneous similitude and successive similitude. This term may not gain wide currency but it is used here to emphasize the fact that induction in a direction opposite to contrast can and does occur. Under some stimulus circumstances, a surround of a particular color and brightness induces not the opposite color and brightness, as in contrast effects, but the same color and brightness, a similitude effect, into the neighboring region. The brightness induction effect has been investigated by Helson (1963) and has been referred to as assimilation or the Bezold spreading effect. Those terms, however, do not make it clear that the effect is opposite to that of contrast; it appears that "similitude" might better make this point.

The stimulus conditions under which one sees simultaneous similitude as opposed to simultaneous contrast have not been completely specified psychophysically. In general it seems that where the inducing surround is small, the induction tends to be in the direction of similitude, whereas when the surround is large, the induction tends to be in the direction of contrast (Helson, 1963).

The stimulus conditions which produce successive similitude have been more thoroughly studied. After one looks at a red light, the predominant immediate after-impression is that of green, a negative after-effect. But, particularly if the preceeding stimulus had been very intense, one might also see a positive after-image, an induction not of the opposite color but of the same color. The former is successive contrast and the latter might be called successive similitude. Comparable effects are also seen in the black-white domain. In fact, what we are calling successive similitude has been shown by BARLOW and SPARROCK (1964) to persist for long periods of time after an intense flash, only gradually dissipating as the receptor photopigments are reconstituted (by which they attempt to explain dark adaptation). One has an alternation, then, after a flash of light, of successive contrast and successive similitude, the latter lasting for long periods of time if the stimulation was intense enough to produce a significant amount of photopigment bleaching.

These spatio-temporal interactions have been examined to some extent in recordings from the goldfish ganglion cells (DAW, 1967) and in recordings from LGN cells in macaque (DE VALOIS and JONES, 1961; DE VALOIS and PEASE, 1971) and in the case of achromatic effects in the LGN and cortex (BAUMGARTNER and HAKAS, 1959; BAUMGARTNER, 1961). An example of successive color contrast can be seen in Fig. 16.

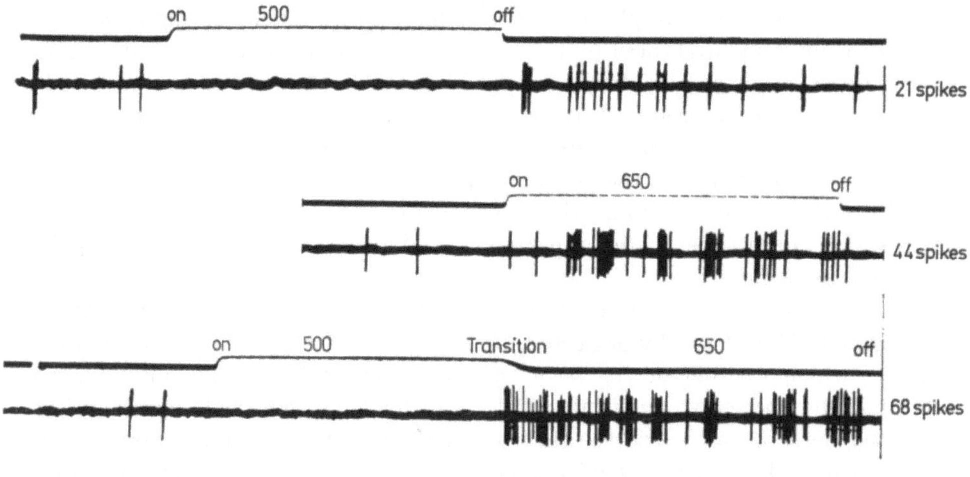

Fig. 16. An example of successive color contrast shown by a spectrally opponent cell. This cell (+ R — G) fires either to a red light, or to the offset of a green light (successive color contrast). Shifting from a green to a red light produces a summation of these two responses. From DE VALOIS and JONES (1961)

Neural rebound is a common property of cells in the central nervous system; at the termination of a stimulus which produces inhibition, a cell shows rebound excitation; and after excitation, rebound inhibition. This, plus the opponent organization of both the color and black-white systems, would appear at least qualitatively to account for both chromatic and achromatic successive contrast effects. Such successive contrast effects are very prominently seen in the responses of monkey LGN cells (DE VALOIS and JONES, 1961).

The extent to which successive similitude is found in either the color (spectrally opponent) or black-white (spectrally non-opponent) systems has not been determined. No neural evidence has been provided for long-lasting positive aftereffects of stimulation, although psychophysical evidence (BARLOW and SPARROCK, 1964; ALPERN and CAMPBELL, 1963; DE VALOIS and WALRAVEN, 1967; GESTRIN and TELLER, 1969) indicated that such long-lasting color and brightness aftereffects are transmitted to the central nervous system.

It is commonly supposed that the center-surround organization of non-opponent cells seen in the retina and LGN accounts for simultaneous brightness contrast. This has been questioned (DE VALOIS and PEASE, 1971) on the basis of the fact that simultaneous brightness contrast operates over very large distances perceptually, whereas the lateral interactions seen at the LGN are found to extend over only very small distances. These lateral interactions are seen as contour-enhancing, rather than as true simultaneous contrast mechanisms. The same would appear to be true for the results in the cat (BAUMGARTNER, 1961), although here the corresponding behavioral evidence is lacking.

The double-opponent organization found by DAW (1967) in the goldfish provides a mechanism by which simultaneous color contrast might occur, although here again the question must be raised of whether the distances over which the physiological responses and the psychophysical effects occur are the same. Spectrally double-opponent cells are not often seen in the LGN of the macaque visual system. It would thus appear that simultaneous brightness and simultaneous color contrast in the primate visual system depend upon interactions that occur past the LGN. This is the one clear instance in which color processing does not appear to be completed by the thalamic level.

The physiological basis for simultaneous color and brightness similitude has not been examined. However, it should be noted that the blur of the retinal image due to diffraction and stray light produces an intensity and wavelength distribution across the retina which, if other factors were not involved, would produce simultaneous similitude. Thus a grey surface on a white background actually has more light on it than the same surface on a black background. Under most circumstances color and brightness contrast presumably overcomes this similitude effect. But the surrounds of cells in the visual path have very low sensitivity per unit area compared to the receptive field centers: large annuli must be used to evoke surround antagonism. It is possible that when the inducing area consists of fine lines (the circumstances under which similitude effects are found; HELSON, 1963) the areas covered are sufficient to activate the receptive field centers (the lines are thus detected), but insufficient to activate receptive field surrounds and thus produce contour enhancement and contrast.

IV. Cortex

The primate LGN projects completely to the striate cortex. As we have seen, in the macaque (and presumably in the human visual system), some 70–80 % of the cells in the LGN are transmitting color information. What happens to that information when it reaches the cortex – where it goes, how it is further processed, what role it plays in other kinds of cortical processes – is virtually unknown at this time, for we have only a few tantalizing hints as to the color organization of the cortex. The main body of our knowledge of visual processing at cortical levels comes from the magnificent studies of cortical units by HUBEL and WIESEL, mainly in the cat, but also in the macaque monkey. But these investigators, for the most part, have not been concerned (rightly so, in the cat) with how the system processes chromatic information. Therefore, we are left with studies of gross evoked potentials, which tell us little of interest about the organization of the system, and a few studies of responses of cells in the cortex, studies which have not been carried to the point of really clarifying the organization.

1. Evoked Potentials

With gross electrodes on the surface of the scalp and the use of averaging techniques, one can reliably record small cortical evoked potentials. These potentials doubtless reflect some aspect of the overall activity of large numbers of cells in the underlying cortical region. We have certainly discovered enough about the way in which many different kinds of information are processed in the nervous system to know that individual cells have very specific functions and that neighboring cells are very often processing quite different kinds of information. Given this, averaging across even two cells, to say nothing about thousands, would obviously result in data which are difficult to interpret. There are some things, however, for which evoked potential recordings can be useful. One is as a kind of gross anatomical device to determine whether a given region is functionally involved with a particular sort of activity – if stimuli are devised which stimulate cells if and only if they are engaged in that activity. Also, correlations can sometimes be made between various psychophysical phenomena and the evoked responses from a particular region of the brain of the same human observer.

Unfortunately, many of the evoked potential studies have merely looked at the responses to flashes of monochromatic light. Such responses by themselves tell us nothing about color processing. Unless the lights are either equated for luminance (not energy), or some response characteristics are found which are distinctive to certain spectral regions, and these characteristics are maintained despite changes in luminance, one cannot even say that the region is involved in color processing. And the fact that one part of the brain may respond differently to flashes of monochromatic light than another can be equally uninformative. Thus, MASSOPUST et al. (1969) report, in the squirrel monkey, that a flash of red light gives larger gross responses from the occipital pole than does a blue, whereas further anterior one obtains larger responses to the blue than to the red. This necessarily need have nothing whatsoever to do with color vision, but merely reflect different photopic-scotopic balances in the cortical projections to the two areas.

Clearly, the most appropriate sort of stimulus to use to detect color processing would be one which could only be detected by the organism on the basis of color differences, i.e., with the different spectral stimuli equated for luminance.

Fairly clear evidence that the visual cortex is involved in color processing comes from a study by SHIPLEY et al. (1965) in which they report that the responses from the occipital region to different wavelengths of light, apparently equated for photopic luminosity, were consistently different in the case of an individual with normal color vision but almost indistinguishable in the case of a dichromat. BURKHARDT and RIGGS (1967) also showed evidence for color responses by the use of a constantly flickering light presented against backgrounds of different wavelengths matched for photopic luminance. They found larger responses when the background was of a wavelength different from that of the flickering stimulus.

In general, however, evoked potentials recorded from the occipital cortex have told us little more than that this region is involved in processing color information, a fact which should be fairly apparent both from the color-vision defects that result from cortical lesions in monkeys (WEISKRANTZ, 1963) and from the nature of the geniculate information arriving there.

2. Unit Recordings

Some of the technical problems involved in recording from single units in the visual cortex and adequately examining their behavior are more difficult than those required for recording at earlier visual stages; a considerable number of the early cortical unit studies thus have little to offer in the way of solid knowledge. The difficulty with holding cells for long periods of time impelled some investigators to use extremely short flashes presented in very rapid succession, a stimulus situation which is extremely difficult to evaluate or to relate to psychophysical data. Also, many anesthetics produce a gross loss in sensitivity of cortical cells. Perhaps the greatest difficulty, however, was that not until HUBEL and WIESEL showed that most cortical cells are specifically responsive to the shapes of objects and not just to flashes of light did investigators present adequate stimuli to cortical cells; they thus found a large number of cortical units to be unresponsive.

Some units in the macaque monkey visual cortex were found (DE VALOIS, 1960b) to respond to different colors in much the same way as do geniculate cells, except that some had binocular inputs and their spontaneous activity was so low that the presence of an inhibitory input was not obvious. MOTOKAWA et al. (1962) reported spectrally opponent cortical units which appear similar to those seen at the LGN. However, LENNOX-BUCHTHAL (1961, 1962) and ANDERSON et al. (1962) reported that the cortical cells they found (in the mangabey monkey *Cercocebus*) responded only to very narrow spectral regions with no response to other spectral regions, see Fig. 17. However, the reports in all these experiments of absence of inhibition from other spectral regions should be looked on with caution, since adequate tests for the presence of inhibition were not performed. The very low spontaneous firing rates found in the cortical cells in the DE VALOIS (1960b) and LENNOX-BUCHTHAL (1961, 1962) experiments would make it very difficult to observe inhibition on the basis of decreases in firing from the spontaneous rate. However if a stimulus from a region that might be inhibitory – that is, from the opposite spectral region – were combined with a stimulus from the area that produced excitation, the presence of inhibition might very well be made apparent. We find such to be invariably the case in LGN cells which give only excitation to limited spectral ranges. Also, in the case of the LENNOX-BUCHTHAL (1961, 1962) experiments, extremely short (8 msec) flashes of light were presented in extremely rapid succession at luminances far above the adaptation level. The use of such short flashes would make it virtually impossible to observe any inhibition during the stimulus, even in the case of a cell with a fairly high spontaneous activity rate, and a short stimulus is quite inadequate to produce rebound excitation at its termination (rebound excitation is seldom observed for stimuli of less than one second duration. The correlated perceptual phenomena of after-images require prolonged prior exposure to the stimulus, except for extremely bright flashes).

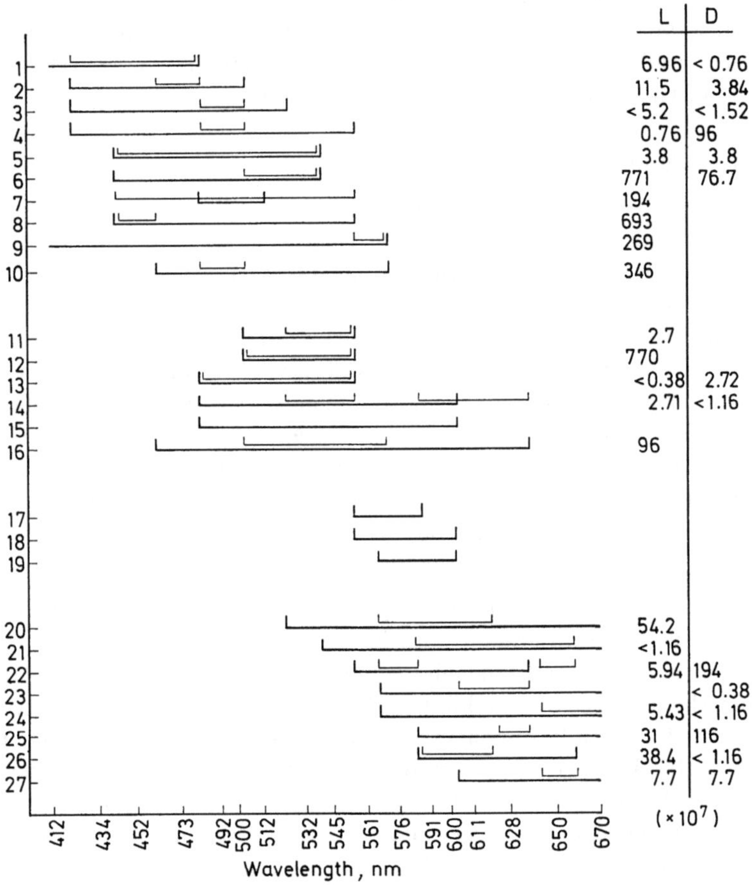

Fig. 17. Narrow spectral response curves obtained from each of 27 cortical cells. Each cell was found to respond to only a limited spectral range when brief (8 msec.) flashes of monochromatic light were given. From ANDERSEN et al. (1962)

In none of these unit recording experiments was an optimally shaped stimulus used. This, as we know from HUBEL and WIESEL's experiments, is a line of a particular width, orientation, etc. for virtually all striate cells. The factors mentioned above, plus the absence of the optimal shape and durations of stimuli, make it difficult to evaluate the significance of the narrow spectral response curves produced by flashes of monochromatic light. One should note that one can often record very narrow spectral response curves from LGN cells too: at spectral extremes under normal circumstances, and at any spectral locus if the sensitivity is depressed so that only the optimal stimulus wavelength evokes a response.

On the other hand, it would not be too surprising if the cortical processing of color information produced a further narrowing, through excitatory-inhibitory interactions, of spectral response curves from those seen at the geniculate level. Clearly, in order to see the numerous colors that we do in the environment, we must have some way of comparing the outputs of different classes of geniculate opponent cells. We see, for instance, orange under those circumstances in which the + Y − B and the + R − G cells are about equally active. There presumably must be some neural system, most likely an individual cell, which makes a comparison of the relative activity rates of these two opponent cell types. If this were the case, one might very well find cells which would respond to narrower bands of the spectrum than one finds at earlier levels. The evidence at this time is not sufficiently compelling, however, to force this conclusion.

Gouras (1970) has examined how many and which cone types feed into each of a number of macaque cortical cells, isolating the individual cone inputs by the use of chromatic adaptation. Virtually every combination of cones was found to feed into one or another of the cells in the sample. Some of the cortical cells were found to have inputs from all three cone types, a situation seldom if ever found at the LGN. Unfortunately, Gouras did not examine how the cells responded to different stimuli under normal adaptation conditions, so it is hard to know what the functions of the various cortical cells were (the intense chromatic adaptation used to isolate a single cone input, by that very act, of course, also eliminates whatever color discrimination might have been present. The fact that a cell has one or two or three cone types feeding into it tells nothing about what it does. It might under any of those circumstances be encoding movement, for instance, not color information). However, most of the cells received the different cone inputs in an opponent manner, so they at least probably were involved with color processing.

Hubel and Wiesel (1968) have examined in some but not all of the macaque cortical cells they have studied the responses to different wavelengths of light. They report that relatively few of the cells in the cortex seem to be concerned with color: about a quarter of the simple cells and less than 10 % of the more complex ones. They classified a cell as being unresponsive to color if it responded to flashes of light of all the different wavelengths, rather than responding to only a narrow band or showing, as geniculate cells do, opponent responses to different spectral regions. It is possible, as will be discussed below, that many cells that in fact receive color information were not so classified by Hubel and Wiesel.

Among the cells that they found to give differential color responses, Hubel and Wiesel (1968) found all the simple cells (only a sample consisting of 6 cells) were + R − G, with a long, narrow central excitatory region responsive only to red light, and inhibitory green flanks. That white light restricted to the center produced no responses indicates a greater specificity of chromatic response than is seen in the LGN. About half the complex and hypercomplex cells with specific color sensitivity were found to respond to white light and to a restricted spectral range; the other half responded only to a restricted range of monochromatic lights.

Several recent studies of cortical cells (BOLES, 1971; POGGIO, MANSFIELD, and SILLITO, 1971; HEPLER, YUND, and DE VALOIS, in preparation) all report far higher percentages of cortical cells responsive to color than did HUBEL and WIESEL (1968). Part of the discrepancy may be that foveal rather than parafoveal areas were studied. A large number of the color-responsive cortical cells show no orientational selectivity but differ from LGN cells in many response characteristics, such as spectral selectivity and binocularity. We find whole columns of cells with similar color characteristics.

HUBEL and WIESEL (1968) report the presence of a very few cortical cells which have a donble-opponent organization similar to that discovered by DAW in the goldfish retina. In these cells there was a far surround of opposite opponent organization to the near center-surround. It is not clear that these double-opponent cells are any more numerous here than in the LGN, where, as was noted earlier, they are very occasionally found. Such an organization might very well be involved in simultaneous color contrast, as DAW hypothesized, but much study of such cells in the primate cortex must be made before the operation of such a system can be specified.

HUBEL and WIESEL (1968) classified as being unconcerned with color those cells that responded similarly to flashes of monochromatic light regardless of wavelength, and also to white light. These cells may indeed be receiving no color information: they may correspond to spectrally non-opponent cells seen at the geniculate, which have been shown in color discrimination tasks (DE VALOIS and MARROCCO, 1971) to be color blind. However, quite another interpretation might be given to the responses of these cortical cells; namely, that they are generalizing across the various colors rather than ignoring them. That is, they are receiving color information but are using this color information to detect form and then are responding to the form independent of its color. For example, one might present a person with a red pencil on a grey background matched with it for luminance, and ask him what it was. He might say that it is a pencil. When presented with a green pencil on the same background, he might also say it is a pencil, and similarly with a blue pencil. One might hastily conclude that such an individual is color blind because he responded similarly to red, green, and blue, but that is, of course, not the case, since if he were color blind he would not have seen any of the pencils. Rather, he was responding not to color but to the form, although he could perceive the form only from color information.

It is conceivable, then, that many of the complex and hypercomplex cortical cells respond to lines of a particular orientation, etc., but that the information upon which they detect a line's presence may be color as well as black-white information, with the color information coming from all of the various opponent cell types (we have indeed found many cells of this type in the monkey cortex).

Strong evidence has been reported (PETTIGREW et al., 1968; HUBEL and WIESEL, 1968) for cells that show this sort of generalization in the black-white domain: cells that respond to either a black bar or a white bar. These cells certainly cannot be said to be blind to black-white differences or they could not detect either the black or white bars. Rather, they discriminate the black bar from its background and the white bar from its background, but then respond to either one.

One sometimes tends to forget that color vision evolved not as an aesthetic mechanism, but because it enables us to see objects better than if we were color-blind and dependent just upon black-white information. Visual information is often largely redundant: we are all acquainted with the fact that one can abolish color information, as in black and white photography, and still be able to see objects, we would also discover, if we abolished luminance differences in a scene, that we can see objects perfectly well purely on the basis of color differences.

It would appear from the work of Hubel and Wiesel (1959, 1962, 1968) that the visual cortex is primarily a form-discriminating network. Lines of varying orientation and size are first detected, then put together into patterns of greater complexity to determine, perhaps at some later stage, whether a chair is present at a particular point in the environment, rather than a tree. This sort of pattern detection would require huge numbers of different cells to detect many different patterns. It would seem unlikely that there should be separate mechanisms for discriminating forms of each different color in each region of the visual field; that is, that there should be one network of cells looking for white chairs, another for red chairs, another for green or black, etc. (we can in fact perceive objects in hundreds of different colors). Rather, it would seem more likely that the color and brightness information may be processed and stored in relatively few cortical cells while most of the cortical cells, receiving information from both LGN black-white (spectrally non-opponent) and color (spectrally opponent) cells, detect the pattern.

Whether or not a cell (or a person) is extracting color information from the environment can be determined if one tests with an appropriate stimulus: one which can be detected only on the basis of color differences, such as are used in conventional color-blindness test plates. A flash of monochromatic light, which consists of both a color and a luminance change, is not such a stimulus.

3. Past the Striate Cortex

It is rather presumtuous even to include such a section, because we know so little about what happens to color information within the striate, to say nothing of the regions to which it projects. Hubel and Wiesel have not reported explorations of areas 18 and 19 in the monkey, as they have in the cat. However, Gross et al. (1970) report on cells even further "up" the path, in the inferotemporal lobe, which appear to be several orders of magnitude more complex than striate cells. Some appear to be selectively sensitive only to certain complex shapes in which color plays a role – i.e., the object must be of a certain color as well as of a certain shape.

V. Summary

With reference to the processing of color information in the visual system (with prime emphasis on higher primates):

1. Color-processing cells have their inputs (in macaque and man) from three types of cones: S cones of λ_{max} of approximately 440 nm, M cones of λ_{max} c.

540 nm, and L cones of λ_{max} of approximately 570 nm; plus, under mesopic conditions outside the fovea, rods with λ_{max} of 510 nm.

2. The absorption curves of the cone photopigments all roughly approximate the Dartnall nomogram in shape. Therefore the L and the M cones absorb all across the spectrum, not just in restricted spectral regions.

3. The outputs of the different cone types feed up many different paths within the geniculo-striate system. Information about the characteristics of the retinal stimulus, such as its color, its brightness, its shape, whether or not it is moving, etc. are all determined by later "feature extractors" which compare the outputs of receptors in different ways.

4. The color of the stimulus striking a group of receptors is detected by ganglion and LGN cells making three types of comparisons: (a) red and green objects are detected by a pair of mirror-image cell types which difference the L and M cone outputs; (b) yellow and blue objects are detected by a differencing of the L and S cone outputs (or perhaps the M and S); (c) white and black objects by summing the L and M cone outputs in one region and subtracting that from the L and M outputs in adjacent regions.

5. Many aspects of color vision can be understood from LGN opponent cell activity. (a) The hue of a spectral light corresponds closely to which of the four opponent cell types is most active; the saturation of the light to the relative activity in the opponent vs. the non-opponent cells. These relations between the LGN cell activity and the color seen hold in chromatic adaptation experiments and in the Bezold-Brücke situation, but not contrast situations or when the stimuli are very small. (b) Wavelength and saturation discrimination can be accounted for quantitatively from opponent cell discrimination behavior.

6. At the striate cortex, four further changes may occur, although the evidence for these is far poorer than that for points 1–5 above. (a) Some unknown interaction must effect the separation of the black-white and color information being carried by the spectrally opponent cells. (b) The outputs of the various spectrally opponent cells may be differenced from each other to produce cells with extremely high spectral selectivity. (c) Cells of a particular spectrally opponent variety but receiving from different retinal areas may be combined to form colored-line detectors. (d) Cells which detect lines of different colors may be combined to form cells which detect lines regardless of color (although it is color information which allows the lines to be detected initially). (e) Interactions over long distances may occur to provide color contrast.

References

Abramov, I.: Further analysis of the responses of LGN cells. J. Opt. Soc. Amer. **58**, 574—579 (1968).

Alpern, M., Campbell, F. W.: The behavior of the pupil during dark-adaptation. J. Physiol. (Lond.) **165**, 5—7 (1963).

Andersen, V. O., Buchmann, B., Lennox-Buchthal, M. A.: Single cortical units with narrow spectral sensitivity in monkey (*Cercocebus torquatus atys*). Vision Res. **2**, 295—307 (1962).

Anderson, K. V., Symmes, D.: The superior colliculus and higher visual functions in the monkey. Brain Res. **13**, 37—52 (1969).

Barlow, H. B.: Summation and inhibition in the frog's retina. J. Physiol. (Lond.) **119**, 69—88 (1953).
— Sparrock, J. M. B.: The role of afterimages in dark adaptation. Science **144**, 1309—1341 (1964).
Baumgartner, G.: Die Reaktionen der Neurone des zentralen visuellen Systems der Katze im simultanen Helligkeitskontrast. In: Jung, R., Kornhuber, H. H. (Eds.): The Visual System: Neurophysiology and Psychophysics. Berlin-Göttingen-Heidelberg: Springer 1961.
— Hakas, P.: Reaktionen einzelner Opticusneurone und corticaler Nervenzellen der Katze im Hell-Dunkel-Grenzfeld (Simultankontrast). Pflügers Arch. ges. Physiol. **270**, 29 (1959).
Blackwell, H. R., Blackwell, O. M.: Rod and cone receptor mechanisms in typical and atypical congenital achromatopsia. Vision Res. **1**, 62—107 (1961).
Boles, J.: Colour and contour detection by cells representing the fovea in monkey striate cortex. Paper presented at first annual meeting, Society for Neuroscience, 1971.
Boynton, R. M., Gordon, J.: Bezold-Brücke hue shift measured by color-naming techniques. J. Opt. Soc. Amer. **55**, 78—86 (1965).
Brindley, G. S.: Physiology of the retina and the visual pathway. London: Edward Arnold 1960.
Brown, R. H.: Bright visibility curve of the rabbit. J. gen. Physiol. **17**, 323—338 (1937).
Burkhardt, D. A., Riggs, L. A.: Modification of the human visual evoked potential by monochromatic backgrounds. Vision Res. **7**, 453—459 (1967).
Clarke, F. J. J.: Further studies of extra-foveal colour metrics. Optica Acta **10**, 257—284 (1963)
Daw, N. W.: Goldfish retina: organization for simultaneous color contrast. Science **158**, 942—944 (1967).
— Colour-coded ganglion cells in the goldfish retina: extensions of their receptive fields by means of new stimuli. J. Physiol. (Lond.) **197**, 567—592 (1968).
DeValois, R. L.: Color vision mechanisms in the monkey. J. gen. Physiol. **43**, 115—128 (1960a).
— In: Mechanisms of Colour Discrimination. London: Pergamon Press 1960b, 111—114.
— Analysis and coding of color vision in the primate visual system. Cold Spr. Harb. Symp. quant. Biol. **30**, 567—579 (1965).
— Physiological basis of color vision. Tagungsbericht Internationale Farbtagung Color 69, Stockholm, 1970.
— Abramov, I., Jacobs, G. H.: Analysis of response patterns of LGN cells. J. Opt. Soc. Amer. **56**, 966—977 (1966).
— — Mead, W. R.: Single cell analysis of wavelength discrimination at the lateral geniculate nucleus in the macaque. J. Neurophysiol. **30**, 415—433 (1967).
— Jacobs, G. H.: Primate color vision. Science **162**, 533—540 (1968).
— — Abramov, I.: Responses of single cells in visual system to shifts in the wavelength of light. Science **146**, 1184—1186 (1964).
— — Jones, A. E.: Responses of single cells in primate red-green color vision system. Optik **20**, 87—98 (1963).
— Jones, A. E.: Single-cell analysis of the organization of the primate color-vision system. In: Jung, R., Kornhuber, H. H. (Eds.): The Visual System: Neurophysiology and Psychophysics. Berlin-Göttingen-Heidelberg: Springer 1961.
— Marrocco, R. T.: Single cell analysis of saturation discrimination in the macaque. Vision Res. **13**, 701—711 (1973).
— Pease, P. L.: Contours and contrast: responses of monkey lateral geniculate nucleus cells to luminance and color figures. Science **171**, 694—696 (1971).
— Smith, C. J., Kitai, S. T., Karoly, A. J.: Responses of single cells in different layers of the primate lateral geniculate nucleus to monochromatic light. Science **127**, 238—239 (1958).
— Walraven, J.: Monocular and binocular aftereffects of chromatic adaptation. Science **155**, 463—465 (1967).
Dowling, J. E.: Structure and function in the all-cone retina of the ground squirrel. In: The physiological basis for form discrimination. Symposium at Brown University, Providence, R. I., 17—23 (1964).
Gestrin, P. J., Teller, D. Y.: Interocular hue shifts and pressure blindness. Vision Res. **9**, 1267—1271 (1969).

GOURAS, P.: Primate retina: duplex function of dark adapted ganglion cells. Science **147**, 1593—1594 (1965).
— Identification of cone mechanisms in monkey ganglion cells. J. Physiol. (Lond.) **199**, 533—547 (1968).
— Trichromatic mechanisms in single cortical neurons. Science **169**, 489—492 (1970).
— LINK, K.: Rod and cone interaction in dark-adapted monkey ganglion cells. J. Physiol. (Lond.) **184**, 499—510 (1966).
GRANIT, R.: Sensory mechanisms of the retina: London: Oxford University Press 1947.
GROSS, C. G., BENDER, D. B., ROCHA-MIRANDA, C. E.: Visual receptive fields of neurons in the inferotemporal cortex of monkey. Science **166**, 1303—1306 (1969).
HELSON, H.: Studies of anomalous contrast and assimilation. J. Opt. Soc. Amer. **53**, 179—184 (1963).
HERING, E.: Zur Lehre vom Lichtsinne. Wien: Karl Gerolds Sohn 1878.
HILL, R. M.: Unit responses of the rabbit lateral geniculate nucleus to monochromatic light on the retina. Science **135**, 98—99 (1962).
— MARG, E.: Single-cell responses of the nucleus of the transpeduncular tract in rabbit to monochromatic light on the retina. J. Neurophysiol. **26**, 249—257 (1963).
HUBEL, D. H., WIESEL, T. N.: Receptive fields of single neurones in the cat's striate cortex. J. Physiol. (Lond.) **148**, 574—591 (1959).
— — Receptive fields of optic nerve fibers in the spider monkey. J. Physiol. (Lond.) **154**, 572—580 (1960).
— — Integrative action in the cat's lateral geniculate body. J. Physiol. (Lond.) **155**, 385—398 (1961).
— — Receptive fields, binocular interaction and functional architecture in the cat's visual cortex. J. Physiol. (Lond). **160**, 106—154 (1962).
— — Receptive fields and functional architecture of monkey striate cortex. J. Physiol. (Lond.) **195**, 215—243 (1968).
HUMPHREY, N. K.: Responses to visual stimuli of units in the superior colliculus of rats and monkeys. Exp. Neurol. **20**, 312—340 (1968).
HURVICH, L. M., JAMESON, D.: An opponent-process theory of color vision. Psychol. Rev. **64**, 384—404 (1957).
JACOBS, G. H.: Spectral sensitivity and color vision of the squirrel monkey. J. comp. physiol. **56**, 616—621 (1963).
— DEVALOIS, R. L.: Chromatic opponent cells in squirrel monkey lateral geniculate nucleus. Nature (Lond.) **206**, 487—489 (1965).
— GAYLORD, H. A.: Effects of chromatic adaptation on color naming. Vision Res. **7**, 645—653 (1967).
— YOLTON, R. L.: Distribution of excitation and inhibition in receptive fields of lateral geniculate neurones. Nature (Lond.) **217**, 187—188 (1968).
— — Visual sensitivity and color vision in ground squirrel. Vision Res. **11**, 511—537 (1971).
JACOBSON, M.: Spectral sensitivity of single units in the optic tectum of the goldfish. Quart. J. exp. Physiol. **49**, 384—393 (1964).
JUDD, D. B.: Color perceptions of deuteranopic and protanopic observers. J. Res. nat. Bur. Stand. **41**, 247—271 (1948).
KADOYA, S., MASSOPUST, L. C., JR.: Unit activity in the extrageniculate striate system of the squirrel monkey. Anat. Rec. **166**, 327 (1970).
KLÜVER, H.: Visual functions after removal at the occipital lobes. J. Psychol. **11**, 23—45 (1941).
KÖNIG, A.: Über „Blaublindheit". Sitzb. Akad. Wiss. Berlin, 718—731 (1897).
KUFFLER, S. W.: Discharge patterns and functional organization of mammalian retina. J. Neurophysiol. **16**, 37—68 (1953).
LENNOX-BUCHTHAL, M. A.: Some findings on central nervous system organization with respect to color. In: JUNG, R., KORNHUBER, H. H. (Eds.): The Visual System: Neurophysiology and Psychophysics. Berlin-Göttingen-Heidelberg: Springer 1961.
— Single units in monkey, *Cercocebus torquatus atys*, cortex with narrow spectral responsiveness. Vision Res. **2**, 1—15 (1962).
LIE, I.: Dark adaptation and the photochromatic interval. Doc. Ophthalm. **17**, 411—510 (1963).

MARKS, W. B.: Visual pigments of single goldfish cones. J. Physiol. (Lond.) **178**, 14—32 (1965).
— DOBELLE, W. H., MACNICHOL, E. F., JR.: Visual pigments of single primate cones. Science **143**, 1181—1183 (1964).
MARROCCO, R. T.: Maintained discharge characteristics and receptive field organization of optic tract fibers and LGN cells in the monkey. In preparation. (1971).
MASSOPUST, L. C., JR., WOLIN, L. R., KADOYA, S.: Differential color responses in the visual cortex of the squirrel monkey. Vision Res. **9**, 465—474 (1969).
MEAD, W. R.: Analysis of the receptive field organization of macaque lateral geniculate nucleus cells. Doctoral dissertation, Indiana University, Bloomington, Indiana, 1967.
MEYER, D. R., MILES, R. C., RATOOSH, P.: Absence of color vision in cat. J. Neurophysiol. **17**, 289—294 (1954).
MICHAEL, C. R.: Receptive fields of single optic nerve fibers in a mammal with an all-cone retina. III: Opponent color units. J. Neurophysiol. **31**, 268—282 (1968).
— Dual opponent-color cells in the lateral geniculate nucleus of the ground squirrel. J. Amer. Physiol. **57**, 254 (1971).
MOTOKAWA, K., TAIRA, N., OKUDA, J.: Spectral responses of single units in the primate visual cortex. Tohoku J. exp. Med. **78**, 320—327 (1962).
PEARLMAN, A. L., DAW, N. W.: Opponent color cells in the cat lateral geniculate nucleus. Science **167**, 84—86 (1970).
PETTIGREW, J. D., NIKARA, T., BISHOP, P. O.: Responses to moving slits by single units in cat striate cortex. Exp. Brain Res. **6**, 373—390 (1968).
PITT, F. H. G.: The nature of normal trichromatic and dichromatic vision. Proc. Roy. Soc. Lond. **132B**, 101—117 (1944).
POGGIO, G. F., MANSFIELD, R. J. W., SILLITO, A. M.: Functional properties of neurons in the striate cortex of the macaque monkey subserving the foveal region of the retina. Paper presented at first annual meeting. Society for Neuroscience 1971.
POLSON, M. C.: Spectral sensitivity and color vision in Tupaia glis. Doctoral dissertation, Indiana University, Bloomington, Indiana (1968).
POLYAK, S.: The retina. Chicago: University of Chicago Press 1941.
— The vertebrate visual system. Chicago: University of Chicago Press 1957.
RODIECK, R. W., STONE, J.: Analysis of receptive fields of cat retinal ganglion cells. J. Neurophysiol. **28**, 833—849 (1965).
RUSHTON, W. A. H.: A cone pigment in the protanope. J. Physiol. (Lond.) **168**, 345—359 (1963).
— A foveal pigment in the deuteranope. J. Physiol. (Lond.) **176**, 24—37 (1965).
SECHZER, J. A., BROWN, J. L.: Color discrimination in the cat. Science **144**, 427—429 (1964).
SHIPLEY, T., JONES, R. W., FRY, A.: Evoked visual potentials and human color vision. Science **150**, 1162—1164 (1965).
SNYDER, M., KILLACKEY, H., DIAMOND, I. T.: Color vision in the tree shrew after removal of the posterior neocortex. J. Neurophysiol. **32**, 554—563.
STILES, W. L.: Colour vision: the approach through increment threshold sensitivity. Proc. nat. Acad. Sci. (Wash.) **45**, 100—114 (1959).
SVAETICHIN, G.: Spectral response curves from single cones. Acta physiol. scand. **39**, (Suppl. 134), 17—46 (1956).
— MACNICHOL, E. F., JR.: Retinal mechanisms for chromatic and achromatic vision. Ann. N.Y. Acad. Sci. **74**, 385—404 (1958).
WAGNER, H. G., MACNICHOL, E. F., WOLBARSHT, M. L.: Functional basis for "on" center and "off" center receptive fields in the retina. J. Opt. Soc. Amer. **53**, 66—70 (1963).
WALD, G.: The receptors of human color vision. Science **145**, 1007—1016 (1964).
WEISKRANTZ, L.: Contour discrimination in a young monkey with striate cortex ablation. Neuropsychologia **1**, 145—164 (1963).
WESTHEIMER, G., CAMPBELL, F. W.: Light distribution in the image formed by the living human eye. J. Opt. Soc. Amer. **52**, 1040—1045 (1962).
WIESEL, T. N.: Receptive fields of ganglion cells in the cat's retina. J. Physiol. (Lond.) **153**, 583—594 (1960).
— HUBEL, D. H.: Spatial and chromatic interactions in the lateral geniculate body of the rhesus monkey. J. Neurophysiol. **29**, 1115—1156 (1966).

Willmer, E. N.: A physiological basis for human color vision in the central fovea. Doc. Ophthalm. **9**. 235—313 (1955).

Wolin, L. R., Massopust, L. C., Jr., Meder, J.: Differential color responses from the superior colliculus of squirrel monkeys. Vision Res. **6**, 637—644 (1966).

Wright, W. D.: Researches on normal and defective colour vision. London: Henry Kempton 1947.

Yund, E. W.: A physiological model of color and brightness contrast. Doctoral dissertation, Northeastern University, Boston. Massachusetta (1970).

Chapter 4

Neurophysiology of Binocular Single Vision and Stereopsis

By

PETER O. BISHOP, Canberra City (Australia)

With 16 Figures

Contents

Introduction

In the last few years observations made from single cells in the striate cortex have revealed neural mechanisms which doubtless underlie many of the phenomena of binocular vision. These new observations will be reviewed and a neurophysiological theory of binocular single vision and depth discrimination, already outlined elsewhere (BARLOW, BLAKEMORE, and PETTIGREW, 1967; NIKARA, BISHOP, and PETTIGREW, 1968; PETTIGREW, NIKARA, and BISHOP, 1968 a and b; JOSHUA and BISHOP, 1969) will be further elaborated below. Before proceeding with this task, it will be instructive to make brief reference to some of the evolutionary aspects of binocular vision and, prompted by our neurophysiological observations, to suggest new interpretations of some features of the evolution of single vision and depth discrimination. In addition it will be necessary to consider the psychophysics of binocular vision since it is the goal of the sensory physiologist to interpret psychophysics in neurological terms. In sharp contrast to form perception and pattern recognition, there is available in relation to binocular vision a well-structured, relatively coherent and widely-accepted body of psychophysical observation and theory, suggesting at once that the problems concerned are more tractable and perhaps less complex than those for form perception. A further matter to be considered is the role of eye movements and their co-ordination in the binocular process since vision is pre-eminently a sensorimotor activity.

I. Properties of Binocular Vision

1. Binocular Single Vision and the Evolution of Stereopsis

Animals with a binocular field of view are to be found throughout the vertebrate series and there has been a constant tendency to enlarge the binocular field at the expense of the monocular. From being initially lateral, a frontal eye position has been independently evolved several times (WALLS, 1942). However binocular singleness of vision, in itself, confers no benefits on the animal; it is a problem to be overcome rather than an advantage to be gained. Binocular singleness of vision is the price that has to be paid for the evolutionary advantage of some form of visual depth discrimination based on binocular parallax. Thus in man, with the exception of stereopsis, seeing with two eyes is marginally, if any, better than seeing with one – absolute threshold, differential threshold and visual acuity are about the same. Of course, binocular vision and singleness of vision are not necessarily equivalent, though it is hard to believe that binocular vision wherever it has evolved, is ever diplopic under normal circumstances. We must conclude therefore, from the repeated appearance of binocular vision in evolution and the absence of benefit from single vision as such, that stereopsis has evolved a number of times but, as we shall see, its neural basis in the submammalian vertebrate is almost certainly of a radically different kind than that in the higher mammals.

All sub-mammalian vertebrates have a total decussation of the optic nerve fibres at the chiasma so that binocular single vision and stereopsis are not dependent upon the type of partial chiasmal decussation to be found in the higher mammals – total decussation is, therefore, compatible with both singleness of vision and binocular depth discrimination. Among the mammals only the carnivores, lower primates and man have developed binocular vision to any extent and, within this group, the redistribution of the optic nerve fibres at the chiasma proceeds in step with the increasingly frontal position of the eyes. As WALLS (1942) points out, such a partial decussation permits the topographical projection onto the visual centres of a system of corresponding retinal points. However WALLS, impressed by the probability that many a lower vertebrate with a total decussation at the chiasma has both binocular single vision and depth discrimination, considered that, in the carnivores and primates, the bi-retinality, as such, plays no particular part in single vision and stereopsis. In his view, the value of partial decussation is to be sought in the realm of motor activity where it might serve to facilitate the precise motor co-ordination of the two eyes required in binocular vision. While not minimizing its importance for motor control, the contrary suggestion will now be made, namely that the partial decussation has evolved primarily in the interests of stereopsis and that the neural mechanisms that permit a fine discrimination of visual depth also provide the basis for the binocular control of eye movements.

Since the recognition of form is not an essential prerequisite for stereopsis, JULESZ (1964, 1965 a) has argued that the neural mechanisms for binocular depth discrimination in man come into play at a very early stage in the central processing that subsequently leads to the final phases of form and pattern perception. As the pathways from the two eyes preserve their identity through the lateral geniculate nucleus, the most probable locus of the neural mechanisms for binocular depth discrimination is the geniculo-cortical arrival layer (Layer IV) in the striate cortex. It will be shown below that the neural mechanisms concerned operate on a mosaic, point-by-point, basis in their assessment of retinal image disparity, again in keeping with the idea that the ultimate discrimination of form has yet to take place. The proposal will also be made that the information from corresponding small retinal areas, on coming together on single cells in the striate cortex, has assigned to it a depth value that is preserved in the subsequent stages of the analysis of form.

The partial decussation at the chiasma is the most efficient arrangement for ensuring that the optic nerve fibres from the corresponding small regions of the two retinas are brought together at the earliest possible stage for the point-by-point data analysis underlying depth discrimination. It is difficult to imagine a comparable alternative based on a total decussation at the chiasma. Furthermore it is possible that the partial decussation has evolved in preference to the total decussation because it is only in this way, on the basis of a mosaic, point-by-point arrangement of corresponding points, that a high level of stereoscopic acuity can be reached in binocular single vision without, at the same time, losing any of the visual acuity that has been achieved for monocular vision. Stereoscopic acuity in man is of the same order of magnitude as monocular vernier acuity. Total decussation at the chiasma suggests that, for the animal who possesses it, the neural analysis of form precedes stereopsis, the singleness of vision involving the conjunction of forms rather than the correlation of points, the quality of the stereopsis being thereby reduced.

2. Psychophysics of Binocular Vision

Both the concept of the horopter and of corresponding retinal points were developed soon after Kepler had established the fact of the retinal image at the beginning of the 17th Century (Boring, 1942), and, ever since that time, the development of these concepts has been the mainstay of our understanding of the phenomena of binocular single vision and stereopsis. In its original definition, the horopter was the locus of all points seen as single in the binocular visual field for a constant position of the eyes and the concept of corresponding retinal points provided the basis for the singleness of vision. Two retinal points, one in each eye, are said to be corresponding when they have a common visual direction and objects whose images fall on the corresponding points will then be seen as single. Objects behind and in front of the horopter will be seen double. The concept of the horopter can be restated as the locus of points whose images fall on corresponding points of the retinas for a constant position of the eyes. One of the first attempts to describe the true form of the horopter was the geometrical construction due to Vieth (1818) and Müller (1826), the horopter of the fixation plane being the circle that passes through the fixation point and the optical centres of the two eyes (Fig. 1A). It was soon appreciated that the empirical horopter does not coincide with the Vieth-Müller circle, though the Hering-Hillebrand horopter deviation was only so-named at a much later date. Our present detailed knowledge of the empirical horopter is of relatively recent origin (cf. Ogle, 1950). For a viewing distance within about 1 metre, the horopter in the fixation plane is concave towards the observer and lies between the Vieth-Müller circle and the fronto-parallel plane through the fixation point (Fig. 1B). However the phenomenon of

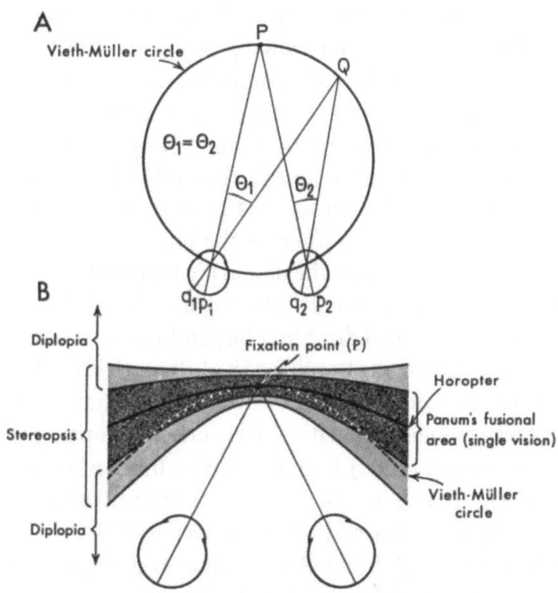

Fig. 1.A The Vieth-Müller horopter circle. The images of points P and Q fall on geometrically corresponding points in the two retinas. B Diagram showing relative positions of the Vieth-Müller circle and the empirical horopter, the overlapping areas for stereopsis and for binocular single vision and the regions where diplopia begins

"Panum's fusional area" introduces a complicating factor since it makes the horopter only a special plane in a three-dimensional region of single vision and we must now discuss this complication in somewhat more detail. The phenomenon of binocular parallax had been known since ancient times, but WHEATSTONE (1838), by his invention of the stereoscope, first recognized that the disparate retinal images provide the cue to binocular depth perception. He also recognised that his observations, by their implication that disparate retinal points are being stimulated, contradicted the theory of corresponding points as a basis for the singleness of binocular vision. To take account of WHEATSTONE's discovery, PANUM (1858) restated the theory of corresponding points by proposing that for any point in one eye there is a small circle or area of points on the other retina, stimulation of which will cause fusion of the inputs (cf. Fig. 5). Although PANUM's original statement was in terms of retinal elements, the name Panum's fusional area subsequently became applied to the area or region in space in which objects are seen singly (Fig. 1B). However, the concept of an area of fusion leaves unresolved the basic issue of how it is that, for a given fixation point, two retinal elements, one in each eye can, at the same time, be both disparate to take account of binocular parallax and yet corresponding in the interests of single vision. A further complication arises from the fact that stereoscopic depth perception is still possible even if the disparate retinal images are not fused. There is a region in space, both in front of and behind Panum's fusional area, in which stereoscopic depth is experienced in association with diplopia (Fig. 1B). Thus fusion of the disparate images is not a prerequisite for stereoscopic depth perception.

Horizontal disparity gives rise to stereopsis, vertical disparity does not. It is perhaps not surprising that a stereoscopic mechanism using vertical disparities has not developed since, with the horizontal disposition of the eyes, these disparities do not occur in the straight-ahead viewing position except at very close range. This directional anisotropy further distinguishes the stereoscopic and fusional mechanisms. Vertical disparities are fused as well as horizontal and, if care is taken to avoid vergence eye movements in the interests of fusion, the width of Panum's area for central vision is the same in the vertical as in horizontal direction, about 15 min arc in each case (BRECHER, 1942; MITCHELL, 1966). The dimensions of Panum's fusional area are taken either as the largest retinal disparity that still gives rise to a single fused image or as the disparity threshold for diplopia. Although vertical disparities do not give rise to stereopsis, the depth experience is still possible in the presence of vertical doubling beyond the vertical extent of Panum's area, i.e. with combined horizontal and vertical disparities. Thus the region over which stereopsis occurs is much larger, in all directions than that for single vision. With increasingly large image disparities, the sense of depth gradually fades and finally ceases to exist. OGLE (1962) found that the limit of disparity permitting a qualitative sense of the depth of two double images was about 15 min arc at the maculas and nearly 3.5° at a retinal eccentricity of 6°.

Though the phenomena of binocular rivalry and suppression must find an important place in any theory of binocular single vision and stereopsis, TREISMAN (1962) has drawn attention to the many gaps in our knowledge of psychophysics in this area. Little is known about what aspects of the stimulus complex are inhibited under different conditions and whether different small features of the stimulus are always suppressed at random in rivalry or whether the input to an eye can be suppressed as a whole. Little is known about the criteria which determine whether the inputs from corresponding regions of the two eyes are sufficiently similar to fuse or sufficiently different to compete.

An important new development in the study of binocular vision has been the use by Julesz (1964) of random-dot targets as stereo pairs. To monocular inspection the stereo targets appear merely as random dots without any evidence of shape or regular pattern and the three dimensional form only emerges in binocular vision after fusion of the stereo pair has taken place. Furthermore both visual texture discrimination and depth perception appeared to use some, at least, of the same basic mechanisms. Thus it seems that both the fusional and the stereoscopic mechanisms belong in the chain of events leading up to pattern recognition, but come into play at a very early stage in the latter process. These observations encourage the neurophysiologist to expect a solution to the problems of binocular single vision and depth discrimination without coming up against the much more difficult problem of form perception. Furthermore an understanding of the means by which particular features of the environment are selected for fusion and analyzed for depth information will doubtless provide valuable clues to the later stages in the brain by which form and pattern are distinguished.

The use of stabilized retinal images continues to be a fruitful source of important new observations that must eventually find a place in our neurophysiological theory. Using binocularly stabilized retinal images of random-dot targets, Fender and Julesz (1967), after first bringing the right and left images within Panum's fusional area, were then able to pull the images slowly and symmetrically apart on the retinas by about 2° in the horizontal direction without loss of stereopsis or fusion.

3. Binocular Eye Movements

The precise control of binocular eye movements is an essential component in the development of stereopsis and the precision of this control is probably one of the limiting factors in the quality of the depth discrimination. Any theory of binocular vision must take into account both the fusional movements that establish single vision and the spontaneous eye movements of steady bi-fixation (Alpern, 1962). Fluctuations in retinal image disparity pose a problem for stereoscopic acuity analogous to that of the eye movements of fixation for monocular visual acuity. Just as the possibility of fusion over small but finite corresponding areas of the retina broadened the concept of corresponding points, a further broadening may be needed to take account of the varying binocular disparity due to the imperfect correlation between the motions of the two eyes.

While Dove's (1841) observation that stereoscopic depth could be experienced during the flash of an electrical spark demonstrated that eye movements are not necessary for stereopsis, nevertheless convergence is needed to bring the retinal images within the critical limits of disparity. Retinal image disparity is the essential stimulus for fusional movements and neither diplopia nor any stereoscopic effect resulting from disparity are necessary. Thus fusional movements occur without a break in single vision when a very weak prism is placed before one eye while binocularly-fixating a given object. The random dot patterns of Julesz

(1964), for which there is no monocular recognition of shape, also demonstrates that the perception of double images is not a prerequisite for fusion. In this case, there is no difference between fusion thresholds and thresholds for stereopsis.

During steady fixation three categories of eye movement are generally recognised: *Flicks*, which are sudden changes in fixation; *Drifts*, which are the slow irregular movements between two successive flicks and *Tremor* which is a relatively high frequency low amplitude oscillation superimposed on the drifts.

In Table 1, modified from PRITCHARD and HERON (1960), the monocular eye movements of fixation in man and cat are compared. Most of the values for the cat are due to PRITCHARD and HERON (1960) though the few observations of HEBBARD and MARG (1960) are in general agreement particularly with respect to the relative infrequency of flicks. In monocular fixation in man, the effect of all involuntary eye movements is to keep the image of the fixation point within a retinal region about 20 min arc in diameter (DITCHBURN and GINSBORG, 1953). Any position within this area is almost equally acceptable. The location of the visual axis is therefore a two-dimensional probability distribution.

The eye movements of binocular fixation have not been studied to the same extent as those for the one eye. In binocular fixation, the flicks always occur in both eyes simultaneously and are of approximately the same direction and magnitude. In general, flicks and drifts occur in opposite directions but the drifts are less-well correlated between the two eyes than are the flicks. Because of inequalities in the motions of the eyes, there is a constantly varying amount of binocular disparity. The flicks tend to correct fixational errors and recently ST-CYR and FENDER (1969) have shown that the drifts also have a corrective function though to a lesser extent than the flicks. ST-CYR and FENDER (1969) further considered whether the correction for binocular disparity could be accounted for by two monocular fixation mechanisms or whether an additional contralateral

Table 1. Comparison of monocular eye movements of fixation in cat[a] and man[b]

Movement	Cat	Human
Flick		
Mean amplitude	35 min arc	6.5 min arc
Interflick period	very large	0.83 sec (mean)
Duration	40 msec	40 msec (maximum)
Mean velocity	13°/sec	10°/sec
Drift		
Mean amplitude	25 min arc approx.	3.5 min arc
Maximal amplitude	2°	9.0 min arc
Mean velocity	30 min arc/sec	5.0 min arc/sec
Tremor		
Mean amplitude	24 sec arc	18 sec arc
Median amplitude	18 sec arc	18 sec arc
Maximal frequency	65 Hz	150 Hz

[a] PRITCHARD and HERON (1960). HEBBARD and MARG (1960).

[b] YARBUS (1967). DITCHBURN and FOLEY-FISHER (1967). FENDER and JULESZ (1967).

(binocular) control was required. They concluded that most of the corrections could be accounted for by two separate monocular mechanisms. There was no evidence that binocular vertical discrepancy of the visual axes triggers corrective movements. On the other hand vergence error is a stimulus for correction by drifts, but not by flicks.

The recent work of FENDER and his colleagues confirms the earlier observations of DITCHBURN and GINSBORG (1953) that the binocular disparities during steady bi-fixation are usually within 15 min arc. Depending upon the nature of the fixation target the mean retinal disparity caused by flicks is about 2 min arc while that for the drifts is slightly larger (ST-CYR and FENDER, 1969). However some flicks may produce much greater disparities extending beyond 20 min arc (FENDER and JULESZ, 1967). The errors of convergence or divergence caused by the drifts develop at slow rates, of the order of 1 min arc/sec.

II. Functional Architecture and Receptive Field Topography

1. Neurophysiology: Basic Concepts

Our present understanding of the neurophysiological basis of binocular single vision and stereopsis began with HUBEL and WIESEL's (1959, 1962) observation that single units in the striate cortex could be driven from both eyes and that it was possible to plot separate receptive fields for each eye. Before discussing their observations in more detail it will be helpful to define some of the terms in common usage in this work and to provide an outline of the experimental procedures that are generally adopted. Most of the observations to which reference will be made have been obtained from the anaesthetized cat or monkey, the general experimental arrangement being shown in Fig. 2B. The animal, rigidly held in a stereotaxic frame, faces a tangent screen placed at 1 to 2 metres in front of the eyes. Total paralysis of the animal and various other procedures are used to reduce eye movements to an acceptable minimum so that the retina can be regarded as fixed with respect to the tangent screen. Single neurons are recorded extracellularly by microelectrodes inserted into the striate cortex. The visual pathways from the two eyes retain their identity through the lateral geniculate nucleus (LGN) and finally converge onto the cells in layer IV of the striate cortex. Although there is now direct evidence for widespread binocular interaction in the LGN, the receptive field for the non-dominant eye being usually only inhibitory (SANDERSON, DARIAN-SMITH, and BISHOP, 1969), its significance for binocular vision is not sufficiently clear to warrant detailed discussion here. One possible function concerns the suppression of weak activity associated with double images such as might occur when receptive fields related to non-corresponding retinal areas or elements are stimulated by the one object.

The receptive field of a visual neuron is the region of the visual field (tangent screen) over which the firing of that cell can be influenced (HUBEL and WIESEL, 1962). Because of the experimental difficulties in determining the limits of influence, the concept of a *minimal response field* (BARLOW, BLAKEMORE, and PETTIGREW, 1967) has been introduced and unless otherwise indicated in this paper, the term receptive field will be used in this sense. The procedure adopted for plotting the minimal response field is dictated by the fact that the great majority of striate neurons respond best to the movement of a line stimulus (slit of light, dark bar or light-dark border) having a particular orientation and a preferred direction of movement (Fig. 2C). Unless otherwise qualified, the term orientation refers to the optimally-orientated slit or line stimulus, the convention being that it is specified by the clockwise angle that the slit or line makes with the zero (vertical) meridian in the visual field (Pettigrew, Nikara, and Bishop, 1968 a). Usually the preferred direction of movement of the stimulus is at right angles to its orientation. In plotting the minimal response field, the stimulus orientation is first established and then two lines are marked on the screen, indicating the beginning and end of the response to an optimally-oriented moving slit, bar or edge, whichever is the most effective. These two lines form the primary borders of the plot.

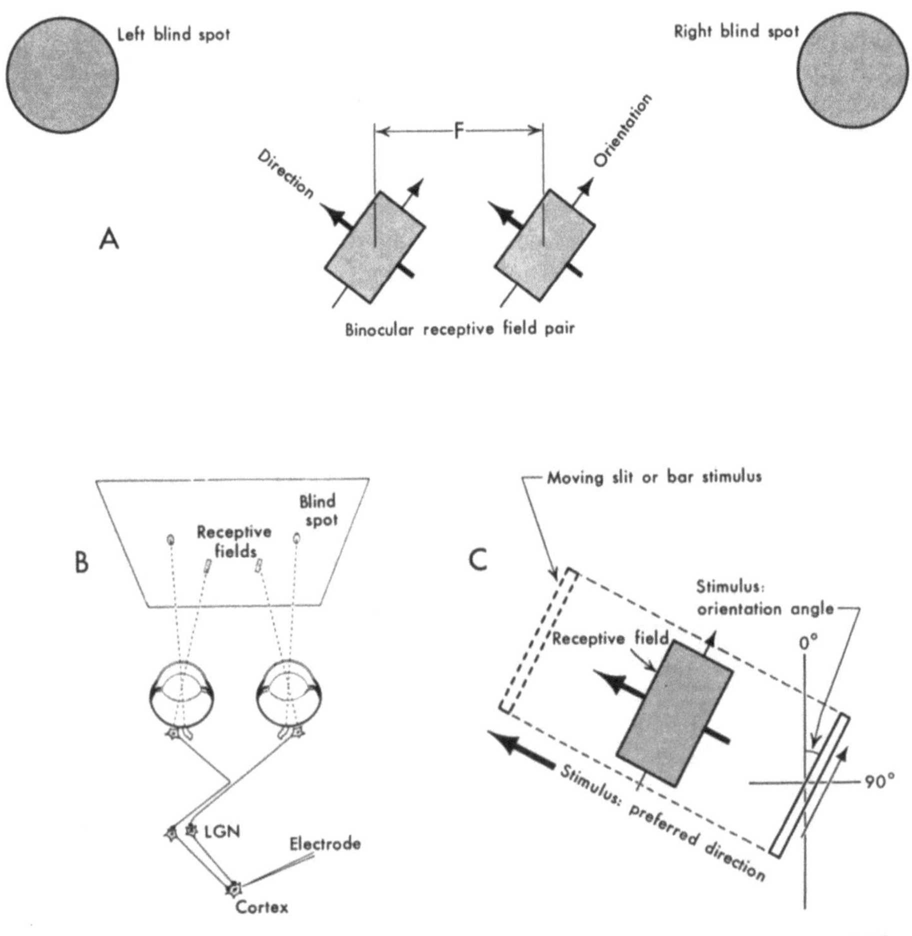

Fig. 2.A Diagram showing typical positions on a tangent screen of the two receptive fields of a binocularly-activated striate neuron in relation to their respective blind spots when the extra-ocular muscles are paralyzed. B Same as for *A* but also showing the neural pathways form the two eyes converging on a single striate neuron. C Diagram to illustrate the definition of the term "orientation" in respect to a line stimulus

Lateral borders, at right angles to the primary borders, are determined by shifting laterally the continuously oscillating stimulus, maintaining its optimal orientation, until the end of the target has moved out of the field and no consistent response can be elicited. Repeating this procedure on the other side completes the plot. The receptive field is probably more likely to be irregularly elliptical in shape, the rectangular outline of the minimal response field being simply a consequence of the method used to determine it. Furthermore it must be stressed that the area of influence, either inhibitory or subthreshold, which properly belongs to the receptive field extends well beyond the minimal response field. However the plots made in this way give a good guide to the centre of the area of influence on the tangent screen, the position of the receptive field being taken as its geometrical centre.

Because of the uncertainty regarding the location of the area centralis in the cat, the position of each receptive field is first measured with respect to the blind spot of the same side (Fig. 2A), the co-ordinates being then transformed to be centred on the estimated position of the fixation point (NIKARA, BISHOP, and PETTIGREW, 1968) and expressed in terms of a horizontal angle (azimuth) to the right and left of the fixation point and a vertical angle (elevation) above and below the fixation point.

When the extraocular muscles are paralyzed, the eyes take up a fixed position in divergence and hence the two receptive fields of a binocularly-activated cortical neuron become horizontally separated on the tangent screen by the distance F in Fig. 2A. Under normal circumstances, the eyes would re-converge and, if the cat were looking at some point on the tangent screen, it is reasonable to suppose that the two receptive fields would then be superimposed. The extent to which this is the case will be considered in some detail below since it is of the utmost importance in binocular vision.

2. Neural Basis of Stereopsis: Essential Requirements

The neural mechanisms underlying binocular depth discrimination must provide two essential steps if they are to make use of the cues contained in the retinal image disparities. The first step is the selection from the retinal images in the two eyes those separate parts that are images of the same feature in the visual field. Thus each separate part of the retinal image in the one eye must be accurately paired with its fellow in the image in the other eye. The evidence of psychophysics suggests that this is done on a mosaic, feature-by-feature basis. The psychophysical evidence also suggests that wherever retinal rivalry occurs there may be active suppression of one or other member of the incongruous pair of image features, the selection again being for specific features on a local, region-by-region basis. The second step in the analysis of image disparities is the estimation of visual depth by assessing the small displacements in the relative positions of the paired image features that result from binocular parallax. A basis for the identification of the paired parts of the two retinal images is provided by the fact that each binocularly-activated neuron has the same specific stimulus requirements (trigger features) for the receptive fields of the two eyes. Furthermore, as BARLOW et al. (1967) have pointed out, it is unlikely that there will be many identical trigger features in any small part of the monocular image so that similar features located in approximately the same region of the image of each eye can be safely assumed to belong to the same feature of the object. The example they give is that a black line of a particular orientation in one image should be paired with the black line of the same orientation at the most nearly corresponding position in the other image because both are likely to be images of the same feature of the object. The phenomenon of receptive field disparity to be described below provides the necessary mechanism for the second step by which retinal image disparities are assessed.

3. Receptive Field Properties of Binocularly-Activated Neurons

Cells which can be independently driven from both eyes always have receptive fields in approximately corresponding positions in the contralateral hemifield. The two fields have the same size and arrangement so that whatever stimulus is the most effective in one eye – in form, orientation, and direction and rate of movement – is also the most effective in the other eye (HUBEL and WIESEL, 1962, 1965 a; BARLOW, BLAKEMORE, and PETTIGREW, 1967; PETTIGREW, NIKARA, and BISHOP, 1968 b). Almost without exception, the only difference between the two fields as judged by hand plotting relates to the phenomenon of eye dominance which has been studied in detail by HUBEL and WIESEL (1962, 1963, 1965 a and b, 1968). For some cells the responses from the two eyes are about equal, but for others one eye tends to dominate. The distribution of ocular dominance ranges from cells driven only by the contralateral eye, through those driven equally by the two eyes to those driven exclusively by the ipsilateral eye.

Cell Columns. Each small area of the cortex can be subdivided, by sets of vertical partitions extending from surface to white matter, into several independent systems of discrete cell aggregations to which the term "column" has been applied (MOUNTCASTLE, 1957; HUBEL and WIESEL, 1962). Besides being grouped into cell columns according to receptive field orientation, cells in the striate cortex are also aggregated according to ocular dominance although the organization in the cat is less clear than it is in the monkey (HUBEL and WIESEL, 1962, 1968). The cortical parcellation on the basis of ocular dominance seems to be quite independent of columns defined by orientation, the two systems apparently having entirely independent borders. Of the two types of columns, those associated with eye dominance seem to be larger, often including several orientation columns. Similarly within an orientation column there may be more than one region defined by ocular dominance.

4. Binocular Receptive Field Disparity

Maps of the topographical projection of the visual field onto the various cortical areas have been considerably refined by the introduction of single-unit recording techniques. The topographical projection is mapped continuously onto the occipital cortex so that a progressive movement over its surface corresponds to a similar movement in the visual field. However HUBEL and WIESEL (1962, 1968) have shown that the detailed topographical representation does not hold at a microscopic level. Without any net change in receptive field position, successively recorded cells in an electrode track perpendicular to the surface of the striate cortex do, nevertheless, show an irregular variation in receptive field position from cell to cell. This random staggering of receptive field positions for the one cortical locus determines the ultimate grain of the so-called point-to-point projection in the visual system. The irregular variation in monocular receptive field position just described is to be carefully distinguished from the random distribution of binocular receptive field disparities to be considered below, although the two phenomena are undoubtedly closely inter-related.

The cell columns described above are concerned with a particular stimulus parameter (orientation or eye preference) but, as we have seen, precise receptive field position is not one of them. In fact a given small region of the visual field must be represented in a number of different columns defined by orientation or eye preference. Although the cell columns are not organized on a strict topographical basis, HUBEL and WIESEL (1962, 1968) considered the range of the random variations in receptive field position to be observed along the length of a column. In the monkey striate cortex they found that the area covered 2–4 times that of the average receptive field. Taking the average receptive field to be 0.5° across this represents a range of 1.5° for receptive field centres. The comparable estimate for the cat striate cortex is probably slightly larger, possibly 2–3° (HUBEL and WIESEL, 1962). This is in keeping with CREUTZFELDT and ITO's (1968) observation that the distribution of field centres of geniculate fibres entering a minute cortical area were distributed within 2–3° around the field centres of the cortical cells recorded from the same electrode track.

The phenomenon of receptive field disparity forms the keystone of the neurophysiological theory of binocular depth discrimination developed simultaneously by BARLOW, BLAKEMORE, and PETTIGREW (1967) and NIKARA, BISHOP, and PETTIGREW (1968), and further elaborated by JOSHUA and BISHOP (1969) and HENRY, BISHOP, and COOMBS (1969). With normal use of the eyes the two receptive fields of a binocularly-activated striate cell are presumably held in register and the term "receptive field disparity" concerns the extent to which, under conditions of perfectly steady binocular fixation, all the receptive field pairs can be in register at the one time in the one plane. Under experimental conditions, paralysis of the extra-ocular muscles diverges the visual axes and hence separates the receptive fields of the two eyes on the tangent screen. The divergence is usually sufficient to lead to the formation of two separate groups of receptive fields, one for each eye, as units are recorded successively along an electrode track perpendicular to the cortical surface. As pointed out above, there is a random scattering within each group of receptive fields, the mean position of the group being characteristic for the particular cortical locus. Though apparently random for the one eye, it is still possible that the members of each pair of receptive fields are so correlated that the separation between them remains constant over the group. If this were the case, and in the absence of disjunctive rolling of the eyes, superimposition of one pair of receptive fields would entail superimposition of them all. That there is no such correlation between the receptive fields of the two eyes is readily seen from Fig. 3 (NIKARA, BISHOP, and PETTIGREW, 1968) where the group of receptive fields for the right eye (CONTRA) has been moved bodily downwards and to the left so that the members of pair 5 lie vertically in line one underneath the other. By arbitrarily aligning pair 5, it is easy to see the horizontal disparities of the remaining receptive field pairs. The latter are out of line by varying amounts – only 2 and 8 are reasonably in line and pair 4 is particularly out of correspondence. In a similar way, vertical disparities would be seen by positioning the two groups so that pair 5 lay on the same horizontal line. It should be stressed that the example in Fig. 3 has been selected for illustrative purposes only, the disparities being atypically large.

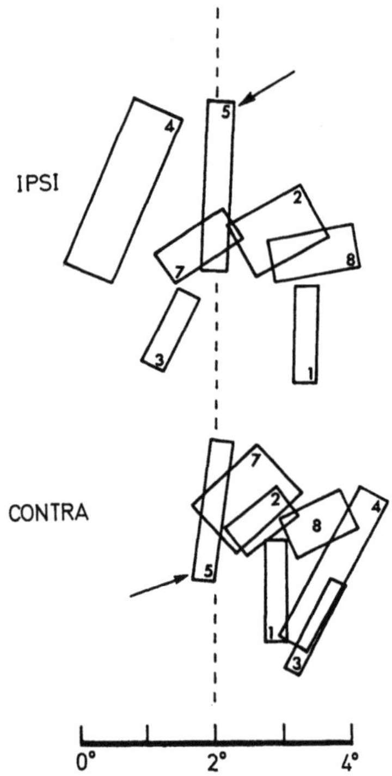

Fig. 3. Plots of receptive field pairs recorded from binocular units in the left striate cortex in the one experiment and arranged so as to illustrate the concept of receptive field disparity. The disparities are atypically large (NIKARA, BISHOP, and PETTIGREW, 1968)

The upper, diagrammatic part of Fig. 4 shows the method used to arrive at a quantitative measure of the receptive field disparities. The receptive fields in the right-eye group have been moved so that their geometrical centres superimpose, the positions of the left-eye members of each pair being adjusted by a similar amount in each case. In this way all the scatter has been transferred to the left-eye group of receptive fields. For diagrammatic purposes, it has been assumed that there is no vertical disparity. The histogram below the left eye in Fig. 4, prepared from data in NIKARA, BISHOP, and PETTIGREW (1968), shows that the horizontal receptive field disparities have an approximately normal distribution about the position of exact correspondence. The vertical receptive field disparities had a similar distribution, the resultant scattergram being shown in Fig. 4B. These observations have been confirmed by JOSHUA and BISHOP (1969) who found that, for the region within 4° of the visual axis, the scatter of the receptive field disparities was about 1.2° (S. D. 0.5°) in both horizontal and vertical directions. In arriving at these estimates careful attention was paid to the various sources of experimental error particularly residual eye movements during paralysis and the reproduceability of repeated plots of the same receptive field.

It should be noted at this stage that the spread of binocular receptive field disparities is of the same order as the spread of the random staggering of receptive field positions for the one cortical locus described above. The suggestion will be made later on that the receptive field disparities arise from the random cross-coupling between the cells in adjacent monocular cell areas in layer IV of the striate cortex.

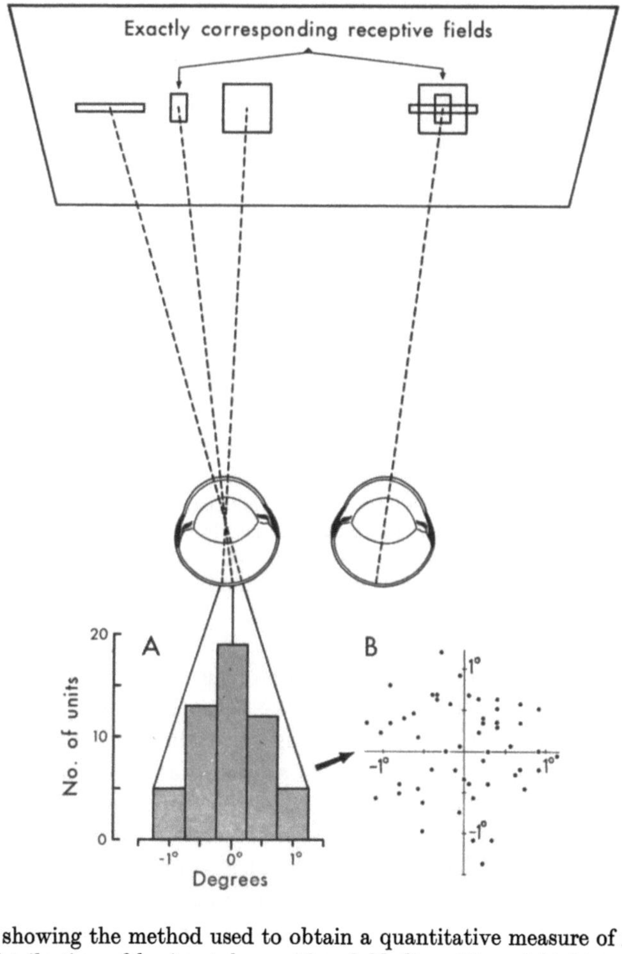

Fig. 4. Diagram showing the method used to obtain a quantitative measure of receptive field disparities. A Distribution of horizontal receptive field disparities of 54 binocular receptive field pairs. B Scattergram of the horizontal and vertical receptive field disparities of the same units as in *A*. Data for A and B from Nikara, Bishop, and Pettigrew, 1968

If a sufficiently large number of receptive field plots had been available, the scattergram in Fig. 4B could have been prepared in another way. A class of binocular receptive fields could have been selected such that all the members for the one eye (right) had the same visual direction and hence fell accurately one on top of the other on the tangent screen. The positions of the receptive fields for the members of the same class but for the other eye (left) could then have been plotted directly to produce the scattergram in Fig. 4B without the need for any adjustments in the receptive field positions. This possibility leads to the general scheme for the binocular receptive field array illustrated in Fig. 5. Five receptive field locations are indicated for the right eye, at each one of which a large series of

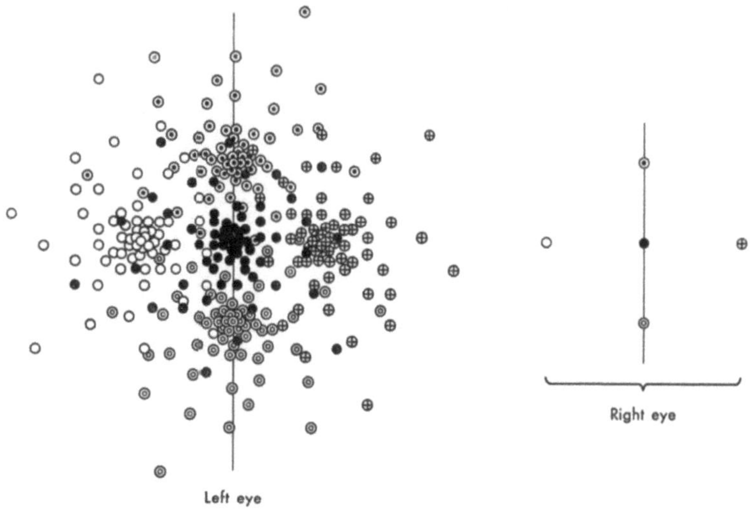

Fig. 5. General scheme (diagrammatic only) for the binocular array of receptive field disparities illustrated by five sets of binocular receptive field pairs with the eyes diverged by paralysis. For the right eye, all the members of each set overlap at a single point in the visual field. For the left eye, the bivariate distributions of the members of each set are centred about points exactly corresponding with their respective points in the hemifield of the right eye

receptive fields are accurately superimposed one on top of the other. The five distributions of the left-eye members of the receptive field pairs are centred on points which are accurately corresponding with their respective points for the right eye. Each distribution presents an approximately normal scattergram for which, in the centre of gaze, the two variables are uncorrelated and the standard deviation of the spread in the horizontal and vertical directions are equal. Each distribution overlaps its neighbours on all sides so extensively that, in nature, in all probability, there is virtually a continuous distribution over the whole visual field though with rapidly declining density as retinal eccentricity increases. Furthermore the two eyes are symmetrically related so that, in the above example, the left eye could have been substituted for the right and vice versa.

The key place of receptive field disparity in the neural basis of binocular depth discrimination is readily appreciated from Fig. 6, the diagrammatic arrangement of receptive fields being the same as those in Fig. 4. In this case, however, convergence of the eyes has enabled the exactly corresponding receptive fields to be superimposed in the middle frontoparallel plane. Once again all the receptive fields for the right eye have the same visual direction. It can be readily seen that the narrow rectangular fields superimpose in the distal fronto-parallel plane whereas the large square fields will fall over one another in the proximal plane. The essential components for the neural mechanism of binocular depth discrimination would be provided if the response from the binocularly-activated striate neuron differed sufficiently depending upon whether its receptive fields were in register or slightly out of register in the plane of the stimulus. This matter will be taken up in detail later on.

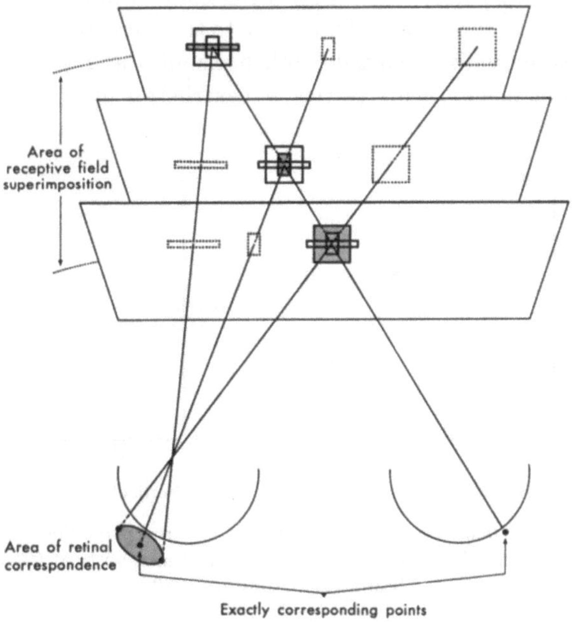

Fig. 6. Diagram showing a rearrangement of the upper part of Fig. 4 to indicate the way in which the phenomenon of receptive field disparity might form the basis of a neural mechanism for binocular depth discrimination

5. Receptive Field Disparity and Retinal Eccentricity

With increasing retinal eccentricity there is a gradual decrease in the separation, F, of the receptive fields of a binocular unit (Fig. 7), as well as progressive changes in the local receptive field disparities. The decrease in F has two important consequences. Firstly, the slope of the change with eccentricity determines the shape of the horopter surface (Fig. 9) and secondly, any distribution of receptive field disparities obtained by pooling over a range of eccentricities will have two components: a systematic element due to the progressive fall in F and a random element for the spread of disparities about the local mean value of F.

It is difficult to plot a sufficient number of receptive field pairs over a range of eccentricities in the one animal so as to obtain a satisfactory relationship between F and eccentricity. Figure 7B, which is based on the pooled data from 19 experiments, shows a decrease in F equivalent to 0.7° of visual angle for a change of 14° from the centre of gaze (JOSHUA and BISHOP, 1969). Beyond 16° of horizontal eccentricity, F decreases rather more rapidly. There is also a decrease in F with eccentricity in the vertical direction (Fig. 7C).

In addition to the change in F, JOSHUA and BISHOP (1969) have shown that horizontal receptive field disparities increase and vertical disparities decrease with increasing horizontal eccentricity. Up to 16° of retinal eccentricity the standard deviation of the spread of horizontal disparities increased from 0.5° to 0.9° while that for the vertical disparities decreased from 0.5° to 0.35°. The reduction in the vertical disparity is the more significant in that it took place despite a significant increase in experimental error due to the increase in size of the receptive fields in the periphery. By contrast, both horizontal and vertical receptive field disparities did not change with vertical eccentricities up to 20° from the visual axis.

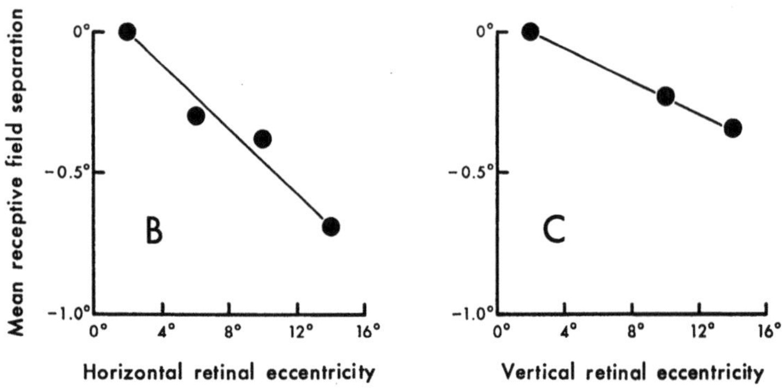

Fig. 7. Graphs showing decrease in mean separation of receptive-field pairs of binocularly-activated striate neurons with increasing horizontal (B) and vertical (C) retinal eccentricity (JOSHUA and BISHOP, 1969)

6. Outline of a Neurophysiological Theory

The following outline is a modified version of a general theory but forward by JOSHUA and BISHOP (1969).

A number of terms need precise definition and must be used with care if confusion is to be avoided. Receptive field pairs can be either corresponding or disparate only if they converge on a single cortical neuron. Furthermore, only a relatively small proportion of binocular receptive field pairs can be exactly corresponding. Consider all those receptive fields for the one eye, say the right, that have a common visual direction: their fellow receptive fields for the other eye will have a two dimensional gaussian distribution (Figs. 4 and 5). However, by definition, only some of these receptive field pairs will be corresponding, namely those whose receptive field for the left eye lie at the centre (mean) of the gaussian distributions. By definition receptive field pairs that are not corresponding show receptive field disparity. It is the functional architecture of the retino-cerebral connections which determine whether or not the two receptive fields of a binocularly activated neuron are corresponding or disparate. Hence, in order to be able to say whether a particular receptive field pair in a given animal is corresponding or disparate it would be necessary to have a sufficiently large distribution of receptive field pairs in the same vicinity. Unfortunately, under experimental conditions, when binocular interaction on single cortical neurons is being studied, this information will usually not be available.

In regard to a binocularly activated neuron, the terms 'corresponding' and 'disparate' should not be used to refer to the locations of the two receptive fields with respect to one another. It is particularly important that the concepts of correspondence and disparity should be distinguished from the idea of receptive fields being in register or out of register (aligned or misaligned). When receptive fields are precisely superimposed on the tangent screen they are said to be in register and when they are offset or separated from one another they are out of register or misaligned. Whether or not they are in register depends upon the direction of gaze of the two eyes and the prism settings when prisms are used. Thus for a given position of the eyes and for a particular depth in space (tangent screen), corresponding receptive fields may be out of register and disparate receptive fields may be in register. The concept of receptive fields being in register is discussed in more detail below (see Part III). It implies a condition of maximal binocular facilitation rather than the coincidence of minimal receptive field centres. The latter, however, provides a good approximation to the position for maximal binocular facilitation.

The visual axis may be defined as the visual direction with the greatest concentration of receptive fields and the two eyes are directed towards a particular fixation point when the maximum number of corresponding receptive field pairs are in register at that point. In normal binocular fixation, the residual eye movements of steady gaze are centred on this position of maximum receptive field alignment. Any disjunctive movements away from this position will lead to a reduction in binocular facilitation and a fall in the spike discharge from striate neurons. The flicks, and to a lesser extent the drifts, tend to restore maximal receptive field alignment. The horopter is defined as the surface in the visual field for which corresponding receptive fields are in precise register. The horopter is therefore the surface which contains the fixation point and for which, irrespective of retinal eccentricity, the greatest number of receptive fields are in register. The condition of maximal receptive field aligment over the retina generally is required for the integrity of the horopter and it is this condition which, in a manner analogous to the direction of gaze, determines the relative orientation of the eyeballs with respect to one another. The horopter surface, determined and maintained in the interests of maximal receptive field alignment, provides a surface of reference in relation to which objects are localized in binocular depth discrimination. Object points which stimulate the one neuron via its two receptive fields will be treated by the brain as single even though the retinal images may be disparate. The single object point will be localized to a position in space at which the two receptive fields are in precise register.

Whether or not a receptive field pair is in register for a particular depth depends upon the binocular alignment of the eyes at the time. Nevertheless the narrow range of the receptive field disparity distributions means that, for any given fixation point, the visual directions of binocular receptive field pairs will intercept only over a relatively restricted range of depths in space centred on the fixation point. This range of depths is determined almost entirely by the *horizon-*

tal receptive field disparity distributions and is clearly related to Panum's fusional area. Receptive field pairs showing vertical disparity may nevertheless have zero horizontal disparity and so become effectively exactly corresponding when random eye motions displace the visual axis of one eye above that of the other. It is proposed that the primary role of vertical receptive field disparities is to offset the effects of the random eye movements of binocular fixation. Thus receptive field pairs each of which are both vertically and horizontally disparate become essential components in the visual mechanisms underlying both binocular single vision and stereopsis.

A clear distinction must also be drawn between the terms "retinal image disparity" and "receptive field disparity". The former concept is purely one of geometrical optics while the latter is mainly concerned with the organisation of the retino-cerebral pathways by which the cortical neurons are influenced. Fig. 8 illustrates the distinction and enables a further elaboration of the neural theory of binocular single vision and stereopsis. *A*, *B*, *C* and *D* in Fig. 8 are the visual projections of retinal *points* or *elements* onto the tangent screen and as such are the geometrical concepts of the psychophysics of binocular vision. *A* and *B* are the projections of

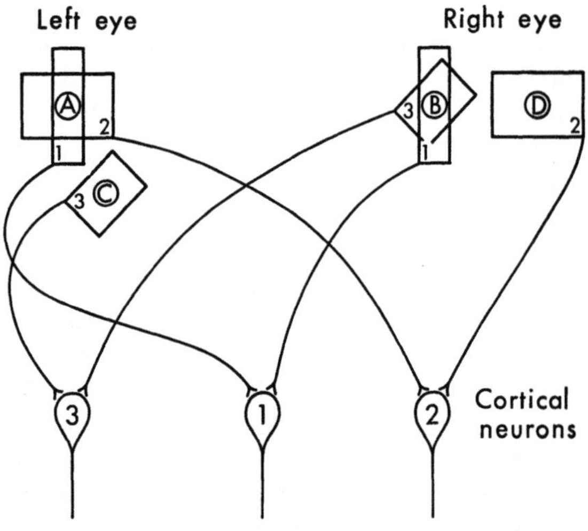

Fig. 8. Diagram illustrating the way in which the phenomenon of receptive field disparity may be used to explain how two retinal elements, one in each eye, can, at the same time and for a given fixation point, be both disparate, to take account of binocular parallax and yet corresponding in the interests of single vision. For details see text (Joshua and Bishop, 1969)

corresponding retinal *elements* and *C* and *D* of disparate *elements*. The receptive fields of striate unit 1, which are corresponding, are accurately in register with *A* and *B*. Units 2 and 3 have disparate receptive fields although one member of each pair is accurately in register with *elements A* and *B* respectively. If we now consider *A* and *B* as coinciding on the tangent screen, *A* can be simultaneously localized both in the direction of *B* by unit 1 and in a direction between *A* and *C* by unit 3. Similarly unit 2 would localize *A* between *B* and *D*. This provides an explanation for the fact that two retinal *elements*, one in each eye, can, at the same time and for a given fixation point, be both disparate, to take account of binocular parallax, and yet corresponding in the interests of single vision. Thus a single visual localization is obtained both by the stimulation of corresponding retinal *elements* (the horopter) and by the stimulation of disparate retinal *elements* (Panum's fusional area and stereopsis).

A binocular unit will be specifically activated from either eye by the same single feature in object space because the two receptive fields have the same highly specific trigger features. The particular neuron will be discharged or have its firing pattern characteristically modified only when the feature having the required stimulus parameters is located in space at the position where the unit's receptive fields superimpose. The stereoscopic assessment of depth depends upon the receptive field disparities of the neurons concerned, the discrimination being made with respect to the horopter as a reference surface.

7. The Horopter

The definition of the horopter in psychophysics depends upon the experimental method used for its determination. Using the nonius method (OGLE, 1962), the horopter may be defined as the locus of object points whose images in the two eyes, by falling on corresponding retinal elements, are localized in the same visual direction. Neurophysiologically, the horopter was defined above as the surface which contains the fixation point and for which, irrespective of retinal eccentricity, the greatest number of receptive fields are in register. By inspection of Fig. 5 it can be seen that the greatest number of receptive fields will be in register in that surface for which each of the series of receptive fields for the right eye superimposes the receptive field partner for the left eye that lies at the centre of its respective distribution. If we assume right-left symmetry, the same horopter surface would provide maximal superimposition of the left-eye series on the centres of the right-eye distributions.

Figure 9A shows the method used for determining the horopter in the cat (JOSHUA and BISHOP, 1969). Two receptive field pairs are shown, the left-eye member of each pair being a uniform gray and the separation, F, for the central pair being greater than the separation, F', for the peripheral pair. Since $F > F'$ in the plane of the tangent screen, the convergence of the eyes required to superimpose the central pair of receptive fields on the fixation point would displace the left-eye member of the peripheral pair to the right of its partner leading to a crossed receptive field disparity. The peripheral pair would then superimpose in a surface somewhat closer to the cat than the fronto-parallel plane through the fixation point. The horopter is determined by considering the mean receptive field separations for groups of receptive fields at each of a series of points of increasing retinal eccentricity (Fig. 7). The mean receptive field separations at each of the points in Fig. 7 are to be regarded as equivalent to the separation in Fig. 5 between a particular series of right-eye receptive fields and the centre of the distribution of the left-eye partners belonging to the same binocular units.

Using the data in Fig. 7, the horopter in Fig. 10 has been drawn to scale for a fixation point at 25 cm. The form of the horopter depends upon the asymmetrical relationship of the visual directions of the receptive fields for the two eyes with respect to retinal eccentricity. In view of the uncertainties regarding the precise relationship between mean receptive field separation and retinal eccentricity however, too much emphasis should not be placed upon the quantitative aspects

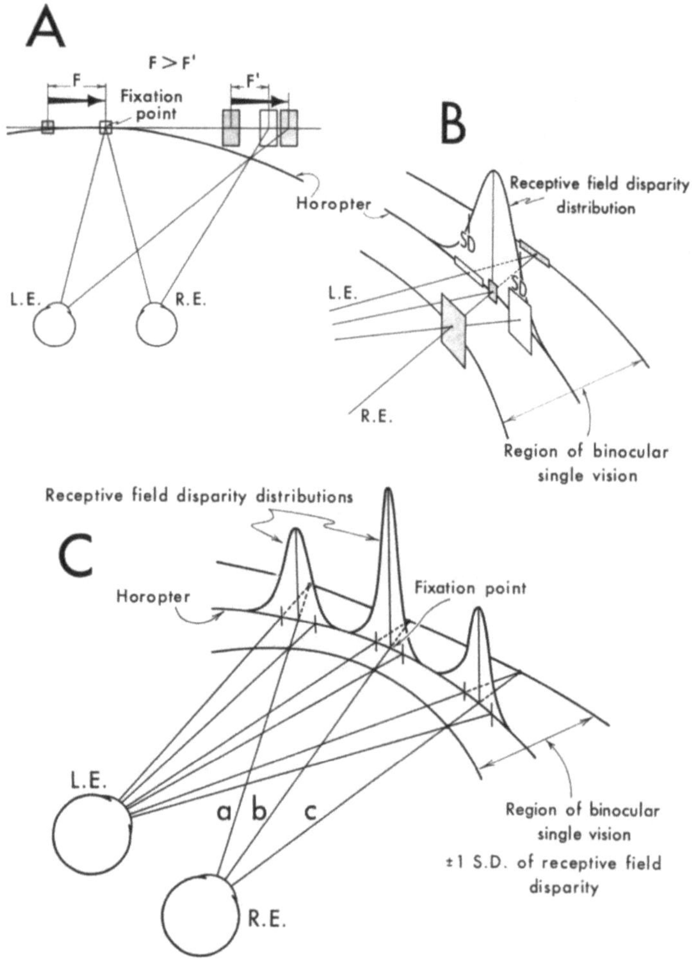

Fig. 9. A Construction for the horopter in the cat. The decrease in mean receptive field separation with increasing retinal eccentricity ($F > F'$) causes the horopter to be concave towards the animal (JOSHUA and BISHOP, 1969). B and C Construction of a region of binocular single vision in the cat analogous to Panum's fusional area in man. The limits of the region of binocular single vision are set by ± one standard deviation of receptive field disparity about a number of selected points on the horopter (JOSHUA and BISHOP, 1969)

of the construction in Fig. 10. In man for viewing distances under about 4 metres, the horopter lies between the Vieth-Müller circle and the objective fronto-parallel plane. In Fig. 10 the horopter lies just inside the Vieth-Müller circle. A reduction in the slope of the curve in Fig. 7 would, however, lead to a flattening of the horopter so that it would then come to lie between the Vieth-Müller circle and the fronto-parallel plane.

The finding that the mean receptive field separation varies with retinal eccentricity is the first neurophysiological evidence for a suggestion put forward many years ago by Hillebrand (1893) to account for the departure of the empirical horopter from the Vieth-Müller circle. The latter departure has been called the

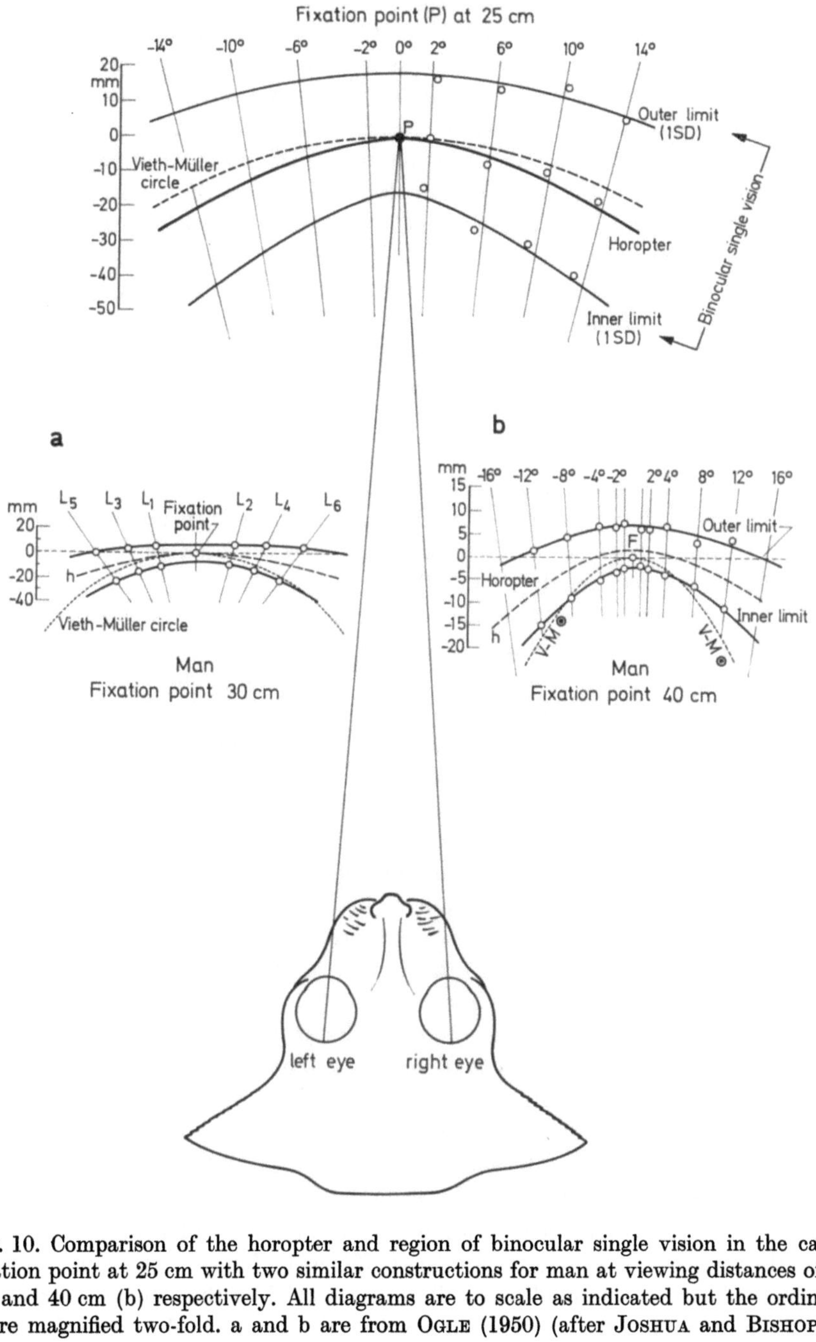

Fig. 10. Comparison of the horopter and region of binocular single vision in the cat for a fixation point at 25 cm with two similar constructions for man at viewing distances of 30 cm (a) and 40 cm (b) respectively. All diagrams are to scale as indicated but the ordinates in b are magnified two-fold. a and b are from Ogle (1950) (after Joshua and Bishop, 1969)

Hering-Hillebrand horopter deviation. HILLEBRAND (1893) proposed that there was an asymmetrical spatial distribution of corresponding points on the retinas of the two eyes and on this basis he was able to provide a qualitative explanation of the change in shape of the horopter with viewing distance. More recently OGLE (1950, 1962) attempted a quantitative description of the deviation in terms of a parameter, H, called the Hering-Hillebrand horopter deviation coefficient. OGLE's analysis lead to the conclusion that HILLEBRAND's qualitative description was not borne out on a quantitative basis and that the spatial distribution of corresponding points on the retinas of the two eyes varies with fixation distance. In neurophysiological terms this would mean that the slope of the curves in Fig. 7 vary with fixation distance in a manner analogous to the change in the parameter, H, described by OGLE (1962). Recently FLOM and ESKRIDGE (1968) re-examined the problem of change in retinal correspondence with viewing distance. Using an after-image technique which is unaffected by changes in dioptrics, they concluded that retinal correspondence is stable within 6 min arc at an eccentricity of 12° for changes in viewing distance from 10 cm to 600 cm. With present neurophysiological techniques and at the level of the striate cortex there is no indication that the directional values of receptive fields are other than stable.

8. Panum's Fusional Area

The attempt will now be made to arrive at a region of binocular single vision in the cat which may be regarded as analogous to Panum's fusional area in man. The term "Panum's fusional area" usually refers to the area or region in space in which objects are seen singly (Fig. 1B).

Fig. 5 has been drawn as if all the receptive field disparities had been transferred to the left eye, the fields in the right eye being arranged in regular array. In a similar manner changes in retinal image disparity might be considered as a variable image disparity in the left eye associated with a fixed image in the right eye. As retinal image disparity increases, the image points in the left eye will gradually move away from the exactly corresponding receptive fields at the centres of the two-dimensional receptive field distributions. There will thus be a gradually decreasing density of receptive fields available for binocular depth analysis until finally, at the limits of the distributions, binocular depth discrimination will no longer be possible. This sequence is paralleled in man by the gradual loss in stereoscopic depth discrimination with increase in retinal image disparity until the stereoscopic experience finally fades and ceases to exist. Panum's fusional area in man lies well within the limits for stereoscopic depth perception. Thus the region for binocular single vision in the cat is likely to be well within the limits of the distribution of receptive field disparities. The same conclusion can also be reached in another way. With increasing retinal image disparity, the disparate image points will stimulate a decreasing proportion of receptive field pairs the members of which converge onto single binocular units in the striate cortex and a steadily increasing proportion of receptive fields that are not binocular pairs, each of the latter receptive fields leading to the discharge of a different striate unit. In other words, if the image points are sufficiently disparate only one member of each binocular receptive field pair will be stimulated by the same feature of the object. Thus at some critical image disparity, the number of units stimulated binocularly may be insufficient to sustain single vision particularly in the face of an increasing number of binocular units stimulated only monocularly. Diplopia is then experienced but the scattered binocular receptive field pairs that are still being stimulated will provide some measure of binocular depth discrimination. Unfortunately no behavioural data regarding binocular depth discrimination is available for the cat.

The method used for the construction of a possible region of binocular single vision in the cat is illustrated in Fig. 9B and C (JOSHUA and BISHOP, 1969). The construction was based upon the four points used to define the horopter and on the horizontal receptive field disparity distributions at each of these four points. The latter points were spaced at successive 4° intervals of retinal eccentricity. The limits of the region of binocular single vision were arbitrarily set at one standard deviation on either side of the mean of the receptive field disparity distributions. The curves were then completed by assuming symmetry about the fixation point, *P*. The completed horopter line and region of binocular single vision is shown in Fig. 10 (JOSHUA and BISHOP, 1969) and by way of comparison two examples of the determination of Panum's fusional area in man (OGLE, 1950).

9. Receptive Field Disparity and Nasotemporal Overlap

The nasotemporal partition of the retina and the partial decussation at the chiasma pose a special problem for binocular depth discrimination which can be readily appreciated from Fig. 11. If there were a sharp vertical line of separation between the two halves of the retina, the situation would be as in Fig. 11A where the two visual axes are shown crossing at the fixation point. Object points between the visual axes, both in front and behind the fixation point, would still project to both cerebral hemispheres, but the one hemisphere would only receive information about any of these points via the one eye. Thus in the vicinity of the fixation point and in the very region where it would be of the greatest value, binocular depth discrimination would not be possible because of the absence of binocular parallax. Retinal disparity information would, however, become available if there were a region of overlapping projection as shown in Fig. 11B. An overlap would, in any case, be needed to compensate for the eye movements of binocular fixation.

There is now clear-cut histological and neurophysiological evidence for a vertical strip of nasotemporal overlap. In the cat retina, STONE (1966) used histological methods to show that the ganglion cells in a median strip of nasotemporal overlap supply both optic tracts. On the neurophysiological side, there is also conclusive evidence that a vertical strip of visual field centred on the midline (zero meridian) has a bilateral representation in the cerebral cortex centred on the 17/18 boundary in each occipital lobe (LEICESTER, 1968; BLAKEMORE, 1968; NIKARA, BISHOP, and PETTIGREW, 1968; JOSHUA and BISHOP, 1969). While this boundary zone receives a projection from the opposite hemisphere via the corpus callosum (HUBEL and WIESEL, 1967; CHOUDHURY, WHITTERIDGE, and WILSON, 1965), there is nonetheless strong evidence that the nasotemporal overlap projects via the optic tract of the same side. An overlap similar to that in the cortex is found in the lateral geniculate nucleus (KINSTON, VADAS, and BISHOP, 1969) and the overlap in the cortex survives section of the corpus callosum (LEICESTER,

1968). Furthermore HUBEL and WIESEL (1967) found that virtually all the single fibres they recorded from the posterior corpus callosum could be driven by visual stimulation and all were binocularly-activated. If the nasotemporal overlap in the cortex was based solely on a transcallosal projection one would expect the callosal fibres to be monocularly-activated. Thus the transcallosal projection reinforces a nasotemporal overlap which exists independently of it.

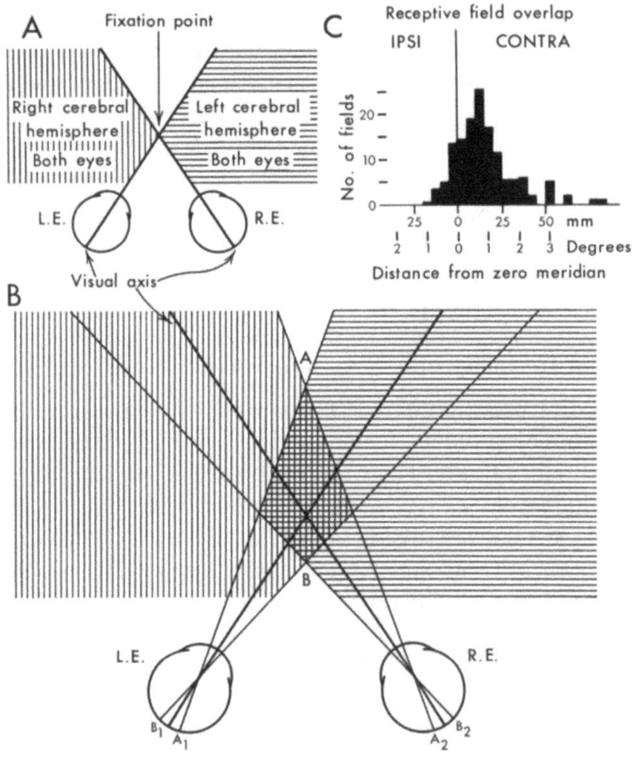

Fig. 11. A and B Diagrams to illustrate the need to include some degree of nasotemporal overlap in any neural mechanism for binocular depth discrimination (modified from BLAKEMORE 1968). C Nasotemporal overlap of the centres of the receptive fields of neurons recorded from the region of the boundary between Areas 17 and 18 and plotted as though all the neurons were located in the left cerebral hemisphere (LEICESTER, 1968)

From the above considerations, the suggestion can be made that the nasotemporal overlap and the phenomenon of receptive field disparity have a common basis. The two vertical lines in Fig. 5 represent corresponding vertical meridians for the two eyes, their separation being due to the paralysis of the eyes. The line for the right eye passes through the three vertical points at each of which there is a set of superimposed receptive fields. Each receptive field for the right eye forms the right-eye member of a binocular pair, the left members of the three sets having a two-dimensional gaussian distribution about points respectively corresponding to those for the right eye. The nasotemporal overlap is then a natural consequence of considering the two vertical lines to be zero meridians.

The three observations, namely: the nasotemporal overlap of retinal ganglion cells, the nasotemporal overlap of cortical receptive fields and the receptive field disparity in the region of the midline are all closely similar in nature and angular magnitude. The histogram in Fig. 5C, taken from Leicester (1968), shows the distribution about the zero meridian of the centres of receptive fields of cells recorded in the vicinity of the 17/18 boundary. All the receptive fields have been considered as though they belonged to cells in the left visual cortex so that any spread to the left of the zero meridian represents a nasotemporal overlap. The distribution of the receptive field centres to the left of the zero meridian has a range and standard deviation which are only slightly less than the corresponding values for the distribution of receptive field disparities for the central region (range 1.2°, S. D. 0.5°; Nikara, Bishop, and Pettigrew, 1968; Joshua and Bishop, 1969). The pooled distribution of the horizontal receptive field disparities contained in Leicester's Table 1 (Leicester, 1968) has a standard deviation of 0.88°. While the latter is somewhat larger than the value for the central region (S. D. 0.5°), a consequence presumably of pooling data from both central and peripheral retinal regions, it suggests that the positions of the receptive fields of the binocular pairs are randomly distributed about the zero meridian. In other words a receptive field to the left of the zero meridian is just as likely to have its partner to the right of the zero meridian as to the left.

The two nasotemporal overlap distributions, retinal and cortical, are monocular while the receptive field disparity distribution is binocular so that it is difficult to make a valid comparison between them. There are grounds for believing that both the ganglion receptive fields and the striate cell receptive fields to which they contribute have common geometrical centre points (Leicester, 1968). Nevertheless it is not easy to make a quantitative comparison between the neurophysiological data from the cortex and Stone's estimate of the nasotemporal overlap of retinal ganglion cells. He described the median strip of overlap as being 0.9° across, but his data shows that the spread is rather larger than this even on the basis of the relatively sharp median edge of the temporal hemiretina. While the median edge of the nasal hemiretina is much more diffuse, the more widely overlapping cells, in this case, apparently do not project to the cortex. Both of the nasotemporal overlap distributions apparently have a slightly smaller effective range than that of the receptive field disparities about their mean as would be expected if the latter arose from a random cross-coupling between the cells involved in the overlap.

It is hard to escape the conclusion that the nasotemporal overlap forms the basis for binocular depth discrimination in the vicinity of the fixation point and that the limits of the receptive field disparity for the central region determine the range over which stereopsis can be experienced.

10. Vertical Receptive Field Disparities

Vertical receptive field disparities are an essential component in the neural mechanisms which are responsible for binocular single vision and stereopsis. In the first place they provide the basis for the vertical extent of Panum's fusional area, both the receptive field disparities and Panum's fusional area having the same relative dimensions in the vertical as in the horizontal direction. Vertically disparate receptive fields may also be horizontally disparate, the latter of course providing the necessary basis for binocular depth discrimination. This explains the persistence of stereopsis in the presence of vertical image disparities such as occurs when object points are viewed at relatively near distances. OGLE (1955) has studied the effect on stereoscopic depth perception arising from a transverse disparity when a vertical disparity is also introduced. Stereoscopic acuity decreased only slowly with increase of vertical disparities up to as much as 25—30 min arc. Stereoscopic perception persisted for some distance after the onset of vertical doubling just as is the case in the horizontal direction (Fig. 1). The relationship of vertical receptive field disparity to the eye movements of binocular fixation will be discussed below.

III. Dynamic Aspects of Binocular Interaction

Having discussed functional anatomy and receptive field topography in relation to binocular vision, we may now consider the more dynamic aspects of binocular interaction and particularly the way in which the cells in the striate cortex indicate by their firing different degrees of receptive field alignment. When the receptive fields are superimposed in the plane of an optimal stimulus, the firing is markedly facilitated (Fig. 12) and when the fields are out of register as they would be if they superimposed in a plane other than that of a single optimal stimulus, they mutually inhibit one another. One of the remarkable observations we have made (PETTIGREW, NIKARA and BISHOP, 1968 b) is that, although the

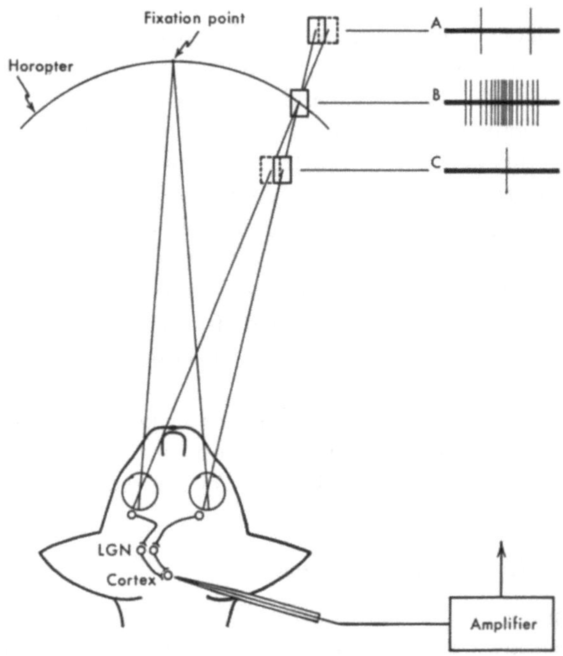

Fig. 12. Diagram illustrating the facilitation of the binocular response from striate neurons when their receptive fields are in register and the inhibition of the response when the fields are out of alignment

receptive fields may be several degrees across, their displacement from exact register by only a few minutes of arc may lead to mutual inhibition. So far we have regarded binocular receptive fields as being in register for a particular optimal stimulus when the geometrical centres of their minimal response fields coincided in the plane of that stimulus. While this serves as a good approximation, it is clear that the concept of being in register implies a condition of maximal binocular facilitation rather than of purely geometrical alignment. Barlow, Blakemore, and Pettigrew (1967) have already drawn attention to the fact that the position of receptive field alignment for maximal facilitation of the binocular response may diverge slightly from one of coincidence of the minimal response fields. Based on positions for maximal facilitation, they introduced the concept of a binocular centre for receptive fields and they used these centres in their analysis of the distribution of horizontal receptive field disparities. Much further work needs to be done to provide a quantitative description of the total receptive field of striate neurons including regions of inhibition and subliminal excitation. Only then will it be possible to arrive at a satisfactory estimate of receptive field disparity based on functional criteria. Work along these lines is in progress in this laboratory.

1. Binocular Interaction on Striate Neurons: Experimental Methods

The method used for studying the interaction on striate neurons of impulses coming from the two eyes is shown in Fig. 13 (Pettigrew, Nikara, and Bishop, 1968 a). The Risley biprisms before each eye were used to manipulate the visual

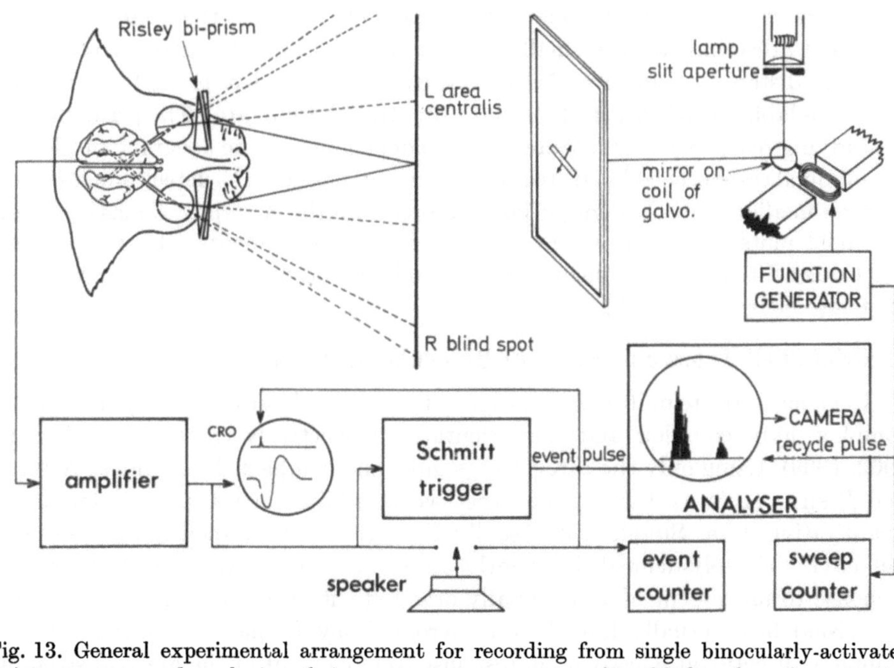

Fig. 13. General experimental arrangement for recording from single binocularly-activated striate neurons and analyzing their responses to a moving slit of light when the receptive fields of the cell, positioned on a translucent screen by means of rotary biprisms, are either in register or in varying degrees of misalignment (Pettigrew, Nikara, and Bishop, 1968 a)

direction of the receptive fields. In this way the divergence due to the paralysis can be overcome and the eyes made to converge on a common fixation point in the plane of the tangent screen or alternatively the receptive fields can be moved, each independently of the other, to any convenient place on the screen. In addition it is possible to move the receptive fields in small steps into and out of exact alignment and to study the firing patterns of the neurons when they are stimulated with their receptive fields out of alignment by varying amounts. Slits of light of adjustable dimensions, orientation and intensity were projected onto the rear of a translucent screen and moved by reflection from a mirror attached to the coil of a moving coil galvanometer. The slit was moved under the control of a function generator backwards and forwards across the receptive fields and the responses of the neuron were recorded as average response histograms (cf. KOZAK, RODIECK, and BISHOP, 1965) by means of a multichannel scaler. Binocular interaction was studied by recording the responses of each eye separately and then when both eyes saw the slit.

By the above technique we were able to examine the mechanisms of binocular interaction under conditions not far removed from normal bifixation of an object in the visual world. With the stimulus moving in the plane of the tangent screen at a fixed distance from the eyes, horizontal shifts in prism setting were equivalent to changes in viewing distance both to the front of and beyond the plane of the tangent screen. This, in turn, is equivalent to the standard psychophysical procedure whereby the viewing distance is fixed and the stimulus plane is set at different distances from the eyes. By examining the way in which binocularly-activated units responded to a single stimulus passing over the receptive fields in various degrees of misalignment it was possible to assess the ability of striate neurons to discriminate the small amounts of retinal image disparity associated with the distance of objects. Prism settings can be graded as finely as desired, those available to us being in steps of 0.1 dioptres (3.4 min arc). The prism method of studying receptive field interactions is independent of the accuracy of plotting the minimal response field and striate neurons can be shown to detect much smaller changes in alignment than might be supposed from the accuracy of making receptive field plots. Recently we have added further refinements (HENRY, BISHOP, and COOMBS, 1969) using a phase-shift method for studying the binocular responses of striate neurons (Fig. 16).

2. Striate Cell Types and Kinds of Binocular Interaction

In general striate units can be assigned to one of at least four categories, simple, complex, hypercomplex and non-orientated respectively (HUBEL and WIESEL, 1962, 1968). Using only the categories, simple and complex, PETTIGREW, NIKARA, and BISHOP (1968 a) distinguished the two in additional ways based largely on quantitative data. Simple cells usually have little or no spontaneous activity, but have a sharply-defined, frequently unimodal, average response histogram in response to movement of an optimally-oriented slit or bar over a small region of the visual field, usually less than 1° across. They frequently show marked or complete directional selectivity in response to movement of the optimal stimulus. Slowly moving stimuli are preferred, around 2°/sec. Simple unimodal units with the smallest receptive fields show the greatest specificity for stimulus orientation with a distinct preference for horizontal and vertical orientations. In the cat, but much more clearly in the monkey, simple cells tend to aggregate in or close to layer IV in the striate cortex so that the great majority of cells in that layer are of the simple type.

Complex cells usually have a relatively active maintained discharge with a median of 8 spikes/sec. The receptive fields are larger than for simple cells, being generally more than 3° across and the units give a brisk sustained discharge while the stimulus is moving over the full traverse of the receptive field. They resemble simple cells in respect to stimulus orientation and directional selectivity. The spike frequency of the discharge is higher than with simple cells, they respond better to somewhat faster stimulus movements (6°/sec) and there is usually a fairly complex multimodal average response histogram.

The striate cells in both cat and monkey have a wide spread of ocular dominance, ranging from cells showing equality in the effects exerted by the two eyes to those which are discharged only by the one or the other eye. Table 2 (data from Hubel and Wiesel, 1962, 1968) shows the percentage distribution of monocular and binocular units among the various cell types in both cat and monkey. About 20% of the cells in the cat striate cortex are discharged only monocularly whereas this becomes about 40% in the monkey. Cells of the simple type show a much greater preference for one or other eye than do complex cells. In the monkey, the great majority of simple cells (88%) are discharged only monocularly and, in the cat, the majority (78%) of monocularly-activated cells are of the simple type (Hubel and Wiesel, 1962, 1968). While Hubel and Wiesel (1962) have recognised the occurrence of binocular influences on cells that were brought to discharge only from the one or the other eye, they nevertheless found many cells (20% in cat and 40% in monkey) which were without any binocular influence. Pettigrew, Nikara, and Bishop (1968 b) made a similar observation for the cat. Using techniques designed to reveal inhibitory and subliminal excitatory binocular effects, Henry, Bishop, and Coombs (1969) have recently shown that, in the cat, most of the cells previously regarded as exclusively monocular do, in fact, have powerful binocular influences upon them (see below).

Pettigrew, Nikara, and Bishop (1968 b) have described the three possible types of binocular interaction, facilitation, summation and occlusion. Later work (Henry, Bishop, and Coombs, 1969) makes it clear that these three effects are normal aspects of the binocular process and that, depending upon the stimulus parameters and the receptive field alignment or misalignment, they are all to be observed in the one cell.

Table 2. Cell types and binocular interaction in cat and monkey

	Cat (Hubel and Wiesel, 1962)		Monkey (Hubel and Wiesel, 1968)	
	Monocular %	Binocular %	Monocular %	Binocular %
Simple cells	21	79	88	12
Complex cells	17	83	34	66
Hypercomplex cells			15	85
Non-orientated cells			88	12
Total cells	20	80	39	61

3. Binocular Interaction on Simple Cells

Fig. 14a shows the averaged responses of a simple unit to a slit that was optimally-oriented at 135° and moved in the preferred direction downward and to the left (cf. Fig. 14b) across the two receptive fields (PETTIGREW, NIKARA, and

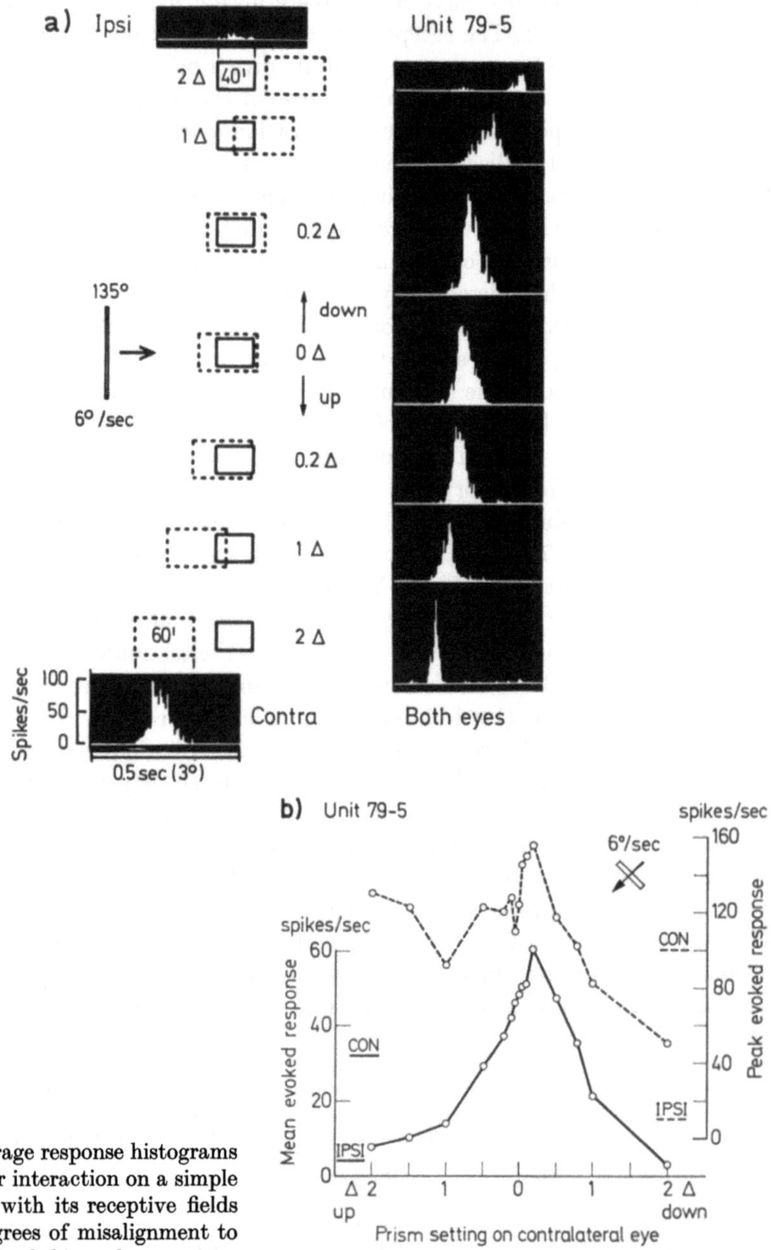

Fig. 14. a Average response histograms of the binocular interaction on a simple striate neuron with its receptive fields in varying degrees of misalignment to show the optimal binocular response when the receptive fields are accurately superimposed in the stimulus plane (PETTIGREW, NIKARA, and BISHOP, 1968 b). b Graph of the averaged binocular response of the unit in a as a function receptive field misalignment. The respective levels of the monocular responses are indicated (PETTIGREW, NIKARA, and BISHOP, 1968 b)

Bishop, 1968 b). Only the histograms for the preferred direction are shown since there was no response in the opposite direction. To monocular stimulation, the unit showed a marked contralateral eye-dominance (Contra), with only a very weak response from the ipsilateral eye (Ipsi). The two receptive fields are represented by the two rectangles and, as is usual with marked eye-dominance, the minimal response field for the weaker eye (continuous line) is smaller than for the dominant eye (broken line). After overcoming, with prisms, the divergence of the eyes due to the paralysis, the position of the ipsilateral receptive field was then held constant. The binocular response histograms in the right hand column (Both eyes) were obtained by sweeping the same slit across the two receptive fields as they were offset with respect to one another in successive steps by moving the contralateral field along the line of the slit movement. Although the preferred direction for stimulus movement was downward and to the left (Fig. 14b), by way of simplification and to give a left to right stimulus sequence, the positions of the receptive fields have been completely rearranged in Fig. 14 as though the preferred direction for stimulus movement was left to right as indicated by the arrow and with appropriate inversion of the prism settings. Hence, going down the binocular responses, the ipsilateral receptive field is first stimulated before the contralateral, then at the same time and finally after the contralateral field. The binocular response is maximal when the two receptive fields are exactly superimposed (0.2 Δ), the initial prism power of 9 dioptres (Δ) used to bring the two receptive fields into register being 0.2 Δ short of this optimal setting. One prism dioptre is equivalent to one centimetre on the tangent screen at one metre or 0.57°. A change in prism power and hence in receptive field position, in either direction, causes the binocular response to fall off. Whenever the receptive fields are out of alignment mutual inhibition occurs, this being particularly the case when the non-dominant eye (Ipsi) is stimulated slightly before the dominant eye. Thus although the non-dominant eye gave only a very weak response to monocular stimulation it was, nevertheless, the dominant eye as far as inhibition was concerned. In general, for simple units, there seems to be a reciprocal relationship with respect to monocular excitation and binocular inhibition: the less dominant the eye in respect to monocular excitation the more powerful its inhibitory effect on the other eye (cf. also Figs. 15 and 16).

Figure 14 b graphs the changes in the binocular response over a range of receptive field alignments of approximately $\pm$ 2 Δ ($\pm$ 1.1°) on either side of exact alignment (0.2 Δ). Two measures of the response are shown: the peak evoked response (spikes/sec) and the mean evoked responses in spikes/sec averaged over the time taken for the stimulus to traverse the two receptive fields. The response levels from stimulation of each eye alone (Ipsi and Con) are shown for comparison. Both measures of the binocular response show a sharply-defined optimum when the contralateral prism setting is 0.2 Δ down and an obvious decrement when the prism setting is 0.2 Δ (7 min arc or 5 min arc horizontal misalignment) in either direction from the optimum.

Another unit deeper in the same electrode track as that used to record the unit in Fig. 14 had the same optimal stimulus orientation (135°) and preferred direction but required quite a different prism setting for optimal binocular stimulation, namely an ipsilateral setting of only 6 Δ rather than 9 Δ and therefore indicative of a relative receptive field disparity of 1.7°. One of the main limiting factors in the analysis of receptive field disparity outlined above was uncertainty regarding the receptive field boundaries and the use of the geometrical centre of the minimal response field. The great advantage of determining receptive field disparity by the variable prism method is that it does not depend upon the accuracy with which the receptive fields can be plotted and it uses a functional measure of receptive field alignment. The method as at present developed is, however, too time-consuming to provide a satisfactory distribution of receptive field disparities in the one animal.

A number of additional features of binocular interaction on simple cells are illustrated by the unit in Fig. 15. In most respects the cell was typical of the simple type. There was virtually no spontaneous activity and it gave a sharply-peaked, almost unimodal, completely directionally-selective average response histogram. There was almost complete left eye (IPSI) dominance with only a very weak response from the contralateral eye. The unit was unusual in that it responded optimally to a horizontal slit moving vertically downwards at quite high speed (31°/sec) and, in keeping with its location at about 10° eccentric to the centre of gaze, it had a large minimal response field (3°). When the receptive fields were optimally superimposed the binocular response was 162% of the monocular response from the dominant eye and this binocular response was reduced to 15% when the receptive field of the weaker eye was offset slightly ahead of the field of the dominant eye. Thus, despite the large size of the receptive fields, they were, nevertheless, still quite sensitive to relatively small misalignments, the binocular response falling from 151% to 100% for a change in alignment of only 17 min arc. However the unit in Fig. 16 (Henry, Bishop, and Coombs, 1969), which had receptive fields close to the visual axis, was much more sensitive to misalignment, the binocular response falling from the peak of facilitation (167%) to absolute inhibition (0%) for a misalignment of less than 10 min arc from the position of optimal superimposition.

4. Layer IV Simple Cells as Binocular Gate Neurons

The observations made by Julesz (1964, 1965 a and b) using random-dot targets as stereo pairs suggest that the mechanisms concerned in binocular depth discrimination come into play at a very early stage in the chain of binocular events. The first major interaction between the pathways from the two eyes takes place when they come together in the striate cortex. On anatomical grounds, it is a reasonable suggestion that the principal neurons concerned in the first stage of the binocular process are the stellate cells in layer IV of Brodmann since the majority of lateral geniculate axons terminate in relation to them. There is now good evidence for the idea that stellate cells are to be identified with neuron receptive fields of the simple type. The stellate neuron is the predominant cell type in layer IV and, in both cat and monkey, the majority of cells recorded in layer IV are of the simple type (Hubel and Wiesel, 1962, 1968). This is particularly the case for layer IV in the cat where complex cells are very rarely found. However linking the first stage of the binocular process with the stellate cells in layer IV immediately encounters the difficulty that the majority of the simple cells in layer IV are apparently only monocularly-discharged. This is particularly

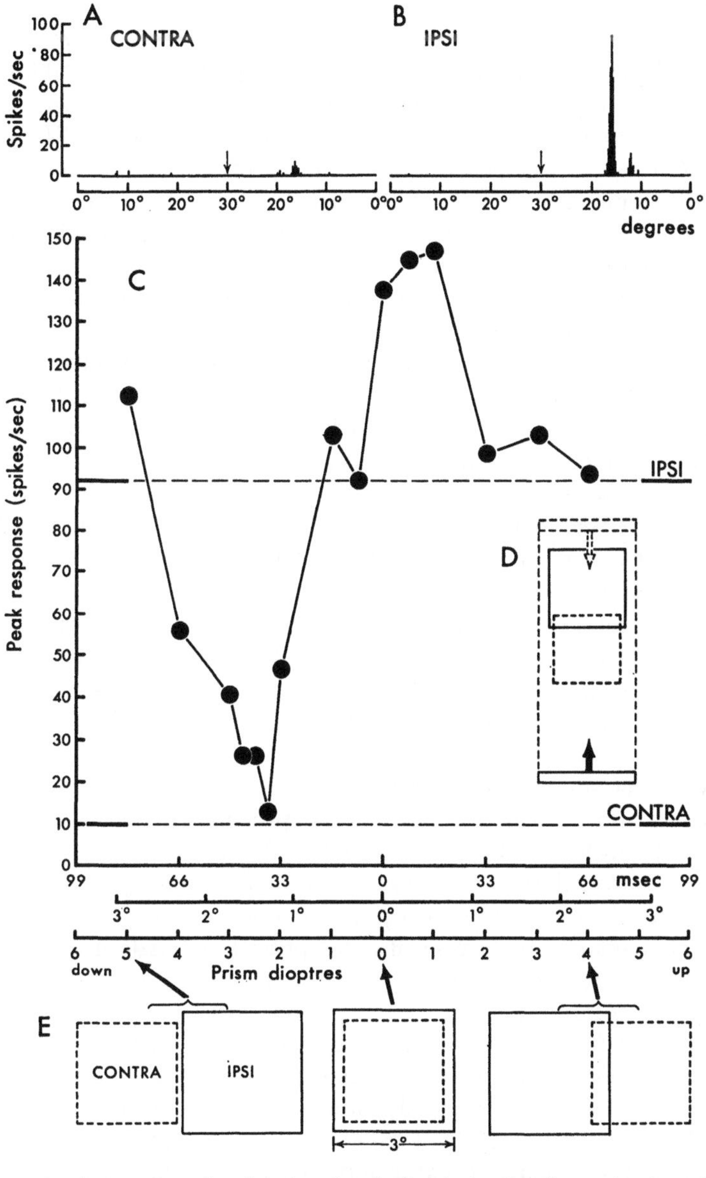

Fig. 15. Binocular interaction showing fine depth discrimination by a simple striate neuron with large receptive fields (3° across) situated about 10° from the visual axis (Joshua and Bishop, 1969). A and B Monocular average response histograms to a slit moving up and down (arrow-turn-round point) across the receptive fields. C Graph of the averaged binocular responses of the unit as a function of receptive field misalignment. Broken lines-levels of the monocular peak responses. D Diagram showing slit (orientation 90°) moving vertically first upward, then downward across the receptive fields with the ipsilateral field displaced above the contralateral roughly corresponding to 4 prism dioptres (Δ) up. E Diagram of receptive field positions at three different prism settings (5 Δ down, 0 Δ and 4 Δ up) on the assumption that the stimulus moved from left to right and back again instead of up and down

the case at the level of layer IVB in the monkey where there is a mosaic of alternating left-eye and right-eye representation, each apparently almost pure (HUBEL and WIESEL, 1968). By the vertical arrangement of the cell columns, the mosaic sub-division of the cortical surface is reproduced through all the cell layers. However even in the binocular columns, which tend to have a larger cross-section, there is still a mosaic of monocularly-activated regions in layer IVB. HUBEL and WIESEL (1968) have developed the idea that, at the level of layer IV, binocular interaction is minimal and that, as the visual input is transmitted over several stages to more complex cells in the upper and lower layers, there is progressively more intermixing between the eyes.

The difficulty regarding the monocular activation of simple cells has now been resolved by the observation that the majority of them are in fact exquisitely binocular (HENRY, BISHOP, and COOMBS, 1969). While these cells may be exclusively discharged from the one eye, it is still the case that a receptive field may be plotted for the non-dominant eye, the field being largely inhibitory with a very small region of subliminal excitation. This observation has been missed previously because of a number of circumstances. Most simple cells have little or no spontaneous activity (PETTIGREW, NIKARA, and BISHOP, 1968 a) so that an inhibitory effect is difficult to recognize. Furthermore plotting a receptive field for the non-dominant eye involves a knowledge of its likely position on the tangent screen, a particularly precise control of the stimulus variables, an artifically-raised level of maintained activity as a background against which to demonstrate inhibition and, finally, analysis of firing patterns by quantitative methods. It is now suggested that most of the monocularly-discharged simple cells in layer IV are acting as binocular gate neurons since they are only discharged when the stimulus variables are exactly the same for both receptive fields and, in addition, only when the stimulating feature is located at a precise depth in space characteristic for the particular cell. Although the simple cell fails to be discharged when the non-dominant eye is stimulated on its own, even by a stimulus judged to be optimal for the dominant eye, nevertheless it provides a gating control over the discharge from the dominant eye by reason of a receptive field made up of a very small sharply-defined region of subliminal excitation associated with a broad, but powerful region of inhibition.

The unit used for Fig. 16 was a simple cell having no spontaneous activity. The average response histogram from the dominant eye on its own showed a sharply-defined, unimodal, completely directionally selective response to a moving slit, there being no response at all from the non-dominant eye using the same, but independently-applied, optimal stimulus (HENRY, BISHOP, and COOMBS, 1969). The details of the receptive field for the non-dominant eye were, however, revealed by stimulating the non-dominant eye with the same stimulus as before, but this time against the background of a maintained discharge produced by the separate, but simultaneously applied repetitive stimulation of the dominant eye. The background discharge from the dominant eye produced the conditions necessary for the stimulus to the non-dominant eye, previously ineffective or apparently so, now to reveal a narrow peak of subliminal excitation (0.3° across) and a broad region of powerful inhibition (2° across). The subliminal excitation has the same precise stimulus requirements as for the response from the dominant eye, but the

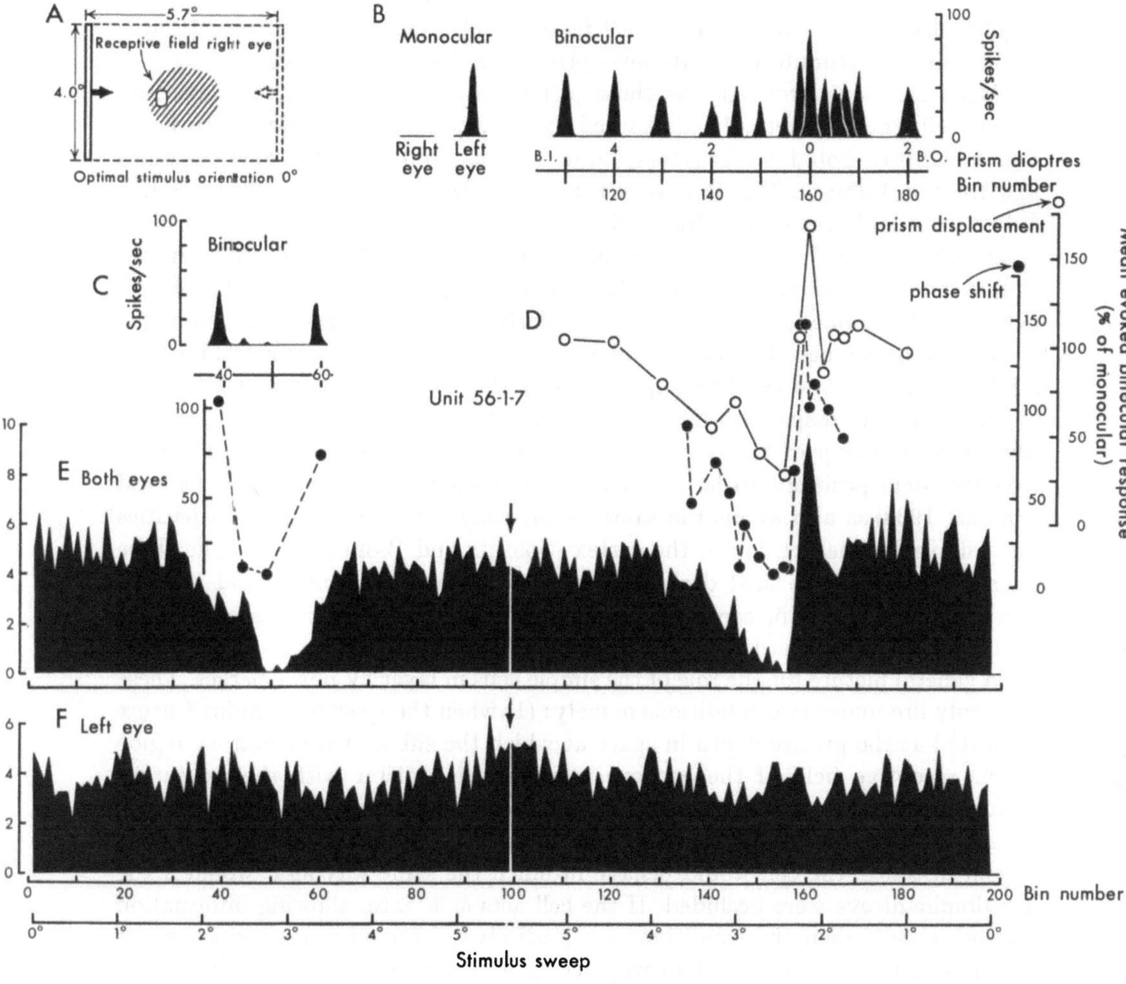

Fig. 16. Details of the demonstration of a receptive field for the non-dominant eye of a striate neuron monocularly-discharged only from the other (dominant) eye (Henry, Bishop, and Coombs, 1969). Partial (*B*, *C*) and complete (*E*, *F*) average response histograms to a 4° long slit of light moved to and fro (traverse 5·7°; arrow, turn-round point) across the two receptive fields. Histograms *B*, *C*: Left eye, unimodal completely directionally-selective response; Right eye, no response. Histogram *F* (dominant eye): Simulated maintained discharge ("activated discharge") produced by slit of light moved to and fro (0.1 Hz) across excitatory region of receptive field of dominant eye. Stimulus asynchronous with multichannel analyzer, hence relatively uniform distribution over 200 bins. Histogram *E* (both eyes): same as *F* but with the non-dominant eye also and separately stimulated by slit of light as in *A* and as for *B* and *C*. Analyzer synchronized to non-dominant eye sweep. Dual stimulation demonstrates for the non-dominant eye a predominantly inhibitory receptive field (*A*-cross-hatched) with a directionally-selective peak of subliminal excitation (*A*-clear region). *B*, *C* and *D*: inhibitory and facilitatory effects by the non-dominant eye demonstrated by the prism and phase shift methods. For details see Henry, Bishop, and Coombs (1969)

inhibition is much more general, occurring on both limbs of the stimulus sweep. While the selectivity of the subliminal excitation and the generality of the inhibition have yet to be fully investigated, it seems to be the case that these cells are highly selective for a particular object feature and exercise a powerful veto over a wide range of stimuli which are not optimal for the dominant eye.

Much the same properties as those just described are, in fact, displayed by simple cells that are actually discharged by the separate stimulation of each eye and it had been noted earlier (Pettigrew, Nikara, and Bishop, 1968 b; Joshua and Bishop, 1969; cf. Fig. 15 above) that even when the non-dominant eye was only very weakly responsive by itself, it nevertheless had a very powerful binocular inhibitory effect on the dominant eye. Thus the monocularly-activated simple cells described above (cf. also Henry, Bishop, and Coombs, 1969) probably belong to the same class as those binocularly-activated simple cells having a strong eye-preference. It has already been observed that the zone of binocular inhibition for the one cell has approximately the same dimensions as the range of receptive field disparities from cell to cell for the particular cortical locus. A further relationship of possible significance in this context is that the range of receptive field positions to be found for units within a cell column (Hubel and Wiesel, 1968) is also about the same as the range of receptive field disparities for a similarly-small region of the cortex (Joshua and Bishop, 1969). The three phenomena, receptive field disparity, zone of binocular inhibition and random receptive field position, are of approximately the same dimensions and might well have a common basis.

A general picture for the role of the simple cells in layer IV now emerges. These cells only fire under two conditions namely: (1) when the specific stimulus feature is located at the precise depth in space at which the subliminal excitatory region in the receptive field of the one eye is in precise register with the excitatory region of the field of the other eye and (2) when the excitatory region for the dominant eye falls outside the inhibitory zone of the non-dominant eye. In the second instance, the cell would behave in much the same way as it would if the non-dominant eye were occluded. If the cell acts as a gate, allowing information through either when the conditions are precisely set for stereopsis or when one eye is occluded, it would explain why seeing with two eyes differs so little from seeing with one, except in respect to the stereoscopic experience. In the intermediate circumstance, when the receptive fields are over-lapping but the excitatory regions are not in register, the response from the unit would be drastically reduced or abolished. Thus over a large part of the range of receptive field alignments the cell would fire only when the conditions are set for the binocular depth mechanism to come into play. This arrangement, which ensures binocular singleness of vision over a significant portion of the spread of receptive field disparities, is presumably to be correlated with Panum's fusional area. This would explain why the zone of inhibition for each cell is of approximately the same order as the spread of receptive field disparities. A further suggestion can also be made namely that the receptive field disparities arise from the random cross-coupling between the cells in adjoining mosaic subdivision of left-eye and right-eye representations in layer IV. Such a cross-coupling, presumably via interneurons, would explain why the random staggering of receptive field positions in a cell column approximates the spread of receptive field disparities. The necessary anatomical basis for the pooling of receptive fields of units across an area of monocular representation in layer IV to provide the broad band of inhibition for cells in an adjoining area of opposite monocularity, is provided by the tangential dendritic spread of the stellate cells described by Colonnier (1964) together with their tangentially running axons.

5. Binocular Rivalry and Receptive Field Inhibitory Zones

The term binocular rivalry or retinal rivalry is usually applied to the situation in which the two eyes are stimulated by patterns which are sufficiently different to make fusion impossible, rivalry being what the subject perceives in the non-fusion situation (Levelt, 1965). In the past, there has been a general tendency to consider the phenomenon as though it belonged among the visual illusions rather than as being part of the normal binocular process. Following Hubel and Wiesel's (1959, 1962) observations on the receptive fields of striate neurons, Levelt (1965) has recently considered the consequences of the assumption of different eye-dominance for each small unit of the visual field. One advantage is that binocular rivalry becomes a normal component in binocular vision. The theory we have put forward above not only comes to the same conclusion but also provides an explicit account of the role of rivalry in the binocular process.

Whenever the optimal stimulus for a particular striate neuron is located at that position in space at which the excitatory regions of the two receptive fields superimpose, the cell fires and the object-feature will be "seen". If, however, the stimulus is optimal in every respect except that it is located in a plane other than that for superimposition of the excitatory regions, the cell will be inhibited and the object-feature may not be seen. Under normal viewing conditions and for planes other than that for optimal superimposition, the excitatory regions of the two receptive fields will be presented, in the environment, with separate features which are likely also to be dissimilar. Inhibition is particularly effective under these circumstances (Joshua and Bishop, 1969; Henry, Bishop, and Coombs – unpublished observations).

The inhibitory action is, however, only locally applied, extending about 2° or so from the position of optimal receptive field superimposition (Figs. 15 and 16). Even for cells with the smallest receptive fields the inhibitory zone extends about 1° on either side of the position for optimal facilitation. As noted by Pettigrew, Nikara, and Bishop (1968 b) the zone of binocular inhibition for any one cell is approximately the same as the range of receptive field disparities from cell to cell. The spread of receptive field disparities for cells in the central region is about 2°, increasing beyond 3° when the receptive field eccentricity is 10° or more (Joshua and Bishop, 1969). The latter observation suggests that, in man, the extent of the suppressive effect in binocular rivalry should be roughly the same as that for qualitative stereopsis.

Kaufman (1963) made a direct psychophysical test of the spread of suppressive action on the part of contours in binocular rivalry. He presented a horizontal line to the left eye and two vertical lines to the right eye, so that the binocular impression was an intersection of a horizontal and two vertical lines. The enclosed single line disappeared and reappeared, the suppression time remaining at a maximal level up to about 14 min arc separation of the vertical lines and thereafter decreasing rapidly and monotonically with angular distance to reach a low level at about 2°. The parallel with the binocular inhibitory effects on striate neurons is obvious. The instability which is such a feature of the suppression in binocular rivalry might result from the changing receptive field disparities consequent upon the eye movements of binocular fixation.

In the case of the receptive field of the non-dominant eye in Fig. 16 the small region of subliminal excitation is eccentrically located in the large inhibitory zone. The position of the small excitatory region within the receptive field varies from cell to cell and the suggestion is made that the receptive fields with the most eccentrically located excitatory regions are those with the greatest disparity, the inhibitory zone being directed inwards towards the horopter surface. In this way binocular rivalry would be virtually confined within the range of receptive field disparities and a stimulus or trigger feature which produces optimal binocular activation of a cell having a particular receptive field disparity, will inhibit all other cells requiring the same stimulus, but having different receptive field disparities.

6. Receptive Field Disparity and the Eye Movements of Binocular Fixation

The limits to which the extraocular muscles can maintain steady binocular fixation have been discussed above. Because the movements of fixation are not perfectly correlated in the two eyes, there is a constantly varying amount of binocular retinal image disparity. The phenomenon of receptive field disparity provides an explanation of the otherwise puzzling fact that the quality of stereopsis is maintained despite the fluctuations of image disparity. Inspection of Fig. 5 shows that depth information will still be available provided the eye movements of fixation keep any given point in the right-eye visual field well within the scatter of the receptive field disparities for the left-eye members of the binocular receptive field pairs concerned. If this is the case the observation is not unexpected that the retinal image disparities that occur during binocular fixation are only rarely greater than the dimensions of Panum's fusional area for central vision (Fender and Julesz, 1967) and probably always much less than the limits of retinal image disparity for the qualitative experience of stereopsis in the presence of diplopia particularly for parafoveal vision (Ogle, 1962).

Stereoscopic acuity continues to improve with increasing duration of exposure of the test-object. Ogle and Weil (1958) reported a four-fold decrease in stereoscopic threshold with increasing exposure-times from 6 msec to 1 sec. Below 6 msec the threshold remained constant. As Rønne (1956) has suggested, it is as though the subject "chooses, by trial and error, the optimal horopter . . . which gives the best approximation to average minimal disparities," analogous to curve fitting to a number of points. The role of involuntary eye movements in stereoscopic acuity is still uncertain but with increasing duration of exposure it is likely that binocular fixation will come to centre on the position of maximal receptive field alignment and hence of maximal stereoscopic acuity. Shortess and Krauskopf (1961), using a stabilized image technique, came to the conclusion that eye movements bear no relationship to stereoscopic acuity. This result can, however, be criticized on the grounds of contact lens slip on the cornea, since the experiment is a particularly severe test for slippage (cf. BARLOW, 1963).

It is suggested that the range of binocular retinal image displacements that occurs during fixation always allows a sufficient number of receptive field pairs to be in register in the plane of the stimulus to provide depth information. The asymmetrical character of the binocular responses, such as those in Figs. 14, 15 and 16, could provide the necessary clue to the direction that the corrective flick must take to bring a greater number of receptive fields into register. The unit in Fig. 15 illustrates the fact that simple cells responding optimally to the vertical movement of a horizontal slit may detect vertical receptive field misalignments as well as other cells detect horizontal misalignments. The detection of vertical misalignments is clearly an essential preliminary for corrective eye motions in the vertical direction.

7. Binocular Interaction on Complex Cells

It is perhaps not surprising that much less is known about binocular interaction on complex cells than on simple cells. In contrast to simple cells, binocular interaction on complex cells tends to be more symmetrical and inhibition is not such an obvious feature. With respect to binocular facilitation there are at least two classes of complex cells.

Facilitation over a Narrow Range of Receptive Field Misalignments. PETTIGREW, NIKARA, and BISHOP (1968 b) described a complex cell which had receptive fields 6° across but which required prism shifts of only 2 Δ in either direction to reduce the binocular response to less than either of the monocular responses and much less than the monocular response of the dominant eye. The two receptive fields had only to be moved out of alignment by about one sixth of their length to produce marked depression of the binocular response. Thus the specificity of some complex units for a narrow range of receptive field misalignments may be comparable to that of a simple cell.

An important characteristic of complex cells is that a very small amplitude movement of a slit or line stimulus, provided it is correctly oriented, will produce a response. The direction of movement and orientation of the slit are important but not the exact position within the receptive field. In other words, the cell has specificity for movement in a particular direction but has generalized for position and commonly also for contrast. If we consider the complex cell having as input a number of simple cells all having the same directional selectivity (HUBEL and WIESEL, 1962) then, in the case of the complex cell described above, we must also suppose that these simple cells all have the same, or very nearly the same, specificity with regard to binocular depth. Under natural conditions such a unit would be optimally activated by a line stimulus over a relatively wide range of visual directions (6°) but only over a much narrower range of depth in space. It would have a fine discrimination for depth but would be generalized for visual direction and contrast.

Facilitation over a Wide Range of Receptive Field Misalignments. Some complex cells have a multimodal binocular response curve and a wide range of binocular facilitation ($> 3\ \Delta$) suggesting that these cells receive an input from a number of simple cells with a range of different optimal stimulus misalignments (PETTIGREW, NIKARA, and BISHOP, 1968 b). For these cells there is stimulus generalization with respect to binocular depth as well as visual direction.

8. Temporal Aspects of Binocular Interaction

Binocular depth discrimination involves both temporal as well as spatial factors although the two are very closely inter-related. When test objects are viewed with light exposures too brief to allow of eye movements, depth discrimination depends upon the spatial relationships of the receptive field pairs that happen to obtain at the time of the flash. However, even under these circumstances a temporal factor can be introduced by placing a filter before one eye (RØNNE, 1956). It is a weakness in the neurophysiological theory that all our analyses have so far been done with moving targets. Stereoscopic perception occurs with stationary retinal images and it will be necessary to show that striate neurons can make depth discriminations when they are stimulated by fixed targets flashed on their receptive fields.

Since the speed of the stimulus is known, prism dioptres in Figs. 14 and 15 can be converted into milliseconds so that the abscissae then represent the time intervals by which the one receptive field is stimulated before or after the other. In the case of Fig. 15, a facilitatory response only occurs when the two receptive fields are stimulated together or when the nondominant eye is stimulated no more than about 33 msec after the dominant eye. When the receptive field of the nondominant eye is stimulated 33 msec before that of the dominant eye the response is almost abolished. Thus with moving targets the receptive fields have to be stimulated in a particular order and in a particular temporal relationship for maximal facilitation to occur. For those units so far studied, the range of receptive field interactions in temporal terms extends over 100–200 msec (JOSHUA and BISHOP, 1969).

With normal steady gaze most of the objects we observe are stationary and the only time differences that can occur between the stimulation of the two receptive fields arise as a result of the eye movements of steady fixation. Of importance for the visual process are the slow drifts which, in the cat, average about 30 min arc/sec. The time difference between the peak of facilitation and maximal depression for the unit in Fig. 15 was about 33 msec. With the eyes moving at 30 min arc/sec, this corresponds to a spatial receptive field separation of 1 min arc. Thus the unit's firing would be drastically altered if the two receptive fields passed over a stationary object with the receptive fields spatially separated by as little as 1 min arc and the total range of the interaction would be exceeded if the separation were greater than about 3 min arc.

The *stereophenomenon described by* PULFRICH (cf. LIT, 1949) provides a test for the neurophysiological theory. If a plumb bob oscillating in a frontal plane is viewed with a filter before one eye, the bob appears to move in an elliptical path which seems to be nearer the subject for movement in one direction and farther away for movement in the opposite direction. While the magnitude of the effect depends primarily on the density of the filter, increasing with increasing density, it increases also with reduction in the absolute levels of retinal illumination of the two eyes. The PULFRICH effect is probably largely due to the difference in the latency at the striate cortex of the impulses from the two eyes. Depending upon the experimental conditions, calculated latency differences range from about 5–25 msec (Lit, 1949, 1960). If Lit's data can be applied to the cat, then under our experimental conditions (Pettigrew, Nikara and Bishop 1968 a) one might expect a latency difference of about 15 msec using a filter of density 1.0 before one eye. In the case of the units in Figs. 14 and 15, a latency difference of 15 msec would have been equivalent to receptive field misalignments of 0.16 Δ (5.4 min arc) and 0.9 Δ (31 min arc) respectively. These misalignments would have been detectable in both cases as a change in the amplitude of the binocular response. A direct test of these expectations is, however, clearly needed.

Conclusion

Hering and Helmholtz conflicted on a number of issues, none more strongly than on problems concerned with binocular vision (Helmholtz, 1911). Helmholtz advocated an "empirical theory of vision" in which even the local signs for the visual direction of each retinal element have their meanings learnt from experience. He thought that the single picture of the external world was produced, not by any "anatomical mechanisms of sensation", but by a mental act. It is perhaps surprising that Helmholtz, a physiologist, should have relied so strongly on mentalistic explanations without attempting to seek underlying neural mechanisms. Indeed he specifically warns against the assumption of a neuroanatomical mechanism whereby nerve fibres proceeding from corresponding points in the two retinas come together in the brain (Helmholtz, 1881). Nearly 60 years later Sherrington (1940) concluded, in terms reminiscent of Helmholtz, that there was no evidence that the nervous paths from two corresponding retinal points reach a common mechanism in the brain, the conjunction of the reports from the two eyes being a mental not a physical event. Even as late as 1959, Ogle (Ogle 1959) found it necessary to put arguments for and against a physiological basis for stereoscopic vision.

Hering presented a complete contrast to Helmholtz by attempting to describe the phenomena of binocular vision in physiological terms. He attributed three local signs to every retinal point – one for vertical position, another for horizontal and a third for binocular depth discrimination. Two corresponding retinal points have identical vertical and horizontal values but opposed depth values. The latter cancel one another so that the object point appears on the horopter. The algebraic sum of the positive and negative depth values of slightly non-corresponding points provided the basis for stereoscopic perception. Hering's theory implied that the binocular perception of depth was based on innate neuroanatomical structures. In more recent years Rønne (1956) has put forward a detailed theory of binocular vision in physiological terms. These accounts show a very penetrating grasp of the probable neural mechanisms but neither Hering nor his successors attempted any direct observations on the brain. Their use of physiological terms is formal and symbolic rather than experimental. The plotting of receptive fields of binocularly-activated single neurons in the visual cortex has made it possible, for the first time, to formulate a detailed neurophysiological theory of binocular single vision and stereopsis based on direct experimental evidence. The theory put forward above accounts for a wide range of binocular phenomena, but many of the details of the development of the theory are only suggestions that need to be put to the test.

Postscript

The above chapter was completed in March 1969 and the developments that have taken place since that time have been the subject of recent reviews (BISHOP and HENRY, 1971; RODIECK, 1971; cf. also BISHOP, 1970 a, b). There has been progress both in psychophysics and neurology but this postscript aims only to draw attention to some of the more important developments on the neurological side.

Some Criticisms of the Neural Theory. The neural theory of stereopsis has yet to receive detailed critical assessment but RODIECK (1971) has drawn attention to a number of difficulties. Most of the experimental evidence relates to the cat. It is true that there is, as yet, no satisfactory behavioural observations to show that the cat possesses stereopsis although evidence in this direction is now becoming available (R. FOX and R. R. BLAKE – personal communication). This difficulty would be resolved by the application to the cat of the Julesz random dot technique that has so convincingly demonstrated stereopsis in the macaque monkey (BOUGH, 1970). A number of authors, including RODIECK, have had difficulty in accepting the presence of vertical receptive field disparity as an essential element in the neural theory of stereopsis. The case for the presence of such disparities is briefly stated below (cf. also BISHOP and HENRY, 1971). A much more cogent criticism by RODIECK is that only cells sensitive to horizontal misalignments of their two receptive fields can make binocular depth discriminations. This view is certainly accepted and has been discussed by BISHOP and HENRY (1971). Furthermore the criticism is valid that BARLOW, BLAKEMORE, and PETTIGREW (1967), NIKARA, BISHOP, and PETTIGREW (1968) and JOSHUA and BISHOP (1970) treated all cortical cells equally whereas one might expect that those cells most sensitive to horizontal misalignment should carry a greater weight. This criticism undoubtedly detracts from the value of the receptive field disparity distributions that have been published so far. New determinations of these distributions will clearly be needed both in the light of the above criticism and to take account of our new knowledge concerning the organization of cortical receptive fields (see below). The necessary experiments would be a formidable undertaking and are probably best deferred until our understanding of the organization of the various types of cortical receptive field is more complete.

Receptive Fields of Simple Cells in Cat Striate Cortex. The neural theory for binocular single vision and stereopsis will clearly need modification and development in the light of the new information that is becoming available concerning the properties of the receptive fields of cortical cells, particularly those of simple cells in the striate cortex (HENRY and BISHOP, 1971; BISHOP, COOMBS, and HENRY, 1971 a, b, 1973). A significant feature of these studies is the greater understanding of the importance of inhibitory components in the receptive field organization. The receptive field region from which a spike discharge is obtained is called a discharge centre (see below). Even for a stimulus contour of optimal orientation moving in the preferred direction there are, for simple cells, relatively wide inhibitory sidebands to one or both sides of the discharge centre. For receptive fields near the centre of gaze, these inhibitory regions are about 2° across. When the stimulus orientation is angled away from the optimal, the cell is inhibited over the whole extent of the receptive field (BISHOP et al., 1973 c).

In binocular interaction studies a new concept has been introduced, namely that of a *binocular interaction field* (BISHOP, HENRY, and SMITH, 1971 d). Binocular interaction fields have now been plotted for many simple cells. Each binocularly-activated cell has two interaction fields, one for each eye. The binocular interaction field for one eye plots the changes in the amplitude of the response from the other eye as the two receptive fields of the binocularly-activated cell are moved across one another, first into and then out of register in the plane of the optimal stimulus (i.e. tangent screen). Binocular interaction fields differ with the various types of simple cell, i.e. cells that are discharged only from the one eye, cells binocularly-discharged but with very weak or absent monocular responses and cells showing binocularly-opposite direction selectivity. Almost without exception simple cells have shown binocular facilitation when the two receptive fields are in register. For the common type of simple cell, the facilitated response is about 75% greater than the sum of the separate monocular responses to the same optimal stimulus. Facilitation switches to depression for very small degrees of receptive field misalignment in a direction at right angles to the optimal stimulus orientation, the gradient of the transition from facilitation to depression being usually extremely steep. With misalignment, the discharge to binocular stimulation is commonly abolished while the discharge centres are still overlapping one another by about half of their extents, the midpoints of the centres being then separated by only about 20 min arc. The fact that the two receptive fields of simple cells mutually inhibit each other only when the misalignment is in a direction at right angles to the preferred stimulus orientation is an important factor which will have to be taken into consideration in the neural theory of stereopsis (BISHOP and HENRY, 1971; cf. also HUBEL and WIESEL, 1970).

More recently, the demonstration that simple cells are discharged only by the edges of a moving slit of light or dark bar and not by the general body of the stimulus has led to the recognition that each of the two types of edge has its own receptive field organization (BISHOP et al., 1971 a). There are, in other words, separate discharge centres for each type of edge, and when a cell responds to both types, the two centres may have differing spatial relations with respect to one another: they may be coincident, partially offset or separated from one another. Again the spatial arrangement of the centres for one direction of stimulus movement may not be the same as that for the other. In addition to a discharge centre (or centres) single edge receptive fields also have inhibitory regions which however, differ from discharge centres in that they are not specific for a the particular edge (SHERMAN, HENRY, and BISHOP – unpublished observations). When two or more spatially separated discharge centres are present for the one direction of stimulus movement the cell is classified as bimodal or multimodal respectively. No systematic study has yet been made of the relative spatial arrangements of the discharge centres and inhibitory regions in the two receptive fields of binocularly-activated simple cells.

The neural theory of stereopsis will also have to incorporate these new complexities of receptive field organization. The concept of receptive field disparity as developed above is probably still valid for the class of unimodal simple cells. In the case of these cells, although the monocular receptive field may have two discharge centres, one for each type of edge, the centres are either spatially coincident or not offset to any marked degree, so that the cells still behave as though they had a single discharge centre. The situation is rather more complex in the case of bimodal and multimodal simple cells where the discharge centres are not only spatially separated but may display various combinations of types of discharge centre. Binocular facilitation to a single edge will probably still occur but only when the two discharge centres that are in register are specific not only for the edge in question but also for the specific direction of edge movement. It is possible that these cells may respond to one pattern at one depth and another pattern at another depth. In other words these cells may not have a unique receptive field disparity value.

Neural Mechanisms for Stereopsis in Monkey. Recently, HUBEL and WIESEL (1970) have found cells sensitive to binocular depth in area 18 of the macaque monkey cortex but failed to find cells with similar properties in area 17. The most common type of binocular depth cell either failed to respond, or responded only weakly, to monocular stimulation whereas appropriate stimulation of the two eyes together produced very brisk responses. Similar depth cells have been studied in cat area 17 (BISHOP et al., 1971 d). HUBEL and WIESEL also observed the phenomenon of receptive field disparity and noted that the displacement of the field for one eye, relative to the field for the other eye, was usually at right angles to the receptive field organization. Thus vertically oriented fields were displaced horizontally, whereas with oblique fields there was a vertical component to the disparity as well. This correlation between receptive field orientation and disparity has not so far been observed in the cat. The most common depth cells in the monkey showed maximal discrimination for depth when the receptive fields were misaligned at right angles to the optimal orientation and relative insensitivity for depth when the misalignment was in the direction of the optimal orientation. BISHOP et al. (1971 d) have shown that the simple cells in the cat striate cortex behave in an exactly analogous manner and they have also revealed the nature of the receptive field organization responsible for this behaviour (cf. HENRY and BISHOP, 1971). An unexpected observation in the monkey was that the proportion of binocular depth cells increased as the electrode moved out into cortex having an increasingly peripheral representation away from the vertical midline (HUBEL and WIESEL, 1970). This is unexpected because stereoscopic acuity in man declines rapidly away from the centre of gaze. A possible explanation for the greater frequency of binocular depth cells in the periphery is that the increasing size of their receptive fields makes their detection the more likely (cf. BISHOP et al., 1971 d).

Constant Depth and Constant Direction Cell Columns. Observing that neighbouring binocular depth cells in area 18 of the monkey cortex often had the same horizontal disparity, HUBEL and WIESEL (1970) have suggested that there might be constant depth cell columns segregated from cells that are not particularly concerned with depth. They further suggested that constant depth columns would be distinct from constant orientation columns. Blakemore (1970 a) has also brought forward evidence for constant depth columns, as well as constant direction columns, in area 17 in the cat cortex. In both cases, however, Blakemore used constant orientation as the principal criterion for the identification of cells as belonging to the one functional cell column. This criterion clearly excluded any possibility of distinguishing separate and distinct columns for depth and orientation.

While there may well be constant depth columns, there are, however, strong arguments against accepting Blakemore's observations as satisfactory evidence for their existence. He identified columns on the basis of an average of 3 cells per column (range 2–4) separated in the cortex by a total distance of up to 2 mm in the case of each column. Brief comment is made below concerning the limitations of these observations but perhaps the most serious objection concerns the fact that the majority of the cells in the columns had receptive fields with horizontal eccentricities from about 5° to more than 20° into the periphery. The difficulty is that the exposed dorsal surface of area 17 (here postlateral gyrus) receives a projection from a vertical strip of visual field that does not extend laterally from the zero vertical meridian more than about 5° (Joshua and Bishop, 1970; Joshua and Bishop – unpublished observations; Bishop et al., 1971 d). The strip of visual field is probably less than 5° wide in the vicinity of the zero horizontal. To record cells with receptive fields beyond 5° azimuth, an electrode directed from the dorsal surface must pass down through the cortex on the medial surface of the hemisphere and hence pass mainly at right angles to the radially oriented cell columns of the suprasplenial and splenial gyri. The situation is further complicated by the changing orientations of the columns in the region of the suprasplenial sulcus. Blakemore gives no indication of the depth beneath the surface of the cortex at which the cells were recorded and appropriate histological checks were not carried out to show that the electrode had, in fact, been directed parallel to the columnar arrangement of the cells and fibres in the cortex. It is difficult to avoid the conclusion that the majority of the "cell columns" studied by Blakemore could not have been derived from anatomically-based cell columns. Indeed, quite apart from the above criticism, it can be argued that, considering groups of 2–4 cells, all the 5 criteria Blakemore used for judging units as belonging to one functional column are quite compatible with the possibility that the cells could have been recorded from two or more anatomically-based cell columns that were parallel to one another and separated from each other by a distance of up to 2 mm.

Furthermore Blakemore gives no indication of the basis for the decision as to the range of receptive field disparity values that was acceptable for cells to belong to a depth column and similarly with respect to directions for a direction column. The statistical analysis used by Blakemore as his principal evidence for separate depth and direction columns is invalidated by the fact that all the cells having properties intermediate between those for depth and those for direction were arbitrarily excluded from the analysis. In the circumstances it is hardly surprising that a highly significant difference between the two populations should emerge. Clearly further work is needed to establish the existence of constant depth columns in the visual cortex.

Nasotemporal Overlap. The concept of receptive field disparity as a basis for stereopsis requires the possibility that receptive fields for one eye that are located on the nasotemporal line (zero vertical meridian) be paired with receptive fields for the other eye that are located across the midline into the "wrong "hemifield. This possiblity, in turn, requires that there be a nasotemporal overlap projected from each eye to the visual cortex via the optic tract and lateral geniculate nucleus of each side. SANDERSON and SHERMAN (1971) have now completed the evidence for such a projection in the cat by showing that a strip of retina straddling the zero vertical meridian projects to the three main laminae (A, A_1 and B) of the dorsal nucleus of the lateral geniculate body and to the adjoining part of the medial interlaminar nucleus. LEICESTER's (1968) evidence for a similar nasotemporal overlap projected to the region of the 17/18 border in the visual cortex has also been confirmed by BLAKEMORE (1969).

Fine and Coarse Stereopsis. From observations made on two subjects, one with traumatic sagittal section of the optic chiasm (BLAKEMORE, 1970 b) and the other with surgical transection of the corpus callosum (MITCHELL and BLAKEMORE, 1970), it was concluded that the nasotemporal overlap that is projected via an optic tract is not an important factor for stereopsis in man. There are, however, compelling arguments against such a conclusion. BISHOP and HENRY (1971) have drawn a distinction between what they have termed *fine* and *coarse* stereopsis. Briefly, fine stereopsis describes the unique experience of "seeing solid". It matches closely similar features of the two retinal images within very localized regions of the visual field and over a very narrow range of spatial disparities. Fine stereopsis is closely linked with the phenomena of binocular single vision and retinal rivalry. By contrast, coarse stereopsis is a much less specific process which can operate on visual configurations quite dissimilar in form and luminance, and separated by several degrees in spatial position (cf. WESTHEIMER and TANZMAN, 1956; WESTHEIMER and MITCHELL, 1969; MITCHELL, 1969, 1970). Coarse stereopsis is a phenomenon of double images and commonly operates beyond the range of retinal rivalry. BISHOP and HENRY (1971) have suggested that BLAKEMORE and MITCHELL's observations concern the phenomenon of coarse stereopsis and have no direct bearing on the problems of fine stereopsis. The demonstrations by AKELAITIS (1941) and GAZZANIGA, BOGEN, and SPERRY (1962) that their split-brain patients possessed stereopsis unfortunately does not resolve this difference in interpretation because there was no guarantee that these patients used retinal disparities that could have involved a nasotemporal overlap.

LINKSZ (1971) also disagrees with BLAKEMORE and MITCHELL's conclusions. He too draws a clear distinction between stereopsis as "a unique experience based on innate and functioning retinal correspondence" on the one hand and what he terms "convergence stereopsis" on the other. He believes that BLAKEMORE and MITCHELL have studied the phenomenon of "diplopic convergence and deconvergence stereopsis" analyzed many years ago by HEINE (1900). LINKSZ points out that this kind of stereopsis requires an entirely different kind of disparity – diplopic disparity – with the diplopia forcing convergence to occur so as to eliminate the disparity.

Global Stereopsis. Perhaps the major problem facing the neural theory of stereopsis concerns the phenomenon of global stereopsis (JULESZ and SPIVACK, 1967) by which form and pattern recognition can arise from purely stereoscopic cues. The nature of this problem is discussed by BISHOP and HENRY (1971).

Acknowledgements

The author is grateful to Mrs. CAROL JACOB for her skill in the preparation of the figures and to Miss ISABEL SHEAFFE for her bibliographic and secretarial assistance.

References

AKELAITIS, A. J.: Studies on the corpus callosum. II. The higher visual functions in each homonymous field following complete section of the corpus callosum. Arch. Neurol. Psychiat. **45**, 788—796 (1941).

ALPERN, M.: Movements of the eyes. In: DAVSON, H. (Ed.): The Eye, Vol. 3, part 1, pp. 3—187. New York-London: Academic Press 1962.

BARLOW, H. B.: Slippage of contact lenses and other artefacts in relation to fading and regeneration of supposedly stable retinal images. Quart. J. exp. Psychol. **15**, 36—51 (1963).

— BLAKEMORE, C., PETTIGREW, J. D.: The neural mechanism of binocular depth discrimination. J. Physiol. (Lond.) **193**, 327—342 (1967).

BISHOP, P. O.: Seeing with two eyes. Aust. J. Sci. **32**, 383—391 (1970 a).

— Beginning of form vision and binocular depth discrimination in cortex. In: SCHMITT, F. O. (Ed.): The Neurosciences: Second Study Program, 471—485. New York, N. Y.: Rockefeller University Press 1970 b.

— COOMBS, J. S., HENRY, G. H.: Responses to visual contours: spatio-temporal aspects of excitation in the receptive fields of simple striate neurones. J. Physiol. (Lond.) **219**, 625—657 (1971 a).

— — — Interaction effects of visual contours on the discharge frequency of simple striate neurones. J. Physiol. (Lond.) **219**, 659—687 (1971 b).

— — — Receptive field of simple cells in the cat striate cortex. J. Physiol. (Lond.) (in press) (1973).

— HENRY, G. H.: Spatial vision. Ann. Rev. Psychol. **22**, 119—160 (1971).

— — SMITH, C. J.: Binocular interaction fields of single units in the cat striate cortex. J. Physiol. (Lond.) **216**, 39—68 (1971 d).

BLAKEMORE, C.: Binocular depth discrimination and the nasotemporal division. J. Physiol. (Lond.) **205**, 471—497 (1969).

— The representation of three-dimensional visual space in the cat's striate cortex. J. Physiol. (Lond.) **209**, 155—178 (1970 a).

— Binocular depth perception and the optic chiasm. Vision Res. **10**, 43—47 (1970 b).

— Binocular interaction in animals and man. Ph. D. thesis, Univ. Calif., Berkeley 1968.

BORING, E. G.: Sensation and perception in the history of experimental psychology. New York: Appleton-Century-Crofts 1942.

BOUGH, E. W.: Stereoscopic vision in the macaque monkey: a behavioural demonstration. Nature (Lond.) **225**, 42—44 (1970).

BRECHER, G. A.: Form und Ausdehnung der Panumschen Areale bei fovealem Sehen. Pflügers Arch. ges. Physiol. **246**, 315—328 (1942).

CHOUDHURY, B. P., WHITTERIDGE, D., WILSON, M. E.: The function of the callosal connections of the visual cortex. Q. J. exp. Physiol. **50**, 214—219 (1965).

COLONNIER, M.: The tangential organization of the visual cortex. J. Anat. **98**, 327—344 (1964).

CREUTZFELDT, O., ITO, M.: Functional synaptic organization of primary visual cortex neurones in the cat. Exp. Brain Res. **6**, 324—352 (1968).

DITCHBURN, R. W., FOLEY-FISHER, J. A.: Assembled data in eye movements. Optica Acta **14**, 113—118 (1967).

— GINSBORG, B. L.: Involuntary eye movements during fixation. J. Physiol. (Lond.) **119**, 1—17 (1953).

DOVE, H. A.: Die Combination der Eindrücke beider Ohren und beider Augen zu einem Eindruck. Ber. preuss. Akad. Wiss. (1841) Cited by Boring, E. G. (1942).

FENDER, D., JULESZ, B.: Extension of Panum's fusional area in binocularly stabilized vision. J. opt. Soc. Amer. **57**, 819—830 (1967).

FLOM, M. C., ESKRIDGE, J. B.: Change in retinal correspondence with viewing distance. J. Amer. opt. Ass. **39**, 1094—1097 (1968).

GAZZANIGA, M. S., BOGEN, J. E., SPERRY, R. W.: Some functional effects of sectioning the cerebral commissures in man. Proc. nat. Acad. Sci. (Wash.) **48**, 1765—1769 (1962).

HEBBARD, F. W., MARG, E.: Physiological nystagmus in the cat. J. opt. Soc. Amer. **50**, 151—155 (1960).

HEINE, L.: Sehschärfe und Tiefenwahrnehmung. Albrecht v. Graefes Arch. Ophthal. **51**, 146—173. Quoted by Mitchell and Blakemore (1970).

HELMHOLTZ, H. v.: Popular lectures on scientific subjects. Translation edited by E. Atkinson. First series. London: Longmans, Green & Co. 1881.

— Handbuch der Physiologischen Optik, third ed. (Ed. by Gullstrand et al.) English translation edited by J. P. C. Southall, Opt. Soc. Amer., New York, 1924—1925; reprinted by Dover, New York, 1962. (1909—1911).

HENRY, G. H., BISHOP, P. O.: Simple cells of the striate cortex. In: NEFF, W. D. (Ed.): Contributions to Sensory Physiology, Vol. 5, pp. 1—46. New York: Academic Press 1971.

— — COOMBS, J. S.: Inhibitory and subliminal excitatory receptive fields of simple units in cat striate cortex. Vision Res. **9**, 1289—1296 (1969).

HILLEBRAND, F.: Die Stabilität der Raumwerte auf der Netzhaut. Z. Psychol. Physiol. Sinnesorg. **5**, 1—60. Cited by OGLE, 1950. (1893).

HUBEL, D. H., WIESEL, T. N.: Receptive fields of single neurones in the cat's striate cortex. J. Physiol. (Lond.) **148**, 574—591 (1959).

— — Receptive fields, binocular interaction and functional architecture in the cat's visual cortex. J. Physiol. (Lond.) **160**, 106—154 (1962).

— — Shape and arrangement of columns in cat's striate cortex. J. Physiol. (Lond.) **165**, 559—568 (1963).

— — Receptive fields and functional architecture in two nonstriate visual areas (18 and 19) of the cat. J. Neurophysiol. **28**, 229—289 (1965 a).

— — Binocular interaction in striate cortex of kittens reared with artificial squint. J. Neurophysiol. **28**, 1041—1059 (1965 b).

— — Cortical and callosal connections concerned with the vertical meridian of visual fields in the cat. J. Neurophysiol. **30**, 1561—1573 (1967).

— — Receptive fields and functional architecture of monkey striate cortex. J. Physiol. (Lond.) **195**, 215—243 (1968).

— — Stereoscopic vision in macaque monkey. Nature (Lond.) **225**, 41—44 (1970).

JOSHUA, D. E., BISHOP, P. O.: Binocular single vision and depth discrimination. Receptive field disparities for central and peripheral vision and binocular interaction on peripheral units in cat striate cortex. Exp. Brain Res. **10**, 389—416 (1970).

JULESZ, B.: Binocular depth perception without familiarity cues. Science **145**, 356—362 (1964).

— Some neurophysiological problems of stereopsis. In: NYE, P. W. (Ed.): Information Processing in Sight Sensory Systems, pp. 135—142. Pasadena, California: Cal. Tech. Press. 1965 a.

— Texture and visual perception. Sci. Amer. **212**, 38—48 (1965 b).

— SPIVACK, G. J.: Stereopsis based on vernier acuity cues alone. Science **157**, 563—565 (1967).

KAUFMAN, L.: On the spread of suppression and binocular rivalry. Vision Res. **3**, 401—415 (1963).

KINSTON, W. J., VADAS, M. A., BISHOP, P. O.: Multiple projection of the visual field onto the medial portion of the dorsal lateral geniculate nucleus and the adjacent nuclei of the thalamus of the cat. J. comp. Neurol. **136**, 295—316 (1969).

KOZAK, W., RODIECK, R. W., BISHOP, P. O.: Response of single units in lateral geniculate nucleus of cat to moving visual patterns. J. Neurophysiol. 28, 19—47 (1965).

LEICESTER, J.: Projection of the visual vertical meridian to cerebral cortex of the cat. J. Neurophysiol. 31, 371—382 (1968).

LEVELT, W. J. M.: On binocular rivalry. p. 1—110. Institute for Perception RVO-TNO, Soesterberg — The Netherlands 1965.

LINKSZ, A.: Comment on BLAKEMORE (1969, 1970 b) and MITCHELL and BLAKEMORE (1970) in Surv. Ophthalm. 15, 348—353 (1971).

LIT, A.: The magnitude of the Pulfrich stereophenomenon as a function of binocular differences of intensity at various levels of illumination. Amer. J. Psychol. 62, 159—181 (1949).

— The magnitude of the Pulfrich stereophenomenon as a function of target velocity. J. exp. Psychol. 59, 165—175 (1960).

MITCHELL, D. E.: Qualitative depth localization with diplopic images of dissimilar shape. Vision Res. 9, 991—994 (1969).

— Properties of stimuli eliciting convergence eye movements and stereopsis. Vision Res. 10, 145—162 (1970).

— Retinal disparity and diplopia. Vision Res. 6, 441—451 (1966).

— BLAKEMORE, C.: Binocular depth perception and the corpus callosum. Vision Res. 10, 49—54 (1970).

MOUNTCASTLE, V. B.: Modality and topographic properties of single neurons of cat's somatic sensory cortex. J. Neurophysiol. 20, 408—434 (1957).

MÜLLER, J.: Beiträge zur vergleichenden Physiologie des Gesichtsinnes. Leipzig: Cnobloch. (1826). Cited by BORING, E. G. 1942.

NIKARA, T., BISHOP, P. O., PETTIGREW, J. D.: Analysis of retinal correspondence by studying receptive fields of binocular single units in cat striate cortex. Exp. Brain Res. 6, 353—372 (1968).

OGLE, K. N.: Researches in binocular vision. Philadelphia: W. B. Saunders 1950.

— Stereopsis and vertical disparity. Arch. Ophthal. 53, 495—504 (1955).

— Theory of stereoscopic vision. In: KOCH, S. (Ed.): Psychology: A Study of a Science. Vol. 1. Sensory, Perceptual and Physiological Formulations, pp. 362—394. New York: McGraw-Hill Book Co. Inc. 1959.

— Spatial localization through binocular vision. In: DAVSON, H. (Ed.): The Eye, Vol. 4, pp. 271—320. New York: Academic Press 1962.

— The optical space sense. In: DAVSON, H. (Ed.): The Eye, Vol. 4, Part II, pp. 211—417. New York-London: Academic Press 1962.

— WEIL, M. P.: Stereoscopic vision and the duration of the stimulus. Arch. Ophthal. 59, 4—17 (1958).

PANUM, P. L.: Physiologische Untersuchungen über das Sehen mit zwei Augen. Kiel (1858). Cited by Boring, E. G. (1942).

PETTIGREW, J. D., NIKARA, T., BISHOP, P. O.: Responses to moving slits by single units in cat striate cortex. Exp. Brain Res. 6, 373—390 (1968 a).

— — — Binocular interaction on single units in cat striate cortex: simultaneous stimulation by single moving slit with receptive fields in correspondence. Exp. Brain Res. 6, 391—410 (1968 b).

PRITCHARD, R. M., HERON, W.: Small eye movements of the cat. Canad. J. Psychol. 14, 131—137 (1960).

RODIECK, R. W.: Central nervous system: afferent mechanisms. Ann. Rev. Physiol. 33, 203—240 (1971).

RØNNE, G.: The physiological basis of sensory fusion. Acta Ophthal. 34, 1—26 (1956).

SANDERSON, K. J., DARIAN-SMITH, I., BISHOP, P. O.: Binocular corresponding receptive fields of single units in the cat dorsal lateral geniculate nucleus. Vision Res. 9, 1297—1303 (1969).

— SHERMAN, S. M.: Nasotemporal overlap in the visual field projected to the lateral geniculate nucleus in the cat. J. Neurophysiol. 34, 453—466 (1971).

SHERRINGTON, C. S.: Man on his Nature. London: Cambridge University Press 1940.

SHORTESS, G. K., KRAUSKOPF, J.: Role of involuntary eye movements in stereoscopic acuity. J. opt. Soc. Amer. 51, 555—559 (1961).

St-Cyr, G. J., Fender, D. H.: The interplay of drifts and flicks in binocular fixation. Vision Res. **9**, 245—266 (1969).

Stone, J.: The naso-temporal division of the cat's retina. J. comp. Neurol. **126**, 585—600 (1966).

Treisman, Anne: Binocular rivalry and stereoscopic depth perception. Quart. J. exp. Psychol. **14**, 23—37 (1962).

Vieth, G. U. A.: Ueber die Richtung der Augen. Ann. Phys. (Lpz.) **58**, 233—255 (1818). Cited by Boring, E. G. (1942).

Walls, G. L.: The vertebrate eye and its adaptive radiation. Bloomfield Hills, Michigan: Cranbrook Press 1942.

Westheimer, G., Mitchell, D. E.: The sensory stimulus for disjunctive eye movements. Vision Res. **9**, 749—755 (1969).

— Tanzman, I. J.: Qualitative depth localization with diplopic images. J. opt. Soc. Amer. **46**, 116—117 (1956).

Wheatstone, C.: Contributions to the Physiology of Vision. — Part the First. On some remarkable, and hitherto unobserved, Phenomena of Binocular Vision. Phil. Trans. roy. Soc. **128**, 371—394 (1838).

Yarbus, A. L.: Eye Movements and Vision. Russian text, Nauka, Moscow, 1965. English translation by B. Haigh and L. A. Riggs, New York: Plenum Press 1967.

Chapter 5

Visual Stability and Voluntary Eye Movements

By

DONALD M. MACKAY, Keele, Staffordshire (Great Britain)

With 4 Figures

Contents

1. The Nature of the Problem

When the eyes or the body are moved so that the retinal image changes, the resulting signals from the retina resemble in some respects those which would have been generated if the eyes had been stationary and part of the visual world had moved. When the image displacement is not due to use of the oculomotor or locomotor system, a strong impression of motion-of-the-visual-scene is created. This is so even when the displacement is produced by the voluntary use of other muscles, for example by tapping the eyelid with a finger. During normal exploratory eye-movement and locomotion, however, no such instability of the visual world is seen. As we visually examine an object or walk towards it, it is perceived as a stationary structure despite the jerky translations, shearing motions or pulsatile expansions that take place in its retinal image. Conversely, if paralysis of extra-ocular muscles (HELMHOLTZ, 1866; MACH, 1906) or the use of optical image-stabilizing apparatus (DITCHBURN and GINSBORG, 1952) prevents the image from changing in the way that would normally match the voluntary movement in question, illusory motion-of-the-visual-world is seen even though the retinal image is stationary. (That this applies not only to stabilization against image translation can easily be verified by looking at a nearby object through high-powered binoculars while walking towards it).

The questions raised by these facts are of two kinds. (A) What sources of information are used by the central nervous system in differentiating between image changes due to objective motion of the visual world and those due to voluntary movements? (B) What use does the system make of the information? What information-processing operations are required to explain the stability of the visual world in normal circumstances?

One well-known theory postulates in answer to question B a "suppressive" mechanism, which is supposed to attenuate transmission of retinal signals during a saccadic eye-movement (Holt, 1903). We shall criticize this theory in detail in Section 7.

A second and more widely-accepted answer to B, which we shall call the "subtractive" theory, is that since no illusory motion is normally seen, the information about eye movements must be used to "cancel out" or "subtract" the effects of voluntary movements of eye or body on the retinal signals (Von Holst and Mittelstaedt, 1950; Sperry, 1950; von Holst, 1954, 1957; Teuber, 1960). To quote von Holst (1954), "... the efference [to the effector] leaves an "image" of itself somewhere in the CNS, to which the reafference (i.e. sensory signs of the resulting movement) compares as the negative of a photograph compares to its print. ... When the efference copy and the reafference exactly compensate one another, nothing further happens. When, however, the reafference is too small or ... too great, ... the difference can either influence the movement itself, or for instance, ascend to a higher centre and produce a perception". The proposed subtractive mechanism, which von Holst later (1957) schematised in the form of Fig. 3a, is quite analogous to that by which the map of the coastline on a ship's radar screen is prevented from rotating when the ship's head turns, by using signals indicative of ship-rotation to subtract an equal and opposite rotation from the input to the display. (The main difference is that in radar practice the information is derived as "inflow" from a gyro-compass, rather than from the "outflow" to the ship's steering gear (compare Sections 2(b) and 2(c) below); but this does not affect the present point). The subtractive theory of perceptual stability is often erroneously attributed to Helmholtz (1866) who did answer question A in terms of what he called the "effort of will "(*Willensanstrengung*) but in fact offered no specific hypothesis in answer to B.

One of the chief aims of this chapter will be to question some presuppositions underlying both of these answers. Our argument will be that although the sensory mechanism cannot properly interpret sensory changes without some information as to the activity of the motor system, from an information-engineering standpoint the need is not for the changes due to voluntary movement to be *eliminated* from the sensory input, but for them to be appropriately *evaluated* by the central mechanism responsible for the organism's "conditional readiness" to reckon with its environment (MacKay, 1957, 1958a, 1961b, 1966). Such an evaluative mechanism, as we shall see, can operate with much less accurate oculomotor information than a subtractive mechanism.

Clearly, correct evaluation implies in practice that if eye movement causes an image displacement Θ on the retina, a target at a bearing of Φ relative to the optic axis should be perceived as having a bearing (Φ—Θ). In this sense perceptual stability tautologously involves "subtraction". But this is a truism, not a theory; and it leaves unanswered our physiological question, as to the signal-processing operations required to account for the normal absence of perceived motion.

It is equally clear that at the level of *sensori-motor coordination* an effectively subtractive neural process (Fig. 1) could be a convenient means of modifying the responsive orientation of the organism towards a retinally-represented target

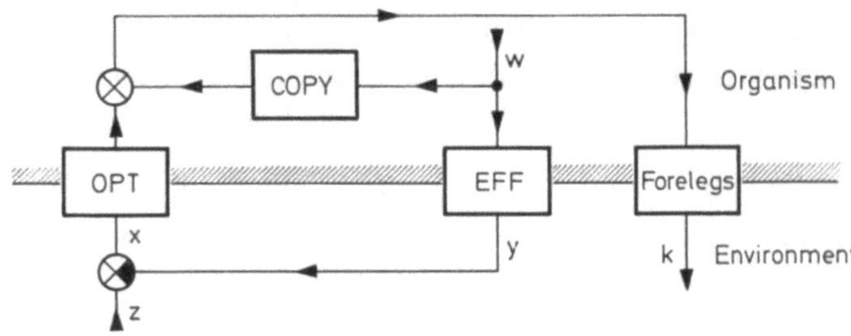

Fig. 1. The model of sensori-motor coordination proposed by von HOLST and MITTELSTAEDT (1950) to allow for the effects of self-produced changes in orientation of the eye. A target in direction z is signalled by the optic system (OPT) at a bearing of x (relative to the eye axis) because of a rotation of the eyes y in response to a command signal w. w is presumed to give rise to a "COPY" signal equal but opposite in sign to that produced by y. This "COPY" signal is added to the optic system output so as to cancel the effect of y, so that the direction of motor action k equals that of the target z (From MITTELSTAEDT, 1962)

so as to take account of the orientation of the retina itself (VON HOLST and MITTELSTAEDT, 1950; SPERRY, 1950). A compensatory process of this kind may be exemplified by the slow changes in the orientation of visual receptive fields in response to bodily tilt, which have been observed in crustaceans (WIERSMA and YAMAGUCHI, 1967) and in cats (HORN and HILL, 1969). What we must not do is to confuse this problem of the stability of sensori-motor coordination with the quite different problem of *perceptual* stability. For the former, VON HOLST's subtractive feedback scheme (sometimes modified to utilize proprioceptive information) seems to have a wide range of application. It is only its use in the latter context that we shall be concerned to question, for reasons that will become clearer in Sections 3 and 4.

2. Possible Sources of Information

As a first step it will be useful to survey the sources from which the CNS could conceivably acquire the necessary information to distinguish the visual consequences of voluntary movement. Three possibilities suggest themselves:

(a) The visual signals themselves might have "telltale" features not present in signals that result from objective motion of the visual scene.

(b) The oculomotor or bodily movements responsible could be indicated by "inflow" from proprioceptors.

(c) The "outflow" of signals to the oculomotor or other motor systems could be monitored at some point in the chain of control.

a) Telltale Features: Motion-signalling Versus Position-Signalling Systems

Gibson (1950, 1966) has been particularly insistent on the differences between transformations of the retinal image that result from voluntary eye movement and those due to objective motion in the visual scene. It is undeniable that few if any naturally occurring objective movements involve rigid displacement of the whole retinal image. On the other hand tapping lightly on the canthus of the eye, or on a small mirror before the pupil, can produce rapid displacements of the whole image quite similar to those in natural saccades; yet they evoke an illusory impression of motion that is unmistakably absent during comparable voluntary saccades. In this case at least it seems implausible to attribute the instability perceived to the presence or absence of "telltale features" in the retinal signal itself.

There is however a class of image movements that can generate telltale retinal signals, in the sense of signals that are not produced by normal saccadic inspection. These are continuous translations of the image of the kind produced by pressing rather than tapping the canthus, or by visually following an object which is moving smoothly over a patterned background (See Section 5 below). In both of these cases a strong impression of illusory motion of the background is created.

A particularly striking "stroboscopic illusion" (MacKay 1958a, 1961a) shows how much more sensitive the visual system is to a continuous motion of the image over the retina than to a discontinuous displacement through the same visual angle. A patterned field is stroboscopically illuminated at a rate well below that for persistence of vision — say at 6 flashes per second. The field contains one or more continuously illuminated objects, such as an indicator lamp or the glowing cathodes of some vacuum tubes. If now the combined image is gently moved over the retina by slight intermittent pressure on the canthus of one eye (with the other closed), for small enough displacements only the steadily lit objects are seen to move.

The interpretation proposed by MacKay was that in addition to its position-signalling mechanism, the visual system had a mechanism sensitive specifically to what might be called the "drifting" of an image continuously over the retina. Such a mechanism could signal velocity-as-such, rather as the reeds in a river-bed could serve specifically to indicate the relative velocity of the water by the torque on their roots. This velocity-information was presumed to be processed in a separate channel from that representing the successive positions of the image (see Section 6 below). Since the continuously moving image could stimulate both mechanisms, whereas the intermittently presented image could stimulate only the position-signalling one, it would be understandable that the movements perceived were not the same. The inference suggested was that the magnitude of perceived displacement depended on the combined "weight" of evidence from both the drift-sensitive and position-sensitive systems (MacKay, 1958a, 1961a).

It was suggested that the same physiological postulate could account for the familiar "waterfall" after-effect, whereby after viewing a continuously moving surface one perceives a stationary surface as paradoxically "moving but not changing its location" in the opposite direction. By analogy with brightness- or colour-adaptation, it seemed plausible that prolonged excitation of a population of motion-sensitive elements by stimuli drifting in one direction would bring about physiological adaptation in the excited elements, so that the "zero-response" of the velocity-signalling system would be shifted in the negative direction.

This interpretation, which seems to be generally accepted among psychologists (see for example SUTHERLAND, 1961; GREGORY and ZANGWILL, 1963; SEKULER and GANZ, 1963) has found support from the discovery in animals by HUBEL (1958), BARLOW and HILL (1963), BAUMGARTNER et al. (1964), and others of visual units preferentially sensitive to motion of a patterned image in one direction, which show transient suppression of their resting discharge as an after-effect of stimulation (see Chapter 6). If we accept it as probable that there are analogous populations of units in man, at least at a post-retinal level, we may take it that our CNS does receive a sensory input from continuous drifting of the image over the retina which is characteristically different from that due to voluntary (saccadic) exploratory movements, without implying that this difference alone accounts for perceptual stability in the latter case. We return to this point in Sections 5 and 6.

b) Proprioceptive "Inflow"

Sir CHARLES SHERRINGTON (1918) was perhaps the most distinguished advocate of the view that "inflowing" proprioceptive information from the extraocular muscles maintained perceptual stability. He had earlier demonstrated the existence of the necessary stretch receptors (SHERRINGTON, 1898), which has been confirmed in both animals and man by more recent work (COOPER and DANIEL, 1949; CHRISTMAN and KUPFER, 1963; BACH-Y-RITA, 1971). Experimental support for the "inflow" theory has been claimed by GUREVITCH (1959), who found signs of "overshoot" and successive approximation ("hunting") in the course of individual attempts to reproduce (in the dark) a previously-learned saccadic movement. Although 40% of movements were as accurate in the dark as with visible targets, his data do not rule out the possibility that the "hunting" resulted from successive comparisons of integrated *outflow* against an internal standard, so they cannot be regarded as conclusive evidence of "inflow" signals at work.

In the case of illusory world-motion seen when attempts are made to move a paralysed eye, it seems particularly unlikely that the discharge of stretch receptors could be responsible. Only those of the unparalysed eye would be activated by such attempts; and in a clinical study by JACKSON and PATON (1909) the observed movements of this eye did not match closely the illusory sensations of motion reported by the patient.

To date, indeed, there seems to be no consensus as to the functional significance of eye-muscle proprioceptors (BACH-Y-RITA, 1971), though it would be consistent with present knowledge (see for example COOPER et al. 1955; FUCHS and KORNHUBER, 1969) to suppose that they serve to stabilize the transfer characteristics of the extraocular muscles, as proposed by LUDVIGH (1952a, b — see Fig. 3b). The hypothesis that they give subjective knowledge of eye position (in man) was conclusively disproved for passive rotations up to 40° by BRINDLEY and Merton (1960), thereby discouraging attempts to explain perceptual stability by reference to the "inflow" they provide.

c) Oculomotor "Outflow"

If the motion of the eyeball were precisely determined by the outflow of impulses to the eye muscles, a "corollary signal"[1] derived from this outflow could provide all the information needed by the CNS as to the position of the eyes. In physiological reality the precision that can be expected is well below that with which the retinal image can be fixated. In an otherwise empty visual field, displacements of the order of 20 sec of arc can be detected (Basler, 1906), so that if exploratory eye movements had to be inferred from "outflow" signals with this degree of precision, the oculomotor transfer function would need to be stable to better than 0.1%. Observations of the accuracy of voluntary eye movements (Young, 1962), as well as physiological common sense, discourage the idea that oculomotor "outflow" could routinely specify eye movements, let alone eye position, with an accuracy of this order.

There is nevertheless increasing evidence that oculomotor information of some kind is available to a number of brain areas usually regarded as sensory. Bizzi (1966, 1967) and Jeannerod and Putkonen (1970) for example have found eye-movement-related discharges in the lateral geniculate nucleus of cats, though not in that of alert monkeys. Wurtz (1969b) emphasises that no sign has been seen of a corollary discharge in the primate visual pathway *up to the striate cortex* during the rapid eye movements of wakefulness. Whether such a discharge may be associated with slow tracking movements remains uncertain. (see Potthoff et al., 1967; Kawamura and Marchiafava. 1968; Bach-y-Rita, 1971). Wurtz and Goldberg (1971) have, however, identified two types of eye-movement-sensitive units in monkey *superior colliculus*. One type responded to either induced following or spontaneous eye movements in light or dark, and to caloric nystagmus. The other did not respond strongly to caloric nystagmus or to spontaneous eye movements, but fired (generally ahead of the ocular EMG) when the eye was made to follow a moving fixation point.

Although it is not always possible to rule out "inflow" from proprioceptors, the results of Goldberg and Wurtz show that in at least some cases corollary discharges exist which are not tied directly to eye movements, and so presumably originate in "outflow" from the central process controlling the movement concerned. The main question to which we now turn concerns the kind of work that may need to be done by such corollary signals if they are necessary for perceptual stability.

3. Perception of Change

Before trying to frame any physiological hypothesis to account for the stability of the visual world during voluntary eye movements, it will be useful to analyse logically what is needed for the perception of *genuine* changes in the environment. From an information-engineering standpoint, the perceptible world

[1] The term "corollary discharge" was introduced by Sperry (1950), who independently of von Holst proposed that such signals might "compensate for the retinal displacement".

can be regarded as a source of constraints on action (internal and external) and the organization of action, particularly locomotion. The basic functions of sensory perception are thus twofold: to keep uptodate both (a) the evaluation and selection of ongoing effector action, and (b) the *conditional* readiness of the organism to take account of the contents of its environment when organising action.

Imagine first, then, an ideal environment in which nothing changed. Here the *only* changes in sensory input would be caused by movements of the sensory surface (in this case the retina) relative to the sensory field (in this case the optical projection of the visual world). In this imaginary situation, the function of the eyes would be simply to offer running feedback whereby the organism could steer its way among the fixed contents of its world. The location of these could be permanently represented in some suitable internal form, which would impose conditional constraints on the planning of locomotion analogous to those that a map of a town imposes on the driver of a car. The physiological details of this internal representation or "map"[2] are not important for the present argument. The point is that once it had been built up, in whatever form, there would be no further need for any coupling to it from the sensory input, as long as the world was stable. It would be pointless to allow the input to play constantly upon the "map", having to be corrected by means of elaborate "compensations" for the changing sensory effects of locomotion and eye movements, yet contributing to it only 100%-redundant information. On the other hand, there would be no question of "suppressing" these sensory changes: they would be fully employed in their primary function of locomotor feedback. In other words, sensory signals in a stable world would have a high selective-information rate with respect to locomotor guidance, but a negligible selective-information rate with respect to internal map-making.

To take an analogy from the tactile domain (MacKay, 1962), we may think of a blind man finding his way around a room full of stationary objects. He uses a variety of muscles to move a sensory surface (his palm or finger-tip) over the sensory field (the surfaces of objects). The result is a constantly changing stream of "rubbing sensations" which are not at all imperceptible; but once he has become familiar with the objects around him these changing signals serve simply as feedback for his motor system, and as clues to the detailed structure of the objects he encounters. The changes in sensory input, however strongly they may be perceived, have *no selective-information-content*, in the technical sense (Shannon, 1948; MacKay, 1969) requiring any *changes* in the blind man's internal "map" of the objects he explores. On the contrary, in the circumstances they only provide him with fresh (and highly-redundant) positive evidence that the disposition of the familiar objects has *not* changed. Here no question of "cancelling" or "suppressing" these changing signals arises; so it is difficult to see why it should be thought necessary in the visual context. In both cases, what is needed is simply a *criterion* (varying both in nature and in precision according to the movement being made) whereby the input signals can be *evaluated* for any "map-changing" content.

[2] For brevity we shall use the term "map" to refer to whatever physiological configuration serves the same organizing function as a map does. As argued below, it need not, and probably should not, be a detailed topographical analogue.

What then if an object in the environment does move? In the first place, the sensory signals generated by its motion must be distinctive in some way, so that they call for a change in the internal representation of its location in the external world: a change whose subjective correlate is presumably the perception-of-its-motion. In the case of our blind man, for example, it is not the presence of sensory change as such, but a sufficiently large departure from the accustomed sequence of sensory changes, that would provide evidence of any changes in his environment. In other words, the perception of environmental change (or stability) depends not on the *detection* (or non-detection) of sensory *change* as such, but on the detection (or non-detection) of significant map-changing *information-content* in the sensory input, i.e. significant according to criteria determined by the ongoing programme of movement. *A change in criterion may therefore produce the same perceptual effect as a change of input.*

Thus if an attempt to explore an object by motion of the sensory surface did *not* generate displacement signals, the absence of these would provide *prima facie* evidence that the object was moving at the same rate as the surface. The same would be expected to hold for a moving object fixated against an unpatterned background, and also for illusory "objects" such as after-images, or the cortically-induced phosphenes studied by Brindley and Lewin (1968). Not surprisingly, all these are perceived as "moving with the eyes" when the subject turns his eyes voluntarily, but not when the eyeballs are passively rotated. The absence of retinal motion-signals in the presence of oculomotor activity is evaluated as evidence of motion in the frame of external space.

It may be noted in passing that the observation by Brindley and Lewin seems to dispose of at least one version of the "subtractive" theory of perceptual stability, according to which the effects of voluntary eye movements upon retinal signals were supposed to be annulled by a corollary discharge to the lateral geniculate nucleus (LGN). In their case the human subject was blind, and phosphenes were induced by direct stimulation through electrodes fixed in relation to the occipital cortex; yet when voluntarily rotating her eyeballs the subject saw the phosphenes "drifting with the eye" just as retinal afterimages would do. In other words, bypassing the LGN made no significant difference to the circumstances in which the phosphenes were seen as stationary or moving — which makes it doubtful that perceptual stability normally depends on keeping the cortical "image" stationary by means of compensatory transformations of input at the LGN level.

Secondly, the evaluative process must be sufficiently detailed to specify, not only *that* a change in the "map" is called for, but also *what* change is wanted; and it should be coupled to the internal representation in such a way as to bring about this change without disturbing the other features represented. To put it conversely, the environment must be internally represented in a way that makes it easy for the location attributed to a moving object to be changed appropriately by the "mismatch signals" from the evaluative process, with the minimum of disruption of its other attributed features (e.g. shape) or those of its neighbours. Unlike a drawing, in which the location of objects can be changed only by rubbing out all details of their form and sketching them afresh elsewhere, what is needed is some kind of representation wherein details of form are laid down separately from details of relative location — as they are in a scale-model with movable components, or in a verbal description in which each item has a changeable "address" (see end of next Section).

Thirdly, the mechanism should take account of the fact that the information-content of the changes actually perceived in the world generally amounts to a minute fraction of the total information represented in the map at any one time. The maximum rate at which new visual information can be absorbed has in fact been psychophysically estimated at about 5 bits per exposure in man (MILLER, 1956), as compared with a potential selective-information content for the visual field of the order of 10^6 bits. In other words, even to deal with a changing environment, our internal perceptual "map" has only a relatively low-capacity informational coupling to the sensory input. It has therefore no alternative but to run on a conservative principle of "informational inertia": the "null-hypothesis" that the map is correct until *sufficient evidence* is received to the contrary. This is also the informationally efficient solution, since it reserves the main capacity of the visual channel for current feedback, and allows changes in the "map" to be prescribed at the much lower information-rate demanded by the typical environment.

Such informational parsimony has, of course, its penalties. Given the opportunity to make an "economical "match, the low-rate updating mechanism may adopt absurdly non-veridical makeshift strategies, leading to striking perceptual illusions. An instructive example is the perceptual impression evoked by a statistically uniform field of dynamic "visual noise", which shows clear signs of an internal matching response to "information overload" that is anything but random, and is matched only with respect to some gross statistical parameters of the stimulus (MACKAY, 1961 b).

Fig. 2. This "Ray" figure (MACKAY, 1957) gives an impression of local distention or contraction of the visual field when fixated half way along one of its rays and then moved to and fro in a direction parallel to that ray. The distortion disappears if a marker is placed at the fixation point so as to generate adequate retinal clues to the motion of the figure

Another is the "rubber sheet" effect seen when a pattern such as Fig. 2 is displaced to and fro over the retina by gentle pressure on the canthus (MacKay, 1960). If gaze is directed at a point halfway along a ray pointing in the direction of motion, the field in its vicinity seems to undergo elastic expansion and contraction, rather than displacement. If, however, a landmark such as an orthogonal line is superposed on the region of fixation, this sensation is immediately abolished and replaced by one of displacement. The implication is that the changes made in the "map" are the most parsimonious that do justice in each locality to the mismatch signals for that locality, regardless of their *intellectual* improbability.

4. Physiological Requirements

What kinds of physiological implications follow from these logical requirements? The first and most significant point is that since it is not the energy but the information-content of sensory signals that determines whether a change in the internal representation of the perceived world is called for, there is no obvious need to postulate that the changes in sensory input resulting from voluntary eye movements are either physiologically *subtracted* by means of an oculomotor "efferenzkopie" (von Holst, 1957, see Fig. 3a) or *suppressed* (Holt, 1903) in order to account for the stability of the perceived world. What is needed is not *physical elimination*, but *informational evaluation* in the light of what the motor system is about. "Evaluation" here does not entail (or deny) conscious mental activity, but refers simply to the kind of computing process, familiar in feedback-guided automata, in which sensory signals are compared against criteria, so as to generate feedback signals indicating whether the states of affairs signalled are "satisfactory" or otherwise, and what corrective or reinforcing actions are required, if any. We can equally well think of this, in the present context, as a process by which the internal "map" is evaluated as "satisfactory" or as "needing revision" on the basis of the evidence from incoming signals. Whichever way we look at it, it is clear that *the detailed criteria of "satisfactoriness" must be determined from moment to moment according to the course and goal of current motor activity*: so there is a need for signal pathways *to the sensory evaluator* from those points in, or even prior to, the motor command system at which the necessary information is available.

What is distinctive in the present analysis, however, is not the postulation of these pathways (since some connection between sensory and motor systems would be needed on almost any theory of perceptual stability), but rather the purpose that the motor information or "corollary discharge" is here required to serve: not to *subtract from* or *cancel* parts of the visual input, but more generally to *set criteria* that determine whether and in what respect that input calls for readjustment of the internal "map". The main physiological implication is that this motor information need not be generated or transmitted with the high precision of the visual input, as it would have to be in a "subtractive" model. On the contrary, the less accurate the control of the retinal image by the motor system in question, the greater must be the tolerance of the evaluator, and the cruder the motor information that will be sufficient to set the criteria of evaluation appropriately.

On this conservative principle, low accuracy in the motor system need not cause perception of spurious motion (as it would on a "subtractive" principle), but only a corresponding inability to recognize as genuine any real environmental motion whose contribution to the sensory signal is too small to transgress the tolerance of the evaluator. We have here to recognize a sharp distinction between the threshold for detection of sensory *change* and the threshold for detection of

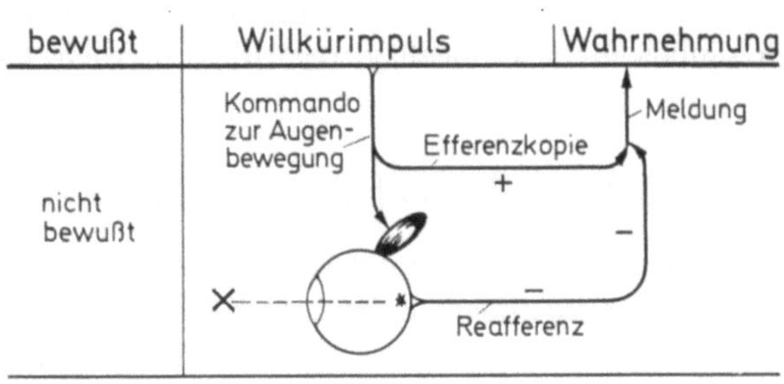

a

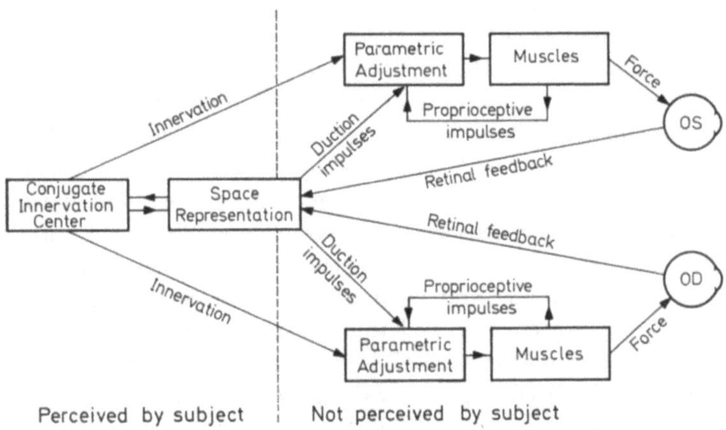

b

Fig. 3. a Von Holst's (1957) diagram illustrating his theory that perceptual stability is maintained by the subtraction of an "*efferenzkopie*" (a corollary discharge derived from the outflow to the eye muscles) from the "reafferenz" signal generated by retinal image displacement in voluntary eye movement (See also v. Holst and Mittelstaedt, 1950). b Ludvigh's (1952b) diagram of the postulated interactions between oculomotor and sensory signals required for the maintenance of perceptual stability

map *mismatch*. The fact that a small change in the relative velocity of a surface in contact with the fingertip may be easily detected under passive conditions is no guarantee that under active exploration the same sensory sequence will give rise to a change in perceived velocity of the surface. Similarly, if a signal to the oculomotor system controls the magnitude of a saccade with an accuracy of $x\%$ and signals from the retina indicate a jump of the retinal image, the question to be settled by the evaluative process is whether this jump of the image implies that the world is stable or demands a change in the "map coordinates". The answer in this case depends on whether the image jump is *significantly* different from that which the saccade was calculated to bring about on the "null hypothesis" of stability, i.e. different by more than $x\%$. Provided that the two "match" within this tolerance, no "map change" need be initiated, and no problem of perceptual instability arises, even though the "map" itself represents the coordinates in question to a far higher precision, and though the first approximate jump may have to be followed by a supplementary corrective one.

This point seems to have been appreciated by LUDVIGH (1952b) whose tentative schematic diagram (Fig. 3b) does not commit him to a "subtractive" theory of perceptual stability as von HOLST's (Fig. 3a) appears to do. LUDVIGH's account, however, does not propose any specific answer to our question B (Sec. 1). He speaks only of the absence or presence of "balancing impulses from the conjugate innervation center" as determining whether or not motion of the environment is perceived.

Although the physiological representation of the perceived visual world (i.e. what we have been calling the "map", as distinct from representations of the visual field) has yet to be understood, there are encouraging hints in the direction indicated at the end of Section 3. INGLE et al. (1968), SCHNEIDER (1969), HUMPHREY (1970) and WEISKRANTZ and COWEY (1970), for example, present evidence that in a number of species from hamster to monkey the ability to identify patterns visually and the ability to localise objects in space depend on quite different departments of the visual system, being selectively affected by ablation of the visual cortex and of the superior colliculi respectively. The picture is not entirely clear cut, and possible differences between species must be reckoned with. DIAMOND (1971) for example reports the tree shrew and ground squirrel to be surprisingly competent in visually guided behaviour even after removal of extra-striate as well as striate cortex. SCHILDER et al. (1971) report that ablation of "area 17, most of 18 and part of 19" left some form discrimination in three out of nine monkeys. Conversely, SPRAGUE and BERLUCCHI (1971) find considerable impairment of form discrimination after lesions of the superior colliculus in cats. But in general it seems fair to conclude that complementary aspects of the visual input – the "what" and the "where" – are dealt with to some extent by separate neural sub-systems. It may be added that cortical analysis of the visual input by "complex feature-sensitive elements" (see Chapter 2 and 21) would lend itself to a mode of representation in which the location attributed to an object could be readily changed without affecting the representation of its other attributes.

Since the primary function of the internal representation is to maintain a conditional readiness to reckon with the state of affairs giving rise to the visual signals, the physiologically natural "mapping" dimensions (degrees of freedom) would be those of the repertoire of action, both internal and external. In other words, neighbouring states of the visual world should be represented by neighbouring states of organization of the repertoire to match them (MACKAY, 1956, 1966). Considerations such as these make it less surprising that something like the *location* of a percept, which makes specific demands primarily on the planning of motor (including oculomotor) activity, should have its neural correlate in quite a different part of the organizing system from attributes such as its form. They also suggest that the evaluative process we have been discussing may have to be sought at levels still more "central" than the cortical projection areas.

5. Contrasts between Visual and Tactile Exploration

The example of tactile exploration establishes the principle that sensory signals indicating image motion need not be suppressed or cancelled from perception in order to preserve perceptual stability; but visual exploration differs from tactile in important respects that must now be considered.

In the first place, although the human eye is capable of smooth rotation in pursuit of a visible target (including even a stabilized near-foveal image: see KOMMERELL, 1971), or in pursuit of an unseen target when this is moved by the subject's own hand (GERTZ, 1916; STEINBACH, 1969), normal exploratory eye-movements are saccadic, so that the retinal input comprises a succession of discrete images, linked by brief phases of rapid image motion lasting from 30–180 msec (HYDE, 1959; ROBINSON, 1968. Both the duration and maximal velocity of a saccade increase with its amplitude). Thus whereas the blind man's palm is "brushed" by his sensory field, the exploring retina normally "hops" over its field at speeds that may reach 600°/sec (WESTHEIMER, 1954). This is well above the upper threshold of psychophysical motion detection, found by BROWN (1958) to be about 35°/sec for short paths. Accordingly, as noted in Section 2a, during saccadic visual exploration we would expect little or no sensation of image drift analogous to the "brushing" we feel in tactile exploration.

Secondly, whereas we have a good proprioceptive system of joint receptors to keep track of exploratory hand movements, human subjects, as already noted, show no comparable awareness of the position of the eyeball if visual clues are removed (BRINDLEY and MERTON, 1960). Wearing a contact lens to which an image source is attached by a stalk, for example (YARBUS, 1957; MACKAY, 1958 b), so that the retinal image is stabilised, one is repeatedly astonished to discover in what direction the stalk is pointing, despite the fact that voluntary *changes* in eye direction are betokened by a subjective motion-of-the-visual-scene in the same direction. This is not to say that no proprioceptive information is available centrally from eye muscles in man, but only that such information is not apparently used to update our internal representation of the direction of gaze. For this we rely mainly on the visual input, much as an air pilot has to do in the absence of dead-reckoning instruments. Given an occasional "fix", eye-position can be estimated (with diminishing accuracy) by the integration of signals indicating the size of intervening saccades. As mentioned above, GUREVITCH (1959) has found what he took to be behavioural evidence of proprioceptive feedback; but even if this interpretation were accepted his data would offer no evidence of proprioceptive signalling of eye *position*, as distinct from *changes* in eye position, to the higher levels of the CNS.

Thirdly, and most significantly, when the retina is *continuously* displaced relative to the visual image the subject has an irresistible impression of motion of the visual world, although no such impression is generated in the case of continuous tactile exploration. Why should this be? The impression is strongest when the smooth eye movement is imposed from without (as by pressure on the canthus); but it is still present, in the foveal region, when the eye muscles themselves are used in smooth pursuit of a moving fixation point, as in the illusion observed by Filehne (1922). So the absence or presence of oculomotor innervation is not the critical factor.

The answer suggested by our analysis is that the visual evaluator, unlike the tactile, is only poorly equipped to evaluate smooth drifting of the sensory stimulus over the sensory surface as positive evidence of stability. This is reasonable, since our foveate eye does not normally explore by continuous motions. Its goal is a (sudden) *displacement* of the sensory field; so it is the appropriate changes in retinal image *position*, rather than continuous image *motion*, that offer confirmatory neural evidence of the stability of the visual world during saccadic eye movements.

Finally, the eye differs from other sense organs in having delicate reflex and compensatory mechanisms to help keep it locked to its visual target during passive or active locomotion, head movements and the like. One of these is responsible for the familiar involuntary drift-and-flick of "railway nystagmus", which keeps one selected region of a passing landscape steady on the fovea despite the multitude of disparate drift signals from retinal regions exposed to the (moving) images of foreground and background (Dichgans and Jung, 1970). Another, controlled by the labyrinth, normally compensates for head rotation, and causes "vestibular nystagmus" when the semicircular canals are abnormally stimulated by continuous rotation or by warming or cooling one ear.

Movement of an organism in a visually perceived surround normally generates simultaneous visual and vestibular inputs. Jung (1947, 1953), using eye-movement recording with human subjects, has demonstrated a delicate interplay of these in the control of eye movements. When they interfere with each other the visual stimuli are usually dominant over the labyrinthine. The complexity of visuo-vestibular integrations has been shown physiologically in rabbits and fishes by Dichgans et al. (1970) and Schmidt et al. (1970), who have observed efferent vestibular nerve discharges from the brainstem modulated in synchrony with the rapid phase of nystagmus.

Since compensatory eye movements are almost entirely involuntary they fall outside our present topic (see Grusser's Chapter in this Volume, and Vol. VI on the Vestibular System, edited by Kornhuber); but they do raise somewhat similar questions for the theory of visual stability. The illusions of movement associated with nystagmus are discussed by Howard and Templeton (1966) and Jung (1968; 1971). On the whole they show the features that would be expected if (a) the slow phase stimulates drift-detectors whose output is normally evaluated as evidence of velocity "mismatch"; (b) the compensatory flicks, though producing equal *displacement* of the image in the opposite direction, are too rapid to excite drift-detectors, and so fail to annul the "mismatch" generated during the slow phase.

6. Possible Functions of Drift-sensitive Elements

Although relatively inactive during saccades, drift-sensitive elements have no lack of likely functions in other contexts. For one thing, they may be expected to play a vital biological role in the detection of *relative* movement between different parts of the visual field, and particularly in signalling the overall expansion or contraction of the retinal image that supplies evidence of movement towards or away from textured surfaces (GIBSON, 1950, 1966). But more important in the present context is their natural suitability as feedback generators in the minimization of retinal image motion during ocular pursuit (OYSTER and BARLOW, 1967; ATKIN, 1969), or in the maintenance of fixation (SPRAGUE and MEIKLE, 1965; MCILWAIN and BUSER, 1968). The ocular grasping and holding of a chosen feature of the visual world on the centre of the fovea requires precisely the sort of control signals that would be offered by a population of elements sensitive to the presence and direction of retinal image *drift*. If coupled in negative feedback to the generator of smooth motion in the oculomotor system with a sufficiently high overall sensitivity, they could readily be arranged to make the retina 'lock on' to the dominant feature of the image in their field of operation, much as the bristles of a hairbrush would keep it "locked" to sufficiently salient features of a rough surface. The fine eye movements observed during fixation (PRITCHARD, 1964) might help to keep this feedback active.

In line with the foregoing hypothesis, it has been shown by RASHBASS (1961) that step changes in the *velocity* of a fixation target (which excite drift detectors) elicit predominantly smooth pursuit motion, whereas step changes in its *position* elicit saccadic corrections. ATKIN (1969) reports that certain non-foveal areas are more effective than the fovea in controlling pursuit velocity. STARR (1967) offers clinical confirmation (in a case of Huntington's Chorea) that the control mechanisms for saccades and for smooth tracking are anatomically as well as functionally distinct. In the frontal eye fields (FEF) of the monkey, BIZZI (1967, 1968) has found much more evidence to the same effect. He identified two quite different kinds of FEF unit associated with eye movements.: Type I, firing during saccadic movements but silent during slow pursuit; and Type II, firing during slow pursuit and discharging steadily when the eyes were oriented in a specific direction. BIZZI and SCHILLER (1970) showed that head turning did not affect these firing patterns, but was related to the activity of still another population of FEF cells (see also WHITTERIDGE, 1960; BRUCHER, 1964; JUNG, 1968, esp. 213-5, ROBINSON, 1968; ROBINSON and FUCHS, 1969). Since during smooth pursuit the retinal image of a target is more or less stationary, our perception of its motion must depend on signals of extra-retinal (presumably oculomotor) origin. BIZZI's Type II signals would be well suited to this purpose.

If the dominant feature to which the fovea is locked moves smoothly in relation to others, these must generate mismatch signals in the drift-detectors over which they pass. FILEHNE's illusion could be regarded as subjective evidence of the resulting disturbance of the updating mechanism. A striking feature of this illusion is that regions of the visual background near the moving object seem to move more rapidly backward than the object seems to move forward. This is related, but not identical, to the well-known AUBERT-FLEISCHL paradox, that if a stationary point is fixated instead of the moving object, the perceived velocity of the object is exaggerated by a factor recently estimated to be about 1.6 (KORNER and DICHGANS, 1967; DICHGANS et al., 1969; DICHGANS and JUNG, 1970). The difference between this latter and FILEHNE's case is that a textured background need not be visible, so that some contrast effects that might complicate the estimation of perceived velocity can be avoided. In both cases, it seems reasonable to conclude that the high sensitivity of the drift detecting system is responsible for the perceptual anomalies. It would be unwarranted, however, to conclude that if the scale factor of the drift detectors were lower by 1.6 the impression of background motion in Filehne's case should disappear, as it does in saccadic exploration. This would depend on whether the visual evaluator could "learn" to evaluate drift signals of an appropriate strength as evidence of stability, in the same way as those from our fingertips are accepted in tactile exploration: which must be left at present an open question.

In summary, then, drift detectors may contribute to motion perception in two ways: (a) by indicating motion of the image over the retina, so giving rise to what Jung (1971) calls "retinal motion perception"; (b) by helping to keep the retina locked to a moving target, the pursuit of which occasions what Jung terms "oculomotor motion perception". The entirely different physiological mechanisms underlying these two modes of motion detection are responsible for a number of psychophysical anomalies (see Jung, loc. cit.).

7. Visual "Suppression" during Saccades

It is well-known that the psychophysical threshold for the perception of faint test-flashes is raised by about half a log. unit if they are presented just before or after the onset of a saccade (Latour, 1962; Volkmann, 1962). Small displacements of the visual field (up to 2° of arc) during a saccade may also fail to be noticed (Sperling and Speelman, 1965; Wallach and Lewis, 1965; Beeler, 1967; but see Yarbus, 1967 for contrary findings).

On the physiological side Bizzi (1966) has found corresponding reductions in photically evoked responses of the cat's LGN during a saccade, and Collewijn (1969) describes similar effects on a late (ca. 250 msec) peak of the photically evoked response in rabbit colliculus and visual cortex. Analogous findings in the case of the human occipital evoked potential have been reported for near-threshold flashes by a number of workers, among them Michael and Stark (1967), and Duffy and Lombroso (1968). Interpretation of these results is complicated, however, by the fact that late components of the evoked response are known to be affected more by psychological factors such as stimulus significance than by stimulus intensity (see Chapter 28).

All of the foregoing observations have tended to be taken as support for "suppression" theories of visual stability, on the presupposition that signals from the retina during a saccade must give an impression of world-movement unless suppressed. The fact that the foveal flash visibility curve may drop over 60 msec *before* onset of eye movement (Volkmann et al., 1968) was taken to confirm that the "suppressive" effect must be due to an inhibitory corollary discharge emitted beforehand by the oculomotor system as part of its preparation for a saccade.

Unfortunately for this view, it turns out that visibility may also be reduced when the eye is moved passively (Richards, 1968). Moreover, if the visual field is displaced in saccadic fashion before a stationary eye, a similar *anticipatory* drop in visibility of a test flash is observed although the eye is immobile (MacKay,

1970 a). This is so even when the test flash and moving field are presented to different eyes (MacKay, 1970 b). Since there is here no possibility of "advance warning" signals from the oculomotor system, it can no longer be argued that the "anticipation" observed in connection with an active saccade requires us to postulate such signals. The implication rather seems to be that in both passive and active cases retroactive reduction in visibility may be caused, at least in part, by central interference with the subsequent processing of the test signal, due to the surge of activity generated in the visual system by the sudden displacement of the visual field (Mitrani et al., 1970). It is known that image-movement in the periphery of the retina can modify the responsiveness of cortical units having fields nearer to the fovea (Macilwain, 1964). Moreover, even the retinal ganglion cell response to a brief flash may last 70 msec or more (Levick and Zacks, 1970), so that "retroactive" interference is readily understandable without invoking a prior corollary discharge. It may also be significant in this connection that Palka (1969), working with the cricket, and Wurtz (1969 a, b) with the monkey, have each recently found single visual units which are suppressed equally by active and passive displacements of a textured retinal image. Wurtz points out that this would not be expected if the suppression were due to a corollary discharge from the oculomotor system (see chapter on Visual Cortical Neurones by Brooks and Jung).

Clearly, if the maintenance of fixation depends even partly on negative feedback from drift detectors, a saccadic change in fixation point requires this feedback to be momentarily suppressed or overridden. (In terms of our hairbrush analogy, this would correspond to lifting the hairbrush out of contact with the salient features before trying to displace it to a new location.) The interesting point is that since physiologically observed drift detectors have a critical image velocity (found to be of the order of 150°/sec in the monkey: Wurtz, 1969 a, b) above which their output falls to zero or becomes inhibitory, no active suppression mechanism should be needed for this purpose with saccadic speeds above the critical velocity. If the first step towards a change of fixation point were a shift in the criterion of equilibrium in the fixation feedback loop, this in itself could elicit a rapid enough eye movement to exceed the critical velocity, so inactivating the motion feedback from the drift detectors and tripping the system into "free" saccadic flight. Motion feedback would automatically come into action and again lock the retina to the image as velocity fell below the critical level for drift-detection towards the end of the saccade.

Even if the transition from "locked" to "free" motion were assisted by active suppression of motion feedback, such suppression would clearly have no necessary connection with the maintenance of perceptual stability, and might not even have any effect on the subjective visibility of a flash during a saccade. Conversely, it seems unjustifiable to regard the small elevations of perceptual threshold found during saccades as significant evidence of active suppression, either for the purpose of "unlocking" fixation or for purposes of perceptual stability.

8. Perceived Location of Stimuli Presented during Saccades

A further visual anomaly which has sometimes been taken to support the "subtractive" theory of perceptual stability is the mislocation of stimuli presented briefly just before or after onset of a voluntary saccade. Sperling and Speelman (1965), and Bischof and Kramer (1968) are among those reporting that the perceived location of a flash stimulus tends to be displaced in the direction of the eye-movement, by an amount varying with the relative timing of the two.

Here again, however, it turns out that displacing the stimulus field when the eye is stationary can produce quite similar perceptual mislocations (Sperling and Speelman, 1965; MacKay, 1970 c). Fig. 4 for example shows that a flash occurring tens of milliseconds before the displacement of a scale (in front of a stationary eye) is mislocated by an amount which is maximal for flashes presented

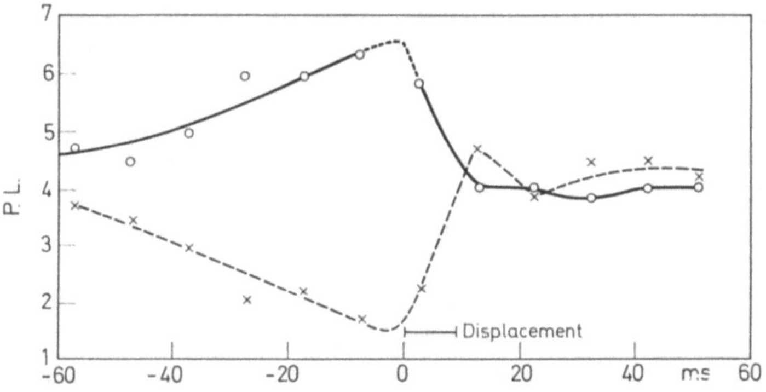

Fig. 4. Perceived location (P. L.) of a brief flash presented before, during and after rapid horizontal displacement of a horizontal scale by approximately $2^1/_2$ scale divisions, before a stationary eye. The flash was physically coincident at all times with the mark "4" on the moving scale. Upper and lower curves show results for displacements in opposite directions, indicating no significant perceptual bias (From MacKay, 1970 c)

just before onset of displacement. To account for this it seems sufficient to postulate that the neural computation of relative location requires integration of evidence over a finite period of time, which on both physiological and psychophysical grounds might be expected to last 50—100 msec or even longer. Thus if the retinal location of the scale changes abruptly a few msec after the flash occurs, the integration process will have two successive and conflicting neural representations of scale position in terms of which to perform its computation. Flashes presented so that their neural signals overlap the neural 'changeover period' are bound to be mislocated to a greater or lesser extent. Assuming that the relative weight given to the earlier scale position diminishes monotonically with increasing time interval between flash and scale displacement, a curve of the general form of Fig. 4 can be predicted. It is therefore doubtful whether any oculomotor "corollary discharge" need be invoked to explain the analogous mislocations of flashes observed when the saccadic image displacement in question is produced by a voluntary eye movement.

This conclusion might at first seem inconsistent with the findings of Matin and Pearce (1965), whose subjects were able to judge the *physical* location of a flash presented halfway through a saccade of 2° 11′ with an accuracy of 10 or 20 min of arc. This led these authors to infer that eye position, even during the course of a saccade, must be "proprioceptively" monitored with astonishing accuracy. Unfortunately, it appears that their flash was triggered " at a precisely known point near the middle of the saccade". There was thus a close correlation between its physical location relative to the initial fixation point and the retinal location of its image, so that the subjects' performance might reflect only their ability to judge the latter, and cannot be taken as conclusive evidence of the accuracy or even the existence of any proprioceptive compensation.

An interesting finding by MATIN and PEARCE was that the perceived location of the mid-saccadic flash relative to a second brief flash presented about 100 msec earlier agreed with their respective *retinal* image positions, showing no signs of any "compensatory shift of coordinates". This is in line with results of an extensive study by STOPER (1967) in which be concluded that during smooth pursuit movements also "the relative position of (flashed) stimuli is determined entirely by the areas of the retina . . . stimulated". He also showed that with a stationary flashing source a sensation of "stroboscopic motion" could be induced while the eye was making smooth pursuit movements, with no sign of cancellation between the oculomotor motion signal and that due to retinal displacement, even though the two were opposite in sign.

On the other hand, it is interesting to note that the perceived motion of an afterimage in darkness is reported to correlate well with the recorded direction of saccadic as well as slow components of eye movement, particularly when the subjects are either "undirected" or simply told to look "straight ahead" (FISCHER and KORNMÜLLER, 1930; MACK and BACHANT, 1969). This suggests that information regarding both saccadic and slow components may be used for re-setting criteria of evaluation, although (according to STOPER's observations) there is no sign of "subtractive" reduction of signals from retinal image motion, either smooth or saccadic, by an oculomotor corollary discharge in smooth *pursuit*. With regard to saccadic eye motion ROCK and EBENHOLTZ (1962) have found that two spatially separated stimuli flashing alternately at a fixed rate can induce a sensation of movement even when the eye is made to jump back and forth from one to the other so that both images have the same retinal location. Conversely, a single stationary flashing stimulus is seen as such (with no apparent movement) even when the eye jumps to and fro so that the image falls alternately on two different retinal points. In other words, veridical perception of the location of a flash stimulus is not interfered with by *saccadic* eye movements, (any more than is the veridical perception of steady stimuli), unless the flash is presented so near to the time of the saccade that its signal overlaps the changeover period mentioned above.

9. Conclusions

Our main conclusion is that the traditional analysis of the problem of perceptual stability during exploratory eye movements requires to be reformulated. From an information-theoretical standpoint, the sensory consequences of such movements are not a disturbance requiring to be eliminated from the visual input; rather they supply positive (if sometimes crude) evidence that the visual world has remained stable during the movement. In order that they should be properly evaluated as such, corollary information as to current eye movements must be available to whatever mechanism updates the internal "map" on the basis of sensory signals; but the example of tactile exploration shows that it is by no means necessary *a priori* that this information be used to "cancel" or "suppress" the sensory changes from perception in order that we should perceive the field of exploration as stable. There is therefore no justification on grounds of perceptual stability for expecting to find physiological mechanisms, driven by the oculomotor system, whose function is to subtract or suppress components from the visual input to *perceptual* centres. All the experiments reviewed which have been thought to point in this direction have turned out to admit of other interpretations. In the collicular opto-motoric sub-system, of course, there is plenty of scope for computing operations, such as the comparison of optical feedback with internally-generated goal-spcifications, which might involve the subtraction of one signal from another; but this is quite another matter.

Fundamental to the clarification of this problem is the fact that relatively slow drifting of an image over the retina can excite a separate signalling system from that which is sensitive to saccadic image displacement. Only the second of these is activated in normal saccadic exploratory movement; eye movements that activate the first "drift-sensitive" system, even in voluntary pursuit of a slowly moving target, normally give rise to illusory sensations of motion. Only when optical drifting results from movement of the body as a whole (e.g. when walking towards a textured surface) is it evaluated as evidence of velocity of approach to a stable object.

We have sharply distinguished our problem of perceptual stability from that of the maintenance of sensory-motor coordination, where von Holst's idea of the stabilization of orienting reflexes by subtraction of an "Efferenzkopie" continues to be found fruitful (Mittelstaedt, 1966). Our main argument has also been independent of whether "outflow" from the oculomotor command system or proprioceptive "inflow" (or a combination of both) supplies the information as to what the eyes are doing, though evidence reviewed seems predominantly to favour "outflow" in this role. The chief question at issue has been the purpose for which the corollary information is required. Our informational analysis has suggested that it is needed to bring about not subtraction or suppression but *informational evaluation* of the visual changes caused by voluntary eye movements. For when properly evaluated, these changes provide positive evidence of the stability of the visual world, which it would be wasteful as well as needless for the system to eliminate or ignore.

Note added in proof: This chapter was substantially complete before the Freiburg Symposium on Cerebral Control of Eye Movements and Motion Perception in July 1971 published by S. Karger (Basel) in 1972 J. Dichgans, E. Bizzi (Eds.). Readers are referred to this Symposium particularly for the most recent developments in the neurophysiology of extrastriate visual projections and their relations with the oculomotor system.

References

Atkin, A.: Shifting fixation to another pursuit target: selective and anticipatory control of ocular pursuit initiation. Exp. Neurol. **23**, 157—173 (1969).

Bach-y-Rita, P.: Neurophysiology of eye movements. In: Bach-y-Rita, P., Collins, C. C. (Eds.): The Control of Eye Movements. Academic Press 1971.

Barlow, H. B., Hill, R. M.: Selective sensitivity to direction of movement in ganglion cells of the rabbit retina. Science **139**, 412—414 (1963).

Basler, A.: Über das Sehen von Bewegungen. I. Die Wahrnehmung kleinster Bewegungen. Pflügers Arch. ges. Physiol. **115**, 582—601 (1906).

Baumgartner, G., Brown, J. C., Schultz, A.: Visual motion detection in the cat. Science **146**, 1070—1071 (1964).

Beeler, J. W., Visual threshold changes resulting from a spontaneous saccadic eye movement. Vision Res. **7**, 769—775 (1967).

Bischof, N., Kramer, E.: Untersuchungen und Überlegungen zur Richtungswahrnehmung bei willkürlichen sakkadischen Augenbewegungen. Psychol. Forsch. **32**, 185—218 (1968).

Bizzi, E.: Changes in the orthodromic and antidromic response of optic tract during the eye movements of sleep. J. Neurophysiol. **29**, 861—870 (1966).

— Discharge of frontal eye field neurones during eye movements in unanaesthetized monkeys. Science **157**, 1588—1590 (1967).

— Discharge of frontal eye field neurones during saccadic and following eye movements in unanaesthetized monkeys. Exp. Brain Res. **6**, 69—80 (1966).

— Schiller, P. H.: Single unit activity in the frontal eye fields of unanaesthetized monkeys during eye and head movement. Exp. Brain Res. **10**, 151—158 (1970).

Brindley, G. S., Merton, P. A.: The absence of position sense in the human eye. J. Physiol. (Lond.) **153**, 127—130 (1960).

— Lewin, W. S.: The sensations produced by electrical stimulation of the visual cortex. J. Physiol (Lond.) **196**, 479—494 (1968).

Brown, R. H.: Influence of stimulus and luminance upon the upper speed threshold for the visual discrimination of movement. J. Opt. Soc. Amer. **48**, 125—128 (1958).

Brucher, J. M.: L'aire Oculogyre Frontale du Singe — ses fonctions et voies afferentes. Bruxelles: Arscia 1964.

Christman, E. H., Kupfer, C.: Proprioception in extra-ocular muscles. Arch. Ophthal. **69**, 824—829 (1963).

Collewijn, H.: Changes in visual evoked responses during the fast phase of optokinetic nystagmus in the rabbit. Vision Res. **9**, 803—814 (1969).

Cooper, S., Daniel, P.: Muscle spindles in human extrinsic eye muscles. Brain **72**, 1—24 (1949).

— — Whitteridge, D.: Muscle spindles and other sensory endings in the extrinsic eye muscles: The physiology and anatomy of these receptors and their connections with the brainstem. Brain **78**, 564—583 (1955).

Diamond, I. T.: Two visual systems in tree shrews (*Tupaia glis*) and squirrels (*Sciurus carolinesis*). Proc. Int. Union of Physiological Sciences, XXV International Congress, Munich, 1971. Vol. VIII, 225—226 (1971).

Dichgans, J., Jung, R.: Attention, eye movements and motion detection: facilitation and selection in optokinetic nystagmus and railway nystagmus. In: Evans, C. R., Mulholland, T. B. (Eds.): Symposium on Attention, pp. 348—375. London: Butterworths 1970.

— Körner, F., Voigt, K.: Vergleichende Skalierung der afferenten und efferenten Bewegungssehens beim Menschen: Lineare Funktionen mit verschiedener Anstiegssteilheit. Psychol. Forsch. **32**, 277—295 (1969).

— Wist, E. R., Schmidt, C. L.: Modulation neuronaler Spontanaktivität im N. vestibularis durch optomotorische Impulse beim Kaninchen. Pflügers Arch. ges. Physiol. **319**, 154 (1970).

Ditchburn, R. W., Ginsborg, B. L.: Vision with a stabilized retinal image. Nature (Lond.) **170**, 36—37 (1952).

Duffy, F. H., Lombroso, C. T.: Electrophysiological evidence for visual suppression prior to the onset of a voluntary saccadic eye movement. Nature (Lond.) **218**, 1074—1075 (1968).

Filehne, W.: Über das optische Wahrnehmen von Bewegungen. Z. Sinnesphysiol. **53**, 134—145 (1922).

Fischer, M. H., Kornmüller, A. E.: Optokinetisch ausgeloste Bewegungswahrnehmungen und optokinetischer Nystagmus. J. Psychol. Neurol. (Lpz). **41**, 273—308 (1930).

Fuchs, A. F., Kornhuber, H. H.: Extra-ocular muscle afferents to the cerebellum of the cat. J. Physiol. (Lond.) **200**, 713—722 (1969).

Gertz, H.: Über die gleitende (langsame) Augenbewegung. Z. Psychol., Physiol., Sinnesorg. Abt. 2: Z. Sinnesphysiol. **49**, 29—58 (1916).

Gibson, J. J.: The Perception of the Visual World. Boston: Houghton Mifflin 1950.

— The Senses considered as Perceptual Systems. Boston: Houghton Mifflin 1966.

Gregory, R. L., Zangwill, O. L.: The origin of the autokinetic effect. Quart. J. exp. Psychol. XV, 252—261 (1963).

Grusser, O.-J., Finkelstein, D., Grusser-Cornehls, U.: The effect of stimulus velocity on the response of movement-sensitive neurones of the frog's retina. Pflügers Arch. ges. Physiol. **300**, 49—66 (1968).

Gurevitch, B. K.: Possible role of higher proprioceptive centres in the perception of visual space. Nature (Lond.) **184**, 1219—1220 (1959).

Helmholtz, H. von: Handbuch der Physiologischen Optik, Vol. 3, Leipzig: Voss 1866.

Holst, E. von: Relations between the central nervous system and the peripheral organs. Animal Behaviour **2**, 89—94 (1954).

— Aktive Leistungen der menschlichen Gesichtswahrnehmung. Studium Generale **10**, 231—243 (1957).

— Mittelstaedt, H.: Das Reafferenzprinzip (Wechselwirkungen zwischen Zentralnervensystem und Peripherie). Naturwissenschaften **37**, 464—476 (1950).

Holt, E. B.: Eye movement and central anaesthesia. Harvard Psychol. Studies **1**. 3—45 (1903).

Horn, G., Hill, R. M.: Modifications of receptive fields of cells in the visual cortex occurring spontaneously and associated with bodily tilt. Nature (Lond.) **221**, 186—188 (1969).

Howard, I., Templeton, W.: Human Spatial Orientation. New York: Wiley 1966.

Hubel, D. H.: Cortical unit responses to visual stimuli in non-anaesthetized cats. Amer. J. Ophthal. **46**, 110—122 (1958).

— Wiesel, T. N.: Shape and arrangement of columns in cat's striate cortex. J. Physiol. (Lond.) **165**, 559—568 (1963).

Humphrey, N. K.: What the frog's eye tells the monkey's brain. Brain, Behav. Evolution **3**, 324—337 (1970).

Hyde, J. E.: Some characteristics of voluntary human ocular movements in the horizontal plane. Amer. J. Ophthal. **48**, 85—94 (1959).

Ingle, D., Schneider, G. E., Trevarthen, C. B., Held, R.: Locating and identifying: Two modes of visual processing. Psychol. Forsch. **31**, Numbers 1 and 4, 42—62 and 299—348.

Jackson, J. Hughlings, Paton, L.: On some abnormalities of ocular movements. Lancet **176**, 900—905 (1909).

Jeannerod, M., Putkonen, P. T. S.: Oculomotor influences on lateral geniculate body neurons. Brain Res. **24**, 125—129 (1970).

Jung, R.: Die Registrierung des postrotatorischen und optokinetischen Nystagmus und die optisch-vestibulare Integration beim Menschen. Acta oto-laryng. (Stockh.) **36**, 199—202 (1947).

JUNG, R.: Zur Physiologie und Pathologie des optisch-vestibularen Systems beim Menschen. In: BERGMANN, G. v., FREY, W., SCHWIEGK, H. (Eds.): Handbuch der Inneren Medizin. 4. Aufl. Bd. V/1, Neurologie I, 1325—1379. Berlin-Göttingen-Heidelberg: Springer 1953.

— Optisch-Vestibulare Regulation der Augenbewegungen, des Bewegungssehens und der Vertikal-Horizontal-Wahrnehmung: ein Beitrag zur Optisch-Vestibularen, Optisch-Oculomotorischen und Optisch-Gravizeptorischen Integration. Brain and Mind Problems, 185—226. Il Pensiero Scientifico 1968.

— Neurophysiological and psychophysical correlates in Vision Research. In: KARCZMAR, A.G., ECCLES, J.C. (Eds.): Brain and Human Behavior. Berlin-Heidelberg-New York: Springer 1971.

KAWAMURA, H., MARCHIAFAVA, P.L.: Excitability changes along visual pathways during eye tracking movements. Arch. Ital. Biol. **106**, 141—156 (1968).

KÖRNER, F., DICHGANS, J.: Bewegungswahrnehmung, optokinetischer Nystagmus und retinale Bildwanderung. (Der Einfluß visueller Aufmerksamkeit auf zwei Mechanismen des Bewegungssehens.) Albrecht v. Graefes Arch. klin. exp. Ophthal. **174**, 34—48 (1967).

KOMMERELL, G.: Extrafoveal after-images and oculomotor control. In: BIZZI, E., DICHGANS, J., (Eds.): Cerebral Control of Eye Movements and Perception of Motion. Basel: S. Karger (in Press) (1972).

LATOUR, P.L.: Visual threshold during eye movements. Vision Res. **2**, 261—262 (1962).

LEVICK, W.R., ZACKS, A.L.: Responses of cat retinal ganglion cells to brief flashes of light. J. Physiol. (Lond.) **206**, 677—700 (1970).

LUDVIGH, E.: Possible role of proprioception in the extra-ocular muscles. Arch. Ophthal. **48**, 436—441 (1952 a).

— Control of ocular movements and visual interpretation of environment. Arch. Ophthal. **48**, 442—448 (1952 b).

MACH, E.: The Analysis of Sensations. Fifth edition. English translation. New York: Dover 1959.

MACK, A., BACHANT, J.: Perceived movement of the afterimage during eye movements. Percept. and Psychophys. **6** (**6** A), 379—384 (1969).

MACKAY, D.M.: Towards an information-flow model of human behaviour. Brit. J. Psychol. **47**, 30—43 (1956).

— The stabilization of perception during voluntary activity. Proc. 15th Int. Congress of Psychology, pp. 284—285. Amsterdam: North-Holland Publishing Co. 1957.

— Perceptual stability of a stroboscopically lit visual field containing self-luminous objects. Nature (Lond.) **181**, 507—508 (1958a).

— Moving visual images produced by regular stationary patterns (II). Nature (Lond.) **181**, 362—363 (1958b).

— Modelling of large-scale nervous activity. S.E.B. Symp.: Models and Analogues in Biology, (J. W. L. BEAMENT, ed.), pp. 192—198. Vol. 14. Cambridge University Press 1960.

— Interactive processes in visual perception. In: ROSENBLITH, W.A. (Ed.): Sensory Communication, pp. 339—355. Boston-New York: M.I.T. and Wiley 1961 a.

— The visual effects of non-redundant stimulation. Nature (Lond.) **192**, 739—740 (1961 b).

— Theoretical models of space perception. In: MUSES, C.A. (Ed.): Aspects of the Theory of Artificial Intelligence, pp. 83—104. New York: Plenum Press 1962.

— Cerebral organization and the conscious control of action. In: ECCLES, J.C. (Ed.): Brain and Conscious Experience, pp. 422—445. Berlin-Heidelberg-New York: Springer 1966.

— Information, Mechanism and Meaning. Boston: M.I.T. Press 1969.

— Elevation of visual threshold by displacement of retinal image, Nature (Lond.) **225**, 90—92 (1970a).

MacKay, D. M.: Interocular transfer of suppressive effects of retinal image displacement. Nature (Lond.) **225**, 872—873 (1970b).

— Mislocation of test flashes during saccadic image displacement. Nature (Lond.) **227**, 731—733 (1970c).

Matin, L., Pearce, D.: Visual perception of direction for stimuli flashed during voluntary saccadic eye movements. Science **148**, 1485—1488 (1965).

McIlwain, J. T.: Receptive fields of optic tract axons and lateral geniculate cells: peripheral extent and barbiturate sensitivity. J. Neurophysiol. **27**, 1154—1173 (1964).

— Buser, P.: Receptive fields of single cells in the cat's superior colliculus. Exp. Brain Res. **5**, 314—325 (1967).

Michael, J. A., Stark, L.: Electrophysiological correlates of saccadic suppression. Exp. Neurol. **17**, 233—246 (1967).

Miller, G. A.: The magical number 7 plus or minus 2. Psychol. Rev. **63**, 81—97 (1965).

Mitrani, L., Mateev, S., Yakimov, N.: Intensity threshold changes during voluntary saccade of the eyes. Proc. Bulgar. Acad. Sci., **23**, 1577—1579 (1970).

Mittelstaedt, H.: Control systems of orientation in insects. Ann. Rev. Entomol. **7**, 177—198 (1962).

— Grundprobleme der Analyse von Orientierungs-Leistungen. Jahrbuch 1966 der Max-Planck-Gesellschaft zur Förderung der Wissenschaften e.V. 121—151, 1966.

Oyster, C. W., Barlow, H. B.: Direction-selective units in rabbit retina. Science **155**, 841—842 (1967).

Palka, J.: Discrimination between movements of eye and object by visual interneurones of crickets. J. Exp. Biol. **50**, 723—732 (1969).

Potthoff, P. C., Richter, H. P., Burandt, H.-R.: Multisensorische Konvergenzen an Hirnstammneuronen der Katze. Arch. Psychiat. Z. ges. Neurologie **210**, 36—60 (1967).

Pritchard, R. M.: Physiological nystagmus and vision. In: Bender, M. B. (Ed.): The Oculomotor System. New York: Harper and Row 1964.

Rashbass, C.: The relationship between saccadic and smooth tracking eye movements. J. Physiol. (Lond.) **159**, 326—338 (1961).

Richards, W. A.: Visual suppression during passive eye movement. J. Opt. Soc. Amer. **58**. 1159—1160 (1968).

Robinson, D. A.: Eye movement control in primates. Science **161**, 1219—1224 (1968).

— Fuchs, A. F.: Eye movements evoked by stimulation of frontal eye fields. J. Neurophysiol. **32**, 637—649 (1969).

Rock, I., Ebenholtz, S.: Stroboscopic movement based on changes of phenomenal rather than retinal location. Amer. J. Psychol. **75**, 193—207 (1962).

Schilder, P., Pasik, P., Pasik, T.: Some evidence for brightness, form and color discrimination by monkeys lacking striate cortex. Proc. Int. Union of Physiological Sciences, XXV International Congress, Munich 1971, Vol. IX, 498.

Schmidt, C. L., Wist, E. R., Dichgans, J.: Alternierender Spontannystagmus, optokinetischer und vestibularer Nystagmus und ihre Beziehungen zu rhythmischen Modulationen der Spontanaktivität im N. vestibularis beim Goldfisch. Pflügers Arch. **319**, R 155 (1970).

Schneider, G. E.: Two visual systems. Science **163**, 895—902 (1969).

Sekuler, R., Ganz, L.: Aftereffect of seen motion with a stabilized retinal image. Science **139**, 419—420 (1963).

Shannon, C. E.: A mathematical theory of communication. Bell Syst. Tech. J., **27**, 379—423; 623—656 (1948).

SHERRINGTON, C. S.: Further note on the sensory nerves of the eye muscles. Proc. Roy. Soc. **64**, 120—121 (1898).

— Observations on the sensual role of the proprioceptive nerve supply of the extrinsic eye muscles. Brain. **41**, 332—343 (1918).

SPERLING, G., SPEELMAN, R.: Visual spatial localization during object motion, apparent object motion and image motion produced by eye movements. Paper presented at the convention of Optical Society of Am., Oct. 7, 1965. Abstracted in J. Opt. Soc. **55**, 1576 (1965).

SPERRY, R. W.: Neural basis of the spontaneous optokinetic response produced by visual inversion. J. Comp. Physiol. Psychol. **43**, 482—489 (1950).

SPRAGUE, J. M., MEIKLE, T. H., JR.: The role of the superior colliculus in visually guided behaviour. Exp. Neurol. **11**, 115—146 (1965).

— BERLUCCHI, G.: Role of the pretectum-superior colliculus in form discrimination in the cat. Proc. Int. Union of Physiological Sciences. XXV International Congress, Munich 1971. Vol. VIII, 223—224 (1971).

STARR, A.: Localization of objects in visual space with abnormal saccadic eye movements. Brain. **90**, 541—545 (1967a).

— A disorder of rapid eye movements in Huntington's chorea. Brain. **90**, 545—564 (1967b).

STEINBACH. M, J.: Eye tracking of self-moved targets: The role of afference. J. Exp. Psychol. **82**, 366—376 (1969).

STOPER, A. W.: Vision during Pursuit Movement: The role of oculomotor information. Ph. D. Thesis, Department of Psychology, Brandeis University, 1967.

SUTHERLAND, S.: Figural after-effects and apparent size. Quart. J. Exp. Psychol. XIII, 222—228 (1961).

TEUBER, H.-L.: Perception. In: FIELD, J. (Ed.): Handbook of Physiology — Neurophysiology III, pp. 1595—1668. Baltimore: Williams and Wilkins 1960.

VOLKMANN, F.: Vision during voluntary saccadic eye movement. J. opt. Soc. Amer. **52**, 571—578 (1962).

— SCHICK, A. M. L., RIGGS, L. A.: Time course of visual inhibition during voluntary saccades. J. opt. Soc. Amer. **58**, 562—569 (1968).

WALLACH, H., LEWIS, C.: The effect of abnormal displacements of the retinal image during eye movements. Percep. and Psychophys. **1**, 25—29 (1965).

WEISKRANTZ, L., COWEY, A.: Filling in the scotoma: A study of residual vision after striate cortex lesions in monkeys. Prog. Physiol. Psychol. **3**, 237—260 (1970).

WESTHEIMER, G.: Eye movement responses to a horizontally moving stimulus. Arch. Ophthal. **52**, 940—941 (1954).

WHITTERIDGE, D.: Central control of eye movements. In: FIELD, J. (Ed.): Handbook of Physiology: Section 1., Neurophysiology, Vol. 2, pp. 1089—1109. Washington, D. C.: American Physiological Society 1959.

WIERSMA, C. A. G., YAMAGUCHI, T.: Integration of visual stimuli by the crayfish central nervous system. J. exp. Biol. **47**, 409—431 (1967).

WURTZ, R. H.: Response of striate cortex neurons to stimuli during rapid eye movements in the monkey. J. Neurophysiol. **32**, 975—986 (1969a).

— Comparison of effects of eye movements and stimulus movements on striate cortex neurons of the monkey. J. Neurophysiol. **32**, 987—994 (1969b).

— GOLDBERG, M. E.: Superior Colliculus cell responses related to eye movements in awake monkeys. Science **171**, 82—84 (1971).

YARBUS, A. L.: A new method of studying the activity of various parts of the retina. Biophysics **2**, 165—167 (1957).

— Eye Movements and Vision. New York: Plenum Press 1967.

YOUNG, L. R.: A Sampled Data Model for Eye Tracking Movements. Sc. D. Thesis, Dept of Aeronautics and Astronautics, M.I.T. 1962.

Chapter 6

Neuronal Mechanisms of Visual Movement Perception and Some Psychophysical and Behavioral Correlations

By

OTTO-JOACHIM GRÜSSER and URSULA GRÜSSER-CORNEHLS, Berlin (Germany)

With 37 Figures

Contents

I. Introduction

Visual movement perception has to be considered a special *perceptual quality* of the visual modality as color, brightness, size etc. are sensory qualities for the perception of the visual world (EXNER, 1875). Movement detecting neuronal systems are probably present in the nervous system of all animals which respond to visual signals more complicated than simple light-dark stimuli.

Prey and enemies of most animals move. The voluntary and involuntary movements of the eyes, the head and the body, on the other hand, continuously shift the image of the stationary visual world across the receptor mosaic of the retina. Thus the development of movement sensitive neuronal systems can be considered as a very useful phylogenetic acquisition. The neurophysiological data about movement detecting neuronal systems, found in different species and at different levels of the visual system, support the view that movement sensitive neuronal systems have developed along similar principles at different times and stages of phylogenesis. With respect to their functional integration, visual movement detecting neurons can be divided into three classes:

(1) Movement sensitive neurons which analyse properties of the stationary visual stimulus pattern, which is continuously shifting on the retina due to voluntary and involuntary eye and head movements.

(2) Movement sensitive neurons which control saccades and pursuit movements of the eye and movements of the head when a moving visual stimulus is observed, or a complex, stationary pattern is explored.

(3) Movement detecting visual neurons which, when activated, form the physical basis of the psychophysical perception of motion of a visual stimulus pattern.

Movement perception might occur when *the stimulus is not physically moving*, just as brightness enhancement is seen in psychophysical experiments using stimulus patterns where no luminance enhancement is present (e.g. Brücke-Bartley effect, simultaneous brightness contrast), or as Fechner-Benham colors are seen in a flickering "white" stimulus, which is perceived as colorless at lower or higher stimulus frequencies. The perception of *apparent motion*, however, in our opinion indicates that movement sensitive neurons (group 3) in the brain are then "paradoxically" activated by the stimulus pattern.

Neurophysiological investigations of movement detecting neuronal systems have been made under different aspects during the last 15 years:

(1) The *qualitative* properties of the different stimulus patterns which activate movement sensitive neurons were investigated.

(2) The *receptive field organization* of movement detecting neurons was explored.

(3) *Quantitative relationships* between different stimulus parameters (angular velocity, size, contrast, position of the moving stimulus, duration of motion etc.) and the neuronal response were established.

(4) Interpretations concerning the *functional integration* of the movement sensitive neuronal systems into the overall performance of the visual system and interpretations with respect to visually guided behavior were presented.

As in the other fields of sensory physiology, psychophysical and behavioral observations are an important guide for integrating the neurophysiological results, obtained with microelectrode recordings, into a general theory of movement perception.

The psychophysical observations described in the next chapter indicate that the analysis of the neuronal mechanism of the sensory *and* the motor part of the visual system, including the visuo-motor coordination in the brain, is a prerequisite for a complete description of the physiological basis of motion perception. We will, however, restrict our review to microelectrode findings in movement detecting neuronal systems of the *sensory part* of the visual system. At the present time, much more neurophysiological data are available from the sensory part than from the motor part of the visual system. For the more recent neurophysiological work about neuronal mechanisms of the eye and head movement control systems, the reader is referred to the publications of Bizzi (1967, 1968, 1969), Bizzi and Schiller (1970), Cohen and Feldmann (1968), Duensing (1961), Duensing and Schaefer (1957), Precht (1969), and to the books edited by Bender (1965) and Bach y Rita (1971). The present report emphasizes the *general principles* of neuronal mechanisms for movement perception, the *different classes* of movement sensitive neurons and the *quantitative* aspects of the response of movement detecting neuronal systems, elicited by a wide variety of stimulus parameters. The review will be restricted to experimental findings in *vertebrates*.

A systematic report of the experimental findings in movement sensitive neuronal systems in different species (including invertebrates) and in different structures of the visual system was given in our recent review article (Grüsser and Grüsser-Cornehls, 1969). A schematic synopsis of the different classes of movement sensitive neurons in different species of vertebrates is presented in Table 1 (p. 352—355).

1. Visual Detection of Motion: A Sensory and Motor Achievement

In psychophysical experiments one has to discriminate between two types of visual motion perception: Perception of real movement and perception of many kinds of apparent motion. A concise review of psychophysical findings concerning visual movement perception will be given in this chapter, in order to provide a general framework for integrating the manifold findings in single visual neurons at a later date.

Real movement perception. The visual detection of motion of an object occurs when an object in the visual world changes its position against a stationary background and the change of position is above the threshold for minimal dislocation and minimal angular velocity (AUBERT, 1884, 1887; BASLER, 1906; KOFFKA, 1931; GRAHAM, 1966). For a quantitative analysis of visual movement perception, the following main parameters of the moving stimuli, which also determine the threshold for movement perception, have to be considered: *angular velocity; size* and *contrast; position and distance of the moving object; duration* of motion; *change in the angular size*, if the object is moving in the z-axis; and general level of light and color adaptation.

The suprathreshold range of velocity can be divided into three parts. At lower stimulus velocities, movement *and* direction of movement are perceived. Above a certain angular velocity only movement can be seen, while discrimination of the direction of the moving stimulus is no longer possible. At higher speeds, only a homogenous impression of the visual stimulus is left, and movement can no longer be detected. The values of angular velocity at which these different perceptual phenomena occur are dependent on the parameters mentioned above. The most important factor seems to be the *temporal* frequency of the pattern stimulating the single photoreceptors (FOSTER, 1969).

The *visual acuity* for moving targets during horizontal pursuit movements decreases linearly with an increase of the angular velocity from 1 to about 50 degrees/sec. A further non-linear decrease in acuity with increasing angular velocity is especially strong with a speed above 80 degrees/sec and might be attributed to the movement of the image across the retina (LUDVIGH, 1949; MILLER and LUDVIGH, 1962; JAEGER and HONEGGER, 1964; METHLING and WERNICKE, 1968; METHLING, 1970). If the eyes are immobile, the visual acuity decreases rapidly with an increase in the angular speed of the stimulus for targets moving faster than 1 degree $\cdot$ sec^{-1}.

In psychophysical experiments with human observers, a power function with an exponent of between 0.79 and 0.81 was found for the *relationship* between the *perceived velocity* of a small moving object and the *angular velocity*. If a periodic pattern of large black and white stripes was used as a moving stimulus, however, a linear relationship between the angular velocity of the motion perceived is valid (MASHHOUR, 1964; DICHGANS et al., 1969; KÖRNER, 1969). The speed of a moving object seems to be about 1.6–1.8 times higher if the eyes are immobile than if they "follow" the moving object, the image of which is then more or less "stabilized" on the retina at a low angular velocity ($<$50 degrees $\cdot$ sec^{-1}) (AUBERT, 1887; FILEHNE, 1922; BROWN, 1931; KORNMÜLLER, 1931; KÖRNER and DICHGANS, 1967).

During the observation of a moving stimulus it is quite possible that the stationary background elicits apparent motion, while the moving visual stimulus seems to be motionless: The moon seen as wandering behind apparently motionless clouds in the darkened sky is a well known example of such an illusion. Under certain stimulus conditions (especially large moving patterns) the moving stimulus is subjectively perceived as stationary, while the observer has the sensation of being moved (circular vection of FISCHER and KORNMÜLLER, 1930).This phenomenon can easily be elicited when the subject is sitting in a large, horizontally turning drum with black-white stripes, as used for the measurements of optokinetic nystagmus, and the optokinetic eye movements are suppressed for a longer period of time by fixation on a small stationary object. Then the observer often has the illusion of being rotated in the direction opposite to that of the moving pattern. If one stands on a bridge across a fast flowing river and looks into the river, one can easily have a similar illusion. The river appears to be flowing more slowly than it actually is, while the observer on the bridge has the sensation of moving against the flow of the river.

Apparent movement. The perception of apparent movement is elicited by a wide variety of visual stimulus patterns. The simplest form of apparent motion is elicited by the movement of the image across the retina due to eye movements. A *passive movement* of the eyeball elicits the perception of movement of the whole visual world. Objects in the central part of the visual field seem to make larger movements than objects in the periphery of the visual field periphery. Paradoxically, however, one does not then see a relative movement between the objects of the visual field (MACKAY, 1958). A good visible after image viewed with closed eyelids or in complete darkness appears motionless, when the eyeball is moved passively. This observation indicates that the muscle receptors of the eye do not contribute signals directly eliciting movement perception.

Apparent motion is seen in the *after-effects* of real movement. After one has looked at a rotating, patterned disc (for example a rotating record disc) or another visual pattern continuously moving in one direction ("waterfall") for some minutes, one will perceive stationary objects moving in the direction opposite to the conditioning stimuli. This movement after-effect is best seen with moving random dot patterns. It is strongly reduced, if one eye is conditioned by the moving stimulus, and the test stimulus thereafter is seen by the other eye.

Apparent motion is also perceived if two stationary visual stimuli are briefly projected onto different parts of the visual field with a temporal delay (φ-movement, EXNER, 1875; WERTHEIMER, 1912, review of current literature AARONS, 1964). φ-movement is also elicited if the two stationary stimuli have different shapes. Then the "moving" stimulus paradoxically changes its shape during the apparent motion.

Another type of apparent motion is seen in stationary visual stimulus patterns whose luminance changes. A simple lecture hall experiment demonstrating this might be mentioned: A ring of light is projected onto a screen. The luminance of the ring is varied sinusoidally at 0.5–3 cps. A stationary spot of light, less bright than the ring, might be projected inside the ring. Apparent expansion and contraction of the spot of light inside the ring occurs synchronous with the sinusoidal change in the luminance of the ring. The movement in the stationary stimulus pattern is probably caused by the change of lateral inhibition, which depends on the luminance of the ring.

φ-movement is also seen if the two successive stimuli are not perceived by the same eye. A simple experiment might demonstrate this. When one reaches out the arm and monocularly fixes the thumb first with one, then with the other eye, movement of the thumb and the hand is seen against the background.

The neuronal basis of this movement perception must be at or beyond the binocular integration in the primary visual cortex, while the movement after-effect phenomena are located, at least in part, at a monocular neuronal system. With binocular stereoscopic stripes composed of monocular random patterns, JULESZ and PAYNE (1968) found φ-movement, which had some properties differing from those of the monocular φ-movement.

Visual perception and eye movements. Involuntary eye movements (nystagmus), caused by vestibular stimulation, intoxication or acute diseases of the central nervous system, will give rise to the perception of motion of visual stimuli, which are stationary in actuality. In this case the perception of movement is induced primarily by the movement of the image on the retina during the slow phase of the nystagmus. No apparent motion of the visual world, however, is perceived by subjects who have a *congenital nystagmus*. If those subjects look into a mirror, they cannot see their own nystagmus. During *voluntary* eye movements (saccades) the well-structured visual world is perceived as stationary. The same is true for the movement of a patterned visual image on the retina during the small involuntary eye movements (eye tremor, flicks and drifts), which occur in addition to the more extensive voluntary eye movements. It was formerly assumed that during saccadic voluntary eye movements, a general suppression of visual perception exists, which is caused by central inhibitory mechanisms (HOLT, 1903; JUNG, 1953). Recent experiments indicate that during the saccadic eye movements only a slight increase of the visual threshold occurs (LATOUR, 1962; ZUBER and STARK, 1966; STARR et al. 1969; MITRANI et al., 1970).

Tachistoscopically projected stationary patterns presented for 50 μsec during a saccade, brought about by voluntary or optokinetic eye movements, are easily recognized and no impairment of contrast or gestalt perception can be found (GRÜSSER unpubl., 1958; RISOS, 1965). We found that during saccadic eye movements perception of motion in the direction of the eye movement is partially suppressed, whereas the perception of movement perpendicular to the direction of the eye movement is less impaired.

The fact that during saccadic eye movements no disturbing perception of movement due to the shift of the retinal image occurs, can be explained, if the temporal stimulus frequency of the photoreceptors is taken into consideration. One can easily calculate that the temporal frequency with which the single receptors are stimulated by a well structured light-dark pattern during a saccade will be above the flicker fusion frequency. Hence, during a saccade, the response at the receptor output level corresponds approximately to that elicited by a short gray stimulus which does not last long enough for perception. A dark interval has to last at least 60–80 msec to reach the threshold of perception (BASLER, 1906). For the same reason, when the eyelids are partially or completely closed quickly during the fast saccades of voluntary or nystagmic eye movements or during nystagmus, the visual perception is not impaired (ADAMS, 1958; HABERICH and FISCHER, 1954). This consideration indicates that central suppression of pattern and movement perception is not necessary at all during saccadic eye movements. The possible central suppression of visual *movement* detection during saccadic eye movements is incomplete, however, because a *single* small light stimulus viewed in a dark room is seen rapidly moving during saccadic eye movements. During the beginning and the end of each saccade, the angular speed of the eye is slow enough that a "smearing" of the contours of the visual pattern should be perceived. This, however, is not the case. The perception of sharp contours during the exploration of a visual pattern by voluntary eye movements might be explained by *metacontrast*. The response elicited by the pattern after the end of each saccade suppresses by means of lateral backward masking the response elicited by the smeared image during the saccade (GRÜSSER, 1972). GERRITS and VENDRIK (1970) explored the effect of different velocities of a moving pattern on the contour perception under otherwise stabilized retinal stimulation. They found that simulation of saccadic displacements alone was not sufficient to restore normal vision, while the simulation of drifts regenerated completely a disappeared pattern.

Apparent motion of a visual stimulus is also seen if one fixes his eyes on a small visual object in darkness and no voluntary eye movements are performed. After some seconds of fixation, a very large and irregular movement of the stimulus pattern is sometimes seen. It is assumed that this "autokinesis" is caused by small drifts which shift the stimulus image from the fovea to adjacent parafoveal areas, whereby movement sensitive neurons are activated. The degree of the drifts and the extension of the apparent movements of the stimulus, however, do not correspond to each other. CRONE and VERDUYN LUNEL (1969) therefore attributed the illusion to the activation of a highly sensitive class of parafoveal movement detecting neurons, which is normally used for the control of slow pursuit movements, when the eyes follow a moving object.

A long lasting after-image appears motionless when it is viewed in complete darkness and the eyes are moved passively. It is seen moving as soon as the eyes move actively, either in saccades or in slow pursuit movements. The latter can be elicited if the after-image is elicited in the parafoveal region and the subject tries to fix the after-image and consequently "follows" the shifting after-image with his eye movements. The perception of movement of an after-image is a direct proof that visual movement perception might occur independently of any change in the position of the retinal stimulus.

Since HELMHOLTZ (1868) it has been assumed that during active eye movements efferent nervous activity interacts with the afferent visual signals (see also v. UEXKÜLL, 1928). This idea was further developed by v. HOLST and MITTELSTAEDT (1951) in their "Reafferenztheorie" and applied for the explanation of perceptual stability of the world during eye movements: The central efference copy of the motor command counteracts the afferent information. For example, if the voluntary command "movement of the eyes to the right" is performed, the image of the visual world on the retina shifts to the right and elicits the perception of movement of the visual world to the left, if not cancelled by the counteracting efference copy. This outflow hypothesis is strongly supported by the casual observation that apparent movement perception is reported without any eye movements by patients with acute paralysis of the eye muscles. For example, if a patient with an acute paralysis of the abducens nerve of the right eye intends to look monocularly with his right eye horizontally to the right, he sees the surrounding world shifting to the right, although his eye is not moving at all (c.f. also KORNMÜLLER, 1931). The explanation of this phenomenon, according to the theory of reafference, is that movement perception is caused by the non-cancelled efference copy in the feedback loop. This theory would also explain the apparent movement of a long lasting after-image during voluntary eye movements.

The experimental findings mentioned in this chapter indicate that for a complete description of the neurophysiological basis of movement perception, in addition to the neuronal mechanisms in the sensory and motor part of the visual system, the effect of vestibular stimulation, the influence of the mechanoreceptors of the neck and the interconnections between voluntary and involuntary eye movement control systems have to be investigated. Many years of neurophysiological investigations will be necessary to provide an experimental basis for a complete neurophysiological description of the neuronal mechanisms, which are the physical correlate of such a simple perception as performed when one observes a flying swallow hunting a butterfly on a warm summer afternoon.

2. Are Movement-Sensitive Neurons always Employed in the Detection of Visual Motion?

A *movement-sensitive* visual neuron may be defined as a neuron in the visual system from which (considering natural stimuli) moving visual patterns elicit a significantly stronger response than stationary ones, even if optimal temporal and spatial frequencies of the stationary stimuli projected to the receptive fields of the neuron are used. A *movement specific* neuron is defined as a unit which is activated or inhibited (again considering natural stimuli) only by moving visual stimuli and not by stationary ones.

Does movement sensitivity or movement-specificity imply that these neurons are employed within the neuronal network of the visual system for the detection of motion in the visual world or even for measurement of the angular velocity of a moving object? We think that such a conclusion cannot necessarily be drawn. As mentioned on p. 334, one has to assume that many of the movement sensitive neurons in the central visual system indeed serve to analyse the visual signals received from the stationary visual world and a second group of movement sensitive visual neurons controls eye and head movements. These arguments are deduced from considerations involving voluntary and involuntary eye movements.

From experiments with stabilized retinal images (DITCHBURN and GINSBORG, 1952; RIGGS et al., 1953; FENDER, 1964), it is well known that the contours of a stabilized image disappears within a few seconds due to local adaptation mechanisms in the afferent visual pathways (HERING). The contours of a stabilized retinal image do not disappear, however, during flicker stimulation (KEESEY, 1965; YARBUS, 1966).

The small involuntary eye movements may therefore compensate for the frequency characteristics of the afferent visual system and counteract local adaptation (Adrian, 1928; c.f. van de GRIND et al. 1972, Chapter 7, in this Volume). The eye movements shift the effective visual stimulus into a temporal frequency range at which retinal neurons are strongly activated. This compensatory mechanism for local adaptation, however, in effect compels the central part of the visual system to analyse information about the stationary visual world from neuronal signals elicited by the continuously shifting image on the retina. Movement-sensitive neurons, especially those in the visual cortex which are also sensitive to the spatial orientation of the contrast border or other more complicated features of the stimulus, seem to be particularly well adapted for the analysis of a structured visual worldthe image of which is continuously moving on the retina. From these considerations, one can expect that only a part of the specialized classes of visual neurons which are found to be movement sensitive are really used as visual motion detectors. It seems to be very probable that only those movement sensitive neurons, for which the neuronal activation is correlated over a wide range with the angular velocity of the moving stimulus, are candidates for motion detectors.

Voluntary and involuntary eye movements which are adjusted to the frequency properties of the physiological processes in the receptor cells are indicators of the first step of sensory-motor coordination, necessary for the perception of the visual world. The movement sensitivity of most of the central visual neurons may be explained as the second stage of adjustment, at which the central analysing units are adapted to the spatio-temporal properties of the input from the retina. The third step of adjustment can be seen by the finding that voluntary and involuntary eye-movements during the perception of a patterned visual stimulus are not at random, but are controlled by the contours and by more complex features of the stimulus (YARBUS, 1966; FENDER, 1964; ST. CYR and FENDER, 1969).

II. Receptive Fields and Different Classes of Movement-Detecting Neurons in the Visual System

1. The Receptive Field (RF)

The concept of the receptive field *(receptive area)* was introduced into neurophysiology by ADRIAN, CATELL and HOAGLAND (1931) for the description of the cutaneous area from which a *single* mechanoreceptive neuron could be stimulated. HARTLINE (1938, 1940) used the term "receptive field" to describe the area in the frog's retina from which he could obtain a response in a single optic nerve fiber. Receptive field (RF) is also an adequate definition to designate the area in the *visual field* which is projected to the receptive field at the retinal surface by the optical system of the vertebrate eye. KUFFLER (1953) demonstrated that the simple concentric receptive fields of the retinal neurons in cats could be subdivided into two classes, according to their excitatory response to small spots of light: on-center and off-periphery *(on-center neurons)* and off-center and on-periphery *(off-center neurons)*. In these neurons with concentric RFs the RF-center and the RF-periphery are mutually antagonistic: in on-center neurons illumination of the RF-center and darkening of the RF-periphery elicits excitation, while illumination of the RF-periphery and darkening of the RF-center elicits inhibition. The contrary is true for off-center neurons.

The receptive fields of movement-sensitive neurons are frequently more complex than the concentric RFs. We have therefore proposed the term "*excitatory receptive field*" (ERF) for the region of the retina or the visual field from which adequate stimulation elicits neuronal excitation (GRÜSSER-CORNEHLS et al., 1963). With the aid of a stimulation technique using small single spots of light, an ERF might be subdivided into more or less regular *on-zones*, *off-zones* and *on-off-zones*, but stimulation of the ERF with a single spot of light can also be completely ineffective, as is the case for some of the neurons having "complex" or "hypercomplex" receptive fields in the visual cortex (HUBEL and WIESEL, 1962, 1965). The region of the visual field from which only inhibition can be elicited by adequate stimulation is called the "*inhibitory receptive field*" (IRF). This concept assumes that within the ERF excitation *and* inhibition might be elicited, while only inhibitory effects appear when the IRF is adequately stimulated. The ERF and IRF together constitute the RF which is now defined as that part of the retina or visual field from which adequate stimulation evokes either an inhibitory or an excitatory effect, or both, in a single visual neuron.

The requirements of an "adequate stimulus" might be highly specific and might include a certain arrangement of contrast borders, direction of movement, angular size and even a special shape of the moving stimulus. For the organization of the RF of movement detecting neurons an ERF surrounded by an IRF seems to be much more common than the reverse or than an irregular distribution of ERF- and IRF-parts within the receptive field. For the functional interpretation of neuronal nets composed of neurons having receptive fields, one has to keep in mind that receptive fields of neurons in the same region of the afferent visual system overlap a hundredfold or even a thousandfold. One stimulus pattern, moved across a small retinal area, will therefore activate some 100 or 1000 central neurons having overlapping ERFs and IRFs.

2. Classes of Movement-Detecting Neurons

Five main classes of movement-detecting neurons can be distinguished:

(a) Neurons which, under *natural stimulus conditions*, respond only or predominantly to moving visual stimuli, *independent* of the direction of movement *(M-neurons)*.

(b) Movement-detecting neurons for which neuronal activation depends on the direction of the moving stimulus *(DS-neurons)*. The direction in which the stimulus has to be moved to elicit an optimal response is called the *preferred direction;* the direction in which no response or the weakest response occurs, the *null-direction*.

Two subclasses can be discriminated. In subclass 1 the preferred direction and the null-direction form a 90 degree angle. In subclass 2 the null-direction and the preferred direction are 180 degrees apart. All directions in which the movement of a stimulus through the ERF elicits a response lie within the "response sector", defined by an angle with respect to the coordinates of the visual field (GUSELNIKOV and VODOLASZKIJ, 1968).

(c) Movement-sensitive neurons which respond better to moving stimulus patterns than to stationary ones, but for which the response also depends on the *spatial orientation* of the contrast borders of the stimulus relative to the axes of the visual field *(OS-neurons)*.

(d) Complex movement detecting neurons (CM-neurons) are neurons which respond in parts of the ERF to motion of a stimulus pattern, while in other parts of the receptive field stationary illumination elicits neuronal activation. Movement sensitive neurons having receptive fields with irregularly arranged ERFs and IRFs will also be included in this class.

(e) Neurons are called Z-neurons if they respond only, or predominantly, to motion in the Z-axis, when a moving object approaches the eye (Z^+-neurons) or moves away from the eye (Z^--neurons). Z-neurons respond only weakly or not at all to movement in a plane perpendicular to the Z-axis between the object and the nodal point of the eye's optical system.

III. The Response of Concentric On-Center and Off-Center Neurons to Moving Stimuli

1. Neurophysiological Data

An analysis of the response of on-center and off-center neurons having concentric RFs (Kuffler, 1953) to moving stimuli might contribute to the understanding of the function of movement sensitive or movement detecting neurons in the higher parts of the visual system. As mentioned in the preceding chapter, movement of a patterned stimulus across the RF corresponds in the time domain to intermittent light stimulation of the photoreceptors. The response to moving contrast stimuli, however, is determined in addition by the *sequential* stimulation of the antagonistic RF-periphery and RF-center. It is therefore of interest to measure the response of visual neurons (which respond well to intermittent, stationary stimulation) to moving light-dark patterns and the dependence of this response on the parameters of the stimuli (luminance, contrast, size, oscillation frequency, oscillation amplitude and position of the oscillating stimulus within the receptive field) (Rodieck and Stone, 1965; Enroth-Cugell and Robson, 1964; Foerster and Grüsser, 1969; Grüsser, Grüsser-Cornehls and Hamasaki, 1971).

From Fig. 1 it is evident that the response enhancement by a simultaneous contrast border, demonstrated first by Baumgartner and Hakas (1959, 1962) with stationary stimulus patterns, also appears in the response of retinal neurons elicited by contrast borders moving slowly across the RF. As the border between RF-center and RF-periphery is approached by the moving contrast border, either entering or leaving the RF-center, the neuronal response increases (Fig. 1, cat retina: Rodieck and Stone, 1965; cat geniculate body: Grüsser and Grüsser-Cornehls, 1961; "Eintrittsaktivierung" and "Austrittsaktivierung"). This border contrast-enhancement of the neuronal response disappears if the moving light or black bars are too narrow (widths less than half the size of the RF-center) or are moved too fast ($> 5-15$ degree $\cdot$ sec^{-1}). At higher stimulus velocities a light bar of $1-5$ degrees diameter elicits *rhythmical activation* of the neurons. This can be explained only in part by the observation that the conduction velocity of lateral inhibition from the RF-periphery is rather slow ($400-480$ degrees $\cdot$ sec^{-1}

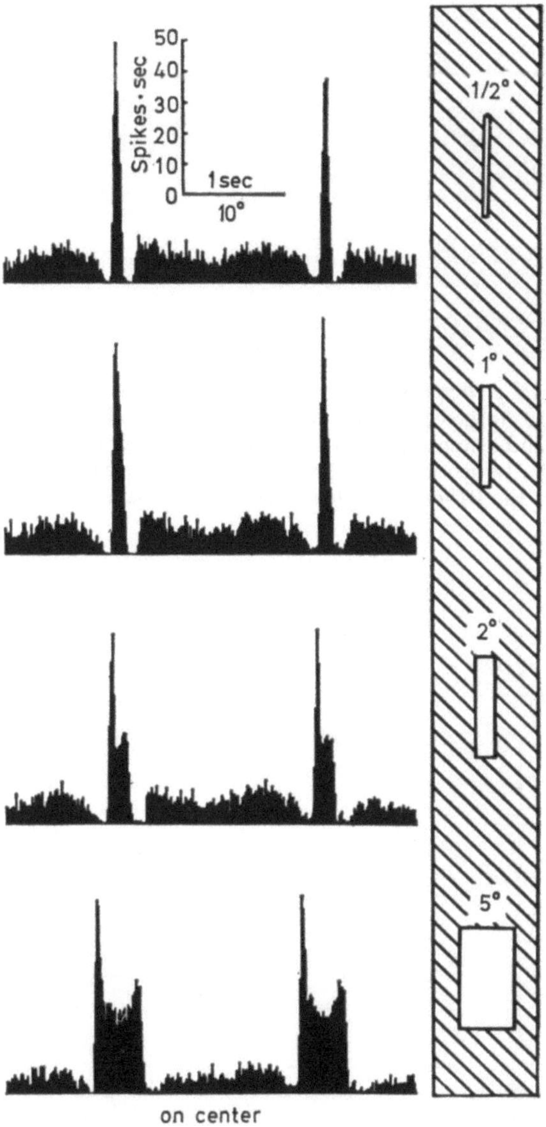

Fig. 1. *Averaged impulse pattern of a cat's retinal on-center neuron*
Movement of a vertical slit of light of 5 degrees width through the receptive field at an angular velocity of 10 degrees · sec^{-1}. As the slit enters or leaves the RF-center, an enhanced border contrast activation is seen in the response. This simultaneous contrast enhancement disappears as the slit of light becomes narrower. Averages of 30 responses; luminance of slit 5 cd · m^{-2} (From Rodieck and Stone, 1965)

Wuttke et al., 1966; Wuttke and Grüsser, 1968). In addition, one type of lateral inhibition has a slower frequency response than excitation, and the lateral inhibition might also have a somewhat longer "local" latency than the excitatory processes in the RF-center (Figs. 2 and 3).

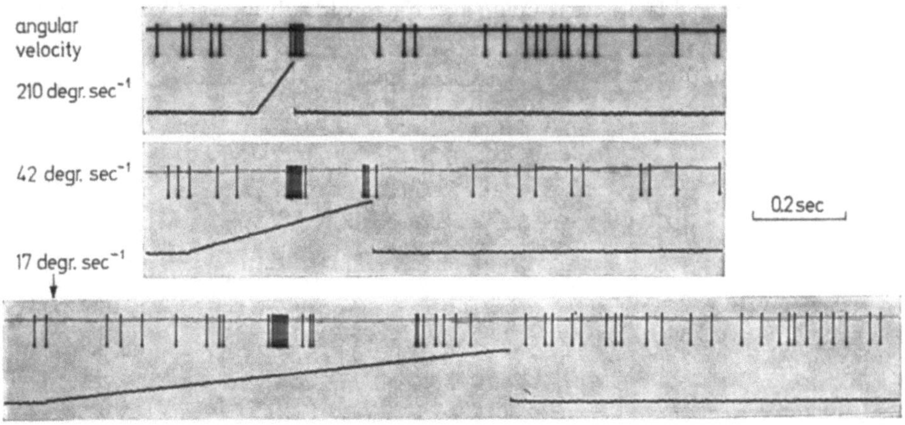

Fig. 2 a. *Impulse pattern of an on-center neuron in the cat's lateral geniculate body*
A vertical slit of light (1×6 degrees) is moved in one direction across the RF at three different angular velocities. A rhythmical impulse pattern is elicited. The primary activation period is preceded by an inhibitory period which is elicited by the initial movement of the stimulus through the RF-periphery (encephale isolé preparation)

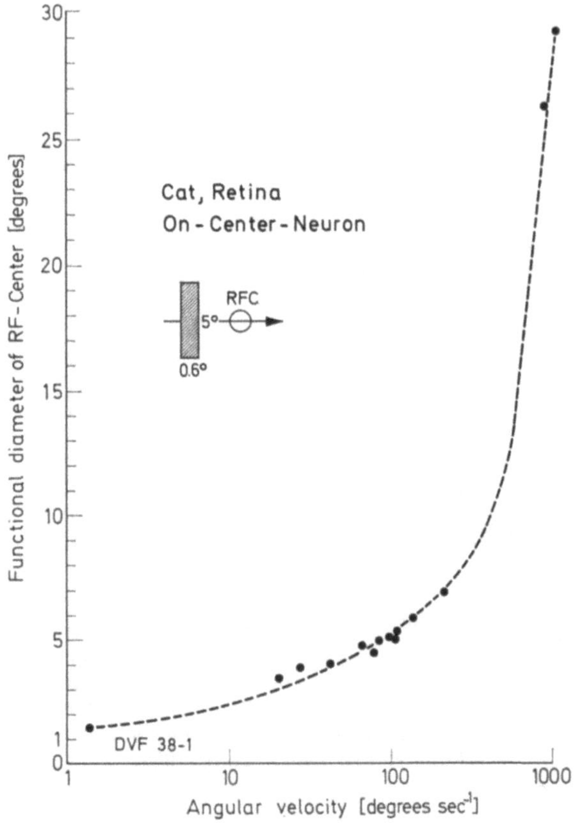

Fig. 2b. *Relationship between angular velocity and the functional diameter of the RF-center of a retinal on-center neuron.*
A vertical slit of light (0.6×5 degrees) was moved horizontally through the RF. The functional size of the RF-center was calculated from the duration of the primary activation period (see Fig. 2a)

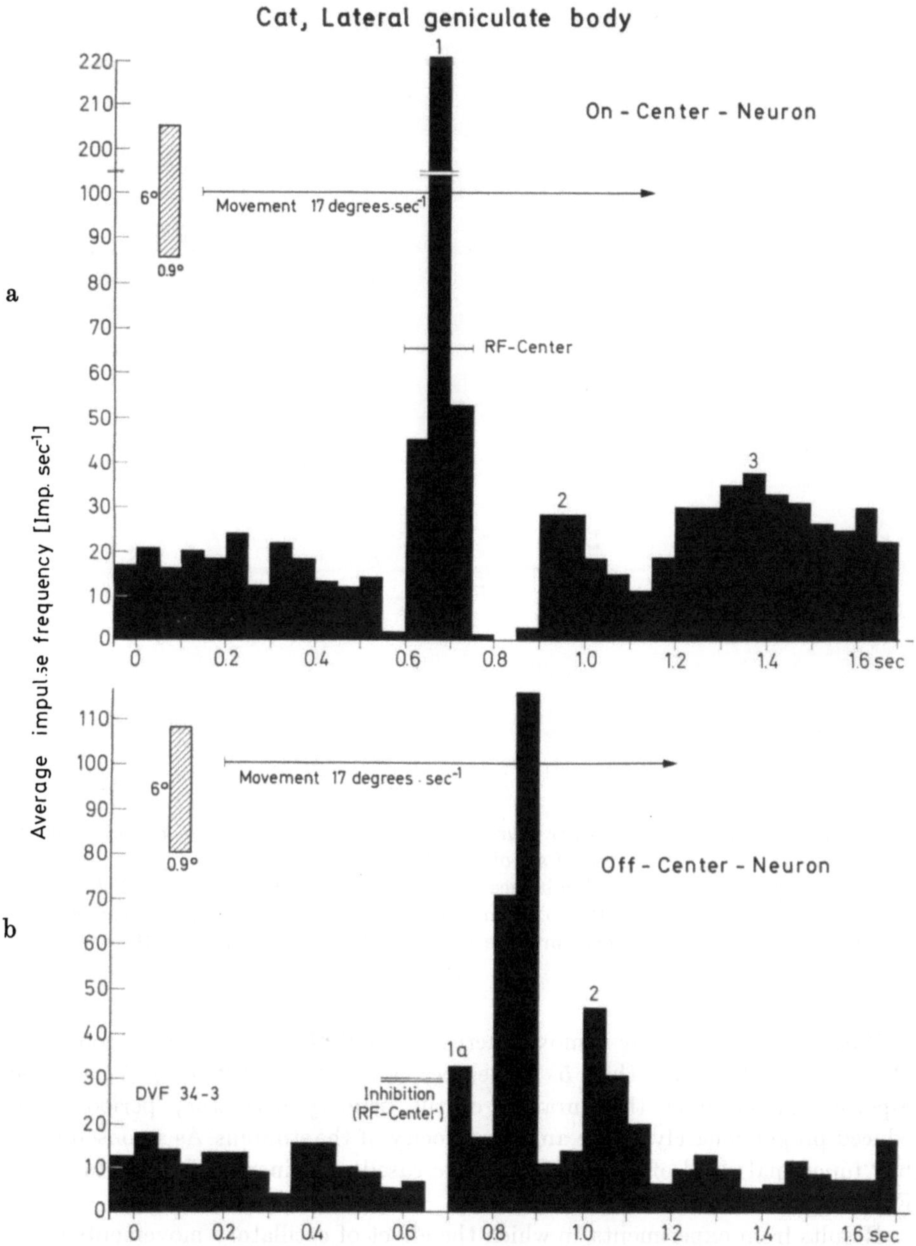

Fig. 3. *Averaged impulse pattern of two neurons in the cat's lateral geniculate body.* A vertical slit of light (0.9 × 6 degrees) was moved horizontally through the RF at an angular velocity of 17 degrees · sec^{-1}. The neuronal response pattern was averaged for 20 successive stimuli occurring every 10 sec. a Response pattern of an on-center neuron. b Response pattern of an off-center neuron, having its RF in about the same region of the visual field as the on-center neuron. Both neurons responded with a grouped discharge pattern. Note the temporal relationship between the excitatory periods in the response pattern of one neuron and the inhibitory periods in the response pattern of the other neuron. Luminance of the moving stimuli 56 asb, background luminance 4 asb

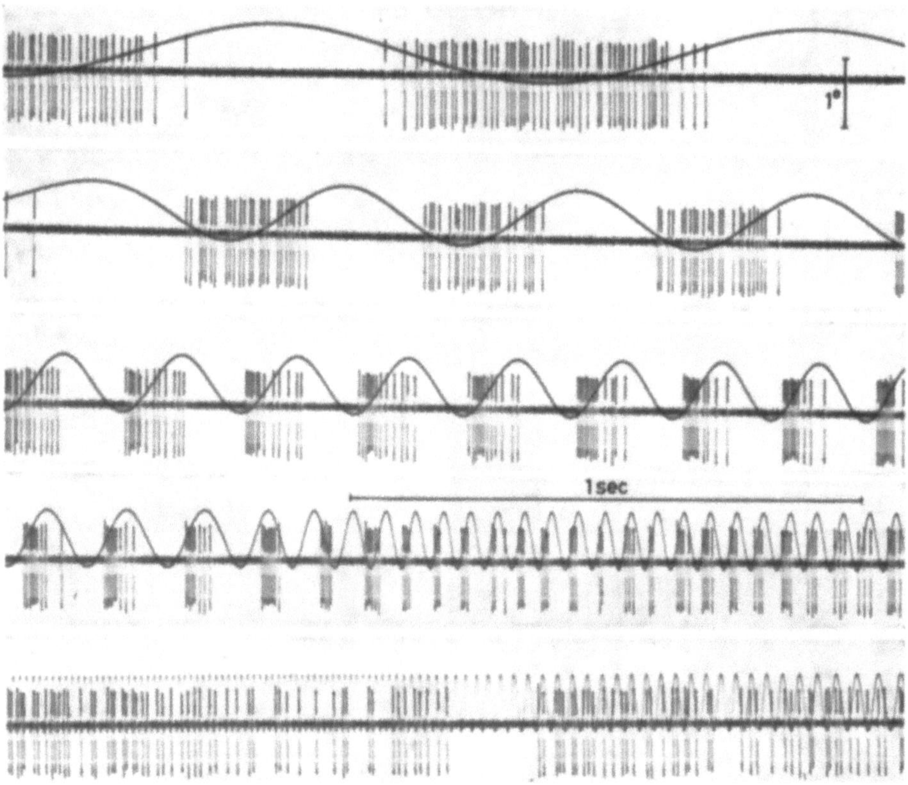

Fig. 4a. *Response of a retinal on-center neuron of the cat to sinusoidal, horizontal oscillation of a vertical slit of light 1 × 15 degrees.*
Oscillation amplitude 0.9 degrees; luminance of the moving stimulus 300 asb, as in Fig. 4b—4d; background illumination about 10 asb. Diameter of the RF-center 0.6 degrees. The light bar moves from the RF-periphery into the RF-center but not through the RF-center

When a small light bar is moved across the RF of neurons having concentric RFs at a speed greater than 5–10 degrees · sec^{-1}, the *duration of the neuronal* response, measured as the duration of the primary activatory period is not reduced proportionately to the angular velocity of the stimulus. As a consequence, the "functional size" of the RF center increased with increasing stimulus speed (Fig. 2b).

Results from experiments, in which the effect of oscillatory movements of the stimulus across the RF of retinal and geniculate neurons in cats was investigated, are shown in Figs. 4 and 5. The frequencies applied included nearly the whole frequency range of the voluntary and involuntary eye movements (Ditchburn and Foley-Fisher, 1967). Oscillatory movements of the stimulus patterns across the receptive field increase the neuronal activation in a similar manner as intermittent stationary stimulation (Grüsser, 1956; Grüsser and Creutzfeldt, 1957; c.f. van de Grind et al., Chapter 7, page 494 of this Volume). The oscillation frequency at which the maximal neuronal activation is reached depends

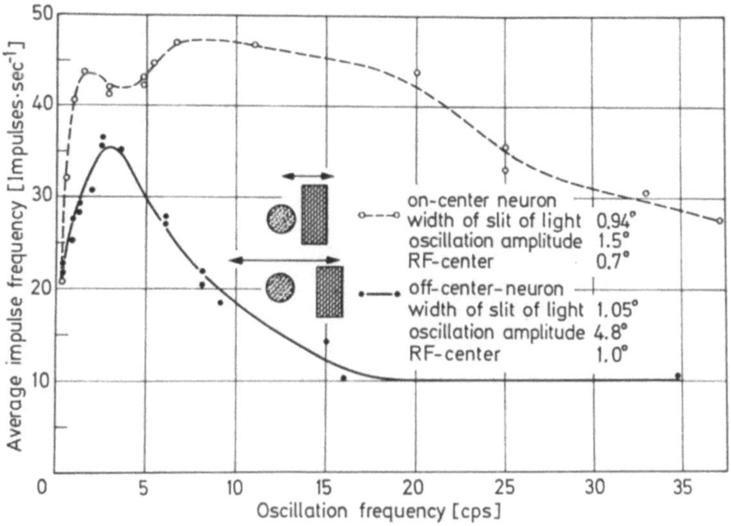

Fig. 4b. *Relationship between average impulse frequency (ordinate, impulses · sec⁻¹) and the sinusoidal oscillation frequency (abscissa).*
Measurements from two retinal neurons in cats. In the on-center neuron, the light bar moves only into the RF-center from one side, while in the off-center neuron the light bar oscillates through the RF-center and the RF-periphery on both sides

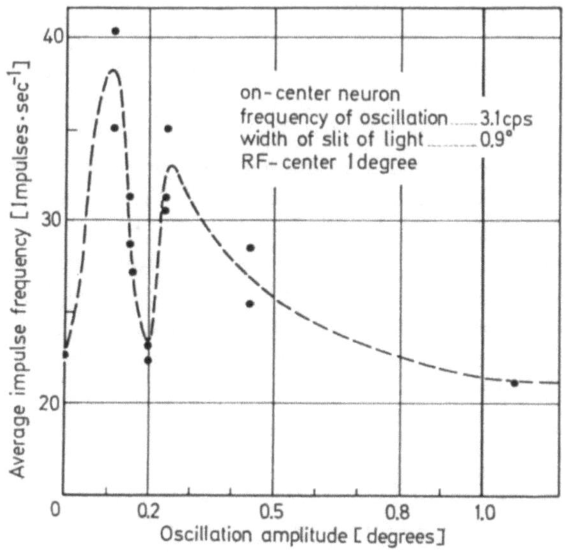

Fig. 4c. *Relationship between the average neuronal impulse frequency of an on-center neuron and the amplitude of the horizontal, sinusoidal oscillation of a bar of light (0.9 × 15 degrees).*
Oscillation frequency 1—3 cps. Diameter of the RF-center about 1 degree. The oscillation is symmetrical with respect to the RF-center

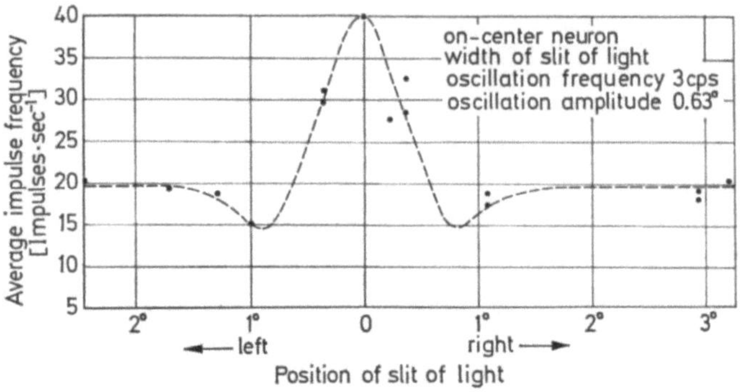

Fig. 4d. *Relationship between the neuronal activation of a cat's retinal on-center neuron (ordinate) and the position of a horizontally oscillating bar of light within the RF.* Oscillation frequency 3 cps, amplitude 0.6 degrees, width of the bar 0.9 × 12 degrees (Fig. 4a—4d from experiments done in cooperation with M. FOERSTER, from GRÜSSER and GRÜSSER-CORNEHLS, 1969)

on the luminance, oscillation amplitude, position and spatial property of the stimulus in relation to the diameter of the RF-center. Due to the antagonistic effect of the RF-center and RF-periphery, a non-monotonic function of the oscillation amplitude and the neuronal activation is obtained (Fig. 4c). The frequency maximum is reached at a lower frequency in off-center neurons than in on-center neurons if similar stimulus conditions are applied. Neurons in the geniculate body reach the response maximum at a somewhat lower oscillation frequency than the corresponding retinal neurons (3–12 cps versus 8–20 cps under photopic stimulus conditions).

Further evidence for the interaction of spatial and temporal signal processing is seen in the response to moving stimulus patterns, as described in the work of ENROTH and ROBSON (1966). They moved sinusoidal gratings, having different spatial frequencies, across the RF of retinal neurons of cat. They stated that the *optimal spatial resolution depends on the temporal frequency of the stimuli*, i.e., in their experiments, on the velocity with which a given spatial grating pattern was moved across the retina.

The response of concentric RF on-center and off-center neurons in the cat's retina and geniculate body is not affected by a change in the direction of the movement of the stimulus pattern through the RF. This indicates that the rotational spatial symmetry found with stationary patterns also holds for the response to moving light-dark borders.

2. Psychophysical Correlations

The data shown in Figs. 4 and 5 indicate that moving a contrast border across the RF enhances the neuronal response at a certain traverse speed or oscillation frequency, similar to the way in which intermittent stimulation enhances the average neuronal activation at a stimulus frequency between about 5 and 25 cps. The maximal neuronal activation elicited by moving stimuli was found, as a rule, at somewhat lower oscillation frequencies. From psychophysiological data it is

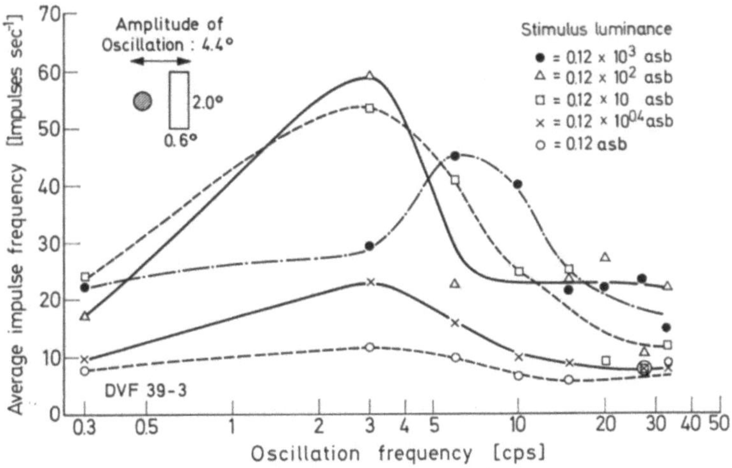

a

Fig. 5. *Neuron of the lateral geniculate body of cat.*
a Relationship between the neuronal activation (average impulse frequency, ordinate) and the frequency of the triangular or sinusoidal oscillation of a bar of light (0.6×2 degrees) at different luminance levels

well known that eye movement (or movement of a "stabilized image") counteracts local adaptation. A stimulus pattern, completely stabilized on the retina, disappears within seconds (Ditchburn et al., 1951; Riggs et al., 1952). The data obtained in single retinal and geniculate units indicate that the primary effect of eye movements counteracting local adaptation occurs at the retinal level: movement of a stimulus across the RFs enhances the neuronal response.

The question whether eye movements improve pattern perception or visual acuity has recently been discussed (Keesey, 1960; Krauskopf, 1952, 1962; van Nes et al., 1967; van Nes, 1968). There is only a small increase, if any at all, in the "dynamic visual acuity" when slowly moving stimulus patterns are used in comparison to the "static" visual acuity (non-stabilized retinal image, pursuit movements, Schober et al., 1968). If the eyes follow the moving object at stimulus velocities above 1—3 degrees $\cdot$ sec^{-1}, the visual acuity and pattern perception rapidly decrease as the velocity of the stimulus increases (c.f. p. 336).

The decrease in visual acuity in such experiments is attributed to incomplete pursuit movements of the eyes. If the eyes are immobile and acuity measurements are made with moving stimuli, the visual acuity is diminished at much lower angular speeds, while the perception of form is strongly, or completely impaired at a stimulus speed above 5—10 degrees $\cdot$ sec^{-1}. These observations with patterned stimuli can be predicted from the experiments performed by Froehlich and his coworkers (1921, 1929). They measured the duration of the primary image and the after-images elicited by a small moving slit of light. As the angular speed of the stimulus exceeds about 5 degrees $\cdot$ sec^{-1} (this limit depends on the stimulus luminance), the "primary image" of the moving stimulus increases in length (Fig. 2). At higher angular velocities, positive and negative after-images, moving

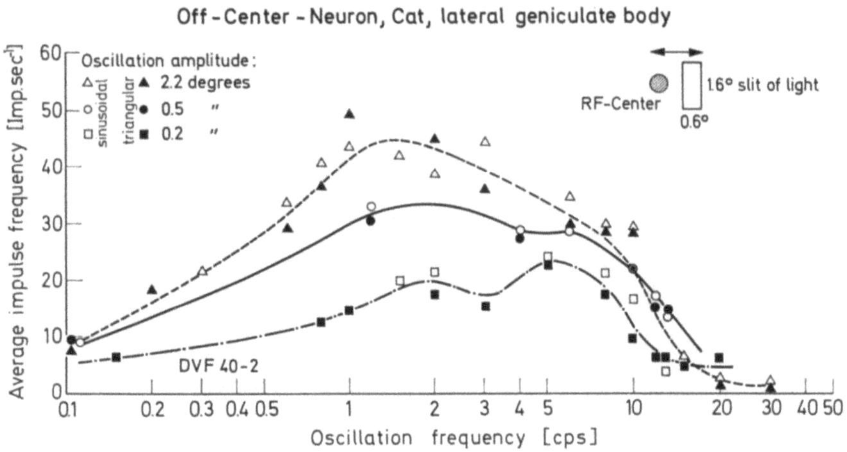

b

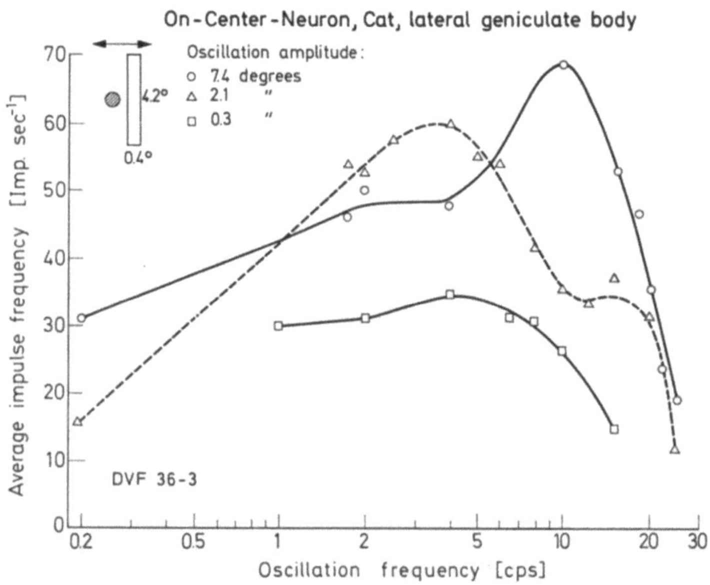

c

Fig. 5 b and c. Relationship between oscillation frequency (abscissa) and average neuronal impulse frequency obtained with three different amplitudes in an on-center neuron (b) and an off-center neuron (c). Note the different effects of the change in amplitude on the "optimal" frequency in the on- and in the off-center neurons. Triangular oscillations. Stimulus luminance in (b) and (c) about 60 asb, background 0.5 asb (From GRÜSSER, GRÜSSER-CORNEHLS, and HAMASAKI, 1970)

behind the primary images, appear. An elongated light band of positive after images, interrupted by dark bands (negative after-images) is then perceived. From these psychophysiological observations one could assume that the neuronal re-

Table 1

Animal	Structure	on-center off-center neurons (on-off-c.)	M-neurons	DS-neurons	OS-neurons	Complex M-neurons	Z-neurons	Authors
Monkey (Macaca mulatta)	superior colliculus		+ ERF: 2—90°			+		Humphrey (1968), Wurtz (1970), Kadoya et al. (1971), Wurtz and Goldberg (1971)
Monkey (spider monkey, Macaca mulatta)	visual cortex (Area 17)	+ color-coded		+	+			Hubel and Wiesel (1968), Wurtz (1968, 1969a, b, c), Spinelli et al. (1970) Creutzfeldt et al. (1971)
Monkey	inferotemporal cortex			+	+	+		Gross et al. (1969, 1972)
Cat	retina	+	(+) re-stricted area ?	+				Kuffler (1952, 1953), Stone and Fabian (1966), Spinelli and Weingarten (1966)
Cat	optic tectum	+ (on-off)	+	+				Marchiafava and Pepeu (1966), Straschill and Taghavy (1966, 1967), Buser and McIlwain (1968), Straschill and Hoffmann (1968, 1970), Sterling and Wickelgren (1969, 1970, 1971), Sprague et al. (1968), Harutiunian-Kozak et al. (1968), Feldon et al. (1970), McJlwain and Fields (1971), Schaefer (1971)
Cat	praetectal area			+			+	Harutiunian-Kozak et al. (1968), Straschill and Hoffmann (1969)
Cat	geniculatum laterale	+		+ (on-off)				Grüsser and Saur (1960), Hubel and Wiesel (1961), Kozak et al. (1965), McIlwain and Creutzfeldt (1968), Meulders and Godfraind (1969)

Cat	pulvinar nucleus lat. thalami	+ (on-off)	+	+	+		GODFRAIND et al. (1969)
Cat	primary visual cortex	+	+	(+)	+		JUNG et al. (1952, 1957), HUBEL and WIESEL (1959, 1962), GRÜSSER and GRÜSSER-CORNEHLS (1960, 1961, 1972), BAUMGARTNER et al. (1963, 1964), DENNEY et al. (1968), PETTIGREW et al. (1968a, b), CREUTZFELDT and ITO (1968), SPINELLI and BARRETT (1969), CREUTZFELDT et al. (1969), EYSEL and GRÜSSER (1970), JONES (1970) NODA et al. (1971)
Cat	secondary and tertiary visual cortex		+	+	+	+	HUBEL and WIESEL (1965), GRÜSSER and HENN (1969), EYSEL and GRÜSSER (1971)
Cat	middle and lateral supra-sylvian gyrus	on-off	+ large ERF	+	+		DOW and DUBNER (1969, 1971), WRIGHT (1969), HUBEL and WIESEL (1969), GRÜSSER, GRÜSSER-CORNEHLS and HAMASAKI (1970)
Rabbit	retina	+	+	+	+	+	BARLOW and HILL (1963), BARLOW et al. (1964), BARLOW and LEVICK (1965), LEVICK (1965), OYSTER and BARLOW (1967), OYSTER (1968)
Rabbit	optic tectum	+ ?	+	+	?		SCHAEFER (1962, 1966, 1967, 1971), HILL (1966), HORN and HILL (1966), MASLAND et al. (1971)
Rabbit	geniculate body	+	+	+	+		HILL (1962), ARDEN (1963), FUSTER et al. (1965), LEVICK et al. (1969)
Rabbit	visual cortex	+ on-off	+	+	+	+	GRÜSSER-CORNEHLS and GRÜSSER (1961), ARDEN et al. (1967), HUGHES (1968)
Rat	retina	+					BROWN and ROJAS (1965)
Rat	geniculate body	+		+ on-off-center			MONTERO and BRUGGE (1969), MONTERO et al. (1968, 1970, 1971)
Rat	optic tectum	+	+				SIMINOFF et al. (1966), HUMPHREY (1967, 1968)

Table 1 (continued)

Animal	Structure	on-center off-center neurons (on-off-c.)	M-neurons	DS-neurons	OS-neurons	Complex M-neurons	Z-neurons	Authors
Rat	pretectal region	on-off +	+	+				SIMINOFF et al. (1967)
Ground squirrel	retina	+ (color +)		+				MICHAEL (1967, 1968)
Ground squirrel	optic tectum		+	+	+			MICHAEL (1969), MICHAEL (1970, 1971)
Ground squirrel	geniculate body	+ (color +)						MICHAEL (1969)
Opossum (Didelphis)	optic tectum	+ mainly on	+	+	+ few			GOODWIN and HILL (1968), CHRISTENSEN and HILL (1969)
Opossum	visual cortex	+ no antagonistic surround	+					CHRISTENSEN and HILL (1969, 1970)
Pigeon	retina	+	+	+	+			MATURANA (1962), MATURANA and FRENK (1962), GUSELNIKOV and VODOLAZSKIJ (1968)
Pigeon	nucleus rotundus		+	+				REVZIN (1969)
Pigeon	optic tectum		+	+				REVZIN (1969), JASSIK-GERSCHENFELDT (1970)
Pigeon	Hyperstriatum		+		+			REVZIN (1969, 1970), O'FLAHERTY (1971)

Anurans (water-frogs, tree-frogs, toads)	retina	off, on-off	+	+ ? (some class 3 neurons			HARTLINE (1940a, b), BARLOW (1953), BYZOV (1958), MATURANA et al. (1960), GRÜSSER-CORNEHLS et al. (1963—1969), GRÜSSER et al. (1963—1972), GAZE and JACOBSON (1963), NORTON et al. (1970), CHUNG et al. (1970), FELDMAN et al.(1971)
Anurans (water-frogs, tree-frogs, toads)	optic tectum diencephalon		+	+	+	+	LETTVIN et al. (1961), GRÜSSER and GRÜSSER-CORNEHLS (1968, 1970), EWERT (1971), SZEKELY (1971), FITE (1969)
Salamander	retina	(+)	+	+	+		GRÜSSER-CORNEHLS and HIMSTEDT (1971)
Necturus	retina	+	(+)	+			WERBLIN (1970), DOWLING (1970), NORTON et al. (1969)
Gecko (gecko gecko)	retina	+	+				GRÜSSER (1963)
Iguana	retina	+ (color ?)	+				GRÜSSER et al. (1969)
Triturus	retina	+	+	+			CRONLY-DILLON (1966, 1968)
Goldfish	retina	+ (color)	+	+			MCNICHOL, WAGNER, and WOLBARSH (1961), GAZE and JACOBSON (1963, 1964), CRONLY-DILLON (1964), DAWN (1968)
Goldfish	tectum		+	+		+	GRAUER (1967), MACKEBEN (1969), HERMANN (1971)
Turtle (Emys blandings)	retina	+ mainly without antagonistic surround		+			LIPETZ and HILL (1970)

sponse in the retina lasts longer than the period during which the slit of light is physically stimulating the corresponding region of the retina. Hence, the central visual system receives signals at the *same* time from areas which were stimulated by the light slit at *different* times. Thus the limited frequency response characteristics of the retinal network is responsible for the rapid decrease in the visual acuity when a stimulus is quickly moved across the retina. Figs. 2 and 3 show that these neurophysiological mechanisms exist exactly as discussed. The periodic activation of on-center neurons corresponds to the positive after images of a moving slit of light, while the periodic activation of off-center neurons is correlated with the dark bands (Grüsser and Grützner, 1958).

The neurophysiological basis for the reduction of visual acuity is demonstrated by Fig. 2b, whereby it is seen that the functional size of the RF-center of a retinal on-center neuron increases with an increase in the stimulus speed. Therefore, neurons in the central nervous system receive impulses *simultaneously* from retinal and geniculate ganglion cells having their receptive fields along a large part of the path of the moving light stimulus. The greater the angular speed of the stimulus, the larger the part of the path.

Since limited temporal and spatial frequency properties exist for the retinal network, the motor control system of the eyes must, on the one hand, prevent the fading of a visual pattern whose image is projected to one part of the retina, and, on the other hand, motor commands should not move the retinal image across the retina at too high a speed (or oscillation frequency). The interplay of fast saccades occurring every 0.3–1.5 sec (during which pattern vision is interrupted because of retinal "smearing") and slow drifts having an angular speed of about 0.05–0.1 degrees · sec^{-1} (during which patterns are well perceived) seems a very good adaptation of the motor control system of eye movements to the properties of the sensory part of the visual system (c.f. Ditchburn and Foley-Fisher, 1967; St. Cyr and Fender, 1969).

IV. Quantitative Investigations of the Response of M-Neurons

M-neurons were found in various species and at different levels of the afferent visual system (Table 1). The effect of different stimulus parameters on the response of M-neurons was quantitatively analysed in the cortical visual association areas and in the optic tectum of the cat, in the optic tectum of the rabbit and rat and in the retina of amphibia (frogs, toads and salamander).

1. Neurons in the Superior Colliculus of Mammals

Humphrey (1968) recorded two classes of M-neurons from the superficial layer of the rat's superior colliculus (up to 1 mm depth). In the upper layer, small field units (ERF-size of 2–15 degrees) dominate. The "optimal speed" of a moving black or white target (2–10 degrees diameter) is 5–10 degrees · sec^{-1}. Faster movement (25 degrees · sec^{-1}) does not produce a response. These units also respond to stationary light stimuli (0.5 degree diameter) projected to the ERF. In the lower layer, M-neurons have an ERF-size of 30–90 degrees, the responsiveness in the ERF might vary locally, and "silent" zones inside of the ERF also occur. Neuronal adaptation is pronounced. The ERFs are frequently elongated and tend to be orientated so that the longest axis is horizontal. Direction sensitivity does not seem to be prominent.

In the rabbit's optic tectum (SCHAEFER, 1966), the size of the ERF of movement sensitive neurons also increases from the superficial to deeper layers, where more neurons are direction sensitive than in the superficial layers. An increase in the size of the ERF with an increase in the depth of the neurons from the surface of the superior colliculus is also reported for cats (STRASCHILL and TAGHAVY, 1966, 1967), but the ERF-size likewise increases with the distance of the ERF from the area centralis in the visual field (STERLING and WICKELGREN, 1969). About three quarters of the movement sensitive neurons in the colliculus superior of the cat are DS-neurons.

Upon investigating the effect of the angular velocity of a light stimulus moving through the ERF of M-neurons in the optic tectum of cats, STRASCHILL and TAGHAVY (1966, 1967) found a power function for the relationship between the average impulse frequency (R) and the angular velocity (v), which had an exponent of 0.67. Measurements of the angular velocity in the data of STRASCHILL and TAGHAVY, however, were recently questioned by MCILLWAIN and BUSER (1968), who found responses to much higher angular velocites in the collicular units of cats than did STRASCHILL and TAGHAVY.

SCHAEFER (pers. communication, 1969) also found a power function to be valid between the neuronal activation and the angular speed of the stimulus for neurons located in the optic tectum of the rabbit.

2. M-Neurons in the Retina of Amphibia

Quantitative studies on the response of amphibian retinal M-neurons were made in urodela (salamander) and in anurans (toads and frogs). In the frog's retina *(Rana pipiens, Rana esculenta)*, three classes (class 1, 2, and 3) out of four different classes of retinal ganglion cells exhibit movement-sensitive (class 1 and 3) or movement specific (class 2) responses, if natural stimulus conditions are considered (MATURANA et al., 1960; LETTVIN et al., 1959, 1961). The nerve fibers of these different classes of retinal ganglion cells end at different levels at the dendrites of the tectal nerve cells. Class 1 neurons are most distal, class 4 neurons most proximal to the cell body (Fig. 6).

Several classes of movement sensitive neurons were found in the retinal ganglion cells of toads *(Bufo bufo, Bufo viridis, Bufo pelobates fuscus)*, tree frogs *(Hyla septentrionalis)* (GRÜSSER-CORNEHLS and LÜDCKE, 1970; GRÜSSER and GRÜSSER-CORNEHLS, 1970) and salamander (GRÜSSER-CORNEHLS and HIMSTEDT, 1970), in the goldfish (JACOBSON, 1962; GAZE and JACOBSON, 1963; CRONLY-DILLON, 1964) and in the *Triturus* (CRONLY-DILLON and GALAND, 1966; CRONLY-DILLON, 1968).

For the ganglion cells of the amphibian retina (frog, toad, salamander) it was found that the following parameters determine the response of movement detecting neurons: *position* of the moving stimulus in the ERF, *contrast* between stimulus and stationary background, *angular velocity*, *area* of the moving stimulus and *time between successive movements* of a stimulus along the same pathway through the receptive field. The quantitative relationship between neuronal response and these parameters has been investigated during the last 9 years in *Rana esculenta*, *Bufo bufo*, *Hyla septentrionalis* and *Salamandra salamandra* in our Berlin laboratory. The main characteristics of the four different classes of neurons found in the frog's retina, which send their axons to the tectal surface, are shown in Table 2 and are summarized in Fig. 6.

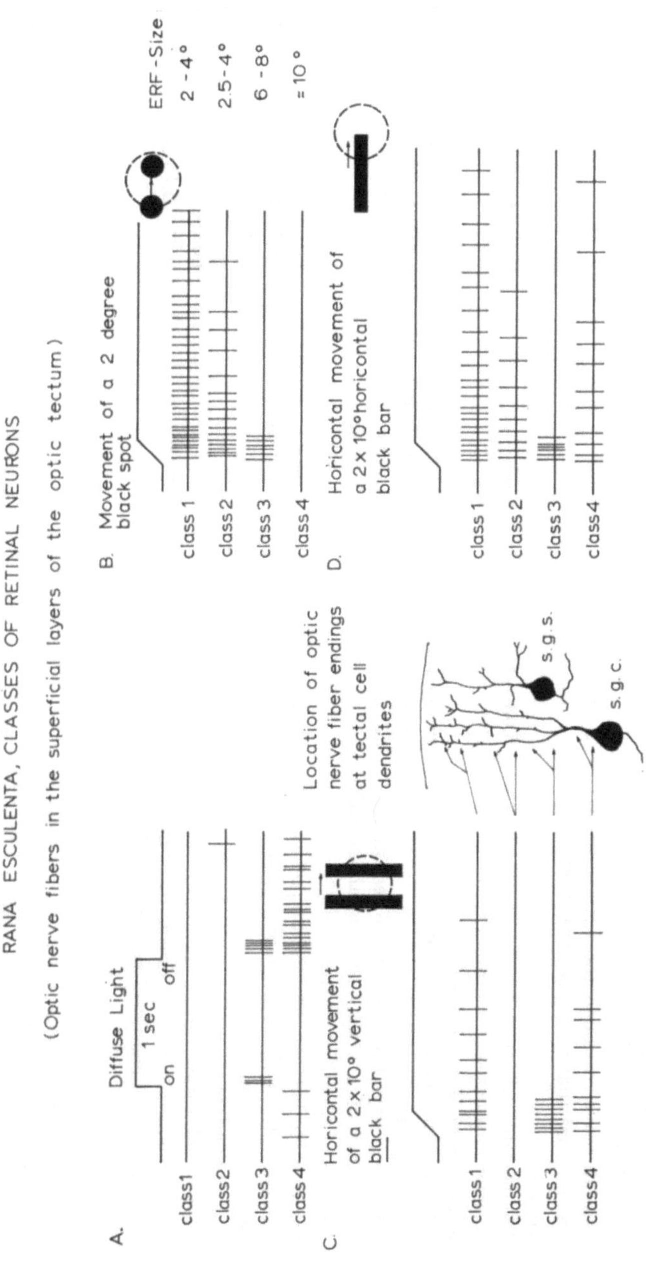

Fig. 6. *Schematic representation of the principal properties of the four main classes of retinal ganglion cells of the frog sending fibers to the optic tectum*

The synaptic endings of the four classes of afferent fibers are located at different levels of the apical dendrites of the tectal nerve cells. Shown are the responses of the four classes of retinal neurons elicited by diffuse illumination, a short horizontal displacement within the ERF of an extended, vertical black bar, a small black stimulus and a horizontally oriented black bar. Three of the neuronal classes (class 1—3) are movement sensitive; class 2 is movement specific (From GRÜSSER and GRÜSSER-CORNEHLS, 1970)

Table 2. Neurons of the frog's retina (Rana esculenta, ganglion cell layer, axons ending in the optic tectum)

	Class 1	Class 2	Class 3	Class 4
ERF-diameter	2—3°	2.5—4°	6—8°	approx. 10°
IRF-diameter	5—6° (min)	6—10°	10—12°	approx. 15°
response to diffuse light	no	no	weak on-off-activation	off-activation
response to small, stationary spots of light projected into the ERF	on-activation	on-, on-off- or off-activation in different zones	short on-off-activation	weak off-activation
response to small, stationary spots of light projected to the IRF	no	no	no	on-off-activation in the border-zone of the IRF
Stimuli moved through the ERF: Optimal angular size (darker or brighter than background)	2—3°	3—5°	6—8°	10° (only darker than background)
Minimal dislocation	0.03—0.05°	0.05—0.1°	0.12—0.25°	?
relationship between neuronal response and angular size of moving object	?	log	log	log
relationship between neuronal activation and angular velocity	power function exponent 0.5	power function exponent 0.7	power function exponent 0.95	linear for large moving stimuli (> 5°)
lower velocity threshold of the response to movement of a black stimulus about half the size of the ERF; degrees · sec^{-1}	$\leqq$ 0.02	$\leqq$ 0.05	0.05—0.1	about 0.1
relationship between impulse frequency and contrast of a dark object moving against a bright background	decrease in neuronal activation with contrast reduction	power function exponent 0.55	power function exponent 0.7 (max. 1.2)	decrease in neuronal activation with reduction of contrast
neuronal adaptation	+	+++	+	very low
directional selectivity	no	no	no (some +)	no
time constant for the decrease in neuronal activity after a change of position inside the ERF	4—10 sec	0.5—2 sec	0.05—0.15 sec	—
response to stroboscopically illuminated patterns. a) stationary	no	no	on-off	off-activation
b) moving (apparent motion)	?	++	+	off-activation
spatial summation of the response to 2 objects moving simultaneously	?	smaller than algebraic sum	smaller than algebraic sum	?
inhibition by stimuli moving through the IRF	+	+++	+	?

Position of the Moving Stimulus within the ERF. The neuronal response is strongest if the stimulus is moved through the ERF-*center;* it decreases, approximating a Gaussian distribution, if the stimulus shifts towards a more peripheral area of the ERF. Fig. 7a shows the spatial distribution of the neuronal impulse

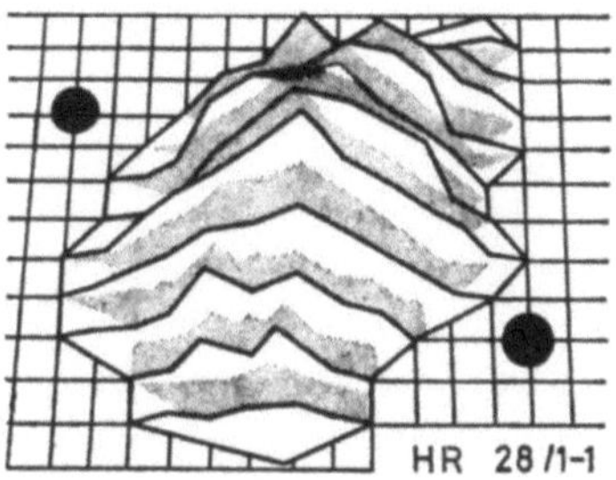

Fig. 7a. *Responses of a class 2 neuron in the retina of Rana esculenta to a spot moving along different paths through the ERF.*
Black spot of 0.5 degrees diameter moving on a white background (contrast 0.97, background luminance about 560 asb). The average neuronal activation is plotted on the ordinate, the position along one path on the abscissa and the position of the paths within the ERF on the z-axis (From HENN and GRÜSSER, 1968)

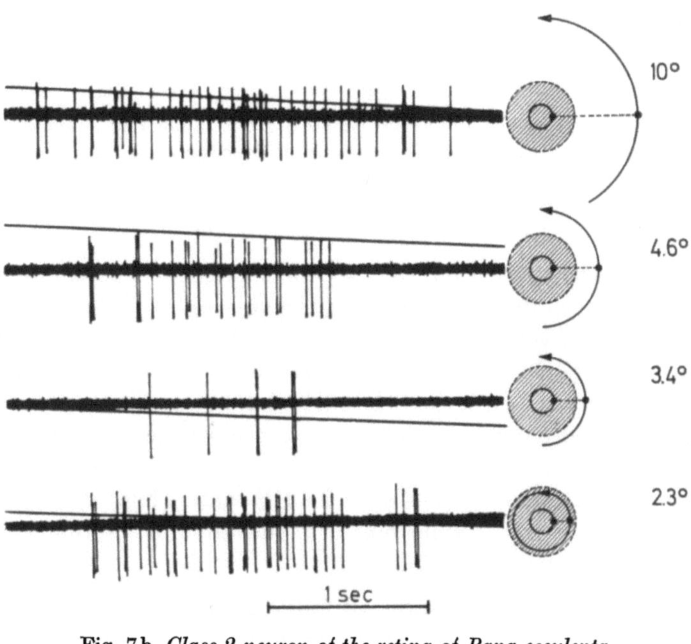

Fig. 7b. *Class 2 neuron of the retina of Rana esculenta.*
Responses to a small, black spot of 1.2 degrees diameter moving on a white background around a circle having a diameter of 2 degrees and located on the ERF. An additional black spot is moving around a circle through the IRF. As the separation between the stimulus in the IRF and the stimulus in the ERF is reduced, the inhibitory effect elicited by the stimulus moving through the IRF becomes stronger and reaches a maximum near the border between ERF and IRF. When the peripheral spot entered a circle of 3.3 degrees diameter, the neuronal response increased (From GRÜSSER and GRÜSSER-CORNEHLS, 1968). Numbers on the right give the distance between the moving stimuli

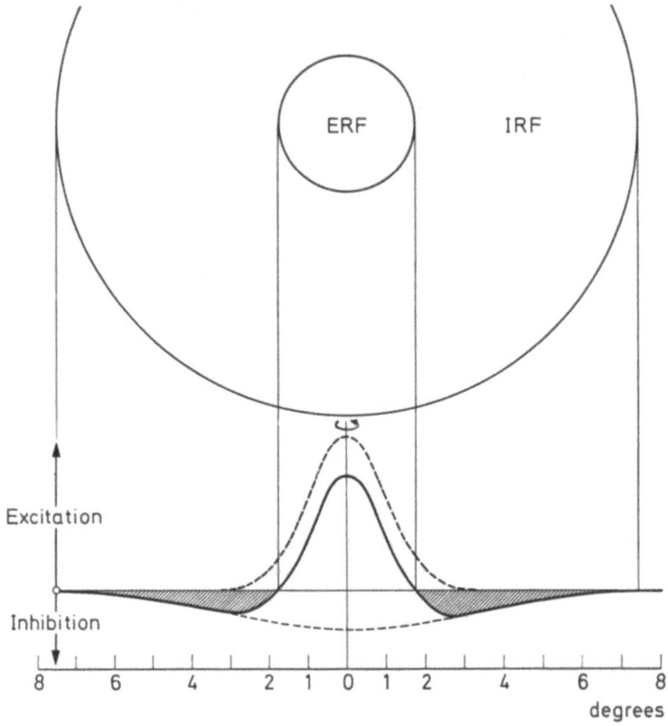

Fig. 7c. *Schematic representation of the distribution of excitation and inhibition elicited by stimuli moving across the ERF and IRF of movement detecting class 2 neurons in the retina of a frog*

frequency elicited within the ERF by a stimulus having a constant velocity, area and contrast. As in concentric RF neurons, the greatest excitability in movement detecting neurons lies within the ERF-center and decreases towards the ERF-periphery. In all three classes of movement sensitive neurons in the retina of the frog, the ERF is surrounded by an IRF. The inhibition elicited by stimuli projected to the IRF is also dependent on the distance between the stimulus and the ERF-center. Inhibition is strongest along the inner-border between ERF and IRF and decreases with the distance from the RF-center (Fig. 7c). Optimal inhibition from the IRF also requires movement of the stimulus. Stationary patterned stimuli elicit no inhibition (Figs. 7b and 10).

Angular Velocity. The response of all four neuronal classes depends on the angular velocity of the moving stimulus. It does not depend, as a rule, on the direction of the movement (Fig. 8). For all four neuronal classes a power function with a constant exponent c was found to be valid for the relationship between the velocity and the average impulse frequency ($\overline{R}$) of the stimulus when crossing the ERF (Fig. 9a), if the area of the moving stimulus was smaller than the ERF:

$$\overline{R} = k \cdot v^c \quad [\text{Impulses} \cdot \text{sec}^{-1}.] \tag{1}$$

The exponent c is significantly different for the three neuronal classes: 0.5 for

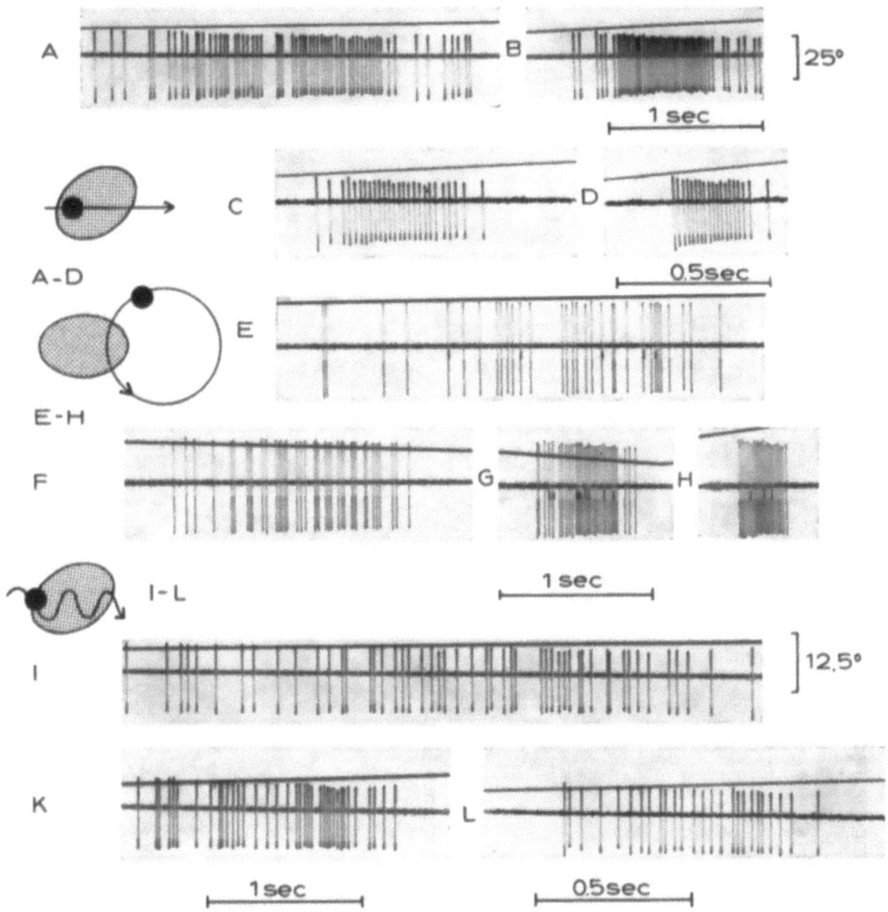

Fig. 8. *Responses of three different class 2 neurons in the retina of Rana esculenta to movement of a dark, round spot through the ERF along a straight, a round and a sinusoidal path at different angular velocities.*
A—D Black spot, 1.5 degrees diameter, white background, 500 asb, contrast 0.97, angular velocity A: 0.9; B: 2.4; C: 4.8; and D: 9.6 [degrees · sec^{-1}]. E—H 1 degree black spot moving along a circle through the ERF and the IRF. Angular velocity in E: 1; F: 2; G: 5; H: 10 [degrees · sec^{-1}]. I—L Movement of a small dark stimulus along a sinusoidal path. Length of the sinusoidal path through the ERF 2.4 degrees, vertical amplitude 1.4 degrees. Average velocity in I: 1.5; K: 3.6; L: 7.2 [degrees · sec^{-1}] (From GRÜSSER, FINKELSTEIN, and GRÜSSER-CORNEHLS, 1968)

class-1 neurons, 0.7 for class-2 neurons and 0.95 for class-3 neurons in *Rana esculenta* (FINKELSTEIN and GRÜSSER, 1965; GRÜSSER et al., 1966). In class-2 neurons with stimuli larger than 5 degrees, the constant c becomes greater than 0.7. This is due to the different effect of stimulus velocity on excitatory and inhibitory mechanisms in the RF (GRÜSSER et al., 1968). For stimuli smaller than 4 degrees, in class-2 neurons the exponent $c = 0.7$ does not vary with a change in the shape of the stimulus, the contrast between the stimulus and background, the position

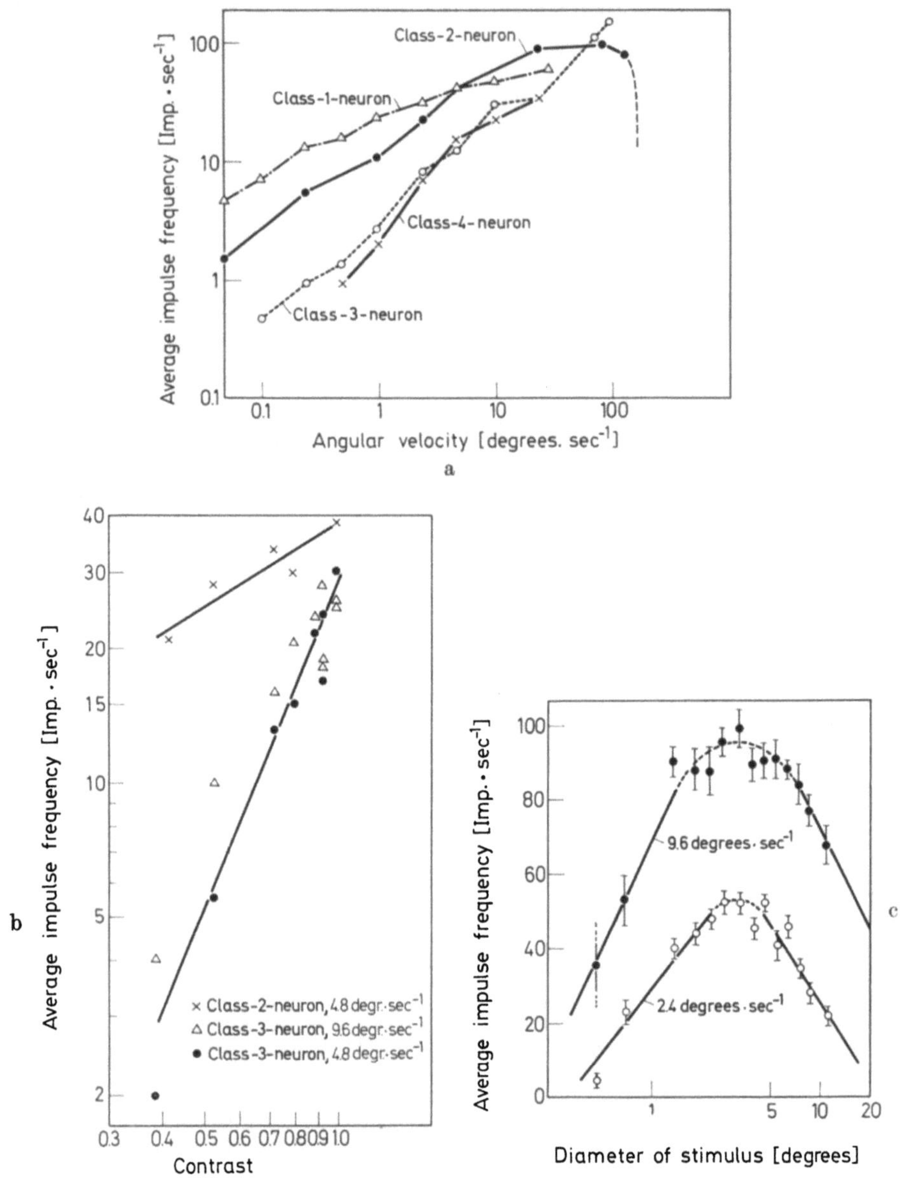

Fig. 9a—c. *The influence of various stimulus parameters on the response of the different classes of neurons in the ganglion cell layer of the frog's retina (Rana esculenta).*

a Relationship between *average impulse frequency* (ordinate) and the *angular velocity* of a stimulus moving through the ERF (abscissa). Black stimuli moving on white background (size: 1.2 degrees for the experiment with class 1 and class 2 neurons, 2.7 degrees for class 3 and 23 degrees for class 4 neurons). The curves obtained in these experiments can best be described by a power function having differing exponents for the various classes of neurons (c.f. Table 2). b Relationship between the *average neuronal impulse frequency* (ordinate) and the *contrast* between a gray spot moving on a white background through the ERF. Background luminance about 560 asb. Stimulus diameter 2.7 degrees. The response was measured for three units. The action potentials of optic nerve fibers were recorded successively from the same area in the optic tectum (From Grüsser and Grüsser-Cornehls, 1969). c *Relationship between the average impulse rate (ordinate)* of 14 class 2 neurons of the frog's retina and *the diameter of the black, round stimuli (abscissa)* moving on a white background through the ERF at two different stimulus velocities (contrast 0.93, background luminance about 560 asb, from Butenandt and Grüsser, 1968)

of the stimulus in the ERF, and the type of movement (linear, sinusoidal, irregular, zig-zag). The response is mainly dependent on the *temporal sequence* with which single receptors within the ERF change their state of excitation.

In different species of amphibia under the same stimulus conditions, the exponent c may differ significantly from species to species for the different classes of retinal neurons (Table 3, GRÜSSER and GRÜSSER-CORNEHLS, 1970; GRÜSSER-CORNEHLS and LÜDCKE, 1970, 1971).

Table 3. Exponent c of velocity function of neurons of the amphibian retina, projecting to the optic tectum. $\overline{R} = K\,V^c$ [Impulses · sec^{-1}]

Species	Class 1	Class 2	Class 3	Class 4
Rana esculenta	0.5	0.7	0.95[a]	1.0[a]
Bufo Bufo	0.7		1.0–1.2[a]	1.1[a]
Hyla septentrionalis	0.5		0.97[a]	1.0[a]
Salamander	superficial fibers 0.7		Layer II 1.4	Layer III 0.85

[a] Valid only for a black stimulus moving on white background.

Table 4. Exponent $\varkappa$ of the contrast function of neurons of the anuran retina, projecting to the optic tectum. $\overline{R} = K \cdot C^{\varkappa}$ [Impulses · sec^{-1}]

Species	Class 1	Class 2	Class 3
Rana esculenta	0.4–0.6	0.55	0.7 (max. 1.2)
Bufo Bufo	0.3–0.5		0.3–0.7[a]
Hyla septentrionalis	0.6		0.2–0.7[a]

[a] Depends on the angular velocity.

Contrast. The response of retinal ganglion cells in all classes of frogs, toads and salamander depends on the contrast between the moving stimulus and the stationary background (GRÜSSER-CORNEHLS et al., 1963; SCHIPPERHEYN, 1964). In class-2 neurons of *Rana* a white stimulus moving on a black background elicits 30–50% less neuronal activation than a black object of the same size moving on a white background through the ERF. In class 1 and 3 of *Rana esculenta* no such difference was found. If the luminance of the background is I_b and that of the stimulus I_s, the contrast may be defined as $C = (I_b - I_s)/(I_b + I_s)$. A power function was found to be valid for the contrast function in retinal neurons of Rana, Hyla, and Bufo:

$$R = k_2 C^{\varkappa} \quad [\text{Impulses} \cdot \text{sec}^{-1}]\,. \tag{2}$$

The average exponent $\varkappa$ is 0.55 in class-2 neurons and 0.75 in class-3 neurons of Rana (Fig. 9, GRÜSSER et al., 1967). The contrast function was found to vary more from unit to unit than the velocity function. This variability might be the reason why no significant differences were found for the same class of retinal neurons in the different species of anurans investigated. A significant difference exists, however, between the contrast function of class-2 and class-3 neurons in *Rana esculenta*. This difference is best seen by comparing responses obtained from class 2 and class 3 neurons from the same animal, the receptive fields of which are located in the same region of the visual field (Fig. 9b). In class 3 neurons of Hyla and Bufo the exponent of the contrast function, however, depends on the angular velocity of the stimulus.

Level of General Illumination. The organization of the concentric RF of neurons in the cat's retina changes as the adaptation level shifts from the scotopic to the photopic range (BARLOW et al., 1957; DOMBERG, 1971). We investigated whether the same is true for the functional organization of the RF of M-neurons in the anuran retina. In class 2 neurons of *Rana esculenta* the ERF size and the average neuronal excitability remain constant when the level of the background luminance is between about 800 asb and a scotopic adaptation level of 0.01 asb. At scotopic levels of illumination, however, there is no longer any difference in the neuronal activation, whether it is elicited by centrifugal or centripetal movement of the stimulus. Under photopic stimulus conditions, *centripetal* movement elicits a stronger neuronal response than *centrifugal* movement (LICKER, 1968). In the toads *(Bufo bufo)* and treefrogs *(Hyla septentrionalis)* the organization of the RFs of movement sensitive class 2 and 3 neurons, however, does change with dark adaptation. The average neuronal activation increases (with constant contrast, angular velocity and stimulus size) at mesopic levels of adaptation, as compared with the responses at photopic adaptation levels. At scotopic levels the size of the ERF of some neurons increases (GRÜSSER-CORNEHLS and LÜDCKE, 1970). The increase in the diameter of the ERF is probably due to a reduction in the strength of the antagonistic inhibitory mechanisms of the surround. Hence, the ERF can reach its "real size", which depends on the anatomical connections only and not on the functional interaction with inhibition (Fig. 7c).

When monochromatic light having different wavelengths is used for the illumination of the stimulus pattern, the response of class 2 neurons to moving dark stimuli remains the same, independent of the color of the background (440nm to 680nm; LICKER, 1968). On the basis of the findings of DONNER (1959), DONNER and RUSHTON (1959), REUTER (1969), and SCHEIBNER and BAUMANN (1970) in single on-off and off-units in the isolated retina of the frog (probably class 3 and class 4 neurons), one can, however, expect that in the retina of anurans *color contrast* between a moving stimulus and the background might also have an effect on the neuronal response of movement detecting neurons.

Size of the Moving Stimulus. The spatial summation of excitation elicited from different parts of the ERF was measured by two methods: either two spots were moved simultaneously through the ERF along two different paths and the responses to single and double spot stimulation were compared, or the relationship between stimulus area and neuronal response was determined (cf. HARTLINE, 1940). With the first experimental technique it was found that the neuronal summation was always smaller than the algebraic sum of the neuronal activity elicited by each spot alone. If the response elicited by one stimulus is called A, that elicited by the other B, the following equation best describes the experimental finding in class 2 neurons (HENN and GRÜSSER, 1968):

$$R_{AB} = \frac{A + B}{1 + k_i (A + B)} \quad [\text{Impulses} \cdot \text{sec}^{-1}]\,. \tag{3}$$

In addition, k_i is a function of the distance between the two stimuli and is lowest at a distance of about 0.7 degrees.

The following relationship between the average impulse frequency and the size of the moving stimulus was found to be valid for the second type of experiments with class 2 neurons in the retina of *Rana esculenta* (BUTENANDT and GRÜSSER, 1968):

$$R = k_3 \left(\log \frac{\text{ERF}}{A^*} - \left|\log \frac{A}{\text{ERF}}\right|\right) \quad [\text{Impulses} \cdot \text{sec}^{-1}]\,. \tag{4}$$

A* is the threshold size (0.2 degrees for class 2 neurons) and "ERF" is the angular size of the excitatory receptive field. The response of class 1 and class 3 neurons is also dependent on the size of the stimulus. In both classes, the relative inhibitory effect of the IRF is less than in class 2 neurons. The maximum of the response is also reached if the area of the moving stimulus is about the same as the ERF.

Similar results were obtained for the area-function of M-neurons in the retinal ganglion cell layer of tree-frogs and toads (GRÜSSER-CORNEHLS and LÜDCKE, 1970, 1971).

For class 2 neurons of *Rana esculenta* the averaged experimental data are also adequately described by a logarithmic, parabolic equation (BUTENANDT, 1969):

$$\overline{R}_2 = a \log A^2 + b \log A + c \quad [\text{Impulses} \cdot \text{sec}^{-1}]\,. \tag{4a}$$

whereby a, b, and c are constants.

Eqs. (3) and (4) are only valid for round or quadratic stimuli. Concerning the relationship between neuronal response and the area of the moving stimulus, an increase in the stimulus size in the direction of the movement changes the neuronal response less than an increase in the stimulus size perpendicular to the direction of movement. A response maximum depending on the area of the stimulus as exhibited in Fig. 9c is only reached if the size of the stimulus exceeds the borders of the ERF in the direction perpendicular to the direction of movement. This finding also indicates that inhibition from the IRF is elicited predominantly by moving stimuli and not by stationary ones.

Neuronal Adaptation. If the same pathway through the receptive field is traversed by two successive moving stimuli with a short period of time between them, class 2 and class 3 neurons are less activated the second time (GRÜSSER-CORNEHLS et al., 1963). A repetitive to and fro-movement through the ERF results in a strong neuronal adaptation. A quantitative relationship was found for the decrease in the response to a second stimulus moved along the same path t seconds after the first. If R_2 is the average impulse frequency elicited by the second moving black stimulus and R_1 is the activation elicited by the first one, the following eq. is approximately valid for class 2 and class 3 neurons of *Rana:*

$$R_2 = R_1 (1 - k^* e^{-t/\tau}) \quad [\text{Impulse} \cdot \text{sec}^{-1}]\,. \tag{5}$$

The time constant τ and k^* in this equation depend on the general metabolism of the frog. τ is between 1.5 and 5 sec for class 2 and class 3 neurons; $k^* \geqq 1$; the better the metabolic condition of the frog, the smaller is k^*,.

Eq. (5) only approximately describes the recovery function of M-neurons after the stimulus has traversed the receptive field. More recent measurements in Rana, Hyla, and Bufo (GRAUER, 1969) indicate that a complete description of the recovery function requires two exponential functions.

The neuronal adaptation definitely does not occur at the ganglion cell output level, because the response elicited by a stimulus moving through the RF is not affected if it is immediately preceded by another stimulus moved along *another* part of the ERF (GRÜSSER-CORNEHLS et al., 1963).

Interdependence of the Effects of the Various Stimulus Parameters. By systematically pairing the stimulus parameters mentioned in the preceding sections, we measured the extent of their interaction in the elicitation of responses from movement detecting class 2 neurons. In class 2 neurons of *Rana esculenta* above a certain threshold level of excitation, a multiplicative combination of Eqs. (1) – (4) was the best way of describing the experimental results within the range of angular velocities from 0.05 to about 25 degrees · sec⁻¹ and for stimuli smaller than the ERF:

$$\bar{R} = k \cdot v^{0.7} \cdot C^{0.55} \cdot \log \frac{A}{A^*} \quad [\text{Impulses} \cdot \text{sec}^{-1}]\,. \tag{6}$$

In class 2 and 3 neurons of Bufo bufo and Hyla septentrionalis the coupling between Eqs. (1) – (4) was more complex. In these retinas the exponent of the contrast function for some units also depends on the angular velocity. This finding is probably due to the effect of threshold mechanisms (GRÜSSER-CORNEHLS and LÜDCKE, 1970, 1971).

3. Movement Sensitivity of the Inhibitory Receptive Field

In the class 1, 2, and 3 neurons in the anuran retina, the response elicited by a stimulus moving through the ERF is strongly reduced by a stimulus moving simultaneously through the IRF. The inhibitory effect is independent of the direction of movement of the stimulus in the ERF (GRÜSSER-CORNEHLS et al., 1963). The diameter of the IRF might be as large as 10 degrees or even larger (GRÜSSER et al., 1964; KEATING and GAZE, 1970). As is evident from experiments, such as are shown in Fig. 7b, the inhibition is strongest near the border between the ERF and IRF. The superposition of two spatial Gaussian distributions with different widths forms the basis of a good quantitative model for the spatial distribution of excitation and inhibition within the RF of anuran M-neurons (Fig. 7c). Elicitation of inhibition from the IRF also requires that the stimulus is moving. This statement is substantiated by the experimental results shown in Fig. 10: A small spot moved through the ERF elicits only a weak response if a regularly structured or random pattern of dots is simultaneously moved in the IRF. As soon as the IRF stimulus becomes stationary, the spot moving in the ERF elicits a strong response.

The inhibitory effect elicited by a stimulus moving through the IRF also undergoes adaptation. This is shown by the following experiment: A contrast stimulus is moved through the ERF, another through the IRF. By comparing the responses to movement of the center spot alone and to the combined movements through the ERF and IRF at the same speed, a quantitative measure of the inhibitory effect from the IRF can be obtained. If the moving stimulus in the IRF is preceded by another stimulus moved along the same path, the inhibitory

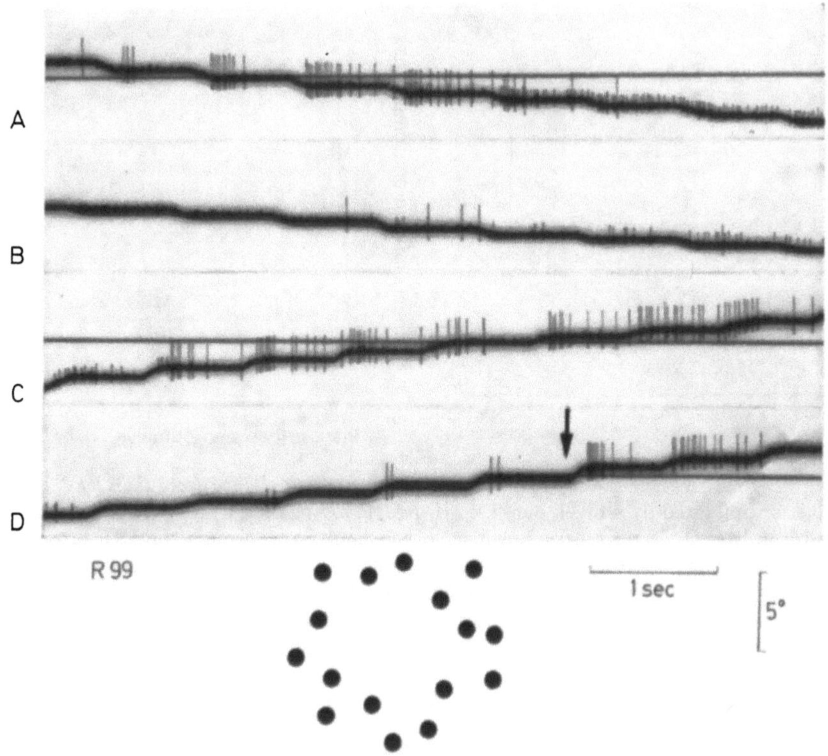

Fig. 10. *Response of a class 2 neuron in the retina of Rana pipiens to a small, black spot of 1.5 degrees diameter, moving in steps through the ERF (A and C) and moving synchronously with 16 spots of the same size, distributed at random within the IRF (C and D).*
The strong inhibition elicited by the stimuli in the IRF immediately disappears when the stimuli cease moving (D, arrow). Inhibition is also elicited when the stimuli in the IRF move in the opposite direction or at right angles to the direction of movement of the spot in the ERF (From GRÜSSER and GRÜSSER-CORNEHLS, 1968a)

effect on the activation elicited by the stimulus moving through the ERF center is reduced (disinhibition). This reduction depends on the time lag between the movement of the conditioning stimulus and the movement of the stimulus pair in the ERF and in the IRF (Fig. 11, GRAUER and GRÜSSER unpubl., 1970).

If the whole stimulus pattern is illuminated stroboscopically, a stationary patterned background does not reduce the response of class 2 neurons of *Rana* to a stimulus moving through the ERF. As soon as the pattern of the background changes its position during the pauses between the flashes, an inhibitory effect elicited from the IRF appears.

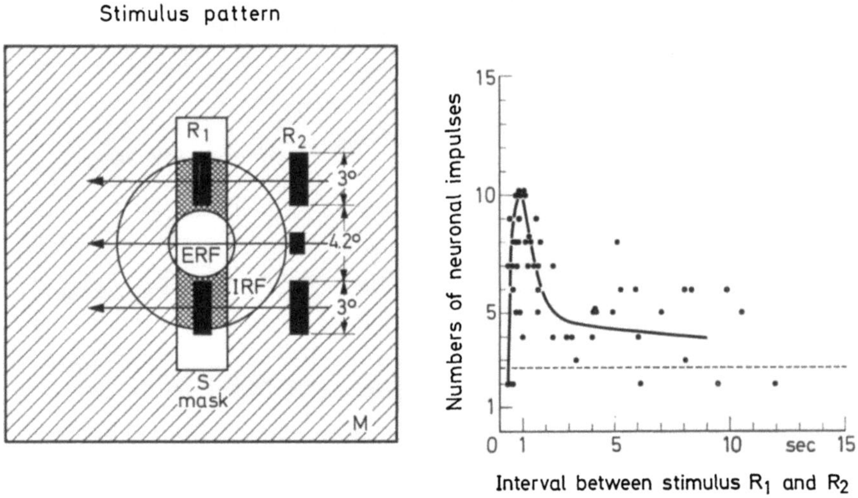

Fig. 11. *Neuronal adaptation of the inhibitory effect elicited from the IRF.* The ERF is stimulated by a black bar (1×2 degrees) moving on a white background; the IRF by two black bars (1×3 degrees) following 2 other black bars through the IRF after a given temporal delay. The inhibitory effect is a function of this temporal delay. The average number of neuronal discharges elicited by this stimulus combination (ordinate) is plotted against the temporal delay. The stimuli appeared within a window in a mask, which covered part of the RF. (From GRAUER and GRÜSSER, unpubl. 1970)

4. Responses to Moving Patterns Illuminated Stroboscopically

The results of experiments, in which the moving contrast pattern is illuminated by very short, repetitive flashes, provide an answer to the question whether *movement or a change in position* of a contrast border within the ERF is the adequate stimulus. For the understanding of the RF organization of M-neurons, it is important to determine whether their responsiveness is dependent upon a temporally linked change in the level of excitation of *adjacent* receptors or on such changes in receptors anywhere in the ERF. With respect to the psychophysical experience (Section I,2), this activation of M-neurons would correspond to apparent motion perception. EWERT (1968) demonstrated that the toad's prey-catching behavior, elicited normally by small moving objects, can also be elicited when the moving object is illuminated with stroboscopic light. The response of movement specific class 2 neurons in the frog's retina to a pattern moving under stroboscopic illumination is shown in Fig. 12 (GRÜSSER-CORNEHLS, 1968). It demonstrates that class 2 neurons do respond to a change in position of the stimulus within the ERF (GRÜSSER-CORNEHLS et al., 1963). The neuronal activation not only depends on the parameters mentioned in Eqs. (1)–(5), but also on the luminance and the frequency of the flashes, which illuminate the moving pattern (Fig. 13). Up to a stimulus frequency of 25–28 flashes per sec, the

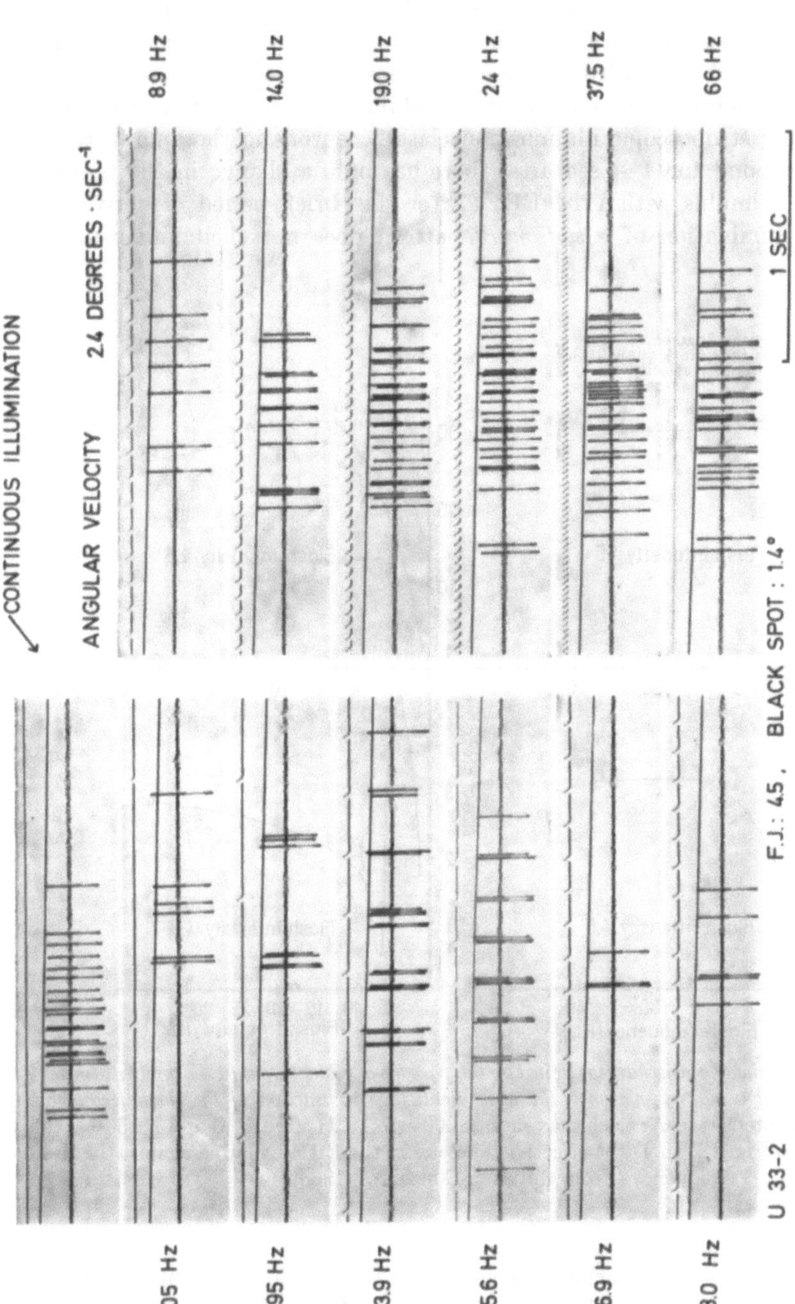

Fig. 12. *Class 2 neuron of the retina of Rana esculenta.* Moving pattern illuminated by short flashes (apparent movement stimulation). Responses to a change in position of a black spot of 1.4 degrees diameter moving on a white background through the ERF at an angular velocity of 2.4 degrees · sec^{-1}. Contrast between the black spot against the white background 0.97. Duration of the flashes illuminating the ERF shorter than 100 μsecs (rise time 14 μsec, exponential decay, time constant 20 μsec), maximal luminance 36000 asb, corresponding to about 180 asb with a flash lasting 20 msec. The neuronal discharges exhibit grouping with frequencies up to 24 flashes per sec (From Grüsser-Cornehls, 1968a)

neuronal response exhibits a rhythmical activity corresponding to the flash frequency.

At higher levels of flash intensity, the neuronal activation is a non-monotonic function of the flash frequency (Fig. 13). A minimal neuronal response is obtained at very low flash frequencies between 3 and 8 flashes per sec. This finding can be explained by the assumption that inhibitory and excitatory mechanisms, located presynaptically to the recorded ganglion cell, have different temporal transfer characteristics, whereby one type of lateral inhibition strongly decreases above 5–8 cps. During stroboscopic illumination class 2 neurons in the anuran retina continue to respond for 1–4 sec after there has been a change in the position of a contrast stimulus within the ERF. After this brief period of activation, stroboscopic illumination of a *stationary* pattern does not elicit any neuronal discharge.

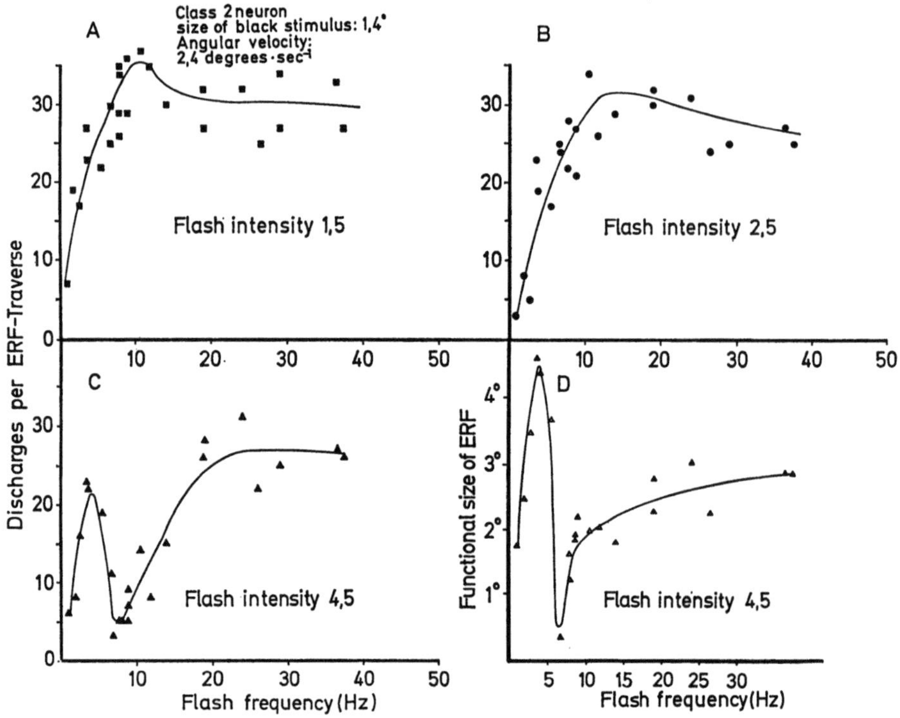

Fig. 13. *The total number of discharges elicited by the passage of a black spot (1.4 degrees diameter) through the ERF of a class 2 neuron (Rana esculenta) as a function of the flash frequency.* 3 different flash intensities. A: flash maximum 6000 asb, B: 11500 asb, C and D: 36000 asb flash duration see Fig. 12. In D the effective diameter of the ERF is plotted against the flash frequency (From Grüsser-Cornehls, 1968a)

The velocity function described by Eq. (1) is also valid for stroboscopically illuminated moving patterns. The exponent ($c = 0.7$) does not depend on the flash frequency, while the multiplicative constant k does. The results described in this section demonstrate that the only requirement for a response from this type of neurons is that within a certain period of time after a stimulus has changed its position within the ERF, another stimulus has to appear in the ERF in a position not necessarily adjacent to the first stimulus. A stimulus suddenly appearing at location B within the ERF after a preceding stimulus has occurred at position A

elicits a stronger activation, as a rule, than if the stimulus at B is not preceded by the stimulus at A. This finding indicates the existence of a lateral facilitatory mechanism within the ERF (Grüsser-Cornehls, 1968).

5. The Ontogenetic Development of Movement Detecting Neurons in the Frog's Retina

A comparative study of the experimental data presented in Table 1 leads to the conclusion that movement detecting neuronal systems developed in various species during *phylogenesis*. The first information about the *ontogenetic* development of M-neurons was provided by the study of Reuter (1969). His investigations were mainly concerned with the change in the color signal processing of the retina as tadpoles mature to adult frogs, but his paper also provides important information concerning M-neurons in the retinal ganglion cell layers of the tadpole and the adult frog. The distribution of ganglion cells in the adult frog, as shown in Table 2, is significantly different from that in the tadpole retina. Class 2 neurons are rarely found in the retina of the tadpole, while class 1 neurons ("sustained edge detectors" of Maturana et al., 1960) are frequently found. The difference between both classes of neurons, in addition to the smaller ERF in class 1 neurons, is mainly characterized by weaker inhibition from the IRF in class 1 neurons. In *Rana* neurons can sometimes be found, which can be characterized as having transitional properties between class 1 and class 2 neurons (Maturana et al., 1960; Grüsser-Cornehls et al., 1963). They may be remnants of the change occurring during metamorphosis. As evidenced by Reuter's work, the different stages in the development of the anuran retina may serve as a good model for further quantitative study of the ontogenetic development of movement detecting neuronal networks.

6. M-Neurons in the Tertiary Visual Cortex and in Cortical Visual Association Areas of the Cat

In the tertiary visual cortex (area 19) and the anterior portion of the middle and the lateral suprasylvian gyri of the cat, movement sensitive neurons belonging to categories M, DS, and OS were found (Hubel and Wiesel, 1965, 1969; Dow and Dubner, 1969, 1971; Grüsser and Grüsser-Cornehls, 1969; Wright, 1969). The visual association area in the suprasylvian gyrus receives fibers from the nerve cells in the primary, secondary and tertiary visual cortex (Wilson, 1969).

In these areas we measured the extent to which the activation of M-neurons depended on various stimulus parameters, such as round spots or slits moving through the ERF at different angular speeds (Grüsser, Grüsser-Cornehls, and Hamasaki, 1970). The response of M-neurons was found to depend on the angular velocity and on the size, contrast and position of the stimulus in the ERF. For some of these neurons, too, a power function gives a good description of the relationship between average neuronal activation and the angular speed of the stimulus. M-neurons in the suprasylvian gyrus frequently give an optimal response at a rather low stimulus speed (5–30 degrees $\cdot$ sec^{-1}) and also exhibit neuronal adaptation.

It was further found that "apparent movement" stimulation, i.e., short light stimuli (5–15 msec duration) projected successively to different locations within the ERF, elicits a strong neuronal activation. The activation depends on the frequency of the stimuli when the angular velocity of the change in position within the ERF is kept constant. The response to intermittently illuminated stationary stimuli exhibits strong neuronal adaptation and ceases within 1–4 sec after a change in the position of the stimulus.

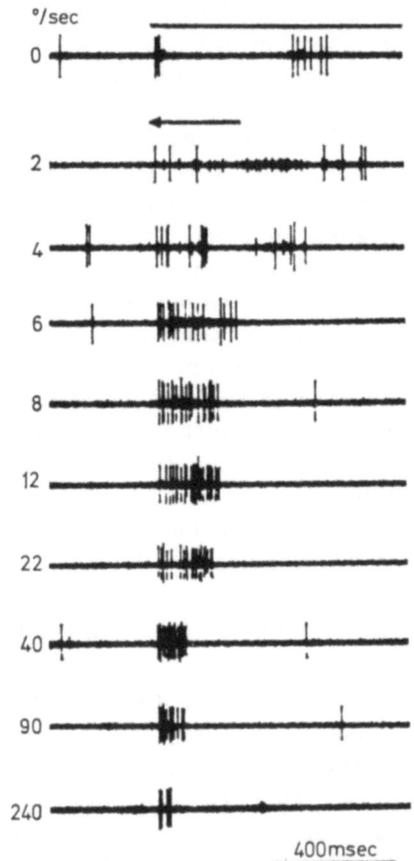

Fig. 14. *Responses of a movement sensitive neuron in the visual cortex of the awake monkey (Macaca mulatta) to movement of a slit of light 0.5 × 2 degrees in size.*
The duration of the neuronal responses decreases as the angular velocity of the stimulus increases. The average neuronal impulse frequency increases with the angular speed and reaches a maximum between 40 and 90 degrees · sec^{-1} (From WURTZ, 1969)

From the data obtained so far, the M-neurons in the central visual system of cats in general exhibit features similar to those of the movement sensitive neurons in the anuran retina. A main difference, however, is the size of the ERF, which in the M-neurons in the visual area of the suprasylvian gyrus might reach values up to 30 degrees.

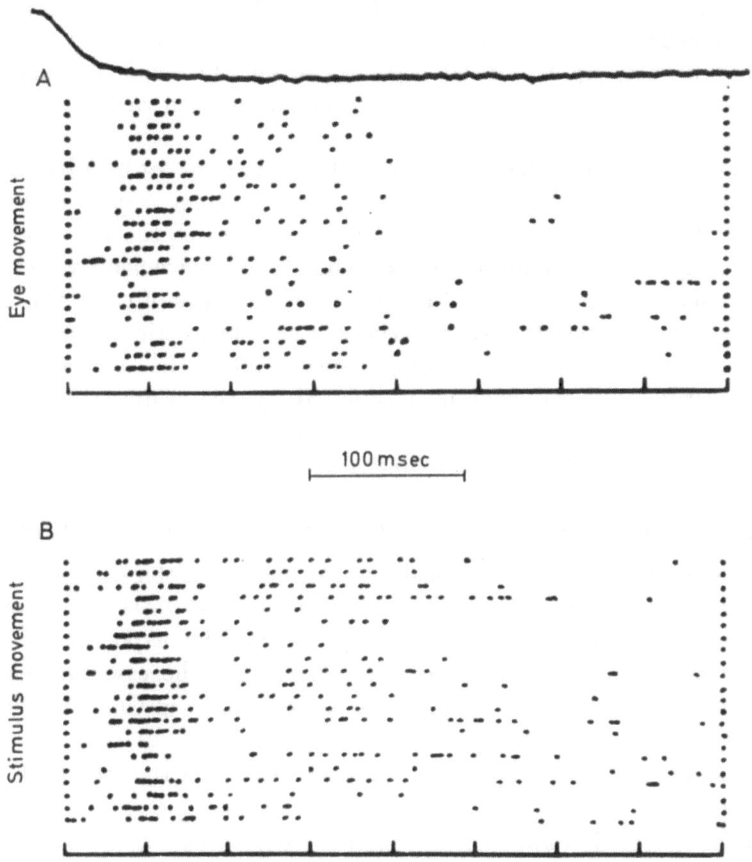

Fig. 15. *Response of a neuron in the primary visual cortex of the awake monkey (Macaca mulatta) to a slit of light which is moving through the ERF either by "voluntary" eye movements (A) or by movement of the stimulus across the stationary retina (B) at an angular velocity of 900 degrees · sec^{-1}, corresponding approximately to the angular speed of the monkey's eye movements* (Wurtz, 1969).
The responses were not significantly different. Either the eye movement (A) or movement of the stimulus triggers the sweep; 23 and 22 successive responses were recorded. Each dot represents a neuronal impulse. Sweep duration 400 msec (From Wurtz, 1969)

7. M-Neurons in the Monkey's Primary Visual Cortex

In the primary visual cortex of monkeys *(Macaca mulatta)* movement sensitive neurons were found (in addition to neurons responding to stationary visual stimuli), which often require a certain orientation of the main contrast border. Some of these neurons belong to the DS-class; the others, however, respond to movement of a spot of light, independent of the direction of motion and exhibit all the essential properties of M-neurons summarized in the next section (Wurtz, 1968, 1969a, b, c; Hubel and Wiesel, 1968). The response of these M-neurons depends on the angular velocity of the stimulus. The optimal velocity according to Wurtz, however, varies from unit to unit. Fig. 14 shows the response of an

M-neuron in the monkey's visual cortex (WURTZ, 1969). In this unit, for example, the average neuronal impulse rate during the passage of the light stimulus through the ERF increases up to 40–90 degrees per sec. The unit still responds at 240 degrees per sec, an angular velocity reached during the saccades of voluntary or involuntary eye movements.

WURTZ did his experiments in unanaesthetized animals which had been conditioned to perform voluntary eye movements when a certain stimulus was presented. Therefore, he could compare the responses to the movement of the pattern's image across the retina that were due to eye movements with those due to pattern movements (Fig. 15). In general, for all types of cortical neurons no difference in the response to these two types of movement stimulation of the ERF was detected. This important finding indicates that the corollary discharge (TEUBER, 1960) of visual neurons postulated by the reafference principle of VON HOLST and MITTELSTAEDT (1950 c.f. Section I,2) exists (if at all) only at a higher level of cortical visual information processing.

8. A Model of the Receptive Field Organization of Movement Sensitive Neurons

The movement sensitive neuronal system in the amphibian retina and the movement sensitive neurons in the optic tectum and the visual cortex of the cat exhibit some common response properties. From the experimental findings described in the preceding sections one can extract the essential features of movement sensitivity in a neuronal network (GRÜSSER and GRÜSSER-CORNEHLS, 1969). Most M-neurons exhibit:

(1) A short, transient response to local stimulation within the ERF at "on" and/or "off".

(2) Fast neuronal adaptation to repetitive stimulation at the same location in the ERF.

(3) Neuronal adaptation to a stimulus repeatedly passed along the *same path* through the ERF.

(4) A pronounced response to stationary stimuli projected successively to different locations within the ERF (apparent motion response).

(5) M-neurons in the amphibian retina (class 1, 2, and 3) have an IRF. If the degree of inhibition from the IRF is relatively strong, then the neuronal responsiveness is limited to the movement of *small* objects through the ERF (class 2 neurons).

From these experimental observations the following conclusions can be drawn concerning the function of the elements forming the RF of M-neurons (Fig. 16):

(a) At a neuronal element between the receptors and the M-neuron, the DC form of the receptor potential to steady illumination or darkening is transformed into transient on-, off- or on-off-responses. According to the intraretinal recordings of BYZOV (1959), DOWLING and WERBLIN (1969), WERBLIN (1970) and NORTON et al. (1970) from amphibian retinas, these bandpass properties (which are obtained by adding high pass filter properties to the low pass filter mechanisms of the late receptor potential) appear in the response of some of the bipolar and amacrine cells.

(b) The *neuronal adaptation* might be due to an increase in the thresholds of the neuronal elements between the receptors and the M-neurons following repeated excitation, or a decrease in the transmitter release or transmitter efficiency. The neuronal adaptation definitely does not occur at the impulse generating site of the M-neurons, because after complete neuronal adaptation, a change in the position of the moving stimulus within the ERF elicits a brisk neuronal response.

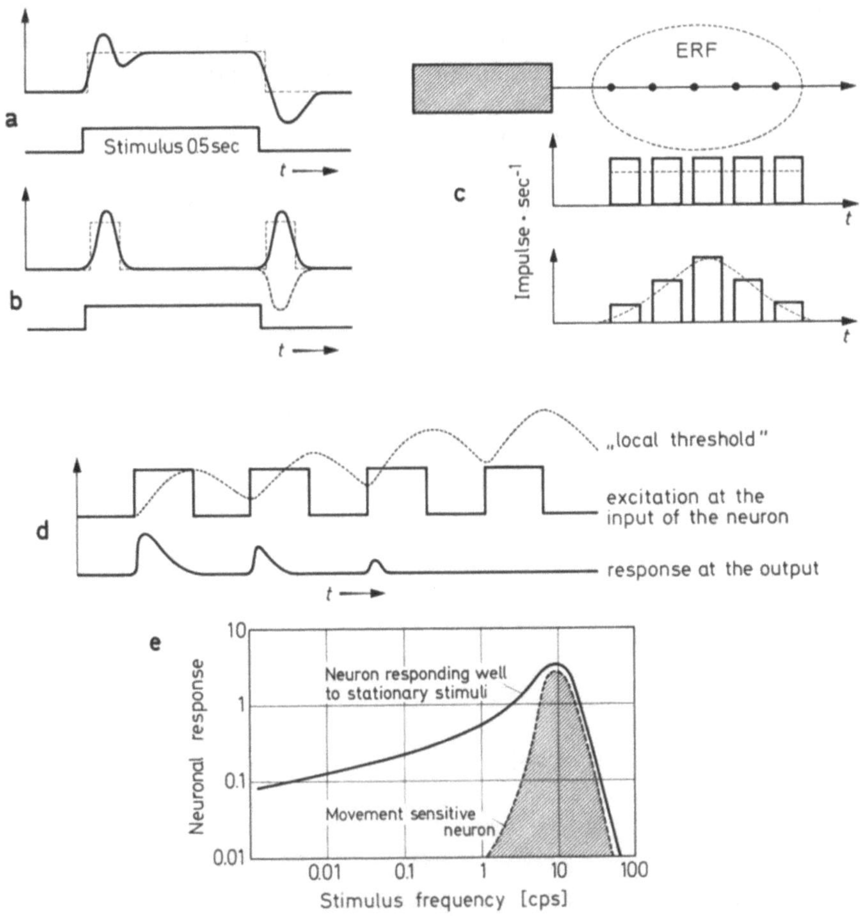

Fig. 16. *Schematic representation of the essential properties of elements in the ERF of M-neurons.* a Response of a receptor to a slit of light moving slowly across its RF. b Response of a neuron between receptor and M-neuron (e.g., bipolar cell in the retina) to the same stimulus. The DC response from the receptor is changed into a transient on- (-off)-response. c Responses of 5 units shown in b to the leading edge of a light bar moving through the ERF of an M-neuron (upper curve) and the input excitation to the M-neuron obtained by weighting of the upper with the spatial excitability distribution in the ERF (lower curve). The RFs of the phasic response units (b) are adjacent to one another. d Change in the synaptic activation of an M-neuron due to the change in the "local threshold" in the ERF when the same part of the ERF is stimulated repeatedly. e Frequency response characteristics of an M-neuron and a visual neuron which respond well to stationary stimuli (From GRÜSSER and GRÜSSER-CORNEHLS, 1969)

(c) The phenomenon we called neuronal adaptation, however, might also be caused by a long lasting, temporal summation of inhibitory processes elicited simultaneously with excitation from the ERF. An experimental indication for the existence of rather slow inhibitory processes in the retinal network of class 2 neurons of the frog's retina are the findings obtained with stroboscopically illuminated moving contrast stimuli (Section IV, 3).

(d) The power functions found to describe the experimental data can easily be explained by the bandpass properties of the signal processing in the network of the RF of M- and DS-neurons. If the conditions for the organization of the perceptive unit of an M-neuron, mentioned in this section, are modelled by means of an electronic model of the vertebrate retina, the main properties of M-neurons including the power function (Eq. 1) are found for movement sensitive model neurons (Eckmiller and Grüsser, 1972).

V. Direction Selective Visual Neurons (DS-Neurons)

In order to determine whether a visual neuron is direction-sensitive, one has to perform certain tests to exclude a *pseudo-direction-sensitivity* caused by fast neuronal adaptation, by incomplete traverses through the ERF or by sensitivity to the *orientation* of a moving contrast border. Only those movement sensitive neurons are designated DS-neurons, for which the direction sensitivity remains constant as the *orientation* of the main contrast borders within the ERF varies.

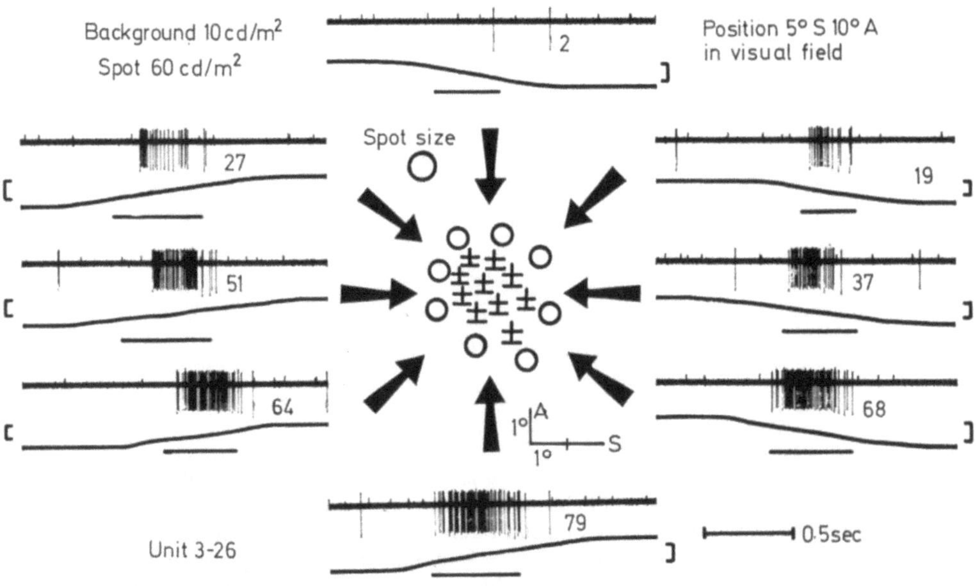

Fig. 17. *Recording of the action potentials of a DS-neuron in the ganglion cell layer in the retina of the rabbit.*

A small spot of light of about 0.8 degrees diameter is moved through the ERF at an angular velocity between 3 and 10 [degrees · sec^{-1}]. Stimulus luminance 60 cd · m^2, background 10 cd · m^2. Preferred direction, vertical upwards; null direction, vertical downwards (From Barlow, Oyster, and Levick, 1964)

The response of DS-neurons, however, might depend on the size and contrast of a moving stimulus. Some of the orientation sensitive neuronal systems described in Section VI exhibit, in addition, direction-sensitivity. These neurons will be included in the category for OS-neurons, since, as a rule, for a strong neuronal activation a stimulus pattern is required of which the predominant feature is that the main contrast borders have a certain orientation.

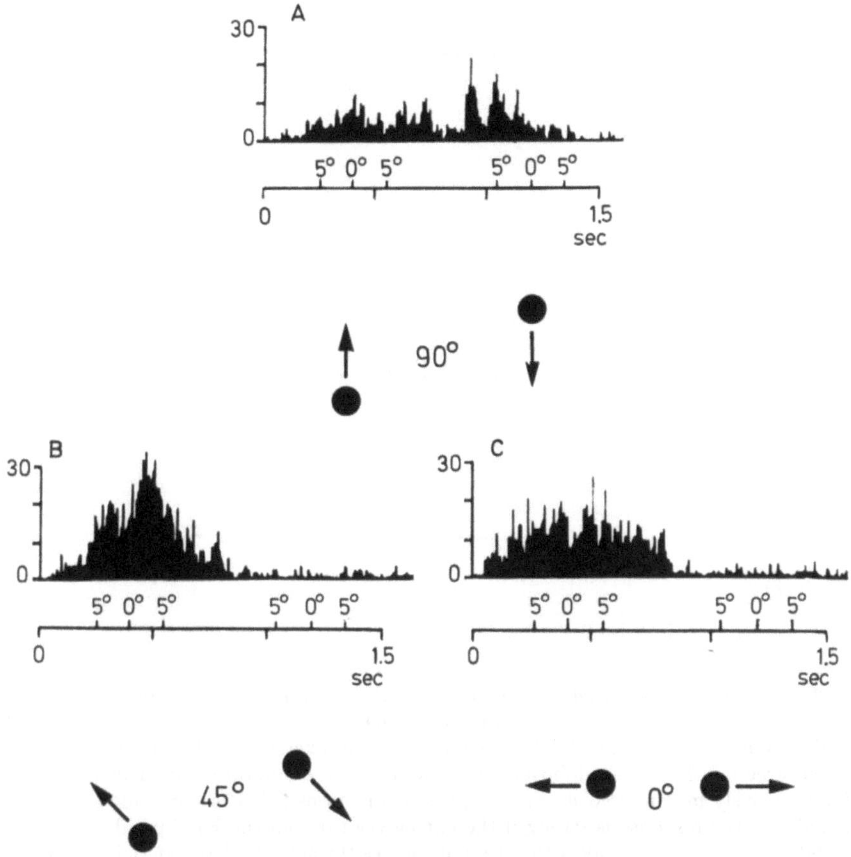

Fig. 18. *Response of a DS-neuron in the lateral geniculate body of the cat to a black spot of 8 degrees diameter moving in different directions through the ERF at 25 [degrees · sec⁻¹]*
Ordinate: averaged instantaneous impulse frequency; abscissa: time or position of the stimulus. The optimal direction was 45 degrees to the left upwards (From Kozak, Rodieck, and Bishop, 1965)

As Table 1 demonstrates, movement sensitive neurons, which are in addition direction-sensitive, can be found in nearly all parts of the afferent visual system from the ganglion cell layer of the retina to the higher visual centers in the cerebral, cortex. One probable exception is the retina of higher mammals, where DS-neurons have only been found in the cat in a restricted central area (Stone and Fabian, 1968). No such cells have as yet been found in the monkey retina (Hubel and

WIESEL, 1968). The nerve cells in the monkey's retina and geniculate body seem to be mainly involved in the processing of color, contrast and light-dark signals (DEVALOIS, 1965), while the more complicated spatial features of a patterned stimulus are extracted by a higher order of visual neurons.

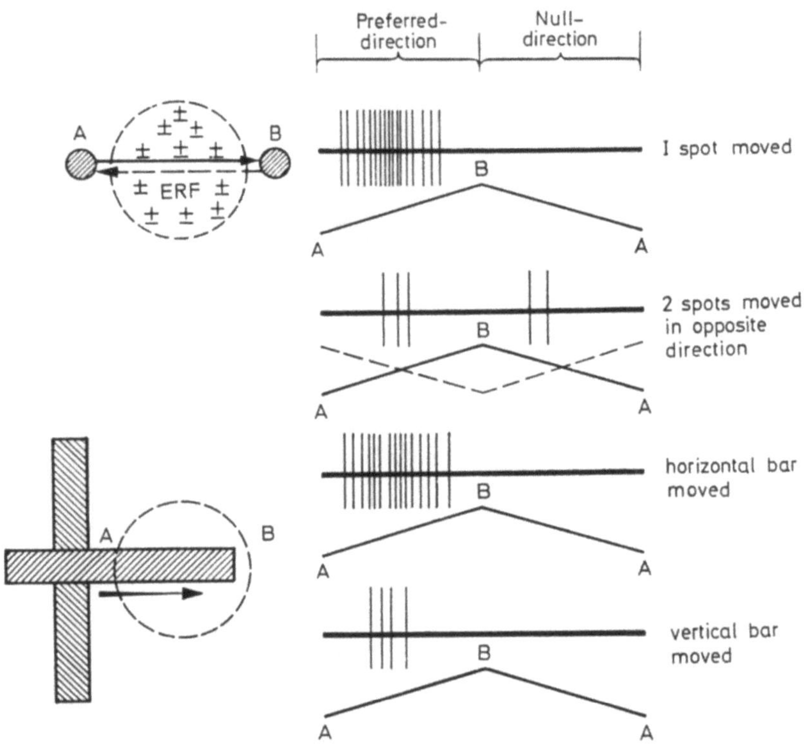

Fig. 19. *Diagram of the findings in DS-neurons (on-off-center) in the squirrel retina* (MICHAEL, 1966).
A small spot moving through the ERF elicits a strong excitation when it is moving in the preferred direction. It elicits inhibition when moving in the null direction. This can be shown by simultaneously moving two stimuli in opposite directions (B). If the moving stimulus is an elongated bar, the response is strong if the bar moves through the ERF parallel to the main axis, while the response is weak if the stimulus moves through the ERF at right angles to the the main axis. This experiment indicates the existence of movement sensitive IRFs

Intraretinal recordings in Necturus and in frogs (WERBLIN, 1970; NORTON et al., 1970) indicate that direction-sensitivity does not exist in neurons of these vertebrate retinae distal to the ganglion cell layer, while an increased sensitivity to moving stimuli, in comparison with stationary ones, exists for the RFs of some of the bipolar and amacrine cells.

Frequently, DS-neurons respond only to stimuli that are not larger than the ERF. This indicates that the ERF is surrounded by a strong IRF, which is also movement sensitive. For the DS-neurons in the superior colliculus of the cat, STERLING and WICKELGREN (1969) demonstrated that the inhibitory processes, which could be elicited by a stimulus moving within the IRF, have the same direction-sensitivity as the excitatory processes elicited from the ERF. Similar findings were reported by MICHAEL (1970) for DS-neurons in the superior colliculus of the ground squirrel. Inhibition within the ERF is most strongly elicited by a spot moving in the null direction and thus ERF-inhibition has a direction sensitivity 180 degrees opposite to the optimal direction of movement to elicit inhibition from the IRF.

Figs. 17 and 18 show examples of the responses of DS-neurons in the retina of the rabbit (BARLOW et al., 1964) and the geniculate body of the cat (KOZAK et al., 1965). Fig. 19 summarizes the results obtained in DS-neurons of the squirrel's retina (MICHAEL, 1966, 1968a, b). These neurons have a movement and direction-sensitive IRF surrounding the ERF. Since movement through the ERF in the null direction elicits inhibition, two spots moving simultaneously through the ERF in opposite directions elicit no or only weak neuronal activation.

1. The Response Sector

The response sector is defined as that angular sector comprising all directions of movement in which a stimulus moving through the ERF will activate the DS-neuron (GUSELNIKOV and VODOLAZSKIJ, 1968). A good graphical representation of the response sector is obtained when the neuronal activation is plotted as the length and the movement direction as the angle of a vector (Fig. 20).

The response sector can be used to further classify the DS-neurons. In the first class of DS-neurons the preferred direction (maximum vector length) in the response sector and the null direction (minimum vector length) are 90 degrees apart. Such neurons were found in the visual cortex of the cat (Figs. 21 and 20, BAUMGARTNER et al., 1964, 1965; HUBEL and WIESEL, 1962; CAMPBELL et al., 1969; COOPER and ROBSON, 1968). Some of the experimental data yielding the type of response sectors shown in Fig. 20c, might result, however, from the fact that the neurons are also sensitive to the *orientation* of the main contrast borders. In most of these experiments, the direction of movement of the bar-shaped stimuli or light-dark stripes was moved perpendicular to the orientation of the main contrast borders.

For the second class of DS-neurons, the preferred and the null-direction are 180 degrees apart (Figs. 18–20). The response sector can be rather wide, as is the case for DS-neurons in the rabbit's retina, or very narrow, as for some DS-neurons in the retina of the pigeon (GUSELNIKOV and VODOLAZSKIJ, 1968).

DS-neurons in the same region of the brain can exhibit quite different response sectors, even when the stimulus conditions are the same. For example, Fig. 15 on page 49 of volume VII/3B shows that for some of the neurons in the superior colliculus of the cat, the neuronal activation is a monotonic function of the direction of movement (φ) (response proportional to $\sin(\varphi/2)$, while for other neurons the neuronal activation changes quite abruptly with a small change in the direction of movement of the stimulus (STRASCHILL and TAGHAVY, 1967, 1968).

In the afferent visual system of the rabbit DS-neurons are found in the retinal ganglion cell layer and in the geniculate body. The response sector is narrower for the higher order of neurons; this "sharpening effect" of the direction sensitivity

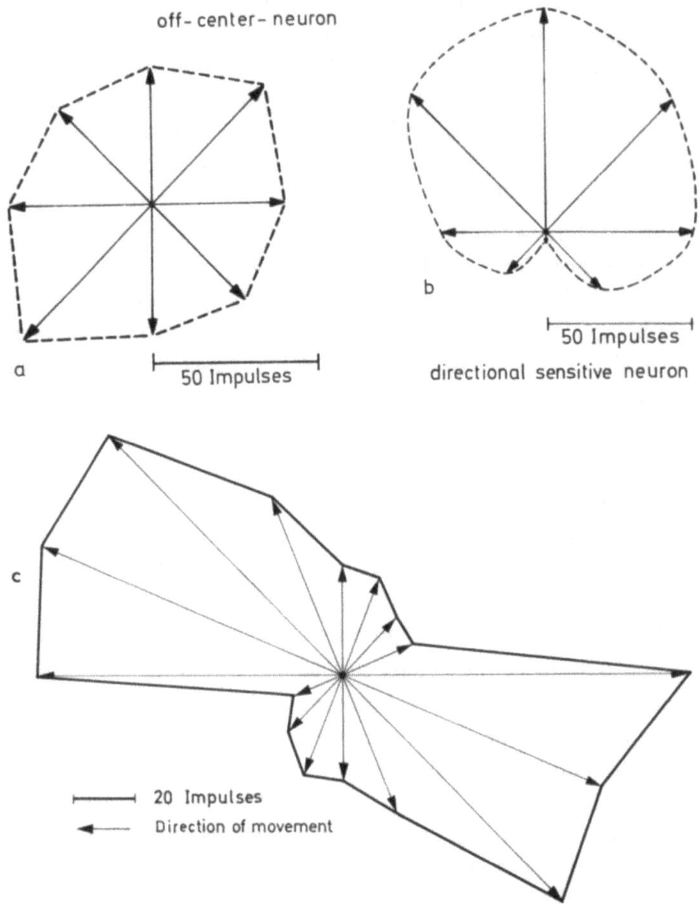

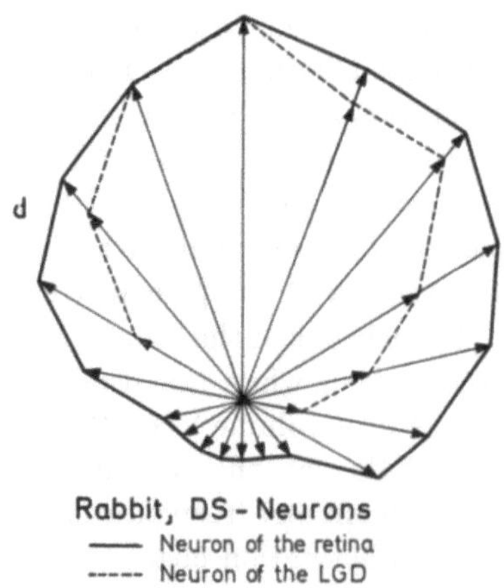

Fig. 20a—d. *Response sectors of DS-neurons.* a Response sector of an off-center neuron in the rabbit retina having no direction-specificity. b Response sector of a DS-neuron in the rabbit retina (on-off-center), which has a preferred direction (vertical upwards). c Response sector of a cortical movement-sensitive neuron in area 17 of the cat. The null direction is at right angles to the preferred direction. Stimuli are black and white stripes oscillating sinusoidally and moving through the ERF in different directions (oscillating frequency 1 cps, as in Fig. 21). d "Sharpening effect" of two layers of DS-neurons connected with each other in the rabbit's retina and lateral geniculate body. Response sector of a retinal and a lateral geniculate body neuron. a and b are measurements from the data published by BARLOW, LEVICK, and OYSTER (1965), c is from BAUMGARTNER (1964) and d is from LEVICK, OYSTER, and TAKAHASHI (1970)

is due to the interaction of synaptic excitation and inhibition at the geniculate cell membrane (Fig. 20d, LEVICK et al., 1969).

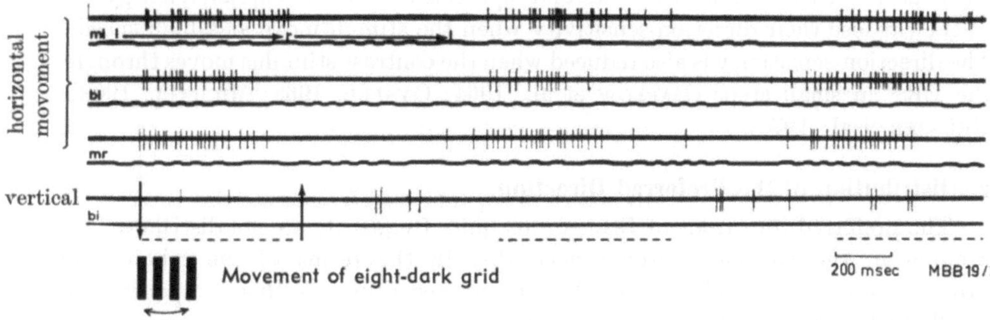

Fig. 21. *Responses of a neuron in the primary visual cortex of the cat to monocular and binocular (bi) stimulation with light-dark stripes moving sinusoidally to and fro through the ERF at 1 cps.* The direction-sensitivity is the same for both eyes. No binocular summation was found. Movement of the pattern in a vertical direction elicits only a weak response. When the pattern is moved in a horizontal direction, the neuronal activation is elicited only by a movement from left to right (From BAUMGARTNER, 1964)

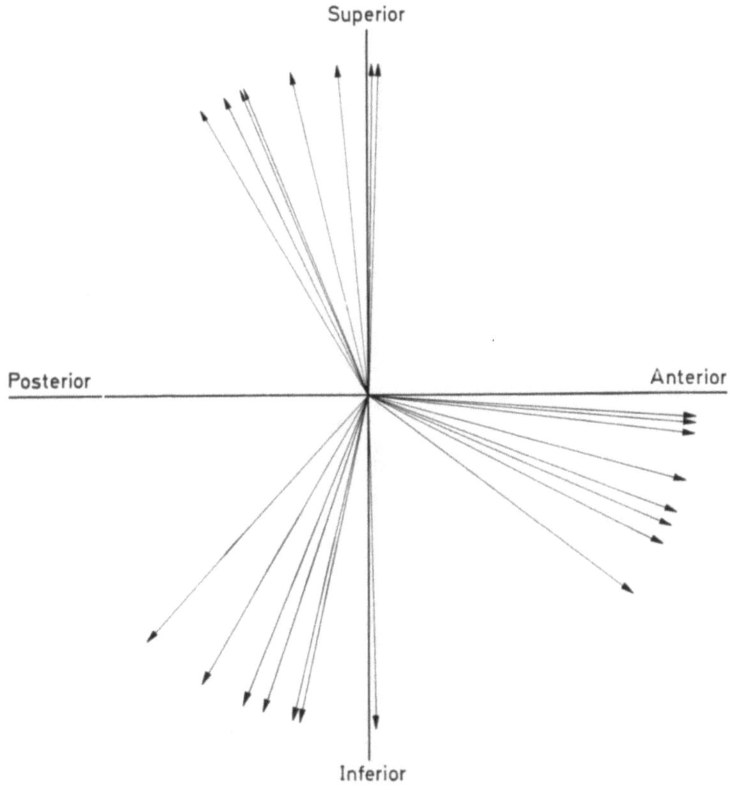

a

Fig. 22a u. b. *Distribution of the preferred directions of DS-neurons in the rabbit's retina.* a. preferred directions in DS-neurons having on-centers

It is not yet known how the shape of the vector diagram changes with the change in the angular velocity or the contrast between the moving stimuli. The DS-neurons in the retina of rabbits and squirrels and in the suprasylvian gyrus of the cat lose their direction-sensitivity when the stimuli move at a high speed. The direction-sensitivity is also reduced when the contrast stimulus moves through the ERF in small steps (Barlow et al., 1964; Oyster, 1968; Michael, 1968; Grüsser et al., 1970).

2. Distribution of the Preferred Direction

The preferred directions of DS-neurons quite frequently are not distributed at random in the various neuronal networks. In the retina of the rabbit one can discriminate two classes of DS-neurons according to their responses to small stationary stimuli projected to their receptive fields: *on*-DS-neurons and *on-off*-DS-neurons (Oyster and Barlow, 1967; Oyster, 1968). The preferred directions of these two different classes of DS-neurons are, first of all, not distributed at random and, secondly, significantly different for each class (Fig. 22).

In the DS-neurons of the rabbit's optic tectum, the preferred direction for units having their RF in the anterior visual field of each eye (binocular part)

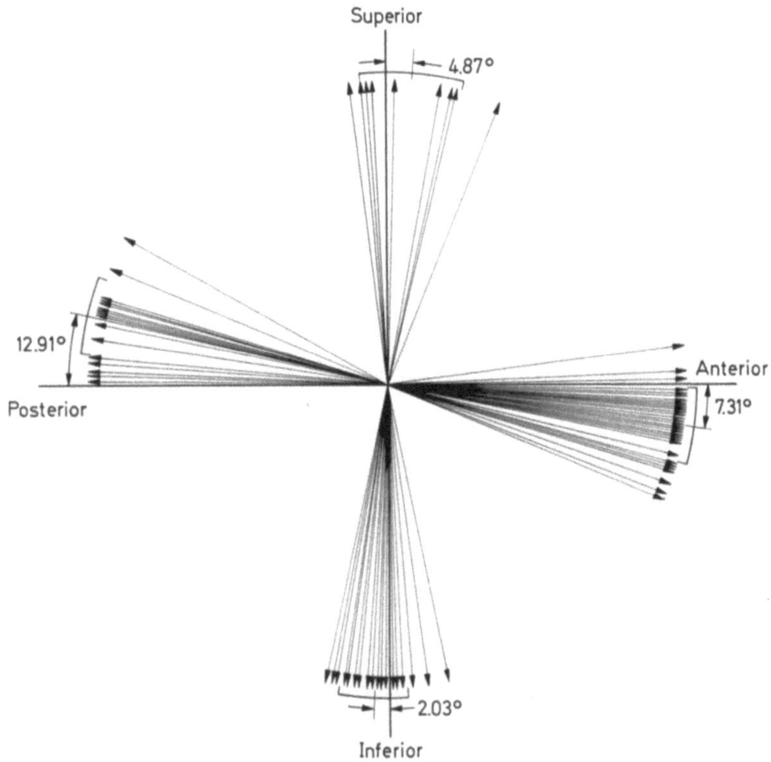

b

b. preferred directions in DS-neurons having on-off-centers (From Oyster and Barlow, 1967)

points towards the posterior part of the visual field. For units having their receptive field in the posterior (caudal) visual field, the preferred direction points towards the anterior part (SCHÄFER, 1966, Fig. 23). In addition, neurons having their receptive fields in the upper part of the visual field show preferred directions pointing upwards, while in units having the receptive field in the lower visual field, the preferred directions point downwards.

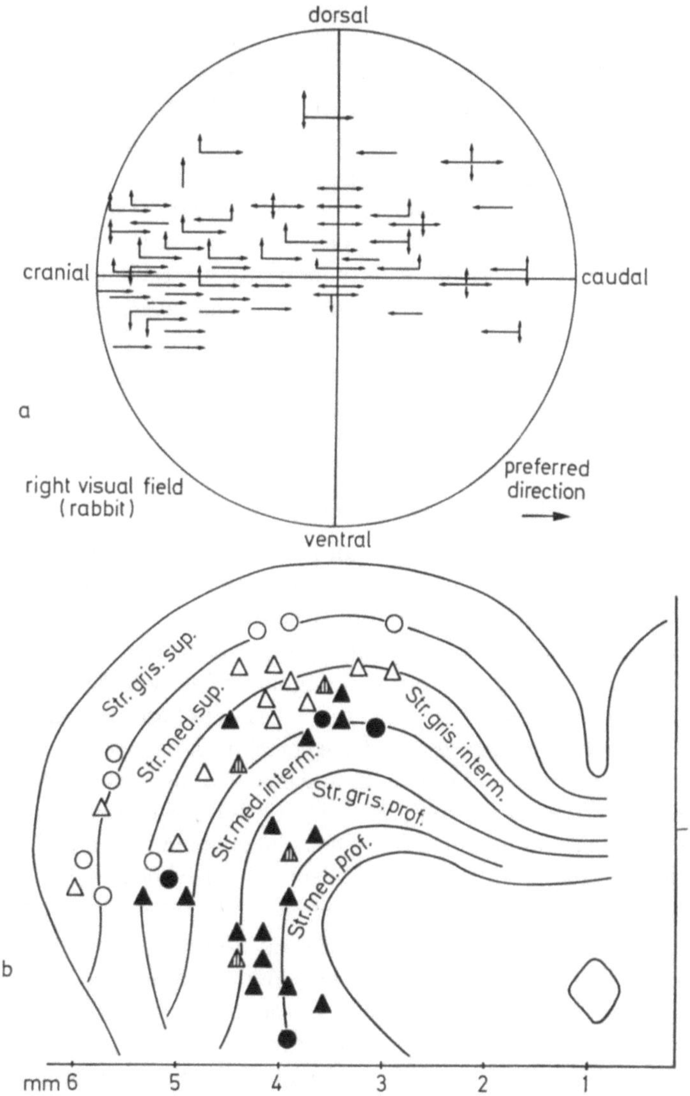

Fig. 23. a *Distribution of the preferred directions of DS-neurons in the superior colliculus of the rabbit.* b *Distribution of the location of the different types of collicular neurons in the layers of the rabbit's colliculus superior.*

open Circles: Neurons responding well to diffuse flashes of light. *Triangles:* DS-Neurons responding to moving stimuli. Filled circles and filled triangles indicate that no response is obtained by diffuse light stimuli. (From SCHAEFER, 1966)

A somewhat different distribution of preferred directions exists for the tectal units of the cat (Straschill and Hoffman, 1969; Sterling and Wickelgren, 1969). This difference might be due to the greater degree of binocular vision in cats in comparison to the rabbit. DS-neurons in the superior colliculus, which frequently can be activated by both eyes, have preferred directions (upwards and horizontally) away from the center of the visual field. Because of the regular projection of the retina into the tectum opticum of the cat (Apter, 1945, 1946) and the rabbit (Schäfer, 1962), the distribution of the preferred directions of DS-neurons is regular with respect to the tectal surface. Units in a given column perpendicular to the tectal surface have similar preferred directions, and their receptive fields are in the same region of the visual field.

In the striate and extrastriate visual cortex of the cat, the preferred directions of neurons within a single column perpendicular to the surface of the visual cortex are also similar (Hubel and Wiesel, 1962, 1965). In the primary visual cortex of the monkey, the nerve cells in one column have either the same preferred direction or preferred directions exactly opposite to each other (Hubel and Wiesel, 1968). Most neurons in the visual cortex, however, are also orientation sensitive. The preferred orientation of the contrast border of an optimal stimulus is probably a more important characteristic of the organization of the vertical columns of the visual cortices than the preferred direction of the movement of the stimulus.

3. The Effect of Stimulus Velocity and Contrast

In the superior colliculus of the cat, an increase of the angular velocity might either increase or decrease the neuronal response of DS-neurons belonging to the same class. In this respect they differ from M-neurons, for which the neuronal activation increases with an increase in the angular velocity of the stimulus (Straschill and Taghavy, 1967). The optimal angular velocity in the cat's tectal DS-units varies between 0.5 and 20 degrees · sec^{-1} (Sterling and Wickelgren, 1969). In the retina of the rabbit there is a positive correlation between the size of the receptive field of DS-neurons and the maximum speed of movement which elicits the greatest response (Barlow et al., 1964). On-center DS-neurons in the rabbit's retina reach their maximum activation when the angular velocity of the stimulus is around 1 degree · sec^{-1}, while the on-off-center DS-neurons reach the maximum neuronal activation between 3 and 6 degrees · sec^{-1} (Oyster, 1968).

DS-neurons in the geniculate body of the cat respond well to targets moving in the preferred direction at 25 degrees · sec^{-1} (Kozak et al., 1964). In the retina of the squirrel, the neuronal activation of DS-neurons increases with an increase of the angular velocity between 0.1 and 30 degrees · sec^{-1} (Michael, 1966, 1968). A quantitative relationship between neuronal activation and angular velocity, however, was not measured. DS-neurons in the visual cortex of the monkey (Wurtz, 1969a, b, c) are also sensitive to the angular velocity of the stimuli. The activation is independent of whether the stimulus is moved in the visual field or whether the eye moves, thereby moving the image of the stationary stimulus across the RF in the retina. In the monkey, the angular velocity of the fast saccades of eye movements exceeds 500 degrees per sec; however, the DS-neurons, as well as the M-neurons, in the visual cortex are evidently able to respond to movement at a higher speed than that reached by fast eye movements.

We measured the relationship between neuronal activation and the angular velocity for the DS-neurons in the visual area 19 and in the suprasylvian gyrus of the cat. As was found for some of the directional, non-selective M-neurons in this cortical area, the neuronal activation of DS-neurons depends on the angular speed of a stimulus and is fairly well described by a power function (Fig. 24, 25a).

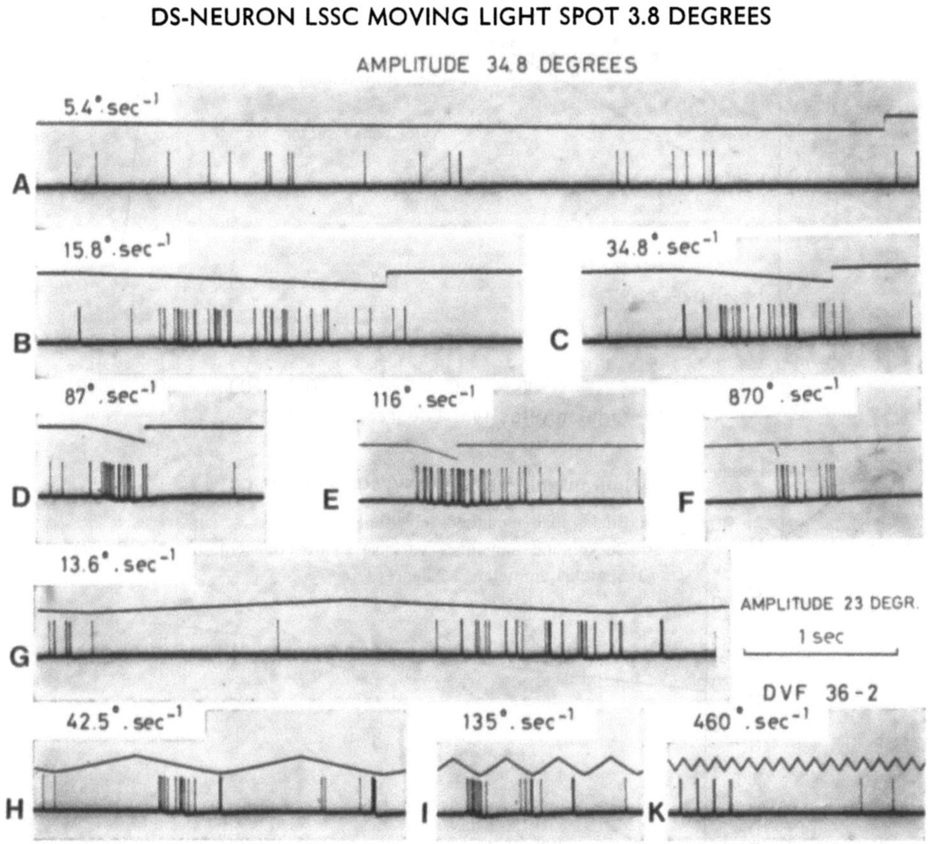

Fig. 24. DS-neuron from the lateral suprasylvian cortex of the cat (encephale isolé preparation). Response to a moving spot of light of 3.8 degrees diameter. Amplitude of movement 34.8 degrees. A—F: Movement through the ERF in the preferred direction at different angular velocity. The neuronal impulse rate increases as the angular velocity increases (A—F) from 5.4 [degrees · sec^{-1}] up to about 100 [degrees · sec^{-1}]. G—K: To and fro movement across the ERF in the null- and in the preferred direction at different oscillation frequencies. Amplitude of movement 23 degrees. As the stimulus reaches an oscillation frequency above 10 Hz (K) only inconstant responses are obtained

The effect of reducing both the contrast and the general illumination of the entire stimulus pattern was investigated for DS-neurons in cortical area 19 and in the suprasylvian gyrus (visual association area) of the cat. When the stimulus luminance is reduced and the background luminance is kept constant (thus the contrast between moving light stimulus and background is reduced), the response sector does not change.

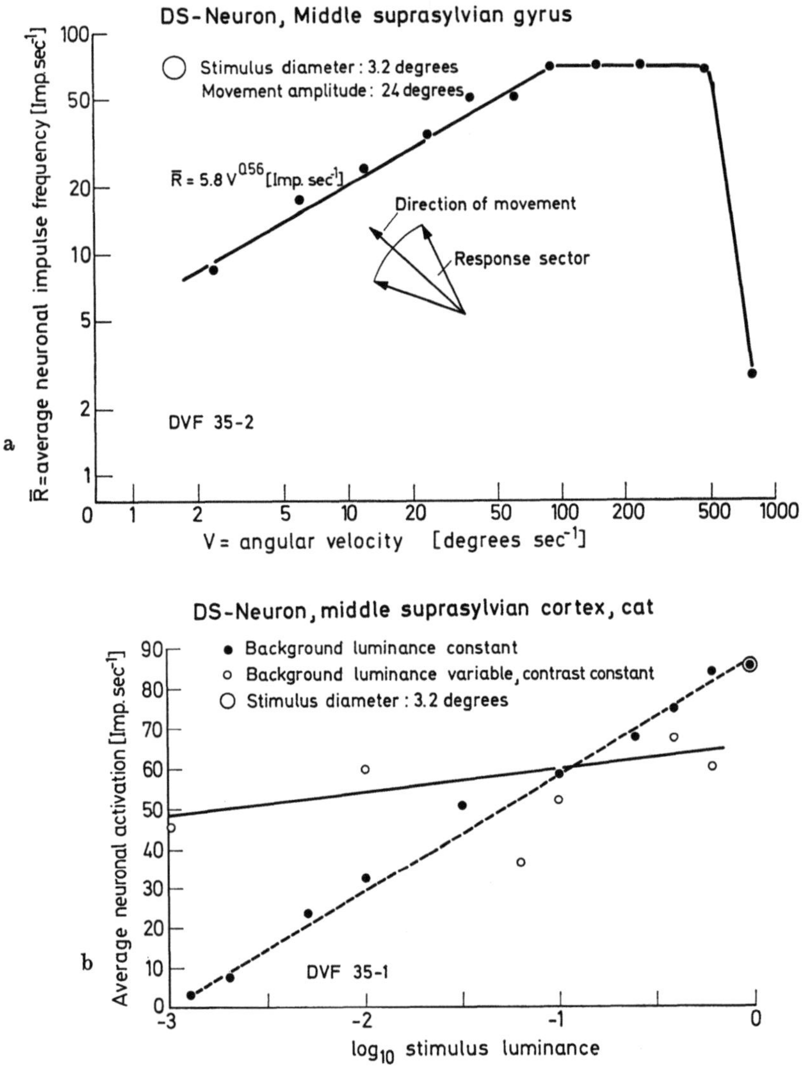

Fig. 25a u. b. *Effect of different stimulus parameters on the response of two DS-neurons in the middle suprasylvian cortex of the cat.*

a Velocity function. Response to a round spot of light (3.2 degrees diameter) moving through the ERF in the preferred direction. A power function $\bar{R} = 5.8 \times v^{0.56}$ (Impulses · sec^{-1}) was found for the relationship between the average neuronal impulse frequency ($\bar{R}$, ordinate) and the angular velocity (v, abscissa). b Dependency of the average neuronal activation upon the luminance of a spot of light of 3.2 degrees diameter, moving in the preferred direction through the ERF at an angular velocity of 5.3 degrees · sec^{-1}. The background illumination was either held constant (dots, 4 asb) or reduced proportionately to the luminance of the moving stimulus. In the latter case the contrast between moving stimulus and background was held constant (circles). The stimulus luminance at $\log I = 0$ is 160 asb

The neuronal activation, however, is approximately a linear function of the logarithm of the luminance of the moving light stimulus (or the contrast between stimulus and background), if the background luminance is kept constant and the stimulus luminance is changed. If the general luminance of the stimulus pattern is reduced (by placing neutral density filters in front of the cat's eye), the neuronal response decreases only slightly and the response sector is not changed. Thus, the *response*, as well as the response sector, is invariant with respect to the level of the general illumination of the stimulus pattern (Fig. 25b).

4. Responses to Spatially Oscillating Stimuli Having Different Frequencies and Amplitudes

We measured the relationship between the frequency of spatially oscillating stimuli (light spots, light bars, etc.) and the responses of DS-neurons in area 19 (tertiary visual cortex) and in the visual association area in the suprasylvian gyrus of cats. The results of one such experiment are demonstrated in Fig. 26 The optimal neuronal activation is obtained at a significantly lower oscillation frequency than in retinal neurons and in most of the geniculate neurons of the same animals (compare Figs. 4, 5, 26, and 35). The relationship between the oscillation frequency and the neuronal response is only slightly changed or not at all if the illumination of both the stimulus pattern and the background is reduced.

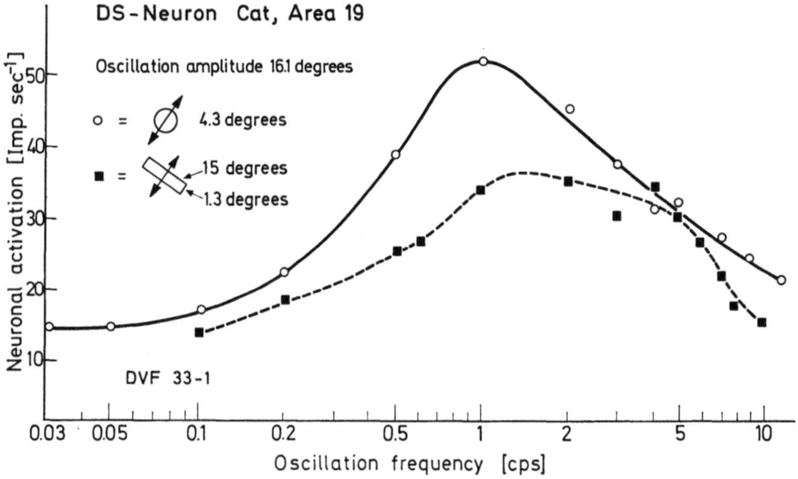

Fig. 26. *Relationship between the oscillation frequency and the average neuronal activation (ordinate) of a DS-neuron in area 19 of the cat's visual cortex.*
Preferred direction 45 degrees upwards to the right. A round spot of light of 4.3 degrees diameter elicits a stronger response than a bar of light (1.3×15 degrees) oriented at right angles to the direction of movement. Oscillation amplitude 16.1 degrees

5. Responses to Apparent-Movement Stimuli

DS-neurons in the retina of the rabbit and the squirrel respond to *stationary* light stimuli projected sequentially to *different parts* of the ERF. The response to a single stationary stimulus is weak. If, however, two light stimuli (*A* and *B*) are projected to two different parts of the ERF with a certain time delay, a strong response is obtained, which indicates the summation of the excitation elicited by each of the two stimuli. The response is strong if the succession of the stimuli is in the preferred direction, while the response to the second stimuli (*B*) is inhibited if the succession of the stimuli is in the null direction. A limited temporal interval for this type of response facilitation indicates that this direction sensitivity to apparent-movement stimulation might be due to an asymmetric conduction of inhibition within the ERF (squirrel retina: MICHAEL, 1966, 1968; rabbit retina:

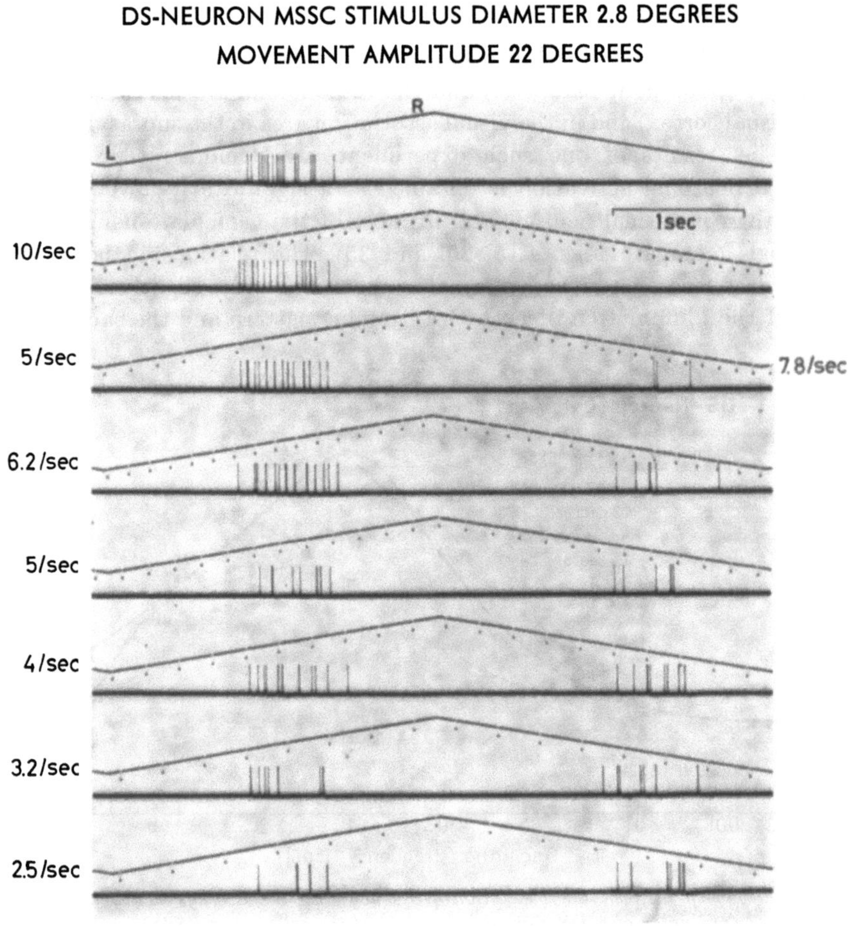

Fig. 27. DS-neuron from the medial suprasylvian cortex of the cat (encephale isolé preparation). Responses to a round spot of light of 2.8 degrees diameter moving across the ERF in the preferred and in the null-direction (amplitude of movements 22 degrees). The neuronal response as the stimulus is moved across the ERF from the left to the right. This direction sensitivity is also visible with movement during intermittent stimulation with flashes of 15 msec duration. At a stimulus frequency below 8 flashes per sec, the direction sensitivity disappears, the neuron responds to movement in both directions. At a flash frequency below 4 flashes per sec the response elicited by the moving stimulus is for both directions about the same

Barlow and Levick, 1965; cat optic tectum: Straschill and Taghavy, 1967). Another possible explanation fo this direction sensitivity would be that lateral excitation is only conducted in the preferred direction and "and-gate" neurons facilitate the unidirectional spread of excitation within the neuronal network.

The experiments with two stationary spots of light indicate that φ-movement in the preferred direction is an adequate stimulus for the activation of DS-neurons, and neighbouring receptors within the ERF need not be stimulated successively for neuronal activation to be elicited. This finding corresponds with the "change in position-response" which is a general property of M-neurons (page 370).

Neurons in area 19 and in the suprasylvian visual cortex of cats also respond with directional selectivity to the apparent movement of stationary stimuli, such as a moving spot or bar of light which is illuminated intermittently with very short flashes (5–15 msec duration). If the light spot changes its position within the ERF in the preferred direction, a rhythmical discharge pattern is then elicited in DS-neurons. The impulse pattern is locked to the flash frequency as long as the frequency is less than 10–15 flashes per sec. The average neuronal response depends on the frequency at which the moving stimulus is illuminated when the other parameters (stimulus size, contrast, adaptation level, etc.) are held constant. In addition, the *directional selectivity* of the neuronal response disappears at low flash frequencies (Fig. 27).

The point at which the neurons no longer show directional selectivity depends not only on the flash frequency, but also on the *angular velocity* of the intermittently illuminated stimulus. From this finding one can conclude that the

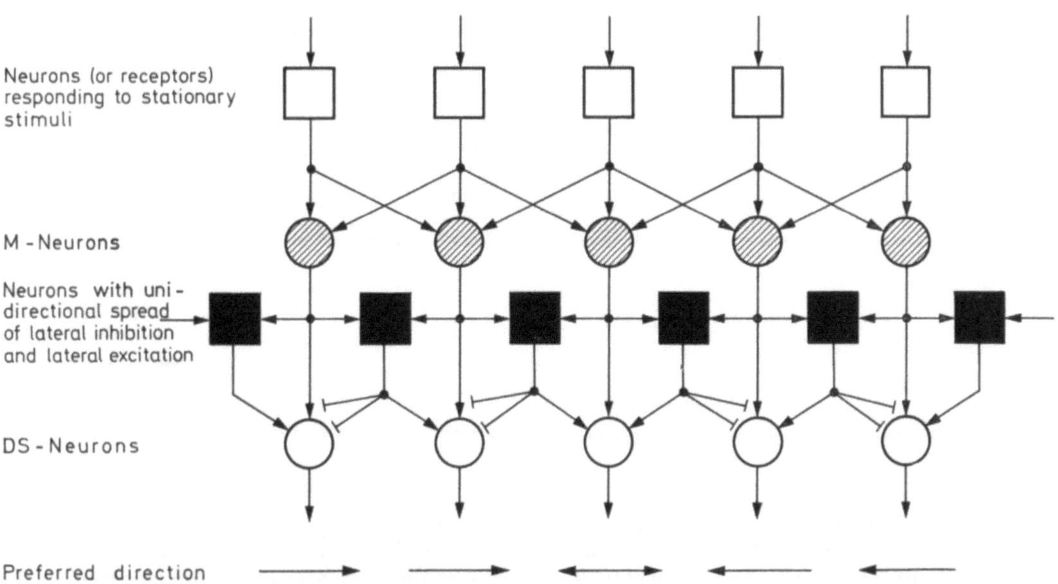

Fig. 28. *Schematic drawing of the neuronal connections of the elements which constitute the RF of a DS-neuron.*
The direction sensitive neurons 1, 2 and 4, 5 exhibit different preferred directions, depending on the direction of the lateral spread of inhibition and excitation of "horizontally" conducting elements in the RF

directional selectivity of the neuronal response is actually a function of the length of time occurring between the stimulation of different parts of the ERF and of the distance between the stimulated parts of the ERF. From these observations one can calculate the conduction speed of the unilateral spread of inhibition, taking place in the ERF of DS-neurons and the extension of the unilateral inhibiting subunits. For the latter we found values of about 0.8—2 degrees for the DS-neurons in the cat's suprasylvian cortex. The temporal and spatial properties of lateral inhibition in the RF of DS-neurons is probably the reason why directional selectivity disappears when continuously illuminated stimuli are moved very rapidly or extremely slowly through the ERF (BARLOW et al., 1964; OYSTER, 1968).

6. Models of the Receptive Field Organization of DS-Neurons

The experimental data mentioned in the preceding sections indicate that the directional selectivity of DS-neurons is invariant with respect to the size of the stimulus (as long as it is smaller than the ERF), the position of the moving stimulus in the ERF, the contrast between stimulus and background and (within a certain range of operation) the angular speed of the stimulus. BARLOW et al. (1964, 1965) demonstrated further that directional selectivity exists also for small changes in the position of the stimulus within the ERF. They thereforepostulated that there are "sub-units" within the ERF which also are direction selective.

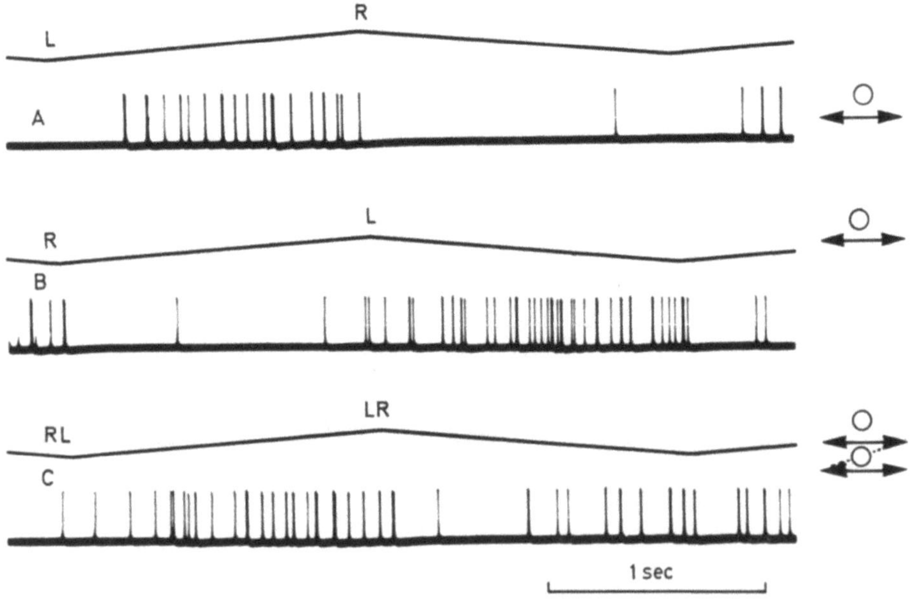

Fig. 29. Impulse pattern of a single cell from the *medial suprasylvian cortex* elicited by a round spot of light (12 degrees diameter) moving through the receptive field. Movement amplitude 23 degrees. The DS-neuron is activated by the movement of the stimulus from left to right and not to the movement from right to left. The two stimuli in A and B move across separate paths along the same direction through the ERF. C: Simultaneous movement of both stimuli, one from left to right, the other from right to left through the ERF. The neuronal response is only a little reduced in comparison to the responses of each stimulus alone (A, B)

The suppression of the neuronal impulses, elicited by a stimulus moving in the preferred direction by another stimulus moving simultaneously across the ERF in the null-direction (Fig. 19), further indicates that directional selectivity is caused by an *active inhibitory process* spreading within the ERF in the *null-direction* (Barlow et al., 1964, 1965). Hence, horizontally conducting neurons transmitting inhibition in the null-direction may be assumed to exist as presynaptic elements of DS-neurons. In addition, there might be a unidirectional lateral spread of excitation in the preferred direction within the same neuronal layer (Fig. 28). Furthermore, one has to assume that either the DS-nerve cells themselves or a layer of neurons between receptors and DS-neurons have M-properties, i.e., that the temporal bandpass characteristics and threshold mechanisms discussed in Section 4.8 are present in these neurons.

In some cortical DS-neurons the spread of lateral inhibition is rather weak, if a stimulus is moved into the null-direction across the ERF. When two stimuli are moved simultaneously through the ERF, one into the preferred direction, the other into the null-direction, the response of these DS-neurons then is only slightly or not at all reduced (Fig. 29).

Some of the DS-neurons in cortical area 19 of the cat and the visual association cortices in the suprasylvian gyrus exhibit strong *neuronal adaptation* to stimuli moved repeatedly along the same path through the ERF. For these neurons one has to assume that the excitatory synaptic processes or the excitation in more distal neurons contributing to the excitation of the DS-neurons show fatigue.

Fig. 28 summarizes schematically the properties one has to assume for the neuronal elements constituting the ERF of a DS-neuron. The IRF of these DS-neurons is probably organized similarly, whereby the preferred direction for the *inhibition* elicited from the IRF is the same as that for the *excitation* elicited from the ERF. Hence, the directional selectivity of the lateral inhibition from the IRF is 180 degrees opposite to the directional selectivity of lateral inhibition effective within the ERF. The simplest explanation for the DS-effect in the IRF is the assumption that DS-neurons whose ERFs are in adjacent regions of the visual field are interconnected by recurrent, inhibitory, collateral axons.

If two layers of DS-neurons are interconnected, the directional selectivity of the neurons in the second layer will be enhanced, assuming that DS-units in the first layer having their preferred direction in one direction are connected by excitatory synapses to the second order DS-neurons, while DS-neurons having preferred directions in other directions send inhibitory contacts to the second order neuron. This type of organization of the synaptic input might explain why the response sector of geniculate body DS-neurons in rabbits is narrower than that of the retinal DS-neurons (Levick et al., 1969, Fig. 20d).

VI. Movement Sensitive Neurons, Responding to the Spatial Orientation of a Contrast Border (OS-Neurons)

1. Subclasses of OS-Neurons in the Mammalian Visual Cortex

The most extensive investigations of visual cortical neurons, of which the responsiveness depends on the orientation of one or more light-dark borders moving through the ERF, was performed by HUBEL and WIESEL (1959, 1962, 1965, 1968), using anaesthetized cats and monkeys.

In the primary visual cortex most of the orientation sensitive neurons also respond when stationary patterns are illuminated sinusoidally, if the border orientation and the frequency of the light stimuli are properly chosen. A large part of these neurons, however, respond only slightly or not at all to diffuse illumination of the retina or to a small, stationary spot of light projected to the ERF. Several subclasses of neurons, sensitive to the orientation of a contrast border projected to the ERF, can be discriminated, since their sensitivity to movement and movement direction varies considerably (Fig. 30).

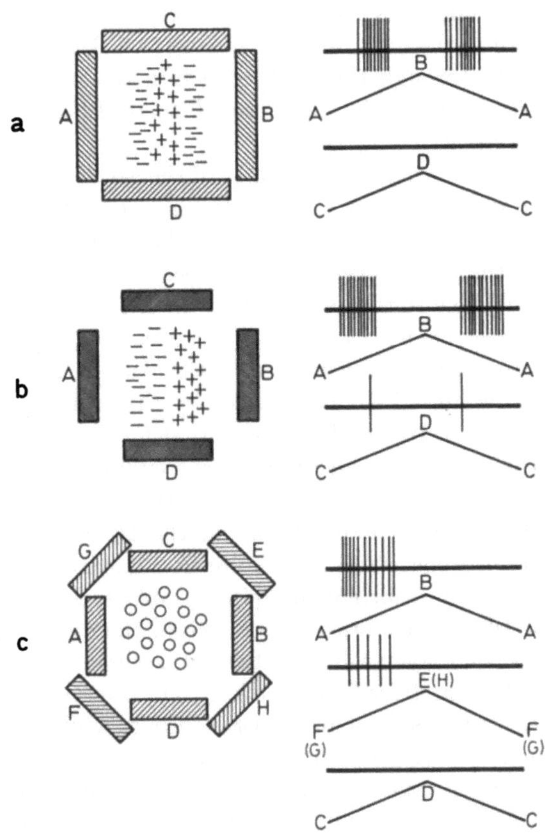

Fig. 30a—c. *Schematic summary of the findings of* HUBEL *and* WIESEL *(1959, 1962) for the response of OS-neurons in the primary and secondary visual cortex of the cat.*

a *Simple RF-neuron* which responds to vertically oriented, horizontally moving slits; no response to horizontally oriented and vertically moving slits of light. The analysis of the RF with stationary light spots (on-off) reveals an elongated on-center (+) flanked by two off-zones (−). b Simple receptive field neuron with elongated on- and off-zone. Responses to vertically oriented, horizontally moving black bar; no response to horizontally oriented, vertically moving bars. c Complex RF-neuron which does not respond (○) to stationary stimulation of the RF but responds strongly to bars moving through the RF. This response also depends on the orientation of the bar and the direction of the movement

Neurons with Simple Receptive Fields. A considerable proportion of the neurons of the cat's and monkey's primary visual cortex (area 17) have simple RFs with a concentric, oval or elongated symmetric structure (Fig. 30). The configuration of the simple receptive fields can be explored by small, intermittently illuminated, stationary spots of light. According to the responses to those stimuli, one can discriminate *on-zones* (+) in the ERF, which are activated by light stimuli and inhibited by dark stimuli, and *off-zones* (−) where the contrary is true. A station-

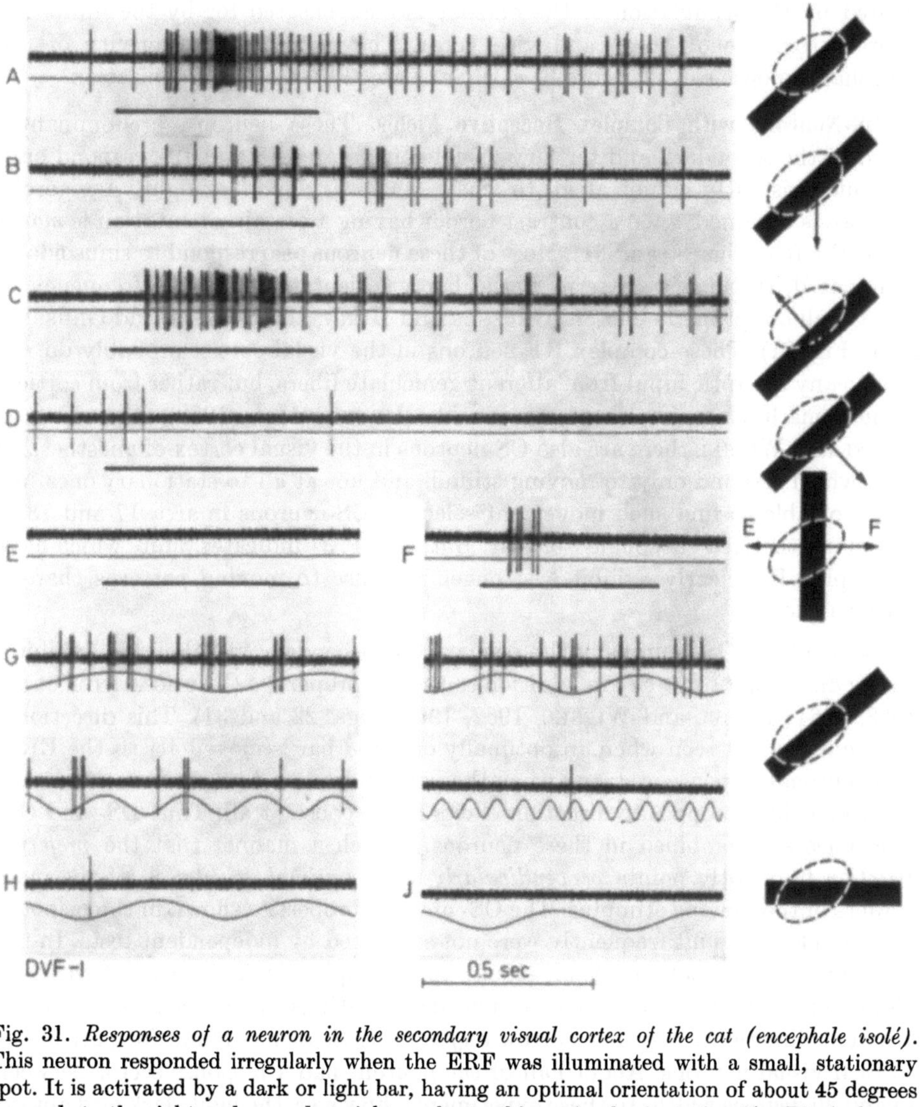

Fig. 31. *Responses of a neuron in the secondary visual cortex of the cat (encephale isolé).* This neuron responded irregularly when the ERF was illuminated with a small, stationary spot. It is activated by a dark or light bar, having an optimal orientation of about 45 degrees upwards to the right and moved at right angles to this optimal orientation (A—D). A change in the orientation of the bar reduces the response (E, F). In D the bar is moved out of the ERF into the IRF; therefore no response is obtained. A strong neuronal response is also obtained when the optimally oriented pattern is illuminated by low frequency sinusoidal light (G). A change in the stimulus frequency (G) or in the orientation of the bar (vertical in H, horizontal in J) abolishes the response (GRÜSSER and HENN, 1968; published in GRÜSSER, 1969)

ary, light-dark border is an optimal stimulus if the light part of the stimulus covers the on-zone and the dark part the off-zone. Likewise, these neurons respond strongly to a moving contrast border when the moving edge is parallel to the RF-axis, and the direction of movement is at right angles to the RF-axis. The activation is slightly stronger for optimally oriented *moving slits* than for *stationary stimuli*, because the slit enters the on-zone and leaves the off-zone simultaneously. If the slit is oriented perpendicular to the RF-axis and moves parallel to it, on- and off-zones are stimulated simultaneously, and the excitation elicited by the stimulation of the on-zone is compensated for by the inhibition elicited from the off-zones and vice versa. The orientation sensitivity of the neuronal response can therefore be easily explained by the RF-organization.

OS-Neurons with Complex Receptive Fields. These neurons, predominantly found in the secondary and tertiary visual cortex (area 18 and 19), respond only very inconsistently or not at all to small, stationary spots of light. A vigorous response is obtained when a contrast border having a certain orientation is *moved* across the RF (Figs. 30 and 31). Most of these neurons also respond to sinusoidally illuminated, stationary patterns if the border-orientation and the frequency of the stimuli are properly chosen (GRÜSSER and HENN, 1969; EYSEL and GRÜSSER, 1971; Fig. 31). These complex RF-neurons in the visual cortex probably do not receive any synaptic input from afferent geniculate fibers, but rather from cortical OS-neurons having simple receptive fields (DENNY et al., 1968). According to HUBEL and WIESEL there are also OS-neurons in the visual cortex of anesthetized cats, which respond only to moving stimuli and not at all to stationary ones. We were not able to find such movement selective OS-neurons in area 17 and 18 of the unanesthetized encephale isolé cat. But, as Fig. 31 indicates, units which have a complex RF clearly exhibit a stronger response to moving patterns than to stationary ones.

Some of the OS-neurons in the primary and secondary visual cortex exhibit a strong *directional selectivity* in addition to the OS-properties (BAUMGARTNER et al., 1963, 1964; HUBEL and WIESEL, 1962, 1965, Figs. 22 and 31). This directional selectivity is best seen when an optimally oriented bar is moved across the ERF. The neuronal response is strong when the pattern is moved in one direction, while movement in the opposite direction elicits no response at all. Thus DS- and OS-properties are combined in these neurons in such a manner that the *preferred direction* frequently points *perpendicularly to the optimal orientation* of the main borders of the moving stimulus. The OS- and DS-properties shown in the response of a complex RF unit frequently were not separated by independent tests. In the experiments by CAMPBELL et al. (1969), for example, single neurons of the cat's visual cortex were stimulated with light-dark gratings moving across the RF in different directions. A rather narrow *angular selectivity* was found for part of the cortical visual neurons. About half of the neurons had a response sector between 14 and 26 degrees width. In these experiments, however, the light-dark grid always moved perpendicular to the striation. Hence, the response depended on the OS- *and* the DS-properties of the receptive field.

In addition to requiring a stimulus whose main borders have a certain orientation and which moves in a certain direction across the ERF, some cortical visual neurons also require a stimulus having a certain size before an optimal response is elicited (Fig. 34). From these findings one can conclude that in these neurons an IRF is either adjacent to or surrounds the ERF.

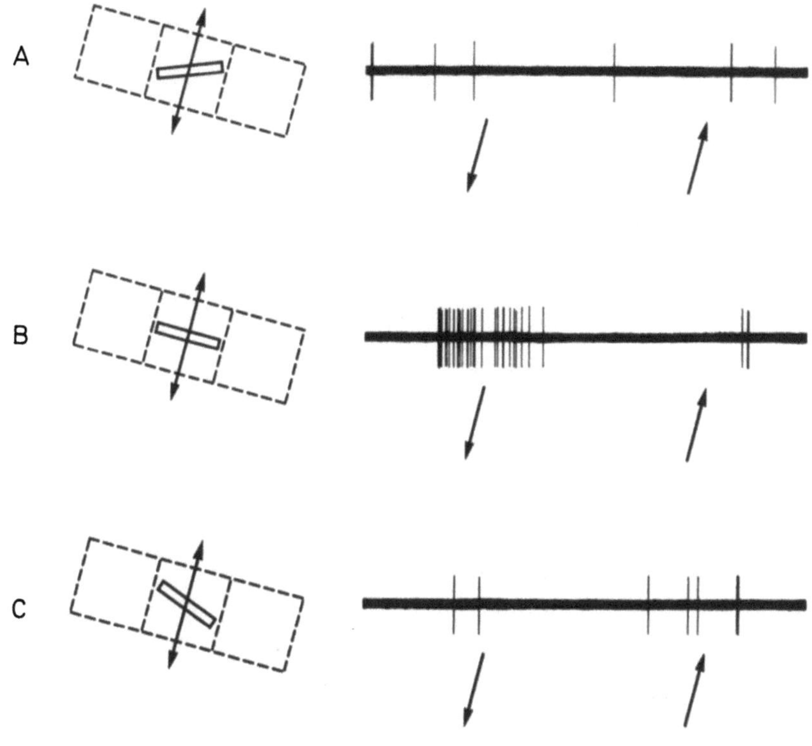

Fig. 32a. Response of a direction sensitive OS-neuron with hypercomplex receptive field from the colliculi superiores of the ground squirrel (by courtesy of Dr. CH. MICHAEL, 1971). A—C: Response to a bar-shaped light stimulus, moving in the optimal direction. A and C: The orientation of the light slit deviates 20 degrees from the optimal orientation. B: a strong response is elicited only if optimal orientation of the bar is obtained. This OS-neuron exhibits strong direction selectivity. Activation only to downward and not to upward movement

As mentioned above, in most of the experiments in which the functional organization of the RF of OS-neurons was explored, no independent tests were performed to find out whether the response is influenced more by DS- or OS-properties. In a very careful study of the response of OS-neurons in the superior colliculus of the squirrel, MICHAEL (1970, 1971) was able to demonstrate that the OS-properties of the ERF are more important for the response than the DS-properties (Fig. 32a). MICHAEL further found that the inhibitory response elicited from the IRF had the same OS-properties as the response elicited by a pattern moving through the ERF. The directional selectivity of the inhibition elicited from the IRF is the same as the directional selectivity of the excitation elicited from the ERF and thus is 180 degrees opposed to the directional selectivity of the lateral inhibition elicited when a pattern is moved through the ERF in the null direction (Fig. 32b).

The synaptic mechanisms, which together with the synaptical connections and the neuronal transfer properties determine these unique properties of the neuronal response, can be further clarified by intracellular recordings. The first measure-

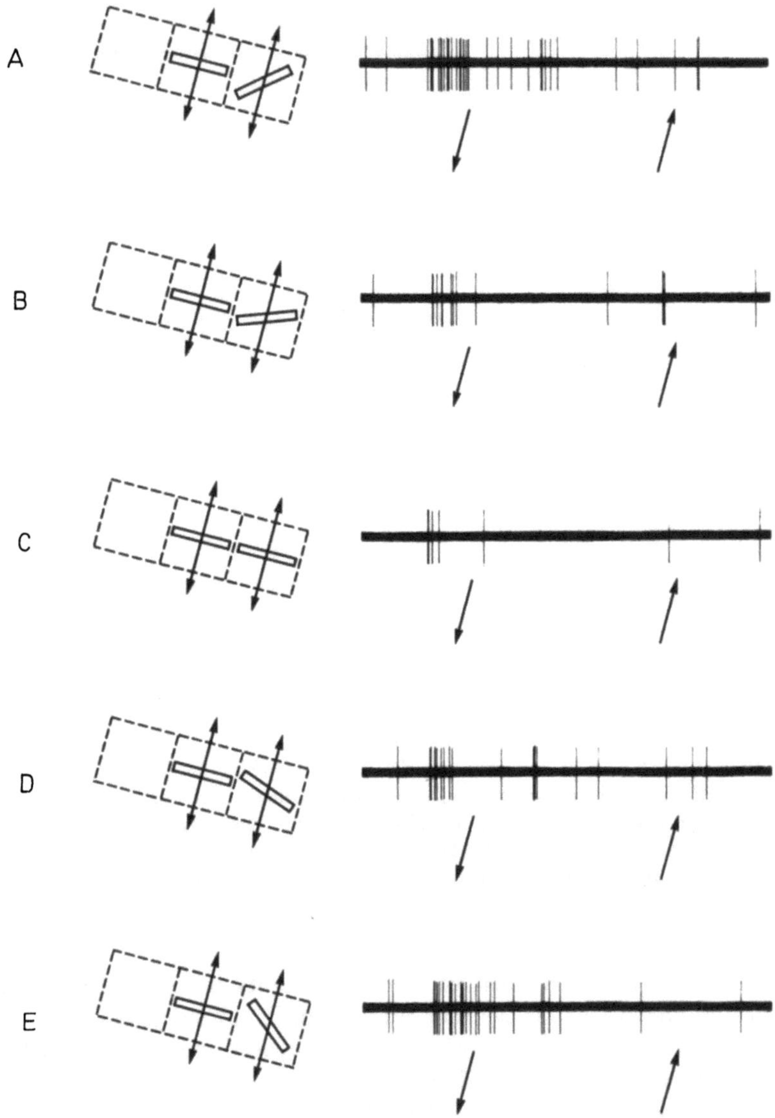

Fig. 32b. Same neuron as in Fig. 32a (MICHAEL 1971). The recording examples demonstrate the orientation sensitivity of the IRF. Simultaneous movement of two slit-shaped light stimuli, one located within the ERF, the other within the IRF. A: The stimulus within the IRF deviates 45 degrees from the optimal orientation. No inhibitory effect (compare with response Fig. 32a; B) The stimulus in the IRF deviates 20 degrees from the optimal orientation. Now an inhibitory effect is visible. C: Optimal orientation for both stimuli, in the ERF and in the IRF, only weak response to downward movement. D, E: The stimulus in the IRF deviates in the other direction from the optimal orientation: decrease of the inhibitory effect. Therefore strong activation to downward movement

ments of the synaptic input of cortical DS- and OS-neurons in area 17 of the cat were made by Creutzfeldt and Ito (1968). Their measurements indicate that the number of geniculate body neurons having excitatory synaptic contacts to cortical neurons might be less than ten. For direction sensitive neurons, movement in the preferred direction elicits EPSPs, while movement in the null-direction elicits IPSPs. This finding indicates that at least part of the direction sensitive inhibitory mechanisms are of the postsynaptic type.

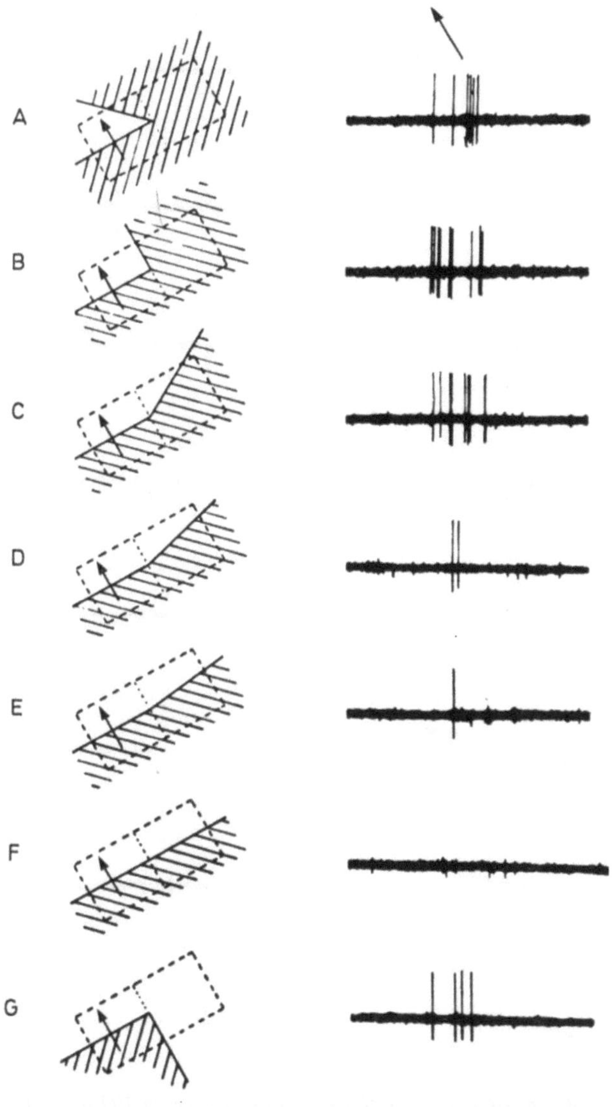

Fig. 33. *Neuron from the secondary visual cortex (area 18) of the cat; hypercomplex receptive field.* The optimal response is obtained if two contrast borders are oriented perpendicular to each other (B) and are moved in a certain direction through the ERF. The orientation of the activating corner is also important for the strength of the response. The neuronal response depends on the angle which is formed by the two contrast borders (Hubel and Wiesel, 1965)

Neurons with Hypercomplex Receptive Field Organization. Neurons with hypercomplex receptive field organization were found in visual area 18 (10%) and area 19 (68%) by HUBEL and WIESEL (1965). A moving stimulus was found to be more effective than a stationary one for all units with hypercomplex RFs. Several subclasses of hypercomplex RFs can be discriminated:

(a) Neurons which respond to two moving contrast edges which intersect (contrast corner). The angle of the intersection, the orientation of the pattern, the

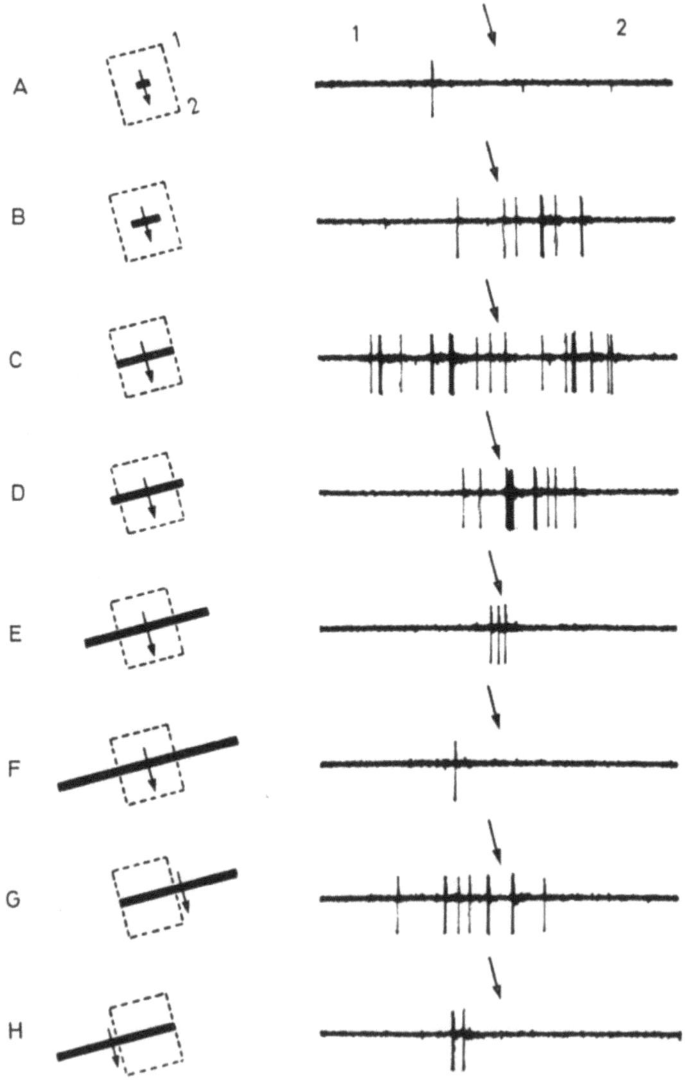

Fig. 34. *Response of a neuron in the tertiary visual cortex (area 19) of the cat, having a hypercomplex receptive field.*
The stimulus was optimal when moved in a certain direction (downwards to the left) and its main border was oriented at right angles to the direction of movement. The moving stimulus elicited a response only if it did not exceed the ERF. This finding indicates the existence of an IRF which flanks the ERF on both sides (From HUBEL and WIESEL, 1965)

position in the RF and the direction of movement are variables influencing the neuronal response. The angular velocity is optimal around 2 degrees $\cdot$ sec^{-1} (Fig. 33).

(b) Neurons which respond to interrupted contours (double-stopped edge), moving through the ERF.

(c) Neurons which respond only to contrast stimuli not exceeding a certain *width and length*, moving in a certain direction through the ERF and having a certain orientation (Fig. 34).

(d) Neurons which require two contrast corners, moving through the ERF in a certain direction to be optimally activated.

(e) Neurons which require the movement of a double-stopped edge ("tongue") of a certain width in order to be strongly activated. In contrast to subclass (b), the effect elicited by increasing the width of the tongue is independent of the position of the stimulus moving through the ERF; "it is as if properties like those of the previous cell (subclass b) had been generalized over a considerable area of the retina" (Hubel and Wiesel, 1965).

The ERF of hypercomplex OS-neurons, as a rule, is larger than the ERF of OS-neurons with a simpler RF-organization. Neuronal adaptation, directional selectivity and a high variability of the response to an optimal stimulus are essential properties of these classes of cortical neurons. Their main functional significance is probably the extraction of certain gestalt properties of the stimulus. The directional and movement sensitivity of these neurons might be interpreted as a useful adaptation to the movement of the retinal image of a stationary object, caused by the voluntary and involuntary eye movements.

2. The Effect of Different Stimulus Parameters on the Response of OS-Neurons

In the preceding section different classes of OS-neurons were described, whereby the parameters used for the classification were the orientation and position of one or more contrast borders of the stimulus moving within the ERF. Not much data are available concerning the effect of other stimulus parameters on the response of simple, complex or hypercomplex RF neurons on the visual cortex. Pettigrew et al. (1968) measured the relationship between the neuronal response and the *angular velocity* of the contrast stimulus moving across the ERF. The highest number of impulses per traverse through the ERF is obtained when the stimulus moves through the ERF at a low angular velocity (<5 degrees $\cdot$ sec^{-1}), while the *instantaneous* impulse frequency may reach a peak at a somewhat higher angular velocity ($20-60$ degrees $\cdot$ sec^{-1}). The average impulse frequency is maximal when the velocity of the stimulus is $10-20$ degrees $\cdot$ sec^{-1}.

In order to determine the mechanisms for pattern recognition it is not only necessary to investigate the velocity of the image, but also the effect of the temporal rate at which a given retinal locus is repeatedly stimulated. With free eye movements this factor depends on the amplitude and sequence of voluntary saccades and slow drifts (p. 336—340). The retinal effect of eye movements can be simulated in the curarized preparation by triangular or sinusoidal spatial oscillation of the stimulus across the RF. Experiments of this type reveal that cortical OS-neurons are highly sensitive to the frequency and the amplitude of pattern oscillation (BURNS et al., 1962). As a rule, the maximal neuronal response is obtained at an oscillation frequency between 0.5–5 cps, i.e., at a significantly lower oscillation frequency than in retinal and geniculate neurons (Fig. 35).

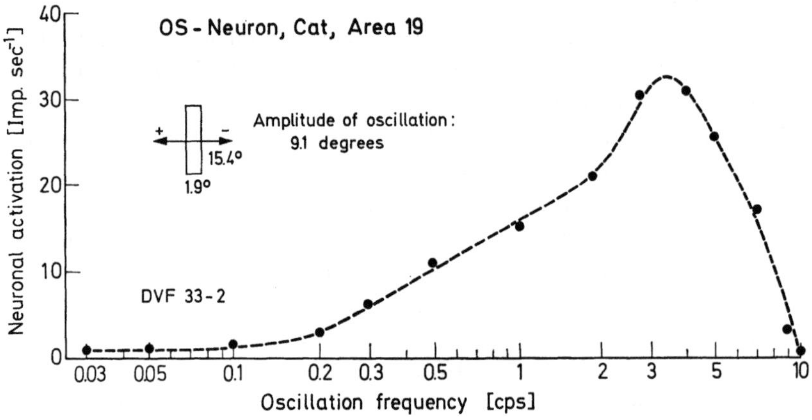

Fig. 35. *Relationship between the neuronal activation (ordinate) of an OS-neuron in area 19 of the cat's visual cortex and the frequency of a horizontal, oscillatory movement (preferred and null-direction) within the ERF.*
Optimal orientation: vertical. Bar of light: 1.9×15.4 degrees, amplitude of oscillation: 9.1 degrees (From GRÜSSER, GRÜSSER-CORNEHLS, and HAMASAKI, 1970b)

The cortical units with simple RF-organization, in contrast, respond to oscillation frequencies similar to those activating geniculate neurons, if a stimulus having the appropriate size, orientation and direction of movement is chosen.

When a *stationary pattern* is illuminated with *sinusoidal light* having different temporal frequencies, neurons with complex and hypercomplex receptive fields in cortical areas 18 and 19 are only activated by a slow change (<3–5 cps) in the luminance of the stimulus (Fig. 31), whereby no clear cut on- or off-response can be discriminated.

The *contrast* between moving stimulus and background affects the neuronal response. From the data obtained so far from complex or hypercomplex RF-neurons in the secondary and tertiary visual cortex, one sees that there is a large decrease in the neuronal response only when the contrast is below 0.5 (GRÜSSER et al., 1970).

Binocular activation is present in most of the movement sensitive complex and hypercomplex RF-neurons of the secondary and tertiary cortex, although ocular dominance might occur (HUBEL and WIESEL, 1965). As a rule, the optimal properties of a stimulus for maximal neuronal activation, i.e., the preferred direction of movements, optimal speed, the optimal location of the stimulus within the visual field, etc., are the same for both eyes. For neurons which might measure the stereoscopic depth of an object in space, the disparity between the images on both retinae may also play a role (BLAKEMORE et al., 1969; P. O. BISHOP et al., 1969). These problems, however, are beyond the scope of this review article.

3. Models of the Receptive Field Organization of OS-Neurons

As mentioned in the preceding sections, many OS-neurons exhibit M- and DS-properties. The general principles for the RF-organization of M- and DS-neurons, as discussed in Sections 4.8 and 5.6, can therefore be applied to a model of the RF-organization of the OS-neurons. In addition, one has to explain the OS-properties of the different specification. A simple model can easily be derived by assuming that simple RF-neurons form the main synaptic input to complex RF-neurons which in turn activate or inhibit hypercomplex RF-neurons (HUBEL and WIESEL, 1965). The geometry of the RF-organization of the presynaptic units together with the geometry of synaptic inputs and the alternative to elicit either excitatory or inhibitory postsynaptic potentials determine the OS-properties of the neurons at which the contacts synapse. For details the reader is referred to the papers by HUBEL and WIESEL and to the diagrams presented in our recent review article (GRÜSSER and GRÜSSER-CORNEHLS, 1969).

VII. Complex Movement Sensitive Neurons

These are neurons which have an even more complicated response characteristic than the movement sensitive OS-neurons described in the preceding chapter.

Part of the neurons in the *visual cortex of the rabbit* (ARDEN et al., 1967), for example, have a complex receptive field which is sensitive to stationary stimuli in one part, while in other parts the stimulus must move before a strong response is elicited. The region that is most responsive to stationary stimuli ("static center") might be located in a different part of the visual field than the region which is most sensitive to moving stimuli ("movement center"). Optimal stimuli for these neurons are objects which move through the movement center to the static center and stop there. Activation of other neurons in the rabbit's visual cortex requires moving stimuli with "complex contours" which, however, were not described in detail by the investigators (ARDEN et al., 1967).

In the nerve cell layers of the *optic tectum* of frogs and toads there are also several classes of neurons which require that an optimal stimulus have a high degree of specificity (LETTVIN et al., 1961; GRÜSSER and GRÜSSER-CORNEHLS, 1969). These neurons probably direct the head and body movement towards prey or away from a dangerous enemy.

In the first investigation of LETTVIN et al. (1961) only two classes of neurons could be discriminated in the tectum opticum (cell layer) of Rana pipiens: *Newness neurons*, responding to moving objects appearing anew in the visual field and *Sameness neurons*, which have a large receptive field and continue to respond to the same moving object as long as that remained in the ERF.

In a qualitative investigation of the neuronal responses in the tectum of *Bufo bufo* and *Rana esculenta* using microelectrodes we found 6 main classes of units which are visually activated and can be discriminated according to the location of the stimulus within the visual field and the specificity of the properties of the moving stimulus. We used the designation T 1 – T 6 for these different classes of nerve cells (GRÜSSER and GRÜSSER-CORNEHLS, 1970).

The action potential of *Class T 1-neurons* is recorded primarily in the anterior and deeper parts of the optic tectum. The ERF has a size of 15–30 degrees and is located in the *binocular visual field*. Many of these neurons respond when each eye is monocularly stimulated with a moving stimulus and exhibit some neuronal summation when both eyes are stimulated by the moving contrast pattern. Some neurons respond only to *binocular* stimulation with a moving object (binocular "and"-gate neurons, c. f. also FITE, 1969).

Class T 2-neurons respond to moving objects of 2–15 degrees diameter, have very large ERFs extending throughout nearly the entire visual field of one eye and are more strongly activated, the more nasal the stimulus is within the ERF. Some of these neurons also exhibit DS-properties.

The properties of *Class* T *3-neurons* are described in the next section.

Class T 4-neurons are similar to class T 2-neurons in that they have very large ERFs, but respond to all moving objects having diameters larger than 1–2 degrees. In addition, these neurons are frequently activated by tactile stimulation of the skin, either of the head or of the body and extremities located ipsilateral to the eye affected. Some of these neurons also respond to the slightest general vibration of the whole animal and to acoustic stimuli.

The RF of *Class T 5-neurons* varies between 5 and 15 degrees in size, might exhibit some DS-properties and often has a "striation" of several ERF and IRF stripes. Thus the response during the movement of a contrast-stimulus through the RF consists of a series of activatory periods interrupted by silent periods (Fig. 8 in GRÜSSER and GRÜSSER-CORNEHLS, 1968).

Class T 6-neurons have their receptive fields in that part of the visual field located above the animal. The ERF extends about 40–60 degrees to both sides (indicating binocular activation) and up to 120 degrees from the caudal part of the visual field to the nasal part. These neurons respond only to moving objects larger than 8 degrees and exhibit very fast neuronal adaptation (corresponding to the "newness" character of the units described by LETTVIN et al., 1961).

Besides responding to tactile, vibratory and acoustic stimulation, some single tectal units also respond to vestibular stimulation (SAMSANOVA, 1965; MKRTYCHEVA, 1966).

As we will discuss in Section X, the functional significance of the response of tectal units in the frog can only be understood at the present time if one assumes that these units serve as control mechanisms for motor responses of the animal. Experiments with chronically implanted microelectrodes will be necessary to obtain further information on this subject.

VIII. Neurons Activated by Movement of a Stimulus in the Z-Axis (Z-Neurons)

1. Z-Neurons in the Cat's Pretectal Region

STRASCHILL and HOFFMANN (1969), working on the pretectum of the cat, discovered units apart from M- and DS-neurons, which responded to an object moving in the midsagittal plane. Some of these units are activated by movement towards the animal (Z^+-neurons), other neurons are activated when the target moves away from the animal (Z^--neurons). When tested with stationary, bright or dark stimuli continuously increasing or decreasing in size, the Z^+-neurons are activated during an increase in the size of the stimulus, while the Z^--neurons respond to a decrease in the area of the stimulus. A change in the luminance of a stationary stimulus or movement in a vertical or horizontal direction does not elicit a strong neuronal activation (Fig. 36).

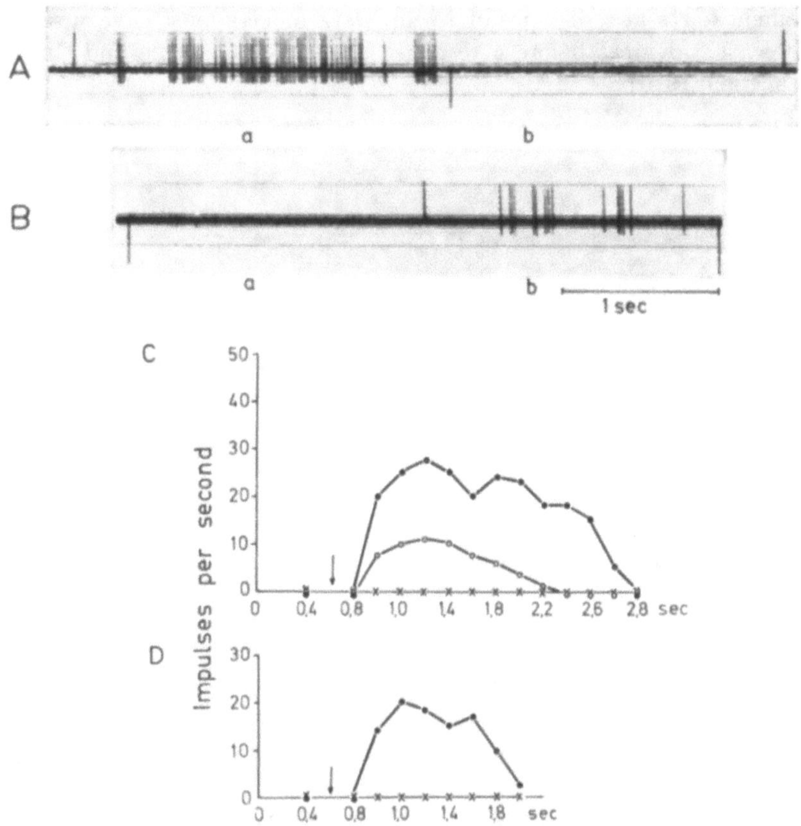

Fig. 36. A *Response of a Z-neuron in the pretectal region of the cat to an object moving in a midsagittal plane towards the animal (a) and away from it (b).* B Response to a continuous decrease (a) and increase in the diameter of a stationary spot of light projected to the ERF. C Average response of the neuron from 10 repeated stimuli of the type shown in A (dots) and to 10 movements in the frontal plane (circles). D Average of ten responses of the stimulus type shown in B (From Straschill and Hoffmann, 1969)

2. Z-Neurons in the Deeper Layers of the Optic Tectum of Anurans

Cells located in the deeper and medial parts of the optic tectum of toads and frogs respond to an object moving in the Z-axis towards the animal (Z^+-neurons), but only weakly or not at all when the object is moved tangentially through the ERF. The ERF is located in the nasal, anterior part of the visual field (Fig. 37) and has a diameter of about 20–30 degrees. No activation was found when the object was moved in the midsagittal plane away from the animal. When two stimuli were moved simultaneously in the Z-axis, one towards and one away from the animal, it was evident from the response that inhibition in the null-direction (away from the animal) is weak, if present at all. Fast neuronal adaptation appears if a contrast object is moved repeatedly along the same path towards the animal. The neuronal activation depends on the velocity at which the object is approaching the animal and reaches a maximum between ·1 and ·3 m · sec^{-1} when the movement starts at a distance of 25 cm. Dark objects must have an angular size of at least 3 degrees before a response is elicited (GRÜSSER and GRÜSSER-CORNEHLS, 1969, 1970).

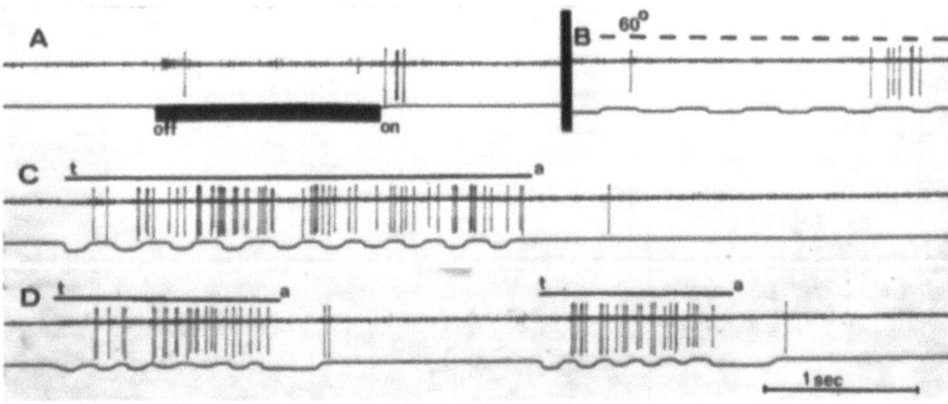

Fig. 37 A—C. *Z-neuron in the deep layer in the middle part of the optic tectum of Rana esculenta.*
A Weak response to diffuse illumination. B Weak response as a black disc of 3 cm diameter is moved along a perimeter having an arc of 60 degrees and at a distance of 25 cm from the eye (velocity about 25 degrees · sec^{-1}). C and D: Movement of the 3 cm black disc on a gray background (contrast about 0.5) towards ($t-a$) and away ($a-t$) from the eye from 25 to 3 cm distance at an average speed of about 7.6 (G) and 15 cm per sec (D). Activation as the stimulus approaches the animal (From GRÜSSER and GRÜSSER-CORNEHLS, 1970)

IX. Neuronal Activation in the Visual Centers during Eye Movements

Eye movements either occur spontaneously or are elicited by visual, vestibular, or acoustic stimuli. In addition to the small amplitude, high frequency eye tremor, two other main types of eye movements could be discriminated in the higher order of mammals: fast saccades and slow pursuit movements (or fast and slow nystagmic movements). Their central control mechanisms are probably located in two different regions of the brain-stem. The rapid eye movements are controlled by neurons in the pontine reticular formation, while the mechanisms for controlling slow eye movements are located in the mesencephalic reticular formation (e.g. COHEN and FELDMANN, 1969; HENN and COHEN, 1971, 1972).

As discussed in Section I,1, the continuous movement of the visual image of the stationary world on the retina is due to the interplay of saccades and slow, drifting eye movements. This is the "peripheral" level of interaction between the sensory and the motor part of the visual system. In addition, at several levels of the sensory visual system there exists a "central" interaction between signals elicited in the motor control system for eye movements and the sensory visual input. According to STRASCHILL and HOFFMANN (1970), some of the neurons in the deeper layer of the superior colliculus of the cat are activated during active eye movements in complete darkness, during active eye movements which shift a stationary, visual stimulus across the retina and by real movement of a contrast pattern across the ERF. Another class of tectal neurons is inhibited by active eye movements occurring in complete darkness or when patterned, stationary stimuli are viewed, while neuronal activation is elicited by the movement of the stimulus pattern without eye movements.

In the superior colliculus of the monkey, WURTZ and GOLDBERG (1971) found units located in the deeper layers which were activated before active eye movements occurred.

Presynaptic inhibition was found in optic nerve terminals in the geniculate body and the superior colliculus of cats during fast eye movements (KAWAMURA and MARCHIAFAVA, 1968).

MONTERO and ROBLES (1970, 1971) found that during vestibular nystagmus single cells in the geniculate body of the rat are maximally or minimally activated about 100–150 msec after the fast phases had begun. In these experiments the nystagmus was recorded in one eye, while the other eye was visually stimulated after it had been immobilized by severing all eye muscles.

Slow potential recordings with macroelectrodes (BIZZI, 1966; COHEN and FELDMANN, 1969) in the geniculate body of cats and monkeys also furnished some evidence that oculomotor and visual signals interact at the geniculate level. This interaction, however, might be due to the effect of efferent cortico-geniculate fibers.

WURTZ (1970) could demonstrate that in the primary visual cortex of monkeys the responses of cortical cells are essentially the same whether a moving stimulus is used or a stationary pattern is shifted across the ERF by voluntary eye movements. This finding makes it rather questionable that any inhibitory interaction between the oculomotor system and the afferent visual system between the retina and the visual cortex exists in these animals. In view of the psychophysical finding mentioned in Section I,2, it is also highly unlikely that such an inhibitory interaction occurs in the human visual afferent system.

In the cat's visual cortex single neurons are activated when a cortical region, located frontal to area 18 and 19, is stimulated electrically. Such a stimulation elicits conjugate eye movements towards the contralateral side and downwards. The long latency (80–200 msec) before neurons in the primary visual cortex are activated can be observed regardless of whether eye movements occur or are suppressed by Curarine or Flaxedil injection (encephale isolé preparation, GRÜSSER and GRÜSSER-CORNEHLS, 1961).

Our knowledge of the interaction between optomotor and visual signals in the visual sensory system is still rather incomplete. Psychophysical observation in humans and microelectrode findings in higher mammals, however, indicate that the old hypothesis of "central anaesthesia" (Holt, 1903) or complete central inhibition of visual signal processing during eye movements can be discarded.

X. Correlation between the Behavioral Responses of Anurans and the Activation of Movement Detecting Visual Neurons

The behavior of frogs and toads is partially determined by visual stimuli. Quantitative data are available concerning the response of frogs and toads to different moving, visual stimuli, measured either in the natural habitat (Grüsser and Grüsser-Cornehls, 1968) or in the laboratory (Ewert, 1969; Ingle, 1968). Three main visually controlled behavioral patterns can be discriminated: locomotion within the habitat, prey-catching and avoidance of moving enemies. Prey-catching and flight responses are probably mainly inborn mechanisms controlled by certain "key stimuli" *(Schlüsselreize)*. The neuronal network controlling the whole behavioral pattern corresponds to the inborn release mechanism *(angeborener Auslösemechanismus)*. In addition, one can assume that some acquired release mechanisms *(erworbener Auslösemechanismus)* play a role in the prey-catching behavior (Schneider, 1954).

The prey-catching behavior of the frog, elicited by a small moving object, consists of the following motor sequences: orientation of movement, binocular fixation, jumping, tongue snapping, swallowing, "cleaning". The response for orientation is elicited maximally by a moving object having an angular size of about 4 degrees (Ingle, 1968), while jumping towards the prey is the most frequent response to stimuli having an angular size of 8–12 degrees (Grüsser and Grüsser-Cornehls, 1968). Larger moving stimuli elicit a flight response.

The quantitative relationship between the different stimulus parameters and the responses of the different neurons in the ganglion cell layer of the retina was described in Section IV,2. From these data, it is evident that the retina performs an important function by filtering the signals elicited by the key stimuli triggering the avoidance and the prey-catching response. A closer inspection of the experimental data, however, reveals that the retinal filter mechanisms definitely are not refined enough that one can characterize the function of a class of retinal neurons as "bug detectors". The response of the 4 classes of optic nerve fibers of the frog, which have their synaptic endings in the optic tectum, to typical natural stimuli is presented in a semi-quantitative manner in Table 5. It is evident from this table that every visual stimulus which elicits a behavioral response activates two or three of the 4 classes of retinal neurons. Therefore, the filtering function of the retinal signal processing is only the first step towards distinguishing prey from enemy or other behaviorally important stimuli in the habitat.

According to Table 5, it could be assumed that the invariants "prey" or "enemy" might be distinguished by a higher class of visual neurons in the optic tectum. From the data described in Section VII, however, we know that this opinion has little support from the experimental findings in tectal units. One has to assume that the average retinal input is "weighted" by the tectal neurons. Their discharge in turn represents *motor commands*. The latter view is supported

Table 5. Correlation between behavioral responses and the activation of retinal neurons of Rana (GRÜSSER and GRÜSSER-CORNEHLS, 1968, 1970)

Stimulus in the experiments	Simulates	Class 1 neurons	Class 2 neurons	Class 3 neurons	Class 4 neurons	Behavioral response
White spot 2 degrees, moved on dark background at 0.5 degrees · sec^{-1}	worm	++	+	0	0	prey-catching
Black spot, 3 degrees, moved on white background, 5 degrees · sec^{-1}	walking fly or bug	++	+++	(+)	0	prey-catching
Black spot, 6 degrees, moved on white background, 20 degrees · sec^{-1}	flying libella	+	++	+++	(+)	prey-catching
Black area 2 degrees, moved on white background, 10 degrees · sec^{-1}	enemy (bird)	(+) borders	0	++ border	++	flight
White area of 20 degrees, moved on dark background, 10 degrees · sec^{-1}	enemy (white stork)	(+) border	0	+ borders	(+) borders	no response or flight
Dark area of fading contours moved with patterned background	dark hiding place for a moving frog	0	0	(+)	++	hiding
Many small, dark spots moved on a white background (0.5—10 degrees · sec^{-1})	swarm of flies	(+)	0	(+)	(+) or 0	no response
Dark-white stripes or more structured contrast pattern; large movements	visual surround during fast body and head movements	(+)	0	+	+	general visual perception of the habitat
Small movements	visual surround during respiratory eye movements	(+)	(+)	+	0	general visual perception

by the findings of RUBINSON (1968) that in the anurans tectal efferent fibers make direct contacts via the tectospinal and tectobulbar tract with motor neurons. If the "motor command function" of the tectum (AKERT, 1949; HESS et al., 1949; BÜERGI, 1957) is assumed to be performed by class T 1–T 6 neurons described in Section VII (p. 402–403), one has to conclude that the behavioral invariants "moving prey" or "enemy" are represented by the function of all complex movement detecting neurons of the optic tectum and *not by one neuronal class alone*. There is evidently a *sequential activation* of neurons of different tectal classes which in turn elicit the motor patterns, described above, in the proper sequence for catching prey. For example, if a flying insect (e.g. a libella) appears in the lateral visual field of a frog, class 1, 2, and 3 neurons in the retina are activated. Their synaptic inputs to the tectum activate class T 5 neurons, which command the movements for orientation. This movement shifts the stimulus across the ERF of class T 2 neurons into the binocular part of the visual field, where class T 1 neurons are activated. Their activation might control the jumping of the frog, which brings the prey to the frog's head, thus activating class T 3 neurons (described in Section VIII). Finally, this activation might trigger the tongue movement and snapping of the prey. EWERT (1967) demonstrated that, in addition, neuronal mechanisms outside of the optic tectum definitely play a role in discriminating between prey and enemy, since toads try to catch moving objects of all sizes after certain pretectal areas have been destroyed (c.f. SZEKELY 1971).

Recent comparative studies concerning retinal signal processing in different anurans and urodeles (GRÜSSER-CORNEHLS and LÜDCKE, 1970, 1971; GRÜSSER-CORNEHLS and HIMSTEDT, 1971) produced substantial evidence that behavioral differences can be linked to some qualitative and quantitative differences in the signal processing of retinal M-neurons (frogs, toads, tree-frogs, salamander). The behavioral differences among the anurans is paralleled by different neuronal filter mechanisms in the retina providing the input to the complex M-neurons of the optic tectum (GRÜSSER-CORNEHLS, 1972).

XI. The Possible Roles of the Different Classes of Movement Sensitive Neuronal Systems in Mammals

In the preceding section we gave a short synopsis of the possible functional importance of movement sensitive neurons in the visual system of anurans having different degrees of complexity. Relatively simple and "automatic" response patterns exist for these animals. This simplifies the attempt to draw general conclusions about the functional significance of the neurophysiological findings. Such a synthesis is more difficult to achieve for the visual neurons in higher mammals.

From the considerations described in Section I, we were tempted to apply some of the following *a priori* hypotheses (A–C) as guidelines for the functional interpretation of the neurophysiological findings:

(A) *Movement detecting neurons extract perceptual information about movement, angular velocity and direction of movement of a visual stimulus pattern.*

An animal or a human observer discriminates between moving (real or apparent) and stationary objects by means of M-neurons. From the psychophysical data mentioned in Section I, it is predictable that at least some of these neurons measure the angular velocity of an object and therefore should have a regular velocity function. This is indeed the case for cortical neurons in the visual association area (Fig. 24, 25). Neurons which are used to make a "judgement" about the *direction* of movement should have a rather narrow response sector. Those neurons exist in the DS-systems of the visual and paravisual cortex.

Under certain circumstances, *apparent movement* can not be discriminated from *real movement*. Therefore, an essential property of the M and DS neuronal systems employed in visual perception should be that they are strongly activated by stimuli showing apparent movement. All cortical M and DS neuronal systems tested so far fulfil this requirement.

The existence of the properties mentioned for M or DS neurons, however, is only a necessary, but not a sufficient condition for classifying them in the category with neurons employed at the "perceptual level". The *location* of the nerve cells within the brain seems to be the other important factor. For example, behavioral experiments by HAMILTON and LUND (1970) have demonstrated that the forebrain in monkeys is necessary for the perception of movement and the direction of movement even though the neurons in the superior colliculus (WURTZ and GOLDBERG, 1970) probably fulfil all the requirements of DS- or M-properties discussed above. On the other hand cortical M- and DS-neurons as well as OS-neurons, employed in the perception of contrast borders and other parameters of visual patterns, also influence eye movements to some extent. This will be discussed below.

(B) *M- and DS-neurons control eye-movements.*

An object moving in the visual world elicits voluntary or involuntary eye movements. The pursuit movements of the eyes are intended to keep the image of the moving object within the center of gaze. For this purpose the eye movements have to be matched with the velocity and the direction of movement of the object. Neurons in the superior colliculi and pretectum probably serve this purpose. In higher mammals, however, their activity is modified and controlled by cortical neurons (visual and paravisual cortex, STERLING and WICKELGREN, 1969; RIZZOLATTI et al., 1970), which also have perceptual functions. This finding is not at all surprising in light of the psychophysical finding that in higher animals pursuit movements depend on the attentiveness. For human observers the eye movements during the observation of a well structured visual pattern also depend on the orientation of the contrast borders in this pattern (c.f. Section I).

In *lower mammals*, it is possible that the DS-neurons in the retinal ganglion cell layer also extract signals from the visual stimulus, which might be used for the control of eye movements. Such an interpretation is in accordance with the experimental finding of the non-random distribution of the preferred directions in retinal DS-neurons of the rabbit (Fig. 22, OYSTER and BARLOW, 1967; OYSTER, 1968).

In addition to eliciting eye movements, moving visual stimuli can also elicit head and body movements. Thus, the M- and DS-neurons in the tectum and the pretectal region might control the motor nervous systems for head and body movements as well as the oculomotor system. This assumption would explain the experimental finding that electrical stimulation of the tectum opticum not only elicits eye movements, but also body movements and turning or upward and downward movements of the head (HESS et al., 1947; SCHAEFER, 1968; MAYER and SCHAEFER, 1969).

(C) *The movement and direction sensitivity of cortical OS-neurons is a functional adaptation to the continuously moving retinal image of the stationary world.*

The psychophysical findings supporting this hypothesis were discussed in Section I,2. This hypothesis is further supported by the experimental findings that the optimal neuronal activation of OS-neurons is within the frequency range of the voluntary and involuntary eye movements within the range of the intervals between saccades. In psychophysical experiments intermittent stimulation of stationary patterns at a low frequency is a partial substitute for eye movements in the recognition of stationary stimuli. In neurophysiological experiments such stimuli are also effective in activating cortical OS-neurons (Fig. 31). The neurophysiological findings in single neurons of the visual cortex of cats and monkeys, however, indicate that a well structured, visual pattern is more readily perceived when explored with free eye-movements than when the retinal image is stabilized. The oculomotor system, controlled by the activity of cortical, movement sensitive OS-neurons, does not move the eyes at random with respect to the patterns, but in such a manner that cortical OS-neurons are optimally activated. Hence, the eye movements during the observation of a structured object in the visual world not only prevent the retinal image from fading and increase the neuronal activation at the lower levels of the afferent visual system, but also optimalize the conditions for sampling signals from the visual world.

Acknowledgement

Our own experiments mentioned in the present review were supported by grants from the Deutsche Forschungsgemeinschaft (Gr 161 and Gr 276) and the National Institute of Health (Research Grant B 07575). The manuscript was prepared while one of us (O.-J. G.) held an International Scholarship from Research to Prevent Blindness Inc., New York at the Bascom Palmer Eye Institute, University of Miami. We thank Prof. E. W. D. NORTON, chairman of the Department of Ophthalmology, University of Miami, for his kind hospitality, Miss B. COOK for her assistance in the English translation and Mrs. U. SAYKAM for her typing of the manuscript.

References

AARONS, L.: Visual apparent movement research. Review 1935—1955, and bibliography 1955—1963. Percept. Motor Skills **18**, 239—279 (1964).

ADAMS, A.: Nystagmographische Untersuchungen über den Lidnystagmus und die physiologische Koordination von Lidschlag und rascher Nystagmusphase. Arch. Ohr.-, Nas.- u. Kehlk.-Heilk. **170**, 543—558 (1957).

ADRIAN, E. D.: The basis of sensation. The action of the sense organs. London: Christophers Ltd. 1928.

— CATTELL, M., HOAGLAND, H.: Sensory discharges in single cutaneous nerve fibers. J. Physiol. (Lond.) **72**, 377—391 (1931).

AKERT, K.: Der visuelle Greifreflex. Helv. physiol. pharmacol. Acta **7**, 112—134 (1949).

AKIMOTO, H., CREUTZFELDT, O.: Reaktionen von Neuronen des optischen Cortex nach elektrischer Reizung unspezifischer Thalamuskerne. Arch. Psychiat. Nervenkr. **196**, 494—519 (1958).

Altman, J.: Some fiber projections to the superior colliculus in the cat. J. Comp. Neurol. **119**, 77–95 (1962).
— Carpenter, M. B.: Fiber projections of the superior colliculus in the cat. J. Comp. Neurol. **116**, 157–178 (1961).
— Malis, L. J.: An electrophysiological study of the superior colliculus and visual cortex. Exp. Neurol. **5**, 233–249 (1962).
Andrew, A. M.: Action potentials from the frog colliculus. J. Physiol. (Lond.) **130**, 25 P (1955).
Apter, J. T.: Projection of the retina on superior colliculus in cats. J. Neurophysiol. **8**, 123–134 (1945).
— Eye movements following strychninization of the superior colliculus of the cat. J. Neurophysiol. **9**, 73–86 (1946).
Arden, G. B.: Types of response and organization of simple receptive fields in cells of the rabbit's lateral geniculate body. J. Physiol. (Lond.) **166**, 449–467 (1963a).
— Complex receptive fields and responses to moving objects in cells of the rabbit's lateral geniculate body. J. Physiol. (Lond.) **166**, 468–488 (1963b).
— Hill, R. M., Ikeda, H.: Receptive fields of rabbit visual cortex. J. Physiol. (Lond.) **189**. 73 P (1967).
— Ikeda, H., Hill, R. M.: Rabbit visual cortex: Reaction of cells to movement and contrast. Nature (Lond.) **214**, 909–912 (1967).
Armstrong, C. M.: Monosynaptic activation of pyramidal cells in area 18 by optic radiation fibers. Exp. Neurol. **21**, 413–428 (1968).
Aubert, H.: Die Bewegungsempfindung. Pflügers Arch. ges. Physiol. **39**, 347–370 (1886).
— Die Bewegungsempfindung. Pflügers Arch. ges. Physiol. **40**, 459–480 (1887).
Barlow, H. B.: Action potentials from the frog's retina. J. Physiol. (Lond.) **119**, 58–68 (1953a).
— Summation and inhibition in the frog's retina. J. Physiol. (Lond.) **119**, 69–88 (1953b).
— Blakemore, C., Pettigrew, J. D.: The neural mechanism of binocular depth discrimination. J. Physiol. (Lond.) **193**, 327–342 (1967).
— Hill, R. M.: Selective sensitivity to direction of movement in ganglion cells of the rabbit retina. Science **139**, 412–414 (1963).
— — Levick, R. W.: Retinal ganglion cells responding selectively to direction and speed of image motion in the rabbit. J. Physiol. (Lond.) **173**, 377–407 (1964).
— Levick, W. R.: The mechanism of directionally selective units in rabbit's retina. J. Physiol. (Lond.) **178**, 477–504 (1965).
Basler, A.: Über das Sehen von Bewegungen. I. Die Wahrnehmung kleinster Bewegungen. Pflügers Arch. ges. Physiol. **115**, 582–601 (1906).
Baumann, Ch.: Receptorpotentiale der Wirbeltiernetzhaut. Pflügers Arch. ges. Physiol. **282**, 92–101 (1965).
Baumgartner, G.: Reaktionen einzelner Neurone im optischen Cortex der Katze nach Lichtblitzen. Pflügers Arch. ges. Physiol. **261**, 457–469 (1955).
— Die Reaktionen der Neurone des zentralen visuellen Systems der Katze im simultanen Helligkeitskontrast. In: Neurophysiologie und Psychophysik des visuellen Systems, S. 296–311. Berlin-Göttingen-Heidelberg: Springer 1961.
— Neuronale Mechanismen des Kontrast- und Bewegungssehens. Ber. dtsch. ophthal. Ges. **66**, 111–125 (1964).
— Brown, J. L., Schulz, A.: Visual motion detection in the cat. Science **146**, 1070–1071 (1964).
— — — Responses of single units of the cat visual system to rectangular stimulus patterns. J. Neurophysiol. **28**, 1–18 (1965).
— Hakas, P.: Reaktionen einzelner Opticusneurone und corticaler Nervenzellen der Katze im Hell-Dunkel-Grenzfeld (Simultankontrast). Pflügers Arch. ges. Physiol. **270**, 29 (1959).
— — Neurophysiologie des simultanen Helligkeitskontrastes. Pflügers Arch. ges. Physiol. **274**, 489–510 (1962).
— Schulz, A., Brown, J. L.: Unterschiedliche Reaktionen auf bewegte Reizmuster bei corticalen und Geniculatumneuronen. Pflügers Arch. ges. Physiol. **278**, 69–70 (1963).
Bender, M. M. (ed.): The oculomotor system. New York: Harper & Row 1964.

BISCHOF, N., KRAMER, E.: Untersuchungen und Überlegungen zur Richtungswahrnehmung bei willkürlichen sakkadischen Augenbewegungen. Psychol. Forsch. **32**, 185–218 (1968).

BISHOP, G. H.: Fibre groups in the optic nerve. Amer. J. Physiol. **106**, 461 (1933).

BISHOP, P. O., KOZAK, W., LEVICK, W. R., VAKKUR, G. J.: The determination of the projection of the visual field on the lateral geniculate nucleus in the cat. J. Physiol. (Lond.) **163**, 503–539 (1962).

BIZZI, E.: Changes in the orthodromic and antidromic response of optic tract during the eye movements of sleep. J. Neurophysiol. **29**, 861–870 (1966).

— Discharge of frontal eye field neurons during eye movements in unanesthetized monkeys, Science **157**, 1588–1590 (1967).

— Discharge of frontal eye field neurons during saccadic and following eye movements in unanesthetized monkeys. Exp. Brain Res. **6**, 69–70 (1968).

— SCHILLER, P. H.: Single unit activity in the frontal eye fields of unanesthetized monkeys during eye and head movement. Exp. Brain Res. **10**, 151–159 (1970).

BLAKE, L.: The effect of lesions of the superior colliculus on brightness and pattern discrimination in the cat. J. comp. Physiol. **57**, 272–278 (1959).

BROWN, J. E., ROJAS, J. A.: Rat retinal ganglion cells: receptive field organization and maintained activity. J. Neurophysiol. **28**, 1073–1090 (1965).

BROWN, J. F.: The visual perception of velocity. Psychol. Forsch. **14**, 199–232 (1931).

BÜTTNER, U., GRÜSSER, O.-J.: Quantitative Untersuchungen der räumlichen Erregungssummation im rezeptiven Feld retinaler Neurone der Katze. I. Reizung mit 2 synchronen Lichtpunkten. Kybernetik **4**, 81–94 (1968).

BÜRGI, S.: Das Tectum opticum. Seine Verbindungen bei der Katze und seine Bedeutung beim Menschen. Dtsch. Z. Nervenheilk. **176**, 701–729 (1957).

BURNS, B. D., HERON, W., PRITCHARD, R.: Physiological excitation of visual cortex in cat's unanesthetized isolated forebrain. J. Neurophysiol. **25**, 165—181 (1962).

— PRITCHARD, R.: Contrast discrimination by neurons in the cat's visual cerebral cortex. J. Physiol. (Lond.) **175**, 445—463 (1964).

— SMITH, G. K.: Transmission of information in the unanesthetized cat's isolated forebrain. J. Physiol. (Lond.) **164**, 238–251 (1962).

BUSER, P., DUSSARDIER, M.: Organisation des projections de la rétine sur le lobe optique, étudiée chez quelques téléostéens. J. Physiol. (Paris) **45**, 57—60 (1953).

BUTENANDT, E.: Diss., Zool. Institut, Freiburg 1969.

— GRÜSSER, O.-J.: The effect of stimulus area on the response of movement detecting neurons in the frog's retina. Pflügers. Arch. ges. Physiol. **300**, 283—293 (1968).

BYZOW, A. L.: Die Dynamik der Labilität einzelner funktioneller Einheiten der Netzhaut des Frosches bei Intensitätsveränderungen von Flimmerlicht (Russ.). Arbeiten aus der Akademie der Wissenschaften UdSSR **105**, 852—855 (1955).

— Lability of single retinal units in some mammals. J. Physiol. USSR **42**, 1011—1020 (1956).

— Analysis of the distribution of potentials and current arising in the retinal response to light stimulation. On the activity of the two types of bipolarse. Biofizika **6**, 689—701 (1959).

— Functional properties of different cells in the retina of cold-blooded vertebrates. Cold Spr. Harb. Symp. quant. Biol. **30**, 547—558 (1965).

— HANITZSCH, R.: Intrazellulär abgeleitete Reaktionen verschiedener Netzhautzellen des Frosches und Axolotl (Russ.). Fiziol. Zh. **52**, 3 (1966).

CAJAL, S. RAMON y: Die Retina der Wirbelthiere. Übers. von GREEFF, Wiesbaden: J. F. Bergmann 1894.

CAMPBELL, F. W., CLELAND, B. G., COOPER, G. F., ENROTH-CUGELL, C.: The angular selectivity of visual cortical cells to moving gratings. J. Physiol. (Lond.) **198**, 237—250 (1968).

— COOPER, G. F., ENROTH-CUGELL, C.: The spatial selectivity of the visual cells of the cat. J. Physiol. (Lond.) **203**, 223—235 (1969).

— — ROBSON, J. G., SACHS, M. B.: The spatial selectivity of visual cells of the cat and the squirrel monkey. Proc. Physiol. Soc. (1969), J. Physiol. (Lond.) **204**, 120—121 (1969).

— KULIKOWSKI, J. J.: Orientational selectivity of the human visual system. J. Physiol. (Lond.) **187**, 437—445 (1966).

CHANG, H. T., CHIANG, C., WU, C.: Sci. Sinica (Peking) 8, 1131 (1959), Zit. nach MKRTYCHEVA, 1965.

Christensen, J. L., Hill, R. M.: A review of the anatomy and neurophysiology of the opossum (Didelphis virginiana) visual system. Amer. J. Ophthal. **46**, 440—446 (1969).

Cohen, B., Feldman, M.: Relationship of electrical activity in pontine reticular formation and lateral geniculate body to rapid eye movements. J. Neurophysiol. **31**, 806—817 (1968).

Collewijn, H.: Changes in visual evoked responses during the first phase of optokinetic nystagmus in the rabbit. Vision Res. **9**, 803—814 (1969).

Cooper, G. F., Robson, J. G.: Directionally selective movement detectors in the retina of the grey squirrel. J. Physiol. (Lond.) **186**, 116—117 P (1966).

Creutzfeldt, O. D., Ito, M.: Functional synaptic organization of primary visual cortex neurones in the cat. Exp. Brain Res. **6**, 324—352 (1968).

Cronly-Dillon, J. R.: Units sensitive to direction of movement in goldfish optic tectum. Nature (Lond.) **203**, 214—215 (1964).

— Pattern of retinotectal connections after retinal regeneration. J. Neurophysiol. **31**, 410—418 (1968).

— Galand, G.: Analyse des résponses visuelles unitaires dans le nerf optique et le tectum du triton, Triturus vulgaris. J. Physiol. (Paris) **58**, 502—513 (1966).

Crosby, E. C., Henderson, J. W.: The mammalian midbrain and isthmus regions. Part II: Fiber connections of the superior colliculus. J. comp. Neurol. **88**, 53—91 (1948).

Daniel, P. M., Kerr, O. I. B., Senevirtane, K. N., Whitteridge, D.: The topographical representation of the visual field on the lateral geniculate nucleus in the cat and monkey. J. Physiol. (Lond.) **159**, 87—88 (1961).

Daw, N. W.: Goldfish retina: Organization for simultaneous color contrast. Science **158**, 942—944 (1967).

Denney, D., Baumgartner, G., Adorjani, C.: Responses of cortical neurones to stimulation of the visual afferent radiation. Exp. Brain Res. **6**, 265—272 (1968).

DeValois, R.: Behavioral and electrophysiological studies in primate vision. In: Neff, W. L. (Ed.): Contributions to Sensory Physiology. New York: Academic Press 1965.

Dichgans, J., Körner, F., Voigt, K.: Vergleichende Skalierung des afferenten und efferenten Bewegungssehens beim Menschen: Lineare Funktionen mit verschiedener Anstiegssteilheit. Psychol. Forsch. **32**, 277—295 (1969).

Ditchburn, R. W., Fender, D. H.: The stabilized retinal image. Optica Acta **2**, 128—133 (1955).

— — Mayne, S.: Vision with controlled movements of the retinal image. J. Physiol. (Lond.) **145**, 98—107 (1959).

— Foley-Fisher, J. A.: Assembled data in eye movements. Optica Acta **14**, 113—118 (1967).

— Ginsborg, B. L.: Vision with a stabilized retinal image. Nature (Lond.) **170**, 36—37 (1952).

— — Involuntary eye movements during fixation. J. Physiol. (Lond.) **119**, 1—17 (1953).

Donner, K. O., Reuter, T.: Visual adaptation of the rhodopsin rods in the frog's retina. J. Physiol. (Lond.) **199**, 59—87 (1968).

— Rushton, W. A. H.: Rod-cone interaction in the frog's retina analyzed by the Stiles-Crawford-effect and by dark adaptation. J. Physiol. (Lond.) **149**, 303—317 (1959).

Dow, B. M., Dubner, R.: Visual receptive fields and responses to movement in an association area of cat cerebral cortex. J. Neurophysiol. **32**, 773—783 (1969).

— — Single-unit responses to moving visual stimuli in middle suprasylvian gyrus of the cat. J. Neurophysiol. **34**, 47—55 (1971).

Dowling, J. E.: Organization of vertebrate retina. Invest. Ophthalm. **9**, 655—680 (1970).

— Synaptic organization of the frog retina: An electron microscopic analysis comparing the retinas of frogs and primates. Proc. roy. Soc. B **170**, 205—228 (1968).

— Werblin, F. S.: Organization of the retina of the mudpuppy, Necturus maculosus. I. Synaptic structure. J. Neurophysiol. **32**, 315—338 (1969).

Dubner, R., Brown, F. J.: Response of cells to restricted visual stimuli in an association area of cat cerebral cortex. Exp. Neurol. **20**, 70—86 (1968).

Duensing, F.: Die Erregungskonstellation im Rautenhirn des Kaninchens bei den Labyrinthstellreflexen (Magnus). Naturwissenschaften **48**, 681—690 (1961).

— Schaefer, K.-P.: Die Neuronenaktivität in der Formatio reticularis des Rhombencephalon beim vestibulären Nystagmus. Arch. Psychiat. Nervenkr. **196**, 265—290 (1957).

DUENSING, F., SCHAEFER, K.-P.: Die Aktivität einzelner Neurone der Formatio reticularis des nicht gefesselten Kaninchens bei Kopfbewegungen und vestibulären Reizen. Arch. Psychiatr. Nervenkr. **201**, 97—122 (1960).
EIBL-EIBESFELDT, I.: Nahrungserwerb und Beuteschema der Erdkröte. Behavior **4**, 1—35 (1951).
ENROTH-CUGELL, CH, ROBSON, J. G.: The contrast sensitivity of retinal ganglion cells of the cat. J. Physiol. (Lond.) **187**, 517—552 (1966).
ERICKSON, R. A.: Visual search performance in a moving structured field. J. Opt. Soc. Amer. **54**, 355—405 (1964).
EWERT, J.-P.: Elektrische Reizung des retinalen Projektionsfeldes im Mittelhirn der Erdkröte (Bufo bufo L.). Pflügers Arch. ges. Physiol. **295**, 90—98 (1967).
— Aktivierung der Verhaltensfolge beim Beutefang der Erdkröte (Bufo bufo L.) durch elektrische Mittelhirnreizung. Z. vergl. Physiol. **54**, 455—481 (1967).
— Verhaltensphysiologische Untersuchungen zum „stroboskopischen Sehen" der Erdkröte (Bufo bufo L.). Pflügers Arch. ges. Physiol. **299**, 158—166 (1968).
— Quantitative Analyse von Reiz-Reaktionsbeziehungen bei visuellem Auslösen der Beutefang-Wendereaktion der Erdkröte (Bufo bufo L.). Pflügers Arch. ges. Physiol. **308**, 225—243 (1969).
EXNER, S.: Über das Sehen von Bewegung und die Theorie des zusammengesetzten Auges. S.-B. Akad. Wiss. Wien, mat.-nat. Kl., Abt. III **72**, 156—190 (1875).
EYSEL, U. T., GRÜSSER, O.-J.: Neurophysiological basis of pattern recognition in the cat's visual system. In: GRÜSSER, O.-J., KLINKE, R. (Hrsg.): Zeichenerkennung in technischen und biologischen Systemen. Berlin-Heidelberg-New York: Springer 1971.
FENDER, D. H.: Control mechanisms of the eye. Sci. Amer. **187**, 1—11 (1964).
— GILBERT, D. S.: Temporal and spatial filtering in the human visual system. Sci. Progr. **54**, 41—59 (1966).
FILEHNE, W.: Über das optische Wahrnehmen von Bewegungen. Z. Sinnesphysiol. **53**, 134—145 (1922).
FINKELSTEIN, D., GRÜSSER, O.-J.: Frog retina: detection of movement. Science **150**, 1050—1051 (1965).
— — REICH-MOTEL, H.: Reaktionen einzelner Retinaneurone des Frosches (Rana esculenta) auf bewegte Reize verschiedener Winkelgeschwindigkeit. Pflügers Arch. ges. Physiol. **283**, R 48—49 (1965).
FISCHER, M. H., KORNMÜLLER, A. E.: Optokinetisch ausgelöste Bewegungswahrnehmungen und optokinetischer Nystagmus. J. Psychol. Neurol. (Lpz.) **41**, 273—308 (1930).
FITE, K. V.: Single unit analysis of binocular neurons in the frog optic tectum. Exp. Neurol. **24**, 475—488 (1969).
FOERSTER, M. H., GRÜSSER, O.-J.: unpublished results (1969).
FOSTER, D. H.: The responses of the human visual system to moving spatially-periodic patterns. Vision Res. **9**, 577—590 (1969).
FRÖHLICH, F. W.: Grundzüge einer Lehre vom Licht- und Farbsinn. Jena: Gustav Fischer 1921.
— Die Empfindungszeit, S. 365. Jena: Gustav Fischer 1929.
FUSTER, J. M., CREUTZFELDT, O., STRASCHILL, M.: Intracellular recording of neuronal activity in the visual system. Z. vergl. Physiol. **49**, 605—622 (1965).
GAEDT, CH., GRÜSSER, O.-J.: The dependence of simultaneous contrast activation of retinal neurons on the temporal frequency of the stimuli. Unpubl. Study 1968.
GAREY, L. Y.: Interrelationship of the visual cortex and superior colliculus in the cat. Nature (Lond.) **207**, 1410—1411 (1965).
GAZE, R. M.: Binocular vision in frogs. J. Physiol. (Lond.) **143**, 20 P (1958a).
— The representation of the retina on the optic lobe of the frog. Quart. J. exp. Physiol. **43**, 209 (1958b).
— JACOBSON, M.: Convexity detectors in the frogs visual system. J. Physiol. (Lond.) **169**, 1—3 (1963a).
— — Types of single unit visual responses from different depths in the optic tectum of the goldfish. J. Physiol. (Lond.) **169**, 92—93 P (1963b).
— — A study of the retinotectal projection during regeneration of the optic nerve in the frog. Proc. roy. Soc. B **157**, 420—448 (1963c).

Gaze, R. M., Keating, M. J.: Receptive field properties of single units from the visual projection to the ipsilateral tectum in the frog. Quart. J. exp. Physiol. **55**, 143—152 (1970).

Gesteland, R. C., Howland, B., Lettvin, J. Y., Pitts, H.: Comments on microelectrodes. Proc. I.R.E. **47**, 1856—1862 (1959).

Godfraind, J. M., Meulders, M.: Effects de la stimulation sensorielle somatique sur le champs visuels de neurones de la région génuille chez le chat anaestesié au chloralose. Exp. Brain Res. **9**, 183—201 (1969).

— — Veraart, C.: Visual receptive fields of neurons in pulvinar, nucleus lateralis posterior and nucleus suprageniculatus thalami of the cat. Brain Res. **15**, 552—555 (1969).

Goodwin, H. E., Hill, R. M.: Receptive fields of marsupial visual system. I. The superior colliculus. Amer. J. Optom. **45**, 358—363 (1968).

Graham, C. H.: Perception of movement. In: Graham, C. H. (Ed.): Vision and visual perception, pp. 575—588. New York: John Wiley & Sons 1965.

Grauer, C.: Mikroelektrodenableitungen aus dem Tectum opticum vom Goldfisch während des optokinetischen Nystagmus. Unpublished Study 1967.

— Neuronale Adaptation in der Froschnetzhaut. Unpublished Study 1969.

Gross, C. G.: Visual functions of inferotemporal cortex. Handbook of Sensory Physiol. **7** (1970).

— Bender, B. D., Rocha-Miranda, C. E.: Visual receptive fields of neurons in the inferotemporal cortex of the monkey. Science **166**, 1303—1306 (1969).

Grüsser, O.-J.: Reaktionen einzelner corticaler und retinaler Neurone der Katze auf Flimmerlicht und ihre Beziehungen zur subjektiven Sinnesphysiologie. Med. Diss. Freiburg i. Br. 1956.

— A quantitative analysis of spatial summation of excitation and inhibition within the receptive field of retinal ganglion cells of cats. Vision Res. Suppl. **3** (1971) (in press).

— Creutzfeldt, O.: Eine neurophysiologische Grundlage des Brücke-Bartley-Effekts: Maxima der Impulsfrequenz retinaler und corticaler Neurone bei Flimmerlicht mittlerer Frequenzen. Pflügers Arch. ges. Physiol. **263**, 668—681 (1957).

— Dannenberg, H.: Eine Perimeter-Apparatur zur Reizung mit bewegten visuellen Mustern. Pflügers Arch. ges. Physiol. **285**, 373—378 (1965).

— Finkelstein, D.: Analyse eines auf „Bewegungswahrnehmung" spezialisierten Neuronensystems in der Froschnetzhaut. In: Kroebel, W. (Hrsg.): Fortschritte der Kybernetik, S. 83—95. München: R. Oldenbourg 1967.

— — Grüsser-Cornehls, U.: The effect of stimulus velocity on the response of movement-sensitive neurons of the frog's retina. Pflügers Arch. ges. Physiol. **300**, 49—66 (1968).

— Grüsser-Cornehls, U.: Mikroelektrodenuntersuchungen zur Konvergenz vestibulärer und retinaler Afferenzen an einzelnen Neuronen des optischen Cortex der Katze. Pflügers Arch. ges. Physiol. **270**, 227—238 (1960).

— — Reaktionsmuster einzelner Neurone im Geniculatum laterale und visuellem Cortex der Katze bei Reizung mit optokinetischen Streifenmustern. In: Jung, R., Kornhuber, H. (Hrsg.): Neurophysiologie und Psychophysik des visuellen Systems, S. 313—324. Berlin-Göttingen-Heidelberg: Springer 1961.

— — Demonstration des inhibitorischen Umfeldeffektes bei bewegungsspezifischen Neuronen der Froschnetzhaut *(Rana pipiens, Rana esculenta)*. Pflügers Arch. ges. Physiol. **291**, 86 R (1966).

— — Neurophysiologische Grundlagen visueller angeborener Auslösemechanismen beim Frosch. Z. vergl. Physiol. **59**, 1—24 (1968a).

— — Die Informationsverarbeitung im visuellen System des Frosches. In: Marko, H., Färber, G. (Hrsg.): Kybernetik 1968, S. 331—360. München: R. Oldenbourg 1968b.

— — Die Neurophysiologie visuell gesteuerter Verhaltensweisen bei Anuren. Verh. Zool. Gesell. **64**, 201—218 (1970).

— — Bullock, Th.: Functional organization of receptive fields of movement detecting neurons in the frog's retina. Pflügers Arch. ges. Physiol. **279**, 88—93 (1964).

— — Finkelstein, D., Henn, V., Patutschnick, M., Butenandt, E.: A quantitative analysis of movement-detecting neurons in the frog's retina. Pflügers Arch. ges. Physiol. **292**, 100—106 (1967).

GRÜSSER, O.-J., GRÜSSER-CORNEHLS, U., HAMASAKI, D.: Responses of neurons of the retina and the geniculate body of cats to moving visual stimuli. (1970; publication in preparation).
— — — The effect of different stimulus parameters on the response of movement and direction sensitive neurons in the visual association areas of the cat's brain. (1970; publication in preparation).
— — LICKER, M.: Further studies on the velocity function of movement-detecting class-2 neurons in the frog retina. Vision Res. 8, 1173—1185 (1968).
— — PATUTSCHNICK, M.: The contrast sensitivity of movement-detecting neurons in the frog's retina. Pflügers Arch. ges. Physiol. (1971, in preparation).
— — SAUR, G.: Reaktionen einzelner Neurone im optischen Cortex der Katze nach elektrischer Polarisation des Labyrinths. Pflügers Arch. ges. Physiol. **269**, 593—612 (1959).
— HENN, V.: Unpubl. Study 1968.
— REICH-MOTEL, H.: Neuronale Adaptation bewegungsspezifischer Neurone der Froschnetzhaut. Unpubl. Study 1965.
— REIDEMEISTER, CH.: Flimmerlichtuntersuchungen an der Katzenretina. II. Off-Neurone und Besprechung der Ergebnisse. Z. Biol. **111**, 254—270 (1959).
— SAUR, G.: Monokulare und binokulare Lichtreizung einzelner Neurone im Geniculatum laterale der Katze. Pflügers Arch. ges. Physiol. **271**, 595—612 (1960).
GRÜSSER-CORNEHLS, U.: Reaktionen bewegungsempfindlicher Neurone der Froschnetzhaut bei stroboskopischer Belichtung des Reizmusters. Pflügers Arch. ges. Physiol. **294**, 65 (1967).
— Response of movement-detecting neurons of the frog's retina to moving patterns under stroboscopic illumination. Pflügers Arch. **303**, 1—13 (1968a).
— GRÜSSER, O.-J.: Reaktionsmuster der Neurone im zentralen visuellen System von Fischen, Kaninchen und Katzen auf monokulare und binokulare Lichtreize. In: JUNG, R., KORNHUBER, H. (Hrsg.): Neurophysiologie und Psychophysik des visuellen Systems, S. 275—286. Berlin-Göttingen-Heidelberg: Springer 1961.
— — BULLOCK, TH.: Reaktionen einzelner Retinaneurone des Frosches *(Rana pipiens)* bei Reizung mit bewegten optischen Mustern. Pflügers Arch. ges. Physiol. **278**, 60—61 R (1963).
— — — Unit response in the frog's tectum to moving and non-moving visual stimuli. Science **141**, 820—822 (1963).
— HIMSTEDT, W.: Responses of retinal and tectal neurons of the salamander *(Salamandra salamandra)* to moving stimuli. (Brain Behav.Evol. in press 1972)
— LÜDCKE, M.: Vergleichende neurophysiologische Untersuchungen zur Signalverarbeitung in der Netzhaut von Anuren. Pflügers Arch. ges. Physiol. **319**, R 148 (1970).
— — A quantitative study of signal processing in the retina of the cuban tree-frog *(Hyla septentrionalis)*. Z. vergl. Physiol. (in preparation, 1971).
GUSELNIKOV, V. J., VODOLAZSKIJ, A. N.: Einige Angaben über den Sehanalysator der Taube (Detektoreigenschaften der Retina) (russ.). Nauč, dokl. vysš. školy. biol. nauki **9**, 45—52 (1968).
HABERICH, F. J., FISCHER, M. H.: Die Bedeutung des Lidschlags für das Sehen beim Umherblicken. Pflügers Arch. ges. Physiol. **267**, 626—635 (1958).
HAMDI, F. A., WHITTERIDGE, D.: The representation of the retina on the optic tectum of the pigeon. Quart. J. exp. Physiol. **39**, 111—119 (1954).
HAMILTON, CH., LUND, J. S.: Visual discrimination of movement midbrain or forebrain. Science **170**, 1428—1430 (1970).
HARTLINE, H. K.: The response of single optic nerve fibres of the vertebrate eye to illumination of the retina. Amer. J. Physiol. **121**, 400—415 (1938).
— The receptive fields of the optic nerve fibres. Amer. J. Physiol. **130**, 690—699 (1940a).
— The effects of spatial summation on the retina on the excitation of the fibres of the optic nerve. Amer. J. Physiol. **130**, 700—711 (1940b).
HARUTIUNIAN-KOZAK, B., KOZAK, W., DEC, K.: Single unit activity in the pretectal region of the cat. Acta Biol. Exp. (Warzawa) **28**, 333—343 (1968).
— — — BALCER, E.: Responses of single cells in the superior colliculus of the cat to diffuse light and moving stimuli. Acta Biol. Exp. (Warzawa) **28**, 317—331 (1968).

HEBBARD, F. W., MARG, E.: Physiological nystagmus in the cat. J. Opt. Soc. Amer. **50**, 151—155 (1960).

HELMHOLTZ, H.: Handbuch der physiologischen Optik. 1. Aufl. Leipzig: Voss 1867.

HENN, V., GRÜSSER, O.-J.: The summation of excitation in the receptive field of movement-sensitive neurons of the frog's retina. Vision Res. **9**, 57—69 (1969).

— REITER, H.: Die Erregungsintegration im excitatorischen rezeptiven Feld bewegungsspezifischer Retinaneurone des Frosches (Rana esculenta). Pflügers Arch. ges. Physiol. **289**, R 86—87 (1966).

HEUSSER, H.: Die Lebensweise der Erdkröte Bufo bufo (L.); das Orientierungsproblem. Rev. Suisse de Zool. **76**, 443—518 (1969).

HILL, R. M.: Unit response of the rabbit lateral geniculate nucleus to monochromatic light on the retina. Science **135**, 98—99 (1962).

— Receptive field properties of the superior colliculus of the rabbit. Nature (Lond.) **211**, 1401—1409 (1966).

HIMSTEDT, W.: Experimentelle Analyse der optischen Sinnesleistungen im Beutefangverhalten der einheimischen Urodelen. Zool. Jb. Physiol. **73**, 281—320 (1967).

— SCHALLER, F.: Versuche zu einer Analyse der Beutefangreaktionen von Urodelen auf optische Reize. Naturwissenschaften **53**, 619 (1966).

HIRSCH, H. V. B., SPINELLI, D. N.: Visual experience modifies distribution of horizontally and vertically oriented receptive fields in cats. Science **168**, 869—871 (1970).

HOLST, E. v., MITTELSTAEDT, H.: Das Reafferenzprinzip (Wechselwirkungen zwischen Zentralnervensystem und Peripherie). Naturwissenschaften **37**, 464—476 (1950).

HOLT, E. B.: Eye movement and central anesthesia. Psychol. Rev. Mon. Suppl. **4**, 3—45 (1903).

HONEGGER, H., SCHÄFER, W. D., JAEGER, W.: Untersuchungen über die Sehschärfe für bewegte Objekte. II. Vergleich von horizontaler und kreisförmiger Projektion der Sehzeichen. Albrecht v. Graefes Arch. Ophthal. **178**, 132—146 (1969).

HORN, G.: The effect of somaesthetic and photic stimuli on the activity of units in the striate cortex of unanesthetized, unrestrained cats. J. Physiol. (Lond.) **179**, 263—277 (1965).

— HILL, R. M.: Responsiveness to sensory stimulation of units in the superior colliculus and subjacent regions of the rabbit. Exp. Neurol. **14**, 199—223 (1966a).

— — Effect of removing the neocortex on the response to repeated sensory stimulation of neurons in the midbrain. Nature (Lond.) **211**, 754—755 (1966b).

— — Modifications of receptive fields of cells in the visual cortex occuring spontaneously and associated with bodily tilt. Nature (Lond.) **221**, 186—188 (1969).

HUBEL, D. H.: Integrative processes in central visual pathways of the cat. J. Opt. Soc. Amer. **53**, 58—66 (1963).

— Single unit activity in striate cortex of unrestrained cats. J. Physiol. (Lond.) **147**, 226—238 (1959).

— Transformation of information in the cat's visual system. In Vol. III, Proc. Int. Union Physiol. Sci. 1962, 160—169.

— WIESEL, T. N.: Receptive fields of single neurones in the cat's striate cortex. J. Physiol. (Lond.) **148**, 574—591 (1959).

— — Receptive fields of optic nerve fibres in the spider monkey. J. Physiol. (Lond.) **154**, 572—580 (1960).

— — Integrative action in the cat's lateral geniculate body. J. Physiol. (Lond.) **155**, 385—398 (1961).

— — Receptive fields, binocular interaction and functional architecture in the cat's visual cortex. J. Physiol. (Lond.) **160**, 106—154 (1962).

— — Shape and arrangement of columns in cat's striate cortex. J. Physiol. (Lond.) **165**, 559—568 (1963).

— — Receptive fields and functional architecture in two non-striate visual areas (18 and 19) of the cat. J. Neurophysiol. **28**, 229—289 (1965).

— — Receptive fields and architecture of monkey striate cortex. J. Physiol. (Lond.) **195**, 215—243 (1968).

— — Visual area of the lateral suprasylvian gyrus (Clare-Bishop area) of the cat. J. Physiol. (Lond.) **202**, 251—260 (1969).

HUBER, G. C., CROSBY, E. C., WOODBURNE, R. T., GILLIAN, L. A., BROWN, J. O., TAMTHAI, B.: The mammalian midbrain and isthmus region. Part I: The nuclear pattern. J. comp. Neurol. **78**, 129—534 (1943).

HUMPHREY, N. K.: The receptive fields of visual units in the superior colliculus of the rat. Proc. Physiol. Soc. **189**, 86—88 P (1967).

INGLE, D.: Visual releasers of prey-catching behavior in frogs and toads. Brain Behav. Evol. **1**, 500—518 (1968).

JACOBS, J. H.: Receptive fields in visual systems. Brain Res. **14**, 553—575 (1969).

JACOBSON, M.: The representation of the retina on the optic tectum of the frog. Correlation between retinotectal magnification factor and retinal ganglion cell count. Quart. J. exp. Physiol. **47**, 170—178 (1962).

— Spectral sensitivity of single units in the optic tectum of the goldfish. Quart. J. exp. Physiol. **49**, 384—393 (1964).

— GAZE, M. R.: Types of visual response from single units in the optic tectum and optic nerve of the goldfish. Quart. J. exp. Physiol. **49**, 199—209 (1964).

JAEGER, W., HONEGGER, H.: Untersuchungen über die Sehschärfe für bewegte Objekte. Albrecht v. Graefes Arch. Ophthalm. **166**, 601—616 (1964).

JASSIK-GERSCHENFELD, D.: Somesthetic and visual responses of the superior colliculus neurones. Nature (Lond.) **208**, 898—900 (1965).

— MINOIS, T., CONDÉ-COURTINE, F.: Receptive field properties of directionally selective units in the pigeon's optic tectum. Brain Res. **24**, 407—421 (1970).

JONES, B. H.: Responses of single neurons in cat visual cortex to a simple and a more complex stimulus. Amer. J. Physiol. **218**, 1102—1108 (1970).

JULESZ, B., PAYNE, R. A.: Differences between monocular and binocular stroboscopic movement perception. Vision Res. 8, 433—444 (1968).

JUNG, R.: Neuronal discharge. Electroenceph. clin. Neurophysiol., Suppl. **4**, 57—71 (1953).

— Nystagmographie: Zur Physiologie und Pathologie des optisch-vestibulären Systems beim Menschen. In: BERGMANN, G. v. (Hrsg.:) Handbuch der Inneren Medizin, 4. Aufl., Bd. V/1, S. 1325—1379. Berlin-Göttingen-Heidelberg: Springer 1953.

— Korrelationen von Neuronentätigkeit und Sehen. In: JUNG, R., KORNHUBER, H. (Hrsg.): Neurophysiologie und Psychophysik des visuellen Systems, S. 410—434. Berlin-Göttingen-Heidelberg: Springer 1961.

— BAUMGARTNER, G.: Neuronenphysiologie der visuellen und paravisuellen Rindenfelder. 8. Int. Congr. Neurol. Wien 1965, Symp. Proc. **3**, S. 47—75.

— BAUMGARTNER, R. v., BAUMGARTNER, G.: Mikroableitungen von einzelnen Nervenzellen im optischen Cortex der Katze: die lichtaktivierten B-Neurone. Arch. Psychiat. Nervenkr. **189**, 521—539 (1952).

KALIL, R. E., CHASE, R.: Corticofugal influence on activity of lateral geniculate neurons in the cat. J. Neurophysiol. **33**, 459—474 (1970).

KANO, C.: Die Wirkung der anschaulichen Größenunterschiede auf die Bewegungsschwelle bei übereinstimmender Größe der gereizten Netzhaut-Areale. Psychol. Forsch. **33**, 242—253 (1970).

KAWAMURA, H., MARCHIAFAVA, P. L.: Excitability changes along visual pathways during eye tracking movements. Arch. ital. Biol. **106**, 141—156 (1968).

KEATING, M. J., GAZE, R. M.: Observations on the surround properties of the receptive fields of frog retinal ganglion cells. Quart. J. exp. Physiol. **55**, 129—142 (1970).

KEESEY, U. T.: Effects of involuntary eye movements on visual acuity. J. Opt. Soc. Amer. **50**, 769—774 (1960).

— Visibility of a stabilized target as a function of rate and amplitude of luminance variation. J. Opt. Soc. Amer. **55**, 1577 (1965).

— RIGGS, L. A.: Visibility of Mach bands with imposed motions of the retinal image. J. Opt. Soc. Amer. **52**, 719—720 (1962).

KENNEDY, J. L.: The natures and physiological basis of visual movement discrimination in animals. Psychol. Rev. **43**, 494—521 (1936).

— The effect of complete and partial occipital lobectomy upon the thresholds of visual real movement discrimination in the cat. J. gen. Psychol. **54**, 119—149 (1939a).

KNAPP, H., SCALIA, F., RISS, W.: The optic tracts of Rana pipiens. Acta neurol. scand. **41**, 325—355 (1965).

KÖRNER, F.: Die Geschwindigkeitsperzeption beim Bewegungssehen. Ber. dtsch. Ophthal. Ges. **69**, 569—1972 (1969).

— DICHGANS, J.: Bewegungswahrnehmung, optokinetischer Nystagmus und retinale Bildwanderung. Der Einfluß visueller Aufmerksamkeit auf zwei Mechanismen des Bewegungssehens. Albrecht v. Graefes Arch. klin. exp. Ophthal. **174**, 34—48 (1967).

KOFFKA, K.: Die Wahrnehmung von Bewegung. In: BETHE, A. u.a. (Hrsg.): Handbuch der normalen und pathologischen Physiologie, Bd. XII/2, S. 1156—1214. Berlin: Springer 1931.

KOMMERELL, G., THIELE, H.: Der optokinetische Kurzreiznystagmus. Albrecht v. Graefes Arch. klin. exp. Ophthal. **179**, 220—234 (1970).

KORNMÜLLER, A. E.: Eine experimentelle Anaesthesie der äußeren Augenmuskeln am Menschen und ihre Auswirkungen. J. Psychol. Neurol. (Lpz.) **41**, 354—366 (1931).

KOSTELYANYETS, N. B.: The influence of the speed of increment of the test-object upon the characteristics of the response of the ganglion off-cell of the frog's retina. Biofizika **157**, 1225—1228 (1964) (Russ.).

— Investigation of receptive off-fields of frog retina by means of dark moving stimuli. Zh. vyssh. nerv. Deyat. Pavlova **15**, 521—524 (1965) (Russ.).

KOZAK, W., RODIECK, R. W., BISHOP, P. O.: Responses of single units in lateral geniculate of nucleus of cat to moving visual patterns. J. Neurophysiol. **28**, 19—47 (1965).

KRAUSKOPF, J.: Effect of retinal image motion on contrast thresholds for maintained vision. J. Opt. Soc. Amer. **47**, 740—741 (1952).

— Effect of target oscillation on contrast resolution. J. Opt. Soc. Amer. **52**, 1306 (1962).

KUFFLER, S. W.: Discharge patterns and functional organization of mammalian retina. J. Neurophysiol. **16**, 37—68 (1953).

LATIES, A. M., SPRAGUE, J. M.: The projection of optic fibers to visual centers of the cat. J. comp. Neurol. **127**, 35—70 (1966).

LATOUR, P. L.: Visual threshold during eye movements. Vision Res. **2**, 261—262 (1962).

LAZAR, G., SZEKELY, G.: Golgi studies on the optic center of the frog. J. Hirnforsch. **9**, 329—344 (1967).

— — Distribution of optic terminals in the different optic centres of the frog. Brain Res. **16**, 1—14 (1969).

LETTVIN, J. Y., MATURANA, H. R., MCCULLOCH, W. S., PITTS, W. H.: What the frog's eye tells the frog's brain. Proc. I.R.E. **47**, 1940—1951 (1959).

— — PITTS, W. H., MCCULLOCH, W. S.: Two remarks on the visual system of the frog. In: ROSENBLITH, W. (Ed.): Sensory Communication, pp. 757—776. Cambridge: M.I.T. Press 1961.

LEVICK, W. R.: Receptive fields of rabbit retinal ganglion cells. Amer. J. Optom. **42**, 337—343 (1965).

— Receptive fields and trigger features of ganglion cells in the visual streak of the rabbit's retina. J. Physiol. (Lond.) 188, 285—307 (1967).

— OYSTER, C. W., TAKAHASHI, E.: Rabbit lateral geniculate nucleus: sharpener of directional information. Science **165**, 712—714 (1969).

LICKER, M.: Changes in receptive field organization of movement detecting neurons of frog retina dependent on adaptation level. Pflügers Arch. ges. Physiol. **294**, 64 (1967).

— PANTEN, B.: Unpubl. Study 1968.

LIPETZ, L. E., HILL, R. M.: Discrimination characteristics of the turtle's retinal ganglion cells. Experientia (Basel) **26**, 373—374 (1970).

LIT, A.: Visual acuity. Ann. Rev. Psychol. **19**, 27—54 (1968).

LÖMO, T., MOLLICA, A.: Activity of single units in the primary optic cortex in the unanesthetized rabbit during visual, acoustic, olfactory and painful stimulation. Arch. ital. Biol. **100**, 86—120 (1962).

LUDVIGH, E.: Visual acuity while one eye is viewing a moving object. Amer. Arch. Ophthal. **42**, 14—22 (1949).

LUND, R. D.: Terminal distribution in the superior colliculus of fibers originating in the visual cortex. Nature (Lond.) **204**, 1283—1285 (1964).

Lunkenheimer, H.-U., Grüsser, O.-J.: Nichtlineare Übertragungseigenschaften retinaler Neurone der Katze. Pflügers Arch. ges. Physiol. **291**, 88 (1966).

MacKay, D.M.: Perceptual stability of stroboscopically lit visual field containing self-luminous objects. Nature (Lond.) **181**, 501—508 (1958).

— Elevation of visual threshold by displacement of retinal image. Nature (Lond.) **225**, 90—92 (1970).

Mandl, G.: Localization of visual patterns by neurons in cerebral cortex of the cat. J. Neurophysiol. **33**, 812—826 (1970).

Marchiafava, P.L., Pepeu, G.: The responses of units in the superior colliculus of the cat to a moving stimulus. Experientia (Basel) **22**, 51—53 (1966).

— — Electrophysiological study of tectal responses to optic nerve volley. Arch. ital. Biol. **104**, 406—420 (1966).

Marg, E., Adams, J.E.: Evidence for a neurological zoom system in vision from angular changes in some receptive fields of single neurons with changes in fixation distance in the human visual cortex. Experientia (Basel) **26**, 270—272 (1970).

Mashhour, M.: Psychophysical relations in the perception of velocity. Acta Universitatis Stockholmiensis, Stockholm Studies in Psychology, Vol. 3, p. 176. Stockholm: Almquist & Wiksell 1964.

Maturana, H.R.: Number of fibers in the optic nerve and the number of ganglion cells in the retina of Anurans. Nature (Lond.) **183**, 1406 —1407 (1959).

— Functional organization of the pigeon retina. Proc. 22nd int. Cong. Physiol. Sci. **3**, 170—178 (1962).

— Frenk, S.: Directional movement and horizontal edge detectors in the pigeon retina. Science **142**, 977—979 (1963).

— Lettvin, J.Y., McCulloch, W.S., Pitts, W.H.: Physiological evidence that cut optic nerve fibres in the frog regenerate to their proper place in the tectum. Science **130**, 1709 (1959).

— — — — Anatomy and physiology of vision in the frog (Rana pipiens). J. gen. Physiol. **43**, 129—175 (1960).

McGill, J.L.: Organization within the central and centrifugal fibre pathway in the avian visual system. Nature (Lond.) **204**, 395—396 (1964).

McIlwain, J.T., Buser, B.: Receptive fields of single cells in the cat's superior colliculus. Exp. Brain Res. **5**, 314—325 (1968).

— Creutzfeldt, O.D.: Microelectrode study of synaptic excitation and inhibition in the lateral geniculate nucleus of the cat. J. Neurophysiol. **30**, 1—21 (1967).

— Fields, H.L.: Superior colliculus: Single unit responses to stimulation of visual cortex in the cat. Science **170**, 1426—1428 (1970).

Meulders, M., Godfraind, J.M.: Influence du reveil d'origine réticulaire sur l'étendue des champs visuels des neurones de la région genuillée chez le chat avec cerveau intact ou avec cerveau isolé. Exp. Brain Res. **9**, 201—220 (1969).

Meyer, D.L., Schott, D., Schaefer, K.-P.: Reizversuche im Tectum opticum frei schwimmender Kabeljaue bzw. Dorsche (Gadus morrhua L.). Pflügers Arch. ges. Physiol. **314**, 240—252 (1970).

Michael, C.R.: Receptive fields of directionally selective units in the optic nerve of ground squirrel. Science **152**, 1092—1094 (1966).

— Receptive fields of opponent color units in the optic nerve of the ground squirrel. Science **152**, 1095—1096 (1966).

— Receptive fields of single optic nerve fibres in a mammal with an all-cone retina. II. Directionally selective units. J. Neurophysiol. **31**, 257—261 (1968).

— Visual response properties and functional organization of cells in the superior colliculus of the ground squirrel. Vision Res. Suppl. 3 (in press, 1971).

Miller, I.W., Ludwigh, E.: The effect of relative motion on visual acuity. Invest. Ophthal. **7**, 83—116 (1962).

Minkowski, M.: Experimentelle Untersuchungen über die Beziehungen der Großhirnrinde und der Netzhaut zu den primären optischen Zentren, besonders zum Corpus geniculatum externum. Arb. hirnanat. Inst. Zürich **7**, 259—362 (1913).

MITRANI, L., MATEFF, ST., YAKIMOFF, N.: Smearing of the retinal image during voluntary saccadic eye movements. Vision Res. **10**, 405—410 (1970).

— — — Temporal and spatial characteristics of visual suppression during voluntary saccadic eye movements. Vision Res. **10**, 417—422 (1970).

MKRTYCHEVA, L. J.: Elements of the functional organization of the visual system in the frog. Zh, vyssh. nerv. Deyat. Pavlova **15**, 513 (1965); Amer. Fed. Proc. Transl. Suppl. **23**, (2), T 373—376 (1966).

MKRTYCHEVA, L. T., SAMSONOVA, V. G.: Sensitivity of neurons of the frog's tectum to changes in the intensity of light stimulus. Vision Res. **6**, 419—426 (1966).

MONTERO, V. M., BRUGGE, J. F.: Direction of movement as the significant stimulus parameter for some lateral geniculate cells in the rat. Vision Res. **9**, 71—88 (1969).

— ROBLES, L.: Oculomotor modulation on lateral geniculate nucleus cell responses. Vision Res. Suppl. 3 (in press, 1971).

MOORE, R. J., GOLDBERG, J. M.: Ascending projections of the inferior colliculus in the cat. J. comp. Neurol. **121**, 109—136 (1963).

MORGAN, C. T.: The visual discrimination of real movement in the cat. Psychol. Bull. **34**, 519—523 (1937).

MUNTZ, W. R. A.: Microelectrode recordings from the diencephalon of the frog (Rana pipiens) and a blue-sensitive system. J. Neurophysiol. **32**, 699—711 (1962).

NORTON, A. C., CLARK, G.: Effects of cortical and collicular lesions on brightness and flicker discrimination in the cat. Vision Res. **3**, 29—44 (1963).

NORTON, A. L., SPEKREIJSE, H., WAGNER, H. G., WOLBARSHT, M. L.: Responses to directional stimuli in retinal preganglionic units. J. Physiol. (Lond.) **206**, 93—107 (1970).

NYBERG-HANSEN, R.: The location and termination of tectospinal fibers in the cat. Exp. Neurol. **9**, 212—227 (1963).

ORBAN, G., WISSAERT, R., CALLENS, M.: Influence of brain stem oculomotor area stimulation on single unit activity in the visual cortex. Mathematical analysis of the results. Brain Res. **17**, 351—355 (1970).

OTSUKA, R., HASSLER, R.: Über Aufbau und Gliederung der corticalen Sehsphäre der Katze. Arch. Psychiat. Nervenkr. **203**, 212—234 (1962).

OYSTER, C. W.: The analysis of image motion by the rabbit retina. J. Physiol. (Lond.) **199**, 613—635 (1968).

— BARLOW, H. B.: Direction selective units in rabbit retina: Distribution of preferred directions. Science **155**, 841—842 (1967).

PALKA, J.: An inhibitory process influencing visual responses in a fibre of the ventral nerve cord of locusts. J. insect Physiol. **13**, 235—248 (1967).

PANTLE, A., SEKULER, R.: Contrast response of human visual mechanisms sensitive to orientation and direction of motion. Vision Res. **9**, 397—406 (1969).

PANTLE, A. J., SEKULER, R. W.: Velocity-sensitive elements in human vision: Initial psychophysical evidence. Vision Res. **8**, 445—450 (1968).

PARTRIDGE, L. D., BROWN, J. E.: Receptive fields of rat retinal ganglion cells. Vision Res. **10**, 455—461 (1970).

PATUTSCHNICK, M., GRÜSSER, O.-J.: Der Einfluß des Reiz-Hintergrundkontrastes auf die Aktivierung bewegungsspezifischer Neurone der Froschnetzhaut (Rana esculenta). Pflügers Arch. ges. Physiol. **291**, R 85 (1966).

PETTIGREW, J. D., NIKARA, T., BISHOP, P. O.: Responses to moving slits by single units in cat striate cortex. Exp. Brain Res. **6**, 333—390 (1968).

— — — Binocular interaction on single units in cat striate cortex: Simultaneous stimulation by single moving slit with receptive fields in correspondence. Exp. Brain Res. **6**, 391—410 (1968).

PICKERING, S.: The extremely long latency response from on-off-retinal ganglion cells: Relationship to dark adaptation. Vision Res. **8**, 383—387 (1968).

— VARJU, D.: Ganglion cells in the frog retina: Inhibitory receptive field and long latency response. Nature (Lond.) **215**, 545—546 (1967).

PRECHT, W., RICHTER, A., GRIPPO, J.: Responses of neurons in cat's abducens nuclei to horizontal angular acceleration. Pflügers Arch. **309**, 285—309 (1969).

PRITCHARD, E. M., HERON, W.: Small eye movement in the cat. Can. J. Psychol. **14**, 131—137 (1960).

RACKENSPERGER, W., REITER, H., WUTTKE, W., SNIGULA, F.: Die Reaktion einzelner Retinaneurone auf sinusförmige Leuchtdichtänderung. Pflügers Arch. ges. Physiol. **283**, R 50 (1965).

RAMON Y CAJAL, P.: Investigaciones de histologica comparado de los centros opticos de distinos vertebrados. Doctoral Thesis. Universidad de Zaragoza 1890. Zit. nach MATURANA et al. 1970.

RASMUSSEN, A. T.: Tractus tectospinalis in the cat. J. comp. Neurol. **63**, 501—525 (1936).

REICH-MOTEL, H., BUTENANDT, E.: Nicht photochemisch bedingte Adaptation in der Netzhaut von Fröschen (Rana esculenta). Pflügers Arch. ges. Physiol. **283**, R 28 (1965).

REIDEMEISTER, CH., GRÜSSER, O.-J.: Flimmerlichtuntersuchungen an der Katzenretina. I. und II. Z. Biol. **111**, 241—270 (1959).

REUTER, T.: Visual pigments and ganglion cell activity in the retina of tadpoles and adult frogs (Rana temporaria L.). Acta Zool. Fenn. **122**, 3—64 (1969).

REVZIN, A. M.: Unit responses to visual stimuli in the nucleus rotundus of the pigeon. Fed. Proc. **26**, 2238 (1969).

— A specific visual projection area in the hyperstriatum of the pigeon (Columba livia). Brain Res. **15**, 246—249 (1969).

RIESEN, A. H., AARONS, L.: Visual movement and intensity discrimination in cats after early deprivation of pattern vision. J. comp. physiol. Psychol. **52**, 142—149 (1959).

RIGGS, L., RATLIFF, A., CORNSWEET, J. C., CORNSWEET, T. N.: The disappearance of steadily fixated visual test objects. J. opt. Soc. Amer. **43**, 495—501 (1953).

RIGGS, L. A., TULUNAY, U.: Visual effects of varying the extent of compensation for eye movements. J. opt. Soc. Amer. **49**, 741—746 (1959).

RISOS, A.: Die visuelle Wahrnehmung während des optokinetischen Nystagmus des Menschen. Pflügers Arch. ges. Physiol. **283**, R 63 (1965).

RIZZOLATTI, G., TRADARDI, V., CAMARDA, R.: Unit responses to visual stimuli in the cat's superior colliculus after removal of the visual cortex. Brain Res. **24**, 336—339 (1970).

RODIECK, R. W., STONE, J.: Response of cat retinal ganglion cells to moving visual patterns. J. Neurophysiol. **28**, 819—832 (1965).

— — Analysis of receptive fields of cat retinal ganglion cells. J. Neurophysiol. **28**, 833—849 (1965).

ROYE, D. B.: Visual pathways in the frog as determined by the Guillery modification of the Nauta-Gygax technique. M.S. Thesis, University of Florida, Gainesville 1966, p.42.

RUBINSON, K.: Projection of the tectum opticum of the frog. Brain Behav. Evol. **1**, 529—561 (1968).

SAMSONOVA, V. G.: Functional organization of different types of neurons in the frog visual center. Zh. vyssh. nerv. Deyat. Pavlova **15**, 491 (1965); Übers. in Fed. Proc. Transl. Suppl. **22**, 384—388 (1965).

SANIDES, F., HOFFMANN, J.: Cyto- and myeloarchitecture of the visual cortex of the cat and of the surrounding integration cortices. J. Hirnforsch. **11**, 79—104 (1969).

SCALIA, F., KNAPP, H., HALPERN, M., RISS, W.: New observations on the retinal projection in the frog. Brain Behav. Evol. **1**, 324—353 (1968).

SCHAEFER, K.-P.: Mikroableitungen vom Tectum opticum. Proc. Int. Union Physiol. Sci. 1962, Vol. 1, part II, p. 496.

— Mikroableitungen im Tectum opticum des frei beweglichen Kaninchens. Arch. Psychiat. Nervenkr. **208**, 120—146 (1966a).

— Experimenteller Beitrag zum Problem des Bewegungssehens. Fortschr. Med. **84**, 65—68 (1966b).

— Neuronale Entladungsmuster im Tectum opticum des Kaninchens bei passiven und aktiven Eigenbewegungen. Arch. Psychiatr. Nervenkr. **209**, 101—125 (1967).

SCHEIBNER, J., BAUMANN, CH.: Properties of the frog's retinal ganglion cells as revealed by substitution of chromatic stimuli. Vision Res. **10**, 829—837 (1970).

SCHIPPERHEYN, J. J.: Contrast detection in frog's retina. Acta Physiol. Pharmacol. Neerl. **13**, 231—277 (1965).

SCHNEIDER, D.: Beitrag zu einer Analyse des Beute- und Fluchtverhaltens einheimischer Anuren. Biol. Zbl. **73**, 225 (1954).

SCHOBER, H., MUNKER, H., GRIMM, W.: Zur Erkennbarkeit bewegter Objekte: Dynamische Sehschärfe. Klin. Mbl. Augenheilk. **151**, 395—402 (1967).

SCHWASSMANN, J. O., KRUGER, L.: Organization of the visual projection upon the optic tectum of some freshwater fish. J. comp. Neurol. **124**, 113—126 (1965).

SEFTON, A. J.: The innervation of the lateral geniculate nucleus and anterior colliculus in the rat. Vision Res. **8**, 867—888 (1968).

— Properties of cells in the lateral geniculate nucleus. Vision Res. Suppl. 3 (in press, 1971).

SEKULER, R. W., RUBIN, E. L., CUSHMAN, W. H.: Selectivities of human visual mechanisms for direction of movement and contour orientation. J. opt. Soc. Amer. **58**, 1146—1150 (1968).

SIEGERT, P.: Raumorientierung „rindenblinder" Katzen. Pflügers Arch. ges. Physiol. **242**, 515—556 (1939).

SIMINOFF, R., SCHWASSMANN, H. O., KRUGER, L.: An electrophysiological study of the visual projection to the superior colliculus of the rat. J. comp. Neurol. **127**, 435—444 (1966).

— — — Unit analysis of the pretectal nuclear group in the rat. J. comp. Neurol. **130**, 329—342 (1968).

SMITH, K. U.: Visually controlled responses under conditions of stimulation associated with apparent movement in the cat. Psychol. Bull. **34**, 537—538 (1937).

— Visual discrimination in the cat: V. The postoperative effects of removal of the striate cortex upon intensity discrimination. J. gen. Psychol. **51**, 329—369 (1937).

— KAPPAUF, W. E.: A neurological study of apparent movement. J. gen. Psychol. **23**, 315—327 (1940).

SPINELLI, D. N.: Visual receptive fields in the cat's retina; complications. Science **152**, 1768—1769 (1966).

— Receptive field organization of ganglion cells in the cat's retina. Exp. Neurol. **19**, 291—315 (1967).

— BARRETT, T. W.: Visual receptive field organization of single units in the cat's visual cortex. Exp. Neurol. **24**, 76—98 (1969).

— WEINGARTEN, M.: Afferent and efferent activity in single units of the cats optic nerve. Exp. Neurol. **15**, 347—362 (1966).

SPRAGUE, J. M., MARCIAFAVA, P. L., RIZOLATTI, G.: Unit responses to visual stimuli in the superior colliculus of the unanesthetized mid.-pont. cat. Arch. ital. Biol. **106**, 169—193 (1968).

— MEIKLE, T. H., JR.: The role of the superior colliculus in visually guided behaviour. Exp. Neurol. **11**, 115—146 (1965).

ST. CYR, G. J., FENDER, D. H.: The interplay of drifts and flicks in binocular fixation. Vision Res. **9**, 245—265 (1969).

— — Non-linearities of the human oculomotor system: time delays. Vision Res. **9**, 1490—1503 (1969).

— — Non-linearities of the human oculomotor system: gain. Vision Res. **9**, 1235—1246 (1969).

STARR, A., ANGEL, R., YEATES, H.: Visual suppression during smooth following and saccadic eye movements. Vision Res. **9**, 195—197 (1969).

STERLING, P.: Receptive fields and synaptic organization in the superficial gray of the cat superior colliculus. Vision Res. Suppl. 3 (in press, 1971).

— WICKELGREN, B. G.: Visual receptive fields in the superior colliculus of the cat. J. Neurophysiol. **32**, 1—23 (1969).

STONE, J., FABIAN, M.: Specialized receptive fields of the cat's retina. Science **152**, 1277—1279 (1966).

STRASCHILL, M., HOFFMANN, K. P.: Response characteristics of movement-detecting neurons in pretectal region of the cat. Exp. Neurol. **25**, 165—176 (1969).

— — Functional aspects of localization in the cat's tectum opticum. Brain Res. **13**, 247—283 (1969).

— — Activity of movement sensitive neurons of the cat's tectum opticum during spontaneous eye movements. Exp. Brain Res. **11**, 318—326 (1970).

STRASCHILL, M., TAGHAVY, A.: Neuronale Reaktionen im Tectum opticum der Katze auf bewegte und stationäre Lichtreize. Exp. Brain Res. **3**, 353—367 (1967).

SVAETICHIN, G.: Spectral response curves from single cones. Acta physiol. scand. **39**, Suppl. 134, 17—46 (1965).

TAGHAVY, A., STRASCHILL, M.: Bewegungsneurone im Tectum opticum der Katze. Pflügers Arch. ges. Physiol. **289**, R 82 (1966).

TEUBER, H.-L.: Some observations on the superior colliculi of the cat by J. ALTMAN. In: JUNG, R., KORNHUBER, H. (Hrsg.): Neurophysiologie und Psychophysik des visuellen Systems, S. 217—221. Berlin-Göttingen-Heidelberg: Springer 1961.

— Perception. In: Handbook of Physiology, Sect. Neurophysiology 1/Vol. III, p. 1595—1686.

THOMSON, L. C.: The localization of function in the rabbit retina. J. Physiol. (Lond.) **119**, 191—209 (1953).

VAN NES, F. L.: Enhanced visibility — by regular motion of retinal images. Amer. J. Physiol. **81**, 367—374 (1968).

— KOENDERINK, J. T., NAS, H., BOUMAN, M. A.: Spatiotemporal modulation transfer in the human eye. J. opt. Soc. Amer. **57**, 1082—1083 (1967).

WAGNER, H. G., MACNICHOL, E. F., WOLGARSHT, M. L.: Functional basis for "one-centre" and "off-centre" receptive fields in the retina. J. opt. Soc. Amer. **53**, 66—70 (1963).

WERBLIN, F.: Responses of retinal cells to moving spots. J. Neurophysiol. **33**, 342—350 (1970).

WERBLIN, F. S., DOWLING, J. E.: Organization of the retina of the mudpuppy. II. Intracellular recording. J. Neurophysiol. **32**, 339—355 (1969).

WERTHEIMER, M.: Experimentelle Studien über das Sehen von Bewegung. Z. Psychol. **61**, 161—265 (1912).

WHITESIDE, T. C. D., GRAYBIEL, A., NIVEN, J. I.: Visual illusion of movement. Brain **88**, 193—210 (1965).

WICKELGREN, B., STERLING, P.: Receptive fields in cat superior colliculus. Physiologist **10**, 344 (1967).

WIESEL, T. N.: Receptive fields of ganglion cells in the cat's retina. J. Physiol. (Lond.) **153**, 583—594 (1960).

WILSON, M. E.: Cortico-cortical connexions of the cat visual areas. J. Anat. **102**, 375—386 (1968).

WRIGHT, M. J.: Visual receptive fields cells in a cortical area remote from the striate cortex in the cat. Nature (Lond.) **223**, 973—975 (1969).

WURTZ, R. H.: Comparison of effects of eye movements and stimulus movements on striate cortex neurons of the monkey. J. Neurophysiol. **32**, 987—994 (1969).

— Response of striate cortex neurons to stimuli during rapid eye movements in the monkey. J. Neurophysiol. **32**, 975—986 (1969).

— Visual receptive fields of striate cortex neurons in awake monkeys. J. Neurophysiol. **32**, 727—742 (1969).

— Visual cortex neurons: response to stimuli during rapid eye movements. Science **162**, 1148—1150 (1968).

— GOLDBERG, M. E.: Superior colliculus cell responses related to eye movements in awake monkeys. Science **171**, 82—83 (1971).

WUTTKE, W., GRÜSSER, O.-J.: The conduction velocity of lateral inhibition in the cat's retina. Pflügers Arch. **304**, 253—257 (1968).

YARBUS, A. L.: Eye movements and vision. New York: Plenum Press 1967.

ZUBER, B. L., STARK, L.: Saccadic suppression: Elevation of visual threshold associated with saccadic eye movements. Exp. Neurol. **16**, 65—79 (1966).

References 1970 – 1972

The article was finished and sent to the editor in December 1970. On the following pages are references given for literature published between 1970 and April 1972. These references are mentioned in the text only if the authors did send us 1970 unpublished manuscripts.

Papers Published 1970—1972

BACH-Y-RITA, P., COLLINS, C.C., HYDE, J.E.: The control of eye movements, 560 p. New York: Academic Press 1971.

BARMACK, N.H.: Dynamic visual acuity as an index of eye movement control. Vision Res. **10**, 1377—1391 (1970).

— Modification of eye movements by instantaneous changes in the velocity of visual targets. Vision Res. **10**, 1431—1441 (1970).

BARTLETT, J.R., DOTY, R.W., CHOUDHURY, B.P.: Modulation of unit activity in striate cortex of squirrel monkeys by stimulation of reticular formation. Fed. Proc. **29**, (1970).

BELEKHOVA, M.G., KOSAREVA, A.A.: Organization of the turtle thalamus: visual, somatic and tectal zones. Brain, Behav. Evol. **4**, 337—375 (1971).

BISHOP, P.O., COOMBS, J.S., HENRY, G.H.: Responses to visual contours: Spatio-temporal aspects of excitation in the receptive fields of simple striate neurones. J. Physiol. (Lond.) **219**, 625—657 (1971).

— — — Interaction effects of visual contours on the discharge frequency of simple striate neurones. J. Physiol. (Lond.) **219**, 659—687 (1971).

— HENRY, G.H., SMITH, C.J.: Binocular interaction fields of single units in the cat striate cortex. J. Physiol. (Lond.) **216**, 39—68 (1971).

BIZZI, E., SCHILLER, P.H.: Single unit activity in the frontal eye fields of unanesthetized monkeys during eye and head movement. Exp. Brain Res. **10**, 151—158 (1970).

BRANDT, T., DICHGANS, J., KÖNIG, E.: Perception of self-rotation (circular-vection) induced by optokinetic stimuli. Pflügers Arch. **322**, R 98 (1972).

CHOW, K.L., MASLAND, R.H., STEWART, D.L.: Receptive field characteristics of striate cortical neurons in the rabbit. Brain Res. **33**, 337—352 (1971).

CHRISTENSEN, J.L., HILL, R.M.: Receptive fields of single cells of a marsupial visual cortex of Didelphis virginiana. Experientia (Basel) **26**, 43 (1970).

CHUNG, S.-H., RAYMOND, S, A., LETTVIN, J.Y.: Multiple meaning in single visual units. Brain Behav. Evol. **3**, 72—101 (1970).

COLLEWIJN, H.: The normal range of horizontal eye movements in the rabbit. Exp. Neurol. **28**, 132—143 (1970).

— An analog model of the rabbit's optokinetic system. Brain Res. **36**, 71—88 (1972).

— Latency and gain of the rabbit's optokinetic reactions to small movements. Brain Res. **36**, 59—70 (1972).

CORAZZA, R., LOMBROSO, C.T.: The neuronal dark discharge during eye movements in awake encephale isolé cats. Brain Res. **34**, 345—359 (1971).

CREUTZFELDT, O., PÖPPL, E., SINGER, W.: Quantitativer Ansatz zur Analyse der funktionellen Organisation des visuellen Cortex (Untersuchungen an Primaten). In: O.-J. GRÜSSER, KLINKE, R. (Hrsg.): Zeichenerkennung durch biologische und technische Systeme, S. 81—96. Berlin-Heidelberg-New York: Springer 1971.

DOWLING, J.E., WERBLIN, F.S.: Synaptic organization of the vertebrate retina. Vision Res. Suppl. **3**, 1—15 (1971).

EBBESSON, S.O.E.: On the organization of central visual pathways in vertebrates. Brain Res. Evol. **3**, 178—194 (1970).

ECKMILLER, R., GRÜSSER, O.-J.: Electronic simulation of the velocity function of movement-detecting neurons. Proceedings from: Cerebral Control of Eye Movements and Perception of Motion in Space (Freiburg 1971). Bibl. ophthal. (Basel) **83**, 486—489 (1972).

EFRON, R., LEE, D.N.: The visual persistence of a moving stroboscopically illuminated object. Amer. J. Psychol. **84**, 365—376 (1971).

ERKE, H., GRÄSER, H.: Reversibility of perceived motion: Selective adaptation of the human visual system to speed, size and orientation. Vision Res. **12**, 69—87 (1972).

EWERT, J.-P.: Aufnahme und Verarbeitung visueller Information im Beutefang- und Fluchtverhalten der Erdkröte Bufo bufo (L.). (Information processing in the visual system of toads and the release of prey catching and escape reactions). Sonderdruck aus Verhandlungsbericht der Deutschen Zoologischen Gesellschaft, 64. Tagung, S. 218—226 (1970).

— Single unit response of the toad's (Bufo americanus) caudal thalamus to visual objects. Z. vergl. Physiol. **74**, 81—102 (1971).

EWERT, J.-P., BORCHERS, H.-W.: Reaktionscharakteristik von Neuronen aus dem Tectum opticum und Subtectum der Erdkröte Bufo bufo (L.). Z. vergl. Physiol. **71**, 165–189 (1971).

EYSEL, U.TH., GRÜSSER, O.-J.: Neurophysiological basis of pattern recognition in the cat's visual system. In: GRÜSSER, O.-J., KLINKE, R. (Hrsg.): Zeichenerkennung durch biologische und technische Systeme, S. 60–80. Berlin-Heidelberg-New York: Springer 1971.

FELDMAN, J.D., GAZE, R.M., KEATING, M.J.: Delayed innervation of the optic tectum during development in Xenopus laevis. Brain Res. **14**, 16–23 (1971).

FELDON, S., FELDON, P., KRUGER, L.: Topography of the retinal projection upon the superior colliculus of the cat. Vision Res. **10**, 135–143 (1970).

FOSTER, D.H.: A model of the human visual system in its response to certain classes of moving stimuli. Kybernetik **8**, 69–84 (1971).

— The response of the human visual system to moving spatially-periodic patterns: further analysis. Vision Res. **11**, 57–81 (1971).

GERRITS, H.J.M., VENDRIK, A.J.H.: Artificial movements of a stabilized image. Vision Res. **10**, 1443–1456 (1970).

GOULD, J.D., PEEPLES, D.R.: Eye movements during visual search and discrimination of meaningless, symbol and object patterns. J. exp. Psychol. **85**, 51–55 (1970).

GROSS, C.G., ROCHA-MIRANDA, C.E., BENDER, D.B.: Visual properties of neurons in inferotemporal cortex of the Macaque. J. Neurophysiol. **35**, 96–111 (1972).

GRÜSSER, O.-J.: Metacontrast and the perception of the visual world. Pflügers Arch. **333**, R 98 (1972).

— GRÜSSER-CORNEHLS, U.: Interaction of vestibular and visual inputs in the visual system. Progr. Brain Res. (in press) (1972).

— — Comparative physiology of movement-detecting neuronal systems in lower vertebrates. (Anura and Urodela). Bibl. ophthal. (Basel) **83**, 456–464 (1972).

— — HAMASAKI, D.: Proc. XXV, Int. Congr. Physiol. Sci., München, Vol. 2, 652 (1971).

GRÜSSER-CORNEHLS, U.: Bewegungsempfindliche Neuronensysteme im visuellen System von Amphibien. Eine vergleichende neurophysiologische Untersuchung. Nova Acta Leopoldina (Halle); (in press, 1972).

HENN, U., COHEN, B.: Einzelzellableitungen aus Augenmuskelmotoneuronen während Sakkaden und Fixationsperioden bei wachen Rhesusaffen. Pflügers Arch. **333**, R 95 (1972).

HERMANN, H.T.: Saccade correlated potentials in optic tectum and cerebellum of Carassius auratus. Brain Res. **26**, 293–304 (1971).

HOFFMANN, K.-P., STONE, J.: Conduction velocity of afferents to cat visual cortex; a correlation with cortical receptive field properties. Brain Res. **32**, 460–466 (1971).

HONEGGER, H., ALEXANDRIDIS, E.: Sehschärfe für bewegte Objekte in Abhängigkeit von der Adaptation. Albrecht v. Graefes Arch. klin. exp. Ophthal. **181**, 1–11 (1970).

INGLE, D.J.: Brain mechanisms and vision: subcortical systems. Science **168**, 1493–1494 (1970).

— Prey-catching behavior of anurans toward moving and stationary objects. Vision Res. **3**, 447–456 (1971).

JONES, B.H.: Responses of single neurons in cat visual cortex to a simple and a more complex stimulus. Amer. J. Physiol. **218**, 1102–1107 (1970).

KADOYA, S., WOLIN, L.R., MASSOPUST, L.C.: Collicular unit responses to monochomatic stimulation in squirrel monkey. Brain Res. **32**, 251–254 (1971).

— — MASSOPUST, L.C., JR.: Photically evoked unit activity in the tectum opticum of the squirrel monkey. J. comp. Neurol. **142**, 495–508 (1971).

KANO, C.: Die Wirkung der anschaulichen Größenunterschiede auf die Bewegungsschwelle bei übereinstimmender Größe der gereizten Netzhaut-Areale. Psychol. Forsch. **33**, 242–253 (1970).

KEATING, M.J., GAZE, R.M.: Rigidity and plasticity in the amphibian visual system. Brain Behav. Evol. **3**, 102–120 (1970).

KRUGER, L.: The topography of the visual projection to the mesencephalon: A comparative survey. Brain Behav. Evol. **3**, 169–177 (1970).

MACKAY, D.M.: Elevation of visual threshold by displacement of retinal image. Nature (Lond.) **225**, 90–92 (1970).

— Mislocation of test flashes during saccadic image displacements. Nature (Lond.) **227**, 731–733 (1970).

MARG, E., ADAMS, J.E.: Evidence for a neurological zoom system in vision from angular changes in some receptive fields of single neurons with changes in fixation distance in the human visual cortex. Experientia (Basel) **26**, 270–271 (1970).

MASLAND, R.H., CHOW, K.L., STEWART, D.L.: Receptive-field characteristics of superior colliculus neurons in the rabbit. J. Neurophysiol. **34**, 148–156 (1971).

MCILWAIN, J.T.: Cortical origin of collicular directional selectivity in the cat. A review of the evidence. Brain Behav. Evol. **3**, 219–221 (1970).

— FIELDS, H.L.: Interactions of cortical and retinal projections on single neurons of the cat's superior colliculus. J. Neurophysiol. **34**, 763–772 (1971).

METHLING, D.: Über die unterschiedliche Erkennbarkeit von Landoltringöffnungen verschiedener Lage bei Augenfolgebewegungen. Vision Res. **10**, 543–548 (1970).

— Sehschärfe bei Augenfolgebewegungen in Abhängigkeit von der Gesichtsfeldleuchtdichte. Vision Res. **10**, 535–541 (1970).

MEULDERS, M.: Intégration centrale des afférences visuelles. J. Physiol. (Lond.) **62**, 61–109 (1970).

MICHAEL, CH.R.: Integration of retinal and cortical information in the superior colliculus of the ground squirrel. Brain Behav. Evol. **3**, 205–209 (1970).

— Visual response properties and functional organization of cells in the superior colliculus of the ground squirrel. Vision Res. Suppl. **11**, 299–308 (1971).

MONTERO, V.M., ROBLES, L.: Saccadic modulation of cell discharges in the lateral geniculate nucleus. Vision Res. Suppl. **11**, 253–268 (1971).

NODA, H., FREEMAN, R.B., GIES, B., CREUTZFELDT, O.D.: Neuronal responses in the visual cortex of awake cats to stationary and moving targets. Exp. Brain Res. **12**, 389–405 (1971).

NOTON, D., STARK, L.: Scanpaths in saccadic eye movements while viewing and recognizing patterns. Vision Res. **11**, 929–942 (1971).

PICKERING, S.G., VARJU, D.: The retinal On-Off components giving rise to the delayed response. Kybernetik 8, 145–150 (1971).

REVZIN, A.M.: Some characteristics of wide-field units in the brain of the pigeon. Brain Behav. Evol. **3**, 195–204 (1970).

RONCHI, L., SCANDIFFIO, E.: Is peripheral movement illusion mediated by directionally selective units? Atti Fond. G. Ronchi **25**, 855–865 (1970).

SCALIA, F., GREGORY, K.: Retinofugal projections in the frog: Location of the postsynaptic neurons. Brain Behav. Evol. **3**, 16–29 (1970).

SCHAEFER, K.P.: Unit analysis and electrical stimulation in the optic tectum of rabbits and cats. Brain Behav. Evol. **3**, 222–240 (1970).

— MEYER, D.L., SCHOTT, D.: Optic and vestibular influences on ear movements. Brain Behav. Evol. **4**, 323–333 (1971).

SCHICK, F., STRASCHILL, M.: Neuronale Reaktionen in einer visuellen Assoziationsarea der Katze während spontaner Augenbewegungen. Pflügers Arch. **333**, R 96 (1972).

SCHILLER, P.H., KOERNER, F.: Discharge characteristics of single units in superior colliculus of the alert rhesus monkey. J. Neurophysiol. **34**, 920–936 (1971).

SCHNEIDER, G.E.: Mechanisms of functional recovery following lesions of visual cortex or superior colliculus in neonate and adult hamsters. Brain Behav. Evol. **3**, 295–323 (1970).

SPINELLI, D.N., PRIBRAM, K.H., BRIDGEMAN, B.: Visual receptive field organization of single units in the visual cortex of monkey. Intern. J. Neuroscience **1**, 67–74 (1970).

SPRAGUE, J.M., BERLUCCHI, G., DI BERARDINO, A.: The superior colliculus and pretectum in visually guided behavior and visual discrimination in the cat. Brain Behav. Evol. **3**, 285–294 (1970).

STERLING, P.: Receptive fields and synaptic organization of the superficial gray layer of the cat superior colliculus. Vision Res. Suppl. **11**, 309–328 (1971).

— WICKELGREN, B. G.: Function of the projection from the visual cortex to the superior colliculus. Brain Behav. Evol. **3**, 210–218 (1970).

STRASCHILL, M., HOFFMANN, K. P.: Activity of movement sensitive neurons of the cat's tectum opticum during spontaneous eye movements. Exp. Brain Res. **11**, 318–326 (1970).

SZEKELY, G.: The mesencephalic and diencephalic optic centres in the frog. Vision Res. Suppl. **11**, 269–279 (1971).

WURTZ, R. H., GOLDBERG, M. E.: Superior colliculus cell responses related to eye movements in awake monkeys. Science **171**, 82–84 (1971).

ZEKI, S. M.: Convergent input from the striate cortex (Area 17) to the cortex of the superior temporal sulcus in the rhesus monkey. Brain Res. **28**, 338–340 (1971).

Chapter 7

Temporal Transfer Properties of the Afferent Visual System

Psychophysical, Neurophysiological and Theoretical Investigations

By

W. A. van de Grind, Amsterdam (Netherlands), O.-J. Grüsser, Berlin (Germany), and H.-U. Lunkenheimer, Lausanne (Switzerland)

With 53 Figures

Contents

I. Introduction

A. Scope of the Review

The afferent visual system (a.v.s.) includes all those stuctures of the retina and the brain which are mainly concerned with the *sensory functions* of vision and not with the polysensory integration of other modalities into the visual system or with visuo-motor control mechanisms. With this definition, the a.v.s. of vertebrates includes the optical apparatus of the eye, the retina, the pathways from the eye to the brain and in mammals the geniculate body, the primary visual cortex and part of the optic tectum. In the present report, we include also the description of the temporal response characteristics of neurons of the secondary and tertiary visual cortex. We do not give, however, a description of the special "feature extracting" properties of some of these cortical neuronal systems which are relevant to the analysis of visual patterns and which are described in another chapter of this volume (Stone and Freeman, Chap. 2, p. 153). Also excluded is the description of responses of tectal neurons, which are mentioned in two other chapters of this handbook (Sprague, Berlucchi, and Rizzolatti, 1973; Grüsser and Grüsser-Cornehls, 1973).

The function of the a.v.s. can be described as a filter for the chromatic, spatial and temporal properties of visual signals. The working range of this filter (intensity function) comprises about three ^{10}log-units, while the overall working range, including all adaptive mechanisms, extends from about 1 photon per receptor per minute to 10^{11} times this value. Above this level, the retina can be damaged by the energy of the absorbed light.

The present report deals mainly with the *temporal* filter functions; the spatial and the chromatic filter functions of the a.v.s. are only considered insofar as they interact with the temporal frequency characteristics of the system. It will be shown that the temporal filter functions of the a.v.s. can be described by means of linear system theory only under certain experimental restrictions. Within a wide range of stimulus conditions, the responses of the visual system to intermittent light stimuli exhibit non-linearities, which cannot be neglected. The best known non-linearity, for example, is the logarithmic relationship between stimulus luminance and brightness perception, which correlates with the logarithmic relationship between stimulus luminance and the average neuronal impulse frequency in neurophysiological experiments.

This review is divided into three parts:

1. Psychophysics of Visual Dynamics (Section II)

In Section II, a review is presented of the psychophysical findings concerning the dynamical characteristics of the visual system. Emphasis is on the study of responses to periodic light stimulation (flicker). We have tried to avoid too great an overlap with the excellent reviews in this field of Piéron (1961) and of J. L. Brown (1965), but in order to make the present study self-contained, we included a brief discussion of some of the previously covered data. Wherever possible we emphasized results obtained since 1964.

2. Neurophysiology of Visual Dynamics (Section III—VI)

In the second part of this paper a description and discussion of corresponding neurophysiological data obtained in single-neuron recordings from the retina, the optic tract, the geniculate body and the visual cortex of vertebrates are given. Neurophysiological data obtained with *macro*electrode recordings (sumpotentials) from the eye or other parts of the visual system are mentioned only if they provide additional important information for the interpretation of single unit data.

Whenever it was possible to correlate psychophysical or behavioral findings with neurophysiological data, this correlation is briefly mentioned in small print. Only very few microelectrode data are available from the visual system of man, so one has to rely on data obtained in animals. Neurophysiological results in higher mammals and primates can be correlated with psychophysical findings in humans only if one accepts the a priori hypothesis that the essential functional properties of the a.v.s. are highly similar in the human and in higher mammals. In addition, this correlation builds a bridge between two forms of experience which cannot be related with each other applying the category of causality. Nevertheless, we think that such a correlation between subjective perception and neuronal activity is very useful in understanding the function of neuronal systems. The mutual predictability of experimental results in both fields supports the pragmatic significance of this procedure, based on a positivistic theory of cognition.

3. Modelling Studies of Visual Dynamics (Section VII—X)

After a review of different models developed from the results of psychophysical experiments (Section VII), we try to analyse the contribution of the different elements of the retina (such as receptors, synapses, neurons) to the temporal transfer properties of the whole system (Section VIII—X). The overall temporal transfer characteristics of the visual system as found in psychophysical experiments can be deduced from the signal transmission properties of the several relay stations within the 5—7 neuron chain between the retinal receptors and the visual cortex. A system description of the whole a.v.s. will be given in a later report.

Of course, the present study is also incomplete to the extent that not all papers on flicker phenomena that have been published are summarized. The remarkable annotated bibliography on flicker by LANDIS (1953), covering the period 1740 to 1952, already contains 1149 references and many papers have appeared since. An important collection of additional material on the subject can further be found in the proceedings of a symposion on flicker held in Amsterdam in honour of H. DE LANGE: HENKES and VAN DER TWEEL (1964). The literature of the 19th and the beginning of the 20th century is well described by the handbook articles of FICK (1879), von HELMHOLTZ (1896), von KRIES (1905) and TSCHERMAK (1929).

B. Systems Theory as a Guide

One can define a system as a set of three sets, respectively called the set of input variables, the set of output variables and the set of systems equations, which describe the relations between input variables on the one hand and output variables on the other.

If all systems equations are linear, the system is called a linear system. In the psychophysical and neurophysiological experiments and theories of present interest, circumstances are usually chosen in such a way that a description can be given in terms of one equation or graph describing the relation between one input variable (the stimulus) and one output variable (the response), where the input and output variables are time functions. In neurophysiological experiments, it is normally possible to measure responses that are indeed time functions, but in psychophysical experiments the input time function is usually meant to generate only once a binary output function ("seen" or "not seen", "flicker" or "no flicker", etc.). One can only correlate these two research domains by *postulating* something about the transformation between a hypothetical internal time variable response and the actually measured binary output in psychophysical experiments. The latter transformation is performed by a hypothetical system called the central *detector* (Section VII A). Thus, if it is, for example, stated that some transfer function "is measured" psychophysically, this means that a transfer function is *deduced* from a set of input time functions, a corresponding set of binary output functions and postulates regarding the detector.

After a detector is postulated, the psychophysicist is in the same situation as the neurophysiologist in that he has a set of input time functions and a set of output time functions from which he wants to deduce an efficient system description. For a "badly" nonlinear system there are no fixed rules to solve this problem, whereas a linear system can be completely described from a few measurements with well chosen stimuli. In studying an unknown system it is therefore rewarding to start with the hypothesis that the system is linear. Stimuli that can (save for an additive constant and/or a constant scaling factor) be described by either of the following three functions, or some combination of them have been especially useful in research on sensory systems:

(a) The unit pulse function or Dirac-pulse which can be defined in somewhat sloppy terms as:

$$\delta(t) = \begin{Bmatrix} \infty, t = 0 \\ 0, t \neq 0 \end{Bmatrix} \qquad \text{and} \quad \int_{-\infty}^{+\infty} \delta(t)\, dt = 1\,. \tag{1}$$

(b) The unit step function which is defined as:

$$H(t) = \begin{Bmatrix} 1, t \geqq 0 \\ 0, t < 0 \end{Bmatrix}. \tag{2}$$

(c) The sine function:

$$\begin{aligned} x(t) &= \sin(\omega t + \varphi) \\ \omega &= \text{radian frequency} \\ \omega &= 2\pi f \\ f &= \text{frequency} \\ \varphi &= \text{phase at } t = 0\,. \end{aligned} \tag{3}$$

Since light intensity cannot be negative, the sine function with intensity-amplitude I on "background" $\bar{I}$ must satisfy the condition $I \leqq \bar{I}$ and the ratio $I/\bar{I}$, called the modulation depth m, then satisfies the condition $0 \leqq m \leqq 1$ (see also Fig. 1C).

With the techniques of Fourier analysis every periodic function can be written as a sum of sine waves (and every nonperiodic function as the integral over an infinite number of sine waves) where the "component" sine waves have characteristic amplitudes, frequencies and phases. The relation between the amplitude and frequency of the Fourier components is summarized in the amplitude frequency spectrum. The convenience of the Fourier analysis gives the sine wave an important place in the study of systems. For linear systems the superposition principle makes it possible to predict the response to every periodic or nonperiodic stimulus from the knowledge of the response to sinewaves of all frequencies. The principle of superposition states that for a linear system, the response to a linear combination of input signals is the same linear combination of responses to the component input signals. Furthermore, the superposition principle does not hold for nonlinear systems. Accordingly a check of its validity can be used to see whether the studied system is linear or not.

Another convenient test of the linearity hypothesis stems from the fact that the sinewave response of a linear system must again be a sinewave of the same frequency, but possibly of other amplitude and phase. In conclusion then, one has all the facts needed to describe a linear system (with its "transfer function"), if an amplitude-frequency and phase-frequency curve are measured. Alternatively, one can calculate these curves for a linear system from the response to a Dirac pulse or unit step function (see any book on systems theory).

Thus, systems theory can be used as a guide in studying neurobiological systems (see e.g. pp. 435 and 546). In the study of visual dynamics, however, much of the basic data were gathered before systems theoretical concepts were generally accepted or even known. Moreover, it was technically difficult to generate good sinewave stimuli, whereas it was easy to generate stimulus wave forms such as those shown in Fig. 1A and B. In the 19th century, rotating black and white

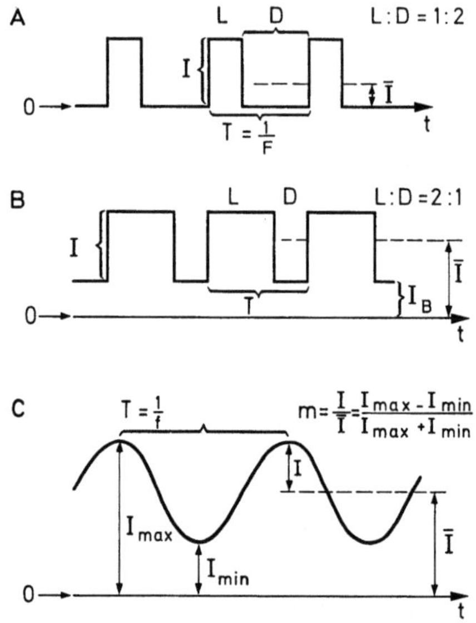

Fig. 1. *Parameters of rectangular (A, B) and sinusoidal (C) flicker stimuli.* Abscissa: time, ordinate: stimulus luminance or retinal stimulus intensity. L light period, D dark period, $\bar{I}$ average stimulus luminance, I_b background luminance, T period of the periodic rectangular or sinusoidal stimuli, I_{max} maximum stimulus luminance, I_{min} minimum stimulus luminance of the sinusoidal stimuli, I amplitude of the sinusoidal stimuli, m depth of modulation. L/T is often called the pulse-to-cycle fraction (PCF) or the light-time fraction. Unfortunately some authors have called it light-dark-ratio (e.g. LANDIS, 1954). Therefore note that we call L/D the light-dark-ratio

discs, either viewed as a whole, or viewed through a hole in a screen directly in front of the spinning discs, were the preferred stimulating devices. In the first half of the 20th century, many experiments were done with electrically or electronically controlled shutters that periodically interrupted a light beam.

A combination of rotating polarizer and fixed analyzer (e.g. KELLY, 1961a), electronically controlled cathode ray tubes or television projection tubes (VAN DER GON et al., 1958), television screens (e.g. VAN NES, 1968b), fluorescent tubes (SPERLING, 1965), modulated Xenon lamps (our experiments) and glow modulators (McDONALD, 1960) are among the more recent stimulating devices.

Despite the fact that it is possible in principle to summarize the classic flicker data obtained with stimulus wave forms like those of Fig. 1A and B ("rectangular" flicker wave form) in sinewave equivalences, the traditional descriptions are followed first, pp. 438–446. This offers just as sound a basis as sinewave data (presented next, pp. 446–467), but in addition it gives some insight into the way of thinking of the many brilliant experimenters working with rectangular flicker in the pre-sinewave age.

Part 1. Psychophysics of Visual Dynamics

II. Perceptual Phenomena Caused by Intermittent Stimulation of the Retina

A. The Critical Flicker Frequency and the Talbot-Plateau-Law

Since Greek antiquity, it is known that an intermittent light stimulus appears to be fused above a certain frequency of intermittence (PTOLEMAEUS of Alexandria, approx. 80–160 A.D., see GOVI, 1885; LEJEUNE, 1956). In psychophysical experiments, the *critical flicker frequency* (CFF) or *flicker fusion frequency* (FFF) represents the frequency of transition from the perception of flicker to steady light. This frequency can be measured with decreasing or increasing stimulus frequencies. Slightly different results might be obtained from measurements with increasing frequency in comparison with measurements with decreasing stimulus frequency. Small differences in results are also seen between the case in which the observer is allowed to find the CFF by himself, and that in which this is done by the investigator. For the general rules described in the following pages, however, the type of determination of the CFF only plays a subordinate role.

One of the first psychophysical measurements of the CFF was performed in the first half of the 18th century by J. SEGNER (1740) at the University of Göttingen. In his invitation to the medical doctoral dissertation of G. G. BIELCK, he showed that intermittent light stimuli appeared to be fused above a certain frequency. He explained this experimental result by the assumption that the fusion of intermittent light is caused by the persistence of the sensation after each light change. However, he used his data to speculate about some *inherent physical* properties of the light, the temporal distribution of photons in the light flux ("raritate luminis"). Since the early 19th century, especially since the work of PURKINJE (1819, 1823) and of PLATEAU (1830, 1834, 1835), the era of systematic investigations of the physiological properties of the eye and the visual system began. Most of the outstanding physiologists and physicists working in the field of vision also studied flicker. For an early review, see VON HELMHOLTZ (1896).

An intermittent light stimulus, having frequency components all of which exceed the CFF, will match in brightness a steady light which has the same time average luminance. This Talbot-Plateau-law (1834, 1835) suggests that above the CFF, the response of the visual system proximal to the location where fusion occurs is proportional to a steady stimulus of which the luminance I_m is equal to the average luminance $\bar{I}$ of the intermittent stimuli (c. f. Fig. 1):

$$I_m \to \bar{I} = \frac{1}{T} \int_0^T I(t)\, dt\,. \tag{4}$$

T is the period of the fundamental frequency of the flicker stimulus $I(t)$.

B. Determinants of the CFF for Stimuli with a Rectangular Wave Form

The CFF depends on the physical characteristics of the stimulus, the mode of presentation, the state of the visual system, experimental procedures and the general condition and personal characteristics of the subject. In the following sections, we concentrate on the first three factors and regard the influence of the others largely as "noise" or experimental variability.

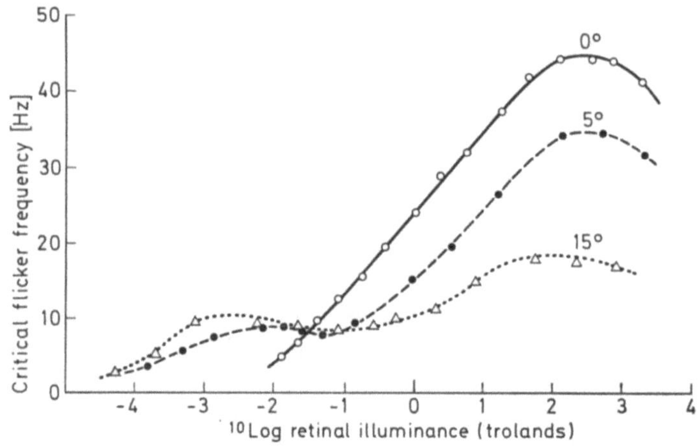

Fig. 2a. *Dependency of the psychophysical CFF (ordinate) on the stimulus intensity* (^{10}log trolands, abscissa) for circular stimulus patterns of 2 degrees diameter and different retinal localization (fovea, 5 and 15 degrees above the fovea, HECHT and VERRIJP, 1933)

1. Stimulus Luminance

At a given level of general adaptation, the CFF depends on the luminance (I) of the intermittent stimuli (BAADER, 1891):

$$\mathrm{CFF} = c_1 \log \frac{I}{I_s} + b_1\ [\mathrm{Hz}]\,, \tag{5}$$

where I_s is the increment threshold on background I_B (Fig. 1) and c_1 and b_1 are constants. This relationship is called the Ferry-Porter-law (1892–1902). The logarithmic relationship is approximately valid for three to four ^{10}log-units of I within the photopic stimulus range. At a higher stimulus intensity, the CFF changes only slightly. The validity of Eq. (5) was tested also with scotopic levels of stimulus intensity in the dark-adapted eye, whereby the intermittent testfields were located outside the fovea. The data of Fig. 2a exhibit a non-monotonic change of the intensity function of the CFF with a kink in the curve within the mesopic range of illumination. For the straight part of the scotopic intensity function of the CFF, the constant c_1 of Eq. (5) is significantly smaller (1.5–5 Hz/^{10}log-unit) than for photopic stimulus intensities (8–14 Hz/^{10}log-unit; PORTER, 1902; PIÉRON, 1922; HECHT, SHLAER and VERRIJP, 1933; HECHT and SMITH, 1936; CROZIER et al., 1937; Fig. 2a). The maximum CFF is reached with large and bright stimuli. Values between 80 and 86 Hz were found for stimuli smaller than 30 degrees (Allen, 1926; Hylkema, 1942c), but, with a stimulus field of 50 degrees diameter, Roehrig (1959a) found a maximum CFF of 107 Hz in a 28 year old subject.

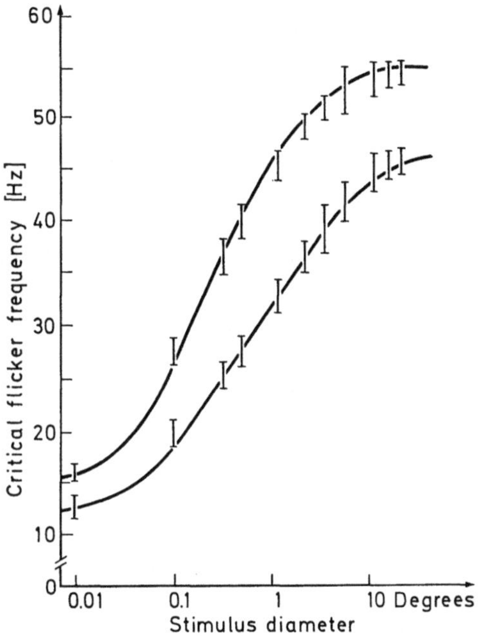

Fig. 2b. *Dependency of the CFF (ordinate) on the stimulus area* (abscissa) for circular stimuli of 2 different average luminances (*L*-*D*-ratio = 1 : 1). On the abscissa the diameter of the stimuli (degrees) is plotted. The data are from measurements made by different authors, combined by Landis (1954). The curves are calculated according to Eq. (13)

2. Stimulus Area

The CFF depends on the angular size of the intermittent stimuli (Fig. 2b, Granit and Harper, 1930; Landis, 1954):

$$\text{CFF} = c_2 \log A + b_2 \text{ [Hz]} \qquad (6)$$

where A is the stimulus area and c_2 and b_2 are constants. This relationship is usually called the Granit-Harper-law. These authors measured the validity of Eq. (6) for a circular stimulus with diameters between 0.98 and 5.0 degrees. An extension of the validity of Eq. (6) up to test field diameters of about 50 degrees was later found by Roehrig (1959). Eq. (6) was confirmed by Piéron (1935), Hecht and Smith (1936), Lloyd (1951), Berger (1953), Foley and Kazdan (1964).

c_2 in Eq. (6) depends also on the shape of the stimuli. As a rule, the CFF is smaller for annular stimuli than for light discs of the same area (Robinson, 1970). However, for a foveally seen testfield, the entire stimulus area is not always effective in determining the CFF. By systematically removing increasingly large central portions from testfields with sizes between 6.9 and 49.6 degrees, Roehrig

(1959b) showed that about 20 % of the smallest and 66 % of the largest testfield can be removed without changing the CFF. Hence, the CFF is not determined by total area but by those parts of the retina that have the highest flicker sensitivity (c.f. p. 443).

The increase of the CFF with increasing stimulus area can also be observed if a stimulus pattern with separated light spots is used (GRANIT, 1930a, b; GRANIT and HARPER, 1930). Small spots of 1 degree diameter, arranged symmetrically on an imaginary circle of 3 degrees and 4 degrees diameter respectively and observed 10 degrees lateral from the fovea, lead to a significantly higher CFF than is found for each spot alone. With foveal observation of the same pattern, the CFF increases only slightly. The same is true if the pattern is viewed with the retinal periphery and the stimulus luminance is high in the photopic range. For foveal vision, the CFF for two adjacent half-fields is significantly higher than the CFF for a single half-field of 2 degrees diameter. As the two half fields are more and more separated, the rise in the CFF for the double half-field relative to that of the single half-field decreases (0–10 degrees; GRANIT and HARPER, 1930). These observations on spatio-temporal summation are early psychophysical indications of spatial-interaction at fusion in the retinal neuronal network.

3. Interaction between Intensity and Area Effects

GRANIT and HARPER also investigated the combined effect of a change of the stimulus luminance and the stimulus area. They found that a multiplicative combination of Eqs. (5) and (6) has to be considered:

$$\mathrm{CFF} = k \log\frac{I}{I_s} \log A + k_1 \log\frac{I}{I_s} + k_2 \log A + b \ [\mathrm{Hz}] \,. \tag{7}$$

WEALE (1957a, b) proposed to reduce Eq. (7) to the following general expression:

$$\mathrm{CFF} = c_0 \log Q + b_0 \ [\mathrm{Hz}] \tag{8}$$

where Q is the number of quanta measured in threshold units:

$$Q = \frac{I}{I_s} \cdot A^p \,. \tag{9}$$

WEALE termed p the "index of spatial summation".

From Eqs. (8) and (9) one obtains:

$$\mathrm{CFF} = c_0 \log\frac{I}{I_s} + p \cdot c_0 \log A + b_0 \ [\mathrm{Hz}] \,. \tag{10}$$

In contrast to Eq. (7), Eq. (10) is the simple additive combination of Eqs. (5) and (6). It was first found to be valid within the fovea (maximum 40 min). WEALE, however, was able to show that when allowance is made for the non-homogeneity of the *cone population* in the parafoveal region of the retina and the approximate number of cones illuminated by the circular stimuli of different areas is considered, Eqs. (8) and (10) are valid up to a stimulus diameter of 7.5 degrees (foveal fixation, data from KUGELMASS and LANDIS, 1955).

Since the rods already have a maximum CFF around 20–22 Hz (p. 444), WEALE's calculations seem to be justified. However, LLOYD (1951) found a slightly higher CFF in a peripheral test area than in a foveal area that according to his estimates contained more cones. On the other hand, with small test fields

(0.2 degree diameter) the CFF decreases from the fovea towards the periphery. The CFF is then positively correlated with the local photopic *visual acuity*, i.e. with the reciprocal value of the diameters of the receptive field centers, which constitute the elementary summing areas. A further quantification of the summation-explanation incorporating ideas regarding the number of cones per summation area as a function of retinal location would be very helpful to find a better basis for the theoretical explanation of the experimental findings. A promising start of such a quantification of these ideas can be found in VAN DOORN et al. (1972).

As one can see from Fig. 2a and 2b, the logarithmic relationship of Eq. (5) through (10) is valid only for a limited range, and in most experiments S-shaped curves were obtained. A more unified description of the experimental data can be reached by the following equation:

$$\mathrm{CFF} = \frac{Q}{1 + k_i Q}\ [\mathrm{Hz}]\,. \tag{11}$$

From Eq. (9) and (11) one obtains:

$$\mathrm{CFF} = \frac{I A^p}{I_s + k_i A^p}\ [\mathrm{Hz}]\,. \tag{12}$$

Eq. (11) or (12) can be transformed into a logarithmic equation of the tanh (c.f. NAKA and RUSHTON, 1966; LIPETZ, 1971, this Handbook, Vol. I, p. 191):

$$\mathrm{CFF} = \frac{1}{2k_i}\left[1 + \tanh\left(\frac{1}{2}\ln k_i Q\right)\right]\ [\mathrm{Hz}]\,. \tag{13}$$

As Fig. 2b demonstrates, the experimental data are described over a wider range by Eq. (12) or (13) than by the simple logarithmic equation (10).

4. The Light-Dark Ratio

The light-dark ratio (defined in Fig. 1) of the flickering stimuli influences the CFF. Under the condition that the Talbot-Plateau-level is kept constant ("compensated stimuli"), the CFF increases with decreasing L-D-ratio to values of approximately 0.005 for the latter. For "uncompensated" flicker stimuli, in which the maximum intensity is kept constant and the light-dark ratio is varied, the maximum CFF is reached for medium stimulus intensities at a L-D-ratio between 1 : 1 and 1 : 3 (PORTER, 1912; PIÉRON, 1928, 1961; COBB, 1934b; BARTLEY, 1937; CROZIER and WOLF, 1941b; GALIFRET and PIÉRON, 1948; ERLICK and LANDIS, 1952; LANDIS, 1954). PORTER (1912) found that for uncompensated flicker stimuli the following equation fitted his experimental data well:

$$\mathrm{CFF} = a + k_3 \log\left[(T - D)\cdot D\right]\ [\mathrm{Hz}]\,. \tag{14}$$

We propose to call this the second law of PORTER. D and T are defined in Fig. 1. For very bright uncompensated flicker stimuli, however, the maximum CFF also increases with a decrease of the light-dark ratio down to 0.005 for an intensity of 10^6 trolands (LLOYD and LANDIS, 1960).

5. Steady Surround

The CFF depends on the contrast of the flicker field against a steadily illuminated surround. In this case, an increase of the surround luminance increases the CFF until the surround luminance reaches 60—100 % of the Talbot-Plateau-level of the intermittent stimuli. Under certain conditions, a log function may also describe this relationship (LYTHGOE and TANSLEY, 1929). A further increase of surround luminance decreases the CFF (FRY and BARTLEY, 1936; BERGER, 1954; RIPPS et al., 1961; HARVEY, 1970a). When the surround luminance is set to the Talbot-Plateau-level [(Eq. (4)], the area (A_b) of this steadily illuminated field surrounding a small foveal flickering test field also influences the CFF:

$$\mathrm{CFF} = k_4 \log A_b + b_4 \,[\mathrm{Hz}]\,, \tag{15}$$

k_4 was found to be about 0.1 for foveal vision (FOLEY, 1961). Data for the CFF measured with a modulated surround are discussed on p. 452.

6. Chromatic Effects

In the light-adapted eye, for monochromatic equal-energy flicker stimuli, the CFF correlates fairly well with the photopic sensitivity curve. This means that monochromatic stimuli of equal subjective brightness and otherwise constant physical parameters (area, light-dark ratio, background luminance, etc.) reach the same CFF. The validity of this observation is the conditio sine qua non for the application of flicker photometry (PORTER, 1902b; ALLEN,1926; HECHT and SHLAER, 1936).

7. Retinal Location

With otherwise constant stimulus parameters, the CFF is dependent on the location of the intermittent stimulus within the visual field (HYLKEMA, 1942b, c; MONJÉ, 1952). With test field diameters above 0.5 degree and mesopic or lower photopic stimulus conditions, the peripheral retina has a higher CFF than the fovea. HYLKEMA found that the larger the test field size, the farther the maximum CFF was shifted into the retinal periphery (Fig. 3). This finding is usually explain-

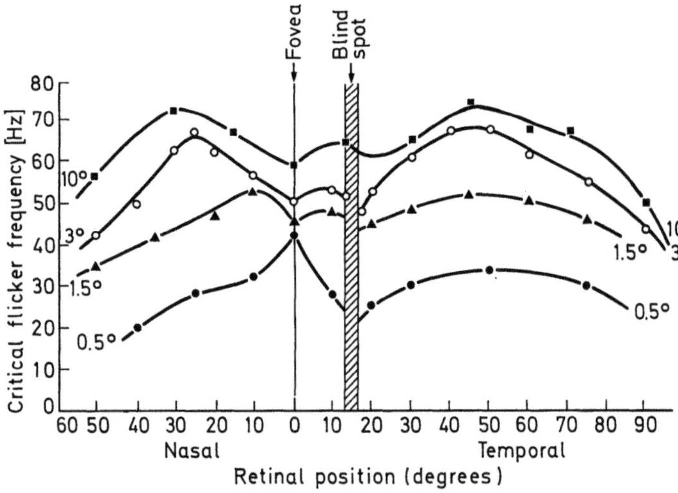

Fig. 3. *Dependence of the psychophysical CFF of round flickering test fields* ($\bar{I}$= 2460 cd/m², *L-D*-ratio = 1 : 1) *of different diameters on the retinal location.* While for small spots the maximum CFF is reached in the fovea centralis, for larger flickering test fields the maximum CFF is reached in the periphery of the visual field (HYLKEMA, 1942)

ed by a larger extension of spatial summation in the visual field periphery (see p. 441).

With small test fields (0.2 degree diameter) or test fields of higher photopic luminance and an area up to 2 degrees diameter, the CFF decreases from the fovea towards the periphery and is, under these conditions, positively correlated with the local visual acuity (c.f. p. 442, CREED and RUCH, 1932; ROSS, 1936; ALPERN and SPENCER, 1953).

8. Binocular Light Stimuli

The CFF of a flickering test field seen with both eyes is slightly higher than the monocular CFF (SHERRINGTON, 1904, 1906; CROZIER and WOLF, 1941a; IRELAND, 1950; BAKER, 1952a, b, c; THOMAS, 1955). The *phase angle* between 2 monocularly seen stimuli, projected on corresponding parts of the retina of each eye ("dichoptic" vision), influences the CFF. The CFF for binocular out-of-phase stimulation is reliably lower (2–9 %) than for binocular in-phase stimulation (SHERRINGTON, 1904; BAKER and BOTT, 1951; THOMAS, 1955). The reduction of the CFF by alternating light stimuli in comparison to the CFF with synchronous binocular stimuli, however, is not strong enough to support the assumption that the neuronal level of flicker fusion is the same as that of binocular fusion (SHERRINGTON, 1904).

The binocular CFF is also reduced with alternating dichoptic flicker stimuli in comparison with synchronous ones if the left half of the fovea of the one eye is stimulated and the right half of the other (BAKER, 1952c). In such experiments, due to the anatomical conditions, the binocular interaction involves cortical areas beyond the primary visual cortex. When the flicker stimuli projected to corresponding areas of each eye have different values of I but are flickering in phase, the binocular CFF is found to be between the CFF-values found for each eye alone. This finding indicates that not only binocular *facilitatory* mechanisms but also *inhibitory* mechanisms determine the CFF. This postulate is supported by the observation that the monocular CFF is reduced temporarily when a steady light stimulus is projected to the corresponding area of the other eye.

9. Dark Adaptation

The CFF of flicker stimuli seen with the peripheral retina is lower in the dark adapted state than in the light adapted state for the same I/I_s values (LYTHGOE and TANSLEY, 1929). However, during the course of dark adaptation for a *given value of* I, the CFF rises proportionately to the negative log of the threshold I_s (ERNST, 1968). This rule is valid whether I_s is changed by the presence of a background illumination or by the temporal course of dark adaptation after preadaptation with a short light stimulus of high intensity. From Fig. 2a, one can draw the conclusion that the maximum fusion frequency obtained at scotopic stimulus luminances in the dark adapted eye is below 12 Hz (SCHATERNIKOFF, 1902). Von KRIES (1902), however, reported that a totally color-blind human observer reached a maximum CFF of 20–22 Hz. This seems to be the upper frequency limit of rod vision.

10. Frequency Adaptation

Granit and von Ammon (1930) found that for peripheral vision the CFF of a 2.5 degrees test field decreases during the first 10 sec of observation. For foveal vision, only a very weak similar effect was described by these authors. They concluded that the decrease of the CFF is caused by the *local adaptation* which is known to be stronger in the periphery of the visual field than in the fovea. Besides local adaptation, however, there exists a specific adaptation effect of *flicker* stimulation, which is different from local adaptation caused by steady light. Exposure to a flickering light above 10 Hz but below the CFF, lasting longer than 10 sec, produces a fall in the CFF (Simonson and Brožek, 1952; Arnold, 1953; Alpern and Sugiyama, 1961; Turner, 1965). Looking at a steady light of the same Talbot luminance for 3 min also reduces the CFF but the reduction is significantly less than that elicited by subfusional flickering light. When the conditioning flicker stimuli have an intermittency above the CFF, a slight *increase* of the CFF of a test stimulus might be seen, if the retina is adapted more than 50 sec to the flickering light (Alpern and Sugiyama, 1961; Turner, 1965). The recovery of the facilitatory or inhibitory effects, elicited by long-lasting flicker adaptation, occurs within 1–5 min. The maximum drop of the CFF is obtained with an adapting frequency of about half the CFF. The relationship between the decrease (f_d) of the initial CFF and the adapting frequency (f_a) is fairly well described by a *u*-shaped parabolic function (Rey and Rey, 1965; average of 4 observers, 1 degree stimulus diameter foveal observation, rectangular flicker stimuli, photopic adaptation):

$$f_d = 22\, x \cdot (1 - x) - 1 \text{ [Hz]} \tag{16}$$

whereby $x = f_a/\text{CFF}$.

Turner (1965) reported that the frequency adaptation effects are also present when the adapting stimulus is viewed with one eye and the induced changes of the CFF are measured with the other eye. Therefore, the changes in the CFF appear to be, at least in part, also of central and not only of retinal origin.

R. A. Smith (1970, 1971) adapted his subjects for 30 sec to a *sinusoidally* modulated stimulus of 2 degrees diameter and measured immediately after this conditioning period the increase of the threshold modulation of the same flicker stimulus at different frequencies. Under such experimental conditions, the adaptation effect is maximum for the adapting frequency, but is also present for the whole frequency range between 4 and 50 Hz. These findings are very important with respect to the measurements of the threshold modulation curves as shown in Fig. 4 and 5. Suppose the subject is asked to set the modulation threshold for a flickering test field at a frequency above 10 Hz. Then the threshold might rise by as much as a factor 2, because of the mentioned adaptation effect, during the time the subject needs to make his setting.

Pantle (1971) measured this change of the de Lange curves (c.f. p. 447) after 4 min of adaptation to rectangular flicker stimuli (*L*-*D*-ratio = 1 : 1, 400 td luminance, 4 degrees diameter). The change of the de Lange-curve depends on the frequency of the adapting stimuli. A maximum rise of threshold is reached for the frequency range below 20 Hz with an adapting flicker frequency of 11 Hz, while for higher frequency ranges of the de Lange-curve, an adapting flicker stimulus of

32 Hz is more effective. PANTLE concluded from his findings that the overall frequency response characteristics of the a.v.s. are determined by parallel neuronal channels, having different frequency optima.

11. Observation Time and Transient Effects

Since the measurements of BASLER (1911), it is known that the CFF for periodic flashes in a long stimulus train is significantly higher than the fusion frequency for two flashes. The CFF of a short train of flashes increases with the time ($t_0 < 1$ sec) one can observe the intermittent stimulus pattern (MARBE, 1903; GRANIT and HAMMOND, 1931; ANDERSON et al., 1966; NELSON, BARTLEY, and HARPER, 1964; BARTLEY, NELSON, and RONNEY, 1961). The relationship between the CFF and the number of rectangular flicker stimuli in a short stimulus train can be approximated in part by exponential functions. These data indicate that the a.v.s. needs a certain time to reach the optimum sensitivity to flicker.

12. The Condition of the Subject

Further factors which influence the CFF are the age, somatic or psychic fatigue, the general metabolic level and the level of arousal of the subject (review: SIMONSON and BROŽEK, 1952). Some data indicate a diurnal variability of the CFF, which exhibits also a slightly positive correlation with the body temperature (LANDIS and HAMWI, 1954, 1956; ARNOLD and WACHOLDER, 1953; SIMONSON et al., 1941; SCHMIDTKE, 1951). Hyperventilation increases the CFF (RUBINSTEIN and THERMAN, 1935), anoxia and increased CO_2-plasma concentration decreases the CFF (SEITZ, 1940; ALPERN and HENDLEY, 1952). Sedative drugs, alcohol and barbiturates lower the CFF (ENZER, SIMONSON, and BALLARD, 1944; BERG, 1949; LANDIS and ZUBIN, 1951; BJERVER and GOLDBERG, 1955; AIBA, 1959; FRÜHAUF, 1971). The CFF increases slightly after excitation with caffeine, pervitine or benzedrine (SIMONSON and ENZER, 1942; BECK, 1951; MUCHER and WENDT, 1951). In general, one can say that the drugs increasing the level of arousal, at least in a small dosage, tend to increase the CFF, while drugs reducing the level of arousal decrease the CFF. The effects are altogether small. A precise and repeated measurement of the CFF with standardized stimulus conditions, however, might be useful as a test for drug effects and other influences which change the level of arousal.

The foveal CFF of an amblyopic eye is significantly reduced in comparison to the CFF obtained with the same stimuli projected to the fovea of the good eye (FEINBERG, 1956).

C. Sinewave Flicker Stimuli at Threshold

Sinusoidally modulated light was used in an experimental study of flicker fusion phenomena as early as 1922 by IVES. He concluded that not the waveform itself, but rather the parameters of the fundamental component (first harmonic) determine the fusion frequency. COBB (1934a, b) reconsidered the results and analysis of IVES and showed that for low frequencies the higher order Fourier components (higher harmonics) also influence the results. Yet it was not until the work of DE LANGE (1952–1961) that sinusoidal flicker stimuli acquired their present dominant role in psychophysical experiments on flicker.

As mentioned in Section I B the application of sinusoidal stimuli has the advantage that only one frequency component is present in the stimulus. A visual sinusoidal stimulus (Fig. 1 c) is generally described as:

$$x(t) = \bar{I}\,(1 + m \sin \omega t)\,. \tag{17}$$

The stimulus is completely described by three parameters, the average luminance $\bar{I}$, the modulation m and the frequency $f = \omega/2\pi$, whereas four parameters are necessary to describe square wave flicker stimuli completely. An important attraction of sinewave data is that they have proved to enable a quite good prediction of the CFF to complex dynamic stimuli via the so-called Fourier equivalence calculation (see pp. 435–438 and the review by KELLY, 1972). Thus from the results obtained with sinusoidal stimuli for different values of $\bar{I}$, a quite efficient description of some of the dynamic properties of the a.v.s. is possible.

1. Sinewave Threshold Curves

Kelly measured (c.f. his Fig. 9 in Vol. VII/4 of this Handbook), like DE LANGE (1952), the threshold modulation depth at the flicker fusion or its inverse, the modulation sensitivity for different frequencies with different levels of $\bar{I}$ (0.06–9300 trolands). These curves, giving on log-log coordinates the relation between threshold modulation and frequency, with $\bar{I}$ as a parameter, are called de Lange-curves (Fig. 4). In Fig. 5a KELLY's data are plotted in a coordinate system in which the absolute amplitudes of the sine waves are considered. It is evident from this figure that the upper frequency range obtained with each $\bar{I}$-value fits into a "master curve". According to KELLY (1969b), this master curve can be very well described by the formula

$$y = 6000 \exp\left[-(\pi f)^{1/2}\right] [\text{trolands}^{-1}] \tag{18}$$

whereas the Ferry-Porter-law would predict:

$$y = c \exp\left[-kf\right] [\text{trolands}^{-1}]\,. \tag{19}$$

In this formula y is the absolute sensitivity defined as $(m^* \cdot \bar{I})^{-1}$, where m^* is the threshold modulation depth. For the limited range of experimental data, the difference between these two functions is rather small, even though it is of theoretical interest (KELLY, 1969b).

In Fig. 5b the data of Fig. 5a are replotted in the traditional way, as in Fig. 2a (KELLY, 1961b). Fig. 5b also shows that the Ferry-Porter-law is approximately valid in a restricted luminance range for sinewave stimulation. The range of validity decreases with a decrease of modulation. It can be seen by comparison with Fig. 2a and b, that Eq. (12) or (13) will also fit the data of Fig. 5b.

From Fig. 4 and 5a, it is apparent that for photopic stimuli the sensitivity of the a.v.s. reaches a maximum at "medium" frequencies. The brighter and the larger the flickering field, the more pronounced is the sensitivity peak or "resonance" phenomenon and also the higher the resonance frequency. Both the resonance phenomenon and the high frequency cutoff have been rich sources of inspiration for modelling studies (Part 3). Data as shown in Fig. 5a have been used in conjunction with Fourier equivalence calculations to predict with considerable success experimental results obtained for the CFF with rectangular stimulus wave forms of different light-dark ratios: de Lange (1954), Kelly (1961d), Gibbins and Howarth (1961). For detailed reviews of this and related work see J. L. Brown (1965) and Kelly (1972).

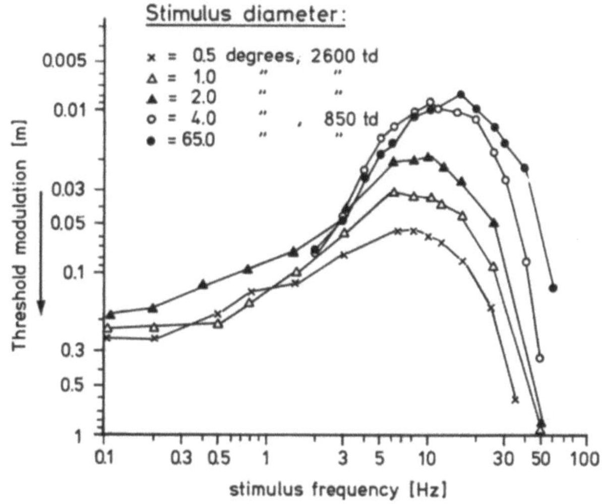

Fig. 4. Threshold modulation curves for sinusoidally flickering test fields of different areas. As the area increases, the frequency at which the optimum sensitivity is reached increases (Keesey, 1970)

2. Area Effects and Spatio-Temporal Interactions

In the de Lange-curve the frequency of maximum flicker sensitivity (resonance frequency) and the cutoff frequency shift to higher values for larger stimulus areas, whereas the low frequency sensitivity decreases (Kelly, 1959, 1964; van der Gon, 1959; van der Tweel, 1961; Keesey, 1970, Fig. 4). It is of interest to know whether this effect is due to the increased area or to the increased border between flickering field and background. Both the total number of stimulated receptive fields ("area effect") and the number of receptive fields covered by the image of the flicker field boundary ("edge effect") depend on the retinal location of the test field and of its area. Moreover, eye movements may play a role because they sweep the border of the test field across the retinal receptive fields. The question now is, which combination of these factors might explain the mentioned changes of the de Lange-curve caused by increasing the stimulus area.

First of all, essentially the same difference between de Lange-curves for small and for large test fields was found in stabilized image experiments by Keesey (1970) reported by Kelly (1964), and West (1968). This rules out the influence of eye movements as an explanation. Kelly (1969a) has recently devised methods to separate edge effects from the area-effect. He concludes that the edge-effect has no influence at high frequencies, where only area counts. Kelly's findings further indicate that, although flicker sensitivity is enhanced at all frequencies by an increase of area, the edge effect is much more important at low frequencies than the area effect. This might be explained by the assumption that lateral inhibition,

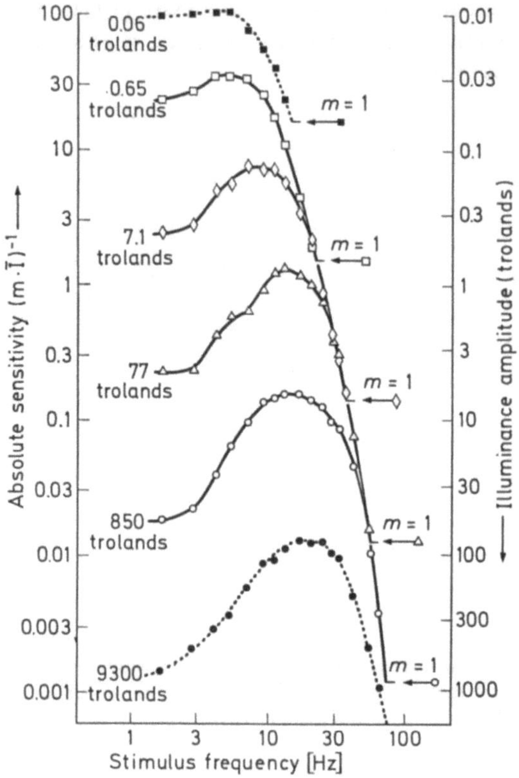

Fig. 5a. *Sine wave threshold amplitude data.* Plots on an absolute sensitivity scale (Ordinate) *rather* than the relative scale used in de Lange-curves (Fig. 4). For $m \to 1$, the results obtained with different $\bar{I}$-values fit into a common "master curve" (KELLY, 1961)

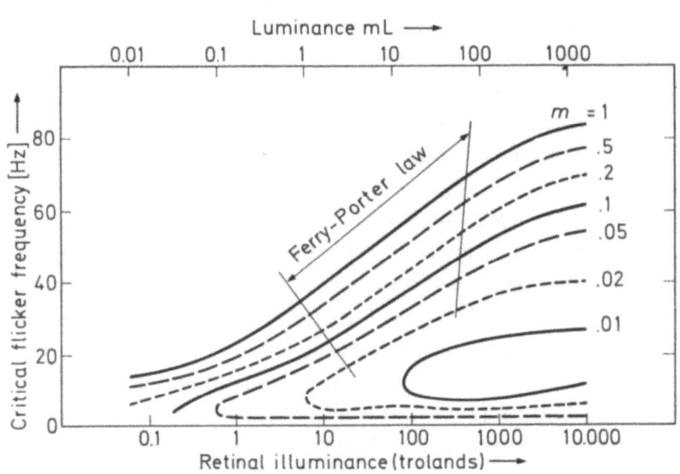

Fig. 5b. The data of Fig. 5a are replotted. The dependence of the CFF on $\bar{I}$ (abscissa, or $m \cdot I$) is plotted for sinusoidal flicker stimuli of different depths of modulation ($m = 1$ to $m = 0.01$). The range of validity of the Porter-law, i.e. the logarithmic relationship between CFF and $\bar{I}$ is reduced as m decreases (KELLY, 1961). The curves found for $m = 0.5$ to $m = 1$ are well approximated by Eq. (13)

which influences the spatial filtering characteristics of the a.v.s., is mainly effective at low temporal frequencies (c.f. p. 508).

Whatever the explanation, the experimental results clearly show that the temporal filtering characteristics depend upon the spatial frequency contents of the stimulus and the reverse has also been shown to be true. Quite a number of studies of these "spatio-temporal interactions" in the a.v.s. have appeared in recent years (Schade, 1956; Kelly, 1960, 1966, 1969a, 1970, 1971b; Robson, 1966; Kulokowski et al., 1966; van Nes, 1968a, b; van Nes et al., 1967; Kulikowski, 1971a, b).

Two types of experimental setups were used in these studies, namely:

(a) *Standing Waves*. A grating of linear stripes of which the luminance changes sinusoidally in a direction perpendicular to the stripes will be called a sinusoidal grating. The spatial frequency f_s of such a grating is usually expressed in cycles per degree of visual angle (cpd). If the luminance of such a grating is modulated sinusoidally in time with temporal frequency f_t, a "standing wave" is generated. Kelly (1966) used circular symmetric Bessel function patterns also modulated sinusoidally in time. These standing waves have nodes and antinodes and at the points of the retina where nodes are imaged, the stimulation does not change with time. This might limit their importance as basic spatio-temporal stimuli.

(b) *Travelling Waves*. Such stimuli are defined as

$$I = \bar{I}\ \{1 + m \sin [2\,\pi\,(f_s \cdot x + f_t \cdot t)]\} \tag{20}$$

where the temporal frequency $f_t = v \cdot f_s$, if v is the angular velocity of the uniformly drifting sinewave grating. Manipulation of f_s and v makes it easily possible to study the responses of the a.v.s. for all sorts of (f_t, f_s)-combinations. Travelling waves have the advantage that all points of the retina receive an equal type of stimulation.

The results obtained with standing and travelling waves are qualitatively in good agreement. Figs. 6 and 7 are representative of the results obtained with travelling waves (van Nes, 1968a): The spatio-temporal sensitivity reaches an optimum between 0.5–3 Hz and 2–3 cpd in the isomodulation contours of Fig. 7 (van Nes, 1968a, foveal vision, 85 td, artificial pupil). In a spatio-temporal amplitude threshold surface published by Kelly (1966, 1972), for standing waves the maximum lies at somewhat higher spatial and temporal frequencies, but these data are for the higher luminance of 1000 td (circular gratings viewed with a natural pupil). One can expect that in the light-adapted eye for parafoveal vision, the temporal optima change only slightly, whereas the spatial optima are shifted to considerably lower values. In other words, for uniformly moving gratings, the point of optimum sensitivity will move away from the fovea when spatial frequency is decreased and this will hold over a wide range of drift velocities (van Doorn et al., 1972).

Psychophysical experiments with spatial *square wave* gratings and *rectangular* temporal stimuli of different durations (Hilz, 1965; Schober and Hilz, 1965; Nachmias, 1966, 1967) confirm the general conclusion from Figs. 6 and 7: The visibility of rectangular gratings („Rechteckgitter") not only increases with the stimulus luminance, but also with the stimulus duration. In the experiments of Schober and Hilz, the minimum contrast threshold of the gratings was reached

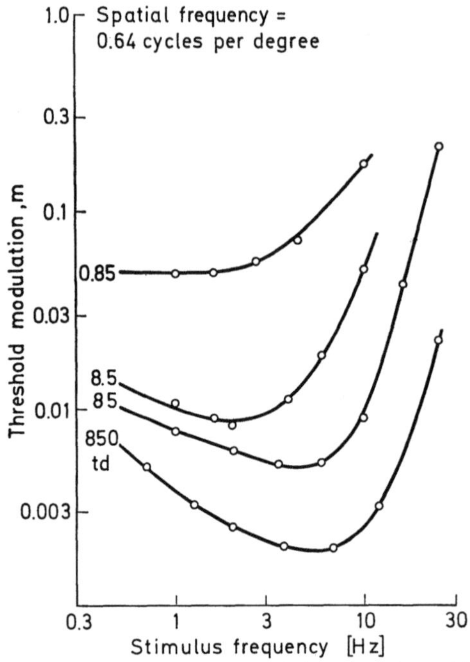

Fig. 6. *Temporal threshold modulation curves for flicker detection with* spatially sinusoidally modulated gratings (spatial frequency 0.64 cycles per degree) drifting at different speeds across the visual field. Measurements were made at 4 different $\bar{I}$-values: 0.85, 8.5, 85 and 850 trolands (VAN NES, 1968)

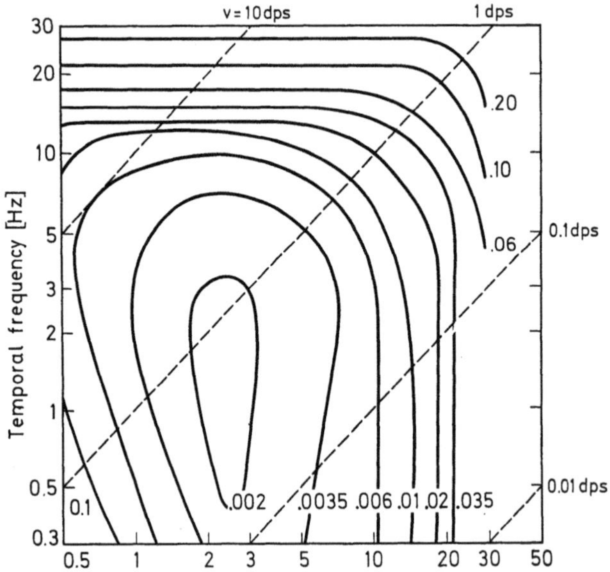

Fig. 7. *Contours of isomodulation lines for the threshold modulation obtained in a light-adapted human observer with sinusoidal gratings of different spatial frequencies* ($\bar{I}$ = 85 td, abscissa), which were moved at different angular velocities across the visual field, thus producing a certain temporal stimulus frequency (ordinate). The non-symmetric distribution of these isomodulation lines for threshold stimulation exhibits the non-linear properties of spatio-temporal interaction at the flicker threshold. The dashed lines are loci of a constant grating velocity (VAN NES, 1968)

between 2.5 and 4.2 periods per degree and increased with a stimulus duration up to 1000 msec.

3. Flicker Thresholds with Phased Surrounds

On p. 442 was discussed how a steady surround of a luminance near the Talbot level increases the CFF. If the surround is also flickered and in phase with the test stimulus (which means a larger flickering area), the CFF proves to decrease relative to the values found with steady surround (LURIA and SPERLING, 1962; FOLEY, 1963). On the other hand, if test field and surround are flickering 180 degrees out of phase, the CFF of the test field increases (1.5 degree center, 5.7 degrees annular surround, LEVINSON, 1964). FIORENTINI and MAFFEI (1970) performed simular experiments with very small sinusoidally modulated patterns (3′ circular center, 9.5′ annular surround) projected to the fovea. The modulation threshold $m\varphi$ is a function of the phase angle φ between center and surround stimulus (Fig. 8). The relative facilitatory (out-of-phase) or inhibitory (in-phase) effects decrease as the stimulus frequency increases above 2–5 Hz (LEVINSON, 1964). Above 12 Hz, φ does not influence the threshold modulation at all. We found that the data published by FIORENTINI and MAFFEI are well described by the following equation for $f \leqq 10$ Hz (Fig. 8):

$$m_\varphi = (m_{\max} - m_{\min}) \sin\left(\frac{\varphi}{2} + \lambda\right) + m_{\min} \tag{21}$$

whereby m_φ is the threshold modulation for the center test field, when the surround sine wave is delayed by the phase angle φ and λ is a constant.

4. Chromaticity Modulation

DE LANGE (1958a) measured a number of amplitude gain characteristics for coloured sinusoidally intensity-modulated test fields of 2 degree diameter with a 60 degree white annular surround for two average luminance values. He found clear differences between the amplitude gain characteristics for different colours and noted that this implied, according to linear filter theory, corresponding phase shift differences. Subsequently, DE LANGE (1958b) experimented with two sinusoidally luminance modulated components of different colours but of equal subjective brightness and with an adjustable phase relation. These components were superimposed on a 2 degree testfield (causing "heterochromatic" flicker) surrounded by a 60 degree annular white surround. With a phase difference of $180° + \psi$ between the components, one has for $\psi = 0°$ exact "counterphase heterochromatic flicker". For low frequencies of alternation one then sees a periodic change from the one color component to the other, but on increasing the frequency, a point will be reached where the hues fuse. This frequency is called the critical color fusion frequency (CCFF). DE LANGE (1958b) pointed out that a brightness flicker remains above the CCFF and he showed that this brightness flicker can be eliminated at the CCFF by an appropriate adjustment of ψ. DE LANGE (1958b) as well as WALRAVEN and LEEBEEK (1964a, b) who extended DE LANGE's heterochromatic flicker measurements and corresponding analysis, explained this finding in terms of a zone theory of color vision (c.f. VON KRIES, 1905) as follows.

For equal brightness of the red and green components (if we concentrate on these two), the response amplitudes in the a.v.s. at the input of a hypothetical flicker detector are also equal when the CCFF is reached. The amplitude of the brightness signal, which in this case is thought to be the algebraic sum of the outputs of the green and the red channel, depends on the differential phase, however. The green and red outputs will exactly nullify each other only when they are in exact counterphase, which will occur if the differential phase shift is compensated at the input. Quantification of this explanation is of course only possible in the context of a specific model of color vision (WALRAVEN and LEEBEEK, 1964b).

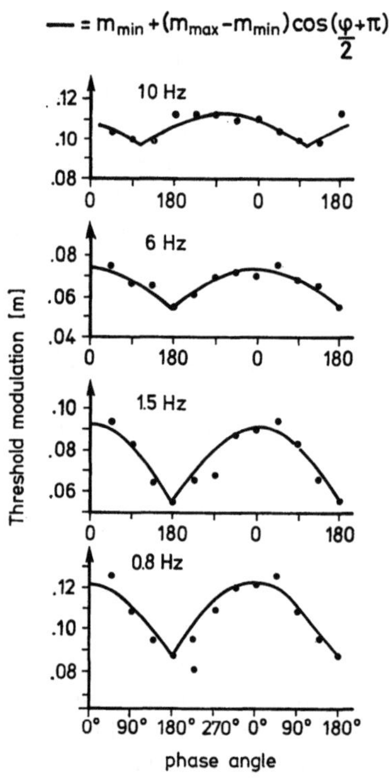

Fig. 8. *Threshold modulation for the flicker perception of a sinusoidally modulated small test field* (3′ diameter, projected to the fovea). The test field is surrounded by a light annulus of 9.3′ outer diameter which is also modulated sinusoidally with the same frequency as the center-stimulus but with different phase angle (abscissa). The center threshold modulation depends on the phase angle φ, but the effect of changing φ decreases as the stimulus frequency increases from 0.8—10 Hz (data from FIORENTINI and MAFFEI, 1970). The curves are calculated from Eq. (21)

From the mentioned differences found by DE LANGE (1958a, b) between gain and phase characteristics for the different colors, it also follows that, if different color channels are supposed to be present in the a.v.s., these channels must have different "time constants". Moreover, these time constants depend upon the state of adaptation of the channels. Further analysis is therefore possible on the basis of experiments in which a flicker stimulus of a given color is projected to an adapting background of another color. KELLY (1962c) measured de Lange-curves for

65 degree test fields under different conditions and found sensitivity peaks around 6, 12 and 24 Hz, which could selectively be suppressed or enhanced by chromatic adaptation. Burckhardt (1966) developed a model consisting of three color channels with different "refractory periods" (about 30 ms for red, 60 ms for green and 120 ms for blue) that might explain Kelly's findings. Different bandpass characteristics of the color sensitive mechanisms of the a.v.s. for blue (430 nm), green (530 nm) and red (648 nm) sinusoidal flicker stimuli, for a foveal test field of 2.5 degrees diameter and 540 trolands, were also found by Green (1969). In con-

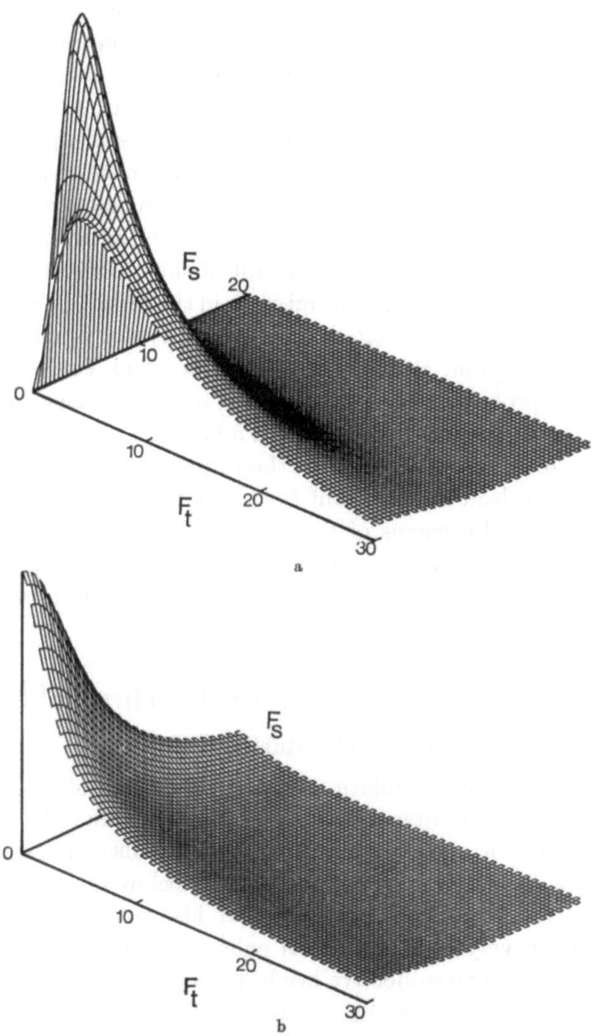

Fig. 9. *Spatio-temporal sensitivity surfaces for sinusoidally luminance* modulated, drifting gratings (top) and for constant luminance chromaticity modulated, drifting gratings (bottom). The surfaces were calculated and plotted by a digital computer according to the data of van Nes (1968) and van der Horst and Bouman (1969) respectively. The a.v.s. apparently functions as a simple, spatio-temporal, low-pass filter for chromaticity information, whereas "tuning" phenomena occur in the luminance transfer characteristics (after Koenderink et al., 1972)

trast to BURCKHARDT'S interpretation of KELLY'S data, GREEN found that the red and green mechanisms have frequency response characteristics with peaks of sensitivity only a few Hz apart,but with different shapes at low and intermediate rates of sinusoidal flicker. The blue-sensitive mechanism has a significantly depressed modulation sensitivity curve relative to the curves for the red and green mechanism, and thus it has a much lower flicker fusion frequency (c.f. BRINDLEY et al., 1966; GREEN, 1969).

VAN DER HORST (1969a, b) used a color television monitor to study heterochromatic flicker , which also allowed him to study spatio-temporal chromaticity interactions (VAN DER HORST and BOUMAN, 1969). These studies clearly show an important difference between the spatio-temporal information processing characteristics of the a.v.s. for luminance patterns on the one hand and chromaticity patterns on the other. Both in space and in time, the constant luminance chromaticity transfer characteristics lack the resonance phenomena which are prominent in the luminance transfer characteristics. This can most clearly be seen from the spatio-temporal sensitivity surfaces in Fig. 9 (KOENDERINK et al., 1972). Interpretation of the heterochromatic flicker data might be complicated by the fact that the chromaticity modulation is the result of simultaneous changes in luminance of the two components of the light mixture in the test field. Therefore "chromaticity changes" have to be calculated, for example as VAN DER HORST did, from a CIE chromaticity diagram. Recent data of REGAN and TYLER (1971) indicate, however, that it is probably not always correct to assume that color mixing laws that hold for the static case also hold when the mixed components have time variant intensities. Therefore, the introduction by these authors of a wavelength modulated light generator might lead to important further developments. The data obtained by REGAN and TYLER by means of this improved stimulus technique, however, also strongly confirm the older findings that the chromaticity mechanisms are attenuated at much lower stimulus frequencies than the brightness mechanisms of the a.v.s.

D. Flicker Phenomena above the Threshold Level

1. Brightness Enhancement (Brücke-Bartley Effect)

Not only the threshold modulation for the perception of flicker, but also the subjective brightness of an intermittent light stimulus above the flicker threshold depends on the frequency. BREWSTER (1834) and BRÜCKE (1864) discovered that intermittent light appears brighter at frequencies below the CFF than above the CFF. EBBECKE (1920a, b) and BARTLEY and his co-workers (1938–1964) investigated quantitatively this brightness enhancement, which is usually called Brücke-Bartley effect: The subjective brightness of intermittent light is enhanced at medium frequencies above the brightness of the steady stimulus. In contrast to BARTLEY'S repeated claim that the brightness enhancement appears at or near 10 Hz, the subjective brightness of flicker stimuli can reach its maximum at any frequency between 3 and 15 Hz for photopic stimulus intensities. The relative amount of brightness enhancement increases with the stimulus luminance ($\bar{I}$) and the size of the stimulus area (A) at photopic levels of retinal adaptation. Moreover, the larger I and A, the higher the resonance frequency at which the maximum

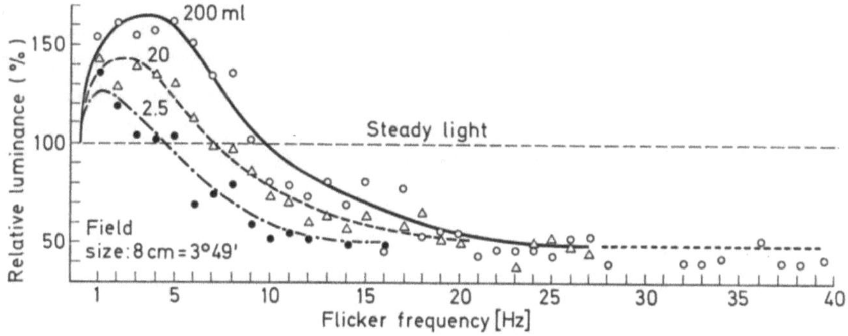

Fig. 10a. *Brücke-Bartley-effect obtained with rectangular flicker stimuli* (*L-D*-ratio 1 : 1). Mesopic level of adaptation. The subjective brightness (relative luminance of the steady light, which matches the brightness of the flicker stimulus) is plotted on the ordinate, the frequency of the flicker stimuli on the abscissa. Stimulus diameter 3°49′. Three different levels of $\bar{I}$ (Rabelo and Grüsser, 1961)

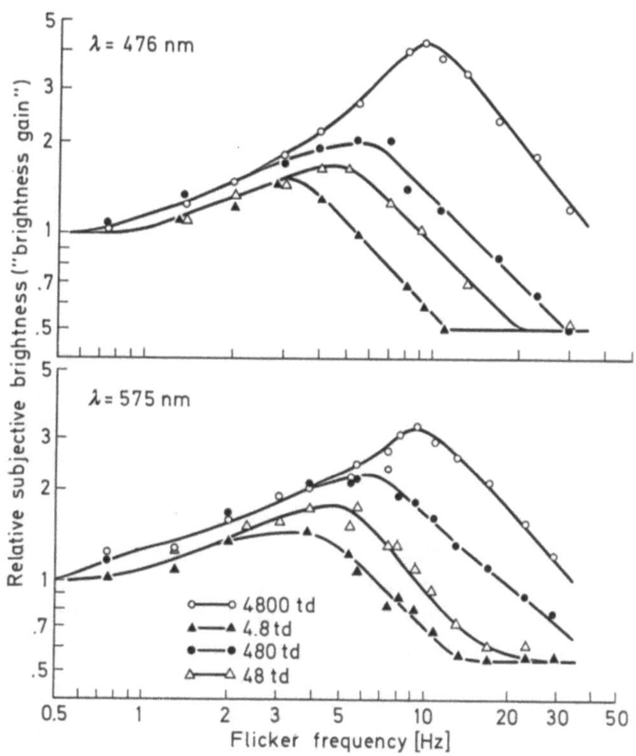

Fig. 10b. *Brücke-Bartley-effect obtained with flicker stimuli of monochromatic light* (476 and 575 nm) at 4 different levels of average retinal illumination (4.8—4800 td). The relative subjective brightness ("brightness gain") is plotted on the ordinate on a logarithmic scale, while the frequency of the sinusoidal stimuli is also plotted on a logarithmic scale on the abscissa. As for white flicker stimuli (Fig. 10a) the relative brightness enhancement increases as the stimulus intensity increases and the resonance frequency shifts into a higher frequency range (van der Horst and Muis, 1969)

brightness enhancement is reached (Fig. 10a, RABELO and GRÜSSER, 1961). When the *L*-*D*-ratio is reduced below 1 : 1 at medium photopic stimulus luminance, the brightness enhancement is reduced and the resonance frequency shifts to lower values, whereas with very bright flicker stimuli, a reduction of the *L*-*D*-ratio to 1 : 2 might increase the Brücke-Bartley effect (GRÜSSER and REIDEMEISTER, 1958; WASSERMAN, 1966b).

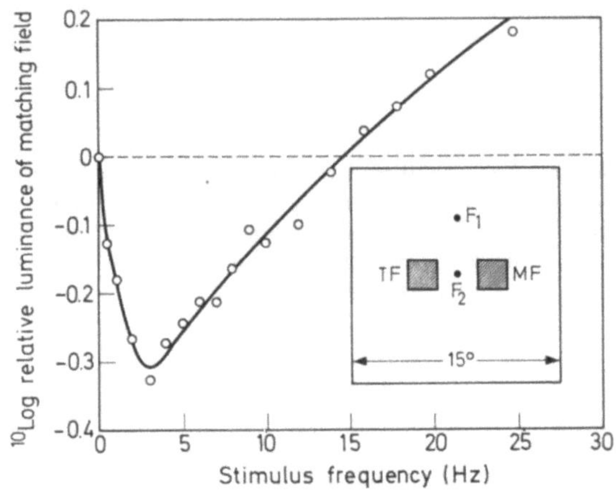

Fig. 10c. *Darkness enhancement.* Luminance of a steady field (MF) to match the low luminance (dark) periods of a rectangular test field (TF) flickering at different frequencies ($L : D = 1 : 1$). The luminance of the TF is maximum 582 cd/m², minimum 64.5 cd/m². 0 on the ordinate is 64.5 cd/m². F1 and F2 are fixation points (GLAD and MAGNUSSEN, 1972)

An objective indication of the subjective brightness enhancement is the increased constriction of the pupil. With sinusoidally modulated light, VARJÚ (1964) found a minimum pupil diameter and a maximum brightness enhancement at about 1–3 Hz in human observers. These low frequency values of the Brücke-Bartley maximum are probably caused by the 15 min dark adaptation preceding the measurements of VARJÚ. For the light-adapted cat, a minimum pupil diameter is reached with bright, rectangular, flicker stimuli around 10 Hz (GRÜSSER, 1957a). The Brücke-Bartley effect is closely related to the Broca-Sulzer-brightness enhancement, seen with single, short, rectangular stimuli in comparison to longer lasting light stimuli. Intermittent illumination "cuts off" the decrease of subjective brightness of a photopic stimulus that lasts longer than 80–200 msec (EXNER, 1868, 1870). This relationship was recognized by BROCA and SULZER, 1902 (c.f. EBBECKE, 1920a, b). WASSERMAN (1965) assumed that both phenomena are caused by completely identical mechanisms. In our opinion, both effects are closely related but not identical. For the measurement of the brightness of periodic rectangular stimuli, the a.v.s. is in the steady state for a periodic response, for the measurement of the brightness of single flashes, this is not the case.

The effect of *different colors* on the brightness enhancement was investigated by several authors. The most careful study of which we know is that of VAN DER HORST and MUIS (1969). These authors investigated the brightness enhancement of monochromatic sinusoidally modulated flicker stimuli of 476, 510 and 575 nm (Fig. 10b).

As with white light, the brightness enhancement and the resonance frequency increase as the luminance of the chromatic stimuli is increased (4.8–4800 td). The relative increase, however, is independent of the wave length. This is not too surprising because the chromatic response functions are, according to the opponent color theory, approximately zero at the three wave lengths mentioned. In one experiment, however, van der Horst and Muis measured the brightness enhancement of 7 different monochromatic light stimuli of 4800 trolands. The deviation of the curves from each other is small with respect to the resonance frequencies. Therefore, van der Horst and Muis concluded that in contradiction to the opinion of Wasserman (1965), Ball and Bartley (1962), the spectral composition of the stimuli has no effect on the Brücke effect. Van der Horst and Muis observed a flicker depending hue shift, which is strongest with monochromatic light of 625 nm. For a retinal stimulus intensity of 4800 td, the maximum hue shift is reached for the longer wave lengths at a slightly higher frequency (12 to 15 Hz) than for the shorter wave lengths. Nilson and Nelson (1971) confirmed these data. For monochromatic light between 425 and 650 nm, the maximum hue shift occurs around 540 and 650 nm. The hue shift is always accompanied by a desaturation effect (c.f. also Bartley and Nelson, 1960; Bleck and Craig, 1965a, b).

It is difficult to locate the physiological mechanism responsible for the brightness enhancement by means of psychophysical experiments alone. Bartley and his co-workers claimed that a cortical mechanism related to the α-rhythm is responsible for this effect, while Grüsser (1956) and Grüsser and Creutzfeldt (1957) postulated on the basis of neurophysiological findings a retinal origin (see p. 495). An important psychophysical contribution to the clarification of this problem was provided by Motokawa et al. (1956). Motokawa and Iwama (1950) demonstrated that illumination of the retina increases the electrical excitability of the eye during the period following the illumination (ζ-effect). Motokawa et al. measured the ζ-effect after intermittent light stimulation. The ζ-effect is largest at frequencies around 10 Hz at which in Motokawa's experiments the brightness enhancement was at a maximum. Motokawa concluded therefore that the brightness enhancement of flickering light is predominantly of retinal origin, a view supported by all neurophysiological evidence mentioned in this review (p. 497).

2. Darkness Enhancement

If one judges the subjective darkness of a grey test field, placed on a white background, both viewed intermittently through a sector disc with light grey sectors, one perceives not only a brightness enhancement of the white background of the stimulus, but also an enhancement of the darkness of the grey field. This darkness enhancement is seen between 1 and 6 Hz and, as a rule, at lower stimulus frequencies than the maximum brightness enhancement (Grüsser unpubl., 1961; mentioned in Rabelo and Grüsser, 1961). In a similar manner, the darkness of the low intensity periods of a flickering test field (rectangular or sine wave stimuli), projected to a background of photopic luminance, appears darkest at frequencies between 2 and 4 Hz. From these observations we concluded that the neuronal "darkness"-system of the a.v.s. (D-system) is activated more strongly by the dark pauses of intermittent light than by steady darkness. Furthermore, the maximum activation of the neuronal darkness system is expected to occur at lower stimulus frequencies than that of the brightness system.

Quantitative measurements about this darkness enhancement were recently published by GLAD and MAGNUSSEN (1972). They used a quadratic flickering test field of 2 degrees size surrounded by a 15 degree background of 582 cd/m². The luminance of the test field was modulated by the rectangular flicker stimuli ($L : D = 1 : 1$) between 64.5 and 582 cd/m². The subject had to set the luminance of a steady field of the same size and on the same background, placed in 3 degrees distance from the test field, to the same "level" of darkness as the flickering test field. The results of measurements of this type are shown in Fig. 10c: At low stimulus frequencies the subjective darkness increases with increasing stimulus frequency, reaches a maximum at about 3 Hz and decreases continuously above this frequency level.

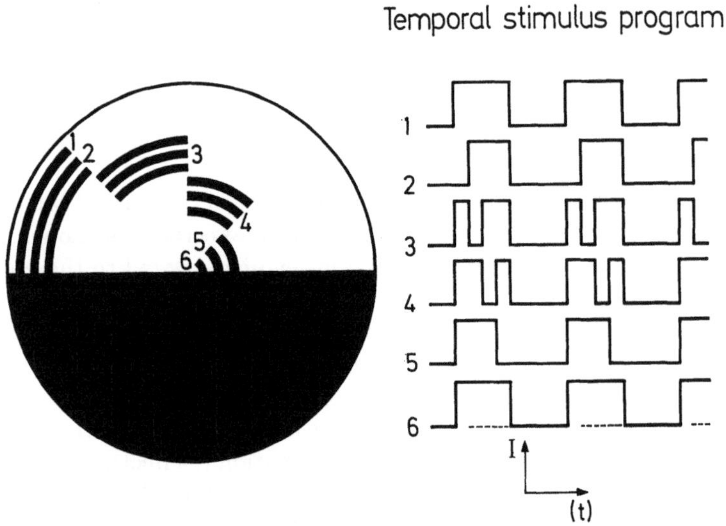

Fig. 11. *Benham top (A) and the temporal stimulus program (B)* of the different sector parts, when the disc is rotated clockwise

3. Fechner-Benham Colors

At flicker frequencies between 3 and 40 Hz, an intermittent white light stimulus might appear colored. The color depends on the area and the frequency of the stimuli and is rather fluctuating and unsaturated. This flicker-induced color perception is enhanced and shifted to a lower frequency region when certain spatially structured stimuli are used. The pattern-induced flicker colors were first described by PREVOST and are usually called Fechner-Benham colors, because both these authors developed a specially designed flicker-disk, the rotation of which induces in the human observer color perception (FECHNER, 1838; BENHAM, 1895; FRY, 1935; review: COHEN and GORDON, 1949). Fig. 11 shows BENHAMS "Top", which is perceived at a certain rotation frequency (3–6 Hz) as colored. For a counter-clockwise rotation, the 3 outer sectors appear as red rings, the next sectors as green rings, the third sectors appear light bluish, while the inner sectors are of a dark violet color. These pattern-induced flicker colors also have a low saturation.

Pattern-induced flicker colors are also seen, if the Benham top is illuminated by monochromatic light (Christian and Haas, 1948).

The moving patterns of Benhams top cause periodical light-dark stimulation, whereby neighbouring retinal areas receive different periodic stimuli, which evidently interact simultaneously. The temporal program of the stimuli at a given point of the retina and the phases between different stimuli of neighbouring retinal districts are of importance for the perceived colors. If the disc rotates clockwise, the sequence of the colors is reversed (red is seen in the inner rings, etc.). The colors also depend on the angular width of the whole pattern: The visual angle at which Benhams top produces optimum pattern-induced flicker colors is smaller in the fovea than in the more peripheral parts of the retina (von Campenhausen, 1968, 1969, 1970). Movement of the pattern is *not* essential to the perception of flicker colors. A stationary pattern of dark and light stimuli leads to the same color perception, if the temporal program of the flicker stimuli is adequately chosen (von Campenhausen, 1970).

Pattern-induced flicker colors are seen at a lower frequency than the color change of a diffuse, intermittently illuminated light stimulus (Welpe, 1970). Flicker colors, however, do not depend on any regular pattern at all. This can be easily observed on a television screen on which "white noise" appears. From an optimum distance, which is different for foveal and parafoveal vision, one sees besides black and white spots, numerous colored spots in a pseudo-brownian movement on the TV-screen. At a low illumination level, a few, irregularly distributed, moving blue spots are seen between the white and black spots. As the average luminance of the TV-screen is increased, green and red-purple spots, both of low saturation, appear in addition. They are more densely distributed than the blue spots, but appear also irregularly. With a further increase of the TV-screen luminance, the black or gray spots disappear and the flickering spots with the three colors are distributed irregularly on a white flickering background. It seems reasonable to assume that if the luminance changes of neighbouring spots on the TV-screen by chance follow each other at a certain frequency and have a certain relative phase, flicker colors appear for the same reason (temporal and spatial sequence) as in the case of the Fechner-Benham top.

4. Visual Acuity

Visual acuity increases with the average luminance of a stimulus pattern. This is a well known daily experience. In reading a newspaper, for example, the higher the brightness, the better the perception of the letters. Some investigators, therefore, hypothesized that the visual acuity would also increase when brightness enhancement of the flicker stimuli is observed. Measurements of the visual acuity at flicker frequencies at which brightness enhancement is seen, indicate, on the contrary, that visual acuity *decreases* when brightness enhancement appears. Usually these measurements are made in comparison with the visual acuity for stimuli of the same Talbot level as the flicker stimuli (Feilchenfeld, 1909; Geratewohl and Taylor, 1953; Nachmias, 1958, 1961; Nelson, Bartley, and Soules, 1963; Bourassa and Bartley, 1965; Bartley and Ball, 1968; Ball and Bartley, 1970). This decrease of visual acuity during the increase of the subjective brightness is not paradoxical. Both phenomena can easily be explained by the same mechanism, if one assumes certain frequency response characteristics for lateral inhibition in the retinal neuronal network (c.f. p. 500).

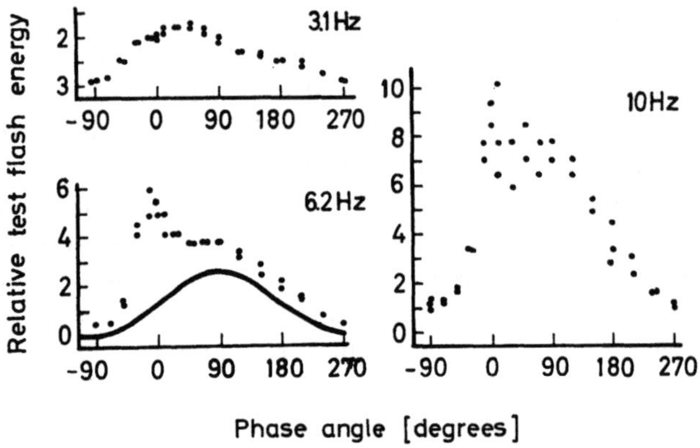

Fig. 12. *Relative intensity (ordinate) of a threshold incremental test stimulus* (2 degrees diameter), projected on a sinusoidally modulated background (22 degrees diameter, 1200 td retinal illuminance at peak, $m > 0.99$). The sine wave at the lower left illustrates the relative amplitudes of test and background stimuli (SHICKMAN, 1970)

5. The Temporal Course of Subjective Brightness of a Slow Flicker Stimulus

Sinusoidally modulated light at low frequencies (< 3 Hz) appears symmetrically modulated only if the depth of modulation is below 0.2 ($\bar{I}$ in the photopic range of adaptation). The larger m, the more the subjective brightness deviates from a symmetrical (linear) temporal modulation. A sinusoidally modulated (0.1–3 Hz) spot of light appears to be asymmetrically modulated: The period of increasing brightness appears shorter than the period of decreasing brightness. At a frequency level above 10–15 Hz, it is not possible to say from the flicker perception whether the temporal change of the stimulus is triangular, sinusoidal or rectangular. Between these frequencies and the CFF, the perception of lightness and darkness of the flicker stimuli is also unequally distributed. Especially for a frequency range below the CFF, the flicker impression is not a dark interruption of the perceived test field but only a grey inconstancy of its brightness. The grey periods seem to be much shorter than the light periods and appear irregularly distributed across a large flickering field.

The apparent depth of modulation decreases monotonically as the stimulus frequency increases above 1 Hz (foveal stimuli of 2.3 degrees diameter, $\bar{I} = 3$ cd/m^2) As $\bar{I}$ increases (30 or 300 cd/m^2) for $m > 0.25$, the apparent depth of modulation increases up to 5–10 Hz. Above this frequency range, the apparent depth of modulation also decreases monotonically with increasing flicker frequency (MARKS, 1970).

BOYNTON, STURR, and IKEDA (1961) used the increment threshold technique to measure the course of excitation elicited by *rectangular* flicker stimuli of different frequencies. Their results indicate that the temporal course of excitation elicited by rectangular flicker stimuli is non-linear in the lower and medium frequency range, while at higher frequencies (30 Hz), only the first harmonic of the rectangular stimuli elicits a response.

SHICKMAN (1970) made the same measurements with sinusoidal stimuli. At low stimulus frequencies (< 3 Hz), the non-sinusoidal temporal change of the threshold corresponds approximately to the temporal course of the subjective brightness of the stimuli (Fig. 12). His threshold measurements also indicate a clearly non-linear change of excitation in the lower frequency range for sinusoidal stimuli ($m = 0.9$).

6. Flash Frequency and Perceived Flash Rate

The number of flashes perceived in a train of suprathreshold flicker stimuli is identical with the real number of flashes only when $f < 3$ Hz. As the stimulus frequency increases above 3 Hz, a linear correlation still exists between the number of flashes in a stimulus train of different duration and the number of flashes perceived. The linear regression coefficient γ is then smaller than 1, i.e. the ratio of the perceived flash rate f_p and the flicker frequency f decreases as the stimulus frequency increases. FORSYTH and CHAPANIS (1958) used a hyperbolic function to describe this relationship, but their data can also be adequately fitted with an exponential function:

$$\gamma = \frac{f_p}{f} = \exp\,[(1 - f)/f^*]\,. \tag{22}$$

The constant f^* is 11–15 and decreases with retinal excentricity.

When the subjective flash rate f_p is measured by setting a click rate subjectively equal to the flash rate, f_p decreases for a fixed f with excentricity (LICHTENSTEIN et al., 1963).

Despite the evident differences between subjective flicker rate and stimulus frequency, another type of psychophysical experiment nevertheless indicates that the stimulus frequency is represented at all frequencies below the CFF by the neuronal impulse patterns in central parts of the a.v.s. Diffuse, dichoptic stimuli of different frequencies (f_1 and f_2), projected to corresponding areas of each retina, cause visual beats. The subject sees transient increments of brightness. The beat rate corresponds to the difference of the frequencies ($f_1 - f_2$), i.e. the physical beat between the stimulus frequencies. Therefore, one can conclude that the frequencies f_1 and f_2 must be accurately represented by the neuronal activity of the a.v.s. beyond the neuronal level of binocular fusion (KARRER, 1967, 1968, 1969). The perception of visual beats is only slightly affected when the luminance of the dichoptic flicker stimuli is unequal. The perceived brightness of the beats decreases as the frequencies f_1 and f_2 increase, it also decreases as the beat frequency increases. KARRER (1969) concluded from these findings that the processing of temporal phenomena within the a.v.s. is at least partially independent of the processing of the information about brightness.

7. Subjective Patterns in a Flickering Homogeneous Field

A homogeneous bright field, flickering above 6–8 Hz and below the CFF, elicits subjective patterns which were extensively described by PURKINJE (1823). The patterns change as the frequency of the intermittent stimuli changes; they are described as inconsistent checkerboards, stars, grids, spirals etc. (SMYTHIES, 1959a, b; WELPE, 1970). Most observers also see rotatory or translatory movements of these subjective patterns at certain stimulus frequencies. The flicker patterns might also appear for some observers slightly colored (c.f. pp. 459–460). When uniform test fields are alternated with fields filled with small stationary random dots, the random dots are seen in "quasi-brownian" motion (FIORENTINI and MACKAY, 1965).

The threshold luminance I_p, at which the subjective patterns appear, depends on the frequency f of intermittency (REMOLE, 1971):

$$\log I_p = \alpha f + \beta \,, \tag{23}$$

$$I_p = b \exp [\alpha f] \,. \tag{24}$$

The threshold I_p varies considerably between different observers (at 6 Hz between 0.1 and 1 cd/m^2 for a test field of 8.5 degrees diameter, $L : D = 1 : 1$). I_p is higher for blue and green stimuli than for red and yellow flicker stimuli of equal luminance. The wave length effects are modified by chromatic preadaptation (REMOLE, 1971).

In our opinion, the appearance of subjective flicker patterns indicates that cortical neurons which respond "normally" only to contours, contour interruptions etc. (see p. 527) are activated by diffuse flicker stimuli of a certain temporal frequency.

8. Difference Limens for Flicker Frequency

Despite the remarkable difference between the stimulus frequency and the perceived flicker frequency (p. 462), the a.v.s. has a very good ability to discriminate slight differences in the frequency of 2 simultaneously observed flickering test fields. The differential limens (Δf) of the light adapted eye are smaller than 1 Hz in the whole frequency range below the CFF. The relative difference limens ($\Delta f/f$) are in most measurements between 0.5 and 5 % (SCHWARZ and WINTZER, 1954; SCHWARZ et al., 1955; MOWBRAY and GEBHARD, 1955, 1960; GEBHARD et al., 1955). The relative differential sensitivity ($f/\Delta f$) decreases linearly with increasing flicker frequency between 5 and 35 Hz, if the stimulus is seen with the peripheral retina, while for foveal vision the maximum relative differential sensitivity is reached between 10 and 25 Hz (SCHWARZ et al., 1955; MOWBRAY and GEBHARD, 1960).

9. Fusion Frequency of Gestalt Properties

An intermittently illuminated pattern, i.e. a flickering triangle, a circle, a cross or a more complex pattern, disappears during the dark pauses, if the flicker frequency is below 2 Hz (rectangular stimuli). The spatial structure of a flickering pattern reaches a persistence at flicker frequencies above 2–6 Hz. Above these frequencies, the pattern does not disappear during the dark pauses. However, its brightness changes according to the flicker stimuli, while the light-dark contours do not seem to change at all. These findings indicate that besides the critical fusion frequency, which is a measurement for the upper temporal resolution of the brightness mechanisms within the a.v.s. a much lower fusion frequency exists, at which the Gestalt properties reach temporal persistence. We call this frequency G.F.F. (Gestalt Fusion Frequency). As Fig. 13 indicates, the G.F.F. is not or only slightly dependent on the stimulus luminance.

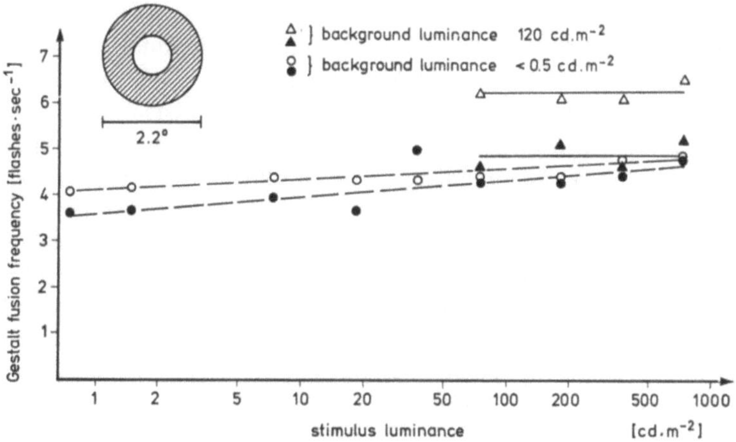

Fig. 13. *Fusion frequency (GFF) of the Gestalt properties of a simple stimulus (annulus).* The luminance of the annulus was modulated rectangularly in time ($L : D = 1 : 1$). Measurements were made at two levels of background luminance (128 cd/m² and < 0.5 cd/m²). Average values from two observers. The subjects were asked to find the frequency (ordinate) at which the intermittently illuminated annulus reaches persistence of its contours. Filled symbols: values from measurements with decreasing stimulus frequency; open symbols for increasing frequency. Data from an experiment done in cooperation with H. QUERFURTH and D. STANGE

E. Visual Phosphenes Elicited by Electrical Stimulation of the Retina

As all other sense organs, the retina responds also to inadequate stimuli. The visual sensations evoked by non-photic stimulation are called phosphenes. According to CLAUSEN (1955), electrically induced phosphenes were first described in 1755 by LEROY. Reviews of the findings of the 18th and 19th century were given by v. HELMHOLTZ (1896) and also by PFLÜGER (1865), who showed that the phosphene threshold follows his laws of electrotonic stimulation. In the context of the present review, phosphenes elicited by alternating electrical current are of interest, because this method provides a second tool for the measurement of the frequency response characteristics of the a.v.s. (reviews: CLAUSEN, 1955; MOTOKAWA, 1970). In the psychophysical experiments with human observers, the phosphenes are usually elicited by current applied via two electrodes, one placed on the forehead, the other on the temporal bone near the eye. Phosphene thresholds are lower, the nearer one electrode is placed to the eye. When sinusoidal or intermittent, rectangular, electrical stimuli of different frequencies are used, the threshold strength θ_p of the phosphene is a function of the stimulus frequency (SCHWARZ, 1940; MOTOKAWA and IWAMA, 1950; ABE, 1951; BOUMAN et al., 1951; CLAUSEN, 1955). Most authors agree that the minimum threshold in the light-adapted eye is reached around 20 Hz. According to the data published by SCHWARZ (1947,

Fig. 14a), an approximate linear relationship between the log of the phosphene threshold θ_p above and below 20 Hz and log f is valid:

$$\log \theta_p = \alpha' \log f + \beta' , \tag{25}$$

$$\theta_p = b' \cdot f^{\alpha'} \; [\mu A] . \tag{26}$$

The constants α' and β' depend on the background illumination. β' increases as the background luminance increases, while $|\alpha'|$ increases as the background luminance is reduced. α' is negative below 20 Hz and positive above 20 Hz. When the subject observes concentric light rings the phosphene flicker threshold for the foveal rings reaches its minimum θ_p-value at 33 Hz, while the peripheral rings have a θ_p-minimum at 20 Hz (MEYER-SCHWICKERATH and MAGUN, 1951). MOTOKAWA and EBE (1953), measuring the threshold-frequency relationship in much smaller steps than SCHWARZ, found for the threshold curve several subminima at about 7.5, 27.5, 35, 42.5 and 77 Hz (Fig. 14b). The maximum stimulus frequency at which flicker phosphenes can be elicited is reached between 100 and 120 Hz. Measurements at higher frequencies would require stimulus intensities at which strong pain sensations are aroused.

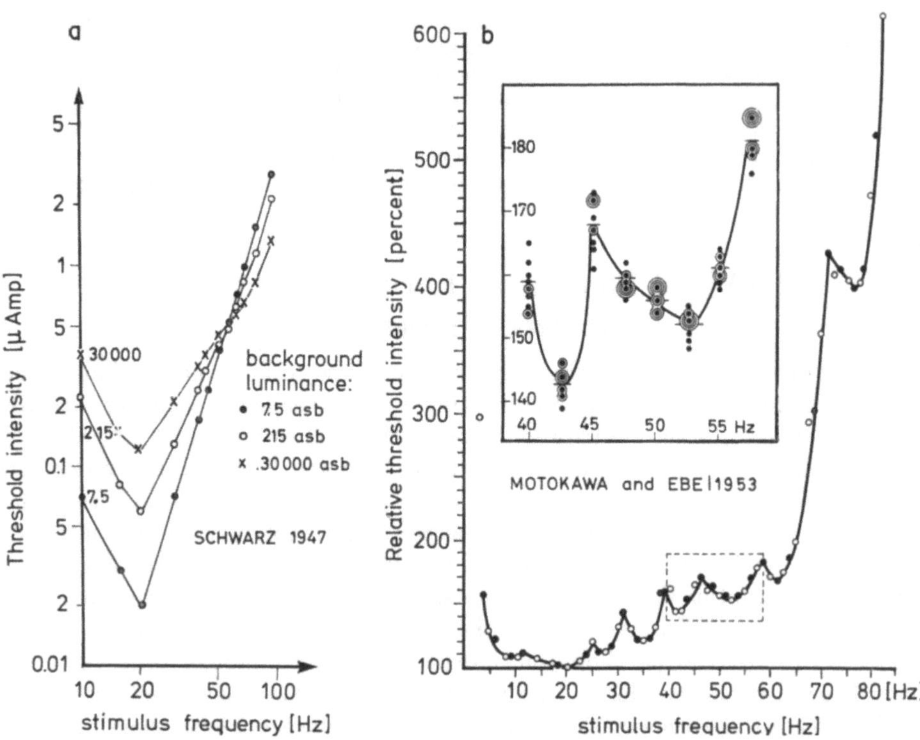

Fig. 14a. *Phosphene threshold intensity (ordinate), obtained with alternating current stimulation of the eye of a human observer* adapted to three different levels of background illumination. Abscissa: frequency of a.c. (SCHWARZ, 1947)

Fig. 14b. *Relative phosphene threshold intensity (ordinate) obtained with alternating current stimulation of human eye.* Abscissa: Frequency of the a.c., linear scale. Points of different marks were obtained from the same subject on different days. Inset: part of the curve between 40 and 60 Hz a.c. enlarged (MOTOKAWA and EBE, 1953; MOTOKAWA, 1970)

In the light adapted eye, the flicker phosphenes are elicited up to higher stimulus frequencies than in the dark adapted eye. The frequency at which the minimum threshold is reached is also lower in the dark adapted eye (3—10 Hz, square wave stimuli or sine wave stimuli, Lohmann, 1940; Bouman and ten Doesschate, 1969).

At stimulus frequencies above 110 Hz, at strengths below flicker threshold, subjective phosphene *patterns* are aroused by the AC current (Lohmann, 1940). These patterns disappear above 220 Hz. Regularly or irregularly oriented grids, star shaped figures etc. are reported, which frequently appear to be colored. The spatial distance between the stripes of the phosphene pattern is smaller, the higher the stimulus frequency.

Motokawa (1970) assumes that the minima found in the threshold frequency curve are caused by different frequency optima of the different color mechanisms in the light adapted eye. This idea is supported by the results of Takahashi et al. (1956) who measured the electrical excitability at 36, 42 and 77 Hz AC and different chromatic adaptation. The electrical excitability of the retina to 36 Hz AC reaches a maximum with blue light. At 42 Hz AC green stimuli and at 77 Hz red stimuli enhance the sensitivity most (c.f. Motokawa and Ebe, 1953). In the dark adapted eye the resonance curves shift to lower frequencies. A maximum enhancement of the electrical excitability is found for chromatic light of 507 nm and 565 nm at about 20 Hz. Barnett (1941) found a second resonance optimum in the dark adapted eye at 7.5 Hz, which is believed to be also caused by rod mechanisms.

In summary, one can conclude from the experimental findings on AC-induced visual phosphenes that the a.v.s., stimulated beyond the input level of the receptors, is able to follow to higher stimulus frequencies than for light stimulation. This finding indicates that *one* limitation of the flicker fusion frequency might be located within the receptor cells. The phosphene data further indicate that the neuronal substrates of different chromatic mechanisms seem to have also different frequency properties. Hence, not only color depending mechanisms located in different types of receptors, but also color depending mechanisms of the retinal neuronal network differ in the color specific frequency transfer properties.

The observations made with simultaneous stimulation of the eye by intermittent electrical and light stimuli clarify the problem of the site of action of the electrical current in the retina. According to the findings of Brindley (1955) this is either the synaptic region between receptors and bipolar cells or the bipolar cells. The visual beats which appear if flicker stimuli and electrical stimuli have a slightly different frequency, are still perceived at light stimulus frequencies far above the CFF. Therefore, Brindley (1962) concluded that, above the psychophysically determined photopic CFF, the overall output signals of the photoreceptors of the human retina still follow the frequency of the light stimulus (< 120 Hz). This conclusion is indeed consistent with neurophysiological findings about the limitation of the flicker threshold of retinal neurons (c.f. p. 477).

Veringa (1964) found that sinusoidally modulated current enhances or suppresses the flicker perception of a sinusoidally modulated test field, if the phase angle φ_c between current and light is appropriately adjusted. φ_c depends on the flicker frequency. It is the larger, the higher the flicker frequency, φ_c reaches under photopic stimulus conditions about 420 degrees at 60 Hz, at scotopic conditions about 800 degrees at 18 Hz (VERINGA and ROELOFS, 1966). VERINGA hypothesized that φ_c is the phase angle by which the foremost process of vision in the receptors is delayed, before it reaches the site of interaction with the AC-current.

Part 2. Neurophysiology of Visual Dynamics

III. Intracellular Recordings of Single Retinal Neurons

A. Receptor Potentials

Intracellular recordings from the inner segments of vertebrate photoreceptors were obtained in both rods and cones by several authors (review see TOMITA, 1970). To our knowledge, however, no systematic investigation of the *flicker response* of single cones or rods was performed. Therefore, the response of the mass potential of the eye (electroretinogram, ERG) has to be used for an approximate analysis of the frequency response properties of retinal receptors. This analysis might be facilitated by occlusion of the arteriae centrales retinae. After electro- or photocoagulation of these vessels, nearly all retinal neurons except the photoreceptors degenerate in the cat retina (STONE, 1969). From such a retina a slow ERG response, probably representing only the rod activity, can be recorded (MAFFEI and POPPELE, 1968). The amplitude of the response of this mass potential, elicited by sinusoidally modulated diffuse light stimuli is attenuated above 2 Hz and reaches its CFF below 25 Hz.

In the normal ERG, the a-wave is probably caused predominantly by the summated receptor potentials of the eye. One can conclude from the latencies and the frequency components of the ERG in the response to single flashes that at a stimulus frequency above 10 Hz, the AC-component of the ERG response probably consists mainly of the a-wave component. Therefore, the measurement of the ERG-CFF might also provide some information about the temporal response characteristics of the photoreceptors. At lower and medium frequency flicker, the photoreceptor component of the ERG is certainly masked by potentials from other sources (c.f. GRANIT, 1947, 1955). Hence, the frequency response characteristics of the ERG have to be interpreted very carefully.

The response of the ERG to flicker stimulation was intensively investigated by numerous authors in many species of vertebrates. The quantitative relationship between the CFF of the ERG and the stimulus luminance is well described by a logarithmic function (Fig. 15b):

$$\mathrm{CFF} = k_e \log \frac{I}{I_s} \,[\mathrm{Hz}] \,. \tag{27}$$

For animals with a pure rod retina, the CFF of the ERG does not exceed 20–22 Hz. In accordance with these experimental findings, DODT et al. (1967) found a maximum CFF of 17 Hz in the ERG of a totally color-blind human observer (rectangular flicker stimuli, 1000 td, $L:D = 1:1$). The constant k_e of Eq. (27), as in psychophysical experiments, is also for the ERG response significantly smaller for rod vision than for cone vision. Therefore, for a mixed cone-rod retina, a kink is

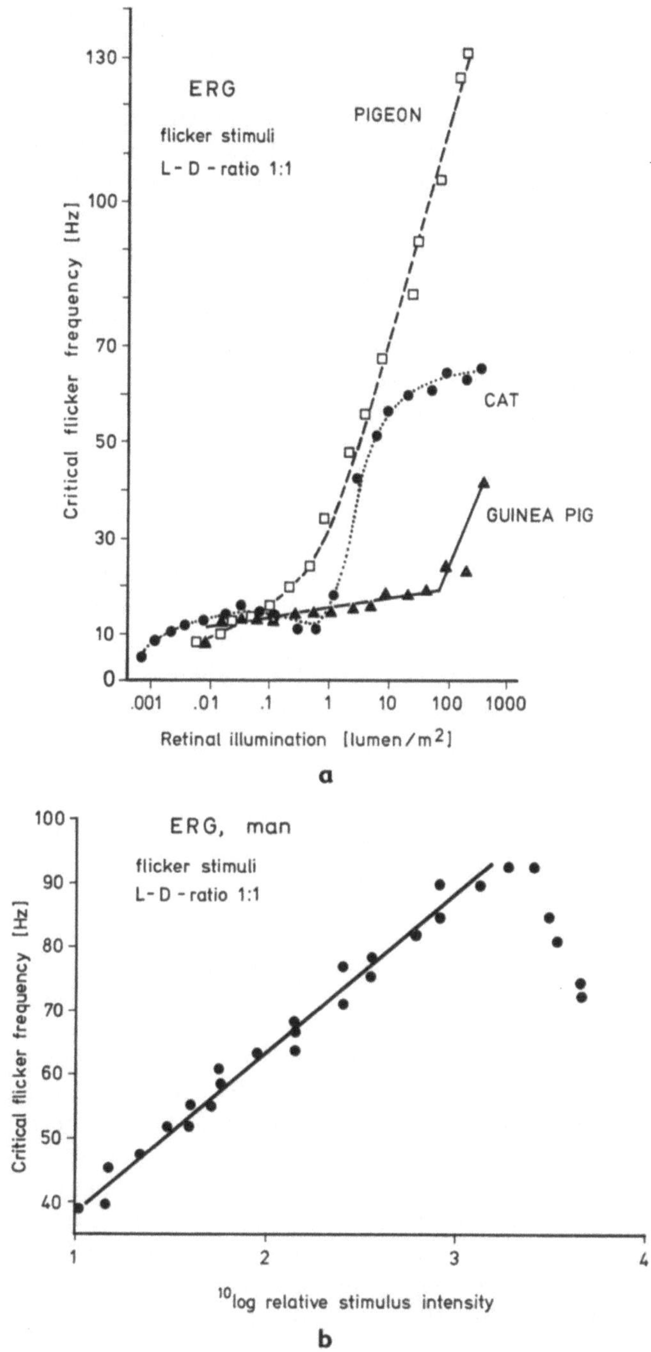

Fig. 15a and b. *Relationship between CFF of ERG (ordinate) and intensity of diffuse rectangular flicker stimuli* ($L:D = 1:1$). a The examples are from an animal with a cone dominated retina (pigeon), an animal with a mixed retina (cat) and an animal with a rod dominated retina (guinea pig). The maximum CFF reached with high intensities depends on the number of cones in the retina. (After HRACHINOVA and B. SCHMIDT, 1968). b Human retina. Photopic flicker stimuli (HECK, 1957)

usually found in the curve of the relationship between CFF and log stimulus luminance at mesopic stimulus intensities (Fig. 15a). The CFF-intensity function of the ERG of a pure cone retina (e.g. European ground squirrel, *Citellus citellus*, or Iguana) does not have this kink (MENEGHINI and HAMASAKI). The maximum CFF in the ERG of pure cone retinae is reached between 80 and 120 Hz. The CFF of the human ERG reaches its maximum values around 90 Hz (DODT, 1953, 1954; DODT and ENROTH, 1954; HECK, 1957; BORNSCHEIN, 1961, 1964; BORNSCHEIN and SZEGVARI, 1958; BORNSCHEIN and LAHODA, 1964; BORNSCHEIN and SCHUBERT, 1953; MENEGHINI and HAMASAKI, 1967; HRACHOVINA and B. SCHMIDT, 1968).

One can conclude from these findings that these frequency response characteristics of the rods are significantly different from those of the cones. The rod response is attenuated at significantly lower frequencies than the cone response. This statement is supported by the comparison of the temporal courses of the *intracellularly recorded receptor potentials*, which are significantly slower for rods in comparison with cones (TOYODA et al., 1970; TOMITA, 1970).

The same conclusion about different frequency response characteristics of rod and cone systems can also be reached by the analysis of the ERG-responses elicited by suprathreshold sinusoidal flicker stimuli. As mentioned, however, the ERG obtained under such stimulus conditions does not only reflect the activity of the receptors. The response maximum for sinusoidal light stimulation is found to be between 2 and 4 Hz for *rod dominated* retinae, while *cone dominated* retinae give a response maximum between 12 and 25 Hz. (REUTER, 1972; CERVETTO, 1968; GOURAS and GUNKEL, 1962, 1964; LEGEIN et al., 1969).

Despite the missing intracellular recordings of flicker responses of single retinal receptors, it seems safe to conclude from the presented ERG-data and from additional psychophysical indications (p. 466) that rods and cones have different frequency transfer properties. The upper frequency limit is significantly lower for rods than for the cones.

B. Bipolar Cell Potentials

Intracellular recordings from bipolar cells were obtained in the retina of fish (KANEKO and HASHIMOTO, 1969; KANEKO, 1970) and of the mudpuppy *Necturus* (WERBLIN and DOWLING, 1969; WERBLIN, 1970). The response of the membrane potential of bipolar cells to light stimulation of the receptors of its receptive field (RF) is either a depolarizing or a hyperpolarizing slow potential. An *antagonistic* organization of RF-center and of the RF-periphery was found. No or only rudimentary action potentials are present. TOYODA (1971, personal communication) investigated in goldfish the responses of bipolar cells, horizontal cells and amacrine cells to sinusoidal light stimuli of different frequencies ($m = 0.15$). Some of his data are shown in Fig. 16. A maximum amplitude of the bipolar cell potential is reached at stimulus frequencies between 2 and 3 Hz. At a stimulus frequency below 1–2 Hz, only a slight temporal delay between the sinusoidal stimuli and the neuronal response appears, while above this frequency level, the negative phase angle of the response increases with increasing stimulus frequency and reaches about 2π at 10 Hz. The frequency response characteristics of the goldfish retinal neurons, however, can be used only with restriction as a model of the mammalian retinal response, because the experiments were done in isolated retinae at room temperature. One can assume that the frequency response characteristics of comparable mammalian retinal neurons are shifted into a higher frequency range.

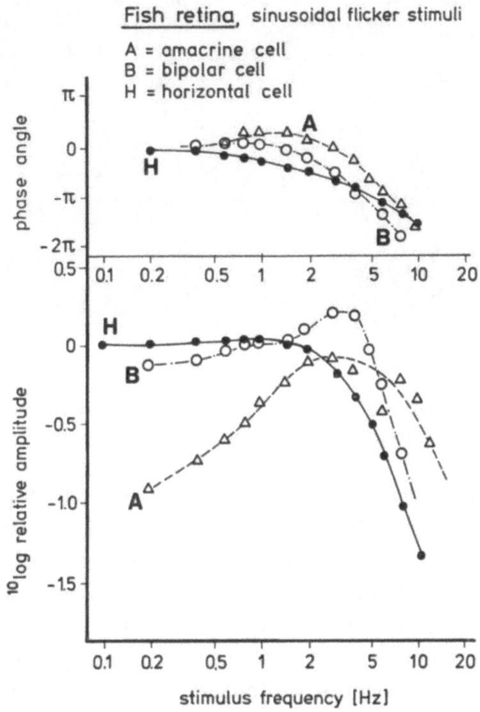

Fig. 16. *Amplitude frequency characteristics and phase response of an amacrine cell* (*A*), bipolar cell (*B*) and a horizontal cell (*H*) of the goldfish retina for diffuse light stimuli $m = 0.15$ (Toyoda, 1971; personal communication)

C. Horizontal Cell Potentials (S-Potentials)

Intracellular recordings from single horizontal cells were first reported by Svaetichin (1956), MacNichol and Svaetichin (1958) and Motokawa et al. (1957) in the fish retina. Similar intracellular responses from horizontal cells were obtained in the cat retina (Motokawa et al., 1957; Grüsser, 1957b, 1960, 1961; Jung et al., 1957; Brown and Wiesel, 1959). In the first papers, these *S*-potentials were referred to as "cone potentials" or "receptor potentials", because depth measurements of the microelectrode tip indicated an origin of these potentials in the synaptic layer between receptors and bipolar cells. The studies of Werblin and Dowling (1969) in the Necturus retina, of Tomita et al. (1969) and of Kaneko (1970, 1971) in the fish retina, and of Steinberg and Schmidt (1970, 1971) in the cat retina provided good evidence by means of intracellular dye injection through the recording microelectrode and histological examination that the sites of recording of the *S*-potentials are the horizontal cells.

In the cat retina, a white light or a monochromatic light of any wave length always leads to a hyperpolarizing response (*L*-type) of the horizontal cells. In the fish retina besides the *L*-type responses *H*-cell potentials can be recorded which exhibit a color dependent "biphasic" or "triphasic" response with hyperpolarization or depolarization depending on the wavelength of the monochromatic stimuli (two *C*-types).

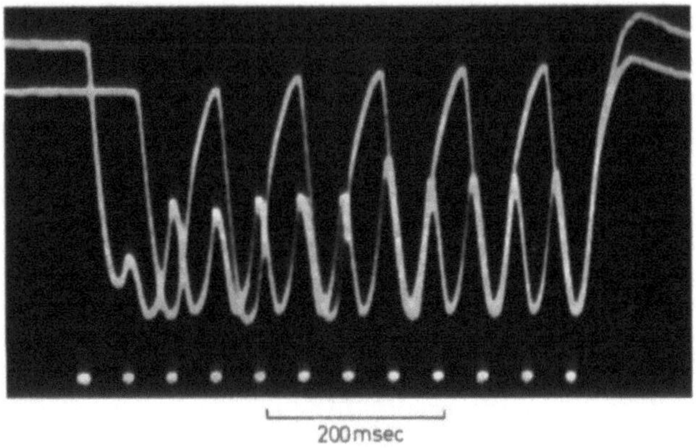

Fig. 17. *Intracellular recordings of a H-cell potential of the fish retina.* Responses to 100 μsec flashes delivered at a frequency of 10 and 20 flashes per sec (SVAETICHIN, 1956)

In his first report, SVAETICHIN (1956) described the *S*-potentials of the fish retina elicited by flickering flashes of constant duration (100 μsec) but different frequencies. He demonstrated that the AC-component of the fish *S*-potential responses is reduced to about one half as the flash frequency increases from 10 to 20 flashes per sec (Fig. 17). With bright flicker stimuli, responses appear as high as 50 Hz. SVAETICHIN pointed out the close relationship between the logarithmic function which relates the stimulus intensity *I* and the amplitude of the *S*-potentials and the logarithmic function which relates the stimulus intensity and the CFF of the *S*-potentials.

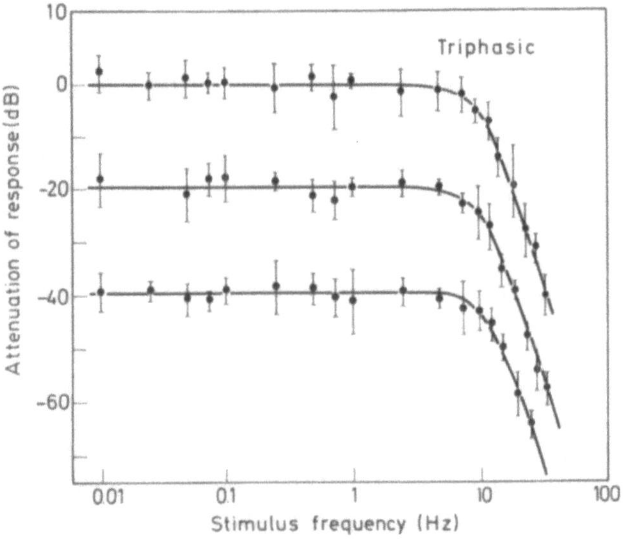

Fig. 18. *Amplitude frequency response characteristics of a (triphasic) color-depending H-cell potential* of the goldfish retina stimulated with monochromatic light of 3 different wave lengths (SPEKREIJSE and NORTON, 1971)

A careful study of the dynamic characteristics of the L-type and C-types of S-potentials in the fish retina was reported by SPEKREIJSE and NORTON (1970, carp retina). According to their findings, the behavior of the S-potentials of the fish can be described fairly well by a linear low pass filter system of the fourth order. The amplitude frequency characteristics of the 3 different types of S-potentials (2 C-types, 1 L-type) are rather similar, independent of the wave length of the stimuli. The phase characteristics of the responses, however, depend on the wave length of the stimuli. The high frequency attenuation amounts to 24 dB/octave and the cut-off frequency is around 11.5 Hz. The following transfer function (solid line in Fig. 18) describes the experimental data very well:

$$T(s) = \frac{\omega_0^4}{\left(s^2 + \frac{\omega_0}{Q}\, s + \omega_0^2\right)^2} \qquad (28)$$

whereby $\omega_0 = 2\,\pi f_0$ and f_0 is the cut-off frequency and $Q = 0.7$ is a constant.

The response of S-potentials in the mammalian retina (cat) to intermittent flicker stimuli (rectangular flicker, $L:D = 1:1$, or short flashes of < 1 msec duration) were investigated by GRÜSSER (1957, 1960). Recording examples from these experiments are shown in Fig. 19.

With rectangular flicker stimuli the amplitude of the H-cell potentials decreases as the stimulus frequency increases above 3 Hz. At the CFF the response "disappears" in the noise of the recording. In the photopic range a logarithmic relationship between the CFF and the stimulus intensity is valid (Eq. [5)]. When flickering light, having a frequency a few Hz below the CFF, is turned on and off the H-cell potentials follow the intermittent flicker stimuli only after a "non flickering part" lasting up to half a second.

In the experiments of GRÜSSER (1957, 1960), the maximum flicker frequency to which the H-cells responded was found to be about 75 Hz for a 200 Lux stimulus intensity and direct retinal illumination. In our recent experiments with much stronger sinusoidal flicker stimuli, single H-cells in the cat retina responded to frequencies up to 115 Hz.

Above 10–15 Hz, the H-cell response to rectangular flicker stimuli is nearly sinusoidal, i.e. the response is mainly determined by the fundamental frequency of the stimuli. Later experiments (GRÜSSER and LUNKENHEIMER, 1966; FOERSTER, VAN DE GRIND and GRÜSSER, 1973) with sinusoidally modulated light stimuli revealed that the response of H-cells in the cat retina clearly exhibit non-linear components at stimulus frequencies below 8 Hz, if $m > 0.4$ (Fig. 20a). At higher frequencies however, the H-cell responses compare fairly well with those of a linear filter. Fig. 20b shows the amplitude and phase response characteristics of H-cell potentials (Fig. 20a) obtained with sinusoidal light stimuli, projected to the receptive field (RF) of the investigated cells. For the attenuation in the amplitude frequency characteristics one can discriminate two different slopes at a frequency range above and below 25–30 Hz. The attenuation in a medium frequency range above 3 Hz is 6 db/octave or less, while the high frequency attenuation is about 36 db/octave (Fig. 20b).

No antagonistic surround is found in the RF of cat's H-cells. The forms of the frequency response characteristics in general do not change very much if the stimulus size is increased, however, the CFF increases significantly. The amplitude of the horizontal cell potentials (R_H) increases with the stimulus area ($\bar{I}$ and m = constant). The following relation fits the data well for a restricted frequency range (FOERSTER et al., 1972/1973):

$$R_H = k_H \cdot A^{0.5}\,[\mathrm{mV}]\,. \tag{29}$$

For the overall transfer properties of the horizontal cells, not only the AC-component of the response is of interest, but also the average DC-response. This is the sum of the hyperpolarizing DC-response and the averaged AC-components. Measurements of these values are obtained by turning on and off, for several seconds, sinusoidal flicker stimuli of different frequencies.

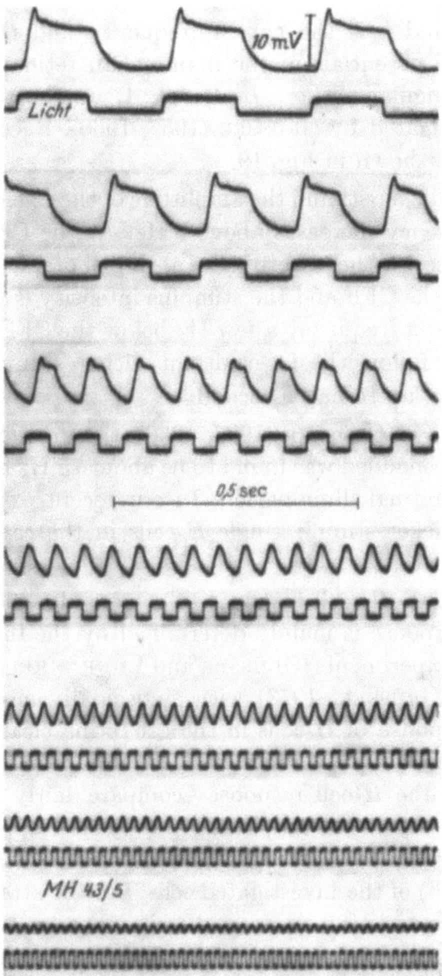

Fig. 19. *Intracellular recording examples from a H-cell potential of the light-adapted cat retina.* Direct retinal illumination, diffuse flicker stimuli 500 Lux (GRÜSSER, 1960). Note that in this recording the positivity of the potential is downward

We found that the time-averaged responses elicited by sinusoidal flicker stimuli change only little for stimulus frequencies above 5 Hz.

The amplitude of the H-cell potential (R_H) increases at constant I as the depth of modulation (m) of the sine wave stimuli increases. The relationship between R_H and m is well described by Eq. (33). At constant values of m and f the amplitude of the H-cell potential increases as I increases. A linear relationship between log I and R_H is valid for part of the experimental data (Fig. 20c). However, a tanh (log I)-function, analog to Eq. (12) predicts the data over a wider range than the simple log function (c.f. Nakas analysis of the H-cell potential of the fish retina, 1971). As Fig. 20c shows, the slope of the intensity function decreases above 5–8 Hz. There exists, in contrast to the findings about the intensity function of retinal ganglion cells (p. 476), no increase of this slope between 1 and 8 Hz.

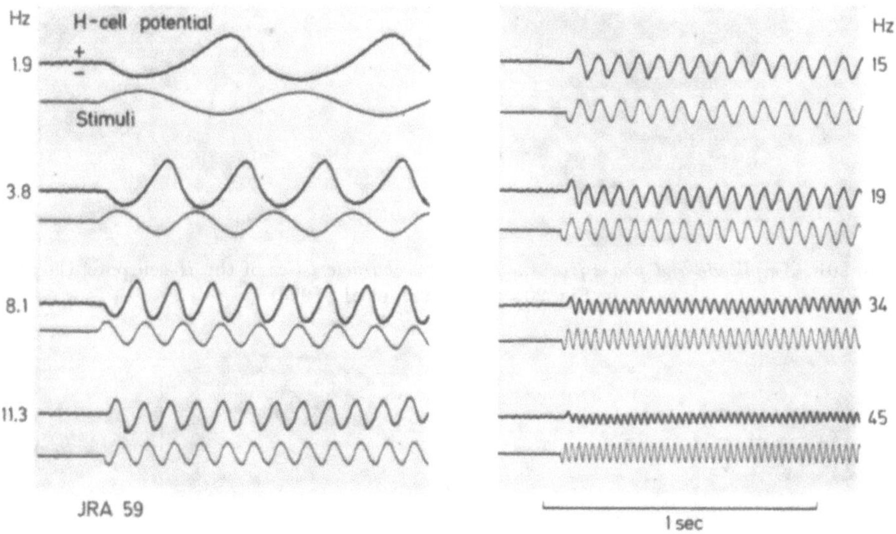

Fig. 20. *H-cell responses from the cat's retina to sinusoidal stimuli*

Fig. 20a. *Intracellular recording of H-cell* potential of the light adapted cat retina. Sinusoidal stimulation ($\bar{I}$ about 1600 cd · m^{-2}, $m = 0.9$, stimulus diameter 4.3 degrees)

The spectral sensitivity curve of light-adapted S-potentials reaches its maximum at about 560 nm and resembles Granits photopic dominator of the cat retina (Granit, 1955; Steinberg, 1969). With dark adaptation and a simultaneous reduction of the stimulus intensity down to scotopic levels, the spectral sensitivity of the S-potentials shifts to a maximum at 505 nm. From these findings, Steinberg concluded that cones *and* rods contribute to the response of the S-potentials in the cat's retina. With dark-adaptation and scotopic stimulus intensities, the H-cell potential becomes significantly slower. Therefore, one can postulate that the frequency response characteristics of H-cell potentials depend on the level of retinal adaptation. In the dark-adapted retina, the responses to single light flashes have frequency characteristics similar to those found for the rod mechanism (p. 469). No systematic measurements of the flicker response of H-cell potentials in the dark adapted eye were performed as yet.

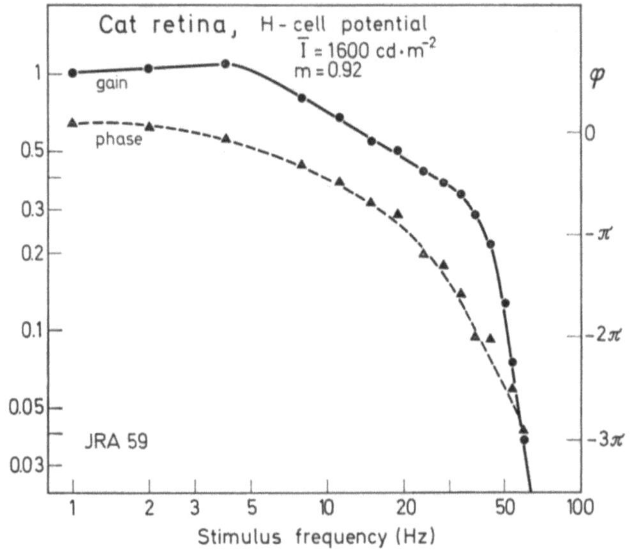

Fig. 20b. *Amplitude and phase frequency response characteristics* of the H-cell potential shown in Fig. 20a (FOERSTER et al., 1973)

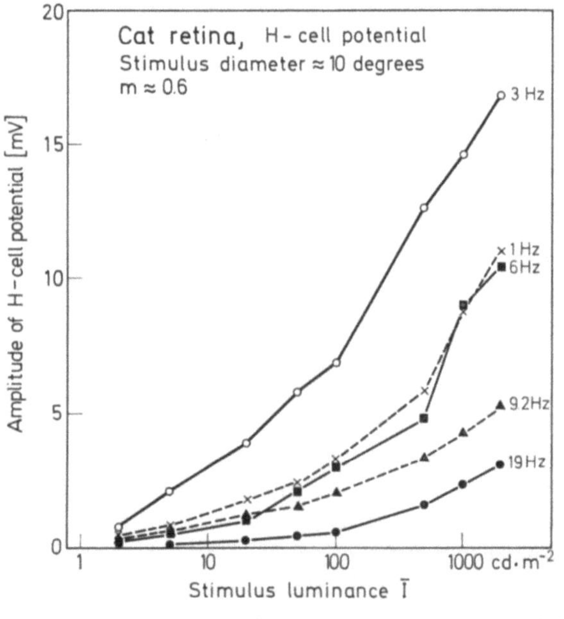

Fig. 20c. *Intensity function* of intracellularly recorded H-cell potential of the cat retina (light adapted). Sinusoidal stimuli of different frequencies, $m \approx 0.6$. Stimulus diameter ≈ 10 degrees (FOERSTER et al., 1973)

D. The Response of Amacrine Cells

Intracellular recordings from amacrine cells were obtained in the retina of the mudpuppy and of the goldfish (WERBLIN and DOWLING, 1969; WERBLIN, 1970, 1972; KANEKO, 1971, 1972). The response of amacrine cells to illumination of the RF-center is characterized by a transient on-off depolarization on which rudimentary short impulses are sometimes superimposed. In the cat retina, FOERSTER and VAN DE GRIND (1973) recorded similar intracellular responses in the layer between ganglion cell and horizontal cell. However, they did not identify the recorded cells by dye injections.

TOYODA (1971, personal communication) studied in the goldfish retina the response of single amacrine cells to sinusoidal stimuli of different frequencies ($m = 0.15$). The results of one of his experiments are shown in Fig. 16. It is evident from these curves that amacrine cells, in contrast to the horizontal cells, show strong tuning phenomena in the frequency range between 2 and 5 Hz. The relative enhancement at the resonance frequency is also higher in the amacrine cells than in bipolar cells.

E. Glial Cell Responses

The glial cells of the retina (Müller-cells) respond to a single flash with a delayed and slow membrane depolarization (*Necturus*, MILLER and DOWLING, 1970). The temporal course of the glial cell response corresponds very well to the *b*-wave of the ERG. This correlation applies also to intermittent light flashes; for both the glial cell responses and the *b*-wave, the CFF is reached at very low stimulus frequencies around 1–3 Hz.

F. Intracellular Recordings from Ganglion Cells

Intracellular recordings from ganglion cells of the cat's retina by WIESEL (1959) proved that the antagonistic organization of the receptive field (RF) first described on the basis of extracellular recording techniques (KUFFLER, 1952, 1953) is also present in the slow membrane potential responses: For on-center neurons light stimuli projected in the RF-center cause a depolarization, while light stimuli projected in the RF-periphery lead to a hyperpolarization of the membrane potential. Light off in the RF-center elicits a hyperpolarization, while light off in the RF-periphery leads to a membrane depolarization accompanied by an increase of the neural impulse frequency. In recent experiments, we measured the frequency response characteristics of the slow membrane potentials of single retinal ganglion cells in cats for rectangular and sinusoidally modulated flicker stimuli, projected to the RF-center (FOERSTER and VAN DE GRIND, 1973, Fig. 21, 22). To obtain a better quantitative measurement of the membrane potential changes a number of response-cycles for the same stimulus frequency were graphically superimposed, whereby the generated impulses were neglected. From the data of such a "superimposogram" one can plot an amplitude frequency response curve (Fig. 22). As one sees from these results, the amplitude of the ganglion cells' slow potential increases at frequencies between 7 and 20 Hz, if photopic flicker stimuli are applied. Above 30 Hz the amplitude gain curve of the postsynaptic potential changes can approximately be fitted by a straight line with a slope of 10–16 dB/octave. Above the stimulus frequency at which the ganglion cells stop to respond to the flicker stimuli with at least 1 impulse per period (CFF), one can see that the subthreshold membrane potential changes still follow the flicker stimulus (Fig. 21). These findings

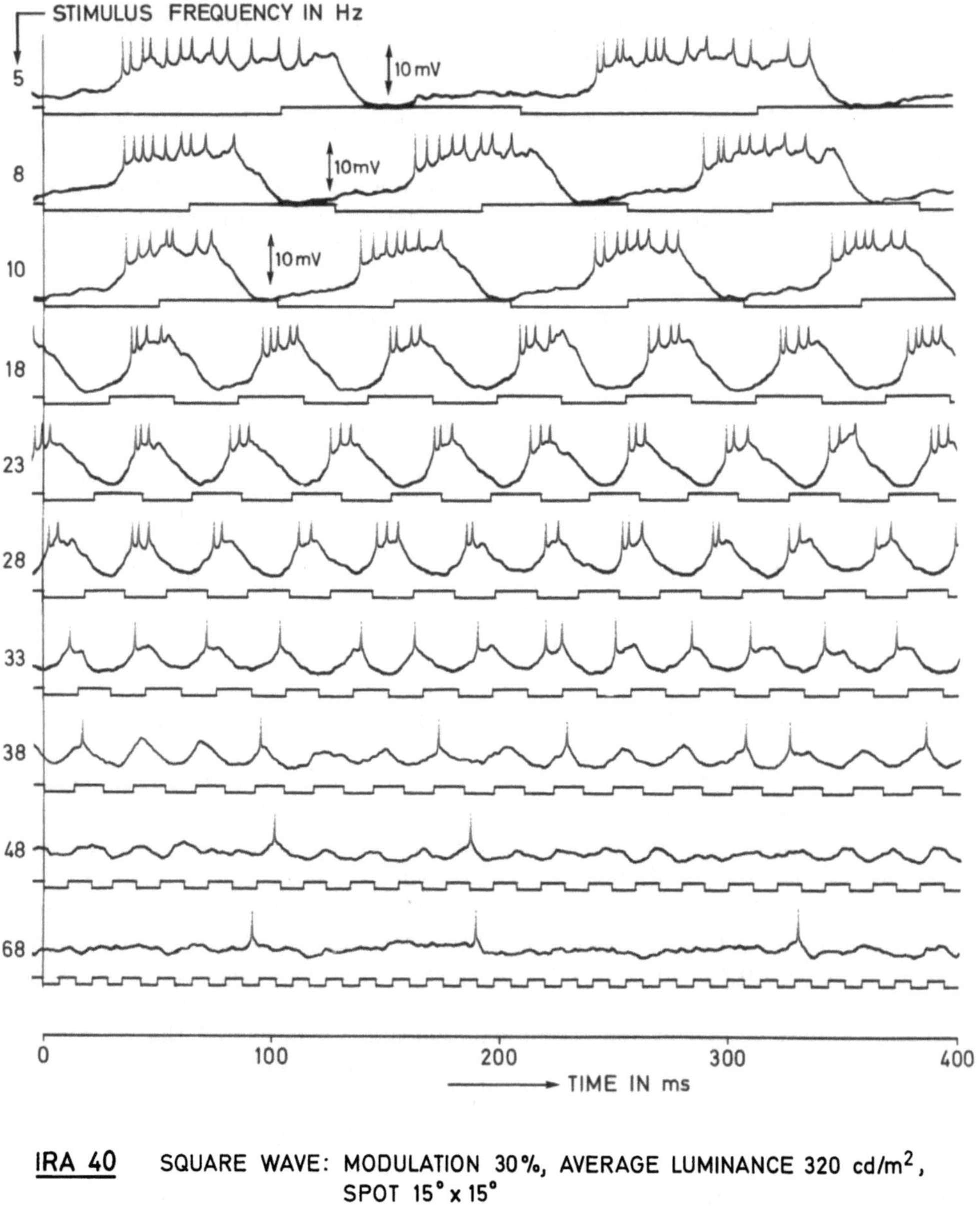

Fig. 21. *Intracellular recording of slow membrane potential changes and impulse sequence elicited by square-wave modulated flicker.* Stimuli of 15° × 15° (square) centered on the RF of a ganglion cell (off-center) in the light-adapted retina. Membrane potential without light stimulation approximately — 70 mV. The action potentials were decreased in amplitude by filtering (0—2.5 kHz) during playback and additional optical clipping during filming in order to emphasize the slow membrane potential. FOERSTER and VAN DE GRIND (1973)

indicate that the mechanisms which determine the CFF of the a.v.s. include the impulse generating processes of the ganglion cell membrane. The attenuation of the *H*-cell potential (Fig. 20b) begins at lower frequencies than that of the slow membrane potential of the ganglion cell (Fig. 22). The initial slope of the *H*-cell potential's gain-frequency curve between 3 and 30 Hz is not very steep, but in the high frequency range it might be 36 dB/octave or more (Foerster, van de Grind, Grüsser, in preparation, 1973).

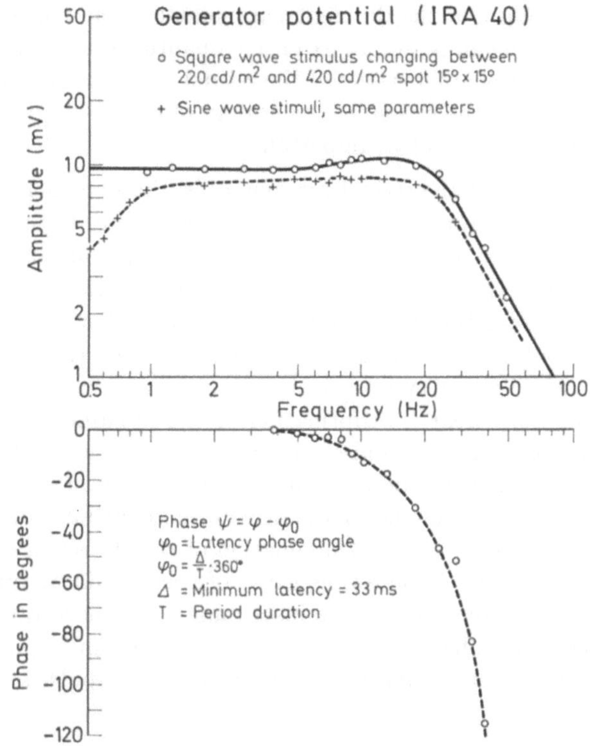

Fig. 22. *Amplitude- and phase-frequency response characteristics of the intracellularly recorded membrane potential* changes of an off center ganglion cell in the light adapted cat retina (c.f. Fig. 21) (Foerster and van de Grind, 1973)

IV. The Output Signals of the Retina

A. Introduction

In the preceding chapter, the intracellular responses to flicker stimuli were described for the different neurons of the retinal network. Some important information was gained by these direct intracellular recordings about the contribution of the different neuronal elements to the flicker response of the neuronal network, which forms the "perceptive unit" of a retinal ganglion cell. At the present time, however, there still exists a technical barrier for the full utilization of the analytical power of intracellular recordings in the mammalian retina: It is difficult to obtain stable intracellular recordings from the mammalian retina for a period longer than a few minutes. The longest time that we were able to keep an ultrafine micropipette intracellularly in a horizontal cell of the cat retina was about 70 min, while for stable intracellular recordings of ganglion cells, 20 min was the maximum except for two cases of 50 and 90 min respectively (FOERSTER and VAN DE GRIND, in prep. 1973). In addition, one probably selects giant ganglion cells for stable intracellular recordings. These short periods of intracellular recordings relative to the painstaking and time consuming preparation are insufficient for a complete investigation of the effect of the different parameters of the flicker stimuli on the neuronal response. Therefore, for an extensive quantitative analysis of the spatial and temporal transfer characteristics of the retinal neuronal network, the analysis of the output signals of the retina, i.e. recordings of the action potentials of single optic nerve fibers, is still of great importance.

At the axon hillock of the retinal ganglion cells, the slow membrane potentials of the ganglion cells are transformed into a sequence of all-or-nothing impulses. This transformation (encoder process) is certainly *not* a simple linear process. A linear relationship between the membrane depolarization and the impulse rate is only approximately valid for the steady state. For relatively fast changes of the membrane potential containing frequency components above 5 Hz, a non-linear transformation of the ganglion cell membrane potential to the instantaneous impulse rate takes place. This non-linear transformation of the dynamic components of the membrane potential into an impulse sequence is probably a general property of many nerve cells; it is also found for the transformation of a receptor potential into an impulse sequence (muscle spindle, EYSEL and GRÜSSER, 1970). In the latter case, it was shown that a temporal summation of the threshold elevations which remain after each impulse is the prime cause of the non-linear transformation of the slow membrane potential into the instantaneous impulse rate. This "self inhibiting" mechanism probably applies generally to the transformation of dynamic changes of the nerve cell membrane potentials into an impulse sequence (see also pp. 542–543).

The action potentials of the ganglion cells can be recorded directly from the retina, as GRANIT and his co-workers and KUFFLER did. With this technique, action potentials of large ganglion cells are selected (RUSHTON, 1949). Another place of access for microelectrodes to the output signals of the retina is the optic disc. A rather easy way to record the impulse patterns of single retinal ganglion cells is provided by microelectrodes which are stereotactically inserted into the optic tract (WIESEL, 1960). This technique was applied in most of the studies in our laboratory during the last 10 years. With this recording technique, however, one probably also selects action potentials of the larger optic tract fibers, i.e. the action potentials of the non-myelinated fibers of the optic tract, which are present in most species, are not recorded by the tungsten microelectrode.

Beginning with the pioneer work of CH. ENROTH (1952), the responses of retinal ganglion cells to flicker stimuli were first investigated with diffuse illumination of the whole retina. During the last 10 years, flicker stimuli with more complex spatial properties (different stimulus sizes, contrast borders. spatial gratings etc.) were applied for the analysis of the temporal and spatial transfer characteristics of the neuronal network of the retina. If not mentioned otherwise, the cat was the experimental animal in all the microelectrode studies mentioned in the remaining sections of this part of the paper.

B. Classes of Retinal Ganglion Cells

In the ganglion cell layer of the vertebrate retina, 3 different types of neuronal responses were found (Hartline, 1938a, b, 1940, frog; Granit, 1947): *on-neurons, off-neurons* and *on-off-neurons*. The area of the retina from which light flashes cause activation or inhibition is called the receptive field (RF) of the ganglion cell. The set of receptors and other elements of the neuronal network which are functionally connected with a single ganglion cell is designated *perceptive unit* (= retinal sampling unit, van de Grind et al., 1970). Kuffler (1952, 1953) discovered the meanwhile "classical" functional organization of the RF of mammalian retinal ganglion cells: The RF can be subdivided into concentrically arranged zones of different functional values. In *on-center neurons*, the RF-center is activated by "light on" and inhibited by "light off", while the RF-periphery is activated by "light off" and inhibited by "light on". The RF-organization of the off-center neurons is the mirror image to that of the on-center neurons (off-activation and on-inhibition in the RF-center, on-activation and off-inhibition in the RF-periphery).

A further subdivision of the retinal on-center and off-center neurons is possible on the basis of their spatio-temporal summing properties (Enroth-Cugell and Robson, 1966; Cleland et al., 1971b, Saito et al., 1970, 1971). The on- or the off-center neurons of class I respond to a longer lasting change of the illumination of the RF-center with transient activation or inhibition, while the class II-neurons exhibit a sustained response to center stimuli. The inhibitory effect caused by stimulation of the RF-periphery is stronger in class II-neurons. The average conduction velocity is higher in class I on-center and off-center neurons than in class II-neurons, i.e. class I-neurons have optic nerve fibers of a larger diameter than class II-neurons. The average RF-center is found to be larger in class I than in class II-neurons. This subdivision of retinal ganglion cell responses, however, depends also on the state of anaesthesia. In encephale isolé preparations the higher the amount of additionally injected barbiturates, the more retinal neurons are found to respond according to the transient type. In addition, a gradual transition from a sustained type response (X-cells of Enroth-Cugell and Robson) to the transient type response (Y-cell) can be observed when the response of the same retinal neuron is recorded in encephale isolé animals without anaesthesia and at 3–8 different levels of pentobarbital anaesthesia (5–50 mg/kg cat, Foerster and Grüsser, 1969).

It seems rather probable that most of the "on-neurons" described in the neurophysiological papers before 1952 (c.f. Granit, 1947, 1950) are *on-center* neurons and the off-neurons are off-center neurons. Part of the on-off neurons of Granit are probably neither on-center nor off-center neurons, but units which have a specialized receptive field, as described in the cat retina recently by Spinelli (1966, 1967) and Stone and Fabian (1966).

With chromatic light stimuli of different wave lengths, WOLBARSHT et al. (1961) explored the RF-organization of ganglion cells in the goldfish retina. They found in some units a color-depending antagonistic organization of the RFs, which might also be present in part of the ganglion cells of the monkey's retina (c.f. GOURAS, 1968, 1971). The response of retinal ganglion cells to chromatic flicker stimuli, however, was not explored in these species. Therefore we refrain from giving an elaborate description of these interesting color-depending organization principles of the RF.

C. Responses of Retinal Ganglion Cells to Diffuse Flicker Stimuli

ENROTH (1952) and DODT and ENROTH (1954) investigated the responses of single retinal ganglion cells in the dark-adapted and light-adapted retina of the cat. Recording examples from these investigations are shown in Fig. 23. ENROTH defined the critical flicker frequency of the ganglion cells as that stimulus frequency, at

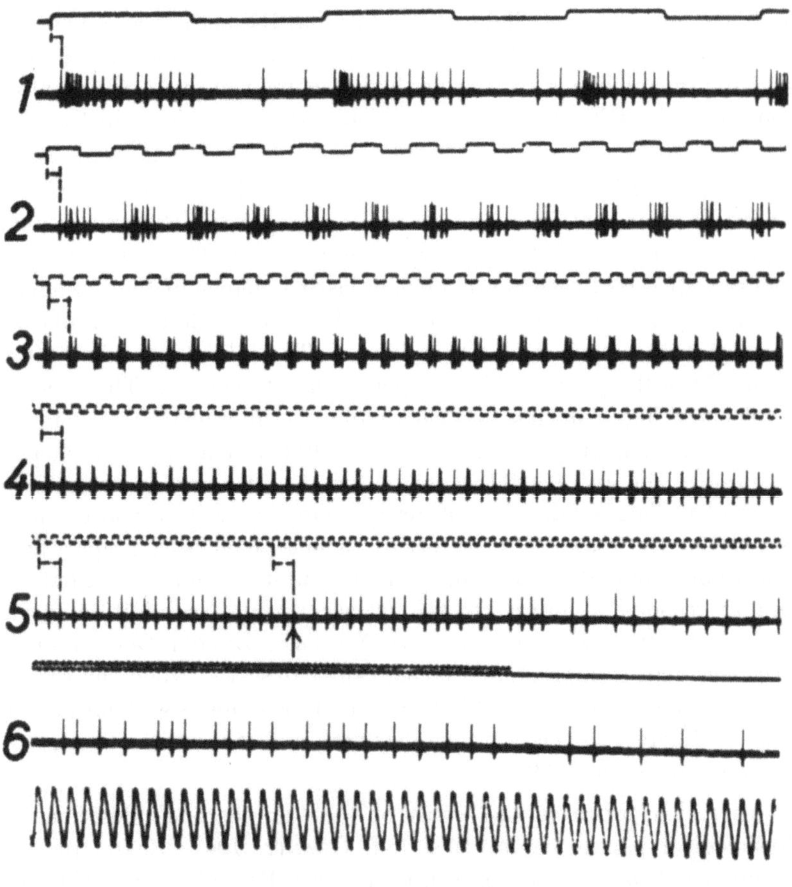

Fig. 23. *Response of an on-center neuron in the light-adapted cat retina to diffuse rectangular flicker stimuli* ($L:D = 1:1$) of increasing frequency. Latency marked by broken lines at the beginning of each record. The CFF (L) is reached at 72 flashes per sec. Direct retinal illumination 5500 Lux intensity (DODT and ENROTH, 1954)

which not every period of the flicker stimulus elicits at least one neuronal impulse. Above the CFF, however, the neuronal impulses are still correlated with the flicker stimuli (1 : 2, 1 : 3, 1 : 4 . . . "scaling", c.f. p. 510). The range of this frequency scaling of the neuronal response extends, as a rule 5—15 Hz above the CFF. The stimulus frequency above which, even an elaborate computer analysis shows no correlation between flicker stimuli and neuronal discharges is further called TFF (Talbot Fusion Frequency). A close positive correlation exists between the frequency values of the CFF and of the TFF; therefore the "laws" described on the following pages for the CFF are also valid for the TFF, if a scaling factor is taken into account.

On-neurons and off-neurons behave differently at stimulus frequencies above the CFF. The *sustained* on-center neurons exhibit an impulse rate which is identical to that elicited by a steady state stimulus of the same luminance as the time-averaged flicker stimuli. This is the neurophysiological correlate of the Talbot-Plateau law (Eq. (4). p. 438). Off-center neurons, however, are mostly inhibited by high frequency flicker stimuli in the same way as they are inhibited by a steady light of the same average luminance. *Transient* on-center neurons discharge at a very low rate above the TFF.

The CFF and the TFF of the retinal ganglion cells depend on the stimulus intensity (Fig. 24) and follow roughly the Ferry-Porter law [Eq. (5)].

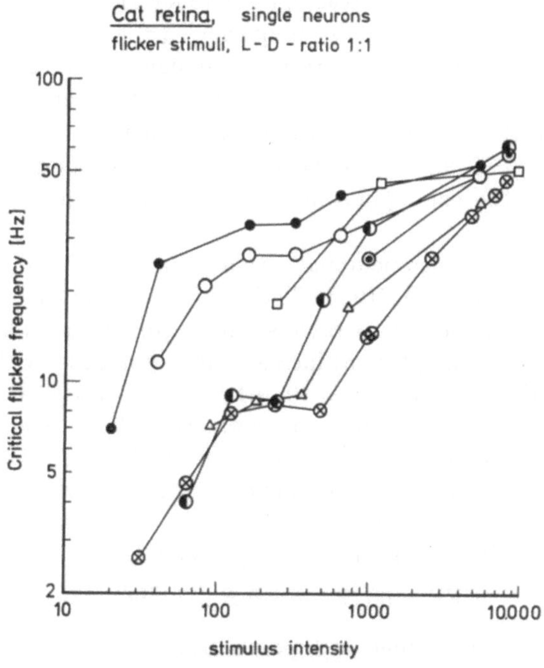

Fig. 24. *Relationship between the stimulus intensity (abscissa) and the CFF* (ordinate) of 7 different ganglion cells in the light-adapted cat retina. Diffuse flicker stimuli, $L:D = 1:1$ (DODT and ENROTH, 1953)

In the cat retina, we found two different general functions for this dependency of the neuronal CFF on the stimulus intensity: Part of the optic nerve fibers reach the highest CFF even at luminances above 500 cd/m² between 20 and 24 Hz. In other neurons, the CFF-intensity function exhibits a kink within the mesopic range of stimulus luminance and the CFF reaches 60–70 Hz for photopic stimulus intensities above 400 cd/m². In the latter type of units, the slope of the CFF-intensity function is significantly less steep in the scotopic and mesopic range than in the photopic range. From these findings, another subdivision of the retinal on-center and off-center neurons into two different classes is possible. The perceptive unit of the one is connected probably only with rods; the perceptive unit of the other processes signals from both receptor types, rods and cones (c.f. also OGAWA et al., 1966).

ENROTH (1952) investigated very carefully the conditions of fusion and demonstrated that the CFF is essentially related to the *latency properties* of the neuronal response. She discriminated between the latency of excitatory and of inhibitory processes and assumed that the CFF is reached as soon as the excitatory and inhibitory processes elicited by the flicker stimuli completely overlap each other. A negative correlation exists between the CFF and the discharge latency of on-center and off-center neurons (linear correlation coefficients between -0.44 and -0.69). These findings were confirmed by GRÜSSER and RABELO (1958) who used intermittent short flashes of constant duration (< 1 msec).

As the stimulus intensity increases, the latency of on- and off-center neurons decreases while the "initial impulse frequency" (= average impulse rate of the first 3 impulses), caused by a slow rectangular light stimulus increases (GRANIT, 1947). As a consequence of this finding and the negative correlation between CFF and latency, a positive correlation exists between the initial impulse frequency and the CFF ($r_{\text{off}} = +0.72$, $r_{\text{on}} = +0.71$). This relationship, first found for the dark-adapted retina of the cat, applies also to the light-adapted retina (DODT and ENROTH, 1954).

The relationship between neuronal activation and CFF was further elaborated by the work of FUKADA et al. (1966). These authors found that the CFF obtained in different neurons under comparable stimulus conditions (diffuse light stimuli, $L:D = 1:1$, photopic adaptation) increases monotonically with the conduction velocity of the optic nerve fiber ($v_n = 10-64 \text{ m} \cdot \text{sec}^{-1}$). The CFF has also a positive correlation with the maximum neuronal activation ($\overline{R}_{\text{max}}$, see p. 495), elicited by the flicker stimuli (GRÜSSER, 1956; FUKADA et al., 1966). For the discussed relationships, FUKADA et al. found exponential functions, which are for on-center neurons:

$$\text{CFF} = 100\,(1 - e^{-0.1}\, v_n + 0.55)\ [\text{Hz}]\,, \tag{30}$$

$$\text{CFF} = 90\,(1 - e^{-0.01}\, \overline{R}_{\text{max}})\ [\text{Hz}]\,. \tag{31}$$

In the light of the later findings of FUKADA and SAITO (1971), these data can be easily understood. According to FUKADA, the size of the RF-center increases as the conduction velocity of the optic nerve fibers becomes larger. Moreover, with diffuse flicker stimuli, the amount of light energy received per perceptive unit depends on the diameter of the RF. Hence, large RFs lead to a stronger flicker activation and therefore to a higher CFF of the ganglion cell than small RFs (see p. 495).

Since the work of HARTLINE and GRANIT, it is well known that the average impulse rate of retinal ganglion cells depends on the stimulus luminance. The neuronal activation also depends on other stimulus parameters. One of these is the stimulus *frequency*. This is shown in Fig. 25 for the cat retina (rectangular stimuli, $L : D = 1 : 1$, GRÜSSER, 1956, 1957; GRÜSSER and CREUTZFELDT, 1957; REIDEMEISTER and GRÜSSER, 1958): The average neuronal impulse rate of individual retinal on-center and off-center neurons first increases as the frequency of the flicker stimuli increases and reaches a maximum between 3 and 20 Hz. For a higher flicker frequency up to the TFF, the average impulse rate decreases again.

The relationship between the average neuronal impulse frequency and the flicker frequency depends on the intensity of the flicker stimuli: The brighter the flickering light, the more the average impulse rate is enhanced at medium stimulus frequencies and the higher the flicker frequency at which the impulse frequency maximum is reached. A positive correlation exists between the CFF and the impulse frequency at the response maximum ($r = 0.62$ for on-neurons, $r = 0.85$ for off-neurons), (GRÜSSER, 1956; GRÜSSER and RABELO, 1958; REIDEMEISTER and GRÜSSER, 1958; GRÜSSER and REIDEMEISTER, 1958; FUKADA et al., 1966, 1971).

The enhancement of the average activation of the on-center neurons corresponds to the psychophysically measured brightness enhancement (Brücke-Bartley effect, p. 455). This explanation is based on the assumption that the average neuronal impulse rate of on-center neurons correlates with subjective brightness. This positive correlation is also suggested by the parallel observation that both neuronal impulse rate and subjective brightness are logarithmic functions of the stimulus intensity.

As Fig. 25 shows, not only the impulse rate of the on-center neurons but also that of the off-center neurons depends on the flicker frequency. On the average,

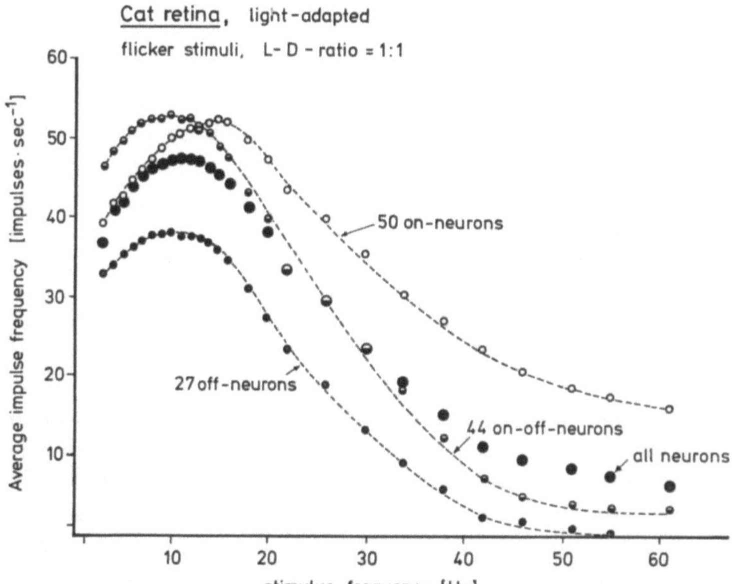

Fig. 25. *Relationship between the average impulse rate and the stimulus frequency* (diffuse flicker stimuli, direct retinal illumination of approximately 5 Lux intensity, $L : D = 1 : 1$, cat retina). The response maximum is reached in off-center neurons (off-neurons and on-off neurons) at a significantly lower stimulus frequency than in on-center neurons (GRÜSSER and REIDEMEISTER, 1959)

however, the increase of off-center neuron activation is less than that of the on-center neurons and the maxima are reached at stimulus frequencies below that for the maxima of the on-center neurons.

This finding probably correlates with the darkness enhancement described for psychophysical studies with flicker stimuli on p. 458. The much lower activation of off-center neurons in comparison with on-center neurons at stimulus frequencies above 15 Hz can be correlated with the flicker perception at this frequency level: The dark periods appear much shorter and rather inconstant in comparison to the bright periods of the flickering spot. The difference in the frequencies at which the response maximum of off-center neurons and of on-center neurons is reached indicates, that darkness enhancement should be observed at a lower stimulus frequency than brightness enhancement. This is indeed the case (p. 459).

The effect of the *light-dark ratio* of diffuse rectangular flicker stimuli on the response of cats' retinal on-center and off-center neurons was investigated by REIDEMEISTER and GRÜSSER (1959). In general, the CFF of on-center neurons decreases with a decrease of the light-dark ratio below 0.5 ("uncompensated" flicker stimuli). In the off-center neurons, however, no such regular relationship holds. Some off-neurons increase, some decrease their CFF, when the light-dark ratio decreases. These findings might be considered as the neurophysiological correlate of the second law of PORTER (p. 442).

The relationship between stimulus frequency and neuronal impulse rate is also significantly influenced by a change of the light-dark ratio (REIDEMEISTER and GRÜSSER, 1959). From the experimental results such as those shown in Fig. 3, of the paper of REIDEMEISTER and GRÜSSER (1959) one can predict that the relative brightness enhancement of the Brücke-Bartley effect should also depend on the light-dark ratio of the stimuli. With uncompensated flicker stimuli, this was indeed found in psychophysical experiments: The brightness enhancement is significantly reduced as the L-D-ratio is changed from 1: 2 to 1: 3 or 1: 7 (see p. 457).

The negative correlation between the neuronal CFF and the impulse latency is also valid for flicker stimuli with a light-dark ratio < 1. The increase of the neuronal latency with increasing flicker frequency is generally valid only in on-center neurons. If the light-dark ratio is smaller than 1 : 5, off-center neurons decrease the latency with an increase of the stimulus frequency (GRÜSSER and REIDEMEISTER, 1959).

D. Responses to Spatial Stimulus Patterns Modulated Sinusoidally in Time

From KUFFLER's results, it was predictable that the flicker data obtained with diffuse light stimuli will provide only incomplete and indirect information about the temporal response properties of the investigated neuronal network. The different frequency transfer properties of the components of the perceptive units cannot be explored with diffuse light stimuli. It was therefore necessary to investigate the response of retinal ganglion cells to flickering stimuli with spatial light-dark patterns adequately adapted to the functional organization of the concentric RF. Fig. 26 shows some stimulus patterns used for quantitative studies of the spatio-temporal transfer properties of the retinal network. Also a useful stimulus for the analysis of the spatio-temporal transfer properties of the retinal network is a flickering or moving spatial sinusoidal grating (see p. 450).

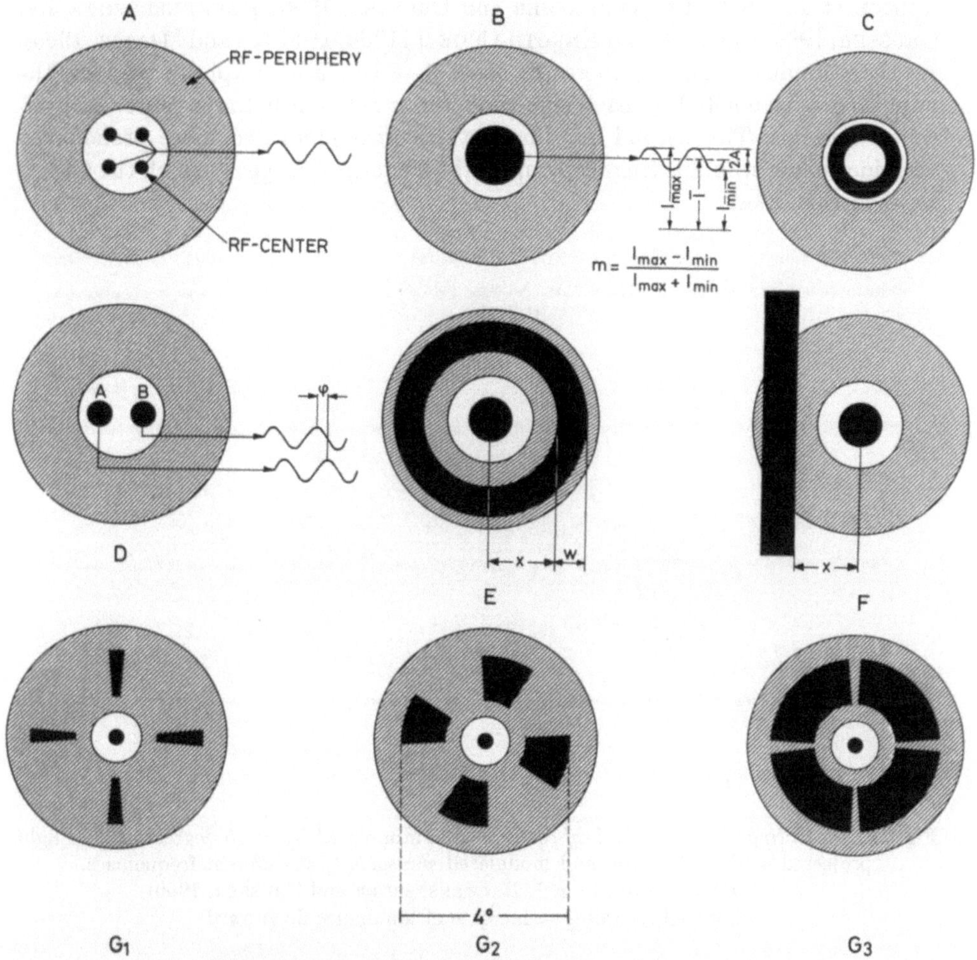

Fig. 26. *Different stimulus patterns used for the exploration of the spatio-temporal response characteristics of retinal ganglion cells in the cat.* RF-center white, RF-periphery shaded, light-stimuli black. The size of the stimulus pattern was variable

1. The Variation of the Depth of Modulation (m); a Test for Response Linearity

The response of on-center and off-center neurons to sinusoidal stimulation of the RF-center is linear only with restricted stimulus conditions: $m < 0.2 - 0.4$, $f < 5$ Hz, stimulus size < RF-center. Under these conditions, the instantaneous impulse rate of most retinal neurons varies sinusoidally in time. As a rule, when $m > 0.4$, two non-linear components appear in the neuronal response: At low stimulus frequencies (< 5 Hz), the instantaneous impulse rate of on-center neurons increases *faster* than the corresponding sine wave, but decreases more slowly. The non-linearity of the response of off-center neurons is the mirror image of the on-neuron's non-linearity: The instantaneous impulse rate rises more slowly than the sine wave and shortly after the maximum off-response is reached, the neuronal impulses cease rather abruptly (Fig. 27a, b, Rackensperger et al., 1965; Lunken-

HEIMER et al., 1966; LUNKENHEIMER and GRÜSSER, 1966; RACKENSPERGER and GRÜSSER, 1966; CLELAND and ENROTH-CUGELL, 1966; HUGHES and MAFFEI, 1966). As the stimulus frequency increases *above* 5–8 Hz, *pauses* appear between the impulse groups elicited by each sine wave for $m > 0.2$ in both on-center and off-center neurons. This second non-linearity becomes more and more pronounced with increasing stimulus frequency up to the frequency range at which the CFF is reached (Fig. 27a).

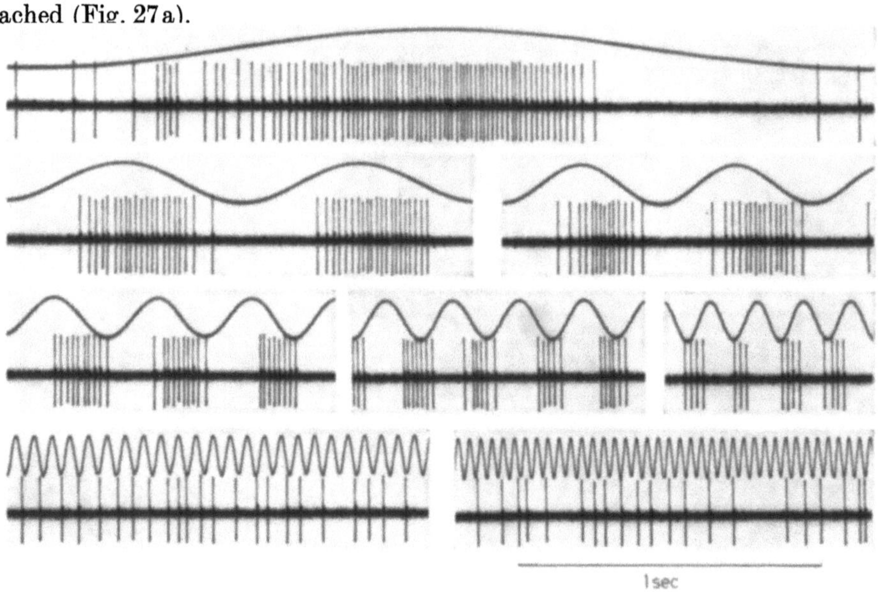

Fig. 27a. *Recordings from a retinal off-center neuron* stimulated by a 2.5 degrees spot of light projected to the RF-center and modulated *sinusoidally* at different frequencies. ($m = 0.9$, $\bar{I} = 190$ cd · m^{-2}, RACKENSPERGER and GRÜSSER, 1966) Photocell recording = increase of luminance downwards

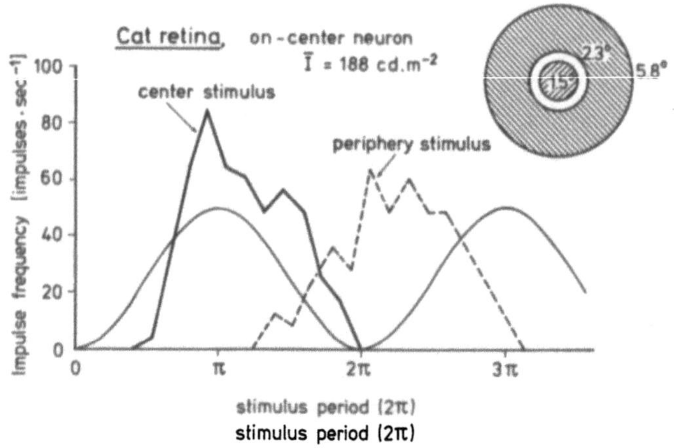

stimulus period (2π)

Fig. 27b. *Response pattern of* an *on-center neuron of the cat's retina to sinusoidal light stimulation of the RF-center* (solid line, 1.5 degrees stimulus diameter) and of the RF-periphery (broken line, light annulus 2.3 degrees inner diameter, 5.8 degrees outer diameter). Average luminance of the stimulus 188 cd · m^{-2}, $m > 0.95$. Frequency of sine wave = 1.3 Hz (RACKENSPERGER and GRÜSSER, 1966)

HUGHES and MAFFEI (1966) used a power function to describe the time course $R(t)$ of the neuronal activation at low stimulus frequencies (0.1–1.0 Hz, diffuse sine wave stimuli):

$$R(t) = S + k_h \bar{I}^b (1 + m \sin \omega t)^b \text{ [Impulses} \cdot \text{sec}^{-1}] , \qquad (32)$$

S is the spontaneous activity in darkness, the constant $b = 0.75$, the constant k_h is variable for different neurons. This equation, however, holds only for low stimulus frequencies.

The asymmetric temporal course of the instantaneous impulse frequency correlates well with the subjective brightness aroused by a slow sinusoidal stimulus as described on p. 461: The subjective brightness increases faster than it decreases.

Another simple test of linearity is the measurement of the average impulse rate obtained with sine wave stimuli of different m or f, but constant $\bar{I}$. If the neuronal response is linear, i.e. the instantaneous impulse rate modulation is sinusoidal for a sine wave stimulus, the average impulse rate ($\bar{R}$) should be independent of m and f. This condition can be observed only rarely in class II on-neurons (sustained type), if $m < 0.2$.

In most neurons the higher m, the more non-linear the neuronal response. If either $\bar{R}$ or the maximum instantaneous impulse frequency during one sine wave (R_{max}) is plotted versus m, the neuronal response can be described by the following non-linear equation Fig. 28a:

$$\bar{R} = \frac{S + \sigma m}{1 + k_i \sigma m} \text{ [Impulses} \cdot \text{sec}^{-1}] . \qquad (33)$$

σ is a scaling constant. The inhibition constant k_i decreases as the stimulus frequency increases. It is also smaller in class II on-center or off-center neurons than

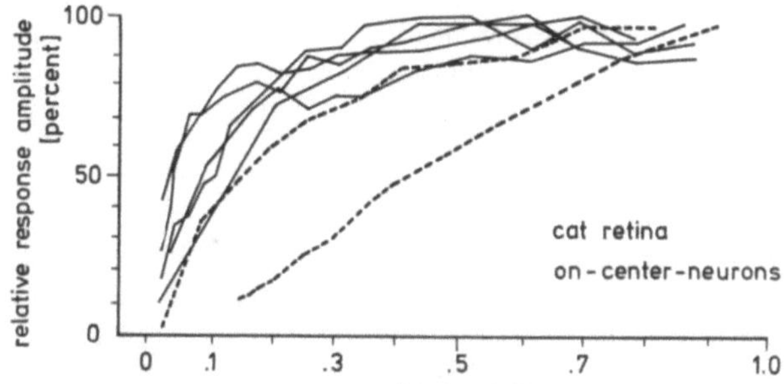

Fig. 28a. *Non-linear relationship of the response amplitude (ordinate) and the depth of modulation (m, abscissa) for different on-center neurons of the cat retina.* Average retinal illuminance caused by sinusoidal stimuli 5000 td, $m = 0.98$ (full line) or 6000 td, $m = 0.99$ (broken line). Test field smaller than RF-center, 8 Hz (CLELAND and ENROTH, 1966)

in class I neurons. According to the findings of CLELAND and ENROTH-CUGELL (1966), k_i is at 8 Hz also smaller in off-center neurons than in on-center neurons. Because of the decrease of k_i, with increasing stimulus frequency, the relationship between $\bar{R}$ (or R_{max}) and m approximates a linear function for higher stimulus frequencies (Fig. 28b). From these findings one can conclude that the network properties that correspond with k_i of Eq. (33) have less influence the higher the stimulus frequencies above 3.5 Hz.

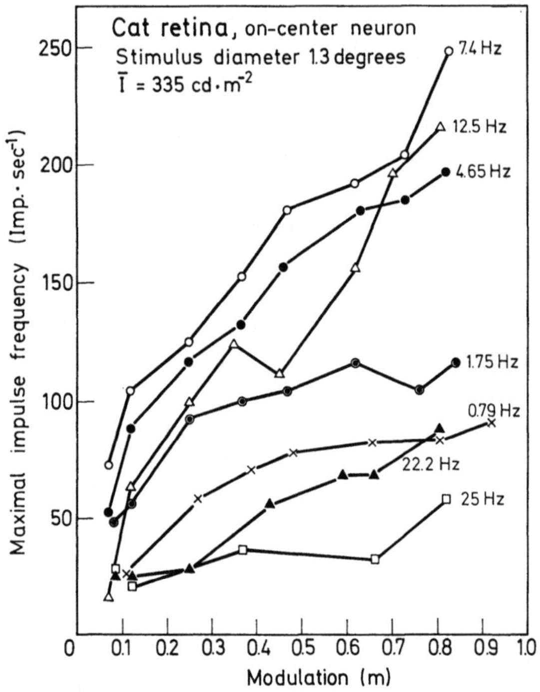

Fig. 28b. *On-center neuron*. The response maximum (= maximum impulse frequency obtained during one bin of the computer measurements, c.f. Fig. 33a) is plotted on the ordinate as a function of m

2. Stimulus Frequency (f)

As with diffuse rectangular flicker stimuli, the neuronal response to small flickering test fields is more and more delayed when the frequency of a *sine wave* stimulus increases. From the average neuronal responses (PST-histograms) obtained at different stimulus frequencies, one can calculate the contribution of the first and higher order harmonics (for details see FOERSTER et al., 1973; SCHELLART and SPEKREIJSE, 1972).

The phase lag φ' of the neuronal response relative to the sine wave stimuli increases as the stimulus frequency increases above 3 Hz and reaches at the CFF values up to 3-4 π (Fig. 28c). At a given stimulus frequency the phase angle between stimulus and responses is smaller the larger the stimulus area and the stimulus intensity. Both the latency of the excitatory process and the latency of the inhibitory processes, elicited from the RF-center, depend on the stimulus frequency (Fig. 28d). As the flicker frequency approaches the CFF the relative duration of the excitatory period decreases.

The neuronal response elicited by the sine wave stimuli (measured as the average impulse rate $\bar{R}$, the difference between $\bar{R}$ and the average impulse frequency above the TFF, the maximal impulse rate R_{max}, reached during one sine wave period (Fig. 33a) and the amplitude of the first harmonic of the response R_1) increase as the stimulus frequency increases above 2—3 Hz and reach a maximum at medium stimulus frequencies. At frequencies above this tuning frequency a

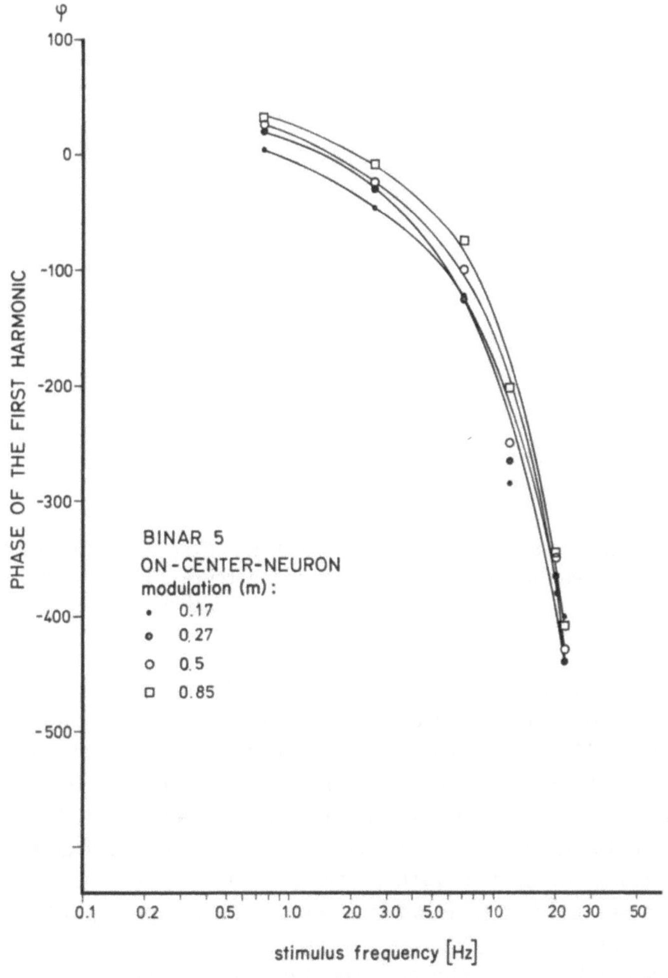

Fig. 28c. *On-center neuron of the light adapted cat retina.* Phase characteristics of the first harmonic of the neuronal response pattern. Sinusoidal stimuli of different depths of modulation. $I = 335\ \mathrm{cd \cdot m^{-2}}$

further increase of the stimulus frequency reduces the neuronal response again. In part of the retinal neurons two response maxima are present. The response enhancement at medium stimulus frequencies, as a rule, is stronger in class I neurons than in class II neurons (p. 480).

As mentioned on p. 484, the neuronal response enhancement depends on m, $\bar{I}$ and A. The tuning frequency is reached at higher stimulus frequencies the larger these values. This observation correlates with the psychophysical findings that the Brücke-Bartley brightness enhancement is stronger with larger m, $\bar{I}$ and A.

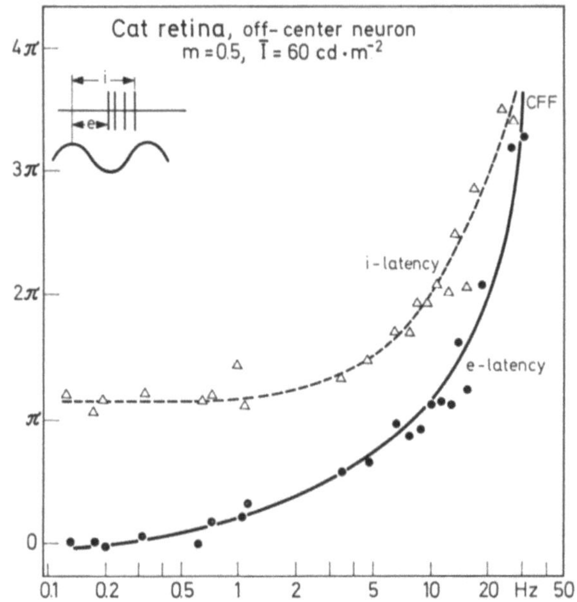

Fig. 28d. *Off-center neuron of the cat retina.* Dependency of the latency of the off-excitation and the on-inhibition on the stimulus frequency (log, abscissa). The latencies are expressed in radians (LUNKENHEIMER, 1968)

In the upper frequency range (> 20 Hz, photopic stimuli) the attenuation of R (off-center neurons and transient on-center neurons), $R - R_{TFF}$, $R_{\max}$ and R_1 increases (Fig. 31, 32). Negative slopes between 30 and 48 db/octave were found (RACKENSPERGER et al., 1965; LUNKENHEIMER and GRÜSSER, 1966; GRÜSSER, 1966; MAFFEI et al., 1968; CLELAND and ENROTH-CUGELL, 1966). SCHELLART and SPEKREIJSE (1972) measured the dynamic characteristics of ganglion cells in the goldfish retina. For high frequencies the amplitude characteristics obtained with different values of $\bar{I}$ join together in a common envelope if plotted on an absolute sensitivity scale.

CLELAND and ENROTH-CUGELL (1966) measured for 5 retinal ganglion cells the dependency of the *threshold modulation* of a spot of light projected to the RF-center on the stimulus frequency. The neuronal sensitivity (inverse of threshold modulation) increases at medium frequencies approximately like the gain-increase described above for suprathreshold stimulation. The correspondence between threshold and suprathreshold stimulation is not exact (Fig. 29). The fact that there is a significant difference between both curves shows that the investigated neuronal network has strong non-linear properties.

The findings of CLELAND and ENROTH-CUGELL can be correlated with the de Lange threshold sensitivity curves obtained in psychophysical measurements. Parallel to the findings reported in Fig. 29 there seems to be a difference between psychophysical de Lange curves for threshold sensitivity and the curves describing the subjective brightness as a function of the frequency of suprathreshold flicker stimuli.

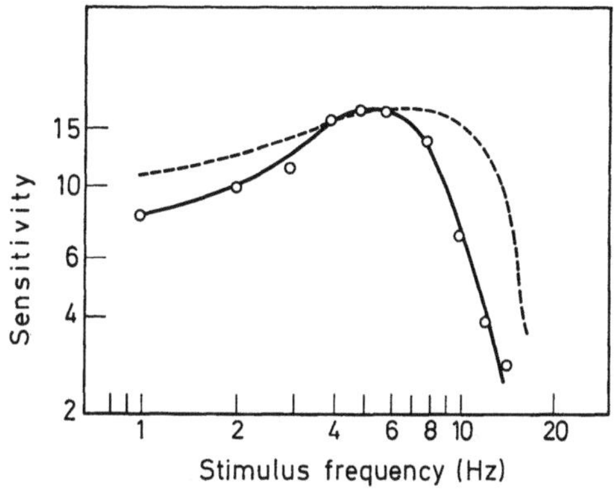

Fig. 29. *On-center neuron of the cat retina.* Relationship between threshold sensitivity (ordinate) and stimulus frequency (abscissa). For each frequency the experimenter determined the depth of modulation at which he could barely hear an impulse frequency modulation synchronous with the luminance fluctuations. The inverse of these values is plotted (open circles). The broken line is the amplitude response from the same cell obtained at a modulation depth of 0.5 (Cleland and Enroth-Cugell, 1966)

3. Stimulus Intensity (I)

If the average luminance of the sine wave stimuli is varied, keeping m and f constant, the *intensity function* of retinal neurons can be measured. The results are well described by a logarithmic function (Fig. 30a):

$$\bar{R} = S + k_I \log \frac{I}{I_s} \text{ [Impulses} \cdot \text{sec}^{-1}] \qquad (34)$$

and

$$R_{\max} = S + k_I^* \log \frac{I}{I_s} \text{ [Impulses} \cdot \text{sec}^{-1}] \,. \qquad (35)$$

As for the S-potentials of the retina a tanh (log I-function) also describes the experimental intensity function over a wider range:

$$\bar{R} = S + \frac{1}{k_i^*} \left[1 + \tanh \left(\frac{1}{2} \ln \frac{k_i^* \bar{I}}{I_s}\right)\right] \text{ [Impulses} \cdot \text{sec}^{-1}] \,. \qquad (36)$$

The *intensity function* of single retinal neurons depends on the stimulus area, the level of adaptation and the stimulus frequency. The gain of the logarithmic intensity function is represented by the slope, i.e. the multiplicative parameter k_I of Eqs. (34), (35). This parameter reaches its highest value at a frequency level between 5 and 15 Hz. The frequency at which the response maximum is reached depends also on the stimulus area and probably on the class (I, II) of the retinal ganglion cell. The dependency of k_I of Eq. (34) on the stimulus frequency is shown in Fig. 30b (stimuli covering about 70% of the RF-center). Eq. (36) is a transformation of the following equation (Naka and Rushton, 1966; c.f. p. 442):

$$\bar{R} = S + \frac{\bar{I}}{I_s + k_i^* \bar{I}} \text{ [Impulses} \cdot \text{sec}^{-1}] \,. \qquad (37)$$

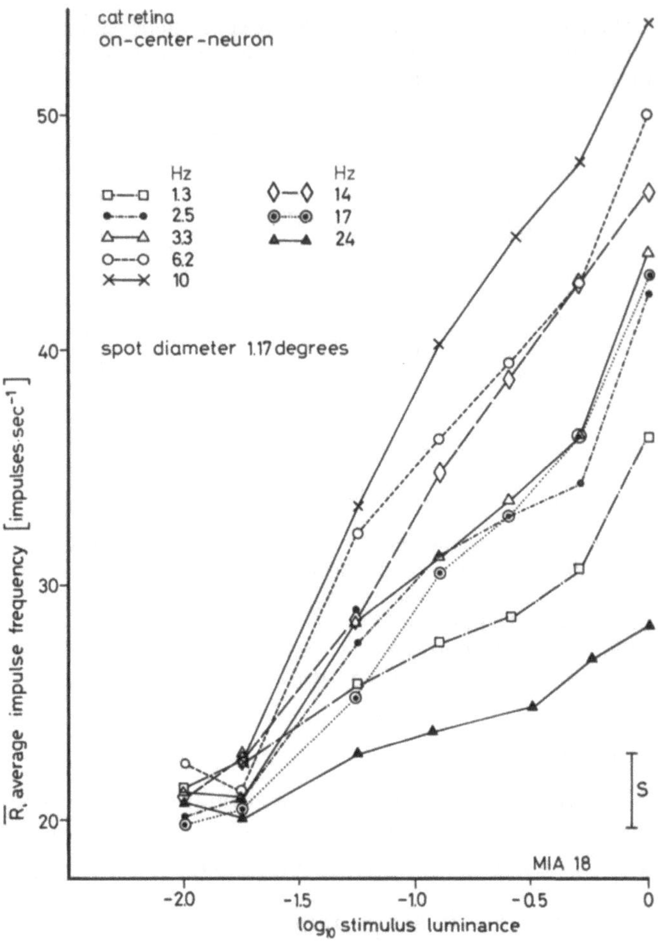

Fig. 30a. *On-center neuron* of the cat retina. Relationship between the average neuronal impulse frequency (ordinate) and the stimulus luminance (^{10}log, ordinate, $0 = 420$ cd · m^{-2}). Diameter of the test field projected to the RF-center = 1.17 degrees. The measurements are made with different stimulus frequencies (1.3 Hz to 24 Hz). A maximum of the slope is reached at about 10 Hz. S = range of spontaneous activity in darkness (GRÜSSER, 1969)

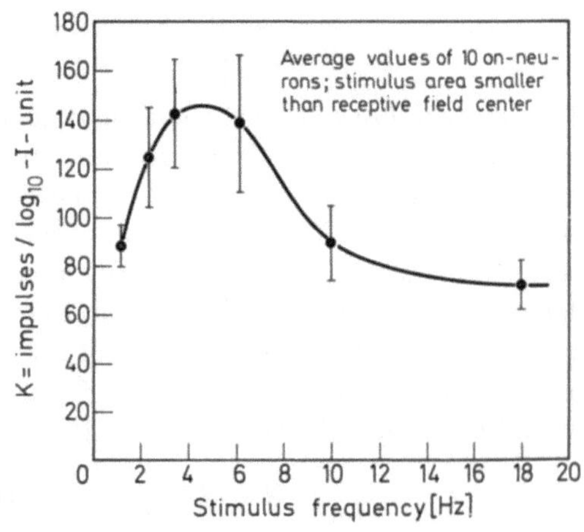

Fig. 30b. *Average values of on-center neurons of the cat retina.* The dependence of the slope of the intensity function (Fig. 30a) on the stimulus frequency (ordinate) is plotted. Stimulus diameter adjusted so that about 70% of the RF-center was illuminated, $m > 0.9$ (GRÜSSER, 1969)

This formula as well as Eq. (12) can be interpreted in terms of the shunting inhibition model proposed by VARJÚ, 1965 and by FURMAN (1965). The inhibition constant k_i^* decreases with increasing stimulus frequency and reaches a minimum between 3 and 12 Hz. It increases again above this value and reaches near the CFF almost the same value that it has below 1 Hz (GRÜSSER, unpubl. 1969).

4. Spatial Summation of Flicker Responses within the RF-Center

The effect of the spatial properties of flickering stimuli on the temporal transfer characteristics was investigated with two different approaches:

(a) Round test fields of different diameters (Fig. 26), centered on the RF-center, are used as stimuli and the relationship between neuronal activation and stimulus area is determined ("area function", see GRÜSSER, 1971). With this traditional method of measuring area effects (HARTLINE, 1940), the results depend not only on the stimulus area, but also on the *distribution of excitability* within the RF. It is known that the neuronal excitability is maximum in the center of the RF-center and decreases towards the RF-periphery (HARTLINE, 1940; KUFFLER, 1952, 1953; WUTTKE et al., 1965; WUTTKE and GRÜSSER, 1966; RODIECK and STONE, 1965; ENROTH et al., 1966; FREUND and GRÜNEWALD, 1969).

The results reported by GRÜSSER (1971) show that the area function depends on the stimulus frequency. The area at which the strongest neuronal response is

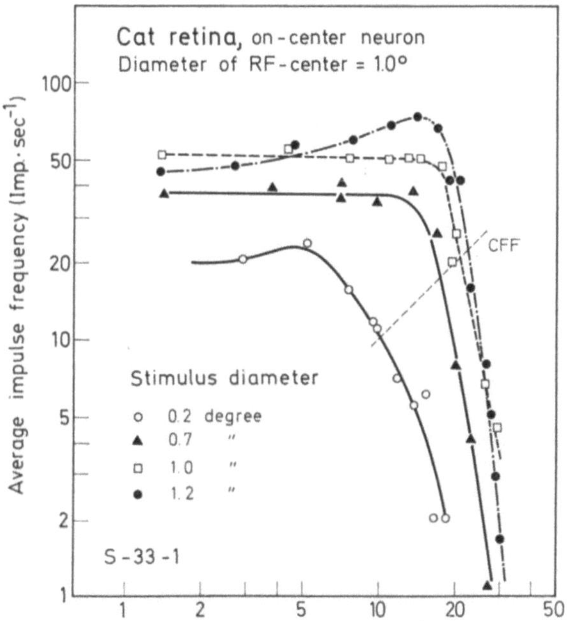

Fig. 31a. *On-center neuron of the cat retina.* Relationship between stimulus frequency (abscissa) and average impulse frequency (ordinate), obtained with flickering test fields ($m = 0.9$) of 4 different areas. Diameter of the RF-center = 1.0 degree. When the stimulus area extends beyond the RF-center, the neuronal activation to lower stimulus frequencies is reduced (GRÜSSER and LUNKENHEIMER, unpubl. 1966)

reached ("functional size" of the RF-center) is smallest at stimulus frequencies below 3 Hz. At higher stimulus frequencies, this functional size of the RF-center increases, i.e. the relative inhibitory effect of the RF-periphery evidently decreases as the stimulus frequency increases above 3–5 Hz. These findings clearly indicate that the overall spatial summing properties of the neuronal network, which forms the perceptive unit of a retinal ganglion cell, depend strongly on the temporal properties of the stimulus.

As Fig. 31a, b demonstrates, the relationship found between the neuronal activation ($\overline{R}$, R_{max}) and the stimulus frequency with diffuse rectangular stimuli (p. 484, Fig. 25), applies also for sinusoidal stimuli of different areas. When $\overline{I}$ and m are held constant, the larger the stimulus area, the higher the frequency at which the retinal ganglion cell reaches the response maximum, the CFF and the TFF. In addition, the relative increase of the neuronal activation is stronger with stimuli larger than the RF-center than with those smaller than the RF-center. As long as the stimulus size is smaller than the RF-center the neuronal response at all

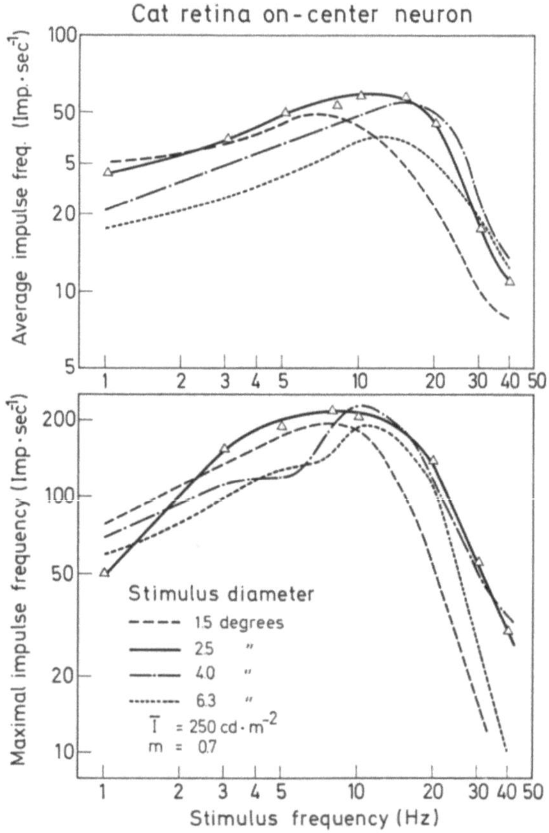

Fig. 31b. *On-center neuron of the cat retina.* Same relationship as in Fig. 31a. Test field smaller or larger than the RF-center (= 2.0 degrees). As the stimulus diameter extends beyond the RF-center, the neuronal response is reduced at low stimulus frequencies below that obtained with test fields smaller than the RF-center. For a test field of 6.4 degrees diameter, however, the neuronal response is also reduced at stimulus frequencies up to 25 Hz (from an experiment done in cooperation with U. EYSEL)

stimulus frequencies becomes larger as the stimulus area increases. If the stimulus area extends beyond the RF-center, the neuronal response is diminished at low stimulus frequencies, but in most neurons it is increased at stimulus frequencies from 8—10 Hz up to the CFF or TFF, when the stimulus area increases beyond the

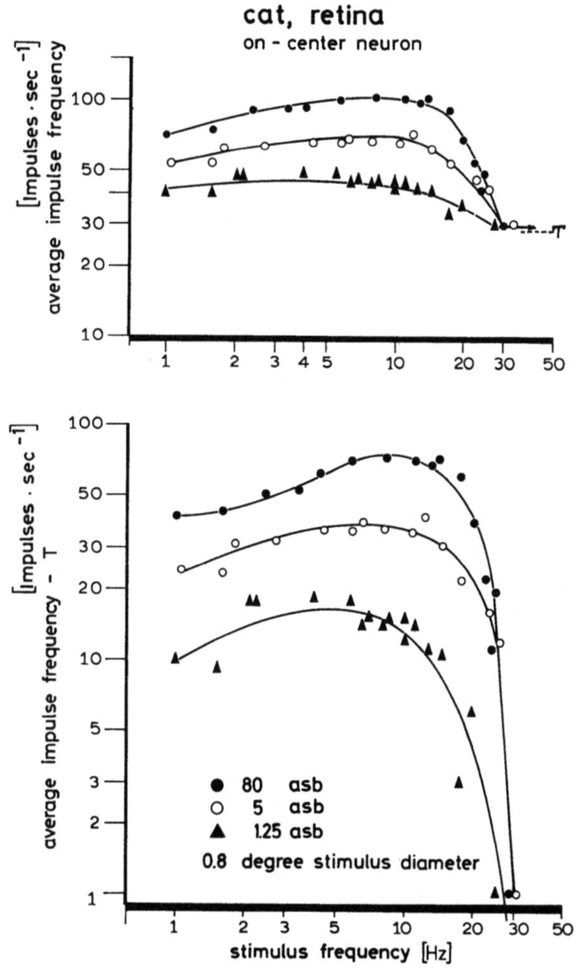

Fig. 32. *On-center neuron of the cat retina.* Same relationship as in Fig. 31a. Sinusoidal stimulation of the RF-center (stimulus diameter 0.8 degrees, $m > 0.9$). Responses obtained at 3 different values of $\bar{I}$ (251 cd · m^{-2}, 16 cd · m^{-2}, 4 cd · m^{-2}). In the lower diagram the difference between the neuronal activation and the neuronal impulse frequency above the CFF is plotted (Grüsser and Lunkenheimer, 1966, from Grüsser, 1971)

RF-center size. This observation again indicates that one form of lateral inhibition elicited from the RF-periphery is attenuated at a lower stimulus frequency than the excitatory processes elicited from both the RF-center and the RF-periphery.

For the relationship between CFF (or TFF) and the stimulus area ($\bar{I}$ and m constant) the neurophysiological data are well summarized, both for on-center and off-center neurons, by Eq. (6). Hence, all these neurophysiological data correlate directly with psychophysical findings.

From the neurophysiological results shown in Fig. 31a, b, one can also conclude that the Brücke-Bartley effect should depend on the area of the flicker stimuli and, more specifically, that the perceived brightness enhancement increases with an increase of the stimulus area. In addition, the tuning frequency should shift into higher frequency ranges as the stimulus area increases, when m and $\bar{I}$ are held constant. This prediction is indeed confirmed by corresponding psychophysical findings in human observers (p. 448).

(b) The spatial summation of excitation within the RF-center can be measured without the complications of a variable excitability within the RF by the following method (Fig. 26A, BÜTTNER and GRÜSSER, 1966, 1968; GRÜSSER et al., 1970, 1971): Small spots of equal size are projected into the RF-center so that the neuronal activation caused by each spot of light alone, is approximately the same for all spots. For concentric RFs, this is the case if the light spots are placed on an imaginary circle centered on the RF-center. With this type of stimulus pattern, the spatial summation of excitation within the RF-center turns out to be non-linear for suprathreshold stimuli and depends on the stimulus frequency. According to CLELAND and ENROTH-CUGELL (1968) a *linear* summation of excitation, elicited by different spots, projected to the RF-center, exists for *threshold* stimuli. As long as the stimuli do not enhance the neuronal activation more than 15–20 impulses/sec above the spontaneous impulse rate, the experimental findings might still be approximated by a linear model of spatial summation. For stimuli that cause a higher activation, however, the majority of the investigated retinal neurons exhibit a clear non-linear component in the spatial summation. The non-linear deviation depends on the distance between the light spots and decreases with the spots' distance as long as they are projected in the RF-center (BÜTTNER and GRÜSSER, unpubl. 1971).

Fig. 33a demonstrates that the deviation from a linear spatial summation at suprathreshold levels of activation also depends in the stimulus frequency. The most satisfactory and simple description of the experimental results is again obtained by the application of FURMAN'S model of multiplicative forward inhibition (BÜTTNER and GRÜSSER, 1968; GRÜSSER et al., 1970). We assumed that *one type* of lateral inhibition acting in the whole RF (center and periphery) is processed by an inhibition pool, formed by one of the horizontal neuronal systems (horizontal cell layer or amacrine cell layer) of the retina. The following equation describes the experimental results fairly well:

$$R_n = \frac{S + \sum_n b_i}{1 + {}_sk_i \sum_n b_i} \text{ [Impulses} \cdot \text{sec}^{-1}\text{]} . \qquad (38)$$

R_n is the average impulse frequency elicited by n spots of light, b_i is the average excitation above the spontaneous activity S elicited by one spot of light alone and ${}_sk_i$ is the average inhibition constant describing the interaction of the assumed lateral inhibition pool and the excitatory processes within the perceptive unit. With different ${}_sk_i$, Eq. (38) is also valid for the maximum neuronal activity (R_{max}) reached during the sinusoidal stimulus period. Computer programs using iterative computation of the Chi-square values were used to find the ${}_sk_i$-value giving a best fit of Eq. (38) to the data obtained with different stimulus frequencies. The average inhibition constant ${}_sk_i$ of Eq. (38) reaches a minimum at a frequency between 8 and 15 Hz (Fig. 33a). A model with a one or two stage interaction of linear (subtractive) lateral inhibition was also tested as a description of the experimental results, but without success. These findings provide additional experimental evidence that a *non-linear* model has to be assumed for the spatial summation of excitation within the RF-center. Each spot of light projected to the RF-center elicits not only excitatory but simultaneously inhibitory (shunting) processes. The assumed mutual lateral inhibition within the RF-center has frequency properties that differ from those of the excitatory mechanisms, that determine the supra-threshold neuronal response.

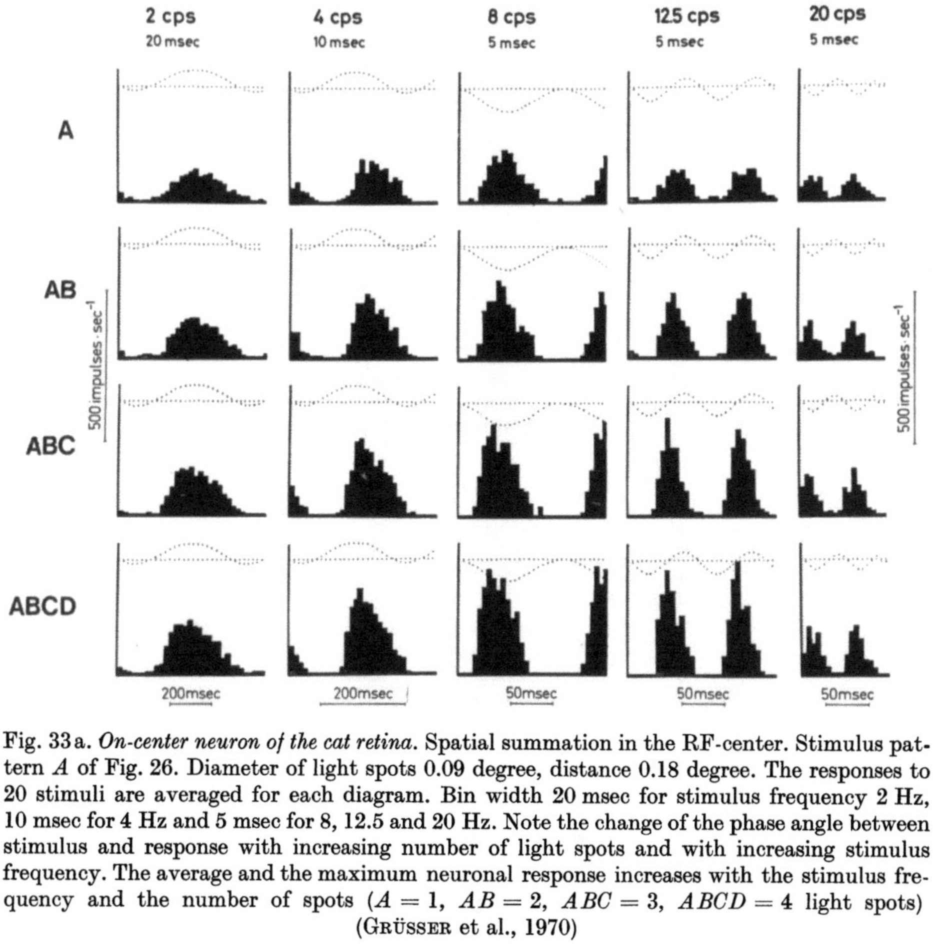

Fig. 33a. *On-center neuron of the cat retina.* Spatial summation in the RF-center. Stimulus pattern A of Fig. 26. Diameter of light spots 0.09 degree, distance 0.18 degree. The responses to 20 stimuli are averaged for each diagram. Bin width 20 msec for stimulus frequency 2 Hz, 10 msec for 4 Hz and 5 msec for 8, 12.5 and 20 Hz. Note the change of the phase angle between stimulus and response with increasing number of light spots and with increasing stimulus frequency. The average and the maximum neuronal response increases with the stimulus frequency and the number of spots ($A = 1$, $AB = 2$, $ABC = 3$, $ABCD = 4$ light spots) (GRÜSSER et al., 1970)

The CFF for both on-center and off-center neurons increases linearily with the log of the number of light spots. This finding correlates with the psychophysical measurements of GRANIT and HARPER (p. 440).

(c) The interaction of spatial and temporal signal processing in the perceptive unit of retinal neurons is also found when measurements are made with sinusoidal gratings moving across the receptive field. With respect to the overall, spatial summing properties of the RF, ENROTH-CUGELL and ROBSON (1966) discriminated two types of retinal on-center or off-center neurons (linear spatial summation of excitation in X-cells, non-linear summation in Y-cells, c.f. p. 480). In both types of retinal ganglion cells, the average response depends on the spatial frequency of the gratings. The optimum spatial frequency of the gratings depends on the functional size of the RF-center and covers therefore a wide range of values. The

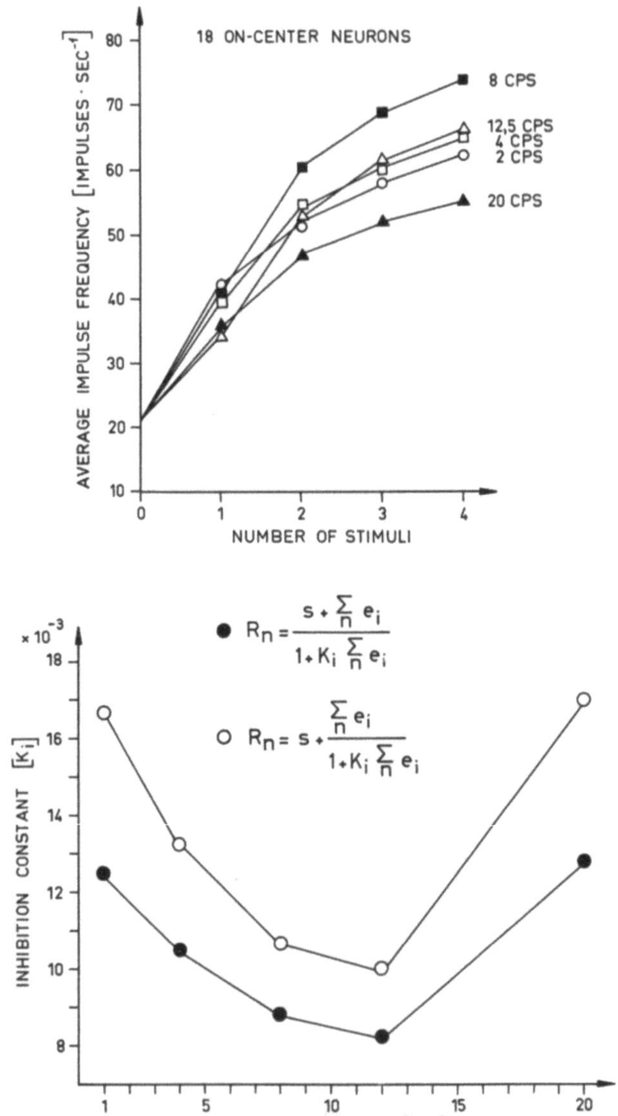

$$R_n = \frac{s + \sum_n e_i}{1 + K_i \sum_n e_i}$$

$$R_n = s + \frac{\sum_n e_i}{1 + K_i \sum_n e_i}$$

Fig. 33b. *On-center neuron of the cat retina.* Quantitative analysis of spatial summation within the RF-center. Top: average impulse frequency of 18 on-center neurons (ordinate) plotted versus the number of light spots illuminated (stimulus pattern A in Fig. 26, abscissa). Sinusoidal stimulation at 2.4, 8, 12.5 and 20 Hz, $m > 0.9$ (Grüsser et al., 1970). Bottom: Dependency of the inhibition constant k_i of Eq. (36) on the stimulus frequency. Computer calculated values (minimum Chi-square sum) from the data of Fig. 33a (Grüsser, 1971)

spatial frequency at which the optimum neuronal response is obtained, however, also depends on the temporal frequency of the stimuli.

Such findings can be interpreted as the first neurophysiological correlates to corresponding psychophysical measurements, the results of which were discussed in II c.

5. Spatial Summation of Periphery-Inhibition as a Function of Stimulus Frequency

Separate measurements of the neuronal activity caused by stimulation of the RF-center alone (R_c) and of the activity caused by simultaneous stimulation of RF-center and RF-periphery (R_{c+p}) provide data from which conclusions about the inhibition I_p exerted by the RF-periphery can be deduced. Good evidence is provided by several research groups (Rackensperger, 1969; Enroth-Cugell and Pinto, 1970, 1971; Amecke-Mönninghof and Buettner, 1971; Grüsser, 1971) that a linear interaction of inhibition elicited from the RF-periphery with the RF-center response can be assumed, as long as the response is well above the threshold level for impulse generation. Under such conditions I_p can be determined as the difference between R_c and R_{c+p}. In order to study the summation properties of the peripheral inhibition I_p, Grüsser et al. (1971) measured R_{c+p} as a function of the area A_p of a peripheral flickering stimulus. Stimulus patterns as $G_1 - G_3$ of Fig. 26 were used in these experiments, i.e. areas of the same relative inhibitory effects were added as the periphery stimulus increased. The following relation was found to describe the data (Fig. 34):

$$I_p = R_c - R_{c+p} = \frac{\alpha A_p}{1 + k_{ip}\,\alpha A_p} \,. \tag{39}$$

As Fig. 34 shows, the relative inhibitory effect obtained from the RF-periphery and its spatial summing properties depends also on the stimulus frequency. The gain of periphery inhibition (α in Eq. (39)) decreases as the stimulus frequency increases above 2 Hz, while the spatial summation becomes more linear, i.e. the inhibition constant k_{ip} of Eq. (39) also decreases as the stimulus frequency increases. From these findings it was concluded that the spatial summing properties of the inhibitory effect exerted by the RF-periphery and its dependence on the stimulus frequency are in principle quite similar to those of the excitatory processes in the RF-center described on pp. 494–500 (Grüsser et al., 1971; Grüsser and Lütgert, 1972).

6. Spatial Summation of Excitation from the RF-Periphery

Stimulation of the RF-periphery elicits also neuronal activation, which at low stimulus frequencies can be attributed in part to temporal disinhibition (off-excitation following on-inhibition in on-center neurons, on-excitation following off-inhibition in off-center neurons). The spatial summation of this periphery excitation was measured with stimulus pattern G of Fig. 26: The neuronal activation increases as the stimulus area in the RF-periphery increases. The relationship between neuronal activation and stimulus area can also be described by Eq. (38). Like the spatial summation of excitation within the RF-center, spatial summation of excitation elicited from the RF-periphery is not only a non-linear process but it also depends in a similar way on the stimulus frequency.

We measured the dependence of the CFF on the area of the periphery stimulus with stimulus pattern G of Fig. 26 ($m > 0.9, \bar{I} > 10\,\text{cd/m}^2$). The CFF increases approximately with the log of the stimulus area. An additional steady illumination of the RF-center of on-center neurons also leads to a higher CFF obtained with the flickering stimulus pattern in the RF-periphery. In off-center neurons this CFF for peripheral stimulation decreases with additional center-stimulation with steady light. The neuronal activation ($\bar{R}$, R_{max}) depends on the frequency of the periphery stimulus. The maximum neuronal response is obtained between 5 and 20 Hz. This maximum response is larger and is reached at a higher stimulus frequency the larger the area and/or the intensity of the flickering stimulus projected to the RF-periphery.

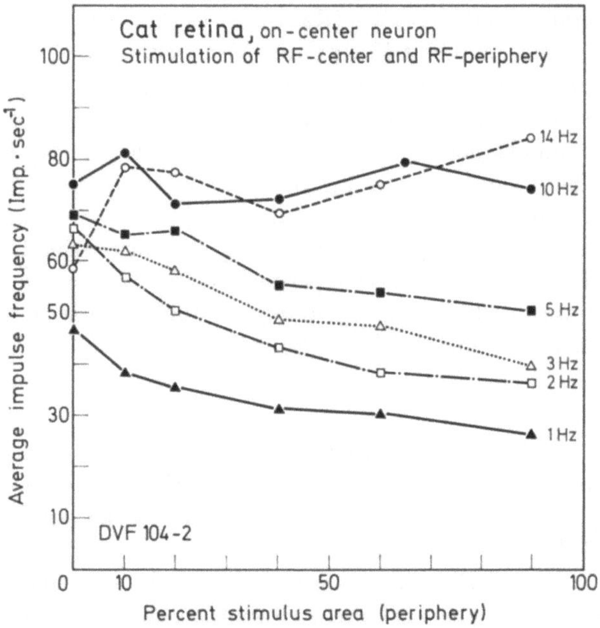

Fig. 34. *Spatial summation of inhibition from the RF-periphery of an on-center neuron of the cat retina.* Stimulus pattern *G* of Fig. 26. Ordinate: average impulse frequency, abscissa: percent area of the stimulus in the RF-periphery (sectors of a ring, pattern *G*, Fig. 26). At stimulus frequencies below 10 Hz the neuronal activation elicited by the combined stimulation of the RF-center and RF-periphery decreases relative to the response obtained from the RF-center alone. The spatial summation of this inhibitory effect is a non-linear function of the stimulus area. The average inhibitory effect decreases, as the stimulus frequency increases. Above 10 Hz the facilitatory periphery effect is stronger than the inhibitory effect (Grüsser and Lütgert, 1972)

We could not confirm the findings of Maffei et al. (1970) who claim that the overall periphery signals are attenuated at lower frequencies than those of the RF-center. This statement holds only for the *inhibitory interaction* of RF-center and RF-periphery. These experimental differences might be due to the relatively smaller area of the stimulus in the RF-periphery used in the experiments of Maffei et al. or more probably by stray light effects (Maffei et al., used dilated natural pupils without artificial pupil).

7. Dynamic Interactions between RF-Center and RF-Periphery

A sinusoidally modulated spot of light, which is larger than the RF-center, causes a smaller neuronal response at low stimulus frequencies than a spot of light which covers only the RF-center. This inhibitory effect of the RF-periphery is less at higher stimulus frequencies (p. 495). In most of the on-center neurons and in many off-center neurons, the neuronal activation caused by a sinusoidal flicker stimulus above 10 Hz, covering the whole RF, is stronger than the response obtained with a stimulus that covers the RF-center alone. This finding indicates a decrease of the periphery inhibition at medium and higher stimulus frequencies. These experimental findings can be explained by two different mechanisms:

(a) With increasing stimulus frequency, the inhibitory effect from the RF-periphery is more strongly attenuated than the excitatory processes from both the RF-center and the RF-periphery.

(b) The phase frequency response of the inhibitory process from the RF-periphery is different from that of the center process. Therefore, at higher frequencies the inhibitory effect from the RF-periphery is out of phase with the excitatory effect elicited from the RF-center.

In order to discriminate between these two possibilities, experiments with the stimulus patterns *E* or *G* in Fig. 26 were performed: Center stimulus and periphery stimulus were modulated with the same frequency, but the phase angle φ between these two was varied in steps of 20 or 40 degrees between 0 and 360 degrees

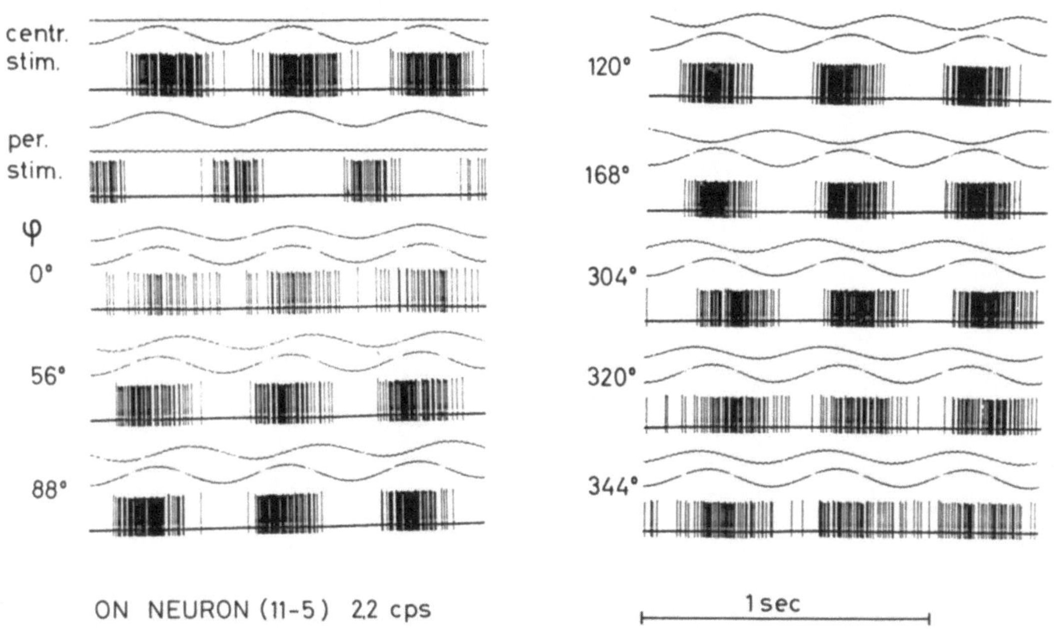

Fig. 35. *Dynamic interaction of the signals caused by stimulation of the RF-center and the RF-periphery of an on-center neuron of the cat retina.* The RF-center was stimulated by a spot of light (1 degree diameter) and the RF-periphery by a light annulus. Both stimuli were modulated sinusoidally at the same frequency (2.2 Hz). The phase angle φ between both is varied. Note the significant change of the response pattern with change of φ (Grüsser and Rackensperger, 1967; from Grüsser, 1971)

(Fig. 35, 36, GRÜSSER and RACKENSPERGER, 1966; GRÜSSER et al., 1971; GRÜSSER, LÜTGERT, and RACKENSPERGER, 1972).

When $R_{c+p}(\varphi)$ is the average neuronal impulse rate obtained at a constant stimulus frequency, but variable phase angle φ, the following function describes well the experimental data:

$$\overline{R}_{c+p}(\varphi) = (\overline{R}_c + \overline{R}_p)\left[\alpha + \beta \sin\left(\frac{\varphi}{2} + \lambda\right)\right] \text{ [Impulses} \cdot \text{sec}^{-1}] \quad (40)$$

$\alpha \leqq 1$ and $\beta < 1$ are constants, $\overline{R}_c$ is the average impulse rate obtained from stimulation of the center alone and $\overline{R}_p$ the average impulse rate obtained by stimulation of the periphery alone. Instead of Eq. (40) one can also write:

$$\overline{R}_{c+p} = \overline{R}_{\min} + (\overline{R}_{\max} - \overline{R}_{\min}) \sin\left(\frac{\varphi}{2} + \lambda\right) \text{ [Impulses} \cdot \text{sec}^{-1}] \quad (41)$$

where $R_{\min}$ and $R_{\max}$ are the minimum and maximum values respectively obtained at a given frequency f (Fig. 36).

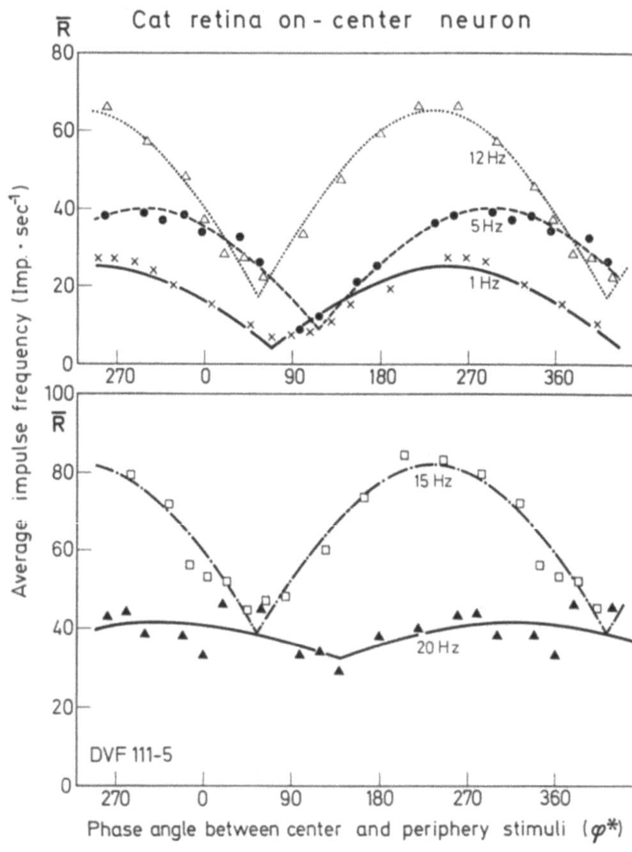

Fig. 36. *On-center neuron of the cat retina.* Stimulus pattern G_3, Fig. 26. Relationship between the phase angle φ^* between center stimulus and periphery stimulus (abscissa) and the average impulse frequency (ordinate). The curves are from Eq. (40) fitted by computer to the data (minimum Chi-square values). $\overline{I} = 600$ cd · m^{-2}, $m = 0.9$ (GRÜSSER, LÜTGERT, and RACKENSPERGER, 1972)

As discussed on 500, the inhibitory processes elicited from the RF-periphery are attenuated more strongly at stimulus frequencies above 3 Hz than the excitatory processes, elicited from both the RF-center and the RF-periphery. However, the inhibitory processes elicited from the RF-center (direct inhibition, see below) have frequency response characteristics similar to those of the excitatory processes. There is also a slight change in the phase relation of inhibitory and excitatory processes which can be seen from the frequency dependent variation of R_{c+p} (Fig. 36). At stimulus frequencies above 20–30 Hz the inhibitory influence becomes independent of the phase angle φ. Therefore $(R_{\max} - R_{\min}) \to 0$. From the data mentioned on p. 500 one can assume that the inhibitory periphery process reaches its fusion frequency at 10–20 Hz and is constant above this frequency, while the inhibitory signals from the RF-center reach the CFF near the CFF of the excitatory signals.

From these experimental findings, we concluded that both mechanisms(a) and (b), mentioned above are present. The attenuation of the lateral inhibitory signals from the RF-periphery, is more important than the slightly different phase frequency response of excitatory and inhibitory periphery signals as a mechanism which causes the increase of the CFF with increasing stimulus area beyond the size of the RF-center.

Eq. (40) and Eq. (41) directly result from the assumption that inhibition and excitation from the RF-center and RF-periphery interact linearly. The constant α is the weighting factor for the summation of center excitation and periphery excitation. For many neurons $\alpha = 1$ at low stimulus frequencies.

As one can see, Eq. (40) and Eq. (41) are identical with the Eq. (21), which describes the psychophysical findings about the interaction of a flickering, small light spot and a surrounding, flickering annulus (Fiorentini and Maffei, 1971). The formal identity again suggests a close correlation between psychophysical and neurophysiological data.

8. The Effect of Steady Illumination of the RF-Periphery on the Flicker Response of the RF-Center

The flicker activation of retinal ganglion cells depends not only on the parameters of the flickering test field but also on the size (A_B) and the luminance (I_B) of the steadily illuminated surround. In on-center neurons, the CFF and the neuronal response to intermittent stimulation of a test field projected to the RF-center decreases as the luminance or area of a steady, peripheral, surround stimulus increases. When the steady surround stimulus extends beyond the RF-periphery, however, its inhibitory effect decreases with a further increase of the surround area (disinhibition).

In the off-center neurons, the reverse effect is observed. A steady illumination of the RF-periphery facilitates the response obtained by a flickering test field in the RF-center. Also the CFF is increased by a steady stimulation of the RF-periphery (Fig. 37). The larger the area of the steady stimulus projected to the RF-periphery, the more the neuronal activation and the CFF increase. The facilitatory effect of the surround stimulus is maximum at a stimulus intensity equal to that of the flickering stimulus in the RF-center. Here again an increase of the surround stimulus beyond the RF-periphery decreases the neuronal response. This inhibitory effect from the far periphery corresponds to the disinhibition observed in on-center neurons (LUNKENHEIMER, 1969).

If the flickering stimulus projected to the RF-center of off-center neurons does not cover the whole RF-center and the surrounding, steady light still covers part of the RF-center, the facilitatory effect from the RF-center is masked by the steady stimulus in the RF-periphery.

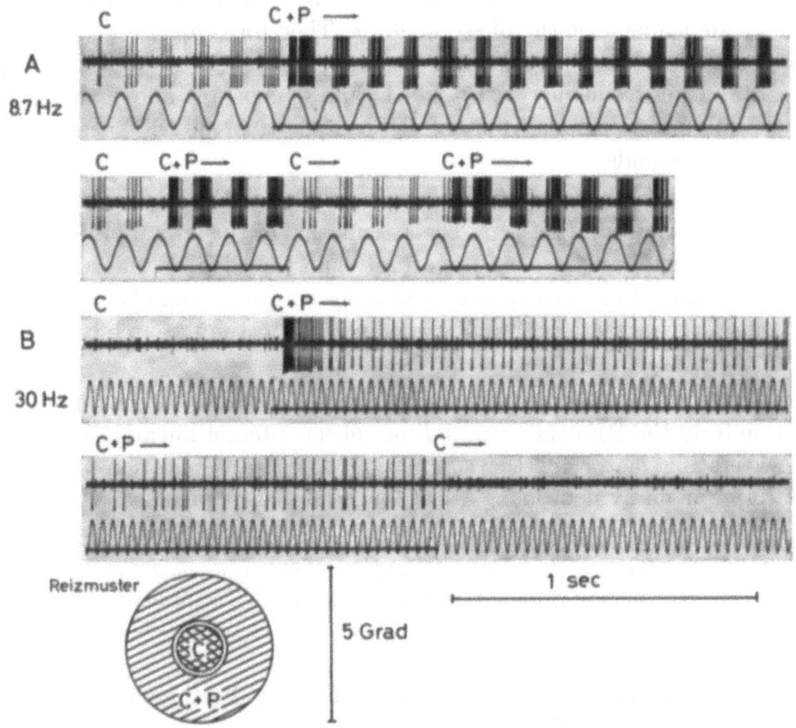

Fig. 37. *Off-center neuron of the cat retina.* Responses to sinusoidal stimulation of the RF-center (C) and to sinusoidal stimulation of the RF-center in combination with steady illumination of the RF-periphery ($C + P$). Average luminance of the sinusoidal stimulus and of the RF-periphery approximately equal ($= 250$ cd $\cdot$ m^{-2}, $m = 0.7$), a: Stimulus frequency of the center stimulus $=$ 8.7 Hz. Response enhancement with additional steady stimulation of the RF-periphery. b: The CFF caused by the stimulus alone is reached at 28 Hz. Intermittent stimulation of the RF-center at 30 Hz elicits no response. Additional steady illumination of the RF-periphery causes activation. The impulses are grouped in the frequency of the center stimulus (GRÜSSER, 1972)

A steady light stimulus projected to the RF-periphery of off-center neurons not only increases the average neuronal activation but it also leads to a "linearization" of the neuronal response pattern. The facilitatory or the inhibitory effects of the steady stimuli projected to the RF-periphery are well described by a logarithmic function of the luminance and the area of the stimulus (GRÜSSER, LICKER, and LUNKENHEIMER, in preparation 1972). The steady light inhibitory or excitatory effects of the RF-periphery can be easily explained if one assumes that the periphery stimulus adds or subtracts a constant value of the average membrane potential of the ganglion cell. Hence, the center response is superimposed on different steady state values of the membrane potential and reaches therefore the impulse threshold under different conditions. The impulse pattern depends on the "distance" between the impulse threshold and the membrane potential and it is this "distance" that is manipulated with the surround stimulation. The average depolarization level of the ganglion cell membrane as well as the temporal course of the threshold decay after each impulse determine whether or not the periodic membrane excursions are transformed into a sequence of impulses, i.e. they determine at which frequency the CFF is reached (Fig. 21).

The effect of the steady illumination of the RF-periphery on the CFF is a good example of how the *temporal* transfer characteristics of a nerve cell, which also depend on the sampling frequency of its threshold mechanism, can be changed by a DC-shift of the average membrane potential (c.f. Grüsser, 1972).

The neurophysiological findings described in this section can be interpreted as the correlate for the dependence of the psychophysically measured CFF of a flickering test field surrounded by a steady light as described on pp. 442—443. If one assumes that at the perceptual level the phenomena correlate with the sum of on-center and off-center neurons' activity, the average CFF of both systems increases, because the CFF of off-center neurons is clearly increased by a steady surround stimulus.

9. Enhancement of the Flicker Response of Retinal Neurons by McIlwain's Periphery Effect

McIlwain (1964, 1966) studied the influence of distant retinal areas on the reponse of retinal ganglion cells and of LGN-neurons of cats, stimulated in the RF-center by liminal intermittent light. Dark objects moved in the far surround of the receptive field of retinal ganglion cells lower the threshold to responses obtained by stimuli projected to the RF-center or to the RF-periphery. This effect can be elicited at a distance up to 90 degrees from the RF-center and is greatly reduced by barbiturate anesthesia. The angular velocity and the amplitude of the movement as well as the size and the contrast of the stimuli moving in the far periphery influence the strength of this facilitatory periphery effect. It seems rather probable that the facilitation is caused by disinhibition, i.e. the excitation elicited by the moving targets is caused by an inhibition of the inhibiting RF-periphery in the immediate surounding of the RF-center. The periphery effect is not caused by stray light (Levick et al., 1965).

10. Interaction of Excitation and Direct Inhibition from the RF-Center

A light spot projected to the RF-center not only causes excitation but also inhibition. We call this inhibitory process, i.e. the off-inhibition from the center of on-center neurons and the on-inhibition from the center of off-center neurons, *direct inhibition*. The summation of direct excitation and direct inhibition elicited from the RF-center and the relative frequency properties of both processes were investigated with stimulus pattern D of Fig. 26: Two small light spots of 0.22 degrees diameter and 0.15 degrees distance are projected to the RF-center. The luminance of the light spots is varied sinusoidally (0.5–35 Hz) and the phase angle between the two sinusoidal stimuli is changed in steps from 0 to 360 degrees. By this method, excitatory periods elicited by the one stimulus are arbitrarily superimposed on inhibitory responses to the other light stimulus and vice versa (Fig. 56). The neuronal response (average impulse rate, $\overline{R}$, maximum impulse rate, (R_{max}) depends in a very simple way on the phase angle φ'. Of course, the neuronal activation is also dependent on the stimulus frequency (p. 489). The relative relationship between φ' and the neuronal activation, however, is independent of the stimulus frequency between 2 and 20 Hz (CH. BÜTTNER et al., 1969, 1971). It can be described by the following functions:

$$\overline{R} = b\,(\overline{R}_a + \overline{R}_b)\cos\left(\frac{\varphi'}{2} + \lambda'\right)\ [\text{Impulses}\cdot\text{sec}^{-1}] \tag{42}$$

and

$$R_{max} = b'\,(R_{a,\,max} + R_{b,\,max})\cos\left(\frac{\varphi'}{2} + \lambda'\right)\ [\text{Impulses}\cdot\text{sec}^{-1}]\,. \tag{43}$$

These two functions were found to fit the data from a large collection of on-center neurons, but they also fitted the data of the few off-center neurons that were tested. However, on the average, the neuronal activation is smaller in the off-center neurons than in the on-center neurons.

From these findings, the conclusion was drawn that the excitatory and the inhibitory signals from the RF-center that probably interact at the ganglion cell membrane are summed linearily at frequencies below 20 Hz (For the arguments why Eqs. (42) and (43) are valid for linear summation of direct inhibition, see CH. BÜTTNER et al., 1971). In addition, one can conclude that the amplitude and the frequency properties of the direct inhibition do not deviate significantly from the frequency characteristics of the direct excitation, if the stimulus frequency is below 20 Hz. Furthermore, the amount of inhibition during off-periods is in this frequency range proportional (but not necessarily equal) to the amount of excitation, i.e. the gain of the direct inhibition is the same as the gain of the direct excitation. The amount of direct inhibition is on the average larger in off-center neurons than in on-center neurons, when the steady background is considerably lower than $\overline{I}$. In on-center neurons the phase relationship between on-excitation and on-inhibition changes at stimulus frequencies above 20 Hz. However, the interaction between excitation and inhibition above 20 Hz and near the CFF can still be described by a simple model of linear summation.

11. The Dependency of the Simultaneous Contrast Activation on the Stimulus Frequency

From the data described on pp. 485—504, it is predictable that the response of retinal ganglion cells to intermittently illuminated contrast borders projected on the RF, also depends on the frequency of the flicker stimuli. BAUMGARTNER and HAKAS (1959, 1960, 1961) described how the response of cats' single optic tract fibers depends on the position of a light-dark border within the RF: If the neuronal activation is plotted against the border's position, the resulting "spatial distribution" of the neuronal activation correlates well with the psychophysical findings on the brightness and darkness enhancement across a contrast border (MACH-bands). The neuronal activation reaches a maximum when the position of the light-dark border corresponds with the border between RF-center and RF-periphery. BAUMGARTNER explains these findings by the well-founded assumption that under this stimulus condition the overall effect of lateral inhibition is at its minimum.

Fig. 38 demonstrates that the contrast enhancement of retinal neurons depends on the stimulus frequency, when the light part of the light-dark pattern is illuminated sinusoidally. The largest relative contrast enhancement is measured at low stimulus frequencies. When the neuronal activation increases at medium stimulus frequencies (4—10 Hz,) the additional response enhancement when the border is at its optimal position becomes less pronounced (Gaedt and Grüsser, 1966; Gaedt and Lunkenheimer, 1966).

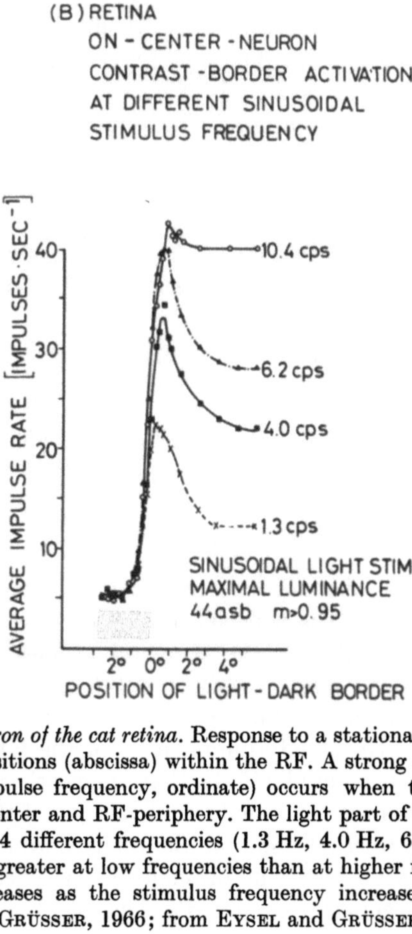

Fig. 38. *On-center neuron of the cat retina.* Response to a stationary light-dark contrast border, placed at different positions (abscissa) within the RF. A strong enhancement of the neuronal response (average impulse frequency, ordinate) occurs when the contrast border is at the border between RF-center and RF-periphery. The light part of the contrast border is illuminated sinusoidally at 4 different frequencies (1.3 Hz, 4.0 Hz, 6.2 Hz, 10.4 Hz). The relative MACH-band-effect is greater at low frequencies than at higher frequencies. The average neuronal activation increases as the stimulus frequency increases up to 10 Hz (Gaedt and Grüsser, 1966; from Eysel and Grüsser, 1971)

These neurophysiological data suggest that at the psychophysical level the increase of the subjective brightness of a flickering contrast stimulus (Brücke-Bartley effect) should be paralleled by a diminished border contrast. The effect of the stimulus frequency on the appearance of MACH-bands was, as far as we know never measured psychophysically. The findings about the reduction of the visual acuity at stimulus frequencies at which brightness enhancement of the flicker stimuli is perceived (p.460), however, might be viewed as a psychophysical correlate of the neurophysiological results described in this section. It is well known from psychophysical experiments that visual acuity is closely related to simultaneous contrast.

12. Neuronal Responses above the CFF

The ratio impulse frequency/flicker frequency above the CFF becomes smaller than 1, but the impulse intervals are still dependent on the flicker stimuli (ENROTH, 1952). We call the flicker frequency, above which no relationship between the intermittent stimuli and the impulse pattern is present, the Talbot flicker frequency (TFF) because above the TFF the Talbot law applies to the neuronal response (p. 438). Between CFF and TFF, the retinal neurons act as scaling devices. The relative downward scaling becomes larger, the more the flicker frequency approaches the TFF.

The properties of the neuronal response between CFF and TFF were further examined by FUKADA et al. (1966), OGAWA et al. (1966), FUKADA and SAITO (1970) and GRÜSSER and VESPER (1970). Fig. 39 shows interval histograms of the impulse sequence of an off-center neuron elicited by flicker stimuli below and above the CFF. The grouping of the intervals around whole number multiples of the flicker frequency is evident. In addition, the analysis of the impulse pattern by calculation of joint interval histograms reveals that successive intervals are not statistically independent of each other.

The local maxima of the distribution of the intervals grouped around whole multiples of the flicker interval follow between the CFF and the TFF an exponential function:

$$H_n = H_1 \exp [- t/\tau^*] \quad [\text{Impulses} \cdot \text{sec}^{-1}] \tag{44}$$

whereby H_1 is the number of intervals in the class $1/f$, H_n the number of the intervals in the class $1/nf$ (GRÜSSER and VESPER, 1970). The time constant τ^* of Eq. (44) was found in off-center neurons to be a linear function of the frequency of the sinusoidal flicker stimuli between CFF and TFF:

$$\tau^* = a_1 f - a_2 \quad [\text{msec}] \tag{45}$$

whereby a_1 and a_2 are constants. From these findings, one can conclude that between the CFF and the TFF the neuronal response pattern depends not only on the flicker frequency but also on the temporal properties of the threshold elevation occurring after each impulse and on internal synaptic "noise". This explanation is consistent with the finding described on pp. 504–506 for the dependency of the CFF on the steady illumination of the RF-periphery. It also explains the findings of FUKADA et al. (1971), that under otherwise equal stimulus conditions the CFF increases with the conduction velocity of the optic nerve fiber. The larger the optic nerve fiber, the shorter the relative and absolute refractory period. Hence, the interaction between threshold decay after each impulse and slow membrane potential occurs more frequently the larger the conduction velocity (and the shorter the refractory period of the impulse generating mechanism).

13. The Effect of Barbiturates and of Ethyl Alcohol on the Neuronal Flicker Response

The data about the spatial and temporal transfer characteristics of retinal neurons were usually gathered by measurements in cats anaesthetized with barbiturates. Therefore the effect of barbiturates on the neuronal response has to be studied (SCHMIDT and CREUTZFELDT, 1968, 1969; FOERSTER and GRÜSSER, 1967,

1969). At low doses (10 mg per kg pentobarbital delivered to encephale isolé preparations) the neuronal spontaneous activity and the flicker response of cats' retinal neurons might be slightly enhanced. Above 15 mg/kg pentobarbital, the

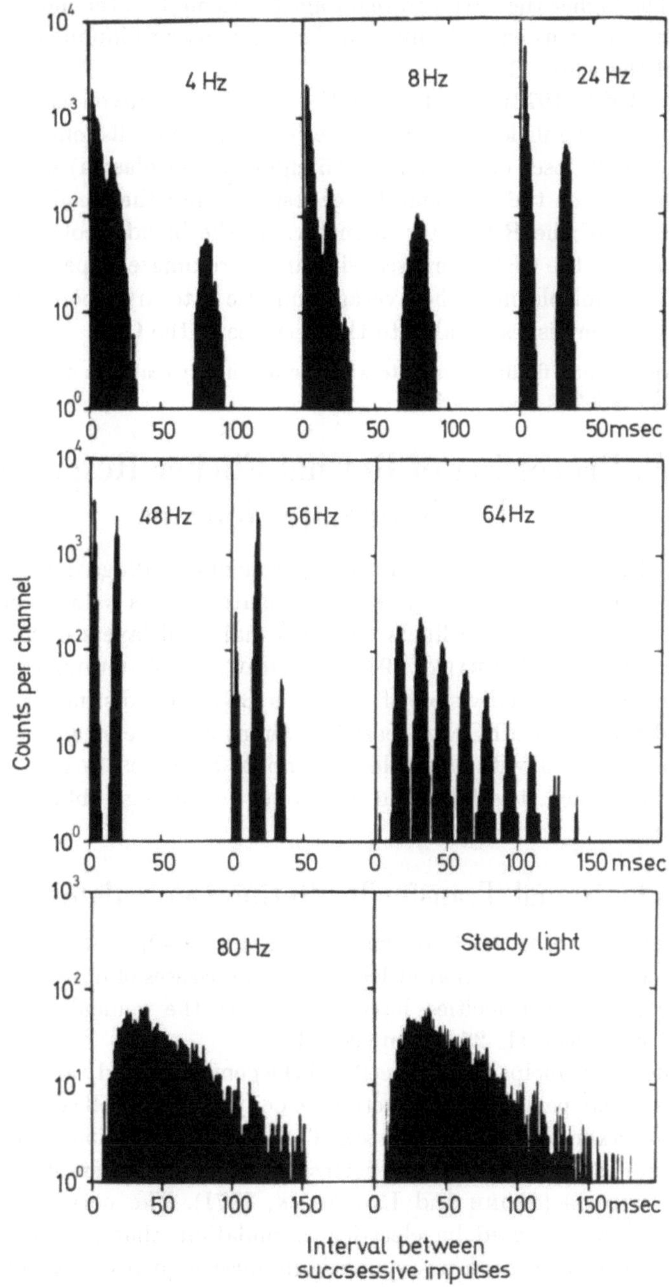

Fig. 39. *Off-center neuron of the cat retina.* Interval histograms of the responses caused by a flickering test field (17.5 degrees diameter, average luminance 8.4 lm · m^{-2}). CFF at about 50 Hz, TFF above 65 Hz (Ogawa et al., 1966)

CFF and the average neuronal activation elicited by sinusoidal stimulation of subfusional frequencies decrease rapidly with increasing level of anaesthesia. The response type (X/Y, sustained/transient) is also affected by the amount of barbiturates. The higher the barbiturate dosage (> 25 mg/kg), the more the on-center and off-center neurons exhibit only transient responses to illumination of the RF-center or of the whole RF.

Stange (1970, 1972) investigated the effect of intravenously administered ethyl-alcohol on the flicker response of retinal ganglion cells (encephale isolé preparations). Small doses of alcohol (< 0.5 mg/ml blood plasma) have only a very small or no effect on the neuronal flicker response, but they lead to a slight functional decrease of the RF-center diameter. As the blood alcohol level increases above 0.8 mg/ml, the CFF decreases with an approximate slope of 5 Hz for every mg alcohol per ml plasma. The average impulse rate in response to subfusional flicker stimuli diminishes parallel to the decrease of the CFF.

These experimental findings correlate with the psychophysical data on the decrease of the CFF caused by sedative drugs (p. 446).

V. Processing of Retinal Flicker Responses by Geniculate Neurons

In this chapter, emphasis is on the experimental findings in single neurons of the corpus geniculatum laterale (lateral geniculate nucleus = LGN) of the cat. The majority of the optic nerve fibers end in 3 main cell layers of the LGN (Minkowski, 1913, 1920; O'Leary, 1940; Hayhow, 1958). Guillery (1966, 1970) proposed recently a subdivision of the cat's LGN into distinct laminae. In the human and the primate LGN at least 6 different cell layers can be discriminated (3 with ipsilateral, 3 with contralateral input, Le Gros Clark, 1942). If not otherwise mentioned, the following data refer to findings obtained in the cat's LGN neurons.

A. The Signal Transfer by Retino-Geniculate Synapses

Since the work of G. H. Bishop and O'Leary (1940), and P. O. Bishop et al. (1953) it is well established that at least two main classes of optic nerve fibers with different conduction velocities have endings at the geniculate cells (class I: $50-80$ m $\cdot$ sec^{-1}, class II: $25-40$ m $\cdot$ sec^{-1}).

As a rule, the principal cells (P-cells) of the geniculate body are excited by optic tract axons of one particular conduction velocity. Each P-cell sends via the visual radiation one axon to the visual cortex. The axon's conduction velocity correlates with the conduction velocity of the corresponding retinal afferents that end at the geniculate neurons (Stone and Hoffmann, 1971). The maximum optic nerve impulse frequency, caused by electrical stimulation, that a single LGN-neuron follows with a synchronous postsynaptic impulse sequence is significantly lower than the maximum frequency the corresponding optic nerve fibers can follow (Cornehls, 1958; Grüsser-Cornehls and Grüsser, 1960; P. O. Bishop et al., 1962; P. O. Bishop, 1964). The signal transfer through LGN-synapses and neurons has at higher input frequencies properties of a "scaling device". This is somewhat surprising because the large synaptic contacts of one optic nerve fiber (SZENTAGOTHAI et al., 1966) might produce an EPSP large enough to reach the postsynaptic threshold for the generation of an impulse (SINGER et al., 1972). The scaling effect of the geniculate synapses is also observed when pre- and postsynaptic action potentials are simultaneously recorded (HUBEL and WIESEL, 1960; CLELAND et al., 1971; ECKHORN and PÖPEL, 1971) or the EPSP-sequence is compared with the impulse pattern by means of intracellular or quasi-intracellular recordings (e.g. MCILWAIN and CREUTZFELD, 1967; CREUTZFELDT, 1968; SINGER and CREUTZFELDT, 1970; ECKHORN and PÖPEL, 1972).

Fig. 40 shows two typical diagrams for the relationship between frequency of the electrical stimuli applied to the optic nerve and the impulse frequency of an optic nerve fiber and of two geniculate neurons (optic radiation fibers), the one belonging to class I, the other to class II. The upper frequency limit for a 1 : 1 input-output relation is significantly lower in class II geniculate neurons than in class I neurons. In addition, longer lasting repetitive synchronized activation of the excitatory optic nerve synapses between optic nerve fiber and geniculate P-cells leads to a change of the scaling factor (posttetanic depression, GRÜSSER and

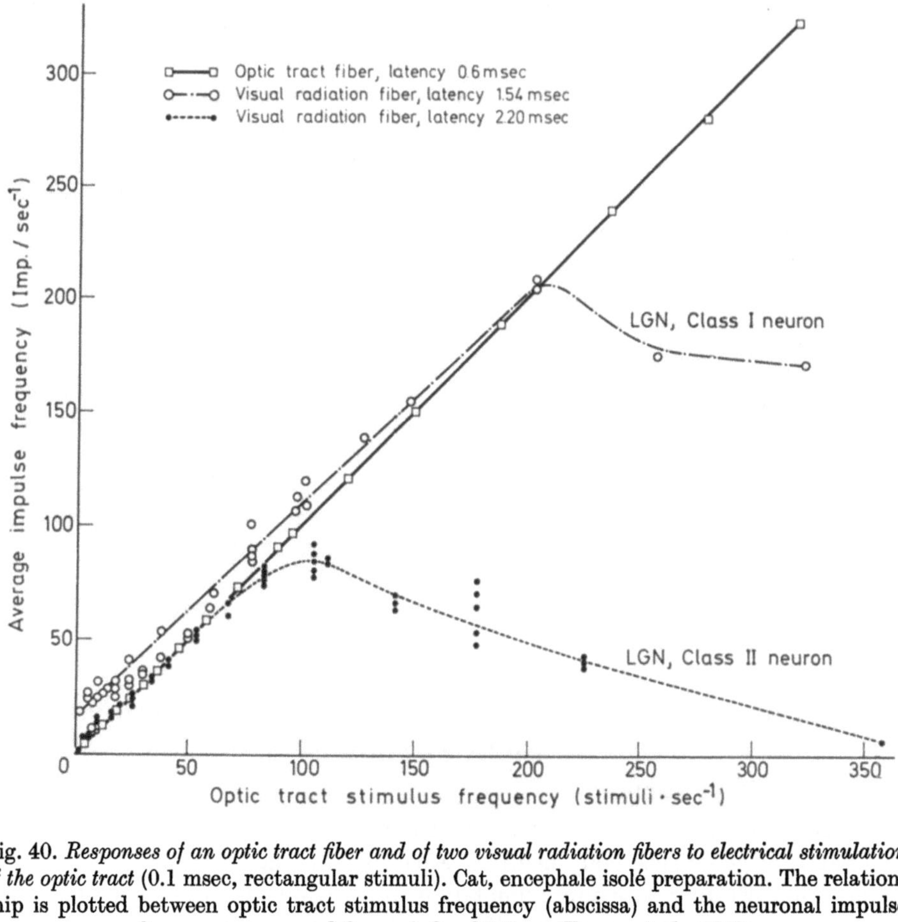

Fig. 40. *Responses of an optic tract fiber and of two visual radiation fibers to electrical stimulation of the optic tract* (0.1 msec, rectangular stimuli). Cat, encephale isolé preparation. The relationship is plotted between optic tract stimulus frequency (abscissa) and the neuronal impulse frequency measured for periods of 0.5 sec (EYSEL et al., 1972)

GRÜTZNER, 1958; EYSEL, GRÜSSER, and PECCI-SAAVEDRA, 1972). The limiting frequency of the signal transfer through geniculate neurons, however, is by far much higher than the CFF obtained with very high intensity flicker stimuli at the level of the optic nerve fibers. Therefore, the impulse generating mechanisms at the LGN-neurons do not constitute any additional basic limitations to time resolution. Nevertheless, the complete neuronal network within the LGN, including recurrent inhibitory mechanisms, might act as an effective frequency filter.

B. Classes of Neurons in the LGN

Besides the *P*-cells in the geniculate nucleus, which send axons to the visual cortex, single unit recordings were obtained from interneurons (*I*-cells), which probably have their synaptic contacts mainly within the LGN (rat, BURKE and SEFTON, 1966; cat, MCILWAIN and CREUTZFELDT, 1967). With respect to the functional organization of their concentric RF, *P*-cells are subdivided into two, about equally frequent classes: On-center neurons and off-center neurons. Another subdivision is possible with respect to the spatio-temporal signal transfer properties of the neurons: "Transient" neurons and "sustained" neurons (CLELAND et al., 1971). With this classification a relationship with the conduction velocity of the afferent optic nerve fibers holds as well: The "transient response" neurons receive their input from the retina via fast optic nerve fibers (class I), the "sustained response" neurons via slower class II optic nerve fibers (CLELAND et al., 1971; STONE and HOFFMANN, 1971). Type I neurons have, on the average, larger RFs than type II neurons. This might be caused mainly by the fact that the ganglion cells of the area centralis have smaller RFs than the ganglion cells in the RF-periphery and have axons of type II (STONE and FREEMAN, 1971). EYSEL and GRÜSSER (1971) classified geniculate neurons according to their functional properties with respect to spatial summation, flicker response and the response to moving contrast borders: "Contrast cells" (*C*-neurons) and "luminosity cells" (*L*-neurons). In both classes, on-center and off-center neurons are found. The *L*-neurons correspond to the "sustained response" type of CLELAND et al., the *C*-neurons to the "transient response" type. The interpretation of the functional significance of this classifiction, however, is different. CLELAND et al. believe that the sustained type serves primarily the signalling of steady local differences of retinal illumination and the transient type constitutes an initial stage of development of specific sensitivity to moving patterns. On the other hand, EYSEL and GRÜSSER assume that contrast detection is the main functional property of the *C*-neurons, while *L*-neurons signal average brightness (c.f. also BAUMGARTNER, 1961; BAUMGARTNER et al., 1965).

C. Responses of Geniculate Neurons to Intermittent Light Stimuli

1. Diffuse Flicker Stimuli

In general, the "laws" described for the CFF of retinal neurons (p. 480) are also valid for geniculate body neurons, especially for the sustained *L*-type:

(a) The CFF increases approximately in proportion with the logarithm of the average luminance of rectangular flicker stimuli (neurophysiological correlate of the Ferry-Porter-law). We found 2 types of neurons: The one type reaches a maximum CFF between 20 and 22 Hz, even if flicker stimuli of higher photopic stimulus intensity are applied. The other type exhibits a *kink* in the CFF-stimulus luminance curve at mesopic stimulus levels and reaches a maximum CFF between 50 and 60 Hz for high stimulus intensities of the flicker light (Fig. 41, Grüsser and Grüsser-Cornehls, unpubl. 1960).

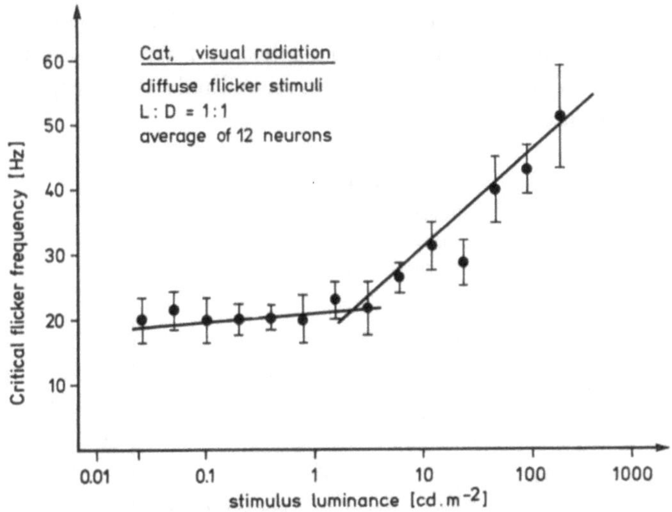

Fig. 41a. Cat, encephale isolé preparations, on-center neurons of the LGN (recordings of action potentials from optic radiation fibers). Dependency of the average CFF (ordinate) on the stimulus luminance (abscissa), diffuse flicker stimuli, light-dark ratio = 1 : 1 (Grüsser and Grüsser-Cornehls, unpubl. 1960)

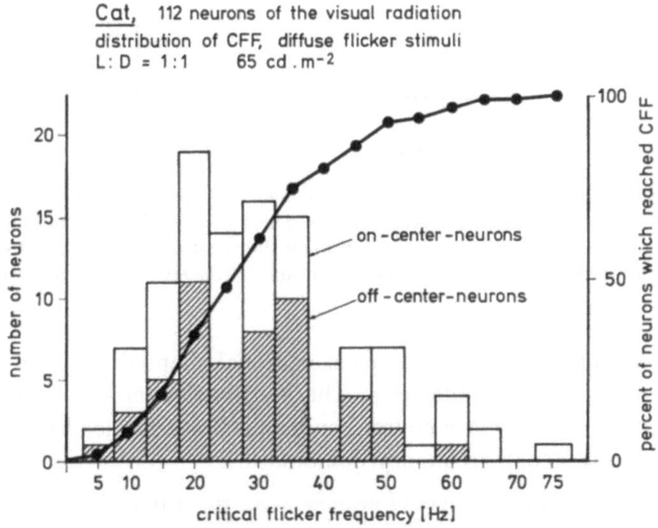

Fig. 41b. Cat, encephale isolé preparations, 112 on-center neurons and off-center neurons of the LGN. Distribution of the CFF at one stimulus luminance (64 cd · m^{-2}). Diffuse flicker stimuli, $L : D = 1 : 1$ (Grüsser and Grüsser-Cornehls, unpubl. 1960)

(b) When the CFF of different neurons of the LGN is measured under comparable stimulus conditions (diffuse flicker stimuli of constant intensity parameters, encephale isolé preparations), the variability of the CFF of different neurons is rather high. With the same flicker stimuli, repeated measurements of the CFF or TFF in the same neurons, however, reveal a rather high constancy (< ± 5 %). The distribution of the CFF in a large collection of geniculate neurons is shown in Fig. 41 b. When the relationship between the neuronal impulse rate of the *P*-cells and the flicker frequency is plotted under comparable stimulus conditions in different neurons, the response optima are also reached at different stimulus frequencies. The neurons with the lower CFF reach correspondingly their response maxima at lower stimulus frequencies.

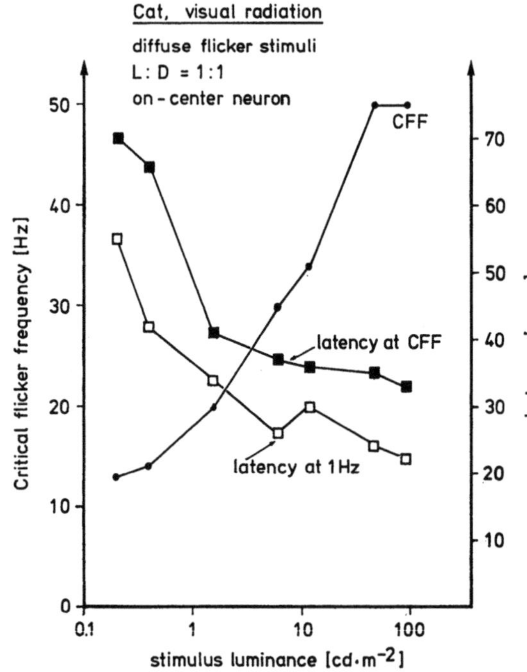

Fig. 41c. *On-center neuron* of the LGN. Cat, encephale isolé preparation (recording of action potentials from the optic radiation axon). Dependency of the CFF, the latency at the CFF and the latency at 1 Hz flicker stimuli (ordinate) on the stimulus luminance (abscissa). Diffuse flicker stimuli, light-dark ratio 1 : 1 (GRÜSSER and GRÜSSER-CORNEHLS, unpubl. 1960)

It has been suggested that there exist neuronal channels in the a.v.s. that are tuned to specific spatial frequencies. The distribution of the CFF and of the response maxima of geniculate neurons suggests that a similar hypothesis can be proposed for the temporal frequency domain.

As Fig. 41b indicates, the number of neurons which respond to the flicker stimulus with rhythmic discharges decreases as the stimulus frequency increases. This observation probably correlates with the psychophysical finding that a large homogeneous flickering test field, above 10—15 Hz, is perceived as rather inhomogeneous; different parts of the test field seem to flicker at different times.

(c) The relationship between flicker frequency and neuronal response latency described for retinal on-center and off-center neurons holds also for the *P*-neurons of the LGN. Thus, the relationship found between response latency and CFF as well as between neuronal activation and CFF for retinal neurons is also valid for the flicker response of geniculate cells (Fig. 41 c, Grüsser and Grüsser-Cornehls, unpubl. 1960).

(d) The relationship between the average or the maximum neuronal impulse rate of the *P*-cells and the flicker frequency is in general the same as in retinal neurons (p. 484, De Valois, 1958; monkey; Arden and Liu, 1960, rabbit; Grüsser and Saur, 1960; Rabelo and Grüsser, 1961, cat). The same rules apply for this relationship as described for the retinal on-center and off-center neurons: The maximum neuronal response is obtained at higher stimulus frequencies, the larger the stimulus area and the higher the stimulus intensity.

The cells of the *Z*-class (probably interneurons) found in the geniculate body, that are strongly inhibited by "light on" as well as by "light off", exhibit the reverse relationship between flicker frequency and impulse rate: The neuronal response reaches a minimum between 5 and 10 Hz. Above the frequency at which the *P*-cells reach flicker fusion, these neurons again increase their impulse rate.

As for the retinal neurons, the relationship between neuronal impulse rate of P-cells and flicker frequency can be considered as the neurophysiological correlate of brightness and of darkness enhancement (c.f. p. 484 and 485).

(e) If action potentials from pre- and postsynaptic neurons, which have their RF in about the same retinal area, are recorded within the LGN of a given preparation, the CFF of the retinal neurons, as a rule, is slightly higher than the CFF of the geniculate neurons. If the animals (cats) are anaesthetized by pentobarbital, the neuronal impulse rate of the LGN neurons decreases much faster with an increasing amount of barbiturates than that of the retinal neurons. In encephale isolé preparations, the neuronal impulse rate of retinal neurons obtained with slow and medium stimulus frequencies is 50–80 % higher than that of geniculate neurons. This difference is larger in the anaesthetized preparation (200–300 %, sinusoidal light stimuli, $m > 0.9$, 1–40 Hz).

2. Binocular Flicker Stimuli

Part of the *P*-cells in the LGN of the cat are activated when the RF in the *dominant* eye is stimulated (on-center or off-center response), while stimulation of the corresponding area in the *non-dominant* eye causes inhibition. With diffuse light stimuli, Grüsser and Saur (1960) found that only 10 % of the LGN-neurons were inhibited by stimulation of the non-dominant eye. In their experiments, however, the visual cortex was removed and therefore possible cortico-geniculate interactions were abolished. When the neuronal activity is recorded from the visual radiation of the intact brain, about 50 % of visual radiation fibers are activated by the dominant eye (on-activation or off-activation) and also inhibited (on-off-inhibition) by diffuse stimulation of the non-dominant eye (cat, encephale isolé preparations, Grüsser and Grüsser-Cornehls, unpubl. 1960). The other 50 % of the visual radiation fibers respond only to stimulation of the *dominant* eye (on-activation or off-activation), while stimuli applied to the *non-dominant* eye are completely ineffective. SANDERSON et al. (1969, 1971) further demonstrated that the inhibitory effect of the stimulation of the non-dominant eye might be increased when instead of diffuse light stimuli, specific stimulus patterns are projected to the corresponding area in the non-dominant eye.

Indeed the flicker response of the LGN-neurons (visual radiation fibers) of the pure monocular type does not depend on stimulation of the non-dominant eye. The CFF of neurons activated by the dominant eye and inhibited by the non-dominant eye, however, is not determined by the dominant eye alone. When the frequency of the flicker stimuli applied to the dominant eye is set near the CFF, synchronous or alternating stimulation of the non-dominant eye reduces the flicker response significantly (Fig. 42a). Therefore, the binocular CFF of these LGN-neurons is

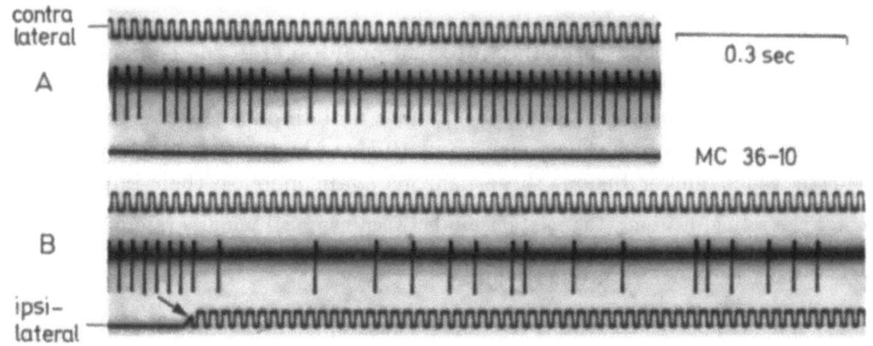

Fig. 42a. *On-center neuron of the cat's LGN* (recordings from the optic radiation axon). *A*: responses to diffuse intermittent light stimulation of the dominant, contralateral eye (48 Hz, 645 cd · m^{-2}). At these stimulus frequencies the CFF is just reached. *B*: additional synchronous stimulation of the ipsilateral eye (same stimulus frequency and luminance) leads to a significant reduction of the neuronal impulse rate and therefore to a reduction of the CFF

slightly lower than the monocular CFF when the RF in the dominant eye is stimulated. Stimulation of the non-dominant eye alone causes an on-off-inhibition, which follows the flicker frequency only up to 5–8 Hz. Above this frequency, intermittent diffuse light stimulation of the *non-dominant* eye leads to a general reduction of the spontaneous activity of the LGN-neurons (GRÜSSER and GRÜSSER-CORNEHLS, unpubl. 1960).

When the dominant and the non-dominant eye are stimulated with different flicker frequencies, one can also see an inhibiting effect of the stimuli applied to the non-dominant eye on the flicker activation, caused by stimulation of the dominant eye. This is shown in Fig. 42b. In this neuron, stimulation of the dominant (ipsilateral) eye causes an off-activation, while 4 Hz light stimulation of the non-dominant eye causes an irregular predominantly inhibitory response. A slight inhibitory effect of the non-dominant eye also exists for steady illumination (Fig. 42b trace *B*). This inhibitory effect is much stronger when the non-dominant eye is intermittently illuminated (4 Hz, Fig. 42b trace *D*). This finding indicates that the dominant eye controls the dynamic signal transfer properties of the LGN-relay station for this class of neurons, but the non-dominant eye modifies these signal transfer characteristics slightly.

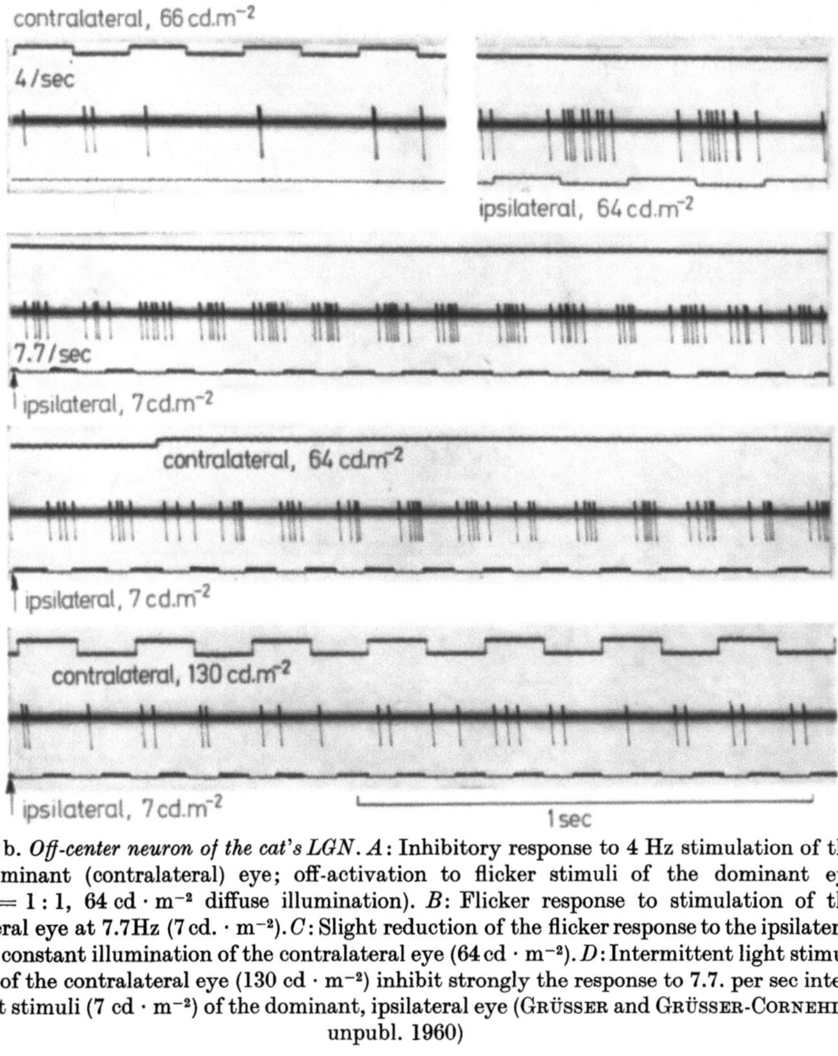

Fig. 42b. *Off-center neuron of the cat's LGN. A*: Inhibitory response to 4 Hz stimulation of the non-dominant (contralateral) eye; off-activation to flicker stimuli of the dominant eye ($L:D = 1:1$, 64 cd · m^{-2} diffuse illumination). *B*: Flicker response to stimulation of the ipsilateral eye at 7.7Hz (7 cd. · m^{-2}). *C*: Slight reduction of the flicker response to the ipsilateral eye by constant illumination of the contralateral eye (64 cd · m^{-2}). *D*: Intermittent light stimuli (4 Hz) of the contralateral eye (130 cd · m^{-2}) inhibit strongly the response to 7.7. per sec intermittent stimuli (7 cd · m^{-2}) of the dominant, ipsilateral eye (Grüsser and Grüsser-Cornehls, unpubl. 1960)

3. Responses of LGN-Neurons to Selective Flicker Stimulation of the RF

(a) *Area and Intensity Function.* The area-response relationship of on-center and off-center neurons of the luminosity type depends similarly on the stimulus frequency as described for retinal neurons: For small flickering test fields centered on the RF the neuronal activation increases as the stimulus area increases. The maximum response is obtained when the stimuli just cover the whole RF-center. When the flicker stimuli extend beyond the borders of the RF-center the neuronal response becomes less. As in retinal neurons this inhibitory effect of the RF-periphery depends also on the stimulus frequency. In luminosity type LGN-neurons the inhibitory periphery effect is maximum at low stimulus frequencies and decreases when the stimulus frequency increases above 3 Hz (Fig. 43a, Rackensperger, unpubl. 1970; Eysel et al., 1971).

For the intensity function and its dependency on the frequency, similar rules apply for LGN-neurons of the luminosity type as for on- and off-center neurons of the cat retina (p. 492).

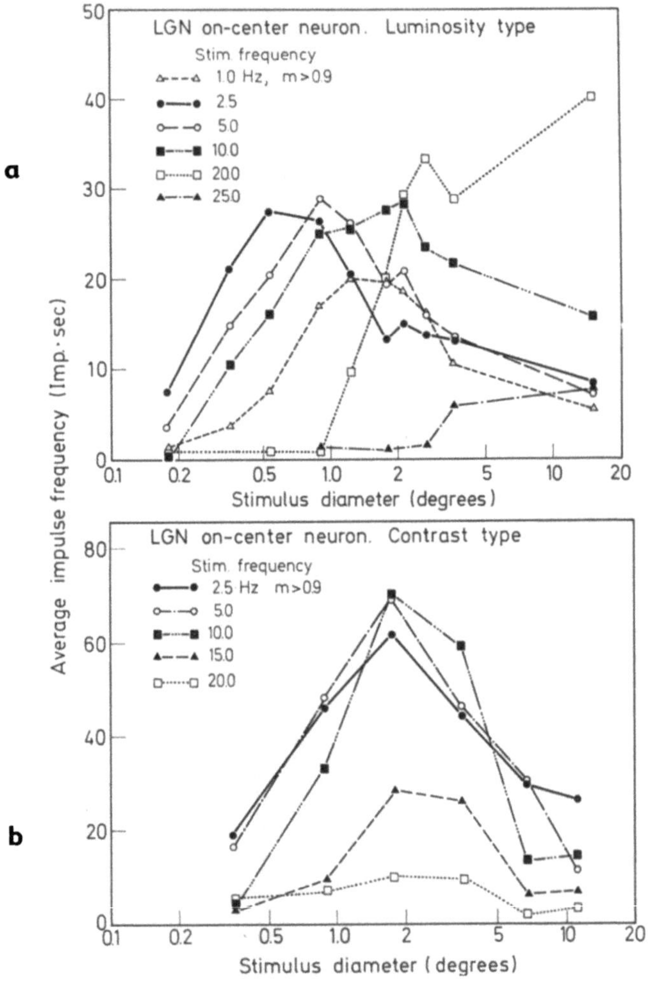

Fig. 43a. *On-center neuron of the cat's LGN.* Luminosity type. Relationship between the average neuronal activation (ordinate) and the diameter of the round flickering test field (abscissa) that is projected to the RF-center. *I* is approximately 200 cd · m^{-2} (RACKENSPERGER, unpubl. 1967)

Fig. 43b. *On-center neuron of the cat's LGN.* Contrast type. Same relationship as in Fig. 43a (RACKENSPERGER, unpubl. 1970)

In the *contrast neurons* of the cat's LGN, the inhibitory effect of light stimulation of the RF-periphery is significantly stronger than in luminosity-neurons. Some of these neurons respond only with a rather weak activation to flicker stimuli that cover the whole RF. This strong periphery inhibition, however, is much less frequency-dependent than in luminosity neurons (Fig. 43b, RACKENSperger, unpubl. 1970). Therefore the response enhancement caused by the simultaneous contrast of a sinusoidally illuminated border is only slightly affected by the frequency of the stimuli (Eysel and Grüsser, 1971). These findings indicate that the frequency response characteristics of the inhibitory process, elicited by the stimulation of the RF-periphery, differ for luminosity neurons and for contrast neurons. At the level of the LGN, a new lateral inhibitory mechanism appears in contrast neurons, which evidently has nearly the same frequency transfer properties as the excitatory process.

The CFF increases in both types of geniculate neurons as the stimulus area increases. This area effect is less strong in contrast neurons than in luminosity neurons.

The intensity functions of contrast and luminosity neurons differ also. While the response of luminosity neurons is in a similar way dependent on the stimulus luminance and stimulus intensity as described on p. 492 for retinal neurons, this is not the case for contrast neurons. These neurons also increase their neuronal activation with the stimulus luminance, but the overall influence of stimulus intensity is less than in the luminosity neurons.

The relationship between $\bar{I}$ and the CFF (or TFF) is well described for both contrast and luminosity neurons by Eq. (5) and Eq. (12). A similar interrelationship of area and intensity effects as found in retinal neurons is observed in the luminosity neurons of the LGN. Moreover, part of the LGN-neurons do not exceed a maximum CFF of 22–25 Hz, even with large and bright flicker stimuli.

(b) *Neuronal Response Enhancement at Medium Flicker Frequencies.* The relationship between stimulus frequency and the neuronal response is in general very similar for both on-center and off-center luminosity neurons of the LGN to that described on p. 489 for the neurons of the retina: The larger the stimulus intensity and the stimulus area, the stronger the relative response enhancement at medium stimulus frequencies and the higher the tuning frequency at which the response maximum is reached.

In on- and off-center neurons of the contrast type the relative response enhancement to flicker stimuli of medium frequency is more pronounced than in the luminosity neurons. The relative effect of stimulus luminance on this relationship, however, is significantly less in contrast neurons than in luminosity neurons.

(c) *Center and Periphery Responses.* Maffei and Fiorentini (1972) examined 24 units of the LGN of the cat in search of differences in the center and periphery responses to sinusoidal stimulation at different frequencies. In the upper frequency range they found similar attenuation characteristics for both the center and the periphery response of LGN-cells. They concluded that the surround mechanisms of LGN receptive fields are controlled by the center response from homolog presynaptic retinal neurons, having their RF-center located in the RF-periphery of the investigated LGN-cell.

(d) *Flicker Threshold Sensitivity Curves.* When the threshold sensitivity or equal response curves of single geniculate neurons are measured at different stimulus frequencies, the curves obtained are very similar to the de Lange threshold curves found under similar conditions for the subjective flicker threshold in human subjects (Fig. 44a, compare with Fig. 5a, Spekreijse et al., 1971, monkey).

These findings indicate the close relationship between neuronal data and psychophysical findings for the frequency response characteristics of the a.v.s. at threshold.

(e) *Linear and Non-Linear Responses to Sinusoidal Stimuli of Different Modulation Depths.* The two non-linearities to sinusoidal stimulation of the RF described in retinal on-center and off-center neurons (p. 488) are also present in the response of LGN-neurons (RACKENSPERGER, 1970; EYSEL and GAEDT, 1971; cat, SPE-

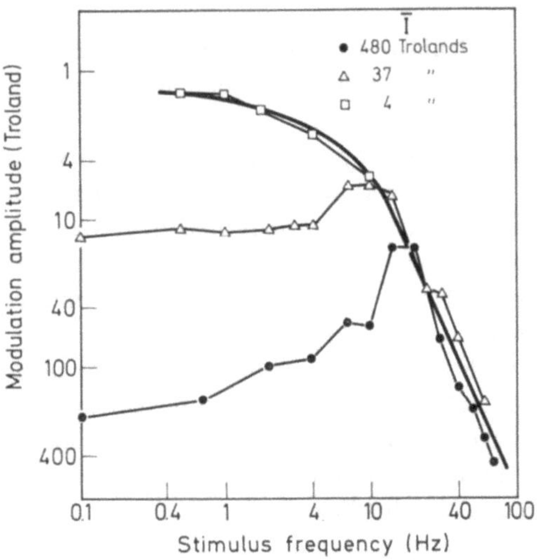

Fig. 44a. Absolute modulation amplitude (ordinate, $m \cdot \bar{I}$) at constant response of a monkey LGN cell for three different adaptation levels $\bar{I}$. Abscissa: frequency of the sinusoidal light stimuli (SPEKREIJSE et al., 1971)

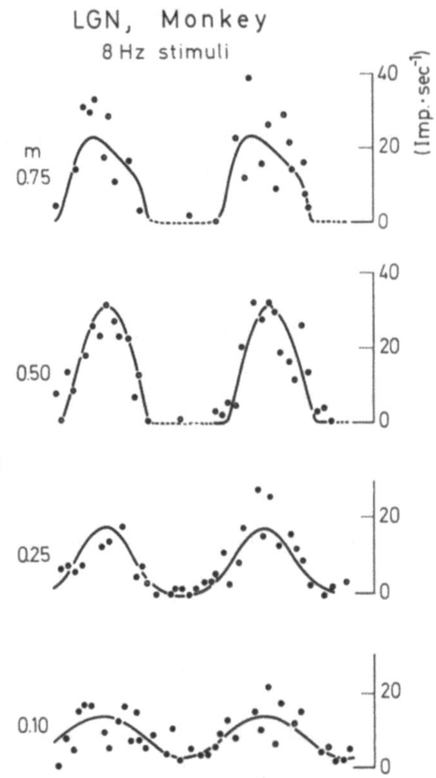

Fig. 44b. Response pattern of a neuron of the LGN of the macaque monkey (green +/red — -cell). Sinusoidal light stimuli projected to the RF-center, 8 Hz, different depth of modulation. Distortion of the response for higher modulation depth (SPEKREIJSE et al., 1971)

KREIJSE et al., 1971, monkey). Both non-linearities are even more pronounced in the contrast neurons, either of the on-center or of the off-center type, than in retinal neurons or LGN-luminosity neurons. Fig. 44b demonstrates these two non-linearities in the response of a chromatic opponent cell (green+/red−) in the LGN of the rhesus monkey. The response non-linearities depend on the stimulus frequency in a similar manner as described for the retinal neurons. In contrast to these findings, MAFFEI et al. (1965a, b) claim that in cat's geniculate cells stimulated with diffuse sinusoidally modulated light stimuli, "the oscillation of the firing rate appears to be a perfect replica of the sine wave". According to our findings, linear responses are obtained only in *L*-neurons, if $f < 2$ Hz, $m < 0.2 - 0.3$ and $A <$ RF-center, for photopic values of $\bar{I}$ (EYSEL and GRÜSSER, unpubl. 1968).

The psychophysical correlations mentioned on p. 488 regarding the subjective brightness of sinusoidally modulated light and the corresponding neuronal activity in retinal on-center and off-center neurons therefore apply also to the responses of LGN-cells.

(f) MAFFEI and RIZZOLATTI (1966) tried to measure the transfer properties of the geniculate cells by means of the comparison of the input and output response patterns. They averaged the amplitude and phase frequency response of 3 retinal ganglion cells and compared the results with the same data measured in 3 geniculate neurons (diffuse illumination of the RF with sinusoidally modulated light). From the difference of these two pairs of curves, plotted on a log-scale, they calculated a function which they called the "transfer function" of the geniculate cell. In the light of the non-linearity of the responses of retinal and geniculate neurons, the different types of neurons (sustained, transient, *L*-type, *C*-type) and the unrelated locations of the RFs of the 6 investigated neurons, the results of MAFFEI and RIZZOLATTI have hardly any significance with respect to the transfer properties of the geniculate cells. The interesting technique of simultaneous recordings of input and output impulse sequences developed by CLELAND et al. (1971) together with the procedure applied by MAFFEI and RIZZOLATTI probably provide a method to measure the contribution of the LGN-neurons to the overall a.v.s. transfer properties.

4. Changes of the Flicker Response of LGN-Neurons by Non-Visual Synaptic Inputs

The flicker response of geniculate neurons is affected by non-visual synaptic activation. The reticular formation of the brain stem is probably one important source of the "non-specific" synaptic inputs of the *P*-cells in the LGN, but certainly there are other sources for non-visual inputs. Despite many non-visual synapses at the LGN-cells (c.f. SZENTAGOTHAI et al., 1968; PECCI-SAAVEDRA et al., 1971), the spontaneous activity of LGN-neurons in cats is mainly determined by the visual input. When the optic nerve is sectioned, the spontaneous activity of cat's LGN-neurons, which is in encephale isolé preparations between 15 and 25 impulses · sec^{-1}, is reduced to 0.5–2 impulses · sec^{-1} (EYSEL et al., 1972).

MAFFEI et al. (1965) found in "midpontine pretrigeminal cats" that a sinusoidal retinal light stimulus modulates the impulse sequence of geniculate neurons during behavioral and EEG wakefulness, while during "synchronous sleep" the correlation between light stimuli and neuronal impulse pattern becomes weak or even disappears totally. The non-specific input might act in part as "noise" for the transfer of specific visual signals through the LGN, but there might also be some inhibitory (presynaptic ?) interaction between visual and non-visual input at the level of the LGN. The relative effect of this non-specific synaptic "noise" on the flicker response of geniculate neurons decreases as the part of the RF-center stimulated by sinusoidally modulated light increases (MAFFEI, 1966; experiments with 1–4 spots of light, projected to the RF-center, similar to stimulus pattern *D* in Fig. 26).

TAIRA and OKUDA (1962) also demonstrated that the transmission of visual signals through LGN-neurons is increased by non-visual sensory, arousal stimuli or electrical stimuli applied to the mesencephalic reticular formation. Most *P*-cells of the LGN (recordings from single fibers in the visual radiation) are facilitated by reticular activation, whereby the response to light stimuli becomes stronger and more irregular. TAIRA and OKUDA observed only in one radiation fiber an inhibitory effect caused by reticular activation.

VI. Responses of Single Neurons of the Visual Cortex to Flicker Stimuli

GRÜSSER (1956, 1957), GRÜSSER and CREUTZFELDT (1957a, b) and GRÜSSER and GRÜSSER-CORNEHLS (1958, 1960) investigated the response of single neurons recorded extracellularly by means of micropipettes in the primary visual cortex (area 17) in cats. They used diffuse flicker stimuli applied monocularly or binocularly with dichoptic stimulus conditions. During the quantitative analysis of the data, GRÜSSER and CREUTZFELDT averaged all responses of neurons which responded to diffuse flicker stimuli. They did not separate recordings from possible presynaptic afferent radiation fibers and postsynaptic elements. Later studies revealed that when one records with micropipettes from the layers of the primary visual cortex, one can easily pick up action potentials of afferent radiation fibers. Hence, part of the data attributed to cortical neurons (c.f. also JUNG et al., 1952; JUNG, 1955; BAUMGARTNER and JUNG, 1955) in the work of GRÜSSER and CREUTZFELDT are probably responses of single visual radiation fibers. During the preparation of the present paper the protocols and film recordings of these old experiments were therefore re-examined (impulse shape of the action potentials, injury discharges as a sign for recordings from cortical cells, etc.). In addition, data obtained by experiments performed during the last few years, in which we used patterned and/or moving light stimuli for the activation of single neurons of area 17 or area 18 in the cat (c.f. GRÜSSER and GRÜSSER-CORNEHLS, this handbook, Chapter 6) were used for the preparation of the following sections.

A. Responses of Cortical Neurons to Diffuse Flicker Stimuli

According to HUBEL and WIESEL (1959, 1962) cortical neurons, even those with simple receptive fields, do not respond to diffuse light stimuli in anaesthetized cats. This finding of HUBEL and WIESEL is in contrast to the findings of JUNG et al. (1952, 1955, 1958) and GRÜSSER and CREUTZFELDT (1957) who reported that in encephale isolé cats part of the cortical neurons also respond to diffuse light stimuli. In recent experiments (1968–1972) several investigators of our laboratory have recorded many neurons in the primary and secondary visual cortex and have investigated the response of these neurons to patterned and diffuse flicker stimuli. We found that one class of units having simple receptive fields (parallel or concentric on- and off-zones) responds also to diffuse flicker stimulation of one or both eyes (encephale isolé cats, fast spontaneous EEG waves). The responses correspond to visual evoked potentials elicited by the flicker stimuli (Fig. 47, c.f. also SCHWARTZ and LINDSLEY, 1964; STURR and SHANSKY, 1971, cat; LOPEZ DA SILVA, 1970, dog). According to depth measurements of the microelectrode tip, the neurons which respond also to diffuse flicker stimuli are mainly located in layer IV of the primary visual cortex. For these units the "laws" described for the flicker response of LGN-neurons are also valid:

(a) The CFF increases within the photopic range approximately according to Eq. (12) or Eq. (13) (Ferry-Porter-law). Behavioral findings (Kappauf, 1938) reveal that the neuronal CFF corresponds fairly well to the CFF measured in behavioral studies.

(b) The CFF increases when the area of the flicker stimuli increases as long as the stimulus is restricted to the excitatory receptive field (ERF). The ERF is that part of the receptive field from which the neuronal activity is increased by adequate stimuli. In many cortical neurons the ERF is surrounded by an inhibitory receptive field (IRF), from which only inhibitory effects are caused by adequate stimuli.

(c) The average neuronal response increases as the flicker frequency increases up to 5–15 Hz, where it reaches a maximum. With further increases of the flicker frequency, the neuronal response decreases again (Grüsser, 1956, 1957; Grüsser and Creutzfeldt, 1957a).

(d) The CFF of different cortical neurons does not vary greatly for a given cortical cell under constant stimulus conditions, but changes considerably from cell to cell. Many cortical units follow flicker stimuli even under bright photopic stimulus conditions only up to 6–12 Hz (Fig. 45, Fig. 48, Grüsser, 1956; Grüsser and

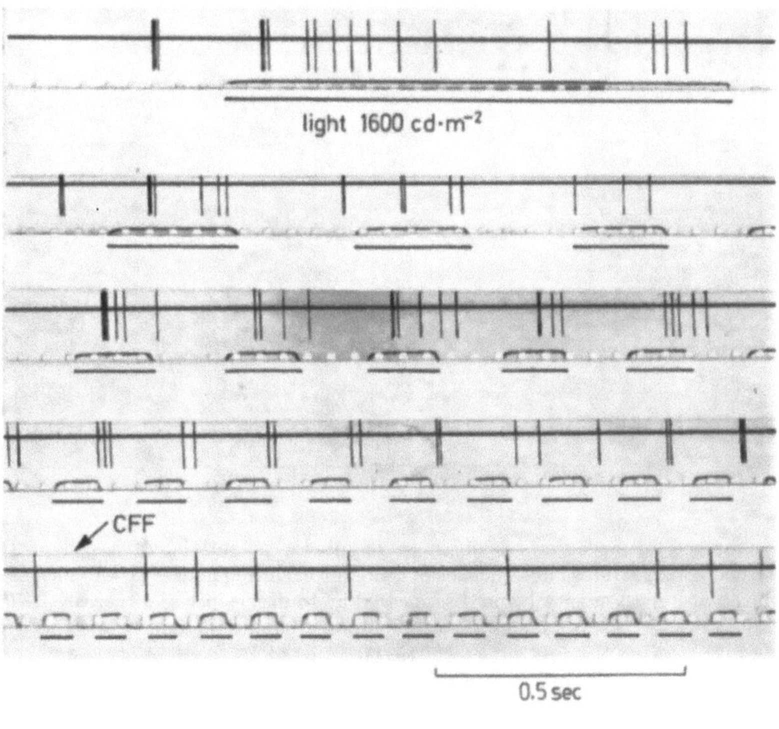

Fig. 45. *On-neuron* of the primary visual cortex of the cat (encephale isolé preparation). Weak response to diffuse illumination of the contralateral (dominant) eye (1600 cd · m^{-2}). The neuron follows intermittent light stimuli and reaches the CFF at a rather low frequency of 7.5 Hz (Grüsser, 1956)

CREUTZFELDT, 1957). A small number of units which are binocularly driven have a CFF comparable to that obtained in LGN-neurons under the same stimulus conditions.

The neurons of area 17, which have complex receptive fields (HUBEL and WIESEL, 1962; P. O. BISHOP and co-workers, 1964–1972; c.f. STONE and FREEMAN, Chapter 2, Vol. VII/3 A, this handbook), respond preferably to contrast borders of specific orientation and limited size moving through the ERF perpendicular to the optimal orientation. Part of these neurons, however, respond also to diffuse flicker stimuli. Quite frequently this response is rather irregular and the CFF is reached at low stimulus frequencies (< 10–15 Hz), even when bright intermittent flicker stimuli are applied. In some units, no response to sinusoidal stimuli is found for low frequency flicker stimuli (< 6 Hz) but a "photic driving" of these units occurs in a frequency range between 8 and 12 Hz. This finding corresponds well with the intracellular recordings of KUHNT and CREUTZFELDT (1971) who found that the averaged subthreshold excitatory postsynaptic potentials (EPSPs) in cats' cortical neurons are enhanced by diffuse light flashes applied with this frequency. Inhibitory postsynaptic potentials (IPSPs) elicited by the same flashes, however, decrease in amplitude, when the flash frequency is above 6 Hz (Fig. 46).

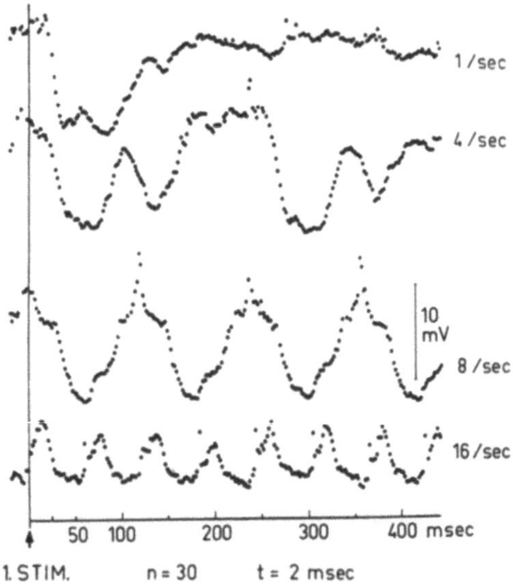

Fig. 46. Averaged intracellular responses (postsynaptic potentials) of a cell of the cat's primary visual cortex. Diffuse light flashes of different flash frequencies (1—16 flashes per sec) IPSP reduced at 8 flashes per sec and disappeared at 16 flashes per sec. Increased EPSP at 8 and 16 flashes per sec. Arrows: spontaneous membrane potential (KUHNT and CREUTZFELDT, 1971)

This "paradoxical" activation of cortical units by medium frequency diffuse flicker stimuli (despite the neuronal specialization to respond to certain spatial stimulus patterns) might be considered as the neurophysiological correlate of the flicker patterns seen in a large homogeneous flickering field (c.f. p. 462).

B. Binocular Flicker Stimuli

GRÜSSER and GRÜSSER-CORNEHLS (1960, 1965) used a specially designed binocular flicker stimulator with which independent dichoptic flicker stimuli could be applied to each eye. Three types of binocular interaction were found by these authors in cortical units:

(a) In part of the cortical neurons the dominant eye causes an activation (on-, off- or on-off-) while the other eye elicits inhibition of the spontaneous impulse sequence. The CFF with binocular synchronous light stimuli is in these neurons significantly lower than the CFF observed with flicker stimulation of the dominant eye (photopic stimulus conditions, rectangular flicker stimuli, $L : D = 1 : 1$).

(b) A rather small part of the recorded cortical neurons (about 5 %) respond only with a weak activation or not at all to diffuse monocular illumination of each eye alone, while binocular *synchronous* flicker stimuli cause a strong on-off-response. This *binocular facilitation* disappears when the dichoptic flicker stimuli are out of phase. The CFF of these neurons is much higher for synchronous binocular flicker stimulation than for alternating binocular or monocular flicker stimuli (Fig. 47, GRÜSSER-CORNEHLS and GRÜSSER, 1961).

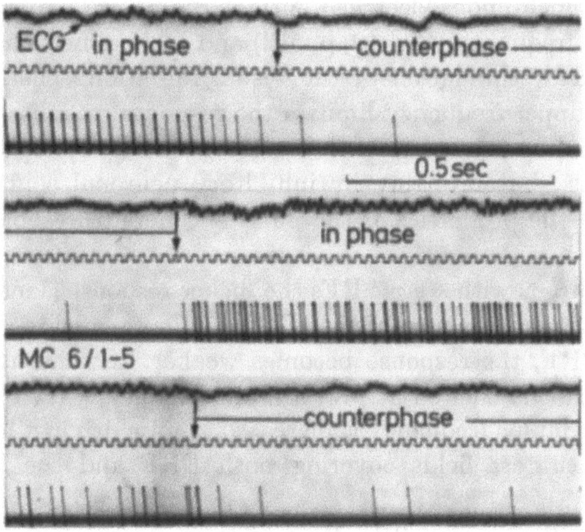

Fig. 47. *Binocular activated neuron* of the primary visual cortex of the cat. Regular responses to synchronous, no response to alternating binocular flicker stimuli. The stimuli are either in-phase or 180 degrees out of phase (GRÜSSER-CORNEHLS and GRÜSSER, 1961)

If in the human visual cortex, part of the binocularly activated neurons respond similarly, this observation might be interpreted as the neurophysiological correlate of the CFF-enhancement, observed in psychophysical experiments with synchronous stimuli in comparison with the CFF obtained with alternating binocular or with monocular flicker stimuli (see p. 444).

(c) A few cortical neurons were found which respond to diffuse flicker stimulation only with an inhibition of the spontaneous activity. The inhibitory effect of binocular flicker stimulation, as a rule, is higher than that elicited by monocular flicker stimuli of each eye alone.

C. Responses of Cortical Neurons to Flickering Stimulus Patterns

(a) *Neurons with Simple RFs.* Neurons of area 17 which have simple RFs (concentric type or parallel on- and off-zones) respond to small flickering test fields projected to the excitatory receptive field (ERF) with a strong activation, when the frequency of the flicker stimuli is below 6–10 Hz. In comparison to LGN-neurons (cat), stimulated with the same stimulus patterns (at the same level of retinal adaptation), the upper frequency limit of the cortical neurons for monocular stimulation of the dominant eye on the average is reduced. In most neurons of this type the CFF is reached between 12 and 20 Hz even when very bright flicker stimuli are used.

For the majority of the neurons of the a.v.s. (retina, LGN, primary visual cortex) the following rule seems to be valid: When a flickering round test field, for example of one degree diameter, is placed into the RF such as to cause optimal activation and the highest CFF, the upper frequency limit (CFF or TFF) decreases with the number of synaptic stations traversed (c.f. Grüsser et al., 1962). For example, if the CFF of a retinal neuron is reached at 40 Hz, that of the corresponding geniculate neuron is between 30 and 35 Hz and the CFF of the cortical neuron driven directly by geniculate fibers is reached between 12 and 20 Hz. Geniculate neurons can by synchronous electrical optic nerve fiber stimulation be driven up to a much higher frequency (Fig. 40, p. 512) and cortical neurons can similarly be driven by electrical stimulation of the visual radiation. Therefore, the overall reduction of the upper frequency limit in the response to light stimulation of the retina is not caused by properties of the single excitatory synapses in the neuronal chain of the a.v.s., but probably by inhibitory neuronal connections (negative feedback) and by the properties of the pulse encoders (thresholds, accomodation etc.).

In cortical neurons with *simple* RFs the flicker response is maximum when the stimulus covers the greater part of the ERF. As the test field extends beyond the borders of the ERF, the response becomes weaker. In some units no rhythmic flicker response can be elicited at low stimulus frequencies with such a stimulus pattern. As the stimulus frequency is increased, photic driving, however, might be observed also with test fields covering both ERF and the surrounding IRF (Fig. 48).

(b) *Neurons with Complex RFs.* Most of these neurons are activated also by stationary, low frequency, sinusoidally modulated stimulus patterns, having an optimal orientation or extension (usually light bars of a certain length and width), projected to the ERF. In many neurons of this type the response to intermittent flicker stimulation ceases at stimulus frequencies above 3–5 Hz (c.f. Chapter 6 of this volume, Grüsser and Grüsser-Cornehls, 1972). Part of the area 17 neurons with complex RFs do not respond per se to stationary intermittent flicker stimuli. However, a *transient* flicker response is obtained in these units when the position of the flickering test field in the ERF is changed. With intermittent illumination in a frequency range below 8–12 Hz and within 1–5 sec after a change of position, these neurons discharge in the rhythm of the flicker stimuli, whereby the neuronal response decreases with time and finally ceases after some seconds, if the flicker stimulus remains in a stationary position.

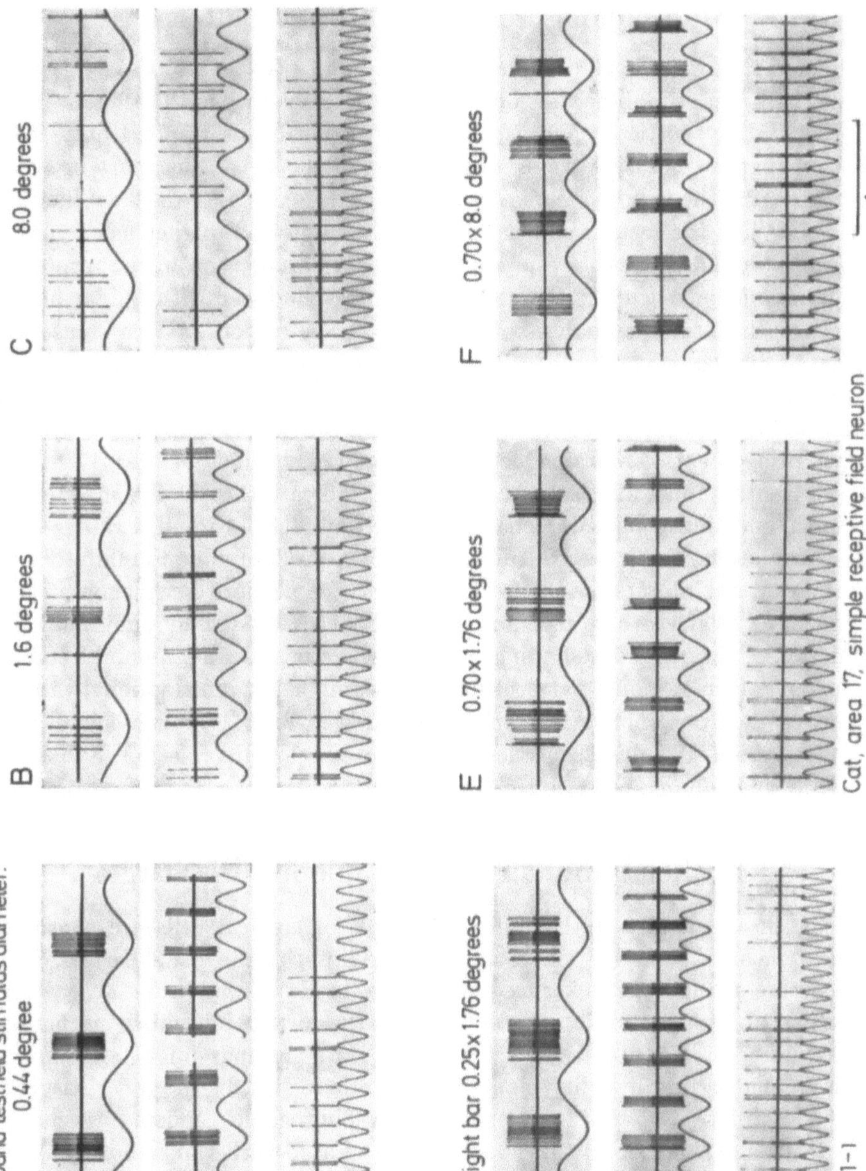

Fig. 48. Cat, encephale isolé preparation. Neuron of the primary visual cortex (simple receptive field, off-dominated) response to sinusoidally modulated test fields ($\bar{I} = 672$ cd · m^{-2}, $m = 0.8$). As the diameter of the round test field (A — C) is increased above 0.5 degrees, the neuron responds less. An 8 degree flickering test field causes a regular flicker response only between 6 and 10 Hz. Sinusoidally modulated light bars of different shape projected to the RF, however, cause a strong response as long as one side of the bar is not larger than 1 degree (from an experiment done in cooperation with R. Saunders and D. Stange, 1972)

D. Responses of Cortical Neurons to Low Frequency Sinusoidally Modulated Flicker Patterns

The non-linearities in the response of retinal and LGN-neurons to sinusoidally modulated flicker stimuli become even more pronounced when the afferent signals are processed by cortical neurons. Many cortical neurons respond only to the increasing part of the sine wave, others only to the decreasing part or with a response during the maximal slope (increasing or decreasing) of the slow sine wave (< 3 Hz). When the instantaneous impulse rate is plotted, higher harmonics are much more dominant in the response of cortical neurons than in the response of retinal or geniculate neurons. These findings correspond well with similar findings obtained with harmonic analysis of the EEG-flicker response. With sinusoidally modulated diffuse light stimuli, higher frequency components are present in the EEG-response (van der Tweel, 1961, 1964; van der Tweel and Verduyn Lunel, 1965; Spekreijse, 1966, man; Lopes da Silva, 1970, dog). The averaged postsynaptic potentials obtained by intracellular recordings of area 17 neurons (cat) by Kuhnt and Creutzfeldt (1971) also indicate that at the subthreshold level of the membrane potential changes elicited by flicker stimuli, a rather strong effect of higher harmonics is present below 20 Hz. Probably due to inhibitory and excitatory lateral interactions, higher harmonics become more conspicuous in cortical responses relative to retinal and LGN-responses caused by sinusoidal light stimulation of the eye. These findings indicate that the precise temporal frequency components of the response patterns have a less important informational value for the cortical neuronal network than for the lower retinal or geniculate levels while the *spatial* stimulus properties play a more important role (c.f. Stone and Freeman, this volume, p. 171—188).

E. Flicker Stimulation and Epileptic Discharges

When the responses of the different classes of cortical cells are considered together, the average activation of the neuronal input to the visual cortex reaches its maximum at medium flicker frequencies (6—15 Hz). In addition, according to the findings of Kuhnt and Creutzfeldt (1971) the cortical inhibitory mechanisms are attenuated strongly at these frequencies. Both findings might explain why there is an increased probability that epileptic brain seizure discharges are triggered by light stimulation of this frequency range ("photogenic" epileptic discharges, c.f. Gastaut and Corriol, 1948). When cats are treated with Eukraton® and reach a lowered seizure threshold, in single neurons of the motor cortex epileptic discharges are triggered by rhythmic diffuse flashes at medium and lower frequencies (Glötzner and Grüsser, 1968).

F. The Effect of Non-Specific Activation on the Flicker Response of Cortical Neurons

Akimoto and Creutzfeldt (1958) found that single neurons in the cat primary visual cortex (encephale isolé preparations) can be activated or inhibited by the electrical stimulation in the non-specific thalamic nuclei. The response to diffuse light flashes is often enhanced by non-specific stimuli. The same is true for the flicker response of cortical neurons. During and shortly after electrical stimulation of intralaminar thalamic nuclei (8–50 electrical stimuli), the CFF of cortical neurons is enhanced and reaches values 5–10 Hz above those without electrical stimulation (Grüsser, 1956; Creutzfeldt and Grüsser, 1957). This observation indicates that the flicker activation is higher at the subthreshold level of cortical neurons than at the suprathreshold level. Similar to the mechanism discussed in connection with Fig. 37, the non-specific activation shifts the average membrane potential nearer to the impulse threshold. Thus the sampling process of the impulse encoder mechanism reaches this threshold more often and neuronal discharges can more frequently be generated by the subthreshold changes of the postsynaptic potentials caused by the input activation. This finding also explains the observation described by Grüsser (1956) that the CFF is higher in cortical neurons, the higher the spontaneous activity during darkness (correlation coefficient $r = +0.5$).

G. Frequency Properties of Single Neurons in the Secondary and Tertiary Visual Cortex

Hubel and Wiesel (1962, 1965) described the functional characteristics of the RFs of single nerve cells in the secondary and tertiary visual cortex of the cat as being "complex" or "hypercomplex" (c.f. Stone and Freeman, Chapter 2 of this volume). These higher order visual cells respond only weakly or not at all to small stationary spots of light projected to their RF, but a vigorous response is obtained when contrast borders of a certain configuration and orientation are moved across the ERF. Diffuse flicker stimuli, as a rule, do not activate these neurons in area 18 and 19 at all or only very weakly. When stationary light stimuli of optimal orientation are placed into the ERF or when specific patterns, projected to the ERF, are illuminated sinusoidally, many of these neurons respond also to stationary stimuli. The optimal stimulus frequency is in all of these neurons reached between 0.5 and 5 Hz (Fig. 20c in Chapter 6 of this handbook, Grüsser and Grüsser-Cornehls, 1972). This low upper frequency limit can also be observed when spatially oscillating stimuli are used for the activation of these neurons (Fig. 23 in Chapter 6 of this handbook, Grüsser et al., 1971, 1972).

In general the data obtained by Grüsser, Grüsser-Cornehls and Hamasaki (1971, 1972) indicate that the more complex the spatial stimulus properties have to be to obtain an optimal activation of a higher order cortical visual neuron, the lower, on the average, is the upper frequency limit of the neuronal response. This finding indicates that with increasing complexity of the perceptive unit not only the spatial band pass properties are more and more narrowed but also the temporal band pass characteristics.

This low upper frequency limit in the response of cortical complex or hypercomplex neurons, which probably "extract" some Gestalt features of the visual stimulus patterns, correlates well with the low fusion frequency which is observed for the contours of simple visual stimulus patterns as shown in Fig. 13.

Part 3. Modelling Studies of Visual Dynamics

VII. Cascade-Filter Models

Most of the models to be reviewed in this section can best be characterized as "descriptive models". Even though they are often presented in the form of electrical or chemical analogs, their main aim is to give a good description of as wide a set of data as possible, without claiming physiological relevance. The analogs in this case serve didactic purposes. To the extent that these models are merely efficient restatements of the data, they can never be falsified, only refined.

At present, some trend exists to search for physiological correlates for the models of the "cascaded-integrator" type as will be discussed in section 7.5. In that case, at least the physiological relevance can be falsified by new experimental data. This tendency to split up the descriptive models at the global level into subsystems with physiological relevance, i.e. to fill the black boxes with "reasonable" hardware, might lead to an iterative modelling process converging to neuronal network models of the a.v.s.

A. Detectors

A "detector" is a system that classifies a continuum of signals into few classes, such as "seen" (supra-threshold, "1") and "not seen" (subthreshold, "0"). Whatever neuronal machinery is responsible for a binary classification like seen/not seen, this final decision-process can be represented by a constant threshold detector (*CT*-detector) with an effectively infinite bandwidth. This is possible because the comparison of one variable $y(t)$ with another variable, the threshold $\theta(t)$ can be represented as the comparison of some transformed signal $F\{y(t)\}$ with a time invariant threshold θ. In this transformation the possible filtering characteristics of the detector are projected in the pre-detector system.

In psychophysically oriented models of visual dynamics the detector is nowadays usually defined as a device classifying all signals of which the *changes* exceed some fixed amplitude $\pm \Delta L$ as "seen" and all other signals as "not seen". This type of detector, which is sometimes called a peak-detector, is designated in this review as "constant amplitude (*CA*-)detector". For waveforms that are asymmetrical relative to the zero amplitude-line this detector is ill-defined, because one should choose whether one-sided threshold crossings suffice or not for a "seen" output. Since the one-sided *CA*-detector (where either only the positive signal excursions are compared with threshold $+\Delta L$, or only the negative excursions with $-\Delta L$) dominates the field, we will simply refer to the one-sided version as "*CA*-detector". In principle this *CA*-detector can be viewed as the series connection of an ideal differentiator (that filters out the *DC*-component) and a *CT*-detector. Obviously in the absence of a *DC*-component in the detector input signal, the *CA*- and *CT*-detector are equivalent.

Ives (1922b) assumed that in order to detect flicker, the rate of change (second derivative) of some signals in the nervous system has to exceed a constant critical value. In other words the Ives-detector consists of a series connection of an ideal second order differentiator and a *CT*-detector.

Data such as those of Fig. 4 and 5 describe by virtue of the presentation method the transfer characteristics of a hypothetical filter F (a "de Lange-filter"), situated between the light input and the input of a hypothetical CA-detector. Correlation of electrophysiological data with psychophysical de Lange curves is only directly possible if this point is born in mind. For example, if amplitudes of a slow-potential are measured, the observer actually "applies a CA-detector to his data". However, if he processes some other features of the signal (latency or number of spikes per cycle or firing frequency etc.), some transformation of the resulting transfer functions might become necessary. This point has not received attention in the literature.

There is one more problem when the CA-detector concept is used in analysing flicker data or constructing theories. The CA-detector in fact only signals "something is seen" (or not), it does not detect periodicity, which is a prerequisite of the concept "flicker". Here again authors have been careless in comparing e.g. flash detection data with flicker detection data, without stating how a detector would discriminate between a periodic and a non-periodic threshold crossing. A theory of realistic, physiologically relevant detectors (rather than a detection theory), therefore appears to be a desirable future development.

B. Linear Cascade-Filter Models

A series connection of filters, such as schematized in Fig. 49A will be called a cascade filter. We will first discuss cascade-filters of which the individual stages are linear time invariant systems. In that case, the gain of the cascade-filter is the product of the gains of the individual stages and the total phase shift is equal to the sum of the stage phase shifts.

(a) DE LANGE (1952) plotted his flicker data in the way shown in Fig. 4 and first noted the formal similarity of the low luminance curve (see Fig. 5a) with the amplitude-frequency characteristics of a cascade-filter consisting of RC-integrator stages (Fig. 49B). For RC-integrator stage i, with input voltage v_{i-1} and output voltage v_i, one can easily write down a differential equation under the simplifying assumption that the stage is almost unloaded ("loose coupling"). Then $v_i = v_{i-1} - R_i I_C$, where I_C is the current through the condensor. Calling $R_i C_i = \tau_i$ one has

$$\tau_i \frac{dv_i}{dt} + v_i = v_{i-1}. \tag{46}$$

Such an RC-integrator acts as a low pass filter and the amplitude gain for a sinewave in the steady state can directly be written down from this differential equation (write $i\omega\tau_i v_i$ for the first term and calculate the modulus of the complex function v_i/v_{i-1}):

$$|G|_i = \frac{1}{(1+\omega^2\tau_i^2)^{1/2}} \tag{47}$$

where ω is the radian frequency, $\omega = 2\pi f$.

Cascading n stages that are loosely coupled and have equal time constants, one finds the amplitude gain for the cascade-filter as

$$|G| = [1 + (2\pi f\tau)^2]^{-n/2}. \tag{48}$$

In the log-log plot of Fig. 4 and 5a, the slope of the de Lange curves in the high frequency range (where the 1 in the latter formula can be neglected) directly provides an estimate of n, whereas the cutoff frequency determines τ. DE LANGE (1952) concluded that an adequate fit for his low luminance data (4.3 troland) is obtained with $n = 5$ and $\tau = 12.7$ msec. In addition, his data for higher background luminances can in the high frequency range (above 15 Hz) also be fitted with Eq. (48) by taking $n = 6$, $\tau = 7.7$ msec at 43 troland and $n = 12$, $\tau = 3.5$ msec at 430 troland. However, the resonance phenomena (the maxima in Fig. 4 and 5a) still have to be taken care of and to this end, DE LANGE (1952) suggests a negative feedback from the output of his RC-chain to the input. Since the phase shift

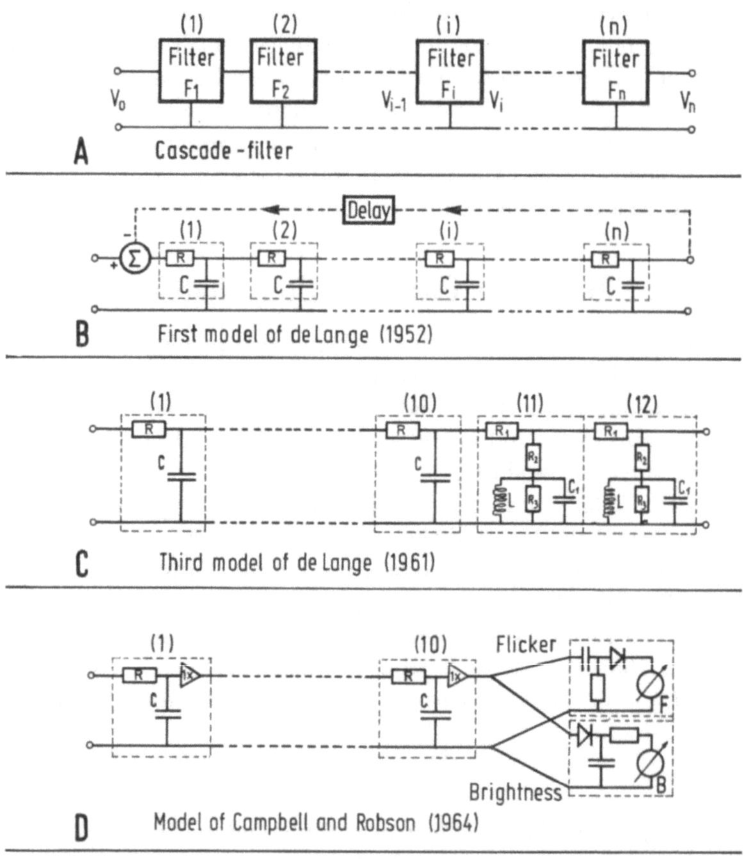

Fig. 49. Linear cascade-filter models of the dynamics of vision. *a* The general construction of a "cascade-filter". If all component filters F_1 — F_n are linear, the system is called a linear cascade-filter. *b* Cascade of leaky integrators proposed by DE LANGE (1952) as a phenomenological model of the behavior of the a.v.s. for high frequency periodic light stimuli. If a delayed feedback is added the model also explains the enhancement of sensitivity (Fig. 4, 5a) at medium stimulation frequencies. *c* Another model by DE LANGE (1961), which also mimics high frequency data and enhancement at medium frequencies, but without feedback. It is in better quantitative agreement with the psychophysical data than the model under *b*. *d* CAMPBELL and ROBSON (1964) noted the necessity to postulate different *detectors* for flicker and for brightness (peak reading device) if a simple linear cascade-filter model after DE LANGE is to explain data on flicker detection as well as on brightness enhancement (Brücke-Bartley effect)

$\varphi = \arctan(2\pi f \tau)$ has to be exactly π at the resonance frequencies (7, 8 and 9 Hz), it then proves necessary to postulate time delays in the feedback loop. Values for this time delay of 14, 12 and 11 msec respectively give acceptable results.

De Lange, however, notes that this theory only provides a first suggestion and not an optimum fit to the data. He regards his *RC*-sections as possible models of physiological subprocesses such as diffusion (Section VII E).

(b) The second model developed by de Lange (1961) mimics for a number of different wave forms the high frequency low luminance flicker detection behavior of the a.v.s. by means of four cascaded integrator stages all having the same time constant τ, but different values of R and C.

(c) The third de Lange model (de Lange, 1961; Fig. 49C) like the first is intended to describe the high frequency data as well as the resonance phenomena of the high luminance curves. It consists of 10 cascaded *RC*-integrators (first order low pass filters) in series with two bandpass stages incorporating an inductance. The fit to the high frequency data is good, but the overall fit is more indicative than exact.

(d) Levinson (1968) was apparently unhappy with *RC*-models and developed a hydraulic analog of the n-stage leaky integrator model as a didactic vehicle. He also developed a method to check with psychophysical experiments whether the simple cascade models describe not only the steady state sine wave gain characteristics of the a.v.s. but also transient effects. For example, if a sine or square wave modulation of up to 600 Hz is switched on, without changing the average illumination, the model predicts a transient response which is much slower than the (invisible) high frequency modulation. Levinson (1968) showed that these slow transients indeed also seem to occur in the human a.v.s., because in the psychophysical experiment the subjects report seeing a flash, both at onset and offset of the modulation of the flickering light, even though the frequency is above the CFF ("pseudoflashes"). Apparently the cascade filter models thus also describe some transient effects reasonably well.

The "pseudoflashes" can also be given repetitively, if a stimulus is used that changes periodically from frequency f_1 to f_2 with both f_1 and f_2 above the CFF. Such an experiment was carried out by Forsyth and Brown (1961) and analyzed by Forsyth and Brown (1962) and Matin (1962). One clear discrepancy between the behavior of a cascade filter model and the psychophysical results obtained by Levinson (1968) is that to the observer, the pseudoflashes at the beginning and end of the modulation look alike, whereas the model predicts a light and a dark flash. Here one might save the cascade filter model by ascribing the confusion of "dark" and "light" flashes to the detector (c.f. final remarks of Section VII A).

(e) Campbell and Robson (1964) proposed a 10 stage *RC*-model with isolation amplifiers between the stages (to guarantee a "loose coupling") and provided with two detectors at the output. They state that their measurements on the Brücke-Bartley-effect as well as on flicker can be explained with this model if a *CA*-detector is used for the flicker detection channel and a peak detector with long time constant for the brightness channel. These authors also schematize a specific electronic implementation of the detectors used in conjunction with the cascade-filter model (Fig. 49D).

C. Non-Linear Adapting Cascade-Filter Models

The principle shortcoming of all linear cascade-filter models, even at a descriptive level, is that they do not include adaptational mechanisms. Moreover, for large stimulus changes at a fixed average adaptation level, such mechanisms will also influence the responses. Thus, the linear models are "small signal steady state" models for fixed adaptation levels and to make them more realistic they have to be extended with non-linearities. It has been noticed in several electrophysiological studies that both the time scale of the response and the sensitivity change with adaptation level. However, the sensitivity changes are much larger than the time scale changes.

(a) FUORTES and HODGKIN (1964) suggest for the transduction from light to receptor potential in the Limulus eye an n-stage leaky integrator model of which the leak resistances of the stages are regulated by the model's output signal (Fig. 50A). Disregarding for a moment this feedback regulation of time constants, we will look once more at the linear cascade filter model with n RC-integrator sections. From Eq (46) one can calculate the pulse response of this linear cascade-filter model [for the result see Section VIIE, Eq. (52)].

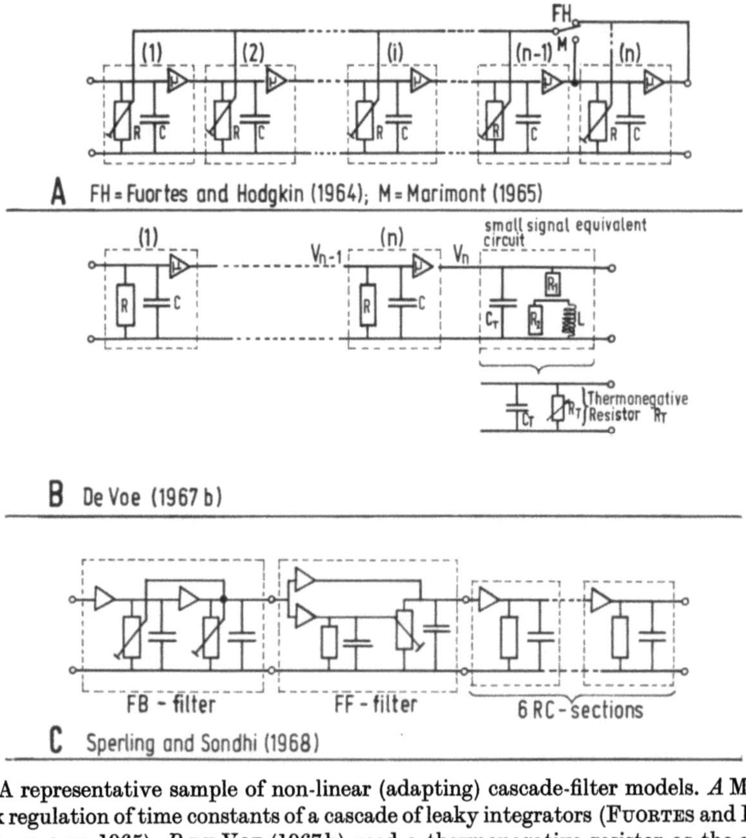

Fig. 50. A representative sample of non-linear (adapting) cascade-filter models. *A* Model with feedback regulation of time constants of a cascade of leaky integrators (FUORTES and HODGKIN, 1964; MARIMONT, 1965). *B* DE VOE (1967b) used a thermonegative resistor as the non-linear stage. For small signals it can be replaced by the linear small signal equivalent circuit shown in this figure. *C* This model of SPERLING and SONDHI (1968) contains two non-linear filters with feedforward and feedback adaptation, followed by a linear 6-stage cascade of leaky integrators

From this pulse response it can then be seen that the time scale of the response is proportional to R, but the sensitivity (maximum amplitude of the pulse response) to R^{n-1}. This fact makes it possible for $n > 2$ to mimic the disproportionate increase of sensitivity relative to response duration with dark adaptation. To this end, it only has to be assumed that dark adaptation causes an increase of R. In other words, the conductances of the receptor membrane $g = 1/R$ have to increase for higher input (and thus output) signals.

Consequently FUORTES and HODGKIN postulated:

$$g = g_0 (1 + \alpha v_n) \tag{49}$$

as the implementation of the idea that an automatic gain control prevents overload and at the same time improves time resolution at higher intensities. This leads to the following set of differential equations relating the output v_i of stage i with its input v_{i-1} for $i \in \{1, 2, \ldots, n\}$

$$C \frac{dv_i}{dt} + v_i g_0 (1 + \alpha_i v_n) = \mu v_{i-1} \tag{50}$$

where μ is the transconductance (ampère/volt) of the isolation amplifiers (Fig. 50a) and α_i the feedback weighting factor for stage i. FUORTES and HODGKIN were mainly concerned with the case $\alpha_i = \alpha$ for all i, but they suggest that in some cases a better fit to their data might be obtained by leaving one or a few sections uncontrolled. That would mean that the corresponding α_i-values would have to be zero.

From curve fitting FUORTES and HODGKIN (1964) conclude that the flash response of Limulus receptor potential to small flashes either on a background or after pre-adaptation to strong light, can be reasonably well described by the linear de Lange -model of Eq. (46). The best fitting values for n differ strongly between cells, however. For example in 9 cells, they find for flashes given to a light adapted eye in the dark, values of n from 7—14 with an average around 10. Next they compare step responses of the complete non-linear model of Eq. (50) with $n = 10$, with Limulus photoreceptor step responses. The fit is reasonable over some $3^1/_2$ decades of step intensities, although the model responses are somewhat too oscillatory. Flash responses (intense flashes) are not compared with the above model but with the following modified version:

$$C \frac{dv_i}{dt} + v_i g_0 e^{\alpha_i v_n} = \mu v_{i-1}. \tag{51}$$

Again with $n = 10$ the model's pulse responses are more sharply peaked than the Limulus photoreceptor flash response. Altogether we have the feeling that the success of the model, considering its complexity, is not impressive.

(b) PINTER (1966) estimates best fit parameters of the Fuortes-Hodgkin-model from the light adapted sinusoidal frequency response of the Limulus photoreceptor membrane potential. With these parameters he predicts the response to large light steps in the dark adapted eye. The predictions prove to be poor. The fit of the Fuortes-Hodgkin model to PINTER's sinewave data is reasonable above 1 Hz. However, for lower frequencies an extra "linear lead network" is necessary at the output of the Fuortes-Hodgkin model to explain the continuing drop of gain with decreasing frequency.

(c) MARIMONT (1965, Fig. 50A, switch on M) has drawn attention to the fact that the Fuortes-Hodgkin-model gave excessive sharpness for the pulse response and an overshoot for the step response due to too great a delay in gain reduction. She proved that a much better fit to the Limulus data of FUORTES and HODGKIN could be obtained by taking the feedback from the stage next-to-last. A translation of the filter model in terms of a chemical compartmental model can also be found in her paper (MARIMONT, 1965; see also BORSELLINO et al., 1965a, and BORSELLINO and FUORTES, 1968).

(d) Of course we are here only interested in the Limulus-models as far as they prove applicable to mammalian visual dynamics. At high frequencies of sinewave stimulation, the Fuortes-Hodgkin-model behaves as a linear cascade-filter model with RC-integrators and it will therefore fit human high frequency flicker data like the de Lange -models. The Fuortes-Hodgkin-model also shows the "resonance" phenomenon (see e.g. Fig. 6 in PINTER, 1966) and can therefore, at least in principle, also be applied to medium frequency human flicker data. MATIN (1968) has applied the Fuortes-Hodgkin-model to human psychophysical data on critical duration, differential luminance threshold, CFF and visual adaptation with encouraging results.

ERNST (1968) discusses the question whether the effect of a background on CFF is comparable with that of the after-effect of a bright adaptation light (cf. p. 444). He concludes that the Fuortes-Hodgkin-model qualitatively explains his psychophysical data, if he assumes in addition that for a given value of R, the C-values must be lower in the presence of a background than in the presence of after-effects of a bright light.

(e) DE VOE (1966, 1967b) proposed for the light-adapted wolf spider eye a non-linear cascade-filter model that contains RC-integrator sections, preceded (DE VOE, 1966) or followed (DE VOE, 1967b, Fig. 50B) by a single non-linear stage. The non-linear stage consists of a capacitance C_T shunted by a time variant thermonegative resistor R_T. For such a thermonegative resistor, an increase in current slowly decreases R_T due to the increase of temperature.

In the literature on such devices, it has been proved that a thermonegative resistor when it is biased by a steady current (in the model due to a steady background intensity), shows small signal linearity and can be replaced by an equivalent circuit consisting of an inductance and resistance in parallel series connected with another resistor (Fig. 50B). With this replacement a transfer characteristic can be calculated and it proves to fit the wolf spider data of DE VOE (1967a, b) quite well.

D. Multi Control-Loop Cascade-Filter Models

SPERLING and SONDHI (1968) developed a cascade-filter model of which the last 6 sections are the familiar RC-integrators, but this time they are preceded by a two-stage Fuortes-Hodgkin-model (which they called an FB-filter) and a feed-forward (FF-)filter: Fig. 50C. The FB-filter compresses the dynamic range of the stimulus into the range that the rest of the filter can handle and is therefore placed directly at the input side. The FF-filter models the Weber-law behavior for low input frequencies and for flashes.

The complete set of differential equations describing the model can easily be written down, but a general analytic solution is not known. Therefore, data- and curve-fitting have to be based upon numerical methods, simulation studies or approximations for limiting cases.

SPERLING and SONDHI chose their parameters so as to give an optimum fit to increment flash detection data of HERRICK and of GRAHAM and KEMP. With the same parameters, the fit to the flicker threshold modulation data of DE LANGE was fairly good. On the whole, the model fits the different sets of data reasonably well, but it is somewhat complicated to handle and a number of parameters have to be quite arbitrarily fixed. This indicates redundancy in the model.

An interesting part of the Sperling-Sondhi-model is the FF-filter, which mimics Weber-law behavior without the use of explicit logarithmic transformations. This is achieved by dividing the present input by the time average of recent inputs. The basic idea of realizing the Weber-law in this way was already described by BARLOW (1965; see VAN DE GRIND et al., 1971b).

E. The Physiological Relevance of Cascade-Filter Models

DE VOE (1967b) has also discussed the physiological relevance of the non-linear stage of his cascade filter model and he argues convincingly that it might represent delayed rectification as for example found in squid giant axons. The cascaded *RC*-filter part of all models discussed above, may, according to DE VOE (1967b), be directly translated into the proenzyme-enzyme model of WALD (1965), the IVES (1922b) diffusion model or its reformulation by RUSHTON (1965) in terms of a leaky cable.

LEVINSON (1966) envisioned a membrane in which an inherent shower of particles with Poisson distribution rains constantly. Photons cause holes or "apertures" in this membrane and each opened aperture immediately starts to gather the inherent Poisson process particles. If a fixed number of particles is caught in a hole, the hole "closes", which is an elementary output event. The *rate* of output events at time t, $R(t)$, is then proportional to M, the total number of holes created by the photons of a short flash, the rate $\bar{r}$ of the inherent Poisson process per hole area ($\bar{r}$ is called $1/\tau$) and the probability of a particular hole having caught $n-1$ particles at that time. Regarding this rate $R(t)$ of the elementary output events as the output signal, LEVINSON thus found for the impulse response:

$$R(t) = M\left(\frac{t}{\tau}\right)^{n-1} \frac{e^{-t/\tau}}{\tau\,(n-1)!} \qquad (52)$$

which is formally equivalent with the pulse response of an n-stage *RC*-filter as can be calculated from Eq. (50) with $\alpha_i = 0$, or from Eq. (46).

In order to include the feedback control of time constant and gain of the Fuortes-Hodgkin-model, LEVINSON postulated that "inactivating" particles are generated at a rate proportional to the output signal. With the additional assumption of the presence of a stream of neutral particles, this leads, as LEVINSON proves, to a formal equivalence with the complete Fuortes-Hodgkin-model. LEVINSON'S single stage stochastic interpretation of the feedback cascade-filter model has the added advantage that it is easier to incorporate the variability found in experimental studies for the best fitting value of n, since n can also be assumed to be a stochastic parameter.

VIII. Models which Include an Impulse Generating Mechanism

In most models of flicker detection it is assumed that flicker fusion is mainly (though certainly not exclusively) a retinal phenomenon. In some early theories, such as the photochemical theory of HECHT (HECHT and VERRIJP, 1933; HECHT, 1937), the fusion locus was even assumed to be the primary phototransduction process. This assumption is in disharmony with modern neurobiological evidence, but one might argue that the resulting formulae retain some value as phenomenological descriptions. However, HECHT's theory also fails to predict modern sine wave data.

Diffusion theories of vision such as those by LASAREFF (1913, 1914, 1922) and IVES (1922b) have also lost their original physiological basis, but they are probably more flexible as phenomenological models. For example KELLY (1969b) recently showed that the VERINGA (1961) diffusion model, which is akin to the above diffusion theories, fits the high frequency sine wave flicker data rather accurately [Eq. (18), p. 447)] and in any case better than the cascade filter models.

Since diffusion equations also often turn up in attempts to describe the pulse generating properties of neurons (c.f. JOHANNESMA, 1968) there is ample room for neuronal reinterpretations of diffusion models of flicker fusion.

Such a reinterpretation is necessary if diffusion theories are to be used to explain the high frequency sine wave data mechanistically, because the data presented in Figs. 21 and 37 clearly indicate that pulse-generating mechanisms play an important role in limiting the time resolution of the visual system. Thus, despite the success of some phenomenological models, it is important to develop models that include pulse-generating mechanisms ("pulse encoders" or briefly "encoders"). A review of such models will therefore be presented below.

A. An Impulse-Generating Mechanism Used as Non-Ideal Peak Detector

In the model proposed by LEVINSON and HARMON (1961) a neuromime (nerve cell model) of the type described by HARMON (1961) is used in conjunction with ENROTH's fusion criterion (p. 481). The neuronal network preceding this detector is simulated by a series connection of 5 *RC*-integrator stages and a non-linear stage that slightly distorts the waveform of the cascade-filter output, so as to have a faster increase and slower decrease. The total model mimics the high frequency envelope of DE LANGE's flicker data reasonably well, but this is mainly due to the 5-stage cascade-filter. On the whole, the influence of the neuromime and non-linear stages on the amplitude frequency characteristics is only a small one. Moreover, they hardly add to phase shift or spike response latency. The neuromime mainly translates the cascade-filter output into a somewhat more physiological signal form (impulse sequence).

The model, as judged from the neuromime impulse output, also gives a fair description of some of the cat ganglion cell latency data found by ENROTH (1952, p. 483). No attempt was made by LEVINSON and HARMON to include "resonance" or adaptation phenomena in the model in such a way that a quantitative comparison with neurophysiological or psychophysical data, other than the "high frequency envelope" of DE LANGE's curves, became possible. In view of data such as those in Fig. 21, it would seem more realistic to postulate a more active role for a neuromime in limiting the high frequency responses.

B. A Refractory Model of Flicker Fusion

KELLY (1961c, 1962a) used a pulse encoder with absolute relative and refractory behavior in his model of flicker fusion phenomena. In this pulse encoder a pulse is generated as soon as the graded potential (slow membrane potential) $g(t)$ equals the value of the threshold function $h(t)$, after which event the threshold $h(t)$ is infinite during the absolute refractory period T_0 and then recovers exponentially to its resting value h_0 according to the following function with the refractory time constant τ:

$$h(t) = h_0 \exp\left[-\frac{t - T_0}{\tau}\right]. \tag{53}$$

The time origin $t = 0$ is taken to be the moment of generation of the most recent spike. The graded potential $g(t)$ is not affected by the firing of the encoder (no "reset"). A problem with KELLY's encoder is that no transfer function is known for it. KELLY therefore sets up an approximate calculation based on the idea that all phase relations between generator potential and firing moments will eventually occur in this non-phaselocked situation. One can then first calculate the input "threshold" amplitude value A_t for which the shortest and longest interpulse interval never differ by more than a fixed value δ, however favorable the phase might be. A similar argument leads to an upper limit of A_t.

Finally the harmonic average of $A_{t\max}$ and $A_{t\min}$ is defined as the threshold amplitude at the input. The gain-frequency function estimated in this way for the encoder proves to be similar to that of a low pass filter with steep high frequency cutoff. The cutoff frequency depends on the average output interval $\overline{T}$ (sampling interval), which can be calculated for a given average input A from $h(\overline{T}) = A$ with Eq. (53). In this way the effect that a higher average luminance leads to a higher cutoff frequency is incorporated in KELLY's model by the relation between sampling frequency and average input signal of the encoder. KELLY states that a hyperbolic function for $h(t)$ would provide better results. The results of Fig. 37 provide evidence that the basic idea of an impulse encoder, viz, that an increase of the average membrane potential leads to an increased sampling rate, and therefore increases the cutoff frequency, is probably correct (GRÜSSER, 1972).

In order to include the low frequency decrease of flicker sensitivity, KELLY has postulated a second order filter (bandpass) preceding the encoder stage. The complete model resulted in a reasonable fit to his data, although the high frequency cutoff caused by the encoder is too steep. This point is typical for (deterministic) pulse encoders and will therefore be discussed in a separate section on the possible use of encoders in models of visual dynamics. However, we will first turn to the problem of analyzing the dynamic behavior of pulse encoders per se.

C. Transfer Functions for Pulse Encoders

If the optic nerve is stimulated electrically, it can follow stimulus frequencies far above those ever observed during light stimulation. Thus absolute refractory phenomena can be neglected in flicker fusion theories. Relative refractoriness, that is the gradual recovery of the threshold from its high immediate postfiring value, might determine the sampling characteristics of the encoder, as was postulated for the encoder of KELLY (1961c, see the preceding section). Let us for the sake of brevity call this principle "refractory sampling". Real neurons indeed have a relative refractory period during which the threshold gradually approaches some resting value. This mechanism constitutes the main sampling principle of the encoding of graded potentials into a neuronal impulse sequence. Temporal summation of threshold elevations, remaining after each impulse, lead to a nonlinearity of this sampling mechanism (c.f. EYSEL and GRÜSSER, 1970b; EYSEL, 1971).

As an alternative encoder mechanism, an "integrate and fire" encoder is often proposed. For such devices it is postulated that the graded potential is integrated until the value of the integral equals the (constant) threshold. Then an impulse is generated, the graded potential is simultaneously reset and integration starts anew. Despite the fact that this model of neuronal impulse encoding is not too realistic with respect to findings with intracellular microelectrode recordings (see VAN DE GRIND and GRÜSSER, 1973; Fig. 21) it has some interesting features which were investigated by several authors. For the formal description of the neuronal encoding mechanisms it is conceivable that a neuron acts at low firing rates as an "integrate and fire" encoder and at higher firing rates as a "refractory encoder" (VAN DE GRIND et al., 1971a). Literature on refractory encoding is scarce (KELLY, 1961c; MUNDIE, 1969; EYSEL, 1971; GRÜSSER, 1972) and except for KELLY's approximation, no transfer functions have been derived. Many authors have discussed "integrate and fire" mechanisms and transfer functions have also been calculated in some cases.

JONES et al. (1961) formulated the concept of an I.P.F.M. ("integral pulse frequency modulator"). In this device a pulse is generated as soon as the integral of the input signal reaches K. Then the integration starts anew (the integrator is "reset"). Since K unit input pulses (or the equivalent area of a continuous signal) are necessary for every output pulse, VAN DE GRIND et al. (1971a) called it a K-scaler: Fig. 51a. As an approximate realization of the K-scaler JONES et al. (1961) proposed an RC-integrator or "leaky integrator" followed by a discharge tube (threshold and reset) and called the resulting system a "relaxation oscillator". (a "KT-scaler" in the terminology of VAN DE GRIND et al., 1970, where K symbolizes the scaling factor and T the leak time constant (Fig. 51a). For $T \to \infty$ the leaky integrator becomes a "perfect integrator", and the KT-scaler a K-scaler.) For a K-scaler with sinusoidal input, BAYLY (1968, 1969) has calculated the output pulse frequency spectrum, KNIGHT (1969) derived a transfer function with a perturbation calculation and KOENDERINK et al. (1971) and KOENDERINK (1972) gave a direct derivation. The transfer function is similar to that of a zero order hold device. The transfer function of the latter device was first proposed by BORSELLINO et al. (1965b) as a phenomenological description of the dynamics of the slowly adapting stretch receptor organ of Crustacea:

$$B(i\omega) = \frac{1 - \exp(i\omega/f_0)}{i\omega/f_0}\,. \tag{54}$$

The result derived by Koenderink et al. (1971), which is very similar, is:

$$\frac{m_{\text{out}}}{m_{\text{in}}} = \left|\frac{\sin(\pi\omega/\omega_0)}{(\pi\omega/\omega_0)}\right| (m_{in} \ll 1) \tag{55}$$

where $\omega_0 = 2\pi\bar{I}/K$ is the mean output frequency. Koenderink (1972) argues that a transfer function of a single K-scaler (I.P.F.M.) is a useful concept only if it is defined on the basis of an averaging process at the output, for example by using a PST-histogram as the output signal. For an ideal K-scaler the output modulation depends on the phase of the discharge moments relative to the stimulus and "beats" might occur as Koenderink (1972) proves (see also Stein and French, 1970). These beats disappear only if uncertainties such as drift and noise are explicitly allowed and the output is defined as an appropriate average over many cycles of a single I.P.F.M.-response or over one cycle of an ensemble of such responses.

D. Pulse Encoders in Models of Visual Dynamics

It is remarkable that only very few models of visual dynamics have been formulated that incorporate a pulse encoder. Yet this would seem to be an urgent task of theorists, because of the following physiological findings:

(1) The CFF is proportional to the mean output impulse rate of retinal and cortical ganglion cells in the cat (Dodt and Enroth, 1954; Grüsser, 1956; Dodt, 1964).

(2) The membrane potential of retinal ganglion cells in the cat still follows periodic light changes above the CFF for its output pulse signal (Fig. 21).

(3) It is possible with certain stimulus configurations that mainly change the average membrane potential to manipulate the CFF for the cat's retinal ganglion cells (Fig. 37).

One reason why pulse encoders are used so seldom in visual modelling studies might be their analytic complexity. Simulation studies could, however, well provide the necessary insight. In simulation studies it is further possible to study more complete encoding schemes that for example include accomodation (c.f. Eysel, 1971; van de Grind et al., 1971a) and more realistic amounts of variability (c.f. Stein and French, 1970; van de Grind et al., 1971a).

The following points are important in developing encoding schemes for models of vision:

(a) Any neuromime with an internal reset (like the K-scaler) raises the question how this reset should be interpreted physiologically. In the intracellular recordings by Foerster and van de Grind (1973) for example, the membrane potential of retinal ganglion cell of the cat was not noticeably changed (reset) by the generation of a nervous impulse (Fig. 21). The variable to be reset in the model encoder should therefore, if anything, represent some neuronal process that is not recorded by the intracellular micropipette.

(b) The slope of the gain-frequency curve at high frequencies is steeper for the pulse encoders studied so far than that found in psychophysical de Lange-curves and also than in most neurophysiologically measured gain characteristics. Extra assumptions are therefore necessary to complement the postulate that encoders determine the high-frequency slope of de Lange-curves. For psychophysical data the usual extra assumption is that several channels, all with different carrier frequencies, are involved simultaneously (ensemble-explanation). For single-cell transfer functions this explanation obviously breaks down and one has to assume either a suitable drift or stochastic changes of the carrier frequencies.

IX. Dynamic Models that Include Inhibition Phenomena

All models discussed above are one-channel models and as such they do not incorporate hypotheses about the influence of spatial factors. KELLY (1959) showed that for low frequencies the flicker sensitivity measured psychophysically is smaller for large test fields than for small ones. This depression of sensitivity might be taken to suggest that lateral inhibition decreases flicker sensitivity at low frequencies for large test fields. There is abundant experimental support for such an interpretation in the electrophysiological literature. Quantitative theories are scarce, however, with the possible exception of theories relating to Limulus.

A. The Principles of Tuning by Inhibition

DE LANGE's method (1952, see VII B) of tuning a cascade-filter with delayed negative feedback can also be applied in multi-channel models. In that case the feedback can be taken from the same and/or from neighbouring channels and it could thus represent self-inhibition and /or lateral inhibition of the recurrent (feedback) type.

RATLIFF et al. (1967) proposed this tuning principle as an explanation of the difference between small field and large field flicker responses in the lateral eye of Limulus. For this eye recurrent lateral and self-inhibition of the subtractive type are known to be present. In the model of the eccentric cell of the lateral eye of Limulus described by RATLIFF et al. (1967, 1969), KNIGHT (1969), DODGE et al. (1970), KNIGHT et al. (1970) and their colleagues (see the review in this handbook by HARTLINE and RATLIFF, 1972) lateral inhibition not only provides tuning, but also causes "amplification" at the tuned frequencies. "Amplification" means in this case that the response of a given ommatidium is larger for a large homogeneous flicker test field (where lateral inhibition is present) than for test fields of the size of a single ommatidium (no lateral inhibition). Lateral inhibition of course always reduces the DC-component, but at the tuned frequencies the AC-components of the inhibition channels are shifted 180° in phase relative to the generator potential and thus they add at the encoder input.

B. Lateral Inhibition in a Psychophysical Theory of Flicker

Imagine a high contrast spatial sinewave or squarewave grating, flickered in such a way that the bright bars at time t are the dark bars at $(t + 1/2\ T)$ and v.v., with $T =$ the flicker period. Such a stimulus can be called a counterphase grating (VAN DER TWEEL and SPEKREIJSE, 1968; KELLY, 1969a). KELLY (1969a) ascribed the difference between the de Lange curves for a uniform field (spatial frequency 0) and for a high spatial frequency counterphase grating to the action of lateral inhibition. In the latter case lateral inhibition was assumed to have no influence in the time domain and the difference (log-log scale) between the former and the latter gain characteristics was therefore postulated to represent the gain characteristics of lateral inhibition. The gain characteristics found in this way for the lateral inhibitory connections are like those for a simple low pass filter of rather low order and there is a slight dependence of this gain curve on adaptation level.

C. Network Models of Spatio-Temporal Visual Signal-Processing

At very low light intensities quantum effects play a noticeable role in perception (for reviews of the so-called "quantum theory of vision" see Bouman, 1961, 1969). Under such circumstances the retinal photoreceptors are known to act as single photon detectors and can therefore be viewed as "all pass" systems. Studies of the dynamics of the a.v.s. under such conditions might therefore reveal the dynamics of post-receptor structures without the obscuring influences of receptor-level adaptation (see below). The "De Vries-Rose law" for human scotopic vision states that the threshold luminance of a small short flash given on a scotopic background luminance $\bar{I}$ is proportional to the square root of $\bar{I}$. This law can be implemented with the help of an adapting "KT-scaler" (Fig. 51 D). A feedback adapting KT-scaler (Fig. 51 A), which together with a CA-detector embodies the de Vries-Rose law, was proposed by van de Grind et al. (1970, 1971 b) who called the device a "VR-machine" (de Vries-Rose machine).

Extending these ideas to photopic vision a "W-machine" (Weber-machine) was proposed by the same authors. This W-machine (Fig. 51 b) again together with a CA-detector, embodies the Weber law, which states that the threshold luminance of a small short flash given on a photopic background of luminance $\bar{I}$ is proportional to $\bar{I}$. It is interesting that this W-machine is in principle a simple feed-forward adapting KT-scaler, making it the mirror image of the VR-machine. The W-machine functions as a logarithmic differentiator of which the steady state output level is kept constant regardless of the average input level. Only changes in the input lead to output changes for high enough background intensities (for further details see Koenderink et al., 1970, 1971).

The pulse-generating part of these machines can be removed without noticeably changing the dynamic behavior. This leads to the continuous versions shown in Fig. 51 C, which are more suited for modelling studies of intraretinal signal processing.

The continuous versions of the machines, the W_c-machine and the VR_c-machine of Fig. 51 c, were proposed by Koenderink et al. as a receptor- (or possibly triadic receptor synapse) model and a bipolar cell model respectively. Both the VR-machine and the W-machine are in fact extended embodiments of the principle of shunting inhibition as proposed and studied by Furman (1965). The former system embodies feedback or recurrent shunting inhibition, the latter feedforward or nonrecurrent shunting inhibition. Furman's formulations were based on the static case, whereas the mentioned machines also contain specific postulates regarding the dynamics of shunting inhibition.

With adequate choices of the parameters (VAN DE GRIND et al., 1970), the W-machines (receptor models) function in the low luminance range as all-pass systems, so that the VR-machines (bipolar cell models) series connected with a pulse encoder (K-scaler as ganglion cell model) determine retinal dynamics. This tandem system has a gain curve like the 0.65 troland de Lange curve in Fig. 5a and also fits for different time constants a Limulus eccentric cell (also a bipolar cell) gain curve. (VAN DE GRIND et al., 1971a. For an extensive discussion of the

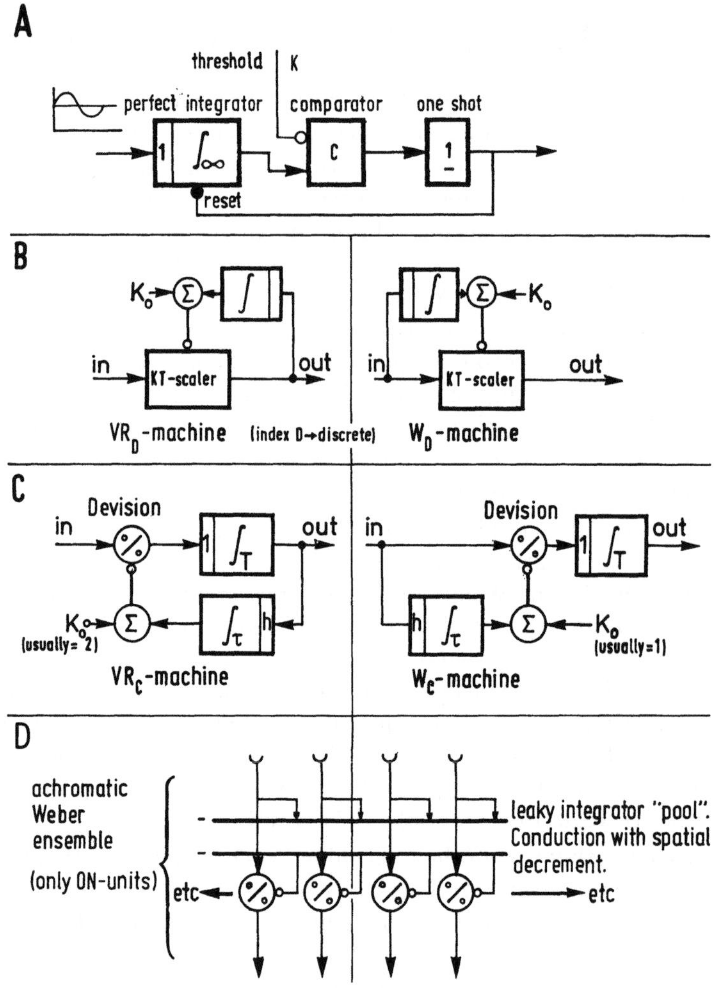

Fig. 51. A KT-scaler. The input signal is "integrated" by the leaky integrator with leak time constant T until threshold K is reached. The device then generates an impulse, the integrator is reset and the process starts anew. Components (B and C) of the network models of the retina and an example (D), developed in a series of papers by KOENDERINK, VAN DE GRIND, and BOUMAN (see text). B A "de Vries-Rose machine" (VR-machine) and a "Weber-machine" (W-machine), which deliver output pulses on receiving suprathreshold input signals of a continuous or discrete (pulse-)form. C Continuous (output signal) versions of the VR-machine and W-machine. These machines embody feedback and feedforward shunting inhibition (FURMAN, 1965) respectively. D "Weber-ensemble model" used as a simplified (photopic but achromatic) model of the retina (KOENDERINK et al., 1971; VAN DOORN et al., 1972)

use of VR-machines in psychophysical quantum vision models, see van de Grind et al., 1971 b.)

At photopic luminance levels the VR-machines almost completely loose their influence on the network dynamics. Therefore Koenderink et al. (1971) and van Doorn et al. (1972) studied the simplified "photopic" model of Fig. 51 d, which they called a "Weber-ensemble model" (achromatic vision), and Koenderink et al. (1972) discussed a chromatic variation.

The achromatic Weber-ensemble model predicted correctly de Lange-curves for various spatial configurations, viz. a blurred edge field, a sharp edge field and a counterphase center-surround pattern, as measured psychophysically by Kelly (1969 a): Koenderink et al. (1971). The data of van Nes (1968 b) obtained with drifting spatial sine wave patterns (Section II C) were also successfully predicted by the model (op. cit.). Further, increment threshold data of Graham and Kemp (1938) were very accurately fitted and an attempt was made to explain constant brightness data.

The chromatic variation of the Weber ensemble model (Koenderink et al., 1972) simulated very well the dynamic response found by de Valois et al. (1967) for spectral opponent cells in the LGN of macaque monkeys.

Considering the basic simplicity of these network models, their scope is amazingly large and it is therefore encouraging to note the close similarity with neurophysiologically inspired models, for example the one discussed in Section X. Svaetichin et al. (1971) have recently, be it in qualitative terms, developed a model of the functioning of the triadic synapse in the receptor pedicles, that is very similar to the W-machine. Finally a high degree of formal similarity also exists with the work of Sperling (1970).

X. Outline of a Neurophysiological Model of Spatio-Temporal Signal Processing in the Retina

In this section we present the outline of a model of the mammalian retina[1] (Grüsser and Lunkenheimer, 1966; Büttner and Grüsser, 1968; Grüsser, 1971). This model is based on neurophysiological data such as those reviewed in Section IV and on neuroanatomical findings (Cajal, 1896; Polyak, 1941; Dowling and Boycott, 1966; Boycott and Dowling, 1969; Dowling and Werblin, 1971, Fig. 52). A hardware simulation of parts of this model was recently described (Eckmiller, 1971; Eckmiller and Grüsser, 1972) and a more extensive description is in preparation. Fig. 52 shows the general structure of the model of the perceptive unit of an on-center ganglion cell of the cat's retina. We will first discuss the typicall "vertical" elements (*R*-, *B*- and *G*-cell models). Then a brief description of the spatial interactions and of the properties of the *H*- and *A*-cell models is given. For the sake of brevity we restrict the discussion to the findings related to on-center ganglion cells.

[1] An outline of the model was also presented by one of the authors (O.-J. G.) at the symposion "Theory of temporal factors in vision and visual perception", June 1966, Rochester, N. Y., USA.

(a) *Receptors*. The absorption of a photon by the visual pigments in the outer segment of a receptor triggers some unknown processes that result in a change of the permeability for specific cations of the receptor membrane. In the case of the vertebrate photoreceptors the result seems to be mainly a decrease of the sodium conductivity of the receptor membrane and this finally causes the hyperpolarizing receptor potential (c.f. TOMITA, 1970). The static relationship between the intensity I of a steady light stimulus and the receptor's hyperpolarizing potential V_R is non-linear (c.f. BAYLOR and FUORTES, 1970) and can be described well with a "shunting inhibition" equation (p. 497):

$$V_R = \frac{C_R\, I}{I_{SR} + K_R\, I}\,\mathrm{mV}\,, \tag{56}$$

where I_{SR} is the receptor's threshold intensity at a given level of adaptation and C_R is the weighting constant of the excitatory pathway and K_R of the inhibitory pathway. Ideas as to why a receptor membrane might behave in this way were presented by ZERBST et al. (1962).

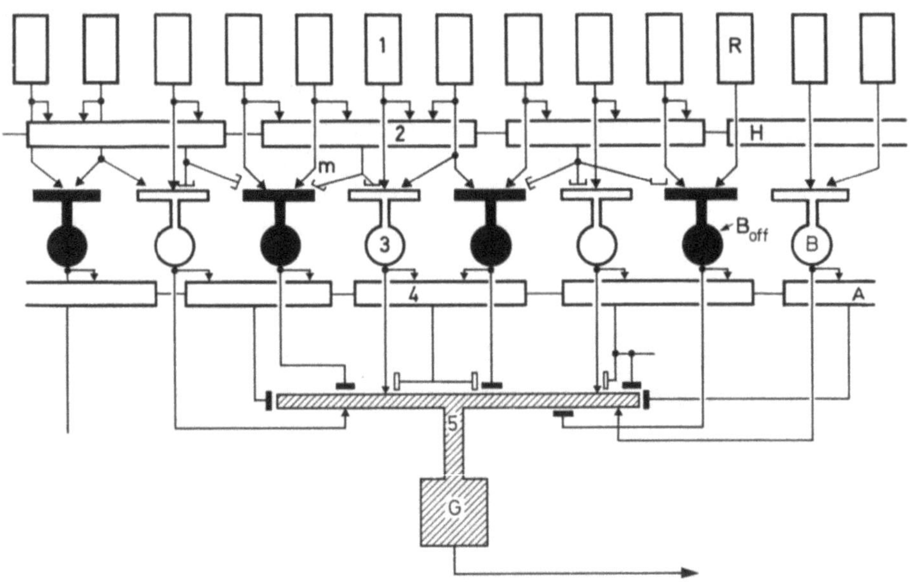

Fig. 52. Block diagram of the elements of the perceptive unit of an on-center ganglion cell (cat retina). *R* receptors, *H* horizontal cells, *B* bipolar cells, *A* amacrine cells, *G* ganglion cell. A possible second layer of amacrine cells is neglected (lateral excitation)

A general dynamic extension of the above static description can be obtained if it is assumed that the shunting inhibition process has dynamic properties that differ from those of the direct excitatory process: Fig. 53.

In Fig. 53 the filters F1–F3 represent the photochemical and chemico-electrical processes in the receptor and can probably be postulated to be simple low (3.–5.) order linear filters e.g. of the cascaded integrator type [Section VII, Eq. (48)]. The complex transfer function of the receptor can then from the block diagram of Fig. 53 directly be written down in terms of the complex transfer functions of the component filters. The static input-output relation [Eq. (56)] then follows as the limiting case for static input signals. The receptor model of Fig. 53 is similar to a "Weber-machine" (Fig. 51 c) and to a model proposed by REICHARDT (1961) to account for the light reactions of *Phycomyces*.

VARJÚ (1965) showed that the response of such a system to sinusoidal inputs shows an increase of the average response that very well explains quantitatively the brightness enhancement (Brücke-Bartley effect) treated in Section II D.

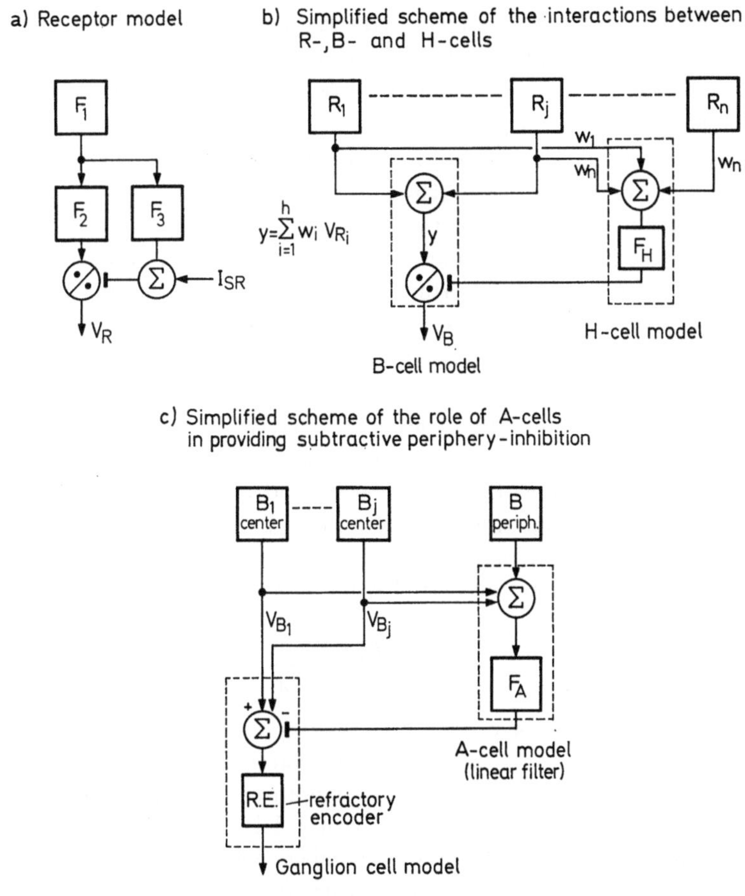

Fig. 53. Schematic diagrams of the different parts of the model explained in Section X

(b) *The Bipolar Cells*. The influence of the bipolar cells on the *dynamics* of the system is assumed to be negligible in the high frequency range at a given photopic adaptation level, while attenuation occurs at low frequencies. Thus in the model the *B*-cells are represented by bandpass systems which add or subtract signals from their receptive fields (antagonistic structure). Two types of bipolar cells are assumed (on-center and off-center, Fig. 52). The on-center bipolars in the RF-center of ganglion cells activate on-center ganglion cells and inhibit off-center ganglion cells, while off-center bipolars have the reverse effect.

(c) *The Ganglion Cells.* The receptor signals from the RF-center of an on-center ganglion cell are assumed to be summed by on-bipolar cells and then arrive at the ganglion cell membrane. The G-cell models are postulated to be pulse encoders that function according to the "refractory sampling" principle discussed in Section VIII C (Here with a hyperbolic threshold function). Moreover it is assumed in our model that the encoder shows "accomodation" phenomena, embodied by a summing of "remaining-threshold-elevations" (EYSEL and GRÜSSER, 1970; EYSEL, 1971).

For a refractory encoder the CFF should be higher for shorter relative refractory periods. This prediction conforms with the findings that there exists a positive correlation between the CFF and conduction velocities of the optic nerve fibers (p. 483), because this implies a negative correlation between CFF and the relative refractory period.

(d) *The Horizontal Cells.* So far we have only considered the simplest situation where only a few receptors of the RF-center of an on-center cell were assumed to be illuminated. To incorporate the function of the horizontal elements of the retina into the model we have to look once more at the experimental data about spatial interactions in the RF. The experiments described on p. 497 and 500 show that both spatial summation of excitation in the RF-center and of inhibition in the RF-periphery can be described with similar non-linear relations [Eq. (38) and Eq. (39)].

BÜTTNER and GRÜSSER (1968) assumed that the output of the receptors is pooled by horizontal cells (2 in Fig. 52) and that the pool's output interacts again with the receptor signals at the bipolar cell dendrites (3 in Fig. 52). The experimental results suggest that this interaction can again be described by FURMAN's (1965) shunting inhibition model (c.f. also VARJÚ, 1965). When n receptors or possibly receptor groups are activated equally, the overall input to the ganglion cell is in the static case:

$$V_B = \frac{C_B n\, V_R}{1 + K_B\, n\, V_R}\,. \tag{57}$$

A dynamic extension can be obtained as follows. Assume that the horizontal cell pool including its synapses on the bipolar cells can be represented by a simple, first-order low pass filter with an upper frequency limit between 3 and 8 Hz (Horizontal conduction of signals in the pool occurs with a delay and with spatial decrement and thus a spatial "weighting" of input signals results). The attenuation of higher frequency signals increases only slowly with frequency for a first order filter. However, when the stimulus frequency reaches the frequency limit of the receptors (filter F1 and F2 in Fig. 53), the slope of the overall gain curve of the H-cell potential should become much steeper. As Fig. 20 shows, these ideas are corroborated by the experimental findings. Because the H-cell pool (schematized as filter F_H in Fig. 53, compare with Fig. 51 D) controls the spatial summation in the RF-center its properties should be reflected at the ganglion cell's output level for the stimulus patterns of Fig. 26c and *d*. The model predicts that the spatial summation should increase and become more linear when the frequency increases above 3 Hz, since the influence of filter F_H (Fig. 53) then decreases. This indeed was found experimentally (p. 494).

(e) *The Amacrine Cells.* These cells are assumed to play a role in the interaction between signals from the center and from the periphery of the ganglion cell's RF. The lateral inhibiting amacrine cells form a pool (again with spatial decrement of horizontally conducted signals) for the bipolar cell output signals and have subtractive inhibitory influence on the ganglion cells. The latter assumption is justified by the findings that in on-center neurons, on-inhibition from the RF-periphery interacts linearly with on-excitation from the RF-center (Amecke-Mönninghof and Buettner, 1970; Enroth and Pinto, 1971, 1972; Grüsser, 1971). From the data described on p. 500 about the frequency properties of this periphery-inhibition process one can conclude that the inhibitory pathway via the *A*-cells has a lower cutoff frequency than the excitatory pathway directly from bipolar to ganglion cell. The principle of the interactions between *B*-, *A*- and *G*-cell models is schematized in Fig. 53. Again at higher frequencies the relative influence of the lateral pathway (F_A in Fig. 53) diminishes. This explains why the response enhancement caused by simultaneous contrast is lower for the range of medium frequencies at which the average neuronal response is enhanced (Fig. 38).

XI. Conclusions

Since the appearance of the annotated bibliography on flicker perception by Landis (1953) and the review by Landis (1954) the research in this area has made considerable progress at least in the following three respects:

(1) The extended use of sine wave flicker stimuli since the work of de Lange (1952) and the accompanying application of Fourier analysis to explain most of the traditional data from fewer sine wave data has brought a theoretical unification and has reduced the bewildering amount of experimental parameters to manageable properties.

(2) Also in 1952, Ch. Enroth introduced flicker stimulation in neurophysiological experiments in which single unit recordings are obtained with microelectrodes. Both the improvement of recording and of stimulation techniques during the last twenty years have led to an enormous extension of the "microelectrophysiological" literature on flicker and visual dynamics.

(3) The development of quantitative flicker theories has also been simplified by the sine wave approach and the concomitant introduction of electrical engineering ideas has unified the modelling concepts (filters, transfer functions, cutoff frequency, black box, etc.). In addition, mechanistic interpretations of formal mathematical theories have become feasible even for a complex system like the a.v.s. by the combination of neurophysiological concepts with "hardware" (electronic) as well as "software" (computer programs) analogs.

The black boxes of earlier psychophysical theories are about to be opened, only to expose as in the box of Pandora a series of new (neurophysiological) black boxes.

Acknowledgements

This work was supported by grants from the Deutsche Forschungsgemeinschaft (Gr 161) and in part by a Public Health Services Research Grant, No. NB-07575 from the National Institute of Neurological Diseases and Blindness, Bethesda, Maryland. Part of the report was prepared while one of us (O-J.G.) was supported by an International Scholarship from Research to Prevent Blindness Inc., New York, at the Bascom Palmer Eye Institute, Department of Ophthalmology, University of Miami (winter semester 1970/1971). He thanks Prof. E. D. W. NORTON, Chairman of the Department of Ophthalmology, University of Miami for his hospitality and the opportunity to work at the Bascom Palmer Eye Institute.

One of the authors (W.A.v.d.G.) was supported by the Alexander von Humboldt Stiftung, Bonn-Bad Godesberg during the time that he held a research fellowship at the Department of Physiology, Freie Universität Berlin (1971/1972).

We thank Mrs. J. DAMES for her assistance in the English translations and Mrs. U. SAYKAM for her accurate typework.

References

ABE, Z.: Influence of adaptation on the strength frequency curve of human eyes, as determined with electrically produced flickering phosphenes. Tohoku J. exp. Med. **54**, 37—44 (1951).

AIBA, S.: The effects of dexamphetamine, sodium amobarbital and meprobamate on critical frequency of flicker under two different surround illuminations. Psychopharmacology **1**, 89—101 (1959).

AKIMOTO, H., CREUTZFELDT, O.: Reaktionen von Neuronen des optischen Cortex nach elektrischer Reizung unspezifischer Thalamuskerne. Arch. Psychiat. Nervenkr. **196**, 496—519 (1958).

ALLEN, F.: The variation of visual sensory reflex action with intensity of stimulation. J. opt. Soc. Amer. **13**, 383—430 (1926).

ALPERN, M., HENDLEY, C. D.: Visual functions as indices of physiological changes in acid-base balance of the blood. Amer. J. Optom. **29**, 301—314 (1952).

— SPENCER, R. W.: Variation of critical flicker frequency in the nasal visual field. Arch. Ophthal. **50**, 50—63 (1953).

— SUGIYAMA, S.: Photic driving of the critical flicker frequency. J. Opt. Soc. Amer. **51**, 1379—1385 (1961).

AMECKE-MÖNNINGHOF, E., BUETTNER, U. W.: Die Wechselwirkung von rezeptivem Feldzentrum und rezeptiver Feldperipherie retinaler Neurone der Katze. I. Phasische Reizung im Zentrum. Pflügers Arch. ges. Physiol. **319**, R 152 (1970).

— — Die Wechselwirkung zwischen Feldzentrum und Feldperipherie von retinalen Neuronen der Katze. II. Phasische Reizung in der Peripherie. Pflügers Arch. ges. Physiol. **319**, R 152 (1970).

ANDERSON, D. A., HUNTINGTON, J., SIMONSON, E.: Critical fusion frequency as a function of exposure time. J. opt. Soc. Amer. **56**, 1607—1611 (1966).

ARDEN, G., LIU, Y.-M.: Some types of response of single cells in the rabbit lateral geniculate body to stimulation of the retina by light and to electrical stimulation of the optic nerve. Acta physiol. scand. **48**, 36—48 (1960a).

— — Some responses of the lateral geniculate body of the rabbit to flickering light stimuli. Acta physiol. scand. **48**, 49—62 (1960b).

ARDUINI, A.: Specific and non-specific components in the activity of a relay nucleus: the lateral geniculate. Arch. ital. Biol. **107**, 715—722 (1969).

ARNOLD, H.: Optische Verschmelzungsfrequenz und ermüdende Beanspruchung. Arbeitsphysiologie **15**, 62—78 (1953).

— WACHOLDER, K.: Weitere Untersuchungen über optische Verschmelzungsfrequenz und ermüdende körperliche Beanspruchung. Arbeitsphysiologie **15**, 139—148 (1953).

BAADER, E. G.: Über die Empfindlichkeit des Auges gegen Lichtwechsel. Med. Diss., Univ. Freiburg i. Br. 1891.

BAKER, C. H.: The dependence of binocular fusion on timing of peripheral stimuli and on central process. Canad. J. Psychol. **6**, 84—91 (1952a).

— The dependence of binocular fusion on timing of peripheral stimuli and on central process. 2. Asymmetrical flicker. Canad. J. Psychol. **6**, 123—130 (1952b).

— The dependence of binocular fusion on timing of peripheral stimuli and on central process. III. Cortical flicker. Canad. J. Psychol. **6**, 151—163 (1952c).

— BOTT, E. A.: Studies on visual flicker and fusion. II. Effects of timing of visual stimuli on binocular fusion and flicker. Canad. J. Psychol. **5**, 9—17 (1951).

Ball, R. J., Bartley, S. H.: Effects of intermittent monochromatic illumination on visual acuity. Amer. J. Optom. **47**, 519—525 (1970).

Barlow, H. B.: Optic nerve impulses and Weber's law. Cold Spr. Harb. Symp. quant. Biol. **30**, 539—546 (1965).

— Fitzhugh, R., Kuffler, S. W.: Change of organization in the receptive fields of the cat's retina during dark adaptation. J. Physiol. (Lond.) **137**, 338—354 (1957).

Barnett, A.: Electrically produced flicker in darkness. Amer. J. Physiol. **133**, 205—206 (1941).

Bartley, S. H.: The neural determination of critical flicker frequency. J. exp. Psychol. **21**, 678—686 (1937).

— Subjective brightness in relation to flash rate and the light-dark ratio. J. exp. Psychol. **23**, 313—319 (1938).

— Some factors in brightness discrimination. Psychol. Rev. **46**, 337—358 (1939).

— Brightness enhancement in relation to target intensity. J. Psychol. **32**, 57—62 (1951).

— Brightness comparison when one eye is stimulated intermittently and the other eye steadily. J. Psychol. **34**, 165—167 (1952).

— Light adaptation and brightness enhancement. Perceptual and Motor Skills **7**, 85—92 (1957).

— Some facts and concepts regarding the neurophysiology of the optic pathway. Arch. Ophthal. **60**, 775—791 (1958).

— Ball, R. J.: Effects of intermittent illumination on visual acuity. Amer. J. Optom. **45**, 458—464 (1968).

— Nelson, T. M.: Certain chromatic and brightness changes associated with rate of intermittency of photo stimulation. J. Psychol. **50**, 323—332 (1960).

— — A further study of pulse-to-cycle fraction and critical flicker frequency. A decisive theoretical test. J. opt. Soc. Amer. **51**, 41—45 (1961).

— — Some relations between sensory end results and neural activity in the optic pathway. J. Psychol. **55**, 121—143 (1963).

— — Ronney, J. E.: The sensory parallel of the reorganization period in the cortical response in intermittent retinal stimulation. J. Psychol. **52**, 137—147 (1961).

— — Soules, E. M.: Visual acuity under conditions of intermittent illumination productive of paradoxical brightness. J. Psychol. **55**, 153—163 (1963).

— Paczewitz, G., Valsi, E.: Brightness enhancement and the stimulus cycle. J. Psychol. **43**, 187—192 (1957).

— Wilkinson, F. R.: Brightness enhancement when entoptic stray light is held constant. J. Psychol. **33**, 301—305 (1952).

Basler, A.: Über die Verschmelzung von zwei nacheinander erfolgenden Lichtreizen. Pflügers Arch. ges. Physiol. **143**, 245—251 (1911).

Baumgartner, G.: Die Reaktionen der Neurone des zentralen visuellen Systems der Katze im simultanen Helligkeitskontrast. In: Jung, R., Kornhuber, H. H. (Hrsg.): Neurophysiologie und Psychophysik des visuellen Systems, S. 296—313. Berlin-Göttingen-Heidelberg: Springer 1961.

— Brown, J. L., Schulz, A.: Responses of single units of the cat visual system to rectangular stimulus patterns. J. Neurophysiol. **28**, 1—18 (1965).

— Hakas, P.: Reaktionen einzelner Opticusneurone und corticaler Nervenzellen der Katze im Hell-Dunkel-Grenzfeld (Simultankontrast). Pflügers Arch. ges. Physiol. **270**, 29 (1959).

— — Die Neurophysiologie des simultanen Helligkeitskontrastes. Reziproke antagonistischer Neuronengruppen des visuellen Systems. Pflügers Arch. ges. Physiol. **274**, 489—510 (1962).

Baylor, D. A., Fuortes, M. G. F.: Electrical responses of single cones in the retina of the turtle. J. Physiol. (Lond.) **207**, 77—92 (1970).

Bayly, E. J.: Spectral analysis of pulse frequency modulation in the nervous system. IEEE Trans. BME-15, 257—265 (1968).

— Spectral analysis of pulse frequency modulation. In: Systems analysis approach to neurophysiological problems. Conf. Proc. Lab. Neurophysiol. Univ. Minnesota, 48—60 (1969).

Beck, A.: Anstieg der Verschmelzungsfrequenz bei Erregung im Elektroretinogramm. Klin. Wschr. **29**, 446—448 (1951).

Benham, C. H.: The artifical spectrum top. Nature (Lond.) **51**, 200 (1895).

BERG, O.: A study of the effect of evipan on the flicker fusion intensity in brain injuries. Acta psychiat. (Kbh.) Suppl. **58**, 1—116 (1949).

BERGER, C.: Area of retinal image and flicker fusion frequency. Acta physiol. scand. **28**, 224—233 (1953).

— Illumination of surrounding field and flicker fusion frequency with foveal images of different sizes. Acta physiol. scand. **30**, 161—170 (1954).

BIELCK, G. G.: De raritate luminis. Medical doctor thesis, 12 p. Introduced by J. A. SEGNER. Göttingen: Vandenhoek 1740.

BISHOP, G. H., O'LEARY, J. L.: Electrical activity of the lateral geniculate of cats following optic nerve stimuli. J. Neurophysiol. **3**, 308—322 (1940).

BISHOP, P. O.: Synaptic transmission. An analysis of the electric activity of the lateral geniculate nucleus in the cat after optic nerve stimulation. Proc. Roy. Soc. **141** B, 362—392 (1953).

— Properties of afferent synapses and sensory neurons in the lateral geniculate nucleus. Int. Rev. Neurobiol. **6**, 191—255 (1964).

— BURKE, W., DAVIS, R.: The identification of single units in central visual pathway. J. Physiol. (Lond.) **162**, 409—431 (1962).

— JEREMY, D., MCLEOD, J. G.: Phenomenon of repetitive firing in lateral geniculate of cat. J. Neurophysiol. **16**, 437—447 (1953).

BJERVER, K., GOLDBERG, L.: Effect of alcohol ingestion on driving ability. Quart. J. Stud. Alcohol **II**, 1—30 (1950).

BLECK, F. C., CRAIG, E. A.: Brightness enhancement and hue: I. The effect of Munsell-hue targets. J. Psychol. **59**, 243—250 (1965).

— — Brightness enhancement and hue: II. Hue shift as a function of steady and intermittent photic stimulation. J. Psychol. **59**, 251—258 (1965).

BORNSCHEIN, H.: Vergleichende Elektrophysiologie der Retina. In: JUNG, R., KORNHUBER, H. (Hrsg.): Neurophysiologie und Psychophysik des visuellen Systems. Berlin-Göttingen-Heidelberg: Springer 1961.

— Physiologische Aspekte des Flimmerelektroretinogramms: Komponenten und Frequenzcharakteristik. Docum. Ophthal. **18**, 85—100 (1964).

— LAHODA, R.: Harmonic analysis of human flicker electroretinogram. Proc. 3rd Int. Conf. Med. Electronics 342—344 (1960).

— SCHUBERT, G.: Das photopische Flimmer-Elektroretinogramm des Menschen. Z. Biol. **106**, 229—238 (1953).

— SZEGVARY, G.: Flimmerelektroretinographische Studie bei einem Säuger mit reiner Zapfennetzhaut *(Citellus citellus)*. Z. Biol. **110**, 285—290 (1958).

BORSELLINO, A., FUORTES, M. G. F.: Responses to single photons in visual cells of Limulus. J. Physiol. (Lond.) **196**, 507—539 (1968).

— — SMITH, T. G.: Visual responses in Limulus. Cold Spr. Harb. Symp. quant. Biol. **30**, 429—443 (1965).

— POPPELE, R. E., TERZUOLO, C. A.: Transfer functions of the slowly adapting stretch receptor organ of Crustacea. Cold Spr. Harb. Symp. quant. Biol. **30**, 581 (1965).

BOUMAN, M. A.: History and present status of quantum theory in vision. In: ROSENBLITH, W. A. (Ed.): Sensory Communication, pp. 377—401. Cambridge, Massachusetts: M.I.T. Press 1961.

— My image of the retina. Quart. Rev. Biophys. **2**, 25—64 (1969).

— DOESSCHATE, J. TEN: Adaptation and the electrical excitability of the eye. Docum. Ophthal. **26**, 240—247 (1969).

— — VELDEN, H. A. VAN DER: Electrical stimulation of the human eye by means of periodical rectangular stimuli. Docum. Ophthal. **5—6**, 151—167 (1951).

BOURASSA, CH. M., BARTLEY, S. H.: Some observations on the manipulation of visual acuity by varying the rate of intermittent stimulation. J. Psychol. **59**, 319—328 (1965).

BOYCOTT, B. B., DOWLING, J. E.: Organization of the primate retina: light microscopy. Proc. roy. Soc. Lond. B **255**, 109—184 (1969).

Boynton, R. M., Sturr, J. F., Ikeda, M.: Study of flicker by increment threshold technique. J. opt. Soc. Amer. **51**, 196—201 (1961).
Brewster, D.: On the influence of successive impulses of light upon the retina. Phil. Mag. (Lond. Edinb.) **4**, 241—245 (1834).
Brindley, G. S.: The site of electrical excitation of the human eye. J. Physiol. (Lond.) **127**, 189—200 (1955).
— Beats produced by simultaneous stimulation of the human eye with intermittent light and intermittent or alternating current. J. Physiol. (Lond.) **164**, 157—167 (1962).
— Physiology of the retina and the visual pathway. London: Edward Arnold 1960.
— du Croz, J. J., Rushton, W. A. H.: The flicker fusion frequency of the blue sensitive mechanism of colour vision. J. Physiol. (Lond.) **183**, 497—500 (1966).
Brink, G. van den, Reijntjes, G. A.: Spatial and temporal facilitation in vision. Vision Res. **6**, 533—551 (1966).
Broca, A., Sulzer, D.: La sensation lumineuse en fonction du temps. C. R. Acad. Sci. (Paris) **134**, 831—834 (1902).
Broekhuijsen, M. L., Veringa, F. T.: Sinusoidal current and perceived brightness II. Vision Res. **12**, 363 (1971).
Brown, J. L.: Harmonic analysis of visual stimuli below fusion frequency. Science **137**, 686—688 (1962).
— Flicker and intermittent stimulation. Chapter 10. In: Graham, C. H. (Ed.): Vision and Visual Perception, p. 251. New York-London-Sydney: John Wiley and Sons 1965.
Brown, K. T., Wiesel, T. N.: Intraretinal recording with micropipette electrodes in the intact cat eye. J. Physiol. (Lond.) **149**, 537—562 (1959).
Brücke, E.: Über den Nutzeffekt intermittierender Netzhautreizungen. Sitzungsber. K. Akad. Wissensch., math.-naturwiss. Klasse. Wien **49**, II, 128—153 (1864).
Büttner, Ch., Büttner, U., Eysel, U., Grüsser, O.-J., Lunkenheimer, H.-U., Schaible, D. Spatial summation in the receptive fields of cat's retinal ganglion cells. I. Summation within the RF-center. Proceedings of the First European Biophysics Congress 1971, Baden, Austria, ed. Broda, E., Locker, A., Springen-Lederer, H. Separatum, Verlag der Wiener Medizinischen Akademie, 257—261 (1971).
— — Grüsser, O.-J.: Die Frequenzeigenschaften der off-Inhibition im rezeptiven Feldzentrum retinaler on-Zentrum Neurone der Katze. Pflügers Arch. ges. Physiol. **312**, 133 (1969).
— — — Interaction of excitation and direct inhibition in the receptive field center of retinal neurons. Pflügers Arch. ges. Physiol. **322**, 1—21 (1971).
— — — Summation of excitation and inhibition in the receptive field center of retinal neurons. Biokybernetik III, Symposion Leipzig 1969, pp. 197—201. Ed. Drischel, H., Tiedt, N. Jena: VEB Gustav Fischer 1971.
— — — Rackensperger, W., Vierkant, J.: Die Summation von zwei unabhängig voneinander ausgelösten Erregungen im rezeptiven Feldzentrum retinaler Neurone der Katze. I. Internat. Symposion Biokybernetik Leipzig (1967). Wiss. Z. Karl-Marx-Universität Leipzig **2**, 178—182 (1968).
Büttner, U., Grüsser, O.-J.: Zeitliche und räumliche Einflüsse auf die Erregungsintegration im rezeptiven Feld retinaler Neurone der Katze. Pflügers Arch. ges. Physiol. **291**, 88 (1966).
— — Quantitative Untersuchungen der räumlichen Erregungssummation im rezeptiven Feld retinaler Neurone der Katze. I. Reizung mit 2 synchronen Lichtpunkten. Kybernetik **3**, Bd. **4**, 81—94 (1968).
— — Spatial summation within the RF-center of retinal neurons. The effect of the distance of two light stimuli. unpubl. (1971).
Burckhardt, Ch. W.: From flicker-fusion to color vision. Biologica Computer Laboratory Report 1—41, Dept. El. Eng., Univ. of Illinois Urbana, Ill. (1966).
Burke, W., Sefton, A. J.: Discharge patterns of principle cells and interneurons in lateral geniculate nucleus of rat. J. Physiol. (Lond.) **187**, 201—212 (1966).
— — Inhibitory mechanisms in lateral geniculate nucleus of rat. J. Physiol. (Lond.) **187**, 231—246 (1966).
Byzov, A. L.: Horizontale Retina-Zellen als Regulatoren der synaptischen Übertragung (russ.). Fiziol. Zh. SSSR **53**, 1115—1124 (1967).
Cajal, S. Ramon y: Die Retina. Translated by Greeff. Wiesbaden: Bergmann 1896.

CAMPBELL, F. W., COOPER, G. F., ENROTH CUGELL CH.: The spatial selectivity of the visual cells of the cat. J. Physiol. (Lond.) **203**, 223—235 (1969).
— ROBSON, J. G.: The attenuation characteristics of the visual system determined by measurements of flicker threshold, brightness and pupilomotor effect of modulated light. Docum. Ophthal. **28**, 83—84 (1964).
CAMPENHAUSEN, CH. VON: Über den Ursprungsort von musterinduzierten Flickerfarben im visuellen System des Menschen. Z. vergl. Physiol. **61**, 355—360 (1968).
— The color of Benham's top under metameric illumination. Vision Res. **9**, 677—682 (1969).
— Musterinduzierte Flickerfarben. Untersuchungen zur Psychophysik des Farbensehens. Verh. dtsch. Zool. Ges. **64**, 227—234 (1970).
CERVETTO, L.: Analysis of the pigeon's electroretinogram. Arch. ital. Biol. **106**, 194—203 (1968)
CHRISTIAN, P., HAAS, R.: Über ein Farbenphänomen (polyphäne Farben). Sitzungsber. Heidelberg, Akad. Wiss. Math.-naturw. Klasse **1948**, 1—28.
CLARK, W. E. LE GROS: The visual centres of the brain and their connexions. Physiol. Rev. **22**, 205—232 (1942).
CLAUSEN, J.: Visual sensations (phosphenes) produced by AC sine wave stimulation, 101 p. Copenhagen (Bergen): Ejnar Munksgaard 1955.
— VANDERBILT, C.: Visual beats caused by simultaneous electrical and photic stimulation. Amer. J. Psychol. **70**, 577—585 (1957).
CLELAND, B. G., DUBIN, M. W., LEVICK, W. R.: Simultaneous recording of input and output of lateral geniculate neurons. Nature, New Biology **231**, 191—192 (1971a).
— — — Sustained and transient neurones in the cat's retina and the lateral geniculate nucleus. J. Physiol. (Lond.) **217**, 473—496 (1971b).
— ENROTH-CUGELL, C.: Cat retinal ganglion cell responses to changing light intensities: Sinusoidal modulation in the time domain. Acta physiol. scand. **68**, 365—381 (1966).
— — Quantitative aspects of sensitivity and summation in the cat retina. J. Physiol. (Lond.) **198**, 17—38 (1968).
— — Quantitative aspects of gain and latency in the cat retina. J. Physiol. (Lond.) **206**, 73—91 (1970).
COBB, P. W.: Some comments on the Ives theory of flicker. J. opt. Soc. Amer. **24**, 91—98 (1934a).
— The dependence of flicker on the dark-light ratio of the stimulus cycle. J. opt. Soc. Amer. **24**, 107—113 (1934b).
COHEN, J., GORDON, D. A.: The Prevost-Fechner-Benham subjective colors. Psychol. Bull. **46**, 97—138 (1949).
CORNEHLS, U.: Reaktionen einzelner Neurone im optischen Cortex der Katze nach elektrischen Doppelreizen des Nervus opticus. Pflügers Arch. ges. Physiol. **268**, 52 (1958).
— GRÜSSER, O.-J.: Ein elektronisch gesteuertes Doppellichtreizgerät. Pflügers Arch. ges. Physiol. **270**, 78—79 (1959).
CREED, R. S., RUCH, T. C.: Regional variations in sensitivity to flicker. J. Physiol. (Lond.) **74**, 407—423 (1932).
CREUTZFELDT, O. D.: Functional synaptic organization in the lateral geniculate body and its implication for information transmission. From: Structure and functions of inhibitory neuronal mechanisms. Proc. of the 4th Internat. Meeting of Neurobiologists, Stockholm, Sept. 1966. Oxford-New York: Pergamon Press 1968.
— AKIMOTO, H.: Konvergenz und gegenseitige Beeinflussung von Impulsen aus der Retina und den unspezifischen Thalamuskernen an einzelenen Neuronen des optischen Cortex. Arch. Psychiat. Nervenkr. **196**, 520—538 (1958).
— GRÜSSER, O.-J.: Beeinflussung der Flimmerreaktion einzelner corticaler Neurone durch elektrische Reize unspezifischer Thalamuskerne. Excerpta med. (Amst.) Int. Congr. Ser. **11**, 148 (1957).
— — Beeinflussung der Flimmerreaktion einzelner corticaler Neurone durch elektrische Reize unspezifischer Thalamuskerne. In: BOGAERT, L. V., RADERMECKER, J. (Eds.): Proc. 1st Int. Congr. Neurol. Sci. Vol. 3, Electroencephalography, Clinical Neurophysiology and Epilepsy, pp. 349—355. London-New York-Paris: Pergamon 1959.
— SAKMANN, B., SCHEICH, H.: Zusammenhang zwischen Struktur und Funktion der Retina. Aus: Kybernetik 1968, Hrsg. MARKO, H., FÄRBER, G. München: Oldenbourg.

Crozier, W. J., Wolf, E.: Theory and measurement of visual mechanisms. IV. Critical intensities for visual flicker, monocular and binocular. J. gen. Physiol. **24**, 505—534 (1941a).
— — Theory and measurement of visual mechanisms. V. Flash duration and critical intensity for response to flicker. J. gen. Physiol. **24**, 635—654 (1941b).
— — Zerrahn-Wolf, G.: Intensity and critical frequency for visual flicker. J. gen. Physiol. **21**, 203—221 (1937).
De Valois, H.: Discussion to Bartley, S. H. Arch. Ophthal. **60**, 775—791 (1958).
De Valois, R. L., Abramow, I., Mead, W. R.: Single cell analysis of wavelength discrimination at the lateral geniculate nucleus of the macaque. J. Neurophysiol. **30**, 415—433 (1967).
De Voe, R. D.: A nonlinear model of sensory adaptation in the eye of the wolf spider. In: Bernhard, C. G. (Ed.): The Functional Organization of the Compound Eye, pp. 309—328. Oxford: Pergamon 1966.
— Nonlinear transient responses from light-adapted wolf spider eyes to changes in background illumination. J. gen. Physiol. **50**, 1961—1992 (1967a).
— A nonlinear model for transient responses from light-adapted wolf spider eyes. J. gen. Physiol. **50**, 1993—2030 (1967b).
Dodge, F. A., Jr, Shapley, R. M., Knight, B. W.: Linear systems analysis of the Limulus retina. Behav. Sci. **15**, 24—36 (1970).
Dodt, E.: Ergebnisse der Flimmerelektroretinographie. Med. Habilitationsschrift, Freiburg i. Br. Dez. 1953.
— Ergebnisse Flimmer-Elektroretinographie. Experientia (Basel) **10**, 330—339 (1954).
— Erregung und Hemmung retinaler Neurone bei intermittierender Belichtung. Docum. Ophthal. **18**, 259—274 (1964).
— Enroth, C.: Retinal flicker response in cat. Acta physiol. scand. **30**, 375—390 (1953).
— Lith, G. M. H. van, Schmidt, B.: Electroretinographic evaluation of the photopic malfunction in a totally color blind. Vision Res. **7**, 231—241 (1967).
— Walther, J. B.: Der photopische Dominator im Flimmer-ERG der Katze. Pflügers Arch. ges. Physiol. **266**, 175—186 (1958).
— Wirth, A.: Differentiation between rods and cones by flicker electroretinography in pigeon and guinea pig. Acta physiol. scand. **30**, 80—89 (1953).
Doorn, A. J. van, Koenderink, J. J., Bouman, M. A.: The influence of the retinal inhomogeneity on the perception of spatial patterns. Kybernetik **10**, 223—230 (1972).
Dowling, J. E.: Organization of vertebrate retinas. Invest. Ophthal. **9**, 655—680 (1970).
— Boycott, B. B.: Neural connections of the retina: fine structure of the inner plexiform layer. Cold Spr. Harb. Symp. quant. Biol. **30**, 393—402 (1965).
— — Organization of the primate retina: Electron microscopy. Proc. roy. Soc. Lond. B **166**, 80—111 (1966).
— Werblin, F. S.: Synaptic organization of the vertebrate retina. Vision Res. Suppl. **3**, 1—15 (1971).
Ebbecke, U.: Über das Augenblicksehen. Pflügers Arch. ges. Physiol. **185**, 181—195 (1920a).
— Über das Sehen im Flimmerlicht. Pflügers Arch. ges. Physiol. **185**, 196—223 (1920b).
Eckhorn, R., Pöpel, B.: A contribution to neuronal network analysis in cat LGB: Simultaneous recordings of maintained activity. Int. J. Neurosci. (in press, 1972).
Eckmiller, R.: Electronic analog models of the retina and the visual system. In: Grüsser, O.-J., Klinke, R. (Eds.): Zeichenerkennung durch biologische und technische Systeme, S. 143—151. Berlin-Heidelberg-New York: Springer 1971.
— Properties of an electronic simulation network of the vertebrate retina and the first layer in the lateral geniculate nucleus. Proc. of the First European Biophysics Congr. 1971, Baden, Austria, Broda, E., Locker, A., Springer-Lederer, H. (Eds.): Separatum, Verlag der Wiener Medizinischen Akademie 267—271 (1971).
— Grüsser, O.-J.: Electronic simulation of the neuronal network of the vertebrate retina. Kybernetik, in prep. 1972/1973.
Enroth, Ch.: Spike frequency and flicker fusion frequency in retinal ganglion cells. J. Physiol. (Lond.) **117**, 18—21 (1952).
— The mechanism of flicker and fusion studied on single retinal elements in the dark-adapted eye of the cat. Acta physiol. scand. **27**, Suppl. 100, 1—67 (1952).

ENROTH-CUGELL, CH., PINTO, L.: Algebraic summation of centre and surround inputs to retinal ganglion cells of the cat. Nature (Lond.) **226**, 458—459 (1970).
— — Properties of the surround response mechanism of cat retinal ganglion cells and centre-surround interaction. J. Physiol. (Lond.) **220**, 403—439 (1972).
— — Pure central responses from off-centre cells and pure surround responses from on-centre cells. J. Physiol. (Lond.) **220**, 441—464 (1972).
— ROBSON, J. G.: The contrast sensitivity of retinal ganglion cells of the cat. J. Physiol. (Lond.) **187**, 517—552 (1966).
ENZER, N., SIMONSON, E., BALLARD, C.: The effect of small doses of alcohol on the central nervous system. Amer. J. Clin. Pathol. **14**, 333—341 (1944).
ERLICK, D., LANDIS, C.: The effect of intensity, light-dark ratio, and age on the flicker-fusion threshold. Amer. J. Psychol. **65**, 375—388 (1952).
ERNST, W.: The dependence of critical flicker frequency and the rod threshold on the state of adaptation of the eye. Vision Res. 8, 889—900 (1968).
EXNER, S.: Über die zu einer Gesichtswahrnehmung nötige Zeit. Sitzungsber. Kaiserl. Akad. Wissensch. **58, II**, 601—631 (1868).
— Bemerkungen über intermittierende Netzhautreizung. Pflügers Arch. ges. Physiol. **3**, 214—240 (1870).
EYSEL, U.: Computer simulation of the impulse pattern of muscle spindle afferents under static and dynamic conditions. Kybernetik 8, 171—179 (1971).
— FLYNN, J. T., GAEDT, CHR.: Spatial summation of excitation and inhibition in receptive fields of neurons in the lateral geniculate body of the cat and the influence of visual deprivation. Pflügers Arch. ges. Physiol. **327**, 82—94 (1971).
— GRÜSSER, O.-J.: The impulse pattern of muscle spindle afferents. A statistical analysis of the response to static and sinusoidal stimulation. Pflügers Arch. ges. Physiol. **315**, 1—26 (1970).
— — Neurophysiological basis of pattern recognition in the cat's visual system. In: GRÜSSER, O.-J., KLINKE, R. (Eds.): Zeichenerkennung durch biologische und technische Systeme, S. 59—80. Berlin-Heidelberg-New York: Springer 1971.
— — PECCI-SAAVEDRA, J.: The signal transmission by degenerating synapses of the cat's lateral geniculate nucleus. Brain Res. (in preparation 1972).
FECHNER, G. T.: Über eine Scheibe zur Erzeugung von subjectiven Farben. Poggendorf Ann. Physik. Chem. **45**, 227—232 (1838).
FEILCHENFELD, H.: Über die Sehschärfe im Flimmerlicht. Z. Psychol. **35**, 1—7 (1904).
FEINBERG, I.: Critical flicker frequency in amblyopia ex anopsia. Amer. J. Ophthal. **42**, 473—481 (1956).
FERRY, E. S.: Persistence of vision. Amer. J. Sci. **44**, 192—207 (1892).
FICK, A.: Die Lehre vom Lichtsinn. In: Hermanns Handbuch der Physiologie der Sinnesorgane, Vol. III/1, pp. 139—234. Leipzig: F. C. W. Vogel 1879.
FIORENTINI, A., MAFFEI, L.: Transfer characteristics of excitation and inhibition in the human visual system. J. Neurophysiol. **23**, 285—292 (1970).
FISCHER, B.: Optische und neuronale Grundlagen der visuellen Bildübertragung: Einheitliche mathematische Behandlung des retinalen Bildes und der Erregbarkeit von retinalen Ganglienzellen mit Hilfe der linearen Systemtheorie. Vision Res. **12**, 1125—1144 (1972).
— FREUND, H.-J.: Eine mathematische Formulierung für Reiz-Reaktionsbeziehungen retinaler Ganglienzellen. Kybernetik **7**, 160—166 (1970).
— MAY, H.-U.: Invarianzen in der Katzenretina: Gesetzmäßige Beziehungen zwischen Empfindlichkeit, Größe und Lage receptiver Felder von Ganglienzellen. Exp. Brain Res. **11**, 448—464 (1970).
FOERSTER, M. H., GRIND, W. A. VAN DE: Intracellularly recorded discrete waves and ganglion cell responses in the intact eye of the cat. Vision Res. in prep. (1973).
— — GRÜSSER, O.-J.: The response of horizontal cells of the cat's retina to flicker stimulation. Vision Res., in preparation (1972/1973).
— GRÜSSER, O.-J.: Der Einfluß der Narkosetiefe auf die Entladungsmuster retinaler Neurone der Katze. Pflügers Arch. ges. Physiol. **294**, 53 (1967).
— — Responses of retinal ganglion cells of the cat at different depths of barbiturate anesthesia. unpubl. work (1969)

Foerster, M. H., Grüsser, O. J., Lunkenheimer, H.-U.: Responses of cat's retinal ganglion cells to sinusoidally modulated light. The effect of the depth of modulation. In preparation (1973).

Foley, P. J.: Interrelationships of background area, target area and target luminance in their effect on the critical flicker frequency of the human fovea. J. opt. Soc. Amer. **51**, 737—740 (1961).

— Critical flicker frequency and phased surrounds. J. opt. Soc. Amer. **53**, 497—498 (1963).

— Kazdan, J.: Area-intensity relations within the fovea for flickering white and part-spectrum targets. J. opt. Soc. Amer. **54**, 547—550 (1964).

Forsyth, D. M.: Use of a Fourier model in describing the fusion of complex visual stimuli. J. opt. Soc. Amer. **50**, 337—341 (1960).

— Brown, C. R.: Nonlinear property of the visual system at fusion. Science **134**, 612—614 (1961).

— — Visual system at fusion. Science **135**, 794—795 (1962).

— Chapanis, A.: Counting repeated light flashes as a function of their number, their rate of presentation and retinal location stimulated. J. exp. Psychol. **56**, 385—391 (1958).

Freund, H.-J., Grünewald, G.: Räumliche Summation und Hemmungsvorgänge im receptiven Feldzentrum von Retinaneuronen der Katze. Exp. Brain Res. **8**, 37—52 (1969).

— — Baumgartner, G.: Räumliche Summation im receptiven Feldzentrum von Neuronen des Geniculatum laterale der Katze. Exp. Brain Res. **8**, 53—65 (1969).

— Lauff, D., Grünewald, G.: Binoculare Interaktion im Corpus geniculatum laterale der Katze. Pflügers Arch. ges. Physiol. **297**, 85 (1967).

Frühauf, A.: Critical flicker fusion during the action of different drugs. I. Caffeine and meprobamate (including a full description of the method). Psychopharmacology **21**, 382—389 (1971).

Fry, G. A.: Color sensations produced by intermittent white light and the three component theory of color vision. Amer. J. Psychol. **47**, 464—469 (1935).

— Bartley, S. H.: The effect of steady stimulation of one part of the retina upon the critical flicker frequency in another. J. exp. Psychol. **19**, 351—356 (1936).

Fukada, Y.: Receptive field organization of cat optic nerve fibers with special reference to conduction velocity. Vision Res. **11**, 209—226 (1971).

— Motokawa, K., Norton, A. C., Tasaki, K.: Functional significance of conduction velocity in the transfer of flicker information in the optic nerve of the cat. J. Neurophysiol. **29**, 698—714 (1966).

— Saito, H.-A.: The relationship between response characteristics to flicker stimulation and receptive field organization in the cat's optic nerve fibers. Vision Res. **11**, 227—240 (1971).

Fuortes, M. G. F., Hodgkin, A. L.: Changes in time scale and sensitivity in the ommatidia of Limulus. J. Physiol. (Lond.) **172**, 239—263 (1964).

Furman, G. G.: Comparison of models for subtractive and shunting lateral inhibition in receptor neuron fields. Kybernetik **2**, 257—274 (1965).

Gaedt, Ch.: Die Abhängigkeit der Kontrastaktivierung retinale Neurone von der Frequenz des Reizlichtes. Med. Dissertation, Physiologisches Institut, Berlin 1968.

— Grüsser, O.-J.: The dependence of simultaneous contrast activation of retinal neurons on the temporal frequency of the stimuli (unpubl. work 1966).

— Lunkenheimer, H.-U.: Die Abhängigkeit der Kontrastaktivierung retinaler Neurone von der Belichtungsfrequenz. Pflügers Arch. ges. Physiol. **291**, 87 (1966).

Galifret, Y., Piéron, H.: Etude des fréquences critiques de fusion pour les stimulations chromatiques intermittentes à brillance constante. Année Psychol. **45/46**, 1—15 (1948).

Gastaut, H., Corriol, J. H.: Sur la forme des ondes induites sur le cortex cérébral par les stimulations lumineuses rhythmées. C. R. Soc. Biol. **142**, 351—353 (1948).

Gebhard, J. W., Duffy, M. M., Mowbray, G. H., Byham, C. L.: Visual sensitivity to the rate of electrically produced intermittence. J. opt. Soc. Amer. **46**, 851—860 (1956).

— Mowbray, G. H.: On discriminating in the rate of visual flicker and auditory flutter. Amer. J. Psychol. **72**, 521—529 (1959).

— — Byham, C. L.: Difference-limens for photic intermittence. Quart. J. exp. Psychol. **7**, 49—55 (1955).

Geratewohl, S. J., Taylor, W. F.: The effect of intermittent light on the readability of printed matter under conditions of decreasing contrast. J. exp. Psychol. **46**, 278—282 (1953).

Gibbins, K., Howarth, C. I.: Prediction of the effect of the light-time fraction on the critical flicker frequency: an insight from Fourier analysis. Nature (Lond.) **190**, 330—331 (1961).

— — The effect of intermittent illumination on the visual acuity threshold. Quart. J. Exp. Psychol. **14**, 167—175 (1962).

Glad, A., Magnussen, S.: Darkness enhancement in intermittent light: An experimental demonstration. Vision Res. **12**, 111—115 (1972).

Glees, P.: The termination of optic fibres in the lateral geniculate body of the cat. J. Anat. (Lond.) **75**, 434—440 (1941).

Glötzner, F., Grüsser, O.-J.: Membranpotential und Entladungsfolgen corticaler Zellen, EEG und corticales DC-Potential bei generalisierten Krampfanfällen. Arch. Psychiatr. Nervenkr. **210**, 313—339 (1968).

— — Tweel, L. H. van der: A source for modulated light. Phys. Med. Biol. **3**, 164—173 (1958).

Gon, J. J., Denier van der, Strackee, J.: Gezichtsscherpte-Een fysisch-fysiologische studie. Thesis Univ. Amsterdam. The Netherlands (1959).

Gouras, P.: Duplex function in the grey squirrel's electroretinogram. Nature (Lond.) **203**, 767—768 (1964).

— Identification of cone mechanisms in monkey ganglion cells. J. Physiol. (Lond.) **199**, 533—547 (1968).

— The function of the midget cell system in primate color vision. Vision Res. Suppl. **3**, 397—410 (1971).

— Gunkel, R. D.: The resonant frequencies of rod and cone electroretinograms. Invest. Ophthal. **1**, 122—126 (1962).

— — The frequency response of normal, rod achromat and nyctalope ERGs to sinusoidally monochromatic light stimulation. Docum. Ophthal. **18**, 137—150 (1964).

Govi, G.: L'ottica di Claudeo Tolomeo. G. B. Paravia 171 p. (1885).

Graham, C. H., Granit, R.: Comparative studies on the peripheral and central retina. VI. Inhibition, summation and synchronization of impulses in the retina. Amer. J. Physiol. **98**, 664—673 (1931).

— Kemp, E. H.: Brightness discrimination as a function of the duration of the increment intensity. J. gen. Physiol. **21**, 635—650 (1938).

Graham, N.: Spatial frequency channels in the human visual system: Effects of luminance and pattern drift rate. Vision Res. **12**, 53—68 (1972).

Granit, R.: Interaction between distant areas in the human eye. J. Physiol. (Lond.) **69**, XVII (1930a).

— Comparative studies on the peripheral and central retina. I. Amer. J. Physiol. **94**, 41—50 (1930b).

— Sensory mechanisms of the retina. London: Oxford University Press 1947.

— The organization of the vertebrate retinal elements. Ergebn. Physiol. **46**, 31—70 (1950).

— Receptors and sensory perception, pp. 280—291. New Haven: Yale University Press 1955.

— The visual pathway. Part III of H. Davson. The eye, Vol. 2, 536—763. New York-London: Academic Press 1962.

— Ammon, W. von: Comparative studies on the peripheral and central retina. III. Amer. J. Physiol. **95**, 229—241 (1930).

— Hammond, E. L.: Comparative studies on the peripheral and central retina V. Amer. J. Physiol. **98**, 654—663 (1931).

— Harper, P.: Comparative studies on the peripheral and central retina II. Amer. J. Physiol. **95**, 211—228 (1930).

Green, D. G.: Sinusoidal flicker characteristics of the color sensitive mechanisms of the eye. Vision Res. **9**, 591—601 (1969).

Grind, W. A. van de, Grüsser, O.-J.: On neuronal pulse encoders (in preparation 1973).

— Koenderink, J. J., Bouman, M. A.: Models of the processing of quantum signals by the human peripheral retina. Kybernetik **6**, 213—227 (1970).

GRIND, W. A. VAN DE, KOENDERINK, J. J., HEYDE, G. L. VAN DER, LANDMAN, H. A. A., BOUMAN, M. A.: Adapting coincidence scalers and neural modelling studies of vision. Kybernetik 8, 85—105 (1971a).

— — LANDMAN, H. A. A., BOUMAN, M. A.: The concepts of scaling and refractoriness in psychophysical theories of vision. Kybernetik 8, 105—122 (1971b).

GRÜSSER, O.-J.: Reaktionen einzelner corticaler und retinaler Neurone der Katze auf Flimmerlicht und ihre Beziehungen zur subjektiven Sinnesphysiologie. Med. Diss. Freiburg i. Br. 1956.

— Lichtreaktionen einzelner Neurone des optischen Systems und ihre Beziehungen zur subjektiven Sinnesphysiologie. Klin. Wschr. **35**, 199 (1957).

— Receptorpotentiale einzelner retinaler Zapfen der Katze. Naturwissenschaften **44**, 522 (1957).

— Rezeptorabhängige Potentiale der Katzenretina und ihre Reaktionen auf Flimmerlicht. Pflügers Arch. ges. Physiol. **271**, 511—525 (1960).

— Rezeptorabhängige R-Potentiale der Katzenretina. In: JUNG, R., KORNHUBER, H. (Hrsg.): Neurophysiologie und Psychophysik des visuellen Systems, S. 56—61. Berlin-Göttingen-Heidelberg: Springer 1961.

— Anatomische und physiologische Grundlagen des Binocularsehens. Habilitationsschrift, Berlin (1963).

— Beispiele für eine systemtheoretische Analyse der Netzhautfunktion. Pflügers Arch. ges. Physiol. **289**, R 85 (1966).

— The frequency response of the retinal ganglion cell responses to sinusoidal light stimulation of the receptive field. Paper given at the Symposion on "Theory of temporal factors in vision and visual perception", June, 1966, Rochester, N. Y.

— Die Intensitätsfunktion retinaler Neurone der Katze. Pflügers Arch. ges. Physiol. **307**, R 143 (1969).

— The intensity function of single neurons of the cat's retina measured with sinusoidally modulated testfields of different area and frequency. Unpubl. work (1969).

— A quantitative analysis of spatial summation of excitation and inhibition within the receptive field of retinal ganglion cells of cata. Vision Res. Suppl. **3**, 103—127 (1971).

— Informationstheorie und die Signalverarbeitung in den Sinnesorganen und im Nervensystem. Naturwissenschaften **59**, 436—447 (1972).

— CREUTZFELDT, O.: Untersuchungen mit Flimmerlicht an einzelnen Neuronen des optischen Cortex. X. Internat. Congress of Physiological Sciences Bruessel, abstr. 388 (1956).

— — Eine neurophysiologische Grundlage des Brücke-Bartley-Effektes: Maxima der Impulsfrequenz retinaler und corticaler Neurone bei Flimmerlicht mittlerer Frequenzen. Pflügers Arch. ges. Physiol. **263**, 668—681 (1957).

— GRÜSSER-CORNEHLS, U.: Microelectrode recordings from single units of the cat's central visual system. In part unpublished work (1958—1960).

— — Mikroelektrodenuntersuchungen zur Konvergenz vestibulärer und retinaler Afferenzen an einzelnen Neuronen des optischen Cortex der Katze. Pflügers Arch. ges. Physiol. **270**, 227—238 (1960).

— — Neuronal discharge and evoked potential in the primary visual cortex of cats. V. Internat. Congr. EEG and Clin. Neurophysiol. 1961, Excerpta Med. Internat. Congr. Ser. **37**, abstr. 6 (1961).

— — Neurophysiologische Grundlagen des Binocularsehens. Arch. Psychiat. Nervenkr. **207**, 296—317 (1965).

— — Neurophysiologie des Bewegungssehens. Bewegungsempfindliche und richtungsspezifische Neurone im visuellen System. Ergebn. Physiol. **61**, 178—265 (1969).

— — Neuronal mechanisms of visual movement perception and some psychophysical and behavioral correlations. This handbook, Vol. VII/3A, p. 334—428 (1973).

— — HAMASAKI, D. I.: Responses of neurons in the cat's visual system to moving light-dark patterns. Proc. Int. Union. Physiol. Sciences, Vol. IX. 25. Int. Congress, Munich 1971, Nr. 652.

— GRÜTZNER, A.: Reaktionen einzelner Neurone des optischen Cortex der Katze nach elektrischen Reizserien des Nervus opticus. Arch. Psychiat. Nervenkr. **197**, 405—432 (1958).

GRÜSSER, O.-J., HELLNER, K. A., GRÜSSER-CORNEHLS, U.: Die Informationsübertragung im afferenten visuellen System. Kybernetik **1**, 175—192 (1962).

— LICKER, M., LUNKENHEIMER, H.-U.: The effect of steady illumination of the RF-periphery on the flicker response from the RF-center of cat's retinal ganglion cells. Vision Res. (in preparation (1972)).

— LÜTTGERT, M.: Spatial summation of inhibition in the RF-periphery of ganglion cells of the cat retina. Pflügers Arch. ges. Physiol. (in preparation) 1972.

— — RACKENSPERGER, W.: Dynamic interaction of signals from the RF-center and the RF-periphery of single retinal ganglion cells of cats. Pflügers Arch. ges. Physiol. (in preparation 1972).

— LUNKENHEIMER, U.: Intracellular responses of cat's S-potentials to diffuse sinusoidally modulated flicker stimuli. unpubl. (1966).

— — Responses of single retinal on-center and off-center ganglion cells of the cat to sinusoidal light stimuli of different frequency modulation, area and background area. Unpubl. work, Berlin 1965—1969.

— — LÜTTGERT, M., RACKENSPERGER, W., WUTTKE, W.: Spatial summation in the receptive fields of cat's retinal ganglion cells. II. Summation within the RF-periphery. Proc. of the First European Biophysics Congress 1971, Baden, Austria, ed. E. BRODA, A. LOCKER, H. SPRINGER-LEDERER, Separatum, pp. 263—266. Verlag der Wiener Med. Akad. 1971.

— RABELO, C.: Reaktionen einzelner retinaler Neurone nach Lichtblitzen. I. Einzelblitze und Lichtblitze wechselnder Frequenz. Pflügers Arch. ges. Physiol. **265**, 501—529 (1958).

— — Die Wirkung von Flimmerreizen mit Lichtblitzen an einzelnen corticalen Neuronen. Electroenc. clin. Neurophysiol. **3**, 371—375 (1959).

— REIDEMEISTER, C.: Flimmerlichtuntersuchungen an der Katzenretina. II. Off-Neurone und Besprechung der Ergebnisse. Z. Biol. **111**, 254—270 (1959).

— SAUR, G.: Monoculare und binoculare Lichtreizung einzelner Neurone im Geniculatum laterale der Katze. Pflügers Arch. ges. Physiol. **271**, 595—612 (1960).

— SCHAIBLE, D., VIERKANT-GLATHE, J.: A quantitative analysis of the spatial summation of excitation within the receptive field centers of retinal neurons. Pflügers Arch. ges. Physiol. **319**, 101—121 (1970).

— VESPER, J.: Responses of retinal ganglion cells at and above the critical flicker frequency. Unpubl. work 1970. Vision Res. (in preparation 1973).

GRÜSSER-CORNEHLS, U., GRÜSSER, O.-J.: Mikroelektrodenuntersuchungen am Geniculatum laterale der Katze: Nervenzell- und Axonentladungen nach elektrischer Opticusreizung. Pflügers Arch. ges. Physiol. **271**, 50—63 (1960).

— — Reaktionsmuster der Neurone im zentralen visuellen System von Fischen, Kaninchen und Katzen auf monoculare und binoculare Lichtreize. In: JUNG, R., KORNHUBER, H. (Hrsg.): Neurophysiologie und Psychophysik des visuellen Systems, S. 275—286. Berlin-Göttingen-Heidelberg: 1971.

GRÜTZNER, A., GRÜSSER, O.-J., BAUMGARTNER, G.: Reaktionen einzelner Neurone im optischen Cortex der Katze nach elektrischer Reizung des Nervus opticus. Arch. Psychiat. **197**, 377—404 (1958).

GUILLERY, R. W.: A study of golgi preparations from the dorsal lateral geniculate nucleus of the adult cat. J. comp. Neurol. **128**, 21—50 (1966).

— The laminar distribution of retinal fibers in the dorsal lateral geniculate nucleus of the cat: a new interpretation. J. comp. Neurol. **138**, 339—368 (1970).

HALSTEAD, W. C.: A note on the Bartley-effect in the estimation of equivalent brightness. J. exp. Psychol. **28**, 524—528 (1941).

HANITZSCH, R., LÜTZOW, A. VON: Das Flimmer-ERG der isolierten Warmblüternetzhaut. Albrecht v. Graefes Arch. klin. exp. Ophthal. **173**, 217—224 (1967).

HARMON, L. D.: Studies with artificial neurons, 1: Properties and functions of an artificial neuron. Kybernetik **1**, 89—101 (1961).

— LEVINSON, J., BERGEIJK, W. A. VAN: Analog models of neural mechanism. IRE Trans. **8**, 107—112 (1962).

HARTER, M. R., WHITE, C. T.: Perceived number and evoked cortical potentials. Science **156**, 406—408 (1967).

HARTLINE, H. K.: Impulses in single optic nerve fibres of the vertebrate retina. Amer. J. Physiol. **113**, 59 (1935).
— The response of single optic nerve fibres of the vertebrate eye to illumination of the retina. Amer. J. Physiol. **121**, 400—415 (1938).
— The receptive fields of the optic nerve fibres. Amer. J. Physiol. **130**, 690—699 (1940a).
— The effects of spatial summation in the retina on the excitation of the fibres of the optic nerve. Amer. J. Physiol. **130**, 700—711 (1940b).
— RATLIFF, F.: Inhibitory interaction in the retina of Limulus. Handbook of Sensory Physiology, Vol. VII/Part 1B. Berlin-Heidelberg-New York: Springer 1972.
HARVEY, L. O.: Flicker sensitivity and apparent brig htness as a function of surround luminance. J. Opt. Soc. Amer. **60**, 860—864 (1970a).
— Critical flicker frequency as a function of viewing distance, stimulus size and luminance. Vision Res. **10**, 55—63 (1970b).
HAYHOW, W. R.: The cytoarchitecture of the lateral geniculate body in the cat in relation to the distribution of crossed and uncrossed optic fibers. J. comp. Neurol. **110**, 1—64 (1958).
HECHT, S.: Rods, cones and the chemical basis of vision. Physiol. Rev. **17**, 239—290 (1937).
— SHLAER, S.: Intermittent stimulation by light. V. The relation between intensity and critical frequency for different parts of the spectrum. J. gen. Physiol. **19**, 965—979 (1936).
— — VERRIJP, C. D.: Intermittent stimulation by light II. The measurement of critical fusion frequency for the human eye. J. gen. Physiol. **17**, 237—249 (1933).
— SMITH, E. L.: Intermittent stimulation by light VI. Area and the relation between critical frequency and intensity. J. gen. Physiol. **19**, 979—991 (1936).
— VERRIJP, C. D.: Intermittent stimulation by light III. The relation between intensity and CFF for different retinal locations. J. gen. Physiol. **17**, 251—265 (1933a).
— — Intermittent stimulation by light. IV. A theoretical interpretation of the quantitative data of flicker. J. gen. Physiol. **17**, 266—282 (1933b).
HECK, J.: The flicker electroretinogram of the human eye. Acta physiol. scand. **39**, 158—166 (1957).
HELMHOLTZ, H. VON: Handbuch der physiologischen Optik. 2nd Ed., 1008 p. Leipzig: L. Voss 1896.
HENKES, H. E., TWEEL, L. H. VAN DER: Flicker. Proceedings of the symposion on the physiology of flicker, September 1963. Docum. ophthal. (Den Haag) **18**, 1—540 (1964).
HILZ, R.: Der Einfluß von Leuchtdichte, Beobachtungsabstand, Darbietungszeit und anderen Parametern auf die Erkennbarkeit von Rechteckgittern. Dissertation, Technische Hochschule München (1965).
HORST, G. J. C. VAN DER: Chromatic flicker. J. opt. Soc. Amer. **59**, 1213—1217 (1969a).
— Fourier analysis and color discrimination. J. opt. Soc. Amer. **59**, 1670—1676 (1969b).
— BOUMAN, M. A.: Spatio temporal chromaticity discrimination. J. opt. Soc. Amer. **59**, 1482—1488 (1969).
— MUIS, W.: Hue shift and brightness enhancement of flickering light. Vision Res. **9**, 953—963 (1969).
HRACHOVINA, V., SCHMIDT, B.: Electroretinogram fusion frequency and retinal illumination of some vertebrate eyes. 6. ISCERG Symposium, pp. 279—282. Leipzig: VEB G. Thieme 1968.
HUBEL, D. H., WIESEL, T. N.: Receptive fields of single neurones in the cat's striate cortex. J. Physiol. (Lond.) **148**, 574—591 (1959).
— — Receptive fields, binocular interaction and functional architecture in the cat's visual cortex. J. Physiol. (Lond.) **160**, 106—154 (1962).
— — Receptive fields and functional architecture in two non-striate visual areas (18 and 19) of the cat. J. Neurophysiol. **28**, 229—289 (1965).
HUGHES, G. W., MAFFEI, L.: Retinal ganglion cell response to sinusoidal light stimulation. J. Neurophysiol. **29**, 333—352 (1966).
HYLKEMA, B. S.: Fusion frequency with intermittent light under various circumstances. Acta ophthal. (Kbh.) **20**, 159—180 (1942a).
— Examination of the visual field by determining the fusion frequency. Acta opthal. (Kbh.) **20**, 181—193 (1942b).
— De versmeltingsfrequentie bij intermitteerend licht (Thesis) Univ. of Amsterdam, the Netherlands. Amsterdam, van Gorcum and Comp. 1942c.

IKEDA, M., BOYNTON, R. M.: Negative flashes, positive flashes and flicker examined by increment threshold technique. J. opt. Soc. Amer. **55**, 560—566 (1965).
— FUJII, T.: Diphasic nature of the visual response as inferred from the summation index of n flashes. J. opt. Soc. Amer. **56**, 1129—1132 (1966).
IRELAND, F. H.: A comparison of critical flicker frequencies under conditions of monocular and binocular stimulation. J. exp. Psychol. **40**, 282—286 (1950).
IVES, H. E.: Critical frequency relations in scotopic vision. J. opt. Soc. Amer. **6**, 254—268 (1922a).
— A theory of intermittent vision. J. opt. Soc. Amer. **6**, 343—361 (1922b).
JAHN, T. L.: Brightness enhancement in flickering light. Psychol. Rev. **51**, 76—84 (1944).
JOHANNESMA, P I. M.: Diffusion models for the stochastic activity of neurons. In: CAIANIELLO, E. R. (Ed.): Neural Networks, pp. 116—144. Berlin-Heidelberg-New York: Springer 1968.
JONES, R. W., LI, C. C., MEYER, A. U., PINTER, R. B.: Pulse modulation in physiological systems, phenomenological aspects. IRE Trans. BME-8, 59—67 (1961).
JUNG, R.: Coordination of specific and nonspecific afferent impulses at single neurons of the visual cortex. In: JASPER, H. H., PROCTOR, L. D., KNIGHTON, R. S., NOSHAY, W. S., COSTELLO R. T. (Eds.): Reticular Formation of the Brain, pp. 423—434. Boston-Toronto: Little, Brown & Co. 1958.
— Mikrophysiologie des optischen Cortex: Koordination der Neuronenentladungen nach optischen, vestibulären und unspezifischen Afferenzen und ihre Bedeutung für die Sinnesphysiologie. Med. Jap. **5**, 693—698 (1959).
— Korrelationen von Neuronentätigkeit und Sehen. In: JUNG, R., KORNHUBER, H. H. (Hrsg.): Neurophysiologie und Psychophysik des visuellen Systems, S. 410—435. Berlin-Göttingen-Heidelberg: Springer 1961.
— BAUMGARTEN, R. VON, BAUMGARTNER, G.: Mikroableitungen von einzelnen Nervenzellen im optischen Cortex der Katze: Die lichtaktivierten B-Neurone. Arch. Psychiat. Nervenkr. **189**, 521—539 (1952).
— BAUMGARTNER, G.: Hemmungsmechanismen und bremsende Stabilisierung an einzelnen Neuronen des optischen Cortex. Ein Beitrag zur Koordination corticaler Erregungsvorgänge. Pflügers Arch. ges. Physiol. **261**, 434—456 (1955).
— CREUTZFELDT, O., GRÜSSER, O.-J.: Die Mikrophysiologie kortikaler Neurone und ihre Bedeutung für die Sinnes- und Hirnfunktionen. Dtsch. med. Wschr. **82**, 1050—1059 (1957).
— — — The microphysiology of cortical neurones. Its significance for sensory and cerebral functions. German Med. Monthly **3**, 269—276 (1958).
KANEKO, A.: Physiological and morphological identification of horizontal, bipolar and amacrine cells in goldfish retina. J. Physiol. (Lond.) **207**, 623—633 (1970).
— Physiological studies of single cells and their morphological identification. Vision Res. Suppl. **3**, 17—26 (1971).
— HASHIMOTO, H.: Recording site of the single cone response determined by an electrode marking technique. Vision Res. **7**, 847—851 (1967).
— — Electrophysiological study of single neurons in the inner nuclear layer of the carp retina. Vision Res. **9**, 37—55 (1969).
KAPPAUF, W. E.: Flicker discrimination in the cat. Psychol. Bull. **33**, 597—598 (1936).
KARRER, R.: Visual beat phenomena as an index to the temporal characteristics of perception. J. exp. Psychol. **75**, 372—378 (1967).
— Visual beats: Phenomenology and preliminary data as a function of age. Psychon. Sci. **11**, 269—270 (1968).
— Visual beats: differential brightness of the stimuli and estimation of brightness. Vision Res. **9**, 429—433 (1969).
— CLAUSEN, J.: Visual beats: preliminary observations of perceived rate as a function of retinal locus stimulated. Percept. Psychophys. **5**, 163—165 (1969).
KATO, H., YAMAMOTO, M., NAKAHAMA, H.: Intracellular recordings from the lateral geniculate neurons of cats. Jap. J. Physiol. **21**, 307—323 (1971).
KEESEY, U. T.: Variables determining flicker sensitivity in small fields. J. opt. Soc. Amer. **60**, 390—398 (1970).
— Comparison of human visual cortical potentials evoked by stabilized and unstabilized targets. Vision Res. **11**, 657—670 (1971).

Kelly, D. H.: Effects of sharp edges in a flickering field. J. opt. Soc. Amer. **49**, 730—732 (1959).
— J_0-stimulus patterns for visual research. J. opt. Soc. Amer. **50**, 1115 (1960).
— Visual signal generator. Rev. Sci. Instr. **32**, 50—55 (1961a).
— Visual responses to time-dependent stimuli. I. Amplitude sensitivity measurements. J. opt. Soc. Amer. **51**, 422—429 (1961b).
— Visual responses to time-dependent stimuli. II. Single channel model of the photopic visual system. J. opt. Soc. Amer. **51**, 747—754 (1961c).
— Flicker fusion and harmonic analysis. J. opt. Soc. Amer. **51**, 917—919 (1961d).
— Information capacity of a single retinal channel. IRE Trans. on I.T. 221—226 (1962a).
— Visual responses to time-dependent stimuli. III. Individual variations. J. opt. Soc. Amer. **52**, 89—95 (1962b).
— Visual responses to time dependent stimuli. IV. Effects of chromatic adaptation. J. opt. Soc. Amer. **52**, 940—947 (1962c).
— Sine waves and flicker fusion. Docum. ophthal. (Den Haag) **18**, 16—35 (1964).
— Frequency doubling in visual responses. J. opt. Soc. Amer. **56**, 1628—1633 (1966).
— Studies of visual perception. Contract DAAKO2-67-C-0146 (1967).
— Flickering patterns and lateral inhibition. J. opt. Soc. Amer. **59**, 1361—1370 (1969a).
— Diffusion model of linear flicker responses. J. opt. Soc. Amer. **59**, 1665—1670 (1969b).
— Effects of sharp edges on the visibility of sinusoidal gratings. J. opt. Soc. Amer. **60**, 98—103 (1970).
— Theory of flicker and transient responses. I. Uniform fields. J. opt. Soc. Amer. **61**, 537—546 (1971a).
— Theory of flicker and transient responses. II. Counterphase gratings. J. opt. Soc. Amer. **61**, 632—640 (1971b).
— Adaptation effects on spatio-temporal sine-wave thresholds. Vision Res. **12**, 89—101 (1972a).
— Flicker. Handbook of Sensory Physiol. Vol. VII/4, pp. 273—302. Berlin-Heidelberg-New York: Springer 1972b.
Knight, B. W.: Frequency response for sampling integrator and for voltage to frequency converter. In: Systems analysis approach to neurophysiological problems. Conf. Proc. Lab. Neurophysiol. Univ. Minnesota, 61—72 (1969).
— Toyoda, J.-I., Dodge, Jr., F. A.: A quantitative description of the dynamics of excitation and inhibition in the eye of Limulus. J. gen. Physiol. **56**, 421—437 (1970).
Knoll, M., Welpe, E.: Vergleich von Anregungsbedingungen, Formklassen und Bewegungsarten optischer und elektrischer Phosphene. Elektromedizin **13**, 128—134 (1968).
Koenderink, J. J.: The concept of the transfer function of an integral pulse frequency modulator. In preparation, 1972.
— Grind, W. A. van de, Bouman, M. A.: Models of retinal signal processing at high luminances. Kybernetik **6**, 227—237 (1970).
— — — Foveal information processing at photopic luminances. Kybernetik **8**, 128—144 (1971).
— — — Opponent color coding: a mechanistic model and a new metric for color space. Kybernetik **10**, 78—98 (1972).
Kohn, H., Salisbury, I.: Electroencephalographic indications of brightness enhancement. Vision Res. **7**, 461—468 (1967).
Korn, A., Scheich, H.: Übertragungseigenschaften der Katzenretina. Kybernetik **8**, 179—188 (1971).
Kries, J. von: Über die Wahrnehmung des Flimmerns durch normale und durch total farbblinde Personen. Z. Sinnesphysiol. **32**, 113—117 (1903).
— Die Gesichtsempfindungen. In: Nagel, W. (Hrsg.): Handbuch der Physiologie des Menschen, Vol. III, S. 105—282. Braunschweig: F. Vieweg u. Sohn 1905.
Kuffler, S. W.: Neurons in the retina: organization, inhibition and excitation problems. Cold Spr. Harb. Symp. quant. Biol. **17**, 281—292 (1952).
— Discharge patterns and functional organization of mammalian retina. J. Neurophysiol. **16**, 37—68 (1953).
Kugelmass, S., Landis, C.: Relation of area and luminance to the threshold for critical flicker fusion. Amer. J. Physiol. **68**, 1—19 (1955).
Kuhnt, U., Creutzfeldt, O. D.: Decreased post-synaptic inhibition in the visual cortex during flicker stimulation. Electroenceph. clin. Neurophysiol. **30**, 79—82 (1971).

KUIPER, J. A., LEUTSCHER-HAZELHOFF, J. T.: Linear and nonlinear responses from the compound eye Calliphora erythrocephala. Cold Spr. Harb. Symp. quant. Biol. **30**, 418—428 (1965).

KULIKOWSKI, J. J.: Some stimulus parameters affecting spatial and temporal resolution of human vision. Vision Res. **11**, 83—93 (1971a).

— Effect of eye movements on the contrast sensitivity of spatio-temporal patterns. Vision Res. **11**, 261—273 (1971b).

— CAMPBELL, F. W., ROBSON, J. G.: Spatial and temporal frequency characteristics of human photopic vision. Proc. of the 2nd Int. Biophys. Congress Vienna 1966.

LANDIS, C.: An annotated bibliography of flicker fusion phenomena covering the period 1740—1952. Armed Forces National Research Council, Vision Committee Secretariat, 3433, Mason Hall, University of Michigan, 130 p. Ann. Arbor Michigan (1953).

— Determinants of the critical flicker-fusion threshold. Physiol. Rev. **34**, 259—286 (1954).

— DILLON, D., LEOPOLD, J., RUTSCHMANN, J.: Changes in the level of blood sugar and sensory and motor performance brought about by insulin coma therapy. J. Psychol. **45**, 275—285 (1958).

— HAMWI, V.: The effect of certain physiological determinants on the flicker fusion threshold. J. appl. Physiol. **6**, 566—572 (1954).

— — Critical flicker frequency, age and intelligence. Amer. J. Psychol. **69**, 459—461 (1956).

— ZUBIN, J.: The effect of thonzylamine hydrochloride and phenobarbital on certain psychological functions. J. Lab. clin. Med. **38**, 873—880 (1951).

LANGE, H. DE: Experiments on flicker and some calculations on an electrical analogue of the foveal systems. Physica **18**, 935—950 (1952).

— Een onderzoek van het flikkerverschijnsel en een mogelijke verklaring van een naar voren gekomen resonantie-effekt T. Ned. Radio Gen. **18**, 1—31 (1953).

— Relationship between CFF and a set of low-frequency characteristics of the eye. J. opt. Soc. Amer. **44**, 380—389 (1954).

— Attenuation characteristics and phase-shift characteristics of the human fovea-cortex systems in relation to flicker-fusion phenomena. Doctoral dissertation, Techn. Univ. Delft. The Netherlands (1957).

— Research into the dynamic nature of the human fovea-cortex systems with intermittent and modulated light. I. Attenuation characteristics with white and colored light. J. opt. Soc. Amer. **48**, 777—784 (1958a).

— Research into the dynamic nature of the human fovea-cortex systems with intermittent and modulated light. II. Phase shift in brightness and delay in color perception. J. opt. Soc. Amer. **48**, 784—789 (1958b).

— Eye's response at flicker fusion to square-wave modulation of a test field surrounded by a large steady field of equal mean luminance. J. opt. Soc. Amer. **51**, 415—421 (1961).

LASAREFF, P.: Theorie der Lichtreizung der Netzhaut beim Dunkelsehen. Pflügers Arch. ges. Physiol. **154**, 459—469 (1913).

— Zur Theorie der Adaptation der Netzhaut beim Dämmerungssehen. Pflügers Arch. ges. Physiol. **155**, 310—317 (1914).

— Untersuchungen über die Ionentheorie der Reizung. IV. Die Theorie der Erscheinungen des Flimmerns beim Dunkelsehen. Pflügers Arch. ges. Physiol. **196**, 177—184 (1922).

LEJEUNE, A.: L'optique de Claude Ptolémée dans la version latinée d'après l'arabe de l'émir Eugène de Sicile. Publ. Univ. de Louvain, Louvain 1956, 358 p.

LENNOX-BUCHTHAL, M. A.: Single unit studies and the mechanism of flicker fusion. Docum. ophthal. (Den Haag) **18**, 245—258 (1964).

LEVICK, W. R., OYSTER, C. W., DAVIS, D. L.: Evidence that McIlwain's periphery effect is not a stray light artifact. J. Neurophysiol. **28**, 555—559 (1965).

LEVINSON, J.: Fusion of complex flicker II. Science **131**, 1438—1440 (1960).

— Nonlinear and spatial effects in the perception of flicker. Docum. ophthal. (Den Haag) **18**, 36—55 (1964).

— One-stage model for visual temporal integration. J. opt. Soc. Amer. **56**, 95—97 (1966).

— Flicker fusion phenomena. Science **160**, 21—28 (1968).

— HARMON, L. D.: Studies with artificial neurons. III. Mechanism of flicker-fusion. Kybernetik **1**, 107—117 (1961).

Lichtenstein, M., White, C. T., Siegfried, J. B.: Apparent rate of flicker at various retinal loci and number of perceived flashes per unit time: a paradox. Percept. Motor Skills **17**, 523—536 (1963).

Lipetz, L. E.: The relation of physiological and psychological aspects of sensory intensity. This handbook Vol. **1**, 191—225 (1971).

Lloyd, V. V.: A comparison of critical fusion frequencies for different areas in the fovea and periphery. Amer. J. Psychol. **65**, 346—357 (1951).

— Landis, C.: Role of the light-dark ratio as a determinant of the flicker-fusion threshold. J. opt. Soc. **50**, 332—336 (1960).

Lohmann, H.: Über die Sichtbarkeitsgrenze und die optische Unterscheidbarkeit sinusförmiger Wechselströme. Z. Sinnesphysiol. **69**, 27—40 (1940).

Lopez da Silva, F. H.: Dynamic characteristics of visual evoked potentials. Inst. Med. Physics. Utrecht, The Netherlands, Report (1970).

Lunkenheimer, H.-U.: Untersuchungen über den Einfluß der Umfeldbelichtung auf die Antwort retinaler Neurone der Katze auf sinusförmige Belichtung der RF-Peripherie (unpubl. manuscript, Berlin 1968).

— Grüsser, O.-J.: Nicht-lineare Übertragungseigenschaften retinaler Neurone der Katze. Pflügers Arch. ges. Physiol. **291**, 88 (1966).

— Rackensperger, W., Schwanz, E., Grüsser, O.-J.: Reaktionen retinaler Neurone der Katze auf Sinuslicht: der Einfluß von Modulationsgrad, Beleuchtungsstärke und Reizfeldgröße. Pflügers Arch. ges. Physiol. **289**, 1282 (1966).

Luria, S. M., Sperling, H. G.: Phase relations in flicker fusion. J. opt. Soc. Amer. **52**, 1051—1057 (1962).

Lythgoe, R. J., Tansley, K.: The relation of the critical frequency of flicker to the adaptation of the eye. Proc. roy. Soc. London **105** B, 60—92 (1929).

MacNichol, E. J., Svaetichin, G.: Electric responses from the isolated retinas of fishes. Amer. J. Ophthal. **46**, 26—46 (1958).

Maffei, L.: Inhibitory and facilitatory spatial interactions in retinal receptive fields. Vision Res. **8**, 1187—1194 (1968).

— Spatial and temporal averages in retinal channels. J. Neurophysiol. **31**, 283—287 (1968).

— Cervetto, L.: Dynamical interactions in retinal receptive fields. Vision Res. **8**, 1299—1303 (1968).

— — Fiorentini, A.: Transfer characteristics of excitation and inhibition in cat retinal ganglion cells. J. Neurophysiol. **33**, 276—284 (1970).

— — Rizzolatti, G.: Transfer properties of the lateral geniculate body. J. Neurophysiol. **30**, 333-340 (1967).

— Fiorentini, A.: Retinogeniculate convergence and analysis of contrast. J. Neurophysiol. **35**, 65—72 (1972).

— Moruzzi, G., Rizzolatti, G.: Geniculate unit responses to sine-wave photic stimulation during wakefulness and sleep. Science **149**, 563—564 (1965a).

— — — Influence of sleep and wakefulness on the response of lateral geniculate units to sinewave photic stimulation. Arch. ital. Biol. **103**, 596—608 (1965b).

— Poppele, R. E.: Transient and steady state electroretinal responses. Vision Res. **8**, 229—246 (1968).

— Rizzolatti, G.: Transfer properties of the lateral geniculate body. J. Neurophysiol. **30**, 333—340 (1967).

Marbe, K.: Tatsachen und Theorie des Talbot'schen Gesetzes. Pflügers Arch. ges. Physiol. **97**, 335—393 (1903).

Marchiafava, P. L.: Binocular reciprocal interaction upon optic fiber endings in the lateral geniculate nucleus of the cat. Brain Res. **2**, 188—192 (1966).

Marimont, R. B.: Numerical studies of the Fuortes-Hodgkin Limulus model. J. Physiol. (Lond.) **179**, 489—497 (1965).

Marks, L. E.: Apparent depth of modulation as a function of frequency and amplitude of temporal modulations of luminance. J. opt. Soc. Amer. **60**, 970—977 (1970).

Matin, L.: Fourier treatment of some experiments in visual flicker. Science **136**, 983—985 (1962).

MATIN, L.: Critical duration, the differential luminance threshold, critical flicker frequency and visual adaptation: A theoretical treatment. J. opt. Soc. Amer. **58**, 404—415 (1968).
McDONALD, H.S.: J. opt. Soc. Amer. **50**, 1128 (1960).
McILWAIN, J.T.: Receptive fields of optic tract axons and lateral geniculate cells: peripheral extent and barbiturate sensitivity. J. Neurophysiol. **27**, 1154—1173 (1964).
— Some evidence concerning the physiological basis of the periphery effect in the cat's retina. Exp. Brain Res. **1**, 265—271 (1966).
— CREUTZFELDT, O.D.: Microelectrode study of synaptic excitation and inhibition in the lateral geniculate nucleus of the cat. J. Neurophysiol. **30**, 1—21 (1967).
MENEGHINI, K.A., HAMASAKI, D.I.: The electroretinogram of the Iguana and Tokay Gecko. Vision Res. **7**, 243—251 (1967).
MEYER, J.J.: Examination of subjects with cranio cerebral trauma using a visual perception test: the de Lange curve. Schweiz. Arch. Neurol. Neurochir. Psychiat. **108**, 213—221 (1971).
MEYER-SCHWICKERATH, G., MAGUN, R.: Über selektive elektrische Erregbarkeit verschiedener Netzhautanteile. Arch. Ophthal. **151**, 693—700 (1951).
MILLER, R.F., DOWLING, J.E.: Intracellular responses of the Müller (Glial) cells of mudpuppy retina: their relation to *b*-wave of the electroretinogram. J. Neurophysiol. **33**, 323—341 (1970).
MINKOWSKI, M.: Experimentelle Untersuchungen über die Beziehungen der Großhirnrinde und der Netzhaut zu den primären optischen Zentren, besonders zum Corpus geniculatum externum. Arb. Hirnanat. Inst. Zürich **7**, 255—362 (1913).
— Über den Verlauf, die Endigung und die zentrale Repräsentation von gekreuzten und ungekreuzten Sehnervenfasern bei einigen Säugetieren und beim Menschen. Schweiz. Arch. Neurol. Psychiat. **6**, 201—252 (1920).
MONJE, M.: Über die regionale Verteilung der Empfindlichkeit in der Netzhaut bei Untersuchung mit intermittierenden Reizen. Pflügers Arch. ges. Physiol. **255**, 499—507 (1952).
MOTOKAWA, K.: Visual function and the electrical excitability of the retina. Tohoku J. exp. Med. **51**, 145—153 (1949).
— Physiology of color and pattern vision, 283 p. Tokyo: Igagu Shoin, Ltd. 1970.
— EBE, M.: Selective stimulation of color receptors with alternating currents. Science **116**, 92—94 (1952).
— — Retinal colour processes caused by intermittent white light. Nature (Lond.) **170**, 79—80 (1952).
— IWAMA, K.: Resonance in electrical stimulation of the eye. Tohoku J. exp. Med. **53**, 201—206 (1950).
— OIKAWA, T., TASAKI, K.: Receptor potential of vertebrate retina. J. Neurophysiol. **20**, 186—199 (1957).
— SUZUKI, E., OOBA, Y.: Retinal responses to intermittent light of subfusional frequencies. Tohoku J. exp. Med. **64**, 161—168 (1956).
MOWBRAY, G.H., GEBHARD, J.W.: Differential sensitivity of the eye to intermittent white light. Science **121**, 173—175 (1955).
— — Differential sensitivity of peripheral retina to intermittent white light. Science **132**, 672—674 (1960).
MUCHER, H., WENDT, H.W.: Gruppenversuch zur Bestimmung der kritischen Verschmelzungsfrequenz beim binokularen Sehen: Änderungen unter Koffein und nach normaler Tagesarbeit. Arch. exp. Path. Pharm. **214**, 29—37 (1951).
MUNDIE, J.R.: Neural calculus. In: PROCTOR, L.D. (Ed.): Biocybernetics of the nervous system, pp. 325—356. London: J. & A. Churchill Ltd. 1969.
NACHMIAS, J.: Brightness and visual acuity with intermittent illumination. J. opt. Soc. Amer. **48**, 726—730 (1958).
— Brightness and acuity with intermittent illumination. J. opt. Soc. Amer. **51**, 805 (1961).
NAKA, K.I., RUSHTON, W.A.H.: *S*-potentials from luminosity units in the retina of fish (cyprinidae). J. Physiol. (Lond.) **185**, 587—599 (1966).
NAQUET, R., KILLAM, K.F., RHODES, J.M.: Flicker stimulation with chimpanzees. Life Sci. **6**, 1575—1578 (1967).
NELSON, TH.M., BARTLEY, S.H., HARPER, E.S.: CFF for short trains of photic stimulation having various temporal distributions and separations. J. Psychol. **58**, 333—341 (1964).

NES, F. L. VAN: Enhanced visibility by regular motion of retinal images. Amer. J. Psychol. 81, 367—374 (1968a).
— Experimental studies in spatio-temporal contrast transfer by the human eye. Doct. Thesis Univ. Utrecht, The Netherlands, 123 p. (1968b).
— KOENDERINK, J. J., NAS, H., BOUMAN, M. A.: Spatio-temporal modulation transfer in the human eye. J. opt. Soc. Amer. 57, 1082—1088 (1967).
NILSSON, T. H., NELSON, T. M.: Hue shifts produced by intermittent stimulation. Vision Res. 11, 697—712 (1971).
OGAWA, T., BISHOP, P. O., LEVICK, W. R.: Temporal characteristics of responses to photic stimulation by single ganglion cells in the unopened eye of the cat. J. Neurophysiol. 29, 1—30 (1966).
O'LEARY, I.: A structural analysis of the lateral geniculate nucleus of the cat. J. comp. Neurol. 73, 405—430 (1940).
PANTLE, A.: Flicker adaptation. I. Effect on visual sensitivity to temporal fluctuations of light intensity. Vision Res. 11, 943—952 (1971).
— Flicker adaptation. II. Effect on the apparent brightness of intermittent lights. Vision Res. 12, 705—715 (1972).
PAUTLER, E. L.: Responses of the isolated mammalian retina to intermittent and steady photic stimulation. Vision Res. 4, 493—498 (1964).
PECCI-SAAVEDRA, J., VACCAREZZA, O. L., READER, T. A., PASQUALINI, E.: Ultrastructural and electrophysiological aspects of denervated synapses in the lateral geniculate nucleus. Vision Res. Suppl. 3, 229—238 (1971).
PFLÜGER, E. F. W.: Untersuchungen aus dem physiologischen Laboratorium zu Bonn, S. 170—171. Berlin: Hirschwald 1865.
PIÉRON, H.: L'influence de l'intensité lumineuse sur la persistance retinienne apparente. Arch. néerl. Physiol. 7, 199—212 (1922).
— Influence du rapport des phases sur la durée d'interruption d'une stimulation lumineuse périodique à la limite du papillotement. C.R.Soc. Biol. 99, 398—400 (1928).
— L'influence de la surface rétinienne en jeu dans une excitation lumineuse intermittente sur la valeur des fréquences critiques de papillotement. C.R. Soc. Biol. 118, 25—28 (1935).
— La vision en lumière intermittente. Monogr. Franç. Psychol. 8, 91 p. (1961).
— Neurophysiological mechanisms of critical flicker frequency and harmonic phenomena. J. opt. Soc. Amer. 52, 475 (1962).
PINTER, R. B.: Sinusoidal and delta function responses of visual cells of the Limulus eye. J. gen. Physiol. 49, 565—593 (1966).
PLATEAU, J.: Über einige Eigenschaften der vom Licht auf das Gesichtsorgan hervorgebrachten Eindrücke. Poggendorf Ann. Physik. Chem. 20, 304—332 (1830).
— Essai d'une théorie générale comprenant l'ensemble des apparences visuelles etc. Mém. de l'Acad. des Sci. et bell.-let., Bruxelles 8, 1—68 (1834).
— Betrachtungen über ein von Hrn. TALBOT vorgeschlagenes photometrisches Princip. Poggendorf Ann. Physik. Chemie 35, 457—468 (1835).
POKORNY, J., SMITH, V. C.: Luminosity and CFF in deuteranopes and protanopes. J. opt. Soc. Amer. 62, 111—117 (1972).
POLYAK, S. L.: The retina. Chicago: Chicago Univ. Press 1941.
PORTER, T.: Contribution to the study of flicker. I. Proc. roy. Soc. London 70 A, 313—329 (1902).
— Contribution to the study of flicker. II. Proc. roy. Soc. London 86 A, 495—513 (1912).
PREVOST, B.: Memories de la Societé de Physique et d'Histoire naturelle de Geneva 3, 121 (1826).
PURKINJE, J. E.: Beiträge zur Kenntnis des Sehens in subjektiver Hinsicht. Prag 1819.
— Beobachtungen und Versuche zur Physiologie der Sinne. Vol. 1, Prag 1823.
RABELO, C., GRÜSSER, O.-J.: Die Abhängigkeit der subjektiven Helligkeit intermittierender Lichtreize von der Flimmerfrequenz (Brücke-Effekt, 'brightness-enhancement'): Untersuchungen bei verschiedener Leuchtdichte und Feldgröße. Psychol. Forsch. 26, 299—312 (1961).

RACKENSPERGER, W.: Two types of LGN-cells in the cat. Their response to stimulation of the receptive field by sinusoidal stimuli of different frequency and area. Unpubl. manuscript (1970).

— GRÜSSER, O.-J.: Sinuslichtreizung der rezeptiven Felder einzelner Retinaneurone. Experientia (Basel) **22**, 192 (1966).

— REITER, H., WUTTKE, W., SNIGULA, F.: Die Reaktion einzelner Retinaneurone auf sinusförmige Leuchtdichteänderung. Pflügers Arch. ges. Physiol. **283**, R 50 (1965).

RANDOLPH, D. I.: Brightness enhancement as a function of frequency, intensity and light-dark ratio. J. opt. Soc. Amer. **54**, 577 (1964).

RATLIFF, F., KNIGHT, B. W., GRAHAM, N.: On tuning and amplification by lateral inhibition. Physiology **62**, 733—740 (1969).

— — TOYODA, J., HARTLINE, H. K.: Enhancement of flicker by lateral inhibition. Science **158**, 292—293 (1967).

REGAN, D.: Some characteristics of average steady-state and transient responses evoked by modulated light. Electroenceph. clin. Neurophysiol. **20**, 238—248 (1966).

— TYLER, C. W.: Wavelength modulated light generator. Vision Res. **11**, 43—56 (1971).

REGAN, S.: Some dynamic features of colour vision. Vision Res. **11**, 1307—1324 (1971).

REICHARDT, W.: Die Lichtreaktion von Phycomyces. Kybernetik **1**, 6—21 (1961).

REIDEMEISTER, C., GRÜSSER, O.-J.: Flimmerlichtuntersuchungen an der Katzenretina. I. On-Neurone und on-off-Neurone. Z. Biol. **11**, 241—253 (1959).

REMOLE, A.: Luminance thresholds for subjective patterns in a flickering field: effect of wavelength. J. opt. Soc. Amer. **61**, 9, 1164—1168 (1971).

REUTER, J. H.: A comparison of flash evoked ERG's and ERG's evoked with sinusoidally modulated light stimuli in a number of rodents. Pflügers Arch. ges. Physiol. **331**, 95—102 (1972).

REY, P., REY, J.-P.: Effect of an intermittent light stimulation on the critical fusion frequency. Ergonomics 8, 173—180 (1965).

RIPPS, H., KAPLAN, I. T., SIEGEL, I. M.: Effect of contrast on CFF and apparent brightness. J. opt. Soc. Amer. **51**, 870—873 (1961).

ROBINSON, D. N.: Critical flicker-fusion of solid and annular stimuli. Science **167**, 207—208 (1970).

ROBSON, J. G.: Spatial and temporal contrast-sensitivity functions of the visual system. J. opt. Soc. Amer. **56**, 1141—1142 (1966).

RODIECK, R. W.: Quantitative analysis of cat retinal ganglion cell response to visual stimuli. Vision Res. **5**, 583—601 (1965).

— STONE, J.: Analysis of receptive fields of cat retinal ganglion cells. J. Neurophysiol. **28**, 833—849 (1965).

ROEHRIG, W. C.: The influence of area on the critical flicker-fusion threshold. J. Psychol. **47**, 317—330 (1959a).

— The influence of the portion of the retina stimulated on the flicker-fusion threshold. J. Psychol. **48**, 57—63 (1959b).

ROSS, R. T.: A comparison of the regional gradients of fusion frequency and visual acuity. Psychol. Monogr. **47**, 306—310 (1936).

ROUFS, J. A. J.: Dynamic properties of vision. I. Experimental relationship between flicker and flash threshold. Vision Res. **12**, 261—278 (1972).

— Dynamic properties of vision. II. Theoretical relationship between flicker and flash thresholds. Vision Res. **12**, 279—292 (1972).

RUBINSTEIN, B., THERMAN, P. O.: The influence of hyperventilation on the fusion frequency of intermittent visual stimuli. Scand. Arch. Physiol. **72**, 26—34 (1935).

RUSHTON, W. A. H.: The structure responsible for action potential spikes in the cat's retina. Nature (Lond.) **164**, 743—744 (1949).

— The Ferrier lecture 1962. Visual adaptation. Proc. roy. Soc. (Lond.) B **162**, 20—46 (1965).

SAITO, H.-A., SHIMHARA, T., FUKADA, Y.: Four types of responses to light and dark spot stimuli in the cat optic nerve. Tohoku J. exp. Med. **102**, 127—133 (1970).

— — — Phasic and tonic responses in the cat optic nerve fibers — stimulus-response relations. Tohoku J. exp. Med. **104**, 313—323 (1971).

Sakmann, B., Creutzfeldt, O.-D.: Scotopic and mesopic light adaptation in the cat's retina. Pflügers Arch. ges. Physiol. **313**, 168—185 (1969).

Sanderson, K. J., Bishop, P. O., Darian-Smith, I.: The properties of the binocular receptive fields of lateral geniculate neurons. Exp. Brain Res. **13**, 178—207 (1971).

— Darian-Smith, I., Bishop, P. O.: Binocular corresponding receptive fields of single units in the cat dorsal lateral geniculate nucleus. Vision Res. **9**, 1297—1303 (1969).

Schade, Sr., O. H.: Optical and photoelectric analog of the eye. J. opt. Soc. Amer. **46**, 721—739 (1956).

Schaternikoff, M.: Über den Einfluß der Adaptation auf die Erscheinung des Flimmerns. Z. Sinnesphysiol. **29**, 241—263 (1902).

Scheich, H., Korn, A.: Timing properties and temporal summation in the retina. Pflügers Arch. ges. Physiol. **327**, 16—36 (1971).

Schellart, N. A. M., Spekreijse, H.: Dynamic characteristics of retinal ganglion cell responses in goldfish. J. gen. Physiol. **59**, 1—21 (1972).

Schmidt, R., Creutzfeldt, O. D.: Veränderungen von Spontanaktivität und Reizantwort retinaler und geniculärer Neurone der Katze bei fraktionierter Injektion von Pentobarbital-Na (Nembutal). Pflügers Arch. ges. Physiol. **300**, 129—147 (1968).

Schmidtke, H.: Über Messung der psychischen Ermüdung mit Hilfe des Flimmertests. Psychol. Forsch. **23**, 409—463 (1951).

Schneider, C. W.: Behavioral determinations of critical flicker frequency in the rabbit. Vision Res. 8, 1227—1234 (1968a).

— Electrophysiological analysis of the mechanisms underlying critical flicker frequency. Vision Res. 8, 1235—1244 (1968b).

Schober, H., Hilz, R.: Contrast sensitivity of the human eye for square wave gratings. J. opt. Soc. Amer. **55**, 1086 (1965).

Schwartz, A. S., Lindsley, D. B.: Critical flicker frequency and photic following in the cat. Bull. Inst. Estudios Med. Biol. **22**, 249—262 (1964).

Schwarz, F.: Über die Wirkung des Wechselstromes auf das Sehorgan. Z. Sinnesphysiol. **67**, 227—244 (1936—1938).

— Über die Reizung des Sehorgans durch niederfrequente elektrische Schwingungen. Z. Sinnesphysiol. **69**, 92—118 (1940).

— Quantitative Untersuchungen über die optische Wirkung sinusförmiger Wechselströme. Z. Sinnesphysiol. **69**, 1—16 (1940).

— Über die Reizung des Sehorgans durch doppelphasige und gleichgerichtete elektrische Schwingungen. Z. Sinnesphysiol. **69**, 158—172 (1941).

— Über die elektrische Reizbarkeit des Auges bei Hell- und Dunkeladaptation. Pflügers Arch. ges. Physiol. **248**, 76—86 (1944).

— Wintzer, H.: Über die Unterscheidung und Beurteilung von Frequenzen rhythmischer Lichtblitze. Pflügers Arch. ges. Physiol. **260**, 74—80 (1954).

— — Langer, H.: Weitere Untersuchungen über die Unterscheidung und Beurteilung von Frequenzen rhythmischer Lichtblitze. Pflügers Arch. ges. Physiol. **261**, 295—301 (1955).

Seitz, C. P.: Effects of anoxia on visual function; a study of critical frequency. Arch. Psychol. **257**, 1—38 (1940).

Sen, T. K.: Visual responses to two alternating trains of high-frequency intermittent stimuli. J. opt. Soc. Amer. **54**, 386—393 (1964).

Senders, V. L.: On reading printed matter with interrupted light. J. exp. Psychol. **47**, 135—136 (1954).

Sherrington, C. S.: On binocular flicker and the correlation of activity of corresponding retinal points. Brit. J. Psychol. **1**, 26—60 (1904).

— The integrative action of the nervous system. New Haven: Yale University Press (1906).

Shickman, G. M.: Visual masking by low-frequency sinusoidally modulated light. J. opt. Soc. Amer. **60**, 107—117 (1970).

Shumake, S. A., Smith, J. C., Taylor, H. L.: Critical fusion frequency in Rhesus monkeys. Psychol. Rec. **18**, 537—542 (1968).

Simonson, E.: Flicker between different brightness levels as determinant of the flicker fusion. J. opt. Soc. Amer. **50**, 328—331 (1960).

SIMONSON, E., BROŽEK, J.: Flicker fusion frequency, background and applications. Physiol. Rev. **32**, 349—378 (1952).
— ENZER, N.: Effect of pervitine (desoxyephedrine) on fatigue of the central nervous system. J. Industr. Hyg. **24**, 205—209 (1942).
— — BLANKSTEIN, S. S.: The influence of age on the fusion frequency of flicker. J. exp. Psychol. **29**, 252—255 (1941).
SINGER, W.: Inhibitory binocular interaction in the lateral geniculate body of the cat. Brain Res. **18**, 165—170 (1970).
— CREUTZFELDT, O. D.: Reciprocal lateral inhibition of on- and off-center neurones in the lateral geniculate body of the cat. Exp. Brain Res. **10**, 311—330 (1970).
— PÖPPEL, E., CREUTZFELDT, O. D.: Inhibitory interaction in the cat's lateral geniculate nucleus. Exp. Brain Res. **14**, 210—226 (1972).
— WÄSSLE, H.: The lateral geniculate body, a multichannel filter for spatial and temporal stimulus parameters. Proc. I. Europ. Congr. Biophys. Vienne 353—357 (1971).
SMITH, R. A., JR.: Adaptation of visual contrast sensitivity to specific temporal frequencies. Vision Res. **10**, 275—279 (1970).
— Studies of temporal frequency adaptation in visual contrast sensitivity. J. Physiol. (Lond.) **216**, 531—552 (1971).
SMYTHIES, J. R.: The stroboscopic patterns. I. The dark phase. Brit. J. Psychol. **50**, 106—116 (1959a).
— The stroboscopic patterns. II. The phenomenology of the bright phase and after-images. Brit. J. Psychol. **50**, 305—324 (1959b).
SPEKREIJSE, H.: Analysis of EEG responses in man evoked by sine wave modulated light. Doctoral dissertation, Univ. of Amsterdam. The Netherlands (1966).
— Rectification in the goldfish retina: analysis by sinusoidal and auxiliary stimulation. Vision Res. **9**, 1461—1472 (1969).
— NORREN, D. VAN, BERG, T. J. T. P. VAN DEN: Flicker responses in monkey lateral geniculate nucleus and human perception of flicker. Proc. nat. Acad. Sci. **68**, 2802—2805 (1971).
— NORTON, A. L.: The dynamic characteristics of color coded *S*-potentials. J. Gen. Physiol. **56**, 1—15 (1970).
SPERLING, G.: Linear theory and the psychophysics of flicker. Docum. Ophthal. (Den Haag) **18**, 3—15 (1964).
— Temporal and spatial visual masking. I. Masking by impulse flashes. J. opt. Soc. Amer. **55**, 541—559 (1965).
— Model of visual adaptation and contrast detection. Percept. and Psychophys. **8**, 143—157 (1970).
— SONDHI, M. M.: Model for visual luminance discrimination and flicker detection. J. opt. Soc. 58, 1133—1145 (1968).
SPIGEL, I. M.: Size-constancy and critical flicker-frequency. Amer. J. Psychol. **77**, 469—471 (1964).
SPINELLI, D. N.: Visual receptive fields in the cat's retina: complications. Science **152**, 1768—1769 (1966).
— Receptive field organization of ganglion cells in the cat's retina. Exp. Neurol. **19**, 291—315 (1967).
SPRAGUE, J. M., BERLUCCHI, G., RIZZOLATTI, G.: this Handbook VIII/3B, 27—101 (1973).
STANGE, D.: Über den Einfluß von Äthylalkohol auf die Funktion einzelner Netzhautneurone der Katze. In: BREITENECKER, L. (Ed.): Beiträge zur gerichtlichen Medizin, S. 327—331. Wien: F. Deuticke 1970.
STEIN, R. B., FRENCH, A. S.: Models for the transmission of information by nerve cells. In: ANDERSON, P., JANSEN, J. K. S. (Eds.): Excitatory Synaptic Mechanisms, pp. 247—257 Oslo: Oslo University Press 1970.
STEINBERG, R. H.: Rod and cone contributions to *S*-potentials from the cat retina. Vision Res. **9**, 1319—1329 (1969).
— The rod after-effect in S-potentials from the cat retina. Vision Res. **9**, 1345—1355 (1969).
— Incremental responses to light recorded from pigment epithelial cells and horizontal cells of the cat retina. J. Physiol. (Lond.) **217**, 93—110 (1971).

STEINBERG, R. H., SCHMIDT, R.: Identification of horizontal cells as S-potentials generators in the cat retina by intracellular dye injection. Vision Res. **10**, 817—820 (1970).
— SCHMIDT, R.: The evidence that horizontal cells generate S-potentials in the cat retina. Vision Res. **11**, 1029—1031 (1971).
STONE, J.: Structure of the cat's retina after occlusion of the retinal circulation. Vision Res. **9**, — 351—356 (1969).
— FABIAN, M.: Specialized receptive fields of the cat's retina. Science **152**, 1277—1279 (1966).
— — Summing properties of the cat's retinal ganglion cell. Vision Res. **8**, 1023—1040 (1968).
— FREEMAN, R. B., JR.: Conduction velocity groups in the cat's optic nerve classified according to their retinal origin. Exp. Brain Res. **13**, 489—497 (1971).
— HOFFMANN, K.-P.: Conduction velocity as parameter in the organisation of the afferent relay in the cat's lateral geniculate nucleus. Brain Res. **32**, 454—459 (1971).
— HOLLÄNDER, H.: Optic nerve axon diameters measured in the cat retina: some functional considerations. Exp. Brain Res. **13**, 498—503 (1971).
STURR, J. F., SHANSKY, M. S.: Cortical and subcortical responses to flicker in cats. Exp. Neurol. **33**, 279—290 (1971).
SUMITOMO, I., IDE, K., IWAMA, K.: Maintained activity and responsiveness to flicker stimulation in rat lateral geniculate neurons. Physiol. and Behavior **3**, 955—959 (1968).
— — — ARIKUNI, T.: Conduction velocity of optic nerve fibers innervating lateral geniculate and superior colliculus in the rat. Exp. Neurol. **25**, 378—392 (1969).
— IWAMA, K., ARIKUNI, T.: A relation between visual field representation of rat lateral geniculate cells and conduction velocities of optic nerve fibers innervating them. Brain Res. **24**, 333—335 (1970).
SVAETICHIN, G.: Spectral response curves from single cones. Acta physiol. scand. **39**, Suppl. 134, 17—46 (1956).
— Receptor mechanisms for flicker and fusion. Acta physiol. scand. **39**, Suppl. 134, 47-54 (1956).
— NEGISHI, K., DRUJAN, B., MURIEL, C.: S-potentials and retinal automatic control systems. In: First Europ. Biophys. Congr. Vienna 1971, Proc. Vol. 5, 77—88 (1971).
SZENTAGOTHAI, J., HAMORI, J., TÖMBÖL, T.: Degeneration and electronmicroscope analysis of the synaptic glomeruli in the lateral geniculate body. Exp. Brain Res. **2**, 283—301 (1966).
TAIRA, N., OKUDA, J.: Sensory transmission in visual pathway in various arousal states of cat. Tohoku J. exp. Med. **78**, 76—97 (1962).
TAKAHASHI, R., MORI, H., YOSHINO, T.: Changes in electrical excitability of the human eye for sinusoidal alternating currents, caused by illumination with white and monochromatic lights. Nichiganshi (Jap.) **60**, 727—734 (1956); cit. after K. MOTOKAWA (1970).
TALBOT, H. F.: Experiments on light. Phil. Mag. **5**, 321—334 (1834).
THOMAS, G. J.: A comparison of uniocular and binocular critical flicker frequencies: Simultaneous and alternate flashes. Amer. J. Psychol. **68**, 37—53 (1955).
TOMITA, T.: Electrical response of single photoreceptor. Proc. IEEE **56**, 1015—1023 (1968).
— Electrical activity of vertebrate photoreceptors. Quart. Rev. Biophys. **3**, 179—222 (1970).
TOYODA, J.: The frequency response of the retinal interneurons: Factors affecting flicker. Unpubl. manuscript (1971).
— HASHIMOTO, H., ANNO, H., TOMITA, T.: The rod response in the frog as studied by intracellular recording. Vision Res. **10**, 1093—1100 (1970).
TOYODA, J.-I., NOSAKI, H., TOMITA, T.: Light-induced resistance changes in single photoreceptors of Necturus and Gekko. Vision Res. **9**, 453—463 (1969).
TSCHERMAK, A.: Licht- und Farbensinn. In: BETHE's Handbuch der normalen und pathologischen Physiologie. Vol. XII, S. 295—501. Berlin: Springer 1929.
TURNER, P.: The modification of critical flicker fusion frequency by an adapting stimulus of flickering light. Vision Res. **5**, 463—470 (1965).
TWEEL, L. H. VAN DER: Some problems in vision regarded with respect to linearity and frequency response. Ann. N. Y. Acad. Sci. **89**, 829—856 (1961).
— Relations between psychophysics and electrophysiology of flicker. Docum. ophthal. (Den Haag) **18**, 287—304 (1964).
— SPEKREIJSE, H.: Visual evoked responses. In: The clinical value of electroretinography. ISCERG Symp. Gent 1966, pp. 83—94. Basel: Karger 1968.

TWEEL, L. H. VAN DER, VERDUYN LUNEL, H. F. E.: Human visual responses to sinusoidally modulated light. EEG Clin. Neurophysiol. **18**, 578—598 (1965).

VARJÚ, D.: Der Einfluß sinusförmiger Leuchtdichteänderungen auf die mittlere Pupillenweite und auf die subjective Helligkeit. Kybernetik **2**, 33—43 (1964).

— Über nichtlineare Analogschaltungen zur Simulierung biologischer Adaptationsvorgänge. Progress in Brain Res. **17**, 74—101. Amsterdam: Elsevier Publ. 1965.

VERINGA, F.: On some properties of nonthreshold flicker. J. opt. Soc. Amer. **48**, 500—502 (1958).

— Enige natuurkundige aspecten van het zien van gemoduleerd licht. Thesis, University of Amsterdam. The Netherlands (1961).

— Phase shifts in the human retina. Nature (Lond.) **197**, 998—999 (1963).

— Electro-optical stimulation of the human retina as a research technique. Docum. opthahl. (Den Haag) **18**, 72—82 (1964).

— Diffusion model of linear flicker responses. J. Opt. Soc. Amer. **60**, 285—286 (1970).

— ROELOFS, J.: Electro-optical interaction in the retina. Nature (Lond.) **211**, 321—322 (1966).

VERNON, M. D.: The binocular perception of flicker. Brit. J. Psychol. **24**, 351—374 (1934).

WALD, G.: Visual excitation and blood clotting. Science **150**, 1028—1030 (1965).

WALRAVEN-P.-L., LEEBEEK, H. J.: Phase shift of alternating coloured stimuli. Docum. ophthal. (Den Haag) **18**, 56—71 (1964a).

— — Phase shift sinusoidally alternating coloured stimuli. J. opt. Soc. Amer. **54**, 78—82 (1964b).

— — BOUMAN, M. A.: Some measurements about the fusion frequency of colors. Optica Acta **5**, 3—7 (1958).

WALTER, W. G.: Colour illusions and aberrations during stimulation by flickering light. Nature (Lond.) **177**, 710 (1956).

WASSERMAN, G. S.: Brightness enhancement and the opponent colour theory. Vision Res. **6**, 689—699 (1966a).

— Brightness enhancement in intermittent light, variation of luminance and light-dark ratio. J. opt. Soc. Amer. **56**, 242—250 (1966b).

WEALE, R. A.: Some observations on the Ferry-Porter law. J. Physiol. (Lond.), P 26—27 (1957).

— The effect of test size and adapting luminance on foveal critical fusion frequency. Symp. on Vis. Problems of Colour (1957).

WELPE, E.: Über die Strukturierung des Gesichtsfeldes bei intermittierender Belichtung des Auges im Frequenzbereich zwischen 33 Hz und der Flimmerverschmelzungsfrequenz. Vision Res. **10**, 1457—1469 (1970).

WERBLIN, F. S.: Response of retinal cells to moving spots: Intracellular recording in Necturus maculosus. J. Neurophysiol. **33**, 342—350 (1970).

— DOWLING, J. E.: Organization of the retina of the mudpuppy Necturus maculosus. II. Intracellular recording. J. Neurophysiol. **32**, 339—355 (1969).

WEST, D. C.: Flicker and the stabilized retinal image. Vision Res. **8**, 719—745 (1968).

WHITE, C. T., LICHTENSTEIN, M.: Some aspects of temporal discrimination. Perceptual and Motor Skills **17**, 471—482 (1963).

WIESEL, T. N.: Recording inhibition and excitation in the cat's retinal ganglion cells with intracellular electrodes. Nature (Lond.) **183**, 264—265 (1959).

— Receptive fields of ganglion cells in cat's retina. J. Physiol. (Lond.) **153**, 583—594 (1960).

WOLBARSHT, M. L., WAGNER, H. G., MACNICHOL, E. F.: The origin of "on" and "off" responses of retinal ganglion cells. In: JUNG, R., KORNHUBER, H. (Eds.): The visual system: Neurophysiology and Psychophysics, pp. 163—170. Berlin-Göttingen-Heidelberg: Springer 1961.

WUTTKE, W., GRÜSSER, O.-J.: Die funktionelle Organisation der rezeptiven Felder von on-Zentrum-Neuronen der Katzenretina. Pflügers Arch. ges. Physiol. **289**, R 83 (1966).

— — The conduction velocity of lateral inhibition in the cat's retina. Pflügers Arch. ges. Physiol. **304**, 253—257 (1968).

— RACKENSPERGER, W., GRÜSSER, O.-J.: Ausbreitungsgeschwindigkeit und räumliches Dekrement der lateralen Hemmung in der Netzhaut. Pflügers Arch. ges. Physiol. **283**, R 49 (1965).

ZERBST, E., DITTBERNER, K.-H., WILLIAM, E.: Über die Nachrichtenaufnahme durch biologische Receptoren. I. Theoretische Untersuchungen zur Ursache der Erregungsbildung. Kybernetik **2**, 160—168 (1965).

Chapter 8

Maintained Discharge in the Visual System and its Role for Information Processing

By

WILLIAM R. LEVICK, Canberra, Australia

With 8 Figures

Contents

I. Retina

1. The Physiological Nature of the Maintained Discharge

A constant feature of recordings from neurones of the sensory side of the nervous system is the occurrence of impulses in the absence of the appropriate stimuli. This activity has sometimes been called "spontaneous" (GRANIT, 1955) but the term while operationally useful becomes gradually less appropriate as more is learned about the causation of the activity. In the visual system, the activity in complete darkness merges gradually with the activity at progressively increasing levels of uniform background illumination and poses the same kind of problem for signalling. One therefore prefers the more generally applicable term, maintained discharge, to unify the description and analysis under a wide variety of conditions. "Ongoing" discharge is usually synonymous, but "background" discharge is frequently employed to describe the low amplitude activity of more distant units accompanying the recording of a single neurone.

The maintained discharge is a delicate and easily perturbed part of a visual neurone's performance: under abnormal conditions (e.g. deep anaesthesia), the maintained discharge may be strongly depressed though responses to flashing lights and moving targets remain. Since recordings from visual neurones inevitably entail surgical, pharmacological and mechanical interference, it is important to review the evidence bearing on the physiological nature of the activity. It appears that a maintained discharge in the absence of stimulation was an anathema to the early investigators: one can sense some uncertainty in their early writings and find remarks conveying their suspicions about its physiological nature (ADRIAN, 1937; HARTLINE, 1938; GRANIT, 1941). Present attitudes are different: a maintained discharge is to be expected. The current problem is to determine to what extent the activity has been disturbed by the arrangements employed to monitor it.

In the earliest investigations the eye, with or without optic nerve etc., was actually removed from the animal (ADRIAN and MATTHEWS, 1927, 1928a, b; HARTLINE and GRAHAM, 1932; HARTLINE, 1934, 1938; GRANIT and THERMAN, 1935) and while these pioneers took particular care of the preparations and produced good evidence to support the physiological nature of the responses they studied, it has to be admitted that the most sensitive properties of the retina may have been disturbed by isolation from an intact circulation and the operative interference. A technique was developed by GRANIT and SVAETICHIN (1939, see GRANIT, 1947) in which the circulation to the eye was preserved and the activity of ganglion cells or their axons recorded by bringing an insulated probe onto the vitreal surface of the retina after removal of cornea, lens and vitreous. The possibilities for abnormal neural behaviour with this preparation have been recognized and discussed (GRANIT, 1941a, b, 1947) and may be summarized as follows: mechanical pressure on the neurone by the probe, movements of the retina due to contraction of muscles. One should also bear in mind the phenomenon of the pressure-phosphene (HELMHOLTZ, 1911) produced by mechanical distortion of the retina; the opened bulb is no longer supported by the intraocular pressure and slight buckling is difficult to avoid. Another significant hazard of the opened eye preparation is the occurrence of a spreading depression (GOURAS, 1958; MARTINS-FERREIRA and DE OLIVEIRA CASTRO, 1966) rather like that affecting the cortex (LEÃO, 1944a, b). The process may be provoked by mechanical stimulation with a needle or occurs spontaneously at the cut edge of the retina; GOURAS showed that it was accompanied by a striking pattern of discharge in retinal ganglion cells (cf. BARLOW, 1953).

Some of the difficulties have been avoided with KUFFLER's development of the unopened eye technique (KUFFLER, 1952, 1953; TALBOT and KUFFLER, 1952). Movement of the retina was controlled by fixing the eyeball to a ring with a cuff of conjunctiva. Surgical interference with the eyeball was reduced to a small puncture at the *ora serrata* which admitted a cannula bearing the glass-insulated platinum wire electrode. The electrode passed across the vitreous and recording was established by placing the tip on the anterior surface of the retina. KUFFLER devoted considerable attention to establishing what structures were being recorded from (cf. RUSHTON, 1949) and whether the electrode was distorting the behaviour. Recordings from single ganglion cell somata and single axons of ganglion cells could be convincingly distinguished. A maintained discharge was observed in both and since it could be abolished in the axons as well as in the cell recordings by appropriate illumination of the receptive field, the electrode could not be held responsible for causing the maintained discharge. Experiments in unanaesthetized decerebrated cats (KUFFLER, 1953; KUFFLER et al., 1957) also revealed brisk maintained discharges which therefore cannot be attributed to the effects of anaesthesia or a possible centrifugal efferent nerve supply to the retina.

A persistent sceptic could still suggest that the puncture wound at the periphery of the retina was the source of the maintained discharge, but this objection was disposed of as soon as recordings were made from axons of the optic nerve or optic tract intracranially. BORNSCHEIN (1958a) combined such recording with a technique for reversible pressure blockade of the eye (cannula in the anterior chamber, connected to a pressure system). He found brisk maintained discharges in optic nerve fibres which must have come from the retina because they could be arrested completely by raising the intraocular pressure; restoration of normal pressure also restored the maintained activity. It scarcely seems worthwhile pursuing the argument further. The conclusion is: in the cat, the animal which has been investigated in greatest detail, the maintained discharge which is observed in most if not all retinal ganglion cells in the absence of photic stimulation is a normal physiological phenomenon.

A maintained discharge of retinal ganglion cells in darkness has been commented on or at least illustrated in a variety of other vertebrates: monkey (HUBEL and WIESEL, 1960), human (WEINSTEIN et al., 1971), rat (GRANIT, 1941a; BROWN and ROJAS, 1965), squirrel (MICHAEL, 1966), guinea pig (GRANIT, 1942), rabbit (BARLOW and LEVICK, unpublished observations), frog (GRANIT, 1941b; MATURANA et al., 1960), tortoise, snake (GRANIT, 1947) and fish (WAGNER et al., 1960). Although the observations have usually been casual and unsystematic, it would be reasonable to suppose that the situation would be at least qualitatively similar to that in the cat.

2. Origin of the Maintained Discharge

Behavioural investigations have indicated that cats have very low absolute thresholds for detection of light (BRIDGMAN and SMITH, 1942; GUNTER, 1951). Indeed, the luminance threshold for cats is about five times smaller than for humans. If the cat retina operated similarly to that of the human, then its rod receptors would also have reached the ultimate in sensitivity: each rod would be capable of signalling the absorption of a single quantum. In view of this possibility, the maintained discharge assumes a major significance in limiting visual performance. Threshold would now be determined by the number of quantal absorptions required to displace the discharge pattern beyond the bounds of its own irregularity. BARLOW (1956, 1957, 1964) has developed a psychophysical theory of visual thresholds which is based upon the idea that the visual system, like all exquisitely sensitive devices, operates in the presence of an intrinsic noise which may possibly be identified with the fuzzy grey sensations experienced in darkness, even when all after-images have faded ("Augenschwarz" of FECHNER, 1860; "Eigengrau" of HERING, 1878; "Eigenlicht" of HELMHOLTZ, 1911). Indirect arguments (BARLOW, 1964) seem to place the source of the noise in the receptors. Since the maintained discharge is the manifestation of the noise, the origin of the maintained discharge is of great interest.

HUGHES and MAFFEI (1965) came to the conclusion "that the dark discharge is the expression of the autochthonous activity of the deafferented ganglion cells or of neural nets not including the bipolar cells or the receptors". Their main evidence is that the time course of change in the maintained discharge rate after extinguishing an adapting light is very much shorter than the change in threshold or sensitivity of the ganglion cell. This was also observed earlier by Barlow et al. (1957a). However, such differences in time scale need not imply the conclusion quoted. The retina handles a very large dynamic range of inputs and must possess mechanisms to adjust sensitivity. If a signal proportional to the output discharge controlled the sensitivity, then the result of Hughes and Maffei would still be obtained. Rodieck (1967) also opposed their conclusion on the ground that correlations can sometimes be found between the firing patterns of adjacent ganglion cells. This certainly argues against the latter being independent generators of maintained discharges, but the result does not exclude the amacrine cells from generating the autochthonous activity.

Certain drugs selectively damage the visual receptors but spare the cells of the inner nuclear and ganglion cell layers. Such drugs are iodoacetic acid (Noell, 1951, 1952) and members of the series of diaminophenoxy alkanes (Edge et al., 1956; Ashton, 1957). Intravenous administration of iodoacetic acid promptly abolishes the ERG, retinal and optic nerve responses to light (Noell, 1951). The maintained discharge recorded from single fibres in the optic nerve (Bornschein, 1958c) greatly increases as the b-wave fails, and then falls to zero. Despite the almost complete absence of "spontaneous activity" in the retina, there was no apparent change in the excitability of the optic nerve 3 weeks after poisoning (Noell, 1952). Similar though more gradual loss of receptor function occurs after administration of the diaminophenoxyalkane, M and B 968A, again coupled with absence of the maintained discharge (Rodieck, 1967; Burke and Hayhow, 1968). All of this points to the receptors as the fundamental source of the maintained discharge in darkness. If, in fact, they are not, we should have to consider a neural origin such as spontaneous release of transmitter at synapses of the kind discussed by Katz (1966), Katz and Miledi (1963) and Blankenship and Kuno (1968).

3. Factors Determining the Maintained Discharge

The aim of this section is to synthesize a picture of what happens to the maintained discharge at various background levels of illumination from zero up. Some type of systematic relation is to be expected for two reasons. Firstly, the size of the pupil depends upon the level of illumination (Fig. 1B; see Kappauf, 1943; Le Grand, 1957); and secondly, the subjective estimate of the brightness of a field is a systematic function of its luminance under the type of conditions considered (Fig. 1A; see Stevens and Stevens, 1963) namely: at equilibrium for a uniformly illuminated area covering as much of the visual field as possible ("Ganzfeld" illumination). Since a systematic relation between maintained discharge and illumination has proved to be rather elusive (Kuffler, 1953; Kuffler et al., 1957; Rodieck, 1967), some possible complications should be discussed.

A significant difficulty is that change of background illumination is generally accompanied by substantial transient changes of maintained discharge which subside with a time constant varying from seconds to minutes in different neurones (Kuffler et al., 1957). There is always some uncertainty that

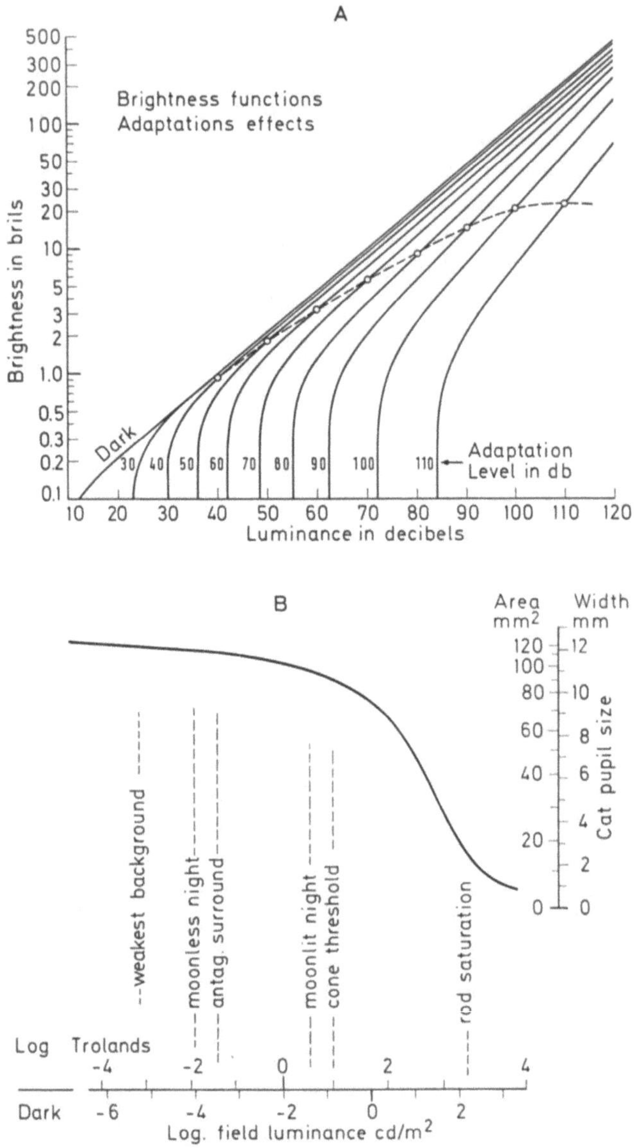

Fig. 1. *Apparent brightness and pupil size in the "Ganzfeld" at difference luminance.* A The dashed line connecting the circles gives the equilibrium relation between *apparent brightness* (ordinate) *of a uniform, steadily exposed field* and its luminance (abscissa) on logarithmic scales: 40 decibels corresponds to 3.18×10^{-3} cd/m² and a shift to the right of 10 decibels multiplies the luminance by 10. The scale of brightness is relative to the apparent brightness of a 5° spot of luminance 3.18×10^{-3} cd/m² flashed for 1 sec on a dark field (1 "bril"). Pupil size not controlled. From STEVENS and STEVENS (1963). B *Size of the cat's pupil* as a function of the luminance of a large uniform field. Redrawn from KAPPAUF (1943) who gave pupil width versus luminance. Conversion from pupil width to pupil area based on unpublished data kindly supplied by WILCOX. This permits the scale of retinal illumination (cat trolands) to be constructed along the horizontal axis. The vertical dashed lines correspond approximately with physiologically significant levels of retinal illumination or field luminance

equilibrium has been reached, especially for large downward steps of illumination which involve slower processes (BARLOW et al., 1957b). The equilibrium output of receptors is probably a simple monotonic function of illumination (cf. HARTLINE and GRAHAM, 1932; RIGGS and GRAHAM, 1940; RUSHTON, 1959) but at least two layers of synapses intervene before the ganglion cell (DOWLING, 1968) and provide opportunities for extensive modification in favour of other specializations of function. Thus particular ganglion cells of the frog signal the presence or movement of variously configured borders within their receptive fields (MATURANA et al., 1960); rabbits (BARLOW et al., 1964) and ground squirrels (MICHAEL, 1968) have cells signalling the direction of movement of targets. Such units specifically disregard the level of illumination in favour of their appropriate stimuli (BARLOW and HILL, 1963; LEVICK, 1965); therefore, one might not find a systematic relation between their maintained discharges and illumination level because their mechanism has specifically removed it.

For orientation, it is helpful to have some idea of the range of illuminations that are appropriate for the visual system (see LE GRAND, 1957). Very approximately, a sheet of white paper has a luminance of: 10^{-4} cd/m² under the night sky without moon; 5×10^{-2} cd/m² in moonlight; 50 cd/m² in a well-lit study; 25000 cd/m² under the midday sun. The effect on the retina of a particular field luminance is given by the resulting retinal illumination which is proportional to the luminance of the surface viewed (B cd/m²) and the apparent pupil area (a mm²). The retinal illumination is conveniently defined by a unit called the troland, td. One troland is the retinal illumination produced when a luminance of 1 cd/m² is viewed through a pupil of 1 mm². In general, $T = B \cdot a$. When the transmission of the ocular media (τ) and the posterior nodal distance (focal length, f) of the eye are known, one can immediately convert from trolands to the more fundamental unit, lumens/m² at the retina: $E = T \cdot \tau/f^2$. The troland is widely used in human work for comparing retinal illuminations in different experiments where τ and f have not been specifically measured. The same definition may be adopted for the cat or any other animal, with the proviso that the specific ocular transmission and focal length must be used when the result is required in lumens/m². Thus 1 human troland is not the same retinal illumination as 1 cat troland, but it does not matter; inter-species comparisons should be based on the fundamental unit, lumens/m². Maxwellian view systems need special treatment (WESTHEIMER, 1966). In the following the troland is used as defined above in relation to the cat and it is given as a scotopic quantity. Numerical aspects of photopic and scotopic units have recently been discussed by SCHEIBNER and BAUMGARDT (1966). The posterior nodal distance of the cat's eye is about 12.5 mm (VAKKUR et al., 1963) and data on the size of cat's pupil as a function of field luminance is shown in Fig. 1A.

In the cat, the weakest background which just begins to elevate thresholds is about 5×10^{-4} td and the antagonistic surround of concentric units appears about 3.5×10^{-2} td (BARLOW and LEVICK, 1969b); Purkinje shift and cone threshold occur at about 10 td (BARLOW and LEVICK, 1968; DAW and PEARLMAN, 1969); rod system saturates at about 2×10^3 td (DAW and PEARLMAN, 1969); 3×10^5 td applied for 10 sec bleaches about 50% of the rhodopsin (calculation based on RUSHTON, 1956). From these facts, it appears that many studies on maintained discharge have employed illuminations only near the top of the range, in some cases producing substantial bleaching of visual pigment. Restoration of equilibrium after bleaching is a slow process in the cat (BARLOW et al., 1957a).

Are the carriers of illumination information the commonly occurring on-centre and off-centre (B-system and D-system of JUNG, 1961a) ganglion cells (JUNG, 1964)? Working over the lower 7 log units of background, BARLOW and LEVICK (1969b) showed that on-centre cells had discharge rates monotonically increasing with background up to 3.5×10^{-2} td (Fig. 2A); at this level the antagonistic surround became detectable. As background was further increased, the firing rate ceased behaving regularly: in different units, it decreased, remained unchanged or fluctuated irregularly. Above 700 td the mean rate again rose.

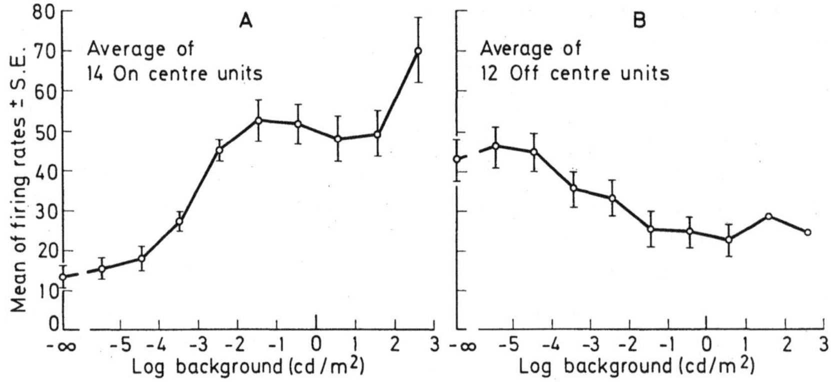

Fig. 2. *Average* (± standard error of the mean) *of the mean discharge rates of 14 on-centre units* (A) *and 12 off-centre units* (B) as functions of the luminance of the uniform background field in the equilibrium condition. Pupil area 7 mm². (From Barlow and Levick, 1969b)

Thus the emergence of the antagonistic surround suspended the monotonic increase of maintained discharge over about 4 log units of increasing background. This has advantages for preserving sensitivity to local luminance changes but information about the background luminance is rendered ambiguous in that range. However, the authors described another rarely encountered type of on-centre concentric unit with distinctly different properties including a monotonic increasing maintained discharge over the range where the ordinary on-centre units signal ambiguously. Taken together, the two classes provide the required information. As a group, the off-centre ganglion cells showed a monotonically decreasing maintained discharge with background, but the slope of the relation was very shallow (Fig. 2B) and there was considerable variation from unit to unit.

Recording from the optic tract, Straschill (1966) followed the relationship over 3 log units, probably at the top of the range. There is some uncertainty in the specification of his stimulus; calculation suggests that his brightest light (~ 200000 td) may have bleached substantial quantities of visual pigment; his weakest light (~ 100 td) probably overlapped 1.5 log units into the range of Barlow and Levick. The higher level monotonic increase in firing rate of on-centre units continued uninterruptedly and off-centre units showed the same shallow reduction as background increased (Fig. 3A, B). There were, however, some complications in this simple picture. Five out of 35 units had a maintained discharge independent of stimulus level and were not included in Fig. 3. More than half of the units were classified as on-off type by virtue of their responses to abrupt change of background. Although they had either on-centres or off-centres the centre did not necessarily control the relation between maintained discharge and background. Thus some on-centre on-off neurones had *decreasing* maintained discharge with stronger backgrounds and some off-centre ones showed an increasing discharge. STRASCHILL suggested that in such cases, the surround determines the relation between maintained discharge and background.

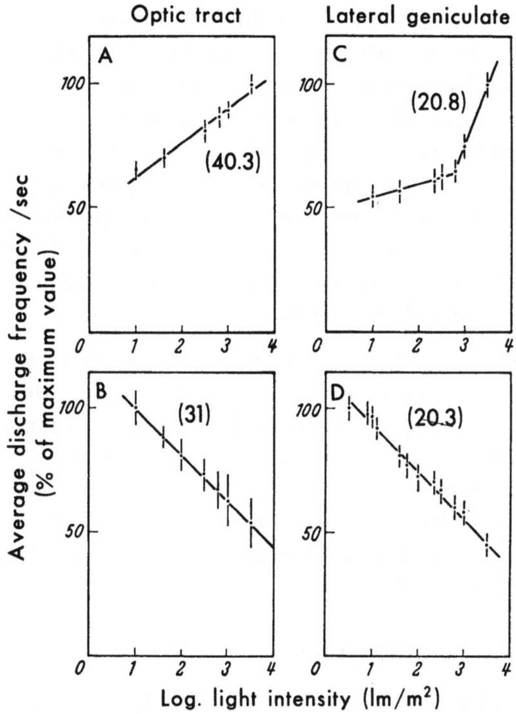

Fig. 3. *Luminance and neuronal discharge rates.*
A, B Continuation to higher backgrounds of the relation between mean discharge rate of retinal ganglion cells and luminance of uniform background. C, D the same for lateral geniculate neurones. In each case, the results from several neurones were pooled (A: 5 units; B: 9 units; C: 6 units; D: 8 units) after expressing each discharge rate as a percentage of the maximum for the particular neurone: the points are the means, the vertical strokes give ± standard deviation. To relate these results to those of Fig. 2, a "light intensity" of 10 lm/m² might be assumed to produce a retinal illumination of about 620 cat trolands (scotopic). The numbers in brackets are the average maximum discharge frequencies (100% point) for each group. (From STRASCHILL, 1966)

It is well-known that the effectiveness of the surround relative to the centre is controlled by adaptation level (BARLOW et al., 1957b; BARLOW and LEVICK, 1969b), but why the effect is more prominent in some units than in others requires further investigation. If at low backgrounds the centres of these units became dominant, the slope of the characteristics would reverse; a particular discharge rate would then correspond to more than one background level. Units of this type could not participate in signalling luminance unless their maintained discharge possessed additional special properties.

By placing a concentric macroelectrode in the optic chiasma ARDUINI and PINNEO (1962) attempted to obtain a measure of the total maintained discharge of all the optic nerve fibres. They recorded the noise of the optic chiasma and used its mean square value (minus the value for the dead preparation) as a quantitity proportional to the total number of impulses/sec. The chiasma was much noisier in the dark, so they inferred that the net retinal activity decreased with illumination. This conclusion does not agree with that of BARLOW and LEVICK (1969b) at the lower levels or of RODIECK (1967) at the higher. STRASCHILL (1966) supported the conclusion because his on-units were outnumbered by the on-off and off-units which had higher discharge rates in darkness than in light. However, some of the illuminations he used probably bleached substantial amounts of visual pigment and may have inadvertently raised the dark discharge. Although detailed experiments are lacking, the discrepancy may be resolved if the maintained discharges of off-centre units are not independent (RODIECK, 1967) or if off-centre units have larger axons (BARLOW and LEVICK, 1969b). Whatever the explanation, it will have very interesting consequences for the mode of action of the retina on the lateral geniculate nucleus.

Summary. Most on-centre ganglion cells show an increasing maintained discharge with background illumination, up to the level at which the antagonistic surround emerges. The firing rate at higher levels depends upon the effectiveness of the surround relative to the centre; it may continue to increase, vary unsystematically or decrease. A rare type of on-centre unit has an increasing maintained discharge with illumination over the range of ambiguous performance. The maintained discharge of most off-centre units decreases with background but the relationship is rather flat.

4. The Problem of Retinal Rhythms

The orderly picture of the maintained discharge described in the preceding section is not always found. On occasions the firing rate of a ganglion cell may be observed to undergo spontaneously a remarkable sequence of oscillations (Fig. 4) which may even drive the discharge frequency beyond the range achieved by steady illumination (Figs. 2, 3). A wide variety of patterns occur (RODIECK and SMITH, 1966) and many investigators have commented on the faster types (e.g. ADRIAN and MATTHEWS, 1928b; GRANIT, 1941b; BARLOW, 1953; KUFFLER et al., 1957; ARDUINI and PINNEO, 1962; DOTY and KIMURA, 1963; ASCOLI and MAFFEI, 1964; BROWN and ROJAS, 1965; LAUFER and VERZEANO, 1967).

The occurrence of patterned discharge in the absence of patterned stimulation naturally raises the question of artifact. Thus KUFFLER et al. (1957) attributed unexpected fluctuations of firing rate to mechanical irritation of the ganglion cell; they also found that pentobarbitone could induce rhythms, as did BROWN and ROJAS (1965). However, RODIECK and SMITH (1967) obtained the rhythms in the absence of anaesthetics and they could exclude mechanical effects by recording rhythms at the optic disc or beyond the eyeball in the optic nerve. Moreover, during the slower rhythms, the activity of a single unit was paralleled by the massed activity in the same optic nerve, suggesting that the pattern was widespread in the retina. This raises the possibility that the rhythms might be driven by possible centrifugal efferents, but severing the optic nerve had no effect on the course of a rhythm. Although similar rhythms occurred simultaneously in each eye, their phases were independent. BARLOW and LEVICK (1969b) found that rhythms appeared when the condition of the eye or the preparation as a whole was in doubt, e.g. towards the end of long experiments. However, fluctuations of firing rate also occurred in apparently normal preparations at levels of background illumination where the surround was just beginning to appear.

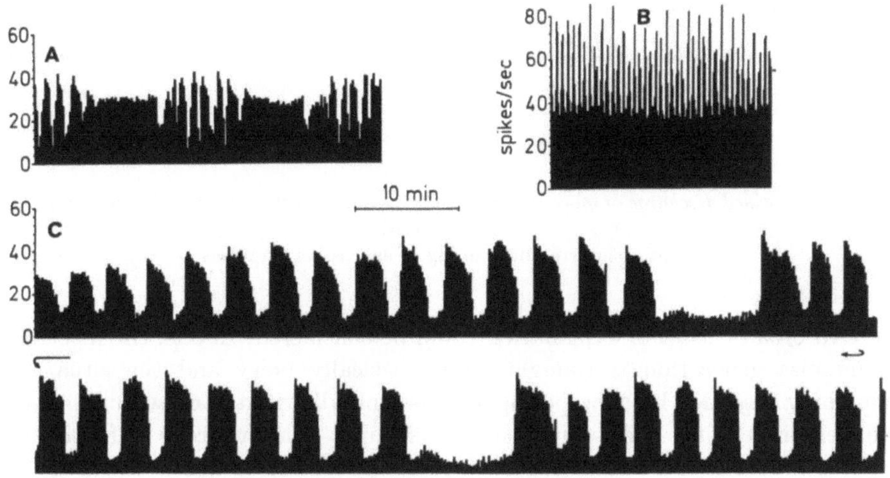

Fig. 4. *Examples of retinal rhythms.*
The upper boundary of the black area traces the firing rate of cat retinal ganglion cells as a function of time in uninterrupted darkness. The time scale refers to all records. Note the variety in the period of the rhythms, the waxing and waning in A and the sustained regularity in C (continued in the lowest record). At the peaks in B, the firing rate goes beyond the values illustrated in Figs. 2 and 3. (From Rodieck, 1967)

The most striking evidence for the physiological nature of the rhythms has been provided by Cavaggioni (1968) who recorded the massed activity in the optic chiasma with chronically implanted macroelectrodes. He often found spontaneously occurring rhythms remarkably like the patterns illustrated by Rodieck and Smith (1967); they occurred in the quiet, unrestrained cat in darkness. At other times an increase in the dark discharge could be related fairly precisely with EEG signs of spontaneous arousal from deep sleep, although the subsequent course of the rhythm could be independent of subsequent EEG changes for a while. Cavaggioni agrees that the form of the rhythm is determined in the retina but suggests that the process may be triggered by arousal. How this occurs is far from clear, but vascular changes (Granit, 1955) and reflex changes of intraocular pressure (Collins, 1967) were suggested.

If rhythms are a normal, physiological occurrence, it is difficult to fit them into the orderly picture of an exquisitely sensitive retina, reliably signalling very weak stimuli by minimum perturbation of a steady maintained discharge.

5. Analysis of the Maintained Discharge

The sequence of impulses in the maintained discharge of retinal ganglion cells is quite unlike the output of the familiar electronic oscillator. Time intervals between successive impulses continually change in apparently haphazard fashion resulting in a more or less irregular discharge (Fig. 5). If signalling is achieved by altering the mean frequency (ADRIAN, 1928), then one might have expected to find a homogeneous carrier frequency in order to obtain the maximum resolution of small differences in the input within a given time. What is the source of the irregularity?

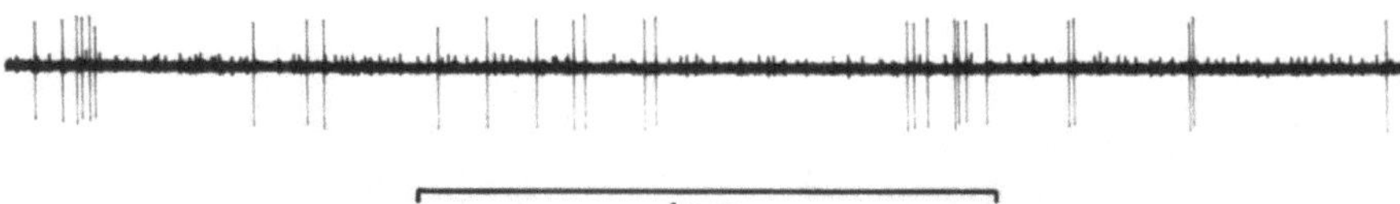

Fig. 5. *Maintained discharge of on-centre retinal ganglion cell in the cat after reaching equilibrium in darkness.*
Note the irregular timing of successive impulses

Two broad classes of explanation could be considered. In one, the mechanism of impulse generation is thought to be basically noisy and the situation is commonly modelled by superposing a Gaussianly distributed noise voltage across the membrane of the spike-generating locus. Input to the neurone shifts the mean level of the noise relative to threshold, thus producing irregular discharge with mean rate controlled by the input (cf. HAGIWARA, 1954; GEISLER and GOLDBERG, 1966). Possibly, the spontaneous miniature synaptic potentials of KATZ and MILEDI (1963) could supply the noise. The other explanation is to suppose that the input to the neurone is itself noisy: impulses arrive on numerous independent input fibres and their effects sum to reach threshold. Irregularity of output is a consequence of independent arrival times. In view of the discussion of the origin of the maintained discharge, the latter alternative seems more likely.

In order to appreciate the problem it is helpful to consider models of stationary random pulse processes (MOORE et al., 1966), the most fundamental of which is the Poisson process. An elementary exposition is not easy to find but helpful discussions are given in FELLER (1957) and COX (1962). Briefly, we may consider a Poisson process as one which with passage of time emits a train of pulses (Poisson time series) such that the probability of a pulse occurring at a particular moment is constant and independent of the past or future history (event-times) of the train. Well-known equations give the distributions of lengths of intervals between pulses (exponential distribution) and of number of pulses per unit time (Poisson distribution). The Poisson model is important because a surprising number of practical situations approximate to it; for example, the series of disintegrations within a lump of radium, the local sequence of telephone calls reaching an exchange, and most important from the present viewpoint, the sequence of quantum absorptions in visual pigment, and the sequence of miniature junction potentials (FATT and KATZ, 1952; BLANKENSHIP and KUNO, 1968) at a synapse.

From the practical point of view, the effects of certain operations upon the Poisson process are worth knowing: 1. When the mean inter-pulse interval is small (high discharge rate), the distribution of the number of pulses per unit of time asymptotically approaches a Gaussian (or "normal") distribution with standard deviation equal to the square root of the mean; 2. When the Poisson process is transduced by a device having a "dead-time" such that each output pulse paralyzes the system for a fixed time τ, the exponential distribution of intervals is displaced bodily to the right by an amount τ and there are no intervals shorter than τ; also, the right-hand tail of the Poisson distribution of number of pulses is progressively squeezed down to the horizontal axis (RICCIARDI and ESPOSITO, 1966). 3. If a device transmits only each n-th pulse of the Poisson pulse train, the distribution of intervals becomes one of the family of gamma distributions (special Erlangian distribution of COX, 1962; Pearson type III distribution of Kuffler et al., 1957; see graphs in Fuortes and Hodgkin, 1964) and the distribution of the number of pulses approximately resembles the shape of the Poisson distribution of the input, but with the horizontal axis scaled by $1/n$ (exact calculation in Cox, 1962, p. 38–39). Another helpful "rule-of-thumb" is that if a device combines a number of independent input channels such that an output pulse occurs whenever a pulse is received from any input channel, the local (short term) properties of the output are asymptotically Poissonian as the number of input channels in increased, regardless of the statistical properties of the individual input processes (Cox, 1962).

The selection of the above details is conditioned by their potential neurophysiological relevance: "dead-time" is equivalent to refractory period; the n-th pulse operation is the idealized version of a neurone requiring n unit steps of input to reach threshold, whereupon the emitted impulse resets the membrane; that is to say, it recognizes the temporal summation of ganglion cells; the combining operation is the embodiment of spatial summation. There are many other important considerations; for example, what are the modifying effects of inhibition? Since the answer depends on particular details (presynaptic or postsynaptic inhibition, or both; degree of dependence of inhibitory process on excitatory action) there are no simple guides. It appears that most of the neurophysiologically significant modifications of the Poisson process lead to mathematically intractable problems if one wants answers in closed form. However, useful advances have been made by dealing with artificial but tractable cases which approximate a particular process (Gerstein and Mandelbrot, 1964; Ten Hoopen, 1966). Penetrating insights have also come from simulating intractable cases on a digital computer (Stein, 1965, 1967a; Perkel et al., 1967).

6. Interval Histogram

By analysis of the distribution of intervals between successive impulses one may test particular hypotheses about the processes giving rise to the maintained discharge. It should be appreciated that the result of the test is to reject a false hypothesis. Since a wide variety of real processes may lead to identical distributions, "getting the expected result" is not always a strong answer on its own to the question of "what is the underlying process".

The interval histograms of retinal ganglion cells show considerable variation from cell to cell and in the same cell at different levels of illumination (Fig. 6A, B, C). Kuffler et al. (1957) found that an exponential distribution could not be adjusted for a satisfactory fit to the experimental histograms even after allowance for dead-time, whereas a gamma distribution could be successfully fitted. However, Herz et al. (1964; see also Fuster et al., 1965b) considered that most of the histograms were of exponential form; the rounded peak with graded reduction near the vertical axis (cf. gamma distribution) were attributed to the modifying effect of the neuronal refractory period. Histograms illustrated by Rodieck (1967) include some that might be of exponential or gamma shape and others which obviously depart from these simple forms. Heiss and Bornschein (1965, 1966) and Heiss (1967) have also seen considerable variety including multimodal types which Rodieck suggested were associated with barbiturate anaesthesia.

Nevertheless, Barlow and Levick (1969b) found that the histograms of on-centre units could be fitted by members of the family of gamma distributions over a wide range of background illuminations, the parameter of the required gamma distribution diminishing systematically with background in each unit. They therefore tested the hypothesis that the ganglion cell was being driven by a Poisson process but transmitted to its output only every nth input pulse where n

is the parameter of the gamma distribution (Fig. 6D). Independent measurement of n from the sensitivity to added light gave moderate agreement at low backgrounds but major discrepancy at high. The hypothesis in its elementary form was therefore ruled out. However, by simulation on a digital computer, STEIN (1965) showed that gamma distributions also arise from a similar model in which input events cause exponentially decaying effects on the neurone membrane rather than the sustained effects implied by simple scaling (Fig. 6E). In this case the gamma

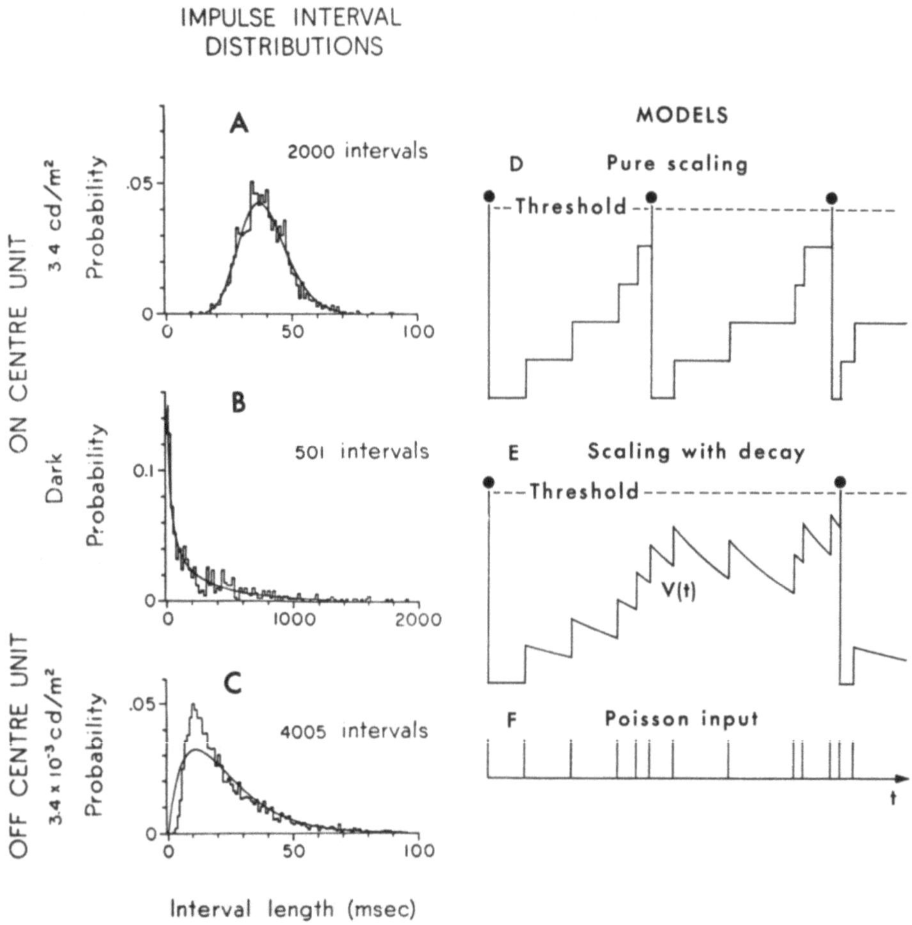

Fig. 6. A, B, C *Interval histograms of the maintained discharge of cat retinal ganglion cells;* A, B from an on-centre unit at a background of 24 cat trolands and in darkness; C from an off-centre unit at 0.024 troland. The smooth curves in each case are best-fitting gamma-distributions. The very different histograms of the on-centre unit are well fitted by suitable choice of the two parameters of the distribution, but the best fit in C is poor. D, E *Hypothetical models* in terms of the membrane potential of the ganglion cell, $V(t)$, the threshold for triggering an impulse (black dot, followed by reset of membrane potential) and an effectively random (Poisson) input train of events (F). The "pure scaling" model is mathematically simple but calls for rather unrealistic membrane behaviour (sustained excitatory post-synaptic potentials); "scaling with decay" is more physiological but mathematically inconvenient (A, B, C from BARLOW and LEVICK, 1969b)

parameter may correspond with a much wider range of neural sensitivities, so there is hope for ultimate agreement. Naturally, the final solution will have to incorporate the modifying effects of the antagonistic surround as well as the time course of recovery of excitability (sub-normal and super-normal phases etc., cf. HODGKIN, 1948; STEIN, 1967b) after each ganglion cell impulse. One cannot argue on the basis of events in motoneurones because retinal ganglion cells lack the complication of a recurrent collateral (CAJAL, 1955).

7. Other Statistical Measures

Joint distributions of successive intervals, serial correlation coefficients of various orders and autocorrelation functions have been investigated (RODIECK, 1967; HERZ et al., 1964). Intriguing facts have been uncovered, but these have not yet been convincingly related to retinal mechanisms. Thus, a small but significant negative correlation between the lengths of adjacent intervals has been found for a proportion of the units. An interjected antidromic impulse is followed by the same average time course of firing (autocorrelogram of RODIECK, 1967) as a naturally occurring impulse. Correlation between the firing patterns of simultaneously recorded units has also been noted.

8. Significance of the Maintained Discharge

It has already been described how the ongoing discharge of on-centre (and to some extent, off-centre) retinal ganglion cells could provide a signal of the level of uniform illumination in the visual field. There are some more general implications of the activity which will be next discussed.

A Limit to the Detection of Weak Stimuli. If there were no maintained discharge, threshold could depend upon the production of a single impulse. One could adopt and use a simple definition of threshold such as the stimulus strength which evoked an impulse on 50% of trials. Sensitivity (or, gradient of response *versus* stimulus strength) would therefore be the sole factor determining threshold. There is a close parallel with the case of the electrical threshold of a nerve fibre. In the presence of an irregular ongoing discharge, the situation is more complex: there is no way of determining with certainty which impulses are associated with a photic stimulus and which are attributable to the maintained discharge. Instead, one must adopt a strategy which seeks to determine the minimum reliably detectable modification of the discharge, and the properties of the maintained discharge therefore enter into the threshold equation. Furthermore, when reliable detection requires several extra impulses, the period over which they are spread becomes important. To see the relation between maintained discharge and threshold, we must consider the relative significance of the new factors introduced.

When one considers the very subtle ways in which messages can be deliberately concealed by coding, the difficulties of providing a general theory for the detection of signals become apparent. In the case of the retina, however, the variability of suprathreshold responses to identical stimuli (LEVICK and ZACKS, 1970) and the absence of special time structure in the maintained discharge argue against an elaborate temporal coding scheme. It has therefore been taken that a brief stimulus is coded as a transient change of average impulse frequency. On this basis, and using the interval distribution of the maintained discharge, FitzHugh (1958) outlined a general method for making optimal judgements of the presence or absence of weak stimuli. The calculation weighed the likelihood that a particular response could have resulted from the presence of a stimulus against the likelihood that it could result from the absence of stimulus. One may see from FitzHugh's equations that lower thresholds are favoured by more regular discharges (larger gamma parameter) or lower mean rates.

Barlow and Levick (1969a) developed a simpler calculation relating threshold and the maintained discharge. In this case the statistical parameter was the variability of the number of impulses occurring per unit time. From the distribution of the number of impulses (pulse number histogram) they determined how many extra impulses would have to be ejected by a stimulus to make the resulting number unlikely to have arisen from the maintained discharge alone (at an assigned level of significance or "false positive rate"). Since the pulse number histograms could be well fitted by Gaussian distributions, the required excess of impulses for reliable detection of a change in the discharge was proportional to the standard deviation of the pulse number histogram. One may see rather more directly how the irregularity of the maintained discharge sets the level of the threshold. It is interesting that thresholds determined in this way agreed with those obtained empirically by listening to the discharge on a loudspeaker or by watching it on an oscilloscope. The approach is a specific realization of the general theory of decision processes in perception (Tanner and Swets, 1954; Swets et al., 1961) in which the noise distribution is explicitly measured.

Improved Bandwidth for Low Amplitude Signals. Granit (1955) suggested that a function of the maintained discharge would be to permit the signalling of inhibitory effects to higher levels. Related to this are the observations of De Valois et al. (1962) who showed that increments and decrements of illumination displace the discharge rate more or less symmetrically about its maintained level. The idea that the maintained discharge is the zero-signal carrier-frequency of a frequency-modulated system has thus gained attention (Hughes and Maffei, 1966). If this is the case, however, the ganglion cell output is quite unlike the conventional FM system where the carrier frequency is usually greater than half the maximum frequency in order to obtain wide dynamic range and linear performance.

There is another difficulty with operation at low carrier frequency. The discharge is characteristically irregular so that relatively long pauses between impulses are not uncommon. For instance, at a mean rate of 30/sec, about 5% of the intervals are longer than 80 msec (gamma parameter = 2). It would therefore take at least 80 msec to infer the presence of a maximum inhibition at the 5% level of significance. On the other hand, ganglion cells can increase their discharge to 800/sec or more, enabling a decision about a maximum excitation to be reached much more rapidly. It seems much more likely that the inhibitory signalling of an on-centre unit is taken over by means of excitation of an off-centre unit with similarly located receptive field. In this way high frequency performance is made available for decreasing as well as increasing illumination (Barlow, 1961).

The significance of the maintained discharge from the above viewpoint is to provide a reasonably high frequency response for low amplitude modulation (near the cross-over point between on-centre and off-centre signalling).

Energizing Function. An aroused state of the cerebral cortex depends upon the integrity and activity of the ascending reticular activating system (Moruzzi and Magoun, 1949) which in turn depends upon incoming activity on the specific afferent pathways. A transection of the brain stem at the mid-pontine level but above the trigeminal nerve leaves the reticular system in connection only with the olfactory and optic inputs. Arduini and Hirao (1959) showed that in the acute version of this preparation, the EEG showed an activation pattern which could reversibly and repeatedly be transformed into the slow-wave sleep pattern if the retinal maintained discharge was shut down by raising the intraocular pressure in both eyes. How much of a role the maintained discharge plays in keeping the normal, intact organism in an alert state is not so clear.

II. Higher Centres

A maintained discharge is also a feature of recordings made from neurones in the lateral geniculate nucleus and visual cortex. Again, there is no reason to believe that the activity is the result of mechanical interference by the electrode; in the case of geniculate neurones, similar firing patterns were found in recordings of their axons in the optic radiation (Hubel and Wiesel, 1961; Levick and Williams, 1964). Transiently raising the intraocular pressure completely blocked or strongly slowed the discharge of the majority of geniculate neurones thus showing that the drive came from the maintained discharge of retinal ganglion cells.

However, the geniculate discharge is quite different from the retinal. Perhaps the most striking difference is that the nature of the geniculate discharge depends upon the state of the rest of the brain. Hubel (1960) observed that in the awake cat, impulses of a geniculate cell occurred at more or less random intervals, the pattern resembling, at least superficially, that described for retinal ganglion cells (Fig. 7A). As the cat became drowsy and finally slept, there developed an increasing tendency to firing in characteristic, brief, high-frequency clusters of impulses. Each cluster consisted of 2–8 spikes at frequencies of up to 500/sec or more (Fig. 7B). Such spontaneous clusters are usually not found in the retinal discharge. Clearly there must be another input to geniculate neurones. Confirming and extending earlier work of Suzuki and Taira (1961), Taira and Okuda (1962) showed that stimulation of a restricted region in the midbrain reticular formation increased the mean dark discharge rate even more effectively than an arousal by auditory or olfactory stimuli. The result held for both on- and off-centre axons of geniculate cells. Optic tract units were unaffected. It is worth noting that the effects of arousal produced opposite effects on the responses of on- and off-units to light, in the sense that off-units became more strongly inhibited by illumination. We might therefore expect more complicated relations between maintained discharge and steady level of illumination than in the case of retinal ganglion cells.

It is now well-established that there are two distinct states of sleep (recently reviewed by Jouvet, 1967): so called "slow-wave sleep" characterized electrophysiologically by large amplitude spindles and slow waves in the EEG; and "rapid-eye-movement (REM) sleep", associated with a higher frequency, low amplitude EEG plus hippocampal theta rhythm, and ponto-geniculo-occipital

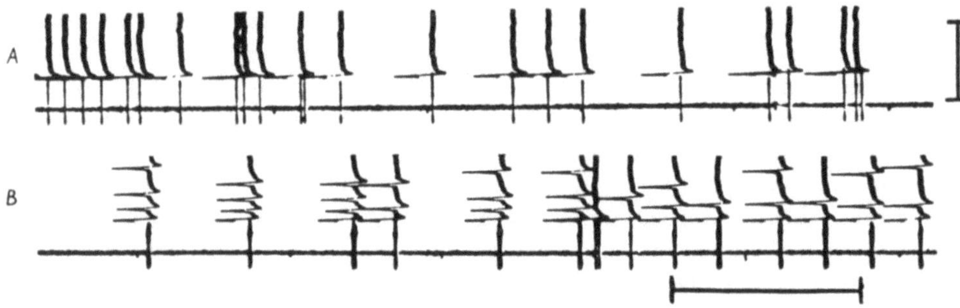

Fig. 7. *Maintained discharge of a lateral geniculate neurone of the cat in the aroused state* (A) *and asleep* (B).
The lower trace of each pair is the conventional spike record photographed on moving film; in order to resolve fine details of the discharge, the second beam of the oscilloscope was arranged to sweep vertically when the time base was triggered by an impulse. In A, the impulses occur singly and irregularly and each triggers a vertical sweep of the upper beam. In B, the impulses are grouped into high-frequency bursts and nearly every vertical sweep displays several impulses. Time scale for vertical records (right), 10 msec; for horizontal records, 1 sec. (From Hubel, 1960)

phasic potentials associated with the rapid eye movements. The two states may alternate with each other at variable intervals. The significance from the present viewpoint is that the dark discharge of geniculate cells is on average greater in REM sleep than in slow-wave sleep and greater even than in the aroused state (Fig. 8) (Sakakura and Iwama, 1967; Sakakura, 1968). Although the clustered firing patterns are considerably de-emphasized in both the aroused state and REM sleep, the discharges still differ: in REM sleep, there tend to be "surges" of impulses lasting for up to 0.5 sec, following the irregularly occurring ponto-geniculo-occipital potentials.

It is therefore clear that the geniculate maintained discharge has to be dealt with along several dimensions: 1) at different levels of illumination and darkness; 2) with respect to the visual properties of the units (on-centre, off-centre etc.); 3) according to the state of the rest of the brain. Species differences may be important because the work of Arden and Söderberg (1961) suggests that in the rabbit, the input from the reticular formation is more effective than that from the retina. Finally the work of Burke and Sefton (1966), confirmed by Suzuki and Kato (1966), Noda and Iwama (1967) and Sakakura and Iwama (1967) has established that geniculate units can be subdivided in yet another way into principal cells and interneurones. Since in no single investigation have all these factors been simultaneously recognized, it is not possible to do more than mention in conclusion some of the measurements that have been made. Since there is a great expansion in the number of distinct visual types of unit in the cortex (Hubel and Wiesel, 1962, 1968; Baumgartner et al., 1964; Barlow et al.,

1967; NIKARA et al., 1968) and multisensory convergence occurs (GRÜSSER et al., 1959; JUNG, 1961; KORNHUBER and DA FONSECA, 1961; HORN, 1963; MURATA et al., 1965), the problems of interpretation are even greater there.

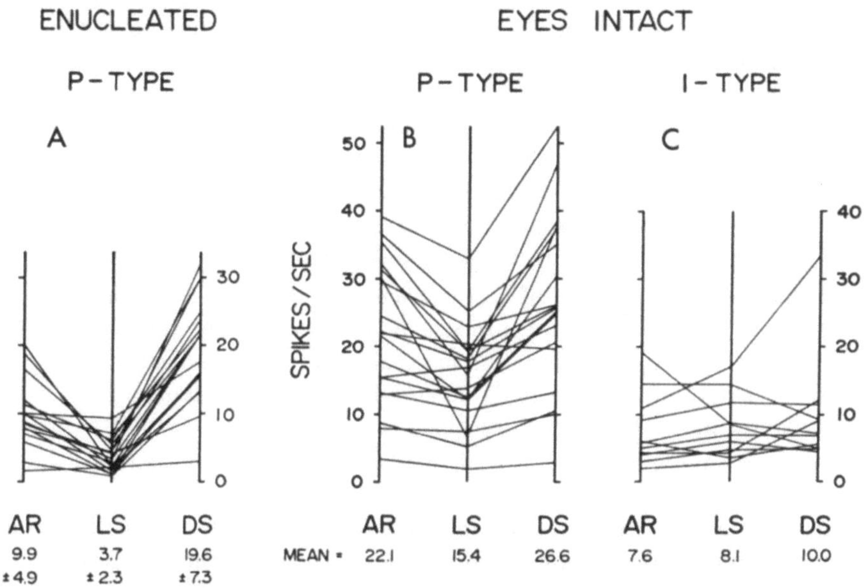

Fig. 8. *Maintained discharge rates of single geniculate neurones of P-type* (presumptive principal cells) *and I-type* (presumptive interneurones) *during three behavioural states* (*AR* = aroused; *LS* = slow-wave sleep; *DS* = rapid-eye-movement-sleep) with intact visual system (B, C) or after chronic enucleation of both eyes (A). Lines connect values from individual neurones, plotted on the three vertical scales of each figure. Mean discharge frequencies for the sample are given under the corresponding scale; standard deviations are also given below in A. Note that for particular *P*-type neurones, the lowest discharge rate usually occurs during slow-wave sleep, the highest during REM-sleep. This relation is preserved despite long-term (2 days +) enucleation which appears to have subtracted about 9 impulses/sec from the average maintained discharge in each state. (A from SAKAKURA, 1968; B and C from SAKAKURA and IWAMA, 1967)

It seems generally agreed that the maintained discharge rate at all levels of illumination including darkness becomes less on average as one goes from retina to geniculate to cortex (Fig. 3C, D) (STRASCHILL, 1966; HERZ et al., 1964). Superimposed upon this is the picture that only neurones of the D-system (off-centre at retina and geniculate) show sustained displacement of their discharge for changes of illumination at all levels of the visual system. Neurones of the B-system in the geniculate and cortex show only transient displacements of the discharge (JUNG, 1953, 1964; BAUMGARTNER et al., 1964). It has therefore been suggested that the brightness of a uniformly illuminated visual field or of background illumination is coded negatively by the maintained discharge of the D-system.

The intrinsic irregularity of the discharge has been investigated in geniculate and cortical neurones by the interval histogram. A wide variety of shapes has been found; attempts at classification are empirically based. Clustered firing of geniculate neurones appears as a peak at the shortest intervals. Intracellular or quasi-intracellular recordings have shown each cluster of spikes to arise from a unitary "delayed depolarizing potential" (McIlwain and Creutzfeldt, 1967; Fuster et al., 1965) which is itself associated with an optic tract synaptic potential. The remainder of the histogram thus corresponds to the intervals between those delayed depolarizing potentials or optic tract synaptic potentials which result in impulse-triggering. Some histograms reveal many peaks (Levick, 1963) which have been explained on the basis of interaction of excitatory and inhibitory afferents (see also Ten Hoopen, 1965). Histograms with less distinct multiple peaks are more common (Bishop et al., 1964; Herz et al., 1964). Radically different histograms are produced by the same unit under the conditions of arousal, slow-wave sleep and REM sleep (Sakakura and Iwama, 1967).

The maintained discharge of cortical neurones does not have the clusters of spikes typical of the geniculate during slow-wave sleep. The interval histogram usually revealed only a single peak; a long tail was typical, indicating long pauses of variable length between periods of activity. The significance of the various firing patterns has not yet been explained.

The overall picture is that the further one goes up the visual pathway, the more successfully do the units cancel the effect of the mean level of illumination. The neurones are designed to transduce more subtle aspects of the visual scene; where these functions are paired and oppositely directed, the significance of the maintained discharge may be to provide reasonable frequency response for low amplitude signals. The functional significance of the differences in maintained discharge in different states of sleep and arousal remains a mystery.

Acknowledgement

The secretarial assistance of Mrs. L. Cowan and Miss B. Ferguson was much appreciated.

References

Adrian, E. D.: The basis of sensation. The action of sense organs. London: Christophers 1928.
— Synchronized reactions in the optic ganglion of *Dytiscus*. J. Physiol. (Lond.) **91**, 66—89 (1937).
— Matthews, R.: The action of light on the eye. I. The discharge of impulses in the optic nerve and its relation to the electric changes in the retina. J. Physiol. (Lond.) **63**, 378—414 (1927).
— — The action of light on the eye. II. The processes involved in retinal excitation. J. Physiol. (Lond.) **64**, 279—301 (1928a).
— — The action of light on the eye. III. The interaction of retinal neurones. J. Physiol. (Lond.) **65**, 273—298 (1928b).
Arden, G. B., Söderberg, U.: The transfer of optic information through the lateral geniculate body of the rabbit. In: Sensory Communication, p. 521—544. New York-London: John Wiley and Sons Inc. 1961.
Arduini, A., Cavaggioni, A.: Transmission of tonic activity through lateral geniculate body and visual cortex. Arch. ital. Biol. **103**, 652—667 (1965).
— Hirao, T.: On the mechanism of the EEG sleep patterns elicited by acute visual deafferentation. Arch. ital. Biol. **94**, 140—155 (1959).
— Pinneo, L. R.: Properties of the retina in response to steady illumination. Arch. ital. Biol. **100**, 425—448 (1962).

ASCOLI, D., MAFFEI, L.: Slow periodicity in the dark discharge of retinal units. Experientia (Basel) **20**, 226—227 (1964).
ASHTON, N.: Degeneration of the retina due to 1 : 5-di(p-aminophenoxy) pentane dihydrochloride. J. Path. Bact. **74**, 103—112 (1957).
BARLOW, H. B.: Action potentials from the frog's retina. J. Physiol. (Lond.) **119**, 58—68 (1953).
— Retinal noise and absolute threshold. J. opt. Soc. Amer. **46**, 634—639 (1956).
— Increment thresholds at low intensities considered as signal/noise discriminations. J. Physiol. (Lond.) **136**, 469—488 (1957).
— Initial remarks. Gruppendiskussion von: Der Informationswert verschiedener Reaktionstypen der Neurone des visuellen Systems. In: Neurophysiologie und Psychophysik des Visuellen Systems. Berlin-Göttingen-Heidelberg: Springer 1961.
— The physical limits of visual discrimination. In: Photophysiology, Vol. 2. New York: Academic Press, Inc. 1964.
— BLAKEMORE, C., PETTIGREW, J. D.: The neural mechanism of binocular depth discrimination. J. Physiol. (Lond.) **193**, 327—342 (1967).
— FITZHUGH, R., KUFFLER, S. W.: Dark adaptation, absolute threshold and Purkinje shift in single units of the cat's retina. J. Physiol. (Lond.) **137**, 327—337 (1957 a).
— — — Change of organization in the receptive fields of the cat's retina during dark adaptation. J. Physiol. (Lond.) **137**, 338—354 (1957 b).
— HILL, R. M.: Selective sensitivity to direction of movement in ganglion cells of the rabbit retina. Science **139**, 412—414 (1963).
— — LEVICK, W. R.: Retinal ganglion cells responding selectively to direction and speed of image motion in the rabbit. J. Physiol. (Lond.) **173**, 377—407 (1964).
— LEVICK, W. R.: The Purkinje shift in the cat retina. J. Physiol. (Lond.) **196**, 2-3 P (1968).
— — Three factors limiting the reliable detection of light by retinal ganglion cells of the cat. J. Physiol. (Lond.) **200**, 1—24 (1969 a).
— — Changes in the maintained discharge with adaptation level in the cat retina. J. Physiol. (Lond.) **202**, 699—718 (1969 b).
BAUMGARTNER, G., BROWN, J. L., SCHULZ, A.: Visual motion detection in the cat. Science **146**, 1070—1071 (1964 a).
— EICHIN, F., SCHULZ, A.: Unterschiede neuronaler Aktivierung im zentralen visuellen System bei langdauernder Verdunkelung und Belichtung des Auges. Pflügers Arch. ges. Physiol. **279**, R 4 (1964 b).
BISHOP, P. O., LEVICK, W. R., WILLIAMS, W. O.: Statistical analysis of the dark discharge of lateral geniculate neurones. J. Physiol. (Lond.) **170**, 598—612 (1964).
BLANKENSHIP, J. E., KUNO, M.: Analysis of spontaneous subthreshold activity in spinal motoneurones of the cat. J. Neurophysiol. **31**, 195—209 (1968).
BORNSCHEIN, H.: Nachweis einer physiologischen Spontanaktivität in Einzelfasern des N. opticus der Katze. Experientia (Basel) **14**, 13—14 (1958 a).
— Spontan- und Belichtungsaktivität in Einzelfasern des N. opticus der Katze. I. Der Einfluß kurzdauernder retinaler Ischämie. Z. Biol. **110**, 210—222 (1958 b).
— Spontan- und Belichtungsaktivität in Einzelfasern des N. opticus der Katze. II. Der Einfluß akuter Jodazetatvergiftung. Z. Biol. **110**, 223—231 (1958 c).
BRIDGMAN, C. S., SMITH, K. U.: The absolute threshold of vision in cat and man with observations on its relation to the optic cortex. Amer. J. Physiol. **136**, 463—466 (1942).
BROWN, J. E., ROJAS, J. A.: Rat retinal ganglion cells: receptive field organization and maintained activity. J. Neurophysiol. **28**, 1073—1090 (1965).
BURKE, W., HAYHOW, W. R.: Disuse in the lateral geniculate nucleus of the cat. J. Physiol. (Lond.) **194**, 495—519 (1968).
— SEFTON, A. J.: Discharge patterns of principal cells and interneurones in lateral geniculate nucleus of rat. J. Physiol. (Lond.) **187**, 201—212 (1966).
CAJAL, S. R. y: Histologie du Système nerveux. Vol. II. French edition Madrid: Consejo superior de Investigaciones cientificas, Instituto Ramon y Cajal 1955.
CAVAGGIONI, A.: The dark-discharge of the eye in the unrestrained cat. Pflügers Arch. ges. Physiol. **304**, 75—80 (1968).
COLLINS, C. C.: Evoked pressure responses in the rabbit eye. Science **155**, 106—108 (1967).

Cox, D. R.: Renewal Theory. London: Methuen 1962.
Daw, N. W., Pearlman, A. L.: Cat colour vision: one cone process or several? J. Physiol. (Lond.) **201**, 745—764 (1969).
De Valois, R. L., Jacobs, G. H., Jones, A. E.: Effects of increments and decrements of light on neural discharge rate. Science **136**, 986—988 (1962).
Doty, R. W., Kimura, D. S.: Oscillatory potentials in the visual system of cats and monkeys. J. Physiol. (Lond.) **168**, 205—218 (1963).
Dowling, J. E.: Synaptic organization of the frog retina: an electron microscopic analysis comparing the retinas of frogs and primates. Proc. roy. Soc. B **170**, 205—228 (1968).
Edge, N. D., Mason, D. F. J., Wien, R., Ashton, N.: Pharmacological effects of certain diaminodiphenoxy alkanes. Nature (Lond.) **178**, 806—807 (1956).
Fatt, P., Katz, B.: Spontaneous subthreshold activity at motor nerve endings. J. Physiol. (Lond.) **117**, 109—128 (1952).
Fechner, G. T.: Elemente der Psychophysik. Leipzig: Breitkopf and Härtel 1860. Translated by H. Adler, edited by E. G. Boring and D. H. Howes: Elements of psychophysics. New York: Holt, Rinehart and Winston 1965.
Feller, W.: An introduction to probability theory and its applications, Vol. I. New York: Wiley 1957.
FitzHugh, R.: A statistical analyzer for optic nerve messages. J. gen. Physiol. **41**, 675—692 (1958).
Fuortes, M. G. F., Hodgkin, A. L.: Changes in time scale and sensitivity in the ommatidia of *Limulus*. J. Physiol. (Lond.) **172**, 239—263 (1964).
Fuster, J. M., Creutzfeldt, O. D., Straschill, M.: Intracellular recording of neuronal activity in the visual system. Z. vergl. Physiol. **49**, 605—622 (1965a).
— Herz, A., Creutzfeldt, O. D.: Interval analysis of cell discharge in spontaneous and optically modulated activity in the visual system. Arch. ital. Biol. **103**, 159—177 (1965b).
Geisler, C. D., Goldberg, J. M.: A stochastic model of the repetitive activity of neurons. Biophys. J. **6**, 53—69 (1966).
Gerstein, G. L., Mandelbrot, B.: Random walk models for the spike activity of a single neuron. Biophys. J. **4**, 41—68 (1964).
Gouras, P.: Spreading depression of activity in amphibian retina. Amer. J. Physiol. **195**, 28—32 (1958).
Granit, R.: Isolation of colour-sensitive elements in a mammalian retina. Acta physiol. scand. **2**, 93—109 (1941a).
— Rotation of activity and spontaneous rhythms in the retina. Acta physiol. scand. **1**, 370—379 (1941b).
— Spectral properties of the visual receptor elements of the guinea pig. Acta physiol. scand. **3**, 318—328 (1942).
— Sensory mechanisms of the retina. London: Oxford University Press 1947.
— Receptors and sensory perception. New Haven: Yale University Press 1955.
— Svaetichin, G.: Principles and technique of the electrophysiological analysis of colour reception with the aid of microelectrodes. Upsala Läk. Fören. För. **45**, 1—4, 161—177 (1939).
— Therman, P. O.: Excitation and inhibition in the retina and in the optic nerve. J. Physiol. (Lond.) **83**, 359—381 (1935).
Grüsser, O.-J., Grüsser-Cornehls, U., Saur, G.: Reaktionen einzelner Neurone im optischen Cortex der Katze nach elektrischer Polarisation des Labyrinths. Pflügers Arch. ges. Physiol. **269**, 593—612 (1959).
Gunter, R.: The absolute threshold for vision in the cat. J. Physiol. (Lond.) **114**, 8—15 (1951).
Hagiwara, S.: Analysis of interval fluctuation of the sensory nerve impulse. Jap. J. Physiol. **4**, 234—240 (1954).
Hartline, H. K.: Intensity and duration in the excitation of single photoreceptor units. J. cell. comp. Physiol. **5**, 229—247 (1934).
— The response of single optic nerve fibers of the vertebrate eye to illumination of the retina. Amer. J. Physiol. **121**, 400—415 (1938).
— Graham, C. H.: Nerve impulses from single receptors in the eye. J. comp. cell. Physiol. **1**, 277—295 (1932).

Heiss, W.-D., Bornschein, H.: Die Impulsverteilung der Daueraktivität von Einzelfasern des N. opticus. Einflüsse von Licht, Ischämie, Strychnin und Barbiturat. Pflügers Arch. ges. Physiol. **286**, 1—18 (1965).

— — Multimodale Intervallhistogramme der Daueraktivität von retinalen Neuronen der Katze. Kybernetik **3**, 187—191 (1966).

Helmholtz, H. von: Handbuch der Physiologischen Optik. II. Bd. Hamburg: Leopold Voss 1911. Translation from 3rd German edition, ed. J.C.P. Southall **1924**. Republished New York: Dover 1962.

Hering, E.: Zur Lehre vom Lichtsinne. Vienna: Carl Gerold's Sohn 1878. Translated by L.M. Hurvich and D. Jameson: Outlines of a theory of the light sense. Cambridge, Mass.: Harvard University Press 1964.

Herz, A., Creutzfeldt, O., Fuster, J.: Statistische Eigenschaften der Neuronaktivität im ascendierenden visuellen System. Kybernetik **2**, 61—71 (1964).

Hodgkin, A. L.: The local electric changes associated with repetitive action in a non-medullated axon. J. Physiol. (Lond.) **107**, 165—181 (1948).

Horn, G.: The response of single units in the striate cortex of unrestrained cats to photic and somaesthetic stimuli. J. Physiol. (Lond.) **165**, 80—81 P (1963).

Hubel, D. H.: Single unit activity in lateral geniculate body and optic tract of unrestrained cats. J. Physiol. (Lond.) **150**, 91—104 (1960).

— Wiesel, T. N.: Receptive fields of optic nerve fibres in the spider monkey. J. Physiol. (Lond.) **154**, 572—580 (1960).

— — Integrative action in the cat's lateral geniculate body. J. Physiol. (Lond.) **155**, 385—398 (1961).

— — Receptive fields, binocular interaction and functional architecture in the cat's visual cortex. J. Physiol. (Lond.) **160**, 106—154 (1962).

— — Receptive fields and functional architecture of monkey striate cortex. J. Physiol. (Lond.) **195**, 215—243 (1968).

Hughes, G. W., Maffei, L.: On the origin of the dark discharge of retinal ganglion cells. Arch. ital. Biol. **103**, 45—59 (1965).

— — Retinal ganglion cell response to sinusoidal light stimulation. J. Neurophysiol. **29**, 333—352 (1966).

Jouvet, M.: Neurophysiology of the states of sleep. In: The Neurosciences: A study program. New York: The Rockefeller University Press 1967.

Jung, R.: Neuronal discharge. EEG. clin. Neurophysiol. Suppl. **4**, 57—71 (1953).

— Neuronal integration in the visual cortex and its significance for visual information. In: Sensory Communication. New York-London: The M.I.T. Press and John Wiley and Sons Inc. 1961a.

— Korrelationen von Neuronentätigkeit und Sehen. In: Neurophysiologie und Psychophysik des visuellen Systems. Berlin-Göttingen-Heidelberg: Springer 1961b.

— Neuronale Grundlagen des Hell-Dunkelsehens und der Farbwahrnehmung. Bericht über die 66. Zusammenkunft der Deutschen Ophthalmologischen Gesellschaft in Heidelberg, 70—111 (1964). München: J. F. Bergmann 1964.

Kappauf, W. E.: Variation in the size of the cat's pupil as a function of stimulus brightness. J. comp. Psychol. **36**, 125—131 (1943).

Katz, B.: Nerve, muscle and synapse. New York: McGraw Hill 1966.

— Miledi, R.: A study of spontaneous miniature potentials in spinal motoneurones. J. Physiol. (Lond.) **168**, 389—422 (1963).

Kornhuber, H. H., da Fonseca, J. S.: Convergence of vestibular, visual and auditory afferents at single neurons of the cat's cortex. Intern. Congr. E.E.G. and Clin. Neurophysiol., 5th, Rome, 1961. Excerpta Medica, Intern. Congr. Ser. 1961.

Kuffler, S. W.: Neurons in the retina: organization, inhibition and excitation problems. Cold Spr. Harb. Symp. quant. Biol. **17**, 281—292 (1952).

— Discharge patterns and functional organization of mammalian retina. J. Neurophysiol. **16**, 37—68 (1953).

— FitzHugh, R., Barlow, H. B.: Maintained activity in the cat's retina in light and darkness. J. gen. Physiol. **40**, 683—702 (1957).

LAUFER, M., VERZEANO, M.: Periodic activity in the visual system of the cat. Vision Res. **7**, 215—229 (1967).
LEÃO, A. A. P.: Spreading depression of activity in the cerebral cortex. J. Neurophysiol. **7**, 359—390 (1944a).
— Pial circulation and spreading depression of activity in the cerebral cortex. J. Neurophysiol. **7**, 391—396 (1944b).
LEGRAND, Y.: Light, colour and vision. London: Chapman & Hall 1957.
LEVICK, W. R.: An interpretation of multimodal interval histograms. J. Physiol. (Lond.) **169**, 110—111 P (1963).
— Pattern abstraction in the rabbit's retina. In: Symposium on Information Processing in Sight Sensory Systems. Pasadena: California Institute of Technology 1965.
— WILLIAMS, W. O.: Maintained activity of lateral geniculate neurones in darkness. J. Physiol. (Lond.) **170**, 582—597 (1964).
— ZACKS, J. L.: Responses of cat retinal ganglion cells to brief flashes of light. J. Physiol. (Lond.) **206**, 677—700 (1970).
MARTINS-FERREIRA, H., DE OLIVEIRA CASTRO, G.: Light-scattering changes accompanying spreading depression in isolated retina. J. Neurophysiol. **29**, 715—726 (1966).
MATURANA, H. R., LETTVIN, J. Y., MCCULLOCH, W. S., PITTS, W. H.: Anatomy and physiology of vision in the frog *(Rana pipiens)*. J. gen. Physiol. **43**, Suppl. 2, 129—175 (1960).
MCILWAIN, J. T., CREUTZFELDT, O. D.: Microelectrode study of synaptic excitation and inhibition in the lateral geniculate nucleus of the cat. J. Neurophysiol. **30**, 1—21 (1967).
MICHAEL, C. R.: Receptive fields of opponent color units in the optic nerve of the ground squirrel. Science **152**, 1095—1097 (1966).
— Receptive fields of single optic nerve fibers in a mammal with an all-cone retina. II: Directional selective units. J. Neurophysiol. **31**, 257—267 (1968).
MOORE, G. P., PERKEL, D. H., SEGUNDO, J. P.: Statistical analysis and functional interpretation of neuronal spike data. Ann. Rev. Physiol. **28**, 493—522 (1966).
MORUZZI, G., MAGOUN, H. W.: Brain stem reticular formation and activation of the EEG. EEG. clin. Neurophysiol. **1**, 455—473 (1949).
MURATA, K., CRAMER, H., BACH-Y-RITA, P.: Neuronal convergence of noxious, acoustic and visual stimuli in the visual cortex of the cat. J. Neurophysiol. **28**, 1223—1239 (1965).
NIKARA, T., BISHOP, P. O., PETTIGREW, J. D.: Analysis of retinal correspondence by studying receptive fields of binocular single units in cat striate cortex. Exp. Brain Res. **6**, 353—372 (1968).
NODA, H., IWAMA, K.: Unitary analysis of retino-geniculate response time in rats. Vision Res. **7**, 205—213 (1967).
NOELL, W. K.: The effect of iodoacetate on the vertebrate retina. J. cell. comp. Physiol. **37**, 283—307 (1951).
— The impairment of visual cell structure by iodoacetate. J. cell. comp. Physiol. **40**, 25—55 (1952).
PERKEL, D. H., GERSTEIN, G. L., MOORE, G. P.: Neuronal spike trains and stochastic point processes. Biophys. J. **7**, 391—418 (1967).
RICCIARDI, L. M., ESPOSITO, F.: On some distribution functions for non-linear switching elements with finite dead time. Kybernetik **3**, 148—152 (1966).
RIGGS, L. A., GRAHAM, C. H.: Some aspects of light adaptation in a single photoreceptor unit. J. cell. comp. Physiol. **16**, 15—23 (1940).
RODIECK, R. W.: Maintained activity of cat retinal ganglion cells. J. Neurophysiol. **30**, 1043—1071 (1967).
— SMITH, P. S.: Slow dark discharge rhythms of cat retinal ganglion cells. J. Neurophysiol. **29**, 942—953 (1966).
RUSHTON, W. A. H.: The structure responsible for action potential spikes in the cat's retina. Nature (Lond.) **164**, 743—744 (1949).
— The difference spectrum and the photosensitivity of rhodopsin in the living human eye. J. Physiol. (Lond.) **134**, 11—29 (1956).
— A theoretical treatment of FUORTES's observations upon eccentric cell activity in *Limulus*. J. Physiol. (Lond.) **148**, 29—38 (1959).

SAKAKURA, H.: Spontaneous and evoked unitary activities of cat lateral geniculate neurons in sleep and wakefulness. Jap. J. Physiol. **18**, 23—42 (1968).
— IWAMA, K.: Effects of bilateral eye enucleation upon single unit activity of the lateral geniculate body in free behaving cats. Brain Res. **6**, 667—678 (1967).
SCHEIBNER, H., BAUMGARDT, E.: Sur l'emploi en optique physiologique des grandeurs scotopiques. Vision Res. **7**, 59—63 (1967).
STEIN, R. B.: A theoretical analysis of neuronal variability. Biophys. J. **5**, 173—194 (1965).
— Some models of neuronal variability. Biophys. J. **7**, 37—68 (1967a).
— The frequency of nerve action potentials generated by applied currents. Proc. roy. Soc. B **167**, 64—86 (1967b).
STEVENS, J. C., STEVENS, S. S.: Brightness function: Effects of adaptation. J. opt. Soc. Amer. **53**, 375—385 (1963).
STRASCHILL, M.: Aktivität von Neuronen im Tractus opticus und Corpus geniculatum laterale bei langdauernden Lichtreizen verschiedener Intensität. Kybernetik **3**, 1—8 (1966).
SUZUKI, H., KATO, E.: Binocular interaction at cat's lateral geniculate body. J. Neurophysiol. **29**, 909—920 (1966).
— TAIRA, N.: Effect of reticular stimulation upon synaptic transmission in cat's lateral geniculate body. Jap. J. Physiol. **11**, 641—655 (1961).
SWETS, J. A., TANNER, W. P., JR., BIRDSALL, T. G.: Decision processes in perception. Psychol. Rev. **68**, 301—340 (1961).
TAIRA, N., OKUDA, J.: Sensory transmission in visual pathway in various arousal states of cat. Tohoku J. exp. Med. **78**, 76—96 (1962).
TALBOT, S. A., KUFFLER, S. W.: A multibeam ophthalmoscope for the study of retinal physiology. J. opt. Soc. Amer. **42**, 931—936 (1952).
TANNER, W. P., JR., SWETS, J. A.: A decision-making theory of visual detection. Psychol. Rev. **61**, 401—409 (1954).
TEN HOOPEN, M.: Multimodal interval distributions. Kybernetik **3**, 17—24 (1965).
— Probabilistic firing of neurons considered as a first passage problem. Biophys. J. **6**, 435—451 (1966).
VAKKUR, G. J., BISHOP, P. O., KOZAK, W.: Visual optics in the cat, including posterior nodal distance and retinal landmarks. Vis. Res. **3**, 289—314 (1963).
WAGNER, H. G., MACNICHOL, E. F., JR., WOLBARSHT, M. L.: The response properties of single ganglion cells in the goldfish retina. J. gen. Physiol. **43**, Suppl. **2**, 45—62 (1960).
WEINSTEIN, G. W., HOBSON, R. R., BAKER, F. H.: Extracellular recordings from human retinal ganglion cells. Science **171**, 1021—1022 (1971).
WESTHEIMER, G.: The Maxwellian view. Vision Res. **6**, 669—682 (1966).

Chapter 9

Neuronal Changes in the Visual System Following Visual Deprivation

By

KAO LIANG CHOW, Palo Alto, California (USA)

With 4 Figures

Contents

I. Introduction

Many studies of the effects of neural function upon neural structure have demonstrated that, within genetically determined constraints, the neural circuitry of the mammalian brain can be modified by the life history of the organism. The role played by sensory stimulation in the development and maintenance of nervous tissue has been actively investigated. The optic centers provide a convenient system to test such afferent effects. The anatomy of the visual pathways is well established, and its functions can be easily assessed by both physiological and behavioral tests. Furthermore, it is possible to prevent visual stimulation completely, a degree of stimulus-control not easily achieved in other senses.

Taking advantage of these facts, investigators as early as the beginning of the century examined in many species the changes in visual centers produced by varying the amount of light stimulation. In lower vertebrates the presence or absence of light stimulation caused reversible changes in retinal structures, such as forward or backward migration of epithelical pigment, contraction or elongation of the receptors, and increased or decreased staining of Nissl material in ganglion cells (DETWILER, 1943). CARLSON (1902) demonstrated in the bird retina that the degree of Nissl staining of bipolar and ganglion cells could be altered by light stimulation. These positive findings are in contrast to the equivocal results obtained from the early studies of mammals. VON GUDDEN (cited in GOODMAN, 1932) and BERGER (1900) were the first to use eyelid-suturing for controlling light input. They reported histological changes in the visual cortex of deprived rabbits, cats, and dogs. GOODMAN (1932) took exception to their results, contending that this method did not exclude all light penetration, and that it could damage the eyeball and hence cause secondary degeneration of the visual centers. The reported changes (i.e. thining of cortex and reduction in number of Nissl granules) were, moreover, slight and not quantitatively studied.

GOODMAN's own study in 1932 attempted both to improve the method of light deprivation and to quantify the histological findings. The brains of six month old rabbits reared from birth in darkness were compared with controls reared in a normal laboratory environment. Other animals, after either eyelid-suturing or enucleation, were also studied. He failed to detect any difference between the brains of the dark-reared and normal animals. Also, there were no group differences in the measurements of the thickness of the cell layers of the retina, superior colliculus, or striate cortex, or in the number of cells per unit area of these structures. In contrast, the superior colliculus and lateral geniculate nucleus of an enucleated animal showed clear degeneration of the optic fiber layers and were smaller than normal. This was secondary degeneration resulting from recision of the end organ and not from the absence or reduction of function. The completely negative findings of GOODMAN probably discouraged further investigation, and this subject lay dormant for the next two decades. Renewed interest in the histology and biochemistry of optic centers following visual deprivation awakened in the fifties, a development correlated with the numerous studies of the effects of visual deprivation on visually guided behavior. RIESEN's (1966) review clearly revealed the stimulating interplay of these two research areas. With the advent of sophisticated cytological, histochemical, and electrophysiological techniques, research on this subject developed rapidly and has reached a new stage in recent years.

II. Methodological Considerations

Contrary to GOODMAN's results, most of the later studies showed that visual deprivation affected the structure and physiology of the visual system. The effects were varied and sometimes contradictory, partly reflecting differences in species, age of the animal, method of light exclusion, and anatomical and physiological techniques used. As will be made clear, many of these variables are relevant, but were not always fully emphasized in interpreting the experimental results.

Workers on mammalian species have used the mouse, rat, rabbit, cat, dog, monkey, and the chimpanzee. The duration of visual deprivation, beginning from birth, varied from a few days to 2 years. Since the degree of maturity of the eye and brain at birth are not the same for all mammals, the lack of stimulation may influence the relatively immature organ of one species more than the comparatively mature one of another. The widely varied life span of the animals also makes it difficult to compare directly the effect of the length of deprivation among species. Is keeping a mouse in a darkroom for one month equivalent to depriving a chimpanzee of light stimulation for 5 years?

Another vexing point is that varying the duration of visual deprivation may reveal confounding interactions, underlying observed functional-structural relationships. The withdrawal of afferent input may retard normal growth, leaving intact the immature tissue, or it may cause atrophy of neural structure. GYLLENSTEIN et al. provided a rare example illustrating this point (1959, 1965). The visual cortical cells (layers 2–4) of mice reared from birth to 2 months of age in darkness had a smaller nuclear diameter and less internuclear material than their normally reared litter mates. However, when the duration in the darkroom was increased to 4 months, similar measurements showed little difference from normal. Presumably, the lack of stimulation initially arrests the normal growth, but later the maturation process overrides this obstacle. If these mice are kept still longer in the darkness, the normalized cells may atrophy.

Other investigators have attempted to produce structural changes of visual centers by keeping adult or young animals in total darkness. Here again, relevant variables are the stage of neural development at which deprivation was initiated and the duration for which it was maintained. Light deprivation may cause irreversible damage in an adult animal, which can be interpreted as reflecting a secondary degenerative process of viable tissue, or it may cause metabolic and enzymatic abnormalities that recover after reinstitution of light stimulation. These two variables, i.e., degree to which the changes are reversible and the level of maturation, are not always clearly controlled in the experimental design. The irreversible changes may represent either an arrest of growth or a regress from a more mature state. The reversible changes, could represent either a normal growth in the process of catching-up after a delay, or the repairing of atrophied tissue.

Various methods have been used to prevent visual inputs. Maintaining animals in a darkroom is convenient and widely used, since there is little possibility of mechanical damage to the eyeball. This technique has the limitation of permitting only binocular deprivation, and thus preventing the intra-subject controls possible with other methods. The method of eyelid suturing permits either monocular or binocular deprivation, but does not exclude all light stimulation (10–20% may penetrate to the retina) although patterned light is prevented. Also, mechanical damage to the eyeball may occur. The advantage of monocular deprivation certainly does not mitigate the additional complication of the peculiar effect that this procedure has on the neural centers. For example, most visual cortical cells are normally excited by both eyes. When one eye is occluded, the resulting imbalance of activity may cause effects quite different from those obtained when stimulation is reduced equally for both eyes.

Most of the studies on structural changes following visual deprivation are based on histological material prepared for light microscopy. The introduction of quantitative measurement and histochemical staining adds a new dimension to these endeavors. The few investigations of changes in ultra-structure and enzymatic activity give only tantalizing hints of the power of these new techniques. In addition to demonstrating an effect, these methods may provide a wedge into the revelation of the molecular mechanisms underlying these changes. Studies of the electrical activity of single neurons or aggregates of neurons are few, and many more are needed. These physiological data, together with the many behavioral effects reported, provide the information needed for a structure-function analysis of the consequences of visual deprivation. The following is a summary and evaluation of the histological, neurochemical, and physiological changes in visual centers that result from visual deprivation. These results are compared with those accompanying transneuronal degeneration after optic enucleation.

III. Effect on Retina and Optic Nerve

1. Histological and Neurochemical Findings

Reports on histological observations of the retina and optic nerve in animals deprived of visual stimuli have been conflicting. Neurochemical studies, on the other hand have generally agreed in demonstrating decreased protein and RNA concentration, as well as changes in enzymatic activity. Table 1 summarizes these studies. Only experiments in which deprivation was initiated at birth are included in Tables 1–3. Studies testing young or adult animals will be discussed separately. Goodman's 1932 paper is the first study listed since it set a standard in experimental design and quantitative method for histological data. This study should not be ignored even though completely negative.

Of the nine studies reporting histological results alone (Table 1, No. 1, 4, 5, 8, 10, 11, 12, 13, 15) or the two with additional neurochemical data (No. 6, 7), six (No. 4, 5, 7, 11, 12, 13) demonstrated structural damage of the retina or optic nerve in the mouse, rat, cat, and chimpanzee. The results included decreases in number and myelinization of optic nerve fibers, density of Müller fibers, thickness of inner plexiform layer, and number of ganglion cells. One of the severest effects appeared in the retina of a chimpanzee that was reared in the dark from birth to 33 months of age (with a few minutes of light exposure daily) and then in a normal environment for another eight years. Fig. 1 b illustrates the temporal side and fovea of the left retina, showing almost complete absence of ganglion cells. It contrasts sharply to the normal retina of another chimpanzee (Fig. 1 a) that was reared, until 7 months old, in the dark except for one and one-half hours daily exposure to diffuse light. The difference in the staining quality is partly due to the different section thicknesses.

The other five studies showed no histological changes. The differences between the studies reporting positive and negative results seem to be due to the selection of experimental animal and the varying lengths of deprivation. Studies yielding negative results generally referred to either a shorter period of visual deprivation (No. 8, 10, 15) or to a particular species, the albino rabbit (No. 1, 6). One may speculate that the rabbit's retina is less easily affected by light deprivation than other species, and that lengthening the deprivation time would eventually cause retinal changes regardless of species. In this regard, the results of Gyllenstein et al. pose a further complication. He (1963, 1966) showed that the optic nerves of mice reared in darkness for 30 days had smaller numbers of fibers and decreased myelinization. By increasing the duration to 4 months these changes were reversed; now the optic nerve recovered to a normal state, but the retina showed

Table 1. Effects of visual deprivation on retina and optic nerve

Author	Animal	Method of deprivation	Length of deprivation	Preparation and methods	Result
1. GOODMAN (1932)	albino rabbit	in darkness	6 months	Nissl and Golgi stain	No change in thickness of the layers or the number and the dendritic branches of the ganglion cells
2. BRATTGÅRD (1952)	albino rabbit	in darkness	2, 4 months	X-ray radiography, ultraviolet absorption	Decrease of protein fraction and RNA concentration of ganglion cell layer
3. HELLSTRÖM and ZOTTERSTRÖM (1956)	cat	in darkness	3—24 days	Nissl stain and sulfhydryl stain	Decreased S—H group staining of retina
4. CHOW et al. (1957)	chimpanzee	in darkness	33 months (few minutes in light daily)	Nissl and protargol fiber stain	Severe degeneration of ganglion cell layer and optic nerve
5. WEISKRANTZ (1958)	cat	in darkness	4 months (30 min in light during last 28 days)	Hematoxylin	Decreased thickness of inner plexiform layer and decreased density of Müller fibers and optic fibers
6. SCHIMKE (1959)	albino rabbit	in darkness	26, 60, 120 days	Retina weighted and homogenized for enzyme essay	Small, significant decrease in activity of malic and lactic dehydrognese, acid phosphatase; increase in glucose 6-phosphate dehydrognese. No change in total weight, total protein, or Nissl stained histology
7. RASCH et al. (1961)	hooded rat, cat, chimpanzee	in darkness	rat, 2, 3 months cat, 3.5—36 months chimpanzee, 33 months	Nissl stain, Azure-B stain and photometric determination of RNA, fast green stain or Millon reaction for protein	Decreased cytoplasmic and nucleolar RNA in cells of all three layers, with decrease protein, smaller ganglion cells and thinner inner plexiform layer

Table 1 (continued)

Author	Animal	Method of deprivation	Length of deprivation	Preparation and methods	Result
8. Terry et al. (1962)	hooded mouse	eyelid closure (unilateral and bilateral)	3—9 weeks	Hematoxylin eosin	No change in general appearances of optic nerve
9. Liberman (1962)	hooded rat	in darkness	17 weeks	AChE in homogenized retina, glycolysis in incubated retina	No change in glycolytic activity, decreased AChE
10. Wiesel and Hubel (1963a)	cat	eyelid closure, contact lens (unilateral)	3 months	Nissl and silver fiber stain	No change in the general appearance of retina and optic nerve, no change in the thickness of retinal layers
11. Gyllensten and Malmfors (1963)	hooded mouse	in darkness	20, 30 days	Osmium	Decreased number and diameter of myelinated fibers
12. Wendell-Smith (1964)	hooded mouse	eyelid closure with tantalum shield (unilateral)	75 days	Potassium permanganate	Optic nerve cross-sectional area decreased 10%
13. Gyllensten et al. (1966)	hooded mouse	in darkness	2, 4 months	Osmium toluidine blue, and hematoxylin	No change in number and diameter of optic fibers, but decrease in thickness of retina and diameter of ganglion cells
14. Maraini et al. (1967)	albino rat	eyelid closure (unilateral)	4—7 months	autoradiography of tritiated leucine	No difference in leucine in corporation in the retinal layers
15. Fox et al. (1968)	dog	eye bandaged (bilateral)	1—35 days	Nissl and myelin stain	No change in the general appearance of retina and optic nerve

slight hypotrophy. Gyllenstein called attention to this reversible effect of prolonged visual deprivation, and suggested that the maturation of the optical nerve was independent of retinal conditions. He also reported that both the numbers and the diameters of the optic nerve fibers in mice that had been kept in darkness from the fourth to the seventh month of age were increased above those of their normal age mates. In this case the absence of function caused an abnormal increase of nerve fibers.

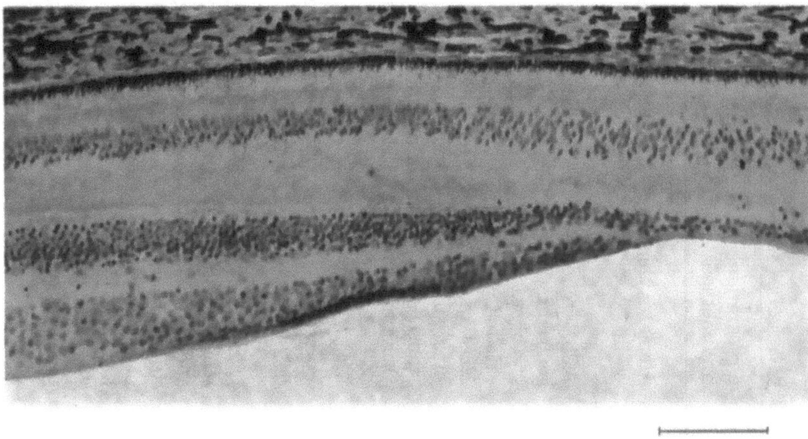

a

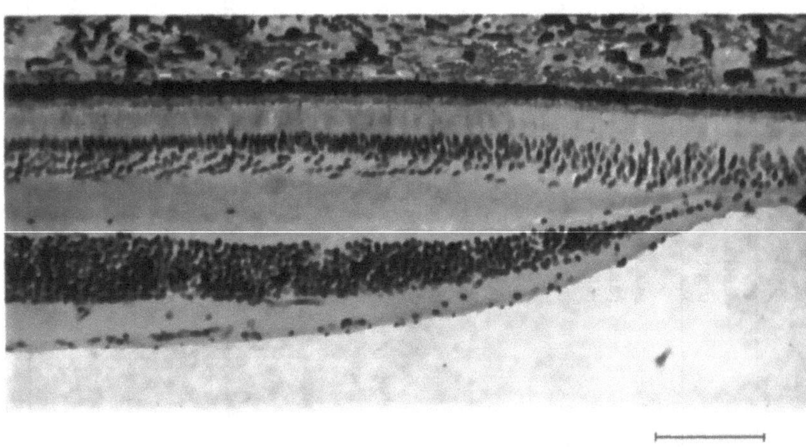

b

Fig. 1 a and b

Fig. 1. *Dark-reared chimpanzee's retina and lateral geniculate body*, compared to normal retina. a The temporal side and fovea of a chimpanzee's *normal*, left retina. 8 μ, hematoxylin and eosin. b The temporal side and fovea of a *dark-reared* chimpanzee's (Snark) left retina, showing the severe depletion of the ganglion cell layer. 12 μ, hematoxylin and eosin. c A cross-section of the left lateral geniculate body of the same *dark-reared* chimpanzee whose retina is shown in b. Note the heavy gliosis and paucity of neurons throughout the medial two-thirds of the nucleus. 50 μ, thionine. Calibration in a and b, 0.1 mm; in c, 1 mm

Five neurochemical studies reported some alterations of retinal composition following visual deprivation (No. 2, 3, 6, 7, 9). Brattgård (1952) introducing modern histochemical techniques to these studies, reported a significant decrease of the protein fraction and RNA concentration of the ganglion cell layer of rabbits reared in darkness. Later, Rasch et al. (1961) confirmed and extended these results to the retinas of the rat, cat, and chimpanzee. They also demonstrated, for the first

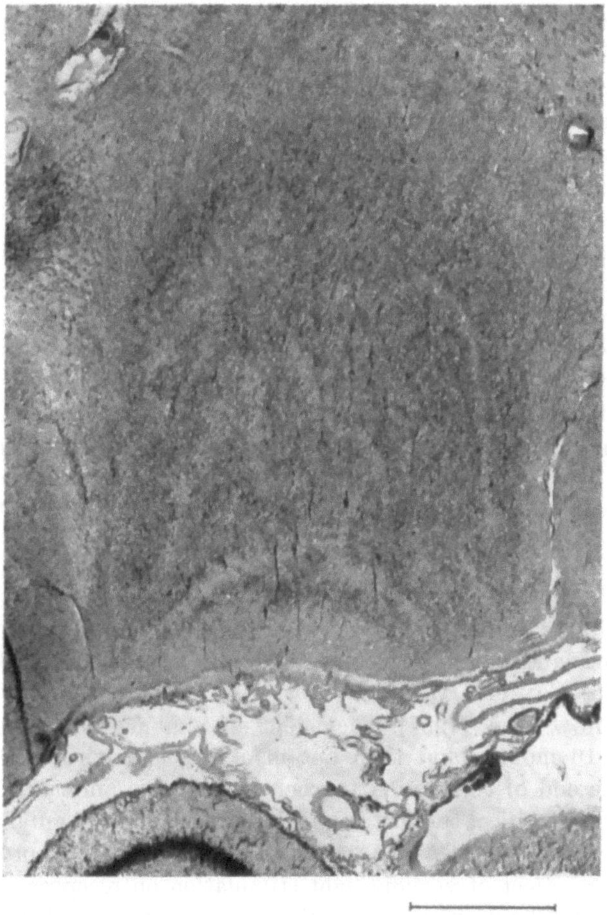

Fig. 1 c

time, changes in the bipolar and receptor cell layer, namely, decreased RNA in these cells of the cat retina. In addition, various enzymatic activities were depressed by visual deprivation including acetylcholinesterase (Liberman, 1962) malic and lactic dehydrogenase, acid phosphatase, (Shimke, 1958) and the sulfhydryl group (Hellström and Zotterström, 1952). Glycolytic activity remained stable (Liberman, 1962), but the activity of glucose 6-phosphate dehydrogenase increased above the normal level (SHIMKE, 1958). Thus, the absence of light stimulation in newborn animals hinders RNA and protein synthesis, and alters the enzymatic activity of the retina. In the majority of cases, this procedure will eventually result in cell death in the ganglion cell layer, but with little or no effect on the bipolar and receptor cells.

Attempts to influence protein production, enzymatic activity or structural integrity of the retinas of older animals either by increasing or decreasing the light stimulation were less successful. That the amount of light stimulation could affect the protein concentration in ganglion cells was demonstrated as early as 1902 by CARLSON. However, this effect was short-term and reversible, signifying the extreme range of metabolic turnover within which a cell can normally function. More recent studies of a similar nature such as that of GIOMIRATO and BAGGIO (1962) show a marked decrease of the nuclear protein mass in rabbit ganglion cells after 3 h of intense light stimulation but only a slight reduction after 3 h in total darkness. These effects were increased when the animal was under barbiturate anesthesia. GLOW and ROSE (1964) reported an inhibition of acetylcholinesterase in rat's ganglion cells by intense light stimulation and a decrease in rate of acetylcholine synthesis after the rat was kept in darkness up to 20 days. BECH (1957) failed to demonstrate any changes of the nucleic acid content of ganglion cells of adult rabbit living 1–14 days in darkness, however, 1 h of light stimulation given to these animals caused a slight but significant increase of nucleic acid concentration of the ganglion cells which was not further potentiated by stronger light. Unlike what he found in the newborn rabbits, SHIMKE (1959) could not observe any change of retinal enzymatic activity in adult albino rabbits after living in darkness for as long as 120 days.

The retinal histology of young and adult animals also proved to be impervious to the lack of light stimuli. Such negative findings have been reported for mouse, rabbit, cat, dog, and monkey (TERRY et al., 1962; SHIMKE, 1959; BURKE and HAYHOW, 1968; FOX et al., 1968; CHOW, 1955). One exceptional case is that reported by CHOW et al. (1957). They kept an 8 month old chimpanzee in the darkroom for 16 months and then out in a normal environment for another 6 years. At the end of this period the behavior of this animal indicated deficit vision, and after sacrifice, the histology of its retina showed an almost complete absence of the ganglion cell layer, but apparently normal receptor layers. It is possible that the effect of varying light stimulation on receptors are subtle and not detectable by light microscopy. KUWABARA (1966) and KUWABARA and GORIN (1968) confirming earlier light microscopy findings (NOELL et al., 1966), showed in electron microscopy that strong light disrupted the flat, laminated discs in the outer segment of photoreceptors of albino rat. After exposure to strong light up to 7 days the two membranes of the discs became separated and distorted into torturous tubular structures. DE ROBERTIS and FRANCHI (1956) reported a decrease in the size of rod and cone synaptic vesicles in albino rabbits kept in darkness from 2–9 days which was not confirmed by MOUNTFORD (1963). CRAGG (1969b) in a carefully designed study showed that retinal receptor terminals increase in width in dark-reared rats but subsequently decrease in width after exposure to day light for only 3 min. In the dark the synaptic vesicles are more widely dispersed in the large terminals than those of light-exposed rats, the total number of vesicles, however, was unchanged. An increase of conventional synapses of amacrine cells was also reported to occur in the inner plexiform layer of light deprived rat retina (SOSULA and GLOW, 1971). But there was no change in the axonal migration rate of protein in retinal ganglion cells of rabbit (KARLSSON and SJOSTRAND, 1971). These few studies suggest that a systematic electronmicroscopic survey of the deprivation effects on retinal structures may help to resolve the manifold results.

In summary, light stimulation regulates the nucleic acid, protein synthesis, and many enzymatic activities of the retina of newborn aminals. Elimination of functional activity disrupts their normal process ultimately leading to cell death. The ganglion cells are more vulnerable than bipolar and receptor cells. The retinal metabolic machinery of adult animals, on the other hand, is much less altered by the amount of light stimulation. Visual deprivation seldom results in irreversible destruction of the retinal cells of mature animals.

2. Neurophysiological Findings

The electroretinogram (ERG) has been widely used as an index of the functional state of the retina. The origin of its components has been assigned to different retinal layers (BROWN, 1968). The *b*-wave of cat ERG was the only one shown to be affected by light deprivation. ZETTERSTRÖM (1956) reported that the *b*-wave first appeared in 2-week old kittens reared in a normal environment, but was delayed to the fourth week in dark-reared kittens. This may be related to a decreased sulfhydryl content of the outer and inner nuclear layer (HELLSTRÖM and ZETTERSTRÖM, 1956). WIESEL and HUBEL (1963b) also obtained normal ERG in one kitten with one lid sutured shut for two months. On the other hand, when the deprivation was lengthened to one year or longer, the cat ERG showed a decrease in *b*-wave amplitude and an increase in the time interval needed for the second of a pair of light flashes to produce a *b*-wave as large as that evoked by the first (BAXTER and RIESEN, 1961; GANZ et al., 1968). This retardation of recovery was quickly restored to normal after a few flash presentations. The *b*-wave amplitude was also decreased when adult cats were kept in darkness for 1 week, and this was reversed after a few days in normal light environment (CORNWELL and SHARPLESS, 1962). The *b*-wave represents the slow-wave activity of the inner nuclear layer and this layer appears intact in histological studies. One may wonder whether the ERG provides a more sensitive measure of retinal function than the histological and neurochemical methods. Ganglion cell activity is not represented in the ERG, therefore the severe defect of this cell layer described in the previous section could not be assessed by this method.

WIESEL and HUBEL (1963) have recorded seven optic tract fibers in a kitten that had had one eye sutured from birth for 2 months. These units (2 were activated from the normal eye, 5 from the formerly closed eye) were completely normal in their response to light stimulation and in their concentric receptive-field organization.

IV. Effect on Lateral Geniculate Nucleus

1. Histological and Neurochemical Findings

The optic fibers terminate primarily on the lateral geniculate nucleus which in turn projects fibers to the visual cortex; other retinal fibers reach the pretectum and superior colliculus. Normal neural activity impinging upon these visual centers can be blocked either by light deprivation or by enucleation. The latter procedure inflicts trauma on the optic nerve causing it to disintegrate, with the eventual atrophy of the postsynaptic cells. Transneuronal change caused by visual deprivation will be compared with the degeneration resulting from enucleation.

The retina provides a nearly exclusive afferent supply to LGN neurons, which rely upon these optic inputs to maintain their integrity. LGN cells undergo transneuronal degeneration after enucleation as was first shown by MINKOWSKI (1920) and LE GROS CLARK (LE GROS CLARK and PENMAN, 1934; GLEES and LE GROS CLARK, 1941) many years ago. The time course, and severity, of degeneration following enucleation varied greatly depending on the animal species and age at the time of deafferentation. For example, HESS (1958) demonstrated atrophy in visual centers of fetal guinea pig (gestation age, 4–6 weeks) a few days after removal of the eye. MATTHEWS et al. (1960) reported a reduction of Nissl substance in neurons and increased gliosis in monkey LGN 4–8 days after enucleation. On the other hand, the LGN of rat, rabbit and cat showed a much slower onset of atrophy, beginning only after weeks or months. Transneuronal degeneration is indicated by decreases in cell size, intensity of Nissl stain, and number of neurons (TSANG, 1937). The amount of reduction in cross-sectional area of atrophied cells has been estimated as 25% in the cat and rabbit (COOK et al., 1951) and as much as 60% in the monkey (MATTHEWS et al., 1960). Except in the monkey (MATTHEWS, 1964) and in man (GOLDBY, 1957), this degenerative process does not result in discernable cell loss in the LGN.

Neurochemical changes were also reported to occur in the LGN following enucleation. These include increased deoxyribonucleic acid content of the cellular nuclei (EECKEN and FAUTREZ, 1965), decreased di- and triphosphopyridine nucleotide diaphorase and acid phosphatase (GAY and SILBERBERG, 1964) and decreased concentration of acetylcholine (MILLER et al., 1969).

Newborn animals deprived of light stimulation show comparable anatomical changes in the LGN. These changes are less severe than those following enucleation, and seldom resulted in detectable neuronal death. Seven of the studies listed in Table 2 (No. 3–6 and 9–11) reported reduced cell size (up to 40%), along with decreased amounts of internuclear material, less cells, or diminished Nissl staining quality. There was also less leucine uptake and acetylcholinesterase concentration in the LGN of deprived animals (No. 7 and 8). The study by KUPFER and PALMER (1964) showed that although the neurons of deprived kittens were small, the enzymatic activity of the cells was similar to that of normal neurons.

Visual deprivation, therefore, may arrest the neuronal growth but not cause degeneration of normal tissue. This concept may be relevant to other studies

Table 2. Effect of visual deprivation on lateral geniculate nucleus

Author	Animal	Method of deprivation	Length of deprivation	Preparation and methods	Result
1. Goodman (1932)	albino rabbit	in darkness	6 months	Nissl and Golgi stain	No change
2. Terry et al. (1962)	hooded mouse	eyelid closure (unilateral and bilateral)	3—9 weeks	Hematoxylin and eosin	No change
3. Wiesel and Hubel (1963a)	cat	eyelid closure (unilateral)	3 months	Nissl stain	Decreased cell cross-sectional area by 40%, pale Nissl stain
4. Kupfer and Palmer (1964)	cat	eyelid closure (unilateral)	1—4 weeks	Nissl stain and cytochemical determination of TPN- and DPN-diaphorase, ChE, dehydrognese of 6-phosphagloconate, succinate and lactate-DPN	Decreased cell size and staining quality but normal enzymatic activity
5. Wiesel and Hubel (1963a)	cat	eyelid closure (unilateral and bilateral)	3 months	Nissl stain	Decreased cell cross-sectional area by 40%, no recovery after 3—15 months in light
6. Gyllensten et al. (1965)	hooded mouse	in darkness	2, 3, 4 months	Gallocyanin	Decreased internuclear materials

Table 2 (continued)

Author	Animal	Method of deprivation	Length of deprivation	Preparation and methods	Result
7. Maraini et al. (1967)	albino rat	eyelid closure (unilateral)	4—7 months	Autoradiography of tritiated leucine	Decreased leucine uptake in the neurons
8. Maletta and Timiras (1967)	hooded rat	in darkness	21 days	Colorimetric determination of AChE and ChE	Decrease of AChE. No change of ChE
9. Fifkova and Hassler (1969)	albino rat	eyelid closure (unilateral)	3 months	Nissl stain	Total volume decreased 5%, cell number decreased 17%
10. Chow and Riesen (unpublished)	chimpanzee	in darkness	33 months (few minutes in light daily	Nissl and Weil stain	Severe decreased neuronal size and increased gliosis in patches throughout the nucleus, pale Nissl stain
11. Chow (unpublished)	cat	eyelid closure (unilateral)	20—24 months	Nissl and Weil stain	Decreased neuronal cross-sectional area by 40%, but recovered to almost normal size after 7—10 months intensive training after eye-opening

listed in the table. Thus the results of Wiesel and Hubel (1965b) were interpreted to indicate more than retarded cell growth. They reported that the reduced cell-size in the LGN of kittens deprived of pattern vision for the first 3 months of life remained even after 3–15 months in a normal light environment. On the other hand, both Dews and Wiesel (1970) and our own data (Chow and Stewart, 1972) showed that some (but not all) cats can regain some vision after visual deprivation from birth if they are subsequently forced to use the deprived eye. The LGN cells of our animals appear to be of normal size. Fig. 2a is a cross-section of a cat's LGN which was unilaterally enucleated at birth and sacrificed 24 months later. Fig. 2b shows the decreased cell size of the LGN in a cat following unilateral eyelid suturing from birth for 23 months. Figs. 3a and 3b show LGN sections of 2 cats who had their left eyelid sutured from birth to 20 and 22 months of age, then their left eye was opened and their right eye closed for 7 and 8 months.

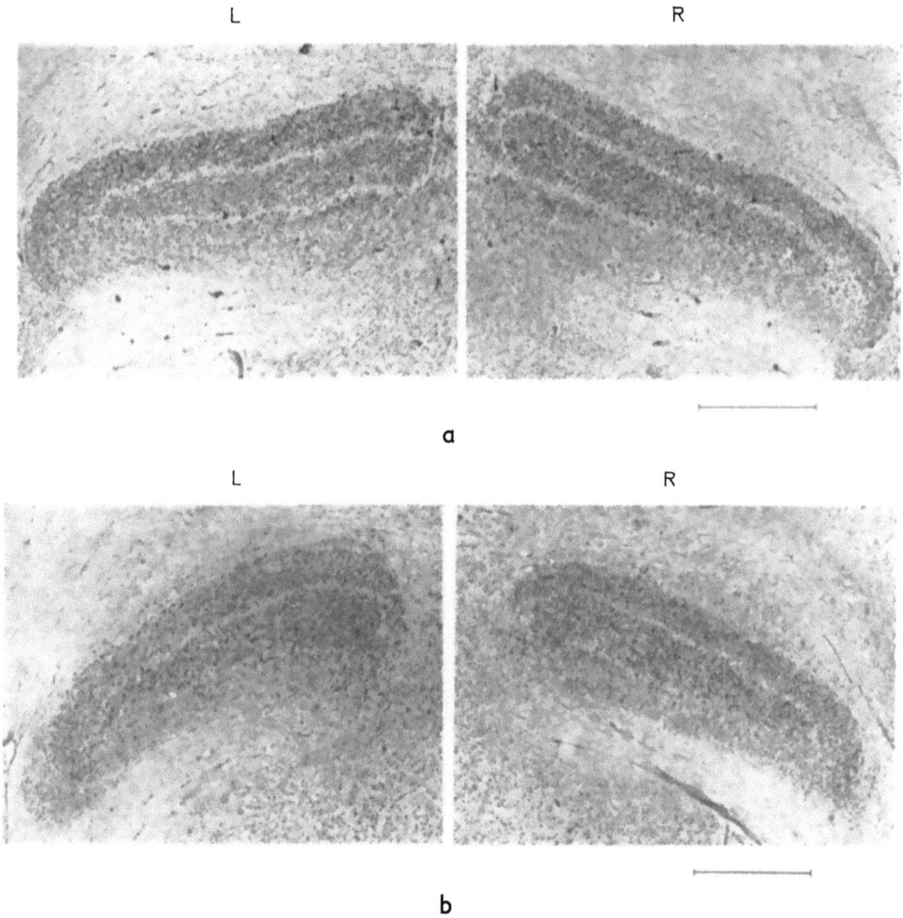

Fig. 2. a *The lateral geniculate body of cat* (ECLG 8) *right eye enucleated at birth*, sacrificed at 23 months of age. On this and the photomicrographs in Fig. 3, the LGN of the two hemispheres were taken from the same thionine stained section; *L*, left LGN; *R*, right LGN; calibration, 1 mm. b The LGN of cat ECLG 7, left eyelid sutured at birth, sacrificed at 24 months of age. Note the degenerated layers receiving the optic inputs from the treated eye

During this time the cats were trained to use their left eye to learn visual pattern discriminations. In contrast to the LGN of cats which were visually deprived from birth and never used the deprived eye, the LGN cells in adjacent layers of the latter 2 cats showed little size differences. Approximately 400 cells in each geniculate of the two animals shown in Figs. 2b and 3b were measured. For the animal shown in Fig. 2b, the cross-sectional areas of cells in the layers receiving optic fibers from the early-deprived left eye (X, 178 μ^2; S.D., 72) did not differ from the early-normal right eye (X, 159 μ^2; S.D., 64). For the LGN cells in Fig. 1b, the comparable values were 115 μ^2 (S.D., 38) and 185 μ^2 (S.D., 82). The 40% reduction in the size of the LGN cells caused by visual deprivation shown in Fig. 1b is comparable to that reported by Wiesel and Hubel (1963a).

The reduction of cell size may be due to either arrest of normal growth or degeneration of mature cells and the subsequent forced use may either renew

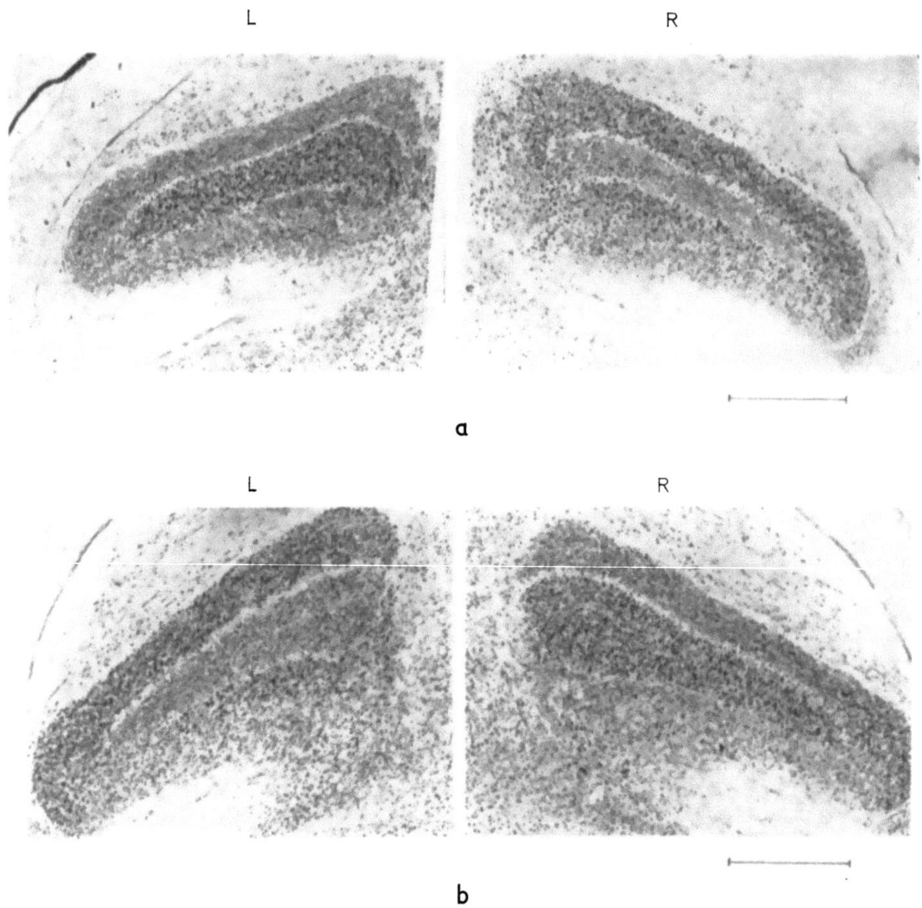

Fig. 3. a *The lateral geniculate body of cat* (ECLG 9) *left eyelid sutured at birth,* at 20 months of age the left eyelid was opened and the right eyelid sutured, sacrificed at 27 months. b *The LGN of cat* (ECLG 12) *left eyelid sutured at birth,* at 22 months of age the left eyelid was opened and the right eyelid sutured, sacrificed at 30 months. Note the lack of clear-cut difference in the appearances of the layers

the growing process or restore the damaged tissue. The difference between these results is probably due to several factors, e.g. the individual differences in the pigmentation of the eyelid, the length of deprivation, and, more importantly, the degree of nervousness and neurotic behavior of the deprived animals. They do not regain vision unless special handling, training, and maintenance conditions are instituted.

Two chimpanzees, one reared in darkness from birth and the other from 8-months to 24-months old showed degeneration of the retinal ganglion cell layer (Chow et al., 1957). In addition, severe degeneration occurred in the LGN. Fig. 1c is a transverse section through the LGN of the chimpanzee whose retinal ganglion cells were severely depleted (Fig. 1b). Widespread degeneration appeared in the LGN, viz. regional gliosis, paucity of neurons, and reduced size of the remaining neurons. It may be significant that the few remaining ganglion cells were scattered uniformly throughout the retina, but the degeneration of LGN occurred in patches. Perhaps the ganglion cells project in large groupings, or the rate of LGN cell degeneration varies widely in different regions.

Visual deprivation instituted after infancy has failed to produce changes in the LGN of the mouse, cat, and monkey (Terry et al., 1962; Wiesel and Hubel, 1963a; Burke and Hayhow, 1968; Chow, 1955).

Cragg (1969) used electron microscopy to examine the ultrastructure of the LGN of albino rats reared in darkness from weaning to 7–8 weeks of age. There was a 15% increase in the diameter of the axonal terminals and a 34% decrease in the density of terminals in the nuclei of deprived animals as compared with their litter-mate controls. Such changes were also present in rats first reared in a light environment up to the weaning time and then put into darkness. This finding suggests that subtle changes of the synaptic knobs may be the primary effect following visual deprivation.

In summary, visual deprivation in newborn animals causes reduction in size and in strength of Nissl staining in LGN neurones. The anatomical results of deprivation are similar, but less severe than those following enucleation. Withdrawal of visual stimulation from young or adult animals has no detrimental effect. However, subtle changes in protein, RNA synthesis and enzymatic activities may be present.

2. Neurophysiological Findings

The striking histological changes of the LGN neurons in kittens with eyelid sutured from birth was not matched by correspondingly severe neurophysiological abnormalities. Wiesel and Hubel (1963a) recorded single cell responses to light stimulation in the LGN of 2 visually deprived, 3-month old kittens and found that 30 out of 34 cells had a normal center-surround field organization. The remaining 4 cells responded sluggishly to light and had extremely large field centers. Our own results confirmed these findings (Chow and Stewart, 1972). However some recent studies using more sophisticated quantitative methods (Eyel *et al.*, 1971; Hamasahi and Pollack, 1972) reported about half the deprived LGN cells displayed a slightly different discharge pattern from normal cells, which may indicate an altered level of surround inhibition of the receptive field. These rather slight electrophysiological effects could mean that the normal functional characteristics of the LGN neurous, which are known to be complete at birth (HUBEL and WIESEL, 1963), were maintained during visual deprivation. Both retinal ganglion cells and LGN neurons show continuous discharges in the absence of light stimulation (LEVICK and WILLIAMS, 1964; BISHOP et al., 1964; ARDUINI, 1961), and persist even after enucleation (BURKE and HAYHOW, 1968). These maintained dark discharges could provide the necessary inputs to keep the neurons functioning normally, although not sufficient to maintain normal size.

The histology of the LGN of adult cats was not affected by visual deprivation, nor was their physiological responsiveness. BURKE and HAYHOW (1968) had kept adult cats in darkness for 56–966 days and obtained normal LGN field responses and multiple-unit discharges to either a single or repeated shocks of the optic nerve. In addition, they used either iodoacetate or 1, 5-di(B-aminophenoxy) pentane dihydrochloride to destroy the retinal receptor cells. This procedure eliminated retinally maintained discharges leaving intact the LGN baseline discharges. Seven to 791 days after the drug treatment the LGN response to single nerve shock was still normal. However, an increased synaptic efficacy was obtained when repeated shocks were applied, as shown by the reduction or elimination of the depression that normally followed such stimulation. This could result from denervation hypersensitivity.

V. Effect on Visual Cortex

1. Histological and Neurochemical Findings

Studies of the visual cortex (area 17) of animals deprived of vision from birth can be classified into 2 groups (Table 3). First are those using Nissl stained material to examine the histological appearance of the cortex. In only two of these studies on the visual cortex of the mouse, rabbit, and cat (No. 1, 3, 4, 5, 6) were there any histological alterations. GYLLENSTEN (1959, 1965) found decreased nuclear diameters and less internuclear material in the visual cortices of mice reared in darkness from birth to 20, 30 days and 2 months of age respectively. These defects, like those of optic nerve fibers, were reversed almost to normal when the deprivation was prolonged to 4 months. In this connection mention should be made of the effect of other sensory inputs on the histology of visual cortex. In a series of studies, KRECH et al. reported that rats reared in a "rich" environment, or trained in learning problems, had thicker, heavier visual cortices with more glial cells than those of normal controls (see BENNETT et al., 1964).

The second group of studies used some variant of the Golgi stain and only one failed to demonstrate some sort of dendritic abnormality of cortical cells in animals visually deprived from birth (No. 1, 7, 8, 11, 12). The type of dendritic defects varied among the several studies. VALVERDE (1967) reported a statistically significant decrease of the number of spines on apical dendrites of layer V pyramidal cells in the visual cortex of mice reared in darkness. But the spines of some but not all apical dendrites recovered to normal number after the animal was allowed to live under normal light enviroment (VALVERDE, 1971). On the other hand, GLOBUS and SCHEIBEL (1967) did not find any changes in the length, branching pattern,

Table 3. Effect of visual deprivation on visual cortex (area 17)

Author	Animal	Method of deprivation	Length of deprivation	Preparation and methods	Result
1. GOODMAN (1932)	albino rabbit	in darkness	6 months	Nissl and Golgi stain	No decrease in cortical thickness or in cell numbers, no change in dendrites
2. SCHIMKE (1959)	albino rabbit	in darkness	26, 60, 120 days	Brain was weighted and homogenized for enzyme essay	No change in histology, total weight, total protein or dehydrogenase activity
3. GYLLENSTEN (1959)	hooded mouse	in darkness	20, 30 days	Gallocyanin	Decreased cell nuclear diameter and internuclear material in layers II—IV
4. TERRY et al. (1962)	hooded mouse	eyelid closure (unilateral and bilateral)	3—9 weeks	Hematoxylin and eosin	No change
5. WIESEL and HUBEL (1963a)	cat	eyelid closure (unilateral)	3 months	Nissl stain	No change
6. GYLLENSTEN et al. (1965)	hooded mouse	in darkness	2, 3, 4 months	Gallocyanin	Decreased cell nuclear diameter and internuclear material in 2 month animal but approaching normal in 4 month animals
7. VALVERDE (1967)	hooded mouse	in darkness	22—25 days	Rapid Golgi	Decreased number of spines in the apical dendrites of layer V pyramidal cells

Table 3 (continued)

Author	Animal	Method of deprivation	Length of deprivation	Preparation and methods	Result
8. Globus and Scheibel (1967)	albino rabbit	in darkness	30 days	Rapid Golgi and Golgi-Cox	No decrease of total number of neurons, no change in the proportion of different types of neurons on the length, branching pattern, and density of spines of stellate cell dendrites, but increased deformation of dendritic spines of pyramids and increased variability of dendritic length of stellate cells
9. Maraini et al. (1967)	albino rat	eye closure (unilateral)	4—7 months	Autoradiography of tritiated leucine	No change
10. Maletta and Timiras (1967)	hooded rat	in darkness	21 days	Frozen and colorimetric determination of AChE and ChE	No change
11. Coleman and Riesen (1968)	cat	in darkness	6 months	Golgi-Cox	Decreased dendritic length and branching of layer IV stellate cells
12. Fox et al. (1968)	dog	eyes bandaged (bilateral)	1—35 days	Rapid Golgi and biochemical determinations	Smaller and less developed dendritic aborizations of Meynert cells of layer V, and increased noradrenalin and dopamine of visual cortex

and spine density in dark-reared rabbits. They observed some peculiarly shaped spines on the central three-fifths of the apical dendrites of pyramidal cells, and an increased variability of dendritic length of the stellate cells. Goodman (1932) also reported completely normal dendrites of the visual cortical cells in dark-reared rabbits. This again suggests that the albino rabbit may be less responsive to the absence of light stimulation. Coleman and Riesen (1968) used newborn kittens and found that 6-months dark-rearing produced shorter dendrites and fewer branchings of the layer IV stellates. These changes were also present in the pyramidal cells of the cingulate gyrus but were absent from the layer V pyramids of the visual cortex. Fox et al. (1968) reported poorly developed dendrites of layer V visual cortical cells in their visually deprived dogs.

The Golgi method thus appears to provide a more sensitive tool capable of revealing subtle morphological alterations not demonstrable either by the Nissl stained material, nor by neurochemical assays (No. 9 and 10). Since the resulting changes were slight, and the Golgi staining was capricious in use, it was almost mandatory to apply statistics to prove the significance of any differences between groups. This in turn points to the importance of adequate sampling. Coleman and Riesen suggested that the results reported by Globus and Scheibel as largely negative is due to the latter's use of small sample size and low power microscopy. On the other hand, Coleman and Riesen's own study was based on data obtained from the dark-reared and normally reared groups of only 3 cats in each case. Since the neuronal dendrites of the visual cortex in one experimental animal were normal, the difference between groups was dependent on the abnormal values obtained from the other 2 cats. Furthermore although it is clear that visual deprivation affects the dendrites of visual cortical cells in newborn animals, it is not clear how to proceed to synthesize the results into a meaningful pattern. For example, one study reported morphological distortions confined to apical dendrites of pyramids, while another found changes only in the dendritic branches of stellates. This would mean that in the first case the visual terminals could directly control the principle output cells, and in the second case that they act primarily upon the internuntials.

Neonatal enucleation caused more severe alterations of cortical morphology than those resulting from visual deprivation. Here again the surgical excision of the end organ had effects not equivalent to complete removal of function. Nuclear size and relative volume of internuclear material decreased in the visual area of mice, which did not recover after long survival times (Gyllensten et al., 1967). Decreases in cortical thickness, weight, neuronal size and number of glial cells were found in the rat and rabbit cortex following enucleation (Tsang, 1937; Bennett et al., 1964; Lindner and Umrath, 1955). Also, damage to the dendritic spines of visual cortical cells was more pronounced than that caused by visual deprivation. Globus and Scheibel (1967) found the spines diminished by as much as 70% of the normal value over the central portion of apical dendrites of pyramidal cells in all layers, though there was no change in the stellate cells. Valverde (1968) however, examined the layer V pyramids of mice and reported decreased spines limited to the shaft segment located in layer IV. He observed abnormal dendritic orientations of those stellate cells with ascending axons that apparently did not send dendrites into layer IV. By demonstrating selective spine degeneration in different dendritic segments, these data may help in elucidating the micro-structure of the specific input-output topology of cortical neurons.

The visual cortex of some adult species was also affected either by removal of the eye or by withdrawal of visual stimulation. These changes included decreased cortical thickness, less cholinesterase concentration, less internuclear material, but increased nuclear deoxyribonucleic acid concentration (TSANG, 1937; BEENETT et al., 1964; GYLLENSTEN et al., 1965; EECKEN and FAUTREZ, 1965).

In summary, visual deprivation produces transneuronal changes in the visual cortex of newborn animals. At a gross level these include a decrease in cortical thickness, cortical weight, nuclear diameter, and internuclear material. More subtle effects revealed by Golgi material were the diminished or distorted spines on the apical dendrites of pyramidals and the abnormal lengths of the stellate dendrites. These derangements of cortical architecture were greater following neonatal enucleation. Selective spine atrophy located on different segments of the pyramidal dendrites were demonstrated to follow different experimental manipulations, such as visual deprivation, enucleation, lesions in lateral geniculate nucleus or corpus callosum. This selectivity presumably reflects the trans-synaptically induced degeneration of the axonal terminals originated from different sources and terminated on different segments of the cortical dendrites.

2. Neurophysiological Findings

The important series of experiments reported by WIESEL and HUBEL (1963b, 1965a) provided the first detailed look at the behavior of cortical neurons following visual deprivation. Microelectrode recording of single cortical cells in kittens monocularly deprived from birth to 2–3 months of age revealed that the great majority of cells had normal receptive field organization and responded to stimulation of the normal eye. Only a few cells could be driven from the previously occluded eye. This very skewed ocular-dominance was in contrast to the large number of cells activated through either eye of normal cats (HUBEL and WIESEL, 1959). Figs. 4a and 4b summarize their data on the differences in the ocular-dominance distribution of cortical cells recorded from normal adult cats and from visually deprived kittens. The deprived animals not only had many more cells that could only be driven through the non-occluded eye, but also had some cells that did not respond to either eye. The few cells that were driven by the deprived eye, usually responded abnormally; such as, some did not have a clear receptive-field orientation, and some failed rapidly to respond after repeated stimulation.

HUBEL and WIESEL (1970) further showed that these cortical effects of visual deprivation appeared during a very restricted period of the kitten's life. Susceptibility to the effects of eyelid-suturing began about the fourth week until about the sixth week. No effect could be demonstrated after 3 months of age. Unilateral eye-suturing in adult cats from 3 months to a year produced no detectable effect. There was also little recovery shown by cortical cells in kittens first monocularly deprived for 3 months and then allowed binocular experiences for as long as 5 years. From this and previously cited work by the same authors it may be concluded that cats show little capacity to recover from the effects of unilateral eye-suturing from birth. GANZ et al. (1968) confirmed this abnormal physiology of single cortical cells including that those cells which could be driven through the deprived eye were less selective in their response to the shape, orientation, and direction of motion of the stimuli.

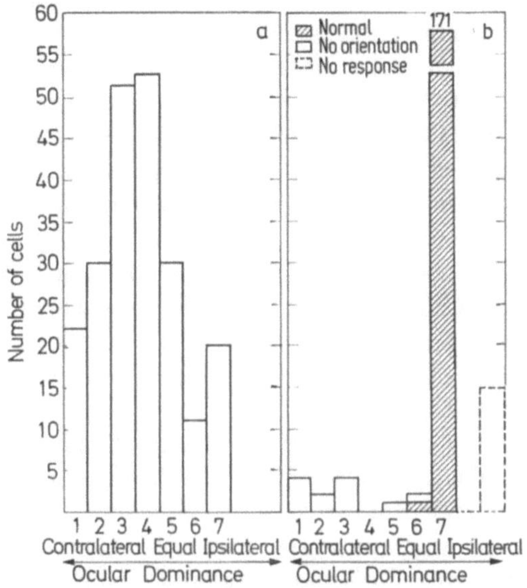

Fig. 4. a *Ocular-dominance distribution of 223 cells recorded from striate cortex of adult cats.* Cells of group 1 were driven only by the contralateral eye and of group 7 only by the ipsilateral eye. Cells in group 4 were driven equally by either eye. For cells in group 3 there were marked dominance of the contralateral eye, for group 3, slight dominance. For cells in group 6 there was marked dominance of the ipsilateral eye, for group 5, slight dominance. b *Ocular-dominance distribution of 199 visual cortical cells recorded from 5 monocularly deprived kittens* (8—14 weeks old). The groups were similarly numbered as in a. Shading indicates cells that had a normal response to visual stimulation; no shading indicates those that had abnormal receptive field orientation. Interrupted lines are cells that failed to respond to either eye. Re-drawn from Wiesel and Hubel (1963b, 1965b). By permission of the authors and the American Physiological Society

In spite of the permanence of affected cortical cell physiology, Dews and Wiesel (1970) reported partial recovery of visually guided behavior, if the normal eye was closed and the animal was forced to use the formerly deprived eye. However, these authors felt that this result does not necessarily indicate a recovery of function of the deprived eye but rather a new process enabling a better utilization of whatever visual information reaches the cortical neurons.

In agreement with these results, we were able to train cats to use their previously occluded eye in visually guided behavior. But in addition, we found that the LGN of these animals likewise showed little atrophy. However, the visual cortical cells in a cat that had almost normal LGN (as shown in Fig. 3b) continued to show the deprivation anomalies. Microelectrode recordings obtained from this animal yielded 23 cells responding to stimulation of the non-occluded eye and 2 cells to stimulation of the previously occluded eye. Furthermore, Wiesel and Hubel (1965a) found that more cortical cells could be driven binocularly in kittens following binocular occlusion than those after unilateral eye-suturing indicating other possible variables not yet taken into account by these experiments.

Several recent reports habe begun to analyze some of the variables. BLAKEMORE and COOPER (1970) reared kittens in darkness except for daily 5 hour exposure to visual environment dominated by either horizontal or vertical black stripes. HIRSH and SPINELLI (1971) exposed kittens to either horizontal or vertical bars painted on eye masks. Later, the receptive field axis of cortical cells recorded from these cats tended to orient according do the rearing conditions; e. g., vertically oriented receptive fields in cats exposed to vertical stripe stimulation. In another line of inquiry, the disparity of the two visual fields along the vertical meridian from the two eyes to the same cortical cells has been considered as the key to stereoscopic vision (BISHOP and HENRY, 1971). This binocular disparity developed during early postnatal period and could be distorted to a limited degree by rearing kittens with prisms in front of their eye which introduced a vertical field disparity (BARLOW and PETTIGREW, 1971; SHLEAR, 1971). These results point to some deprivation conditions being selective, affecting a specific set of cortical synapses but not others.

Aside from these long-term deprivation effects, the physiology of cortical cells is also affected by transient deafferentation. When the intraocular pressure was raised to block completely the optic nerve discharge (BORNSCHEIN, 1958), the LGN neurons first showed an initial increase and then a decrease rate of spontaneous and light-evoked discharges (ARDEN and SODERBERG, 1959; SUZUKI and ICHIJO, 1967). At the beginning or retinal ischemia, some visual cortical cells of cats showed an initial high frequency burst and others an initial inhibition. As the ischemia continued, cell discharges slowed down in both encephalé isolé and cerveau isolé preparations, but more so in the latter which might reach complete silence. The light-evoked responses also increased after about 1 min. This complete neuronal silence was not achieved by other deprivation procedures. Even though this method can be used only for short-term deafferentation, its uniqueness may offer a useful tool for further exploration (BAUMGARTER et al., 1961).

In agreement with the severe disturbance of the single-cell response, the evoked potentials recorded through gross electrodes on the visual cortex were also affected by visual deprivation. However, the reported effects were varied and sometimes contradictory. An increased initial latency and a decreased amplitude of the late response of the evoked potentials were obtained in eye-sutured kitten and dark-reared rabbit (WIESEL and HUBEL, 1963b; SCHERRER and FOURMENT, 1964). In the rabbit, these alterations resembled more the potentials evoked from newborn than from adult animals (MARTY, 1962). A comparable result was the finding that the visual cortex of dark-reared cats showed poor following responses to flicker stimulation (BAXTER, 1966), indicating a less efficient response system. The evoked potentials of visually deprived dog pups also showed great variability in amplitude, and had a long-duration negative wave typical of that in normal pups aged 2–10 days (FOX et al., 1968). A seemingly improved response was reported by LINDSLEY et al. (1964), in dark-reared monkeys that received 1 h of diffuse light per day. Both the "on" and "off" response evoked by light in the visual cortex of these animals were enhanced as compared with normally reared controls. This increased efficiency could be secondary to the earlier-cited potentiation of the LGN response after elimination of the retinal dark-discharges. These alterations of the evoked potentials usually reversed to a normal level after a few days exposure to light stimulation. It is therefore difficult to relate these transient defects of gross evoked potentials to the more permanent change of single-cell responses in the visual cortex of deprived animals.

VI. Effect on the Superior Colliculus

A few studies cited above included observations of the effects of either visual deprivation or enucleation on the histological structure of superior colliculus. Visual inspection of Nissl-stained material revealed no atrophic changes of this nucleus in the dark-reared cat (Wiesel and Hubel, 1965a), rabbit (Goodman, 1932), or mouse (Terry et al., 1962). However, a significant thinning of its layers was generally found after neonatal enucleation in these animals (Goodman, 1932; Tsang, 1937; Bennett et al., 1964; Terry et al., 1962). Again, Gyllensten et al. (1965, 1967) reported a consistent decrease in the relative volume of internuclear material in the superior colliculus of the mouse after rearing in darkness or after removal of the eye from birth, the latter procedure resulting in more volume decrease.

In common with the lateral geniculate body, the superior colliculus receives direct afferent fibers from the retina. Visual deprivation seems to be less effective in altering the structural integrity of the superior colliculus than of the LGN. These few studies (cited above) do not permit an evaluation of the significance of this difference. Similarly, the significance of the studies reporting an increase of noradrenalin and dopamine in the superior colliculus of visually deprived dogs, and a decrease of its total acetylcholinesterase activity in blind or light deprived rats remains to be evaluated (Fox et al., 1968; Bennett et al., 1964; Maletta and Timiras, 1967).

VII. Conclusions

Normal maturation of the structure and function of the visual apparatus of newborn animals depends on an adequate amount of light stimulation. Withdrawal of all visual inputs from birth affects the normal protein synthesis and enzymatic activities of the retina and visual centers, i.e. it leads to reduced total protein mass, decreased concentration of RNA, and lowered activity of acetylcholinesterase and other respiratory and oxidative enzymes. Presumably if they were not corrected these defects would eventually lead to cell death. Instances such as the elimination of retinal ganglion cells (chimpanzee), the thinning of retinal inner plexiform layer, the reduced neuronal cross-sectional areas of the LGN cells (cat), and the decreased total volume and internuclear material of visual cortex (rat, mouse) exemplify the end result of such processes.

Some physiological manifestations are concomitant, though not necessarily correlated with these histological and neurochemical alterations. Microelectrode recordings from unilaterally eyelid-sutured kittens reveal normal single-cell physiology of retinal ganglion cells and LGN cells but a greatly diminished distribution of visual cortical cells that could be driven through the deprived eye.

These findings cannot be easily understood in terms of the normal histological appearances of the retina and visual cortex but with concurrent abnormal LGN neurons. Field potentials obtained from the deprived animals (i.e. ERG and evoked potentials of visual cortex) showed only minor and transient abnormalities. However, these animals were behaviorally almost completely blind when first introduced into a daylight environment. Some of them could regain vision and learn to perform visual discriminations. An integrative evaluation of these somewhat incongruous findings in histology, physiology, and behavior cannot be attempted until the influences of several relevant variables are better controlled.

First, both the age and the growth rate of the animal are important variables. When visual deprivation procedures were initiated from birth, the severity of its effects varied as a function of the maturation state of the visual structures at birth and the rate of its subsequent growth. For these reasons, the same deprivation schedule would affect the visual system of the mouse more than that of the cat. Second, the length of deprivation must be taken into consideration. Presumably, the longer the deprivation time, the more disruptions it produces in the visual system. However, to complicate this issue there are the instances where increasing the length of deprivation time would partially reverse the trend, allowing the visual centers to recover to a nearly normal state. Third, species differences are undoubtedly involved. The visual systems of various animals could be intrinsically unequal in their responses to the disruptive influences of the absence of normal stimulation. After reviewing the literature, one has the impression that the visual system of the albino rabbit is more resistant to the effect of visual deprivation than, say, the hooded mouse. Fourth, the successive stations of the visual system show variable reactions to visual deprivation. Thus, the retinal ganglion cells become completely atrophied in dark-reared chimpanzees leaving their presynaptic cells, the bipolar cell layer, nearly intact. Or a normal optic nerve may persist even though the internuclear material of the LGN was diminished (mouse). In generalizing about these results, the particular level of neurons studied has to be taken into account, together with the fact that the deprivation effects do not always occur in successive steps from the periphery to the center.

The functional state of nerve cells of many regions of the brain has been shown to depend on the activity level of their inputs (HYDÉN, 1943) but the fundamental question of how the neuronal discharges regulate the metabolic activity and maintain the structure of cells has scarcely been studied. In the visual system, the retinal ganglion cells and the LGN cells continue to discharge in total darkness. The abnormal state of the visual centers of deprived animals could not be the result of a simple disuse comparable to the lack of exercise causing muscle atrophy. It is conceivable that these dark-discharges are quantitatively insufficient to provide the required level of metabolic activity to maintain a neuron. On the other hand, qualitative differences between the activity patterns during dark discharge and normal visual experience may produce alterations in synaptic transmission. In this regard, it may be suggested that the more severe effects of neonatal enucleation may be due to the added disappearance of some trophic influences normally supplied by the now degenerated axonal terminals.

Finally, there is the puzzling fact of recovery of function from the adverse effect of visual deprivation. Animals reared in total darkness generally could regain good vision after being removed to a normal light-environment (Riesen, 1966). A few studies have been specifically directed to the possibility of recovery of the structural alterations, but with little result. Reports of the recovery of protein synthesis and enzymatic activities are conflicting. This dissociation of function and structure, and the variable degree of recovery from deprivation point to our ignorance of the role of the genetic and environmental as well as of the structural and functional determinants for the maintenance of a viable visual system. If visual deprivation arrests normal growth, then the recovery would indicate a resumption of the maturation process. If deprivation causes neural atrophy, then recovery would represent either a restitution of the degenerated tissue or a substitution of normal tissue. If both processes are found to contribute to the recovery, then no unitary mechanism can be postulated, and the process underlying recovery must be established for each visual center, in each animal species, and for different age groups.

Acknowledgement

I thank Dr. David Stewart for his critical reading of the manuscript and checking the references.

The preparation of this article was supported by USPHS Grants NB 3816 and 5K6 NB 18,512 from the National Institute of Neurological Diseases and Blindness.

References

Arden, G. B., Söderberg, U.: The relationship of lateral geniculate activity to the electrocorticogram in the presence or absence of the optic tract input. Experientia (Basel) **15**, 163—164 (1959).

Arduini, A.: Influence of visual deafferentiation and continuous retinal illumination on the excitability of geniculate neurons. In: Jung, R., Kornhuber, H. (Eds.): The Visual System: Neurophysiology and Psychophysics, pp. 117—124. Berlin-Göttingen-Heidelberg: Springer 1961.

Barlow, H. B., Pettigrew, J. D.: Lack of specificity of neurons in the visual cortex of young kittens. J. Physiol. (Lond.) **218**, 98—100 (1971).

Baumgarter, G., Creutzfeldt, O., Jung, R.: Microphysiology of cortical neurons in acute anoxia and retinal ischemia. In: Gastaut, H., Meyer, J. S. (Eds.): Cerebral Anoxia and the Electroencephalogram, pp. 5—34. Illinois: Charles C. Thomas 1961.

Baxter, B. L.: Effect of visual deprivation during postnatal maturation on the electroencephalogram of the cat. Exp. Neurol. **14**, 224—237 (1966).

— Riesen, A. H.: Electroretinogram of the visually deprived cat. Science **132**, 1626—1627 (1961).

Bech, K.: The basophilic substances in the retinal ganglion cells and the physiological activity changes in these cells: An experimental histological study. Acta Ophth. Suppl. **46**, 1—105 (1957).

Bennett, E. L., Diamond, M. C., Krech, D., Rosenzweig, M. R.: Chemical and anatomical plasticity of brain. Science **145**, 610—619 (1964).

Berger, H.: Experimentell-anatomische Studien über die durch den Mangel optischer Reize veranlaßten Entwicklungshemmungen im Occipitallappen des Hundes und der Katze. Arch. Psychiat. Nervenkr. **33**, 521—567 (1900).

Bishop, P. O., Henry, G. H.: Spatial vision. Ann. Rev. Psychol. **22**, 119—160 (1971).

— Levick, W. R., Williams, W. O.: Statistical analysis of the dark discharge of lateral geniculate neurons. J. Physiol. (Lond.) **170**, 598—612 (1964).

Blakemore, C., Cooper, G. F.: Development of the brain depends on the visual environment. Nature (Lond.) **228**, 447—448 (1970).

BORNSCHEIN, H.: Spontan- und Belichtungsaktivität in Einzelfasern des N. opticus der Katze. I. Der Einfluß kurzdauernder retinaler Ischamie. Z. Biol. **110**, 210—222 (1958).

BRAATGÅRD, S. O.: The importance of adequate stimulation for the chemical composition of retinal ganglion cells during early postnatal development. Acta Radiol. Suppl. **96**, 1—80 (1952).

BROWN, K. T.: The electroretinogram: Its components and their origins. Vision Res. **8**, 633—677 (1968).

BURKE, W., HAYHOW, W. R.: Disuse in the lateral geniculate nucleus of the cat. J. Physiol. (Lond.) **194**, 495—519 (1968).

CARLSON, A. J.: Changes in the Nissl's substance of the ganglion and the bipolar cells of the retina of the Brandt Cormorant *Phalacrocorax penicillatus* during prolonged normal stimulation. Amer. J. Anat. **2**, 341—347 (1902—1903).

CHOW, K. L.: Failure to demonstrate changes in the visual system of monkeys kept in darkness or in colored lights. J. comp. Neurol. **102**, 597—606 (1955).

— RIESEN, A. H., NEWELL, F. W.: Degeneration of retinal ganglion cells in infant chimpanzees reared in darkness. J. comp. Neurol. **107**, 27—42 (1957).

— STEWART, D. L.: Reversal of structural and functional effects of long-term visual deprivation in cats. Exp. Neurol. **34**, 409—433 (1972).

CLARK, W. E. LE GROS, PENMAN, G.: The projection of the retina in the lateral geniculate body. Proc. roy. Soc. B **114**, 291—313 (1934).

COLEMAN, P. D., RIESEN, A. H.: Environmental effects on cortical dendritic fields: I. Rearing in the dark. J. Anat. **102**, 363—374 (1968).

COOK, W. H., WALKER, J. H., BARR, M. L.: A cytological study of transneuronal atrophy in the cat and rabbit. J. comp. Neurol. **94**, 267—292 (1951).

CORNWELL, A. C., SHARPLESS, S. K.: Electrophysiological retinal changes and visual deprivation. Vision Res. **8**, 389—401 (1968).

CRAGG, B. G.: The effects of vision and dark-rearing on the size and density of synapses in the lateral geniculate nucleus measured by electron microscopy. Brain Res. **13**, 53—67 (1969a).

— Structural changes in naive retinal synapses detectable within minutes of first exposure to daylight. Brain Res. **15**, 79—96 (1969b).

DE ROBERTIS, E., FRANCHI, C. M.: Electron microscope observations on synaptic vesicles in synapses of the retinal rods and cones. J. biophys. biochem. Cytol. **2**, 307—318 (1956).

DETWILER, S. R.: Vertebrate Photoreceptors. New York: The Macmillan Co. (1943).

DEWS, P. B., WIESEL, T. N.: Consequences of monocular deprivation on visual behavior in kittens. J. Physiol. (Lond.) **206**, 437—455 (1970).

EECKEN, H. V., FAUTREZ, J.: Cellular modifications in the rabbit's visual system after unilateral enucleation. Nature (Lond.) **206**, 423—424 (1965).

EYSEL, U. TH., FLYNN, J. T., GAEDT, CHR.: Spatial summation of excitation and inhibition in receptive fields of neurons in the LGN of the cat and the influence of visual deprivation. Pflügers Arch. **327**, 82—94 (1971).

FIFKOVÁ, E., HASSLER, R.: Quantitative morphological changes in visual centers in rats after unilateral deprivation. J. comp. Neurol. **135**, 167—178 (1969).

FOX, M. W., INMAN, O., GLISSON, S.: Age differences in central nervous effects of visual deprivation in the dog. Develop. Psychobiol. **1**, 48—54 (1968).

GANZ, L., FITCH, M., SATTERBERG, J. A.: The selective effect of visual deprivation on receptive field shape determined neurophysiologically. Exp. Neurol. **22**, 614—637 (1968).

GAY, A., SILBERBERG, D. H.: Histochemical correlates of transynaptic degeneration. Arch. Neurol. **10**, 85—90 (1964).

GLEES, P., LE GROS CLARK, W. E.: The termination of optic fibers in the lateral geniculate body of the monkey. J. Anat. **75**, 295—308 (1941).

GLOBUS, A., SCHEIBEL, A. B.: Synaptic loci on visual cortical neurons of rabbit: The specific afferent radiation. Exp. Neurol. **18**, 116—131 (1967).

— — The effect of visual deprivation on cortical neurons — A Golgi study. Exp. Neurol. **19**, 331—345 (1967).

GLOW, P. H., ROSE, S.: Effects of light and dark on the acetylcholinesterase activity of the retina. Nature (Lond.) **202**, 422—423 (1964).

Goldby, F.: A note on transneuronal atrophy in the human lateral geniculate body. J. Neurol. Neurosurg. Psychiat. **22**, 202—207 (1957).

Gomirato, G., Baggio, G.: Metabolic relations between the neurons of the optic pathway and various functional conditions. J. Neuropath. exp. Neurol. **21**, 634—644 (1962).

Goodman, L.: Effect of total absence of function on the optic system of rabbits. Amer. J. Physiol. **100**, 46—63 (1932).

Gyllensten, L.: Postnatal development of the visual cortex in darkness (mice). Acta. morphol. neerl-scand. **2**, 331—345 (1959).

— Malmfors, T.: Myelinization of the optic nerve and its dependence on visual function — a quantitative investigation in mice. J. Embryol. exp. Morph. **11**, 255—266 (1963).

— — Norrlin, M. L.: Effects of visual deprivation on the optic centers of growing and adult mice. J. comp. Neurol. **124**, 149—160 (1965).

— — Norrlin-Grettue, M. L.: Development and functional alterations in the fiber composition of the optic nerve in visually deprived mice. J. comp. Neurol. **128**, 413—418 (1966).

— — — Visual and non-visual factors in the centripetal stimulation of postnatal growth of the visual centers in mice. J. comp. Neurol. **131**, 549—558 (1967).

Hamasahi, D. I., Rackensperger, W., Vesper, J.: Spatial organization of normal and visually deprived units in the LGN of the cat. Vision Res. **12**, no. 5, 843—854 (1972).

Hellström, B., Zetterström, B.: The effect of light on the manifestation of the electroretinogram and on histochemically demonstrable SH-groups in the retina. Exp. Cell. Res. **10**, 248—251 (1956).

Hess, A.: Optic centers and pathways after eye removal in fetal guinea pigs. J. comp. Neurol. **109**, 91—115 (1958).

Hirsch, H. V. B., Spinelli, D. N.: Modification of the distribution of receptive field orientation in cats by selective visual exposure during development. Exp. Brain Res. **13**, 509—521 (1971).

Hubel, D. H., Wiesel, T. N.: Receptive fields of single neurones in the cat's striate cortex. J. Physiol. **148**, 574—591 (1959).

— — Receptive fields of cells in striate cortex of very young, visually inexperienced kittens. J. Neurophysiol. **26**, 994—1002 (1963).

— — The period of susceptibility to the physiological effects of unilateral eye closure in kittens. J. Physiol. (Lond.) **206**, 417—436 (1970).

Hyden, H.: Protein metabolism in the nerve cell during growth and function. Acta physiol. Scand. **6**, Suppl. **17**, 3—136 (1943).

Karlsson, J.-O., Sjostrand, J.: Effect of deprivation of light on axonal transport in retinal ganglion cells of the rabbit. Brain Res. **29**, 315—320 (1971).

Kupfer, C., Palmer, P.: Lateral geniculate nucleus: Histological and cytochemical changes following afferent denervation and visual deprivation Exp. Neurol. **9**, 400—409 (1964).

Kuwabara, T.: Membrane transformation of photoreceptive organs by light. Electron Microscopy (6th International Congr.) **2**, 501—502 (1966).

— Gorn, R. A.: Retinal damage by visible light. Arch. Ophthal. **79**, 69—78 (1968).

Levick, W. M., Williams, W. O.: Maintained activity of lateral geniculate neurons in darkness. J. Physiol. (Lond.) **170**, 582—597 (1964).

Liberman, R.: Retinal cholinesterase and glycolysis in rats reared in darkness. Science **135**, 372—373 (1962).

Lindner, I., Umrath, K.: Veränderungen der Sehsphäre I und II in ihrem monokularen und binokularen Teil nach Extirpation eines Auges beim Kaninchen. Dtsch. Z. Nervenheilk. **172**, 495—525 (1955).

Lindsley, D. B., Wendt, R. H., Lindsley, D. F., Fox, S. S., Howell, J., Adey, W. R.: Diurnal activity, behavior and EEG responses in visually deprived monkeys. Ann. N.Y. Acad. Sci. **117**, 564—587 (1964).

Maletta, G. J., Timiras, P. S.: Acetylcholinesterase activity in optic structures after complete light deprivation from birth. Exp. Neurol. **19**, 513—518 (1967).

Maraini, G., Carta, F., Franguelli, R., Santori, M.: Effect of monocular light-deprivation on leucine uptake in the retina and the optic centers of the newborn rat. Exp. Eye Res. **6**, 299—302 (1967).

Marty, R.: Development post-natal des réponses sensorielles du cortex cerebral chez le chat et le lapin. Arch. Anat. Microscop. Morphol. Exp. **51**, 129—264 (1962).

MATTHEWS, M. R.: Further observations on transneuronal degeneration in the lateral geniculate nucleus of the macaque monkey. J. Anat. **98**, 255—263 (1969).
— COWAN, W. M., POWELL, T. P. S.: Transneuronal cell degeneration in the lateral geniculate nucleus of the macaque monkey. J. Anat. **74**, 145—169 (1960).
MILLER, E., HELLER, A., MOORE, R. Y.: Acetylcholine in rabbit visual system nuclei after enucleation and visual cortex ablation. J. Pharm. exp. Therapeut. **165**, 117—125 (1969).
MINKOWSKI, M.: Über den Verlauf, die Endigung und die zentrale Repräsentation von gekreuzten und ungekreuzten Sehnervenfasern bei einigen Säugetieren und bei Menschen. Schweiz. Arch. Neurol. Neurochir., Psychiat. **6**, 201—252, **7**, 268—303 (1920).
MOUNTFORD, S.: Effect of light and dark adaptation on the vesicle population of receptor-bipolar synapses. J. Ultrastruct. Res. **9**, 403—418 (1963).
NOELL, W. K., WALKER, V. S., KANG, B. S., BERMAN, S.: Retinal damage by light in rats. Invest. Ophthal. **5**, 450—473 (1966).
RASCH, E., SWIFT, H., RIESEN, A. H., CHOW, K. L.: Altered structure and composition of retinal cells in dark-reared mammals. Exp. cell Res. **25**, 348—363 (1961).
RIESEN, A. H.: Sensory deprivation. In: STELLAR, E., SPRAGUE, J. M. (Eds.): Progress in Physiological Psychology, Vol. 1, pp. 117—147. New York: Academic Press 1966.
SCHERRER, J., FOURMENT, A.: Electrocortical effects of sensory deprivation during development. In: HIMWICH, W. A., HIMWICH, H. E. (Eds.): Progress in Brain Research, Vol. 9, pp. 103—112. The Developing Brain. New York: Elsevier 1964.
SCHIMKE, R. T.: Effects of prolonged light deprivation on the development of retinal enzymes in the rabbit. J. biol. Chem. **234**, 700—703 (1959).
SHLEAR, R.: Shift in binocular disparity causes compensatory change in the cortical structure of kittens. Science **173**, 638—641 (1971).
SOSULA, L., GLOW, P. H.: Increase in number of synapses in the inner plexiform layer of light deprived rat retina: quantitative EM. J. Comp. Neurol. **141**, 427—452 (1971).
SUZUKI, H., ICHIJO, M.: Tonic inhibition in cat lateral geniculate nucleus maintained by retinal spontaneous discharge. Japan. J. Physiol. **17**, 599—612 (1967).
TERRY, R. J., ROLAND, A. L., RACE, J.: Effect of eye enucleation and eyelid closure upon the brain and associated visual structures in the mouse. J. exp. Zool. **150**, 165—183 (1962).
TSANG, Y. C.: Visual centers in blinded rats. J. comp. Neurol. **66**, 211—261 (1937).
VALVERDE, F.: Apical dendritic spines of the visual cortex and light deprivation in the mouse. Exp. Brain Res. **3**, 337—352 (1967).
— Structural changes in the area striata of the mouse after enucleation. Exp. Brain Res. **5**, 274—292 (1968).
— Rate and extent of recovery from dark rearing in the visual cortex of the mouse. Brain Res. **33**, 1—12 (1971).
WEISKRANTZ, L.: Sensory deprivation and the cat's optic nervous system. Nature (Lond.) **181**, 1047—1050 (1958).
WENDELL-SMITH, C. P.: Effect of light deprivation on the postnatal development of the optic nerve. Nature (Lond.) **200**, 707 (1964).
WIESEL, T. N., HUBEL, D. H.: Effects of visual deprivation on morphology and physiology of cells in the cat's lateral geniculate body. J. Neurophysiol. **26**, 978—993 (1963a).
— — Single-cell responses in striate cortex of kittens deprived of vision in one eye. J. Neurophysiol. **26**, 1003—1017 (1963b).
— — Comparison of the effects of unilateral and bilateral eye closure on cortical unit responses in kittens. J. Neurophysiol. **28**, 1029—1040 (1965a).
— — Extent of recovery from the effects of visual deprivation in kittens. J. Neurophysiol. **28**, 1060—1072 (1965b).
ZETTERSTRÖM, B.: The effect of light on the appearance and development of the electroretinogram in newborn kittens. Acta physiol. scand. **35**, 272—279 (1956).

Comparative Data

Chapter 10

Principles of the Mosaic Organisation in the Visual System's Neuropil of Musca domestica L

By

VALENTIN BRAITENBERG and NICHOLAS J. STRAUSFELD, Tübingen (Germany)

With 19 Figures

Contents

Introduction

As a short preface to this article we feel it is appropriate to quote, briefly, from a passage in CAJAL's autobiography (1937). He writes: "The complexity of the insect retina is something stupendous, disconcerting and without precedent in other animals. When one considers the inextricable thicket of the compound eye . . . . . when one discovers not one chiasma as in vertebrates but three chiasmas of enigmatic significance besides the inexhaustible supply of amacrine cells and centrifugal fibres; when one meditates, finally, on the infinite number and exquisite adjustment of all these histological factors, so delicate that the highest power of the microscope hardly brings them under observation, one is completely overwhelmed".

These remarks were written after he had already completed extensive studies on both the vertebrate (1888, 1892) and cephalopod (1917) visual system and after a lengthy survey, with SANCHEZ (1915) of the neuronal components in insect optic lobes. But perhaps one should not be too surprised that this passage represented his emotive reaction to arthropod visual neuropil, for it is precisely this impression of fantastic and beautiful complexities that one initially recieves from histological preparations of the insect brain.

Our main purpose in this account is to outline some of the fundamental patterns of neuronal arrangement which contribute to the "inextricable thicket" and to show that after all there is a basic simplicity of organisation in this neuropil, more evident, perhaps, than in any other piece of central nervous tissue.

There are three main synaptic regions in the insect optic lobes (Fig. 1) which receive relays from the compound eyes. Golgi preparations show that these contain

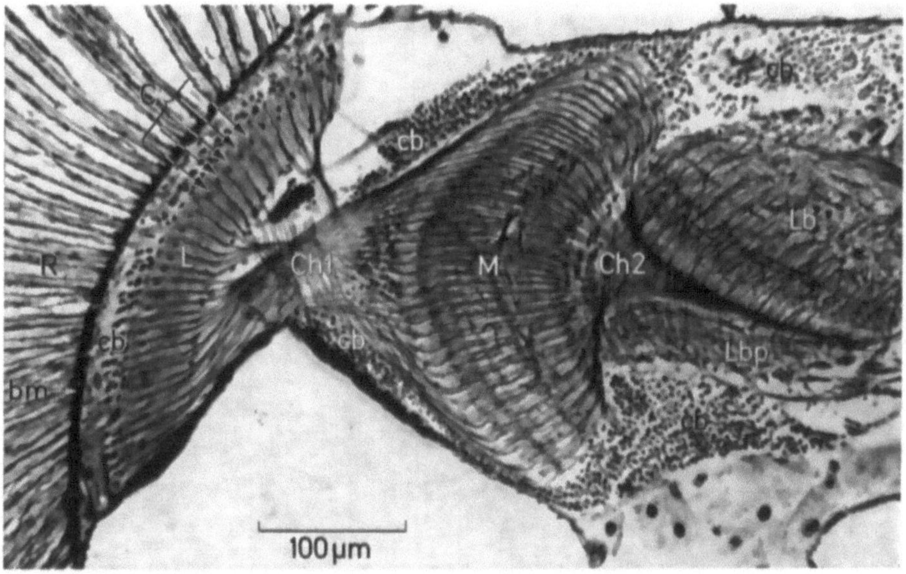

Fig. 1. The three visual ganglia of the fly, connected by the two chiasmata. A portion of retina (*R*) can be seen, separated from the lamina (*L*) by the basement membrane (*bm*). The columnar arrangement of the cartridges (*c*) in the lamina can be clearly made out. Fibres from the cartridges project to the medulla (*M*) via the first optic chiasma (*Ch1*). The decussation of fibres in the chiasma simply results in an antero-posterior cross over of fibres but does not establish a dorsoventral cross over. The structure of the neuropil in the medulla (*M*) is columnar *and* stratified, the striation reflecting mainly layers of tangential and amacrine fibres. Elements from the medullar columns feed into the second optic chiasma (*Ch2*), through which they reach the two regions of the lobular complex, the anterior lobule (*Lb*) and the posterior lobular plate (*Lbp*). Fibres connecting the latter two are also contained in the second optic chiasma. The neural processes which form the neuropil regions of the optic ganglia are derived from cell bodies situated outside the synaptic regions proper (*cb*)

Fig. 2. The "inextricable thicket of fibres" (CAJAL) viewed through the eyes of those dealing with Golgi preparations of the medulla of *Musca* and abstracted from investigations by CAMPOS-ORTEGA and STRAUSFELD, 1972; STRAUSFELD, CAMPOS-ORTEGA, in preparation. The elements that have been drawn here (to the left) include a) some terminals derived from the lamina (L_1, L_2, L_3) and the retina (R_7); b) cells that have perpendicularly oriented main fibres (with respect to the outer and inner face of the region) and are intrinsic to the medulla (*I*); c) cells that are intrinsic to the medulla but have tangentially oriented fields (amacrine cells: *A*, *Am*); d) neurons that project from the medulla to the lobula complex (transmedullary cells (*tm*) to the lobule (*lb*), *T*-cells (*T*) to the lobule or lobula plate (*lbp*), *Y*-cells (*Y*) to the lobule *and* lobula plate); e) tangential cells (*Tan*) that link the medulla to the mid-brain and/or contralateral optic lobe; c) centrifugal and medulla-to-lamina *T* cells (c and *t*, respectively) that link the medulla to the lamina. The diagram to the right outlines the main forms of cells,

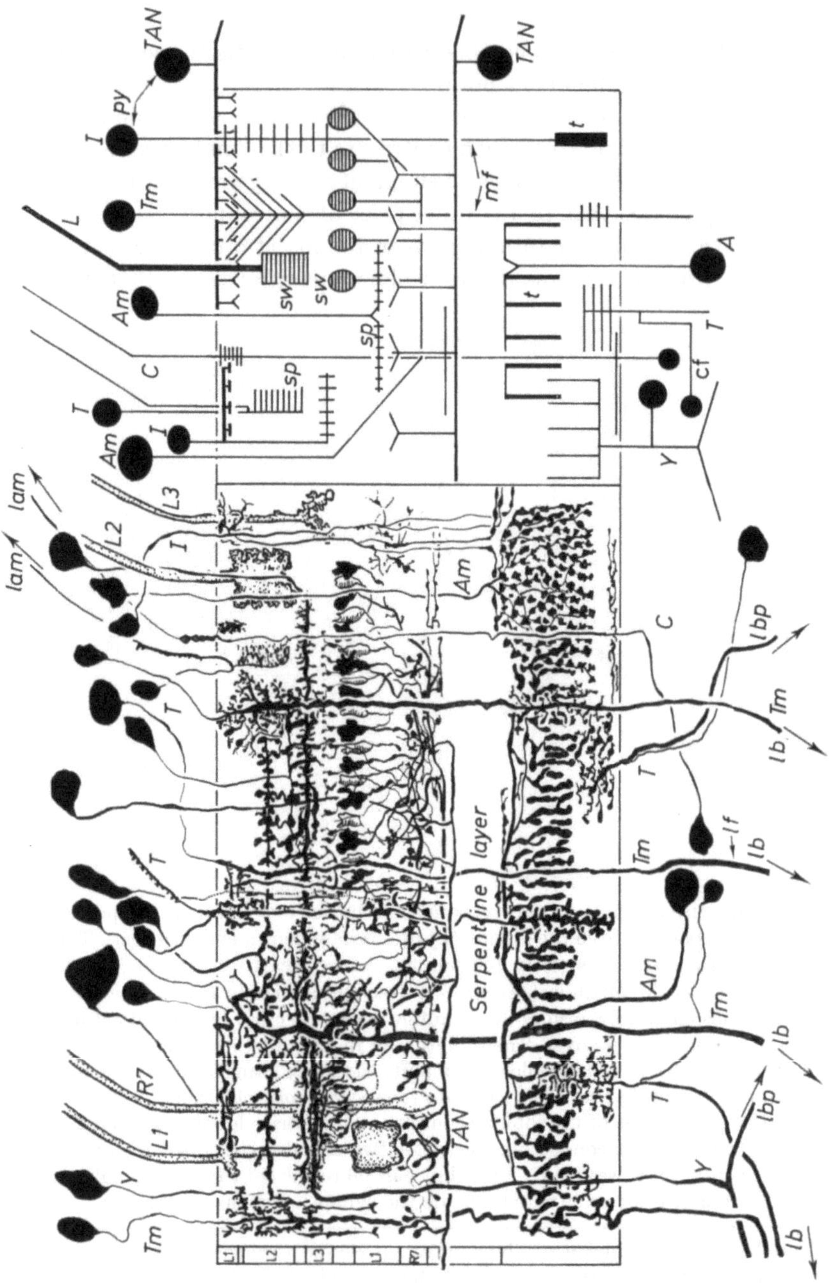

Fig. 2

classified according to STRAUSFELD and BLEST, 1970. Class I, or perpendicular cells include all elements whose long axis is oriented at right angles to the surfaces of the medulla and which serve to link this region to the lobula complex (*tm*, *T* and *Y*-cells). They have perikarya (*py*) above or below this region. Class II cells consist of forms of tangential arborisations. Class III cells include amacrines and intrinsic neurons. The specialisations from cell fibres are either spiny (*sp*), taberous or blebbed (*t*) or large swollen outgrowths (*sw*). *lf* = linking fibre (or connecting fibre) between a medullary arborisation and the corresponding component of the same cell elsewhere in the optic lobes or brain. *mf* = mother fibre of a cell within a neuropil region. *cf* = cell body fibre, connecting the perikaryon with the main pattern of arborization (or the linking fibre). The levels of terminations from the lamina are shown extreme left

within them a surprising variety of neuronal forms (CAJAL and SANCHEZ, 1915; STRAUSFELD and BLEST, 1970; STRAUSFELD, 1970a, b) each of which has a characteristic sillhouette, both in long and cross section, and each has sets of lateral processes disposed at characteristic levels in the neuropil. These contribute to field shapes that are typical of that and only that neuron (Fig. 2). In *Musca domestica*, the housefly, there are over one hundred distinct forms of neural elements in the lobes.

Reconstructions of cell relationships (STRAUSFELD and BLEST, 1970; STRAUSFELD, 1970a), derived from Golgi impregnated elements, give the impression of a complexity of neuroarchitecture which, in terms of its finer anatomical and physiological comprehension, seems analytically almost self-defeating. On the other hand, different histological procedures, namely reduced silver techniques, show up particular features of whole populations of some types of neurons (Fig. 6 and 7) and have the singular advantage of informing us whether or not some elements are arranged with constant spatial relationships to each other or whether they are, so to speak, randomly spaced throughout the layers of neuropil.

Accounts by POWER (1943), MEYER (1951), VIGIER (1907–1908), TRUJILLO-CENÓZ (1965, 1969), TRUJILLO-CENÓZ and MELAMED (1966, 1970), MELAMED and TRUJILLO-CENÓZ (1968), and most recently by BRAITENBERG (1967, 1970, 1972), BOSCHEK (1971), STRAUSFELD (1971a, b), CAMPOS Ortega and STRAUSFELD (1972), BRAITENBERG and HAUSER (1972) and others (see bibliography), illustrate that at least the first synaptic region, the lamina, appears to be rigidly organised into discrete columns and bears little relationship with the tangle of fibers that so amazed CAJAL. The medulla (the second synaptic region) is also clearly columnar and consist of regular arrays of fibre-groups including afferent, efferent and intrinsic (amacrine) neurons (CAMPOS-ORTEGA and STRAUSFELD, 1972b).

Our purpose in this article is to bring together some of the available information from two sources; namely, the morphological taxonomy of neuronal types derived from Golgi analysis on the one hand and, on the other, the symmetry and periodicity of the networks formed by these neurons, as we see them in the reduced silver preparations. To use one method without the other is only of limited value if one wants to discover the anatomical substrate of the computations as revealed by behavioural experiments; and both are incompetent when questions are asked about the detailed synaptology, for which the electron microscope has to be used. We shall consider here only the well established results of E. M. analysis already available in the literature (FERNANDEZ MORAN, 1958; TRUJILLO-CENÓZ, 1965; TRUJILLO-CENÓZ and MELAMED, 1970; TRUJILLO-CENÓZ and MELAMED, 1970; BOSCHEK, 1971; CAMPOS-ORTEGA and STRAUSFELD, 1972a; STRAUSFELD and CAMPOS-ORTEGA, 1972) and will have to postpone the detailed and quite tedious correlations between the electron-and photon-microscopical studies which we are at present undertaking in our institute.

The Mosaic Arrangement of Lenses of the Cornea

There are about 3 000 lenses on the surface of each eye of *Musca*. These vary (by almost a factor 2) in size; the largest being in the medial-anterior part of the retina and the smallest in the back. However, in spite of these variations, the lenses

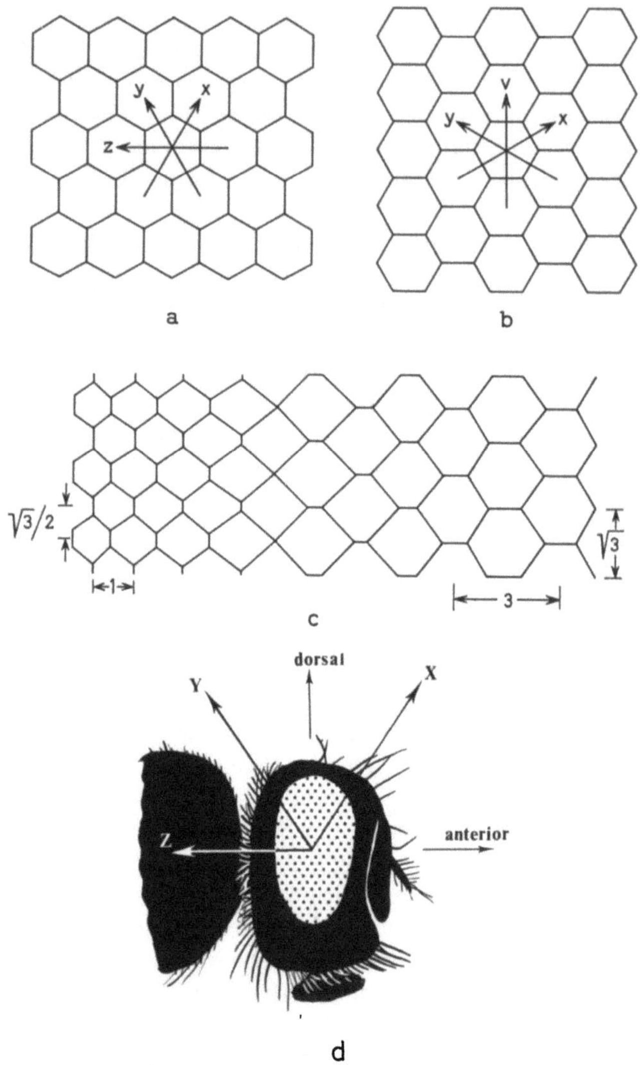

Fig. 3a, b, c, and d. *Arrangement of the facets (lenses) in the cornea.* a Hexagonal lenses in a hexagonal array as in the bee's compound eye. One horizontal and two oblique sets of rows of lenses, x, y and z are the coordinates of this array: z points backward, x obliquely upward and forward, y obliquely upward and backward. b hexagonal lenses in a hexagonal array as in the eye of *Drosophila:* one vertical and two oblique sets of rows of lenses. The vertical direction is marked v. c arrangement of the lenses in the house-fly's eye: there is a transformation of the pattern from the bee-type as in (a) to the *Drosophila*-type as in (b) through a progressive stretching of the array from back to front (left to right in the drawing). The stretching only involves the horizontal dimension, the distances between horizontal rows, $\sqrt{3}/2$ remaining unchanged from back to front. The stretching enlarges the horizontal distances between lenses from 1—3. At the same time the diameters of the lenses are enlarged from $1-\sqrt{3}$. While the enlargement of the lenses from back to front, and the transformation of one type of hexagonal array into another, correspond well to real measurements of the fly's eye, the drawing is an exaggeration insofar as it shows the transformation within a much shorter row (7 ommatidia) than it actually occurs in the fly's eye (horizontal rows of about 36 ommatidia). d to illustrate the position of our coordinates x, y and z with respect to the fly's eye and head. (Modified from BRAITENBERG, 1970)

and underlying receptor elements are arranged in distinct hexagonal coordinates (Fig. 3).

The three axes of the array tend to be related to the macroscopical outline of the compound eye in a regular way. Thus in *Drosophila* one of the axes coincides with the long vertical axis of the elliptical eye and in the bee with the short horizontal one. Even when, as in the house fly, there is a considerable variation of lense size, ranging from small lenses in the back of the eye to ones twice as large in front (Fig. 3c), the horizontal rows of ommatidia appear to be kept fairly parallel to each other and parallel to the short axis of the compound eye, a feat which is accomplished by a distortion of the whole array (Braitenberg, 1967, 1970). Also, the varying interommatidial angles, which could be expected on the basis of the variation of lense size, are compensated and even overcompensated by an increase of the curvature from the front to the back of the eye. In fact, measurements of the angles between the optical axes of neighbouring ommatidia seem to indicate that the smaller angles are found in front where the lenses are largest. Whether this implies better resolution, as well as greater light gathering capacity in the frontal region of the visual field, is a question which we will be in a position to answer only after we have considered the role of the set of the multiple sensory elements, the so-called retinula, with which each ommatidium is equipped.

The Arrangement of Rhabdomeres in the Retinula

Cross-sections of the ommatidia in the retina reveal that for each ommatidium there is a crown of seven elongated cells, connected lengthwise with seven thin rods, the so-called rhabdomeres. There is ample evidence that these are the site of the primary photochemical processes reponsible for vision in insects. In the fly they are arranged in a peculiar asymmetric configuration in the so-called retinula, a feature that was observed long ago by Vigier (1907a) and Dietrich (1909). Six rhabdomeres are arranged around a central seventh and of the six, one is displaced somewhat outside the crown formed by the other five (Fig. 4). All seven points can be fitted into a hexagonal dot pattern. Another interesting feature is that the six cells (numbered according to Dietrich, 1909) have their microvilli oriented so that cells 1 and 4 share one alignment, 2 and 5 a second and 3 and 6 a third. The axis of the microvilli of 7 have a fourth alignment, and those of cell 8 (which lies directly beneath 7 and is geometrically contiguous and optically continuous with it) are arranged at right angles to that of 7 (Trujillo-Cenóz and Melamed, 1966).

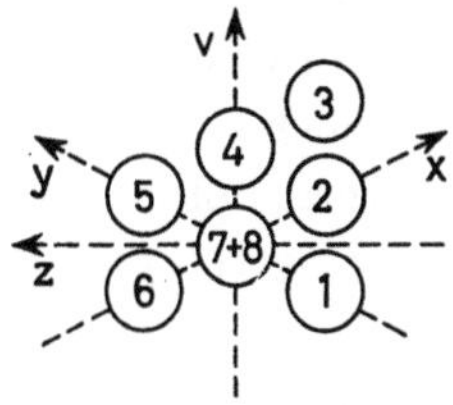

Fig. 4. The pattern of the rhabdomeres in the retinula in the focal plane of the ommatidium to illustrate their positions with respect to the coordinates of the ommatidial array. (From Braitenberg, 1970)

Vigier's Conjecture on the Composition of the Image in the Fly's Eye

The optics of the compound eye of Diptera, the implications of the inverting optics of the individual ommatidia superimposed on the non-inverting optics of the whole compound eye, and the explanation of the retina-lamina projections were expounded with astonishing clarity by VIGIER in 1907 and 1908. It pays to translate most of his small paper which rarely appears in the bibliography of recent authors who deal with the same facts.

He had shown in some previous papers (VIGIER, 1907a, b, c) "1. That the ommatidia of Diptera receive small inverted images; 2. that this fact is in contradiction with the classical theory of mosaic vision; 3. that one of the principal difficulties in the combination of these multiple images or partical photographs into a continuous photograph resides in the inversion of these little images which cannot fuse in one image no matter how closely they are put together; 4. that it is in the ganglionic apparatus of the animal and not in the compound eye that this synthesis is achieved. ..."

He then proudly announces that he was able to follow the course of the fibers which connect the photoreceptors to the visual ganglia and to show how the patterns of excitation, corresponding to the small inverted images seen by individual ommatidia, are combined in a unitary picture.

"The retinula of each ommatidium in Diptera comprises seven photosensitive elements, the rhabdomeres, whose peripheral tips coincide with the focal plane onto which an image is projected by the cornea lens. The rhabdomeres form a narrow fascicle, with one in the center and the other six in the periphery. Since the axes of neighbouring ommatidia diverge (only) very slightly, their visual fields must be partially overlapping. Let us consider a group of seven ommatidia. The central ommatidium receives on its light sensitive surface a small inverted image, which produces seven distinct excitations on the fascicle of seven independent rhabdomeres. Because of the partial overlap of their visual fields the central point of this image appears also in six neighbouring ommatidia, but in each it occupies a different position; it is shifted from the center because of the divergence of the optical axes and, because of the inversion of the little images, the rhabdomeres which are hit by the identical point (of the visual environment) are the following: the axial rhabdomere in the central ommatidium and the rhabdomere farthest from the central ommatidium in each of the peripheral ommatidia of the group."

"The rhabdomeres which correspond to identical points of the ommatidial images and which, therefore, receive the same light intensity are *physiologically homologous*. While they are distinguished topographically, they are equivalent from the point of view of geometrical optics and physiology. Their position is determined precisely and invariably by the regular hexagonal array of ommatidia ...".

He then proceeds to answer the question as to the recomposition of the whole image.

As the fibers of the retinular cells leave the retina at its base, "each bundle of fibers derived from one ommatidium" and "composed of fibers which transmit different excitation", as he had explained earlier, "is dissociated in a very curious way: the bundle is twisted 180 degrees around its own axis, then the fibers of the bundle separate and associate with other fibers stemming from neighbouring ommatidia in order to form together with them those peculiar structures within the periopticum (the lamina) which VIALLANES has christened neuroommatidia (the cartridges of this paper), but whose significance has not yet been recognized. I have shown earlier (VIGIER, 1907b) that it is within these structures that the connexions between the first elements of the chain which represents the optic nerve of the compound eye are established by contiguity".

"The fibers which are thus brought together in a neuroommatidium are precisely the seven[1] fibers which carry identical information derived from neighbouring ommatidia. This is achieved by the convergence toward the axial fiber of the central ommatidium of the group of the homologous fibers of surrounding ommatidia, i.e. of the fibers in each ommatidium having a position farthest away from the central one ... *Fibers stemming from homologous rhabdomeres unite in isodynamic groups, the neuroommatidia.*

At this level the excitations mix: they are in fact transmitted to the *perioptic neurons,* (our neurons L_1 and L_2) which are generally two in number and are situated in the axis of each neuroommatidium from where they reach the epiopticum (the medulla) through the external chiasma."

Thus, he says, a continuous, upright global image of an object is reestablished as a "perioptic neurophotograph".

What is left for us, after this illuminating paper of 1907 is relatively pedestrian labour.

Recent Confirmation of Vigier

There are three points which are left open – or are slurred over in VIGIER's account. Each of them has received an answer in recent years, whereby his idea has been completely confirmed.

I. VIGIER writes that he established the contacts between retinular cell axons and second order neurons of the lamina by means of Golgi and reduced silver techniques. These old histological methods are of course able to give no more than a suggestion of contiguity that may be indicative of synaptic contact but give absolutely no certainty of such contacts. Theoretically the electron microscopist is hardly in a better position since he too must rely on an act of faith in the correlation of those membrane specializations which he calls synapses with the places at which the influences of one neuron onto another probably take place as we know them to occur from electrophysiological evidence. But by all the criteria which are by now standard in electron microscopy, the contacts between the axons of the retinular cells and the two second order neurons situated in the core of each cartridge, as shown in the electronmicrographs of TRUJILLO-CENÓZ (1965), have to be taken as synapses. Moreover, TRUJILLO-CENÓZ attempts to answers a question which VIGIER did not even ask, namely whether the information transmitted to each of the two second order fibers is the same or different. TRUJILLO-CENÓZ claims that

[1] In fact it is only 6; 7 and 8 bypass in close proximity the other six endings in the lamina.

each of the six retinular cell axons entering a cartridge makes contacts of the same kind with each of the two second order fibers. This has been confirmed by Boschek (1971). Furthermore it seems that L1 and L2 probably have identical connectivities with elements in the lamina derived from the medulla (Strausfeld and Campos-Ortega, 1972).

We shall return later to the question of the possible contacts between the retinular cell fibers and those second order neurons which are located outside the core of the cartridge and which are considered neither by Vigier nor by Trujillo-Cenóz.

II. Vigier states that the array of photoreceptor endings in the retinula is "narrow" and that the optical axes of the ommatidia diverge "slightly" but he does not treat the fact that this explanation rests entirely in the precise quantitative geometrical relationships in the angles involved. In other words, Vigier's theory is correct only if the angle between the central axes of neighbouring ommatidia, the "interommatidial angle", has the same magnitude as the angle between the lines of sight of neighbouring rhabdomeres of one ommatidium. Only then, and in addition, only if the hexagonal array of the elements of the retinula has the same orientation as the hexagonal array of the ommatidia in the eye (this Vigier already answered in the affirmative) can it happen that each of the seven lines of sight of one ommatidium is exactly parallel to the central line of sight of another ommatidium. The measurement of these angles was undertaken by Kirschfeld (1967) and the result was exactly as had been implicitly assumed by Vigier.

III. Vigier's argument, as we have translated it, seems to imply (albeit carefully) that the six peripheral rhabdomeres of an ommatidium have lines of sight corresponding to the axes of the six nearest neighbours among the surrounding ommatidia. This implies a rotational symmetry which is in contrast with the array of elements in the retinula as we have illustrated it in Fig. 4 and as Vigier himself has drawn it very neatly in one of his earlier papers (Vigier, 1907a). If one observes that the rhabdomere number 3, the only one which in terms of the hexagonal arrangement of the retinula is not an immediate neighbour of the central rhabdomere number 7, is displaced to a position which continues the regular hexagonal spacing of points one may still save Vigier's idea provided that the fibers between the retina and the lamina do not go from each ommatidium to the six neuroommatidia (cartridges) which are its nearest neighbours but to a set of cartridges which are arranged around the position of that ommatidium in a pattern which repeats, after a rotation around 180 degrees, the pattern of the rhabdomeres in the retinula. That the fibers of one ommatidium are distributed asymmetrically onto the lamina was indeed found by Trujillo-Cenóz and Melamed (1966) on the basis of reconstructions from serial ultrathin sections and by Braitenberg (1967) in silver preparations. The latter author, at that time ignorant of Vigier, rediscovered Vigier's explanation of the projection pattern and showed that when the distortions of the ommatidial array are taken into account, the pattern of the retina-lamina projection exactly annuls the complicated pattern formed by the diverging lines of sight of the rhabdomeres in the ommatidia so that the array of cartridges in the lamina receives nothing but an orderly representation of the visual field, with the definition corresponding to that of the ensemble of the ommatidia, *as if* each ommatidium was simply seeing one point of the environment (Fig. 5).

There are some more minor flaws in VIGIER's account. His implicit assumption that the fiber of rhabdomere number 7, the one occupying the central position in the retinula, joins the other 6 fibres "which carry identical information" in the cartridge, is in fact wrong. We now know that rhabdomere 7 is a double rhabdomere (MELAMED and TRUJILLO-CENÓZ, 1967) and the two elements which compose it deserve a special treatment of their own for various reasons.

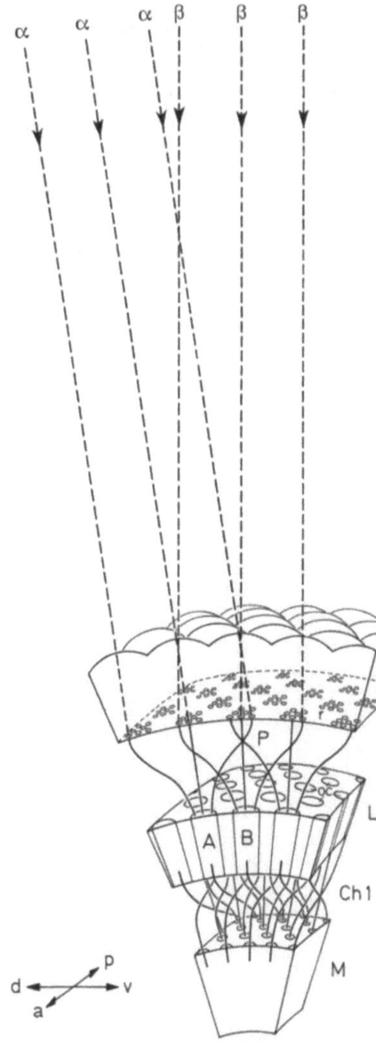

Fig. 5. Diagram of the projection of the visual environment through the cornea onto the retina and hence, via nerve fibres, onto the lamina and the medulla. *C*: array of lenses in the cornea, *R*: array of retinulae (*r*) in the focal plane of the lenses, *p*: projection of each retinula onto 7 optic cartridges (*o. c.*) of the lamina (only some of the fibres are shown), *L:* lamina with cartridges. (A, B . . . .) *Ch*1: first chiasma, *M*: medulla, *a*: anterior, *p*: posterior, *v*: ventral, *d*: dorsal. Optical information from a direction α is conveyed through nervous pathways to cartridge *A*; information from β to cartridge *B* etc.

Fibers of Retinular Cells 7 and 8

We know from the work of MELAMED and TRUJILLO-CENÓZ (1968) that the seventh light guide of the retinula, the one occupying a central position in the cross-section of the ommatidium, is in fact made up of two receptor cells, one (cell 7) lying directly above the other (cell 8). TRUJILLO-CENÓZ has shown that the microvilli which make up the fine structure of the rhabdomeres are oriented in cell 8 at right angles to those in cell 7. CAJAL and SANCHEZ (1915) recognized that at least one fiber from each ommatidium projects *through* the lamina with a termination in the second synaptic region, the medulla. In fact it is now known from TRUJILLO-CENÓZ and MELAMED that the axonal prolongations of retinular cells 7 and 8 project as a pair, keeping in close proximity all the way through the lamina. Golgi studies (STRAUSFELD, 1971a) have shown that both components of this pair end in the medulla; fibers 7 and 8 do not enter the pattern of decussation of retinular cell fibers 1—6 described in the previous chapter. Degeneration studies (CAMPOS ORTEGA and STRAUSFELD, 1972a) have shown R7 and R8 terminating at two levels in the medulla and map the retina-lamina mosaic onto the 2nd synaptic region.

Second Order Neurons in the Lamina

There are at least five different kinds of neurons, whose cell bodies lie between the external plexiform layer of the lamina and the base of the retina, which can be followed through the lamina, across the first optic chiasma and into the second sy-

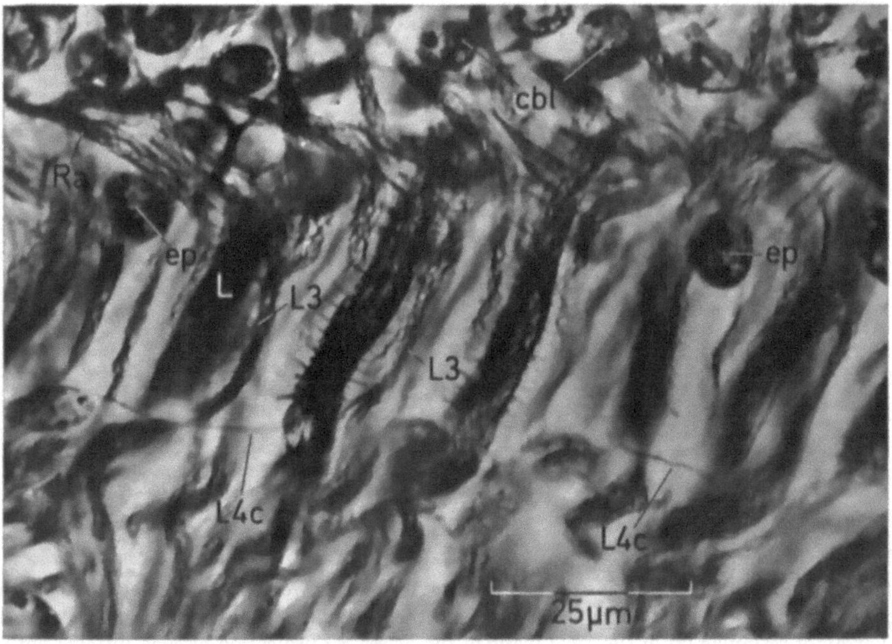

Fig. 6. Section through the lamina, oriented between the *v*- and *x*-axis of Fig. 5, showing monopolar cell bodies (*cbl*), retinular fibres (*ra*), the pairs of L_1 and L_2 (L), nuclei of so-called epithelial cells (glia elements situated between triplets of cartridges [*ep*]), some collaterals of L_4 (L_4c) and, finally, an L_3 element (L_3)

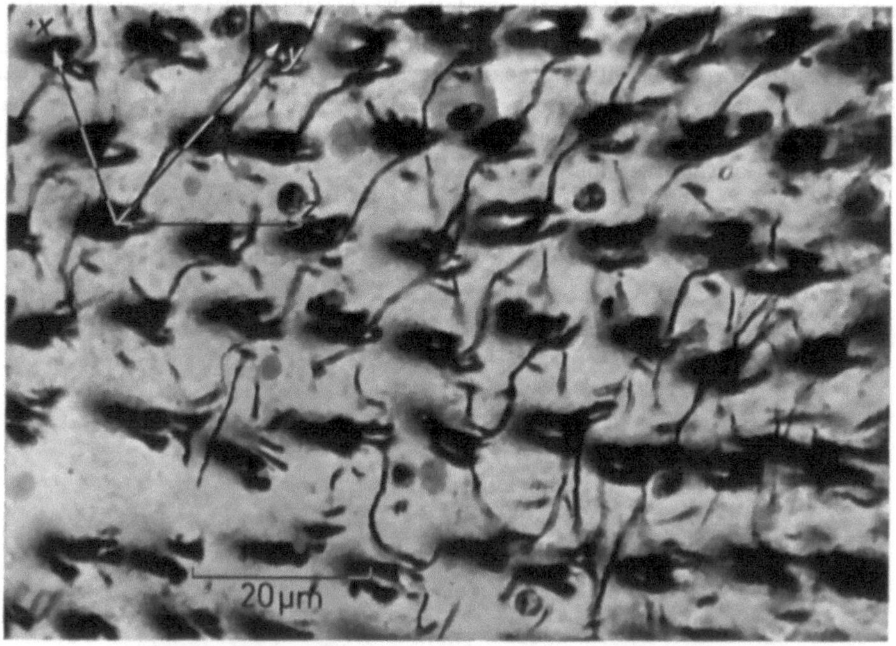

Fig. 7. Tangential section through a layer just beneath the lamina neuropil, showing the network of the L_4-collaterals. The three axes of the eye, *x*, *y and z*, have been indicated in order to show the regular orientation of these elements

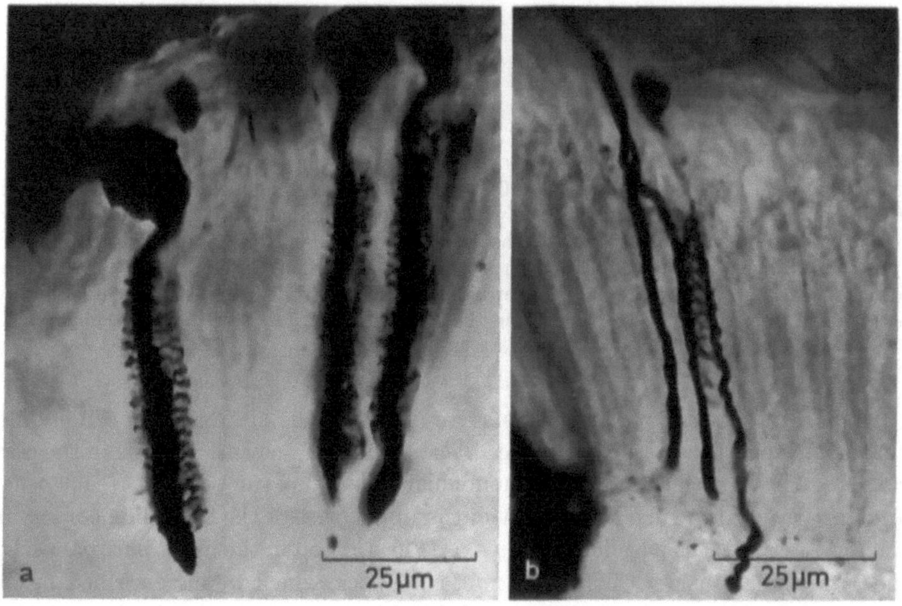

Fig. 8a and b. Laminae of *Musca* and *Calliphora*. a 3 Golgi stained L_1 components in the lamina of *Musca*. b An L_3-neuron in the lamina of *Calliphora*, with two retinular cell fibres that invade two adjacent cartridges. As far as can be seen in this Golgi preparation the right hand fibre (either R_2 or R_3) appears to be in direct contact with L_3

naptic region of the optic lobes, the medulla. Each cell type has a very distinctive silhouette in Golgi impregnated material and features of all of them can be clearly distinguished in reduced silver preparations (Figs. 6, 7, 8, 12). From this we can tell that these five "monopolar" cells of the lamina have a periodicity equal to that of the optic cartridges or neurommatidia, i.e. each cartridge has associated with it one of each of the five monopolars.

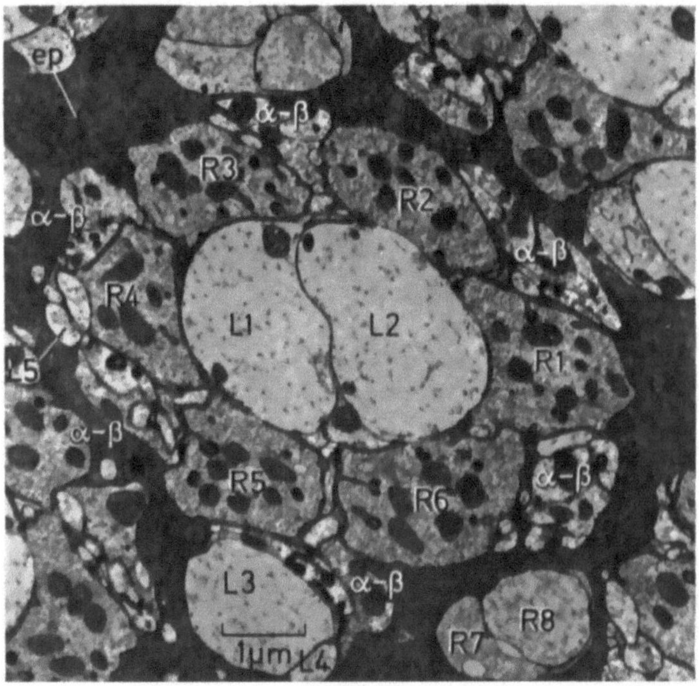

Fig. 9. *Cartridge cross section.* An electron micrograph showing a cross section of a cartridge: the crown of retinula cell profiles (*R*1—*R*6) surround the axial monopolars, *L*1 and *L*2. *L*3 and *L*4 are, at this level (about one half the way down the cartridge), situated outside the crown beside *R*5 and *R*6. *L*5 lies beside *R*4. The pairs of fibres labelled α—β are derived from one type of medulla-to-lamina cell Tl and an intrinsic or amacrine cell (see. Fig. 13). α—β fibres form a basket of processes around the crown of retinula cells. The long visual fibres *R*7 and *R*8 lie separate from the cartridge. They, and the *R* and *L* fibre, are embedded in a glial matrix (darkly stained: ep)

A pair of the monopolar cells is surrounded by the six retinular cell endings that come together in one cartridge. Because of their central position in the cartridge they have been termed axial monopolar cells. They run parallel to each other, and are arranged in such a way that in the upper part of the eye a line conjoining their cross-sections is parallel to the y axis (Fig. 10) and in the lower part parallel to the x-axis. We consider them as distinct cell types, which we have termed L_1 and L_2, for convenience, not only because of this difference of position in the cartridge, but also because they show different staining properties and, more important, because one of the two ends more deeply than the other. One of the pair, L_1, situated with respect to its fellow in the direction of the positive y-axis in the upper hemisphere (and in the direction of the negative x-axis in the lower hemisphere) has, in some of our preparations, a smaller cross-section and is more darkly stained than L_2. These two neurons, as they appear in reduced silver preparations, can be correlated with two different branching patterns as we know them from Golgi material. According to TRUJILLO-CENÓZ (1965) each of these two fibres makes the same type of synaptic contacts with each of the six retinula cell axons by means of lobated extensions (Figs. 9 and 11). The difference of the cross-sections of L_1, and L_2 and the variation of this difference in different regions of the eye, has recently

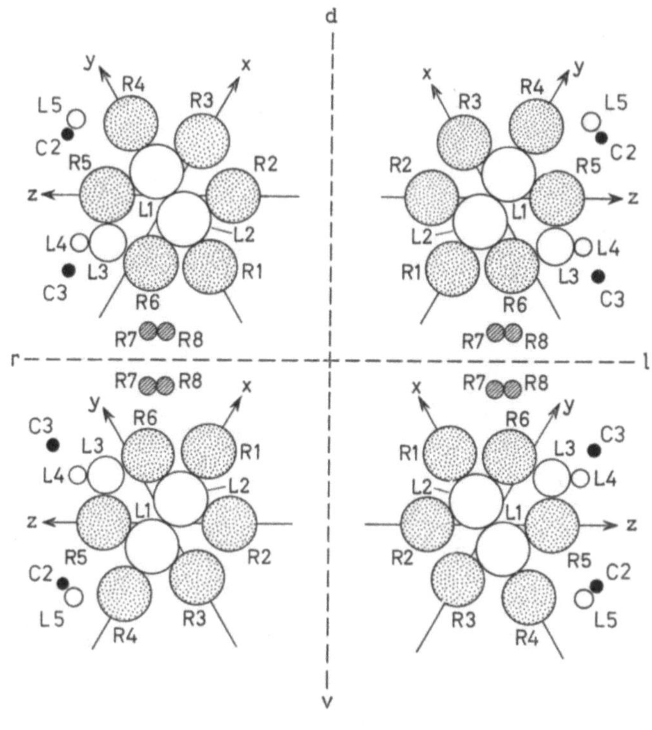

Fig. 10. Arrangement of the elements of the cartridges in the four quadrants of the two laminas. *d:* dorsal, *v:* ventral, *l:* left, *r:* right. The two dotted lines represent the two planes of mirror symmetry of all elements of the visual system of the fly: the mid-sagittal plane and the so-called equatorial plane, corresponding to about the 55th horizontal row of ommatidia. A kaleidoscope with two mirrors in these planes would show the whole visual system if one half of one eye (or one quadrant of a histological preparation through the visual ganglia) was viewed through it. *R*1—*R*6; fibres of retinula cells 1—6 (numbered as in Fig. 5), the "short visual fibres". Their arrangement is counterclockwise in the upper half of the right eye, clockwise in the lower half etc. *R*7 and *R*8: fibres of retinula cells 7 and 8 (the "long visual fibres"). Their position is equatorial with respect to the cartridge to which they belong (HORRIDGE AND MEINERTZHAGEN, 1970). *L*1, *L*2: the two second order fibres forming the core of the cartridge. They are arranged along the y-coordinate in the upper half of the eye, along the x-coordinate in the lower half. L_3, L_4: the two second order fibres lying between the retinula-cell-axons R_5 and R_6. L_4 gives off collaterals (not shown here) which form the regular net illustrated in Fig. 7. L_5: The fifth monopolar cell form, lying between R_5 and R_4. C_2 and C_3: the positions of the terminals of the medulla-to-lamina "centrifugal" cells. The α—β fibres (not diagramed here) lie between the retinula cell profiles. (Data from BRAITENBERG, 1967, 1970; BOSCHEK, 1971; STRAUSFELD, 1971a; CAMPOS-ORTEGA and STRAUSFELD, 1972a; STRAUSFELD and CAMPOS-ORTEGA, 1972)

been the subject of a quantitative electronmicroscopical study and a possible explanation in terms of their role in the mechanism of movement perception has been proposed (Braitenberg and Hauser-Holschuh, 1972).

Another monopolar cell whose perikaryon is situated with those of L_1 and L_2 above the cartridge has its fibre so closely opposed to the fibres of the other two

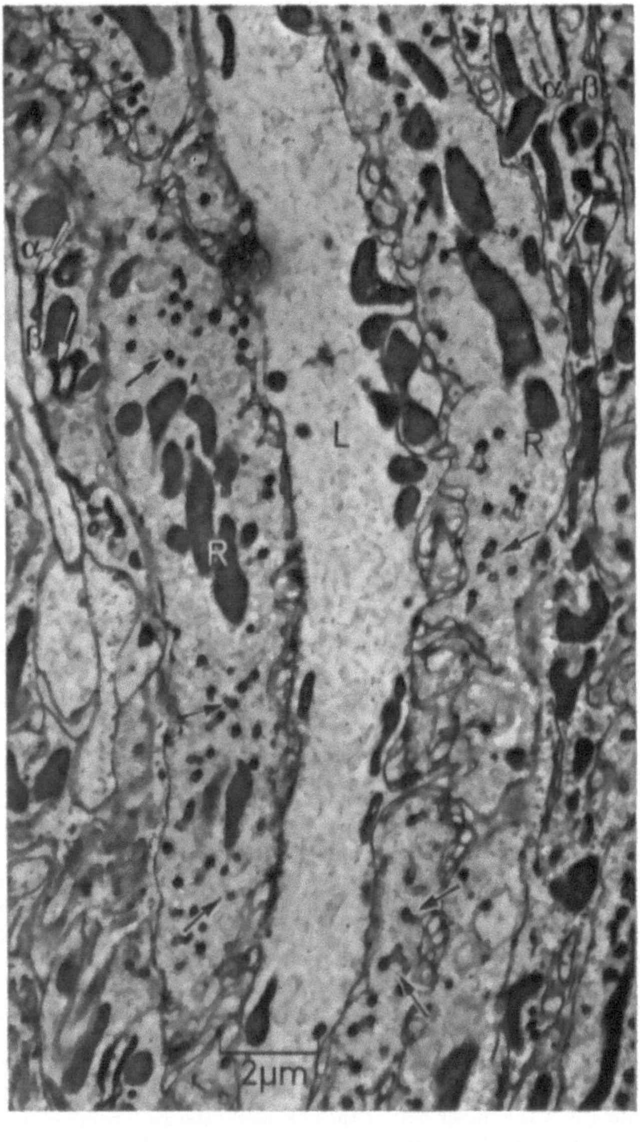

Fig. 11. Section cut parallel to the long axis of a cartridge showing positions of retinula cell terminals (R) either side of an L_1 (or L_2) fibre (L). Adjacent to the retinula cell axon there are portions of the α, β-fibres which are part of the basket terminals of T_1 and arborisations of lamina intrinsic cells (cf. Figs. 9, 13). Peculiar membrane specialisations (arrows) having the form of invaginations can be seen both in the β-fibres and in the retinular terminals

at the outer surface of the lamina as to be almost indistinguishable from them in the light microscope. But further down the cartridge this cell takes a very characteristic course: it projects between R_5 and R_6 to a position posterior and outside the crown of retinula cell terminals and runs down through the lamina. It finally joins the other axons just beneath the inner margin of the lamina. This neuron, which we call L_3, is clearly recognisable in Golgi preparations as well as in reduced silver

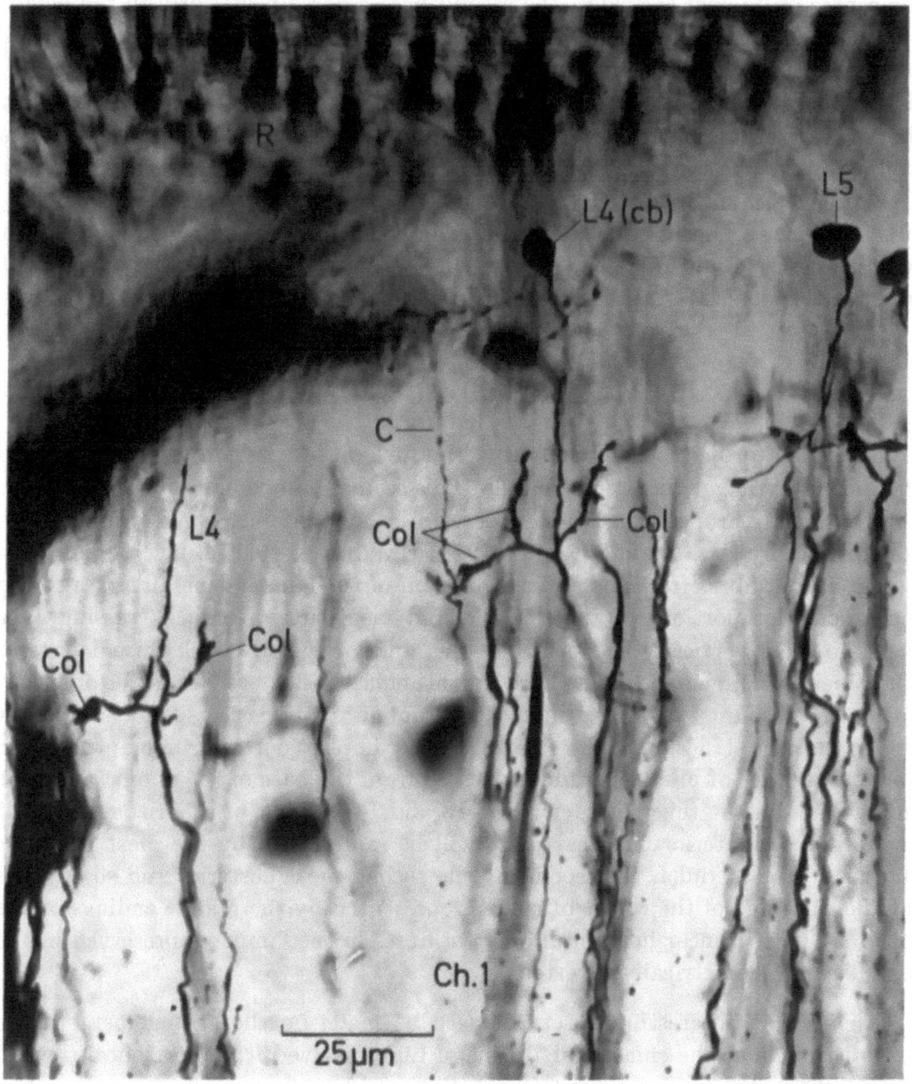

Fig. 12. L_4-neurons (one connected to its cell body *cb*), as seen in a Golgi preparation, and a fifth form of monopolar cell classified as L_5 in Fig. 13. The arrangement of L_4 collaterals should be compared with the network of collaterals as seen in Bodian preparations (Fig. 7). *R:* retina; *c:* centrifugal terminal from the medulla; *col:* L_4-collaterals; Ch_1: first optic chiasma

preparations on the basis of its characteristic posteriorly kinked appearance and asymmetrical arrangement of processes (Fig. 6, Fig. 8b, Fig. 13).

The fourth neuron, L_4, is also situated posteriorly with respect to the cartridge, with its fibre placed alongside that of L_3 for most of its length between the cell-bodies and inner delineation of the lamina. This fourth fibre (Fig. 12) is distinguished from the others by a remarkable branching pattern: just beneath the inner face of the lamina it gives rise to 3 branches, one of which reaches anteriorly into the centre of its own cartridge, the second passes dorso-posteriorly to the neighbouring cartridge lying above it, and the third passes ventro-posteriorly to the cartridge beneath it. In horizontal sections of the lamina, stained with reduced silver, the branches of these cells are seen to form a characteristic network pattern beneath the whole of the lamina's inner face (Fig. 7). Counts of cell-bodies above the lamina as well as direct observations of Golgi impregnated material reveal that there is a fifth type of monopolar cell (L_5) present in all optic cartridges (STRAUSFELD, 1971 a). It passes through the lamina outside the retinula cell crown beside retinula cell terminals R4 and R5. The positions of L1–L5 with the cyclic arrangement of receptor terminals R1–R6 is shown in Fig. 10.

The First Optic Chiasma

There is a projection of fibres from the first optic ganglion, the lamina, into the second optic ganglion, the medulla. This seems to have one thing in common with all the arthropods for which a description of the optic ganglia is available (see reviews by CAJAL and SANCHEZ, 1915; HANSTRÖM, 1928; BULLOCK and HORRIDGE, 1965): fibres from the anterior portion of the lamina generally pass to the most posterior portions of the medulla and vice versa. No such crossing over seems to occur vertically between the upper and lower parts of the visual system. The precise arrangement of the fibres in this chiasma has been a matter of speculation (see account by BRAITENBERG, 1968) but one of us (STRAUSFELD, 1971 b) has shown that the structure of the chiasma is the simplest one imaginable: the linear horizontal order of bundles of fibres derived from the base of the lamina arrive at the medulla in the reverse order (Figs. 13 and 14) so that of a horizontal row of cartridges the anterior-most cartridge will be represented by the most posterior endings in the medulla, the second by the second-most posterior and so on. The middle cartridge of the row is of course represented by the middle endings of the medulla. Thus a linear horizontal order of fibres in the lamina is preserved in the medulla but geometrically inverted.

There is an interesting observation to be made on the arrangement of the fibres throughout the chiasma. Ribbons of fibres derived from horizontal rows of ommatidia in the lamina twist through 180° and terminate as horizontal rows in the medulla. In the upper half of the right eye each ribbon has a right handed twist. (Following the ribbon from the lamina to the medulla or vice versa, one would see a clockwise rotation). In the lower half of the right eye the twist is left handed. In the left eye, of course, the situation is a mirror image (BRAITENBERG, 1970).

Exactly the same local, half hemispheric, variation of the direction of twist has been seen in the bundles leading from the ommatidia to the cartridges of the lamina.

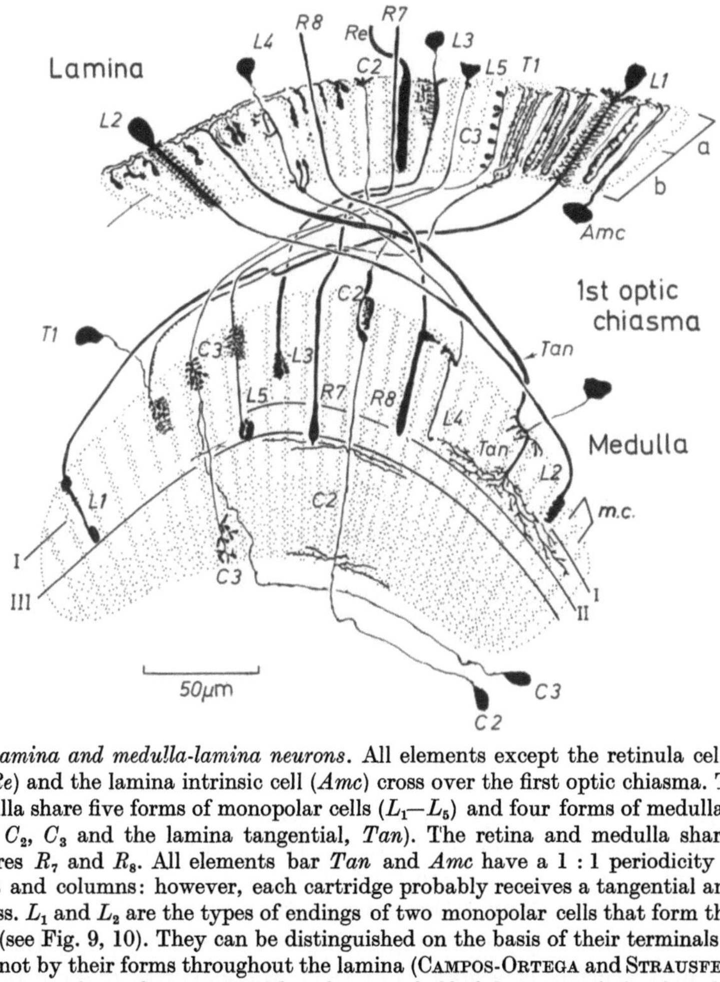

Fig. 13. *Lamina and medulla-lamina neurons*. All elements except the retinula cell terminals R_1—R_6 (*Re*) and the lamina intrinsic cell (*Amc*) cross over the first optic chiasma. The lamina and medulla share five forms of monopolar cells (L_1—L_5) and four forms of medulla-to-lamina cells (T_1, C_2, C_3 and the lamina tangential, *Tan*). The retina and medulla share the long visual fibres R_7 and R_8. All elements bar *Tan* and *Amc* have a 1 : 1 periodicity with optic cartridges and columns: however, each cartridge probably receives a tangential and intrinsic cell process. L_1 and L_2 are the types of endings of two monopolar cells that form the core of a cartridge (see Fig. 9, 10). They can be distinguished on the basis of their terminals in the medulla but not by their forms throughout the lamina (CAMPOS-ORTEGA and STRAUSFELD, 1972a). L_3 has a unilateral set of processes within the outer half of the external plexiform layer of the lamina (levels a, b). Its ending in the medulla lies between the terminals of the two neurons that form the core of the cartridge. L_4 is characterised by its collaterals at the inner face of the lamina and the set of processes at level a in the plexiform layer. The collateral pattern is repeated in the medulla, but with a reverse polarity. L_5, or the "midget" monopolar cell, has one or two lateral processes in the lamina's outer surface: its terminal in the medulla is complex in form and lies at two levels. In addition to the monopolar neurons, whose perikarya lie above the lamina, there are three small-field elements per cartridge derived from cell bodies above or beneath the medulla. T_1 terminates as a basket of fibres in the lamina. It is derived from a set processes at the level of L_2's ending. C_2 and C_3 have distinct forms of endings in the lamina. Their complex wide-field arborizations in the medulla are also easily identifiable. In the lamina a and b indicate the two main layers defined by the presence or absence of tangential processes. In the medulla levels I, II and III indicate the relative depths of the L_2, L_3 and L_1 endings, respectively. *m. c.* indicates the extent of one medullary column. Posterior on this diagram is to the right

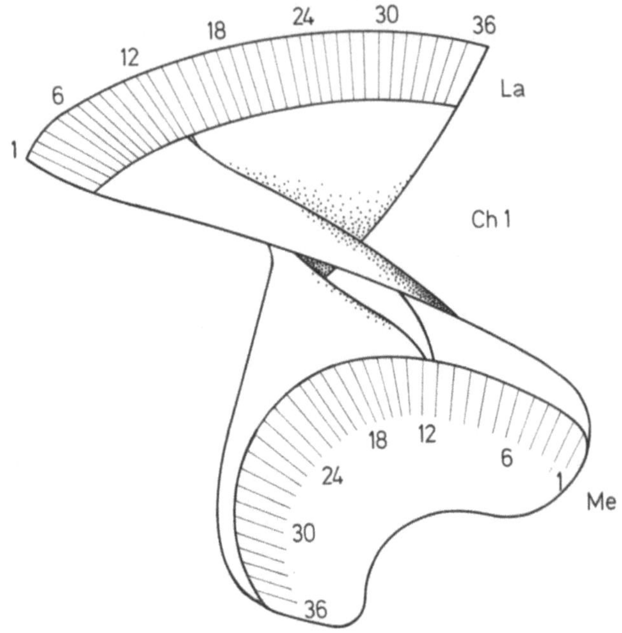

Fig. 14. Diagram to show the topography of cross-over of the sheets formed by the fibre bundles of one layer of the first optic chiasma, *La*, *Ch*1 and *Me* indicate lamina, first optic chiasma and medulla, respectively. The cartridges of the lamina are numbered from front to back. Posterior-most and anterior-most fibres cross over near the medulla's outer face whereas the most lateral fibre bundles, 18—24, cross over very near the inner face of the lamina (STRAUSFELD, 1971b). This is clearly shown in Fig. 1

The Medulla

We can state, with some confidence, the elemental features of the lamina and retina; however, we would prefer to be relatively cautious about the medulla, since, although we know much about the forms of neurons in this region (see Fig. 2 and STRAUSFELD, 1970; CAMPOS-ORTEGA and STRAUSFELD, 1972b) we have little knowledge about how they are arranged with each other throughout the neuropil.

The medulla's surface is perforated by the hexagonal array of input bundles arriving from the lamina. Each of these bundles contains the terminal portions of five monopolar cells from the lamina and the two long visual fibres (R_7 and R_8) from the retina. This means that there are at least seven terminals at a single locus of the medulla's surface which represent a single optic cartridge. There are, in addition, 3 types of cells from the medulla which seem to suggest conduction from each column to the corresponding cartridge in the lamina. This means that each bundle between these two region contains at least 10 fibres (STRAUSFELD, 1971a, b). Contiguous with each of these groups of fibres are at least nine second order interneurons that project through the medulla's depth and finally terminate in the lobula complex via the second optic chiasma (CAMPOS-ORTEGA and STRAUSFELD, 1972b). These fibres seem to be the main constituents of one medullary column. The arrangement of columns implies that at any level of the medulla there is a representation of the retina-lamina mosaic; thus each column can be associated with a single point of the visual environment (Fig. 5). At the base of each column, just above the inner face of the medulla, there are two more cell types which project to the lobule and lobular plate (Fig. 2). Thus there are at least eleven relays to the innermost synaptic regions of the optic lobes which convey the point to point

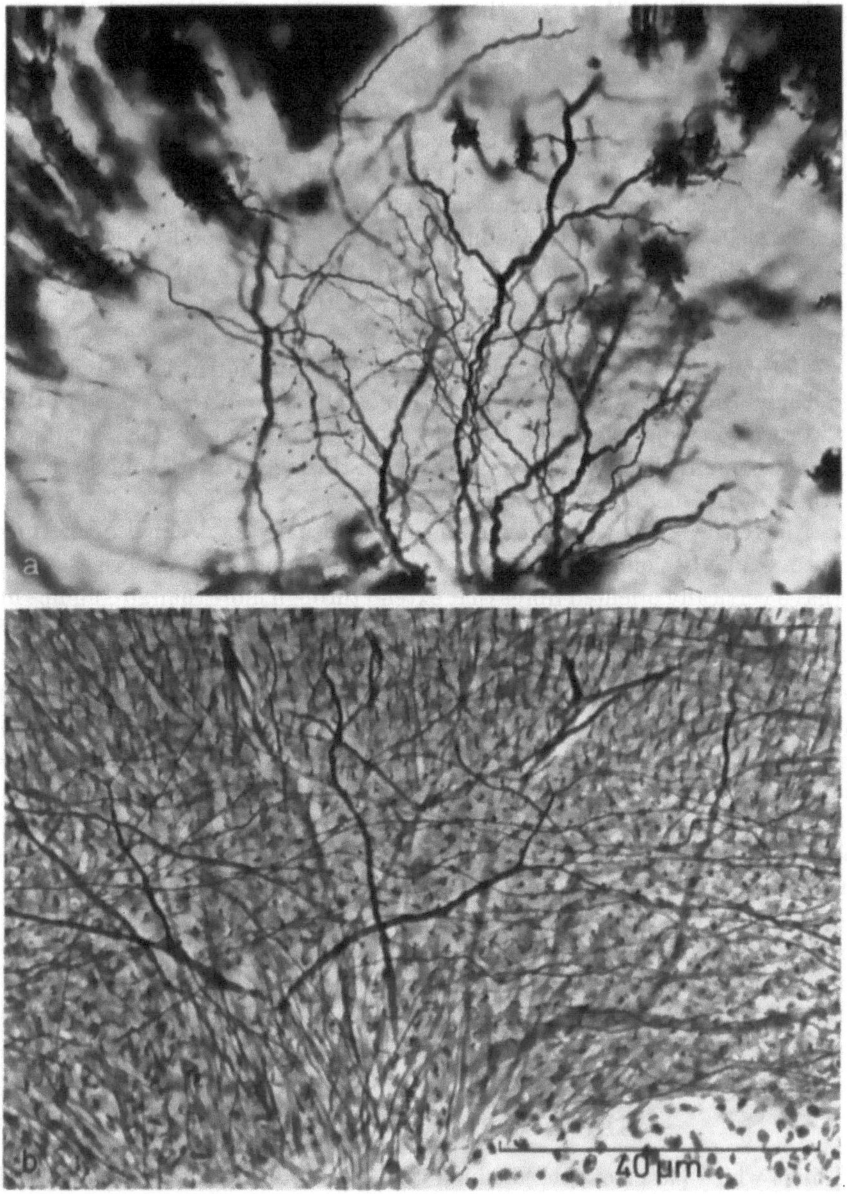

Fig. 15a and b. Tangential fibres in the serpentine layer of the medulla. a As seen in a Golgi preparation of the Blow-fly, *Calliphora erythrocephala*. b The same cell as seen in a Bodian reduced silver preparation

representation of the retinal map onto the two surfaces of the lobular complex (Fig. 1).

In summary then, of the great many forms of small field second order neurons in the medulla (at least 40 types in *Musca*) eleven have so far been related to the periodic arrangement of columns. These cell types have their long axes oriented at right angles to the medulla's surface and a characteristic pattern of their different branching patterns may be related to the different ways they collect signals from the input fibres. Moreover, the processes of the medulla-to-lobula-complex neurons appear to be as strictly oriented along the x, y, z axes of the neuropil as the L_4 collaterals in the lamina (STRAUSFELD and BRAITENBERG, 1970; STRAUSFELD, 1971b).

There are other fibre types in the medulla that have their long axes oriented parallel with the medulla's surfaces. These elements are either intrinsic to the medulla (and are termed amacrine cells) or are derived from neurons, termed tangential cells, that link the medulla with the contralateral optic lobe and/or ipsi-or contralateral mid-brain regions (STRAUSFELD and BLEST, 1970). Tangential cells (Fig. 15) probably mediate interactions within large aggregates of medulla elements (or lamina-retina input groups) themselves representing large regions of the visual field. The salient features of tangential and intrinsic organisation are listed below (from STRAUSFELD, 1970; CAMPOS-ORTEGA and STRAUSFELD, 1972b).

1. Processes from the fields of eight types of tangential cells invade every medullary column. The tangential fields spread through strips of columns or across the whole of the medulla's lateral extent.

2. Above the serpentine layer intrinsic cells usually have small fields through between 1 and 10 columns. There is, though, a set of "line amacrines" (Fig. 2) that have vertically oriented fibres that extend through more than 20 columns in the one direction and are restricted to single columns in the other, horizontally. These fibres form a dense stratum which exhibits an anteroposterior gradient, there being more fibres posteriorly than anteriorly (BRAITENBERG and HAUSER HOLSCHUH, 1972). Two other sets of amacrines have wide oval fields at this level. Beneath the serpentine layer intrinsic cells have large oval fields, densely packed collaterals, and serve vertically arranged strips of columns.

3. There are also intrinsic cells that serve to link one level of the medulla with another and yet others with asymmetric fields that connect widely separated sets of columns at the same level by long axon-like fibres.

4. The second order cells and similarly oriented intrinsic cells mark the columns of a three dimensional lattice. The tangential and intrinsic cell processes arranged at discrete levels form the cross-connections. The geometry of the medulla is in contrast to that of the lamina, which is primarily columnar, and the lobule where cells are arranged supraperiodically.

The Second Optic Chiasma

Fibres which pass from the medulla to the subsequent two ganglia, the lobule and the lobular plate, are collectively known as the second optic chiasma. The pattern formed by this structure (Fig. 1) is at first sight very complex, but can be depicted by the following simple rules;

a) the fibres between the medulla and lobular plate do not undergo any cross over; thus the order of their exit from the medulla, horizontally and vertically, is preserved: the most lateral portions of the lobular plate are connected to the

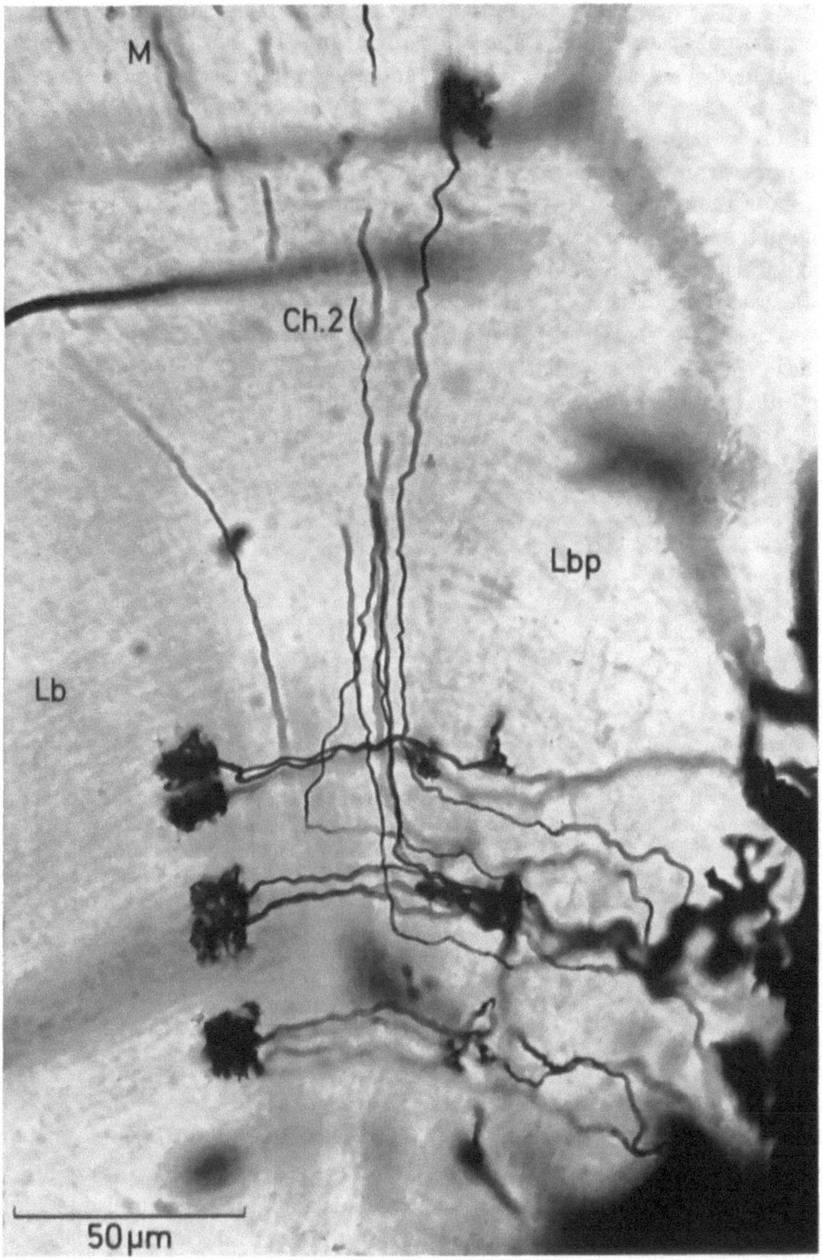

Fig. 16. Fibres in the second optic chiasma (*Calliphora erythrocephala*). This Golgi preparation shows some elements (*T*-cells) that connect the medulla (*M*) to the lobula plate (*Lpb*) and the obule (*Lb*) to the lobula plate. *Ch2:* second optic chiasma

most posterior portions of the medulla and therefore, through the lamina, represent a structural projection of fibres that relay information from the most anterior portions of the visual field.

b) Fibres between medulla and lobule cross in a manner analogous to those described for the first optic chiasma. Anterior fibres from the medulla end in the medial portion of the lobule and thus represent the most posterior portion of the visual field. Posterior fibres from the medulla end in the lateral portion of the lobule.

c) Thus, portions of the lobule and lobular plate lying opposite each other receive fibres from the same locus in the medulla and therefore are related to the same area of the represented visual field. It comes as no surprise then that

d) lobule and lobular plate are connected by a set of parallel fibres (Figs. 16, 18). Like in the first optic chiasma there seems to be no cross-over of fibres in the vertical direction but there is an important difference in the number of layers in the two chiasmata: between the lamina and the medulla each horizontal row of columns in the outermost region gives rise to one layer of the chiasma. But between the medulla and lobule pairs of adjacent horizontal rows of columns in the medulla give rise to single layers of fibres that either cross-over to the lobule or stretch to the lobular plate (Fig. 17).

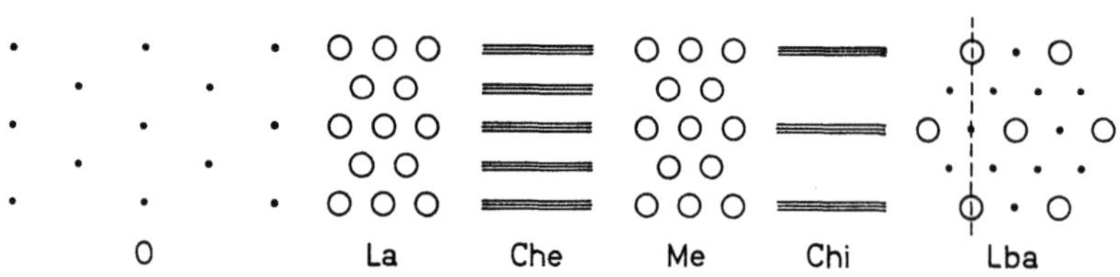

Fig. 17. Arrangement of ommatidial axes in the visual field (0), cartridges in the lamina (*La*), layers of the first (or external) optic chiasma (*Che*), columns in the Medulla (*Me*), layers of the second (or internal) optic chiasma (*Chi*) and of some neuronal elements in the Lobula (*Lba*). The mapping of the elements of O onto *La* is achieved by the transformation already described in the legend of Fig. 5. Each row of elements in *La* corresponds to one layer of *Che*, and to one row of elements in *Me*. However, each layer of *Chi* corresponds to two rows in O, *La* and *Me*. In *Lba* some large neurons (circles) are arranged in a hexagonal array derived from the original one by occupying only every second position in every second row. On vertical cuts (dotted line) these elements appear spaced at distances 4 times the distance of the original horizontal rows (From BRAITENBERG, 1970)

The Lobular Complex

We call the two synaptic regions proximal to the medulla, the (anterior) lobule and (posterior) lobular plate, the lobular complex, since in most other orders of insects there is only one synaptic region at this position, the undivided lobule. We have already outlined the projection of fibres from the medulla to these two regions and it is clear from this that incoming fibres to the lobular plate are located exactly opposite incoming fibres to the lobule when both sets are derived from the same locus in the medulla (Fig. 16). Tangential sections of the lobular plate do not reveal the precise hexagonal arrangement of fibre-cross sections, a pattern which is typical for the first two synaptic regions. Nevertheless the unscrambled arrangement of at least a proportion of the fibres allows the geometric order of the first two synaptic regions to be carried into the third and fourth regions.

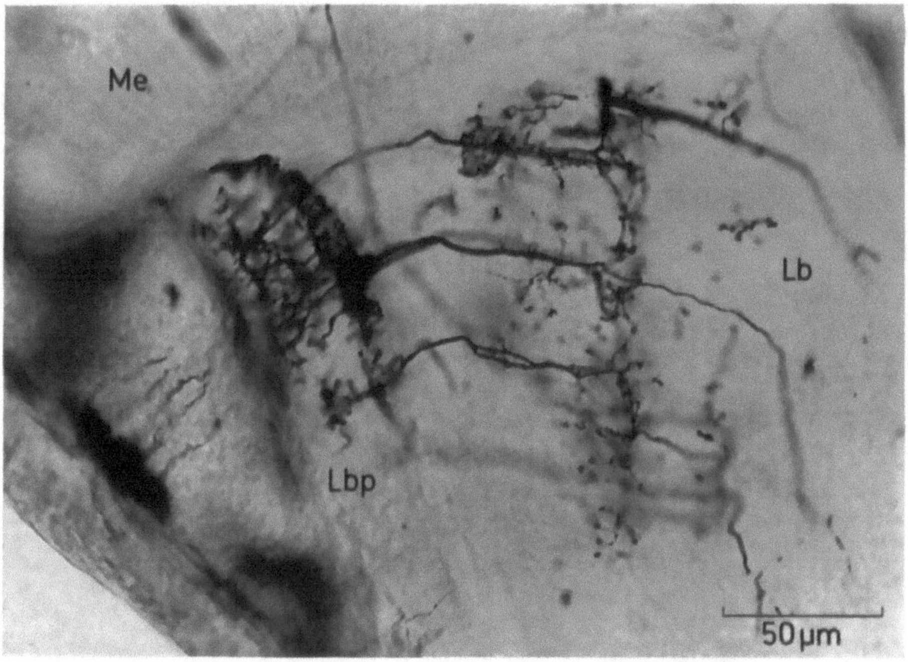

Fig. 18. *Lobular complex.* Golgi preparation of fibres which invade both regions of the lobular complex and connect them to the midbrain. *Me:* medulla; *Lbp:* lobular plate; *Lb:* lobule

The Lobule

There is clearly a periodic arrangement of some of the types of neurons within this region whose population is arranged in a hexagonal array even if the period of the array is equivalent to about two of the unitary periods in the retina, lamina and medulla. A tentative explanation of this supraperiodic arrangement is given in Fig. 17. The diameter of the field spreads of these cells is somewhat larger than their separation, so that there is some overlap between the fields of immediate neighbours.

This supraperiodic arrangement is not solely exemplified by these cells; there are other elements in the lobular complex which are arranged in a similar fashion: for instance to mention only one type, elements which invade both the lobule and lobular plate (Fig. 18). The lobule contains layers of tangential cells, like some of those found in the medulla; however, it seems almost if not wholly devoid of amacrine or intrinsic cells. Both regions of the lobular complex have predominantly horizontally and vertically arranged tangentials which lead to regions in the midbrain (Fig. 19). Of these cells, there are four types (two in the lobular plate and two in the outermost stratum of the lobule) which by virtue of their lateral spread can only interact with narrow horizontal strips of the relayed medullary mosaic (Fig. 19b). Finally, there are some neurons in the lobular complex, whose arborizations spread over the whole of the surface of either of these ganglia and which could serve to relay information from the whole of the visual field.

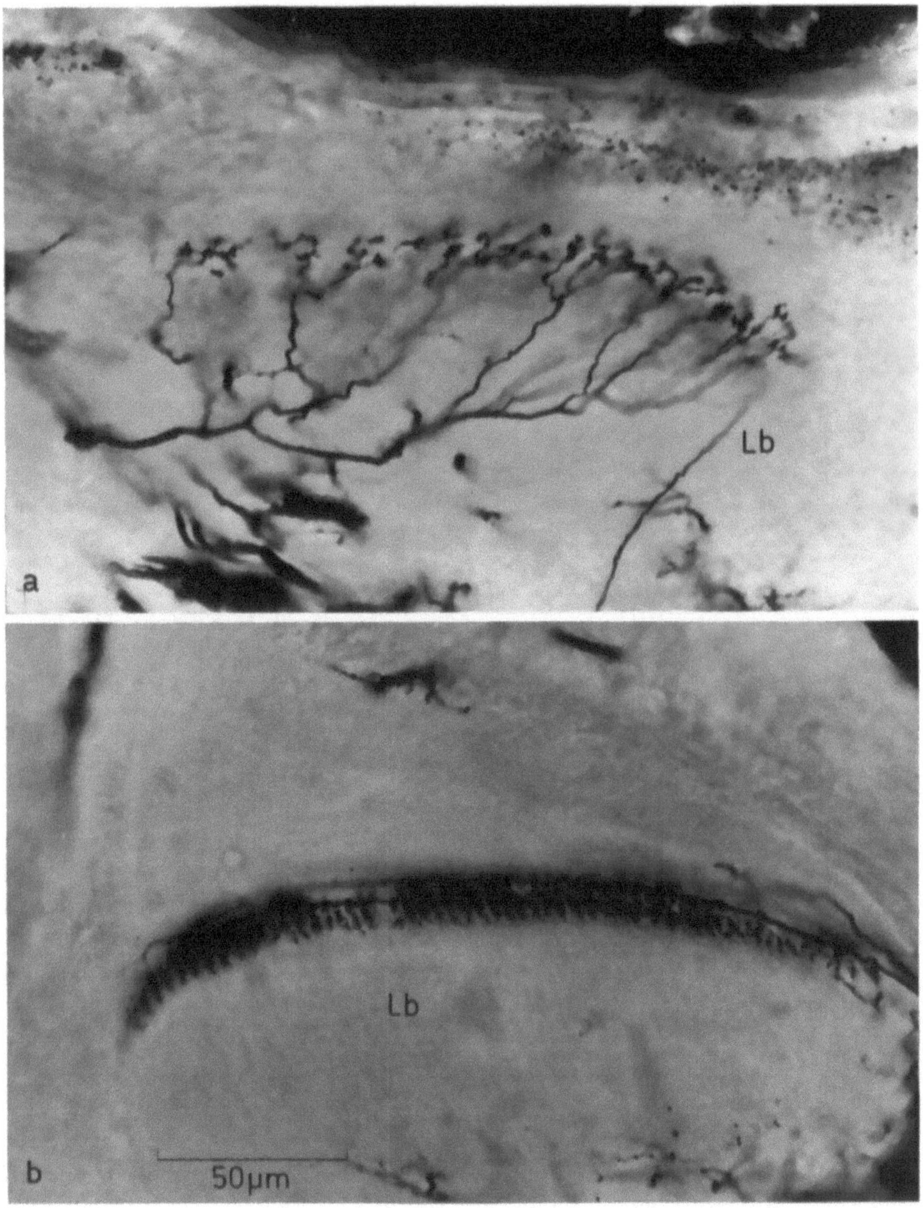

Fig. 19a and b. Two different types of "strip-field-neurons" (Strausfeld, 1970) which invade two different layers in the lobula (*Lb*) and lead to the midbrain

The majority of the types of small-field supraperiodic neurons terminate in glomerular regions of neuropil, such as the optic tubercle (Strausfeld, 1970), in the ipsilateral protocerebrum. There endings are arranged in tangles: all spatial representation of the retina-lamina mosaic is thereby lost. Thus it seems that the lobular complex is the last station to preserve spatial arrays of neurons that correspond to the geometry of the visual field.

Discussion

It has long been recognized that insects can detect movement in their optical environment and react to it in various ways. The best known of these effects are the so-called optomotor reactions, experimentally produced by movement of a constant angular velocity within a large part of the visual field (typically a rotating cylinder) which elicits a rotation (or a tendency to rotate) of the animal about an axis parallel to the axis of rotation of the optical surround. This may be reasonably interpreted as a mechanism having the purpose of stabilizing the flight (or the gait) of the animal with respect to accidental environmental perturbations. In fact, perception of movement of the visual environment to one side, such as might be produced by the effect of a deviation of the animal to the opposite side, will elicit as an optomotor reaction a rotation of the animal to the same side, so as to counteract the deviation. Experiments on the beetle *Chlorophanus* (Hassenstein, 1959) which find their counterpart in similar observations on the bee (Kunze, 1961) the housefly (Fermi and Reichardt, 1963; McCann and MacGinitie, 1965) and the fruitfly (Götz, 1964) indicate that the intensity of the reaction is computed on the basis of a summation over the output of a large number of elementary movement perceiving devices distributed across the whole eye. The individual movement detectors incorporate the principle of cross-correlation of the input derived from just two ommatidia which lie adjacent to each other.

The orientation of the movement detectors has been analysed from behavioral experiments by Götz (1968) for *Drosophila* and *Musca*. He found that the influence of the turning tendency around the "yaw"-axis was maximal when the perceived movement was in the horizontal direction and decreased cosine-fashion to zero for vertical movement. On the contrary, however, if the influence of movement in the visual environment on the forward thrust of the animal was determined (and therefore the influence on the uplift, since the force vector during flight seems to keep a constant upward pitch), it was found that the influence is maximal with movement in the vertical direction and zero with horizontal movement. These experiments would induce one to assumme two very similar mechanisms in the fly's brain, oriented at right angles to each other, one horizontal and one vertical: however, as Götz points out, an even more economical model, having the same properties, could be constructed on the basis of two obliquely oriented sets of movement detectors, whose outputs, when added, would measure the horizontal component and, when substracted, the vertical component of the movement.

There is another set of observations implying that the comparison of velocities perceived in different regions of the visual field could possibly provide the basis for various kinds of abstractions. A fly when it approaches some surface on which it will land "puts out its landing gear", in other words it stretches out its legs toward the landing surface, while during flight the legs are folded and kept tucked close to the body. This landing reaction (GOODMAN, 1960) can be elicited by means of the artificial stimulus provided by a rotating spiral in front of the animal (BRAITENBERG and TADDEI, 1966) which will produce in part of the visual field a streaming motion emanating from one point, apparently reminiscent, to the fly's brain, of the enlargement of the image of a landing surface during the approach manoeuvres.

With this previous knowledge, we approach the histology of the fly's visual ganglia armed with a number of specific questions about the patterns of fibres connections that may be responsible for the operations performed on the visual input.

1) From experiments on the beetle *Chlorophanus*, which are not contradicted by anything in the visual physiology of dipterous insects, it would appear that movement is perceived as a result of a computation between neighbouring ommatidia or between two ommatidia separated by just one ommatidium but not farther apart. Where are the fibres which, in the optic ganglia, span exactly two or three columns?

2) Are there any fibres oriented only in the horizontal direction in the visual ganglia (i.e. parallel to our z-axis), matched by a similar set of fibres oriented vertically? Two such sets of fibres may be inferred from GÖTZ's (1968) observations, or alternatively:

3) Are there two sets of similar, short fibre connections, oriented in two oblique directions, mirror symmetrical to the z-axis? Such oblique fibres could be postulated on the basis of GÖTZ's second model.

4) Where are the large fibers which are responsible for the summation of the outputs of all the individual movement detectors?

5) Where are the fibers which mediate the interactions of various quadrants of the visual field, which one must assume for the perception of an expanding image such as may elicit the landing reaction?

The answer to the second of these questions is in the negative. Nowhere did we find, in our survey of the visual ganglia, identical connections in the vertical and the horizontal direction. On the contrary, questions (1) and (3) may find an answer in one of the oblique sets of connections described, such as the collaterals of the L_4-fibres at the base of the lamina, or the oblique short side branches of neurons in the upper layer of the medulla which can be found in silver preparations, or, possibly, sets of amacrine cells which, as far as can be determined from Golgi analysis span only two or three columns along the x and y axes. Large neurons responsible for integration over large parts of the visual field [questions (4) and (5)] do not seem to be present in the lamina even though this region contains a complicated arrangement of vertically oriented tangential fibres each spanning a limited extent. Large field fibres are, though, found in the medulla and are particularly prominent in the lobular complex. It will take more detailed topographical mapping, both by behavioural experiments and from the histological preparations, before the histological substrate responsible for the various kinds of computation performed by the visual system can be suggested with sufficient plausibility. The main purpose of our contribution has been to provide some of the coordinates and neural landmarks in the visual system through which further attempts to match the geometry of the neuropil with its capacity to analyse features of the visual environment can be reliably carried through.

Acknowledgement

We are grateful to Dr J. A. Campos-Ortega for reading and commenting on the manuscript and also for data concerning levels of *L* terminals and medulla stratification. We would also like to thank Mr. E. Freiberg for the preparation of Figs. 3, 4, 5, 10 and 17 and Mrs K. Schwertfeger for preparation of the manuscript.

References

Boschek, C. B.: On the fine structure of the peripheral retina and lamina ganglionaris of the fly, Musca domestica. Z. Zellforsch. **118**, 369—409 (1971).

Braitenberg, V.: Patterns of projection in the visual system of the fly. I. Retina-lamina projections. Exp. Brain Res. **3**, 271—298 (1967).

— On Chiasms: In: Caianello, E. R. (Ed.): Neural Networks (proceedings of the school on Neural Networks, June 1967 in Ravello). Berlin-Heidelberg-New York: Springer 1968.

— Ordnung und Orientierung der Elemente im Sehsystem der Fliege. Kybernetik **7**, 235—242 (1970).

— Periodic structures and structural gradients in the visual ganglia of the fly. In: Wehner, R. (Ed.): Proc. of the European symposium "Information processing in the visual system of Arthropods", Zürich: March 1972. Berlin-Heidelberg-New York: Springer 1972.

— Holschuh-Hauser: Patterns of projection in the visual system of the fly. 2. Quantitative aspects of lamina-medulla neurons in relation to models of movement perception. Exp. Brain Res. **16**, 184—209 (1972).

— Taddei Ferretti, C.: Landing reaction of *Musca domestica* induced by visual stimuli. Naturwissenschaften **53**, 155 (1966).

Bullock, T. H., Horridge, G. A.: Structure and function in the nervous systems of invertebrates. San Francisco-London: Freeman, W. H., 1965.

Cajal, S. R.: Morfologia y conexiones de los elementos de la retina de los aves. Revista trimestral Histologia normal y patologica (1888).

— Contribucion al conocimiento de la retina y centros opticos de los cefalopodos. Trab. Lab. Invest. biol. Univ. Madr. **15**, 1—83 (1917).

— La rétine des vertébrés. La Cellule **11** (1892).

— Recollections of my life (Recuerdos De Mi Vida). (trans Horne, E. Cragie and Juan Cano). Cambridge-London: M.I.T. 1937.

— Sanchez, D.: Contribucion al conocimiento de los centros nerviosos de los insectos. Parte 1. Retina y centros opticos. Trab. Lab. Invest. Biol. Univ. Madr. **13**, 1—168 (1915).

Campos-Ortega, J. A., Strausfeld, N. J.: The columnar organization of the second synaptic region of the visual system of *Musca domestica* L. Z. Zellforsch. **124**, 561—585 (1972a).

— — Columns and layers in the second synaptic region of the fly's visual system: the case for two superimposed neuronal architectures. In: proc. of the European symposium "Information processing in the visual system of Athropods", Zürich: March 1972. Ed. R. Wehner. Berlin-Heidelberg-New York: Springer 1972.

Dietrich, W.: Die Facettenaugen der Dipteren. Z. wiss. Zool. **92**, 465—539 (1909).

Fermi, G., Reichardt, W.: Optomotorische Reaktion der Fliege *Musca domestica*. Kybernetik **2**, 15—28 (1963).

Fernandez Moran, H.: Fine structure of the light receptors in the compound eye of the insects. Exp. Cell Res. Suppl. **5**, 586—644 (1958).

Götz, K. G.: Optomotorische Untersuchung des visuellen Systems einiger Augenmutanten der Fruchtfliege *Drosophila*. Kybernetik **2**, 77—92 (1964).

— Flight control in *Drosophila* by visual perception of motion. Kybernetik **4**, 199—208 (1968).

GOODMAN, L. J.: The landing responses of insects. I. The landing response of the fly, *Lucilia sericata* and other Calliphorinae. J. exp. Biol. **37**, 854—878 (1960).

HANSTRÖM, B.: Vergleichende Anatomie des Nervensystems der wirbellosen Tiere. Berlin: Springer 1928.

HASSENSTEIN, B.: Optokinetische Wirksamkeit bewegter periodischer Muster (nach Messungen am Rüsselkäfer, *Chlorophanus viridis*). Z. Naturforsch. **14b**, 659—689 (1959).

HORRIDGE, G. A., MEINERTZHAGEN, I. A.: The accuracy of the patterns of connexions of the first- and second-order neurons of the visual system of *Calliphora*. Proc. roy. Soc. B **175**, 69—82 (1970).

KIRSCHFELD, K.: Die Projektion der optischen Umwelt auf das Raster der Rhabdomere im Komplexauge von *Musca*. Exp. Brain Res. **3**, 248—270 (1967).

KUNZE, P.: Untersuchung des Bewegungssehens fixiert fliegender Bienen. Z. vergl. Physiol. **44**, 656—684. (1961).

MCCANN, G. D., MACGINITIE, G. F.: Optomotor response studies of insect vision. Proc. Roy. Soc. B **163**, 369—401 (1965).

MELAMED, J., TRUJILLO-CENÓZ, O.: The fine structure of the central cells in the ommatidia of Dipterans. J. Ultrastruct. Res. **21**, 313—334 (1968).

MEYER, G.: Versuch einer Darstellung von Neurofibrillen im zentralen Nervensystem verschiedener Insekten. Zool. Jahrb. (Anat.) **71**, 413—431 (1951).

POWER, M. E.: The brain of *Drosophila melanogaster*. J. Morph. **72**, 517—559 (1943).

STRAUSFELD, N. J.: Golgi studies on insects. Part II: The optic lobes of Diptera. Phil. Trans. B **258**, 135—223 (1970a).

— Variations and invariants of cell arrangements in the nervous systems of insects (a review of neuronal arrangements in the visual system and corpora pedunculata). Verh. dtsch. Zool. Ges. **64**, 97—108 (1970b).

— The organization of the insect visual system (light microscopy) I. projections and arrangements of neurons in the Lamina ganglionaris of Diptera. Z. Zellforsch. **121**, 377—441 (1971a).

— The organization of the insect visual system (light microscopy) II. The projections of fibres across the first optic chiasma. Z. Zellforsch. **121**, 442—454 (1971b).

— BLEST, A. D.: Golgi studies on insects. Part 1. The optic lobes of Lepidoptera. Phil. Trans. B. **258**, 81—134 (1970).

— BRAITENBERG, V.: The compound eye of the fly (*Musca domestica*): connections between the cartridges of the lamina ganglionaris. Z. vergl. Physiol. **70**, 95—104 (1970).

— CAMPOS-ORTEGA, J. A.: Some interrelationships between the first and second synaptic regions of the fly's (*Musca domestica*) visual system. In: WEHNER, R. (Ed.): Proc. of the European symposium "Information processing in the visual system of Arthropods", Zürich: March 1972. Berlin-Heidelberg-New York: Springer 1972.

TRUJILLO-CENÓZ, O.: Some aspects of the structural organization of the intermediate retina of Dipterans. J. Ultrastruct. Res. **13**, 1—33 (1965).

— Some aspects of the structural organization of the medulla in Muscoid flies. J. Ultrastruct. Res. **27**, 533—553 (1969).

— MELAMED, J.: Electron microscope observations on the peripheral and intermediate retinas of Dipterans. In: BERNHARD, C. G. (Ed.): The functional organization of the compound eye. London: Pergamon Press 1966.

— — Light and electronmicroscope study of one of the systems of centrifugal fibers found in the lamina of Muscoid flies. Z. Zellforsch. **110**, 336—349 (1970).

VIGIER, P.: Mécanisme de la synthése des impressions lumineuses recueillies par les yeux composés des Diptères. C. R. Acad. Sci. (Paris) **63**, 122—124 (1907a).

— Sur les terminations photoréceptrices dans les yeux composès des Muscides. C. R. Acad. Sci. (Paris) **63**, 532—536 (1907b).

— Sur la reception de l'excitant lumineux dans les yeux composes des Insectes, en particulier chez les Muscides. C. R. Acad. Sci. (Paris), **63**, 633—636 (1907c).

— Sur l'existence réelle et le rôle des appendices piriformes des neurons. La neurone perioptique des dipteres. C. R. Soc. Biol. (Paris) **64**, 959—961 (1908).

Chapter 11

Comparative Physiology of Colour Vision in Animals

By

HANSJOCHEM AUTRUM and INGEBORG THOMAS, München (Germany)

With 18 Figures

Contents

First of all the question of whether animals can see colour demands precision, as nature never answers more intelligently than she is asked. It was the ophthalmologist, CARL VON HESS, who designed the first experiment which gave clear and understandable results about animals' colour discrimination (see AUTRUM, 1963). He asked whether animals could differentiate a coloured object from other equal but very different light objects. Also it was important to prove whether animals recognize colour as a special quality and not just its brightness.

To answer this question, one can train animals to a certain colour, and then test it critically by offering the colour along with colourless materials having degrees of brightness from deep black to pure white. If the animal can see colour, it will discover the trained colour from among the colourless objects.

Such research was conducted first by VON FRISCH (1914) on the bee. Numerous other animals were investigated using similar or other methods.

I. Methods

When spectral lights are used to analyse behaviour reactions to colours, it can always be proven that there is colour vision when the reaction (behaviour) to spectral light of wavelength λ_1 and intensity I_1 is different from the reaction to light of wavelength λ_2 and intensity I_2, and with any variation of intensities I_1 and I_2. This holds also for mixtures of colours. If pigment colours are used for testing colour vision one must be sure that they can be distinguished from "uncoloured" ("white") pigments of different degrees of brightness — that is, from white through increasingly grey steps to black[1]. With this method there is at least in principle the danger that pigments or lights which are white or neutral grey to man are coloured to animals. This can occur, when pigments reflect wavelengths which are not visible to the human eye[2]. This type of reaction is shown by bees (Kühn, 1927; Lutz, 1933; Hertz, 1937; Engländer, 1941; Daumer, 1956): Sheets of paper that appear equally white to us although they reflect ultraviolet light to different degrees are perceived by bees as different "whites". On the other hand bees cannot differentiate a "white" without any ultraviolet from blue-green (Daumer, 1956).

Colour vision in animals has been tested with the following methods:

A. Behavioural Studies

a) Training in numerous variations, adapted to the species and problem under investigation.

b) Registration of the optomotoric reactions. These were first applied by Schlieper (1927) in a rotating cylinder with striped patterns of different colours. The responses can be very different: Running or flying after the rotating pattern, compensatory movements of the head or the whole body, nystagmus of eyes or antennae. An apparatus for optomotorical investigations which also permits the measurement of reactions to ultraviolet light has been described by Kaiser (1968). The torque which is produced by the optomotoric reactions of flying insects can be measured electronically (Kunze, 1961) or mechanically.

c) Investigation of the inborn behavioural reaction to objects of a certain colour, e.g. in *Vanessa polychloros* (Ilse, 1928), *Pieris brassicae* (Ilse, 1937), caterpillars of *V. urticae* and *V. io* (Süffert and Götz, 1936), in *Eristalomyia* (Ilse, 1949; Kugler, 1950), and in aphids (Moericke, 1950, 1952, 1954).

d) Selective adaptation to certain colours, e.g. in Daphniae (v. Frisch and Kupelwieser, 1913; Koehler, 1924a) and *Drosophila* (Hamilton, 1922). After the selective adaptation of the eye to certain wavelengths the sensitivity was not uniformly changed over the whole spectral range. Therefore, it is probable that there are several receptor systems each with a different spectral sensitivity. This is an indication, although not a proof that colour vision occurs in these animals (Walther, 1958 a, b; Goldsmith, 1960).

e) Specific fatigue of the component of a taxis in prey-catching action (in Amphibia: Birukow and Meng, 1955). Toads (*Bufo bufo*) remember very many details of size, colour and shape of dummies and rapidly show fatigue when the

[1] The number of grey steps that lie between the brightest white and the darkest black is 20 to maximally 30 in man (Ranke, 1952; Kern, 1952) and is to a large degree independent of the state of adaptation.

[2] Each deviation of the spectral sensitivity of an animal eye from that of the human eye can lead to a different "uncoloured".

same dummy is offered repeatedly. This fatigue is immediately terminated when details of the dummy are changed. By this means it was proven that *Bufo bufo* has colour vision.

f) Investigation of the adaptation of body pigmentation to the colours of the environment (KÜHN and HEBERDEY, 1929; KÜHN, 1950, in cephalopods.)

g) Measurement of phototactic reactions (Daphniae: v. FRISCH and KUPELWIESER, 1913; KOEHLER, 1924a; HEBERDEY, 1936a; bees: BERTHOLF, 1928, 1931; *Drosophila:* FINGERMAN and BROWN, 1953).

B. Heterochromatic Flickering

Heterochromatic flickering was used in connection with electrophysiological observation of action potentials in the eye (*Calliphora:* AUTRUM and STUMPF, 1953; frog, turtle: FORBES et al., 1955).

C. Direct Measurement of the Spectral Sensitivity of Single Visual Cells

We know little about the spectral sensitivity, i.e. the input, of single visual cells. Two methods for measuring the spectral properties of single visual cells directly have been applied:

a) The measurement of the absorption spectrum of the receptor by means of a microspectrophotometer (MARKS, 1963, 1965 a, b; MARKS et al., 1964; BROWN and WALD, 1963, 1964; LANGER and THORELL, 1966; see LIEBMAN, this Handbook, Vol. VII/1, Chapter 12).

b) The electrophysiological investigation of the receptor potential (in insects: AUTRUM and BURKHARDT, 1961; AUTRUM and VON ZWEHL, 1964). TOMITA et al. (1967) recorded intracellular potentials from the fish retina. NAKA and RUSHTON (1966 a, b, c) determined the action spectra of cones indirectly. DE VALOIS et al. have found colour specific neurons in the Corpus geniculatum laterale of monkeys (see Chapter 3). These investigations (see B and C) yield insights into the physiological basis of colour vision and with supplementary behaviour experiments the actual ability to see colours can be determined.

One characteristic distinguishes the receptors of vertebrates from those of invertebrates: The excitation of the photoreceptors in invertebrates results in depolarization (HARTLINE et al., 1952 in *Limulus*; NAKA and EGUCHI, 1962 in *Apis*; BURKHARDT, 1962 in *Calliphora*; AUTRUM and VON ZWEHL, 1962, 1963, 1964 in *Apis*; HAGINS, 1965 in cephalopods). The single cone response of the fish retina is hyperpolarizing (TOMITA, 1965; KANEKO and HASHIMOTO, 1967; TOMITA et al., 1967). Hyperpolarization in response to the light stimulus has also been found by WERBLIN and DOWLING (1969) and TOYODA et al. (1969) in *Necturus* and *Gekko*. The hyperpolarizing potentials of receptor cells in vertebrates fall within a limited range of intensity (about 2 log units). In insects the range of intensity dependence is larger (AUTRUM and VON ZWEHL, 1962). In invertebrates the distal tip of the receptor cell becomes negative (HAGINS, 1965; ZETTLER, 1967), in vertebrates, however, it probably becomes positive (BROWN et al., 1965). In contrast to invertebrate visual cells, the resistance of mudpuppy cones and gekko rods (and some L-type S-potential sites of the mudpuppy) increases upon illumination and decreases upon darkening (TOYODA et al., 1969). — For the chromaticity units (C-units) in the retina of vertebrates see Vol. VII/2, chapter 15.

II. Results

A. Honeybee

1. Behavioural Responses

Of all the animals bees have been investigated most extensively and successfully. In 1914 v. FRISCH using critical methods proved for the first time that bees have a genuine colour sense. The range of wavelengths, within the spectrum, for which the eye of the bee is sensitive, is 650—300 nm; therefore, it goes further into the ultraviolet than does the range for the human eye (KÜHN, 1927; BERTHOLF, 1928, 1931). Within this range bees can distinguish five regions from one another: 1) 650—500 nm (yellow; it comprises our colour qualities of orange, yellow and green). 2) 500—480 nm (blue-green; for us a transitional region). 3) 480—440 nm (blue; our colour quality blue). 4) The transitional region between blue and ultraviolet (440—360 nm, violet; DAUMER, 1956). 5) The region between 360 and 300 nm (ultraviolet), which to us is visible only under certain conditions (high intensity, aphacic[3] persons). By mixing the wavelengths of the ends of the spectrum visible for bees (ultraviolet and yellow) a new region of bee-purple-colours arises (see below; DAUMER, 1956). Therefore, six main colour qualities can be distinguished by bees.

Many important advances beyond the results of earlier authors were made by DAUMER (1956) who worked with a device which permits the mixing of spectral colours and so can be used to investigate bees colorimetrically. Within the above mentioned main regions bees can distinguish still other colours with certainty. The ability to distinguish is weakest in the yellow region and becomes stronger at its green end (530—500 nm). The maximum sensitivity for differences lies in the blue-green region (510—480 nm). Therefore, the colour quality here changes quite rapidly with wavelength. A second maximum lies at the transition from blue to ultraviolet. Even in the ultraviolet region colours are distinguished (DAUMER tested 375 nm against 360 nm).

In the regions of weakest colour differentiation the colours appear more saturated[4] than they do in the transitional regions. Saturation is lowest in the blue-green; the spectral colours of the yellow region have a moderate, about equally strong saturation. Violet and particularly ultraviolet are most strongly saturated. The degree to which colours were saturated was determined by measuring the lowest intensity of the colour that had to be added to a bee-white (see p. 665) surface in order that it would appear just perceptibly coloured to the bees.

Mixing the colours of both ends of the visible spectrum (red and violet) produces for man a new colour quality purple which is not included in the spectrum.

[3] Persons, whose lens has been surgically removed, are called aphacic. The lens absorbs ultraviolet light to a high degree.

[4] Colour perceptions are characterized by three Helmholtz coordinates: hue, brightness and saturation. The more saturated a colour is the less it is mixed with white.

The situation is analogous in bees: Mixing ultraviolet and yellow results in a mixed colour (called bee-purple by DAUMER), which is not confused with any colour of the visible spectrum. Because of the high value of saturation of ultraviolet a mixture of 2% ultraviolet and 98% yellow[5] can be distinguished from pure yellow. On the other hand a yellow-ultraviolet-mixture must be at least 50% yellow in order to be distinguished from ultraviolet. Within the purple region a bee can distinguish several colours. Therefore, the colour circle of bees, like that of man, is closed by hues that result from the mixing of the ends of the visible spectrum (Fig. 1).

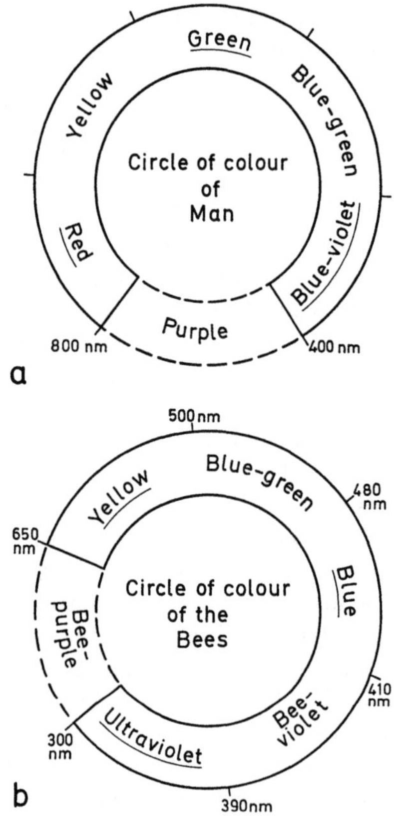

Fig. 1. *Circle of colour (a) of man; (b) of the bee (diagrammatic).* The three basic colours are underlined. By mixing them, the intermediary colours can be produced. Complementary colours stand opposite each other in the picture (From DAUMER, 1956; modified)

When for man the three basic colours red, green and blue are mixed in suitable proportions, the result is "white". For bees the situation is analogous: their basic colours are yellow, blue and ultraviolet. The mixture (55% yellow of 588 nm + 30% blue-violet of 440 nm + 15% ultraviolet of 360 nm) results in a "neutral" hue, which is confused with the white light of a Xenon high pressure arc (the

[5] The percentages give the physical portion of energy of the total amount of irradiation.

spectral energy distribution coincides almost exactly with that of the direct sun light) but with no other colour. Thus, bee-white contains appreciable amounts of ultraviolet; if they are lacking the resulting colour, which appears "white" to our eyes, is confused with blue-green. The following complementary colours result (Fig. 2):

Ultraviolet of 360 nm complementary to 65% yellow of 588 nm + 35% blue-violet of 440 nm (blue-green);

blue-violet of 440 nm complementary to 78,7% yellow of 588 nm + 21,3% ultraviolet of 360 nm (bee-purple);

yellow of 588 nm complementary to 66,5% blue-violet of 440 nm + 33,5% ultraviolet of 360 nm (bee-violet).

For mixtures of colours the same Grassmann laws hold as in man.

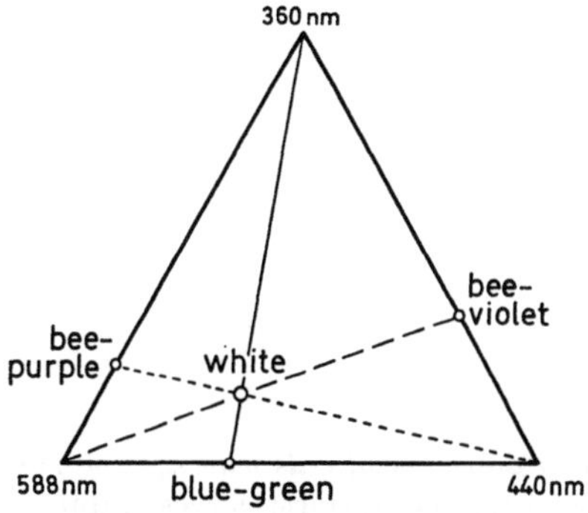

Fig. 2. *Colour triangle of the bee with the three complementary spectral stimuli:* yellow 588 nm, blue-violet 440 nm and ultraviolet 360 nm. The colour location of the white-point has been determined by the center-of-gravity-rule (From Daumer, 1956)

On the whole the rules for colour vision are the same for man and for bees. The decisive difference is the shift of the region of sensitivity to the short wave side of the spectrum.

From these facts, using the analogy of man, one can conclude that there is a receptor system consisting of at least three elements each with a different spectral sensitivity (trichromatic system).

2. Electrophysiology

Three receptor types have been shown to exist in bees by Goldsmith (1958, 1961) using the electroretinogram and by Autrum and von Zwehl (1964) by determination of the spectral sensitivity of single visual cells.

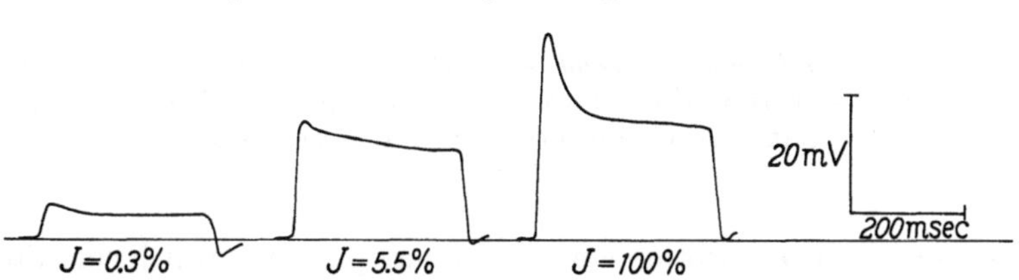

Fig. 3. *Receptor potentials from single visual cells of worker bees.* I = relative intensity of stimulus (From Autrum, 1968)

As soon as the tip of a microelectrode enters a visual cell, the resting potential appears. It lies between 45 and 80 mV. — The receptor potentials of bees are always monophasic and depolarizing (Fig. 3). Within the first 20 msec after the onset of illumination they reach a maximum depolarization which drops off to a plateau. All measurements are based on the maximum. In the dorsal area of the drone eye visual cells were found with a maximum sensitivity at 340 nm together with others with a maximum at 450 nm (Figs. 4, 5). One can compare these sensi-

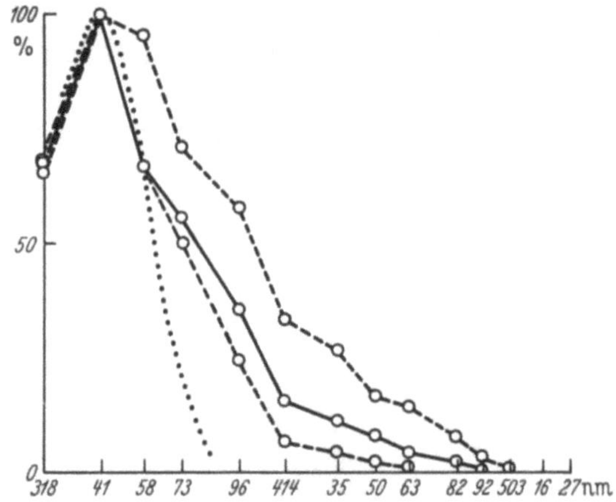

Fig. 4. *Sensitivity curves from drone visual cells with λ_{max} of 340 nm.* —— sensitivity curves, average of 15 cells; - - - narrowest and broadest curves of different single cells; . . . . . DARTNALLs nomogram curve for λ_{max} of 340 nm (From AUTRUM, 1968)

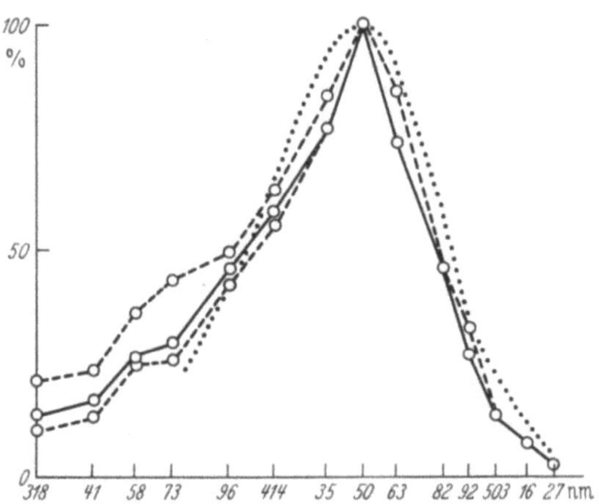

Fig. 5. *Sensitivity curves from drone visual cells with λ_{max} of 450 nm.* —— sensitivity curves, average of 16 cells; - - - narrowest and broadest curves of different single cells; . . . . . DARTNALLs nomogram curve (From AUTRUM, 1968)

tivity curves with the absorption curves of visual pigments which contain retinal, that is with the absorption curves of rhodopsins. DARTNALL (1953) published a procedure for constructing the resonance or absorption curve of any rhodopsin once the maximum of the absorption curve is known. The absorption curve for retinal pigment with a maximum at 450 nm obtained from the nomogram of DARTNALL is included for comparison (dotted curve). The agreement between the sensitivity curve of the drone's single visual cell and the absorption curve for retinal pigment of the same maximum is impressive. In the extreme ventral area of the drone eye — and so far only there — visual cells have been found with a maximum sensitivity at 530 nm (Fig. 6). Here, too, the correlation between the nomogram curve and the sensitivity curve is high.

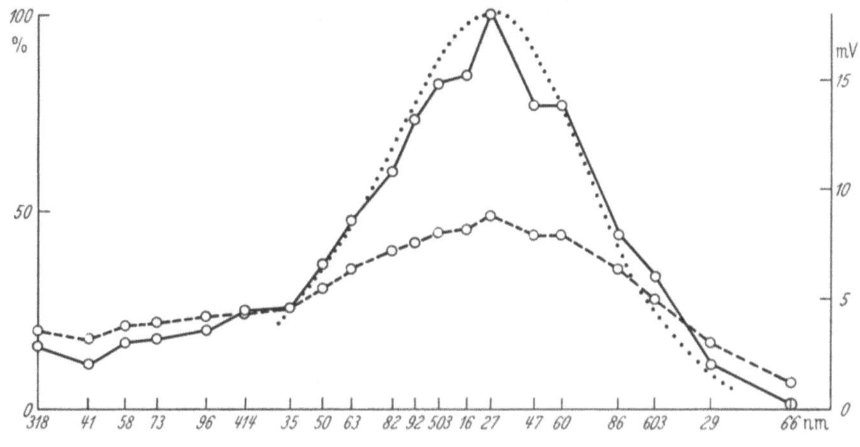

Fig. 6. *Efficiency curve and sensitivity curves from drone visual cells with* λ_{max} *of 530 nm.* - - - efficiency curve (mV on righthand side); —— sensitivity curve (relative values on left-hand side); . . . . . DARTNALLs nomogram curve (From AUTRUM, 1968)

It is harder to insert microelectrodes successfully into the visual cells of worker bees. Cells with a maximum sensitivity at 530 nm can be found throughout the entire compound eye of the female workers (Fig. 7). The nomogram curve for a retinal pigment with the same absorption maximum is also given for comparison.

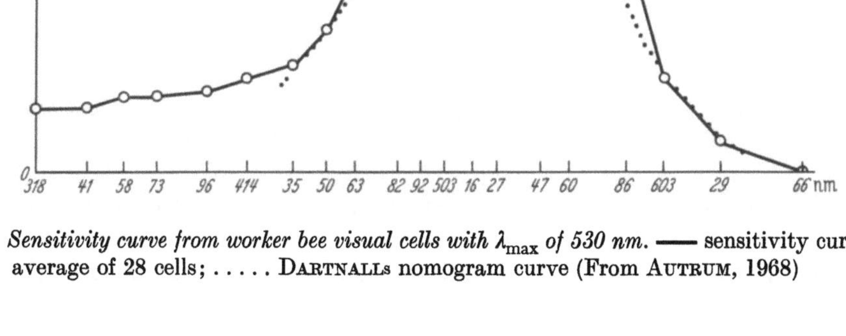

Fig. 7. *Sensitivity curve from worker bee visual cells with* λ_{max} *of 530 nm.* —— sensitivity curve, average of 28 cells; . . . . . DARTNALLs nomogram curve (From AUTRUM, 1968)

The visual cells of both drones and workers show the same spectral characteristics. The sensitivity of the 530 nm receptor extends into the long-wave side of the spectrum to about 650 nm, that is, into the orange. In that region of the spectrum which our eyes see as red, the receptors of the bee's eyes fail to react. On the short-wave side the sensitivity extends into the ultraviolet. However, there is no second maximum as in the case of rhodopsin.

Autrum and von Zwehl did not find a receptor in the workers with a maximum at 450 nm as they did in drones. Instead they found two significantly different receptors with maxima at 430 nm and 460 nm. Figs. 8,9 show the sensitivity curves for these two receptors and the corresponding nomograms.

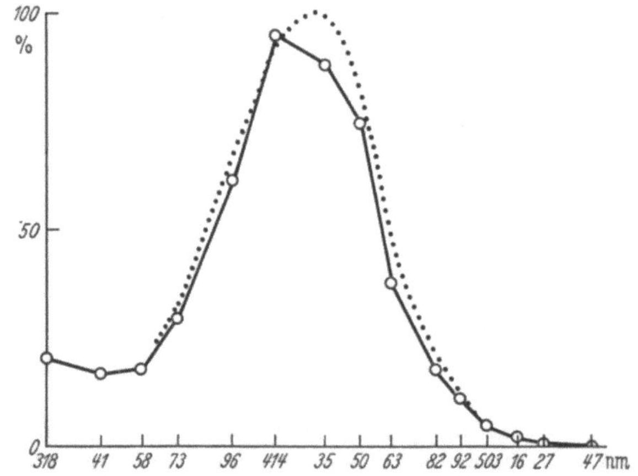

Fig. 8. *Sensitivity curve from worker bee visual cells with λ_{max} of 430 nm.* —— sensitivity curve, average of 3 cells; ..... Dartnalls nomogram curve (From Autrum, 1968)

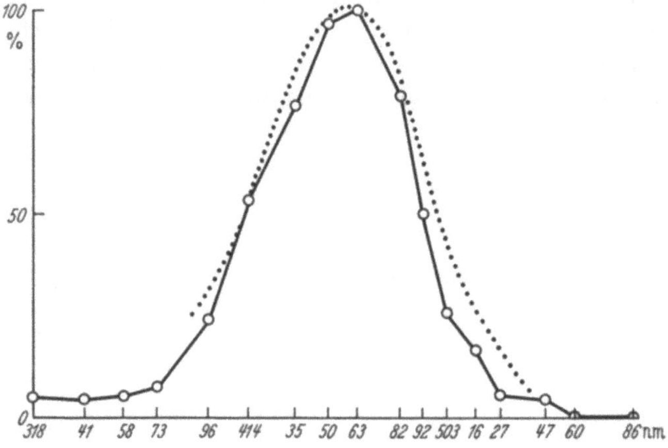

Fig. 9. *Sensitivity curve from worker bee visual cells with λ_{max} of 460 nm.* —— sensitivity curve of one cell; ..... Dartnalls nomogram curve (From Autrum, 1968)

There is complete agreement between the results of the investigations of single visual receptors and the behaviour experiments. The receptor with the maximum at 530 nm has no significant sensitivity beyond 650 nm (Fig. 10).

On the other hand, bees have a special receptor for ultraviolet with a maximum sensitivity at 340 nm which is still at 70% of maximum at 318 nm. A visual pigment with these properties has yet to be identified in bees (GOGALA et al. succeeded in isolating an UV visual pigment from *Ascalaphus* eyes; see p. 674). The ultraviolet sensitivity of the human eye results from a secondary maximum of rhodopsin, which also lies at 340 nm.

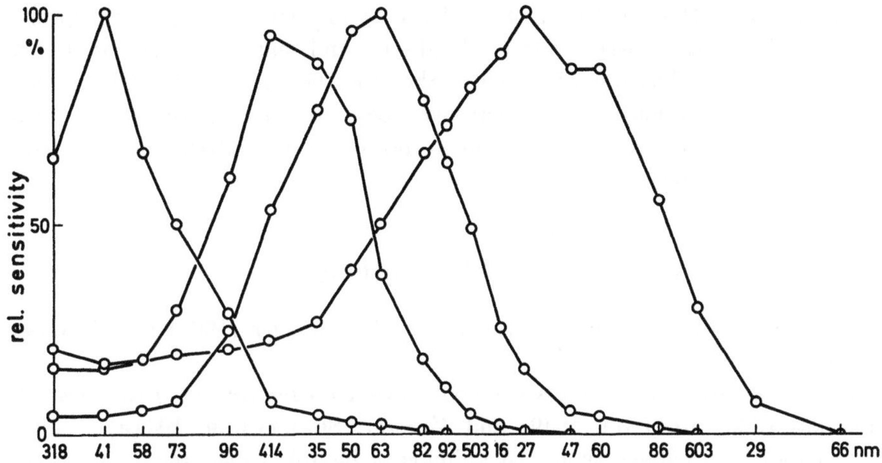

Fig. 10. *Sensitivity curves of the receptor cells of the worker bee eye with maximal responses at 340, 430, 460 and 530 nm* (From AUTRUM, 1968)

Bees have a hard time distinguishing colours in the "yellow region" between 650 nm and about 530 nm. Yellow and orange look very much alike to them, if not fully the same (see p. 664). This conclusion corresponds very closely to the sensitivity curves (Fig. 10). Down to about 610 nm only one receptor responds. Then a second receptor starts to come in weakly. Bees trained to monochromatic colours also confuse the 530—510 nm segment of the spectrum with neighbouring longer and shorter wavelengths. They begin to perceive a change in colour towards the neighbouring region of shorter wavelengths. Bees trained to orient to 520 nm often fly to either the neighbouring yellow or blue-green. According to KÜHN (1927) and DAUMER (1956) the "yellow region" connects with a transitional region which extends from about 520 nm to about 480 nm. This is the "blue-green region" in which bees can distinguish very slight changes in wavelength. Now the ability to distinguish spectral colours is going to be best developed where the slope of sensitivity in different receptors is of opposite sign. In the "blue-green region" the sensitivity of the green receptor begins to fall off. Simultaneously the sensitivity of the 460 nm and also of the 430 nm receptor starts to rise. One is again confronted with the remarkable agreement between the course of the sensitivity curves and the results of behaviour experiments. In the adjoining "blue region" from about 480 nm to around 410 nm, behaviour experiments show some ability to distinguish colours, though not so great as in the transition area between 530 nm and 480 nm. The characteristics of the visual cells indicate that the ability to distinguish colours should drop off considerably at the short-wave end of the "blue region", for the active receptors here drop off together almost in parallel. — Behaviour experiments show a second transition area between 410 nm and 360 nm, where slight differences in colour are discernible. Here the sensitivity of the 340 nm receptor begins and rises sharply. DAUMER (1956) found that the addition of only a small percentage of ultraviolet to spectral blue-violet (440 nm) was enough to make the resulting mixed colour different from blue-violet to bees.

Accordingly, the results of the behaviour experiments agree exceptionally well with the characteristics of the individual visual cells which give rise to colour perception. The behaviour of bees can therefore be correlated with the characteristics of single visual cells and can even be deduced from them. The data on the bee's ability to distinguish colour is, however, not sufficiently extensive to permit exact comparison.

B. Other Insects

For review of insect colour vision see BURKHARDT (1964) and GOLDSMITH (1964).

Hymenoptera: Wasps can see colours (*Vespa vulgaris, V. germanica:* ARMBRUSTER, 1922; MOLITOR, 1939; and *V. rufa*: SCHREMMER, 1941a). While the first two authors believe their test animals to be able to see red, *V. rufa* is said to be red-blind. In investigations on ants TSUNEKI (1950, 1953 on *Camponotus* and *Leptothorax*) and MOLLER-RACKE (1952 on *Lasius*) had negative results. KIEPENHEUER (1968) was able to prove that *Formica polyctena* can distinguish ultraviolet and yellow-green.

Coleoptera: These species are capable of seeing colours: *Chrysomela, Agelastica, Geotrupes* (SCHLEGTENDAL, 1934), *Tropinota hirta* (MOLLER-RACKE, 1952). A clear-cut reaction to brightness in the rotating cylinder is shown by *Ips typographus, Cetonia aurata, Hydrous piceus, Dytiscus marginalis, Attagenus spellio, Aphodius inquinatus* and *Melasoma populi* (MOLLER-RACKE, 1952). In contrast to these results SCHÖNE (1953) reports that the imagines of *Dytiscus* and *Acilius* have a certain capability for distinguishing colours (green-blue, green, yellow-green, orange each from red, orange from green; however, blue-violet and ice-blue cannot be distingushed from orange or red respectively at equal degrees of brightness in the rotating cylinder). For larvae of *Dytiscus* and *Acilius* SCHÖNE could not show any colour specific reactions.

Diptera: In all investigated species colour sense has been demonstrated: *Drosophila* (HAMILTON, 1922; FINGERMAN and BROWN, 1953), *Bombylius* (KNOLL, 1921), *Fannia canicularis* (SCHLEGTENDAL, 1934), *Eristalomyia* (KUGLER, 1950; ILSE, 1949), *Calliphora* (AUTRUM and STUMPF, 1953). In *Calliphora* colour vision has been investigated with the aid of an analysis of the illumination potentials at heterochromatic flickering with monochromatic lights of equal effects. The eye of *Calliphora* must possess at least two types of receptors each with a different spectral sensitivity. MAZOKHIN-PORSHNYAKOV (1960a, on *Calliphora*; 1960b, on *Musca*) worked with a similar method. AUTRUM et al. (1961) found in the electroretinogram of *Calliphora* components that are dependent upon the spectral colour.

Burkhardt (review 1962) succeeded in directly determining the spectral sensitivity of single receptors electrophysiologically. He found three types of receptors (Fig. 11). Langer (1967) was able to prove the existence of two of the electrophysiologically found receptor types by microspectrophotometric measurements

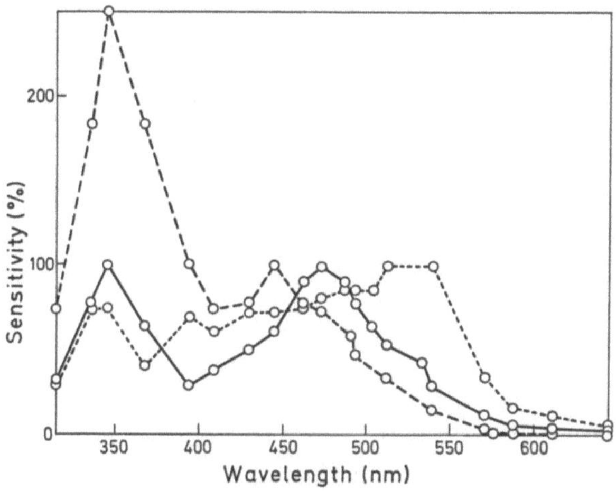

Fig. 11. *Spectral sensitivity of typical individual cells in the compound eye of Calliphora erythrocephala.* Sensitivity maxima in the visible range set at 100%. Solid line: green-type receptor cell; broken line: blue-type; dotted line: yellow-green-type (From Burkhardt, 1962; modified)

on single rhabdomeres of *Calliphora* (Fig. 12). From all this one can conclude that there is colour vision in *Calliphora*. Kaiser (1968) constructed an apparatus in which the effect of rotating striped patterns in spectral colours (ultraviolet included) on the optomotoric reaction of the fly *Phormia regina* could be investigated. He found no clear-cut reactions, but does not conclude that the flies are colour blind.

Lepidoptera: The investigations of Knoll (1922, 1925, 1927), Kühn and Ilse (1925), Ilse (1928, 1937, 1965), Schlegtendal (1934), Schremmer (1941b) and Swihart and Swihart (1970) demonstrated a well developed colour sense and the presence of a simultaneous coloured contrast (Ilse, 1934). The brightness values (effectivity of the stimulus) of different colours vary considerably from species to species. They are, however, very similar for the species within one family. All butterflies see yellow very brightly; red and orange appear dark to Satyridae and light to Nymphalidae, Pieridae and Lycaenidae (v. Buddenbrock and Moller-Racke, 1952). Of special interest is the proof that *Deilephila livornica* can recognize colours even in faint twilight (Knoll, 1925); this is beyond the limit of the capability for colour vision in man. Caterpillars of *Vanessa urticae* and *V. io* (Süffert and Götz, 1936) and of night-moths (Hundertmark, 1936) can probably (in contrast to the larvae of beetles; see above Schöne, 1953) recognize at least a colour region "red to green" and a second one, "blue", as coloured qualities.

Hemiptera: While SCHLEGTENDAL (1934) could not detect any colour specific reactions in *Troilus luridus* using the rotating cylinder (*Cimex* and several Capsidae do not show any optomotoric reactions at all), ROKOHL (1942) was successful in demonstrating colour vision in *Notonecta glauca*. Although it was possible to find for each pigment colour a grey of equal brightness, alternating colours of equal brightness in the striped cylinder resulted in clear reactions which were the

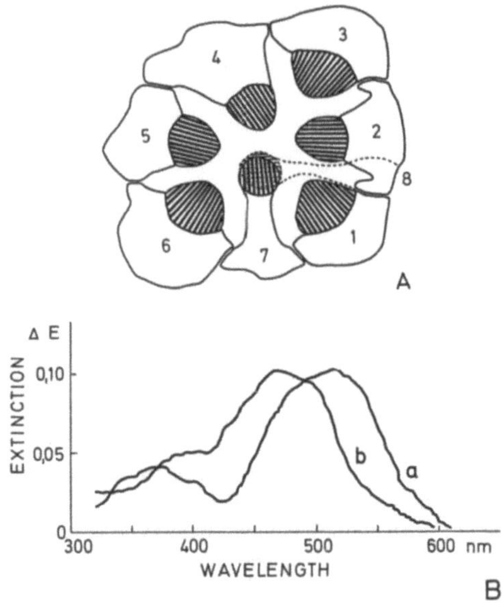

Fig. 12 A. *Order of the 8 visual cells in the ommatidium of the fly Calliphora,* transverse section. Each retinula cell (1—8) forms a rhabdomere. The rhabdomeres 7 and 8 lie within the axis one behind the other. B *Extinction curves of rhabdomeres of single retinula cells in the eye of the fly Calliphora*: (a) of rhabdomere No. 5; (b) of rhabdomeres 7 and 8. For numbering see Fig. 12A. The extinction of rhabdomeres 1—4 and 6 corresponds to that of No. 5; the extinction curves are not significantly different from one another (From LANGER and THORELL, 1966)

stronger the farther apart the colours are within the colour circle. It is interesting that it was possible to demonstrate a zonal colour blindness of the eye of *Notonecta glauca*: Only ommatidia in the dorsoposterior region of the eye are fully capable of colour vision. According to their position and number these correspond to ommatidia that have been formed in the first larval instar. The ommatidia which have been added as late as the imago, those of the ventral and anterior eye portions, are colour blind. The values for brightness of colours are different for both parts of the eye (RESCH, 1954); in the colour blind ventral part of the eye green, orange, yellow (and white) appear considerably lighter than in the dorsal region which is capable of colour vision. The subjective values for brightness of colours change during the development of individuals (OHNESORGE, 1955). The imago sees red, orange and yellow brighter than does the first larval instar and green and blue darker (ROKOHL, 1942, was not able to verify this). Obviously, the physiological properties of the ommatidia change during development of the imago stage. BRUCKMOSER (1968) found three types of receptor cells in the dorsal portion of the eye of *Notonecta glauca* by determining the spectral sensitivity of single visual cells ($\lambda_{max} = 350$, 414 and 567 nm) (see also BENNETT and RUCK, 1969). — Colour vision also exists in *Graphosoma lineatum* and *Mesocerus marginatus* (MOLLER-RACKE, 1952). *Lygus rugulipennis* can distinguish ultraviolet-blueviolet from yellow-yellowgreen (BECH, 1965).

Homoptera: MOERICKE (1950, 1952) found colour specific reactions for numerous aphids: They land preferably on yellow, less frequently on orange, yellow-green and green, hardly at all on red and not at all on blue-green, blue, violet, white, grey and black. Colours also contribute, independent of their intensity, to triggering the puncturing of leaves (puncture into colours between 600 and 500 nm, no puncture at shorter wavelengths and at white-grey-black). Also in aphids a successiv coloured contrast seems to exist. They sting into grey after having been exposed to blue. Species differ particularly with respect to the effect of ultraviolet. While most Homoptera approach saturated yellow most strongly, *Hyaliopterus pruni*, for example, is only attracted by a yellow surface which also reflects ultraviolet (MOERICKE, 1954). The reflection of ultraviolet in the vicinity of attractive landing surfaces decreases the frequency of landing of Homoptera (and Diptera) strongly (bare ground reflects small quantities of ultraviolet only).

Neuroptera: Larvae of dragon-flies (*Aeschna cyanea*) can see colours (KOEHLER, 1924b). AUTRUM and KOLB (1968) determined the spectral sensitivities of single visual cells in the eyes of larvae and imagoes of *Aeschna cyanea* and *A. mixta*, HORRIDGE (1968) of *Anax* and *Libellula*. They found up to four different types of receptors. MAZOKHIN-PORSHNYAKOV (1959) demonstrated the presence of colour vision in imagoes and larvae of *Libellula* with the colorimetric technique (ERG) of AUTRUM and STUMPF (1953) (see also RUCK, 1965). — GOGALA (1967) found electrophysiologically (ERG and single cell recordings) maximum sensitivity to 350 nm in both parts of the divided eye of *Ascalaphus* and a second maximum to 530 nm in the lateral eye. In addition, GOGALA et al. (1970) were able to extract an UV-visual pigment ($\lambda_{max} = 345$ nm) from the frontal part of the *Ascalaphus* eye.

Orthoptera: *Dixippus morosus* behaves in critical tests in the rotating cylinder like a colour blind animal (SCHLEGTENDAL, 1934). However, from this negative result one cannot conclude that this species is colour blind. For *Periplaneta americana* WALTHER (1958a, b) concludes from electrophysiological investigations on selective adaptation that there are at least two systems of receptors with different spectral sensitivities in the upper region of the eye, while there is only one receptor system in the lower region. MOTE and GOLDSMITH (1970) confirmed these results by intracellular recordings from single retinular cells of *Periplaneta*. In the dorsal part of the eye two classes of cells with maximum sensitivity at 365 nm and 507 nm respectively are present. From this it seems at least possible that in *Periplaneta* as in *Notonecta* (ROKOHL, 1942) there is a colour blind section in the eye next to a section capable of colour vision.

In summary, one can say that all of the orders of insects tested so far have colour vision. The sensitivities are different from species to species particularly with respect to the long wave end of the spectrum. Ultraviolet can probably be seen by all insects and by many of them as a colour. That colour blindness actually exists [*Phormia* (KAISER, 1968); *Troilus, Dixippus* (SCHLEGTENDAL, 1934); limited eye regions of *Notonecta* (ROKOHL, 1942) and *Periplaneta* (WALTHER, 1958a, b)], is possible, but has not been proven conclusively.

C. Invertebrates

Among the other invertebrates colour vision has been demonstrated in *Daphniae* (v. FRISCH and KUPELWIESER, 1913; EWALD, 1914; BECHER, 1921; KOEHLER, 1924a; HEBERDEY, 1936a) and *Copepods* (HEBERDEY, 1936b). Later authors had gained negative results with the same methods. But v. FRISCH and KUPELWIESER as well as KOEHLER and HEBERDEY learned from experience that some populations react to colours while others do not. This could explain the occasional negative results. *Daphnia* and *Diaptomus* can distinguish between at least two colour qualities, one long-wave and one short-wave. The assumption of three distinguishable colour qualities by HEBERDEY (1936a, b) is not based on conclusive evidence. According to SMITH and BAYLOR (1953) the negative phototactic reaction to short-wave light disappears when animals are cooled down to 10° C or less. Under such circumstances *Daphnia magna* reacts in the same way to the whole spectrum as it does to long-wave light alone. The ecological significance of this behaviour is probably the following: Daphniae do not intrude into deeper and cooler layers of water, but instead come to a halt when approaching a threshold at which the deeper layer is 10° C or lower. Furthermore, Cladocera (*Daphnia, Ceriodaphnia, Moina* and *Bosmina*) under the influence of short-wave light show a strong component of movement which is vertical to the direction of the propagation of the light. Under normal circumstances, they show a strong component of movement, which is horizontal to the direction of the propagation of the light. In long-wave light the main component of movement lies in the direction of the propagation of the light. Ecologically seen, this behaviour leeds to an accumulation of the animals in plankton-rich zones, since chlorophyll absorbs short-wave light. Therefore, in such zones horizontal movements are observed lesser frequently than in zones poor in plankton and, consequently, rich in short-wave light.

In the median eye (median ocellus) of the horseshoe crab (*Limulus polyphemus*) NOLTE and BROWN (1969, 1970) found two types of receptor cells: one type of visual cells with a maximum sensitivity at 360 nm and a second type with a peak at 530 nm.

The decapods *Crangon* (KOLLER, 1927; SCHLEGTENDAL, 1934), *Leander adspersus* (SCHLEGTENDAL, 1934), *Carcinus maenas* (v. BUDDENBROCK and FRIEDRICH, 1933) and Eupagurids (KOLLER, 1928) can see colours. Using intracellular recording NOSAKI (1969) found two different receptors in the eye of the crab *Procambarus* with maximum sensitivities at 460 nm and 560 nm (in winter) resp. 600—620 nm (in summer). Isolated dark-adapted rhabdoms from the eye of the spider crab *Libinia emarginata* were examined by microspectrophotometry. They contain a single pigment with $\lambda_{max} = 493$ nm (HAYS and GOLDSMITH, 1969).

In the jumping spider *Evarcha falcata* KAESTNER (1950) has demonstrated colour vision with optomotorical reactions: orange and blue can reliably be distinguished from all shades of grey. — In the wolf spider (*Lycosa miami*) secondary eyes have maximum relative sensitivities at 505—510 nm which are unchanged by chromatic adaptation. In contrast, the principal eyes have ultraviolet sensitivities which are 10—100 times greater at 380 nm than at 505 nm (DEVOE et al., 1969, using the ERG). Chromatic adaptations to both wavelengths in principal eyes hardly changed relative spectral sensitivities. According to DEVOE et al., there are two obvious explanations for KAESTNERs behavioural observations: One is that spiders are colour blind and that in KAESTNERs experiments the coloured stripes which the spiders always distinguished readily from all grey stripes gave high contrasts to the grey stripes at ultraviolet wavelengths. A second possible explanation of KAESTNERs results is that the differences in spectral sensitivities between principal and secondary eyes could form the basis of colour vision.

KÜHN and HEBERDEY (1929) and KÜHN (1933, 1950) were able to find a colour sense in *Sepia* and *Octopus* by investigating the adaptation of body pigmentation to the background. A coloured background affects *Sepia* differently than an uncoloured one. On green, yellow and red backgrounds the animals are more strongly (saturatedly) coloured. Depending upon the background they are coloured green or yellow. On a red background they are darker yellow than on a yellow background. Adaptation to blue is impossible; however, against a blue background the colour is strongly veiled with grey. Training experiments with *Octopus* proved that spectral yellow and blue are recognized as colours. These behaviour analyses, however, require verification. ORLOV and BYZOV (1961) found colorimetrically only *one* spectral sensitivity curve in *Octopus* and *Eledone*. This curve with its maximum at 485 nm coincides with the absorption spectrum of cephalopod rhodopsin. — *Pecten* is colour-blind in the rotating cylinder (v. BUDDENBROCK and MOLLER-RACKE, 1953). However, the spectral sensitivity of *Pecten maximus* has two maxima, at 475—480 nm and at 540 nm (CRONLY-DILLON, 1966; behaviour experiments with the shadow reflex: at a sudden decrease of the light intensity *Pecten* closes the shell). Using intracellular recordings MCREYNOLDS and GORMAN (1970) found only one cell type with maximum sensitivity at 500 nm in the eye of *Pecten*. — The investigations by LICHE (1934) on *Limnaea stagnalis* and by MUNDHENKE (1955) on *Limnaea* and *Copea* are methodically insufficient and do not allow any conclusions on colour vision or colour blindness.

For other invertebrates no investigations are available.

D. Vertebrates

The investigations on colour vision in vertebrates have been compiled in tables by AUTRUM (1958). The investigations of recent years have shown that colour vision among vertebrates is much more widely distributed than had been thought earlier.

Fishes: The principle features of colour vision in fish have been elucidated by v. FRISCH (1912, 1913) (summary in v. BUDDENBROCK, 1952). Their colour vision and their ability to distinguish colours are very similar to those of man. In numerous bony fish a simultaneous coloured contrast could be demonstrated (HERTER, 1950). For colour vision in goldfish see Vol. VII/4, Chapter 28. — Microspectrophotometric studies on single cones of goldfish (MARKS, 1963, 1965a, b; MACNICHOL, 1964) have shown that they have three types of cones. The frequencies of the individual types are still uncertain, especially since not all regions of the retina have been investigated and since the distribution within the retina can be quite different (MARKS, 1965a, b). To measure the spectral distribution of the receptor pigments in numerous species of fish SVAETICHIN et al. (1965) have used spectrophotometric methods. They found two different types of cones, three types at the maximum (Fig. 13).

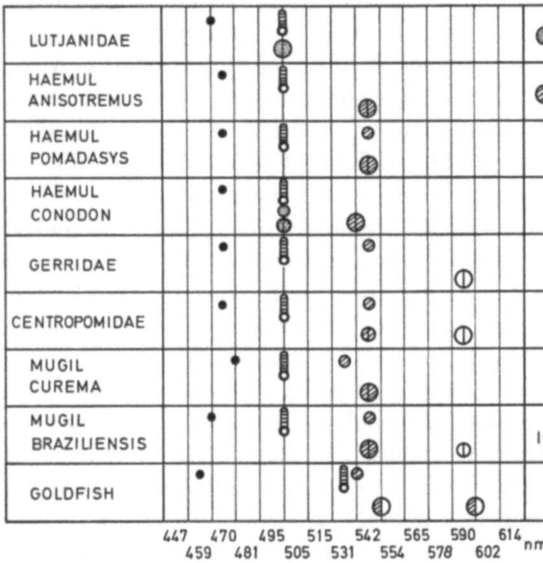

Fig. 13. *Maxima of the spectral absorption of the different types of cone observed in nine different fish species.* The maxima of all the additional short single cone pigments (filled circles) are grouped around 465 nm. The maxima of the rods (indicated by rod-like bars) are at 500 nm, except the 531 pigment rod of the goldfish. The central long single cones (dotted or striated circles) showed maximal absorption either in the 500 nm region or at about 540 nm. The large and small equal double cone absorptions are distributed in the regions 500, 540 and 590 nm, while the pigments of the two members of the unequal goldfish double cones are located at 540 and 590 nm respectively (double cones indicated by divided circles). The different mosaic patterns formed by cones in the different fish species are indicated to the right (From SVAETICHIN et al., 1965)

Using electrophysiological methods three receptor types (Fig. 14) have also been found in fish (*Cyprinus carpio*) by intracellular recordings (TOMITA et al., 1967). Marking experiments have reliably shown that the cone inner segment was the recording site of the single cone response. — Some of the extracellular recordings from cones in the fish retina (MITARAI et al., 1961; SVAETICHIN, et al., 1965) were in fact recordings from cones, while most of the recordings by SVAETICHIN (1953) were of a neuronal origin.

Amphibia: Of special interest is the kind of colour vision in Amphibia. *Salamandra maculosa, Triturus h. helveticus, Bombina variegata* (BIRUKOW, 1950), *Bombina bombina* (THOMAS, 1955), *Bufo bufo* (MENG, 1958), *Rana esculenta, R. pipiens* and *R. clamitans* (THOMAS, 1955) have a trichromatic colour sense. In contrast to these Amphibia that are capable of seeing colours in the whole spectral region between red and blue, there are other species which cannot distinguish extensive regions of the colour circle from grey (Fig. 15). In *Discoglossus pictus* the yellow region, in *Rana latastei* the yellow and yellow-green, in *Rhacophorus buergeri* and *Leptodactylus ocellatus* the orange, yellow and yellow-green, and in *Ceratophrys americana* from the orange to the blue-green regions were equal to grey in optomotoric reactions (THOMAS, 1955). *Limnodynastes tasmaniensis* shows clear-cut colour-specific reactions only for blue. For *Rana temporaria* (in contrast to

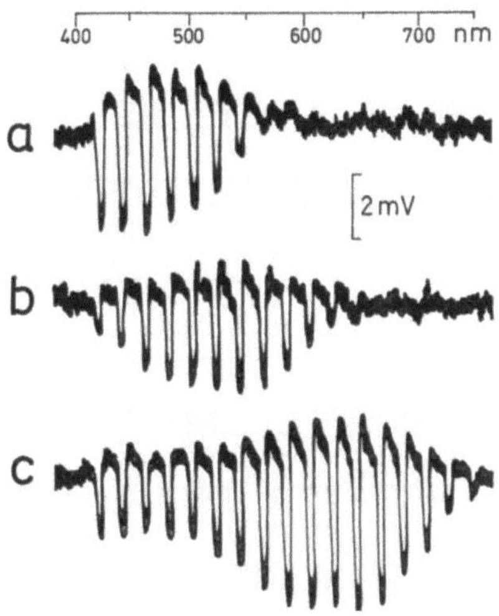

Fig. 14. *Sample recordings from single cones, demonstrating the three types of spectral responses found in the carp retina.* Scanning of the spectrum is made in steps of 20 nm with monochromatic light adjusted to equal quanta (2×10^5 photons per μ^2 sec), and with a duration of light of 0,3 sec at each wavelength followed by an intermission of 0,6 sec. A downward deflection indicates negativity. Recording is made with a C-R coupled amplifier having a time constant of 0,5 sec. The spectral scale is given in terms of nanometers at the top of the figure. A dominant peaking occurs at blue in *a*, green in *b*, and red in *c* (From TOMITA et al., 1967)

Rana esculenta) grey-equalities can be made up for yellow and green (BIRUKOW, 1950). Similar regions with respect to colour vision are claimed for *Triturus* by BIRUKOW (1950). *Triturus helveticus* has a trichromatic colour sense. In *T. vulgaris* and *T. alpestris* there are colour gaps in the yellow, yellow-green and green, and *T. cristatus* responds only to blue-green and blue as qualities. Numerous Amphibia proved to be colour blind: *Xenopus laevis* (BURGERS, 1952), *Alytes obstetricans* (BIRUKOW, 1950), the *Pelobates* species, *Bufo viridis, calamita, arenarum, cognatus, coccifer, regularis* (THOMAS, 1955), *Bufo calamita* (BIRUKOW, 1950), *Hyla arborea, regilla, versicolor, cinerea* and *Adelotus brevis* (THOMAS, 1955). Just how uncertain negative results in investigations on the ability to see colours can be is shown especially in Amphibia. In training and in the rotating cylinder colour vision is not demonstrable in *Bufo bufo*. Using the fatiguability of the taxis component in prey-catching, however, BIRUKOW and MENG (1955) and MENG (1958) could prove that this toad is capable of seeing colours.

The photopic values of the colours are identical in all Amphibia investigated. Only in the red region are differences found between species. The brightness values of the colours in twilight vision are, however, rather different from species to species. Particularly in the newt the brightness maximum of twilight vision is shifted especially far into the long-wave region. The most extreme form is *Triturus helveticus*; its maximum for brightness in twilight vision lies at yellow to yellow-green (BIRUKOW, 1950). During ontogeny of *Rana temporaria*, the brightness for colours characteristic for daylight vision occurs first.

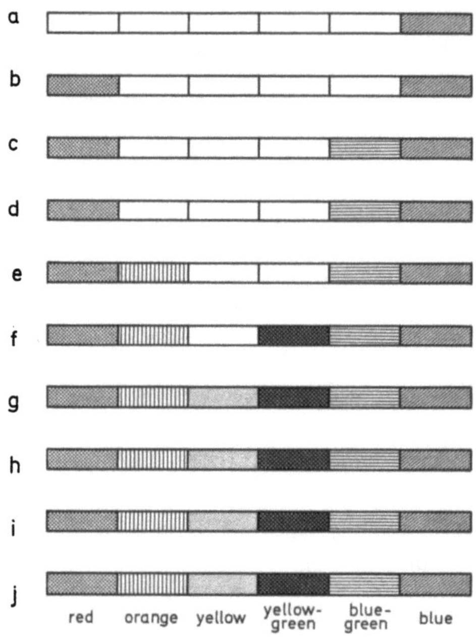

Fig. 15. *Schematic representation of colours to which single Amphibia species respond optomotorically.* The colours not marked (empty spaces) are confused with grey of equal brightness. *a Limnodynastes tasmaniensis*; *b Ceratophrys americana*; *c Leptodactylus ocellatus*; *d Rhacophorus buergeri*; *e Rana latastei*; *f Discoglossus pictus*; *g Bombina bombina*; *h Rana esculenta*; *i Rana pipiens*; *j Rana clamitans* (From THOMAS, 1955)

In the eye of the frog *Rana temporaria* the on-off units are of two types as regards their photopic spectral sensitivity (Fig. 16). Type 1: both the on and the off thresholds throughout the spectrum are determined by cones. Type 2: the off thresholds are determined by cones (GRANITS photopic dominator), the on thresholds by green rods (REUTER, 1969). In this second type the neurons respond in different ways in different parts of the spectrum ("chromatic" units). Between the on and off responses exists a mutual inhibition. Inhibitory interactions between the on and off components of the chromatic units are sometimes observed to produce modulator-like spectral sensitivities. However, specific excitatory interactions may also give rise to narrow sensitivity peaks. The spectral outputs from the visual cells can be sharpened up before the ganglion cell output stage by specific excitatory and inhibitory interactions between the signals from varying types of receptors (DONNER, 1958; DEVALOIS and JONES, 1961; REUTER, 1969).

In the frog's retina, therefore, there are processes with opponent colour responses of much the same type as those observed by DEVALOIS et al. (1958) in the primate lateral geniculate nucleus. Analogous phenomena were observed by DODT and HEERD (1962) in the pineal organ of *Rana temporaria*: inhibition occurred with stimuli of shorter wavelengths (maximally at 355 nm) and excitation with stimuli of green light.

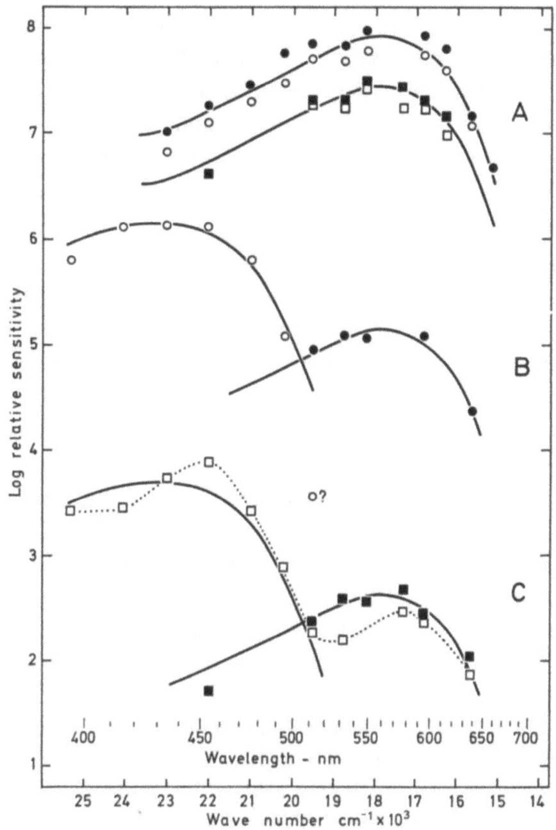

Fig. 16. *Spectral sensitivity (reciprocal of the quantum intensity at threshold) of three on-off units (A, B and C) measured against yellow or red background illuminations in the retinae of adult frogs.* Open symbols: on thresholds. Filled symbols: off thresholds. The continuous lines represent GRANITs photopic dominator of the frog (max. at 560 nm) and the absorption spectrum of visual pigment 433_1 (density 0.3). The experiments are vertically displaced in relation to each other. In experiment A the thresholds are first measured against a 577 nm (circles) background (10^{12} quanta $\times$ mm^{-2} $\times$ sec^{-1} reaching the retina surface) and thereafter against a 640 nm (squares) background (about 2×10^{13} quanta $\times$ mm^{-2} $\times$ sec^{-1}). In experiments B and C the thresholds are measured against a 564 nm background (5×10^{10} quanta $\times$ mm^{-2} $\times$ sec^{-1}). The dotted line is drawn to fit the open squares (From REUTER, 1969)

In the chromatic units of *Rana* the responses are irregular and short in the 495—533 nm region (typically only one impulse with a long latency). In the receptive field of a special tadpole neuron the on and off components were most sensitive in the centre of the receptive field, but their thresholds did not rise in the same way in different directions towards the periphery.

The chromatic on-off units have much the same colour-discriminating properties as the dichromatic optomotor (behavioural) response of the adult frog (BIRUKOW, 1939). In a rotating striped cylinder *Rana temporaria* discriminates between stripes of blue and grey, and of red and grey, irrespective of the relative light intensities. A match was possible, however, between grey and green (or yellow) stripes. Correspondingly the violet-sensitive on components and the red-sensitive off components give strong responses to the change from violet resp. red to grey. Green or white light stimulates both components about equally; then, the chromatic units give only short on-off responses suppressed by mutual inhibition. REUTER (1969) suggests that excitation is mediated by green rods and in some cases by cones as well (Fig. 16), and that an inhibition is mediated by cones sensitive in the 500—600 nm region.

Reptiles: Capable of seeing colours are: *Chrysemys picta* and *Pseudemys scripta elegans* (FORBES et al., 1955), *Clemmys caspica* (WOJTUSIAK, 1933, 1947), *Testudo gigantea* and *T. elephantopus vicina* (QUARANTA, 1952), *Anolis carolinensis* (MUSOLFF, 1955) and *Lacerta agilis* (WAGNER, 1932). *Emys orbicularis* and *Geoclemys reevesi* can be trained to the degree of saturation of colours (BARTKOWIAK, 1949). Among reptiles *Hemidactylus turcicus* and *Anguis fragilis* are colour blind (MUSOLFF, 1955).

Birds: All investigated birds possess colour vision similar to that of man: hummingbirds (POLEY, 1968), fowl (BÄSSLER, 1926), pigeon (HAMILTON and COLEman, 1933; see also Vol. VII/4, Chapter 28), *Melopsittacus undulatus* (PLATH, 1935), *Turdus ericetorum* (ECK, 1939) and *Athene noctua vidalli* (MEYKNECHT, 1941). *Strix aluco aluco* can distinguish colours, but perhaps less reliably than other birds (FERENS, 1948). The connection between plumage colour and colour vision in birds was investigated by PEIPONEN (1963) in the bluethroat (*Luscinia svecica*) and the robin (*Erithacus rubecula*). The significance of coloured oil droplets in the cones of birds has been much discussed. The reflection spectrum of the feathers of the bluethroat, the robin and the yellow wagtail (*Motacilla fl. flava*) probably corresponds to the transmission spectrum of the oil droplets in the dorsal centrum of the retinae of these birds (SMITH, 1942; PEIPONEN, 1963).

Mammals: Capable of seeing colours are the following species: among the insectivores: *Erinaceus europaeus* (HERTER and SGONINA, 1933; HERTER, 1934); among Rodentia: *Sciurus vulgaris* (SÄLZLE, 1936; MEYER-OEHME, 1957), *Spermophilus* (*Citellus*) *citellus* (KOLOSVÁRY, 1934; BONAVENTURE, 1959), *Clethrionomys glareolus* (SÄLZLE, 1936), *Cavia cobaya* (SGONINA, 1936; TRINCKER and BERNDT, 1957). MILES et al. (1956) consider the guinea-pig colour blind (see below). Training experiments on ground squirrels revealed that the colour vision of *Citellus tridecemlineatus* and *C. mexicanus* is dichromatic, i.e. is identical to that of human protanopic dichromacy. The neutral point lies at a certain wave length of the spectrum; this wave length is confused with white (Fig. 17). For both above mentioned species the neutral point lies at 505 nm (JACOBS and YOLTON, 1969). Microspectrophotometric measurements of retinal photopigments (DOWLING, 1964) and electrophysiological indices of visual response (MICHAEL,

1968) suggest that the colour vision of the ground squirrel is based on two cone photopigments (~ 523 nm, ~ 460 nm respectively; see Vol. VII/1, Chapter 8). This pigment arrangement is similar to the situation usually supposed to underlie human protanopic dichromacy (RUSHTON, 1963). For the colour vision of *Citellus leucurus* see CRESCITELLI and POLLACK (1965); for the non-albino rabbit see p. 683. — Among the carnivores *Viverricula indica* and *Mungos ichneumon* (DÜCKER, 1957) and *Suricatta suricatta* (BERNAU, 1969) are capable of seeing colours, also the wolf (*Canis lupus*; EISFELD, 1966) and the domestic dog (ROSENGREN, 1969). For colour vision of the domestic cat see below. Colour vision has been proven,

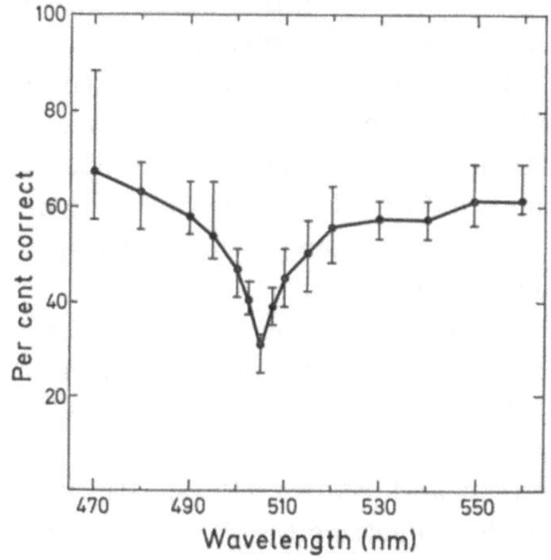

Fig. 17. *Discrimination performance of ground squirrels in the neutral point test.* Each plotted point represents the mean percentage of correct responses for six animals. The vertical lines enclose the total range of performance at each test wavelength (From JACOBS and YOLTON, 1969)

partly in contrast to earlier opinions, for the horse (GRZIMEK, 1952) and for the Zebu (HOFFMANN, 1952). Orange, yellow and green belong to a single colour quality in the Zebu. The tree shrew (*Tupaia glis*) can distinguish colours very well (TIGGES, 1962b; SHRIVER and NOBACK, 1967). The same is true for the squirrel monkey (*Saimiri sciureus*) in the yellow-green, grey and blue regions; however, it cannot distinguish colours in the long-wave part of the spectrum. Its colour vision can be compared with that of protanomalous humans (MILES, 1958b; JACOBS 1963; DEVALOIS and JACOBS, 1968; see Chapter 3). The marmoset (*Callithrix*) is able to see colours in the whole spectrum, even in the red (MILES, 1958a). In *Cebus* colour vision corresponds to that of protanomal humans (GRETHER, 1939; MALMO and GRETHER, 1947). In rhesus monkey (*Macaca rhesus*), in the crab-eating monkey (*Macaca fascicularis*) (TRENDELENBURG and SCHMIDT, 1930), in the macaque (*M. irus*, *M. nemestrina*) (DEVALOIS and JACOBS, 1968) and in the chimpanzee (GRETHER, 1940) the sensitivity for colour differences in the spectrum is equal to or even better than in man; even a simultaneous coloured contrast has been demonstrated in the chimpanzee (GRETHER, 1942). The white-handed gibbon (*Hylobates lar*) and the orang-outang (*Pongo pygmaeus*) also can distinguish colours (TIGGES, 1963a). For colour discrimination in monkey (*Macaca*) see also Chapter 3 and Vol. VII/4, Chapter 28. — By microspectrophotometry three types of cones have been found in the retinae of *Macaca* and of man (MARKS et al., 1964; BROWN and WALD, 1963, 1964).

Among mammals the following species are probably colour blind: *Apodemus sylvaticus* (SÄLZLE, 1936), *Mus musculus* (TRINCKER and BERNDT, 1957), *Mesocricetus auratus* (KNOOP, 1954), the rat (MUNN, 1932; COLEMAN and HAMILTON, 1933; MUNN and COLLINS, 1936; MUENZINGER and REYNOLDS, 1936; in the papers of WALTON, 1933; WALTON and BORNEMEYER, 1939; CAIN and EXTREMET, 1954 it was claimed that the rat and the albino rat can see colours; however, the methods are somewhat open to question), *Genetta suahelica* (DÜCKER, 1957), *Canis latrans*, *C. aureus* and *Alopex lagopus* (EISFELD, 1966) and *Lemur mongoz* (BIERENS DE HAAN and FRIMA, 1930). SÄLZLE (1936) considered the opossum (*Didelphys virginiana*) colour blind, while FRIEDMAN (1967) in more critical investigations was able to demonstrate a well developed colour vision. The difficulties one can expect are shown in the proof of colour vision in the guinea-pig (TRINCKER and BERNDT, 1957): They were able to demonstrate accurate reactions in only a few of a great number of selected animals.

Special problems are posed by the domestic cat. Earlier authors were unable to prove any colour vision in training experiments (DEVOSS and GANSON, 1915; GREGG et al., 1929; MEYER et al., 1954; GUNTER, 1954; DÜCKER, 1957). In recent investigations BONAVENTURE (1961, 1962, 1964), MELLO and PETERSON (1964), SECHZER and BROWN (1964), MEYER and ANDERSON (1965) and MELLO (1968) succeeded in getting the domestic cat to react (usually weakly) to colours in behavioural experiments. This reaction came only after long training which first had to eliminate the preferred distinction according to light intensity. Cats trained to discriminate colours in the mesopic range, where both rods and cones are functioning, retain the ability to distinguish colours at light intensities sufficient to saturate the rods (PEARLMAN and DAW, 1970; DAW and PEARLMAN, 1970). Electrophysiological investigations on the central nervous system (MADSEN and LENNOX, 1955; LENNOX, 1956, 1958a, b; COHN, 1956; OKUDA et al., 1962) confirm these results with respect to the units that had been found to react to colours. They are not nearly as responsive to different colours as they are to brightness. In the Corpus geniculatum laterale and in the optic tract of the cat there are neurons that have a maximum of sensitivity at 500 nm in low light intensities and a maximum at 556 nm at higher intensities (DAW and PEARLMAN, 1969). In addition there is an infrequent type of lateral geniculate cell in the most ventral layer (B) with "on" centre responses to short wavelengths and "off" centre responses to long wavelengths (PEARLMAN and DAW, 1970; Fig. 18). The short wavelength responses are mediated by cones with peak sensitivity at about 450 nm, the long wavelength responses by cones with a sensitivity at 556 nm. On the basis of these findings, colour vision in the cat may be considered dichromatic. In a region of the spectrum near the neutral point (at 500 nm) of the spectral sensitivity curve, monochromatic light should be indistinguishable from white. Furthermore, the cat should be unable to discriminate monochromatic lights of greater than 540 nm, since these would stimulate only one of the two pigments in the cones (PEARLMAN and DAW, 1970).

Among mammals there are some with a *pure cone retina*, the souslik (*Citellus citellus*) and the grey squirrel (*Sciurus carolinensis leucotis*). In the souslik one mm² contains 200000 cones (KARLI, 1951; VILTER, 1954; ARDEN and TANSLEY, 1955b); the number of ganglia cells is, particularly in the region near the Fasciculus opticus, almost equal to the number of cones. In the grey squirrel, for each ganglion

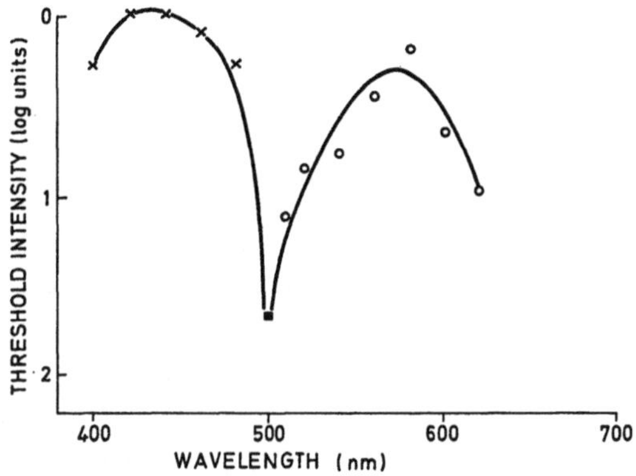

Fig. 18. *Spectral sensitivity for test spots of monochromatic light of various wavelengths illuminating the receptive field centre of a lateral geniculate cell of the cat.* Crosses indicate the threshold for "on" responses and open circles indicate the threshold for "off" responses against white background of 20 cd/m² (From PEARLMAN and DAW, 1970)

cell there are about 2—4 cones (ARDEN and TANSLEY, 1955a). In both these mammals two kinds of cones are found; they lie in different layers and have diffferent shapes. The threshold for the electroretinogram is high compared to that of the rabbit. The shape of the ERG and the spectral sensitivity do not change with adaptation. The sensitivity curve has its maximum at 530 nm; a Purkinje-effect is missing (BORNSCHEIN, 1954; ARDEN and TANSLEY, 1955a, b). In the souslik the curve representing the dependency of the fusion frequency on the light intensity has a bend which is characteristic for mixed retinas (BORNSCHEIN and SZEGVARI, 1958).

Acknowledgement

The authors wish to thank Drs. G. L. and H. J. BECKER for assistance with the translation.

References

ARDEN, G. B., TANSLEY, K.: The spectral sensitivity of the pure cone retina of the grey squirrel (*Sciurus carolinensis leucotis*). J. Physiol. (Lond.) **127**, 592—602 (1955a).

— — The spectral sensitivity of the pure cone retina of the souslik (*Citellus citellus*). J. Physiol. (Lond.) **130**, 225—232 (1955b).

ARMBRUSTER, L.: Über das Farbensehen bei Wespen. Naturwiss. Wochenschr. N. F. **21**, 419—422 (1922).

AUTRUM, H.: Farbensehen der Wirbeltiere. Tabul. biol. ('s-Grav.) **22**, Pars 3, 33—42 (1958).

— Carl v. Hess und der Aufbruch der vergleichenden Physiologie: Farbensinn und Farbensehen. v. Graefes Arch. Ophthalmol. **166**, 3—18 (1963).

— Colour vision in man and animals. Naturwissenschaften **55**, 10—18 (1968).

— AUTRUM, I., HOFFMANN, C.: Komponenten im Retinogramm von *Calliphora* und ihre Abhängigkeit von der Spektralfarbe. Biol. Zbl. **80**, 513—547 (1961).

AUTRUM, H., BURKHARDT, D.: Spectral sensitivity of single visual cells. Nature (Lond.) **190**, 639 (1961).
— KOLB, G.: Spektrale Empfindlichkeit einzelner Sehzellen der Aeschniden. Z. vergl. Physiol. **60**, 450—477 (1968).
— STUMPF, H.: Elektrophysiologische Untersuchungen über das Farbensehen von *Calliphora*. Z. vergl. Physiol. **35**, 71—104 (1953).
— v. ZWEHL, V.: Zur spektralen Empfindlichkeit einzelner Sehzellen der Drohne (*Apis mellifica* ♂). Z. vergl. Physiol. **46**, 8—12 (1962).
— — Ein Grünrezeptor im Drohnenauge (*Apis mellifica* ♂). Naturwissenschaften **50**, 698 (1963).
— — Die spektrale Empfindlichkeit einzelner Sehzellen des Bienenauges. Z. vergl. Physiol. **48**, 357—384 (1964).
BARTKOWIAK, W.: The ability of tortoises to discriminate colour saturations. Bull. Acad. Polon. Cl. Sci. Math. Nat. (B II) **1949**, 21—58 (1949).
BECH, R.: Licht- und Farbreaktionen der *Lygus*-Arten. Biol. Zbl. **84**, 637—640 (1965).
BENNETT, R. R., RUCK, P.: Spectral sensitivities of dark- and lightadapted *Notonecta* compound eyes. J. Insect Physiol. **16**, 83—88 (1970).
BERNAU, U.: Versuche über das Farbsehvermögen der Erdmännchen (*Suricata suricatta*). Z. Säugetierkunde **34**, 223—226 (1969).
BERTHOLF, L. M.: Reactions of the honeybee to light. I. Extent of the spectrum for the honeybee and the distribution of its stimulative efficiency. J. Agricult. Res. **42**, 379—419 (1928).
— The distribution of stimulative efficiency in the ultra-violet spectrum for the honeybee. J. Agricult. Res. **43**, 703—713 (1931).
BIERENS DE HAAN, J. A., FRIMA, M. J.: Versuche über den Farbensinn der Lemuren. Z. vergl. Physiol. **12**, 603—631 (1930).
BIRUKOW, G.: Purkinjesches Phänomen und Farbensehen beim Grasfrosch (*Rana temporaria*). Z. vergl. Physiol. **27**, 41—79 (1939).
— Vergleichende Untersuchungen über das Helligkeits- und Farbensehen bei Amphibien. Z. vergl. Physiol. **32**, 348—382 (1950).
— MENG, M.: Eine neue Methode zur Prüfung des Gesichtssinnes bei Amphibien. Naturwissenschaften **42**, 652—653 (1955).
BLÄSSER, A.: Die partielle relative Farbenblindheit der Hühner. Zool. Jahrb. (Physiol.) **43**, 69—120 (1926).
BONAVENTURE, N.: Sur la sensibilité spectrale et la vision des couleurs chez le Spermophile (*Citellus citellus* L.). C. R. Soc. Biol. (Paris) **153**, 1594—1597 (1959).
— La vision des couleurs chez le chat. Rev. Psychol. Française **6**, 1—10 (1961).
— Sensibilité spectrale et vision des couleurs chez le chat. Rev. Psychol. Française **7**, 75—82 (1962).
— La vision chromatique du chat. C. R. Acad. Sci. Paris **259**, 2012—2015 (1964).
BORNSCHEIN, H.: Elektrophysiologischer Nachweis einer I-Retina bei einem Säuger (*Citellus citellus* L.). Naturwissenschaften **41**, 435—436 (1954).
— SZEGVARI, G.: Flimmerelektroretinographische Studie bei einem Säuger mit reiner Zapfennetzhaut (*Citellus citellus* L.). Z. Biol. **110**, 285—290 (1958).
BROWN, K. T., WATANABE, K., MURAKAMI, M.: Early and late receptor potentials of monkey cones and rods. Cold Spr. Harb. Symp. quant. Biol. **30**, 457—482 (1965).
BROWN, P. K., WALD, G.: Visual pigments in human and monkey retinas. Nature (Lond.) **200**, 37—43 (1963).
— — Visual pigments in single rods and cones of the human retina. Science **144**, 45—52 (1964).
BRUCKMOSER, P.: Die spektrale Empfindlichkeit einzelner Sehzellen des Rückenschwimmers *Notonecta glauca* L. (Heteroptera). Z. vergl. Physiol. **59**, 187—204 (1968).
BUDDENBROCK, W. VON: Vergleichende Physiologie I: Sinnesphysiologie. Basel: Birkhäuser 1952.
— FRIEDRICH, H.: Neue Beobachtungen über die kompensatorischen Augenbewegungen und den Farbensinn der Taschenkrabben (*Carcinus maenas*). Z. vergl. Physiol. **19**, 747—761 (1933).
— MOLLER-RACKE, I.: Neue Beobachtungen über den Farbensinn der Insekten. Experientia (Basel) **8**, 62—66 (1952).
— — Über den Lichtsinn von *Pecten*. Pubb. Staz. Zool. Napoli **24**, 217—245 (1953).

Burgers, A. C. J.: Optomotor reactions of *Xenopus laevis*. Physiol. comp. Oecol. **2**, 272—281 (1952).

Burkhardt, D.: Spectral sensitivity and other response characteristics of single visual cells in the Arthropod eye. Symp. Soc. exp. Biol., Nr. 16: Receptor Mechanisms, pp. 86—109 (1962).

— Colour discrimination in insects. In: Advances in Insect Physiology. Vol. **2**, pp. 131—173. Ed. by Beament, Treherne and Wigglesworth. London-New York: Academic Press 1964.

Cain, J., Extremet, J.: Présentation critique d'une méthode d'apprentissage discriminatif à la couleur chez le rat albinos. C. R. Soc. Biol. (Paris) **148**, 115—117 (1954).

Cohn, R.: A contribution to the study of color vision in cat. J. Neurophysiol. **19**, 416—423 (1956).

Coleman, R. B., Hamilton, W. F.: Color blindness in the rat. J. comp. Psychol. **15**, 177—182 (1933).

Crescitelli, F., Pollack, J. D.: Color vision in the antelope ground squirrel. Science **150**, 1316—1318 (1965).

Cronly-Dillon, J. R.: Spectral sensitivity of the scallop *Pecten maximus*. Science **151**, 345—346 (1966).

Dartnall, H. J. A.: The interpretation of spectral sensitivity curves. Brit. med. Bull. **9**, 24—30 (1953).

Daumer, K.: Reizmetrische Untersuchung des Farbensehens der Bienen. Z. vergl. Physiol. **38**, 413—478 (1956).

Daw, N. W., Pearlman, A. L.: Cat color vision: one cone process or several? J. Physiol. (London) **201**, 745—764 (1969).

— — Cat colour vision: evidence for more than one cone process. J. Physiol. (Lond.) **211**, 125—137 (1970).

DeValois, R. L., Jacobs, G. H.: Primate color vision. Science **162**, 533—540 (1968).

— Jones, A. E.: Single-cell analysis of the organization of the primate color-vision system. In: Jung, R., Kornhuber, H. (Eds.): The Visual System: Neurophysiology and Psychophysics, pp. 178—191. Heidelberg-Berlin-New York: Springer 1961.

— Smith, C. J., Kitai, S. T., Karoly, A. J.: Response of single cells in monkey lateral geniculate nucleus to monochromatic light. Science **127**, 238—239 (1958).

DeVoe, R. D., Small, R. S. W., Zvargulis, J. E.: Spectral sensitivities of wolf spider eyes. J. gen. Physiol. **54**, 1—32 (1969).

DeVoss, J. C., Ganson, R.: Colorblindness of cats. J. Anim. Behav. **5**, 115—139 (1915).

Dodt, E., Heerd, E.: Mode of action of pineal nerve fibers in frogs. J. Neurophysiol. **25**, 405—429 (1962).

Donner, K. O.: The spectral sensitivity of vertebrate retinal elements. Symposium on Visual Problems of Colour; National Physical Laboratory, pp. 541—563. London: H. M. Stationary Office (1958).

Dowling, J. E.: Symp. Physiol. Basis for Form Discrimination, at Brown University, Providence 1964, p. 17.

Dücker, G.: Farb- und Helligkeitssehen und Instinkte bei Viverriden und Feliden. Zool. Beitr. N. F. **3**, 25—99 (1957).

Eck, P. J. van: Farbensehen und Zapfenfunktion bei der Singdrossel, *Turdus e. ericetorum* Turton. Arch. néerl. Zool. **3**, 450—499 (1939).

Eisfeldt, D.: Untersuchungen über das Farbsehvermögen einiger Wildcaniden. Z. wiss. Zool. **174**, 177—225 (1967).

Engländer, H.: Die Bedeutung der weißen Farbe für die Orientierung der Bienen am Stand. Arch. Bienenkunde **22**, 516—549 (1941).

Ewald, L. M.: Versuche zur Analyse der Licht- und Farbenreaktionen eines Wirbellosen (*Daphnia pulex*). Z. Sinnesphysiol. **48**, 285—323 (1914).

Ferens, B.: On the ability of colour discrimination of the Tawny Owl (*Strix aluco aluco* L.). Bull. Acad. Polon. Cl. Sci. Math. Nat. (B II) **1947**, 309—336 (1948).

Fingerman, M., Brown, F. A., Jr.: Color discrimination and physiological duplicity of *Drosophila* vision. Physiol. Zool. **26**, 59—67 (1953).

Forbes, A., Burleigh, S., Meyland, M.: Electric responses to color shift in frog and turtle retina. J. Neurophysiol. **18**, 517—535 (1955).

FRIEDMAN, H.: Colour vision in the Virginia opossum. Nature (Lond.) **213**, 835—836 (1967).
FRISCH, K. v.: Sind die Fische farbenblind? Zool. Jahrb., Abt. allg. Zool. Physiol. **33**, 107—126 (1912).
— Weitere Untersuchungen über den Farbensinn der Fische. Zool. Jahrb., Abt. allg. Zool. Physiol. **34**, 43—68 (1913).
— Der Farbensinn und Formensinn der Bienen. Zool. Jahrb., Abt. allg. Zool. Physiol. **35**, 1—182 (1914).
— KUPELWIESER, H.: Über den Einfluß der Lichtfarbe auf die phototaktischen Reaktionen niederer Krebse. Biol. Zbl. **33**, 517—552 (1913).
GOGALA, M.: Die spektrale Empfindlichkeit der Doppelaugen von *Ascalaphus macaronius* Scop. (Neuroptera, Ascalaphidae). Z. vergl. Physiol. **57**, 232—243 (1967).
— HAMDORF, K., SCHWEMER, J.: UV-Sehfarbstoff bei Insekten. Z. vergl. Physiol. **70**, 410—413 (1970).
GOLDSMITH, T. H.: The visual system of the honeybee. Proc. nat. Acad. Sci. (Wash.) **44**, 123—126 (1958).
— The nature of retinal action potential, and the spectral sensitivities of ultra-violet and green receptor systems of the compound eye of the worker honeybee. J. gen. Physiol. **43**, 775—799 (1960).
— The physiological basis of wave-length discrimination in the eye of the honeybee. In: ROSENBLITH, W. A. (Ed.): Sensory Communication, Chap. 20, pp. 357—375. New York-London: M. I. T. Press & J. Wiley & Sons, Inc. 1961.
— The visual system of insects. In: ROCKSTEIN, M. (Ed.): The Physiology of Insecta, Vol. I, pp. 443—451. New York-London: Academic Press 1964.
GREGG, F. M., JAMISON, E., WILKIE, R., RADINSKY, TH.: Are dogs, cats, and racoons color blind? J. comp. Psychol. **9**, 379—395 (1929).
GRETHER, W. F.: Color vision and color blindness in monkeys. Comp. Psychol. Monogr. **15**, 1—38 (1939).
— Chimpanzee color vision. J. comp. Psychol. **29**, 167 (1940).
— The magnitude of simultaneous color contrast and simultaneous brightness contrast for chimpanzee and man. J. exp. Psychol. **30**, 69 (1942).
GRZIMEK, B.: Versuche über das Farbsehen von Pflanzenessern. 1. Das farbige Sehen (und die Sehschärfe) von Pferden. Z. Tierpsychol. **9**, 23—29 (1952).
GUNTER, R.: The spectral sensitivity of light-adapted cats. J. Physiol. (Lond.) **123**, 409—415 (1954).
HAGINS, W. A.: Electrical signs of information flow in photoreceptors. Cold Spr. Harb. Symp. quant. Biol. **30**, 403—418 (1965).
HAMILTON, W. F.: A direct method of testing color vision in lower animals. Proc. nat. Acad. Sci. (Wash.) 8, 350—353 (1922).
— COLEMAN, T. B.: Trichromatic vision in the pigeon as illustrated by the spectral line discrimination curve. J. comp. Psychol. **15**, 183—192 (1953).
HARTLINE, H. K., WAGNER, H. G., MACNICHOL, E. F., JR.: The peripheral origin of nervous activity in the visual system. Cold Spr. Harb. Symp. quant. Biol. **17**, 125—141 (1952).
HAYS, D., GOLDSMITH, T. H.: Microspectrophotometry of the visual pigment of the spider crab, *Libinia emarginata*. Z. vergl. Physiol. **65**, 218—232 (1969).
HEBERDEY, R. F.: Der Farbensinn helladaptierter Daphnien. Biol. Zbl. **56**, 207—216 (1936a).
— Neue Untersuchungen über den Farbensinn niederer Krebse. Verh. dtsch. Zool. Ges. **1936**, 118—125 (1936b).
HERTER, K.: Dressurversuche mit Igeln (II. Form-, Helligkeitsdressuren, Farbenunterscheidung, Labyrinthversuche, Rhythmus- und Selbstdressuren). Z. vergl. Physiol. **21**, 450—462 (1934).
— Über simultanen Farbkontrast bei Fischen. Biol. Zbl. **69**, 283—300 (1950).
— SGONINA, K.: Dressurversuche mit Igeln (I. Orts-, Helligkeits- und Farbendressuren). Z. vergl. Physiol. **18**, 481—515 (1933).
HERTZ, M.: Beitrag zum Farbensinn und Formensinn der Biene. Z. vergl. Physiol. **24**, 413—421 (1937).
HOFFMANN, G.: Untersuchungen über das Farbsehvermögen des Zebu. Z. Tierpsychol. **9**, 470—479 (1952).

Horridge, G. A.: Unit studies on the retina of dragonflies. Z. vergl. Physiol. **62**, 1—37 (1969).
Hundertmark, A.: Helligkeits- und Farbenunterscheidungsvermögen der Eiraupen der Nonne (*Lymantria monacha* L.). Z. vergl. Physiol. **24**, 42—57 (1936).
Ilse, D.: Über den Farbensinn der Tagfalter. Z. vergl. Physiol. 8, 658—691 (1928).
— Über das Sehen der Insekten, besonders der Tagfalter. Sitzungsber. Ges. naturforsch. Freunde Berlin 1/3, 1—16 (1934).
— New observations on responses to colours in egg-laying butterflies. Nature (Lond.) **140**, 544 (1937).
— Colour discrimination in the dronefly, *Eristalis tenax*. Nature (Lond.) **163**, 255 (1949).
— Versuche zur Orientierung von Tagfaltern. Verh. dtsch. Zool. Ges. **1965**, 306—319 (1965).
Jacobs, G. H.: Spectral sensitivity and color vision of the squirrel monkey. J. comp. physiol. Psychol. **56**, 616—621 (1963).
— Yolton, R. L.: Dichromacy in the ground squirrel. Nature (Lond.) **223**, 414—415 (1969).
Kaestner, A.: Über den Farbsinn der Spinnen. Naturwiss. Rdsch. 8, **1950**, 357—360.
Kaiser, W.: Zur Frage des Unterscheidungsvermögens für Spektralfarben: Eine Untersuchung der Optomotorik der Königlichen Glanzfliege *Phormia regina* Meig. Z. vergl. Physiol. **61**, 71—102 (1968).
Kaneko, A., Hashimoto, H.: Recording site of the single cone response determined by an electrode marking technique. Vision Res. **7**, 847—851 (1967).
Karli, P.: Sur la structure de la rétine du spermophile (*Citellus citellus* L.). C. R. Soc. Biol. (Paris) **145**, 1376 (1951).
Kern, E.: Der Bereich der Unterschiedsempfindlichkeit des Auges bei festgehaltenem Adaptationszustand. Z. Biol. **105**, 237—245 (1952).
Kiepenheuer, J.: Farbunterscheidungsvermögen bei der roten Waldameise *Formica polyctena* Förster. Z. vergl. Physiol. **57**, 409—411 (1968).
Knoll, F.: Insekten und Blumen. Experimentelle Arbeiten zur Vertiefung unserer Kenntnisse über die Wechselbeziehungen zwischen Pflanzen und Tieren. Abh. Zool. Bot. Ges. Wien **12**, 1—116 (1921).
— Insekten und Blumen. II. Lichtsinn und Blumenbesuch des Falters *Macroglossum stellatarum*. Abh. Zool. Bot. Ges. Wien **12**, (1922).
— Lichtsinn und Blütenbesuch des Falters *Deilephila livornica*. Z. vergl. Physiol. **2**, 329—380 (1925).
— Über Abendschwärmer und Schwärmerblumen. Ber. dtsch. Bot. Ges. **45**, 510—518 (1927).
Knoop, I.: Untersuchungen über das Farben- und Formensehen bei Goldhamstern (*Mesocricetus auratus* Waterh.) mit Hilfe der Dressurmethode. Zool. Beitr., N. F. **1**, 219—239 (1954).
Koehler, O.: Über das Farbensehen von *Daphnia magna* Strauss. Z. vergl. Physiol. **1**, 84—174 (1924a).
— Sinnesphysiologische Untersuchungen an Libellenlarven. Verh. dtsch. Zool. Ges. **29**, 83—90 (1924b).
Koller, G.: Über Chromatophorensystem, Farbensinn und Farbwechsel bei *Crangon vulgaris*. Z. vergl. Physiol. **5**, 191—246 (1927).
— Versuche über den Farbensinn der Eupaguriden. Z. vergl. Physiol. 8, 337—353 (1928).
Kolosvary, G.: A study of color vision in the mouse (*Mus musculus*) and the souslik (*Citellus citellus* L.). J. genet. Psychol. **44**, 473 (1934).
Kugler, H.: Der Blütenbesuch der Schlammfliege (*Eristalomyia tenax*). Z. vergl. Physiol. **32**, 328—347 (1950).
Kühn, A.: Über den Farbensinn der Bienen. Z. vergl. Physiol. **5**, 762—800 (1927).
— Über Farbensinn und Anpassung der Körperfarbe an die Umgebung bei Tintenfischen. Nachr. Ges. Wiss. Göttingen, Math.-physik. Kl. Biol. **1** (1933).
— Über Farbwechsel und Farbensinn von Cephalopoden. Z. vergl. Physiol. **32**, 572—598 (1950).
— Heberdey, R. F.: Über die Anpassung von *Sepia officinalis* L. an Helligkeit und Farbton der Umgebung. Verh. dtsch. Zool. Ges. **1929**, 231—237 (1929).
— Ilse, D.: Anlockung von Tagfaltern durch Pigmentfarben. Biol. Zbl. **45**, 144—149 (1925).
Kunze, P.: Untersuchung des Bewegungssehens fixiert fliegender Bienen. Z. vergl. Physiol. **44**, 656—684 (1961).

LANGER, H., THORELL, B.: Microspectrophotometry of single rhabdomeres in the insect eye. Exp. Cell Res. **41**, 673—677 (1966).

LENNOX, M. A.: Geniculate and cortical responses to coloured light flash in cat. J. Neurophysiol. **19**, 271—279 (1956).

— Single fiber responses to electrical stimulation in cat's optic tract. J. Neurophysiol. **21**, 62—69 (1958a).

— The on response to coloured flash in single optic tract fibers of cat: correlation with conduction velocity. J. Neurophysiol. **21**, 70—84 (1958b).

LICHE, H.: Über die photischen Reaktionen bei der Schlammschnecke *Limnaea stagnalis* L. Bull. Acad. Polon. Cl. Sci. Math. nat. (B II) **1943**, 233—249 (1943).

LUTZ, F. E.: Experiments with "stingless bees" (*Trigonia cressoni*) concerning their ability to distinguish ultraviolet patterns. Amer. Mus. Novitates **1933**, 641 (1933).

MACNICHOL, E. F., JR.: Retinal mechanisms of colour vision. Vision Res. **4**, 119—133 (1964).

MADSEN, A., LENNOX, M. A.: Response to coloured light flash from different areas of optic cortex and from retina in anesthetized cat. J. Neurophysiol. **28**, 574—582 (1955).

MALMO, R. B., GRETHER, W. F.: Further evidence of redblindness (protanopia) in Cebus monkey. J. comp. physiol. Psychol. **40**, 143—147 (1947).

MARKS, W. B.: Difference spectra of the visual pigment in single goldfish cones. Ph. D. Dissert. Johns Hopkins Univ. 1963.

— Visual pigments of single cones. In: DEREUCK, A. V. S., KNIGHT, J. (Eds.): Colour Vision. Physiology and Experimental Psychology. Ciba Found. Symp. pp. 208—216. London: J. & A. Churchill Ltd. 1965a.

— Visual pigments of single goldfish cones. J. Physiol. (Lond.) **178**, 14—32 (1965b).

— DOBELLE, W. H., MACNICHOL, E. F., JR.: Visual pigments of single primate cones. Science **143**, 1181—1183 (1964).

MAZOKHIN-PORSHNYAKOV, G. A.: Colorimetric study of colour vision in the dragon-fly. Akad. Nauk SSSR, Biofižika **4**, 427—436 (1959).

— The system of colour vision in *Calliphora*. Akad. Nauk SSSR, Biofižika **5**, 697—703 (1960a).

— Kolorimetrische Untersuchung der Eigenschaften des Farbensehens der Insekten am Beispiel der Stubenfliege. Akad. Nauk SSSR, Biofižika **5**, 295—303 (1960b) (russisch).

MCREYNOLDS, J. S., GORMAN, A. L. F.: Membrane conductances and spectral sensitivities of Pecten photoreceptors. J. gen. Physiol. **56**, 392—406 (1970).

MELLO, N. K.: Color generalization in cat following discrimination training on achromatic intensity and on wavelength. Neuropsychologia (Oxford) **6**, 341—354 (1968).

— PETERSON, N. J.: Behavioural evidence for color discrimination in cat. J. Neurophysiol. **27**, 323—333 (1964).

MENG, M.: Untersuchungen zum Farben- und Formensehen der Erdkröte (*Bufo bufo* L.). Zool. Beitr. N. F. **3**, 313—363 (1958).

MEYER, D. R., ANDERSON, R. A.: Colour discrimination in cats. In: DEREUCK, A. V. S., KNIGHT, J. (Eds.): Colour Vision. Physiology and Experimental Psychology. Ciba Found. Symp., pp. 325—339. London: J. & A. Churchill Ltd 1965.

— MILES, R. C., RATOOSH, P.: Absence of color vision in cat. J. Neurophysiol. **17**, 289—294 (1954).

MEYER-OEHME, D.: Dressurversuche an Eichhörnchen zur Frage ihres Helligkeits- und Farbensehens. Z. Tierpschol. **14**, 473—509 (1957).

MEYKNECHT, J.: Farbensehen und Helligkeitsunterscheidung beim Steinkauz. Ardea **30**, 129—169 (1941).

MICHAEL, C. R.: Receptive fields of single optic nerve fibers in a mammal with an all-cone retina. III. Opponent color units. J. Neurophysiol. **31**, 268—282 (1968).

MILES, R. C.: Color vision in the marmoset. J. comp. physiol. Psychol. **51**, 152—154 (1958a).

— Color vision in the squirrel monkey. J. comp. physiol. Psychol. **51**, 328—331 (1958b).

— RATOOSH, P., MEYER, D. R.: Absence of color vision in guinea pig. J. Neurophysiol. **19**, 254—258 (1956).

MITARAI, G., SVAETICHIN, G., VALLECALLE, E., FATEHCHAND, R., VILLEGAS, J., LAUFER, M.: Glia-neuron interactions and adaptational mechanisms of the retina. In: JUNG, R., KORNHUBER, H. (Eds.): The Visual System. Neurophysiology and Psychophysics., pp. 463—481. Heidelberg-Berlin-New York: Springer 1961.

MOERICKE, V.: Über das Farbsehen der Pfirsichblattlaus (*Myzodes persicae* Sulz.). Z. Tierpsychol. **7**, 265—274 (1950).
— Farben als Landereize für geflügelte Blattläuse (Aphidoidea). Z. Naturforsch. **7b**, 304—309 (1952).
— Neue Untersuchungen über das Farbsehen der Homopteren. Proc. of the Second Conference on Potato Virus Diseases, Lisse-Wageningen 1954, pp. 55—69.
MOLITOR, A.: Zum Farbensehen der Faltenwespen. Zool. Anz. **126**, 259—264 (1939).
MOTE, M. I., GOLDSMITH, T. H.: Spectral sensitivities of color receptors in the compound eye of the cockroach *Periplaneta*. J. exp. Zool. **173**, 137—146 (1970).
MÖLLER-RACKE, I.: Farbensinn und Farbenblindheit bei Insekten. Zool. Jahrb., Physiol. **63**, 153—324 (1952).
MUENZINGER, K. F., REYNOLDS, H. E.: Color vision in white rats. I. Sensitivity to red. J. genet. Psychol. **48**, 58 (1936).
MUNDHENKE, G.: Untersuchungen zur Frage des Farbsehens einiger euthyneurer Schnecken. Zool. Jahrb., Physiol. **66**, 33—61 (1955).
MUNN, N. L.: J. genet. Psychol. **40**, 351 (1932) (see TRINCKER, BERNDT, 1957).
— COLLINS, M.: Discrimination of red by white rats. J. genet. Psychol. **48**, 72 (1936).
MUSOLFF, W.: Untersuchungen über den Farbensinn und Purkinjesches Phänomen bei drei ökologisch verschiedenen Typen der Echsen (Lacertilia) mit Hilfe der optomotorischen Reaktion. Zool. Beitr., N. F. **1**, 399—426 (1955).
NAKA, K.-I., EGUCHI, E.: Spike potentials recorded from the insect photoreceptor. J. gen. Physiol. **45**, 663—680 (1962).
— RUSHTON, W. A. H.: S-potentials from colour units in the retina of fish (Cyprinidae). J. Physiol. (Lond.) **185**, 536—555 (1966a).
— — An attempt to analyse colour reception by electrophysiology. J. Physiol. (Lond.) **185**, 556—586 (1966b).
— — S-potentials from luminosity units in the retina of fish (Cyprinidae). J. Physiol. (Lond.) **185**, 587—599 (1966c).
NEGISHI, K., FATEHCHAND, R.: Cellular mechanisms of a Young-Hering visual system. In: DEREUCK, A. V. S., KNIGHT, J. (Eds.): Colour Vision. Physiology and Experimental Psychology. Ciba Found. Symp., pp. 178—205. London: J. & A. Churchill, Ldt. 1965.
NOLTE, J., BROWN, J. E.: The spectral sensitivities of single cells in the median ocellus of *Limulus*. J. gen. Physiol. **54**, 636—649 (1969).
— — The spectral sensitivities of single receptor cells in the lateral, median and ventral eyes of normal and white-eyed *Limulus*. J. gen. Physiol. **55**, 787—801 (1970).
NOSAKI, H.: Electrophysiological study of color encoding in the compound eye of the crayfish, *Procambarus clarkii*. Z. vergl. Physiol. **64**, 318—323 (1969).
OHNESORGE, E.: Untersuchungen über die Entwicklung des Sehvermögens beim Rückenschwimmer *Notonecta glauca* L. Z. vergl. Physiol. **37**, 424—438 (1955).
OKUDA, J., TAIRA, N., MOTOKAWA, K.: Spectral response curves of postgeniculate neurons in the cat. Tohuko J. exp. Med. **78**, 147—157 (1962).
ORLOV, O. Y., BYZOV, A. L.: Colorimetric investigation of sight in Molluscs, Cephalopoda. Dokl. Akad. Nauk **139**, 723—725 (1961) (russisch).
PEARLMAN, A. L., DAW, N. W.: Opponent color cells in the cat lateral geniculate nucleus. Science **167**, 84—86 (1970).
PEIPONEN, V. A.: Experimentelle Untersuchungen über das Farbensehen beim Blaukehlchen, *Luscinia svecica* (L.) und Rotkehlchen, *Erithacus rubecula* (L.). I. Ann. Zool. Soc. zool. bot. fenn. „Vanamo" **24**, 1—49 (1963).
PLATH, M.: Über das Farbenunterscheidungsvermögen des Wellensittichs. Z. vergl. Physiol. **22**, 691—708 (1935).
POLEY, D.: Experimentelle Untersuchungen zur Nahrungssuche und Nahrungsaufnahme der Kolibris. Bonner Zool. Beitr. **19**, 111—156 (1968).
QUARANTA, J. V.: An experimental study of the color vision of the giant tortoise. Zoologica (N. Y.) **37**, 295—312 (1952).
RANKE, O. F.: Die optische Simultanschwelle als Gegenbeweis gegen das Fechnersche Gesetz. Z. Biol. **105**, 224—231 (1952).

RESCH, B.: Untersuchungen über das Farbensehen von *Notonecta glauca*. Z. vergl. Physiol. **36**, 27—40 (1954).
REUTER, T.: Visual pigments and ganglion cell activity in the retinae of tadpoles and frogs (*Rana temporaria* L.). Acta zool. fenn. **122**, 1—64 (1969).
ROKOHL, R.: Über die regionale Verschiedenheit der Farbentüchtigkeit im zusammengesetzten Auge von *Notonecta glauca*. Z. vergl. Physiol. **29**, 638—676 (1942).
ROSENGREN, A.: Experiments in colour discrimination in dogs. Acta zool. fenn. **121**, 3—19 (1969).
RUCK, P.: The components of the visual system of a dragonfly. J. gen. Physiol. **49**, 289—307 (1965).
RUSHTON, W. A. H.: A cone pigment in the protanope. J. Physiol. (Lond.) **168**, 345—359 (1963).
SÄLZLE, K.: Untersuchungen über das Farbsehvermögen von Opossum, Waldmäusen, Rötelmäusen und Eichhörnchen. Z. Säugetierkunde **11**, 105—148 (1936).
SCHLEGTENDAL, H.: Beitrag zum Farbensinn der Arthropoden. Z. vergl. Physiol. **20**, 545—583 (1934).
SCHLIEPER, C.: Farbensinn der Tiere und optomotorische Reaktionen der Tiere. Z. vergl. Physiol. **6**, 453—472 (1927).
SCHÖNE, H.: Farbhelligkeit und Farbunterscheidung bei den Wasserkäfern *Dytiscus marginalis*, *Acilius sulcatus* und ihren Larven. Z. vergl. Physiol. **35**, 27—35 (1953).
SCHREMMER, F.: Versuche zum Nachweis der Rotblindheit von *Vespa rufa* L. Z. vergl. Physiol. **28**, 457—466 (1941a).
— Sinnesphysiologie und Blumenbesuch des Falters *Plusia gamma* L. Zool. Jahrb., System. **74**, 361—522 (1941b).
SECHZER, J. A., BROWN, J. L.: Color discrimination in the cat. Science **144**, 427—429 (1964).
SGONINA, K.: Über das Farben- und Helligkeitssehen des Meerschweinchens. Z. wiss. Zool. **148**, 350—363 (1936).
SHRIVER, J. E., NOBACK, C. R.: Color vision in the tree shrew (*Tupaia glis*). Folia primat. (Basel) **6**, 161—169 (1967).
SMITH, F. E., BAYLOR, E. R.: Color responses in the Cladocera and their ecological significance. Amer. Naturalist **87**, 49—55 (1953).
SMITH, S.: Response to colour in birds. Nature (Lond.) **150**, 376—377 (1942).
SÜFFERT, F., GÖTZ, B.: Verhalten von Schmetterlingsraupen gegenüber farbigen Flächen. Naturwissenschaften **24**, 815 (1936).
SVAETICHIN, G.: The cone action potential. Acta physiol. scand. **29**, Suppl. 106, 565—600 (1953).
SWIHART, C. A., SWIHART, S. L.: Colour selection and learned feeding preferences in the butterfly, *Heliconius charitonius* Linn. Animal Behav. **18**, 60—64 (1970).
THOMAS, E.: Untersuchungen über den Helligkeits- und Farbensinn der Anuren. Zool. Jahrb., Physiol. **66**, 129—178 (1955).
TIGGES, J.: Untersuchungen über den Farbensinn von *Tupaia glis* (Diard, 1820). Z. Morph. Anthrop. **53**, 109—123 (1963a).
— On color vision in Gibbon and Orang-utan. Folia primat. (Basel) **1**, 188—198 (1963b).
TOMITA, T.: Electrophysiological study of the mechanisms subserving color coding in the fish retina. Cold Spr. Harb. Symp. quant. Biol. **30**, 559—566 (1965).
— KANEKO, A., MURAKAMI, M., PAUTLER, E. L.: Spectral response curves of single cones in the carp. Vision Res. **7**, 519—531 (1967).
TOYODA, J.-I., NOSAKI, H., TOMITA, T.: Light-induced resistance changes in single photoreceptors of *Necturus* and *Gekko*. Vision Res. **9**, 453—463 (1969).
TRENDELENBURG, W., SCHMIDT, I.: Untersuchungen über das Farbensystem des Affen. Z. vergl. Physiol. **12**, 249—278 (1930).
TRINCKER, D., BERNDT, P.: Optomotorische Reaktionen und Farbensinn beim Meerschweinchen. Z. vergl. Physiol. **39**, 607—623 (1957).
TSUNEKI, K.: Some experiments on colour vision in ants. J. Facult. Sci. Hokkaido Univ., Ser. VI, Zool. **10**, 77—86 (1950).
— On colour vision in two species of ants, with special emphasis on their relative sensitivity to various monochromatic lights. Jap. J. Zool. **11**, 187—221 (1953).
VILTER, V.: Histologie et activité électrique de la rétine d'un Mammifère strictement diurne, le Spermophile (*Citellus citellus*). C. R. Soc. Biol. (Paris) **148**, 1768—1771 (1954).

WAGNER, H.: Über den Farbensinn der Eidechsen. Z. vergl. Physiol. **18**, 378—392 (1932).
WALTHER, J. B.: Untersuchungen am Belichtungspotential des Komplexauges von *Periplaneta* mit farbigen Reizen und selektiver Adaptation. Biol. Zbl. **77**, 63—104 (1958a).
— Changes induced in spectral sensitivity and form of retinal action potential of the cockroach eye by selective adaptation. J. Insect Physiol. **2**, 142—151 (1958b).
WALTON, W. E.: Color vision and color preference in the albino rat. I. Historial summary and criticism. II. Experiments and results. J. comp. Psychol. **15**, 359—394 (1933).
— BORNEMEYER, W.: Color discrimination in rats. J. comp. Psychol. **28**, 417 (1939).
WERBLIN, F. S., DOWLING, J. E.: Organization of the retina of the mudpuppy, *Necturus maculosus*. II. Intracellular recording. J. Neurophysiol. **32**, 339—355 (1969).
WOJTUSIAK, R. J.: Über den Farbensinn der Schildkröten. Z. vergl. Physiol. **18**, 393—436 (1933).
— Investigations on the vision of infra-red in animals. I. Investigations on water-tortoises. Bull. Acad. Polon. Cl. Sci. Math. Nat. (B II) **1947**, 43—61 (1947).
ZETTLER, F.: Analyse der Belichtungspotentiale der Sehzellen von *Calliphora erythrocephala* Meig. Z. vergl. Physiol. **56**, 129—141 (1967).

References 1971—1972

The article was finished and sent to the editor in September 1970. The following references are given for literature published 1971 and 1972. These references are not mentioned in the text.

BEIER, W., MENZEL, R.: Untersuchungen über den Farbensinn der deutschen Wespe (*Paravespula germanica* F., Hymenoptera, Vespidae): Verhaltensphysiologischer Nachweis des Farbensehens. Zool. Jahrb. Physiol. **76**, 441—454 (1972).
DIETZ, M.: Erdkröten können UV-Licht sehen. Naturwissenschaften **59**, 316 (1972).
HELVERSEN, O. VON: Zur spektralen Unterschiedsempfindlichkeit der Honigbiene. J. comp. Physiol. **80**, 439—472 (1972).
HIMSTEDT, W.: Untersuchungen zum Farbensehen von Urodelen. J. comp. Physiol. **81**, 229—238 (1972).
HUTH, H. H., BURKHARDT, D.: Der spektrale Sehbereich eines Violettohr-Kolibris. Naturwissenschaften **59**, 650 (1972).
JAEGER, R. G., HAILMAN, J. P.: Two types of phototatic behaviour in anuran amphibians. Nature (Lond.) **230**, 189—190 (1971).
REUTER, T., VIRTANEN, K.: Border and colour coding in the retina of the frog. Nature (Lond.) **239**, 260—263 (1972).
SNYDER, A. W., MILLER, W. H.: Fly colour vision. Vision Res. **12**, 1389—1396 (1972).
STRUWE, G.: Spectral sensitivity of single photoreceptors in the compound eye of a tropical butterfly (*Heliconius numata*). J. comp. Physiol. **79**, 197—201 (1972).
TIEMANN, G.: Untersuchungen über das Farbsehvermögen einiger tagaktiver Echsen und Schildkröten, sowie der Blindschleiche. Z. Tierpsychol. **31**, 337—347 (1972).
TAMURA, T., HANYU, I., NIWA, H.: Spectral sensitivity and color vision in shipjak tuna and related species. Bull. Jap. Soc. Sci. Fish. **38**, 799 (1972).
WEHNER, R., TOGGWEILER, F.: Verhaltensphysiologischer Nachweis des Farbensehens bei *Cataglyphis bicolor* (Formicidae, Hymenoptera). J. comp. Physiol. **77**, 239—255 (1972).

Chapter 12

The Evolution of Mammalian Visual Mechanisms

By

MARVIN SNYDER, Bethesda, Maryland (USA)

With 7 Figures

Contents

The evolution of the structure and function of the mammalian visual system has not, until recently, been studied intensively, and historically these studies have mainly been concerned with the geniculo-striate system. The importance of the superior colliculus (SC) has been reemphasized in recent studies [**66**, **73**, **74**], and a functional pathway from the superior colliculus–lateral posterior nucleus (LP)–neocortex has been demonstrated [1, 72]. In this paper, I will attempt to trace the evolutionary development of some of the visual systems present in the mammals, with particular attention being devoted to the primates, and then to relate these developments to behavior. It will be shown that the visual behavior patterns that have been studied after central visual lesions can be explained on the basis of three mechanisms, lateral geniculate body (GL)–striate cortex, SC–LP–neocortex, and SC–spinal cord.

To conduct a phylogenetic study, it is necessary to employ the brains of contemporary animals. The neural structures of these animals may be more or less well developed or specialized than the ancestral forms that gave rise to them, but it is only through careful study and extrapolation of such material that insight may be gained into the evolution of the mammalian brain.

This study will begin with a review of the visual systems of the opossum and the hedgehog, two primitive mammals whose brains are widely held to be representative of the very earliest mammals. By defining the primitive organization of the visual system in terms of the structures found in these animals, the changes that occurred in evolution can be specified more accurately.

Primitive Mammals

Thalamic Relay Nuclei. The posterolateral portion of the thalamus is highly undifferentiated in both the opossum and the hedgehog [16, 26] as compared to the more commonly studied mammals. In the hedgehog, it is particularly difficult to make a sharp distinction between GL and LP. The GL receives a direct retinal projection [11,28]. Although no laminae are seen, the pattern of terminal degeneration indicates, even at this early evolutionary stage, a central core receiving ipsilateral input and a peripheral zone receiving contralateral input. A pathway has also been demonstrated from the SC to LP [28]. In brief, the posterolateral dorsal thalamus of the hedgehog seems to possess at least two visual subdivisions: GL receiving direct retinal input and LP receiving input from the SC. Thus, there is evidence for a mechanism consisting of at least two independent visual projections at this thalamic level. Possibly a third visual subdivision has also been identified. HALL and EBNER [29] found that a portion of what had been termed LP on the basis of cytoarchitectonic evidence does not receive input from the SC but does receive a direct retinal input. Whether this dorsomedial extension of GL is of special importance or whether a separate representation of the visual field is present must await further experimentation.

Thalamo-Cortical Projections. To investigate the relationship between these thalamic nuclei and the neocortex, the technique of retrograde degeneration has primarily been used [16, 26]. In the hedgehog, lesions restricted to the striate cortex cause slight to moderate degeneration in GL, depending on the size of the lesion. The degenerated zones contain many neurons in various stages of chromatolysis. However, when the lesions are made larger so as to include more of the non-striate posterior neocortex, the degeneration in GL becomes progressively more severe, and, at the same time, degeneration is seen for the first time in LP. Degeneration in LP follows a similar pattern; i.e., it increases when the striate area in addition to the peristriate cortex is damaged. The results of similar experiments on the opossum by DIAMOND and UTLEY [16] are essentially the same. According to the interpretation of these authors, both GL and LP send collateral fibers outside of their essential targets in primitive mammals (Fig. 1A)[1]. Alternatively, the mechanism responsible for the various stages of degeneration observed may involve a collateral to adjacent thalamic rather than to adjacent cortical regions (Fig. 1B [65]), or the severe cell loss after large cortical lesions may be a function of both anterograde and retrograde changes [61]. It is possible that early in mammalian evolution the efferent pathway between the striate cortex and GL was nonexistent or poorly developed as compared to this corticofugal pathway in the

[1] Recent studies employing anterograde techniques have not revealed any degenerating fibers terminating outside of the striate area in either the opossum or hedgehog after lesions in GL [4,29]. These results, although not conclusive, cast some doubt on the validity of this interpretation.

more advanced mammals. Thus, the fiber pathways in the hedgehog may be such that the efferent fibers to GL originate primarily in the peristriate cortex, and, consequently, both striate and peristriate cortex must be removed to produce severe cell loss (Fig. 1C). Further studies with anterograde degeneration techniques may help to elucidate the nature of these pathways and the mechanism responsible for the retrograde cell changes seen after large cortical lesions.

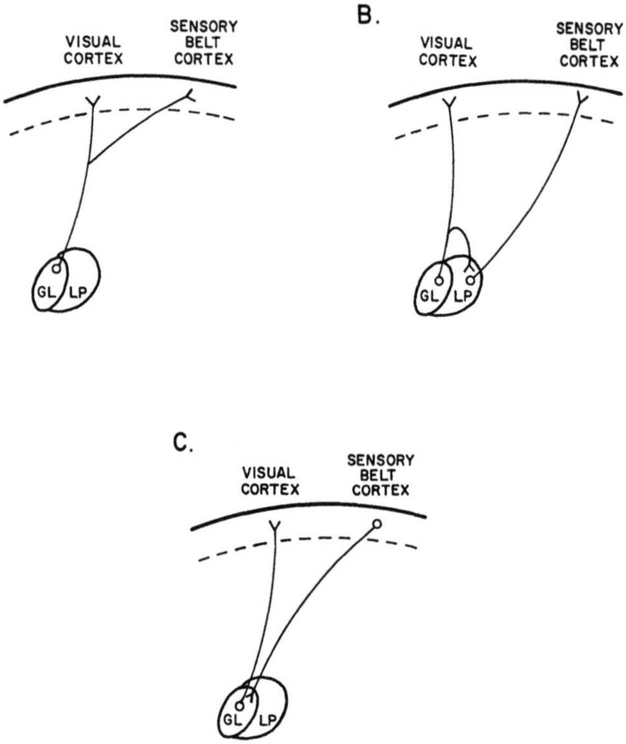

Fig. 1. Schematic representation of possible visual projections in primitive mammals. In (A), the removal of visual and sensory belt cortex leads to degeneration of the thalamic neuron since both the main fiber and its collateral have been destroyed. In (B), it is hypothesized that the removal of sensory belt cortex leads to retrograde transneuronal degeneration of the collateral from GL. Therefore, when striate and sensory belt cortex are removed, the thalamic neuron degenerates for the same reason given above. In (C), it is suggested that the thalamic neuron will degenerate completely only when it is directly affected by reciprocal retrograde and transneuronal degeneration. LP = Lateral posterior nucleus of the thalamus

Visual Physiology. Electrophysiological studies of the hedgehog neocortex indicate that there are at least two visual areas, one in the striate cortex and another within the belt just immediately lateral to the striate area [17]. At present, the underlying mechanism for this second visual representation is not understood; i.e., is it a function of cortico-cortical fibers, of collaterals from GL and/or the dorsomedial GL region, or of fibers from LP? It is tempting to speculate that the organization of V II in the hedgehog is dependent on the interaction of several or all of these possible visual inputs.

With this view of the organization of the primitive mammalian visual system in mind, the brain of the tree shrew will be considered next.

Prosimii

Tree Shrew

The arguments concerning the taxonomic status of the tree shrew are well known [10, 38, 54, 70] and will not be discussed here, except to note that for present purposes it is of little importance whether this animal is the most advanced living insectivore or the most primitive living primate. The brain of the tree shrew is a good approximation of the neural structures thought to have existed in the ancestral primate stock.

Thalamic Relay Nuclei. In contrast to the hedgehog, the GL of the tree shrew is a single, highly differentiated, laminated nucleus. There is further elaboration on the organization of GL seen in the hedgehog, layers 2 and 4 receiving the ipsilateral input and layers 1 and 5 receiving the contralateral input from the retina [11, 22, 76].[2] No evidence for a separate subdivision of the dorsal GL has been found. Just medial to GL is a nucleus which was previously termed pulvinar [72]. As in the hedgehog, this nucleus [1] receives input from the SC, and thus the name LP is now considered more appropriate. The LP in the tree shrew is enormously enlarged as compared to this nucleus in the hedgehog. There is also a dramatic increase in the size of the SC, and, as noted, there is a prominent pathway from the SC to LP.

Thalamo-Cortical Projections. The GL projects in a very specific, topographic fashion to a highly differentiated striate cortex [18]. There is no evidence that the GL projects outside of the striate area in the tree shrew. Fig. 2 shows the projection areas of GL and LP as determined by retrograde degeneration techniques [18]. The cortical targets of these two nuclei are separate, and no evidence is seen for any overlap in the projections. The character of the projections of these nuclei is dramatically different. Small lesions in the striate area produce restricted areas of total degeneration in GL. In marked contrast, lesions in the temporal cortex can produce various stages of degeneration in LP, depending on the size and location of the lesions. Thus, it appears that the major changes between the tree shrew and the hedgehog are in the pattern of projection of GL and in the non-overlap of the thalamic projections to the posterior neocortex.

Visual Physiology. Electrophysiological studies in the tree shrew demonstrate that V I is identical in extent to the anatomically defined striate cortex [17]. A cytoarchitectonically distinct area between the striate cortex and the temporal cortex has been identified in the tree shrew (Fig. 2). This area, which in the tree shrew occupies the same topologic relationship to the striate cortex as V II does to striate cortex in the hedgehog, does not appear to receive input from either GL or LP. Further, V II in the tree shrew is coterminus with this small cytoarchitectonically defined area just lateral to the striate cortex. The source of input, if any, from the thalamus to this V II region in the tree shrew is not known. On the other hand, a cortico-cortical pathway from the striate area to V II has been demonstrated [2].

[2] Layer 3 apparently receives little or no retinal input [11, 22].

A third visual area, recognized by the fact that its neurons have large receptive fields, is found in the projection zone of LP, i.e., lateral to V II. These responses are quite unlike those seen in V II and are more like the responses recorded in the monkey temporal cortex under similar anesthetic conditions [24]. It is suggested that the responses seen in V II of the tree shrew are mediated through cortico-cortical input alone and that in this possible early primate stock the pattern that has been demonstrated in the more advanced primates [14, 36] is already present, namely, the V II responses are mediated entirely through cortico-cortical mechanisms from the striate area.

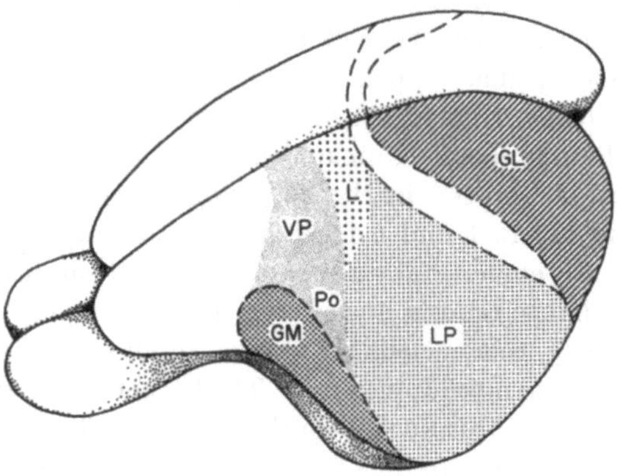

Fig. 2. Projection areas of thalamic nuclei in the tree shrew as determined by retrograde degeneration techniques

These findings do not imply that the responses in the temporal cortex lateral to V II are also mediated only through cortical mechanisms; in fact, it will be shown later that ablation experiments give us every reason to suspect that an important visual pathway exists between the SC and LP and that visual information is transmitted from there to the temporal cortex.

Comparison of Tree Shrew and Hedgehog. It thus appears that early in mammalian evolution a series of dramatic changes occurred in the posterolateral thalamus of the mammals. When comparing the hedgehog and the tree shrew, it is seen that the GL has changed from a poorly defined, diffusely projecting, unlaminated structure to a highly differentiated, topographically precise, multilayered nucleus. Corresponding to these changes, the striate area has undergone extensive elaboration, and a well developed inner granular layer is present as well as a possible homologue to the stripe of Gennari [52]. In addition, no evidence has been found in the tree shrew of any subdivisions of GL which project outside of the striate area. In a similar manner, the widely projecting LP seen in the hedgehog has changed, in the tree shrew, to a nucleus with a rather restricted cortical target. This leads to the possibility that, whereas V II in the hedgehog receives visual input from both cortical and subcortical structures, the V II of the tree shrew, although a homologue of the hedgehog V II, is not activated by input either from LP or GL but depends for its organization on afferents from the striate cortex. This would establish a pattern of serial processing of visual information between V I and V II that may be unique to primates. Finally, the size and differentiation of SC and its pathway to LP in the tree shrew have increased enormously as compared to the hedgehog.

Lemur and Loris

The living intermediate forms between the tree shrew and the monkey, the lemurs and lorises, have not been intensively studied. Although most of these animals are highly specialized, study of their nervous systems should still provide some elucidation of the development of the early primate visual system. The brain of the mouse lemur, *Microcebus*, seems to contain few specializations and consequently does not seem to depart from the ancestral condition as much as the brains of other animals in these groups. Through the courtesy of Sir WILFRED LE GROS CLARK, a brain of the mouse lemur was made available for study. The most pertinent observation made is that the LP is greatly enlarged over that seen in the tree shrew. It is unfortunate that there are no studies of the mouse lemur in which the pathway from SC to LP has been traced. It may be that in this primate one can make the distinction between LP and true pulvinar; i.e., to be consistent with the terminology proposed for the hedgehog and the tree shrew, that part of the posterolateral nuclear complex that receives direct input from the SC should be termed LP and the remainder, the truly intrinsic [64] portion, at least in terms of the afferent pathway, should be termed pulvinar. Since the size of the SC is decreased in the mouse lemur relative to the size of this structure in the tree shrew, it is expected that the total area of the posterolateral complex receiving direct input from the SC would be reduced and a large part of this area is probably "pulvinar." Thus, the first great elaboration of the posterolateral complex, as seen in the tree shrew, seems to have resulted from input from the SC. Further development of the posterolateral area into the pulvinar seems to have resulted from stimuli that came from within the thalamus itself, possibly from LP, from the cortex, or from both.

The only closely related prosimian that has been studied electrophysiologically is the *Galago*. In this animal, both V I and V II have been described [36]. V II is immediately anterior to the striate area, and an analysis of the latencies of the visual responses indicates that V II receives its input from the striate cortex, i.e., V I. In comparing this animal to the others that have been described, it would be most interesting to know what thalamic input, if any, there is to V II in the *Galago*.

Anthropoidea

In the monkey, the last animal in this series, the evidence suggests that, as in the tree shrew, the entire neocortical projection from the GL is restricted to the striate cortex [60, 81]. The posterior thalamus and cortex of the monkey are characterized by a marked increase in the size and differentiation of the pulvinar and its cortical targets. At present, there has been no refined analysis of the afferent and efferent connections of the pulvinar. A pathway has been described from the

SC to a small portion of the inferior pulvinar just medial to the GL [39, 58]. Unfortunately, the cortical target of this region is not precisely known, although it would appear to project to at least a portion of the lunate and inferior occipital sulci. In the nomenclature of this paper, the name LP is now suggested for this subdivision, which was previously included as part of the inferior pulvinar [59]. It is further proposed that what has previously been termed LP in the monkey is homologous to the lateral nucleus (L) in the tree shrew. The evidence for this homology is that both nuclei have similar topological relationships to other thalamic structures, especially to the ventrobasal complex, and both nuclei project to the parietal cortex. Furthermore, anterograde studies of the pretectal region in the monkey will determine whether the pretectum, which projects to L in primitive mammals [5], projects to what has previously been termed LP in the monkey. (Henceforth structures in parentheses indicate the suggested nomenclature.)

Embryological Studies. Further support of the view that a part of the monkey pulvinar is homologous to LP in more primitive forms may be found in embryological studies of the developing thalamus. It has been demonstrated [62] that, in the human, many neurons in the pulvinar are of telencephalic rather than diencephalic origin. Unfortunately, it is difficult to be certain what is meant by the term pulvinar, i.e., how carefully the development of each of the subdivisions was or could be followed. In fact, it is possible that a posterolateral thalamic target of the SC may be absent in man. However, since part of the pulvinar (LP) in the monkey receives input from the SC, it would be most intriguing to learn whether all of the subdivisions of the monkey pulvinar possess neurons of telencephalic origin, or whether the inferior pulvinar (LP) is different, i.e., phylogenetically older and derived entirely from the diencephalon.

Visual Physiology. Electrophysiological studies in the squirrel monkey have indicated that V II is activated via a cortico-cortical pathway from V I [14]. Studies of V II in the rhesus monkey have not been reported, although it is expected that this area will be found immediately surrounding the striate cortex. On the basis of the proposed homologies between the tree shrew and monkey and on the basis of the electrophysiological results, it is suggested that the inferior pulvinar (LP) does not project to V II. However, it is clear from the behavioral work that the cortical visual system in the monkey is composed of many functionally distinct areas [37, 55], and it appears that some of these areas may derive their visual input from the cortex alone, whereas other areas may be sites for interaction of cortical and subcortical input.

Cortical Homologies between the Tree Shrew and Monkey. In establishing homologies between the tree shrew and the monkey, it should also be pointed out that the temporal cortex of the tree shrew is not a homogeneous area and that it is composed of several distinct subdivisions. Likewise, it is not clear that the entire LP nucleus receives input from the SC. Recent experimental results indicate that there is an intimate functional relationship between the prestriate and inferotemporal cortex in the monkey [55]. Thus, it is proposed that the temporal area of the tree shrew is homologous to the prestriate-inferotemporal complex in the monkey, exclusive of V II.

A general view of the evolution of the primate brain, as described here, is shown in Fig. 3.

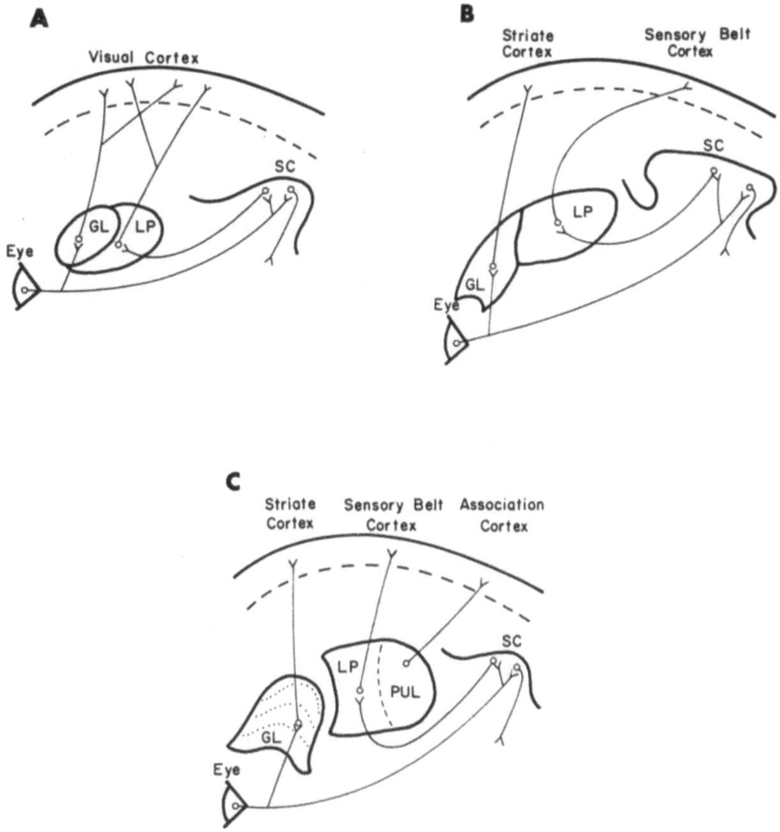

Fig. 3. Schematic representation of the main steps in the evolution of thalamic and cortical visual areas in the primates

Other Mammalian Forms

Cat. In examining forms other than those just described, it appears that the cortical mechanisms for vision developed at different rates and along different lines. For example, the cat is an animal with a highly differentiated, laminated GL. Nonetheless, GL seems to project to the neocortex in a manner more similar to the hedgehog than to the tree shrew [20, 75]. Moreover, the cat, possibly like the hedgehog, seems to have at least two divisions of the dorsal GL, the main body and the medial interlaminar nucleus, i.e., MIN. The main body of GL shows increasing degeneration as posterior neocortical lesions are made larger, and the anterograde studies [22, 42, 81] indicate that this nucleus projects, at least, to areas 17, 18, and the lateral part of the suprasylvian gyrus. The MIN projects to areas 18, 19, and the lateral part of the suprasylvian gyrus. The question of a homology between MIN and the suggested dorsomedial extension of GL in the hedgehog must remain for future analysis. Visual input may also reach the posterior neocortex of the cat from the posterolateral complex (LP), this nucleus being activated by a pathway from the SC [3, 57].

Further information regarding these pathways is provided by electrophysiological studies of the cat. In marked contrast to primates, evoked potentials in V II are not eliminated by lesions of the striate cortex [8], indicating that, although a cortico-cortical pathway may play a part in mediating V II responses [32], it is not the complete mechanism by which V II is activated. It remains for future studies to determine the nature of the contribution of the various subcortical nuclei (GL, MIN, LP) to the visual responses seen in V II. Similarly, the visually evoked responses seen in the suprasylvian gyrus may be mediated by a combination of inputs from the cortex, from GL, from MIN, and from LP [6, 9, 22, 35, 42, 81].

Rodents. Although it is usually accepted that GL projects exclusively to the striate cortex in the rat [47], this conclusion is open to question. LASHLEY reported that the degeneration resulting from a striate only lesion differs from that resulting from a hemidecortication; i.e., after lesions restricted to the striate cortex, many chromatolytic neurons remain in GL [47], whereas after a hemidecortication there are no neurons in GL [48]. These findings may be interpreted, in the light of the foregoing discussion, as indicating that GL in the rat sends collaterals outside of the striate cortex. There is also some evidence in the rat for a second retinal projection to the GL [56], possibly again the homologue of the suggested dorsomedial extension of GL in the hedgehog. In addition, the rat has several visual areas [63], and these areas may be activated by a variety of cortico-cortical and subcortico-cortical interaction.

In another rodent, the squirrel, a more elaborate visual system is seen, including a three-layered GL and a highly differentiated striate cortex. Although there is severe degeneration in GL after localized striate lesions, many neurons still remain in the affected region. As in the other animals described, a strong pathway has been shown from the SC to LP [1]. Although several visual areas have been mapped on the cortex of the squirrel [40], the mechanism underlying these visual responses is not presently known.

Lastly, in the golden hamster, SCHNEIDER [67] reported findings almost identical to those of HALL in the hedgehog in terms of the pattern of the geniculo-cortical projections. SCHNEIDER also found a retinal projection [68] to a dorso-medial extension of GL in the hamster and a pathway from the SC to LP [66, 67].

From this brief review, the following points emerge. (a) The restriction of the geniculo-cortical projection to the striate area appears to be a specialization of the primates. (b) The evidence suggests that in mammals other than the primates there is a separate subdivision of the dorsal GL (also see [12, 30]). (c) There is a visual pathway from the SC to LP to neocortex in all of the groups studied. (d) Although at least two representations of the visual field are seen in the cortex, the mechanisms underlying these responses may be different for different animal groups. Specifically, it appears that a serial processing of information is present between V I and V II in the primates, whereas the nonprimates have a mechanism for both serial and parallel processing of visual information at this level.

Behavioral Considerations

An analysis of the mechanisms underlying the remaining visual function after striate cortex lesions is made difficult by the presence of various visual subsystems. For example, if, in the cat, striate cortex ablations produce little or no visual deficit but lesions that include all of areas 17, 18, 19, and the suprasylvian gyrus produce a severe deficit, does the increased incapacity of the animal result from the removal of (a) collaterals from GL terminating outside of the striate area, (b) fibers from MIN, or (c) visual input from the superior colliculus via LP? In an animal such as the cat, further analysis of this problem may be technically unfeasible, since all of these pathways may overlap at their cortical targets. However, an answer may be provided by studying an animal in which some of these systems are well developed and yet are independent. The tree shrew, as has been shown, fulfills this requirement, since GL projects only to the striate cortex and LP projects only to the temporal region and there are no other retinal targets in the dorsal thalamus.

Posterior Neocortical Ablations in the Tree Shrew. The problem of visual function after striate lesions has been investigated in the tree shrew [72, 73]. To accomplish this, tree shrews were tested in a Yerkes-type, three-door, discrimination apparatus. The stimuli, consisting of either (a) horizontal or vertical stripes or (b) upright or inverted triangles, were transilluminated from the rear. After learning to discriminate these groups of stimuli, as well as an intensity problem, lesions of various sizes were made in the posterior neocortex. Fig. 4 shows a representative case, Tupaia 110, with a lesion that is restricted to the striate area, and Fig. 5 indicates the behavior of this animal. It is clear that Tupaia 110 had lost the preoperatively acquired habits but was able to relearn them in about the same time necessary for the original learning. When allowed free access to a strange room with obstacles placed randomly throughout, this animal could not be discriminated from a normal animal. Fig. 6, on the other hand, shows a case in which almost the entire posterior half of the neocortex had been removed. The degeneration in the thalamus includes not only all of the geniculo-striate system, as was the case with Tupaia 110, but almost all of the LP. Tupaia 103, as shown in Fig. 7, and other animals with similar lesions were severely impaired in relearning the visual problems. Nevertheless, these animals could still catch insects that were swung in front of them and did not bump into objects placed in their path. Moreover, they had no difficulty in making color discriminations [73], although they did fail to climb and jump in strange surroundings and, in fact, fell frequently from a modified visual cliff. It is possible that this residual visual ability is a function of the small number of degenerated neurons that remained scattered through the posterior LP, but it is more tempting to speculate that the SC itself, at least in the three shrew, is capable of mediating spatial, color, and even some pattern perception. From these series of experiments, it was concluded that there are at least three mechanisms of visual analysis in the tree shrew: GL–striate cortex, SC–LP–temporal cortex, and SC–spinal cord.

It is, therefore, necessary to investigate whether the behavior of other species after cortical lesions can be explained on the basis of these three systems.

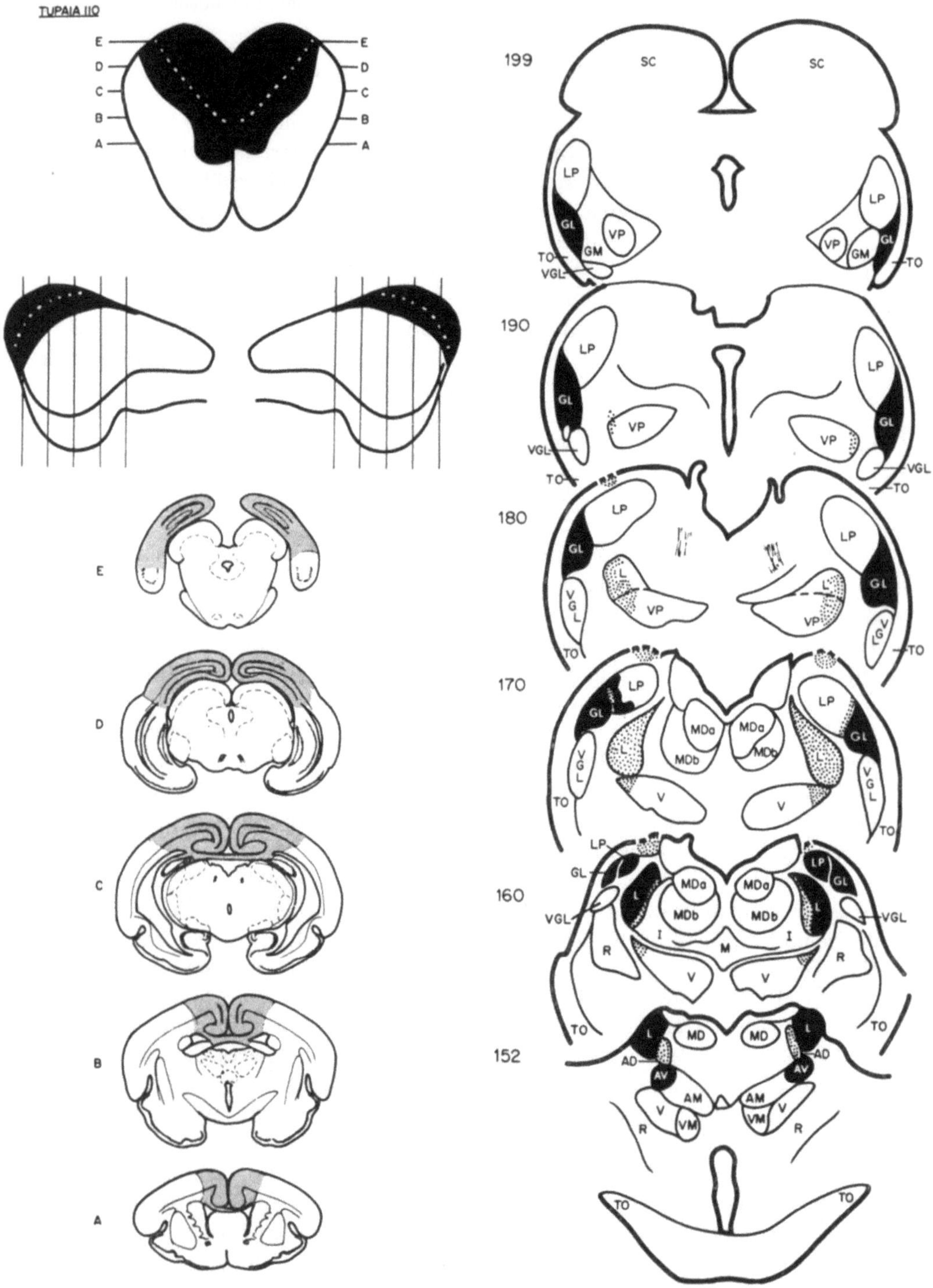

Fig. 4. Cortical lesion and thalamic degeneration for Tupaia 110. In this and the following anatomical figure, the extent of the cortical lesion is shown by a black region upon the dorsal and lateral views of the cortex. The broken white line on the cortical surface marks the boundary of the striate area. In the five standard frontal sections, the lesion is shown by the shaded zone. Representative sections through the thalamus are shown on the right. Severe degeneration is indicated by black and partial degeneration by dots

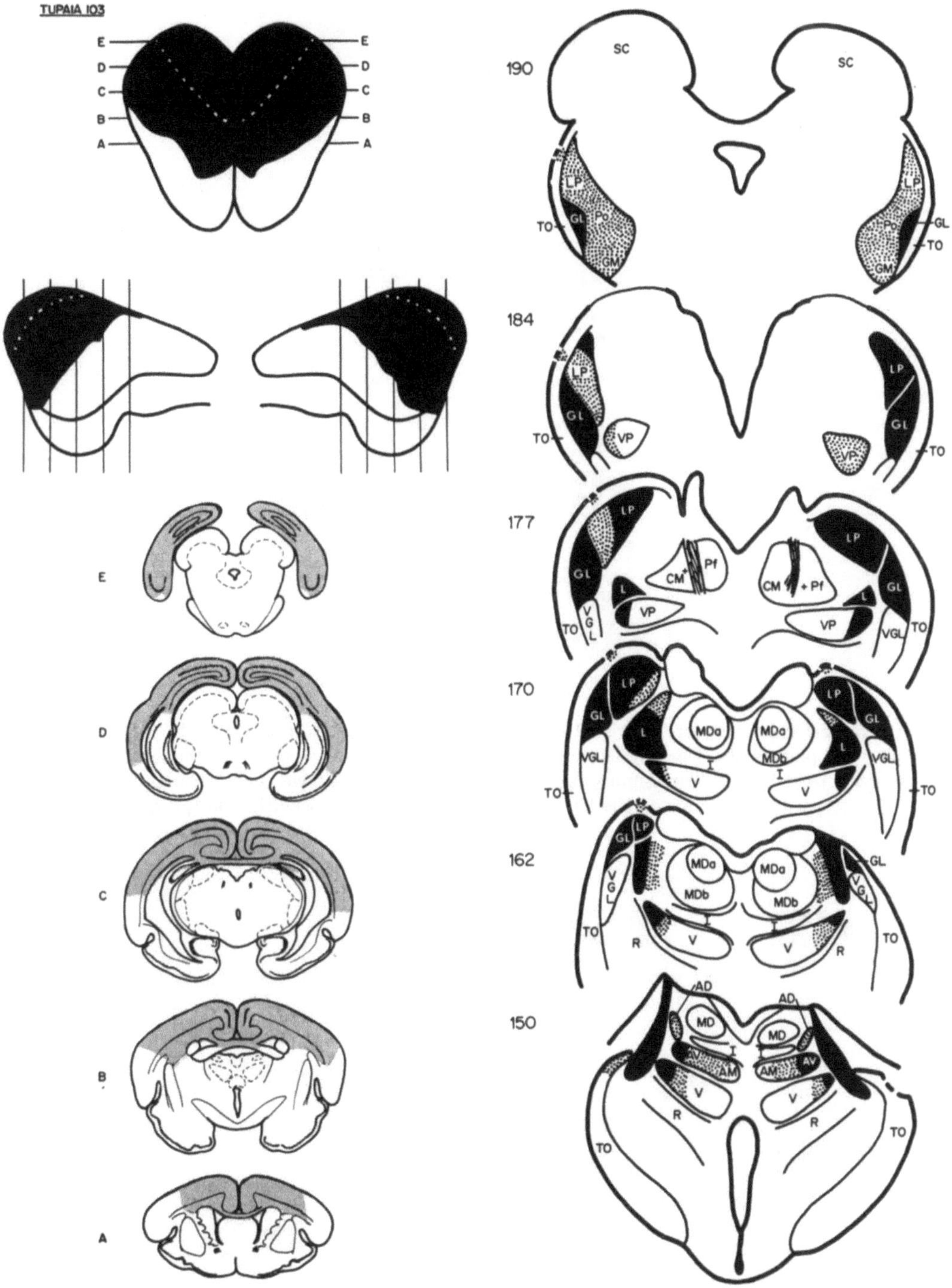

Fig. 6. Cortical lesion and thalamic degeneration for Tupaia 103

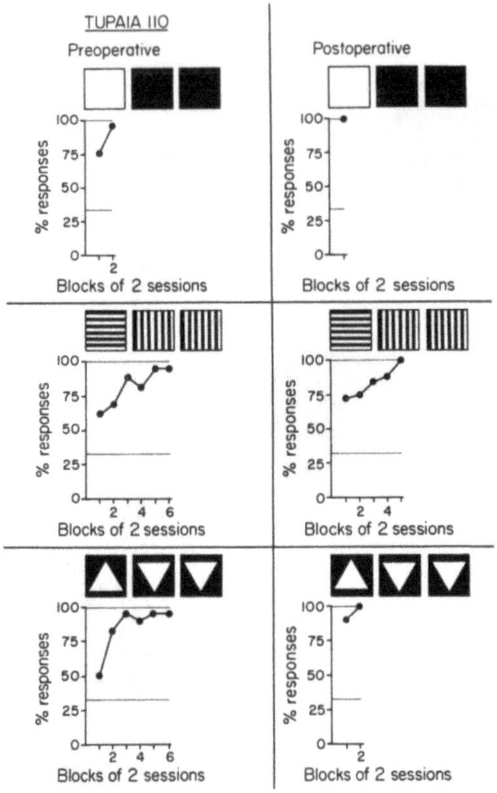

Fig. 5. Performance record of visual discrimination training for Tupaia 110. In this and the following picture of a learning curve, the record before surgery is shown to the left and after surgery, to the right. The stimuli discriminated are pictured above each learning curve with the positive stimulus on the left. Chance level is indicated as either $33^1/_3\%$ or 50%

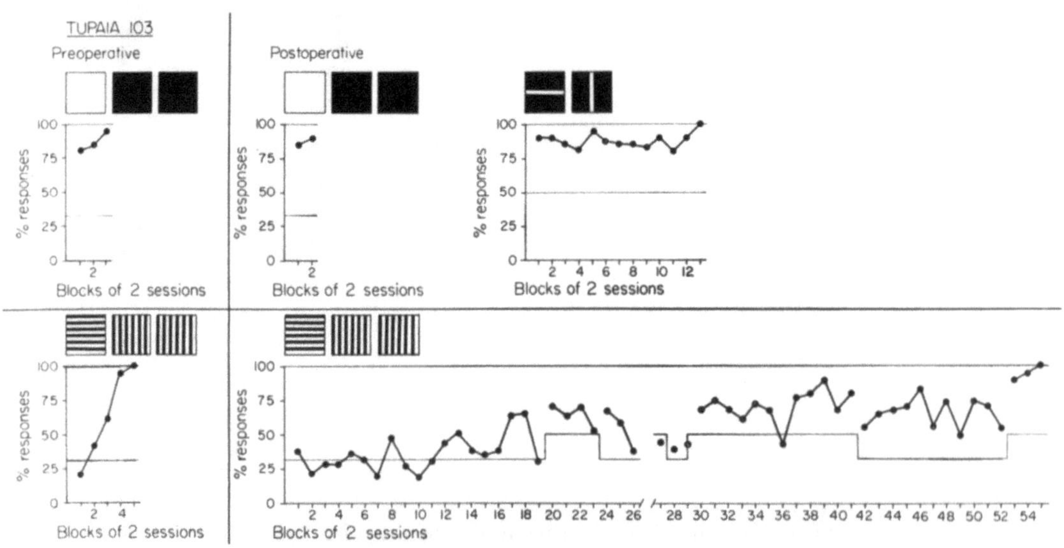

Fig. 7. Performance record of visual discrimination for Tupaia 103

Ablation Studies in Other Mammals

Rat. In LASHLEY's classic studies of the effects of posterior neocortical ablations in the rat, the lesion was not restricted to the striate cortex in any of the cases in which the absence of pattern vision was reported [45, 50]. As pointed out by DOTY [19], LASHLEY's summary of the lesions that failed to produce an impairment in pattern vision is an almost perfect outline around area p, which includes the cortical target of LP. Such animals would be expected to be more like the tree shrews with damage to both the geniculo-striate and the SC–LP–cortex systems. The differences in the behavior of the rat and the tree shrew may be attributable to the extreme development of the SC in the tree shrew and to the fact that LASHLEY's animals were required to jump to the stimuli. Further evidence for the importance for visual function of the LP projection is provided by a comparison of some of LASHLEY's cases. One comparison in particular should be noted. Animal no. 7 [46] with 700 neurons remaining in GL was unique not only in the preservation of this small remnant but also in that the projection area of LP was spared. This now famous case was able to discriminate patterns. Animal no. 15 [49] had the identical portion of the geniculate spared, within an even larger preserved sector. In addition, this animal had an intact area remaining in the contralateral GL, including once again the projection of the temporal retina, an area to which Lashley attributed special importance. The major difference, outside of GL, between these two cases is that case no. 15 had severe damage to the LP projection area. This animal failed all tests for pattern vision. Thus, animal no. 7 with a tiny portion of one geniculate intact but with no damage to the LP projection area could discriminate visual patterns, whereas case no. 15 with a much larger portion of both geniculates remaining but with damage to the cortical target of LP failed on a pattern discrimination.

LASHLEY himself pointed out the differences in visual behavior between animals with lesions that are relatively restricted to the striate area and those cases with deep, large lesions that invade the optic radiations as they leave the thalamus. The former were able to jump gaps as large as 50 cm and orient correctly in space, whereas the latter were unable to perform almost any visual task except intensity discriminations. LASHLEY concluded that "there are enough visual projection fibers going to other regions than the anatomical area striata to mediate vision for position and luminosity of objects" [45]. The undercutting lesions probably also destroyed all of the LP projection fibers as well as the optic radiations. That the LP projection, and not any extrastriate GL fibers, was responsible for the ability to perform on the orienting problems is supported by the fact that some animals with intact striate cortex but severe damage to the LP projection were unable to jump correctly and also by LASHLEY's discovery that lesions in the SC (a finding recently confirmed in the golden hamster [66]) abolish the ability to orient correctly. These results imply that, in the rat, the anterior projections of the SC are essential for the visual behaviors studied. It is not presently known what function the SC–spinal cord pathway mediates in the rat.

The question of whether the rat would be capable of making visual pattern discriminations after lesions restricted to the striate cortex remains an empirical one, but it appears that a visual system from the SC–LP–cortex plays an important role in the visual capacities of this animal.

Cat. A similar pattern of results is seen in the cat. Destruction of area 17 alone produces little or no visual impairment [74, 75]. Only when the lesions are made larger and include the projection area of LP do visual deficits begin to appear [19, 75]. Even after these large lesions, cats may relearn pattern discriminations and recover other visual abilities [82]. Some of the functions that are recovered may be mediated through the SC–spinal system, but the role of GL fibers terminating outside of the striate cortex cannot be ruled out, although, by implication from the data on the tree shrew and the rat, this pathway does not play a major role in the visual behavior patterns that have been studied.

Hedgehog and Golden Hamster. There have been two other recent reports bearing on this problem of extrastriate visual function. HALL [27] reported that the hedgehog is severely impaired on relearning visual pattern discriminations after striate lesions, but even here it was noted that increasing the damage to the projection area of LP increased the visual deficit. SCHNEIDER, in his recent report [66], stressed the lack of ability of the golden hamster to relearn visual pattern discriminations after striate lesions, but again it is unclear from this work whether the projection area of LP was also damaged. Moreover, these animals had the capacity to orient towards objects and snatch food rewards. This was interpreted as a function of the SC alone. It is possible that, if the cortical lesions were enlarged in the hamster so as to include all of the LP projection zone, this capacity would be reduced or abolished. The results of another paper [69] further emphasize the importance of the LP–cortical pathway. Lesions were made in the SC of infant hamsters. When these animals matured, there were direct retinal inputs to the LP, inputs that are not seen in the normal animal. Moreover, the animals receiving lesions in infancy did not show the behavioral deficits at maturity normally seen after SC lesions. This seems to emphasize that an essential pathway to the cortex has been kept intact and somewhat diminishes the argument for the importance of SC alone in mediating these behaviors.

Monkey. The last species to be considered is the monkey. In this animal, as has been shown, the geniculo-striate system is enormously developed, whereas the system from the SC to inferior pulvinar (LP) is reduced in size. Therefore, it would not be suprising to find that monkeys show an extreme visual deficit after lesions in the striate cortex. Although this seemed be the case for many years, it now appears that even in the monkey the visual capacity after striate cortex lesions is much greater than had been thought [33, 34]. In KLUVER'S [43] classic studies of the geniculo-striate system, the lesions in almost all cases must have destroyed not only area 17 but also areas 18 and 19, including the projection area of the inferior pulvinar (LP) (see [78], pp. 153–154). Thus, the severe visual deficit that Kluver observed may be interpreted as a result of damage to both the geniculo-striate and SC–LP–cortex systems. DENNY-BROWN and CHAMBERS [15] reported cases of remaining visual function (movement and contour detection) after lesions restricted to the striate cortex; this capacity could, however, be abolished by lesions invading the projection area of the inferior pulvinar (LP). Similarly, WEISKRANTZ and HUMPHREY [33, 34, 80] have reported remaining visual capacities in the rhesus monkey after lesions relatively restricted to the striate cortex. It was suggested that these visual abilities may be a function of the SC alone. However, it seems more likely that, in light of our discussion and the results of DENNY-BROWN and CHAMBERS, the additional removal of the remaining projection field of the inferior pulvinar (LP) would abolish them.

Mechanisms of Temporal Lobe Function

It thus appears that the visual behavior of mammals after posterior neocortical lesions can be fairly well understood in terms of the three mechanisms that have been described — GL–striate cortex, SC–LP–temporal cortex, SC–spinal cord. But experiments on the tree shrew raise some very interesting problems concerning the mechanism of temporal lobe function in the higher primates. The behavior of tree shrews after removal of both striate and temporal cortex [72] resembles that seen in the monkey after temporal lobe lesions. Therefore, the effects of a temporal lesion alone were studied in the tree shrew. The temporal cortex was removed in several animals [41], and these cases appeared to have many features in common with inferotemporal monkeys. The tree shrews exhibited a loss of visual learning set and relearned easy pattern discriminations in essentially the same time as it had taken them preoperatively, but they became progressively more impaired as the problems were made more difficult. These results were not found when animals with only striate lesions were tested at the same time on the same tasks. In view of this finding, it seems reasonable to assume that the deficit seen in the tree shrew after temporal lesions is mediated, at least in part, by fibers from LP and not via a cortico-cortical pathway as indicated in the monkey by MISHKIN [55]. Of course, removing the temporal cortex of the tree shrew simultaneously eliminated cortical and subcortical input to this area, and the observed deficits may have been the result of this combination. In collaboration with MISHKIN, an attempt was made to reopen the investigation of the effects of pulvinar lesions in the monkey [13] to examine whether such lesions make any contribution to the temporal lobe deficit. The results indicate that, with lesions varying between 75 and 98 % complete [even greater in the inferior pulvinar (LP)], monkeys show no deficit on postoperative retention of visual patterns, on original acquisition of an object learning set, or on the concurrent learning of eight different object discriminations. These tasks are known to be specific for inferotemporal lesions in the monkey [37]. To determine whether it is LP input alone or a combination of LP and cortical input ablation that is responsible for the deficit seen in the tree shrew after temporal lesions, it will be necessary to study the effects of LP lesions on visual behavior in this species.

If no deficit is shown by the tree shrew after lesions in LP, then it would indicate that combined damage to cortico-cortical and subcortico-cortical input is responsible for the deficit seen after temporal lesions and that damage to neither alone is sufficient. Thus, with either the cortico-cortical or subcortical pathway intact, the animal would not show the temporal lobe deficit. It may then be necessary to reevaluate those studies that indicated the essential nature of the cortico-cortical pathway in the monkey [55].

On the other hand, if the LP lesions do produce a deficit in the tree shrew, then the possibility exists that, although the behavioral effects seen after prestriate-inferotemporal lesions are similar in the tree shrew and monkey, they are similar by virtue of convergent rather than parallel evolution.

Conclusion

In the first section of this paper, it was shown that all mammals have two independent visual pathways to the neocortex, one from the GL and the other from the LP. Most of the behavior that has been observed in mammals after damage to central visual structures can be explained on the basis of these two pathways, plus an independently functioning SC–spinal cord system. However, there are other pathways carrying visual information to subcortical and cortical areas. For example, it is well known that retinal fibers project to the pretectum [31, 71], ventral GL [31, 32, 71], and accessory optic nuclei [21, 77]. In addition, there may be at least two other functional pathways to the neocortex: GL—nonstriate cortex and pretectum–L–parietal cortex. Presumably these systems subserve visual functions that have not been determined because of the incomplete nature of our behavioral analyses. Therefore, the question arises as to the significance of these relatively independent systems. It is suggested that these pathways carry information regarding different parameters of the visual world, the total pattern of which indicates that a particular visual event is occurring. Areas of convergence would then represent integrative sites for information dealing with different parameters of a given function, e.g., the abstraction pattern vision. In other words, to use the analogy of WALLS [79], a given external region can be mapped on the basis of many different features, such as climate, topography, and agriculture. However, a complete description would require that all of these parameters be considered.

To extend this analogy to visual function, a comparison between the two systems discussed in the tree shrew, GL–striate cortex and SC–LP–temporal cortex, may be made. That information underlying visual perception is present independently in the striate and temporal cortices is clear from the fact that the tree shrews have to relearn the pattern discriminations after either striate or temporal cortex lesions and also from the fact that the removal of both of these areas produces a more severe deficit than the removal of either alone. Further evidence that the visual information in the SC—LP—temporal cortex system is different from that information in the GL—striate cortex system has been reported by KILLACKEY et al. [41], who showed that a reversal learning set could be reacquired by tree shrews after striate cortex lesions but not after temporal cortex lesions. In addition, animals with striate cortex lesions relearned a triangle discrimination in about the same number of trials as it had taken them preoperatively, but those animals with temporal cortex lesions showed a profound deficit on this problem. However, when these same triangles were embedded within a circle, the animals with striate cortex lesions, although able to relearn the problem, were more impaired than those animals with temporal cortex lesions. These studies provide only a starting point in our attempts to understand the exact nature of the parameters encoded in the various visual systems.

Acknowledgements

The author wishes to thank Dr. M. MISHKIN and Dr. W. C. HALL for their helpful and constructive criticism of this manuscript.

References

1. ABPLANALP, P. L.: Some subcortical connections of the visual system in tree shrews and squirrels. Paper presented at the Conference on Subcortical Visual Systems, M.I.T., 1969.
2. — NAUTA, W. J. H.: Personal communication.
3. ALTMAN, J., CARPENTER, M. B.: Fiber projections of the superior colliculus in the cat. J. comp. Neurol. **116**, 157—178 (1961).
4. BENEVENTO, L. A., EBNER, F. F.: Lateral geniculate nucleus projections to neocortex in the opossum. Anat. Rec. **163**, 294 (1969).
5. — — Prectectal, tectal, retinal, and cortical projections to thalamic nuclei of the opossum in stereotaxic coordinates. Brain Res. **18**, 171—175 (1970).
6. BIGNALL, K. E., IMBERT, M., BUSER, P.: Optic projections to non-striate visual cortex in the cat. J. Neurophysiol. **29**, 396—409 (1966).
7. BISHOP, G. H.: The relation between nerve fiver size and sensory modality: Phylogenetic implications of the afferent innervation of cortex. J. nerv. ment. Dis. **128**, 89—114 (1959).
8. BERKELY, M., WOLF, E., GLICKSTEIN, M.: Photic evoked potentials in the cat: Evidence for a direct geniculate input to visual II. Exp. Neurol. **19**, 188—198 (1967).
9. BUSER, P., BORENSTEIN, P., BRUNER, J.: Etude des systémes "associatifs" visuels et auditifs chez le chat anesthesie au chloralose. Electroenceph. clin. Neurophysiol. **11**, 305—324 (1959).
10. CAMPBELL, C. B. G.: The relationships of the tree shrews: the evidence of the nervous system. Evolution **20**, 276—281 (1966).
11. — JANE, J. A., YASHON, D.: The retinal projections of the tree shrew and hedgehog. Brain Res. **5**, 406—418 (1967).
12. CHOUDHURY, B. P., WHITTERIDGE, D.: Visual field projection on the dorsal nucleus of the lateral geniculate body in the rabbit. Quart. J. exp. Physiol. **50**, 104—112 (1965).
13. CHOW, K. L.: Lack of behavioral effects following destruction of some thalamic association nuclei in monkey. Arch. Neurol. Psychiat. **71**, 762—771 (1954).
14. COWEY, A.: Projection of the retinal on to the striate and prestriate cortex in the squirrel monkey, *Saimiri sciureus*. J. Neurophysiol. **27**, 366—393 (1964).
15. DENNY-BROWN, D., CHAMBERS, R. W.: Visuo-motor function in the cerebral cortex. J. nerv. ment. Dis. **121**, 288—289 (1955).
16. DIAMOND, I. T., UTLEY, J. D.: Thalamic retrograde degeneration study of sensory cortex in opossum. J. comp. Neurol. **120**, 129—160 (1963).
17. — HALL, W. C.: Evolution of neocortex. Science **164**, 251—262 (1969).
18. — SNYDER, M., KILLACKEY, H., JANE, J., HALL, W. C.: Thalamo-cortical projections in the tree shrew. J. comp. Neurol. **139**, 273—306 (1970).
19. DOTY, R.: Functional significance of the topographical aspects of the retino-cortical projection. In: JUNG, R., KORNHUBER, H. (Eds.): The Visual System. Neurophysiology and Psychophysics, pp. 228—245. Berlin-Göttingen-Heidelberg: Springer 1961.
20. GARY, L. J., POWELL, T. P. S.: The projection of the lateral geniculate nucleus upon the cortex in the cat. Proc. roy. Soc. B **169**, 107—126 (1967).
21. GIOLLI, R. A.: An experimental study of the accessory optic system in the Cynomolgus monkey. J. comp. Neurol. **121**, 89—108 (1963).
22. GLICKSTEIN, M.: Laminar structure of the dorsal lateral geniculate nucleus in the tree shrew (Tupaia glis). J. comp. Neurol. **131**, 93—102 (1967).
23. — KING, R. A., MILLER, J., BERKLEY, M.: Cortical projections from the dorsal lateral geniculate nucleus of cats. J. comp. Neurol. **130**, 55—75 (1967).
24. GROSS, C. G., SCHILLER, P. H., WELLS, C., GERSTEIN, G. L.: Single unit activity in temporal association cortex of the monkey. J. Neurophysiol. **30**, 833—843 (1967).
25. HALL, W. C.: Visual pathways from the thalamus to the telencephalon in the turtle and hedgehog. Anat. Rec. **166**, 313 (1970).

26. — DIAMOND, I. T.: Organization and function of the visual cortex in hedgehog. I. Cortical cytoarchitecture and thalamic retrograde degeneration. Brain Behav. Evol. **1**, 181—214 (1968).
27. — — Organization and function of the visual cortex in hedgehog. II. An ablation study of pattern discrimination. Brain Behav. Evol. **1**, 215—243 (1968).
28. — EBNER, F. F.: Parallels in the visual afferent projections to the thalamus in the hedgehog *(Paraechinus hypomelas)* and the turtle *(Pseudemys scripta)*. Brain Behav. Evol. **3**, 135—154 (1970).
29. — — Personal communication.
30. HAYHOW, W. R.: The lateral geniculate nucleus of the marsupial phalanger, *Trichosurus vulpecula*. J. comp. Neurol. **131**, 571—604 (1967).
31. — SEFTON, A., WEBB, C.: Primary optic centers of the rat in relation to the terminal distribution of the crossed and uncrossed optic nerve fibers. J. comp. Neurol. **118**, 295—322 (1962).
32. HUBEL, D. H., WIESEL, T. N.: Receptive fields and functional architecture in two non-striate visual areas (18—19) of the cat. J. Neurophysiol. **28**, 229—289 (1965).
33. HUMPHREY, N. K.: Extrastriate vision in monkeys. Paper presented at the Conference on Subcortical Visual Systems, M.I.T., 1969.
34. — WEISKRANTZ, L.: Vision in monkeys after the removal of the striate cortex. Nature (Lond.) **215**, 595—597 (1967).
35. IMBERT, M., BIGNALL, K. E., BUSER, P.: Neocortical interconnections in the cat. J. Neurophysiol. **29**, 382—395 (1966).
36. IONESCU, D. A.: Post-primary flash-evoked responses in unanesthetized night-and-day-active monkeys. Exp. Brain Res. **7**, 275—298 (1969).
37. IWAI, E.: Two visual foci in the temporal lobe of monkeys. Paper presented at Japan-U. S. joint seminar on Neurophysiological Basis of Learning and Behavior, 1968.
38. JANE, J. A., CAMPBELL, C. B. G., YASHON, D.: Pyramidal tract: A comparison of two primates. Science **147**, 153—155 (1965).
39. JOHNSON, T., SNYDER, M., MISHKIN, M.: Unpublished experiments.
40. KAAS, J., DIAMOND, I. T., HALL, W. C., KILLACKEY, H.: Topographic representation of the visual field in the neocortex of the squirrel. Anat. Rec. **163**, 207 (1969).
41. KILLACKEY, H. P., SNYDER, M., DIAMOND, I. T.: Function of striate and temporal cortex in the tree shrew. J. comp. physiol. Psychol. **74**, 1—29 (1971).
42. KINSTON, W. J., VADAS, M. A., BISHOP, P. O.: Multiple projections of the visual field to the medial portion of the dorsal lateral geniculate nucleus and the adjacent nuclei of the thalamus of the cat. J. comp. Neurol. **136**, 295—316 (1969).
43. KLUVER, H.: Functional significance of the geniculo-striate system. Biol. Symp. **7**, 253—299 (1942).
44. KUYPERS, H. G. T. M., SWARCBART, M. K., MISHKIN, M., ROSVOLD, H. E.: Occipitotemporal cortical-cortical connections in the rhesus monkey. Exp. Neurol. **11**, 245—262 (1965).
45. LASHLEY, K. S.: The mechanism of vision. IV. The cerebral areas necessary for pattern vision in the rat. J. comp. Neurol. **53**, 419—478 (1931).
46. — The functioning of small remnants of the visual cortex. J. comp. Neurol. **70**, 45—67 (1939).
47. — The mechanism of vision. VIII. The projection of the retina upon the cerebral cortex of the rat. J. comp. Neurol. **60**, 57—79 (1934).
48. — Thalamo-cortical connections of the rat's brain. J. comp. Neurol. **75**, 67—121 (1941).
49. — The mechanism of vision. XVII. Autonomy of the visual cortex. J. gen. Psychol. **60**, 197—221 (1942).
50. — FRANK, M.: The mechanism of vision. X. Postoperative disturbances of habits based on detail vision in the rat after lesions in the cerebral visual areas. J. comp. Psychol. **17**, 355—391 (1934).
51. LATIES, A. M., SPRAGUE, J. M.: The projections of optic fibers to the visual centers in the cat. J. comp. Neurol. T**27**, 35—70 (1966).
52. LE GROS CLARK, W. E.: The visual cortex of primates. J. Anat. **59**, 350—357 (1925).
53. — The thalamus of *Tupaia minor*. J. Anat. **63**, 177—206 (1929).
54. — The Antecedents of Man. Edinburgh: Edinburgh University Press 1959.

55. MISHKIN, M.: Visual mechanisms beyond the striate cortex. In: RUSSELL, R.W. (Ed.): Frontiers in Physiological Psychology. New York: Academic Press 1966.
56. MONTERO, V.M., BRUGGE, J.F., BEITEL, R.E.: Relation of the visual field to the lateral geniculate body in the albino rat. J. Neurophysiol. **31**, 221—236 (1968).
57. MOREST, D.K.: Identification of homologous neurons in the posterolateral thalamus of the cat. Anat. Rec. **151**, 390 (1965).
58. MYERS, R.E.: Efferent connections of the superior colliculus in the monkey. Unpublished manuscript.
59. OLSZEWSKI, J.: The Thalamus of Macaca mulatta. Basel: S. Karger 1952.
60. POLYAK, S.: In: KLUVER, H. (Ed.): The Vertebrate Visual System. Chicago: University of Chicago Press 1957.
61. POWELL, T.P.S., COWAN, W.M.: The interpretation of degenerative changes in the intralaminar nuclei of the thalamus. J. Neurol. Psychiat. **30**, 140—153 (1967).
62. RAKIC, P., SIDMAN, R.L.: Telencephalic origin of pulvinar neurons in the fetal human brain. Z. Anat. Entwickl.-Gesch. **129**, 53—82 (1969).
63. ROJAS, J.A., MONTERO, V.M., ROBLES, L.: Organizacion functional de la corteza visual de la rata. Congr. Assoc. Latino-Am. Ciencias Fisiol., Resum. Communic. Libres. Vina del Mar, Chile 1964, p. 98.
64. ROSE, J.E., WOOLSEY, C.N.: Organization of the mammalian thalamus and its relationship to the cerebral cortex. Electroenceph. clin. Neurophysiol. **1**, 391—403 (1949).
65. — — Cortical connections and functional organization of the thalamic auditory system of the cat. In: HARLOW, H.F., WOOLSEY, C.N. (Eds.). Biological and Biochemical Bases of Behavior, pp. 128—150. Madison: University of Wisconsin Press 1958.
66. SCHNEIDER, G.E.: Contrasting visumotor functions of tectum and cortex in the golden hamster. Psychol. Forsch. **31**, 52—62 (1967).
67. — Two visual systems. Science **163**, 895—902 (1969).
68. — Personal communication.
69. — NAUTA, W.J.H.: Formation of anomalous retinal projections after removal of the optic tectum in the neonate hamster. Anat. Rec. **163**, 258 (1969).
70. SIMPSON, G.G.: The principles of classification and a classification of mammals. Bull. Amer. Mus. Nat. Hist. **85**, 1—307 (1945).
71. SINGLETON, M.C., PEELE, T.L.: Distribution of optic fibers in the cat. J. comp. Neurol. **125**, 303—328 (1965).
72. SNYDER, M., DIAMOND, I.T.: The organization and function of the visual cortex in the tree shrew. Brain Behav. Evol. **1**, 244—288 (1968).
73. — KILLACKEY, W., DIAMOND, I.T.: Color vision in the tree shrew after removal of posterior neocortex. J. Neurophysiol. **32**, 554—563 (1969).
74. SPRAGUE, J.M.: Interaction of cortex and superior colliculus in mediation of visually guided behavior in the cat. Science **153**, 1544—1547 (1966).
75. — Visual, acoustic, and somesthetic deficits in the cat after cortical and midbrain lesions. In: PURPURA, D.P., YAHRS, M.D. (Eds.). The Thalamus, pp. 391—417. New York: Columbia University Press 1966.
76. TIGGES, J.: Ein experimenteller Beitrag zum subcortikalen optischen System von Tupaia glis. Folia Primat. **4**, 103—123 (1966).
77. — TIGGES, M.: The accessory optic system in *Erinaceus* (Insectivora) and *Galago* (Primates). J. comp. Neurol. **137**, 59—70 (1969).
78. WALKER, A.E.: The Primate Thalamus. Chicago: University of Chicago Press 1938.
79. WALLS, G.L.: The lateral geniculate nucleus and visual histophysiology. University of California Pub. Physiol. **9**, 1—106 (1953).
80. WEISKRANTZ, L.: Contour discrimination in a young monkey with striate cortex ablation. Neuropsychology **56**, 225—231 (1963).
81. WILSON, M.E., CRAGG, B.G.: Projections from the lateral geniculate nucleus in the cat and monkey. J. Anat. **101**, 677—692 (1967).
82. WINANS, S.: Visual form discrimination after removal of the visual cortex in cats. Science **158**, 944—946 (1967).

Author Index

Page numbers in *italics* refer to the bibliography. Numbers shown in square brackets are the numbers of the references in the bibliography

Subject Index

Reprint from

Handbook of Sensory Physiology

Volume VII/3 A

Central Processing of Visual Information Part A

Edited by R. Jung

Springer-Verlag Berlin · Heidelberg · New York 1973
Printed in Germany

Not in Circulation

Visual Perception and Neurophysiology

By

R. Jung

With 16 Figures

Related Titles

Information Processing in the Nervous System

Proceedings of a Symposium held at the State University of New York at Buffalo 21-24 October 1968.
Editor: K. N. Leibovic
69 figures. XVII, 373 pages. 1969.
Cloth DM 79,20; US $ 35.70
ISBN 3-540-04885-5

This Symposium volume presents a series of papers with the objective of correlating neuronal "machinery" with psychophysiological phenomena. The "theoretical" and "experimental" contributions are thus an attempt to reflect the biological unity, which is often dismembered conceptually into anatomy, physiology, and psychology.

Information Processing in the Visual Systems of Arthropods

Editor: R. Wehner
263 figures. XI, 334 pages. 1972.
DM 36,—; US $ 16.20
ISBN 3-540-06020-0

Complementing the vast literature of vertebrate visual systems this book presents the recent progress in neuroanatomical, neurophysiological, and behavioral work on compound eye systems.

K. Motokawa: Physiology of Color and Pattern Vision

137 figures. 26 tables. 283 pages. 1970.
Cloth DM 72,—; US $ 32.40
ISBN 3-540-04977-0

Published by Igaku Shoin Ltd., Tokyo. Distribution rights for all countries excluding East Asia: Springer-Verlag

The author has specialized in sensory physiology, and his work in this field has attracted international acclaim. This book is arranged as a monograph of his contributions to research in this field over more than 30 years.

Springer-Verlag
Berlin
Heidelberg
New York

Reprint from

Handbook of Sensory Physiology

Volume VII/3 A

Central Processing of Visual Information Part A

Edited by R. Jung

Springer-Verlag Berlin · Heidelberg · New York 1973
Printed in Germany

Not in Circulation

Neurophysiological Mechanisms in the Visual Discrimination of Form

By

J. Stone and R. B. Freeman

With 9 Figures

Reprint from

Handbook of Sensory Physiology

Volume VII/3 A

Central Processing of Visual Information Part A

Edited by R. Jung

Springer-Verlag Berlin · Heidelberg · New York 1973
Printed in Germany

Not in Circulation

Neurophysiology of Binocular Single Vision and Stereopsis

By

P. O. Bishop

With 16 Figures

Reprint from

Handbook of Sensory Physiology

Volume VII/3 A

Central Processing of Visual Information Part A

Edited by R. Jung

Springer-Verlag Berlin · Heidelberg · New York 1973
Printed in Germany

Visual Stability and Voluntary Eye Movements

By

D. M. MacKay

With 4 Figures

Reprint from

Handbook of Sensory Physiology

Volume VII/3 A

Central Processing of Visual Information Part A

Edited by R. Jung

Springer-Verlag Berlin · Heidelberg · New York 1973
Printed in Germany

Not in Circulation

Neuronal Mechanisms of Visual Movement Perception and Some Psychophysical and Behavioral Correlations

By

O.-J. Grüsser and U. Grüsser-Cornehls

With 36 Figures

Reprint from

Handbook of Sensory Physiology

Volume VII/3 A

Central Processing of Visual Information Part A

Edited by R. Jung

Springer-Verlag Berlin · Heidelberg · New York 1973
Printed in Germany

Not in Circulation

Temporal Transfer Properties of the Afferent Visual System. Psychophysical, Neurophysiological and Theoretical Investigations

By

W. A. van de Grind, O.-J. Grüsser and H. U. Lunkenheimer

With 53 Figures

Reprint from

Handbook of Sensory Physiology

Volume VII/3 A

Central Processing of Visual Information Part A

Edited by R. Jung

Springer-Verlag Berlin · Heidelberg · New York 1973
Printed in Germany

Not in Circulation

Maintained Discharge in the Visual System and its Role for Information Processing

By

W. R. Levick

With 8 Figures

Reprint from

Handbook of Sensory Physiology

Volume VII/3 A

Central Processing of Visual Information Part A

Edited by R. Jung

Springer-Verlag Berlin · Heidelberg · New York 1973
Printed in Germany

Not in Circulation

Neuronal Changes in the Visual System Following Visual Deprivation

By

K. L. Chow

With 4 Figures

Reprint from

Handbook of Sensory Physiology

Editorial Board

H. Autrum · R. Jung · W. R. Loewenstein · D. M. McKay · H. L. Teuber

Volume VII/3 A

Central Processing of Visual Information Part A

Edited by R. Jung

Springer-Verlag Berlin · Heidelberg · New York 1973

Printed in Germany

Not in Circulation

Principles of the Mosaic Organisation in the Visual System's Neuropil of Musca domestica L

By

V. Braitenberg and N. J. Strausfeld

With 19 Figures

Reprint from

Handbook of Sensory Physiology

Editorial Board

H. Autrum · R. Jung · W. R. Loewenstein · D. M. McKay · H. L. Teuber

Volume VII/3 A

Central Processing of Visual Information Part A

Edited by R. Jung

Springer-Verlag Berlin · Heidelberg · New York 1973

Printed in Germany

Not in Circulation

Comparative Physiology of Colour Vision in Animals

By

H. Autrum and I. Thomas

With 18 Figures

Reprint from

Handbook of Sensory Physiology

Volume VII/3 A

Central Processing of Visual Information Part A

Edited by R. Jung

Springer-Verlag Berlin · Heidelberg · New York 1973
Printed in Germany

Not in Circulation

The Evolution of Mammalian Visual Mechanisms

By

M. Snyder

With 7 Figures

MIX
Papier aus verantwortungsvollen Quellen
Paper from responsible sources
FSC® C105338

If you have any concerns about our products,
you can contact us on
ProductSafety@springernature.com

In case Publisher is established outside the EU,
the EU authorized representative is:
Springer Nature Customer Service Center GmbH
Europaplatz 3, 69115 Heidelberg, Germany

Printed by Libri Plureos GmbH
in Hamburg, Germany